GLACIERS & GLACIATION

GLACIERS & GLACIATION

DOUGLAS I. BENN

DEPARTMENT OF GEOGRAPHY, UNIVERSITY OF ABERDEEN, UK

AND

DAVID J.A. EVANS

DEPARTMENT OF GEOGRAPHY AND TOPOGRAPHIC SCIENCE, UNIVERSITY OF GLASGOW, UK

A member of the Hodder Headline Group
LONDON

First published in Great Britain in 1998 by
Arnold, a member of the Hodder Headline Group
338 Euston Road, London NW1 3BH

http:\\www.arnoldpublishers.com

Co-published in the United States of America by,
Oxford University Press Inc.,
198 Madison Avenue, New York, NY 10016

British Library Cataloguing in Publication Data
A catalogue entry for this book is available from the British Library

Library of Congress Cataloging-in-Publication Data
A catalog record for this book is available from the Library of Congress

ISBN 0 340 58431 9 (pb) 0 340 65303 5 (hb)

4 5 6 7 8 9 10

Production Editor: Julie Delf
Production Controller: Sarah Kett
Cover Design: T. Griffiths

Composition by J&L Composition Ltd, Filey, North Yorkshire
Printed and bound in Great Britain by The Bath Press, Bath

CONTENTS

PREFACE

For most people living in temperate countries, ice is usually encountered only in small quantities on cold days or as small blocks in drinks. In this everyday form, ice appears to be a rigid, brittle and usually slippery material, but in glaciers and ice sheets ice can exhibit a wide variety of surprising and fascinating behaviour. It can flow plastically, like toothpaste, mould itself around hills and valleys, and creep and slide from high snowfields down towards lower ground. It can carve out huge troughs and alter entire landscapes, scouring away soil and other surficial material or blanketing large tracts with glacial deposits. Glaciers and ice sheets can alter the Earth's climate, chilling the atmosphere and the oceans, and profoundly affecting the global hydrological cycle. They also preserve valuable records of past climate change locked up in ice crystals and air bubbles, and the oscillations of glacier margins provide direct indicators of recent climatic shifts. Furthermore, glaciers can store and suddenly release immense quantities of meltwater, producing catastrophic floods such as those which occurred in Iceland in late 1996 when subglacial volcanic activity melted the base of a portion of Vatnajökull. In heavily populated regions, huge volumes of glacigenic material are continuously excavated by people to use in buildings and roads, and many of the holes left behind are being filled by the ever-increasing mountains of domestic and industrial waste. Thus, either directly or indirectly, glaciers and glaciation impact upon a large proportion of the planet's population.

As students in the 1970s and 1980s we were influenced by the glacial textbooks by Clifford Embleton and Cuchlaine King (*Glacial Geomorphology*, 1975) and David Sugden and Brian John (*Glaciers and Landscape*, 1976), and as researchers and teachers of the 1990s we became increasingly frustrated by the lack of up-to-date versions of these fine volumes. As subscribers to the axiom 'if you want something done, do it yourself', we embarked upon *Glaciers and Glaciation* in 1993. Our aim was to produce a modern synthesis, covering all important aspects of glaciers and their effects. While the focus of the book is on glacial geomorphology and sedimentology, we have also tried to encompass important work in related fields such as glaciology and Quaternary studies. Glaciology as a discipline has expanded enormously in the past few decades, and its importance to glacial geologists and geomorphologists is increasingly clear. It has also become a very rigorous, mathematical subject which can be rather impenetrable to those without a background in physics, and students attempting to explore the research literature can rapidly find themselves lost in a forest of frightening equations. In response, we have tried to present the more important results of modern glaciology in an accessible way, with a minimum of mathematics. Some equations have been included, however, as a means of expressing fundamental relationships in a concise way and as a gentle introduction to the more technical aspects of the literature. Contemplation of these relations should be rewarded by increased insight, although those who feel numerically paralysed should be able to skim them and still grasp the essentials from the text.

We hope that students will find *Glaciers and Glaciation* as interesting and stimulating to read as it was for us to write. It is intended for this book to provide substance for all levels of undergraduate training and to be utilized by postgraduate researchers and university teachers. Given the recent prolific production rate of glacier-related literature, we found it difficult to be comprehensive in our approach, and it is inevitable that some recent research papers will not figure in the text by the time it reaches the bookstores. We may also have missed something or given insufficient space to subjects that you regard as fundamental to the field. If so, please let us know and we will endeavour to put it right in following editions.

ACKNOWLEDGEMENTS

A book reflects an author's personal interests and support. I have been fortunate to enjoy many enchanting landscapes in my career and these have fuelled a passion for glaciers and their products, a passion that developed when I was at school working my way through *The Principles of Physical Geography* by F.J. Monkhouse. I have many people to thank for their support, encouragement and companionship since that time. My mother, father and sister have seen little of me since 1979 when I started my undergraduate programme, but their understanding constitutes the foundation of my support network. Mike Walker provided the irresistible evidence and encouragement that convinced me I would flourish in the North American postgraduate system (he was certainly correct). Bob Rogerson shared with me the wonders of Labrador, and John England supported my PhD research at the University of Alberta. Moreover, John provided me with the opportunity to visit Ellesmere Island on numerous occasions and continues to be a source of inspiration, bestowed through a good-humoured and humble, but always rigorous, approach to research. Keep your stick on the ice scientist! I have managed to visit many exotic locations with the generous support of several institutions, the most significant being the Polar Continental Shelf Project, the Geological Survey of Canada, the Carnegie Trust for the Universities of Scotland, the Royal Society, Norsk Polarinstitutt and the University of Glasgow. A colossal and difficult task was the compilation of the figures for this book, and I am eternally grateful to Les Hill, Yvonne Wilson and Ian Gerrard of the Department of Geography, University of Glasgow, for persevering with me and the eternal 'one last set of figures'. We would both like to thank Laura McKelvie at Arnold, who has been extraordinarily patient with us and our tardy completion rate. Finally, and most significantly, my wife Tessa has given unfailing support, especially during the writing of this book. During the later stages of its writing Tessa has entertained our daughter Tara, and has always given me room to keep the production rate ticking over even though she has a career of her own. I couldn't ask for more.

Dave Evans

The idea that I could make a living by studying glaciers and glaciation came as a revelation one afternoon when I was rock-climbing in the English Peak District. The catalyst was Danny McCarroll, now a lecturer at the University of Wales, Swansea, who was then an undergraduate at Sheffield and manifestly enjoying a better life than mine on the dole. For setting me on my path and for his continuing friendship, I owe him much gratitude. I also wish to thank my teachers John Rice and Colin Ballantyne, who helped me to focus my raw enthusiasm and develop a more measured approach to research. Colin has become one of my most valued friends and collaborators, and my contribution to this book owes much to his influence. My thinking has also been strongly influenced by colleagues who have shared adventures in far-off places, especially Lewis Owen and Chalmers Clapperton, who introduced me to the Himalayas and the Andes and infected me with their passion for these wild and beautiful mountains. My work there and in other areas has been funded by the Natural Environment Research Council, the Carnegie Trust and the University of Aberdeen. I would like to thank Jim Livingstone and Alison Sanderson of the Department of Geography, University of Aberdeen, for help with diagrams, and Laura McKelvie of Arnold for her patience and encouragement during the final stages of writing this book. Most of all I wish to thank my parents and family for their love and support over the years, especially Sue and Andy, who have maintained a happy home while I have been engrossed in my work.

Doug Benn

PART
1
GLACIERS

Glaciers calving into the head of Jokel Fiord, Ellesmere Island. (Photo: D.J.A. Evans)

PART 1
GLACIERS

Glaciers calving into the head of Jokel Fiord, Ellesmere Island. (Photo: D.J.A. Evans)

CHAPTER

1

GLACIER SYSTEMS

1.1 THE GLACIER AS A SYSTEM

Glaciers and ice sheets are among the most fascinating and spectacular phenomena on Earth, covering approximately 10 per cent of the Earth's surface, and locking up over 33 million km^3 of fresh water, enough to raise sea-level over the whole globe by some 70 m. At times during the Quaternary Period of the past 2 million years, global ice coverage was even more extensive, covering approximately one-third of the Earth's surface. The great waxings and wanings of the world's glaciers and ice sheets have profoundly shaped the landscape of huge areas, scouring out rock and sediment and redepositing them as thick accumulations of glacial debris.

Glaciers are also sensitive barometers of climate change, growing and wasting in response to changes in temperature and snowfall, so the fluctuations of past and present glaciers serve as a valuable source of information on the workings of the global climate system. In turn, glaciers and ice sheets exert their own influence on local and global climate, altering pressure systems and wind directions, and serving to keep vast areas locked in perpetual cold.

Why, where and when do glaciers form, and how do they affect the landscape, global climate and the oceans? To answer these questions, it is helpful to visualize glaciers as *systems*, with inputs and outputs, and interactions with other systems, such as the atmosphere, the oceans, rivers and the landscape (see Chorley and Kennedy, 1971; White *et al.*, 1984). Figure 1.1 shows the basic components of a simplified glacier system. Mass and energy enter the system in the form of precipitation, rock debris, gravity, solar radiation and geothermal heat, and leave in the form of water vapour, water, ice, rock debris and heat. This mass and energy is transferred through the system at a variety of rates, with intervening periods of storage within or beneath the glacier.

The most important input to glacier systems is snow and ice derived from direct snowfall, blown snow, and avalanching from slopes above the glacier surface. These inputs, known collectively as *accumulation*, will be present provided prevailing temperatures are low enough. This snow and ice is then transferred downvalley by glacier movement until it reaches areas where it is lost to the system, either by melting and evaporation or by the breakaway of ice blocks or icebergs, known collectively as *ablation* (Fig. 1.2). Glaciers grow where climatic and topographic conditions allow the inputs to exceed the losses (i.e. where accumulation exceeds ablation), and glaciers recede where the outputs are greater than the inputs. The work of glaciers in modifying the landscape by eroding, transporting and depositing rock debris can also be viewed from this system perspective, with debris forming another type of input and output.

1.1.1 Snow and ice

Solid precipitation (snow and ice) forms the primary input of mass for glacier systems. In most areas this differs from the annual precipitation total, because in the warm season much precipitation may fall as rain, so that in glaciological terms the most important factor is the *effective precipitation*. For example, precipitation at the centre of the Antarctic ice sheet is among the lowest on Earth but, because the precipitation falls as snow, it is more effective than in moist areas that lie on the threshold of glacierization. In order to quantify this precipitation effectiveness, Tricart (1969) derived a *nivometric coefficient*, which is the ratio of snowfall to total annual precipitation; a ratio of 1.0 indicates that precipitation is composed entirely of snow.

By using the nivometric coefficient in conjunction with precipitation totals, Sugden and John (1976) provided a guide to glacierization potential. Well-

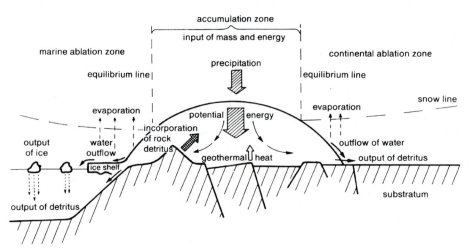

Fig. 1.1 An idealized cross-section through an ice sheet or ice cap when viewed as a system, showing the inputs and outputs of mass and energy. (From Brodzikowski and van Loon, 1991. Reprinted by permission of Elsevier)

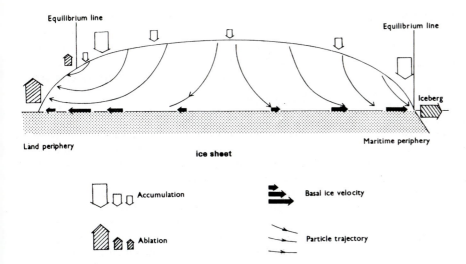

Equilibrium line

Equilibrium line

Iceberg

Land periphery

Maritime periphery

ice sheet

☐☐☐ Accumulation

▰▰ Basal ice velocity

◣◢◢ Ablation

Particle trajectory

Fig. 1.2 A simplified model of an ice sheet system showing distribution of mass input and output and related flow characteristics. Basal sliding is assumed to occur, and is at a maximum in the vicinity of the equilibrium line. (From Sugden and John, 1976)

nourished glaciers will exist in areas where a high precipitation total is matched by a high nivometric coefficient, as may occur in humid high mountain environments such as the Himalaya, or in high-latitude maritime locations such as Svalbard. Regions that are characterized by a high nivometric coefficient but low precipitation totals, such as the Antarctic ice sheet interior and Canadian High Arctic, will be suitable for long-term glacier survival even though vigorous glacier growth is not possible. Clearly these aspects of precipitation are crucial to the glacierization potential of an area, but glacier survival depends on other factors apart from the gross accumulation; the production of glaciers is most often the result of a complex interplay between precipitation, altitude and latitude.

The development of glaciers depends on the survival of snow and ice from one year to the next, and this is crucially dependent on both climatic and topographic factors. Climatically, the most important factor is the amount of energy available to melt ice, which in most areas is closely related to air temperature. Mean annual temperatures provide very little information on the degree of glacierization, however, but there is a close relationship between ablation and the number of days when temperatures are greater than 0°C (*positive degree-days*). The more positive degree-days, the less the likelihood of ice persisting throughout the year. Topographic factors influencing glacier survival are local slope and the presence of deep water at the glacier margin. Snow accumulating on or above steep slopes may avalanche down to lower altitudes where ablation rates are higher, and glaciers terminating in deep water lose large amounts of mass by the breakaway of icebergs, a process known as *calving*.

Most glaciers can be subdivided into two zones, an inner or upper zone where annual accumulation exceeds losses by ablation, and an outer or lower zone where ablation exceeds accumulation. These two zones are known as the *accumulation zone* and the *ablation zone* respectively and are separated by the *equilibrium line*, where annual accumulation and ablation are equal (Fig. 1.2). The positioning of the equilibrium line (the *equilibrium line altitude* or *ELA*) is dictated by local and regional climate and topography. The processes of accumulation and ablation are reviewed in detail in Chapter 2.

1.1.2 Meltwater

Meltwater is an extremely important component of glacier systems. Outputs of water from melting glaciers exert a very strong influence on the hydrology of proglacial areas, and feed water into other parts of the global hydrological system, including the oceans and the atmosphere. As it flows from glacier margins towards the sea, meltwater shapes the land, carving out gorges and depositing broad spreads of gravel, sand and silt. Within and beneath glaciers and ice sheets, liquid water profoundly affects glacier behaviour, controlling rates of glacier flow and influencing the processes and rates of erosion and deposition. The behaviour and work of meltwater are described in Chapter 3.

1.1.3 Glacier motion

Snow and ice is transferred from areas of accumulation to areas of ablation by glacier flow. Flow takes place by a variety of processes, which can be grouped together as *sliding*, *deformation of the ice*, and *deformation of the glacier bed*. By one or more of these processes, glaciers move slowly across the landscape, delivering snow and ice to areas where ablation exceeds accumulation, and meltwater and icebergs can leave the system as outputs. Glacier motion occurs when the forces exerted by the weight and surface slope of the ice overcome the strength of the glacier or its bed, allowing the ice to slide past obstructions on the bed, or the ice to deform like slow, semi-molten lava. When the whole system is in balance, rates of glacier flow match the rates at which snow and ice are added and lost in the accumulation and ablation areas, so that the inputs, throughputs and outputs exist in a state of dynamic equilibrium. However, various factors can throw the system out of balance, causing glaciers and ice sheets to exhibit all kinds of surprising behaviour, such as rapid advances or retreats which are unrelated to climate. The processes and patterns of glacier motion form the topic of Chapter 4.

1.1.4 Erosion, transport and deposition

As glaciers and meltwater flow across the landscape, rock detritus is picked up, transported and redeposited. Glaciers are among the most effective agents of erosion on Earth, excavating impressive troughs and fjord basins and scouring broad areas clear of soil and debris. They are also very efficient transporters of debris, carrying vast amounts of silt, sand, gravel and boulders up to several hundreds of kilometres from their source areas. This debris is then deposited in many types of environment, ranging from the glacier sole to the ocean floor many hundreds of kilometres from the ice margin itself. Erosion, transport and deposition in the subglacial environment form the subject of Chapter 5, and the input and throughput of debris at the ice surface and within the glacier are described in Chapter 6. Deposition at the ice margin and in deep water is discussed in Chapters 7 and 8, respectively.

1.1.5 Lakes and the oceans

The presence of glaciers and ice sheets exerts a strong influence on the distribution of lakes and regional and global sea-levels. Glacier margins dam up temporary lakes, and glacial erosion scours out deep and permanent lake basins. The expansion of ice sheets removes water from the global hydrological system to the extent that the world's oceans were some 150 m lower at the last glacial maximum than at present. Counteracting this, ice sheet growth pushes down the underlying Earth's crust, allowing ocean waters to flood in around the ice sheet margins. Many of the world's glaciers and ice sheets terminate in deep lakes or the sea. Water-terminating glaciers are subject to mass loss by calving, which renders them inherently less stable than land-terminating glaciers, a fact that has important implications for the stability of the West Antarctic ice sheet, the base of which is mostly below sea-level. The interactions between glaciers and deep

water bodies, and processes of subaqueous sediment transport and deposition, are described in Chapter 8.

1.1.6 Glacial sediments, landforms and landscapes

In the mid- and high latitudes, the most obvious legacy of glaciers and ice sheets is the range of sediments and landforms left behind after deglaciation, including corries, troughs, drumlins and moraines. These sediments and landforms can be used to reconstruct the extent and behaviour of former ice masses, and provide vital clues to the past and present workings of the global climate system. Glacial sediments and landforms are also important from an engineering point of view, because so much human activity takes place in terrain affected by glacier erosion or deposition (N. Eyles, 1983a). Part Two of this book is concerned with glacial sediments, landforms and landscapes: Chapter 9 deals with landforms and landscapes of erosion, Chapter 10 with types of sedimentary deposit, Chapter 11 with depositional landforms, and Chapter 12 zooms out to the widest scale to describe the overall impact of glaciation on whole landscapes.

The remainder of this Chapter reviews the large-scale interactions between glacier systems, climate and topography as a basis for understanding the morphology, extent, and past and present distribution of glaciers and ice sheets.

1.2 INTERACTIONS WITH CLIMATE

1.2.1 Glacier formation and decay

The past 2 million years or so of Earth history (the Quaternary Period) have been characterized by periodic climatic fluctuations between cold conditions, when glaciers and ice sheets have become more extensive than today, and warmer conditions when ice cover is much reduced (Fig. 1.3). One of the major achievements of Earth science in recent years has been to show that this sequence of *glacial* and *interglacial* periods is primarily driven by cyclical changes in the Earth's orbit around the Sun (Imbrie and Imbrie, 1979; Berger, 1988; Imbrie *et al.*, 1992, 1993). Three orbital cycles have been identified (Fig. 1.4). First, the shape or *eccentricity* of the Earth's orbit fluctuates, becoming more or less elliptical on a cycle of about 100,000 years; second, the tilt or *obliquity* of the Earth's axis relative to the orbital plane fluctuates over a 41,000-year cycle; and third, the direction of tilt of the Earth's axis relative to the fixed stars undergoes a 23,000-year cycle (*precession*) which alters the timing and variability of the seasons. Taken together, these cycles cause variations in the amount of solar radiation received throughout the year on different parts of the Earth's surface, thus altering the most fundamental input to the Earth's climate system. This external forcing mechanism brings about responses and chain reactions in the Earth's internal elements (notably the atmosphere, the oceans, the hydrological cycle, vegetation cover, and glaciers and ice sheets) which act as a linked system (Bradley, 1985; Imbrie *et al.*, 1992, 1993). Changes in one part of the Earth system in response to an external change can bring about responses in all the other elements because they are coupled or connected. This can set up feedback loops which can amplify the original signal, causing a non-linear response of the global climate system. Glaciers are thought to play an important role in global climate change, serving to amplify or dampen climatic variations. Therefore, glacier growth may be both a consequence and a cause of climatic change, complicating any attempt to understand the timing and severity of glacial cycles (Fig. 1.5; Imbrie *et al.*, 1993).

Although it is generally agreed that ice sheets take longer to grow than to decay, considerably more information is available on the decay of ice sheets than on their inception. This is largely because geomorphic and stratigraphic evidence of glacier recession has a far better preservation potential than any evidence pertaining to glacier buildup (Andrews and Barry, 1978; Dyke and Prest, 1987; Ruddiman and Wright, 1987; Mangerud, 1991a, b; Clark and Lea, 1992; Clark *et al.*, 1993). On any one glacier or ice sheet, ablation rates are usually three or four times the accumulation rates and, therefore, decay times should be one-third to one-quarter of growth times (Paterson, 1994). Because of the significant impact of iceberg calving at marine margins and proglacial lakes during deglaciation, decay rates can be even faster than this (Section 8.2.4). Evidence of global sea-level changes of the order of >10 mm yr^{-1}, which

provide an indicator of rates of change of terrestrial ice sheets, indicate that some inception periods may also be rapid but short-lived (Ruddiman and McIntyre, 1979; Cronin, 1983; Chappell and Shackleton, 1986; Shackleton, 1987; Mix, 1992; Muhs, 1992), constituting a series of 'glacial buildup periods' and culminating in a glacial maximum (Andrews and Barry, 1978). Similarly, rapid ice-sheet decay is recorded in rising global sea-levels (e.g. Fairbanks, 1989).

There is now compelling evidence from both land and oceans to suggest that the large northern hemisphere ice sheets were characterized by relative instability during the last glacial cycle. For example, the stratigraphy of the Hudson Bay area has been interpreted as evidence for more than one growth and decay phase of the Laurentide ice sheet during the last (Wisconsinan) glacial cycle (Shilts, 1982b, 1984b; Andrews *et al.*, 1983; Laymon 1991; Thorleifson *et al.*, 1992; Clark *et al.*, 1993). Furthermore, sediment cores from the North Atlantic have been found to contain several layers of lithic fragments (*Heinrich layers*) dating from the last glacial cycle and interpreted as ice-rafted debris carried by successive 'armadas' of icebergs (Heinrich, 1988; Grousset *et al.*, 1993; Bond and Lotti, 1995). These ice-rafting events are thought to record episodes of advance and breakup of the eastern margins of the Laurentide ice sheet, as the result either of changes in ice sheet dynamics or some unknown climatic forcing mechanism (e.g. Bond *et al.*, 1992, 1993; MacAyeal, 1993a, b). Such evidence indicates that even large ice sheets are dynamic systems, responding relatively rapidly to climatic forcing and internal instabilities, and undergoing huge changes in volume over thousands of years. These rapid volume changes have profound implications for glacial geomorphology, best illustrated in features such as ice flow directional indicators, which record complex changes in ice divides, ice dispersal centres and zones of fast glacier flow (Section 12.4).

Because of the lack of data pertaining to the process of glacier formation and growth (*glacieriza-*

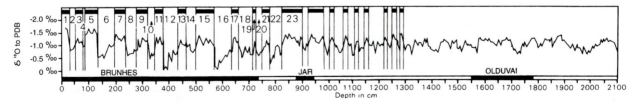

Fig. 1.3 Earth's Quaternary glacial record as shown in the oxygen isotope record from the ocean core V28–239. Glacial periods are given even numbers, and interglacials are given odd numbers and marked by black bars. The lower part of the diagram shows the palaeomagnetic timescale: Brunhes, Jar (Jaramillo) and Olduvai are periods of normal polarity of the Earth's magnetic field. The lower boundary of the Brunhes event is at 0.73 Myr, the Jaramillo is at 0.90–0.97 Myr, and the Olduvai is at 1.67–1.87 Myr. (From Lowe and Walker, 1984. Reproduced by permission of Longman)

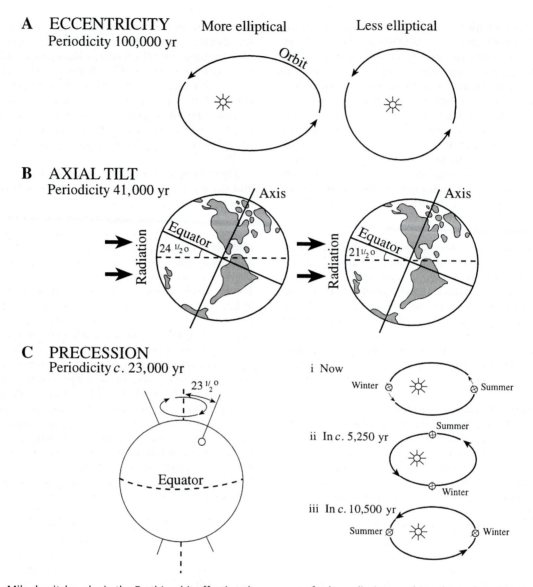

A ECCENTRICITY
Periodicity 100,000 yr

More elliptical

Less elliptical

Orbit

B AXIAL TILT
Periodicity 41,000 yr

Axis

Equator

Radiation

24 ½°

Axis

Equator

Radiation

21½°

C PRECESSION
Periodicity *c*. 23,000 yr

23 ½°

Equator

i Now

Winter   Summer

ii In *c*. 5,250 yr

Summer

Winter

iii In *c*. 10,500 yr

Summer   Winter

Fig. 1.4 Milankovitch cycles in the Earth's orbit affecting the amount of solar radiation reaching the surface. (Modified from Lowe and Walker, 1984)

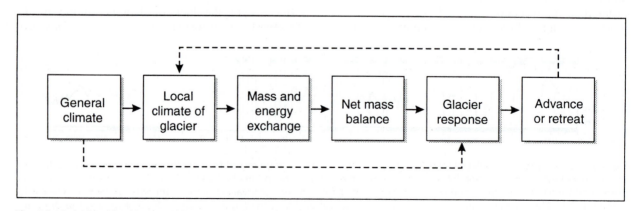

| General climate | → | Local climate of glacier | → | Mass and energy exchange | → | Net mass balance | → | Glacier response | → | Advance or retreat |

Fig. 1.5 Relationships between climate and glacier response. (Modified from Meier, 1965)

tion), models of ice sheet inception are largely theoretically based. Glacierization models have been applied to the inception of the former Laurentide ice sheet of North America, an early example being that of Flint (1943, 1971), who proposed the *highland origin, windward growth* model (Fig. 1.6). This model emphasizes the role of precipitation in mountain areas in glacier growth and can be summarized in four phases (Ives *et al.*, 1975). (a) Snow from storm systems, as they climb over mountain masses, produces ice sheets in high mountains (e.g. the Torngat Mountains in North America). (b) Ice fields grow fastest in the directions that bring the most precipitation. In the case of the Laurentide ice sheet this produced a westward migration of the ice divide and a reversal of flow across the western flanks of the Torngat Mountains. (c) Icefield growth results in total inundation of all high summits. (d) Deglacial thinning reverses the process, and the ice divide migrates back to the mountains. This results in the location of the last ice remnants in the very place where they were initiated.

Alternative models emphasize the coupling of regional-scale topography and climate. Lamb and Woodroffe (1970) identified high-latitude areas as being particularly susceptible to glacierization because they were on the threshold of hosting permanent glacier ice. Their postulates were expanded theoretically by Ives *et al.* (1975), who refined a model referred to as *instantaneous glacierization* (Ives, 1957) for the growth of the Laurentide and Fennoscandian ice sheets. According to this model, snow accumulating on plateau areas first produces small plateau glaciers and then expands and coalesces to produce a multi-domed ice sheet. The buildup of permanent snowfields on Baffin Island during the Little Ice Age has been used as an analogue for continental ice sheet inception by Barry *et al.* (1975). They concluded that a widespread and rapid lowering of regional snow lines on upland plateaux in the eastern Canadian Arctic was a response to only minor summer cooling and that this supported the concept of instantaneous glacierization. The climatic feedback mechanisms inherent in this model were expanded upon by Williams (1978, 1979), who provided data suggesting that late-lying snow cover across the whole of the Canadian Arctic effectively reduced mean air temperatures and increased precipitation over an 8-year instrumented period. Therefore, although astronomical forcing appears to be necessary to initiate long-term climate change, it can be seen that once an ice cover is established, it provides a climatic feedback to further ice sheet growth. In fact, Williams (1979) concluded that the reduced temperatures produced by the snow/ice coverage are more critical to glacierization than precipitation. For example, a lowering of temperature by 10–12°C is required to initiate glacierization of the Keewatin and Labrador sectors of the Canadian Shield. This concept was expanded by Koerner (1980), who suggested that an extension of the average period of annual snow cover and therefore an overall increase in the amount of solar radiation reflected back into space (albedo) is most effective as a climatic feedback mechanism at the start of a glacial period, rather than the growth and feedback of ice caps themselves.

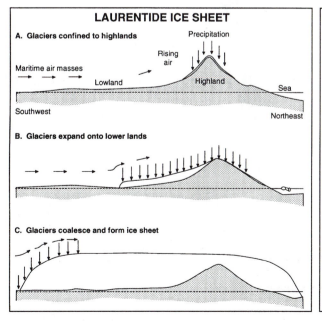

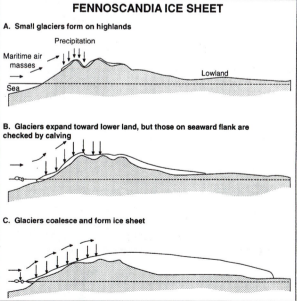

Fig. 1.6 The highland origin, windward growth model of ice sheet inception. (Modified from Ives *et al.*, 1975, after Flint, 1943, 1971)

The growth of ice sheets with large marine-based components was explained by Denton and Hughes (1981a) and Hughes (1986) by the *marine ice transgression hypothesis* (MITH). In this model, glacierization is triggered by the extension of permanent sea-ice into inter-island channels and large marine embayments. The sea-ice then develops into *fast ice*. The ice cover reflects radiation out to space, thus reducing regional temperatures so that the snow line can decline to sea-level. Snow accumulation helps the fast ice to thicken into ice shelves, which eventually ground to form a marine ice dome. Because the marine ice domes 'plug' all the surrounding watersheds, sea-levels drop and continental ice sheets grow through instantaneous glacierization. Hughes (1986) claims that the MITH incorporates the essentials of both instantaneous glacierization (lowering plateau snow-lines) and highland origin, windward growth (ice domes triggering convective snowfall). The MITH glacierization sequence is compared with those of the highland origin, windward growth and instantaneous glacierization models in Fig. 1.7, which illustrates the vast differences in size of possible initiation areas. The larger initiation areas proposed by the MITH are seen as critical to the rapid and uniform development of ice sheets over large areas (Hughes, 1986). This is indeed an important attribute of any model in the light of the evidence for ice sheet instability discussed above.

1.2.2 Effects of ice sheets on climate

Once ice sheets are established, they have considerable, large-scale impacts on regional climate and can often regulate their own existence (Oerlemans and van der Veen, 1984; Paterson, 1994). For example, an ice sheet can exist for some time in a state of disequilibrium with prevailing climate, because its sheer size not only produces a lag in response time but also provides its own environmental buffer to outside forcing mechanisms. Ice masses at high latitudes are

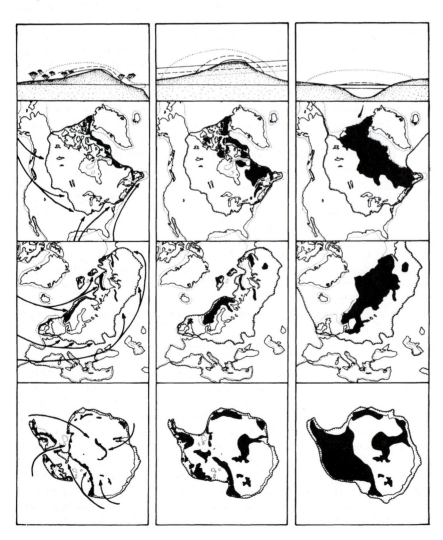

Fig. 1.7 A comparison of the various hypotheses for the formation of the Quaternary ice sheets of North America, north-west Europe and Antarctica. Left – highland origin, windward growth; centre – instantaneous glacierization; right – marine ice transgression. Areas of initiation are black, stages in growth are solid, dashed and dotted profiles (top diagrams) and the maximum extent of ice is the hachured line. Arrows show the routes of the major storm tracks supplying snow to the growing ice sheets. (From Hughes, 1987. Reprinted by permission of Scandinavian University Press)

particularly difficult to remove from the landscape because, even though accumulation may have dropped off considerably, ablation rates are extremely low and the ice can modify its local climate to some extent. The corollary is that some glaciers are essentially relics of former climates (e.g. Bradley and Serrez, 1987).

Once an ice sheet has grown, it modifies its own surface temperature and precipitation patterns. Where an ice sheet forms a dome up to 50 km wide, precipitation can increase with increasing altitude all the way to the summit. Where an ice sheet expands to form a dome of continental proportions, however, the increase in precipitation with altitude occurs only at the outer margins of the ice sheet, and nearer to the summit the precipitation, and thus ice discharge, drops off considerably. For example, the accumulation rate for the Antarctic ice sheet is 10 times greater at most parts of the coast than at the ice sheet centre. The high albedo and high elevation of ice-sheet centres produce very low temperature values, especially during the winter months; winter values plunge to $-70°C$ over Antarctica and to $-40°C$ over Greenland. The sheer sizes of these ice bodies ensure that they exert a large influence over global as well as local-scale climate patterns.

The impact of ice sheet growth on global climate is manifest in the equator–pole temperature gradient. Because ice sheets reduce the absorption of solar radiation, they will accentuate the thermal contrasts that already exist between polar and equatorial regions. Global atmospheric circulation will always react to the increased equator–pole temperature gradient created during glaciations, especially in the northern hemisphere. This it does by increasing the strength of east–west (zonal) and north–south (meridional) circulation.

The climatic impacts of ice sheets have been modelled by a number of researchers and are thought to influence changes in the configuration and marginal oscillations of the ice sheets themselves. Important parameters in such models are the impacts of the reflection of solar radiation by a growing ice sheet (albedo feedback), precipitation changes induced by the interruption of regional climates by ice sheet growth, and the vertical movement of the crust in response to ice loading/unloading (see Sections 1.3 and 1.5; Budd and Smith, 1981; Manabe and Broccoli, 1985; Broccoli and Manabe, 1987a).

The responses by different margins of an ice sheet to climatic change are dictated by their boundary conditions. For example, the West Antarctic ice sheet, which is largely grounded below sea-level and terminates in large floating ice shelves in the Ross and Weddell Seas, responds strongly to changes in sea-level and is very susceptible to global-scale glacial–interglacial cycles and their attendant sea-

level fluctuations. In contrast, the East Antarctic ice sheet is mainly grounded above sea-level, and responds at a much slower rate to direct changes in snow accumulation and temperature (Peixoto and Oort, 1992; Fig. 1.8).

Some glacier bodies appear to be in long-term equilibrium with the present climate of the Earth. The Antarctic and Greenland ice sheets are regarded as more or less permanent features of the Earth's present surface configuration of land and ocean masses. The Antarctic ice sheet as a whole has remained relatively stable since its inception in the Miocene Epoch, more than 20 million years ago (Denton *et al.*, 1993; Section 1.7.2.1). Changes in global climate produce only minor oscillations at the margins of the Antarctic ice sheet compared with the growth and decay of the northern hemisphere Laurentide and Fennoscandian ice sheets. The Greenland ice sheet, by contrast, probably fluctuates to a far greater degree than the Antarctic ice sheet in response to glacial/interglacial climates. Evidence for this comes from interpretations of ice cores from the Canadian Arctic ice caps and the Camp Century and Dye-3 cores on Greenland (Koerner, 1989). It was suggested by Koerner that the Greenland ice sheet may have completely disappeared during the last interglacial, accounting for a 6 m rise in global sea-level at that time (Chappell and Shackleton, 1986). This questions any collapse of the West Antarctic ice sheet during the last interglacial (e.g. Hughes, 1972, 1973; Mercer, 1978; Sugden, 1988), which combined with the melting of a large part of the Greenland ice sheet, would have created far higher global sea-levels than those recorded: global sea-level rises are of the order of 8 m for both the Greenland and West Antarctic ice sheets producing a total of 16 m for a complete ice sheet melting scenario (Barry, 1985). Furthermore, evidence suggests that the ice shelves of the West Antarctic ice sheet are largely in equilibrium with present interglacial conditions (e.g. Fastook, 1984; Lingle, 1984), and so even during the warmer parts of interglacials the marine margins of the ice sheet are not necessarily prone to massive collapse.

The profound influences of former ice sheets on global climate have been modelled using *general circulation models* (GCMs), which are computer simulations of global atmospheric circulation (e.g. Lamb and Woodroffe, 1970; Williams *et al.*, 1974; CLIMAP, 1976; Gates, 1976a, b; Manabe and Broccoli, 1984a, b, 1985; Kutzbach and Guetter, 1986; Broccoli and Manabe, 1987a). Such models have shown that the Laurentide ice sheet acted as a topographic barrier to atmospheric circulation, splitting the mid-latitude westerly jet streams. At the surface, the result was the deflection of the mid-latitude westerlies around the northern and southern margins of

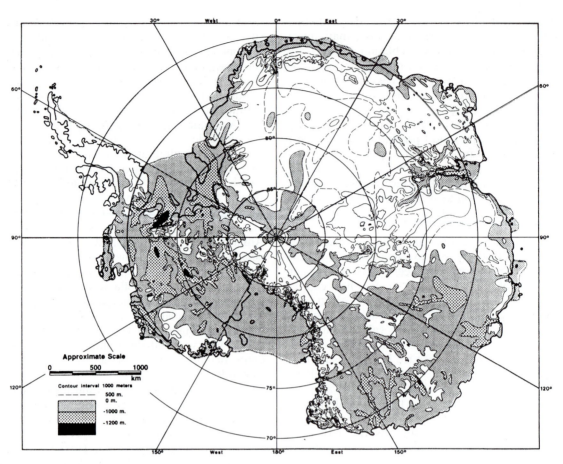

Fig. 1.8 The extent of areas presently below sea-level beneath the Antarctic ice sheet. (From Denton *et al.*, 1993. Reproduced by permission of Scandinavian University Press)

the ice sheet. Because it flowed across an area where the intense insolation reflection from the ice sheet was producing a cold air mass, the flow of cold air contributed to the maintenance of a thick sea-ice cover on the North Atlantic Ocean. At the same time, the western portion of the ice sheet essentially became a permanent blocking anticyclone. The presence of the ice sheet also severely affected precipitation values during full glacial conditions. A large decrease in evaporation and sublimation over the Laurentide ice sheet reduced the amount of moisture available to the atmosphere, prompting Broccoli and Manabe (1987a) to refer to the ice sheet as a *moisture sink*. As a result of this, the regions surrounding the ice sheet suffered from large decreases in precipitation, especially in the summer. The exception to this rule was the area immediately to the south-east of the ice sheet, where precipitation increased. This was part of a transcontinental band of higher precipitation values which stretched across the North Atlantic to the southern edge of the Fennoscandian ice sheet, thought to be associated with the strengthening of the south Laurentide ice sheet westerlies and increased

storminess (Broccoli and Manabe, 1987b). It is suggested by Broccoli and Manabe (1987b) that this ice sheet-induced storm track forms part of a self-sustaining mechanism for ice sheet growth and maintenance, particularly at the southern limits of the last northern hemisphere ice sheets.

Obviously an abundance of precipitation is critical to the growth and maintenance of ice sheets, but the westerlies that fed moisture to the Laurentide ice sheet would be exhausted by the time they reached its eastern margins. Herein lies the partial validity of Flint's (1943, 1971) windward growth model for ice sheets. Can this be applied in any way to the known growth of the Laurentide ice sheet?

Certainly the north-eastern sector of the ice sheet appears to have stood at its last glaciation maximum position at a much later date than any other margin (Miller and Dyke, 1974; Dyke and Prest, 1987; Clark *et al.*, 1993). Opinions on the asynchronous advances of different margins of the Laurentide ice sheet during a single glacial cycle have evolved since the work of Prest (1969; Fig. 1.9A), and are summarized by Dyke and Prest (1987). Research in the 1970s indicated that oscillations of the north-eastern margin of the ice sheet on Baffin Island were out of phase with oscillations of the southern margins in the United States. Furthermore, the north-eastern margins underwent progressively less extensive advances throughout the last glacial cycle, in contrast to the progressively more extensive advances of the southern margins (Fig. 1.9B). A simple climatic theory was put forward to explain such asynchrony. The more extensive advances on Baffin Island were thought to be a function of early ice sheet inception. Later advances were more restricted, owing to the blocking of storm tracks by the developing southern mass of the ice sheet, which led to severe reductions in accumulation in the north-east sector. Further work on the dating of ice margins rendered this simple climatic theory largely untenable (Fig. 1.9C), as advances at both southern and northern margins were seen to be more synchronous. The most recent phase of research has established that some asynchrony does indeed exist and that the south-east and north-

east margins of the ice sheet were out of phase with the north-west, south-west and south margins (Dyke and Prest, 1987; Clark *et al.*, 1993; Fig. 1.9D). Such asynchrony is less obvious or even non-existent in smaller ice sheets (e.g. the North American Cordilleran ice sheet; Fig. 1.10), where ice volume changes in various sectors are not affected by regional-scale climate variability.

1.3 INTERACTIONS WITH THE LANDSCAPE

Although climate is of fundamental importance to the inception and growth of glaciers, the form of the landscape dictates the threshold conditions for glacier occurrence and determines glacier morphology. Given the necessary climatic thresholds for glacierization, a glacier will form, but certain landscapes will produce glaciers of specific size, shape and longevity. Landscapes, particularly those of high relief, also provide the potential energy for the glacier system and are critical to the inputs of rock debris.

At the largest scale, ice sheets such as the former Laurentide or present Antarctic and Greenland ice sheets will accumulate in large continental interiors where temperatures are cold enough to allow snow and ice to persist throughout the year. Mass is lost from such ice sheets largely through calving into marine embayments (Section 8.2.4). Marine embayments may be created by the ice sheets themselves, either through erosion or by the isostatic downwarping of the crust under the weight of the ice. Ice loading has resulted in vast areas of the land surface beneath the Antarctic and Greenland ice sheets being depressed below present sea-level (Fig. 1.8). Accelerated calving at marine-based sectors of ice sheets can drain large amounts of ice from continental interiors, and large ice bodies with a considerable marine-based component, such as the former Laurentide and present West Antarctic ice sheets, have to supply large amounts of ice to their marine margins in order to prevent collapse and disintegration. Such large-scale collapse is thought to have characterized the retreat of parts of the Laurentide ice sheet at the end of the last glaciation, for example in the area of Hudson Strait, although some of the most persistent sectors of the ice sheet (Hudson Bay and Foxe Basin) were marine-based and located on crust that was glacioisostatically depressed hundreds of metres below sea-level (Andrews, 1987a). It is thought that some ice sheets, like the former Barents Shelf ice sheet, possess a self-regulating mechanism whereby ice buildup on the continental shelf ultimately produces massive calving once glacioisostatic loading

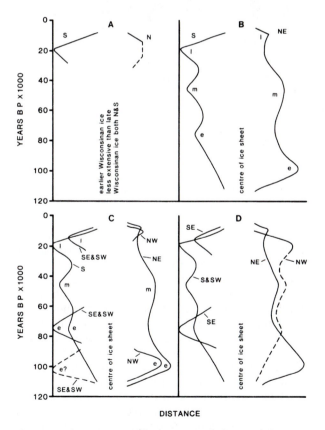

Fig. 1.9 Changing interpretations of the asynchronous marginal oscillations of the Wisconsinan (last glaciation) Laurentide ice sheet. Interpretations A–D are explained in the text. The letters e, m and l refer to the early, middle and late Wisconsinan advances. (From Dyke and Prest 1987. Reproduced by permission of Les Presses de l'Université de Montréal)

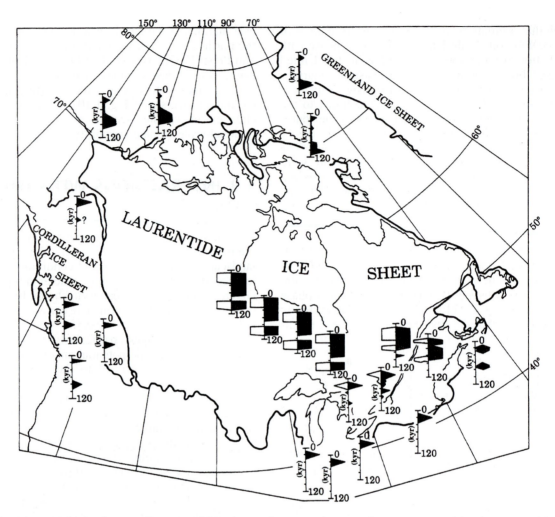

Fig. 1.10 Schematic time–distance diagrams of ice sheet advance and retreat in areas covered by the Laurentide and Cordilleran ice sheets. Alternative interpretations are shown for some areas. (From Clark *et al.*, 1993. Reproduced by permission of Elsevier)

has taken full effect (Weertman, 1961a). A similar model has been presented by N. Eyles and McCabe (1989a) for the Irish Sea portion of the former British ice sheet. Large quantities of mass may also be lost to terrestrial deep water bodies where ice sheets block the natural drainage of a region. For example, the south and south-western margins of the Laurentide ice sheet began to retreat rapidly once proglacial lakes Agassiz and McConnell were at their maximum extents and thus inducing massive-scale ice-front calving (Hughes, 1987; Teller, 1987). Smaller scale but nonetheless important calving of mass was induced by glacial Lake Naskaupi, dammed between the eastern margin of the Laurentide ice sheet and the Torngat Mountains of Labrador and Quebec (Vincent, 1989).

Wherever glaciers occur, they interrupt the flow towards the surface of heat generated by radioactive decay in the Earth's interior (*geothermal heat flux*).

A large ice mass occupying a considerable part of the Earth's surface will slow down the release of this heat to the atmosphere. Furthermore, geothermal heat produces a significant amount of subglacial melting and, combined with glacier thickness, will determine the thermal characteristics of the basal ice. A thick polar ice cap or ice sheet can trap enough geothermal heat to ensure that large areas of the glacier sole are constantly melting. The meltwater so produced may either percolate through the subglacial sediments or form a large subglacial lake, both of which occur beneath the Antarctic ice sheet at various locations. Alternatively, the meltwater may migrate some distance in the direction of the ice sheet margins. It has been suggested by Oerlemans (1982) that such meltwater migration may initiate a large-scale surge when it reaches the margin, draining large amounts of ice from the interior. This surge would reduce the ice thickness and allow geothermal heat to escape, thereby reducing basal temperatures below the melting point and allowing ice to build up once more, thus allowing the cycle to be repeated. The binge–purge cycles resulting from this mechanism have been invoked by MacAyeal (1993a, b) as a possible cause for cyclic growth and decay of the Hudson Bay sector of the Laurentide ice sheet during the last glacial cycle (Section 4.8.3).

Certain topographic and geologic settings are critical to the formation and persistence of large glacier bodies. Using a numerical ice sheet model for the Loch Lomond Stadial ice caps of Scotland, Payne and Sugden (1990) demonstrated that topography plays a critical role in the dynamics of ice sheet growth. For example, the combined effects of various temperature depressions and topography on ice sheet volume increases over time are presented in Fig. 1.11, which illustrates the responses of progressively growing (unstable; A) and steady-state (B) ice caps. During cooler conditions, steady-state ice caps are produced when ice flows down to lower altitudes, where ablation can balance accumulation. Conversely, relatively unstable ice caps are produced when ice flows into lowland areas surrounded by highland terrain, where ablation is unable to balance accumulation and ice begins to thicken; a feedback loop is set up as the ice surface rises and increases the accumulation further. Such a situation is thought to have arisen during ice sheet buildup at Rannoch Moor, Scotland, which is surrounded by high mountains and lies 400–600 m above sea-level.

At smaller scales, glaciers may exist in areas where topography and climate conspire to produce anomalous snow accumulation patterns. Good examples of this are cirque glaciers produced by wind-blown snow accumulating in sheltered lee slopes (e.g. Dolgushin, 1961, for the Urals). Similarly, in Norway, *fall glaciers*, which are a form of glacieret (see Section 1.4.2.6), accumulate at the base of steep rock walls in troughs and fjords, where they are fed by ice avalanching from plateau ice caps at higher altitudes (e.g. Gellatly *et al.*, 1986a, b; Evans and Fisher, 1987). In both cases glaciers may exist below the regional equilibrium line altitude.

1.4 GLACIER MORPHOLOGY

Overall glacier form is a function of climate and topography, and the morphology of any one glacier is unique to its location on the Earth's surface. Consequently, a wide variety of glacier morphologies exist, and clearly these form a continuum from the smallest

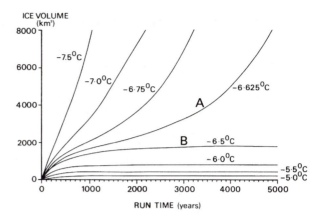

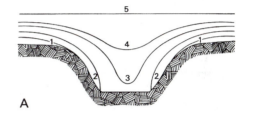

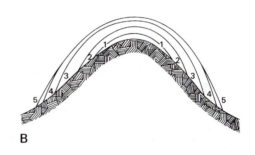

Fig. 1.11 Ice sheet volumes produced by stepped, constant-temperature depressions for intermontaine basins (A) and isolated upland areas (B). The bifurcation in behaviour is highlighted by the contrast between curves A and B and cross-sections of growth stages associated with the two types of topography. (From Payne and Sugden, 1990. Reprinted by permission of John Wiley & Sons)

niche glacier to the largest ice sheet. However, we can adopt a general classification scheme which will serve as a reference point throughout subsequent discussion. The scheme used in this book (Table 1.1) is based on the classifications developed by Ommanney (1969), Østrem (1974), Armstrong *et al.* (1973) and Sugden and John (1976).

1.4.1 Ice sheets and ice caps

Ice sheets and ice caps submerge the landscape, at least in their central portions, and are unconstrained

Table 1.1 Classification scheme for glacier morphology

First-order classification	Second-order classification
Ice sheet and ice cap unconstrained by topography	Ice dome
	Ice stream
	Outlet glacier
Glaciers constrained or controlled by topography	Icefield
	Valley glacier
	Transection glacier
	Cirque glacier
	Piedmont lobe
	Niche glacier
	Glacieret
	Ice apron
	Ice fringe
Marine glaciers	Ice rise
	Glacier ice shelf
	Sea-ice ice shelf

by topography to such an extent that ice flow is largely independent of undulations in the bed. However, at the margins of ice sheets and ice caps, faster-moving ice streams and outlet glaciers are located over depressions in the glacier bed. A size of 50,000 km^2 is adopted as the threshold between an ice cap and an ice sheet. Thus the ice masses presently covering Antarctica and Greenland and those formerly covering Scandinavia, North America, the British Isles and the Barents Shelf and northern Siberia are designated as ice sheets, whereas the ice masses over the highlands of Svalbard, Ellesmere Island, Baffin Island and Iceland are referred to as ice caps (Figs. 1.12 and 1.13 (Plates 1 and 2)).

Ice sheets and ice caps can be subdivided into ice domes, which are high areas of relatively slowly moving ice, and ice streams and outlet glaciers, which are more rapidly flowing zones through which most of the ice is discharged towards the periphery.

1.4.1.1 ICE DOMES

An ice dome is a broad, approximately symmetrical upstanding area of an ice sheet or ice cap. The underlying land surface may be either a topographic high (e.g. East Antarctica) or a vast topographic low (e.g. the centres of the former Laurentide and Scandinavian ice sheets). Ice thicknesses can often exceed 3000 m in ice sheet domes, but beneath ice cap domes thicknesses are only several hundreds of metres.

Irregularities in the subglacial topography can manifest themselves in glacier surface profiles and the location of ice domes. Budd (1970) and Budd and Carter (1971) have demonstrated that local bedrock highs at the base of the Antarctic ice sheet produce locally steepened profiles on the ice surface. Although such surface irregularities may appear severe to the observer on the surface, especially when associated with crevassing, they are insignificant when viewed as part of the total profile length.

1.4.1.2 OUTLET GLACIERS AND ICE STREAMS

The movement of ice within an ice sheet can be subdivided into sheet flow, which occurs within the ice dome areas, and stream flow, which occurs in outlet glaciers and ice streams. Outlet glaciers and ice streams are rapidly moving, channelled ice radiating out from the interiors of ice sheets and ice caps (Fig. 1.14). Outlet glaciers are bounded by ice-free ground at the ice sheet or ice cap margins, whereas ice streams are flanked by slowly moving ice (Bentley, 1987). Some fast-moving parts of ice sheets exhibit characteristics of both ice streams and outlet glaciers, such as the Rutford ice stream in Antarctica, which is bounded by the Ellsworth Mountains on one side and a low-lying part of the ice sheet on the other. The world's longest glacier, the 700-km Lambert Glacier, Antarctica, and the world's fastest glacier, Jacobshavn Glacier, Greenland, are both ice streams along some parts of their length and outlet glaciers along others. Ice streams can be differentiated from the surrounding ice by the presence of heavily crevassed zones along their margins, where rapidly streaming ice shears past sluggish surrounding ice (Fig. 1.15).

Outlet glaciers and ice streams commonly occupy depressions in the glacier bed, which can form deep troughs. The heads of ice streams, at least in some cases, coincide with a step in the subglacial topography separating thin, slowly moving ice from thicker, rapidly streaming ice (McIntyre, 1985). The gradients of outlet glaciers tend to be less steep than those of ice domes (Bentley, 1987). The flow characteristics and morphology of ice streams and outlet glaciers result from conditions at the glacier bed which influence the ice flow mechanisms. In particular, the presence of pressurized water at the bed is thought to be crucial in maintaining rapid sliding and/or flowage of soft, subglacial sediment. Alley *et al.* (1986, 1987a, b, c) have highlighted the importance of deforming till beneath Antarctic ice streams. The mechanisms of glacier flow are discussed in detail in Chapter 4.

Rapid flow through outlet glaciers and ice streams accounts for most of the discharge from ice sheets, and they are responsible for the majority of icebergs and ice-rafted debris reaching the world's oceans. Their importance in draining ice sheets, taken together with their dynamic nature, means that ice streams and outlet glaciers exert a considerable influence on the behaviour and stability of ice sheets, and much research has been undertaken recently on ice

Fig. 1.14 Oblique aerial photograph of Scott Glacier, Queen Maud Mountains, looking upglacier. The glacier is about 12 km wide in the centre of the picture. (US Navy. From Swithinbank, 1988)

Fig. 1.15 Crevasses in the transition zone between faster-moving ice stream and surrounding sluggish ice, Ice Stream B, West Antarctica. (US Navy. From Swithinbank, 1988)

stream dynamics, particularly in the potentially unstable West Antarctic ice sheet (e.g. Bentley, 1987; Bindschadler, 1993). Ice stream dynamics are also thought to have been critical in determining the stability and breakup of the great mid-latitude ice sheets in the past (e.g. Hughes *et al.*, 1985).

1.4.1.3 ICE DIVIDES, DISPERSAL CENTRES AND FLOW LINES

Once an ice sheet or ice cap has developed, either as a single dome or as a complex of coalescent domes (instantaneous glacierization), mass is evacuated from the central accumulation area or areas along flowlines (Fig. 1.16 (Plate 3)). The central accumulation area or areas of an ice sheet are referred to as dispersal centres with respect to glacier flow. In a dispersal centre, an ice divide simply marks the zone where the horizontal velocity is zero. Ice flow is directed in two opposing directions away from the ice divide, rendering the basal ice below the divide almost stagnant (Nye, 1951). This basal zone of little or no basal flow may be up to 100 km wide. Because ice sheets are rarely simple domes, they usually contain several divides from which ice flows at right-angles to the surface contours (Budd, 1969). This horizontal flow component is combined with a vertical flow produced by progressive accumulation, compaction, ice deformation and basal melting or freezing. Ice flow patterns are also influenced by the basal topography and irregular patterns of ice accumulation. This has been demonstrated by Haefeli (1963) for ice flow trajectories in the Jungfraujoch, Switzerland. Haefili used the concept of a *flow divide* to join the points where the horizontal velocity component is zero (Fig. 1.17).

Ice divides will change position as an ice sheet develops and decays. Consequently, the pattern of ice flow will change considerably, and ice directional indicators such as drumlins and flutings will overprint or even destroy each other. The changing ice divides and flow patterns of the former Laurentide ice sheet during the last glaciation have been reconstructed in a series of maps by Dyke and Prest (1987). These mega-scale imprints of glaciation are discussed in detail in Section 12.4.4.

1.4.2 Glaciers constrained or controlled by topography

A number of terms have been introduced to describe the forms of glaciers constrained or controlled by topography, and these are often not associated with any specific size range. In order to provide a comprehensive review, the terms previously provided in Østrem (1974), Embleton and King (1975) and Sugden and John (1976) are discussed here. It must be remembered that there are often no clear distinctions between types of ice mass and that we are always dealing with both spatial and temporal continuums of forms.

1.4.2.1 ICEFIELDS

Icefields differ from ice caps in that they do not possess a domelike surface and their flow is influenced by the underlying topography (Fig. 1.18 (Plate 4)). Icefields will develop in any area with generally gentle but locally fretted topography and at an altitude sufficient for ice accumulation. Good examples of such ice masses are the Columbia icefields in the Canadian Rocky Mountains (Denton, 1975), the icefields of the St Elias Mountains in the Canadian Yukon Territory and Alaska (Field, 1975), the Tien Shan/Kunlun Shan icefields in China (Lehr and Horvath, 1975; Shih Ya-feng *et al.*, 1980), and the Patagonian icefields (Warren and Sugden, 1993), all of which are drained by large valley glaciers.

1.4.2.2 VALLEY GLACIERS

Wherever ice is discharged from an icefield or cirque into a deep bedrock valley it forms a valley glacier (Fig. 1.19). Such glaciers may possess a simple, single-branched planform, or form dendritic networks similar to those of fluvial systems. Like rivers, valley glaciers can be classified according to their position in the drainage basin hierarchy (Horton, 1945). The form of valley glacier networks is often strongly influenced by bedrock lithology and structure. Bedrock slopes beneath valley glaciers are often steep, and so the altitudinal ranges of the glaciers are often very large. Net accumulation increases with altitude, producing a high turnover rate.

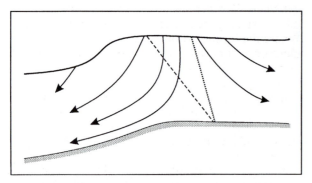

Fig. 1.17 Flow paths related to an ice divide (dotted line) at Jungfraujoch, Switzerland. The ice divide separates flow from the left and right, whereas the flow divide (dashed line) joins the points where the horizontal velocity component is zero. (After Haefeli, 1963)

Fig. 1.19 Valley glacier: the upper Tasman Glacier, Mount Cook massif, South Island, New Zealand. (Photo: M. Kirkbride)

The most distinctive characteristic of valley glaciers is the presence of ice-free slopes overlooking the glacier surface (Fig. 1.19). Such slopes are important sources of snow and ice accumulation in the form of avalanches, and of rock debris falling on to the glacier surface (Sections 2.2 and 6.3).

1.4.2.3 TRANSECTION GLACIERS

Where mountain landscapes are too dissected to support an ice cap and tend to evacuate ice too effectively for it to produce a dome-like profile, they often produce transection glaciers, or interconnected systems of valley glaciers. In such situations, glaciers flow down from several directions into a system of radiating valleys, but the ice overspills the pre-existing drainage divides (Fig. 1.20). Transection complexes form web-like patterns, with ice *diffluences*, where ice flow splits and sends branches down two or more channels, and *confluences*, where ice flow from two or more tributaries converges to form a single unit.

1.4.2.4 CIRQUE GLACIERS

The form of a cirque glacier is dictated by the armchair-shaped bedrock hollow, which acts as an accumulation basin, especially for wind-driven snow. The various sizes and forms of cirque glaciers constitute a continuum from glaciers that are entirely confined by their hosting bedrock hollows (Figs. 1.21 and 1.22) to glaciers that form the heads of larger valley glaciers in more heavily glacierized terrain. Considerable research on cirque glaciers followed on from the highly influential work of Lewis (1960), and a whole chapter is devoted to cirques in Embleton and King (1975). Owing to the plan shape of cirque glaciers, which shows a widening of the basin in the

upper reaches, the glacier flow can be subdivided into converging (accelerating) and diverging (decelerating) components above and below the equilibrium line respectively.

1.4.2.5 PIEDMONT GLACIERS

Wherever valley glaciers debouch on to lowland areas after travelling through bedrock troughs, they will form piedmont glaciers or lobes. Piedmont glaciers such as the Malaspina Glacier, Alaska, and Skeidararjökull, Iceland, are characterized by having large areas below the equilibrium line altitude. The Malaspina Glacier (Fig. 1.23 (Plate 5)) lies entirely in the ablation zone of the Lower Seward Glacier, is 600 m thick, occupies a vast depression up to 250 m below sea-level, contains numerous crumpled medial moraines derived from the feeder valley glaciers, and is fronted by a large area of stagnant ice upon which spruce trees over 100 years old are growing. Numerous smaller examples exist in the Canadian High Arctic, where subpolar glaciers debouch from plateau icefields on to the lowlands of U-shaped valleys (Fig. 1.24). Clearly, the discharge of mass from the accumulation zones of the feeder icefields is sufficient to maintain relatively large glacier surface areas below the equilibrium line.

1.4.2.6 NICHE GLACIERS, GLACIERETS, ICE APRONS AND ICE FRINGES

The smallest ice masses, *ice aprons*, are thin snow and ice accumulations adhering to mountainsides. Similar accumulations occupying small depressions along coasts are referred to as *ice fringes*. Thin ice patches occupying depressions on less precipitous terrain are often referred to as *glacierets* and are produced by snow drifting and avalanching. Glacierets have been called *fall glaciers* where they exist below the ELA as a result of ice avalanching from ice falls at steep plateau edges (Gellatly *et al.*, 1986a). Where the location of an ice body is controlled by a niche or rock bench in a mountain or valley side it is termed a *niche glacier* (Fig. 1.25; Groom, 1959). Niche glaciers and ice aprons differ from large snowpatches in that they undergo significant movement as the result of internal deformation or basal sliding. Glacier motion is initiated when the forces imposed by the weight and surface gradient of the ice overcome the internal resistance. Theoretical considerations suggest that this threshold will be reached when a snowpatch extends between 30 m and 70 m from the base of its backwall (Ballantyne and Benn, 1994a).

Fig. 1.20 Landsat image of Kong Christian IX Land, east Greenland (Canada Centre for Remote Sensing), showing transection glaciers crossing the regional drainage divide (dashed line). (Image provided by Richard Williams, USGS)

1.4.3 Marine glaciers

Wherever glacier ice is forced to float by deeper water or where sea-ice thickens by surface accumulation and bottom accretion, *ice shelves* are produced. Lemmen *et al.* (1988) proposed a classification scheme which reflects the possible different origins of ice shelves, and that scheme is utilized here. *Glacier ice shelves* result from the flotation of glacier tongues; *sea-ice ice shelves* evolve from frozen surface sea-water; and *composite ice shelves* have their origins in both glacier ice and sea-ice. Further discussion of the processes associated with ice shelves is presented in Section 8.2.3.

1.4.3.1 GLACIER ICE SHELVES

Wherever a polar or subpolar glacier terminates in a deep water body, it will float and produce a glacier ice shelf or floating glacier tongue (Fig. 1.26). The

(a)

(b)

Fig. 1.21 Cirque glaciers. (a) Glaciers in deeply incised cirques in the Lahul Himalaya, India (photo: D.J.A. Evans); (b) tropical glaciers in shallow cirques on Nevado Illimani, Cordillera Real, Bolivia (photo: D. Payne)

Fig. 1.22 Slettmarkbreen, a small cirque glacier in Norway. (Photo: D.I. Benn)

Fig. 1.24 Subpolar piedmont lobes fed by a plateau ice cap near Dobbin Bay, south-east Ellesmere Island, Arctic Canada. (Photo: D.J.A. Evans)

Fig. 1.25 A niche glacier on the northern flank of Botnaf-jellet, near Sandane, southern Norway. Note the Little Ice Age moraine and the rock glacier at the base of the slope. (Photo: D.J.A. Evans)

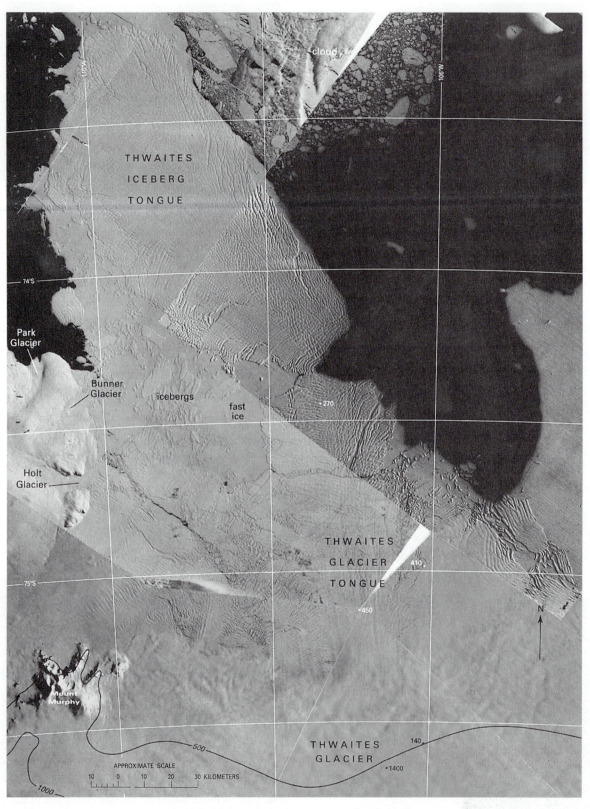

Fig. 1.26 Landsat image mosaic (1972–73) of Thwaites Glacier and Thwaites Iceberg Tongue, Antarctica. Note that the iceberg tongue has become detached from the floating glacier snout and has been rotated at its fracture point. (Image provided by Richard Williams, USGS)

largest and most numerous examples are found around the margins of the Antarctic ice sheet, such as the Amery ice shelf, which is fed by the Lambert Glacier. Smaller examples exist in the Canadian and Greenland High Arctic (e.g. the Milne ice shelf), where perennial sea-ice or re-entrants help to protect the smaller and less vigorous examples from breakup and calving (Lemmen *et al.*, 1988; Jeffries, 1986, 1987).

The ice shelves of Antarctica constitute approximately 7 per cent of the surface area of the ice sheet, make up 44 per cent of the coastline and, through calving, probably account for up to 80 per cent of the ablation total. The two largest ice shelves, the Ronne–Filchner and the Ross (Fig. 1.27), are fed by outlet glaciers from both the East and West Antarctic ice sheets (Swithinbank, 1964) but, because of their size, also accumulate snow on their surfaces and may increase in thickness at the base by bottom freezing in some areas (Section 8.2.3). The Ross, Ronne–Filchner and Amery ice shelves drain a combined area of approximately 62 per cent of the Antarctic continent, flowing at rates of 0.8–2.6 km per year. These high discharges balance the periodic large-scale calving of tabular icebergs around their perimeter, thus ensuring continued survival of the ice shelves (cf. Budd, 1966; Swithinbank, 1969, 1988; Fig. 1.28).

The seaward margins of the Antarctic ice shelves are marked by cliffs rising up to 30 m above sea-level, explaining the name of the Great Barrier given to the outer edge of the Ross ice shelf by the early explorers approaching from the sea (Swithinbank and Zumberge, 1965). The overall thickness at the seaward edges of the Antarctic ice shelves approaches 200 m, which is regarded as a minimum thickness in an unconstrained ice shelf capable of expanding in all directions, because below 200 m thickness the ice cannot effectively creep. As the ice shelves are constrained throughout most of their length, thicknesses increase in a landward direction as well as in places where ice streams enter the floating ice mass from terrestrial-outlet glaciers. Excellent satellite images and aerial photographs together with a full discussion of Antarctic ice shelf characteristics are available in Swithinbank (1988).

1.4.3.2 SEA-ICE ICE SHELVES

Sea-ice ice shelves (Fig. 1.29a) are produced by a combination of thickening by surface snow accumulation and bottom freezing of sea-water (Koenig *et al.*, 1952; Hattersley-Smith, 1955; Crary, 1960; Hattersley-Smith and Serson, 1970; Lyons *et al.*, 1971; see Section 8.2.3). Movement or deformation in a sea-ice ice shelf is induced entirely by its own weight and therefore is determined by the creep rate of the ice in question. Sea-ice ice shelves can exist only where annual temperatures are severe enough for the sea to freeze over and where embayments or a series of offshore islands provide enough anchor points for

Fig. 1.27 Oblique aerial photograph of the eastern edge of the Ross ice shelf on 22 October 1961. Large fractures are releasing icebergs and producing an ice front 20 m high. (Image provided by Richard Williams, USGS)

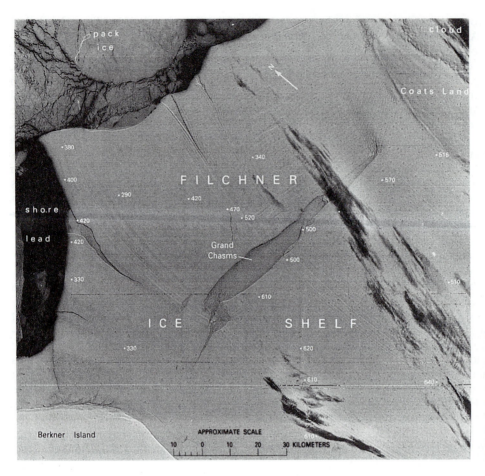

Fig. 1.28 Landsat image (11 November 1973) of the Filchner ice shelf, showing the fracture known as the Grand Chasms. The development of such fractures leads to the release of huge tabular icebergs. (Image provided by Richard Williams, USGS)

sea-ice to thicken without being disrupted by waves and currents.

The sea-ice ice shelves of the Canadian and Greenland High Arctic are recognizable by their characteristic surface 'rolls' (Hattersley-Smith, 1957a, b), which are thought to be produced by wind scouring of the surface. The rolls are made up of a series of elongate ridges and troughs which contain water during warm summers. It is these striking ridges and troughs, with their long axes aligned parallel with the predominant wind patterns, that support the wind deflation origin. The floating *ice islands*, which have been tracked during their passage within the Arctic Ocean currents since the 1950s, possess these surface rolls and originate from the Arctic ice shelves (Hattersley-Smith, 1957a; Crary, 1958; D.D. Smith, 1964; Jeffries, 1992; Fig. 1.29b).

1.4.3.3 COMPOSITE ICE SHELVES

Where floating glaciers and thickening sea-ice are protected from currents and open leads they may combine to produce a composite ice shelf. An excellent example of such an ice shelf is that at Cape Alfred Ernest on the coast of north-west Ellesmere Island in the Canadian Arctic (Lemmen *et al.*, 1988; Jeffries, 1992; Fig. 1.29c). Ice islands on the Arctic Ocean have been traced back to the Cape Alfred Ernest ice shelf through surface rock samples which originated as supraglacial debris on the valley glacier surfaces (D.D. Smith, 1964; Jeffries, 1992). A number of sea-ice and composite ice shelves exist in Antarctica but are generally less well known than their glacially fed counterparts (e.g. the Larsen ice shelf, the Wilkins ice shelf and the ice shelf occupying Prince Gustav Channel; Reece, 1950; Swithinbank, 1988). It must be remembered that the large amount of bottom freezing of sea-water and surface accumulation of snow on the outer margins of Antarctic ice shelves often disguises their glacial origins.

1.4.3.4 ICE RISES

Wherever floating ice shelves thicken to the point where they can ground on offshore shoals, the margins of islands or coastal shelves, they produce *ice rises* (Thomas, 1979b). Where the ice shelf does not possess a sufficient surface gradient to move over an obstacle, ice rises are produced by local thickening due to surface accumulation and not by overriding by

(a)

(b)

(c)

Fig. 1.29 Ice shelves and ice islands of the Canadian Arctic. (a) The Ward Hunt ice shelf on the north coast of Ellesmere Island, as it appeared in 1950. Large calving events have since reduced the size of the ice shelf by half. Note the ridge and trough pattern (rolls) picked out by surface meltwater in the troughs. (Aerial photograph by Energy, Mines and Resources, Canada) (b) An ice island floating on the Arctic Ocean and surrounded by sea-ice. This early spring view was taken before meltwater had collected in the troughs. The north coast of Ellesmere Island is visible in the distance. (Photograph by Polar Continental Shelf Project, EMR, Canada) (c) The Cape Alfred Ernest ice shelf on the Wootton Peninsula, northwest coast of Ellesmere Island, as it appeared in 1950. Large calving events have since reduced the size of the ice shelf considerably. Note the various sea-ice (largely bottom left) and glacier ice components of this composite ice shelf. (Aerial photograph by Energy, Mines and Resources, Canada)

the ice shelf. In such situations an ice rise will possess its own radial flow pattern similar to that of an ice dome, and this will be independent of the general flow direction of the ice shelf that surrounds it. Numerous examples exist within the Antarctic ice shelves, including Roosevelt Island in the Ross ice shelf and Gipps Ice Rise in the Larsen ice shelf (Fig. 1.30). An example of an ice rise in a sea-ice ice shelf is the Cape Discovery Ice Rise, which exists at the landward edge of the Ward Hunt ice shelf, Ellesmere Island, Canada (Lyons *et al.*, 1972). Wherever glacier ice shelves possess sufficient momentum to override an offshore shoal they become heavily crevassed, prompting the term *ice rumples* (Armstrong *et al.*, 1977; Fig. 1.31).

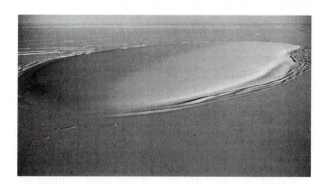

Fig. 1.30 The Gipps Ice Rise in the Larsen ice shelf, Antarctica, 1 January 1972. At that time the ice rise measured 9 × 18 km and was 300 m high. (Photo: Charles Swithinbank)

Fig. 1.31 Ice rises and ice rumples in the Wordie ice shelf, Graham Land, Antarctica (foreground). (Photograph taken on 28 November 1966 by US Navy and provided by Richard Williams, USGS)

1.5 GLACIERS AND SEA-LEVELS

1.5.1 Principles of sea-level change

The growth and decay of glaciers and ice sheets has a profound effect on global sea-level, causing very large regional and global sea-level changes over the course of glacial cycles. Sea-level rise (transgression) and fall (regression) can be caused by glaciers and ice sheets in four main ways (Figs 1.32 and 1.33):

1. *Glacioeustasy*. During glacier build-up, moisture is removed from the atmosphere in the form of snow and locked up as glacier ice. The atmospheric moisture is replaced by evaporation from the oceans, reducing global sea-level. Glacioeustasy refers to the removal of water from the oceans by glacier expansion, and its return by glacier melting. Because of the large volume of continental ice sheets during glaciations, a vast amount of water is taken out of the Earth's hydrological cycle, and global sea-levels drop by a substantial amount at such times.

2. *Glacioisostasy*. In regions occupied by ice sheets, the load placed by the ice on the Earth's crust causes it to sink down into the underlying mantle, depressing the land surface relative to sea-level. Geophysicists have been able to learn much about the characteristics of the lithosphere from its responses to ice-sheet loading and unloading (e.g. Wu and Peltier, 1983; Peltier, 1987; Lambeck, 1990, 1993a), although inferences made about mantle viscosity vary according to interpretations and availability of the glacial and sea-level data being used to constrain ice sheet reconstructions (cf. Mörner, 1980a; Peltier and Andrews, 1983).

3. *Hydroisostasy*. The force balance on the Earth's crust is also influenced by the weight of water in the oceans. Changes in the amount of water in the oceans during glacial cycles affect patterns of crustal loading in the oceans, causing regional and global sea-level changes.

4. *Geoidal eustasy*. The surface of the world's oceans is not a smooth sphere or ellipsoid, but is an equipotential surface with bulges and troughs produced by local variations in the Earth's gravitational field (Fig. 1.34). Bulges and troughs in the geoid change position and shape as the Earth's gravitational field adjusts to changing ice sheet configurations (Mörner, 1976, 1977, 1987). In addition, local isostatic adjustment may occur in response to crustal loading by large

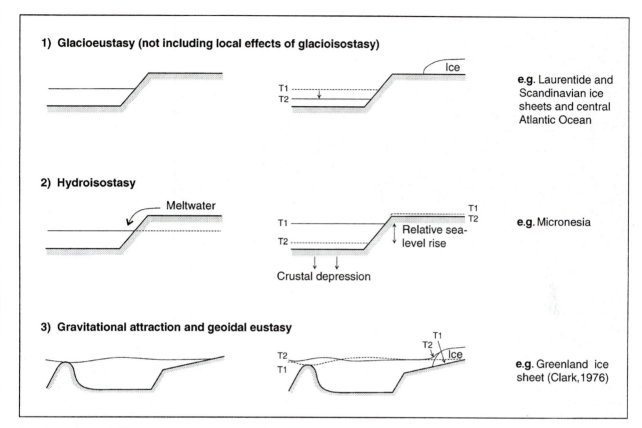

Fig. 1.32 Schematic illustrations of glacioeustasy, hydroisostasy and geoidal eustasy. (Modified from Dawson, 1992)

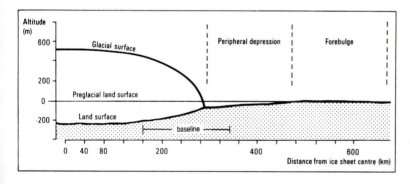

Fig. 1.33 The principle of glacioisostasy, showing the depression of the crust below an ice sheet. The baseline locates the position of the schematic shoreline diagram in Fig. 1.36b

accumulations of sediment. For example, deposition of large deltas by the Mississippi and Ganges has caused local depression of the continental shelf.

Another mechanism of sea-level change is the increase or decrease in density of water due to temperature changes (known as *steric* sea-level change). Water becomes less dense and expands as it is heated above 4°C, so that ocean warming will cause sea-level rise. This mechanism has received very little attention as a cause of Quaternary sea-level changes, but is known to be an important contributor to sea-

level rise in the twentieth century (Warrick *et al.*, 1996).

Therefore, the sea-level history of a particular site during a glacial cycle results from a complex interplay between eustatic and isostatic controls. A considerable literature now deals specifically with sea-levels and glaciation, and good overviews have been provided by Guilcher (1969), Andrews (1974), Mörner (1980b), D.E. Smith and Dawson (1983), Peltier (1987) and Tooley and Shennan (1987). In this section we outline the major principles behind sea-levels and glaciation, and provide some examples of sea-level histories controlled by glacial–interglacial cycles.

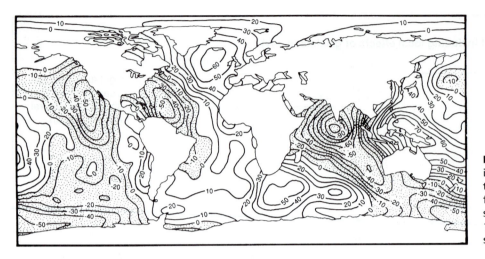

Fig. 1.34 Global geoid map in metres above and below the best-fit ellipsoidal surface. Negative areas shown shaded. (From Dawson, 1992. Reproduced by permission of Routledge)

1.5.1.1 GLACIOEUSTASY

Although regional complications are introduced by glacioisostasy, hydroisostasy and changes in the geoid (Mörner, 1976; Clark, 1980), the main control on global ocean levels on Quaternary timescales is the release and take-up of moisture by evolving ice sheets over glacial cycles. Thus, changes in the global ocean level through Quaternary time allow changes in the volume of ice sheets to be estimated (Donn *et al.*, 1962; Fairbanks, 1989; Bard *et al.*, 1990). Records of glacioeustatic sea-level change have been produced for a number of regions located at distances far enough away from former ice sheets to be unaffected by glacioisostasy (e.g. Shepard, 1960, 1963; Fairbridge, 1961; Bloom, 1967, 1970; Hopley, 1978; Cronin, 1983). It is now widely recognized, however, that global sea-level histories should be based on data from many regions, to eliminate anomalies introduced by geoidal variations, hydroisostasy, and tectonic uplift and depression (Walcott, 1972; Chappell, 1974a; Clark, 1976, 1980; Farrell and Clark, 1976; Clark *et al.*, 1978; Mörner, 1980c, 1987).

To overcome the limitations of local sea-level records, some researchers have turned to the oxygen isotope record from deep ocean sediments. Fluctuations in the concentration of the heavy isotope of oxygen (^{18}O) in foraminifera contained within deep ocean sediments have been linked to the changes in the storage of light oxygen isotopes in glaciers and ice sheets (Bradley, 1985; Section 2.4.2). Thus the oxygen isotope record provides a proxy for ice sheet volume and glacioeustatic sea-level change (Shackleton and Opdyke, 1973; Emiliani, 1978; Shackleton, 1987). However, although the curves of oxygen isotope variation through time give an impression of ice sheet and ocean volume changes, absolute water depth changes are difficult to calculate. Examples of glacioeustatic sea-level histories, which inevitably include geoidal eustatic components, are presented in Section 1.5.2.2.

1.5.1.2 GLACIOISOSTASY

The principles of glacioisostasy were first proposed by Jamieson (1865, 1882) and Shaler (1874), and other perceptive early papers on the subject were presented by De Geer (1892), Goldthwait (1908) and Nansen (1921). Considerable advances in understanding were made during the 1960s and 1970s, when numerous research programmes were undertaken on the sea-level histories of formerly glaciated regions (e.g. Andrews, 1970; Sissons, 1983; Smith and Dawson, 1983). Since that time both geomorphologists and geophysicists have been working on glacioisostasy for different but closely linked reasons. The geophysicists wish to constrain their models of lithospheric deformation using empirical evidence of recent crustal movements, and the geomorphologists want to reconstruct dispersal centres of former ice sheets. An extensive literature now exists on the technical aspects of isostatic research; a selection of important recent papers may be found in volumes edited by Mörner (1980b), D.E. Smith and Dawson (1983) and Sabadini *et al.* (1991).

The basic idea of isostasy is that the Earth's crust (with a mean density of *c.* 2800 kg m^{-3}) is floating on the underlying plastic mantle (with a mean density of *c.* 3300 kg m^{-3}), somewhat like a boat on water. The addition of mass to part of the crust should cause it to sink into the mantle, just as a boat does when cargo is loaded. This principle is reflected in the origin of the word 'isostasy', which is derived from the Greek *isostasios*, meaning 'in equipoise'. The amount of crustal depression resulting from ice sheet loading is a function of ice sheet thickness and the ratio between the densities of ice and rock. Since the density of ice is approximately one-third that of the crust, the depression created beneath an ice sheet is approximately 0.3 times the ice thickness. The amount of depression decreases from the centre of the ice sheet towards the ice sheet margins, where ice is thinner. This simple picture, however, is complicated because the lithosphere has rigidity, and the load of an ice sheet is partly borne beyond the margin. Thus, a *peripheral depression* continues beyond the ice sheet margin up to a distance of between 150 and 180 km (Fig. 1.33). This means that the sea can flood coastal sites in unglaciated areas, owing to crustal loading by an adjacent ice sheet. The resulting higher relative sea-level beyond the ice sheet margins was referred to as the *full glacial sea* by England (1983), because it is a new sea-level relating to full-glacial conditions. The magnitude of sea-level change in peripheral depressions depends partly on global glacioeustatic sea-levels.

The lateral displacement of mantle material from below the centre of ice sheet loading results in the formation of a *forebulge* or area of positive vertical displacement beyond the peripheral depression (Fig. 1.33). The distance between the ice sheet edge to the forebulge depends on the *flexural parameter* of the crust, which represents the amplitude of bending of the lithosphere and is calculated from lithospheric density, thickness, elasticity and other factors (Walcott, 1970). The vertical displacement in the region of the forebulge is not large (approximately one-hundredth the ice sheet thickness), but it does impact on the sea-level history of locations beyond the ice sheet margin. The forebulge beyond the Laurentide ice sheet is thought to have been approximately 18 m high. Because the flexural parameter is not affected by ice sheet thickness, the centre of the forebulge is located between 250 and 280 km from the ice margin regardless of ice sheet size. During deglaciation and unloading, the lithosphere rebounds beneath the ice sheet, and the forebulge collapses and migrates slowly towards the former ice centre (Brotchie and Sylvester, 1969; Walcott, 1970; Clark *et al.*, 1978; Quinlan and Beaumont, 1981). The impacts of rebound and forebulge migration on sea-level change during deglaciation are reviewed in Section 1.5.2.

Very little is known about crustal depression during ice sheet buildup, but in many glaciated regions, raised beaches and glacial deposits provide a detailed record of crustal unloading and rebound. The lithosphere responds to loading in a matter of several thousand years, although it takes in excess of 10,000 years to attain a new equilibrium following ice loading or removal. Uplift or glacioisostatic recovery in response to deglaciation has been subdivided into three periods by Andrews (1970; Fig. 1.35). (a) *Restrained rebound* takes place beneath a thinning ice sheet. Because of occupancy by ice, this period is not recorded by sea-level histories but must be inferred using geophysical models (e.g. Andrews and

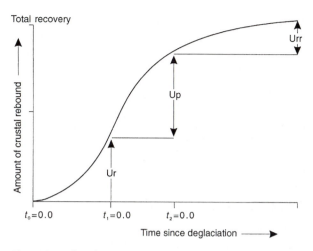

Fig. 1.35 The three sequential periods of glacioisostatic recovery. Initial restrained rebound (Ur) is followed by an accelerated period of postglacial rebound (Up). Future residual rebound (Urr) is expected to decrease through time. Initial deglaciation and crustal unloading occurs at t_0 but marine limits are not recorded in glaciated basins until t_1. (Modified from Andrews, 1970)

Peltier, 1976; Peltier and Andrews, 1983). (b) *Postglacial uplift* is the phase of rebound recorded in an area once ice withdraws. Transgressions and regressions of the sea are recorded by geomorphological and sedimentological evidence, so that sea-level histories can be reconstructed. (c) *Residual uplift* is the rebound still to take place. Because of the long response time of lithospheric recovery, some regions are still rising today in response to the disappearance of the great Pleistocene ice sheets. For example, central Scandinavia and the Hudson Bay region of Canada are still rising by as much as 1 cm per year (e.g. Eronen, 1983). More than 150 m of recovery has still to take place in Hudson Bay, implying that complete recovery may not take place between one glaciation and another.

1.5.1.3 HYDROISOSTASY

Just as the buildup of glacier ice causes depression of the underlying lithosphere, so the changing water loads in the world's oceans during glacial cycles also bring about a crustal response (Bloom, 1967; Walcott, 1972; Chappell, 1974b; Cathles, 1975). This concept of hydroisostasy suggests that oceanic crustal uplift may occur during glaciations when large volumes of water are locked up in continental ice sheets, and that the return of this water during deglaciation will result in the redepression of the oceanic crust. Very few data are available on the amount and rate of hydroisostatic responses to glacial cycles, and the concept of sea floor rebound due to water unloading has been challenged by

Mörner (1987). Some studies have inferred post-glacial hydroisostatic depression of the crust (e.g. Hopley, 1983, for Queensland, Australia), although estimates of the amount of ocean floor depression vary considerably, which is not surprising given the vast differences in the characteristics of the widely scattered study sites. Given the paucity of data, it is of no surprise that estimates of hydroisostatic impacts on global sea-level histories are mostly available only in geophysical models (e.g. Clark *et al.*, 1978; Clark, 1980).

1.5.2 Sea-level records

1.5.2.1 RECONSTRUCTING SEA-LEVEL CHANGE

A wide range of landforms and sediments can be used to reconstruct sea-level histories. Geomorphological features such as deltas, beaches, shingle ridges and erosional platforms mark former shorelines, and may be either submerged or raised relative to present sea-level. The altitude of such features does not coincide exactly with the contemporary sea-level, however. Shingle ridges, for example, may form several metres above sea-level, piled up by extreme storm events, and tend to be consistently higher than estuarine mudflats of the same age. More accurate sea-level data can be obtained from the distribution of marine sediments, which can be identified using microfossils such as diatoms. Estuarine areas and closed basins reached by the sea only at times of high sea-level are particularly suitable areas for stratigraphical studies of sea-level change (e.g. D.E. Smith *et al.*, 1985; Shennan, 1989; Shennan *et al.*, 1993). On many coasts, however, fossiliferous marine sediments are absent, and sea-level reconstructions must rely on geomorphological evidence. Submerged forests and peats also provide evidence for sea-level rise (e.g. Heyworth, 1986). The most important sea-level indicators at low latitudes are coral reefs (Hopley, 1983), which have been submerged by as much as 120 m in Barbados since the last glaciation (Fairbanks, 1989). Whatever evidence is used, the establishment of sea-level chronologies depends on the presence of datable materials. Some of the problems associated with the dating of ancient shorelines were discussed by Sutherland (1986).

In formerly glaciated regions, shorelines of a particular age are typically tilted because they have been uplifted by differing amounts depending on proximity to the centre of ice loading. The tilt of such shorelines can be represented on an *isobase map*, which shows spatial variations in the altitude of contemporaneous shorelines. An isobase, therefore, is a contour joining points of equal uplift over an equal amount of time, and represents the differential emergence of a shoreline towards the area of former maximum ice load. This concept can be illustrated using the example of an ice sheet covering an island (Fig. 1.36; D.J.A. Evans, 1991b). During deglaciation, the sea can flood isostatically depressed parts of the island, forming coastal sediments and landforms. As the land rebounds, these features are raised and tilted, with the area of maximum uplift being located where the ice sheet was thickest. The isobases shown in Fig. 1.36a are those for the 10 kyr BP sea-level: they have been interpolated for the central part of the island, which was occupied by ice at that time. Note that the isobases smooth out the irregular coverage of the ice sheet, reflecting the rigidity of the crust.

As deglaciation and uplift proceed, new shorelines are formed at progressively lower altitudes, then abandoned as relative sea-level continues to fall. Younger shorelines are tilted less steeply than older shorelines, owing to decreasing amounts of differential uplift through time. This is clearly illustrated on *equidistant shoreline diagrams*, which plot the elevations of dated shorelines against distance on a baseline aligned at right angles to the isobases (Fig. 1.36b; Andrews, 1970). Notice how progressive retreat of the ice margin allows later, lower sea-levels to penetrate further inland. This means that although the altitude of a raised shoreline of a particular age increases towards the ice sheet centre, it does not penetrate as far towards the loading centre as a younger and lower shoreline because of late deglaciation. The maximum elevation attained by the sea on any particular stretch of coast is known as the *marine limit* (Andrews, 1970). Two types of marine limit can be defined: the *regional marine limit*, which refers to the highest level reached by the sea over a large region; and the *local marine limit*, which is the maximum sea-level attained in local areas. The local marine limit may be well below the regional marine limit in localities where ice cover persists during regional isostatic uplift and relative sea-level fall. As a result, the altitude and age of the marine limit decrease as one moves from the peripheral depression into the deglaciated terrain, because a large amount of restrained rebound will have taken place inside the ice margin before the sea can invade the coast and produce a shoreline.

Sea-level histories for particular localities can be represented by *relative sea-level curves*, which are graphs showing sea-level relative to today plotted against age. A hypothetical example for a glaciated coast is shown in Fig. 1.36c, which illustrates local sea-level history for a delta formed at the ice sheet limit (site D). Eight radiocarbon dates, on shells in the delta sediments and on driftwood from beaches developed on the delta front as sea-level fell, are used

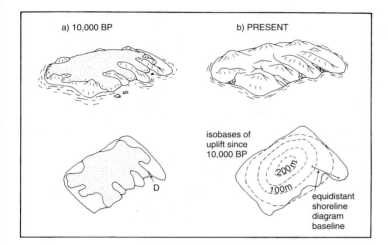

a) 10,000 BP

b) PRESENT

isobases of uplift since 10,000 BP

D

200m

100m

equidistant shoreline diagram baseline

Fig. 1.36a The glacioisostatic signature of a hypothetical island ice sheet. (a) Ice sheet maximum at 10 kyr BP. (b) After complete deglaciation and glacioisostatic rebound. The isobases of crustal recovery, which can be reconstructed using raised marine features, show greatest amounts of uplift near the centre of the island. The site marked D is the former ice-contact delta shown in Fig. 1.36c. The baseline for the equidistant shoreline diagram in Fig. 1.36b is also marked

Fig. 1.36b A hypothetical equidistant shoreline diagram reconstructed along the baseline in Figs 1.33 and 1.36a. This is drawn orthogonal to isobases with the elevations of dated shorelines plotted against distance. This shows the extent and rate of crustal rebound but may be produced only if shorelines are well dated at progressively lower elevations. This diagram has been reconstructed between 160 and 340 km from the centre of the hypothetical island ice sheet, which is a typical distance over which shorelines of different ages can be traced within a study area. D = ice-contact delta in Figure 1.36c

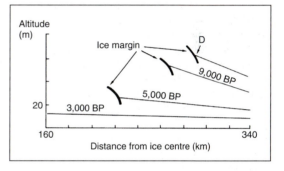

Altitude (m)

Ice margin

D

9,000 BP

5,000 BP

20

3,000 BP

160

340

Distance from ice centre (km)

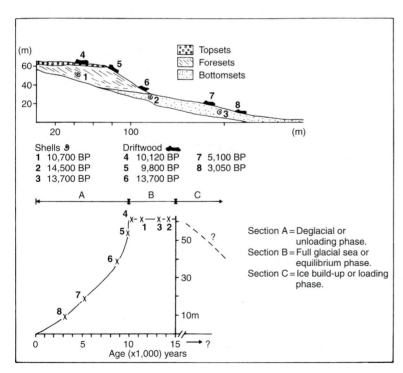

(m)

60

40

20

4

5

1

6

2

7

8

3

20

100

(m)

Topsets
Foresets
Bottomsets

Shells   Driftwood
1 10,700 BP   4 10,120 BP   7 5,100 BP
2 14,500 BP   5 9,800 BP   8 3,050 BP
3 13,700 BP   6 13,700 BP

A   B   C

4
5
1 3 2
50
?
6
30
7
8
10m

0   5   10   15   ?
Age (x1,000) years

Section A = Deglacial or unloading phase.
Section B = Full glacial sea or equilibrium phase.
Section C = Ice build-up or loading phase.

Fig. 1.36c Schematic cross-section through an ice-contact delta located at site D on Fig. 1.36a and the relative sea-level curve that can be reconstructed based upon radiocarbon-dated samples from the site. Shell samples 1–3 were collected from the foreset and bottomset beds of the delta, and relate to the sea-level recorded by the delta surface (marine limit = 61 m). The samples 4–8 date sea-level at the time driftwood was stranded

to date sea-level regression resulting from crustal rebound. The complete response of sea-level at the locality can be subdivided into three segments on the sea-level curve. (a) The *ice build-up* or *loading phase* occurs during ice sheet growth. Crustal depression causes relative sea-level rise. (b) The *equilibrium phase* is recorded by radiocarbon dates on samples 1–4, which show a period of stable high sea-level when the ice sheet was at its maximum thickness and sea-level was in equilibrium with the ice load. The range of these dates will record the length of time for which the ice sheet was at its full-glacial size, and the sea was occupying the full glacial marine limit. (c) The *deglacial* or *unloading phase* is marked by progressive relative sea-level fall, as recorded by radiocarbon dates from successively lower raised beaches on the delta front. This records crustal rebound consequent on ice sheet thinning and deglaciation.

The history of sea-level change at a given site is influenced by global eustatic and local isostatic factors, plus hydroisostatic, geoidal and tectonic influences. The relative importance of eustasy and isostasy depends upon the location of the site, particularly its proximity to ice sheet loading centres. Far from ice sheet margins, sea-level change will mainly reflect eustatic changes, whereas within ice sheet limits the isostatic signal will be dominant. Clark *et al.* (1978) and Clark (1980) subdivided the Earth's surface into six *sea-level zones*, based on characteristic postglacial relative sea-level curves (Fig. 1.37).

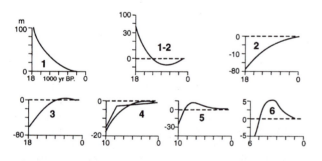

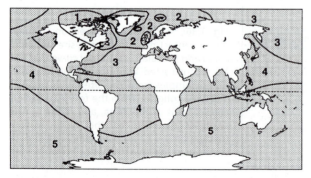

Fig. 1.37 Distribution of sea-level zones and typical relative sea-level curves. (Modified from Clark *et al.*, 1978)

Zone 1 occurs within the limits of the great Pleistocene ice sheets, and has a sea-level history of continuous emergence (marine regression) due to glacioisostatic unloading and uplift. *Zone 1–2* is a transitional zone located close to the ice sheet limits, and is characterized by initial emergence due to glacioisostatic unloading followed by submergence due to the collapse of the forebulge. *Zone 2* is located further from the limits of the great ice sheets, where sea-level history is influenced by eustatic submergence modified by forebulge collapse. *Zone 3* is still further from the ice-sheet limits, and is characterized by initial eustatic submergence followed by transitory emergence several thousand years after deglaciation. The emergence event is a delayed response to distant crustal unloading, and reflects the shape of the migrating forebulge. *Zone 4* is in the tropics and subtropics, and is characterized by continuous eustatic emergence. Sea-level changes are commonly of smaller magnitude than in Zone 2, owing to the absence of forebulge effects. *Zone 5* covers the southern oceans, where sea-level is initially controlled by eustatic submergence. Once meltwater stops flowing into the oceans (*c.* 5 kyr BP), slight emergence takes over owing to hydroisostatic effects. *Zone 6* includes all continental margins except those lying in Zone 2 and is characterized by slight emergence after meltwater ceases to flow into the oceans. This is due to the displacement of mass away from the water-loaded ocean basins to the continents. The sea-level curves in all zones may be altered by local tectonic uplift or subsidence.

A very large number of sea-level curves have been compiled over the past few decades, and good recent reviews of the extensive literature have been provided by van der Plassche (1986) and Pirazzoli (1991). In the following sections, we briefly review characteristic sea-level curves for areas where eustatic factors are predominant (Zones 3–6) and areas strongly influenced by isostatic adjustments (Zones 1, 1–2 and 2).

1.5.2.2 GLACIOEUSTATICALLY DOMINATED SEA-LEVEL HISTORIES

The *World Atlas of Holocene Sea-Level Changes* (Pirazzoli, 1991) contains numerous examples of glacioeustatically controlled sea-level histories from all over the world. We have chosen a few of those examples to illustrate the global influence of glacial cycles on ocean levels. Long sea-level curves spanning the whole of the last glacial cycle have been constructed from emerged coral reefs around the Huon Peninsula, New Guinea (Bloom *et al.*, 1974; Chappell, 1974b), and oxygen isotope concentrations in the foraminifera of deep ocean cores (Shackleton, 1987). The two curves are strikingly similar, and

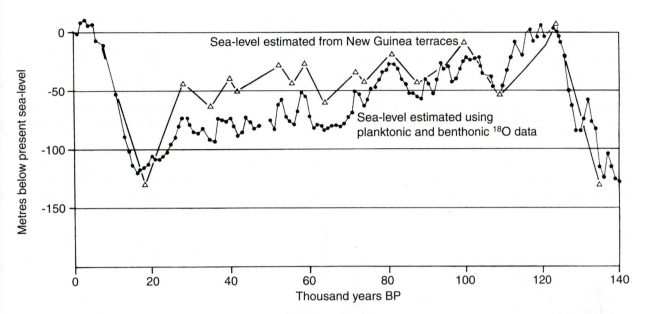

Fig. 1.38 Reconstructed eustatic sea-level histories based on coral terraces in New Guinea and oxygen isotopes from marine microfossils. (From Dawson, 1992. Reproduced by permission of Routledge)

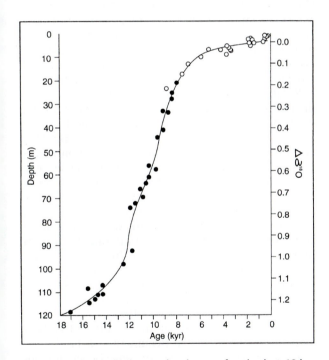

Fig. 1.39 The Barbados sea-level curve for the last 18 kyr. (Modified from Fairbanks, 1989)

show that the global sea-level varies by *c*. 120 m over the course of a glacial cycle (Fig. 1.38).

Figure 1.39 shows the sea-level curve for Barbados since the onset of the last deglaciation 18,000 years ago (Fairbanks, 1989). The curve is typical of regions located at great distances from ice sheet mar-

gins (Zones 3 and 4, Fig. 1.37), and documents submergence of 120 m since deglaciation began. Sea-level rise was fastest between *c*. 12 kyr BP and *c*. 7 kyr BP, when the mid-latitude ice sheets were rapidly receding and large amounts of water were being returned to the oceans. Since *c*. 5 kyr BP, sea-level rise has been slight.

1.5.2.3 GLACIOISOSTATICALLY DOMINATED SEA-LEVEL HISTORIES

Within areas occupied by the great Pleistocene ice sheets (Zones 1 and 1–2), sea-level change since deglaciation has been dominated by regression caused by isostatic uplift. Patterns of uplift can be determined from isobase maps, showing the altitudes of contemporaneous shoreline features. Isobase altitudes do not, however, show absolute values of uplift since shoreline formation, but the total amount of uplift minus the glacioeustatic component of sea-level change. Thus, if a shoreline at an altitude of 20 m is dated to 10,000 yrs BP (when global eustatic sea-level was *c*. 60 m lower than at present), it implies a total uplift of *c*. 80 m since shoreline formation. This approach is, of course, simplistic because it ignores hydroisostatic and geoidal components of sea-level change, but it does show that the amount of uplift recorded by raised shorelines can be very large. In addition, isobase maps can show only the amount of rebound since deglaciation and occupation of an area by the sea. The amount of restrained rebound which occurred before deglaciation cannot be measured directly, and must be reconstructed using geophysical models (e.g. Andrews and Peltier, 1976; Peltier, 1987). For example, sea-level curves in Sweden record up to 250 m of rebound since 9 kyr BP (e.g. Bergsten, 1954; Mörner, 1979; Gronlie, 1981), although the total amount of uplift was probably in excess of 700 m (Mörner, 1980a; Fig. 1.40B).

Isobase maps of formerly glaciated regions provide valuable information on the loading pattern of the crust, and therefore the positions of ice sheet dispersal centres during glaciations. Some generalized isobase maps for eastern Canada, Scandinavia and Scotland are shown in Fig. 1.40. The pattern of uplift in Canada indicates at least three loading centres, reflecting the presence of multiple domes in the

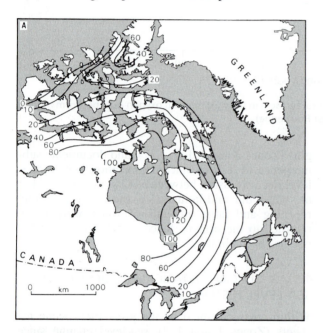

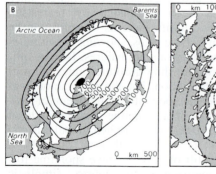

Fig. 1.40 Generalized regional isobase maps. (A) Isobase map showing shoreline emergence in eastern Canada since c. 6000 yr BP. Note the evidence for multiple loading centres. (B) Patterns of absolute uplift in Scandinavia during the Holocene. (C) Generalized isobases for the Main Postglacial Shoreline (dated to c. 7000–6000 yr BP) in Scotland. All contours in metres. (From Lowe and Walker, 1984, after Andrews, 1970, and Mörner, 1980b. Reproduced by permission of Longman)

Laurentide ice sheet. In contrast, rebound patterns in Scandinavia and Scotland are much less complex, probably because of the smaller size of the former ice sheets in these regions (Mörner, 1980b; Eronen, 1983). The domed pattern of uplift in Scandinavia is also clearly visible on the equidistant shoreline diagram in Fig. 1.41, which was constructed using marine and lake shoreline fragments (Donner, 1980).

In detail, however, isobase maps are often considerably more complex, and show that uplift patterns rarely form simple domes. Isobases may exhibit abrupt discontinuities, suggesting that rates of uplift were different on either side of fault zones. An example from Arctic Canada is shown in Fig. 1.42, where isobase patterns point to the presence of a fault in the channel between Prince of Wales Island and Somerset Island (Peel Sound), and a block of untilted terrain in northern Prince of Wales Island (Dyke *et al.*, 1991, 1992; Dyke, 1993a). Further evidence for differential uplift and faulting during isostatic recovery has been discovered in the Forth Valley and the Glen Roy area, Scotland (Sissons, 1972, 1983; Sissons and Cornish, 1982a, b). The degree of displacement across the faults is generally only a metre or so, indicating that faulting and shoreline displacement may go unnoticed unless very accurate surveys are undertaken. Indeed, faulting may be the norm where large blocks of the Earth's crust undergo rapid uplift.

Relative sea-level curves for particular sites within the limits of the Pleistocene ice sheets vary according to the position of the site relative to ice sheet loading centres (Fig. 1.43). Curves from Hudson Bay and north-east Sweden document continuous uplift since deglaciation, reflecting the location of these areas well within the limits of the Laurentide and Scandinavian ice sheets, respectively. Continuous uplift is also recorded by relative sea-level curves from Devon and Ellesmere islands in Arctic Canada, but with smaller magnitude and uplift rates, reflecting limited ice cover in these areas. In Newfoundland, sea-level curves vary from continuous emergence in the north of the island to a combination of emergence and submergence in the south and east (Grant, 1980; Quinlan and Beaumont, 1981, 1982). The influence of forebulge migration on these curves has been modelled by Quinlan and Beaumont (1981).

The varying influence of glacioeustatic and glacioisostatic factors is also evident in sea-level curves from the British Isles (Fig. 1.44 and Fig. 1.45 (Plate 6); Lambeck, 1991, 1993b, c). In the south of the country, sea-level change has been dominated by glacioeustatic submergence resulting from the melting of the large ice sheets. Although isostatic rebound did occur in this area, it was exceeded by eustatic sea-level rise, resulting in net sea-level rise. In Scotland, where the ice sheet was thickest, the

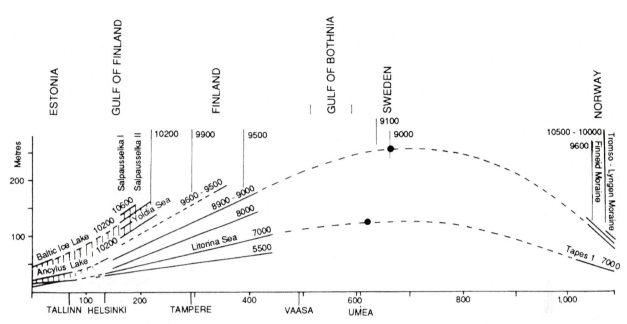

Fig. 1.41 A shoreline diagram across Fennoscandia, showing the extent of surveyed and dated shorelines. These shorelines are continued as broken lines to show the domed uplift produced by the wastage of the Scandinavian ice sheet. Some dated ice margins are included in addition to the positions of the Baltic Ice Lake and Ancylus Lake (vertical lines). (From Dawson, 1992, after Donner, 1980. Reproduced by permission of Routledge)

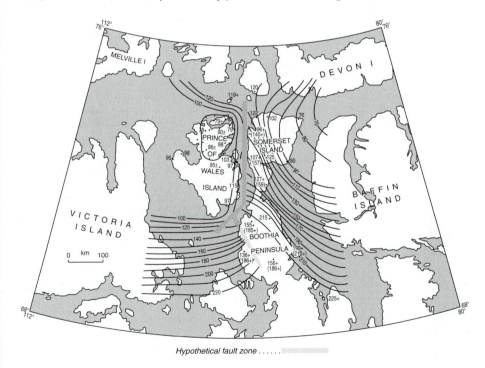

Fig. 1.42 Isobases for the 9.3 kyr BP shoreline in the central Canadian Arctic, showing the hypothetical fault zone between Prince of Wales Island and Somerset Island, and the Prince of Wales Island 'isobase plateau', which is an area that has undergone glacio-isostatic recovery without significant tilting. (From Dyke *et al.*, 1991. Reproduced by permission of the Geological Survey of Canada)

early parts of the sea-level curves record emergence because local isostatic uplift exceeded global eustatic sea-level rise at that time. By *c.* 8000 yr BP, however, isostatic uplift rates had fallen to low values, but global eustatic sea-level was rising rapidly (Fig.

1.39). As a result, a period of relative sea-level rise occurred around the coasts of Scotland, an event known as the Main Postglacial Transgression (Sissons, 1983). By *c.* 5000 yr BP, the great ice sheets had vanished and the glacioeustatic contribution to sea-

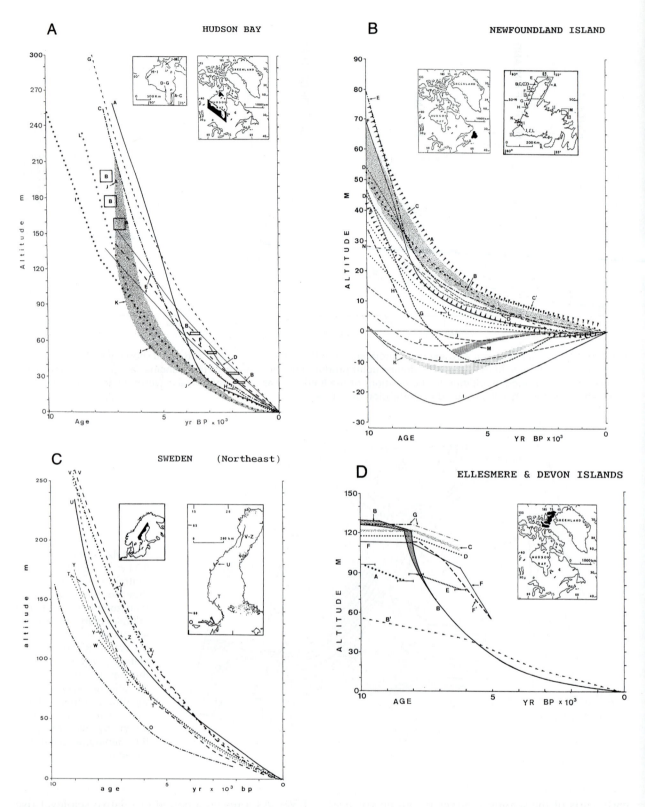

Fig. 1.43 Compilations of sea-level curves from regions affected by glacioisostatic and glacioeustatic sea-level histories during postglacial time: (A) Examples of continuous emergence from Hudson Bay (Zone 1). (B) Family of curves documenting continuous emergence to continuous submergence from Newfoundland (Zone 1–2 transition). (C) Continuous emergence of north-east Sweden (Zone 1). (D) Partial and complete curves from around Ellesmere and Devon Islands, Arctic Canada. (From Pirazzoli, 1991. Reproduced by permission of Elsevier)

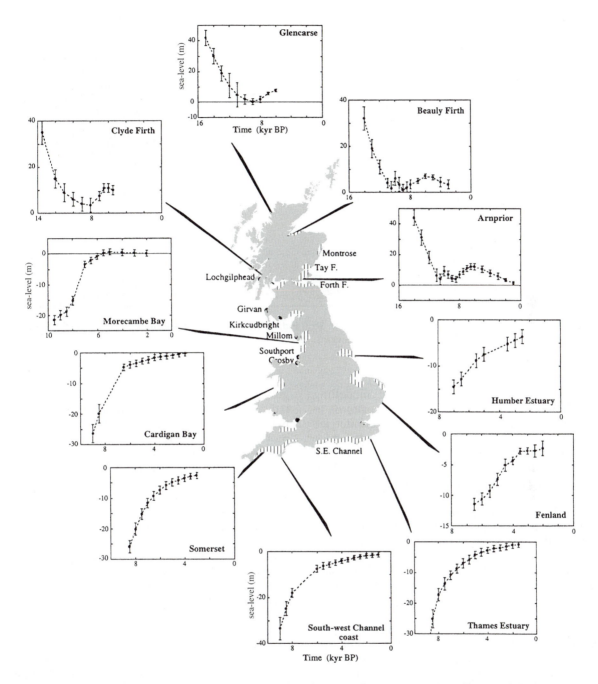

Fig. 1.44a Postglacial sea-level curves from around the coast of Britain, demonstrating the influence of glacioisostasy in the north (emergence) and glacioeustasy in the south (submergence). Although southern sites did undergo glacioisostatic rebound at the close of the last glaciation, it was not recorded above present sea-level because the sea did not rise fast enough to produce raised shorelines. (From Lambeck, 1995. Reprinted by permission of the Geological Society of London)

level change was very small. Isostatic uplift (which had continued throughout the Main Postglacial Transgression) was then able to take over once more as the main control on sea-level change. Thus, over the past few thousand years, sea-level has fallen around the coasts of Scotland, a trend that continues

today. Detailed accounts of late-glacial and post-glacial sea-level change in Britain have been provided by Donner (1970), Sissons and Brooks (1971), Sissons (1983), Firth and Haggart (1989), Haggart (1989), Shennan (1989) and Lambeck (1991, 1993b, c).

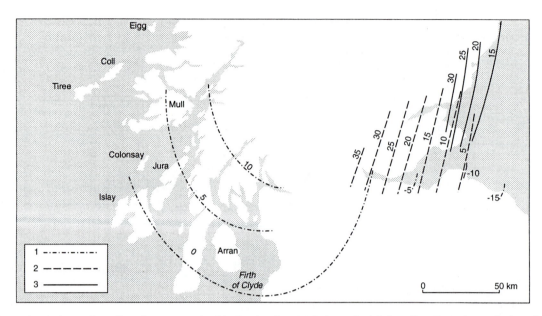

Fig. 1.44b Isobases for various shorelines recognized in Scotland. 1, Main late-glacial shoreline; 2, main Perth shoreline; 3, early shoreline in eastern Scotland. (Modified from Sissons, 1983)

Recent developments in geophysical modelling have allowed researchers to use isobase patterns to arrive at independent reconstructions of former ice sheets. For example, Lambeck (1995) has argued for the existence of large Pleistocene ice sheets over the Barents and Kara seas and the Svalbard archipelago from the presence of raised shorelines in the region. Lambeck's study extends earlier geophysical modelling work (Tushingham and Peltier, 1991), and is compatible with independent reconstructions of ice extent in the region based on glacigenic landforms and sediments (e.g. Vorren and Kristoffersen, 1986; Vorren *et al.*, 1989, 1990; Solheim *et al.*, 1990; Saettem, 1994). Theoretical isobases for the British Isles based on modelling by Lambeck are shown in Fig. 1.45 (Plate 6).

The amount of residual uplift in a region (i.e. the amount of isostatic uplift that has still to occur) can be estimated by the mapping of *free air gravity anomalies*, which are regional variations in the force of gravity (Wu and Peltier, 1983). Negative gravity values indicate that the lithosphere is undercompensated and therefore is still undergoing rebound after ice unloading. Gravity anomaly maps therefore record imprints of former ice loading centres. They commonly echo the patterns of past uplift obtained from sea-level evidence, and point to multiple loading centres below the Laurentide ice sheet but a single centre below Scandinavia (Innes *et al.*, 1968; Walcott, 1970; Balling, 1980).

The sea-level history of a region during the loading phase can be inferred from data on ice sheet growth, although reconstructed histories will differ radically according to the assumptions made regarding ice sheet buildup. For example, England (1992) provided two alternative hypotheses to account for measured amounts of crustal loading in the Canadian High Arctic (Fig. 1.46). Both hypotheses interpret sea-level history in terms of isostatic loading and the glacioeustatic sea-level curves of Fairbanks (1989) and Tushingham and Peltier (1991). Hypothesis A assumes that the ice sheet reached its maximum extent in this region by 18 kyr BP and remained constant until 8.8 kyr BP, when unloading commenced. Relative sea-level rose slowly during this interval because water was being added to the oceans by melting of the southern portions of the Laurentide ice sheet, and other ice sheets. Hypothesis B assumes that glaciers reached their maximum limits after 14 kyr BP following rapid advances from margins similar to those of the present. In this case, isostatic loading occurred later, so that relative sea-level rise must have been more rapid. In both cases, the late date of deglaciation means that eustatic sea-level rise took place before glacial unloading, resulting in maximization of the marine limit. These two hypotheses, therefore, account for the very large amounts of postglacial uplift recorded in the High Arctic without invoking extensive Pleistocene ice cover as proposed by Blake (1970, 1975, 1992b, 1993) and Tushingham (1991). Limited late Pleistocene ice cover in this region is indicated by abundant geological evidence and 'full-glacial' radiocarbon dates from raised shorelines (e.g. England, 1985, 1990; Bednarski,

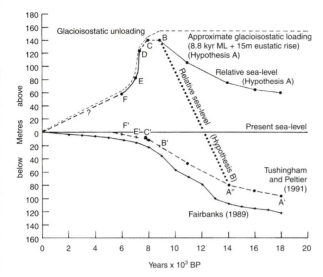

Fig. 1.46 Postglacial relative sea-level curve (C–F) and alternative hypothetical loading-phase curves (A–B and A″–B) for Greely Fiord, central Ellesmere Island, Arctic Canada. Hypothesis A (segment A–B) assumes that glacial loading was fully established by 18 kyr BP and remained constant until 8.8 kyr BP, when unloading commenced. Relative sea-level rose slowly during this interval owing to eustatic flooding (curve A′–F′). Hypothesis B (segment A″–B) assumes that glaciers advanced rapidly to the last ice limit after 14 kyr BP from margins similar to the present. This involved a more rapid transgression because glacial loading and eustatic rise occur in concert. (From England, 1992. Reproduced by permission of the *Canadian Journal of Earth Sciences*)

1986; Lemmen, 1989; D.J.A. Evans, 1990b; England *et al.*, 1991), although recent reinterpretations of the expanding regional sea-level database invoke more extensive ice coverage (England, unpublished).

1.6 PRESENT DISTRIBUTION OF GLACIERS

Glacier ice presently covers approximately 10 per cent or almost 16 million km² of the Earth's surface, most of which is contained within the Antarctic (13.5 million km²) and Greenland (2 million km²) ice sheets. The remaining 3 per cent or 500,000 km² exists as ice caps and small glaciers located at high latitudes or in mountainous regions around the world. These are located predominantly in the northern hemisphere, specifically around the Arctic Ocean basin, in mountainous maritime localities like Norway, Alaska and the Andes, and in continental high mountain terrains like the Alps and the Himalaya. The distribution of glacier ice is documented in more detail in Flint (1971), Hattersley-Smith (1974), Field

(1975), Williams and Ferrigno, (1988, 1989, 1991, 1993) and the World Glacier Monitoring Service (1989). Calculating the volume of glacier ice is more difficult, but some figures have been made available by the radio-echo sounding of some glacier masses and the reconstruction of the bed and surface profiles, providing us with another reminder of the dominance of the Antarctic ice sheet in global ice volumes (Table 1.2).

The present distribution of glaciers reflects the interplay between effective precipitation, temperature and topography described in Section 1.1. These factors vary systematically around the globe, most notably with latitude, altitude, and distance from a moisture source.

1.6.1 Influence of latitude and altitude

Other factors being equal, glaciers will be more extensive closer to the poles because the low solar angle at high latitudes means that less energy is available to melt snow and ice. Similarly, for any given latitude, the likelihood of glacier survival increases with altitude because the thinner air at high altitudes is less efficient at holding heat energy than denser, lower air. The interaction of latitudinal and altitudinal factors creates a broad global pattern of glaciation in which in equatorial regions, glaciers exist only at high altitudes, but occupy progressively lower and lower altitudes towards the poles. For example, glaciers exist only above 4000–5000 m in the tropical mountains of Irian Jaya, Indonesia (Allison and Peterson, 1989), central Africa (Young and Hastenrath, 1991) and Ecuador (Hastenrath, 1981), but glaciers can form at sea-level in high-latitude regions such as the Canadian and Greenland High Arctic (Miller *et al.*, 1975). This pattern clearly emerges in Fig. 1.47 (Plate 7), which shows present and past glacier equilibrium-line altitudes in a north–south transect through the North and South American Cordilleras. Glaciers' ELAs rise from both poles towards the equator, reaching a maximum in the subtropical Mexican Highlands and Bolivian Andes. The simple rise towards the equator is

Table 1.2 Ice volumes (water equivalent) held in the Antarctic and Greenland ice sheets, and all other ice masses

Ice mass	Volume (water equivalent)
Antarctic ice sheet	30.1 million km³
Greenland ice sheet	2.38 million km³
All other ice	180,000 km³
Vatnajökull	3100 km³

Source: Figures from Sugden and John (1976) and Drewry (1983).
Note: The volume of Vatnajökull, Iceland, is shown for comparison.

complicated by the need for effective precipitation, so that in areas with high snowfall, such as southern Alaska, the volcanic peaks of Ecuador, and the Patagonian icefields, glacier ELAs are lower than expected, whereas in more arid areas, such as the Bolivian Andes, glacier ELAs are higher.

In tectonically active terrains like the South American Andes and New Zealand, the process of mountain-building increases the area lying above the regional ELA over time. So, even if the regional climate remains quasi-stable over time, the area of land that lies at higher altitude and therefore in colder local climates increases. Uplift rates in South Island, New Zealand, are between 3.2 m and 7.8 m kyr^{-1} (Bull and Cooper, 1986), and in Ecuador the Andes have been uplifted by approximately 150 m over the last 50 kyr (Clapperton, 1987). Over longer timescales, uplift of Tibet and the Himalaya is thought to have been instrumental in global cooling and the onset of Quaternary glacial cycles (Ruddiman and Kutzbach, 1991; Raymo and Ruddiman, 1992; Beck *et al.*, 1995).

1.6.2 Influences of aspect, relief and distance from a moisture source

Each of these three variables is influential in the production of glaciers at various scales, and they interact with the larger-scale factors of latitude and altitude to bring about particular glacier–landscape relationships (Section 1.3). Obviously, mountain landscapes such as the Himalaya, the European Alps, the North American Rockies and Coast Ranges, the New Zealand Southern Alps and the South American Andes are at sufficient altitude and are large enough

to host considerable mountain icefields, but at local scales mountain shape, aspect and continentality as well as size need to be taken into account.

Although the role of aspect in glacierization is not particularly obvious at regional and continental scales (Chorlton and Lister, 1971), it plays a significant part in the production of glaciers at local scales, especially in marginal settings where the regional snowline is located just below mountain summits (I.S. Evans, 1969). Snowline altitudes may vary by several hundreds of metres in high-relief terrain, owing to the differential receipt of both solar radiation and precipitation, particularly at mid- to high latitudes. The lowest snowlines in many northern hemisphere ranges are in north-east-facing basins, owing to the fact that north-facing slopes receive the least solar radiation and north-east-facing slopes are in the lee of prevailing south-westerly winds, thereby acting as snow traps. Even in areas where the regional snowline lies above the highest summits, glaciers can be produced by snow blow (e.g. Dolgushin, 1961).

In most glaciated regions, precipitous peaks often remain free of snow despite the fact that adjacent highland plateaux at lower elevations host large icefields or ice caps (Fig. 1.48). This is a function of relief which involves summit area and shape as well as altitude. Generally, mountains that are higher in altitude will be able to accumulate enough snow to form a glacier, but narrow summits with steep sides will not have sufficient space for snow accumulation. The concept of critical summit breadth was outlined by Manley (1955), who suggested that a summit 1 km wide will host an ice dome if it rises 200 m above the local firn line, whereas a summit only

Fig. 1.48 The Dent du Géant, Mont Blanc massif, Alps. Steep mountain-tops such as this remain ice-free even though they stand above the glaciation limit. (Photo: D.I. Benn)

100 m wide will need to be up to 700 m above the firn line before snow will accumulate. In some situations mountain slopes will be too steep to hold snow, and therefore valley glaciers will be nourished by avalanche and will exist at altitudes well below the regional ELA (e.g. Andrews *et al.*, 1970; see Section 1.4.2.6). The degree of glacierization in the Home Bay and Okoa Bay areas of Baffin Island, Canada, has been used by Andrews *et al.* (1970) to devise a *shape ratio*, which is defined as the ratio of elevation to area. This ratio increases as mountains become more precipitous or more peaked. The suitability of local topography for glacier inception is best illustrated by the *glaciation level* concept (see Section 1.6.3).

Even though a particular landscape may be suitable for glacier inception, it also may lie at a considerable distance from the nearest moisture source. This means that continentality must be taken into account when assessing glacierization potential. Although continental interiors are characterized by extremely cold temperatures during the winter, they receive very small amounts of precipitation. This means that glacier mass budgets are controlled by summer temperatures in inland settings and by winter precipitation in coastal locations (e.g. Pelto, 1989; Letreguilly, 1988; Chapter 2). An example of the influence of continentality is given in Fig. 1.49, which shows that, for any given latitude, glaciers in the humid Cascade Mountains and Sierra Nevada

occupy lower altitudes than those in the drier Rocky Mountains to the east. Similarly, the rise in glaciation level in Norway is a function of distance from the ocean (Chorlton and Lister, 1971).

At high latitudes, precipitation totals may be insufficient to maintain even mountain icefields, and so the proximity to open water may be critical to glacierization. A good example of this is the distribution of permanent glacier ice in the Canadian Arctic, where large icefields on south-east Ellesmere Island and Devon Island are nourished by the open 'north water' of Baffin Bay, but ice thicknesses and glacier cover diminish in a westerly direction owing to moisture starvation (Koerner, 1977b, 1979). At an ice sheet scale, the importance of distance to moisture source has been stressed by Chorlton and Lister (1968) for the Antarctic ice sheet, where the pattern of snow accumulation appears to be influenced mostly by the nearest available open ocean water.

1.6.3 Glaciation levels or glaciation thresholds

First introduced by Parsch (1882) and Bruckner (1887), the *glaciation level* or *threshold* is a theoretical surface in a glacierized terrain which separates ice-free and ice-covered summits. In a review of glaciation levels and ELAs in the Canadian High Arctic, Miller *et al.* (1975) summarized the importance of the elevation of the glaciation level as a

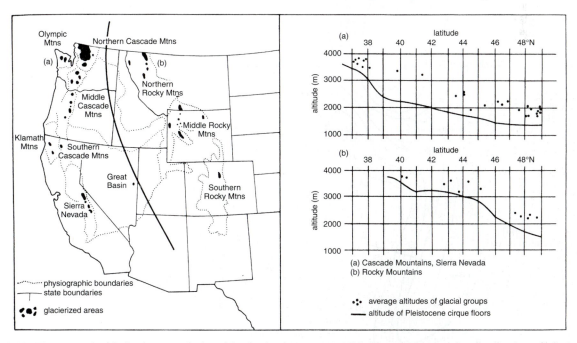

Fig. 1.49 The control of latitude over glacier altitudes in the western USA. The map shows the distribution of glaciers in two regions, separated by the heavy line. The graphs indicate that there is a steeper gradient (b) for the inland region of the Rocky Mountains than for the Cascade Mountains and Sierra Nevada of the continental periphery (a). (From Sugden and John, 1976, after Meier, 1960)

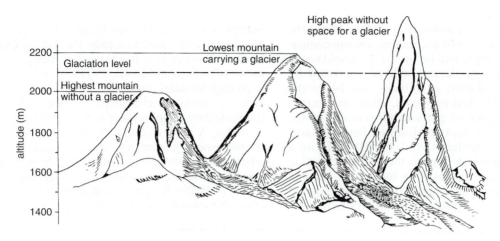

Fig. 1.50 A schematic depiction of the glaciation level in an area containing mountain summits of various shapes and altitudes. The regional glaciation level lies between the lowest suitable summit hosting a glacier and the highest suitable summit without a glacier. (From Sugden and John, 1976, after Østrem, 1966)

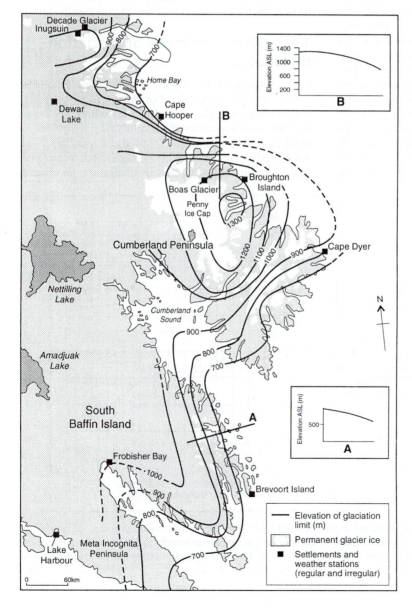

Fig. 1.51 Regional glaciation levels in southern Baffin Island. (Modified from Andrews and Miller, 1972)

provider (a) of an integrated regional climatic index, (b) of a measure of the state of glacierization of a region, and (c) of an indication of the sensitivity of an area to changes in mass balance or climate change. Østrem (1966) suggests that the glaciation level occurs at 100–400 m above the ELA.

Glaciation levels are usually derived by calculating the average difference between the highest unglacierized and lowest glacierized summits (Østrem, 1966; Fig. 1.50). Because regional variations exist, glaciation levels are calculated for localized areas or at the largest scale possible and then contoured (*isoglacihypses*) at a regional or smaller scale (e.g. Østrem, 1966; Andrews and Miller 1972; Miller *et al.*, 1975; Østrem *et al.*, 1981). The mapping of glaciation levels using the summit method is relatively quick and simple, and can provide an overview of local and regional trends, which in turn reflect the interaction of all climatic and topographic variables (Figs 1.51 and 1.52 (Plate 8)). For example, Østrem (1966) demonstrated a rise in the regional glaciation level with distance from temperate west coasts. This relationship simply reflects the turnover rate of mass in more temperate glaciers whereby the large inputs of precipitation on west coasts result in the extension of glacier snouts into very low altitudes to enable melting. In a reassessment of the glaciation level of the southern Coast Mountains of British Columbia, Canada, I.S. Evans (1990) provides evidence for an *all-sided glaciation level* which lies up to 300 m higher than local glaciation levels. The all-sided glaciation level defines the altitude at which glaciers will be formed on all aspects rather than just the northerly slopes of mountains.

1.7 PAST DISTRIBUTIONS OF GLACIERS

During glaciations, glacier ice may cover around 30 per cent of the Earth's surface, and this has had profound effects in shaping the surface features of vast areas that now lie well beyond glacierized terrains. Although reconstructing palaeoenvironments and explaining climate change are not aims of this book, this section will provide a brief overview of past distributions and fluctuations of glacier ice in order to provide a context for following chapters on ancient landforms and sediments.

1.7.1 The pre-Quaternary

A general temperature curve for the Earth (Fig. 1.53) shows that conditions have been favourable for glaciation during several geological periods prior to the Quaternary (Frakes, 1979; Young, 1991). In fact,

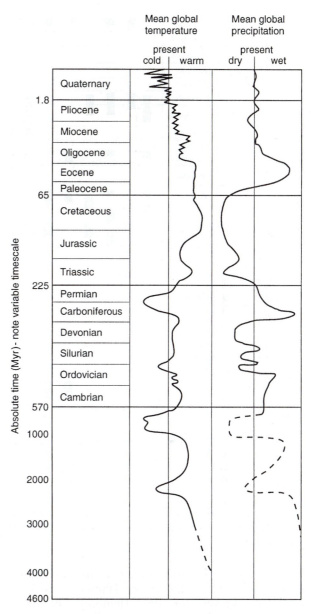

Fig. 1.53 General temperature and precipitation curves for the Earth. (Modified from Bradley, 1985)

the Earth has probably continually experienced some form of glaciation throughout the whole of its history, but styles of glaciation have ranged from continental ice coverage to restricted mountain ice caps (Fig. 1.54). Unlike in the case of the glaciations of the Quaternary Period, we have to look more to geological changes, such as continental drift and mountain-building, than to external forcing mechanisms, such as orbital parameters or solar variability, in order to explain the timing of major glaciations during older geological periods. Numerous studies have been undertaken on glacial sediments dating to various geological periods, and several reviews of the

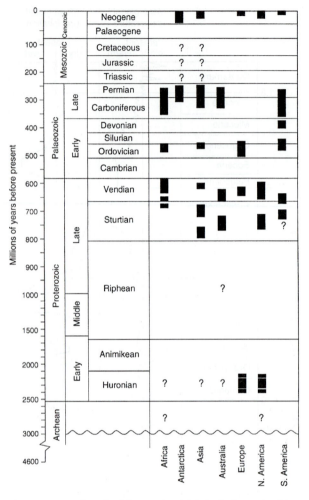

Fig. 1.54 Earth's glacial record, which includes only those deposits that are unequivocally glacial in origin. (Modified from N. Eyles, 1993)

pre-Quaternary glacial record are available (Harland and Herod, 1975; John, 1979a; Hambrey and Harland, 1981; Frakes *et al.*, 1992; Hambrey, 1992; Deynoux *et al.*, 1993; N. Eyles, 1993). Here, we briefly review the Earth's pre-Quaternary glacial record with an emphasis on glacier distributions. Good reviews on the geological controls on glaciation are provided by N. Eyles and Young (1993) and N. Eyles (1993).

Evidence for glaciation in the pre-Quaternary rock record is dominated by sediments referred to as *tillites*, *mixtites* or *diamictites*, which are lithified versions of poorly sorted glacial sediments (Harland *et al.*, 1966; Boulton and Deynoux, 1981; Fig. 1.55). Recent sedimentological work on such rocks has shown that they have a variety of origins, and were deposited both beneath the former glaciers and beyond the margins, often in deep water (e.g. Lindsey, 1971; Hambrey, 1982; Visser, 1983a, b, 1994; Dowdeswell *et al.*, 1985; Visser *et al.*, 1987; C.H.

Eyles, 1988a, b; Frakes and Francis, 1988; C.H. Eyles and Lagoe, 1990; Matsch and Ojakangas, 1991; Young and Gostin, 1991; see Chapters 8, 10 and 11). Sediments deposited by glaciers into the sea (*glacimarine sediments*) provide a particularly valuable record of ancient glaciations because they tend to have higher preservation potential than terrestrial sediments (Gravenor *et al.*, 1984; C.H. Eyles *et al.*, 1985).

The oldest evidence for glaciation dates to the Precambrian Eon (>4600–570 Myr BP). Within the Precambrian sequence the *Huronian Glaciation* dates to the Lower Proterozoic and is represented by three formations of glacial sediments, which are approximately 12,000 m thick and located in the Lake Huron region of Canada. The uppermost formation (*Gowganda Formation*) is the most extensive of the sediments equating to this period and covers more than 120,000 km^2 (Fairbairn *et al.*, 1969; Lindsey, 1971; Young and Nesbitt, 1985). Possible ice coverage at this time is depicted in Fig. 1.56a, showing an ice sheet on the Canadian Shield, which was located at high latitudes during the glaciation, as well as ice masses of unknown extent in southern Africa and western Australia. A recent review of the Gowganda Formation by C.H. Eyles *et al.* (1985) suggests that all the glacial sediments contained therein are subaqueous in origin (e.g. Miall, 1983a, 1985b), a viewpoint that questions the extent of the ice sheet in Fig. 1.56a, since glacimarine sediments may extend much further than the ice itself. Evidence for Late Precambrian or Upper Proterozoic glaciation is available in nearly every one of the Earth's present continents (Harland and Herod, 1975; Spencer, 1975; Hambrey and Harland, 1981; Hambrey, 1983; N. Eyles and Young, 1993; Fig. 1.56b). Some of the most exhaustively studied rocks relating to this period are those of the *Port Askaig Formation* in Scotland (Bjorlykke, 1969; Spencer, 1971, 1981; C.H. Eyles and N. Eyles, 1983a; C.H. Eyles, 1988a) and the *Gaskiers Formation* of Newfoundland, Canada (Williams and King, 1979; Gravenor, 1980), which are again interpreted as being predominantly subaqueous in origin. Terrestrial subglacial tillites dating to the Upper Proterozoic, often associated with subaqueous deposits, occur in Brazil (Pflug and Scholl, 1975; Rocha-Campos and Hasui, 1981; Gravenor and Monteiro, 1983) and other ancient continental settings, but studies and debate on their origin continue (Schermerhorn, 1974, 1975; Deynoux and Trompette, 1976; N. Eyles and Miall, 1984; Gravenor *et al.*, 1984; C.H. Eyles *et al.*, 1985). There is uncertainty about the exact distribution of the Upper Proterozoic ice bodies, but there is evidence for several glaciations, the most recent of which may have spanned the Precambrian–Cambrian boundary (Bjorlykke *et al.*, 1967; Harland and Herod, 1975). Although ice sheets may have per-

AGE	Trench	Forearc	Backarc	Foreland	Intracratonic/ Aulacogenic	Passive Margin
L. CENOZOIC <36 Myr		Gulf of Alaska	Bransfield Strait		Kleszczow Basin Poland, North Sea Rift, Ross Sea/Weddell Sea Rift, Alaskan Interior	Eastern Canadian Continental margin / N.W. European Continental margin
L. PALAEOZOIC 350-250 Myr		Palaeo-Pacific margin of Gondwana: Eastern Australia Antarctica		Karoo Basin S. Africa	Kalahari Basin, Arabian Peninsula, Parana Basin, Brazil, Indian Basins, Australian Interior Basins	
ORDOVICIAN c.400 Myr				West Africa? Central Saudi Arabia		
L. PROTEROZOIC 800-550 Myr		Damara mobile belt, Arabian shield, Paraguay-Araguaia fold belt, Tiddiline basin, North Africa	Gaskiers F'M. NFLD Boston Bay Group	Bakoye GP Jbeliat GP West Africa		Palaeo-Atlantic margin of Laurentia, Paleo-Pacific margin of Laurentia
E. PROTEROZOIC 2,100-1,800 Myr						Huronian supergroup: Gowganda FM
ARCHEAN >2,500 Myr				Witwatersrand Basin S. Africa		

AGE / TECTONIC SETTING OF BASIN

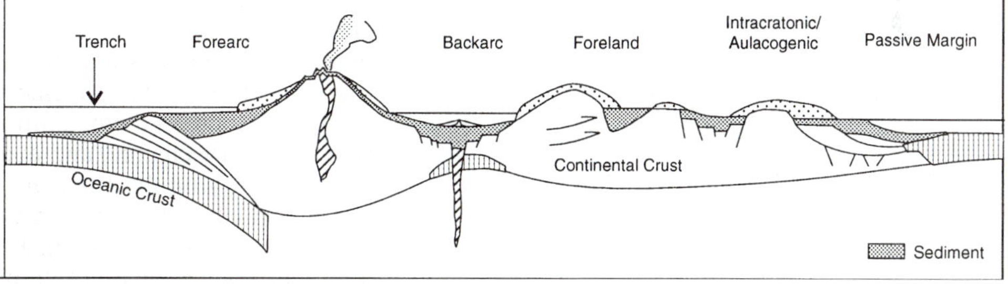

Trench — Forearc — Backarc — Foreland — Intracratonic/Aulacogenic — Passive Margin

Oceanic Crust — Continental Crust — Sediment

Fig. 1.55 The tectonic setting of Earth's glacial record. Glacial successions are placed according to age and tectonic setting. (Modified from N. Eyles, 1993)

sisted in polar regions for a large part of the Upper Proterozoic, continental drift and mountain-building episodes brought about the inception and decay of numerous mountain icefields and continental ice sheets at various times; early notions of a single ice sheet of global extent are not likely to be accurate given the large changes in palaeogeography that took place.

Abundant evidence exists for glaciations throughout the Palaeozoic era (570 Myr–230 Myr BP), specifically during Ordovician, Devonian and Permo-Carboniferous times. The most convincing evidence for Ordovician glaciation lies in Africa, which was centred on the South Pole during Ordovician times (Fairbridge, 1970, 1974, 1979; Bannacef et al., 1971; Harland, 1972; Allen, 1975; Biju-Duval et al., 1981) and hosted a large polar ice sheet consisting of coalescent domes; the ice sheet complex may have been 20 million km² in area, which is some 6 million km² larger than the present Antarctic ice sheet (Fig. 1.57). Further evidence for Ordovician glaciation has been reported from South America, southern Africa, Scotland and North America (Harland and Herod, 1975), suggesting that glacial condi-

tions were not restricted to the south polar region. Evidence for glaciation during the Devonian occurs in South America and southern Africa but is fragmentary, telling us very little about the distribution of glacier ice at this time. Probably the most impact made by any of the pre-Quaternary glaciations was that of the Permo-Carboniferous Gondwanaland ice sheet. This vast ice mass was centred on what is now southern Africa and Antarctica, and glacier flow was directed radially over what are now the coasts of eastern South America, southern Africa, southern Australia and southern India (Fig. 1.58); smaller ice masses were located at the margins of Gondwanaland. The glacier body probably comprised a series of coalescent domes, which at their maximum covered approximately twice the area covered by the present Antarctic ice sheet. The stratigraphic record suggests that the ice coverage of the Permo-Carboniferous fluctuated, and the glacial sediments on the various segments of the former Gondwanaland were probably not laid down at the same time (Adie, 1975). Hundreds of metres of glacial sediments dating to the Permo-Carboniferous have been studied in regions as diverse as South Africa (*Dwyka tillite*;

Hamilton and Krinsley, 1967; Stratten, 1969, 1971; Visser, 1983a, b, 1991; Visser and Hall, 1985; Visser *et al.*, 1987), South America (Frakes and Crowell, 1969; Frakes *et al.*, 1969), India (A.J. Smith, 1963; Casshyap and Qidwai, 1974), Australia (Harris, 1981) and Antarctica (Lindsey, 1971; Matsch and Ojakangas, 1991), recording both terrestrial and sub-aqueous deposition (N. Eyles, 1993) in association with an ice sheet complex that behaved much like the Quaternary and recent continental ice masses (e.g. Hollin, 1969) but whose marine margins changed configuration as Gondwanaland drifted to more northerly latitudes (John, 1979b).

Although we often view the present glacial period as belonging to the Quaternary, the development of our present glacier ice cover began during the Tertiary Period (Denton and Armstrong, 1969; Turekian, 1971), perhaps as a result of tectonic uplift of mid-latitude regions such as the Tibetan Plateau, the Himalaya and the western Cordillera of the Americas (Ruddiman *et al.*, 1989; Kutzbach *et al.*, 1989; Behrendt and Cooper, 1991). Uplift may have initiated increased weathering and erosion and the release of minerals capable of taking up CO_2, thus creating an anti-greenhouse effect. This tectonic model has been questioned, however, by Molnar and England (1990), who suggest that the evidence for uplift (e.g. increased erosion) may actually be a function of climate change, and hence there is a problem of misidentifying the chicken and the egg.

The growth of the Antarctic ice sheet heralded the beginning of Cenozoic glacier growth on the planet after a prolonged phase of global warmth throughout the Mesozoic Era and early Tertiary Period, but the

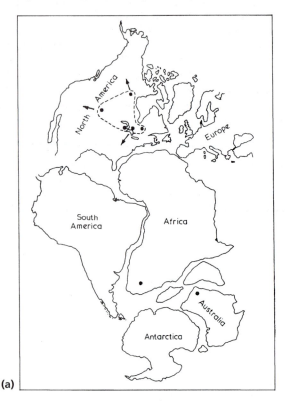

(a)

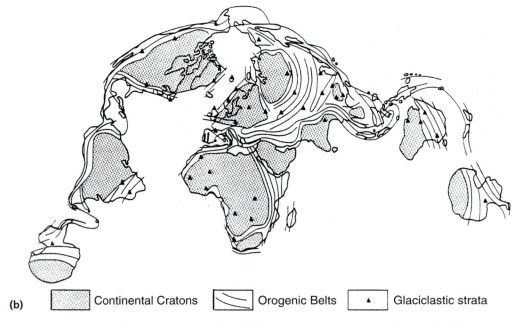

(b)   Continental Cratons   Orogenic Belts   ▲ Glaciclastic strata

Fig. 1.56 Evidence of Precambrian glaciation. (a) The continental arrangement of the Lower Proterozoic, showing the location of glacial sediments (black dots) and a possible North American ice sheet. (From John, 1979. Reprinted by permission of David & Charles) (b) The distribution of Upper Proterozoic glacial sediments. (From N. Eyles 1993. Reproduced by permission of Elsevier)

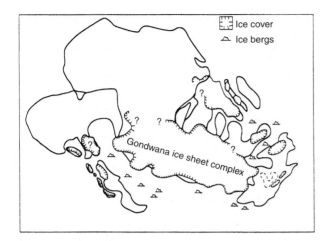

Fig. 1.57 A reconstruction of the Late Ordovician ice sheet. (From N. Eyles, 1993. Reproduced by permission of Elsevier)

development history of the ice sheet is a subject of considerable debate. The offshore glacigenic sediment record has been interpreted by some researchers as an indicator of almost continuous glaciation of Antarctica for the past 40 Myr (e.g. Harwood, 1986; Barrett, 1989; Hambrey *et al.*, 1989, 1992), whereas others prefer a shorter history involving only the past 10–5 Myr (e.g. Shackleton and Kennett, 1975; Drewry, 1978; Leg 113 Shipboard Scientific Party, 1987). These two viewpoints imply a maximum early Tertiary (Eocene) age and a minimum late Miocene/early Pliocene age for ice sheet development.

Since its early development, the volume of the Antarctic ice has certainly fluctuated, and some researchers suggest that it may have disappeared completely at times. Collapse is certainly easier to envisage for the West Antarctic ice sheet than for the East Antarctic ice sheet. This is because of the large

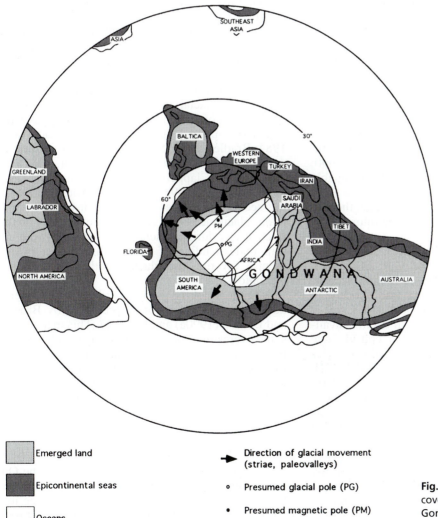

Emerged land

Epicontinental seas

Oceans

Ice-cap

→ Direction of glacial movement (striae, paleovalleys)

∘ Presumed glacial pole (PG)

• Presumed magnetic pole (PM)

Fig. 1.58 A reconstruction of the coverage of the Permo-Carboniferous Gondwanaland ice sheet and glacially influenced marine basins. (From N. Eyles, 1993. Reproduced by permission of Elsevier)

marine-based components of the West Antarctic ice sheet, which make it vulnerable to sea-level rise (e.g. Hughes, 1973; Sugden, 1988). However, collapse of the East Antarctic ice sheet during Pliocene warm intervals has recently been suggested on the basis of interpretations of the Sirius Group glacial sediments, which crop out at high elevations in the Transantarctic Mountains. The Sirius Group sediments contain reworked marine diatoms of Pliocene age, and their mountain summit locations have been explained by glacial transport from sediment-filled depressions. These depressions may have acted as seaways prior to 3 Myr ago, implying that the East Antarctic ice sheet had disappeared at that time (Webb *et al.*, 1984; Webb and Harwood, 1991; Barrett *et al.*, 1992). This interpretation has been challenged in a series of recent papers (Denton *et al.*, 1993; Hall *et al.*, 1993; Huybrechts, 1993; Kennett and Hodell, 1993; Marchant *et al.*, 1993a, b; Sugden *et al.*, 1993; Wilch *et al.*, 1993) which provide geomorphological, geochronological and palaeoclimatological evidence for East Antarctic ice sheet stability since 4.4 million years ago. Denton *et al.* (1993) go further to suggest that, since its inundation of the Dry Valleys in the mid-Miocene Epoch, the East Antarctic ice sheet underwent only minor oscillations throughout the Pliocene, and that presently its Dry Valley outlets occupy a landscape which is inherited from an ancient period of semi-arid erosion. The settlement of this debate between the 'dynamicists' and 'stabilists' (reviewed by Clapperton and Sugden, 1990; Wilson, 1995) would certainly provide invaluable information on the stability of the Antarctic ice sheet during future warm phases of global climate, thereby making a very clear justification for the future study of glaciers and palaeoclimate!

The most impressive stratigraphic record of late Tertiary glaciation is that of the Gulf of Alaska, where the 5 km thick Yakataga formation records glacimarine sedimentation from the late Miocene to the Quaternary (Plafker and Addicott, 1976; Eyles *et al.*, 1991). Excellent sedimentological detail is available throughout the sequence, and a considerable amount of information on glacial depositional styles has been collected from it (Armentrout, 1983; C.H. Eyles, 1987, 1988b; C.H. Eyles and Lagoe, 1990). Glacier growth on the Gulf of Alaska coast has been a consequence of local tectonic uplift, which has been a product of the convergence between the Pacific and North American plates; regional climatic cooling since the late Tertiary has also been effective in the development of permanent ice in the freshly uplifted terrain (Plafker, 1987; Eyles *et al.*, 1991).

Evidence from Alaska and Antarctica, therefore, demonstrates that our present glacier cover began to develop well before the Quaternary. Furthermore, although we can demonstrate that the past 2 Myr of

Earth history has been dominated by glacial–interglacial cycles, glacier ice has been a regular feature of the Earth's surface since the Precambrian; the relative stability of the present Antarctic ice sheet is an example of glacier–climate interactions on a long timescale (millions of years), and although it is affected by external or *exogenic* climate forcing mechanisms, its inception and longevity have been the product of internal or *endogenic* processes, specifically continental drift, for the past 55 Myr (Kennett, 1978).

1.7.2 The Quaternary

From interpretations of ocean cores and long sequences of loess (wind-blown dust) in China, it is apparent that as many as 21 glacial cycles may have affected the Earth's surface during the Quaternary (Dansgaard and Tauber, 1969; van Donk, 1976; Kukla, 1987; Ding *et al.*, 1993; Imbrie *et al.*, 1993; Fig. 1.3). Unfortunately, most terrestrial evidence for the majority of these glaciations has been destroyed by subsequent events, and our geomorphological and stratigraphic record for glaciation is usually restricted to less than four events. It is beyond the scope of this book to present all of the evidence for Quaternary glacial–interglacial cycles, and exhaustive reviews of such material are available elsewhere (e.g. Denton and Hughes, 1981; Lowe and Walker, 1984; Bradley, 1985; Sibrava *et al.*, 1986; Fulton and Andrews, 1987; Ruddiman and Wright, 1987; Fulton, 1989; Clapperton, 1990a; Dawson, 1992; Williams *et al.*, 1993; Calkin and Young, 1995). In this section we concentrate on the distribution of the major glacier ice bodies and comment briefly on their fluctuations during glacial cycles. Clearly, our present understanding of glacier distribution is more precise for the last glaciation than any previous glaciations.

A summary of the global distribution of glacier ice during the last glaciation is presented by Broecker and Denton (1990a) and Fig. 1.47. This map includes some overestimations of ice coverage, and is not supported by field evidence of glacier margins dating to the last glaciation in some areas (e.g. the Canadian High Arctic). However, the map does provide an impression of typical glacier coverage during a glacial cycle.

1.7.2.1 ANTARCTICA

It is most appropriate to open a section on the Quaternary glacier ice cover of the Earth with reference to the Antarctic ice sheet, because it is an ice mass whose history spans the Pliocene–Quaternary boundary. Radiometric dates on nested moraine systems indicate that the East Antarctic ice sheet has fluctuated by only a modest amount during the Quaternary. For example, the moraines and drift in the area of the Ross Embayment and Transantarctic Mountains record maximum expansions of certain outlet glaciers by less than 5 km since the Pliocene (Mercer, 1972; Hendy *et al.*, 1979; Bockheim *et al.*, 1989; Denton *et al.*, 1989a, b; Marchant *et al.*, 1993b; Fig. 1.59). Some glaciers terminating on dry land, like the Taylor and Hart glaciers, appear to have advanced during global interglacials and interstadials, because they drain the polar plateau or occupy parts of the McMurdo Dry Valleys where precipitation increases during warmer periods. Indeed, some glaciers in the area are advancing at the present day in response to the higher precipitation totals of the present inter-

glacial (Denton *et al.*, 1989a). In contrast, the major outlet glaciers which float in the sea have responded in phase with global glacial–interglacial cycles. This is thought to be a response to sea-level fluctuations rather than direct temperature/precipitation controls on the ice sheet (Clapperton and Sugden, 1990). During global glacial maxima, sea-levels drop, allowing Antarctic outlet glaciers to stabilize and expand, whereas higher sea-levels during interglacials destabilize the outlets, causing them to retreat rapidly by calving (Hollin, 1962; Wilson, 1978; Stuiver *et al.*, 1981; Section 8.2). This *glacioeustatic* control on the marine margins of the Antarctic ice sheet may have allowed it to advance to the edge of the continental shelf in most areas (Denton and Hughes, 1981b), but reconstructions of the exact extent of grounded ice during successive Quaternary glaciations vary according to the interpretations of offshore sediment assemblages (cf. Drewry, 1979; Kellogg *et al.*, 1979; Hambrey *et al.*, 1989, 1992; Colhoun, 1991).

1.7.2.2 NORTHERN HEMISPHERE ICE SHEETS

In the northern hemisphere, the most significant glacial events have been the growth and decay of the large mid-latitude ice sheets in Scandinavia, northern Asia and North America. Although we often assume that the Greenland ice sheet is a permanent feature on the Earth's surface, there is now evidence to suggest that it may have almost completely disappeared during the last interglacial at 130 kyr BP (Koerner, 1989). Considerably less is known about the oscillations of mountain icefields in the Himalaya and Tibet, but glacier ice volume changes have been substantial in these regions also.

Maximum estimates of the size of the North American Laurentide ice sheet during glaciations suggest that it grows to be the largest ice body on the planet's surface, constituting more than one-third of the world's glacier cover during any one glaciation and having a volume of anywhere between 18 and 35 million km^3 (Hughes *et al.*, 1981; Fisher *et al.*, 1985). Therefore, its inception and disintegration both have considerable impacts on the global climate system and global sea-levels. Because of the paucity of data pertaining to older glaciations, we here present reconstructions of the ice sheet only for the *Wisconsinan* or last glaciation. The ice sheet may have begun its development at the close of the Sangamonian (last) interglacial, and its coverage may have been more extensive during the early Wisconsinan than at any subsequent time (Vincent and Prest, 1987). However, ice sheet extent during various stages of the Wisconsinan glaciation remains a contentious issue owing to the ambiguity of local and regional databases. Nonetheless, the reconstruction of Dyke and Prest (1987) represents

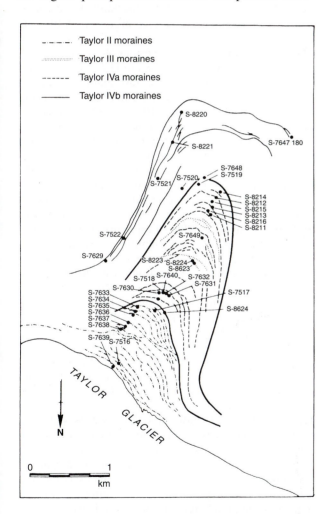

Fig. 1.59 Moraines of the Taylor Glacier in Arena Valley, Ross Embayment, Antarctica. The Taylor IV Drift is thought to be of early oxygen isotope stage 7 or earlier age, the Taylor III Moraines date to oxygen isotope stage 7, and the Taylor II Drift is of oxygen isotope stage 5 age. Note that each glacial depositional phase dates to a global interglacial period. (From Marchant *et al.*, 1993b. Reproduced by permission of Scandinavian University Press)

a commendable effort to reconcile these ambiguities, and has been widely cited.

Dyke and Prest's (1987) reconstruction for 18 kyr BP (Fig. 1.60a (Plate 9)) depicts a Laurentide ice sheet that was separated from the High Arctic Franklin Ice Complex (England, 1976a) by the Lancaster Sound ice shelf. Further ice shelves buttressed the ice margin along the Labrador and Maine coasts and Amundsen Gulf and Melville Sound. The ice sheet possessed three sectors or areas of inception, referred to as the *Keewatin Sector* (comprising *Plains* and *Keewatin Ice*), the *Labrador Sector* (comprising *Hudson* and *Labrador Ice*) and the *Baffin Sector* (comprising *Foxe, Amadjuak* and *Penny Ice*; Dyke *et al.*, 1989). The shrinkage of the ice sheet during the late Quaternary and early Holocene can be followed on Fig. 1.60b–d (Plate 9). The maximum extent of the ice sheet was not synchronous throughout, indicating the time-transgressive response of the ice sheet to the climatic forcing of the last glacial cycle (Fig. 1.10). Of critical significance to the glacial geology and geomorphology of North America is the positioning and migration of the major ice divides and saddles throughout the growth and recession phases of the Laurentide ice sheet. This changing ice sheet geometry created a complex assemblage of landforms, discussed in detail in Section 12.4.4. It has also been suggested that the heart of the Laurentide ice sheet in the Hudson Bay area was somewhat unstable during the Wisconsinan glaciation, in that it underwent a number of collapse stages, owing to either climatic warming or an inherent tendency to periodic unstable advances or *surges* (Andrews *et al.*, 1983; Dredge and Thorleifson, 1987; Clark *et al.*, 1993; MacAyeal, 1993a, b).

The extent of the North American Cordilleran ice sheet is also depicted in Fig. 1.60a (Plate 9), where it is shown as being coalescent with the Laurentide ice sheet along a large portion of its eastern margin. In the west, the ice sheet fed fjord glaciers, which covered large areas of the continental shelf and advanced into the Puget lowlands and Strait of Juan de Fuca in Washington State (Clague, 1989; Easterbrook, 1992). To the south, large glacier lobes also penetrated into Idaho and Montana (Waitt and Thorson, 1983; Richmond, 1986). In the north, the Cordilleran ice sheet coalesced with the expanded mountain icefields of Alaska, which had extended on to the continental shelf (Hamilton and Thorson, 1983), and smaller and largely independent mountain icefields developed in the Mackenzie Mountains of the Northwest Territories (Duk-Rodkin and Hughes, 1992), where the deposits of four separate glaciations have been documented (Hughes *et al.*, 1989). The extent of the Cordilleran ice sheet shown in Fig. 1.60a (Plate 9) remains a contentious issue, however, because some researchers maintain that a corridor

existed for a considerable distance between the two ice sheets during at least the last glaciation (Reeves, 1973; Stalker, 1980; Rutter, 1980, 1984). Such a corridor could conceivably be a function of the differential response rates of the two ice sheets. Differential responses have been demonstrated stratigraphically in southern Alberta, where Laurentide tills consistently overlie Cordilleran tills (Alley, 1973; Stalker, 1976b; Stalker and Harrison, 1977), but non-coalescence of the two ice masses is difficult to prove geomorphologically and geochronologically (MacDonald *et al.*, 1987; Bobrowsky and Rutter, 1992). Although multiple Laurentide and Cordilleran till units have been reported by numerous researchers (e.g. Tharin, 1969; Harris and Boydell, 1972; Alley, 1973; Roed, 1975; Stalker, 1976b; Stalker and Harrison, 1977), recent studies have argued for only one extensive Laurentide glaciation of the westernmost plains in central Alberta (Liverman *et al.*, 1989).

In the Canadian and Greenland High Arctic, maximum reconstructions of the last glaciation depict coalescent Laurentide, Queen Elizabeth Island and Greenland ice (e.g. Blake, 1977, 1992b; Hughes *et al.*, 1977; Denton and Hughes, 1981a; Hughes, 1986). Although Greenland and Ellesmere Island ice has been coalescent during older glaciations (England *et al.*, 1981; England and Bednarski, 1989), the field evidence supports a more restricted or minimum reconstruction of glacier cover during the last glaciation (e.g. England, 1978, 1985, 1990; Hodgson, 1985; Bednarski, 1986; Lemmen, 1989; Evans, 1990), and such a reconstruction has been used by Dyke and Prest (1987) in their maps (Fig. 1.60a (Plate 9)). Such modest responses by high-latitude ice masses is a predictable one, given the aridity of the polar deserts in which they are located, and may be compared to the oscillations of the McMurdo Dry Valley glaciers discussed earlier in this section.

Although the oscillations of the Greenland ice sheet may be large over glacial–interglacial cycles (Koerner, 1989), its margins cannot advance far from their present positions on to the continental shelf before becoming destabilized by deeper water. Extensive moraines thought to date to the last glaciation have been found just offshore around most of the coast of Greenland (Funder, 1989), but reconstructions of the last glaciation in the arid north part of Greenland invoke only modest advances (England, 1985) because of precipitation starvation, as for the adjacent Canadian Arctic archipelago. During an earlier glaciation, the Greenland ice sheet did manage to cross Nares Strait and inundate the north-east corner of Ellesmere Island (England *et al.*, 1981; Retelle, 1986a; England and Bednarski, 1989); glacial erosional forms on the islands in Smith Sound, between Ellesmere Island and Greenland, and shell-bearing till on Carey Oer in Baffin Bay have been ascribed by

Blake (1992) to an ice stream in the coalescence zone of a Canadian High Arctic (*Innuitian*) ice sheet and an expanded Greenland ice sheet.

In Eurasia, ice sheets developed over Britain, Scandinavia, the Barents and Kara Shelves, and parts of northern Siberia. The maximum extent of these ice masses remains contentious, however, some researchers arguing that all were confluent, and others preferring more restricted ice cover consisting of independent ice masses (e.g. Denton and Hughes, 1983; Hughes *et al.*, 1985; Hughes, 1986; Dawson, 1992; Grosswald and Hughes, 1995). Abundant evidence exists for the expansion of the Scandinavian ice sheet into northern Germany, northern Poland and the Baltic states of Estonia, Latvia and Lithuania (Andersen, 1981), but the limits further east remain uncertain. Marine ice sheets developed on the Barents and Kara Shelves, which dammed the northward-draining rivers of the west Siberian lowlands to produce proglacial lakes covering up to 1.5 million

km^2 (Grosswald, 1980; Arkhipov *et al.*, 1986a, b; Baker *et al.*, 1993; Rudoy and Baker, 1993; Fig. 1.61). The Eurasian ice cover also reversed the drainage of the main central European rivers so that they began draining to the Caspian, Black and Mediterranean Seas (Arkhipov *et al.*, 1995). Evidence of former glacier flow emanating from the New Siberian Islands and moving on land in eastern Siberia is used by Grosswald (1984, 1988) and Grosswald and Hughes (1995) to suggest that the Barents/Kara Shelf ice sheet complex was coalescent with a Laptev/East Siberian Sea ice sheet and that this ice extended across to the Alaskan coast. Geochronological constraints on ice-marginal deposits in a number of regions do not support such maximum models for the last glaciation, and debates on ice coverage at that time continue (cf. Grosswald, 1980, 1984; Velichko *et al.*, 1984, 1989; Grosswald and Hughes, 1995; Rutter, 1995). However, compelling geomorphological evidence, largely in the

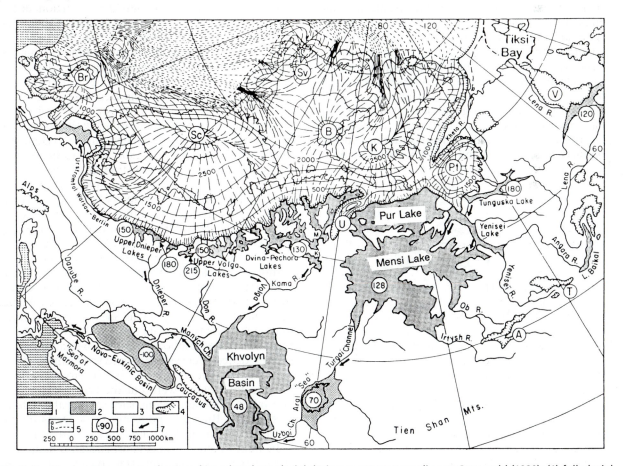

Fig. 1.61 The last Eurasian ice sheet and its related proglacial drainage system according to Grosswald (1980): (1) full glacial sea surface; (2) ice-dammed and other lakes; (3) ice-free land; (4) boundaries of ice sheets and mountain icefields; (5a) lines of ice flow on terrestrial and grounded marine ice; (5b) lines of ice flow on floating glacier ice; (6) altitude of surface of inland lakes in metres; (7) directions of stream flow; (M) Mylva Channel; (K) Keltma Channel. The ice domes and mountain icefields are: (Br) British; (Sc) Scandinavian; (B) Barents; (Sv) Svalbard; (K) Kara; (U) Uralian; (Pt) Putorana Plateau; (V) Verkhoyansk Mountains; (T) Tuva-Sayan Mountains; (A) Altai Mountains. (From Rutter 1995. Reproduced by permission of Elsevier)

form of extensive large end-moraine systems (e.g. Punkari, 1995), does attest to extensive ice coverage during older glaciations, and the palaeogeography reconstructed in Fig. 1.62 must be viewed as an all-time maximum margin comprising diachronous segments.

In Britain, unequivocal geomorphological evidence exists for only two Quaternary glaciations: the *Anglian*, when ice diverted the course of the River Thames southwards (Rose, 1983; Ehlers and Gibbard, 1991), and the *Devensian* (Rose, 1989a; Jones and Keen, 1993). Evidence for an intervening *Wolstonian* glaciation is presently disputed (e.g. Shotton, 1983; Rose, 1987; Rice and Douglas, 1991; Gibbard *et al.*, 1992; Jones and Keen, 1993), even though an extensive glaciation at that time (the *Saalian*) is known to have inundated large areas of northern Europe and, just as the Anglian glaciation diverted the Thames, brought about significant changes in the courses of the north-west European rivers (Gibbard, 1988; Ehlers, 1990). Considerable debate has surrounded the question of coalescence of the British and Scandinavian ice sheets during the last (Dimlington Stadial) glaciation. Certainly the British and Scandinavian ice sheets grew independently from the amalgamation of highland ice masses during the early stages of glacial cycles, but alternative interpretations of the glacial geological/geomorphological record advocate different full-glacial reconstructions. Maximum reconstructions depict confluent Scandinavian and British ice with a confluence zone positioned over the northern North Sea (Boulton *et al.*, 1977, 1985). The occurrence of Scandinavian erratics in British tills and offshore moraine belts certainly lends support to such reconstructions. Furthermore, deeply incised subglacial meltwater channels on the floor of the North Sea have been assigned a last glaciation age by Ehlers and Wingfield (1991), suggesting that the two ice sheets did coalesce. Conversely, a number of recent studies indicate that ice sheet coalescence occurred prior to the last British ice sheet maximum at *c.* 18 kyr BP. Quaternary core stratigraphies from the North Sea indicate that coalescence occurred around 130–200 kyr BP, but later a large area of dry continental shelf, exposed by the glacioeustatic sea-level drop, existed between the Scandinavian and British ice sheets (Sejrup *et al.*, 1987). This is supported by the ice sheet profile reconstructions of Nesje *et al.* (1988), Nesje and Sejrup (1988) and Nesje and Dahl (1990), based upon blockfield distributions in Norway, as well as the ice-marginal positions identified by Andersen (1979), Sutherland (1984a), Cameron *et al.* (1987), Long *et al.* (1988) and Balson and Jeffery (1991) based upon offshore sediment–landform assemblages (Figs 1.63 and 1.64).

1.7.2.3 MOUNTAIN GLACIERS

In addition to the continental-scale ice sheets, numerous mountain glaciers expanded during glacial periods to produce large icefields. Such icefields were located in all the major mountain ranges of the world, including the European Alps (Andersen, 1981; Schlüchter, 1986), the Brooks Range, northern Alaska (Hollin and Schilling, 1981), the Andes (Clapperton, 1993), the Himalaya (Shroder *et al.*, 1993), and the Yenisei–Taymyr region of eastern Russia (Andersen, 1981; Arkhipov *et al.*, 1986a). In addition, icefields on the upland plateau of Iceland coalesced to form a large ice cap which extended to the shelf edge at its maximum configuration (Andersen, 1981). Claims by Kuhle (1985, 1986, 1988b) that large areas of the Tibetan Plateau were covered by an ice cap conflict with most interpretations of the field evidence (Derbyshire *et al.*, 1991; Rutter, 1995), and most researchers agree that Pleistocene glaciations in this region were confined to the Himalaya and other mountain chains surrounding the Tibetan Plateau (e.g. Burbank and Kang, 1991; Derbyshire *et al.*, 1991; Kalvoda, 1992; Owen *et al.*, 1996a).

In South America and New Zealand, some evidence exists for the initiation of glaciers in the Pliocene Epoch (Gage, 1961; Mercer, 1976; Rabassa and Clapperton, 1990; Suggate, 1990). During the Quaternary Period, glaciers have expanded and contracted largely in phase with their larger northern hemisphere counterparts. The most substantial

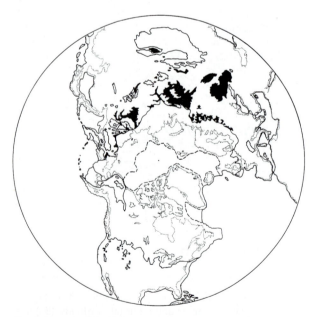

Fig. 1.62 Possible Pleistocene maximum of glacier ice cover in the northern hemisphere. The eastern part of this reconstruction is highly controversial. (Modified from Grosswald and Hughes, 1995)

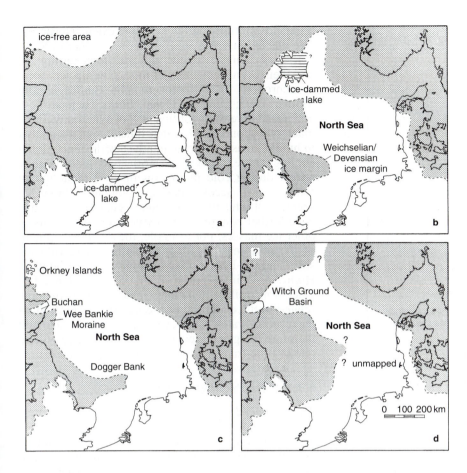

ice-free area

ice-dammed lake

ice-dammed lake

North Sea

Weichselian/ Devensian ice margin

Orkney Islands

Buchan
Wee Bankie Moraine
North Sea

Dogger Bank

a

b

c

Witch Ground Basin

North Sea

? unmapped

0   100   200 km

d

Fig. 1.63 Alternative interpretations of the Weichselian/Devensian glacier margins in the North Sea: (a) Valentin (1957); (b) Jansen (1976); (c) Long et al. (1988); (d) Ehlers and Wingfield (1991). (Reproduced by permission of John Wiley & Sons)

expansions have involved the icefields of the New Zealand Southern Alps and Patagonia. In parts of Patagonia the Quaternary sequence includes interbedded glacial sediments and lavas, providing particularly accurate reconstructions of former glacier fluctuations (Feruglio, 1944). Extensive and well-preserved moraine systems also provide clearly defined margins of former glaciers, especially on the eastern side of the Andes, where glaciers terminated on land rather than as tidewater fronts (Caldenius, 1932; Mercer, 1976; Porter, 1981a; Clapperton, 1993; Clapperton et al., 1995). The extent of the Quaternary glaciations in southernmost South America is depicted in Fig. 1.65, which illustrates the importance of the Andean mountain chain to the accumulation of glacier ice during cold climate stages. The buildup of the Patagonian icefield during the last glaciation was modelled by Hulton et al. (1994), who identified a steep rise in the palaeo-ELA of 4 m km^{-1} from west to east and a more restricted ice advance on the east of the Andes, interpreted as a product of the reduction in precipitation as westerly winds crossed the mountains. The New Zealand Southern Alps provide a physiographic/climatic setting similar to that of the Andes in that they are a young, tectonically active mountain chain receiving

high precipitation totals from the moisture-laden westerlies. The limits of five separate glaciations have been identified on the Quaternary geology map of New Zealand (New Zealand Geological Survey, 1973); however, it is the limits of the last *Otiran glaciation* that are most prominent and appear to document several depositional phases. The palaeo-ELAs of the Tasman and nearby glaciers during these phases were reconstructed by Porter (1975b) and compared with historical (1970–72) ELAs for the same area (Fig. 1.66). Restricted glacial coverage in other areas of the southern hemisphere is reviewed by Colhoun and Fitzsimons (1990) and Galloway (1963) for Australia, by Hall (1990) and Clapperton (1990c) for the islands of the Southern Ocean, and by Rosqvist (1990) for Africa.

In all the glaciated mountain regions of the world, evidence of progressively younger ice advances is partially preserved in the shape of inset moraine sequences and associated landform/sediment assemblages (e.g. Shroder et al., 1993, for the Himalaya and Clapperton, 1993, for the Andes). This may represent the preservation of only part of the landform record in that evidence for restricted early glacial advances has been destroyed by later extensive advances. Alternatively, landscape evolution, particu-

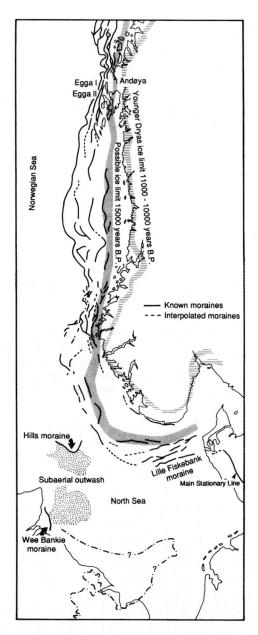

Fig. 1.64 Major moraine systems documenting the former margin of the Scandinavian ice sheet. (From Dawson, 1992. Reproduced by permission of Routledge)

highlands, initially producing more extensive successive glaciations.

Recent dating evidence indicates that, during the late Pleistocene, advances of mountain glaciers were asynchronous in different regions (Gillespie and Molnar, 1995; Benn and Owen, 1997). There are several possible reasons for this lack of synchrony. First, glaciers respond to different climatic signals, so that in humid areas glaciers may advance during a global cooling phase, whereas in arid areas glaciers may retreat during cold phases owing to a reduction in precipitation. Second, the development of large ice sheets will alter regional climate patterns and introduce further complications in the response signals of other ice masses. For example, the late development of the north-eastern margins of the Laurentide ice sheet and the Franklin Ice Complex of the Canadian High Arctic in response to the cooling of the last glaciation reflect changing precipitation patterns related to the evolution of the ice sheet. Third, glacier response rates vary according to glacier size and therefore the glacial record of climate change will be one of time-transgressive response.

1.7.2.4 THE YOUNGER DRYAS

The *Younger Dryas* is the name given to a cold interval (or stade) that occurred near the end of the last glaciation, approximately 11–10 kyr ago, immediately prior to the beginning of the present interglacial period (the *Holocene*). The name comes from *Dryas octopetala* or mountain avens, an attractive white and yellow flower related to the strawberry, which is prominent in the pollen records of those times for lowland Scandinavia. The Younger Dryas was particularly marked around the North Atlantic region, where glaciers readvanced from the restricted locations they had occupied following the last glacial maximum.

It has been argued by Wright (1989) that the Younger Dryas cooling was caused by the release of large volumes of cold meltwater to the North Atlantic from icebergs from the Scandinavian and Siberian ice sheets as well as from the southern margins of the Laurentide ice sheet. Recently, much research has been focused on the role played by drainage of Lake Agassiz, a lake ponded up along the southern edge of the Laurentide ice sheet (e.g. Broecker *et al.*, 1989; Broecker and Denton, 1990a, b; Teller, 1990; Fig. 1.60b (Plate 9)). Initially, the overspill from the lake drained southwards via the Mississippi into the Gulf of Mexico, but about 11,000 radiocarbon years ago the retreating ice allowed a new escape route to open up to the east, through the Great Lakes region and along the St Lawrence River. It has been argued that huge discharges from the lake into the North Atlantic effectively diverted the North Atlantic Drift to a more

larly tectonic uplift, may be capable of producing sequentially less extensive glacier advances through time. For example, a reduction in the areal extent of ice in the Himalayan region through time may be a function of its remarkable uplift of 3000 m during the Quaternary, which has probably acted as a barrier to the summer monsoon, reducing precipitation in the continental interior. This situation is the end result of an evolutionary sequence which probably started off with the uplift of the Himalayan/Tibetan

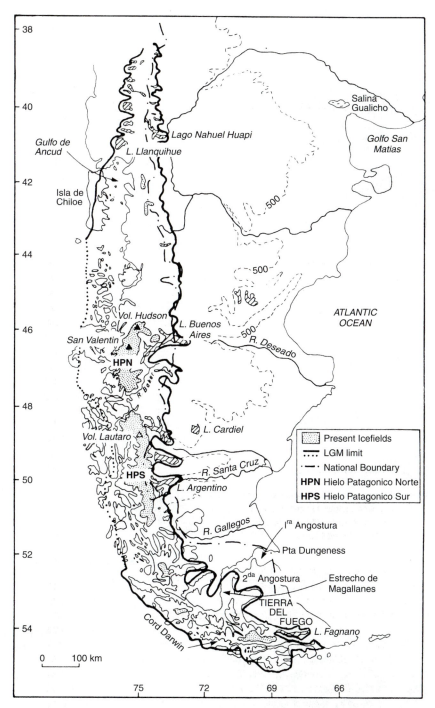

Fig. 1.65 Glacial limits in southern South America. (From Clapperton, 1993. Reproduced by permission of Elsevier)

southerly location, thus initiating a cooling of North Atlantic ocean water, which was transmitted by the atmosphere to the region as a whole. A readvance of the Laurentide ice sheet after 10.4 kyr BP (Teller, 1987) rediverted the Lake Agassiz drainage to the Gulf of Mexico, thereby shutting off the influx of cold water to the North Atlantic and contributing to the close of the Younger Dryas. The reduction in icebergs from the margins of the Scandinavian and Siberian ice sheets, which had retreated from the marine environment, may also have been influential. Although the Gulf of St Lawrence drainage route

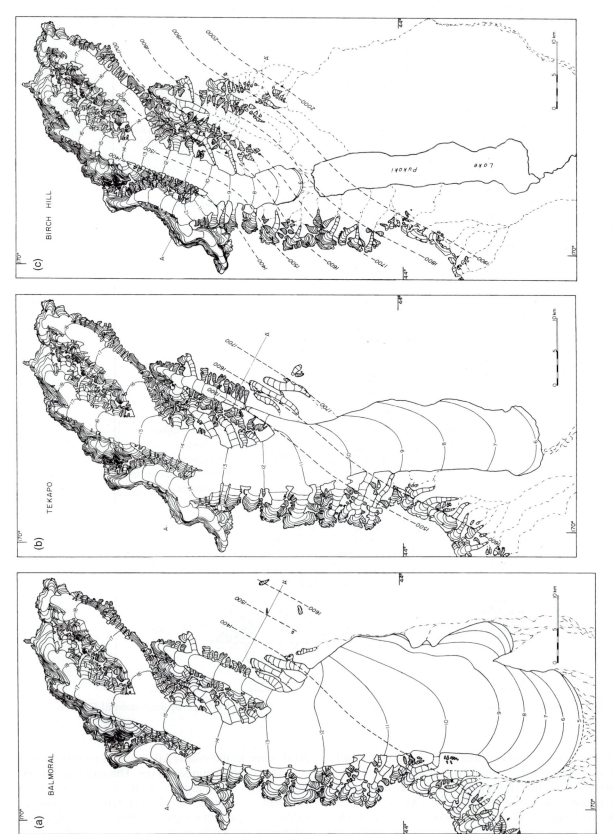

Fig. 1.66 Reconstructions of glacier extent, surface topography and palaeo-ELAs for the Tasman River – Lake Pukaki drainage basin, South Island, New Zealand. Dashed contours represent ELAs (metres). (a) Balmoral Advance (60–28 kyr BP). (b) Tekapo Advance (14.5–13 kyr BP). (c) Birch Hill Advance (12–8 kyr BP). (From Porter,

was reopened owing to ice retreat by 9.5 kyr BP, discharges were perhaps too low to reproduce the diversion of the North Atlantic Drift.

The Younger Dryas shares many characteristics with earlier short-lived cold episodes that punctuated the last glaciation in the North Atlantic region, suggesting that they may share a common origin. The Greenland ice core record for the last glaciation shows that colder than average conditions, known as *Dansgaard–Oeschger events*, occurred every 2000–3000 years during the last glaciation. Each cold event consisted of a gradual cooling followed by a rapid warming, and several culminated in massive discharges of icebergs (Heinrich events) into the North Atlantic, thought to reflect the breakup of advancing ice sheet margins (Section 1.2; Bond *et al.*, 1993; Bond and Lotti, 1995). The Dansgaard–Oeschger events have been attributed to flow instabilities, or surges, of the North American ice sheet, but recent research suggests that ice margins throughout the region advanced simultaneously during many of the events, and that an oceanic or climatic trigger is more likely. Possibilities include oscillations of the 'Atlantic conveyor' due to ocean–ice sheet interactions, or some hitherto unknown climatic cycle, possibly driven by changes in solar radiation. It is therefore possible that the Younger Dryas is part of a larger, little-understood pattern, rather than a unique event requiring a special explanation.

In the British Isles, upland glaciers and icefields readvanced or re-formed during the Younger Dryas (locally known as the *Loch Lomond Stade*; Sissons, 1979a; Gray and Coxon, 1991). During this period some of the most extensive terrestrial moraine belts in the British Isles were constructed at the margins of retreating cirque and icefield outlet glaciers. The largest icefield at this time was the Highland glacier complex in Scotland (e.g. Sissons, 1977a; Thorp, 1986; Bennett and Boulton, 1993a), but smaller icefields and glaciers occupied the mountainous terrain of the Hebridean islands (e.g. Ballantyne, 1989; Benn *et al.*, 1992; Fig. 1.67) and Cairngorms (Sissons, 1980a) of Scotland, the English Lake District (Sissons, 1979b, 1980b) and Pennines (Mitchell, 1991), Snowdonia (Gray, 1982a) and the Brecon Beacons (Shakesby and Matthews, 1993) in Wales, and the Mourne, Wicklow, Donegal and Connemara Mountains and Macgillicuddy's Reeks in Ireland (Gray and Coxon, 1991).

In Scandinavia, the ice sheet margins readvanced, forming the extensive and well-developed Salpausselkä moraines of southern Finland and the Ra moraines of south-western Norway (Donner, 1969, 1978, 1982; Ignatius *et al.*, 1980; Eronen, 1983; Bjorck and Digerfeldt, 1986). Although the Younger Dryas was originally thought to be a strictly European event, evidence is now mounting for glacier

reactivation elsewhere. Grant and King (1984), Mott *et al.* (1986) and Stea and Mott (1989) have argued that glaciers readvanced in the Maritime Provinces of North America, and Dubois and Dionne (1985) concluded that the Quebec North Shore moraine was formed at the south-east margin of the Laurentide ice sheet at this time. Evidence of Younger Dryas ice advances has also been found in the mountains of western North America (Clague, 1975; Porter, 1978; Armstrong, 1981), South America (Mercer and Palacios, 1977; Clapperton, 1985; Clapperton and McEwan, 1985) and New Zealand (Porter, 1975b), suggesting that it may have been a global climatic shift rather than one restricted to areas around the North Atlantic (cf. Rind *et al.*, 1986). However, the causes of such a global climatic event remain obscure, and await further research.

1.7.2.5 RETREAT OF THE LAURENTIDE AND SCANDINAVIAN ICE SHEETS

By the end of the Younger Dryas at 10 kyr BP, the Laurentide ice sheet had retreated to a position of great instability caused by the increasing size and depth of water bodies at the ice sheet margins (Fig. 1.60b (Plate 9)). At approximately 8–9 kyr BP, an extensive moraine system, referred to as the Cockburn Moraines, was deposited by the Laurentide ice sheet after it had largely retreated to terrestrial positions along the Keewatin, Baffin Island and Labrador coasts (Fig. 1.60d (Plate 9); Falconer *et al.*, 1965), and within 1000 years the ice sheet had been reduced to mere remnants in Keewatin, Labrador and Quebec, and Baffin Island. This very rapid demise was due to massive calving within Hudson Strait and Hudson Bay (Dyke and Prest, 1987; Andrews, 1989; Dyke and Dredge, 1989; Vincent, 1989). Remnants of the Laurentide ice sheet still exist today in the form of the Barnes and Penny Ice Caps on Baffin Island, and continue to undergo retreat in our present interglacial climate.

In Europe, the retreat of the Scandinavian ice sheet was also rapid, owing to the presence of various large proglacial lakes and high sea-levels in the area of the Baltic and Gulf of Bothnia. During the Younger Dryas, a large freshwater lake (the *Baltic Ice Lake*) was dammed between the southern ice margin and the north European mainland (Fig. 1.68a; Donner, 1969, 1978, 1982; Ignatius *et al.*, 1980; Eronen, 1983; Bjorck and Digerfeldt, 1986). By 10,000 yr BP (the *Yoldia Stage*), the ice sheet had retreated far enough to allow the flow of marine waters into the Baltic (Fig. 1.68b). Rapid glacier retreat between 10,000 and 9000 yr BP reduced the ice sheet to a remnant dome in northern Sweden, and an extensive freshwater lake, *Ancylus Lake*, had formed in the area of the Baltic and

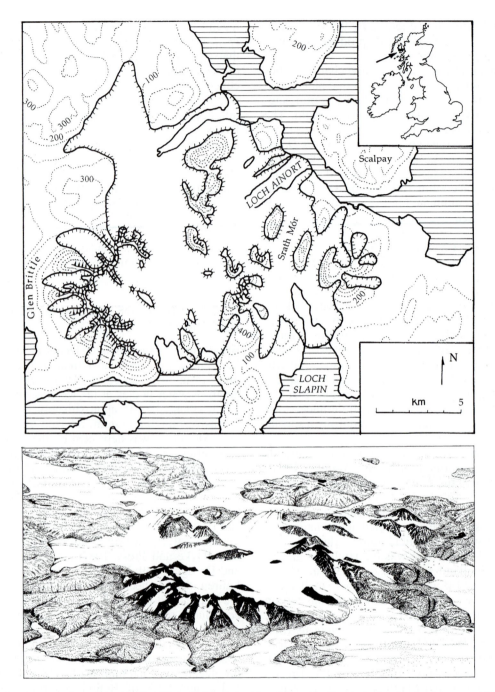

Fig. 1.67 A reconstruction of the Cuillin Hills icefield on the Isle of Skye, Scotland, during the Loch Lomond Stadial (Younger Dryas). The lower diagram is an oblique view over Skye as it would have appeared during the Loch Lomond stadial. (From Ballantyne, 1989. Reproduced by permission of John Wiley & Sons)

Gulf of Bothnia (Fig. 1.68c). The reversion to lake conditions in the Baltic was probably caused by the closing of its marine connection channel in central Sweden due to glacioisostatic rebound; the Baltic did not revert back to marine conditions again until a new outlet was developed in the Danish Sound (*Litorina Sea*; Fig. 1.68d; Eronen, 1983).

1.7.2.6 NEOGLACIATION

As we saw in Section 1.7.2.5, retreat of the great northern hemisphere ice sheets continued into the Holocene Epoch, which is the current phase of the Quaternary Period and began approximately 10,000 years ago. The term *neoglaciation* was introduced by

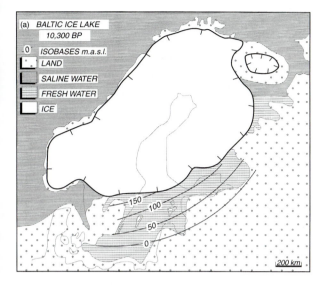

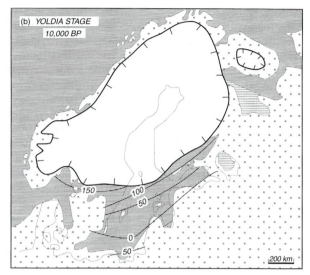

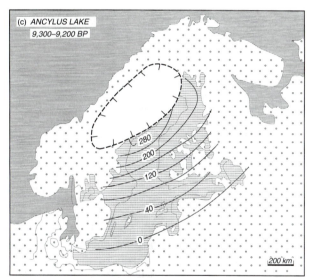

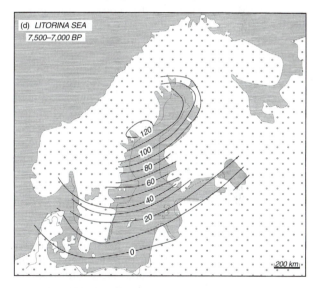

Fig. 1.68 Stages in the recession of the Scandinavian ice sheet and associated marine and lake events: (a) Baltic Ice Lake; (b) Yoldia Stage; (c) Ancylus Lake; (d) Litorina Sea. (Reproduced by kind permission of Matti Eronen)

Porter and Denton (1967) to refer to the regrowth of glaciers after their minimum extent in the early Holocene. Neoglaciation started as early as 8 kyr BP in some regions, but much later in others (Denton and Karlen, 1973).

In North America, ice margins on Baffin Island were behind their current positions by 5 kyr BP, but readvanced briefly before approximately 3.2 kyr BP (Miller, 1973, 1976; Andrews and Barnett, 1979; Dyke, 1979; Dyke *et al.*, 1982a). Neoglacial advances of North American mountain glaciers, dating to various times throughout the Holocene, are reported from the Canadian Cordillera (Denton and Karlen, 1977; Luckman and Osborn, 1979; Osborn, 1985, 1986; Ryder and Thomson, 1986; Ryder, 1987, 1989; Desloges and Ryder, 1990), the US Rocky Mountains (Richmond, 1965; Benedict, 1973a; Currey, 1974), the Alaskan Brooks Range (Ellis and Calkin, 1979, 1984; Calkin and Ellis, 1982; Calkin *et al.*, 1985) and the Torngat Mountains of Labrador (D.J.A. Evans and Rogerson, 1986). The chronologies of these neoglacial advances have been reconstructed from lateral moraine stratigraphies where glaciers overrode previous deposits (e.g. Osborn and Karlstrom, 1989; Desloges and Ryder, 1990) and from inset moraine sequences. An excellent review of the Holocene evolution of the Laurentide and Greenland ice sheets and the Canadian mountain glaciers is available in Fulton (1989).

At arid higher latitudes in North America and in Greenland the responses of glaciers to climate change were very different. The buildup of ice in response to the last glacial cycle was far less intense and slower than that of the Laurentide ice sheet, and maximum glacier limits date to the Holocene (Figs 1.60a and 1.60d (Plate 9)) and deglaciation began anywhere between 10 and 7 kyr BP according to site-specific characteristics (England, 1978, 1985, 1990, 1992; Hodgson, 1985; Bednarski, 1986; Lemmen, 1989; D.J.A. Evans, 1990b; Lemmen *et al.*, 1994a). The Canadian and Greenland High Arctic warmed until the mid-Holocene, when colder conditions gave rise to the formation of the northern Ellesmere Island ice shelves and the readvance of glaciers to their maximum neoglacial positions (Blake, 1981; Bradley, 1990; D.J.A. Evans and England, 1992). Although the ice shelves are now breaking up (Jeffries, 1987), many of the larger glacial systems continue to respond to the mid-Holocene cold phase, testifying to the greatly different response rates of various glacier systems (Stewart and England, 1983; Evans and England, 1992).

Glacier advances have taken place in many parts of Scandinavia during the Holocene (Østrem, 1964; Karlen, 1973; Andersen, 1980). Recent research on the Jostedalsbreen region by Nesje and Kvamme (1991) has suggested that the ice cap Jostedalsbreen disappeared completely during the early Holocene. They further suggest that prominent moraines located just outside the Little Ice Age glacier limits document major readvances by mountain glaciers during the Pre-Boreal–Boreal transition at 9 kyr BP, when the Scandinavian ice sheet was at its Ancylus Lake stage (*Erdalen Event*; Fig. 1.69).

In the southern hemisphere, a large amount of work has been undertaken on Holocene glacial chronologies in New Zealand. Here stratigraphic sequences in overridden lateral moraines provide a pattern of glacier oscillations that closely resembles those of Europe (e.g. Gellatly *et al.*, 1985). The New Zealand Holocene chronology bears some resemblance to the pattern of glacier activity in South America (e.g. Mercer, 1970), although considerable work is still required on neoglacial moraines in South America. The Holocene history of the Antarctic ice sheet is characterized by relatively minor oscillations of certain margins on a timescale that is similar to that of the Arctic, although much research is still required. For example, the ice shelf of the Antarctic Peninsula had disappeared by approximately 6.5 kyr BP but formed again sometime in the later half of the Holocene, when local glaciers and glaciers on the islands of the Southern Ocean also advanced (Clapperton, 1990c). The higher sea-levels of our present interglacial have largely resulted in the recession of the large outlet glaciers of the ice sheet, owing to

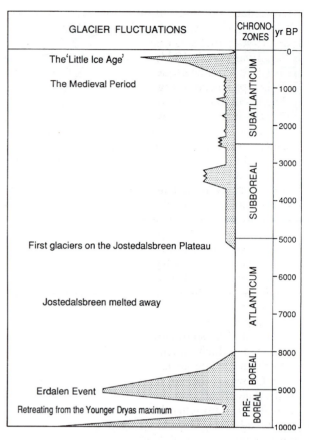

Fig. 1.69 The retreat of the Scandinavian ice sheet and subsequent Holocene glacier fluctuations according to Nesje and Kvamme (1991). (Reproduced by permission of the Geological Society of America)

grounding-line retreat (Section 8.2). However, some terrestrial glacier snouts have advanced in response to the higher precipitation levels of the interglacial (Denton *et al.*, 1989a; Clapperton and Sugden, 1990).

Abundant historical documentation is available for the most recent of the Earth's glacial events, the *Little Ice Age* of the late sixteenth to early twentieth centuries. This was a period of lower temperatures on a global scale which brought about the advance of glaciers by an average of several kilometres at high latitudes and altitudes. An excellent review of the evidence for climate change and glacier response pertaining to the Little Ice Age was provided by Grove (1988). Numerous studies of individual post-Little Ice Age glacier retreat histories, based upon lichenometric and other relative dating techniques, have been reported from around the world. In Iceland and Norway, historical accounts of glacier fluctuations complement the dating techniques employed on recessional moraine sequences. In Norway, the advance and retreat histories of different glaciers appear to vary according to distance from moisture source, latitude and altitude (cf. Matthews and Shakesby, 1984; Erikstad and Sollid, 1986; Ballantyne, 1990a; Bickerton and Matthews, 1992, 1993; D.J.A. Evans *et al.*, 1994). An example of a post-Little Ice Age moraine sequence for Nigardsbreen, Norway, is reproduced in Fig. 1.70.

The historical fluctuations of Icelandic glaciers are presented in Fig. 1.71, which is based largely upon direct observations (e.g. Rist, 1984; Grove, 1988). This shows that some glaciers are still retreating at a considerable rate from their Little Ice Age maximum positions, while others have reached a position of stability or are even readvancing. Whereas the retreats are in response to climate change, stability and readvances are related to a number of factors such as surging behaviour, completion of retreat from lowlands on to plateau surfaces where ablation rates are lower, glacier hypsometry and ice-divide migration.

Some of the best-dated Little Ice Age moraine sequences in North America are in the Rockies and west-coast mountains, where written records and photographs have been used in conjunction with tree-ring and lichenometric dating to obtain ages on landforms which document glacier fluctuations throughout the Holocene (e.g. Luckman and Osborn, 1979; Grove, 1988). An interesting aspect of the Little Ice Age advance in the southern Canadian Rockies is that it represents the most extensive Holocene glacier advance, indicating that conditions were more favourable for glacier growth during the past few hundred years than at any other time since the last glacial period.

Records of glacier responses to the Little Ice Age cooling trend in the Himalaya and China are far more patchy but generally show patterns of retreat since the late nineteenth century (Grove, 1988). Superimposed on the retreat patterns of some glaciers are numerous readvances of various ages, which are thought to relate to fluctuations in temperature and monsoonal air flow (e.g. Shih Ya-feng and Wang Jingtai, 1979; Mayewski *et al.*, 1980; Grove, 1988).

Some glacier snout fluctuations since the Little Ice Age have been documented by direct observation (e.g. Harper, 1993). A climate-change monitoring site has been established at the northern margin of the Barnes ice cap, Baffin Island (Jacobs *et al.*, 1993), in order to link climatic variables with post-Little Ice Age recession rates, which have been studied since the 1960s (e.g. Andrews and Barnett, 1979). Rather than showing continuous retreat since the Little Ice Age, some studies have identified more complex historical glacier fluctuations. For example, short-term precipitation controls are thought to be responsible for recent advances of glacier snouts on Mount Baker, Washington (Harper, 1993), in the Premier Range of British Columbia (Luckman *et al.*, 1987), and on the Franz Josef Glacier, New Zealand (Chinn,

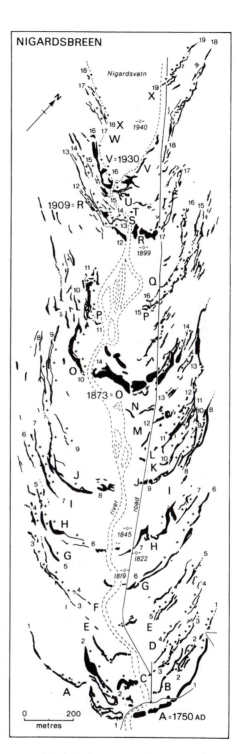

Fig. 1.70 The moraine ridge sequence documenting the retreat of Nigardsbreen, Norway, from its Little Ice Age maximum position. This is typical of the forelands of many actively receding glacier snouts around the world. Moraines are labelled A–X and numbered 1–19 on the north-east and 1–18 on the south-west sides of the foreland. The historical dates are based upon lichenometry which was undertaken at sites marked (+). (From Bickerton and Matthews, 1992)

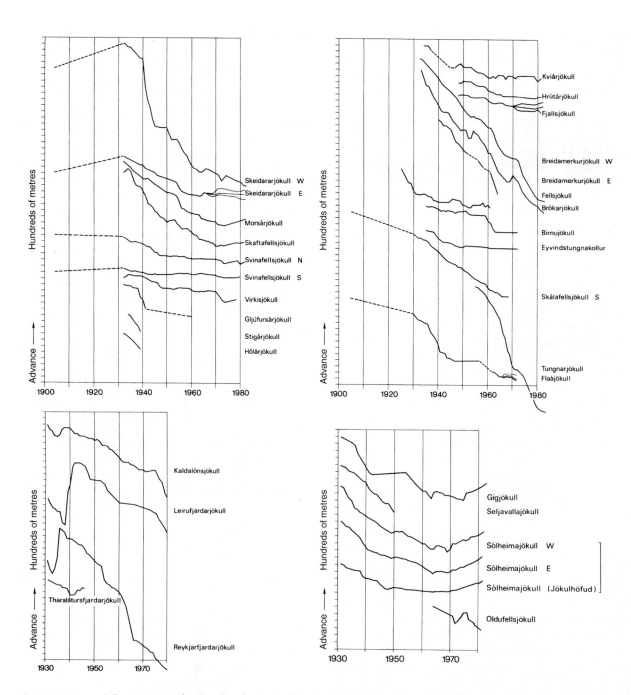

Fig. 1.71 Frontal fluctuations of Icelandic glaciers in historical times showing post-Little Ice Age response. (From Grove, 1988. Reproduced by permission of Routledge)

1989; Brazier *et al.*, 1992), the latter having undergone several readvances during overall retreat. Similarly, many of the glaciers of Europe have undergone readvances over the past few decades after large-scale retreat from their Little Ice Age maximum positions, and although the pattern is complex, the fluctuations appear to be broadly synchronous (Fig. 1.72; Grove, 1988). Such regional synchrony in

Europe has been linked to fluctuations in summer temperature and precipitation (e.g. Reynaud, 1980, 1983) and persistent synoptic situations (Hoinkes, 1968).

At first glance the patterns of Holocene glacier fluctuations based upon numerous studies from around the world appear to be only weakly correlated. However, Röthlisberger (1986) compared the

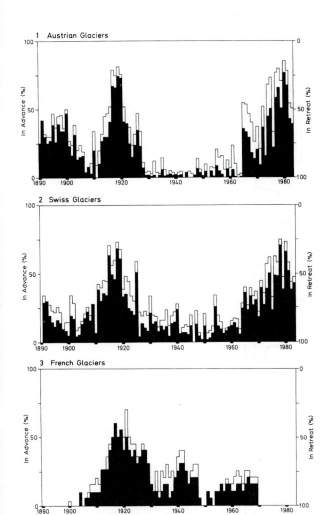

Fig. 1.72 A comparison of the behaviour of Austrian, Swiss and French glaciers since 1890 in terms of the percentage of those found to be retreating, advancing or stationary each year. (From Grove, 1988. Reproduced by permission of Routledge)

moraine sequences only of glaciers of approximately the same size, and collected all data himself in order to achieve compatibility. He produced a series of curves for glacier oscillations in both hemispheres which displays a remarkable synchrony for global glacier responses to Holocene climate change (Fig. 1.73). Such correlations provide a valuable insight into the long-term behaviour of glacial systems worldwide, and help to identify the causes of global climate change.

1.7.2.7 'AVERAGE' GLACIAL CONDITIONS

Although knowledge of glacier distribution is most detailed for the period during and following the last glacial maximum, this period is not necessarily representative of the bulk of Quaternary time. Porter

(1989) has argued convincingly that for most of the Quaternary, glacier extent has been intermediate between the full-glacial maxima and interglacial minima experienced during the Holocene, and that such intermediate conditions are the most relevant to long-term analyses of environmental change and landscape evolution. Using long palaeoclimatic records from ocean cores, Porter argued that in Britain and Iceland average ice cover was similar to that during the Younger Dryas, whereas in Scandinavia ice cover consisted of large mountain icefields covering almost all of Norway and the highlands of Sweden. The Laurentide ice sheet probably consisted of three coalescent domes centred over Keewatin, Labrador and Baffin Island, covering approximately the same area as at 9000 yr BP (cf. Fig. 1.60 (Plate 9)). The Greenland and Antarctic ice sheets were slightly larger than at present, whereas glaciers in major mid- to high-latitude mountain ranges, such as the Southern Alps of New Zealand, the Himalaya, the southern Andes, the European Alps and the Cascade, Brooks and Alaska Ranges, probably extended a few kilometres to a few tens of kilometres beyond their present margins.

According to Porter, this average ice cover represents the principal boundary conditions for much of Quaternary landscape evolution, including the erosion of classic glacial mountain landscapes of cirques and fjords, the development of erosional shore platforms and tropical atolls (because of the correlation between ice cover and global sea-level), and long-term sediment yields.

1.8 THE FUTURE

To what extent are the post-Little Ice Age retreat tendencies an indicator of future glacier behaviour and distribution? In order to address this question we have to assess the influences of long-term climate change cycles and unprecedented human-induced global warming (Warren, 1995). This chapter has demonstrated that different glaciers can respond in completely different ways to the same climate trigger, and so predictions of future climate trends and glacier behaviour must avoid generalizations about recent glacier–climate linkages and look to the longer-timescale Quaternary record of climate and glaciation.

Global warming scenarios (e.g. Leggett, 1990) have tended to invoke images of higher global sea-levels, increased melting of the world's ice bodies, and accelerated calving rates at the marine margins of the Antarctic ice sheet followed by ice sheet collapse (e.g. Hughes, 1972, 1973; Mercer, 1978; Sugden, 1988). The various climate–glacier relationships

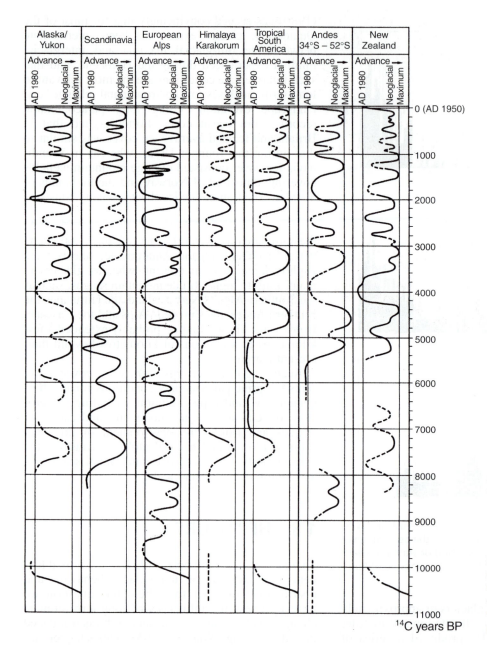

Fig. 1.73 Fluctuations of glaciers in the major glaciated regions of both hemispheres during the Holocene. (From Grove, 1988. Reproduced by permission of Routledge)

reviewed in this chapter clearly show that a warmer world is not necessarily a non-glacierized world. Indeed, glaciers of high-latitude and arid environments, including the great ice sheets of Antarctica and Greenland, may well expand, owing to the increased precipitation totals of a warmer environment.

As we are presently in an interglacial, we should perhaps look to the Quaternary record to understand exactly how extensive previous interglacial ice cover was. We would then know what to expect of our present interglacial in terms of further glacier recession. In Section 1.2.2 we discussed the stability of the Greenland ice sheet and presented the interpretation of ice core records by Koerner (1989), who suggests that the ice sheet may have disappeared completely during the last interglacial. Even if this is an overestimate of the melting total, this observation alone implies that we may not yet have arrived at a natural interglacial equilibrium.

CHAPTER 2

SNOW, ICE AND CLIMATE

Fig. (Sketch No. 2).—*Cliff-view, Cape York.*   Lat. 76°.   Length about 10 miles.

N.N.W.   S.S.E.

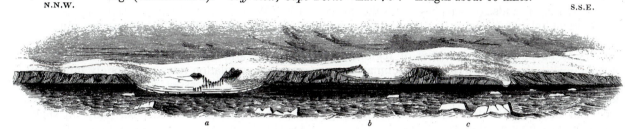

2.1 INTRODUCTION

Glaciers and ice sheets are large, dynamic stores of water, constantly exchanging mass with other parts of the global hydrological system. Glaciers grow by the input of snow and other forms of ice accumulating on their surfaces, and lose mass by melting, the breakaway of icebergs, and other processes. The difference between the gains and losses over a given period of time is the *mass balance*, which is positive if the glacier is gaining mass and negative if it is losing mass. Glacier mass balance reflects the climate of the region in which the glacier is situated, together with site-specific glacier morphology and local topographic setting. The idea of mass balance is therefore an important link between climatic inputs and glacier behaviour, allowing the advance and retreat of many glaciers to be understood in terms of regional or global climatic change. As a result, past climates can be reconstructed from historical or geological evidence for glacier fluctuations, and future water balance can be predicted for catchments containing advancing or retreating glaciers.

Glaciers also contain a direct record of past climates in the ice itself. Successive layers of snow and ice laid down each year contain trapped air bubbles that represent samples of the ancient atmosphere, and the water molecules making up the ice provide a means of reconstructing former air temperatures. Additionally, the layers may contain impurities such as volcanic ejecta, wind-blown dust or radioactive fallout, providing important records of the past.

In this chapter we review the processes by which glaciers gain and lose snow and ice, and examine the links between climate, glacier mass balance and glacier behaviour. The record of past climates contained in glacier ice is described, and we conclude with a discussion of the links between past and present climates and glacier temperature.

2.2 GAINS AND LOSSES

With reference to the glacier system described in Chapter 1, the mass balance of a glacier is the balance between the inputs and the outputs of snow, ice, water and vapour, or between accumulation and ablation. Both accumulation and ablation may take place along the whole length of a glacier during the winter and summer respectively. However, on most valley glaciers and some ice sheets the amount of accumulation will decline downglacier and the amount of ablation will decline upglacier, producing a point on the glacier where ablation equals accumulation: the equilibrium line.

2.2.1 Processes of accumulation

Glacier ice accumulates by the precipitation of snow and other forms of ice at the glacier surface. The most important primary source of ice on most glaciers is snowfall, although amounts vary a great deal from place to place and throughout the year (Fig. 2.1). The highest rates occur in mountainous maritime regions with frequent onshore winds, such as the Coast Range of Alaska, southern Iceland, western Patagonia, and the west coast of New Zealand's South Island. Snowfall accumulation rates as high as $8\,\mathrm{m\,yr^{-1}}$ (water equivalent) have been reported for parts of the Patagonian icefields (Escobar *et al.*, 1992; Warren and Sugden, 1993). Conversely, snowfall is lowest far from any oceanic moisture source, such as the interior of Antarctica, which is one of the most arid regions on Earth. On a more local scale, accumulation rates are strongly influenced by redistribution processes such as wind-blowing or avalanching. High winds can very effectively scour exposed surfaces and redeposit snow in sheltered lee-side locations, while in steep, mountainous areas avalanches can transfer large quantities of snow on to glacier surfaces from the overlying slopes (Vivian, 1975).

Ice crystals may also form on glacier surfaces by the freezing of supercooled vapour or water droplets carried by the wind. The most important type of ice formed by this process is *rime ice*, which can assume strange plant-like shapes built up by the eddying moisture-laden air. Rime ice accumulation is most rapid in cool, humid conditions on surfaces which are most exposed to the wind (Sugden and John, 1976). Some direct accumulation may occur by the freezing of rain-water or groundwater that comes

Fig. 2.1 Snow accumulation layers on Zongo Glacier, Cordillera Real, Bolivia. Successive summer ablation horizons are picked out by dark dust layers. Note the unconformity near the top of the left-hand ice cliff, where ablation has removed some layers. The refreezing of surface meltwater to form icicles is also clearly shown. (Photo: D.I. Benn)

into contact with a glacier, although most liquid water on or in glaciers is meltwater derived from snow or ice. Therefore, freezing is most important as a process for altering the state of the glacier, rather than adding to it.

Glacier accumulation zones have been subdivided according to the nature of melting and refreezing (Benson, 1961; Müller, 1962; Paterson, 1994; Fig. 2.2). Not all the zones are necessarily represented on any one glacier. The highest *dry snow zone* is everywhere below 0°C and is therefore not affected by meltwater. The *dry snow line* separates the dry snow zone from the *percolation zone*, which is characterized by some surface melting. Water in this zone can percolate down through the snow, where it refreezes as horizontal ice lenses or vertical ice glands. The depth of percolation of meltwater increases with decreasing elevation until a point on the glacier (the *wet snow line*) marking the upper limit of the *wet snow zone*, where all the snow deposited during the previous winter period has been raised to 0°C. At lower elevations the refrozen meltwater is in such abundance that it forms a continuous mass of superimposed ice rather than lenses and glands. This

defines the *superimposed ice zone*, and the dividing line between this zone and the wet snow zone on the glacier surface is the *annual snow line* or *firn line*. The *equilibrium line* marks the point where the annual accumulation is exactly balanced by losses from the system: this may be the lower limit of the superimposed ice zone or, if a superimposed ice zone is absent (such as in temperate localities), the lower limit of the wet snow zone. In the latter case, the equilibrium and snow lines coincide. Dry snow zones occur only in the interiors of the Greenland and Antarctic ice sheets, where temperatures are less than −25°C, and at high altitudes (Benson, 1961).

2.2.2 Transformation of snow to ice

If the yearly total of snow or ice accumulating on a glacier surface exceeds local losses by ablation, net accumulation occurs. Year after year, successive accumulation layers are built up, and the deeper layers eventually turn to glacier ice (Fig. 2.1). The transformation of snow to ice occurs as the volume of air-filled pores is reduced and the material increases in density. Freshly fallen snow has a density of

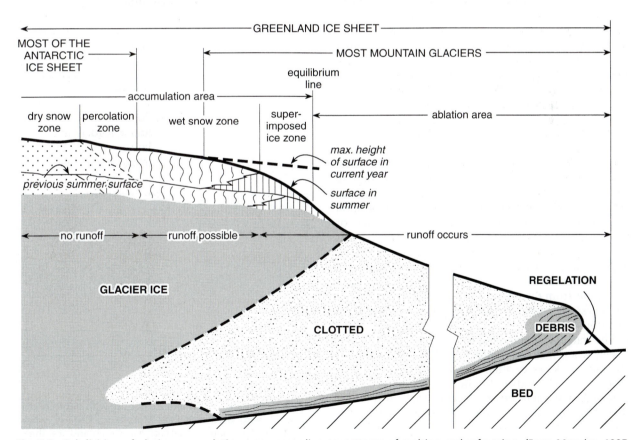

Fig. 2.2 Subdivision of glacier accumulation areas according to patterns of melting and refreezing. (From Menzies, 1995. Reproduced by permission of Butterworth-Heinemann)

50–200 kg m^{-3}, compared with 830–910 kg m^{-3} for glacier ice, and 1000 kg m^{-3} for pure liquid water at 0°C. Pure ice has a density of 917 kg m^{-3}. Snow that has survived one melt season and has begun this transformation is known as *firn,* and has a density of 400–830 kg m^{-3} (Paterson, 1994). The transition between firn and ice occurs when interconnected air passages become sealed off, isolating air in separate bubbles. Additional increases in density beyond this point are achieved by compression of the bubbles, placing the enclosed air under pressure.

The processes by which snow is transformed into ice, and the time taken for the transformation to occur, depend on climate. In cold polar regions and at high altitudes, where melting is unimportant, the principal mechanisms leading to an increase in density are (a) restructuring by wind, (b) the movement of crystals relative to one another, (c) changes in crystal size and shape, and (d) internal deformation of crystals. Winds commonly blow snow across glacier surfaces, either during or after snowfall. This has the effect of breaking up snowflakes into smaller ice crystals, and redepositing them in drifts which have much higher density than snow deposited in still air (mountaineers refer to compact, wind-blown snow as *windslab*). Once deposited, snow increases in density owing to the pressure exerted by overlying snow, which causes crystals to move relative to one another and adopt stronger, more stable packings. As a result, compaction increases with increasing depth of burial. This process is encouraged by progressive changes in the size and shape of crystals in the snowpack, because denser packings are possible for spherical particles than for other shapes. Fresh snow crystals have elaborate, complex shapes with many branches and re-entrants, but the crystals gradually assume more spherical shapes because molecules migrate from one part of the crystal to another, so as to reduce energy gradients across the surface. Migration may be over the surface, within the crystal lattice, or through the air as vapour (Paterson, 1994). Changes in shape are most important in the early stages of transformation, when overburden pressures are low and void spaces are common. Further volume change can result from the internal deformation, or *creep*, of ice crystals when overburden pressures increase (Section 4.3.1).

The transformation of snow to ice is greatly accelerated when melting occurs at the surface or within the snowpack, as is often the case at relatively low altitudes in low and mid-latitudes. Water percolates downwards through the snow, where it will refreeze if it comes into contact with cold snow or ice, forming horizontal lenses or vertical glands of *superimposed ice* (Section 2.2.1). Because freezing of water releases heat (Section 2.2.3.2), the formation of superimposed ice results in warming of the surrounding snow, making it easier for further melting to occur in the future. As a result, progressively deeper layers of snow can be raised to the melting point if melting is accompanied by refreezing at depth. If the surrounding snow or ice is already at the melting point, water may remain in pore spaces until it is refrozen when the weather or season turns colder. The formation of superimposed ice is particularly important in areas where snowfall accumulation is low, allowing the penetration of seasonal 'cold waves' from the surface. For example, about 90 per cent of the ice of the Meighen Ice Cap, Northwest Territories, Canada, was formed by the transformation of snow to superimposed ice (Koerner, 1970).

The contrast in time taken for snow to transform into glacier ice in the presence and absence of melting is well illustrated by the variations in density with depth for the Upper Seward Glacier, Yukon, Canada, and part of the Greenland ice sheet (Fig. 2.3). The Seward Glacier profile is representative of low altitudes in temperate areas, and shows transformation from firn to ice (830 kg m^{-3}) at only 13 m below the surface, or within 3–5 years of burial. In contrast, in the Greenland profile, from a cold continental setting, this change does not occur until a

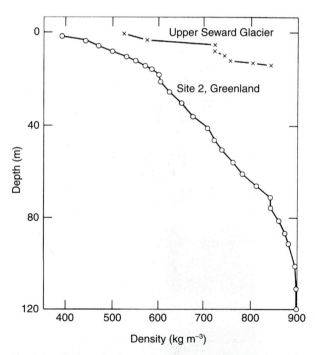

Fig. 2.3 Changes in firn and ice density with depth on the Upper Seward Glacier, St Elias Mountains, Canada, where melting occurs, and on part of the Greenland ice sheet, where there is little or no melting. (From Paterson, 1994, after Sharp, 1951, and Langway, 1967. Reproduced by permission of Pergamon)

depth of 66 m, equivalent to more than 100 years since burial (Paterson, 1994).

2.2.3 Ablation

2.2.3.1 PROCESSES OF ABLATION

Ablation refers to the totality of the processes by which snow and ice are lost from a glacier, and includes wind ablation, the avalanching of ice blocks from terminal ice cliffs, the calving of icebergs into water, melting followed by runoff, evaporation and sublimation.

Wind ablation refers to wind scouring of snow that results in the removal of mass from a glacier. It can be a very important process near the margins of the polar ice sheets, where severe katabatic winds blow out from the continental interior towards the coast. It can also be of local importance on valley glaciers, depending on glacier aspect relative to the prevailing winds.

Avalanching or *dry calving* is an important process of mass loss in mountain areas where glaciers terminate on or above steep slopes. Ice breaking from the glacier front falls down to lower altitudes, where it melts *in situ* (Fig. 2.4). However, if the melt rate is exceeded by the rate of avalanching, glacier ice will re-form at the base of the slope, forming a lower glacier tongue. In this case, the upper ice source and the lower tongue form two parts of a single glacier system. Dry calving is also important at the margins of cold, polar ice masses which commonly form steep ice cliffs. Blocks toppling from the cliffs accumulate as a jumble of icy rubble at the base, which will ablate *in situ* unless overridden by the glacier during a readvance (Section 5.4.3).

Iceberg calving refers to mass loss at the margins of water-terminating glaciers through the breakaway of floating icebergs. Individual calving events vary enormously in scale, from the release of small blocks to huge events involving hundreds or thousands of cubic metres of ice (Fig. 2.5). An extremely large calving event occurred during a rare breakup of the sea-ice that usually surrounds the tongue of Erebus Glacier, Antarctica. In March 1990, a 3.5 km long portion of the floating glacier tongue broke off, producing an iceberg weighing approximately 100 million tonnes. This huge mass of ice was last seen floating past McMurdo Base carrying a hut used by a research team, who fortunately were not in residence at the time (Robinson and Haskell, 1992). More recently, in January 1995, a large portion of the Larsen ice shelf in the Antarctic Peninsula broke up, causing the ice margin to retreat by up to 2 km in only five days (Rott *et al.*, 1996). A spectacular series of photographs of a large calving event from Jacobshavns Glacier, west Greenland, has been published by Epprecht (1987). Globally, calving is an extremely important ablation mechanism, because the Antarctic ice sheet, which contains most of the world's glacier ice, mostly terminates in the sea and exists in an environment which is almost everywhere too cold for surface melting. The mechanisms of iceberg calving have been reviewed by Warren (1992) and are described in detail in Section 8.2.4.

Melting, *evaporation* and *sublimation* are the transformation of ice to water, water to vapour, and ice to vapour, respectively (Fig. 2.6). Mass loss by these processes occurs at the glacier surface, and within pore spaces and other voids below the surface (internal ablation; Ambach, 1955). Internal ablation can be important within snow, but also

Fig. 2.4 Dry calving at the margin of a hanging glacier, Angel Glacier, Jasper National Park, Canada. Ice blocks from the glacier front avalanche down the steep rock slabs and feed the Edith Cavell Glacier below. (Photo: D.J.A. Evans)

Fig. 2.5 Iceberg calving at a tidewater glacier margin, Osbornebreen, Spitsbergen. (Photo: D.J.A. Evans)

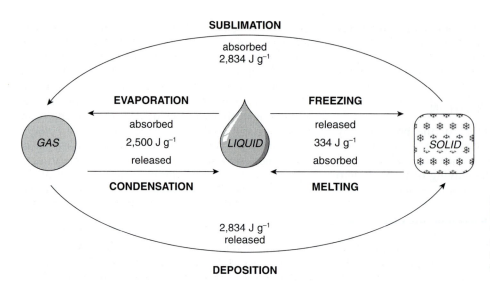

Fig. 2.6 Phase changes between ice, water and vapour, showing the amount of latent heat energy consumed and released by the transformation

occurs below *weathering crusts* on ice surfaces. These crusts are produced by differential absorption of radiant energy along grain boundaries, and consist of porous ice with loosely interlocking crystals. During the formation of weathering crusts, more ice is being melted than would be detected by surface lowering measurements of ablation (Müller and Keeler, 1969).

Melting is the dominant process of ablation on many glaciers, particularly where daytime temperatures exceed 0°C for at least part of the year. Sublimation can predominate, however, in cold continental settings where the air is very dry (the humidity is low), such as the Dry Valleys of Antarctica. Melting, evaporation and sublimation all require the input of radiation and/or thermal energy, which can come from various sources. The energy balance at glacier surfaces, and its implications for ablation, are described in the following section.

2.2.3.2 ENERGY BALANCE AND ABLATION

Ablation by melting, evaporation and sublimation will occur if there is a net surplus of energy at the glacier surface after the ice has been raised to the melting point. Conversely, a net deficit in energy at the surface can chill the ice or lead to ice accumulation through condensation of vapour or freezing of liquid water. The surplus or deficit of energy over time is known as the *energy balance*, and is an important factor when considering glacier ablation rates (Kuhn, 1979; Gruell and Oerlemans, 1986; Paterson, 1994). Components of the energy balance on glacier surfaces are (a) *solar radiation*, (b) *long-wave radiation*, (c) *sensible heat* from the atmosphere, and (d) *latent heat* transferred during melting, freezing, condensation, evaporation and sublimation (collectively termed *phase changes*).

Solar radiation

Solar radiation provides the primary source of energy for the world's climate system. Most solar radiation reaching the lower atmosphere is in the short wavelengths between *c.* 0.2 and 4.0 μm, peaking in the visible part of the spectrum (0.4–0.7 μm) (Barry and Chorley, 1992). This radiation can reach the surface as direct sunshine or as diffuse radiation scattered through the atmosphere. On cloudy days, almost all incoming short-wave radiation is diffuse. Part of the solar radiation is reflected back off the surface, but the remainder can provide energy for heating, melting, evaporation and sublimation. The percentage of radiation that is reflected from a surface is called the *albedo*. Albedo values are high for fresh snow surfaces and low for bare and dirty ice surfaces (Ambach, 1985; Röthlisberger and Lang, 1987; Kuhn, 1989a; Paterson, 1994; Table 2.1). The values in Table 2.1 contrast with values of 2–4 per cent for water (with high sun angles), 10–20 per cent for grass and 35–45 per cent for dry sand (Briggs and Smithson, 1985). Snow surfaces may be responsible for reducing the absorption of short-wave radiation by a glacier surface by over 80 per cent. Therefore, less energy is available for ablation on snow surfaces than on debris-covered ice. This means that the amount of

Table 2.1 Albedos (per cent) for snow and ice

	Range	Mean
Dry snow	80–97	84
Melting snow	66–88	74
Firn	43–69	53
Clean ice	34–51	40
Slightly dirty ice	26–33	29
Dirty ice	15–25	21
Debris-covered ice	10–15	12

Source: Paterson (1994)

ablation will accelerate during the ablation season on the lower parts of mid- and high-latitude glaciers, as winter snow cover gives way to bare ice surfaces. Estimations of albedo values are crucial to the construction of energy balance models over large ice sheets (e.g. Braithwaite and Olesen, 1990a; van de Wal and Oerlemans, 1994).

The amount of short-wave radiation received by a surface also depends on its aspect relative to the position of the Sun. Radiation receipts are highest when the Sun's rays make an angle of 90° with the surface, and lowest when the angle is low (Fig. 2.7). The low solar angle in the high latitudes, and in the mid-latitudes during the winter, means that overall radiation receipts are much lower than in the tropics. At a local level, this global pattern is modified by slope gradient and aspect, and topographic shading by mountains and other barriers, which can result in very different rates of ablation on different parts of a glacier. In the northern hemisphere, solar radiation is at a minimum on north-facing slopes in December and January; in the southern hemisphere the opposite is true.

Long-wave radiation

Energy also reaches glacier surfaces in the form of long-wave (infrared) radiation (Barry and Chorley, 1992). This energy is emitted from the atmosphere, rocks or other surfaces when they have been heated, mainly by incoming short-wave radiation. The ability of the lower atmosphere to trap long-wave radiation is known as the greenhouse effect, a natural phenomenon which is being enhanced by certain types of atmospheric pollution (Houghton *et al.*, 1990; Barry and Chorley, 1992; Kemp, 1994). Water vapour is a particularly efficient greenhouse gas, retaining energy in the atmosphere and making it available for heating and glacier ablation. Therefore, long-wave radiation can be a very important component of the energy

budget of glaciers when the air is humid, such as during hazy or overcast weather. Clear, dry air has a much lower ability to trap long-wave radiation, resulting in lower long-wave receipts during the day and very rapid reductions in air temperature at sunset, when incoming short-wave radiation is cut off. Long-wave radiation is also important at the margins of valley glaciers overlooked by rock walls, where radiation emitted from dark rock surfaces can accelerate ablation (Fig. 2.8).

Sensible heat transfer

Thermal energy exchanged at the interface between the atmosphere and a glacier surface is known as *sensible heat* because it can be felt with the senses. Sensible heat is transported by warm air masses. These can be associated with local air circulation, such as warm *valley winds* that blow up mountain valleys during the day or *Föhn*-type winds on the lee side of mountain ranges (Barry, 1992), or form part of large-scale low-pressure systems such as the Atlantic depressions that commonly affect north-west Europe, and the summer monsoon in the Himalaya. Sensible heat transfer is most efficient when the atmosphere is much warmer than the snow or ice surface, and strong winds and a rough glacier surface encourage turbulence and rapid heat exchange (Kraus, 1973; Paterson, 1994).

Latent heat transfer

Changes of state between ice, water and vapour consume or give up energy, known as latent heat. Melting consumes 334 joules of energy per gram of ice melted $(J g^{-1})$, whereas evaporation consumes $2500 J g^{-1}$, over eight times as much. Freezing and condensation release equivalent amounts of energy, respectively (Fig. 2.6). Therefore, the freezing of rain-water or the condensation of water vapour on a glacier surface can yield large amounts of energy, part of which can

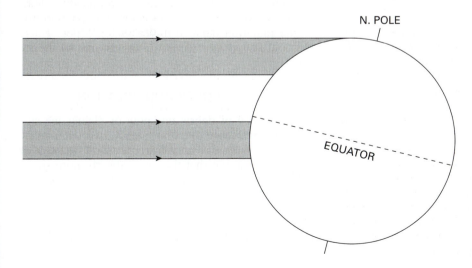

Fig. 2.7 Relationship between latitude and receipt of solar radiation. Near the equator, where the midday sun is high in the sky, solar energy is distributed over a smaller area than near the poles. As a result, more energy is available for heating the surface and atmosphere per unit area at low latitudes

Fig. 2.8 Preferential ablation at the margin of Storbreen, Norway, due to long-wave radiation emitted by warm rock surfaces. (Photo: D.I. Benn)

raise the temperature of the surrounding ice and increase the likelihood of future ablation.

The energy balance

The energy balance at the surface of a glacier is the sum of the individual energy components, and may be expressed as

$$Q_m = Q_s + Q_1 + Q_h + Q_e \qquad (2.1)$$

where Q_m is energy available to melt ice, Q_s is net short-wave radiation flux, Q_1 is net long-wave radiation flux, Q_h is sensible heat transfer, and Q_e is latent heat transfer.

Several variations of this relationship have been published in the literature, in which various components are combined or subdivided according to the source of data or the purpose of the study (e.g. Kuhn, 1979, 1989b; Braithwaite, 1981; Gruell and Oerlemans, 1986; Hay and Fitzharris, 1988; Munro, 1990). The relative importance of each component varies from place to place and through time (Paterson, 1994). In general, net radiation (short-wave and long-wave) provides the most important components, with the highest proportions of short-wave radiation being associated with clear-sky conditions. In energy balance studies of continental climates, net radiation has been calculated to account for approximately 66 per cent of ablation energy (Lang, 1968; Braithwaite, 1981; Braithwaite and Olesen, 1990a). However, in more maritime climates this figure is reduced to bet-

ween 10 and 50 per cent, because the meteorological factors associated with warmer and moist air masses are more important (Ahlmann and Thorarinsson, 1938; Gudmundsson and Sigbjarnarsson, 1972). Short-wave receipts are particularly high on low-latitude, high-altitude glaciers, such as in the northern and central Andes and the Himalaya, where the solar angle is high and the filtering and scattering effect of the lower atmosphere is at a minimum. The long-wave share of the radiation budget will be highest in humid conditions and adjacent to rock surfaces and valley sides. Energy provided by sensible heat transfer and the latent heat of condensation and freezing is most important where warm, humid air is transferred over glacier surfaces by local or regional air circulation. Examples include the western seaboard of North America, southern Iceland, western Norway, and the Pacific coast of Chile, which are strongly influenced by humid maritime air streams (Laumann and Reeh, 1993), and the glaciers of equatorial Africa (Hastenrath and Kruss, 1992). The influence of changing weather conditions on the energy balance components on the Devon Island ice cap, Arctic Canada, has been studied by Holmgren (1971) and Alt (1978).

In their energy balance model for the Greenland ice sheet, van de Wal and Oerlemans (1994) identified a pattern of absorbed short-wave radiation which is dictated by ice sheet albedo and atmospheric transmissivity (Fig. 2.9 (Plate 10)). Absorbed short-wave radiation reaches a maximum at the ice sheet margin, owing to the low albedo there. The minimum absorbed short-wave radiation occurs around the equilibrium line, but receipts increase again towards the centre of the ice sheet, owing to the decreasing cloud cover and increasing atmospheric transmissivity over the ice sheet interior. The model indicates that energy for ablation is derived from turbulent heat flux (40 per cent) and the radiation balance (short-wave and long-wave, 60 per cent). The contribution of the radiation balance increases to 85 per cent in the cold and cloud-free north-eastern part of the ice sheet.

2.2.3.3 DEBRIS COVER AND ABLATION

The presence of debris on snow or glacier ice surfaces can influence ablation rates in two main ways. First, rock surfaces have a lower albedo than snow or bare ice, meaning that they will absorb more short-wave radiation and heat up. They will then re-emit long-wave radiation, so that the adjacent ice will experience greater ablation than it would in the absence of a debris cover. Second, debris can protect underlying snow or ice from ablation by shielding it from incoming radiation or atmospheric heat, thus reducing ablation (Østrem, 1959; Loomis, 1970; Nakawo and Young, 1981, 1982). This will occur if the debris cover is thick enough to prevent heat from the surface being conducted through to the ice during the course of the daily temperature cycle. The albedo effect is greatest for thin or patchy debris cover, whereas the insulation effect dominates for thick, continuous debris, so that debris may enhance or reduce ablation depending on the quantity and distribution. The cross-over point, when ablation is at a maximum, occurs when the debris is *c.* 0.5–1.0 cm thick, and above this point ablation rates rapidly fall below those for clean ice (Fig. 2.10). A debris cover will not only influence the net effect of the heat flow to the glacier surface, but also produce a lag in the response of the underlying glacier surface to short-term changes in air temperatures. At night, when temperatures fall, heat is still stored in the debris cover, so that melting can take place at the debris–ice interface even though the temperature at the debris–atmosphere interface drops below 0°C.

Where the debris cover on glacier tongues is thick and continuous, such as in many high mountain environments, ablation can be negligible. In such cases, ablation is effectively confined to slopes too steep to support superficial debris, such as the walls of open crevasses and other holes on the glacier surface, and steep marginal areas (Fig. 2.11). Ablation proceeds by the preferential melting and retreat of such slopes, in a process known as *backwasting* (N. Eyles, 1979). The rate of backwasting depends on the area of exposed ice slopes, so can accelerate when holes on a glacier surface enlarge and increase their perimeter (Kirkbride, 1993; Section 6.5.4).

Fig. 2.11 Steep bare-ice slopes at the margins of a debris-covered glacier, Lahul Himalaya, India. Ablation is negligible on the easier-angled debris-covered glacier surface, and most mass loss occurs by backwasting of slopes too steep to support a debris cover. (Photo: D.I. Benn)

2.2.3.4 SMALL-SCALE CLIMATE–GLACIER INTERACTIONS

Because of its location, morphology and local topographic setting, each glacier will possess specific microclimatic characteristics. These will reflect the general climate of the region in which the glacier is situated and will determine the response by the glacier's mass balance (Meier, 1965). Changing weather patterns also directly affect heat budgets on a macroscale, as demonstrated by Holmgren (1971) and Alt (1978) for the Devon Island ice cap in Arctic Canada and Brazel *et al.* (1992) for the West Gulkana Glacier in Alaska. Such studies reveal that different synoptic conditions, which either vary throughout or dominate a whole season, will produce vastly different mass balance characteristics for the same glacier over periods of a few years. Examples of heat budgets from different glacier–climate settings are presented in more detail in Paterson (1994).

Measurements of ice ablation have been used to elucidate the relationship between glacier melt rates and temperature by Braithwaite (1981) and Braithwaite and Olesen (1985, 1989) for Canadian Arctic glaciers and the south-western margins of the Greenland ice sheet. Such data have been utilized in energy balance models (e.g. Braithwaite and Olesen, 1990a, b), which demonstrate that a simple relationship exists between ice ablation and air temperature but complications arise when snow cover, albedo effects and wind are being considered. The effects of the various temperature and energy sources on the ablation of Nordbogletscher and Qamanarssup sermia on the Greenland ice sheet are presented in Fig. 2.12. The large amount of ablation energy provided by short-wave radiation even at 0°C is essentially offset by the other sources, of which the sensible heat flux

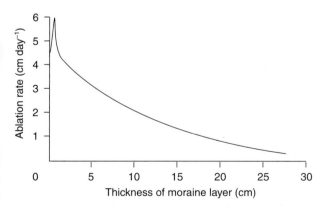

Fig. 2.10 Effect of increasing debris cover on ablation as measured on Isfallsglaciären during July and August 1956. Maximum ablation values occur when the debris cover is 0.5–1 cm thick, at which point the effects of decreased albedo exceed the insulating effect. (Redrawn from Østrem, 1959)

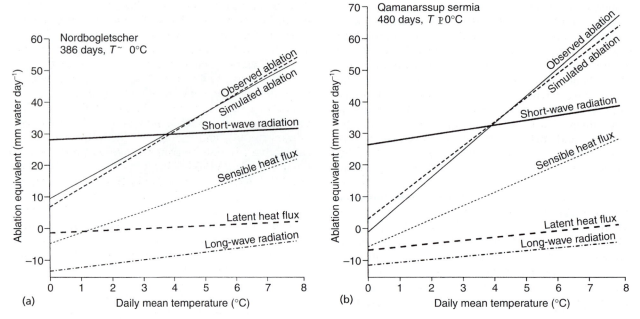

Fig. 2.12 Relationships between air temperature, various components of the local energy balance, and ablation totals for (a) Nordbogletscher and (b) Qamanarssup sermia, Greenland ice sheet. (Modified from Braithwaite and Olesen, 1990a)

is seen to be the most sensitive. Thus Braithwaite and Olesen conclude that the greatest contributor to ablation at the south-west corner of the Greenland ice sheet is sensible heat flux, with smaller contributions from radiation and latent heat flux.

The energy balance models produced by Braithwaite and Olesen (1990a, b) recognize but do not adequately quantify the impact of wind over the glacier surface. Glacier ice masses in the landscape have large impacts on wind patterns in the atmospheric boundary layer. Obviously, the larger the ice mass the more profound these impacts are on local, regional and hemispheric (Section 1.2.2) scales. The surface roughness of an ice body is usually low, and consequently turbulence in the boundary layer is reduced. This has the effect of reducing sensible and latent heat fluxes to the atmosphere. However, high ablation rates of 100–150 mm of water per day are associated with the high temperatures and speeds of föhn winds over the Greenland ice sheet, where an energy balance model calculates average ablation at <40 mm of water per day (Braithwaite and Olesen, 1990b).

The presence of glacier ice will also considerably affect the wind flow over adjacent land surfaces. Cold-air drainage from the ice mass sets up a strong katabatic or glacier wind (Mather and Miller, 1967), which may reach gale force by the time it reaches the margin of a large ice sheet. Such wind flows are strengthened by the differences in radiative and thermal properties and hence the energy balances of ice and tundra, especially during the summer (e.g.

Ohmura, 1982; Duynkerke and van den Broeke, 1994; van den Broeke et al., 1994).

The examples used above are clearly site-specific but can be used as general indicators of climate–glacier relationships. Studies of small-scale interactions between a single glacier and climate are referred to by Marcus et al. (1992) as *representative stations*. They emphasize the importance of filtering for spatial and temporal variability and the linkage of microscale observations to mesoscale and synoptic levels when using such representative stations in assessments of general climate–glacier relationships. Letreguilly and Reynaud (1989) have demonstrated that although different glaciers in a region may show different mass balances for a given year, their mass balance variation over time, centred around the mean, is similar. This confirms that any one glacier can be used as a representative station for glacier–climate interactions in its surrounding region, and provides support for Lliboutry's (1974) concept of a simplified linear model in which it is shown that mass balance variations with time at different sites are similar.

2.3 MASS BALANCE

2.3.1 Definitions

The mass balance of a glacier is defined as the difference between gains and losses (expressed in terms of water equivalent) measured over a specified time period, usually one year. The net change in glacier mass between the same date in successive years defines the *annual mass balance* for the *calendar* or *measurement year*. However, some glaciologists prefer to use the *balance year*, which is measured between two successive annual minima in the mass of the glacier. Because of weather variations between years, this varies in length, but provides a better overall measure of annual changes in mass storage (Paterson, 1994). The annual mass balance is the sum of the *annual accumulation* and the *annual ablation*, which are the gross amounts of mass gained and lost, respectively, over the calendar or balance year. In areas with pronounced seasonal variations in snowfall and temperature, it is useful to subdivide annual accumulation and ablation into *winter* and *summer* totals. The *specific annual mass balance* is the net change in mass on one particular point on a glacier.

2.3.2 Measurement of mass balance

2.3.2.1 DIRECT MEASUREMENTS

Annual accumulation is usually measured in pits excavated into the snowpack or from cores. Annual or seasonal increments of snow are identified from changes in density or grain size, or by a dirt layer representing a summer ablation or non-accumulation surface (Fig. 2.1). The thickness of each layer, divided by average density of constituent snow, allows the annual change in mass to be determined. Seasonal accumulation totals can be obtained using the same method, except that measurements are made at the end of the accumulation season instead of the end of the balance year. Annual and seasonal accumulation often vary widely across a glacier owing to altitudinal variations in snowfall, the influence of topography on wind drifting, and other factors. Therefore, in order to be representative, measurements of accumulation need to be made at several locations.

Annual ablation measurements are made using networks of stakes fixed into the ice. The stakes are set into holes drilled in the ice and are used to measure the drop in the ice surface between the beginning and the end of the balance year. Detailed measurements of ablation over shorter time periods are made using an *ablatometer*, consisting of a freely sliding rod mounted on a cross-arm between two poles (Lewkowicz, 1985; Munro, 1990). Glaciers which lose mass by calving necessitate measurements of velocity and thickness along a cross-section near the terminus, thus facilitating the calculation of total calving over the year.

All these methods obtain the net balance at points on the glacier surface, and so contour maps need to be constructed before the net balance of the whole

glacier can be assessed. Even then, some large errors might be introduced if insufficient sampling points have been selected or inaccurate thickness and density measurements made. A thorough guide to mass balance measurement techniques is available in Østrem and Brugman (1991), which provides details on project design, equipment and analysis of data.

Early attempts to estimate the mass balance of single glaciers are those of Ahlmann (1935, 1946) and Wallen (1948), and a considerable number of new studies were initiated during the International Hydrological Decade (1965–74; e.g. Østrem and Stanley, 1969; Young, 1981). The World Glacier Monitoring Service, organized by the International Association of Hydrological Science (IAHS), the United Nations Environmental Programme (UNEP) and the United Nations Educational, Scientific and Cultural Organization (UNESCO), now provides information on the mass balance characteristics of glaciers. However, few glaciers have been monitored for more than 30 years, which is the period of instrumented time proposed by the World Meteorological Organization (WMO) as being sufficient to make a climatic classification for a region. Two of the most complete mass balance records available are those for Storglaciären in Sweden (started in 1946; Schytt, 1962; Østrem *et al.*, 1973) and Storbreen in Norway (started in 1949; Østrem and Haakensen, 1993), which are often used to illustrate yearly fluctuations in mass balance (Fig. 2.13). In North America the longest mass balance studies are those undertaken on the South Cascade and Blue Glaciers in Washington State, USA, since 1956 and on Peyto, Place and Sentinel glaciers in western Canada since 1965 (Letreguilly and Reynaud, 1989). In the European Arctic, mass balance measurements have been collected for Bröggerbreen, Lövenbreen and Voringbreen on Svalbard since 1966/67 (Hagen and Liestøl, 1990; Hagen *et al.*, 1991; Lefauconnier and Hagen, 1990).

Fig. 2.13 Storbreen, central Norway, where annual mass balance meaurements were begun in 1949. (Photo: D.I. Benn)

Representative mass balance measurements are difficult to obtain for large ice sheets, owing to their sheer size. Modern accumulation patterns on the Greenland and Antarctic ice sheets have been mapped (e.g. Ohmura and Reeh, 1991; Giovinetto and Bentley, 1985), and short-term average values are available. Estimates of the annual average accumulation for the Greenland ice sheet range from 100 mm water-equivalent in the north-central area, to 1500 mm on the south-east coast, reflecting the moisture availability. Estimates of ablation for the Antarctic ice sheet are particularly difficult, and, besides the predominant loss of mass by calving, must take melting below 500 m into account as well as the small amounts of evaporation and wind-driven snow. Estimating the total lost by calving from floating glaciers and ice shelves is complicated by the fact that the floating ice may either melt or accrete at the base, according to local conditions (Section 8.2.3.2). This problem is eliminated if the ice sheet boundary is considered to be at the grounding line or the point at which flotation starts. The mass lost by the ice sheet to a floating margin can then be quantified if the ice thickness and velocity at the grounding line can be calculated.

The inaccuracy of ablation estimates for the Antarctic and Greenland ice sheets is manifest in the various mass balance figures that have been produced. These are summarized by Paterson (1994), who provides ranges of +290 to −2090 Gt for the Antarctic ice sheet, suggesting that both growing and shrinking scenarios can be entertained on the basis of present data (1 Gt = 1016 billion kg). Annual mass balance estimates of the Greenland ice sheet from several studies have been averaged by Weidick (1985), who reports an accumulation total of 500 ± 100 Gt, a calving total of 205 ± 60 Gt, and a melt and runoff total of 295 ± 100.

2.3.2.2 REMOTE SENSING METHODS

Glacier mass balance can also be determined from remote sensing data in the form of aerial photographs or satellite images obtained on successive years or over longer intervals. Changes in the altitude of the glacier surface are measured, the difference providing the change in glacier volume, which can be converted into mass by estimating the average densities of snow, firn and ice on different parts of the glacier. Because of the small changes in mass that are involved on most glaciers, a very high level of accuracy is required in data collection and analysis, and suitable aerial photographs and images for the required times are generally expensive to obtain. However, such data open up the possibility of determining glacier mass balance in very remote areas and for the continent-scale ice sheets of Greenland and Antarctica (Williams, 1986; Williams and Hall, 1993). The problems of determining the mass balance of the Greenland and Antarctic ice sheets are immense because of their sheer size and because they are still responding to climatic events that took place hundreds or even thousands of years ago. Zwally *et al.* (1989) and Zwally (1989) used satellite radar altimeter data to argue that the Greenland ice sheet thickened by up to 20 cm yr^{-1} between 1978 and 1986, a trend that they attributed to increased snowfall carried in by warm moisture-bearing winds. However, this conclusion has been questioned because radar altimeter data are subject to large errors, and the suggested accumulation rates appear to be unreasonably high (Douglas *et al.*, 1990; Oerlemans, 1993). Resolution of these difficulties must await data from more accurate techniques such as laser altimetry. Aircraft-borne laser systems provide extremely accurate ice surface data, and are capable of resolving even small-scale topographic detail. Satellite laser altimeter data for the polar ice sheets should become available around the turn of the century, following the launch of the Earth Observing System (EOS), providing accurate measurements of ice sheet mass balance for the first time. Airborne geophysical and ice radar surveys also provide accurate information from which to reconstruct combined ice sheet surface profiles and subglacial bedrock topography (e.g. Björnsson, 1981; Drewry, 1983; Dowdeswell *et al.*, 1984a, b, 1986).

Satellite images have been particularly useful in documenting the loss of glacier mass by calving from the larger floating portions of ice bodies in both Antarctica (e.g. Jacobs *et al.*, 1986, 1992; Ferrigno and Gould, 1987; Williams and Ferrigno, 1988) and the Arctic (e.g. Jeffries *et al.*, 1992; Jeffries and Shaw, 1993). Such methods of quantification of calving are inherently more accurate than those involving estimates based upon iceberg reports from ships, which have largely formed the basis of calving estimates for the Antarctic ice sheet (e.g. Orheim, 1985). Excellent examples of the use of satellite imagery and aerial photographs in the study of glacier fluctuations are available in the series of US Geological Survey Professional Papers 1386A–K, edited by Williams and Ferrigno.

2.3.2.3 HYDROLOGICAL METHODS

Glacier mass balance can also be estimated by measuring other aspects of the water balance of a glaciated catchment, such as precipitation and runoff. In general terms, the net balance of a glacier can be expressed as

$$B_n = P - R - E \tag{2.2}$$

where B_n is the net balance, P is precipitation, R is runoff and E is evaporation.

More sophisticated versions of this relationship are provided by Collins (1984). To calculate the net balance, or the change in storage, the annual precipitation, runoff and evaporation must be measured or estimated for the whole drainage basin. Such measurements have been made for a large number of glaciers from each of the world's major glaciated ranges (Collins, 1984; Ribstein *et al.*, 1995), although most are for glaciers smaller than 10 km^2 in area, owing to sampling problems in larger catchments. This method cannot distinguish changes in snow and ice volume from changes in storage of liquid water in the glacier system, and so yields results which differ from those derived by direct measurements of mass balance at the glacier surface. Information gained by hydrological methods, however, can provide results more useful for water resource prediction and management. Glacier hydrology is discussed in detail in Chapter 3.

2.3.2.4 CLIMATIC CALCULATIONS

Estimates of glacier mass balance can also be obtained from meteorological data, using measurements or estimates of precipitation, radiation flux, temperature and other factors in energy balance calculations, as described in Section 2.2.3.2. Calculation of the energy balance can provide a powerful method of predicting or reconstructing ablation rates, provided the various sources of energy can be measured or estimated (Müller and Keeler, 1969; Hay and Fitzharris, 1988; Munro, 1990; Braithwaite and Olesen, 1990a; Brazel *et al.*, 1992). Where detailed meteorological data are not available, however, it may be more appropriate to adopt a simpler, statistical approach to estimating ablation rates using mean annual temperatures or *positive degree-days*, defined as the sum of the mean daily temperature (T_i) for all days in which $T_i > 0°C$ (Braithwaite, 1985; Braithwaite and Olesen, 1989; Reeh, 1991; Laumann and Reeh, 1993). For a sample of Norwegian glaciers, Laumann and Reeh (1993) found melt rates of 3.5–5.6 mm H_2O per positive degree-day (mm K^{-1} day^{-1}) for snow, and 5.5–7.5 mm K^{-1} day^{-1} for ice (the difference is due to the higher albedo of snow surfaces, with correspondingly less energy available for melting). Melt rates per degree-day are higher for more maritime locations. This is because higher wind speeds and humidity mean more melting at any given temperature because of sensible heat transfer and the latent heat of condensation.

Using six years of data on Nordbogletscher in Greenland, Braithwaite and Olesen (1989) found a very strong correlation between annual ice ablation and melting degree-days, and a strong correlation between annual ice ablation and mean summer temperatures. Similarly, the annual mass balance of the Barnes ice cap on Baffin Island, Canada, is strongly correlated with mean summer temperature (Hanson, 1987). The Brøggerbreen net balance data also correlate well with estimated net balance figures which were calculated using temperature data (melting degree-days) from Longyearbyen and Ny Alesund (Fig. 2.14; Hagen *et al.*, 1991).

2.3.3 Mass balance gradients

On many glaciers, the amounts of annual ablation and accumulation vary systematically with altitude, although this simple pattern is often complicated by local influences. The rates at which annual ablation and accumulation change with altitude are termed the *ablation gradient* and the *accumulation gradient*, respectively. Taken together, they define the *mass balance gradient*. Several examples of mass balance gradients for North American glaciers are illustrated in Fig. 2.15.

The mass balance gradient links the climatic controls on ablation and accumulation with glacier behaviour, particularly rates of motion, and is therefore a very important measure of glacier activity. Steep mass balance gradients result from heavy snowfall in the upper accumulation area and high ablation rates near the snout, and correspond with high rates of ice throughput (Andrews, 1970; Kuhn, 1984). Such high-activity glaciers occur in moist, mid-latitude areas such as southern Alaska, western Norway, New Zealand and Patagonia. In contrast, shallow mass balance gradients indicate small differences in specific mass balance across the altitudinal range of a glacier, and are associated with slow-moving, low-activity glaciers. Examples of this type are the Devon Ice Cap and White Glacier in Arctic Canada, and the glaciers around the Dry Valleys, Antarctica (Fig. 2.16).

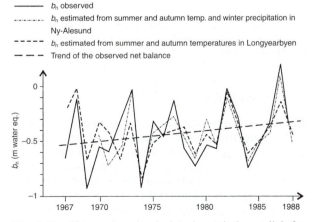

Fig. 2.14 Observed and calculated net balance (b_n) for Brøggerbreen, Svalbard, for the period 1967–88. (Modified from Hagen *et al.*, 1991)

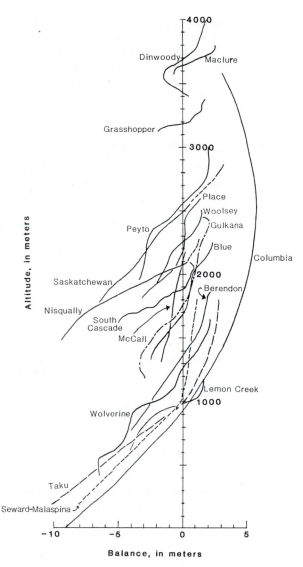

Fig. 2.15 Annual mass balance gradients for glaciers in western North America. The curves show the mean annual amount of net ablation (negative balance) and accumulation (positive balance) for the altitudinal range of each glacier. (From Mayo, 1984. Reproduced by permission of Scandinavian University Press)

2.3.3.1 ABLATION GRADIENTS

In the lower atmosphere, air temperature decreases with barometric pressure, because the thermal energy in a mass of air is proportional to its density. As a result, temperature declines with higher altitude, with average *lapse rates* of about 0.65°C per 100 m, but varying between 0.4 and 0.9°C per 100 m depending on air-mass characteristics (Barry and Chorley, 1992). Thus, ablation generally varies approximately linearly with altitude, being highest at the snout and decreasing at higher altitudes (Schytt, 1967). Ablation

gradients are steepest where the air temperature is frequently above 0°C near the snout, and falling to lower values higher up. This is often the case on mid-latitude glaciers in the summer, but the very steepest ablation gradients are found in the tropics, where ablation occurs throughout the year (Kaser, 1995). Shallow ablation gradients are typical of cold regions, where the main control on ablation is short-wave radiation, which is only weakly dependent on altitude (Kuhn, 1984).

Non-linear ablation gradients can result from altitudinal variations in cloudiness and humidity (which influence radiation receipts, latent and sensible heat transfer, and rates of evaporation and sublimation), proximity to rock walls (which influence amounts of available long-wave radiation and sensible heat transfer), the amount of shading, and glacier aspect (Fig. 2.15; Whittow, 1960; Kruss and Hastenrath, 1987, 1990; Oerlemans, 1989). The commonest cause of non-linear gradients, however, is the presence of debris on the lower parts of glaciers. Ablation on debris-covered glacier snouts may be negligible, and the highest rates may occur in the upper part of the ablation area, where debris cover is thin. In this case, ablation will first increase, then decrease with altitude (Inoue, 1977; Fig. 2.17).

2.3.3.2 ACCUMULATION GRADIENTS

On many cirque and valley glaciers, the amount of net annual accumulation generally increases with altitude, rising from zero at the equilibrium line (Fig. 2.15). This is because the passage of moisture-bearing winds over mountain barriers causes uplift, cooling the air and increasing precipitation, and because the overall proportion of precipitation that falls as snow increases at cooler, higher altitudes. On some mountain ranges, however, the amount of snowfall does not increase with height, and may even decrease (see Columbia Glacier, Fig. 2.15). This tends to be the case in cold high-latitude or high-altitude environments, because very cold air masses cannot carry much moisture, and are therefore inefficient transporters of precipitation. Snowfall can also decrease with altitude in certain high mountain environments where the upper slopes protrude above the level of the principal snow-bearing weather systems. For example, Whittow (1960) showed that in the Ruwenzori Mountains in central Africa, converging moist air masses during the June/July wet season tend to occur at relatively low altitudes, while dry, clear air prevails at the level of the highest summits. Snow accumulation totals may also decrease with altitude on large ice caps and ice sheets, where horizontal distance from moisture sources is an important factor. This is particularly striking on the Antarctic

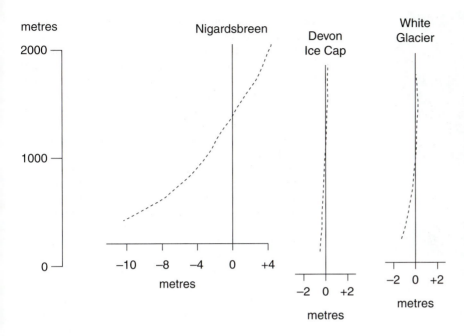

Fig. 2.16 Mass balance gradients for Nigardsbreen, western Norway, and Devon Island ice cap and White Glacier, Arctic Canada. Mass balance on Nigardsbreen exhibits large changes with altitude and the glacier is said to have a steep mass balance gradient. The Arctic glaciers show only small mass balance variation with altitude, and have shallow mass balance gradients. Note that steep mass balance gradients are represented by easier-angled lines on the diagram. (Data from Kuhn, 1984)

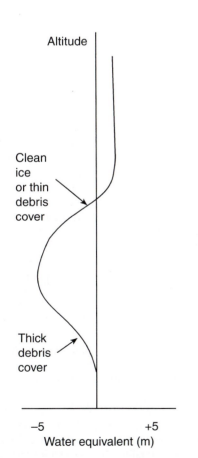

Fig. 2.17 Schematic mass balance gradient for the Khumbu Glacier, Nepal Himalaya, showing the effect of debris cover on the ablation gradient. (Redrawn from Inoue, 1977)

ice sheet, where many low-lying coastal areas have higher precipitation than the high continental interior (Section 1.2.2).

The shape of accumulation gradients can also be strongly influenced by topography. In particular, the redistribution of snowfall by avalanching from steep slopes, and wind scouring from exposed areas, can result in accumulation patterns that differ markedly from original climatically controlled distributions. This is strikingly demonstrated in high mountain environments such as the Himalaya, where many glaciers gain most mass in huge avalanche cones located close to the equilibrium line (Fig. 2.18).

2.3.3.3 THE BALANCE RATIO

Because they are controlled by different climatic variables, the accumulation and ablation gradients on any one glacier generally have different values, with the ablation gradient somewhat steeper than the accumulation gradient. Therefore, most mass balance curves show an inflection at the equilibrium line altitude (Fig. 2.15). The ratio between the two gradients is known as the *balance ratio*, defined as:

$$\text{BR} = b_{\text{nb}}/b_{\text{nc}} \tag{2.3}$$

where b_{nb} and b_{nc} are the mass balance gradients in the ablation area and accumulation area, respectively (Furbish and Andrews, 1984).

The balance ratio ignores any non-linearity that may exist in the respective mass balance gradients Sections 2.3.3.1 and 2.3.3.2), but it remains a useful

Fig. 2.18 A snow avalanche on Taboche, Khumbu Himal, Nepal. Avalanches such as this provide an important input of accumulation on high-altitude glaciers, and strongly influence the form of the mass balance gradient. (Photo: L.A. Owen)

generalization that summarizes the overall mass balance curve of a glacier. For a sample of 22 glaciers in Alaska, Furbish and Andrews (1984) found that balance ratios average about 1.8, indicating that the vertical change in mass balance is 1.8 times as large in the glacier ablation areas as it is in their accumulation areas. The value of 1.8 is slightly lower than that of 2.2 measured for the well-known South Cascade Glacier, Washington State (Fig. 2.19; Meier and Tangborn, 1965; Furbish and Andrews, 1984), and an overall value of 2 may be representative of maritime mid-latitude glaciers in general. In contrast, the balance ratios of tropical glaciers are much higher because of year-round intense melting in their ablation zones and very poor dependence of accumulation with altitude. Analysis of data presented by Francou *et al.* (1995) for the Zongo Glacier, Bolivia, indicates a balance ratio of about 25 (Fig. 2.20).

Glaciers with high balance ratios have small ablation areas compared with the area of the glacier, because only a small area of ablating ice is required

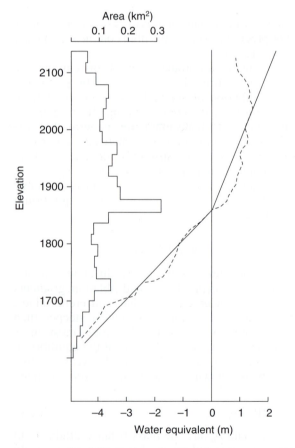

Fig. 2.19 Steady-state mass balance gradient for South Cascade Glacier for the years 1957/58–1963/64 (dashed line), and its representation as linear ablation and accumulation gradients (solid lines). (Redrawn from Meier and Tangborn, 1965)

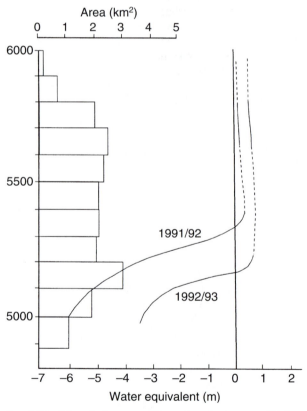

Fig. 2.20 Mass balance gradient for the Zongo Glacier, Cordillera Real, Bolivia, for two successive years. Dashed lines represent interpolated values. During 1992/93 the glacier as a whole was approximately in balance, whereas in 1991/92 it was negative. The marked inflection in the mass balance gradient at the equilibrium line altitude is typical of tropical glaciers where ablation occurs throughout the year. (Redrawn from Francou *et al.*, 1995)

to balance inputs from snowfall higher up on the glacier. In contrast, glaciers with lower balance ratios require larger ablation areas to balance inputs, although the ablation area will still generally be less than half of the whole area of the glacier. Hence, tropical glaciers tend to have smaller ablation areas than mid-latitude glaciers with similar geometry, and their equilibrium line altitudes will be correspondingly closer to the snout.

2.3.4 The equilibrium line

2.3.4.1 CLIMATIC SIGNIFICANCE

Because the equilibrium line is the place where annual accumulation totals exactly balance ablation totals, the equilibrium line altitude (ELA) is very closely connected with local climate, particularly precipitation and air temperatures. The ELA is sensitive to perturbations in either of these two variables, and rises in response to decreasing snowfall and/or increasing frequency of positive air temperatures, and vice versa. Perhaps the best illustration of glacier–climate interactions is the relationship between the net balance and the equilibrium line altitude. When the annual mass balance of the glacier as a whole is negative, the ELA rises, and when the balance is positive, the ELA falls. The ELA value associated with zero annual mass balance is known as the steady-state ELA, because the glacier mass and geometry are in equilibrium with climate. Variations in the altitude of the equilibrium line on a particular glacier, therefore, can be used as an indicator of climatic

fluctuations (Kuhn, 1981). For example, 20 years of mass balance calculations (1967–88) have been used by Hagen *et al.* (1991) to illustrate the impact of climate on the equilibrium line altitude (Fig. 2.21). They used linear regression analysis on the mass balance figures and equilibrium line altitudes for Brøggerbreen and Lovenbreen, Svalbard, over the period 1967–88, which yield ELAs for zero net balance of 246 m and 284 m for the respective glaciers. Changes in the mass balance and ELA of the glaciers can result from changes in the number of positive degree-days or variations in effective precipitation. Variations in ELA with net balance have also been measured for Storbreen in Norway (Liestøl, 1967), Storglaciären in Sweden (Schytt, 1981), Peyto Glacier in Canada (Young, 1981), and other sites.

Fluctuations in the ELA therefore provide an important indicator of glacier response to climate change, and allow reconstructions of former climates and the prediction of future glacier behaviour. Although the ELA is determined by local weather conditions, it is a good indicator of regional climate, because glacier mass balance fluctuations are often strongly correlated over distances of *c.* 500 km (Letreguilly and Reynaud, 1989). However, the area over which glacier ELAs accurately reflect climate will vary according to the steepness of climatic gradients, and will be smallest in the vicinity of major topographic barriers to airflow.

It is useful to specify the climate at the ELA as some unique combination of precipitation and temperature. Figure 2.22 shows the relationship between mean summer temperatures and the mass lost at the

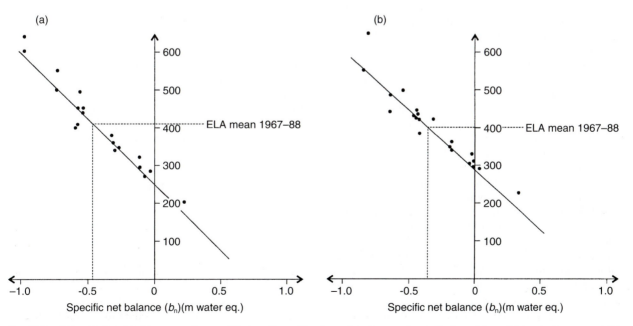

Fig. 2.21 The relationship between the equilibrium line altitude and net annual mass balance for (a) Brøggerbreen and (b) Lovenbreen, Svalbard, for the period 1967–88. (Modified from Hagen *et al.*, 1991)

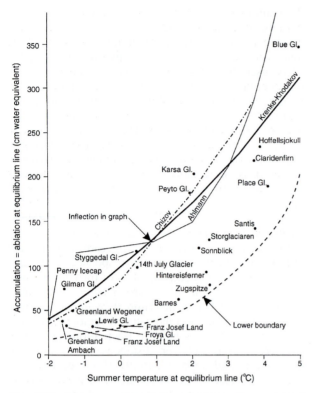

Fig. 2.22 Mean summer temperatures and amount of water lost at the equilibrium line altitude for selected glaciers, and showing curves presented by Ahlmann (1948), Chizov (1964), and Krenke and Khodakov (1966). (Modified from Loewe, 1971)

ELA. The positive correlation between these two variables for a wide range of glaciers reflects the fact that higher levels of mass turnover at the ELA require higher ablation (and hence summer temperatures) to balance the annual specific mass budget. Thus, if the temperature at a glacier ELA is known, the winter precipitation can be estimated. This relationship also explains the regional rise in glacier ELAs with distance from moisture sources in many parts of the world, including western Norway, the Andes and the Himalaya. In areas of lower precipitation, the temperatures required to melt the annual accumulation at the ELA need not be so high as in areas of high precipitation. Thus ELAs will tend to be at higher altitudes, where the temperatures are lower. Recently, Ohmura *et al.* (1992) have correlated mean summer air temperature at the ELA with total annual precipitation for 70 mid- and high-latitude glaciers (Fig. 2.23). Annual precipitation was used rather than annual turnover at the ELA because the former is more readily available for many areas of the world. The predictive power of this relationship was further improved by also considering the amount of incoming short-wave and net long-wave radiation at the ELA. Glaciers with higher radiation receipts

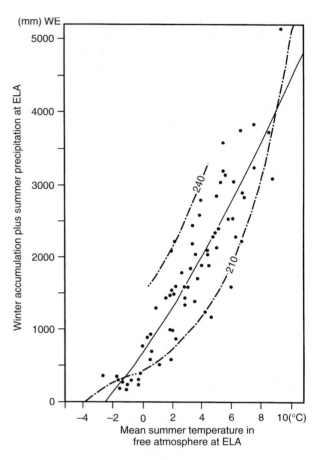

Fig. 2.23 Annual total precipitation and the free-atmospheric temperature observed at the equilibrium line altitudes for 70 glaciers. The dotted lines indicate the best-fit curves for summer radiation of 240 and 210 W m⁻² on the glaciers. (Redrawn from Ohmura *et al.*, 1992)

(i.e. those at lower latitudes or facing towards the equator) will experience greater amounts of ablation relative to air temperature than those with low receipts, and therefore require larger precipitation totals to balance the annual mass budget. Ohmura *et al.* (1992) did not consider tropical glaciers, although a broadly similar sort of relationship should be expected.

2.3.4.2 DETERMINING THE EQUILIBRIUM LINE ALTITUDE

The most rigorous method for determining glacier ELAs is by measurement of glacier mass balance. This allows identification of the annual ELA and, when conducted over several years, the steady-state ELA for equilibrium conditions of zero mass balance (Fig. 2.21; Section 2.3.4.1). However, this approach is very costly and time-consuming, and accurate long-term measurements are available for only a small proportion of the world's glaciers. Fortunately, other methods are available for identifying or estimating glacier ELAs. The ELA for any given year can be identified by observing the distribution of snow and ice on the glacier surface at the time of year when glacier mass is at a minimum. This information can be obtained by field survey or from aerial photographs, observations made during low-level flights over the area, or satellite images (LaChapelle, 1962; Meier and Post, 1962; Østrem, 1975a). On temperate glaciers, the ELA coincides with the *transient snow line*, defined as the lower limit of snow on the glacier surface at the end of the ablation season, and is identifiable in the field and on air photographs and satellite images as the boundary between white snow cover and the off-white glacier ice or firn (Fig. 2.24 (Plate 11); Østrem, 1973, 1975a; Dowdeswell and Drewry, 1989; Kulkarni, 1992). It is also possible to identify signs that indicate the longer-term health of the glacier. For example, Paterson (1994) lists the following situations:

1. a low snow line possessing a sharp boundary with bare ice indicates a positive net balance;
2. a similar boundary but at a high altitude indicates that a period of accumulation has interrupted a long period of negative balance;
3. a snow line separated from bare ice by an area of old firn indicates a more negative balance than the preceding few years.

Locating the equilibrium line on glaciers in polar regions is more difficult, because it is usually at a lower elevation than the snow line, at the lower limit of the superimposed ice zone (Section 2.2.1). Superimposed ice represents accumulation, but can be difficult to distinguish from old glacier ice on air photographs and satellite images.

Air surveys are particularly useful in remote regions, and for giving an impression of the mass balance status of several glaciers in a region. The results are most reliable when checked against direct measurements of glacier mass balance for one or more examples in the same area.

2.3.4.3 RECONSTRUCTING FORMER EQUILIBRIUM LINE ALTITUDES

A variety of methods have been devised to estimate the steady-state ELAs of vanished glaciers, to provide a means of reconstructing former climates in glaciated regions. The most rigorous of these methods are based on the three-dimensional form of the glacier surface, combined with assumed mass balance–altitude relationships. For these methods, a contour map of the former glacier is required, based on geomorphological evidence such as lateral and terminal moraines and periglacial trimlines (Section 12.6.1.2).

Contours of the former glacier surface are drawn joining points of equal altitude at the glacier margins and trending at right angles to any indicators of ice flow direction, such as striae, drumlins and fluted moraines. This exercise is most straightforward for former valley glaciers, where the gradients of the glacier margins are readily established, and where contours are likely to be convex downvalley near the snout and concave in the upper part. The steady-state ELA can then be estimated by the *accumulation area ratio method*, which is based on the assumption that the accumulation area of the glacier (i.e. the area above the ELA) occupies some fixed proportion of the total glacier area (Meier and Post, 1962; Porter, 1975b; Torsnes *et al.*, 1993). If the accumulation area ratio (AAR) is assumed to be 0.6, for example, the ELA is set as the altitude of the surface contour that lies below 0.6 (60 per cent) of the total area (Fig. 2.25). The method is implicitly based on the likely shape of the former mass balance curve, particularly the balance ratio (Section 2.3.3.3). For a rectangular, slab-shaped glacier, the balance ratio (b_{nb}/b_{nc}) equals 1/AAR, so for an AAR of 0.6, BR = 1/0.6 = 1.67. For modern mid- and high-latitude glaciers, steady-state AARs generally lie in the range 0.5–0.8 (Meier and Post, 1962; Hawkins, 1985), with typical values lying

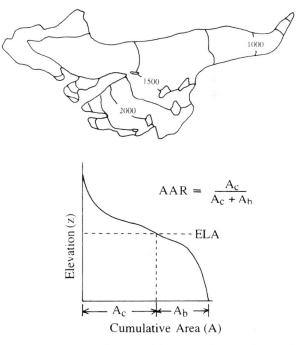

Fig. 2.25 Contoured map and hypsometric curve for a valley glacier. These data can be used in conjunction with an assumed value of the accumulation area ratio (AAR) to derive an estimate of the former steady-state equilibrium line altitude. For example, when the AAR is set at 0.6, the hypsometric curve is used to read off the altitude that lies below 60 per cent of the former glacier area. (Modified from Furbish and Andrews, 1984)

in the range 0.55–0.65 (Porter, 1975b). Glaciers with debris-covered snouts will have rather lower AARs (<0.4), owing to the effect of debris on lowering ablation and increasing the relative size of the ablation area (Müller, 1980; Kulkarni, 1992). ELAs have been estimated by calculating the area-weighted mean altitude or median elevation of glaciers (MEG: Manley, 1959; Sissons, 1974c, 1977a; Sissons and Sutherland, 1976; Meierding, 1982). Sutherland (1984b) found that, for a sample of modern Norwegian glaciers, this method consistently overestimated ELAs, suggesting that an AAR of around 0.6 would be more realistic for glaciers on the eastern seaboard of the North Atlantic. The MEG method is probably best suited to regular-shaped glaciers whose area and altitude distributions do not have a wide range of variation (Meier and Post, 1962; Porter, 1981b; Nesje, 1992).

One shortcoming of the AAR and MEG methods is that they take little account of variations in glacier shape, particularly the distribution of glacier area over its altitudinal range, or *hypsometry*. Accumulation area ratios on modern glaciers are influenced by glacier hypsometry so that, for example, a glacier with a wide accumulation area and a narrow snout will have a different AAR as compared with one with a narrow accumulation basin and a broad snout, even if the ELAs are the same (Furbish and Andrews, 1984). Therefore, former glacier ELAs based on a uniform assumed AAR value may be subject to significant errors if there is a wide range of glacier types and shapes in the area under consideration. To overcome this difficulty, Furbish and Andrews (1984) developed the *balance ratio method*, which takes account of both glacier hypsometry and the shape of the mass balance curve. The method is based on the

fact that, for equilibrium conditions, the total annual accumulation above the ELA must exactly balance the total annual ablation below the ELA. This can be expressed in terms of the areas above and below the ELA multiplied by the average accumulation and ablation, respectively:

$$\bar{d}_b A_b = \bar{d}_c A_c \qquad (2.4)$$

where $\bar{d}_b$ is the average net annual ablation in the ablation area, $\bar{d}_c$ is the average net annual accumulation in the accumulation area, A_b is the area of the ablation area, and A_c is the area of the accumulation area.

If the ablation and accumulation gradients are assumed to be linear, $\bar{d}_b$ and $\bar{d}_c$ are equal to the ablation and accumulation at the area-weighted mean altitudes of the ablation area ($\bar{z}_b$) and the accumulation area ($\bar{z}_c$), respectively. By convention, $\bar{z}_b$ and $\bar{z}_c$ are measured positively from the ELA, and are shown as the heights associated with ½ A_b and ½ A_c on Fig. 2.26. For equilibrium conditions, the altitudes of $\bar{z}_b$ and $\bar{z}_c$ are determined by the balance ratio, according to the relationship:

$$b_{nb}/b_{nc} = \bar{z}_c A_c / \bar{z}_b A_b \qquad (2.5)$$

The steady-state ELA is then defined as the altitude that satisfies equation (2.5) for the balance ratio representative of the area under study (Section 2.3.3.3). The ELA can be found by drawing the hypsometric curve for the glacier, choosing trial values of the ELA and measuring off values of $\bar{z}_c$, $\bar{z}_b$, A_b and A_c, and testing whether b_{nb}/b_{nc} is equal to the desired value. The trial ELA can then be raised or lowered until equation (2.5) is satisfied. Alternatively, the procedure can be fully computerized, making appli-

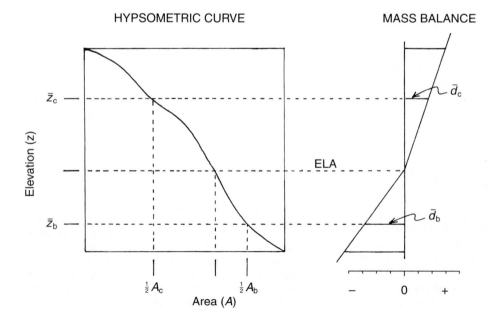

HYPSOMETRIC CURVE

MASS BALANCE

Elevation (z)

$\bar{z}_c$

$\bar{z}_b$

ELA

$\bar{d}_c$

$\bar{d}_b$

½ A_c ½ A_b

Area (A)

−   0   +

Fig. 2.26 Graphical representation of the components of the balance ratio method of calculating glacier equilibrium line altitudes (ELAs); ½ A_b and ½ A_c denote the mid-area values of the ablation area and accumulation area, respectively, and $\bar{z}_b$ and $\bar{z}_c$ are the associated altitudes, measured relative to the ELA. At these altitudes, the net ablation ($\bar{d}_b$) and net accumulation ($\bar{d}_c$) are taken to be equal to the mean values for the ablation and accumulation areas. (Modified from Furbish and Andrews, 1984)

cation of the balance ratio method very quick and easy (Benn and Gemmell, 1997). The influence of hypsometry on glacier response to changes in the ELA is discussed in Section 2.3.6.

Other methods of approximating the steady-state ELA of former valley glaciers have been reviewed by Meierding (1982), Nesje (1992) and Torsnes *et al.* (1993), and include mapping the maximum elevation of lateral moraines (MELM), and setting the ELA at some fixed ratio between the toe of the glacier and the top of the valley headwall (toe-to-headwall altitude ratio or THAR). The *maximum altitude of lateral moraines* provides a rough guide to the former ELA because the flow of the glacier from its accumulation zone to its ablation zone results in the net transfer of debris out of the accumulation zone along the centre line of the glacier and then out towards the margin of the ablation zone, and lateral moraines are deposited only below the ELA (Andrews, 1975). The corollary is that the uppermost altitude of an abandoned lateral moraine marks the palaeo-ELA. A problem arises here when glacier retreat is slow, and continuous provision of debris results in the incremental deposition of lateral moraines in an upslope direction. However, glacier thinning will eventually result in inset moraines (Fig. 2.27). A further problem is that lateral moraines can degrade rapidly on deglaciation and disappear from steep slopes, thus giving a spuriously low ELA. However, the moraine method may be the best approach in areas where former glaciers were debris-covered, and the shape of the mass balance gradient is poorly known.

The THAR method is very crude, as it takes no account of glacier hypsometry or climatic considerations, although it provides a quick method of estimating former ELAs in remote regions where topographic maps are unavailable or unreliable. THARs of 0.35–0.4 have been used by Meierding (1982).

2.3.5 Annual mass balance cycles

The amount of snow and ice stored in glaciers undergoes systematic changes throughout the year, following cycles of gain and loss depending on the seasonal distribution of accumulation and ablation. Several types of cycle are possible depending on the climatic regime, especially the timing of warm and cold seasons, maximum precipitation, and variations in the proportion of precipitation falling as snow (Ageta and Higuchi, 1984; Kaser, 1995). Three types of cycle are discussed here: (a) winter accumulation type, with a well-defined winter accumulation season and summer ablation season; (b) summer accumulation type, with maxima in accumulation and ablation occurring simultaneously during the summer months; and (c) year-round ablation type, with one or two accumulation maxima coinciding with wet seasons.

2.3.5.1 WINTER ACCUMULATION TYPE

The winter accumulation type is the most familiar form of glacier mass balance cycle, and is characteristic of mid- and high-latitude glaciers such as those in the European Alps, Scandinavia, North America and New Zealand (Pytte and Østrem, 1965; Meier and Tangborn, 1965). In these areas, there are pronounced seasonal variations in temperature, resulting in distinct winter accumulation and summer ablation seasons (Fig. 2.28a). Accumulation occurs mainly during the winter, when most precipitation falls as snow, and ablation is at a minimum. As a result, the mass of the glacier as a whole increases during the winter, and is at a maximum in the spring, just prior to the onset of the ablation season. During the summer months (June to September in the northern hemisphere), melting predominates over accumulation, particularly on the lower part of the glacier. Much of the precipitation may fall as rain, although significant snowfalls may still occur at higher altitudes. The mass of the glacier is at a minimum at the end of the ablation season, just prior to the first snows of winter. It should be noted, however, that on some low-altitude glacier snouts, such as those in southern Iceland, ablation may also occur in the winter during the passage of warm fronts.

2.3.5.2 SUMMER ACCUMULATION TYPE

In the summer accumulation type of mass balance cycle, maxima in accumulation and ablation occur

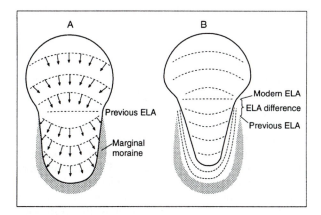

Fig. 2.27 The principle of calculating the depression of equilibrium line altitude on a glacier based on the maximum elevation of lateral moraines (MELM). A previous extent (A) is compared with the modern extent (B) of an idealized glacier. Recessional moraines document ice recession. Dashed lines are contours and arrows indicate direction of flow lines and glacier velocity. (Modified from Nesje, 1992)

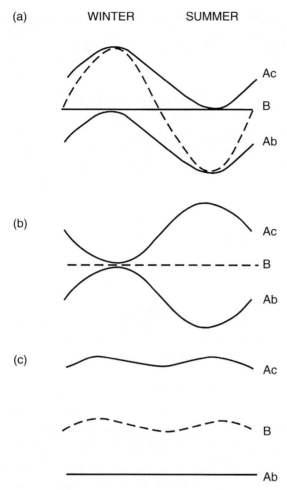

(a)   WINTER   SUMMER

Ac

B

Ab

(b)

Ac

B

Ab

(c)

Ac

B

Ab

Fig. 2.28 Annual mass balance cycles for glaciers of (a) winter accumulation type, (b) summer accumulation type, and (c) year-round ablation type. Ac: accumulation; Ab: ablation; B: balance. (Modified from Ageta and Higuchi, 1984)

season precipitation totals, and the annual net balance may vary widely from one year to the next (Kaser *et al.*, 1990; Francou *et al.*, 1995).

2.3.5.3 YEAR-ROUND ABLATION TYPE

The year-round ablation type of cycle is characterized by ablation throughout the year on the lower parts of glaciers, and is typical of the inner tropics, where seasonal variations in temperature are much less than diurnal variations (Ageta and Higuchi, 1984; Kaser, 1995). Cyclic variation in the amount of ablation can occur, however, as a result of seasonal changes in cloudiness, affecting radiation balance (Whittow, 1960). Accumulation can occur throughout the year on the upper parts of this type of glacier, although distinct peaks may occur coincident with one or two wet seasons (Fig. 2.28c). Double wet-season peaks affect the high-altitude glaciers in the Ruwenzori Mountains and Mount Kenya, east-central Africa, where snowfall is highest during the northward and southward passage of the Inter-Tropical Convergence Zone in March–June and September–December, respectively (Whittow, 1960; Whittow *et al.*, 1963; Hastenrath, 1984). One precipitation peak may be larger than the other, although inter-annual variability can be high. Single accumulation peaks occur on the glaciers of Mount Jaya (New Guinea), Mount Kilimanjaro (East Africa), and on Ecuadorian volcanic peaks (Hope *et al.*, 1976; Hastenrath, 1981, 1984). Again, precipitation totals can vary substantially from year to year.

2.3.6 Long-term variations in mass balance

Annual accumulation and ablation rarely exactly balance, leading to net gains or losses in mass over the year. Such gains and losses may average out over a number of years, in which case there is no long-term change in glacier mass. However, if net negative or positive mass balances are sustained for several years, the glacier will undergo significant changes in mass, resulting in retreat/advance or thinning/thickening. Advance or retreat of the glacier snout in response to mass balance changes will be delayed by a varying amount depending on the rate of mass throughput, glacier geometry and other factors. This characteristic lag is known as the *response time* (Section 7.2.1).

The mass budget or health of a glacier is measured by the *net balance*, which is the difference between two consecutive annual mass balance minima. On all types of glacier, this will usually fall just before the onset of the main accumulation season. Positive net balances (net gains in mass) result from increases in accumulation and/or reductions in ablation, and lead to expansion of the accumulation

more or less simultaneously during the summer (Fig. 2.28b; Ageta and Higuchi, 1984; Kaser, 1995). Glaciers of this type occur in high-altitude areas with a pronounced summer precipitation maximum and a cold, dry winter, such as the monsoon-dominated parts of the Himalayan chain in India and Nepal, and the high Andes of Peru and Bolivia (Yasunari and Inoue, 1978; Higuchi *et al.*, 1982; Ageta and Higuchi, 1984; Francou *et al.*, 1995; Kaser, 1995). In these areas, there is little precipitation in winter, and summer precipitation falls mainly as snow on the upper parts of the glaciers. Ablation is also at a maximum during the summer, when temperatures are highest. As a result, when ablation and accumulation totals are comparable, the total mass of the glacier undergoes little or no change throughout the year, unlike the large cyclic changes typical of winter accumulation type glaciers. Summer accumulation type glaciers are, however, sensitively dependent on wet-zone and lowering of the ELA. Conversely, negative net balances result from an excess of ablation over accumulation, reducing the size of the accumulation area and raising the ELA. In extreme cases, net ablation may occur over the entire surface so that the whole of the glacier lies below the ELA. This situation has occurred on numerous occasions in recent years on Lewis Glacier, Mount Kenya, leading to very rapid retreat (Hastenrath, 1984, 1989; Hastenrath and Kruss, 1992). Long-term trends in glacier mass balance are measured by the *cumulative net balance*, or the running total of annual net balance (Fig. 2.29).

The amount of mass gained or lost by a glacier in response to a given change in the ELA depends on the hypsometry of the glacier. For example, if a glacier has a large proportion of its area close to the ELA, a raising or lowering of the ELA will produce large changes in mass. Conversely, if a small proportion of a glacier's area is near the ELA, the same raising or lowering would have comparatively little effect. Thus, glaciers with large areas of snow and ice near the ELA will exhibit the largest responses to climate change (Fig. 2.30).

Numerous studies have examined the relationship between changes in glacier mass and climatic inputs. For example, Chen and Funk (1990) correlated mass changes of the Rhônegletscher, Switzerland, with climatic records for the period 1882–1987, and showed that most of the mass loss of the glacier is attributable to temperature increases, particularly since the 1940s. These increases form part of a general warming trend since the end of the Little Ice Age in the early nineteenth century. Chen and Funk (1990) suggest that, as a general rule, summer temperatures are more important than precipitation in governing mass changes of mountain glaciers in humid maritime areas. However, Nesje (1989) has shown that changes in both temperature and precipitation correlate with recent fluctuations of outlets of Jostedalsbreen, a maritime ice cap in south-central

Norway. In the New Zealand Alps, retreat of the Stocking Glacier has been positively correlated with monthly temperatures (including a 2-year lag) by Salinger *et al.* (1983), whereas the fluctuations of the Franz Josef Glacier were correlated with precipitation

(b)

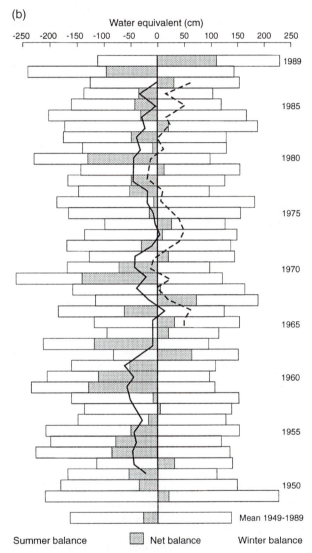

(a)

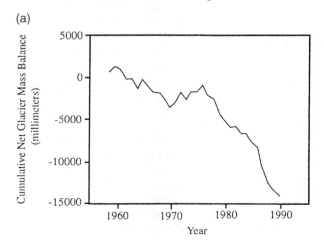

Fig. 2.29 (a) Cumulative net mass balance of South Cascade Glacier, USA, for 1959–90. The graph shows that between 1976 and 1990 the glacier lost *c.* 13 m of water equivalent, averaged over the whole glacier surface. This mass loss was accompanied by dramatic retreat of the glacier margin. (From McCabe and Fountain, 1995. Reproduced by permission of the University of Colorado) (b) Mass balance for Storbreen, Norway (from Østrem and Haakensen, 1993). The five-year running mean is indicated by the solid line. The five year running mean for Nigardsbreen is also indicated (by the dashed line). The total mass lost from Storbreen for a 41-year period (1949–89) is 10.5 m water equivalent. The average net balance for Nigardsbreen is approximately 0.6 m higher than that for Storbreen. (Redrawn from a figure provided by Richard Williams, USGS)

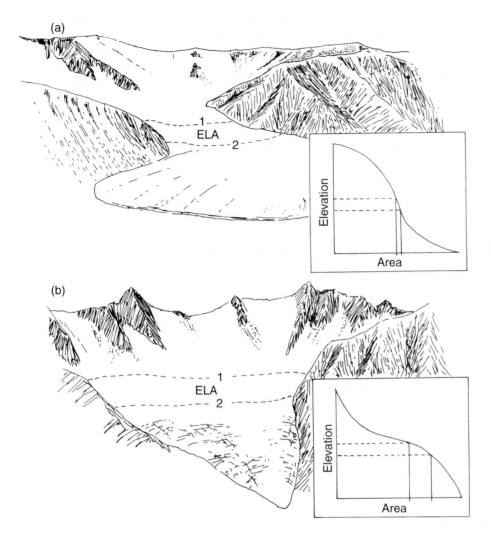

Fig. 2.30 Effects of glacier hypsometry on glacier response to a given change in the equilibrium line altitude (ELA). Glacier A has a relatively small area close to the ELA, whereas glacier B has a large area near the ELA. A fall in ELA from position 1 to position 2 will have a larger effect on the mass balance of glacier B than glacier A

by Hessell (1983) and Brazier *et al.* (1992). The fact that these two glaciers lie on different sides of the main divide may provide some explanation for the inconsistency: the high precipitation values on the west side of the divide, where the Franz Josef Glacier is located, are probably large enough to overprint other climatological variables. A similar situation was reported by Pelto (1989) and Letreguilly (1988) for glaciers in the Coast Ranges of western North America, where accumulation season precipitation is the dominant control of mass balance. Letreguilly also reported that the mass balance of Peyto Glacier, located inland in the Rocky Mountains, is controlled by summer temperature. Problems in isolating a single variable responsible for glacier fluctuations in New Zealand have been overcome by using energy balance models (e.g. Owens *et al.*, 1984; Marcus *et al.*, 1985; Hay and Fitzharris, 1988; Sections 2.2.3.2 and 2.3.2.4).

McCabe and Fountain (1995) demonstrated that negative cumulative mass balance of the South Cascade Glacier, Washington, USA, reflects reduced winter snowfall associated with large-scale shifts in atmospheric circulation over the North Pacific Ocean and northern North America. They showed that, since the 1970s, greater persistence of the Aleutian Low over the north-west Pacific and the associated occurrence of a high-pressure system over western Canada have led to frequent dry north or north-easterly airflow over the glacier, leading to anomalously low snowfall. Links between large-scale atmospheric and oceanic circulation and glacier net balance have also been identified by Thompson *et al.* (1984) and Francou *et al.* (1995), who showed that years of low mass balance on Peruvian and Bolivian glaciers coincide with the occurrence of El Niño, a periodic eastward flow of warm surface waters in the eastern Pacific Ocean associated with large-scale air pressure oscillations (Fig. 2.31). Factors other than temperature and precipitation may also affect net balance. Hastenrath and Kruss (1992) argued that most of the recent negative cumulative balance of Lewis Glacier,

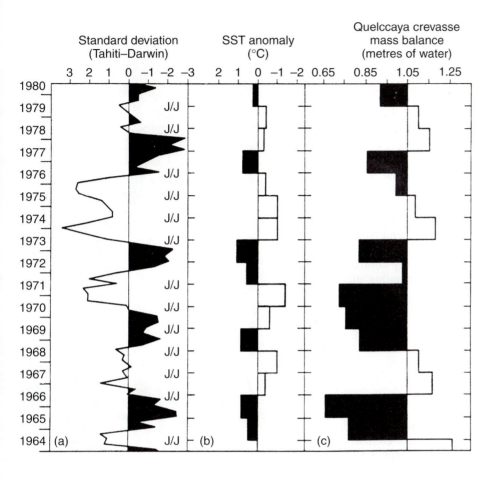

Fig. 2.31 The correlation between accumulation on the Quelccaya ice cap, Peru, and El Niño years. (a) Southern Oscillation Index, showing large-scale pressure differences across the Pacific Ocean. Negative values are associated with El Niño years. (b) Sea surface temperature (SST) anomalies off Peru. Temperatures are higher than normal during El Niño years. (c) Accumulation on Quelccaya ice cap. Lower than normal accumulation occurs during El Niño years. (From Thompson *et al.*, 1984. Reproduced by permission of *Science*)

Mount Kenya, is attributable to the effect of air humidity changes on the surface energy balance.

For vanished glaciers, where climatic records are unavailable, other methods need to be employed to investigate the controls on glacier net balance. Changes in glacier mass can be reconstructed from moraine sequences, and these changes interpreted with reference to palaeoclimatic data or theoretical models. Benn *et al.* (1992) used a combination of geomorphological evidence and pollen stratigraphy to argue that Late-glacial icefields and valley glaciers on the Isle of Skye, Scotland, underwent a two-stage pattern of deglaciation driven by different controls. They argued that initial glacier retreat resulted from a reduction in precipitation towards the end of the Younger Dryas Stade (*c.* 11,000–10,000 radiocarbon years BP), whereas final deglaciation was caused by very rapid temperature increases associated with northward migration of the North Atlantic Oceanic Polar Front at the beginning of the Holocene (cf. Ruddiman and McIntyre, 1981; Atkinson *et al.*, 1987). In South America, glacier advances around the margins of the Bolivian Altiplano during the Late Pleistocene have been correlated with substantial rises in lake levels, strongly suggesting that glacier fluctuations in this region are closely linked to precipitation changes

(Hastenrath and Kutzbach, 1985; Clapperton *et al.*, 1996a). Then, as now, glacier ELAs were lowest in the north-east of the region, indicating that temporal increases in precipitation were probably associated with the westward penetration of moist air masses from the Amazon basin.

The links between climate and glacier mass changes have also been investigated by using mathematical models to find the climatic combinations required to 'grow' or 'melt' glaciers of a known size in a specified time frame. The most detailed models calculate accumulation and ablation totals by considering changes in all contributions to glacier input and energy balance through time (e.g. Gruell and Oerlemans, 1986; Kuhn, 1989b; Braithwaite and Olesen, 1990b). However, this is very time-consuming, and many models integrate the climatic controls into two or more easily calculated parameters such as annual effective precipitation, mean temperature during the ablation season, or ablation–altitude relationships (e.g. Budd and Smith, 1981; Reeh, 1991; Laumann and Reeh, 1993). Glacier–climate models have been developed for many areas and time periods, ranging from Last Glacial icefields in the Scottish Highlands and Patagonia (Payne and Sugden, 1990; Hulton *et al.*, 1994) to twentieth century glacier recession in

Irian Jaya, Indonesia (Allison and Kruss, 1977). While such modelling exercises usefully identify possible causes of glacier fluctuations, the results contain a large degree of in-built uncertainty because the models cannot resolve the unique combination of past climatic conditions, and are also dependent on the accuracy of the laws used to calculate glacier flow rates (Section 4.3).

More rarely, positive mass balances can result from non-climatic causes, such as landsliding on to the glacier surface. In such situations the ablation rate may be reduced by the landslide debris cover, and therefore the mass balance will be positive for a short, climatically unrelated period (Post, 1967; Gardner and Hewitt, 1990; Kirkbride and Sugden, 1992; Fig. 2.17). Non-climatic changes in mass balance can also result from topographically controlled changes in calving rates of water-terminating glaciers, as described in Section 8.2.

2.4 PAST CLIMATES: THE ICE CORE RECORD

2.4.1 Ice cores

Palaeoclimatic records can be obtained from *ice cores* drilled out from glaciers, which contain an annual record of snowfall and other material which is deposited at the ice surface. In the upper layers of cores, annual accumulation layers often stand out clearly as alternate bands of clear and bubbly ice which can be dated by counting down from the surface, like tree rings. For deeper, older parts of cores, however, individual layers may not be apparent, and dating is achieved by indirect means (Bradley, 1985). Short cores can be drilled by hand (Fig. 2.32), but longer cores require mechanical drilling equipment. The longest ice core records are from the interiors of the polar ice sheets, and have yielded immensely valuable data on environmental change in both polar regions. For example, the Vostok core from Antarctica extends back about 160,000 years and the GRIP and GISP cores from Greenland span almost 100,000 years, providing continuous high-resolution records spanning the whole of the last glacial cycle (Lorius *et al.*, 1990; Jouzel *et al.*, 1993; Grootes *et al.*, 1993; McManus *et al.*, 1994). Shorter, but no less valuable, records have also been obtained from ice caps and valley glaciers at lower latitudes, including the Dunde ice cap in Tibet and the Quelccaya ice cap in Peru (Thompson *et al.*, 1986, 1989, 1990; Thompson, 1995).

Glacier ice consists mainly of water, although it also contains impurities such as air bubbles, dissolved ions and solid particles. Analyses of these impurities,

Fig. 2.32 Ice coring by hand. (Photo: N. Spedding)

as well as the molecular composition of the ice itself, have become very important techniques over the past two decades, revealing detailed pictures of ice mass evolution and past and present climatic change. Excellent reviews of ice core analysis are available in Bradley (1985) and Paterson (1994).

2.4.2 Oxygen and hydrogen isotopes

2.4.2.1 ISOTOPIC COMPOSITION OF ICE

In common with most other elements, the atoms of oxygen and hydrogen that make up pure glacier ice can occur as different *isotopes*. Isotopes are alternative forms of an element that result from variations in atomic mass, or the combined number of protons and neutrons in each atomic nucleus. For atoms of each element the number of protons is constant, so variations in mass (and therefore, different isotopes) always result from variations in the number of neutrons in the atom. Oxygen atoms always have eight protons, but may have eight, nine or ten neutrons, yielding three isotopes with atomic masses of 16, 17 and 18 respectively (^{16}O, ^{17}O and ^{18}O). Hydrogen atoms always have one proton, but may have no or one neutron, resulting in two isotopes, ^{1}H and ^{2}H (^{2}H is also known as deuterium, D). Water molecules may therefore consist of any one of nine possible combinations of these five isotopes, ranging from molecules with three light isotopes (^{1}H$_2$^{16}O) to those with three heavy isotopes (^{2}D^{18}O). However, only three of these combinations are common: ^{1}H$_2$^{16}O, ^{1}HD^{16}O, and ^{1}HD^{18}O (Hubbard and Sharp, 1989). In nature, the relative abundance of oxygen isotopes is 99.76 per cent ^{18}O, 0.04 per cent ^{17}O and 0.2 per cent ^{16}O; for hydrogen the figures are 99.984 per cent ^{1}H and 0.016 per cent D. The isotopic composition of glacier ice is known to vary systematically with several factors, the most important of which are the composition of the precipitation from which the ice was formed, and the history of melting and refreezing within the ice. Isotope analyses can therefore yield important insights into past climatic conditions and glacial processes.

2.4.2.2 ISOTOPES AND PALAEOCLIMATE

The isotopic composition of the precipitation falling on a glacier depends on its history of evaporation and condensation as part of the hydrological cycle. During evaporation, water molecules composed of light isotopes turn to vapour more easily than those composed of heavy isotopes, a process known as *fractionation*. The resulting vapour is relatively depleted in deuterium and ^{18}O compared with the initial water. In equilibrium conditions, atmospheric water vapour contains 1 per cent less ^{18}O and 10 per cent less deuterium than average ocean water. Conversely, when water vapour condenses, molecules containing heavy isotopes tend to pass from the vapour to the liquid state more readily than 'light' molecules, resulting in precipitation that is enriched in heavy isotopes compared with the remaining vapour, which will be depleted in heavy isotopes (Dansgaard, 1961). As condensation proceeds, more of the remaining heavy isotopes will be removed, and the vapour will become increasingly depleted in ^{18}O and deuterium. Therefore, progressive cooling of water vapour (e.g. as moist air passes from a warm ocean on to a cold land mass) will result in precipitation with increasingly lighter isotopic composition, as the parent vapour becomes more and more depleted in heavy isotopes. Thus the isotopic composition of precipitation reflects the temperature at which condensation occurs. Although many other factors influence the isotopic composition of precipitation (Bradley, 1985), the effect of temperature is remarkably predictable (Fig. 2.33). This means that the isotopic composition of glacier ice can be used to reconstruct the temperature of the original precipitation, successive annual ice accumulation layers

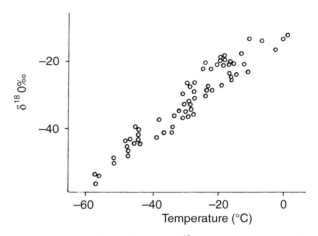

Fig. 2.33 Relationship between ∂^{18}O and mean annual air temperatures, based on values in firn 10–12 m below the surface of the Greenland and Antarctic ice sheets. Note that colder temperatures equate with more negative ∂^{18}O values. (Data from Dansgaard *et al.*, 1973)

yielding a continuous record of temperatures for the lifetime of a glacier or ice sheet.

The isotopic composition of the ice from successive levels is determined by melting the ice in the laboratory and analysing it with a mass spectrometer. Variations in the proportions of ^{16}O and ^{18}O are expressed as ∂^{18}O (delta ^{18}O) values, which measure the difference between the observed ^{18}O/^{16}O ratio and that in a standard water sample, known as Standard Mean Ocean Water (SMOW). Similarly, different proportions of ^{1}H and ^{2}H (D) are expressed as ∂D values, compared with the same standard. Low ∂^{18}O and ∂D values correspond to low palaeotemperatures. Usually, palaeoclimatic studies concentrate on ∂^{18}O values, as in Fig. 2.34, which shows the oxygen isotope record obtained from the GRIP ice core from Greenland, spanning the last glacial/interglacial cycle (Dansgaard *et al.*, 1982). The diagram clearly shows the onset of full glacial conditions at *c.* 120,000 years BP, several warmer interludes (interstadials) that punctuated the last glaciation, and the rapid termination of glacial conditions at *c.* 10,000 years BP. Isotope variations thus provide a priceless record of the severity, timing and pace of global climatic changes (e.g. Bond *et al.*, 1993; Jouzel *et al.*, 1993).

2.4.3 Ancient atmospheres: the gas content of glacier ice

During the transformation of snow into ice below the surface of a glacier, air bubbles become isolated and trapped. These bubbles represent 'fossil air', preserving samples of the atmosphere from the time the ice was formed. Analysis of the composition of such bubbles in ice cores thus provides a means of tracing changes in atmospheric chemistry over many thou-

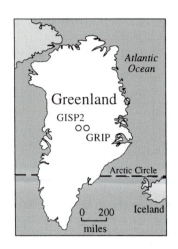

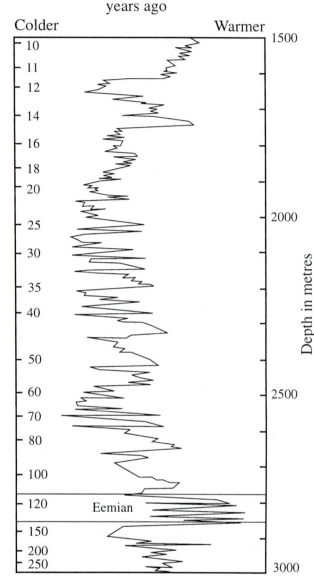

Fig. 2.34 The oxygen isotope record from the GRIP ice core, Greenland. (Redrawn from Dansgaard *et al.*, 1982)

sands of years. Measurements are made by placing a thin slice of a core in a vacuum chamber, cracking the ice and analysing the escaped gases by chromatographic or spectrometric methods. Particular attention has been paid in recent years to the record of variations in 'greenhouse gases' such as carbon dioxide (CO_2) and methane (CH_4), because of their regulating effects on global temperature. Figure 2.35 shows variations in CO_2 and CH_4 and palaeotemperatures derived from $\partial^{18}O$ values for the last glacial–interglacial cycle, obtained from a core drilled at Vostok, Antarctica (Lorius *et al.*, 1990; Jouzel *et al.*, 1993). Oscillations in the greenhouse gas concentrations mirror the climatic fluctuations to a remarkable degree, peak concentrations occurring during warm periods, and low concentrations marking cold

episodes. The parallelism between climate change and greenhouse gas concentrations is thought to reflect feedback mechanisms between glacial, oceanic, atmospheric and biological systems, whereby small climate changes initiate changes in the production and sequestering of trace gases by oceanic and terrestrial plants, which then forces further climatic changes. The ice core record thus provides an extremely important insight into the patterns and mechanisms of climatic change.

2.4.4 Other evidence

Impurities contained within glacier ice also provide records of former environmental conditions. Wind-blown dust and soluble ions can yield data on the

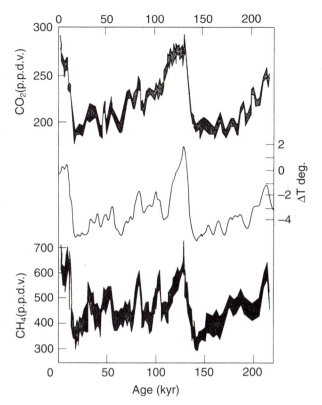

Fig. 2.35 Atmospheric concentrations of the greenhouse gases carbon dioxide (CO_2) and methane (CH_4), and reconstructed atmospheric temperature relative to present for the Vostok ice core, Antarctica. The shaded areas represent the uncertainty in the gas concentrations. (From Jouzel *et al.*, 1993. Reproduced by permission of Macmillan Magazines Ltd)

strength and direction of atmospheric circulation (e.g. Thompson *et al.*, 1990), and tephra horizons and peaks in acidity record the deposition of aerosols injected into the atmosphere by volcanic eruptions (e.g. Hammer 1977, Hammer *et al.*, 1980). Figure 2.36 shows the mean acidity of annual layers from AD 1500 to AD 1972 in the ice core from Crete, central Greenland. Acidity peaks reflect the production of sulphur dioxide by major eruptions, and its transformation to sulphuric acid aerosols in the atmosphere. Most of the acidity peaks can be matched with historical eruptions, including Tambora and Krakatoa in the East Indies. Particularly prominent is the peak associated with the 1783 Laki fissure eruption, Iceland, which caused the death of 90 per cent of the livestock on the island and brought widespread famine (Francis, 1993).

Ice cores worldwide contain radioactive fallout from atmospheric nuclear bomb tests in the 1950s and early 1960s and the Chernobyl reactor accident in 1986. Apart from acting as a reminder of the folly of detonating nuclear weapons, such fallout provides useful stratigraphic marker horizons (e.g.

Crozaz *et al.*, 1966; Koerner and Taniguchi, 1976). For example, the mass balance of the glacier complex Kongsvegen/Sveabreen on Spitsbergen has been reconstructed by Lefauconnier *et al.* (1994) using radioactive fallout from Russian nuclear tests in 1961/62 and the Chernobyl accident.

2.5 ICE TEMPERATURE

Glaciers are not uniformly cold; their temperatures vary in space and time. Many glaciers have temperatures at or close to 0°C, whereas parts of the Antarctic ice sheet are as cold as −40°C. This temperature variation has profound implications for glacier behaviour. In particular, a fundamental distinction exists between ice that is at the melting point and colder ice. Efficient glacier motion can occur only where the ice is at or close to the melting point, and the processes of subglacial erosion, transport and deposition are all sensitively dependent upon temperature. Consequently, it is very important to consider how and why temperature varies in glacier ice.

2.5.1 The melting point of ice

The temperature at which ice melts is not constant at 0°C, but decreases as the ice is placed under increasing pressure at a rate of 0.072°C per million pascals (MPa; 1 pascal = 1 newton per square metre). For example, the pressure at the base of a glacier 2000 m thick is *c.* 17.6 MPa, enough to lower the melting point to −1.27°C. Consequently, the melting point of ice is referred to as the *pressure melting point*. This is actually a slight misnomer, as the melting point of ice is also influenced by the presence of impurities such as solutes. The freezing point of seawater depends on its salinity, but is typically around −2°C at atmospheric pressure. Because the melting point of ice is such an important physical threshold, it is usual to consider ice temperature relative to the local pressure melting point. The importance of pressure melting in regulating glacier motion is examined in Sections 4.5.2 and 4.5.3.

2.5.2 Controls on ice temperature

The temperature of the snow that falls on a glacier surface provides a convenient starting point for considering ice temperature. Following deposition in the form of snow, the initial temperature of glacier ice is modified by three main controls: (a) heat exchange with the atmosphere; (b) the geothermal heat flux; and (c) frictional heat generated by flow (Sugden and John, 1976).

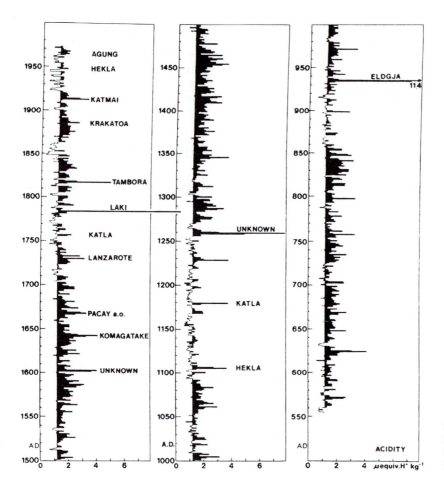

Fig. 2.36 Mean acidity of annual layers from AD 1500 to AD 1972 in the ice core from Crete, central Greenland. (From Hammer *et al.*, 1980)

Heat may be exchanged with the atmosphere by conduction (i.e. the direct effect of air temperature and radiation), the advection of heat in percolating meltwater, and the transfer of latent heat by freezing of water (Section 2.2.3.2). The release of latent heat by the refreezing of meltwater is a particularly important process in glacier accumulation areas in temperate regions, and serves to raise ice temperatures from the cold values typical of winter snow up to or close to the pressure melting point. Where the air temperature is lower than that of the surface ice, heat may be lost to the atmosphere, thus cooling the ice. Therefore, in areas where there is pronounced seasonal air temperature variation, the surface firn of a glacier will experience marked annual changes in temperature, with the firn becoming colder in winter as a *cold wave* penetrates down from the surface, and approaching the pressure melting point in summer owing to latent heat release during the refreezing of meltwater (Sverdrup, 1935; Paterson, 1994). This temperature fluctuation does not affect the deeper ice, which remains at a constant temperature at or close to the pressure melting point throughout the year. In cold regions, such as the interior of Antarc-

tica, the temperature of ice close to the surface is close to the mean annual air temperature.

The geothermal heat flux is the upward transfer of heat resulting from the decay of radioactive isotopes within the Earth. Values of the heat flux are dependent on the tectonic setting; rates lower than $0.04\,\mathrm{W\,m^{-2}}$ (watts per square metre) are typical for continental shield areas which have old, thick crust, and rates in excess of $0.09\,\mathrm{W\,m^{-2}}$ may be recorded in recent volcanic terrains. Considerably higher heat fluxes occur above active volcanoes, and subglacial eruptions can melt extremely large volumes of glacier ice. Obviously, glaciers can continue to exist only in areas where such eruptions are infrequent!

The geothermal heat flux raises the temperature of basal ice, especially where ice is thick and heat is produced faster than it can be conducted towards the surface. Once the ice has been raised to the pressure melting point, excess heat is used to melt ice. The global average rate of $0.0599\,\mathrm{W\,m^{-2}}$ is enough to melt a 6 mm thickness of ice at its pressure melting point each year (Paterson, 1994).

Frictional heat is derived from differential movement (shear) within the glacier, sliding at the base, and the flow of water. In general, heat from this source is significant only close to the glacier bed, where most of the movement occurs. Sliding rates of 20 m yr^{-1} can generate as much heat as the average geothermal heat flux, whereas for fast-sliding outlet glaciers, heat production can be over 100 times as large (Paterson, 1981; Drewry, 1986). The processes of shear and glacier motion are discussed in detail in Chapter 4.

2.5.3 Temperature profiles

The actual temperature of a portion of glacier ice depends on the balance of heat from all these sources. As noted in Section 2.5.2, the temperature of surface firn in glacier accumulation areas is strongly influenced by annual air temperature fluctuations and near-surface processes such as water percolation and refreezing, whereas at slightly greater depths, ice temperatures tend to be similar to mean annual air temperatures. At greater depths, ice temperatures are raised by deep processes, including ice deformation, basal sliding and the geothermal heat flux, so that temperature profiles typically show an increase with depth (Fig. 2.37). Thus, even in cold regions where surface ice is at temperatures well below 0°C, the basal ice may be at the pressure melting point. The relatively high temperatures below thick ice and in places where frictional heat is generated by flow mean that melting temperatures are more likely at the base of rapidly flowing ice streams that occupy bedrock depressions than below thinner, more slowly moving ice to either side. Melting temperatures below ice streams promote rapid ice motion and subglacial erosion (Sections 4.5, 5.3 and 5.6), serving to maintain the ice stream as a rapidly moving, deeply incised feature, so that a self-reinforcing relationship exists between ice thickness, velocity, temperature and landscape evolution.

Ice temperatures are also influenced by the thermal and deformation history of the ice over long timescales. At any one site, ice at progressively greater depths will have originated as snow further and further upglacier prior to its transport to the site by glacier flow. Since the deepest, furthest-travelled ice will have formed at the highest elevations, it should also have the lowest temperatures, so that the temperature profile should exhibit a decrease with depth. Such *negative temperature gradients* have indeed been observed in many shallow boreholes in Greenland and Antarctica (Robin, 1955; Paterson, 1994). Climate change also plays a role in determining ice temperature profiles, and in deep ice cores it is possible to distinguish cold ice deposited as snow during the last glacial cycle from relatively warm ice formed during the present interglacial period (Dahl-Jenssen and Johnsen,

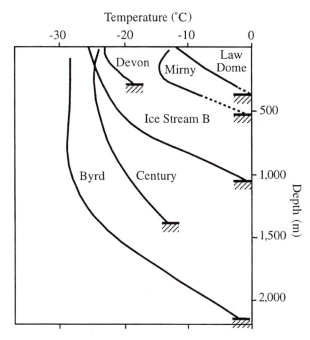

Fig. 2.37 Selected temperature profiles from the Greenland and Antarctic ice sheets. (Modified from Paterson, 1994)

1986). A recent comprehensive review of the controls on ice temperature has been provided by Paterson (1994).

2.5.4 Thermal classification of glacier ice

The temperature of ice exerts a very important control on a wide variety of glacial processes, including glacier motion, meltwater flow and subglacial erosion and deposition. In particular, effective sliding and erosion are critically dependent on the basal ice being at the pressure melting point (Sections 4.5.4 and 5.3.5). For this reason an important distinction is made between *temperate ice*, which is at the pressure melting point, and *cold ice*, which is below the pressure melting point. This classification has been extended to glaciers as a whole, and three classes of glacier have been defined: (a) *temperate glaciers*, which are everywhere at the melting point except for a surface layer a few metres thick which is subject to seasonal temperature cycles; (b) *polar glaciers*, which are below the melting point throughout and are frozen to their beds; and (c) *subpolar glaciers*, which are temperate in their inner regions but have cold-based margins.

However, it is now recognized that this classification is highly simplistic, and that the thermal conditions of individual glaciers may vary both spatially and temporally (Boulton, 1972a; Sugden, 1977; Denton and Hughes, 1981b). Glacier beds may form temperature mosaics, consisting of predominantly melting ice with a few cold patches, or predominantly cold ice with thawed patches, as well as entirely cold or temperate beds (Robin, 1976; Paterson, 1994). Furthermore, zones of frozen and melting basal conditions may migrate through time. For these reasons, whole glaciers should not be placed in a single category, and thermal classification should properly be applied to local conditions. In this book we use the terms *temperate* or *wet-based ice* to refer to ice at the pressure melting point, and *cold* or *cold-based ice* for frozen ice.

CHAPTER

3

GLACIERS AND MELTWATER

3.1 INTRODUCTION

Surface ablation and basal melting of glacier ice can produce large volumes of meltwater (Sections 2.2.3 and 2.5.2). Flowing water in glacial environments exerts an important influence on glacier behaviour and geomorphological processes, and presents both benefits and hazards for human populations. In glacierized regions with low summer rainfall, glacier melt is an important source of water during the growing season, allowing cultivation of valleys that would otherwise be too arid for agriculture. Conversely, the sudden release of stored water in catastrophic floods constitutes a serious and recurrent threat in regions such as the high Andes, Himalaya and Iceland. Recently, glacial meltwater has been exploited as a source of hydroelectric power, particularly in Scandinavia and the European Alps, and the successful building and running of such schemes demands a clear understanding of the principles of water flow and sediment transport. This can be illustrated by attempts to tap into the subglacial drainage of the Glacier d'Argentière in the Mont Blanc Massif by the French–Swiss hydroelectric Emosson Project. After the subglacial intake was put in place, the water disappeared on several occasions during the 1960s and 1970s, often for several weeks at a time. In April 1976 it disappeared, apparently permanently, from its original course, and was not found again for almost three years, a shift that had expensive consequences for the power company (Hantz and Lliboutry, 1983).

The rate of motion of glaciers and ice sheets is sensitively dependent on the pressure and distribution of water at the bed, and some mechanisms of flow, such as basal sliding, are virtually impossible in the absence of liquid water. Water also contributes substantially to glacial erosion, debris transport and deposition, both as a direct agent and in conjunction with ice. In this chapter we examine the flow of glacial meltwater, its transport over, through, below and beyond glaciers, and the processes of glacifluvial erosion and transport. The influence of water on glacier motion is examined in Chapter 4, and the sediments and landforms produced by flowing water are described in Chapters 9, 10 and 11.

3.2 WATER SOURCES, FLOW AND STORAGE

3.2.1 Sources of water

Water may enter glacial drainage systems from melting ice, snow melt, rainfall, dew, runoff from ice-free slopes, and the release of stored water (Fig. 3.1). Surface melting of ice and snow exhibits pronounced daily, annual and random variations according to the local energy balance (Section 2.2.3), causing marked fluctuations in the runoff from glacierized catchments. Rates of surface melting increase with air temperature, radiation flux and rainfall (which carries heat from the atmosphere). Rainfall is particularly effective in melting snow, leading to greater rates of runoff than would occur from similar amounts of precipitation on a snow-free catchment. Water from surface melting and the atmosphere can enter glaciers and melt ice from the walls of englacial and subglacial passageways through the dissipation of sensible heat and mechanical energy. Englacial and subglacial melting can also occur as a result of frictional heat generated by deforming or sliding ice, geothermal heat, and pressure melting associated with glacier flow around bumps (Section 4.5.2). Overall, the amount of melting at glacier beds tends to be more constant through time than melting at the surface, and contributes little to variations in runoff from glacierized catchments.

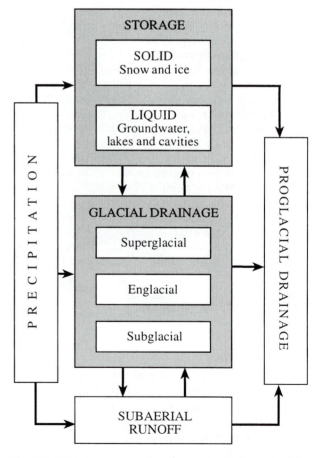

Fig. 3.1 Water sources and pathways in glacierized catchments

The buildup and release of stored water exerts a very strong influence on runoff in glacierized areas, and can cause very high-magnitude flood events. Water can be stored in subglacial, supraglacial, englacial or proglacial lakes, as described in Section 3.2.4.

3.2.2 The permeability of ice

The drainage of water over and through a glacier reflects the permeability of the ice at two scales: *primary permeability*, or the permeability of intact ice and snow, and *secondary permeability*, related to the distribution of tunnels and other passageways. In this sense, glacier drainage is analogous to water flow through limestone karst terrain, which can take place through pore spaces in the rock or tunnels and fissures (Sugden and John, 1976). Primary permeability may be high for snow and firn in which air spaces between ice crystals are linked, but tends to be very low for ice, where bubbles are isolated from one another. For ice at the pressure melting point, however, water can flow through a system of interconnected thin lenses and veins between ice crystals, particularly where pressure gradients are high (Nye and Frank, 1973; Raymond and Harrison, 1975; Hantz and Lliboutry, 1983; Nye, 1989; Mader, 1992). Intact ice below the pressure melting point is impermeable. The bulk of the water draining through a glacier relates to the secondary permeability, and follows a system of conduits ranging in diameter between millimetres and several metres (Nye, 1976; Lliboutry, 1983; Pohjola, 1994). Englacial conduits are maintained by melting of their walls, and tend to be a characteristic of ice at or close to the pressure melting point. Parts of glaciers where the ice is below the pressure melting point, therefore, tend not to have englacial conduit systems, and so surface water will tend to drain supraglacially. However, crevasses can penetrate cold surface ice and take meltwater down to lower levels, where ice may be at the pressure melting point (Holmlund, 1988a; Seaberg *et al.*, 1988).

3.2.3 Water flow in glaciers: basic theory

3.2.3.1 HYDRAULIC POTENTIAL

The flow of water is governed by variations in *hydraulic potential*, a measure of the available energy at a particular time and place. For surface streams, including those flowing at the surface of a glacier, the potential depends only on elevation, and water will always flow directly downslope. For water flowing within or at the base of a glacier, however, the situation is more complex because the hydraulic potential depends on both elevation and water pressure. The idea of hydraulic potential is central to an understanding of water flow within glaciers, so it is worth examining in some detail. The following discussion is a simplified version of that presented by Shreve (1972) and Paterson (1994).

The hydraulic potential $\emptyset$ in an englacial or subglacial conduit can be expressed by the simple relation

$$\emptyset = \emptyset_0 + \emptyset_e + P_w \qquad (3.1)$$

where $\emptyset_0$ is a constant depending on the shape and size of the conduit, $\emptyset_e$ is the potential due to elevation, and P_w is the water pressure.

The elevation potential $\emptyset_e$ is simply the product of the weight of the water and its elevation:

$$\emptyset_e = \rho_w \, gz \qquad (3.2)$$

where ρ_w is the density of water (1000 kg m^{-3}), g is gravitational acceleration (9.81 m s^{-2}), and z is the elevation.

In natural conditions within and below a glacier, the water pressure P_w can vary between *atmospheric pressure* (the pressure of the open air) and *cryostatic pressure* (the pressure exerted by the weight of the overlying ice). The cryostatic pressure is the product of the weight and the thickness of the ice:

$$P_i = \rho_i \, g(H-z) \qquad (3.3)$$

where P_i is the ice pressure, ρ_i is the density of ice (c. 900 kg m^{-3}), g is gravitational acceleration, H is the altitude of the ice surface, and z is the elevation of the point in question.

The water pressure will be atmospheric if the conduit is open to the air and not completely filled with water; in this case the potential depends only on elevation, as for surface streams. If P_w equals the cryostatic pressure, the water is able to support the whole of the weight of the ice, and can lift the glacier off its bed. This can happen on a local scale where ice pressures are lower than average (such as the downglacier sides of obstacles), or on a large scale, forming subglacial lakes (Section 3.2.4.1). The difference between water pressure and ice pressure is known as the *effective pressure*, N, which is written

$$N = P_i - P_w \qquad (3.4)$$

The effective pressure therefore depends on the magnitude of P_w relative to P_i. When P_w is zero, the effective pressure is the same as the total ice pressure, and N is at a maximum; when $P_w = P_i$, the effective pressure is zero and the ice is supported entirely by the pressurized water. The effective pressure is an extremely important concept, which exerts a profound influence on the character of sub- and englacial drainage, glacier motion (Chapter 4) and processes of subglacial erosion and deposition (Chapter 5).

3.2.3.2 PRESSURE AND SIZE OF CONDUITS

For water-filled englacial or subglacial conduits under steady-state conditions, the water pressure is controlled by the balance between two processes. First, water flow produces frictional heat, which melts some of the surrounding ice, tending to enlarge the passage. Second, the ice forming the passage deforms in response to the pressure gradient between the ice and the passage (the effective pressure); this effect tends to contract the passage (Shreve, 1972; Röthlisberger, 1972). For equilibrium conditions, the water pressure will therefore equal the ice pressure plus or minus a pressure change due to decreases or increases in the size of the passage:

$$P_w = P_i + P_m \qquad (3.5)$$

where P_m is the pressure change due to melting or contraction of the walls of the passage. Negative P_m reflects a pressure drop due to efficient melting. Melt rates increase (and, therefore, water pressures decrease) with increasing passage radius, because large channels carry more water and dissipate more heat relative to the area of their walls. Additionally, melt rates are high for downward-sloping passages and are low or negative (i.e. freezing occurs at the walls) for upward-sloping passages, owing to changes in elevation potential and the pressure melting point of ice (Röthlisberger, 1972; Shreve, 1972; Lliboutry, 1983). Conversely, ice deformation rates increase with increasing effective pressure: large pressure differences between the ice and an englacial passage will lead to rapid rates of closure. Two important implications follow from the balance between wall melting and passage closure by ice deformation:

1. Melting and deformation processes allow passages to expand or contract in response to increases or decreases in water pressure. If water pressures increase, the effective pressure falls, reducing tunnel closure rates. Therefore, passages can become larger through more effective melting. Conversely, if water pressures fall, the increased pressure gradient between the ice and the water accelerates passage closure until a new equilibrium is reached. Passages can adjust by melting in a matter of hours, whereas contraction due to ice deformation may take days or even weeks.
2. Because melt rates increase with passage radius, there is an inverse relationship between water pressure and passage radius, and the largest pas-

sages will have the lowest water pressures. Therefore, water will flow towards larger channels following the gradient in pressure, causing them to grow at the expense of smaller ones. As a result, water-filled englacial passages will form a branching network, with large numbers of small tributaries converging on a few large conduits (Fig. 3.2). This statement applies to some, but not all, subglacial drainage systems, for reasons explained in Section 3.4.

3.2.3.3 DIRECTION OF FLOW

The direction of flow in a branching englacial and subglacial drainage network is controlled by variations in hydraulic potential. Water will flow from regions of high hydraulic potential towards regions of low potential, following the steepest *hydraulic gradient*. In other words, water will flow from high elevations to low elevations, and from places where water pressure is high to places where it is low (Fig. 3.2). The precise direction of flow depends on the balance between the elevation and pressure components of the total potential, and their respective rates of change. If we combine equations (3.1), (3.2), (3.3) and (3.5), the potential in a water-filled steady-state passage will be given by

$$\emptyset = \emptyset_0 + \rho_w gz + \rho_i g(H-z) + P_m \qquad (3.6)$$

where $\emptyset_0$ depends on the size and shape of the passage, $\rho_w gz$ is the gravitational potential due to elevation, $\rho_i g(H-z)$ represents the pressure exerted by the weight of the overlying ice, and P_m is the pressure change associated with passage enlargement or contraction. Equation (3.6) shows that the hydraulic potential varies with (a) the elevation of the overlying glacier surface, (b) the elevation of the conduit, and (c) conduit radius (through its influence on P_m). If we ignore for the moment the influence of passage

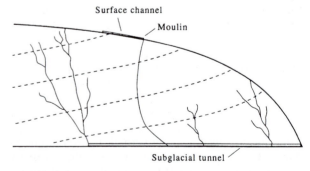

Fig. 3.2 The relationship between equipotential surfaces (broken lines) and englacial and subglacial drainage. Small englacial channels are oriented at right angles to the equipotential surfaces, but the location of the moulin is controlled by a crevasse. (From Paterson, 1994. Reproduced by permission of Pergamon)

size, the rate of change in hydraulic potential, or hydraulic gradient, is given by:

$$\text{grad } \phi = \rho_i g \{\text{grad } H +$$
$$[(\rho_w - \rho_i)/\rho_w] \text{ grad } z\}$$
$$= \rho_i g[\text{grad } H + (0.1 \text{ grad } z)] \qquad (3.7)$$

where grad ϕ is the gradient in hydraulic potential, grad H is the ice surface gradient, and grad z is the gradient of the englacial passage.

The factor 0.1 reflects the relative densities of ice and water. The full derivation of equation (3.6) can be followed in Shreve (1972) and Paterson (1981). Equation (3.7) shows that the potential gradient depends on the ice surface slope and, to a much lesser extent, the slope of the water-filled passage. Planes connecting points where the potential is the same, or *equipotential surfaces,* will therefore rise downglacier, with a gradient approximately 10 times that of the ice surface. If water can drain freely through the glacier, it will flow through the ice following the steepest potential gradient, which lies at right angles to the equipotential surfaces (Fig. 3.2). Once at the bed, water will flow towards the snout at right angles to the *equipotential contours,* which are defined by the intersections of the equipotential surfaces with the bed.

3.2.3.4 NON-STEADY-STATE CONDITIONS

The formula for water pressure given as equation (3.5) is founded on the assumption that the internal plumbing of glaciers is in a steady, equilibrium state. This assumption is reasonable when water discharges do not undergo rapid fluctuations, and the system has time to adjust to new conditions. Although this may be the case where basal melting is the most important source of meltwater, the assumption of steady-state conditions will certainly not hold if large amounts of the water originate at the surface. Rates of surface meltwater production undergo rapid daily variations during the summer ablation season. Therefore, discharges and water pressures in conduits fed from the surface will tend to be out of equilibrium with the pressure in the surrounding ice. High water pressures will occur during periods of rapidly rising discharge, when water backs up the system faster than conduits can enlarge by melting (Seaberg *et al.*, 1988). Conversely, low water pressures will prevail during rapidly falling discharge, when water leaves the system faster than conduits can contract by ice deformation. During periods of rapidly fluctuating discharge, therefore, high discharges will be associated with high water pressures, which is the opposite of the inverse relationship between discharge and pressure predicted

for steady-state conduit systems (Röthlisberger, 1972; Sharp *et al.*, 1989b).

Lliboutry (1983) and Hooke (1984) have pointed out that, because conduits can enlarge fairly rapidly by melting but close much more slowly by ice deformation, the most common condition in systems fed from the surface may be one of low water pressure. Indeed, water pressures may be at or close to atmospheric pressure much of the time during the ablation season, particularly below thin ice (<200 m) where tunnel closure rates are slow. In such cases, the water pressure component of the hydraulic potential becomes zero or very small, and potential is primarily controlled by elevation. In such cases, englacial water will tend to drain vertically downwards through the glacier, and subglacial drainage will tend to follow the local slope of the bed (Hooke, 1984; Holmlund, 1988a).

3.2.4 Storage

Water can be stored in subglacial, supraglacial, englacial or ice-dammed lakes and ponds if discharge is prevented by some form of barrier, related either to ice permeability or to pressure gradients. Storage may be controlled by the glacier itself or by a combination of glacier and local topography, according to local drainage characteristics. Ice-contact water bodies are temporary features, some expanding and contracting in response to glacier fluctuations over tens to thousands of years, and others filling and emptying relatively rapidly in response to glacier dynamics or volcanic activity over periods of weeks to years.

3.2.4.1 SUBGLACIAL LAKES AND PONDS

Ponded subglacial water bodies can vary greatly in size. At the smallest scale, water-filled cavities a few millimetres to several metres across commonly occur beneath temperate glaciers, particularly where the bed is rough. At the opposite extreme, large reservoirs up to 8000 km^2 in area are known from radio-echo sounding to exist beneath the Antarctic ice sheet (Oswald and Robin, 1973; Morgan and Budd, 1975; Oswald, 1975; Robin *et al.*, 1977; Cudlip and McIntyre, 1987; Ridley *et al.*, 1993). It has been suggested that large subglacial lakes may have existed below the Laurentide ice sheet during the last glacial cycle, and that they may have been influential in ice-marginal surging and flooding events (Wright, 1973; Shoemaker, 1991).

Water can accumulate beneath glaciers where regions with relatively low hydraulic potential are surrounded by regions where hydraulic potential is relatively high. In such cases, the hydraulic gradients will cause water to flow towards the subglacial lake or pond, rather than away from it. Two possible situations where large-scale ponding can occur are illustrated in Fig. 3.3 (Björnsson, 1975, 1976; Nye, 1976). In the first case, water can be ponded within a bed depression located beneath an ice dome. Equipotential surfaces beneath the ice dome form concave-up parabolas, reflecting the form of the ice surface. If a bed depression is shallower than the form of the equipotential contours, ponding cannot occur because hydraulic potential decreases away from the centre of the depression, allowing water to flow towards the ice margin (Fig. 3.3b). In contrast, if the bed depression is deeper than the trace of the equipotential contours, ponding will occur because hydraulic potential increases towards the edges of the depression owing to increasing bed elevation (Fig. 3.3a). At equilibrium, the water body will have a concave-up surface along which the pressure is everywhere equal to that at the edges of the basin. The reservoir cannot grow any larger than this, and any further flow of water into the depression will result in equal outward flow.

The second situation where water can be ponded is shown in Fig. 3.3c. In this case, the glacier bed is flat and the ice surface consists of a central depression circled by thicker ice. Hydraulic potential beneath the depression is lower than in the surrounding areas, as shown by the equipotential surfaces, which form convex-up parabolas below the depression. Water flow towards the bed beneath the depression will form a *water cupola*, with a domed upper surface lying along an equipotential surface. This water is prevented from escaping by regions where the hydraulic potential at the bed exceeds that at the surface of the reservoir. As the reservoir expands, these regions will shrink in width, until a hydraulic connection can be made between the reservoir and subglacial drainage pathways beyond. Once water begins to escape from the reservoir, it can enlarge its drainage pathways, and catastrophic drainage can ensue (Björnsson, 1975; Nye, 1976; Section 3.5). Water cupolas are thus less stable than reservoirs located in deep bed depressions.

A particularly effective mechanism for the production of subglacial lakes is ice-melting by volcanic heat flow, such as below the western part of Vatnajökull, Iceland, which is located in one of the most active sections of the mid-Atlantic spreading ridge (Thorarinsson, 1953, 1957; Björnsson, 1974, 1975). Ice centred on the Grimsvötn caldera melts to produce a large subglacial cupola, eventually causing localized supraglacial depressions into which surface meltwater begins to drain (Björnsson, 1976; Fig. 3.4). The lake system at Grimsvötn empties itself catastrophically approximately every six years (Section 3.5), and re-forms in response to the high geothermal heat flows from the caldera (Gudmundsson *et al.*, 1995). Periods of more intense volcanic activity are responsible for substantial changes in the subglacial environment, forming large, unstable lakes. Recent examples from Iceland include the eruption of Katla below Myrdalsjökull in 1918 and the eruption below Vatnajökull in 1996, both of which triggered large floods.

Small-scale water-filled cavities form on the downglacier side of bumps on rough glacier beds, where local ice pressures are low (Section 3.4.4; Fig. 3.19). Below thin or crevassed ice, where subglacial

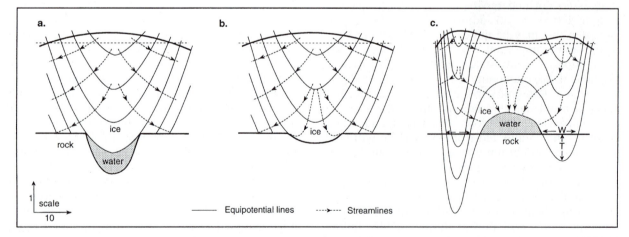

Fig. 3.3 Subglacial lakes. (a) Lake formed in a deep depression beneath an ice dome. The surface of the lake is defined by an equipotential surface. Water cannot flow out of the depression against the potential gradient. (b) Shallow depression beneath an ice dome. Water flowing into the depression can escape along the bed along the potential gradient. (c) Subglacial cupola formed beneath a depression on the ice surface. Water beneath the depression will flow into the cupola, and is prevented from escaping by an ice dam of width W where the ice pressure exceeds water pressure in the cupola. Lake expansion will cause the ice dam to narrow: when $W = 0$, drainage will occur. (Modified from Nye, 1976)

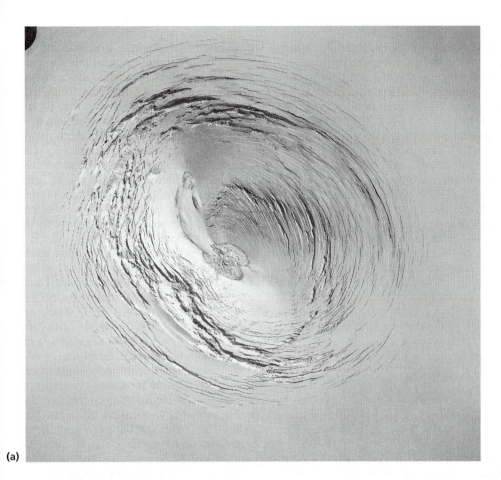

(a)

(b)

Fig. 3.4 (a) An ice cauldron beginning to form on the surface of Vatnajökull, Iceland, as a result of geothermal heating in the Grimsvötn caldera in August 1972. (Aerial photo supplied by Landmaelingar Islands) (b) Final stages of cauldron formation above Grimsvötn. (Photo: S. Thorarinsson)

water may be at atmospheric pressure, air-filled cavities can develop if sliding ice bridges over the lee side of bumps or rock steps (Kamb and LaChapelle, 1964; Vivian and Bocquet, 1973). Where subglacial water is pressurized, however, water-filled cavities form where water pressure (P_w) exceeds the pressure exerted by the ice (P_i) on the bed. Higher water pressures lead to more extensive cavities, as P_w exceeds P_i over larger and larger areas of the bed. Two basic types of cavity can be recognized: *step cavities*, where the downglacier side of the controlling obstacle has an abrupt step, and *wave cavities*, where the bed has a smooth profile (Kamb, 1987). The mechanisms of formation of both types of cavity are essentially the same, but the threshold pressure conditions for cavity formation and growth differ. The mechanics of cavity formation are discussed in detail by Lliboutry (1979), Weertman (1979) and Fowler (1987a).

3.2.4.2 SUPRAGLACIAL AND ENGLACIAL LAKES

Englacial and supraglacial lakes are important temporary storage sites for surface meltwater. Pockets of water in englacial positions are predominantly small and are usually the closed-off remnants of crevasses or meltwater tunnels (Paterson and Savage, 1970).

Supraglacial lakes are formed during the ablation season in all but the coldest of polar settings. On temperate glaciers, supraglacial ponds usually form early in the melt season, but drain away as the englacial drainage network opens up. In contrast, where ice is below the pressure melting point, supraglacial ponds can persist throughout the melt season, owing to the inability of meltwater to penetrate below the uppermost ice layers (Section 3.2.2). On glaciers with large amounts of supraglacial debris, differential melting causes the widespread development of water-filled hollows and sinkholes, or glacier karst (Clayton, 1964; Section 6.5.4.2). As glacier melting proceeds, supraglacial ponds become more numerous and larger, and will persist if englacial drainage is poorly developed or if the englacial water table is high. Partially supraglacial lakes may form if the margins of a glacier become submerged by lake waters dammed between a topographic obstacle or a moraine and the glacier. Often it is difficult to determine whether or not such lakes do overlie glacier ice. As mentioned in Section 3.2.4.1, supraglacial lakes or *ice cauldrons* can also form in depressions created by geothermal melting and ice subsidence (Fig. 3.5; Björnsson, 1975).

3.2.4.3 ICE-DAMMED LAKES

Wherever glacier ice forms a barrier to local or regional drainage, water will be ponded to form an ice-dammed lake. Such lakes are often small, but may cover thousands of square kilometres, or may be as small as <1 km^2 where a trunk glacier dams a small tributary valley. In mountainous terrain, ice-dammed lakes form in one of three locations (Thorarinsson, 1939; Liestøl, 1955; Stone, 1963; Hewitt, 1964, 1982; Maag, 1969): (a) in ice-free side valleys blocked by a glacier in the trunk valley; (b) in trunk valleys where side valley glaciers have blocked the drainage; and (c) at the junction between two valley

glaciers. Marginal ice-dammed lakes are particularly common beside polar and subpolar glaciers, largely because the ice is frozen to the bed and there are no well-developed englacial and subglacial drainage networks through which lakes could drain (Fig. 3.6). In lowland areas, ice-dammed lakes may form wherever glaciers advance into or retreat from terrain that slopes down towards the ice margin. Although many ice-dammed lakes are often referred to as 'proglacial', they owe their existence to ice damming of the drainage and are therefore by definition 'ice-dammed'. The depth of ice-dammed lakes may be controlled by topographic spillways or by the ice itself, depending on the relationship between the local terrain and ice thickness.

Two of the largest examples of ice-dammed lakes are *Glacial Lake Agassiz* which formed during the last glacial cycle when rivers draining towards Hudson Bay were dammed against the margins of the retreating Laurentide ice sheet (2,000,000 km^2 in area; Teller and Clayton, 1983; Teller, 1987, 1995; Kehew and Teller, 1994), and *Glacial Lake McConnell* (215,000 km^2 in area; Lemmen *et al.*, 1994; Fig. 3.7 (Plate 12)). Similarly, large ice-dammed lakes were produced in Siberia as a result of the damming of northward-draining rivers by the Eurasian ice sheet (Rudoy and Baker, 1993; Hughes and Grosswald, 1995). The shorelines of such large lakes are extensive, and therefore can be used to reconstruct patterns of glacioisostatic rebound (Section 1.5), which would normally be unrecorded in continental interiors where there is no raised marine record.

3.2.4.4 PROGLACIAL LAKES

Lakes ponded between frontal glacier margins and topographic high points are referred to as *proglacial* or *frontal lakes* when the lake owes its existence to the topographic barrier rather than the presence of

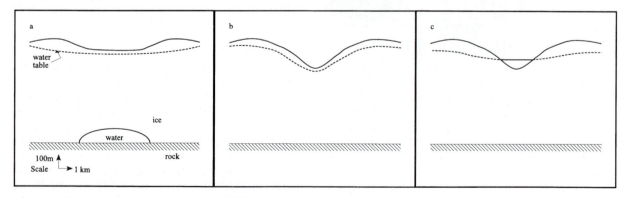

Fig. 3.5 Stages in the formation of a supraglacial cauldron and lake above a geothermal area. (a) The production of a subglacial water cupola beneath an ice surface depression. (b) Formation of a cauldron due to the drainage of the subglacial cupola. (c) Production of a supraglacial lake. (Modified from Björnsson, 1976)

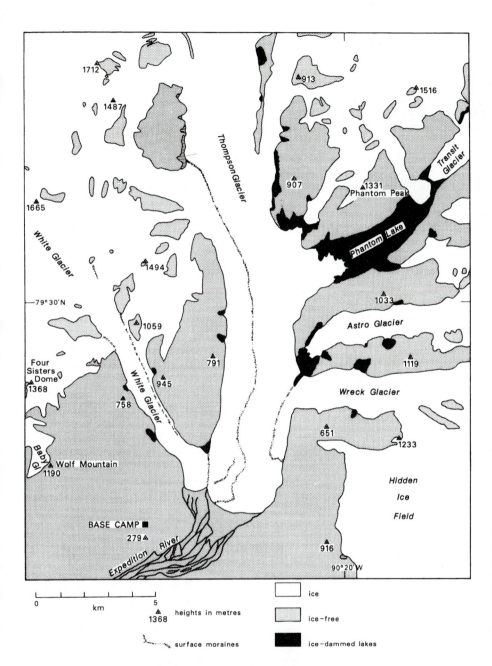

Fig. 3.6 Ice-dammed lakes of various types associated with the Thompson Glacier and adjacent glaciers on Axel Heiberg Island, Arctic Canada. (From Hambrey, 1994. Reproduced by permission of University College London Press)

the ice, as in the case of ice-dammed lakes. A typical example, at the margin of Heinabergsjökull, Iceland, is shown in Fig. 3.8. Moraine-dammed proglacial lakes are common wherever large amounts of debris are available for moraine construction, and such lakes may drain catastrophically if the moraine dam is eventually breached. Such lakes can be dammed by ice-cored sediment accumulations, which eventually collapse as a result of melt-out and therefore release the water from storage. Glacier retreat may also uncover a lower outlet along which some of the water may decant. Excellent examples of evolving

proglacial lakes are found along the margins of Breidamerkurjökull and Fjallsjökull in Iceland (Fig. 3.9 (Plate 13)). These lakes are ponded behind moraines and accumulations of outwash, and have changed size and shape throughout the retreat of the two glaciers from their Little Ice Age maxima (Howarth and Price, 1969; Price and Howarth, 1970; Price, 1982).

Because they are impounded by bedrock and sediment rather than ice, proglacial lakes tend to be more stable than other types of ice-contact water body. However, lake levels may fluctuate in response to catastrophic influxes of water from

Fig. 3.8 Proglacial lake at the snout of Heinabergsjökull, Iceland. The drainage outlet can be seen in the middle distance and the mouth of the river is visible top right. Large icebergs calve from the glacier margin on a regular basis. (Photo: D.J.A. Evans)

sources elsewhere. The proglacial lake of Hein-abergsjökull, for example, fluctuates in response to frequent dumping of water via subglacial tunnels from the ice-dammed lake of Vatnsdalslon at the east margin of the glacier (Thorarinsson, 1939; Björns-son, 1976). Unstable proglacial lakes impounded by latero-frontal moraines are common in many high mountain areas, such as the Cordillera Blanca, Peru, where glacier calving is known to be responsible for the production of large waves erode the restraining moraine and cause catastrophic drainage (Section 3.5; Lliboutry *et al.*, 1977).

3.3 SUPRAGLACIAL AND ENGLACIAL DRAINAGE SYSTEMS

3.3.1 Supraglacial drainage

As noted in Section 2.2.2, surface melt of snow results in the downward percolation of water through the snowpack, where it may refreeze. Where melting exceeds refreezing rates, water will accumulate within the snow, and in areas of low relief may form extensive areas of *slush swamps*. Where snow lies on a slope, however, water will drain laterally through the snowpack, eventually forming rills which join up to form a dendritic drainage network. Such networks can exist entirely within the snow, the only surface expression being slight depressions along the line of subsurface flow (Higuchi and Tanaka, 1982). If discharge is high enough, surface channels will form. These can enlarge rapidly by the entrainment of rafts of snow from the banks, which melt in transit, adding to the fluid discharge. Surface stream networks are often well developed on ice surfaces in glacier ablation

areas, owing to the low primary permeability of glacier ice. Channels are typically a few millimetres to a few metres in depth, and have smooth sides that offer little resistance to water flow (Fig. 3.10). Surface meltwater streams commonly meander, and can exhibit great regularity of wavelength and amplitude (Knighton, 1972). Where structural control is strong, such as where cracks, crevasses or foliation are present, straight channels predominate (Sugden and John, 1976).

On ice masses with large ablation areas, extensive drainage networks can evolve, which may cover several hundred square kilometres (Fig. 3.11). Such networks share many characteristics with drainage networks developed on rock or sediment, but with some important differences (Sugden and John, 1976). (a) Drainage density is very high, and networks tend to lack well-developed trunk streams. This appears to reflect high rates of water supply and the dynamic character of the ice surface, where ice motion and ablation inhibit the development of integrated drainage patterns. (b) Drainage patterns contain many subparallel elements, reflecting structural control and the consistent slope of most glacier surfaces, which contrasts with the incised nature of normal river valleys. (c) The density of streams decreases upglacier, unlike that of normal stream networks, mainly because the amount of melting and water production decreases with altitude. (d) Channel patterns are highly changeable and unstable, because glacier ablation tends constantly to alter the ice surface topography, and crevasses may open up, truncating parts of the network. On small valley glaciers with markedly convex snouts, radial drainage patterns commonly develop, carrying surface drainage towards the glacier margins.

3.3.2 Englacial drainage

Only a small component of the water carried by englacial drainage networks is derived from internal

Fig. 3.10 Surface meltwater stream, Axel Heiberg Island. Note the meanders. (Photo: D.J.A. Evans)

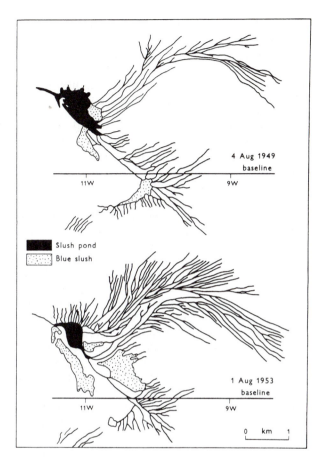

Fig. **3.11** Evolution of a supraglacial stream network on the Greenland ice sheet. (From Sugden and John, 1976; after Holmes, 1955)

Fig. **3.12** A moulin on the Glacier d'Arolla, Switzerland. Dye is being poured into the moulin as part of a tracing experiment. (Photo: I. Willis)

melting of the glacier, and most originates at the surface. Water from supraglacial streams often plunges down vertical or steeply dipping holes in the ice called *moulins* (from the French word meaning 'mill'). Moulins, which are analogous to sink holes in karst, tend to exploit structural weaknesses such as crevasses (Fig. 3.12). New moulins are formed when crevasses open up across the line of a supraglacial stream, capturing the drainage, and can persist after the rest of the crevasse has closed (Stenborg, 1973). Glacier flow can isolate moulins from their supraglacial water supply, particularly if new crevasses are created further upstream, sometimes creating lines of old, abandoned moulins (Holmlund, 1988a). The strong influence exerted by crevasse distribution on the location of moulins means that englacial drainage can be routed in directions very different from those predicted by theory, flow being determined by glacier structure (Stenborg, 1973; Lliboutry, 1983; Seaberg *et al.*, 1988; Fountain, 1993). Changes in crevasse distribution through time can dramatically alter the supraglacial catchments of

englacial streams, with consequences for the amount of water draining through the system.

Water levels in moulins tend to fluctuate through time, and will be highest during periods of rapid melt or rainstorms, when more water enters the system than can be discharged through the englacial and subglacial network. Because the conduit network cannot adjust rapidly enough by melting to cope with high discharges, water backs up the system, raising water levels and water pressures at depth (Holmlund and Hooke, 1983). Conversely, water levels will be low at times of low discharge, before conduits have adjusted to the new conditions by creep closure. At such times, water pressure throughout much of the system may be atmospheric (Lliboutry, 1983; Hooke, 1984). Water level fluctuations have also been observed in boreholes that have connected with englacial or subglacial drainage systems, for similar reasons (Hodge, 1976, 1979; Kamb *et al.*, 1985; Holmlund, 1988a).

From time to time, adventurous researchers have made descents of moulins when the water level is low, such as in winter (Vallot, 1898; Dewart, 1966; Reynaud, 1987; Holmlund and Hooke, 1983; Holmlund, 1988a). The hazards of this questionable practice are made clear in this quotation from Holmlund and Hooke (1983, p. 20): 'While one of us was down in a moulin measuring its geometry, two loud cracks were heard. Almost immediately, the water level in the moulin ... began to rise at a rate of about 7 meters per hour.' A rather safer, though much more time-consuming, approach is to make repeated maps of conduits as they melt out at the glacier surface (e.g. Holmlund, 1988a). If such maps are made over several seasons, a three-dimensional plan of the former englacial network can be built up (Fig. 3.13). One major disadvantage of this method is that the form of conduits can radically alter with time as a result of ice deformation, particularly once they become inactive, so that maps constructed over several years may not give an accurate view of a conduit system as it existed at any one time. Investigations of moulins have shown that the uppermost part is usually straight and vertical, sometimes interrupted by steps or plunge pools, and corkscrew-shaped moulins have been observed in some cases (Dewart, 1966). At greater depths below the surface, moulins tend to change from vertical to downglacier-dipping passageways (Fig. 3.13; Reynaud, 1987), a transition that reflects the depth of crevasses and changing pressure conditions. At depth, moulins tend to be water-filled for much of the time, so that water pressure and hydraulic gradients tend to be controlled by the configuration of the ice, as described in Section 3.2.3.3, resulting in downglacier-dipping passageways. Closer to the surface, moulins are rarely completely filled with water (usually only during peak flow events), so that pressure is often atmospheric and hydraulic potential is controlled by elevation alone. Combined with the presence of open crevasses, this means that water will continue to flow

vertically downwards (Holmlund and Hooke, 1983; Fountain, 1993).

In addition to moulins, englacial drainage is carried by systems of small veins and passages through the ice. Raymond and Harrison (1975) observed a number of tubular passages, suggestive of conduits, in ice cores extracted from Blue Glacier, Washington State, USA. The passages were a few millimetres in diameter, and in two or three cases branched upwards, suggesting a network of englacial conduits as envisaged by Shreve (1972). More recently, Pohjola (1994) used a remote video camera to observe tubular voids in the sides of boreholes drilled in Storglaciären, Sweden, which were interpreted as parts of an englacial channel system. Additionally, several workers have noted that, during drilling of boreholes through ice, the drill occasionally falls freely for distances of up to 1 m, even at depths of 200 m or more, indicating the presence of water-filled englacial passages (Paterson, 1981). It is uncertain how systems of small, branching conduits can develop within glaciers. Pohjola (1994) suggests that such systems may originate in the accumulation area of the glacier, either as drainage networks in the snowpack or along the lines of crevasses. He argues that in both cases, the conduit systems are perpetuated following the transformation of snow to ice, and maintained by the flow of water under pressure.

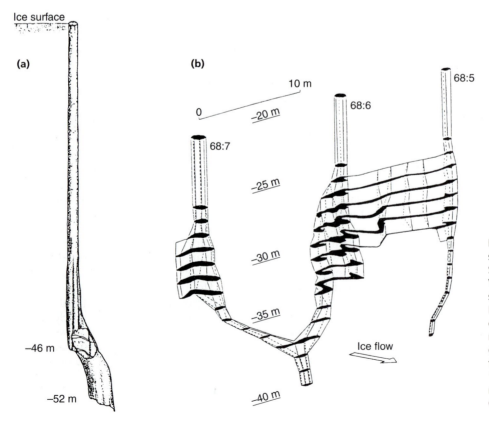

Fig. 3.13 Three-dimensional form of moulins on Storglaciären, Sweden. (a) Vertical upper shaft and sloping lower section. (b) Network of moulins and englacial conduits determined by successive surveys of an ablating glacier surface. (From Holmlund, 1988a. Reproduced by permission of the International Glaciological Society)

Additional information on the character of englacial drainage networks has been obtained by monitoring the flow of water through glaciers in dye-tracing studies. On the basis of the transit time and diffusion of dye injected into moulins in Storglaciären, Sweden, Hooke *et al.* (1988) suggested that part of the englacial drainage of the glacier was through braided conduits rather than a branching network.

3.4 SUBGLACIAL DRAINAGE SYSTEMS

3.4.1 Introduction

Subglacial drainage is now recognized as one of the most important branches of glaciology, because of the profound influence it exerts on ice velocities, glacier stability, and sediment erosion, transport and deposition. Great advances in understanding have been made in recent years, mainly as a result of creative interactions between theoretical analyses and carefully designed field experiments. Several types of subglacial drainage system are now recognized, the distribution of which depends on water discharge, temperature distribution at the ice–bed interface, and the permeability, topography and rigidity of the bed. A fundamental distinction is made between *discrete* and *distributed* systems of subglacial drainage. In the former, water is confined to a few channels or conduits, whereas distributed systems transport water over the whole of, or a large proportion of, the bed. The main types of channelized and distributed subglacial drainage are shown diagrammatically in Fig. 3.14, and are as follows.

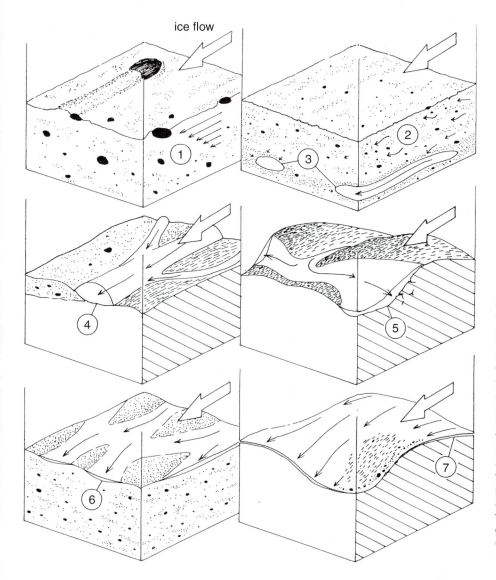

Fig. 3.14 Different types of subglacial drainage system for rigid and deformable beds. 1, Bulk movement with deforming till; 2, Darcian porewater flow; 3, pipe flow; 4, dendritic channel network; 5, linked cavity system; 6, braided canal network; 7, thin film at the ice–rock interface. Types 1, 2, 5, 6 and 7 are distributed systems where drainage takes place over much of the bed, and type 4 is a discrete system in which drainage is carried by a few channels. (From Benn and Evans, 1996)

1. Discrete systems:
   (a) Röthlisberger channels (R-channels) – incised up into the ice;
   (b) Nye channels (N-channels) – incised down into rock or sediment;
   (c) tunnel valleys – incised down into rock or sediment;
2. Distributed systems:
   (a) water film – between ice and bedrock;
   (b) linked-cavity network – between ice and bedrock;
   (c) braided canal network – between ice and sediment;
   (d) porewater flow (Darcian flow) – within subglacial sediment.

Discrete drainage systems are efficient transporters of meltwater, allowing rapid flow through well-connected channel systems. In contrast, distributed systems are inefficient, and meltwater follows more tortuous routes through poorly connected networks. The type of drainage has important implications for glacier motion, by controlling water pressure conditions at the bed (Sections 4.4.2 and 4.5.3).

3.4.2 Channel systems

'*Channel systems*' refers to branching, tree-like networks of tunnels that form efficient drainage pathways capable of carrying large amounts of throughflow. Two fundamental types of subglacial channel are recognized: *Röthlisberger* or *R-channels*, incised upwards into the ice, and *Nye* or *N-channels*, cut down into the glacier bed (see above).

R-channels are similar in many respects to englacial conduits, the principal difference being that they are floored by rock rather than being completely surrounded by ice (Fig. 3.15; Röthlisberger, 1972). R-channels are kept open by melting of the tunnel walls by frictional heat, and tend to contract by ice creep driven by pressure differences between the ice and the tunnel (Section 3.2.3.2; Röthlisberger, 1972; Shreve, 1972). For steady-state conditions, water pressures in R-channels are lower than pressures in the surrounding ice owing to the pressure drop associated with melting of the walls, a tendency that increases with tunnel diameter. Accordingly, large R-channels tend to capture water from smaller ones, leading to the development of a branching (arborescent) network. It should be remembered, however, that for channels fed from the surface, steady-state conditions may be unusual and the pressure within R-channels may be at or close to atmospheric much of the time, particularly close to glacier margins (Röthlisberger, 1972; Lliboutry, 1983; Hooke, 1984). Certainly, the openings, or *portals*, of R-channels at glacier snouts frequently contain only small amounts

Fig. 3.15 Portal at the margin of Eyjabakkajökull, Iceland. (Photo: D.J.A. Evans)

of water relative to their size, particularly late in the melt season (Fig. 3.15). This effect, however, may be partly due to enhanced melting of the walls by warm outside air during the day.

The path taken by R-channels is governed by the hydraulic gradient at the bed. As discussed in Section 3.2.3.3, for equilibrium conditions most of the hydraulic gradient is due to the surface slope of the ice and only a small component is contributed by the gradient of the bed. For ice sheets and ice caps, water flow is approximately parallel to ice flow, but with minor deviations associated with the underlying topography (Shreve, 1985a). The weak influence of elevation on subglacial hydraulic potential, however, means that R-channels need not follow the slope of the bed, but can travel across the slope or even uphill. This is an important difference between subglacial channels and surface streams, the flow of which is controlled only by differences in elevation. Below valley glaciers, parts of the bed are considerably steeper than the surface of the ice, so the influence of subglacial topography on patterns of water flow is correspondingly greater. The fact that valley beds are lower in the middle than at the sides tends to favour the development of a central stream at the deepest point, fed by tributaries from the sides. In the snout area, however, the situation may differ. Near glacier margins, ice surface slopes are relatively large and therefore exert a stronger influence on the hydraulic gradient than the slope of the bed. Glacier snouts usually have convex cross-profiles, which tend to drive subglacial water flow away from the centre-line and towards the margins, a tendency that will be most pronounced where the lateral slope of the ice surface is large and the slope of the valley sides is small. Evidence for this effect exists for many valley glaciers, where subglacial streams have been observed emerging from marginal positions (Fig. 3.16; Stenborg, 1969; Willis *et al.*, 1990; Fountain, 1992).

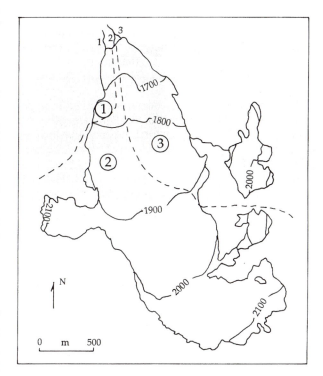

Fig. 3.16 The catchment areas of three melt streams emerging from the South Cascade Glacier, Washington State, USA. Note how meltwater from catchments 1 and 3 emerges from the lateral margins of the glacier owing to the convexity of the ice-surface contours. (Redrawn from Fountain, 1993)

The cross-sections of R-channels vary according to the balance between tunnel enlargement by melting and tunnel closure by ice creep and water freezing on to the walls. Where efficient melting occurs, R-channels should be steeply arched because frictional heating, and hence the melting rate, is greatest above the centre of the channel where the water flow is greatest. Where freezing occurs, however, R-channels are expected to be wide and low (Shreve, 1985a). These predictions imply that there should be systematic variations in the cross-section of an R-channel along its length as it passes through local zones of melting and freezing. In particular, channels flowing downhill should be sharply arched because melting is most efficient for descending flow (Shreve, 1972; Röthlisberger, 1972). Conversely, channels flowing uphill should be wide and low, for two reasons. First, the gain in potential energy associated with increased channel elevation means that less energy is available for melting of the walls. Second, the gain in height means that the water moves into regions where ice pressure is lower and the melting point correspondingly higher. Together, these two effects reduce the amount of ice that can be melted from the tunnel walls, and may cause freezing on to

the walls of steeply ascending tunnels. Shreve (1985a) calculated that freezing should occur for channels ascending with a gradient more than 1.7 times the surface gradient of the ice. Analysis of the cross-profiles of eskers, which represent infilled R-channels, lends some support to this model (Shreve, 1985a).

The second major group of channels is known as Nye channels or N-channels, and these are incised into the substratum (Nye, 1973). The existence of such channels is clearly demonstrated by incised river channels exposed by the retreat of modern glaciers, and abandoned channels cut in bedrock in areas occupied by former ice sheets (Fig. 3.17; Drewry, 1986; Sharp *et al.*, 1989b). N-channels can occur as single isolated features or in braided networks covering large areas of the bed (Walder and Hallet, 1979; Hallet and Anderson, 1982; Sharp *et al.*, 1989b). The latter represent former distributed drainage networks, which are discussed in Section 3.4.4. The presence of N-channels implies that erosion, and hence water flow, is consistently focused along the same route. Such consistency in the direction of water flow is most likely to occur where bedrock topography exerts a strong control on the hydraulic gradient, such as steep-sided valleys or rough glacier beds. This does appear to be the case, as subglacial meltwater channels cut in bedrock do frequently occur along the axes of valleys, or wind between pronounced high points such as roches moutonnées, or cut across ridges via cols (Sissons, 1960a, 1961a, b; Sugden and John, 1976; Walder and Hallet, 1979; Hallet and Anderson, 1982; Sharp *et al.*, 1989b; Section 9.3.4.1).

Fig. 3.17 Nye channel exposed by glacier retreat. (Photo: B. Hubbard)

Branching channel systems can also be incised into subglacial sediments, and can be very large in the case of *tunnel valleys* (Section 9.3.4.2; Boulton and Hindmarsh, 1987; Piotrowski, 1994). The shape of the channel floor is controlled by the erosion and evacuation of sediment, whereas the top of the channel is composed of ice and enlarges or contracts in response to the balance between melting and effective pressure, as for R-channels. Shoemaker (1986a), Shoemaker and Leung (1987), and Boulton and Hindmarsh (1987) have argued that such channels develop to allow the efficient drainage of subglacial aquifers, which would otherwise become unstable as a result of high porewater pressures. The spacing of the channels is dependent on the amount of surface water or basal meltwater to be discharged, and the permeability of the sediment. The pressure conditions and stability of such channels have been considered further by Alley (1992b).

The efficiency of water flow through R-channels and single N-channels has been demonstrated in several dye-tracing experiments on modern glaciers. Where flow is through channels, dye poured down moulins tends to emerge as a single concentrated pulse, indicating little dispersal of the dye during flow, and no temporary storage in backwaters or side channels (Fig. 3.20). Reported water transit velocities fall within the range reported for proglacial outwash streams, and vary between 0.08 m s^{-1} and 0.7 m s^{-1} (e.g. Stenborg, 1969; Kamb *et al.*, 1985; Seaberg *et al.*, 1988; Willis *et al.*, 1990). Actual velocities may be higher than these values if the channel systems are sinuous.

3.4.3 Water films

There is abundant evidence that, where rock beds are overlain by glacier ice at the pressure melting point, the ice and rock are separated by a thin film of meltwater (Weertman, 1972; Hallet, 1979b). Weertman (1964, 1969) supposed that such films constitute the principal drainage system of glaciers, carrying most of the water produced at the surface, interior and bed of the ice. However, it is now widely recognized that subglacial water films cannot be much more than a millimetre thick without breaking up, forming channels at low points on the bed or filling cavities in the lee of obstacles (Lliboutry, 1968; Nye, 1976; Walder, 1982). Therefore, water films have only a very limited ability to transport water, and are believed to form a minor and largely independent drainage component transporting only water produced by local subglacial melting. The bulk of throughflow over hard rock beds, including surface and englacial waters as well as a subglacial component, is probably carried by systems of channels (Section 3.4.2) and/or linked cavity systems (Section 3.4.4). Evidence in support of this view is provided by boreholes to the base of temperate glaciers, which often encounter areas of the bed that are not, at least initially, hydraulically connected to the main subglacial drainage system (Hodge, 1976, 1979; Fountain, 1994).

Subglacial melting can occur over large areas of glacier beds if the amount of heat from geothermal and frictional sources cannot be evacuated through the ice. Most of the water in films below temperate glaciers, however, appears to originate by local pressure melting (Hanshaw and Hallet, 1978; Hallet *et al.*, 1978; Souchez and Lemmens, 1985; Sharp *et al.*, 1990; Hubbard and Sharp, 1993). Pressure melting occurs on the upglacier side of bumps on the bed, where ice-contact pressures are high and the melting point is low. Conversely, at the downglacier side of bumps, ice-contact pressures are low and the melting point is increased, encouraging freezing, a process known as *regelation* (Section 4.5.2.1; Weertman, 1957b; Hubbard and Sharp, 1993). The principal role of the subglacial water film under many glaciers is to transfer water downglacier from areas of pressure melting to areas of regelation; for this reason the water film is sometimes known as the *regelation water film*. The amounts of melting and freezing may not exactly balance, however. Isotopic evidence suggests that, where there is net water production, the excess probably enters channels or a linked-cavity network and is lost to the water film system (Sharp *et al.*, 1990; Hubbard and Sharp, 1993). Conversely, water films may 'soak up' excess water at times of high water pressures.

The thickness of subglacial water films is extremely difficult to establish by direct measurement, because it is impossible to observe such a delicate system without disturbing and altering it. Indirect methods, therefore, must be used. Vivian (1975) noted a deficiency of particles smaller than 200 µm at the ice–rock interface below Alpine glaciers, and reasoned that the finer particles had been carried away in the subglacial water film, which must therefore be roughly 200 µm (0.2 mm) thick. This approach has been refined by Hallet (1979b), who inferred the nature of former subglacial water films from the characteristics of calcite deposits on recently deglaciated limestone surfaces in front of Blackfoot Glacier in the USA. Large areas of the former glacier bed are covered with such deposits, which consist of calcite ($CaCO_3$) precipitated from subglacial waters, together with varying amounts of fine rock particles. The particles appear to have been incorporated in the deposits when calcite precipitated out of the subglacial water film during regelation above the lee side of bumps. Hallet argued that the particles are most likely to have been transported in the water film, and that the size distribution of the particles can be used to reconstruct the film thickness. The coarsest particles are *c*. 110 μm in diameter, although very few exceed 50 μm and most are smaller than 30 μm (Fig. 3.18). Hallet concluded that in most cases the film must be thinner than a few micrometres, with localized and occasional thickening to several tens or even hundreds of micrometres in exceptional cases. The distribution of the coarsest particles in lenses suggests that, where the film was thick, flow was concentrated in irregular microchannels separated by projections of the glacier bed, as predicted by theory (Walder, 1982).

3.4.4 Linked-cavity systems

Lliboutry (1968, 1976, 1979) was the first to suggest that subglacial water can drain through systems of cavities developed between the base of the ice and the underlying bedrock. Such networks consist of numerous cavities linked by narrow connections or *orifices* (Fig. 3.19). Water transit velocities are very low, because of throttling of water flow in orifices, the tortuous nature of the network, and temporary storage in poorly connected cavities. Indeed, not all water-filled cavities need be connected with the main drainage network at any one time, and some may connect only with the subglacial water film. The extent of the active network will vary with water discharge and pressure, as cavities increase or decrease in size and the number of active orifices changes (Kamb, 1987; Sharp *et al.*, 1989b). Striking evidence of former linked-cavity drainage networks is present on recently deglaciated limestone surfaces, where former water-filled cavities and channels are readily identifiable from abundant solutional features (Section 9.3.4; Walder and Hallet, 1979; Hallet and Anderson, 1980; Sharp *et al.*, 1989b).

Dye-tracing experiments below existing glaciers provide indirect but persuasive evidence for linked-cavity drainage systems. In several studies, dye poured down moulins has emerged at the glacier snout in diffuse, multi-peaked patterns, which are very different from the single, sharp-peaked plug of dye characteristic of tunnel systems (Fig. 3.20; Kamb *et al.*, 1985; Seaberg *et al.*, 1988; Willis *et al.*, 1990). Such patterns indicate splitting and rejoining of flow paths, and the temporary storage of water beneath the glacier, compatible with the expected geometry of linked-cavity networks. The time lag between dye injection and its appearance at the snout indicates throughflow velocities of only 0.01–0.086 m s^{-1}, an order of magnitude less than those expected for channels carrying similar discharges.

The mechanics of linked-cavity systems results in very different relationships between water discharge and pressure as compared with those for tunnel systems, with important consequences for drainage network evolution (Walder, 1986; Kamb, 1987; Fowler, 1987a). For small water discharges, ice-melting tends to be a relatively unimportant process of cavity enlargement, because of the large surface area of the drainage system compared with its volume. As we saw in Section 3.2.3, the size of small cavities is mainly determined by the mechanics of ice–bedrock separation, and cavities form where water pressure (P_w) exceeds the pressure exerted by the ice (P_i) on the bed. Therefore, increases in water pressure result in increased carrying capacity. This positive relationship between discharge and pressure is precisely opposite to the steady-state behaviour of R-channels. Whereas large, low-pressure R-channels tend to 'capture' meltwater from smaller, high-pressure ones, no such tendency exists for water-filled cavities at low discharges. Up to a point, the pressure in larger cavities will be greater than in smaller ones, and there will be no tendency for such cavities to capture drainage and undergo unstable growth at the

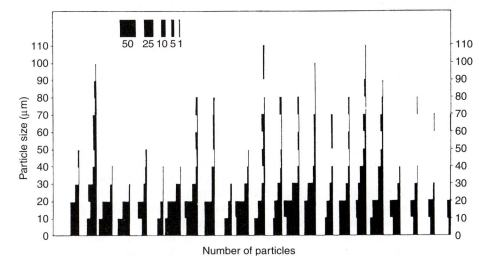

Fig. 3.18 Size distribution of rock fragments in subglacial carbonate deposits. (From Drewry, 1986, after Hallet, 1979b)

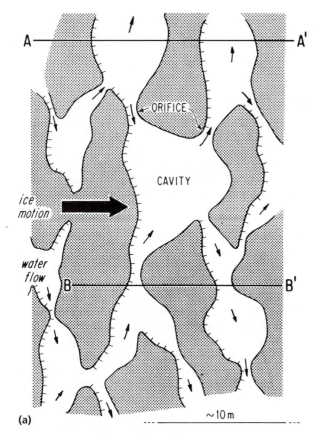

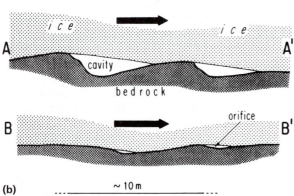

(b)

Fig. 3.19 Schematic diagrams of a linked cavity network. (a) Plan view. (b) Cross-sections through cavities (A–A') and orifices (B–B'). (From Kamb, 1987. Reproduced by permission of the American Geophysical Union)

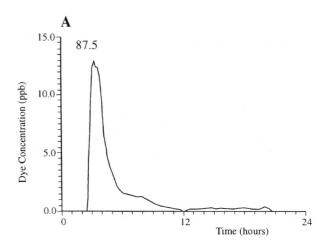

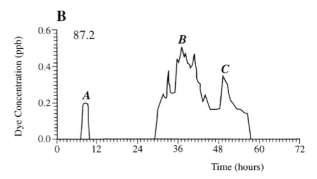

Fig. 3.20 Results of dye-tracing experiments on Midtdalsbreen, Norway, highlighting the difference between discrete and distributed subglacial drainage. (A) A single pulse of dye emerging from the glacier after a short lag, indicating efficient transport through a well-connected conduit. (B) Several dye peaks with lags of up to 48 hours, probably the result of slow transit and temporary storage in a tortuous linked-cavity network. Note the different timescales. (Adapted from Willis *et al.*, 1990)

expense of others. Within limits, therefore, linked-cavity systems tend to be stable features, with no tendency to evolve into branching networks of R-channels.

This relationship begins to break down at high discharges, when ice-melting becomes more important than water pressure in maintaining cavities (Kamb, 1987). As for R-channels, melting of the ice roof offsets some of the ice pressure that would otherwise

tend to close the cavity, and P_w can fall below P_i. Ice melt rates increase with discharge, so increasing water discharges through a cavity result in a greater pressure drop. Therefore, large, low-pressure cavities will 'capture' drainage from smaller, high-pressure cavities, and will tend to grow. With continued cavity growth, drainage will increasingly be carried by a few large conduits, and the linked-cavity network will convert into a system of R-channels. The instability of linked-cavity systems at high discharges provides a mechanism whereby subglacial drainage can 'switch' to a more efficient mode, allowing large amounts of water to be evacuated from the bed.

Important evidence in support of this model of subglacial drainage evolution has been provided by dye-tracing experiments (Kamb *et al.*, 1985; Seaberg *et al.*, 1988; Willis *et al.*, 1990). Studies at Storglaciären, Sweden, and Mitdalsbreen, Norway, have demonstrated that parts of the subglacial drainage systems evolve from distributed, braided systems into efficient conduit systems on a seasonal basis, as discharges increase during the melt season. The timing of the transition is dependent on the supply of water to the system, which in turn reflects the particular weather conditions experienced on the glacier. At Mitdalsbreen, the critical discharge at which transition occurs was estimated at $1 \text{ m}^3 \text{ s}^{-1}$ (total discharge per subglacial catchment). During the winter months, discharges at Storglaciären and Mitdalsbreen drop to zero. It is therefore likely that the drainage network developed during the previous summer will tend to close down in response to ice deformation, and will be reinitiated at the onset of the next ablation season. Kamb *et al.* (1985) found that switches from linked-cavity to channelized drainage were of fundamental importance in halting unstable acceleration during a surge at Variegated Glacier, Alaska. The influence of drainage evolution on the dynamics of this and other glaciers will be examined in Sections 4.5.3 and 4.8.2.

3.4.5 Porewater movement

Groundwater flow through hard rock beds is generally insignificant, owing to the low permeability of most rocks. Subglacial meltwater can, however, enter cave systems developed in limestones, such as Castleguard Cave, below part of the Columbia Icefield, Alberta (Smart, 1986). For glacier beds composed of unconsolidated sediments, however, groundwater flow can be significant. There are two basic mechanisms for porewater movement (Fig. 3.14). First, if the sediment undergoes shear as the result of glacially imposed stresses, *bulk movement* of water will occur as it is carried along with the mineral grains (Clarke, 1987c). Second, porewater can flow relative to mineral grains if driven by a hydraulic gradient (Boulton *et al.*, 1974; Boulton and Jones, 1979; Murray and Dowdeswell, 1992). This process is called *Darcian flow* after the nineteenth-century civil engineer H. Darcy, who conducted early experiments on water flow in soils. Darcy determined that the discharge of a fluid through a porous medium is inversely proportional to the viscosity of the fluid and directly proportional to the pressure gradient and a quantity known as the hydraulic conductivity:

$$Q = -(kA/\eta)(\Delta p/d) \tag{3.8}$$

where k is the hydraulic conductivity, A is the sample cross-sectional area at right angles to flow, η is the fluid viscosity, and $\Delta p/d$ is the pore-water pressure gradient.

With the exception of well-sorted, coarse gravels, the permeability of sediments is low (Table 3.1), so

Table 3.1 Hydraulic conductivities of selected sediments

	$k \, (m \, s^{-1})$		$k \, (m \, s^{-1})$
Clay	$< 10^{-9}$	Fine sand	10^{-7} to 10^{-5}
Silts	10^{-9} to 10^{-7}	Coarse sand	10^{-5} to 10^{-2}
Till	10^{-12} to 10^{-6}	Gravel	10^{-2} to 10^{-0}

Source: Freeze and Cherry (1979)

discharges of water will be low unless pressure gradients are very high. Darcian flow is therefore an inefficient process for the discharge of meltwater, and is thought to be insufficient on its own to drain water from most ice masses (Alley, 1989). Accordingly, Darcian flow probably usually occurs alongside other forms of subglacial drainage, with more efficient drainage systems developing in response to rising porewater pressures in the sediment. Shoemaker (1986a) and Boulton and Hindmarsh (1987) argue that channels evolve by extending headward from regions of highest porewater pressure, and that development of low-pressure channels may therefore be an important mechanism for reducing porewater pressures in saturated sediments. This has implications for glacier stability, because high subglacial porewater pressures reduce the strength of sediments, making them more likely to deform under stress. Details of this process, and its implications for glacier motion, are discussed in Section 4.4.2.

Hubbard *et al.* (1995) have presented evidence for coupled porewater flow and channelized drainage below the Haut Glacier d'Arolla, Switzerland. Using water pressure data from a borehole array connected with a subglacial channel and its surroundings, they found that high water pressures during the day forced water out of the channel into a layer of subglacial till. At night, when water pressures in the channel were low, the flow of water was reversed, driving water from the till back towards the channel. This process apparently results in the flushing of fine debris out of the till into the subglacial drainage system. Boulton and Hindmarsh (1987) found that water pressure variations in subglacial till below Breidamerkurjökull, Iceland, followed a diurnal cycle, also indicating a connection between porewater flow and conduits fed from the surface.

Boulton *et al.* (1995) have modelled meltwater flow through subglacial aquifers below the Saalian ice sheet in the Netherlands. They concluded that throughflow of water was an order of magnitude faster during glacial conditions than during interglacial conditions, owing to large pressure gradients, and that at the ice sheet maximum, porewater will have welled upwards beyond the margin. Such upwelling may have been responsible for restructuring of sediments, forming dewatering pipes and diapirs (upward bulges of remobilized sediment) at depth and extrusion features at the surface.

3.4.6 Braided canal systems

As mentioned in Section 3.4.5, Shoemaker (1986a) and Boulton and Hindmarsh (1987) have argued that branching, low-pressure channels will develop within subglacial aquifers when Darcian flow cannot evacuate all the available meltwater and porewater pressures in the sediment rise. Modelling by Walder and Fowler (1994), however, predicts that dendritic channels will be stable only where the substratum is stiff or rigid, or the hydraulic gradient is large. They propose that where subglacial sediments are soft and prone to deformation, excess water will collect at the ice–sediment interface in widespread braided systems consisting of interconnected broad, shallow *canals* (Fig. 3.14). Walder and Fowler proposed that the relations between discharge and water pressure for braided canal systems will be analogous to those for linked-cavity systems, and will contrast with discharge–pressure relationships for branching channels. For braided canals, greater discharges will be associated with higher water pressures, whereas in branching channels, water pressure decreases with increasing discharge. This means that in canal systems, drainage will tend to remain distributed over much of the glacier bed because large canals will not tend to capture water and grow at the expense of smaller ones.

Clark and Walder (1994) argued that canal systems that form in association with deforming beds should be recognizable in the geologic record as broad lenses of sorted sediments with concave-up lower contacts and nearly planar upper contacts within till units. Such lenses within till sequences have been reported by N. Eyles *et al.* (1982b), Brown *et al.* (1987), Shaw (1987), D.J.A. Evans *et al.* (1995) and Benn and Evans (1996), and are discussed in Section 11.2.1.2. Although the possibility of canal systems has gained support, it must be recognized that as yet there is no direct evidence for their existence from modern glacier beds.

3.4.7 Modelling subglacial drainage systems

One of the most exciting recent developments in subglacial hydrology has been the development of mathematical models of subglacial drainage networks, which allow field observations to be interpreted in terms of physical processes. Equations have been developed that relate water transit times to key characteristics of the drainage system, such as conduit diameter and shape, channel wall roughness, hydraulic gradient, and the existence of storage reservoirs. Using such equations, it is possible to use data from dye-tracing experiments and borehole water-level fluctuations to determine the probable style of subglacial drainage, and to establish whether flow is likely to be through dendritic conduits, linked-cavity networks, Darcian porewater flow or other types of system (e.g. Seaberg *et al.*, 1988; Willis *et al.*, 1990; Hubbard *et al.*, 1995).

Clarke (1996a, b) has developed an original and powerful approach to modelling subglacial drainage networks, known as *lumped-element analysis*. This approach likens drainage networks to electrical circuits consisting of resistors, storage elements and switches. Different types of drainage, such as R- or N-channels, linked cavity systems or Darcian porewater flow, offer varying resistance to flow, and can therefore be modelled as flow resistors. Storage elements can be characterized as differently shaped vessels, which may be open or closed, allowing temporal changes in storage capacity and overflow to be modelled. Finally, different types of switches can be used to simulate changes in flow routeing, which can be triggered by upstream changes in flow volume or capacity. This simple but efficient hydrological model can replicate many observed characteristics of subglacial drainage networks, such as the tendency for large R-channels to capture water from smaller ones, switching between distributed and channelized networks, and the difference between summer and winter subglacial drainage systems. In addition, solute and suspended sediment dynamics have been incorporated in the model, yielding results that compare well with observations (Clarke, 1996a).

Recently, integrated models of glacial drainage have been developed, which allow spatial and temporal variations in runoff and sediment transport to be predicted from catchment topography and meteorological inputs. Richards *et al.* (1996) have conducted an impressive study of catchment hydrology on the Haut Glacier d'Arolla, Switzerland, which combined a detailed measurement programme with comprehensive theoretical analysis. This work allows inputs of meltwater to the system to be predicted from the energy balance at the glacier surface (Section 2.2.3.2), and allows subsequent transport and storage of water and sediment within the glacier to be reconstructed in detail. Richards *et al.* showed that seasonal retreat of the snow line up the glacier alters the volume and spatial pattern of water input to the system, causing a reorganization of the basal drainage network from a distributed system to a conduit system as the melt season progresses. This reorganization results in changes to the subglacial weathering environment and the availability of solutes and suspended sediment, which are reflected in sediment erosion and transport rates. The Arolla project has far-reaching implications for predicting river runoff in glacierized catchments under changing climatic regimes, and highlights the need for similar work to be carried out in other basins.

3.5 GLACIER OUTBURST FLOODS (JÖKULHLAUPS)

3.5.1 Introduction

The hydrology of some glacier systems is characterized by the periodic or occasional release of large amounts of stored water in catastrophic outburst floods. These are widely referred to by the Icelandic term *jökulhlaup* ('glacier-flood'), but are also known as *débâcles* in mainland Europe and *aluviones* in South America (Lliboutry, 1971, 1977; Vivian, 1974). Jökulhlaups may be triggered by (a) the sudden drainage of an ice-dammed lake below or through the ice dam; (b) lake water overflow and rapid fluvial incision of ice, bedrock or sediment barriers; or (c) the growth and collapse of subglacial reservoirs. The periodic nature of jökulhlaups in many areas means that they can often be predicted with some precision, but they are nevertheless severe geomorphological hazards, and their floodwaters can wreak havoc on all human structures located in their path (Mason, 1929, 1930a, 1935; Thorarinsson, 1939; Stone, 1963; Richardson, 1968; Jackson, 1979; Hewitt, 1982; Tufnell, 1984; Grove, 1988). The sudden impact of jökulhlaups is often best described in historical and eyewitness accounts. For example, Thorarinsson (1953) describes the 1934 flood of the Skeidera River, Iceland, following the release of floodwater from Grimsvötn (see Fig. 3.4):

> The river Skeidera, which in March and April normally has its smallest discharge, started to rise on March 22nd. The rise was slow at first, but on the 24th it had reached approximately the summer high water level. On the 28th the water started forcing its way out from under the glacier at several places, breaking up its border. On the morning of March 31st, the glacier burst reached its climax. 40–50,000 m^3 of muddy grey water plunged forth every second from under the glacier border bringing with it icebergs as big as three-storeyed houses. Almost the whole of the sandur, some 1000 km^2 in area, was flooded. At 17.30 hr the same day the burst suddenly started to abate, and by the following morning the discharge of the Skeidera was normal.

It is only during jökulhlaups that the vast Icelandic sandur plains of Skeiderarsandur and Myrdalssandur actually become almost totally flooded. Once the water abates after a flood, channel patterns have often undergone radical changes, icebergs lie stranded on the sandur surface, and roads and bridges

are often either severely damaged or completely destroyed (Fig. 3.21). Such events are not restricted to the volcanic areas of Iceland, and indeed have taken a far heavier toll on humans and their settlements in mainland Europe, North and South America, Asia and New Zealand, where engineers have been attempting to reduce and repair the effects of glacially induced floods for decades. Although jökulhlaups usually occur during the ablation season when meltwater is at a maximum and glacial drainage networks are at their most developed, they have been known to occur during the winter (Liestøl, 1955; Vivian, 1974). In addition, their occurrence is not always cyclic and may reflect changing glaciological conditions associated with glacier snout oscillations (Stone, 1963; Lliboutry *et al.*, 1977; Desloges *et al.*, 1989). Glacier recession or advance–retreat cycles in high-relief terrain are often characterized by two phases of jökulhlaup hazard: an early phase of increasing hazard as lakes build up, and a late phase as lakes drain below and around retreating glacier margins (Fig. 3.22; S.G. Evans and Clague, 1994).

Some of the most violent and destructive jökulhlaups originate in subglacial cupolas in volcanically active areas (Section 3.2.4.1; Thorarinsson, 1956b, 1957; Tryggvason, 1960; Björnsson, 1975, 1976, 1992). The most celebrated example is that of Grimsvötn beneath the Vatnajökull ice cap, which empties approximately every six years, involving the release of up to 4.5 km^3 of water at peak discharges of up to 50,000 m^3 s^{-1} (Fig. 3.23; Björnsson, 1974, 1975, 1992; Gudmundsson *et al.*, 1995). Once meltwater is released by the ice barrier it travels 50 km beneath the ice to emerge on Skeidararsandur, and the release of the water pressure associated with the emptying of the subglacial reservoir is thought to be responsible for setting off volcanic eruptions in the

Fig. 3.21 Jökulhlaup from Summit Lake at Ninemile, Alaska, on 17 September 1967. The discharge was nearly 3000 m^3 s^{-1}, one hundred times larger than the mean discharge of the stream during normal flow conditions. The steel bridge was washed away a few hours after the photograph was taken. (Photo: J.J. Plummer)

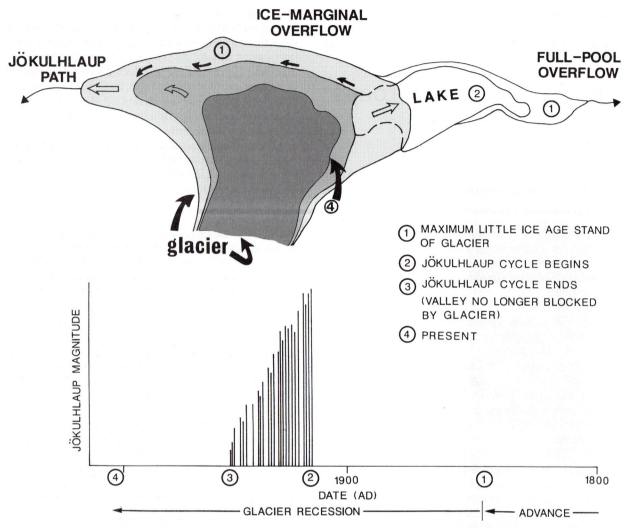

Fig. 3.22 Hypothetical glacier snout positions (open arrows indicate ice flow) and associated ice-dammed lake formation and jökulhlaup activity (top) and a plot of jökulhlaup magnitude against time (bottom), based upon studies in the Canadian cordillera. At time 1, a lake impounded by the glacier is stable and drains via a stable overflow channel. Between times 1 and 2, the lake overflows along the margin of the glacier (solid arrows). At time 2, the first jökulhlaup occurs and sporadic or cyclic outburst floods continue until time 3 when the glacier has retreated to a position where it no longer dams a lake. The glacier then retreats to the present (time 4). (Modified from Clague and Evans, 1994)

Grimsvötn caldera (Thorarinsson, 1953, 1956b). The regularity of the Grimsvötn jökulhlaups means that events can be predicted, and their impact upon roads, bridges and other structures on Skeidararsandur can be anticipated. Since the largest jökulhlaups of the 1930s, flood volumes have decreased in the 1980s and 1990s to 0.6–1.45 km^3 in response to declining volcanic activity and geothermal heating (Gudmundsson *et al.*, 1995). Large jökulhlaups can also be triggered by subglacial volcanic eruptions, such as the 1918 eruption of Katla below Myrdalsjökull (Tómasson, 1996) and the 1996 eruption below Vatnajökull.

Jökulhlaup discharges can exceed ablation-related flows by several orders of magnitude. In Baffin Island, Church (1972) recorded jökulhlaup flows of about 200 m^3 s^{-1} compared with typical peak flows of only 20 m^3 s^{-1}. One of the highest historic jökulhlaup flows followed a subglacial eruption of Katla, below the Myrdalsjökull ice cap in Iceland, and peaked at 1,500,000 m^3 s^{-1} (Maizels, 1995). Although the power of these historical jökulhlaups is indeed awesome, their discharges and geomorphic impacts are minor when compared with the floods released during the last glaciation. Floods from Glacial Lake Missoula, for example, may have reached peak discharges of 21,000,000 m^3 s^{-1} and certainly exceeded 3,000,000 m^3 s^{-1} (Section 9.3.4.4).

Much of the damage created during jökulhlaups is associated with the large amounts of debris that accompany the floodwaters (Fig. 3.24). For example,

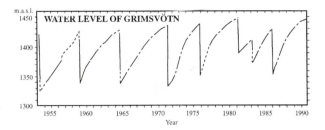

Fig. 3.23 Variations in the water level (metres above sea-level) of Grimsvötn, Vatnajökull, Iceland, showing the results of cyclic lake drainage. (Redrawn from Bjørnsson, 1992)

Fig. 3.24 Moraine-dammed Nostetuko Lake in British Columbia, Canada, which drained catastrophically in 1983 when the Little Ice Age moraine was breached as a result of a glacier avalanche from the Cumberland Glacier in the distance. The scar in the moraine attests to the removal of approximately 1.5×10^6 m³ of material, much of which was redeposited in the debris fan in the foreground. (From Evans and Clague, 1994. Reproduced by permission of Elsevier)

Lliboutry (1971) reported that the 200,000 m³ of water released by the French Tête Rousse débâcle in 1892 brought with it 800,000 m³ of sediment, and Lliboutry *et al.* (1977) reported that the Peruvian Jancarurish flood in 1950 discharged 2,000,000 m³ of water and 3,000,000 m³ of sediment after the failure of a moraine impounding a proglacial lake. S.G. Evans and Clague (1994) reported several examples of catastrophic proglacial lake drainage in British Columbia where debris flows up to 20 m thick, rather

than floodwaters, travel up to 20 km downvalley. Many jökulhlaups also transport large numbers of icebergs. The 1918 flood from Katla, for example, was heavily freighted with icebergs broken off from the destabilized margin of Myrdalsjökull, and left blocks of ice up to 60 m high stranded on the sandur surface when it receded (Tómasson, 1996). In places, icebergs were so numerous that the water itself could not be seen, except where large currents swept forward. At one point, a very large floe was broken from the glacier and carried by the flood, and was said by eyewitnesses to resemble 'whole hillocks covered in snow' (Tómasson, 1996).

Damage to settlements and farmland can take place at very great distances from the outburst source. For example, Mason (1929) reported that the 1926 flood released by the Himalayan Shyok glacier destroyed the village of Abadan and the surrounding cultivated land even though it was at a distance of 400 km from the outburst source. Similarly, the release of the Marjelensee through the Aletsch Glacier, Switzerland, has been responsible for repeated flooding of the Rhône valley as far downstream as Lake Geneva. The completion of a drainage tunnel in 1896 has since allowed water to be slowly released without periods of storage and catastrophic release (Preller, 1896; Collett, 1926; Aubert, 1980).

3.5.2 Jökulhlaup mechanisms

The mechanisms by which jökulhlaups are triggered vary with the position of the water body and the nature of the dam holding back the waters. In particular, different trigger and drainage mechanisms operate for moraine-dammed, ice-dammed and subglacial lakes.

Moraine-dammed lakes generally drain by rapid incision of the sediment barrier by outpouring waters (Fig. 3.25). Once incision begins, the focusing of water flow through the outlet can accelerate erosion and enlargement of the outlet, setting off a catastrophic positive feedback process resulting in the rapid release of huge amounts of sediment-laden water. The onset of rapid incision of the barrier can be triggered by waves generated by glacier calving or ice avalanching, or by an increase in water level asso-

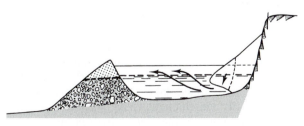

Fig. 3.25 Mechanism of moraine dam failure. (Redrawn from Lliboutry *et al.*, 1977)

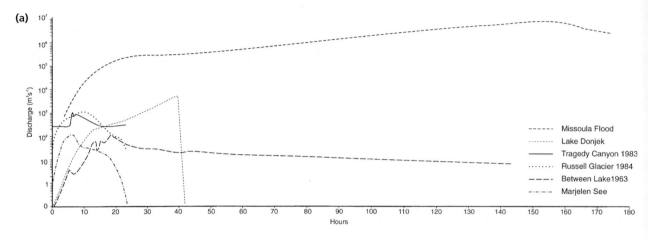

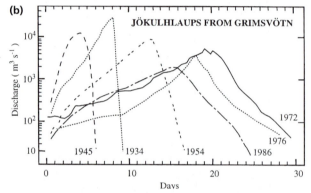

Fig. 3.26 Jökulhlaup hydrographs. (a) Ice-dammed lake outbursts: late Pleistocene flood from Lake Missoula, Washington (estimated); Lake Donjek, Alaska; Tragedy Canyon, BC; Russell Glacier, Greenland; Between Lake, Axel Heiberg Island; Marjelen See, Switzerland. (b) Grimsvötn subglacial lake, Iceland. (Redrawn from Björnsson, 1992)

ciated with glacier advance. For example, an ice avalanche from the Langmoche glacier, Nepal, in August 1985 triggered the drainage of a moraine-dammed lake, and the resulting flood destroyed houses, bridges and a small hydroelectric plant (Vuichard and Zimmerman, 1986, 1987). In Peru, outburst floods from moraine-dammed lakes have been responsible for the devastation of downvalley settlements and huge loss of life (Lliboutry et al., 1977), and spectacular examples of moraine breaching and downstream flooding from proglacial lakes in British Columbia have been reported by Clague et al. (1985) and S.G. Evans and Clague (1993, 1994). The risks of outbursts can be dramatically reduced by the installation of outflow pipes, allowing the waters to drain away continually, although this may not be economically viable in remote areas in poor countries.

Ice-dammed lakes may also drain following failure of the dam, or when water erodes or melts an overflow channel into the dam surface. Maag (1969) reported that lakes dammed by subpolar glaciers on Axel Heiberg Island in the Canadian Arctic emptied by drainage over the glacier surface, along the ice margin or through neighbouring cols. The discharge curves of such lakes display rapid increases up to a maximum followed by a gradual decrease (Fig. 3.26a). Some ice-dammed lakes drain slowly, such as that at the margin of Sydgletscher, south Greenland, which took 14 days to empty in 1981 at an average rate of 200 m^3 s^{-1} (Dawson, 1983). This can be compared with the peak discharge of Graenalon, the largest ice-dammed lake in Iceland and located at the margin of Skeidararjökull, which usually takes approximately nine days to empty and attains peak flows of 6000 m^3 s^{-1} (Thorarinsson, 1939, 1956b).

Most ice-dammed lakes and subglacial cupolas drain rapidly via channels at or near the bed as the result of changes in the glacier hydrological system or changing relationships between pressure conditions in the water reservoir and the surrounding ice.

Nye (1976) developed a mathematical model of ice-dammed and subglacial lake drainage, which has been modified by Spring and Hutter (1981) and Clarke (1982). According to this model, drainage beneath the glacier will be initiated if the hydraulic potential along a possible drainage pathway beneath the ice dam is less than that in the reservoir. Water will then be driven from the reservoir beneath the ice along the hydraulic gradient. This theoretical condition will be met when the lake attains a critical depth, approximately 90 per cent of the thickness of the ice dam, at which hydrostatic pressure at the lake bottom is just enough to support the weight of the ice dam at its thinnest point. The threshold condition for drainage initiation for the case of a subglacial cupola can be visualized by examining Fig. 3.3c. Drainage of the cupola will commence when lake growth reduces the width of the ice dam (W) to zero. At that point, water will begin to flow out of the reservoir along the hydraulic gradient towards the glacier margin.

Once water begins to escape from the reservoir, it can enlarge the subglacial drainage pathway by melting the surrounding ice, either by the mechanical energy of the rushing waters or by thermal energy (Björnsson, 1975; Nye, 1976). Thermal energy may be particularly important in geothermal areas, where lake waters may have been warmed well above the melting point (Björnsson, 1992). The enlargement of the drainage pathway allows more water to escape, which accelerates the melting of the tunnel walls, thus setting up a catastrophic positive feedback process allowing most of the water to drain from the lake. Lake drainage will cease either when the lake is empty or when the tunnel is closed off by ice creep (see Section 3.2.3.2). Creep closure is most effective when there is a large difference between the ice overburden pressure and the water pressure within the tunnel.

This model compares well with observations of jökulhlaup flood hydrographs from a wide variety of ice-dammed and subglacial lake drainage events. Hydrographs typically rise rapidly to a peak discharge and then terminate abruptly (Fig. 3.26b; Rist, 1955; Thorarinsson, 1957; Larsen, 1959; Whalley, 1971; Mathews, 1973; Mottershead, 1975; Theakstone, 1978; Clarke and Waldron, 1984; Sugden *et al.*, 1985; Russell, 1989; Björnsson, 1992). The rapid rise in discharge documents the progressive enlargement of ice tunnels by the draining water, and the abrupt termination in jökulhlaup hydrographs marks either emptying of the lake or creep closure of the drainage tunnel. Jökulhlaup hydrographs contrast markedly with subaerial stream flood hydrographs, which rise rapidly in response to rainfall and then tail off slowly after the rainfall event, owing to the slower evacuation of groundwater. Where lake drainage is complete, tunnels may remain open for some time, delaying refilling of the lake. This is particularly common where the ice dam is thin, and creep closure rates correspondingly low (e.g. Mottershead and Collin, 1976; Russell, 1989). Observations on Norwegian temperate glaciers by Liestøl (1955) suggest that subglacial outlet channels will remain permanently open once the ice dam thickness drops below *c.* 50 m.

The Nye (1976) and Clarke (1982) model has been tested by Björnsson (1992) using empirical data from several ice-dammed and subglacial lakes in Iceland. The model yields good approximations to the observed sequence of events for many jökulhlaups, but with some important differences:

1. Lake drainage can commence before the water level reaches the critical threshold required to overcome the theoretical ice overburden pressures below the ice dam. Instead, drainage begins when the effective overburden pressure below the ice

dam is still greater than the hydrostatic pressure in the reservoir. This indicates that such jökulhlaups are not triggered by simple flotation of the ice dam, but probably also involve pre-existing weaknesses in the ice or reorganizations of the subglacial drainage system in the area surrounding the reservoir (Röthlisberger and Lang, 1987; Björnsson, 1992). Fowler and Ng (1996) have argued that the trigger mechanism is a switch in the subglacial hydrology below the dam from Darcian porewater flow through porous media on the bed, to a distributed canal system at the ice–bed interface. This switch is thought to occur when the effective pressure falls to a critical low, but non-zero, value. Once the switch has occurred, it is like opening a tap, and the flood proceeds. Elaboration of this intriguing model awaits further study.

2. Measured hydrographs tail off much more rapidly from peak discharges than predicted by the theory, indicating that the jökulhlaups are not terminated simply by the creep closure of cylindrical tunnels. Björnsson (1992) suggested that rapid flood termination may reflect the lowering of a flat floating ice lever back on to the bed. Fowler and Ng (1996) modelled flood hydrographs from Grimsvötn for the case of broad, shallow channels, and found a much better fit with the data than for a model based on assumed cylindrical channels.

3. Model simulations fail entirely to describe hydrographs of some jökulhlaups, such as the 1974 flood from the east Skafta cauldron beneath Vatnajökull. In some cases, the empirical data suggest that drainage is not by a discrete tunnel system, as assumed by the model, but in a combination of discrete and distributed drainage at the bed. Shoemaker (1992a) has argued that distributed drainage systems play an important role in some jökulhlaup events, although his ideas have been criticized by Walder (1994). It is clear, however, that many of the details of subglacial drainage are poorly known, and require more research.

Walder and Driedger (1995) have proposed a new jökulhlaup trigger mechanism based on observations of outburst floods from below South Tahoma Glacier, Mount Rainier, USA. On several occasions, destructive debris flows have emerged from the glacier, triggered by the release of meltwater from subglacial storage. The flows usually occur during periods of atypically hot or rainy weather in summer or early autumn, when large amounts of water reach the bed from the surface. Walder and Driedger argued that this sudden water input destabilizes a linked-cavity drainage system at the glacier bed, causing a switch to an efficient conduit system, which rapidly evacuates stored water. The occurrence of this mechanism at South Tahoma Glacier is thought to relate to the geometry of the bed, which may be particularly conducive to the formation of large subglacial cavities.

3.5.3 Jökulhlaup discharges

Because jökulhlaups pose a severe threat to humans and human structures, it is important to be able to estimate the likely magnitude of future floods. Several methods have been devised to predict *peak discharges*, which are the most erosive and destructive phases of floods. Clague and Mathews (1973) calculated a relationship between the volume of water released from ice-dammed lakes and peak flood discharges, based on a sample of 10 lakes. This relationship has the form

$$Q_{max} = 75(V_0/10^6)^{0.67} \qquad (3.9a)$$

where Q_{max} is the peak flood discharge ($m^3 s^{-1}$) and V_0 is the total volume (m^3) of water drained from the ice-dammed lake.

This formula was applied to the August 1979 discharge of glacier-dammed Flood Lake, British Columbia, by Clarke and Waldron (1984), yielding a theoretical peak discharge of $2150 \ m^3 s^{-1}$. This is considerably larger than the maximum discharge of $1200 \ m^3 s^{-1}$ measured at a gauging station 90 km downstream. The Clague–Mathews formula has been modified by several researchers. For example, Costa (1988) suggested the form

$$Q_{max} = 113(V_0/10^6)^{0.64} \qquad (3.9b)$$

and Desloges *et al.* (1989) proposed the form

$$Q_{max} = 179(V_0/10^6)^{0.64} \qquad (3.9c)$$

This method of discharge prediction is not based on any physical mechanism and, according to Clarke (1986), 'confounds understanding but seems to give reasonable results'.

Observations of several outburst floods in North America, Iceland and Scandinavia have shown that peak flood discharges are between two and six times higher than the mean discharge for the whole event (Fig. 3.27; Desloges *et al.*, 1989). Thus, if the volume of water released by a flood and the flood duration are known, the mean and peak discharges can be calculated. Clearly, this method cannot be used to determine the magnitude of future floods, since their duration is unknown. A more physically based way of calculating peak discharges is the *slope area method*, which is based on measurements of the dimensions and slope of channels during peak flood conditions, either from direct observations or geomorphological evidence. First, the peak velocity of the flood is calculated using the Gauckler–Manning formula (Williams, 1988):

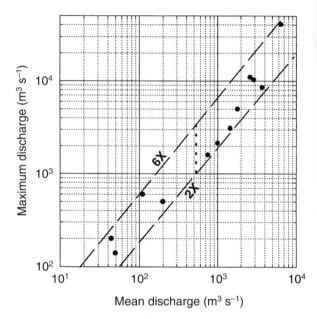

Fig. 3.27 Plot of mean discharge against maximum discharge for several jökulhlaups. The discriminating lines show the maximum discharge as two and six times the mean discharge. The vertical dotted line indicates the estimated mean discharge for the 1984 jökulhlaup at Ape Lake, British Columbia. (Redrawn from Desloges *et al.*, 1989)

$$v = r^{0.67}S^{0.50}/n \qquad (3.10)$$

where v is peak velocity, r is the hydraulic radius of the channel, S is bed slope for a 100 m channel reach, and n is a roughness coefficient known as Manning's n. The hydraulic radius is equal to the cross-sectional area of the channel (A) divided by the wetted perimeter, or the perimeter of the channel that is under water (p):

$$r = A/p \qquad (3.11)$$

Realistic values of Manning's n are difficult to obtain (Williams, 1988). For sediment-floored channels, bed roughness is mainly a function of bed material, particle size and bedform shape, and Manning's n can be estimated from the empirical relation

$$n = 0.038D^{0.167} \qquad (3.12)$$

where D is the average intermediate axis (in metres) of the 10 largest particles on the channel floor (Desloges *et al.*, 1989).

Once the flow velocity has been calculated, the discharge can be found from the continuity equation

$$Q_{max} = vA \qquad (3.13)$$

where Q_{max} is peak discharge, v is peak velocity and A is cross-sectional area of the channel.

Desloges *et al.* (1989) compared the results of the Clague and Mathews formula, the mean versus maximum discharge method and the slope area method for jökulhlaups from ice-dammed Ape Lake, British Columbia. All the methods gave broadly comparable results. The Clague–Matthews method gave a calculated peak discharge of 1680 ± 380 m^3 s^{-1}, the Q_{max} v. Q_{mean} method gave 1080–3240 m^3 s^{-1} and the slope area method gave 1534 and 1155 m^3 s^{-1} at distances of 1 and 12 km from the outlet respectively. Thus it seems that the Clague–Mathews method will yield reasonable predictions of the magnitude of future floods. The other methods cannot be used for prediction, because neither flood duration nor flood depth can be known in advance.

3.6 MECHANICAL EROSION BY MELTWATER

Glacial meltwater can be a very effective agent of erosion, in both subglacial and proglacial settings. The high sediment load carried by many glacial streams, in combination with rapid, turbulent flow, can achieve very high rates of erosion on both hard rock and sediment-floored channels. The processes of erosion can be subdivided into *abrasive wear*, *cavitation erosion*, *fluid stressing* and *particle entrainment from cohesionless beds*, which are discussed in turn below. Erosional landforms created by fluvial action are described in Section 9.3.4.

3.6.1 Abrasive wear (corrasion)

Rapidly flowing water freighted with suspended particles and/or bedload can cause considerable damage to bedrock and cohesive beds by *abrasive wear* or *corrasion*. During this process, flakes of the bed are gouged out by the impact of particles forced against the bed, resulting in pitting, striation and grooving of the bed. Drewry (1986) has presented a detailed, quantitative review of abrasive wear based on models developed by engineers to study the damage caused by flowing slurries in pipelines. The main factors identified by Drewry in causing abrasive wear are the angle at which particles hit the bed, flow velocity, particle concentration, particle and bed lithology, and particle size.

Angle of incidence
The rate of wear is strongly dependent on the angle at which particles hit the bed. For brittle materials, such as rock at low temperatures, the average wear rate per impact is greatest when the angle of incidence is 90°, and declines as particle trajectories make smaller angles with the bed (Tilly, 1969; Allen, 1982a; Drewry, 1986). This means that wear will be encouraged by turbulent flow and winding channels, both of which result in components of flow directed towards the bed. Wear is commonly concentrated at bends and hollows in channels, where flow is forced to change direction, setting up a positive feedback in which such features become more pronounced though time. This explains the focusing of abrasive wear responsible for deep, bowl-shaped potholes by subaerial and subglacial streams.

Flow velocity
Many experimental studies have shown that wear rate increases exponentially with flow velocity. This relationship probably reflects the greater number of impact events per unit time, and the greater amount of energy involved in each event.

Particle concentration
Wear rates increase with the concentration of particles suspended in the flow. At concentrations typical of glacial meltwater (<1 per cent by weight), the relationship is exponential, although at very high concentrations (>20 per cent) the erosion rate begins to fall off, probably owing to the suppression of turbulence. This indicates that flowing slurries are less effective agents of erosion than 'normal' sediment-laden meltwater.

Particle hardness and surface hardness
As for subglacial abrasion, erosion rates will increase with increasing hardness of suspended particles over channel wall hardness. The effects of this relationship can be readily observed in channels cut into rocks of variable hardness. Softer lithologies, such as shales, are invariably preferentially eroded in favour of more resistant materials such as massive sandstones and igneous intrusions.

Particle size
Average wear per impact event increases with the size of the impacting particles, presumably because, for constant impact velocity, larger particles carry a greater amount of energy. The precise relationship is unclear, however, and must be to some extent dependent on the behaviour of different particle sizes at any given velocity (i.e. whether the particles are suspended, saltate, roll or slide).

3.6.2 Cavitation erosion

In fluid flows, *cavitation* is the appearance of bubbles in the flow as a consequence of variations in the flow. This usage of the term 'cavitation' should not be confused with its use by glaciologists to describe the formation of cavities at the bed of a sliding glacier (Sections 3.2.4.1 and 4.5.3). Turbulent water flow over rough beds results in pressure fluctuations, and some regions of the flow experience higher or lower than average pressures. In low-pressure regions, pressure may fall low enough for air dissolved in the water to come out of solution to form bubbles (this process is analogous to the formation of bubbles in a bottle of beer when the top is removed and the vapour pressure in the beer falls to atmospheric). In streams, bubbles formed in low-pressure regions will be carried along by the flow until they reach a region of higher pressure, whereupon they may collapse. If this occurs near channel walls, high impact forces are produced by short-lived but intense pressure and thermal shock waves (Barnes, 1956; Drewry, 1986). Cavitation erosion is a well-known process in hydraulic engineering, and can cause severe damage to pipelines and installations in the form of pitting and etching.

A mathematical discussion of cavitation has been given by Drewry (1986). In general, erosion rates will increase with bed roughness and flow velocity, and decline with the hardness of the channel walls. Rates of erosion may accelerate through time because walls pitted by cavitation damage present greater resistance to flow, causing greater turbulence and pressure fluctuations, thus encouraging bubble formation and collapse. Erosion rates as high as 20 mm hr^{-1} have been reported for artificial structures (Barnes, 1956). In glacial environments, however, it is doubtful whether erosion by this mechanism ever occurs in the absence of corrasion, because glacial streams with velocities great enough to allow cavitation (several metres per second) invariably contain significant amounts of suspended sediment.

3.6.3 Fluid stressing

Fluid stressing refers to the erosion of rock channels or cohesive sediment-walled channels by forces imparted directly by flowing water (Allen, 1982a). Hydraulic stresses can prise off joint-bounded fragments of resistant rocks or cause plastic deformation and failure of muds. Mud beds can be moulded into transverse or streamlined ridges, from which shreds and larger masses of mud can be torn. Alternatively, flakes of mud can be lifted, then peeled off and rolled downcurrent like a carpet.

3.6.4 Particle entrainment from cohesionless beds

As water flows over cohesionless silt, sand or gravel particles with increasing velocity, a critical velocity is reached at which particles of a given size are set in motion. The velocity at which sediment transport begins is known as the *threshold of particle motion* (Allen, 1982a). In essence, entrainment occurs when the drag forces imposed by the flowing water exceed the resisting forces holding the particle in place

(Allen, 1982a, 1985). The resisting forces are proportional to the buoyant weight of the particle, and increase with particle mass, whereas the drag forces increase with particle surface area. This means that larger particles, which have greater mass relative to their surface area, are more difficult to set in motion than small ones. Consequently, small particles will be set in motion by gentle flows, whereas large boulders require exceptionally high velocities before they are entrained. Flume experiments have allowed the construction of diagrams plotting particle size against the threshold of particle motion (Fig. 3.28).

This simple picture is complicated by several factors, two of which are discussed here. First, if the flowing water is turbulent, unsteady flow conditions can act to lift particles from the bed temporarily, making them easier to entrain. Second, the shape and configuration of particles on the bed exerts an important control on the threshold of particle motion, through its effect on both the drag and resisting forces. For example, disc-shaped particles aligned with their *a–b* planes at right angles to the flow are subject to the maximum possible drag forces, and so are relatively easily entrained. The most stable bed configuration, which is most resistant to particle entrainment, is one in which particles are arranged like roof tiles, with their *a–b* planes dipping at a shallow angle upflow (Fig. 3.29). This bed configuration, known as *imbrication*, is commonly found in gravel-bed rivers. The

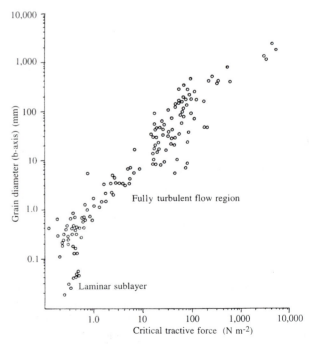

Fig. 3.28 Particle entrainment conditions: relationship between tractive force at the stream bed and particle size for incipient motion. (Adapted from Church and Gilbert, 1975)

Fig. 3.29 Imbricated gravels on a bar surface, Glen Etive, Scotland. Water flow is from left to right. (Photo: D.J.A. Evans)

process by which grains on the bed of a river assume stable orientations is known as *bed armouring*.

3.7 CHEMICAL EROSION AND DEPOSITION

Chemical processes are increasingly recognized as very important agents of erosion in glacial environments (Souchez and Lemmens, 1987; Sharp *et al.*, 1995b). Many mineral species can be dissolved by meltwater flowing over fresh bedrock surfaces or percolating through fine-grained sediments, and carried out of the catchment. Solutes can also be redeposited by meltwater within a glacial catchment, forming precipitate crusts on bedrock or sedimentary particles.

3.7.1 Dissolution and precipitation

Dissolution is the process by which ions (negatively or positively charged atoms or molecules) are released from a solid material and become bonded to water molecules to create *hydrated ions*. The dissolved ions can then be transported as part of the fluid, and may come out of solution and be deposited at a later stage. Positively charged ions (*cations*) form strong bonds with water molecules, creating stable molecules, whereas negatively charged ions (*anions*) form rather weaker bonds with water. Many common ions are soluble in water, including calcium (Ca^{2+}), sodium (Na^+), potassium (K^+), magnesium (Mg^{2+}) and, to a lesser extent, ions of iron (Fe), aluminium (Al) and silica (Si). These ions occur in many common rock-forming minerals such as calcite, feldspar and clay minerals.

When water comes into contact with a soluble material, dissolution proceeds until the concentration of hydrated ions in the solution reaches equilibrium with the prevailing temperature and pressure. The capacity of water to hold ions in solution rises with falling temperatures and increasing pressures. The maximum equilibrium concentration for each combination of pressure and temperature is known as the *saturation point*. Ions will be precipitated out of solution if the solute concentration rises above saturation point, which can happen if water is lost by evaporation or freezing, if temperature rises, pressure falls, or additional solute load is added.

The rate of solute loss from rock surfaces is also influenced by the action of other physical and chemical weathering processes, which can weaken the rock or transform minerals into more readily soluble form. Rock fracture by frost weathering, subglacial abrasion and crushing, and fluvial erosion creates fresh rock surfaces providing new sources of soluble ions. Conversely, selective solution of certain minerals can weaken the rock as a whole, facilitating future rock fracture and the exposure of fresh chemical reaction surfaces. Thus, physical and chemical weathering processes can reinforce one another, leading to higher overall rates of weathering than the sum of each acting in isolation. The composition and solubility of some minerals can be altered by reactions with water (*hydration* and *hydrolysis*) and oxygen (*oxidation* and *reduction*). For example, iron in certain silicate minerals can oxidize, weakening the mineral structure and increasing the likelihood of ions from the mineral passing into solution in surrounding water.

The solubility of minerals is also controlled by the presence of other substances in the water. For instance, the concentration of free hydrogen cations (H^+) is an important factor. Hydrogen ion concentration is measured on the pH (potential hydrogen) scale; distilled water has a pH of 7, acidic water (with a higher concentration of hydrogen ions) has lower pH values, and alkaline water (with a lower hydrogen concentration) has higher pH values. The relationship between solubility and pH for different minerals is not simple, however. Silica is slightly soluble at all commonly occurring pH values, whereas alumina (Al_2O_3) is readily soluble only in water below pH 4 and above pH 9.

The presence of dissolved carbon dioxide (CO_2) in meltwater exerts an important control on the dissolution of calcium carbonate ($CaCO_3$). Calcium carbonate, in the mineral form calcite, is the major constituent of limestone, marble and chalk, and occurs as a secondary mineral in some sandstones and shales. It is readily soluble in the presence of CO_2, yielding free calcium ions in the following reaction:

$$CaCO_3 + H_2O + CO_2 \rightarrow Ca + 2HCO_3$$

Glacial meltwaters may contain very high concentrations of dissolved CO_2 because of high turbulence (which aerates the water), low temperatures, high pressures, and the release of CO_2 from pressurized air bubbles trapped in ice. Consequently, dissolution of calcium carbonate (or carbonation) is a very important process in glaciated areas underlain by calcite-rich rocks.

The abundance of fine-grained sediment in many subglacial and proglacial environments plays a very important role in chemical weathering processes. Fine-grained sediments have a very large surface area, providing extensive reaction surfaces and sources of ions. Combined with an abundance of undersaturated meltwater, the presence of fresh, unweathered rock debris means that there is very high potential for dissolution in glacial environments. In a study of water chemistry at the Glacier Tsidjiore Nouve, Switzerland, Lemmens and Roger (1978) found that supraglacial meltwater which was initially low in solutes approached saturation in sodium and potassium after only 30 m of flow over proglacial sediments. Similarly, experimental work by Sharp *et al.* (1995a) showed that solute concentrations in deionized water rise rapidly on contact with fine rock debris, reaching high values within 2–3 hours (Fig. 3.30). Suspended sediment in turbulent meltwater streams can also be an important source of solutes, providing a large reaction surface area and allowing efficient weathering. In saturated waters, suspended particles can also act as surfaces on which precipitates can collect, thus allowing transport of solutes by meltwater over and above what can be carried in solution.

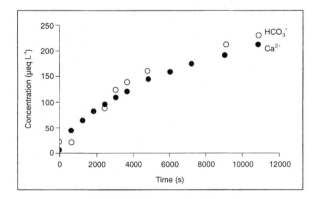

Fig. 3.30 Concentrations of Ca^{2+} and HCO_3^- ions as a function of time in a laboratory dissolution experiment. Fine rock particles were stirred continuously in deionized water in free contact with the atmosphere. (From Sharp *et al.*, 1995a. Reproduced by permission of the International Glaciological Society)

3.7.2 Dissolution and precipitation in subglacial environments

Dissolution and precipitation processes can be particularly important in subglacial environments, where water is in contact with large areas of the bed, there is an abundance of fresh rock surfaces, and large pressure fluctuations occur on short spatial and time scales. Hallet (1976a, 1979b) and Hallet *et al.* (1978) showed that, where wet-based ice is underlain by limestones, dissolution and precipitation of calcium carbonate are intimately associated with the regelation sliding process. Pressure melting of basal ice on the upstream (stoss) side of obstructions produces a film of meltwater which is commonly rich in dissolved CO_2 and low in solutes. As this water comes into contact with fresh rock surfaces and fine-grained debris, it rapidly becomes enriched in calcium ions owing to efficient carbonation. The water flows along the bed to the low-pressure downstream (lee) side of the obstruction, where freezing occurs (Fig. 3.31). Solutes are not readily incorporated in ice, so they accumulate in the remaining unfrozen water. If the concentration of solutes reaches saturation point, precipitation occurs, forming thin layers of calcium carbonate on rock surfaces. Such calcite deposits are commonly finely laminated, recording multiple refreezing events, and may exhibit a variety of textures and crystal forms depending on the chemical history of the water (Hallet, 1976a; Sharp *et al.*, 1990).

Similar processes of solution and precipitation operate for other ions in subglacial meltwaters. For example, on the forelands of many Norwegian glaciers, brown staining commonly occurs in lee-side depressions on gneiss bedrock bumps, and probably represents the precipitation of iron oxides derived from iron-rich minerals such as magnetite and pyrite (Fig. 3.32; Drewry, 1986). Silica and manganese precipitates have also been reported (Andersen and Sollid, 1971).

3.7.3 Chemically based mixing models

Chemically based mixing models have been widely used to derive inferences about glacial drainage systems from the solute content of glacial meltwaters (e.g. Collins, 1979a, b, c; Gurnell and Fenn, 1985; Tranter and Raiswell, 1991; Gurnell *et al.*, 1992). These models assume that water passing through subglacial and englacial components of a glacial drainage system acquire distinct differences in solute content: 'subglacial' water is expected to have a high solute content owing to prolonged contact with fresh bedrock and fine rock particles, whereas 'englacial' water is expected to be low in solutes owing to its origin as debris-poor supraglacial or englacial ice.

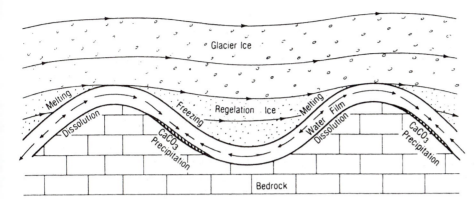

Fig. **3.31** Subglacial dissolution and precipitation of calcium carbonate ($CaCO_3$) in association with glacier sliding over a wavy bed. Dissolution occurs where ice melts in the high-pressure upstream sides of bumps, and precipitation occurs when water is refrozen at the low-pressure downstream sides. (From Hanshaw and Hallet, 1978. Reproduced by permission of *Science*)

Fig. **3.32** Dark iron oxide precipitate deposits in former subglacial cavities, ice-scoured gneiss, Jotunheimen, Norway. Former ice flow from left to right. (Photo: D.I. Benn)

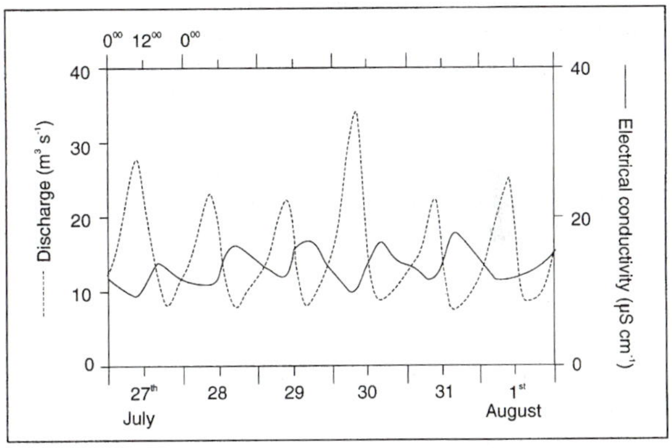

Fig. **3.33** Temporal variations in electrical conductivity (a measure of solute concentration) and discharge for the proglacial stream of Gornergletscher, Switzerland. (Redrawn from Collins, 1979b)

Meltwaters emerging at the glacier margin are assumed to be mixtures of 'subglacial' and 'englacial' water, with a solute content intermediate between the two components. If the solute content of the two meltwater components is known, variations in the solute content in the marginal stream can be used to determine their relative contribution to the mixture from the relation

$$Q_t C_t = Q_{sub} C_{sub} + Q_{en} C_{en} \qquad (3.14)$$

where Q_t, Q_{sub} and Q_{en} are the discharge of the total flow and subglacial and englacial flow components, and C_t, C_{sub} and C_{en} are the solute concentrations of the total, subglacial and englacial flow components (Collins, 1978, 1979a).

Measured solute contents of emerging meltwaters do show systematic variations with discharge, indicating that flow component mixing does occur. Figure 3.33 shows variations in electrical conductivity (an index of solute content) and discharge for the meltwater stream emerging from Gornergletscher, Switzerland, for a six-day period in 1975. Solute content is high at times of low discharge (base flow), when the meltwater is thought to con-

sist mainly of 'subglacial' water, which is rich in solutes. Conversely, solute concentrations are low during peak flow during the afternoons, when solute-poor surface meltwater floods into the system, diluting the base flow component. Similar patterns occur on annual cycles. Solute concentrations tend to be high in winter when there is little dilution of 'subglacial' water by supraglacial meltwater, total discharges are low and residence times are long. Sharp *et al.* (1995b) reported that solute concentrations in the stream draining the Haut Glacier d'Arolla, Switzerland, are three to seven times higher in winter than in summer. However, the total stream discharge is very small in winter, so that despite high concentrations, the total solute flux in winter is low. Metcalf (1986) estimated that for Gornergletscher, less than 10 per cent of the annual total solute load leaves the glacier between September and May.

Although chemically based mixing models yield useful insights into englacial and subglacial drainage dynamics, their usefulness as precise quantitative tools is limited. In particular, Sharp *et al.* (1995a) have argued that the following assumptions of the model are questionable:

1. *Two drainage components.* Subdivision of the internal plumbing of a glacier into only two ('subglacial' and 'englacial') components is simplistic. As we have seen (Sections 3.4.2–3.4.6), there are several possible types of subglacial drainage which are characterized by different rates of throughflow and intermediate storage. Furthermore, the type of drainage may change at different times of the year (Section 3.4.4).
2. *Unique component chemistries.* Most applications of chemical mixing models assume that the 'subglacial' and 'englacial' drainage components have constant solute concentrations, and that all variations in streams at the snout reflect mixing variations. However, solute concentrations of meltwater within glaciers are known to vary in space and time, depending on the storage history, flow velocity and initial concentration (Johannessen *et al.*, 1977; Brown *et al.*, 1994).
3. *Conservative mixing.* Equation (3.14) assumes that the total mass of solutes remains unchanged after mixing of the two drainage components. This is unlikely to be true, however, particularly since mixing may occur some distance from the snout, and bulk meltwaters are commonly undersaturated with respect to abundant reactive minerals.

Sharp *et al.* (1995a) suggest that the problems associated with the standard mixing model could be overcome by conducting detailed analyses of the different solute species present in meltwater rather than using electrical conductivity measurements, which provide only a bulk measure of solute content. Distinctive species with known sources and behaviour (such as SO_4^{2-}, which is mainly derived from the atmosphere) can then be used in model calculations (Tranter and Raiswell, 1991).

3.7.4 Rates of chemical erosion

The abundance of undersaturated meltwater, extensive fresh, reactive rock surfaces and large areas of water–rock contact mean that rates of chemical erosion tend to be high in glacierized catchments. Solute concentrations in the proglacial meltwater stream of the Haut Glacier d'Arolla, Switzerland, indicate that surface lowering rates by chemical weathering are about 14–16 mm kyr^{-1} (Sharp *et al.*, 1995b). These rates are about 25 per cent higher than estimates for the adjacent Glacier de Tsidjiore Nouve and comparable to those for the Gornergletscher catchment (Metcalf, 1986; Souchez and Lemmens, 1987). Even higher rates have been reported from the South Cascade and Berendon Glacier catchments in North America, where runoff is higher than in Switzerland (Reynolds and Johnson, 1972; N. Eyles *et al.*, 1982a). Although surface lowering rates arising from

chemical weathering are much lower than estimated subglacial mechanical erosion rates, they are still very high compared with non-glacierized catchments. Chemical weathering rates in the North American glacier-covered catchments are about three times higher than estimates of the global average (Summerfield, 1991; Sharp *et al.*, 1995b). Thus, glacierization of a catchment can be expected to increase the chemical weathering rate by increasing the throughput of undersaturated, turbulent water and ensuring that meltwaters do not stay in chemical equilibrium with available reactive sediment.

High glacial rates of chemical weathering have important implications for climate change. Sharp *et al.* (1995b) have shown that chemical weathering processes act as a significant sink for atmospheric carbon dioxide, particularly during peaks of meltwater runoff. Thus, peaks in runoff during major deglacial events could remove considerable amounts of CO_2 from the atmosphere, which would tend to counteract any trend in global climatic warming at that time. Drawdown of CO_2 by chemical weathering of the rapidly uplifting Himalayan chain and Tibetan Plateau has been proposed as a major mechanism of global climatic cooling during the late Tertiary (Raymo and Ruddiman, 1992).

3.8 SUSPENDED SEDIMENT AND BEDLOAD

In addition to their load of dissolved ions, glacial rivers also transport sediment as *suspended load* held within the flow, and *bedload* carried along the channel floor. Fast-flowing glacial rivers are capable of transporting huge amounts of suspended load and bedload, contributing to high erosion and deposition rates and profoundly altering the landscape.

3.8.1 Suspended sediment

Suspended sediment refers to grains maintained in transport above the bed. In stagnant water, grains will settle under gravity at a rate proportional to their diameter, so for grains to remain in suspension in flowing water, the settling velocity must be balanced by a vertical component of water flow lifting the grains away from the bed. Suspended sediment is therefore a characteristic of turbulent flows, in which some of the flow is directed upward. The maximum size of particles that can remain in suspension will increase with the turbulence and velocity of the flow, but will generally be finer than fine sand or coarse silt.

Suspended sediment concentrations in glacial streams are not simply a function of flow conditions, however, but also reflect the availability of sediment. For individual flood events, peaks in suspended sediment concentrations often occur before peak discharges, and concentrations are often lower during the falling stage than for similar discharges during the rising stage (Fig. 3.34; Gurnell, 1982). This phenomenon is thought to reflect the rapid flushing out of available fine-grained sediment from the glacier bed and channel floor, and the subsequent depletion of the sediment supply (Østrem, 1975b; Drewry, 1986). Alternatively, meltstreams may display brief pulses or 'slugs' of suspended sediment that are independent of discharge fluctuations, but which may reflect sudden release of sediment from storage due to the collapse of a channel wall or changes in channel patterns. The effects of flushing can also be recognized in seasonal variations in suspended sediment concentration (Vivian and Zumstein, 1973; Drewry, 1986). In winter, when meltwater discharges are low, suspended sediment concentrations are also low, typically only a few milligrams per litre. Sediment concentrations rise rapidly with increasing discharges in the spring, and may reach a few tens of grams per litre. Concentrations tend to fall off later in the ablation season, even though meltwater discharges can remain high, owing to the reduction in available fine-grained sediments.

Where meltwater discharges are high, and loose sediment is readily available, suspended sediment concentrations can become very high, forming *hyperconcentrated flows* (Saunderson, 1977; Lord and Kehew, 1987; Todd, 1989). The suspended sediment is usually of mixed grain size, up to and including gravel, and water content is about 20–60 per cent by weight (Beverage and Culbertson, 1964; Pierson and Costa, 1987). Hyperconcentrated flows are therefore transitional between debris flows and normal stream flows (Section 10.4.9). Particles are maintained in suspension by various mechanisms, including fluid turbulence, but where clay is present,

turbulence is suppressed and cohesive strength may play an important role. Flows may be homogeneous or consist of two layers, with an upper, low-concentration component and a lower, coarse-grained, high-concentration 'carpet' (Todd, 1989). Deposition results from falling shear stresses and flow velocities associated with a reduction in stream gradient and/or flow depth, and occurs by partial or wholesale 'freezing' of the suspension.

3.8.2 Bedload

Fluvial bedload refers to grains swept along close to the bed of a stream, in continuous or intermittent contact with the bed. Three basic modes of bedload transport are possible: sliding, rolling and saltation (Allen, 1982a, 1985). *Sliding* particles retain continuous contact with the bed, but undergo negligible net rotation. Sliding is most important for disc-shaped gravel particles. *Rolling* particles are also in continuous contact with the bed, but move by rotating like a ball or wheel. *Saltation* (from the Latin *saltare*: to leap) involves particles taking relatively long jumps along the bed, and touching the bed only at the start and finish of each jump. The movement of grains in streams may involve all three of these processes occurring together, but in general a sequence of dominant transport modes from sliding to rolling to saltation occurs with increasing driving force. When forces tending to lift the particle exceed the gravitational forces causing settling, contact with the bed is lost and particles enter suspension. The forces imposed on particles by a flowing fluid are primarily a function of fluid velocity, although fluid viscosity and the presence or absence of turbulent flow are also important factors. A mathematical treatment of bedload transport is presented by Allen (1982a).

The grain size and quantity of sediment that can be maintained in motion as bedload increases with dri-

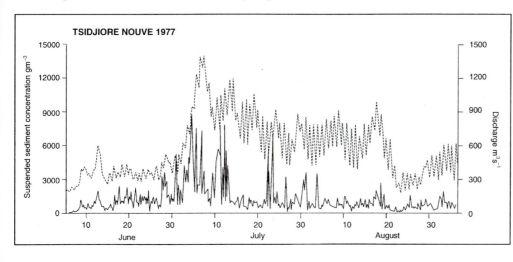

Fig. 3.34 Suspended sediment concentrations (solid line) and discharge (dashed line) for the proglacial stream of glacier Tsidjiore Nouve, Switzerland. (Modified from Gurnell, 1982)

ving stress, and varies between small quantities of silt for low flows through to heavy loads of all grain sizes up to large boulders for extreme flood events (Fig. 3.35). A study by Ashworth and Ferguson (1986) at Lyngsdalselva, Norway, showed that bedload transport was an order of magnitude higher during flows of $10\,\mathrm{m^3\,s^{-1}}$ than for flows of $5-8\,\mathrm{m^3\,s^{-1}}$. The high discharges often associated with glacial melt streams therefore lead to a huge capacity to transport sediment. For example, Østrem (1975b) found that the meltstream of Nigardsbreen, an outlet of the ice cap Jostedalsbre, Norway, transported *c.* 400 tonnes of bedload during a 27-day period in the summer of 1969. A more subjective impression of this capacity can be gained by standing beside a glacial stream in full flood and listening to the deep rumbling of boulders as they crash together and sweep past.

3.8.3 Bedforms

A characteristic feature of stream bedload is that it tends to be organized into distinct patterns or *bedforms*, which represent a dynamic equilibrium response of the bed to prevailing flow conditions. Four basic bedform types can be recognized (Allen, 1983, 1985; Ashley, 1990): plane beds, ripples, dunes and antidunes.

Plane beds are subhorizontal surfaces that develop in sand under two contrasting flow conditions: in slowly moving shallow water (lower flow regime) and more rapidly moving deeper water (upper flow regime). For the lower flow regime, plane beds develop in coarse to very coarse sand (0.8–2.0 mm), and are associated with relatively low rates of sediment transport. In contrast, under upper flow regime

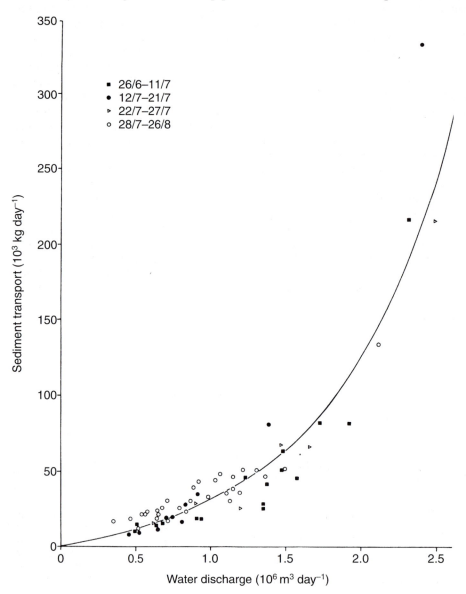

Fig. 3.35 Sediment transport 'rating curve' obtained for the meltstream of a small glacier (Trollbergdalsbreen) in northern Norway, based upon measured daily totals during the summer of 1970. (From Østrem, 1975b. Reproduced by permission of SEPM)

conditions, plane beds form in fine to medium-grained sands (0.0625–1.0 mm) which are mobilized as a *traction carpet* with continuous particle movement over the whole bed. Under these conditions, fluid motion close to the bed is characterized by elongate spiral eddies aligned parallel to flow, called boundary-layer streaks (Kline *et al.*, 1967; Grass, 1971; Allen, 1982a, 1985). These streaks deform the bed into thin, linear grooves and ridges aligned parallel to the former flow direction, called *parting lineation*.

Ripples are distinctive small-scale bedforms that form in sand finer than about 0.6 mm. The ripple crests are oriented at right angles to the flow, and may be long and relatively straight or have short, curving crests, depending on flow conditions. Ripple wavelength varies between 0.1 and 0.6 m, and amplitude (height) up to about 0.04 m (Allen, 1982a, 1985). Ripples are generally strongly asymmetrical, with long, convex-up upstream (stoss) faces and short, steeper lee sides. Ripples migrate downcurrent as the current erodes material from the stoss sides and redeposits it on the lee sides, and undergo vertical accretion if there is an additional input of sediment from suspension (Jopling and Walker, 1968; Allen, 1970).

In popular usage, the word 'dune' refers to large sand ridges formed by wind action, but to sedimentologists it also applies to large transverse sand or gravel bedforms developed below water currents. Certain types of these large bedforms have been variously classified as *dunes*, *bars*, *megaripples* and *sand waves*, but it is now accepted that the general term *dune* should be applied to all large transverse bedforms, in recognition of their genetic affinities (Ashley, 1990; Miall, 1992). Dune wavelength ranges from 1 m to over 1 km, while amplitude

ranges from a few centimetres to several metres. In common with ripples, dunes are repetitive structures and can have long, straight or slightly sinuous crests (*two-dimensional* or *2–D dunes*) or short, curving crests (*three-dimensional* or *3–D dunes*), depending on flow conditions and sediment size (Fig. 3.36b; Allen, 1985; Ashley, 1990). Dunes also migrate downcurrent in response to sediment transport over their stoss surfaces and deposition on their lee faces, and can undergo vertical accretion if sediment supply is sufficient to cause an excess of deposition over erosion. Dunes can have ripples or plane beds superimposed on their stoss surfaces.

Antidunes are bedforms that develop in phase with surface waves in rivers. The surface wave presents its steepest face upstream, forming a phenomenon known by white-water canoeists as a 'stopper' because of the effect it has on downstream progress. Antidunes are oriented transverse to flow, and may consist of long-crested 2–D forms or short-crested 3–D forms. A distinctive type of 3–D antidune was described by Shaw and Kellerhals (1977) and consists of paired mounds joined by a saddle and oriented transverse to flow. These antidunes are arranged in chains parallel to flow, and form under trains of peaked-up, conical surface waves called 'rooster tails'. Antidunes can migrate slowly upstream or downstream.

Experimental studies have demonstrated that each bedform type is associated with particular combinations of particle size and flow conditions, expressed in terms either of shear stress or of flow velocity (Fig. 3.37). The boundaries of such *existence fields* can be sharp or gradational. Sharp boundaries exist at the thresholds separating ripples from upper-stage plane beds, dunes and lower-stage plane beds, meaning that as flow conditions

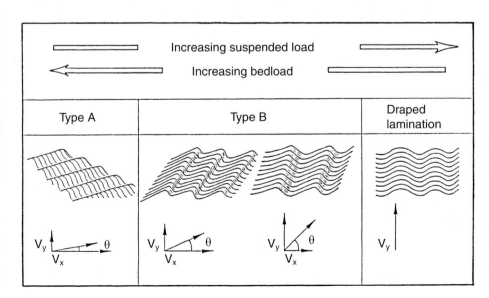

Fig. 3.36a The formation of various types of climbing ripple drift according to the migration rate and the aggradation rate. Ripples climb at the angle θ whose tangent is the mean aggradation rate V_y divided by the downstream migration rate V_x. (From Ashley *et al.*, 1982. Reproduced by permission of Blackwell Scientific)

change there is a sudden jump between the types of bedforms that are stable for any given particle size. In contrast, gradational boundaries such as that between upper-stage plane beds and dunes indicate a gradual transition beween bedforms as flow conditions change.

3.9 PROGLACIAL FLUVIAL SYSTEMS

3.9.1 Discharges and hydrographs

Large variations in meltwater supply combined with fluctuations in rainfall and the release of stored water mean that rivers draining glaciated catchments show widely varying discharge on several timescales. Discharge variations occur on daily and annual cycles, on an irregular basis because of the passage of weather systems, and over long time periods, reflecting changes in glacier mass balance.

The basic daily pattern of water discharge from glaciers into proglacial rivers is driven by the diurnal temperature cycle. Discharge variation consists of a cycle of rising and falling flow superimposed on *base flow*, or minimum daily discharge (Fig. 3.38). Base flow comes from various sources, including sub-

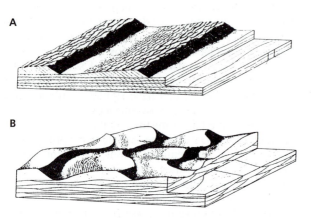

Fig. 3.36b Dune morphology. (A) Two-dimensional dunes; (B) three-dimensional dunes. (From Ashley, 1990. Reproduced by permission of the *Journal of Sedimentary Research*)

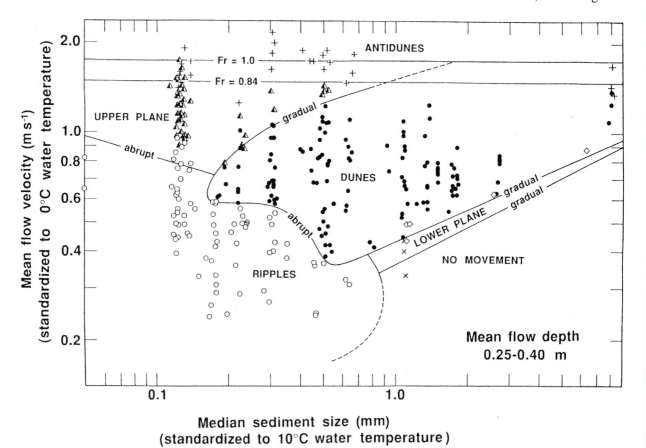

Fig. 3.37 Bedform existence fields, shown as a function of median sediment size and mean flow velocity. (From Ashley, 1990. Reproduced by permission of the *Journal of Sedimentary Petrology*)

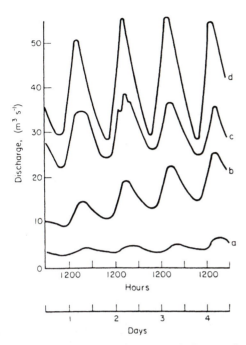

Fig. 3.38 Diurnal variations in discharge from the proglacial stream of Gornergletscher, Switzerland, showing progressive increase in base flow and diurnal cycle amplitude during the ablation season of 1959: (a) 17–20 May, (b) 14–17 June, (c) 23–26 June, (d) 19–22 July. (From Paterson, 1994, after Elliston, 1973. Reproduced by permission of Pergamon)

glacial meltwater, water stored in cavities, meltwater percolating through snow (including firn in the glacier accumulation area), and groundwater (Röthlisberger and Lang, 1987). These components vary little on a day-to-day basis. The superimposed daily cycle consists of those components of the day's meltwater that drain rapidly through the glacier system, including meltwater draining from the glacier ablation area through short routeways into main subglacial conduits, and meltwater that drains entirely supraglacially. Daily discharge peaks lag behind the time of maximum melting on the glacier by a few hours, the length of delay depending on the distance the water has to travel through and below the glacier, and the configuration of the internal drainage network. Lag times are long for distributed drainage systems such as linked-cavity networks, and short for efficient, well-developed channel systems. As the ablation season progresses, the development of subglacial drainage networks from distributed systems to channelized systems can be reflected in a shift towards shorter lag times (Röthlisberger and Lang, 1987; Willis *et al.*, 1990; Hock and Hooke, 1993).

Most glaciers exhibit strong seasonal variations in water runoff, following annual fluctuations in glacier surface ablation due to changes in incoming solar radiation and air temperature. Exceptions are pro-

vided by some high-altitude equatorial glaciers, where significant melting occurs throughout the year. For mid- and high-latitude glaciers, discharges are very low during winter when surface melting is unimportant, and rise rapidly with the onset of snow and ice ablation in spring. The first flood of the melt season has been termed the *spring event* by Röthlisberger and Lang (1987), and is associated with the re-establishment of a connected drainage network through and below the glacier, which allows the efficient discharge of meltwater, rainfall and stored water. High discharges can be expected throughout the summer, particularly during warm, wet weather, when snow melt is at a maximum (Fig. 3.39). Discharges decline in the late summer, when supplies of stored englacial and subglacial water and low-altitude snow beds have dwindled. Runoff falls rapidly to low winter levels with the onset of cold weather in the autumn.

Irregular flood events are associated with (a) individual weather systems, which determine intensity of ablation and rainfall, and (b) changes in the glacial drainage system (Röthlisberger and Lang, 1987). Periods of rapid ablation are usually associated with clear skies in summer, which bring maximum insolation and high air temperatures. During sustained periods of hot weather, proglacial discharge hydrographs exhibit a steady rise in base flow, reflecting the gradual release of water from the glacier accumulation area. Superimposed on this trend are larger than usual daily variations, reflecting daily melting on the lower part of the glacier (Fig. 3.39). The highest weather-related discharge events, however, tend to be associated with high rainfall during summer and autumn storms. Rainfall adds directly to runoff in the basin, and also contributes to ice and snow ablation by efficiently transferring heat from the atmosphere. Such flood events can have considerable geomorphological impacts and cause large amounts of damage to property (Maizels, 1995).

Floods due to changes in the glacial drainage system include those due to the collapse of englacial or subglacial conduits, proglacial ice jams and sediment jams, and drainage of large subglacial and proglacial lakes. The collapse of conduits may result from progressive melting of the walls, glacier flow, or sudden influxes of water from other sources (Warburton and Fenn, 1994), and can initiate the temporary storage and/or release of stored water. Proglacial ice jams are particularly common in alpine areas during spring ice breakup along river courses. Hewitt (1982) has suggested that especially large and destructive floods resulting from ice dams tend to occur in the southern Alaska and Yukon Ranges in North America and the Karakoram Mountains. The largest of all glacier floods are those associated with catastrophic outbursts from subglacial and ice-dammed lakes. The

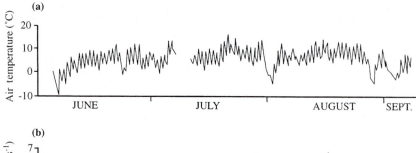

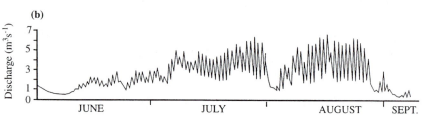

Fig. 3.39 Variations in (a) air temperature and (b) proglacial stream discharge for Haut Glacier d'Arolla, Switzerland, for the ablation season of 1989. Note the low discharges during a cold weather spell in early August. (From Gurnell et al., 1992)

causes and consequences of outburst floods are described in Section 3.5.

Long-term variations can often be related to climatic change and associated changes in water stored as snow and ice. Meltwater discharges are relatively low during cold periods, when storage increases and glaciers advance, and relatively high during warm periods when storage decreases and glaciers retreat (Röthlisberger and Lang, 1987). Discharges will be highest when deglaciation is rapid, and decline as glaciers dwindle and the supply of stored water is exhausted. The reduction in size of glaciers reduces the potential for meltwater yield, as shown by Kasser (1973), who found that a 19 per cent decrease in the area of French glaciers between 1916 and 1968 led to a 16 per cent decrease in mean summer runoff. If glaciers continue to retreat as the result of greenhouse gas-induced global warming, the corresponding reduction in meltwater potential may have far-reaching consequences for communities dependent on this source of water in the summer months. Figure 3.40 shows field systems in the Karakoram Mountains, Pakistan, once irrigated by glacier meltwater but since abandoned following the disappearance of glaciers from the catchment.

3.9.2 Proglacial channel networks

Proglacial channel networks have very distinctive morphologies, because of their highly variable discharges and generally high throughputs of sediment. Where little sediment is available, proglacial rivers may occupy rock-floored channels, but more usually they form networks of shifting sediment-floored channels developed on outwash plains. Outwash plains are often referred to by the Icelandic term *sandar* (singular: *sandur),* and may be hundreds of metres to several kilometres wide. Sandar tend to exhibit systematic downstream changes in morphol-

Fig. 3.40 Disused field systems in the Hunza valley, Pakistan. The presence of glaciers is crucial to human activity in many catchments by providing water for irrigation during the summer months. Agriculture was abandoned in this area when a small glacier in the catchment disappeared. (Photo: L.A. Owen)

ogy, and have been subdivided by Krigstrom (1962) into three zones (Fig. 3.41).

1. *Proximal zone*: here meltwater is confined to a few deep and narrow major sediment-floored channels. These channels may be the extensions of Nye channels developed beneath the margins of the glacier.
2. *Intermediate zone*: flow is in a complex network of wide and shallow braided channels, which shift position frequently and many contain meltwater only during periods of high discharge. Some parts of

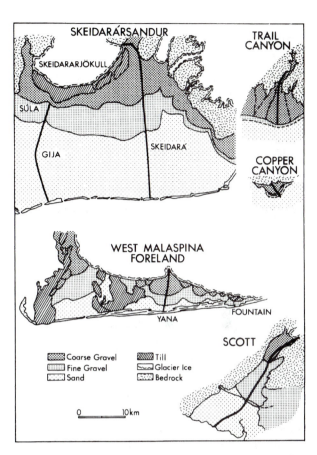

Fig. 3.41 Sandar and proglacial channels. (a) Proximal outwash surface, Stordalen, Norway – Stegholtbreen in background; (b) intermediate channels on incised sandur, Phillips Inlet, Ellesmere Island. (Photos: D.I. Benn and D.J.A. Evans)

Fig. 3.42 Some sandar, with their typical grain sizes. Skeidararsandur is in southern Iceland, the others are in Alaska. (From Boothroyd and Nummendal, 1978. Reproduced by permission of the Canadian Society of Petroleum Geologists)

the sandur surface may be abandoned and inactive, because incision or channel switching has diverted water away. Relict, dry channels are referred to as *palaeochannels*.

3. *Distal zone*: channels are very shallow and ill-defined, and often merge to produce sheet flow during periods of high discharge.

These zones are based upon Icelandic sandar, but can also be recognized elsewhere, such as North America (e.g. Fahnestock, 1969; Church, 1972; Fahnestock and Bradley, 1973; Gustavson, 1974).

Generally, the long profiles of sandar are concave, exhibiting decreasing slope angles in a downstream direction. However, complexities may be introduced where tributaries join the main reach, or valley constrictions or moraines interrupt sandur development (e.g. Church, 1972). The downflow decrease in gradient is associated with a reduction in sediment grain size (Fig. 3.42). Gravels predominate in the proximal zone, giving way to sand or silt in the distal zone. In detail, however, there tend to be only weak correlations between channel slope and clast size (e.g. Fahnestock, 1963; Boothroyd and Ashley, 1975;

Boothroyd and Nummedal, 1978). Possible reasons for this are the role of jökulhlaups and other floods in transporting larger grain sizes over longer distances than during periods of normal meltwater discharge, and the reintroduction of coarse material to channels by bank collapse.

The channel patterns of sandar comprise complex networks of shifting channels and bars arranged in a braided pattern, although meandering and anastomosing patterns have been reported for distal reaches where banks are more stable (Boothroyd and Ashley, 1975; D.G. Smith, 1976, 1983; D.G. Smith and N.D. Smith, 1980). In braided reaches, channels are typically highly unstable and evolve rapidly in response to changes in stream discharge and sediment availability (Fig. 3.43). The presence of non-cohesive, easily erodible bank material, a lack of stabilizing vegetation, and high sandur gradients ($2–50$ m km^{-1}) tend to produce channels with high width/depth ratios and low sinuosity. Channels are separated by *bars*, which become active and subject to erosion and deposition only at peak flows. At other times they

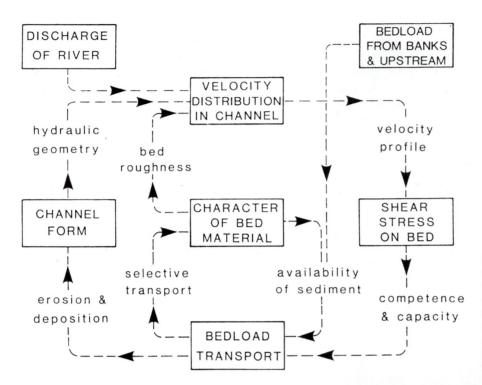

Fig. 3.43 System diagram showing relationships and feedbacks between channel configuration, flow and sediment availability in gravel-bed rivers. (From Ferguson and Ashworth, 1992. Reproduced by permission of John Wiley & Sons)

stand above the active channels and act as stores for large quantities of gravel and sand. In streams where sediment loads are highly variable, the channel and bar forms may change positions over the course of 24 hours, and rapid sediment throughput ensures that steady state is never reached. Only large bar forms produced by the highest-flow events (perhaps jökulhlaups) may become stable over periods of years. As the water stage lowers after bar deposition, the surfaces and margins of bar forms may be dissected by stream erosion and thus may be complex forms produced during a number of cycles of high- and low-water stages.

Successive stages of channel abandonment, due to lateral migration or vertical downcutting, are responsible for producing a series of topographic levels on sandur surfaces. These were first documented on the Donjek River by Williams and Rust (1969), who subdivided the sandur surface into four levels (Fig. 3.44):

- level 1 – main channel level and principal sediment transport route, with little or no vegetation cover and bars exposed only at low flows;
- level 2 – active only during flood stages, with few active channels at other times and sparse vegetation cover;
- level 3 – only low-energy flow during flood stages with moderate vegetation cover in humid areas;
- level 4 – dry islands and interfluves with either a dense vegetation cover in humid areas or aeolian deflation and dune migration.

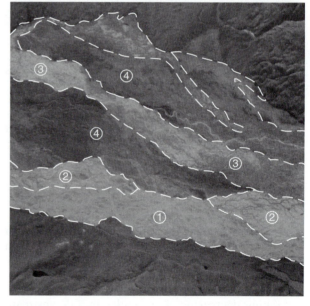

Fig. 3.44 Aerial photograph of part of the Donjek River, Yukon Territory, Canada, showing the four topographic levels of the sandur surface identified by Williams and Rust (1969). (Aerial photograph reproduced by permission of Energy, Mines and Resources, Canada)

Other sandur surfaces, of course, may exhibit more or fewer topographic levels, depending on the history of aggradation and incision at the site.

3.9.3 Bar formation and migration

Bars are dynamic stores of sediment interact with the water flow through the system. Bar growth occurs when initially small irregularities in a channel are amplified by the deposition of sand or gravel, creating conditions favourable to continued accumulation of sediment. Conversely, bar erosion will occur when the bar form is no longer in equilibrium with flow conditions and sediment is removed from the system. Erosion and deposition can occur simultaneously, with sediment being lost from some parts of the bar and gained in others. Erosion and deposition can result in *downstream bar migration*, where the bar travels in a downstream direction by erosion of the upstream face and deposition on an advancing avalanche face on the downstream side; or *lateral accretion*, where sediment deposition occurs on one side of the bar.

Bar formation can occur by deposition of material in midstream, or by erosional dissection of pre-existing topographic highs (Ashmore, 1991; Ferguson, 1993; Maizels, 1995). *Bar initiation* by deposition appears to be a natural consequence of flow within wide, shallow channels, where the shear stress and transport capacity of the flow are low. The stream will deposit sediment preferentially in areas where the shear stress is lowest, forming an obstruction in mid-channel or against one bank. The presence of the obstruction will divert and concentrate the flow around the incipient bar, thus enlarging the channels to either side and isolating and encouraging further growth of the bar. Scouring of the floors and margins of channels remobilizes sediment, providing more material for bar growth in regions of low shear. Two main mechanisms of bar initiation have been recog-

nized (Fig. 3.45; Ashmore, 1991; Ferguson, 1993). The first is *central bar deposition*. This is the classic model of bar formation, in which elongated, migrating sheets of bedload stall in the centre of wide, shallow channels (Leopold and Wolman, 1957). Although central bar deposition is often regarded as the main process of bar initiation, Ashmore (1991) found that it was uncommon in flume experiments, and occurred only when shear stresses were close to the threshold for bedload transport. The second mechanism is *transverse bar conversion*. In this case, bar initiation occurs in areas of flow divergence, where bedload eroded from an upstream scour pool forms a large *lobe* with prograding downstream avalanche faces. Thin bedload sheets stall on top of the lobe, which eventually emerges. Flow is then deflected laterally off the edges of the lobe, forming a symmetrical central bar. This process was originally recognized for small-scale, ephemeral pulses of sediment within channels (Southard *et al.*, 1984), and is now thought by many researchers (e.g. Ashmore, 1991; Bridge, 1993; Ferguson, 1993) to be the main process of bar accretion.

Erosional incision modifies existing bar forms, and is an important mechanism of bar evolution (Schumm and Khan, 1972; Rundle, 1985; Ashmore, 1991; Ferguson, 1993; Maizels, 1995). At high stages, part of the flow may take a short-cut across the bar surface, forming minor channels or *splays*. These splays have a steeper gradient than the main channels to either side, and tend to be preferentially enlarged to form *chutes* and can eventually form new major channels between bar remnants. Sediment eroded by splays and chutes can be redeposited in lobate forms up to several metres wide (known as lobes), which form the nucleus of new bars (Southard

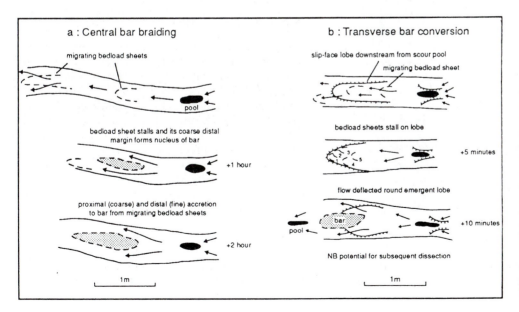

Fig. 3.45 Bar initiation by midchannel deposition. (a) Central bar deposition, in which bedload stalls in areas where shear stress is low. Black patches are pools and arrows show flow direction. (b) Transverse bar conversion. (From Ferguson, 1993. Reproduced by permission of the Geological Society of London)

et al., 1984). Patterns of incision associated with the development of splays and chutes on the surface of mid-channel, lateral and diagonal bars are shown in Fig. 3.46.

As river discharges fall from high to low flows, bars are first eroded and then draped by finer-grained sediments. Their largely sand-covered surfaces may then protrude from the stream flow. During the lower-stage flows, sand accumulation occurs on the downcurrent margins of gravel bars in the form of wedge-shaped cross-stratified deposits. Abandoned channels formed during late-stage incision of the bar may also be filled with sand and/or silt, and small-scale sand bedforms often occur on the bar surface

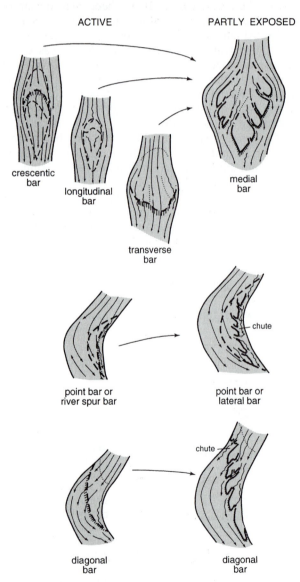

Fig. 3.46 Bar evolution by the erosional dissection of mid-channel, lateral and diagonal bars. (Redrawn from N.D. Smith, 1985)

(Williams and Rust, 1969; McDonald and Banerjee, 1971; Gustavson, 1974; Bluck, 1974).

The processes of sediment deposition, dissection and migration in dynamic channel systems give rise to a wide variety of bar forms, which is reflected in a wide variety of alternative classification schemes (e.g. Collinson, 1986; Ashley, 1990). The most fundamental distinction is between *mid-channel bars*, which form within channels and cause bifurcation of flow, and *bank-attached bars*, which are deposited at channel margins and increase channel curvature. Such bars can be simple forms known as *unit bars*, with simple depositional histories controlled by local flow conditions, or larger, complex forms known as *compound bars* or *braid–bar complexes*, which have multi-stage histories of deposition and erosion (N.D. Smith, 1974, 1985; Ashley, 1990).

Mid-channel unit bars can be subdivided into *crescentic bars*, *longitudinal bars*, and *transverse bars* (Fig. 3.46; N.D. Smith, 1985). Crescentic bars are probably embryonic stages of longitudinal bars. These three bar types develop during high flows when the whole channel is submerged, and are dissected during lower flows to form compound *medial bars*. Bank-attached bars can take the form of *point bars,* which are produced on the insides of channel bends and are separated from the channel bank by a chute, and *diagonal bars* oriented oblique to river flow. During low flows both types become dissected by chutes. Because proglacial outwash environments are very dynamic and individual stream channels are subject to constant changes in water discharges, most bars in sandar and valley trains are the products of multiple depositional and erosional events, and are therefore of complex type (e.g. Church, 1972; Bluck, 1974, 1979).

3.9.4 Polar sandar

Only a small amount of research has been undertaken on the polar desert sandar of Antarctica, even though they have been evolving as proglacial systems for at least 500,000 years (Speden, 1960; Calkin, 1964, 1971; Nichols, 1965, 1971; Rains *et al.*, 1980; Mosley, 1988). The evolution of such sandar is a very slow process owing to the low meltwater discharges and negligible precipitation. Transport rates of $3400 \, t \, yr^{-1}$ have been estimated for the Onyx River in the Wright Valley by Mosley (1988), a rate that is two orders of magnitude less than for Arctic and alpine glacifluvial systems. The morphologies of polar desert sandar have been categorized by Rains *et al.* (1980) as follows: (a) small outwash fans below cirque and hanging valley thresholds; (b) small and discontinuous valley trains interrupted by moraines; (c) small sandar fans in front of glacier snouts or ice-marginal debris aprons; and (d) continuous, thick sandar and valley trains, which are very rare (e.g. the Onyx River in the Wright Valley; Shaw and Healy, 1980; Mosley, 1988).

Process work on the Antarctic sandar indicates that they are produced by processes similar to those that produce Arctic and alpine glacifluvial systems. However, polar desert sandar may take thousands to tens of thousands of years to develop into forms comparable in size to those of Arctic and alpine glacial environments, which have taken only decades to accumulate.

The sediments that comprise polar desert sandar are dominated by sands to fine pebbles; larger gravel sizes are restricted to short stretches of river beds where the glacifluvial modification of moraines has produced a localized lag deposit (Rains et al., 1980) or where occasional higher discharges have transported freshly weathered material from bedrock out-crops (Shaw and Healy, 1980; Mosley, 1988). Despite the small amounts of precipitation in polar deserts, it is presumably possible that lag deposits may evolve into tightly packed mosaics or clast pavements by a combination of summer fluvial activity and winter snow-packing in a manner similar to that described by Davies et al. (1990) for alpine Norway.

Because large-scale and frequent switches of channel position do not occur on polar desert sandar and valley trains, older terraces tend to be covered in dense networks of ice wedge polygons whose growth is indicative of relatively stable, old surfaces. Although such permafrost features are most prominent on the Antarctic glacifluvial systems (e.g. Black, 1973; Shaw and Healy, 1980), they also occur in High Arctic locations where postglacial isostatic uplift has led to outwash incision and terrace abandonment (e.g. D.J.A. Evans and England, 1993).

CHAPTER 4

GLACIER MOTION

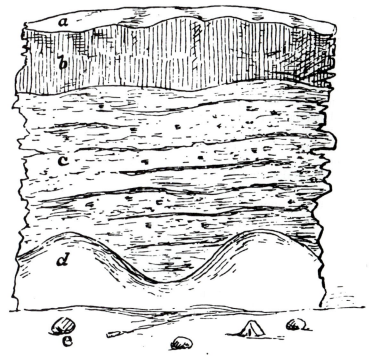

Ella Valley Glacier.—Section of Face of Glacier.

a. Edge of top overhanging in places as it appears from below.

b. Pure ice of a beautiful green color.

c. Ice white and chocolate color, full of small stones and streaks of mud or earth.

d. Apparent snow bank but probably moraine covered with snow.

e. Level of valley, scattered boulders and lumps of ice.

4.1 Introduction
4.2 Stress and strain
4.3 Deformation of ice
4.4 Deformation of the bed
4.5 Sliding
4.6 Spatial patterns of glacier motion
4.7 Temporal variations in glacier motion
4.8 Surging glaciers

4.1 INTRODUCTION

One of the most fundamental characteristics of glaciers and ice sheets is their ability to move. The deformation and sliding of glaciers under the force of gravity slowly transfers snow and ice from high-accumulation areas and continental interiors to areas of ablation, and allows glacial erosion and debris transport to take place. The motion of glaciers was not recognized by the scientific community until the nineteenth century, although it must have been common knowledge to the inhabitants of many glaciated areas, particularly those whose farms and houses were overrun by advancing glaciers during the Little Ice Age (Grove, 1988). The first scientific observation of glacier flow was by Franz Josef Hugi, a professor of physics and natural history, who measured the progress of a boulder along Unteraargletscher, Switzerland, between 1827 and 1836. His results were not universally accepted, however, and sceptical commentators suggested that the boulder had slid along the surface of the ice while the glacier beneath remained motionless! Subsequent observations firmly established the reality of glacier flow (e.g. Forbes, 1846; Agassiz, 1847; Tyndall and Huxley, 1857), and the character of and physical basis for glacier motion have remained an important area of research ever since (Clarke, 1987a).

4.1.1 Balance velocities

Averaged over long time periods, ice flow rates are governed by the climatic inputs to the glacier and the geometry of the catchment. For an ideal glacier of constant size and shape, ice flow through a cross-section must exactly balance the accumulation and ablation taking place upglacier. This idea can be simply illustrated using the 'wedge' concept, which represents the mass lost and gained in the glacier as two wedges (Fig. 4.1; Sugden and John, 1976). The ablation totals increase from zero at the equilibrium line to a maximum at the lower elevations of the snout, producing a wedge of ablation. Similarly, accumulation totals increase from zero at the equilibrium line towards higher elevations, producing a wedge of accumulation. In order to maintain a steady state, the glacier must rectify the loss of the ablation wedge by transferring mass through the equilibrium line. The mass of each wedge is controlled by the snow or ice density, the mass balance gradient, and the width of the wedge (see Section 2.3). Ice discharge through any cross-section on the glacier should equal the combined mass of the wedges located upglacier, with the accumulation wedge taken as positive mass and the ablation wedge as negative mass. This can be expressed quantitatively as:

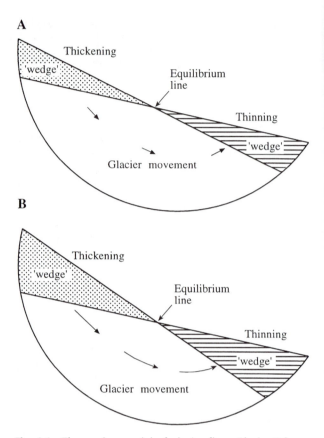

Fig. 4.1 The wedge model of glacier flow. Glacier B has a steeper mass balance gradient than glacier A, so requires higher ice velocities to balance the mass gained and lost in the two wedges. (Modified from Sugden and John, 1976)

$$Q(x) = \Sigma(w_x b_x) \qquad (4.1)$$

where $Q(x)$ is the discharge through a cross-section at a distance x from the highest point on the glacier, Σ represents 'the sum of', and w_x and b_x are the widths and specific net balances, respectively, of successive points upglacier from x.

The average velocity through a cross-section will be given by the equation

$$v(x) = Q(x)/A(x) \qquad (4.2)$$

where $v(x)$ is the average velocity through cross-section x, and $A(x)$ is the cross-section area.

Because of their dependence on mass balance, velocities calculated in this way are known as *balance velocities* (Clarke, 1987b). Two important implications stem from this simple model:

1. Ice discharges increase from the head of the glacier to the equilibrium line, then decrease towards the snout. For a glacier of constant cross-sectional area, average velocities will follow the same pattern.

2. Discharges and velocities will be highest on glaciers with steep mass balance gradients, or where the accumulation from wide basins is focused into narrow channels. The influence of mass balance gradient can be understood by considering two glaciers of identical shape but differing mass balance gradients (Fig. 4.1). Where the mass balance gradient is steeper, the mass of the ablation and accumulation wedges will be greater, thus requiring greater rates of throughflow to maintain equilibrium.

Several authors have used the mass balance gradient at the equilibrium line as a surrogate for the *energy of glacierization* or *glacier activity* (Ahlmann, 1948; Shumsky, 1950; Meier, 1961; Meier and Tangborn, 1965; Pelto, 1988). Regional trends in mass balance gradients indicate that glaciers in humid, maritime areas should flow faster than glaciers in arid, cold climates, and that glacier activity decreases with increasing latitude and greater continentality. This idea was taken further by Andrews (1972), who computed regional patterns in 'glacier power' and erosion potential. However, because of the influence of glacier geometry, there is no simple relationship between mass balance gradients and ice discharge. Indeed, some of the highest glacier velocities in the world have been recorded on Antarctic ice streams, where mass balance gradients are very low but ice from huge catchments is funnelled through relatively narrow outlets (Bentley, 1987; Clarke, 1987b).

Measured velocities on glaciers commonly differ from calculated balance velocities by varying amounts (Fig. 4.2; Clarke, 1987b). The differences may be small, representing temporary deviations from the average on a timescale of hours, days, weeks or months (Willis, 1995), or large, representing major departures from equilibrium conditions over years, decades or even centuries. A distinct class of *surging glaciers* exhibit periodic velocity cycles, with measured velocities oscillating either side of the balance velocity on a timescale of a few to several years (Section 4.8; Meier and Post, 1969; Raymond, 1987; Clarke, 1987b; Sharp, 1988a). In contrast, Jakobshavns Glacier, west Greenland, has very similar balance and measured velocities, indicating that it is close to equilibrium (Bindschadler, 1984; Clarke, 1987b).

Deviations from the balance velocity occur because, in the short term, glacier flow rates depend upon the balance between *driving* and *resisting forces*, which may be out of phase with changes in glacier mass (Fig. 4.3). The driving forces consist of the stresses generated by the surface slope and weight of the ice, and are therefore influenced by gains and losses of mass. The resisting forces result from the strength of the glacier ice, the contact between the glacier and its bed, and the bed itself. Over long periods of time, the driving and resisting forces are kept in a natural and dynamic balance by glacier flow, and the glacier will adopt a surface configuration that generates driving forces sufficient to overcome the resisting forces and balance the mass budget of the glacier. The resisting forces, however, may undergo temporary and rapid changes, mainly because of variations in water input and storage at

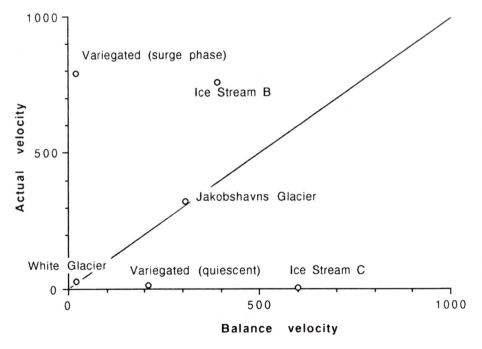

Fig. 4.2 Comparison of balance velocities and measured velocities for selected glaciers (in m yr^{-1}). Jakobshavns Glacier and White Glacier plot close to the diagonal line (where balance and measured velocities are equal) and are thus almost in equilibrium with climate. Ice Stream B is flowing faster than expected, and Ice Stream C is much slower than expected. The surging Variegated Glacier oscillates from one part of the diagram to another on a cycle of several years. (Data from Clarke, 1987b)

(a) (b) (c)

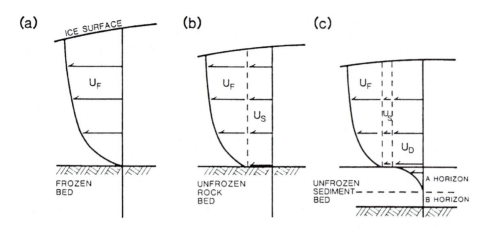

Fig. 4.3 Schematic diagrams showing the vertical distribution of velocity for glaciers moving by different mechanisms: (a) ice deformation only; (b) ice deformation and basal sliding; and (c) ice deformation, basal sliding and deformation of subglacial sediments. (From Boulton, 1996a. Reproduced by permission of the International Glaciological Society)

the bed. As a result, the glacier may speed up or slow down over varying timescales, leading to velocities that differ from those predicted from mass balance alone. Velocity fluctuations and surges and their causes are examined in detail in Sections 4.7 and 4.8. First, the character of driving and resisting forces, and the processes of glacier motion, are discussed in Sections 4.2–4.5. Spatial variations in flow rates are examined in Section 4.6.

4.2 STRESS AND STRAIN

The concepts of stress and strain are fundamental to a clear understanding of why and how glaciers move, and of many aspects of their interaction with the landscape. The words 'stress' and 'strain' are often used interchangeably in everyday language, but in scientific usage their meanings are quite distinct. *Stress* is a measure of how hard a material is being pushed or pulled as the result of external forces, whereas *strain* measures the amount of deformation that occurs as the result of stress. A familiar example of this distinction is what happens when a tube of toothpaste is squeezed. The toothpaste comes out because of deformation (strain) resulting from squeezing of the surface of the tube (the stress). A complete description of the stress and strain in part of a glacier (or toothpaste tube) requires consideration of the forces acting in all directions on the material, and the whole range of resulting deformations. The approach adopted here, however, considers only those components of stress and strain which are central to an understanding of glacier behaviour. Good descriptions of stress and strain in ice and rocks are provided by Paterson (1994) and Twiss and Moores (1992), while a very detailed analysis is presented by Hutter (1983). Also recommended are two highly readable books by Gordon (1976, 1978), which give lively accounts of

consequences of stress and strain in a wide variety of familiar materials and structures.

4.2.1 Stress

To understand stress, it is useful to begin with the concept of force. *Force* can be thought of as the physical influences which change, or tend to change, the state of rest or uniform motion of a mass, and is defined as mass times acceleration. For example, the downward force at the base of a block of ice resting on a surface is the product of the mass of the block and the acceleration due to gravity. The unit of mass used in this book is the kilogram (kg) and that of acceleration is metres per second per second (m s^{-2}), so the unit of force is kg m s^{-2}, or the newton (N). Now, when a force acts on a surface, the intensity of its action depends on the area over which it is distributed. This distributed force is the *stress*, which is defined as force per unit area. A given force acting on a small area results in a larger stress than the same force acting over a large area (Fig. 4.4). For example,

Fig. 4.4 Monty the Mammoth is safely supported by a block with cross-sectional area A_1, but breaks a table leg with the much lower cross-sectional area A_2. Because Monty's weight must be supported by a smaller area, he exerts a greater stress and thus falls to the floor. (With acknowledgements to Twiss and Moores, 1992)

a 1 kg block of ice with a basal area of 10 cm^2 exerts a larger stress on the surface on which it rests than the same mass of ice with a basal area of 100 cm^2. The same effect explains why stiletto heels cause more damage to floors than flat-soled shoes. A commonly used unit of stress is the pascal (1 Pa = 1 N m^{-2}), although for the high stresses encountered at the base of glaciers it is more useful to employ the kilopascal (1 kPa = 1000 Pa). Another unit of stress which is often used in glaciological studies is the bar (1 bar = 100 kPa).

The stress on a surface can be resolved into two components: the stress acting at right angles to the surface (the *normal stress*) and that acting parallel to the surface (the *shear stress*; Fig. 4.5). For equilibrium conditions, the stress on a surface consists of two equal and opposite *tractions* acting on opposite sides of the surface. In the case of normal stresses, the two opposing tractions either press the material together across the surface (*compressive stress*), or tend to pull the material apart across the surface (*tensile stress*). For shear stresses, the tractions are parallel, but act in opposite directions.

At the base of a glacier, the normal stress acting on the bed is mainly the result of the weight of the overlying ice. If the slope of the bed is small, the normal stress (which is compressive) may be approximated as

$$\sigma = \rho_i g h \qquad (4.3)$$

where ρ_i is the density of the ice (c. 900 kg m^{-3}), g is gravitational acceleration (9.81 m s^{-2}), and h is the ice thickness (m). For h = 100 m, the normal stress would be c. 882.9 kPa.

In simple terms, the average shear stress at the base of a glacier is due to the weight of the overlying ice and the slope of the ice surface. For small bed slopes, the shear stress can be approximated as

$$\tau = \rho_i g h \sin a \qquad (4.4)$$

where a is the surface slope of the ice. For h = 100 m and a = 6°, the basal shear stress τ_b would be c. 92 kPa.

An important point to note about these simple formulae is that in both cases, the stress increases with ice thickness (h). This means that the normal

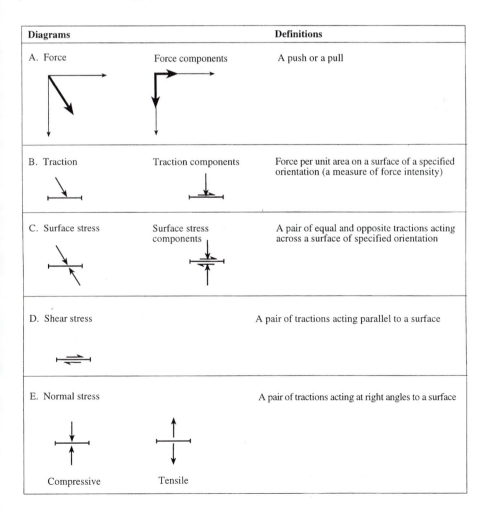

Diagrams		Definitions
A. Force	Force components	A push or a pull
B. Traction	Traction components	Force per unit area on a surface of a specified orientation (a measure of force intensity)
C. Surface stress	Surface stress components	A pair of equal and opposite tractions acting across a surface of specified orientation
D. Shear stress		A pair of tractions acting parallel to a surface
E. Normal stress		A pair of tractions acting at right angles to a surface
Compressive	Tensile	

Fig. 4.5 The concept of stress, including force, stress components, traction, and normal and shear stresses (compressive and tensile). (Modified from Twiss and Moores, 1992, Table 8.1)

and shear stresses in a glacier increase linearly from the surface towards the base, a point that has profound implications for glacier behaviour, as we shall see.

In reality, the stresses within and below glaciers are liable to differ from those given by the above formulae. Two factors are of particular importance. First, undulations of the glacier bed will result in marked stress variations, because the magnitudes of the normal and shear stresses acting across a surface depend on its orientation relative to the applied force. Therefore, for a rough bed there will be areas where stresses are higher than average and others where they are lower than average (Fig. 4.6). Areas with higher than average stresses are known as *stress concentrations*, and have important implications for processes of subglacial erosion and deposition (Chapter 5). The second major factor that can modify the stresses at a given point in a glacier is the pushing or pulling effect of the upstream and downstream ice. This effect is known as the *longitudinal stress*, which is compressive if the ice is slowing down and tensile if the ice is accelerating. Because the longitudinal stress is, by definition, approximately parallel to glacier flow and the surface slope, its principal effect is to modify the shear stress (where bed roughness is high, the longitudinal stress will also significantly affect the normal stress). A theoretical method for calculating longitudinal stresses is given by Nye (1969b) and Paterson (1994). Empirical studies have shown that the influence of longitudinal stress can be incorporated in equation (4.4) if glacier thickness (*h*) and slope (*a*) are averaged over distances of 8 to 16 times the glacier thickness (Bindschadler *et al.*, 1977).

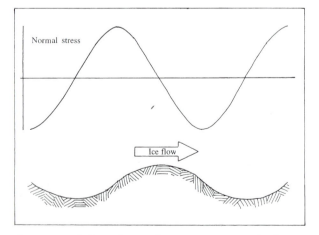

Fig. 4.6 Variation of normal stress over a wavy bed. Normal stress fluctuates around the average value, and is greatest on the upstream sides of bumps and lowest on the downstream sides

4.2.2 Strain

As already noted, *strain* is the change in shape (and size) of a material due to stress. Different materials – such as ice, water, sediments and rocks – exhibit a wide variety of strain responses to a given stress: some will flow quickly, some will flow slowly, some will break, and yet others may appear to be unaffected. In fact, all materials undergo strain of one type or another when placed under stress, no matter how small. It is useful to divide strains into two basic types: recoverable, or *elastic*, strains, and irrecoverable, or *permanent*, strains. Elastic strain refers to a temporary change in shape of a material, which lasts only as long as stress is applied. Once the stress is removed, the material goes back to its original shape. This type of strain is not confined to familiar 'springy' materials such as rubber or the spring in a shock absorber, but is also a characteristic of apparently rigid materials, such as a concrete pavement or bedrock at the base of a glacier. Although the change in shape of a material may be very small, the applied stress causes the compression or stretching of molecular bonds, resulting in the storage of *strain energy*. This stored energy provides the traction to balance that exerted by the external force, which is why you do not fall through the pavement when you walk along the street (Gordon, 1976). At some critical level of stress, which varies widely between different materials, the stored strain energy will be released, resulting in *permanent deformation* or *failure*. The stress above which permanent deformation occurs is the yield stress (which may be zero for some materials). Permanent deformation may take the form of *brittle failure*, where the material breaks along a fracture, or *ductile deformation*, where the material undergoes flow or creep (Fig. 4.7). The processes of brittle failure and ductile deformation are of great importance for glacier flow, erosion and deposition, and are discussed in detail in later sections of this book.

Deformation of a material may or may not involve changes in its volume. Deformation with volume change is known as *dilation*, which may consist of expansion or contraction. Deformation without volume change is termed *constant-volume deformation*. As noted in Section 2.2.2, the transformation of snow to ice in the accumulation area of a glacier results in a decrease in volume and an increase in density of the snow. Where melting and refreezing are negligible, this contraction (or consolidation) occurs in response to the normal stresses imposed by the weight of the overlying snow. Once formed, glacier ice is essentially incompressible; therefore any deformation will be constant-volume strain.

The amount of strain experienced by a material is measured by comparing its shape and size before and after deformation. The patterns of deformation associated with glacier flow may be highly irregular, but are often relatively simple, particularly if small time intervals or volumes of material are considered. Two types of strain commonly associated with glacier flow are *pure shear* and *simple shear* (Fig. 4.7). Pure shear involves flattening or stretching a material under compressive or tensile stresses. As shown in Fig. 4.7, the orientation of lines drawn parallel and normal to the applied stress remains constant as the material is flattened or stretched. In contrast, deformation by simple shear resembles the shearing of a pack of cards, in which the rectangular side of the deck is changed to a parallelogram. This type of deformation occurs in response to shear stresses, and involves the rotation of all lines drawn through a material except those parallel to the shear plane. The type of strain occurring near the beds of glaciers often closely resembles simple shear.

The method of measuring the amount of strain in deformed materials differs for pure and simple shear. Strain resulting from pure shear is measured using changes in the length of lines, whereas strain from simple shear is measured using changes in angles. When studying glacier movement, we also need to know the amount of strain that occurs per unit of time, termed the *strain rate*. The strain rate of most naturally occurring materials rises with increasing stress, in either a linear or a non-linear fashion. Mate-

rials can be classified according to how their rate of strain varies with increasing stress: some examples are shown in Fig. 4.8. The net amount of strain occurring in a given time interval is known as the *cumulative strain*. It is important to note that the cumulative strain depends on both the rate and the duration of strain. For example, high cumulative strains can result from low strain rates sustained over a long period or high strain rates lasting only a short time.

4.2.3 Processes of glacier motion

The movement of glaciers results from permanent strain of the ice and the glacier bed in response to stress. Strain may occur by (a) deformation of the ice; (b) deformation of the bed underlying the glacier; or (c) sliding at the ice–bed interface. Movement at the surface of a glacier is the cumulative effect of these processes acting singly or in combination (Fig. 4.3). Recent research has resulted in a huge increase in knowledge of the processes of glacier motion, but many questions remain about the precise conditions that control strain rates and hence rates of glacier movement. This is partly because resistance to flow depends on several interacting factors – such as temperature, debris content of the ice, bed roughness and water pressure – but mainly because data on actual conditions within and below glaciers are

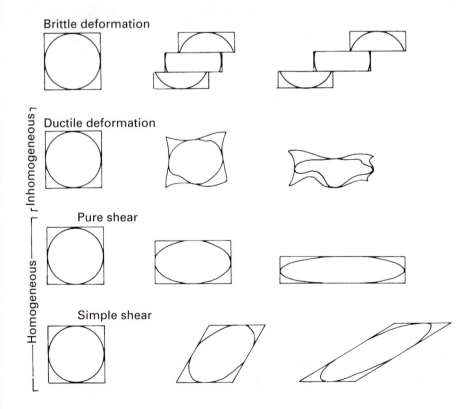

Fig. 4.7 Idealized styles of deformation. (From Benn and Evans, 1996)

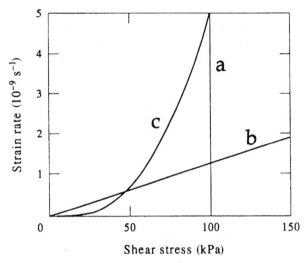

Fig. 4.8 Stress–strain relationships for different types of material. (a) Perfectly plastic material, which remains rigid until the shear stress reaches the yield stress (100 kPa in this case), when it deforms instantaneously; (b) Newtonian viscous material, for which the strain rate is linearly proportional to shear stress; (c) non-linearly viscous material, such as ice. (Modified from Paterson, 1994)

extremely difficult to obtain. Present knowledge on glacier movement has resulted from mathematical and numerical modelling, experimentation with blocks of ice in the laboratory, measurements made in boreholes through modern glaciers, geophysical studies, limited direct observations in natural or artificial subglacial cavities, and interpretation of sediments and landforms in formerly glaciated areas.

4.3 DEFORMATION OF ICE

Deformation of ice refers to changes in the shape of a region of glacier ice in response to stress, and can consist of *creep* or *fracture*.

4.3.1 Creep

Creep is deformation that results from movement within or between individual ice crystals. Ice creep resembles the deformation of metals at temperatures close to their melting point, and results from the interaction of several complex processes (Weertman, 1983; Alley, 1992a). Briefly, movement within crystals may occur by gliding along *cleavage planes* (lines of weakness related to the molecular structure of the crystal) or movement along crystal *defects*, whereas movement between crystals involves changes in shape or size by *recrystallization* at grain boundaries. The effects of creep are strikingly illustrated by the fold structures often exposed on glacier surfaces, which

record substantial changes in the shape of layered ice during flow (Fig. 4.9; Hudleston, 1976; Hambrey and Milnes, 1977). Relationships describing the response of ice to stress, or *flow laws*, have been established in numerous laboratory studies. The most widely used flow law is known as Glen's law (Glen, 1955), which was first adapted for glaciers by Nye (1957). Glen's law can be written as

$$\varepsilon = A\tau^n \qquad (4.5)$$

where ε represents the strain rate, A and n are constants, and τ is the shear stress. The first parameter, A, decreases dramatically with ice temperature (Fig. 4.10), showing that the colder the ice, the less readily it deforms. This is because creep processes are most effective when melting occurs at grain boundaries. The flow law exponent n also varies, but is usually close to 3, which means that the strain rate is proportional to the cube of the shear stress. If the

Fig. 4.9 Fold structures, such as these exposed in the basal ice of a subpolar glacier, Phillips Inlet, Ellesmere Island, provide striking evidence for the deformation of ice. (Photo: D.J.A. Evans)

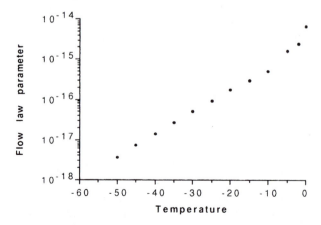

Fig. 4.10 Variation of flow law parameter A (s^{-1} kPa^{-3}) with ice temperature. Note the logarithmic scale of the flow law parameter. Note how the log-linear relationship breaks down near the melting point. (Data from Paterson, 1994)

shear stress is doubled, for example, the strain rate increases eightfold (Fig. 4.8). In practice, it is unrealistic to think in terms of a single flow law for ice, because several processes contribute to creep, and their relative importance changes in space and time. Any general flow law would have to be unmanageably complicated, and it is therefore necessary to formulate different 'flow laws' depending on the dominant processes of creep (Alley, 1992a).

In glaciers, two major factors may cause strain rates to differ from values predicted from laboratory experiments: (a) the orientation of crystals in the ice, and (b) the presence of impurities. Ice crystals exhibit a tendency to split in one particular direction, owing to the presence of parallel cleavage planes controlled by the molecular structure. In fact, ice crystals will deform along such cleavage planes 100 to 1000 times more easily than in any other direction (Weertman, 1973). Therefore, in polycrystalline ice, strain rates are sensitively dependent on the aggregate orientation of ice crystals, a property known as the *ice crystal fabric*. A sample of polycrystalline ice in which large numbers of crystals have cleavage planes aligned parallel to the shear stress (i.e. with a strong fabric) will deform more easily than ice with randomly oriented crystals (a weak fabric; Lile, 1978; Russell-Head and Budd, 1979; Duval, 1981). Indeed, Lile (1978) found that strain rates in ice with a well-developed crystal fabric may be up to 10 times faster than those predicted by Glen's law. Favourable crystal fabrics may develop in ice under steady applied stresses as the result of crystal rotation and recrystallization (Alley, 1992a; Paterson, 1994). In addition, crystal defects may grow (propagate) parallel to the direction of shear, adding to the progressive weakening of ice or *work softening*. Because different parts of a glacier will have been deformed at different rates and at different times, crystal fabrics and strain responses often vary markedly from place to place. Some studies have noted that strong crystal fabrics are characteristic of ice at depth, where stresses are highest (e.g. Gow and Williamson, 1976; Blankenship and Bentley, 1987), although a general model of fabric development and ice deformation remains elusive (Alley, 1992a).

Impurities are known to exert a strong influence on the strain response of glacier ice, although precise flow laws are very difficult to establish. Three main kinds of impurities occur in natural glacier ice: dissolved ions (solutes), gas bubbles and solids (rock particles). The influence of solutes is variable, some tending to harden the ice, some to soften it, and others having no effect (Nakamura and Jones, 1973). Concentrations of gas bubbles usually soften ice, by creating lines of weakness (Hooke, 1973a). The influence of rock particles on the flow characteristics of ice is of great importance, owing to the widespread occurrence of debris in the basal layers of glaciers (Section 5.4). However, different studies have reached conflicting conclusions, some indicating that strain rates increase with debris content, others suggesting that they decrease (Hubbard and Sharp, 1989). Laboratory studies by Nickling and Bennett (1984) concluded that the strength of debris–ice mixtures increases as debris contents increase from zero to 75 per cent, then falls sharply at higher debris contents. Mixtures with more than c. 85 per cent rock debris were found to be weaker than pure ice. This pattern was thought to result from the varying influence of friction between rock particles and the cohesive effect of the ice. As debris content increases from zero to 75 per cent, contacts between particles will be more frequent, increasing the internal friction and resistance to deformation. However, mixtures with very high debris contents contain ice only in the pores between particles, with a consequent loss of cohesive strength. These conclusions conflict with observations made by Echelmeyer and Wang (1987) below the frozen margin of Urumqi Glacier 1 in China. They found that frozen debris at the base of the glacier, with a debris content of 61–79 per cent, was deforming up to 100 times more rapidly than clean ice would at similar temperatures and stresses. Approximately 60 per cent of the motion of the glacier resulted from the deformation of a subglacial debris layer only 36 cm thick. The difficulties in establishing the effect of debris on ice deformation probably arise because so many factors are involved, including the grain size distribution of the debris, temperature-related melting effects around rock particles, and the aggregate orientation (fabric) of the debris and ice crystals.

4.3.2 Fracture

Fracture, or brittle failure, occurs when ice cannot creep fast enough to allow a glacier to adjust its shape under stress. Crevasses are striking examples of fractures formed where ice is pulled apart by tensile stresses (Section 6.2.1). Such tensional fractures tend to be fairly superficial features, because at depth creep rates are higher and ice will flow faster than it can split. On temperate glaciers, where ice is relatively soft, crevasses rarely extend deeper than 30 m or so, whereas on polar glaciers they may be much deeper. In heavily crevassed areas such as ice falls, blocks of ice between crevasses may form spectacular broken pinnacles known as *séracs*, which often collapse under the influence of glacier movement and ablation, adding to the downglacier transfer of ice. Fractures can also develop where ice is undergoing compression. In this case, movement takes place along a shear plane, which is analogous to a fault in rocks. Examples of fractures resulting from compressive stresses are thrust planes at glacier margins, where ice is sheared over more slowly moving ice below. Shear planes may traverse clean or debris-rich ice (Echelmeyer and Wang, 1987).

4.4 DEFORMATION OF THE BED

4.4.1 Observations

In some circumstances, the rocks or sediments beneath a glacier may undergo permanent strain in response to stresses imposed by the ice. Such *subsole deformation* accounts for a substantial share of the forward movement of some glaciers, because strain rates may be equal to or greater than peak strain rates in the overlying ice. The processes of subsole deformation have received a great deal of attention in recent years because they provide possible mechanisms for certain types of unstable glacier behaviour, and could account for the fast flow of some ice streams and outlet glaciers (e.g. Boulton and Jones, 1979; Clarke *et al.*, 1984a; Clarke, 1987b; Alley *et al.*, 1987b; Humphrey *et al.*, 1993). Additionally, deformation of the bed is an extremely important geomorphological process, one which can result in very high rates of subglacial erosion and deposition (Boulton and Hindmarsh, 1987; Boulton, 1996a). In this section we shall consider the evidence for sub-sole deformation and its contribution to the motion of modern glaciers. The implications of bed deformation for debris erosion, transport and deposition are discussed in Section 5.6.

The most striking evidence for deformation of subglacial sediment comes from a series of experiments conducted by G.S. Boulton and co-workers below Breidamerkurjökull, an outlet tongue of the icecap Vatnajökull in southern Iceland (Fig. 4.11). Gaining access to the glacier bed via tunnels cut in from the margin, Boulton inserted segmented rods into the unfrozen till underlying the ice. When excavated several days later, the rod segments had been displaced downglacier by varying amounts, clearly demonstrating that the till was deforming (Fig. 4.12; Boulton, 1979; Boulton and Hindmarsh, 1987). The till has a two-tiered structure, consisting of a porous upper layer some 0.5 m thick (A, Fig. 4.12) and a dense lower layer (B). Most of the observed deformation was by ductile flow of the upper till, accounting for 80–95 per cent of the forward movement of the glacier. During some experiments, a limited amount of deformation also occurred in the lower till, probably by brittle failure along subhorizontal joints. The low strength of the upper till is partly due to its very low density: only 40–45 per cent of the volume of the till was occupied by mineral grains, the rest consisting of water-filled pores. The pressure of the porewater was generally high, and fluctuated on a daily basis owing to variations in the amount of sur-

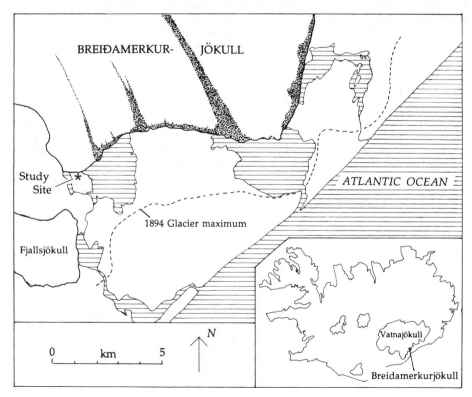

Fig. 4.11

(b)

Fig. 4.11 Breidamerkurjökull, Iceland. (a) Location map showing the location of Boulton's field experiments. (From Benn, 1995) (b) The glacier margin looking east from near the study site. The deposits in the foreground consist mainly of subglacially deformed sediments. (Photo: D.I. Benn)

face meltwater reaching the bed via moulins. Measurements made over a 10-day period demonstrated a strong relationship between porewater pressures and strain rates in the deforming till, with peak strain rates lagging behind peaks in porewater pressure by a few hours. The fact that the upper till was the weakest part of the local glacier–bed system was dramatically proven during one of Boulton's experiments. Water was being pumped out of the till below one of the tunnels prior to excavation, when a plug of partially dried till was forced up through the access hole, rapidly followed by a mass of fluid till. About 13 m³ of fluid till poured into the tunnel, causing widespread collapse of the tunnel floor and acute concern among those who were in the tunnel at the time. This spectacular accident also goes a long way towards explaining why no one has felt moved to repeat Boulton's experiments beneath other glaciers.

A safer approach to studying subglacial deformation below modern glaciers involves placing intruments down boreholes. This approach has several additional advantages over that adopted by Boulton. First, data can be obtained from sites anywhere on a glacier (except dangerously crevassed areas), and not just close to the margin, where conditions may be unrepresentative. Second, continuous records of deformation can be obtained over many months. Third, these methods involve minimal disturbance of the bed, and can yield important data on shear strength *in situ* under natural conditions. On the downside, instrumenting glacier beds requires expensive equipment, and so far no methods are available for measuring absolute velocities rather than strain gradients (Willis, 1995). Measurements of subsole deformation have been made below Trapridge Glacier, Yukon, and Storglaciären, Sweden (Blake *et al.*, 1992; Iverson *et al.*, 1995). Complex patterns of subsole till deformation have been revealed, and rates and directions of strain vary on an hourly basis. Instantaneous strain rates often exceed mean strain rates by an order of magnitude, and short periods of reversed motion have been observed. The great potential of borehole data was highlighted by a lucky accident reported by Humphrey *et al.* (1993), in which a drill stem became jammed in basal sediments below Columbia Glacier, Alaska, and was dragged over the glacier bed for five days before it was recovered. The shape of the striated and bent stem indicated that the uppermost 0.65 m of the bed consisted of deforming sediment, and subsequent analysis allowed an estimate of the strength of the till *in situ* to be calculated. Such estimates provide much-needed data on stress–strain relationships for subglacial till. Engelhardt *et al.* (1978) and Kamb *et al.* (1979) took series of photographs of the bed of Blue Glacier, USA, using a remote-controlled camera placed at the bottom of boreholes. The photographs recorded the motion of unfrozen coarse-grained till, termed *active subsole drift*, which was clearly mechanically distinct from the overlying debris-laden ice. More recently, video techniques have been tried, but to date results have been limited

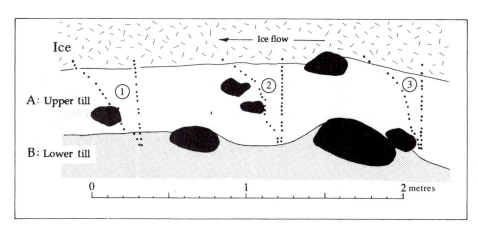

Fig. 4.12 The displacement of segmented rods placed into the till below Breidamerkurjökull, showing subsole deformation. (From Benn, 1995, adapted from Boulton and Hindmarsh, 1987)

owing to instability of the camera and poor visibility (Pohjola, 1993).

The presence of weak, deforming subglacial sediment below some glaciers has also been inferred using seismic sounding techniques. In a series of very influential papers, R.B. Alley and others (Blankenship *et al.*, 1986, 1987; Alley *et al.*, 1986, 1987a, b, c; Alley, 1989) used seismic data to argue that unfrozen deforming till underlies Ice Stream B, one of several ice streams that discharge into the Ross ice shelf in western Antarctica (Fig. 1.15). According to their interpretation, large parts of the ice stream are underlain by deforming till, which contributes an important component of the total ice velocity. Support for this view was provided when porous unfrozen till was recovered from the base of Ice Stream B via boreholes (Engelhardt *et al.*, 1990). Laboratory tests on the till indicated that it was very weak, and would indeed deform at the stresses found beneath the ice stream (Kamb, 1991). Seismic evidence was also used by Richards (1988) to argue that a weak till layer existed beneath Variegated Glacier during its surge in 1982/83 (see Section 4.8). Finally, electrical resistivity and borehole deformation experiments indicate that saturated, deformable till exists beneath part of Storglaciären, in northern Sweden (Hooke *et al.*, 1988).

4.4.2 Processes

Because so few direct observations have been made, current understanding of subsole deformation and its influence on glacier movement is mainly based on theoretical modelling. Theoretical analyses aimed at predicting the strain response of subglacial sediments and rocks to stresses have been presented by Boulton and Hindmarsh (1987), Clarke (1987c), Alley (1989) and Kamb (1991). Although there is considerable difference in detail between these analyses, most agree that subglacial materials will not undergo permanent deformation unless subjected to some threshold stress, known as the *yield stress* or *critical shear stress*. The value of the critical shear stress varies with a number of factors, the most important being the effective normal pressure exerted on the materials by the overlying ice and sediment. Above the critical shear stress the materials will deform, with strain rates varying in a non-linear fashion with the shear stress and the confining pressure. A major aim of these theoretical models is to develop *flow laws* to predict the amount of deformation that will occur under given conditions.

It is worth examining the factors controlling subglacial deformation in some detail, not only as a means of understanding glacier motion, but also because they are of vital importance in processes of subglacial erosion and deposition (Section 5.6).

4.4.2.1 CRITICAL SHEAR STRESS

For a given material, the critical shear stress is the minimum required to overcome the resisting forces that hold the material together. The critical shear stress is therefore equivalent to the shear strength of the material at the onset of permanent deformation. This shear strength is measured in kilopascals or bars. Material strength can be regarded as the sum of two properties: cohesion and intergranular friction.

Cohesion refers to the forces binding a material together, including the chemical bonds between grains and electrostatic forces between charged clay particles. For unlithified materials such as subglacial till, chemical bonding between grains is usually unimportant and cohesion is often due entirely to electrostatic forces. Electrostatic attraction is negligibly small for particles larger than about 1 μm in diameter, and therefore cohesion contributes towards sediment strength only when significant amounts of clay are present. The cohesive strengths of selected sediments and rocks are shown in Table 4.1.

Frictional strength reflects several factors, including the resistance of grains to sliding past each other and resistance to grain crushing. Because frictional resistance is due to interactions between grains, frictional strength is directly proportional to the forces pressing grain surfaces together; that is, the normal stresses. An everyday example of this effect is the friction between sandpaper and the surface of a piece of wood: the harder the sandpaper is pressed down (normal stress), the harder it is to slide along because of the increase in frictional resistance (frictional shear strength). For each material, the specific frictional strength is measured by the ratio between the shear stress and the corresponding normal force at the onset of permanent deformation. As shown in Fig. 4.13, this ratio can be represented as the tangent of an angle if the shear and normal stress vectors are drawn as two sides of a right-angled triangle.

Table 4.1 Typical cohesion values and friction angles for some geological materials

Material	Cohesion (kPa)	Friction angle (°)
Dense sand (well sorted)	0	32–40
Dense sand (poorly sorted)	0	38–46
Gravel (well sorted)	0	34–37
Gravel (poorly sorted)	0	45–48
Bentonite clay	10–20	7–13
Soft glacial clay	30–70	27–32
Stiff glacial clay	70–150	30–32
Till (mixed grain size)	150–250	32–35
Soft sedimentary rock	1,000–20,000	25–35
Igneous rock	35,000–55,000	35–45

Source: Selby (1982)

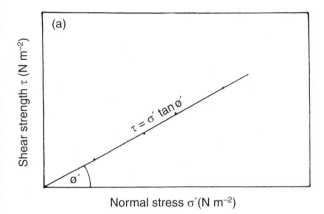

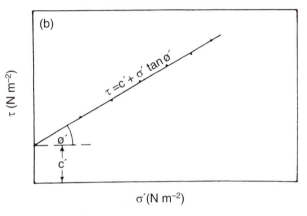

Fig. 4.13 Graphs showing the relationship between normal stress and shear strength for (a) cohesionless, and (b) cohesive materials. The angle of internal friction is the slope of the line. (From Selby, 1983. Reproduced by permission of Oxford University Press)

This ratio is known as the coefficient of friction and written as tan ø, where ø is the *angle of internal friction*. It has a unique value for each material and is influenced by the arrangement, size, shape and resistance to crushing of the constituent grains. The angle of internal friction for some common geological materials is shown in Table 4.1.

4.4.2.2 THE COULOMB EQUATION

The contribution of cohesion and friction to the shear strength of materials was first formulated by the French engineer Charles Augustin de Coulomb in 1776, in what is now known as the Coulomb equation:

$$\tau^* = c + N \tan ø_i \tag{4.6}$$

This equation states that the total shear strength of a material (τ^*) is given by the cohesion (c) plus the product of the normal stress (N) and the coefficient of friction (tan ø$_i$). The Coulomb equation is repre-

sented graphically in Fig. 4.13. Note that whereas the frictional component of strength depends upon the magnitude of the normal stress, cohesion is regarded as a constant, independent of $N(\sigma'$ in Fig. 4.13). For lithified materials this is a reasonable approximation, but where cohesion is due to electrostatic forces (such as for clay-rich tills), cohesion actually decreases with increasing distance between particles. The average inter-particle distance is related to the proportion of the material occupied by air- or water-filled voids, or sediment *porosity*. The influence of porosity on cohesive strength has been incorporated in the till deformation model of Clarke (1987c).

The frictional strength of a granular material will be modified by the presence of water in the pore spaces between grains. At low water contents, surface tension at wet grain contacts serves to pull grains together, increasing normal stresses (N) and causing a rise in frictional strength. This important effect, known as *porewater tension* or negative porewater pressure, explains, among other things, why damp sand is better than dry for building sandcastles. At higher water contents surface tension forces disappear, and for saturated materials part of the weight of the overlying material is transferred from the grains to the porewater. If the water cannot drain away freely, positive porewater pressures develop, and these support part of the normal stress acting between grains. The effective normal stress is therefore reduced, and equals the ice overburden pressure minus the porewater pressure:

$$N = p_i - P_w \tag{4.7}$$

where N is the effective pressure, p_i is the ice overburden pressure and P_w is the water pressure. Note that this formula is identical to that given in Section 3.2.3.1 for the effective pressure at the walls of an englacial or subglacial tunnel (equation (3.4)).

Rising porewater pressures cause a reduction in shear strength because confined water can support compressive stresses (and hence reduce N), but cannot withstand a shear stress. This is why saturated sand is weaker than either damp or dry sand and tends to flow as a slurry when on a slope. The effect of porewater pressure on material strength is incorporated into the Coulomb equation thus:

$$\tau^* = c + (p_i - P_w).\tan ø_i \tag{4.8}$$

where τ^* is the shear strength or critical stress at any point in the material, c is the cohesion, p_i is the normal stress imposed by the weight of the ice, rock particles and water above the point, P_w is the porewater pressure, and ø$_i$ is the angle of internal friction.

4.4.2.3 STRAIN ABOVE THE CRITICAL SHEAR STRESS

Several flow laws have been proposed to describe the strain response of subglacial materials at shear stresses above the critical stress τ^*. To date, the most widely used type of flow law has the form

$$E_s = K(\tau - \tau^*)^a/N^b \qquad (\tau > \tau^*) \qquad (4.9)$$

where E_s is the strain rate, τ is the shear stress, τ^* is the critical stress, N is the effective normal stress, and K, a and b are constants dependent on the properties of the material (Boulton and Hindmarsh, 1987; Boulton, 1987; Alley, 1989; Willis, 1995). This relationship simply states that the strain rate rises as the shear stress becomes increasingly higher than the critical stress, and decreases with the effective normal stress. The precise form of the relationship depends on the values of the constants K, a and b. Note that this flow law is similar to Glen's flow law for ice, but with the added term N^b. The lack of this term in Glen's law reflects the fact that the strain rate of ice, unlike that of unfrozen sediment, is independent of the normal stress.

Together, equations (4.8) and (4.9) explain many of the observed characteristics of deforming subglacial sediment:

1. The deforming layer is confined to the uppermost part of the bed because the normal stress, and hence frictional strength, increases downwards owing to the increasing weight of overlying sediment. As the normal stress increases with depth, a point will be reached when the strength of the sediment will exceed the shear stress, and deformation will cease. It follows that changes in the shear stress or in porewater pressure will result in a thickening or thinning of the deforming layer (Boulton and Hindmarsh, 1987; Alley, 1989; Hart *et al.*, 1990).
2. Sediment strain rates increase upwards, and are highest at the top of the deforming layer where

normal stress is at a minimum. Velocity results from the cumulative effect of strain in the underlying material, and is at a maximum at the top of the layer (Fig. 4.3c).

3. Strain rates increase with porewater pressure, owing to the influence of P_w on effective normal stress and intergranular friction. The dependence of strain rates on P_w implies that most deformation will occur where meltwater cannot be removed rapidly by the subglacial drainage system, allowing pore pressures to build up. Deformation will therefore be encouraged by inefficient drainage systems such as Darcian porewater flow and braided canal systems (Sections 3.4.5 and 3.4.6), and little or no deformation will occur where drainage is by efficient channel systems (Section 3.4.2).

Although equations (4.8) and (4.9) describe many of the characteristics of subsole deformation, there are several additional factors that are thought to be important (Boulton and Hindmarsh, 1987; Hart *et al.*, 1990; Kamb, 1991), six of which will be considered here.

4.4.2.4 DILATANCY

The uppermost 0.5 m of subglacial till at Breidamerkurjökull has very low density and high porosity (Fig. 4.14a; Boulton *et al.*, 1974; Boulton and Hindmarsh, 1987; Benn, 1995). These characteristics, which are believed to be typical of many subglacially deformed sediments, are attributed to a volume increase or *dilation* of the sediment during shear. Dilation has been demonstrated in many laboratory experiments (e.g. Murray and Dowdeswell, 1992) and occurs because grains must 'climb over' each other during shear (Fig. 4.14b). Alley (1991) has suggested that a critical strain rate is required to sustain dilation, and that if the strain rate falls below this threshold, the sediment will collapse back into a denser state (although it will not necessarily regain its pre-deformation density). Because a dilatant sedi-

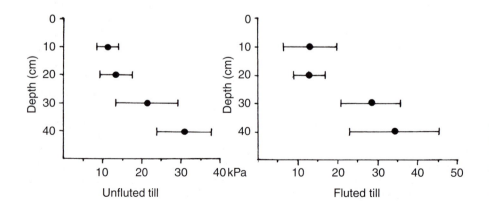

Fig. 4.14a Shear strength of Breidamerkurjökull till. The uppermost layers have low shear strength associated with high porosity. (From Benn, 1995)

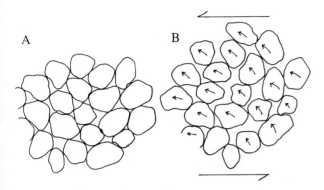

Fig. 4.14b Schematic diagram showing till dilation resulting from shear. The tendency for grains to climb over one another causes an increase in volume

ment is weaker than a dense sediment, Alley argued that two positive feedback mechanisms may operate during sediment dilation and collapse. First, as strain rates rise, dilation occurs, resulting in weaker sediment, which in turn leads to yet higher strain rates. Conversely, as strain rates fall, the sediment collapses and the sediment becomes stronger, leading to still lower strains. He proposed that these feedbacks account for the rapid vertical transition between soft, dilated upper till and the stiff, dense lower till observed at Breidamerkurjökull, and that the threshold between dilatant and collapsed states marks an important jump in the 'flow law' of the till.

4.4.2.5 GRAIN CRUSHING

Hooke and Iverson (1995) and Iverson *et al.* (1996) have suggested that grains in subglacial sediments can shear past each other by breaking, rather than climbing over each other. For this mechanism to be effective, the shear stresses must be greater than the strengths of the grains, and *stress concentrations* due to particular arrangements of grains are thought to be important in generating high enough stresses. This mechanism may well be important in certain types of stiff, non-dilatant material where stresses are high. Where porewater pressures are high and effective normal stresses are low, however, inter-particle contact forces may not be high enough to cause grain fracture, and crushing is likely to be insignificant. Grain size distributions resulting from subglacial shear are considered in Section 5.7.2.

4.4.2.6 SEDIMENT GRAIN SIZE

The strength and strain response of sediments are also influenced by grain size and sorting. Part of the influence of grain size is implicitly included in equation (4.8), because coarse and poorly sorted sediments have high coefficients of friction, whereas fine-grained and well-sorted sediments have low coefficients of friction. The strength of clays depends to a great extent on cohesion (Table 4.1; Selby, 1982). An additional important effect of grain size is its influence on water flow and porewater pressures. Sediment permeability depends to a large extent on the size of the pore spaces through which water can flow; therefore water can flow much more freely through sands and gravels than through clays (Table 3.1). This means that, where glaciers are underlain by fine-grained sediments, the flow of meltwater through the bed is retarded. As a result, porewater pressures build up and may approach the overburden pressure. Thus effective pressures fall, reducing sediment strength and increasing the likelihood of deformation (Boulton, 1987; Boulton and Hindmarsh, 1987).

4.4.2.7 THERMAL PROCESSES

In till flow models to date, no account has been made of thermal processes, such as the freezing of porewater, ice infiltration at the upper boundary of the deforming layer, and the effects of ice melting on the flow and pressure of porewater. One may expect glacier bed materials to encompass a complete spectrum between unfrozen sediment and debris-rich ice, through various ice–water–sediment mixtures, which will certainly exhibit a wide range of flow behaviour. The study of straining ice–water–sediment mixtures is still in its infancy, but preliminary results (Iverson, 1993) suggest that the processes of deformation will be very complex indeed, and that flow laws will be very difficult to establish (see Section 4.3.1).

4.4.2.8 SPATIAL VARIATIONS IN BED STRENGTH

Observations on some modern glaciers and ice streams indicate that the strength of glacier bed materials varies from place to place, consisting of 'sticky patches' within otherwise low-strength beds. High-strength regions will tend to support a disproportionate share of the drag at the bed, thus exerting an important control on glacier velocity. Large boulders in the till beneath Breidamerkurjökull appear to behave in this way by 'bridging' through the upper deforming layer and lodging in the stiffer, lower till (Fig. 4.15; Benn and Evans, 1996). Similarly, Kamb (1991) and Alley (1993) have suggested that where Antarctic ice streams are underlain by deforming till, maximum ice velocities may be controlled by sticky patches at the bed rather than the strain response of the till itself. Kamb has argued that in the absence of such patches, the ice streams would be susceptible to catastrophic acceleration, owing to the highly non-linear till rheology and positive feedback

Fig. 4.15 Boulders lodged at the base of Breidamerkur-jökull, Iceland. (Photo: D.I. Benn)

mechanisms involving the production of meltwater. Therefore, attempts to model glacier velocities using till 'flow laws' may be subject to error if variations in bed strength are ignored. Lodged boulders and other stiff regions in glacier beds are considered to influence the formation of flutings, drumlins and other streamlined forms; study of such features therefore has the potential to reveal conditions below former glaciers and ice streams, and yield insights into ice flow mechanisms (see Chapter 11).

4.4.2.9 DECOUPLING

Recent measurements of till deformation beneath Storglaciären, Sweden, by Iverson *et al.* (1995) indicate that, while till strain rates increase with water pressure up to a point, at very high basal water pressures the strain rates *decrease*, although glacier velocities remain high. This behaviour was attributed to *decoupling* of the glacier from its bed owing to the development of a distributed water system at the ice–till interface. This system may take the form of an anastomosing canal network, as postulated by Walder and Fowler (1994) (Section 3.4.6). The presence of pressurized water at the glacier sole dramatically reduces the shear stress applied to the underlying till, causing till deformation rates to fall. Glacier velocity is not reduced, however, but occurs by enhanced basal sliding, which is favoured by a well-lubricated bed (Section 4.5.3). This model suggests that, for glaciers resting on a bed of deformable till, the following sequence of events should occur as porewater pressure increases:

1. At low porewater pressures, high sediment strength will discourage deformation. Ice will tend to infiltrate the bed at the glacier sole, coupling the glacier to the bed (Boulton and Hindmarsh, 1987). Glacier motion will occur by a combination of processes, including 'ploughing' of debris held in

the basal ice through the upper layers of the till (Brown *et al.*, 1987), limited brittle shear of the till (Boulton and Hindmarsh, 1987; Benn and Evans, 1996), and sliding over large clasts (Benn, 1994a).

2. At higher porewater pressures, till strength is reduced, encouraging deformation. At some critical strain rate, dilation and a transition to pervasive, ductile flow will occur (Alley, 1989). Rising porewater pressures reduce the tendency of basal ice to infiltrate the bed, thus reducing coupling (Boulton and Hindmarsh, 1987).

3. At very high porewater pressures, the development of a distributed drainage system at the ice–till interface causes extensive decoupling and reduction of basal deformation rates (Iverson *et al.*, 1995). Fast flow is achieved by efficient sliding.

Further field experiments are necessary to test the validity of this appealing model for fast-flowing glaciers and ice streams such as Columbia Glacier, Alaska, and Ice Stream B, Antarctica.

4.5 SLIDING

Sliding is the name given to slip between a glacier and its bed. Everyday experience tells us that ice can be extremely slippery, and that the friction between wet ice and a smooth surface (such as the sole of a shoe) is very low indeed. Similarly, if glacier beds consisted of wet ice resting on perfectly smooth surfaces, sliding would be so efficient that glacier movement would be catastrophically fast. Thus, the most important question concerning glacier sliding is not 'how does sliding occur?' but 'how is sliding controlled?'. The fact that glaciers do not run away with themselves is due to a number of interdependent factors that increase the drag between the ice and the bed. The most important factors controlling drag (and hence rate of basal sliding) are (a) adhesion due to freezing of ice to the bed, (b) bed roughness, (c) the quantity and distribution of water at the bed, and (d) the amount of rock debris held in the base of the ice. Drag due to bed roughness is called *form drag*, and that due to rock–rock friction is *frictional drag*.

4.5.1 Adhesion

Effective sliding requires that the basal ice is at or close to the pressure melting point (Section 2.5.1), because the adhesive strength of frozen glacier beds is very high, even where the bed is smooth (this is why it is so difficult to scrape the ice from a car windscreen on a cold morning). Additionally, low temperatures severely reduce the ability of glacier ice to flow around obstacles on the bed, owing to the temperature dependence of creep rates (Section 4.3.1). For glacier beds below the pressure melting point, therefore, the bed can often support larger shear stresses than the ice itself, and movement more readily occurs by ice creep above the bed. Indeed, until recently it was thought that sliding of frozen glacier beds could not occur. However, recent laboratory experiments and theoretical work (Shreve, 1984; Fowler, 1986) predicted that the normal processes of sliding should occur at subfreezing temperatures, albeit at very low speeds. This prediction was strikingly confirmed by measurements made by Echelmeyer and Wang (1987) below Urumqi Glacier 1, China, which demonstrated sliding rates of 0.01 mm day^{-1} (less than 4 mm yr^{-1}) at $-5°C$. Earlier studies of the movement of cold-based glaciers (e.g. Goldthwait, 1960; Holdsworth and Bull, 1970) found no evidence for any sliding, although this may reflect the precision of the methods of measurement.

As noted in Section 2.5.4, frozen and melting conditions may occur as discontinuous mosaics over glacier beds, consisting of frozen patches within mainly thawed beds or thawed patches on mainly frozen beds. In such mosaics, frozen conditions will occur where normal pressures are lower, and the melting point higher than average, such as on the lee side of obstacles (Robin, 1976). Where the bed is frozen, the drag will be higher than average, owing to the adhesive strength of the contact between ice and bed, and frozen patches may therefore exert an important influence on the overall drag at the base of the glacier. Thus, frozen regions can be regarded as another form of 'sticky patch', as discussed in Section 4.4.2.8.

4.5.2 Glacier sliding over rough beds

Real glacier beds are not smooth, but exhibit irregularities, or roughness, at different scales. In order to slide over an irregular bed, ice must be transferred past or round obstacles by some mechanism. The resistance to such transfer around obstacles is known as *form drag* and is a very important factor limiting the rate of glacier sliding. Two mechanisms for transferring ice past obstacles are widely recognized: regelation sliding and enhanced creep.

4.5.2.1 REGELATION SLIDING

The word 'regelation' means 'to freeze again'. In the present context, it refers to processes which allow glacier ice to slide over rough beds by melting on the upglacier side of obstacles and refreezing on the downglacier side. Regelation sliding occurs because most resistance to glacier movement is provided by the upstream sides of obstacles, resulting in locally

high pressures and a consequent lowering of the pressure melting point. This encourages the melting of ice immediately upglacier of obstacles. The resulting meltwater migrates to the low-pressure downglacier side of the obstacle, where it refreezes because the pressure melting point is higher (Fig. 4.16). The ice therefore bypasses obstacles by temporarily turning to water and back again.

Standard theory indicates that the regelation process will be most effective when the latent heat released by freezing on the downstream side of obstacles can be advected through the obstacle to assist in melting on the upstream side (Weertman, 1957b, 1964). Consideration of the heat balance indicates that regelation should be ineffective around obstacles longer than 1 m or so (Kamb, 1970). The flow of meltwater associated with regelation sliding is usually thought to occur in a thin film between the ice and its bed (Weertman, 1964, 1979; Hallet, 1976a, 1979b; Section 3.4.3), although Lliboutry (1993) has argued that, in temperate ice, most water actually flows through a capillary network within a basal ice layer some 20 cm thick. According to Lliboutry, pressure melting at the upstream sides of obstacles occurs at grain boundaries *within* the ice as well as at the ice–bed interface. The associated refreezing is thought to occur in a thin layer (~3.5 cm thick) of the ice above the low-pressure downstream sides of obstacles. This view implies that regelation sliding is closely related to ice creep processes, and indicates that the relations between driving stresses, bed roughness and glacier sliding rates may be very complex indeed. Further complications of the regelation process arise as the result of chemical interactions between the glacier and its bed. Basal ice can entrain and transport ions dissolved from the bed (Section 3.7.1; Hallet, 1976a, b; Souchez and Lorrain, 1978), thus lowering the melting temperature and influencing the regelation process.

The importance of the regelation process in glacier sliding has been recognized for many years on theoretical grounds (e.g. Deeley and Parr, 1914; Clarke, 1987a), and has been confirmed many times by direct

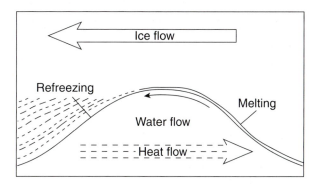

Fig. 4.16 Regelation sliding mechanism

observation in subglacial cavities (e.g. Kamb and LaChapelle, 1964; Hallet *et al.*, 1978; Hubbard and Sharp, 1993). The role of the regelation process in forming basal debris-rich ice, and the characteristics of regelation ice, are described in Section 5.4.

4.5.2.2 ENHANCED CREEP

A second mechanism which contributes to glacier sliding is enhanced creep of the ice in the vicinity of obstacles. It will be remembered that the strain rate of ice varies approximately with the third power of the shear stress (equation (4.5)). Therefore, stress concentrations around the upstream sides of obstacles result in locally high strain rates, causing the ice to accelerate around and (to a lesser extent) over the bump. The basal ice thus continually modifies its shape to allow continued sliding. As noted above, recent theoretical work by Lliboutry (1993) suggests that, in temperate ice, the mechanism of enhanced creep may involve an element of regelation, indicating that ice flow mechanisms may be strongly interrelated. A form of enhanced creep is also important at the bases of cold-based glaciers, allowing the ice to circumvent basal obstacles without sliding. It should be remembered, however, that creep rates are much lower for cold ice, owing to the temperature dependence of the flow law (Glen, 1955).

Enhanced creep is most effective around large obstacles, because larger areas of the basal ice experience enhanced shear stresses and higher strain rates. Because regelation is most effective around small obstacles, it follows that enhanced creep and regelation will vary in relative importance according to the size of obstacles on the bed. For small obstacles, regelation will be the dominant process, whereas enhanced creep will dominate for larger ones. Thus there should be some intermediate *controlling obstacle size* for which neither mechanism is efficient, and which represents maximum resistance to sliding (Fig. 4.17). Theoretical analyses and observations suggest that this critical obstacle size is in the region of 0.05–0.5 m (Kamb and LaChapelle, 1964; Nye, 1969a, 1970; Kamb, 1970). This concept of a controlling obstacle size emphasizes the influence of bed roughness on glacier sliding, and will be of importance when we consider some theories of subglacial erosion and deposition in Chapter 5.

4.5.3 The effect of water pressure on sliding

As we have seen, the presence of liquid water at a glacier bed is a fundamental requirement for effective sliding, because of the lack of adhesion at a thawed bed and the role played by water in the regelation process. Water also contributes to the

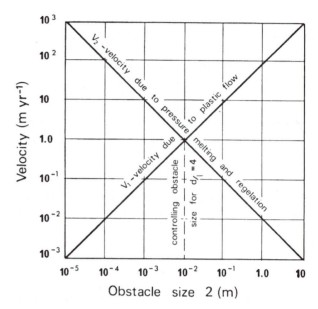

Fig. 4.17 Graph showing the relative effectiveness of regelation and creep for different obstacle sizes. (From Boulton, 1975. Reproduced by permission of Seal House Press)

effectiveness of sliding in other ways, and the distribution and pressure of water at glacier beds are now recognized as the most important factors in regulating short-term velocity fluctuations and glacier surge cycles. Several studies have demonstrated a strong association between periods of increased glacier motion and elevated basal water pressure, as determined from the level of water in boreholes that have made a connection with the subglacial drainage system (e.g. Müller and Iken, 1973; Iken *et al.*, 1983; Kamb and Engelhardt, 1987; Hooke *et al.*, 1989; Meier *et al.*, 1995; Jansson, 1995). Accordingly, basal water pressure forms an important component of modern sliding theory (Lliboutry, 1968, 1987; Fowler, 1987a, b; Kamb, 1987; Schweizer and Iken, 1992; Willis, 1995).

For glaciers flowing over rock beds, the presence of a water film (Section 3.4.3) may enhance sliding by submerging minor obstacles (of the order of millimetres), effectively creating a smoother bed and reducing the importance of the regelation process as a limiting factor in sliding (Fig. 4.18). At a larger scale, bed roughness is reduced by the presence of pressurized water in cavities. In Sections 3.2.4.1 and 3.4.4 we saw that water-filled cavities will form at the lee side of obstacles on the bed when the water pressure (P_w) exceeds the local normal pressure exerted by the ice (p_i) on the bed. The extent of cavity formation is directly linked to water pressure, because as P_w increases it exceeds p_i over larger and larger areas of the bed. Water-filled cavities can occur as isolated features or form part of a linked-cavity network connected by narrow orifices. Pressure

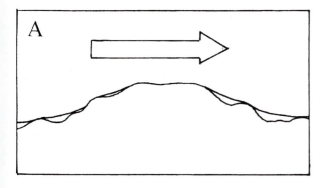

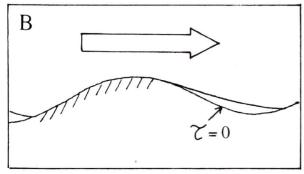

$\tau = 0$

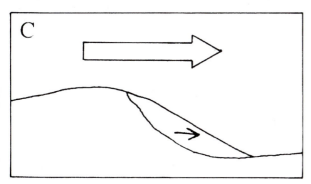

Fig. 4.18 Ways in which water at the bed can influence sliding velocity. (A) Submergence of small bed roughness elements; (B) increasing the local stress in areas of ice–bed contact, thus encouraging enhanced plastic deformation; (C) the 'hydraulic jack' mechanism, in which pressurized water exerts a force against upglacier-facing ice

fluctuations in a linked-cavity network, and the extent of cavity formation, generally reflect variations in the supply of meltwater from the surface (Hooke *et al.*, 1989; Willis *et al.*, 1990; Willis, 1995; Kamb *et al.*, 1995). As well as reducing the effective bed roughness, the presence of water-filled cavities affects sliding behaviour in other ways (Fig. 4.18). The most important is the effect of cavity formation on the distribution of stress at the glacier bed. Because water cannot support a shear stress, the basal shear stress is zero above water-filled cavities. This means that the shear stresses exerted by the glacier must be supported by those parts of the bed remaining in contact

(i.e. the crests and stoss sides of bumps). The increase in stress at these locations increases the efficiency of regelation by reducing the pressure melting point, and enhances plastic deformation, thus increasing the sliding rate (Iken, 1981; Willis, 1995). Another way in which water-filled cavities are thought to influence sliding rates is through the downglacier-directed component of water pressure. Pressurized water in cavities can exert a downglacier traction force on upglacier-facing ice surfaces in cavities, increasing the downglacier driving forces and increasing sliding (Röthlisberger and Iken, 1981). This is known as the 'hydraulic jack' mechanism.

There is growing empirical evidence that the growth of water-filled cavities causes increases in sliding velocity on some glaciers. Short-lived periods of increased velocity on Unteraargletscher and Findelengletscher (Switzerland), Variegated Glacier (Alaska), Storglaciären (Sweden) and Midtdalsbreen (Norway) have been associated with uplift of the glacier surface (Iken *et al.*, 1983; Hooke *et al.*, 1983; Iken and Bindschadler, 1986; Kamb and Engelhardt, 1987; Jansson, 1995; Willis, 1995). These uplift events are thought to be due to 'hydraulic jacking' by pressurized water in lee-side cavities, which lifts the glacier off the bed (Fig. 4.18). Not all glacier uplift events, however, are thought to be due to the hydraulic jack mechanism, and some uplift can be attributed to variations in strain rate associated with velocity variations elsewhere on the glacier and changes in local longitudinal stresses (Balise and Raymond, 1985; Hooke *et al.*, 1989).

The water pressure at which cavities begin to form is dependent on bed roughness, because the size and shape of obstacles govern spatial variations in the normal pressure exerted by the ice, and thus determine the extent and magnitude of low-pressure zones in lee-side locations. The threshold water pressure for cavity formation is known as the *separation pressure*, and for a wavy bed can be defined as

$$P_s = p_i - (\lambda\tau/\pi a) \qquad (4.10)$$

where P_s is separation pressure, p_i is the normal pressure exerted by the ice, λ is bump wavelength, τ is basal shear stress, π has the value 3.1427, and a is bump amplitude.

This relation shows that the separation pressure is high for short-wavelength, high-amplitude bumps, and low for long-wavelength, low-amplitude bumps (Iken, 1981; Schweizer and Iken, 1992; Willis, 1995). Thus, sliding rates will be more sensitive to rising water pressures on smooth beds than on rougher ones.

Enhanced sliding caused by increasing cavity water pressures generates additional meltwater at the bed, because of the increased generation of viscous heat and pressure melting. At a certain point, this creates a positive feedback process by significantly increasing the water pressure in cavities, thus further increasing the sliding rate, which in turn leads to the generation of more meltwater (Schweizer and Iken, 1992). This will lead to unstable, accelerating sliding of the glacier, where velocity increases catastrophically. This will continue until the subglacial drainage system 'switches' to a more efficient channelized system which will evacuate the excess water, reduce the areal extent and pressure of water at the bed, and reduce sliding rates to sustainable levels (see Section 3.4.4). Schweizer and Iken (1992) calculated the *critical pressure*, at which sliding becomes unstable, to be given by

$$P_c = p_i - (\lambda \tau / 2\pi a) \qquad (4.11)$$

This value is halfway between the ice overburden pressure and the separation pressure:

$$P_c = 0.5(p_i + P_s) \qquad (4.12)$$

The transition from stable to unstable sliding as rising water pressure crosses the critical pressure is thought to account for the rapidly accelerating ice flow during glacier surges and some short-term velocity increases on surging and non-surging glaciers (Section 4.8; Kamb, 1987; Schweizer and Iken, 1992; Willis, 1995). According to Schweizer and Iken (1992), the relationship between basal water pressure and sliding rates may take the form:

$$u_b \approx (\tau_b / P_c - P_s)^n \quad \text{for } P_w \leq P_s \qquad (4.13a)$$

$$u_b \approx (\tau_b / P_c - P_w)^n \quad \text{for } P_w > P_s \qquad (4.13b)$$

where u_b is basal sliding velocity, τ_b is basal shear stress, P_w is water pressure, P_c is the critical water pressure, P_s is separation pressure and n is a constant.

The relationships described by these equations are illustrated in Fig. 4.19. Curve A shows the effect of increasing water pressure past P_s and P_c in turn. When P_w is less than P_s, sliding velocity is independent of water pressure and is constant for any given shear stress. When P_w is greater than P_s, but less than P_c, sliding increases non-linearly with water pressure. Above P_c, unstable, accelerating sliding occurs, and velocity increases at an infinite rate. Curve B illustrates the case where the separation pressure and the critical pressure have the same value, i.e. $P_c = P_s$. In this case, sliding motion is uniform until P_s is reached, when the glacier switches instantaneously to a state of unstable sliding.

Although equation (4.13) provides important insights into the sliding behaviour of glaciers during water pressure fluctuations, it is difficult to apply in practice because it assumes an idealized form of the glacier bed. Moreover, for real glaciers the detailed form of the bed is not known, and it is therefore not possible to determine the values of P_s and P_c. For practical purposes, therefore, it is often preferable to use simpler empirical sliding relations, which describe direct relationships between basal water pressure and sliding velocity (Budd *et al.*, 1979). Such relationships have been worked out for a number of glaciers, and generally take the form

$$u_b = \tau^p N^{-q} \qquad (4.14)$$

where u_b is basal sliding velocity, τ is basal shear stress, N = effective normal pressure ($p_i - P_w$), and p and q are empirically determined constants.

This relation indicates that sliding rate increases with basal shear stress and is inversely proportional to the effective pressure (i.e. sliding increases with P_w). Note that the exponential relationship between

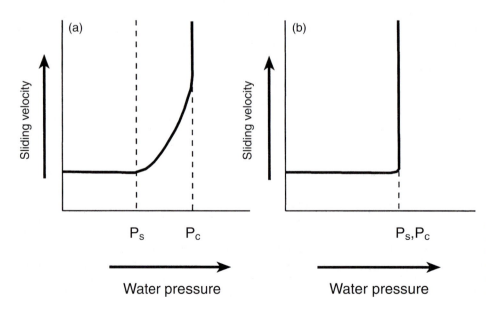

Fig. 4.19 Relationship between sliding velocity and basal water pressure, as modelled by Schweizer and Iken. (Modified from Schweizer and Iken, 1992)

sliding and effective pressure indicates small changes in sliding with water pressure for low values of P_w, and large changes in sliding with water pressure for high P_w, and therefore yields results approximately similar to those of equation (4.13) (Fig. 4.20). On the basis of laboratory experiments, Budd *et al.* (1979) proposed that values of $p = 3$ and $q = 1$ yield the best results. Very similar values were obtained by Bindschadler (1982) using data from Variegated Glacier for summer 1973, although winter velocities for this glacier over the period 1973–81 are best described by equation (4.14) when $p = 5$ and $q = 1$ (Raymond and Harrison, 1987).

Jansson (1995) used velocity data from Storglaciären to derive the empirical velocity relation

$$u_s = 30N^{-0.40} \qquad (4.15)$$

where u_s is surface velocity, and N is effective normal pressure. He also reanalysed data obtained by Iken and Bindschadler (1986) from Findelengletscher, and found that

$$u_s = 371N^{-0.40} \qquad (4.16)$$

The similarity of the exponent in the two relationships led Jansson (1995) to conclude that the sliding

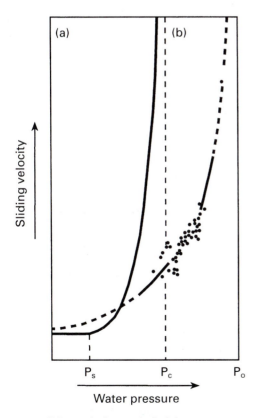

Fig. 4.20 Sliding velocity v. subglacial water pressure showing the discrepancy between theory (a) and observation (b). (Redrawn from Schweizer and Iken, 1992)

mechanism is similar on the two glaciers. It should be clear, however, that empirical power laws such as equations (4.15) and (4.16) are of limited usefulness as general descriptions of glacier sliding. The complexity of the sliding process means that all-purpose sliding laws remain elusive.

4.5.4 Friction and sliding

So far we have considered sliding only for the case of clean ice. In nature, however, basal ice often contains rock debris, and if this debris is in contact with the bed, the resulting *frictional drag* will exert a strong influence on basal sliding. In general terms, the larger the drag, the lower the sliding velocity. Early workers recognized that the drag below particles held in basal ice will be proportional to the force pressing them against the bed (Gilbert, 1910; McCall, 1960). The physics of sliding with friction, however, is very complex and there is still no complete theory that describes how frictional drag and sliding rates vary with applied stresses. Current understanding relies heavily on theoretical considerations and a small number of important laboratory experiments (e.g. Hallet, 1981; Iverson, 1990, 1995; Schweizer and Iken, 1992). Three basic models of subglacial friction have been proposed, known respectively as the *Coulomb*, *Hallet* and *sandpaper* friction models (Schweizer and Iken, 1992). The implications of these three models for processes of subglacial erosion, transport and deposition are discussed in Section 5.2.2.

4.5.4.1 COULOMB FRICTION MODEL

The Coulomb friction model is based on the Coulomb equation (equations (4.6) and (4.8)), and assumes that the friction between rock particles in basal ice and a rigid bed is proportional to the normal pressure N pressing the surfaces together. This model was first developed for subglacial situations in a series of influential papers by G.S. Boulton (1974, 1975, 1979), and it has since been employed in some studies of subglacial sediment deformation (Boulton and Hindmarsh, 1987; Alley, 1989; Section 4.4.2). The Coulomb model predicts that basal friction will increase with ice thickness and is inversely proportional to basal water pressure:

$$\tau_f = (p_i - P_w).\tan \phi_i \qquad (4.17)$$

where τ_f is basal friction, p_i is the ice overburden pressure, P_w is the basal water pressure; and ϕ_i is the angle of internal friction, a value dependent on the nature of the interacting rock surfaces.

For realistic values of the friction coefficient, however, sliding should be possible only at very high basal water pressures (close to ice overburden) or on a steep surface slope (Schweizer and Iken, 1992). As a general model of sliding, the Coulomb model is unrealistic, because it is applicable only for friction between rigid bodies, whereas basal ice is deformable and can mould itself to the contours of the bedrock. It may, however, be a reasonable approximation for (a) the base of rigid slabs of debris-rich ice (Schweizer and Iken, 1992), and (b) subglacial deforming layers, where interstitial ice is absent (Boulton and Hindmarsh, 1987; Section 4.4.2).

4.5.4.2 HALLET FRICTION MODEL

Bernard Hallet (1979a, 1981) developed a friction model based on the idea that ice will deform completely around subglacial particles, and that contact forces will be independent of ice thickness. Hallet reasoned that the contact force between a particle and the bed will be the sum of two components: the buoyant weight of the particle, and drag force resulting from ice flow towards the bed. The buoyant weight is the weight of the rock particle minus the weight of the same volume of ice, and is therefore proportional to the difference between the densities of ice and rock. Ice flow towards the bed exerts a drag force on the particle, thus pressing it against the bed. The faster the flow towards the bed, the higher the drag force, and the higher the resulting contact force. Ice flow towards the bed can result from a combination of three sets of processes (Drewry, 1986): (a) melting due to geothermal heat and sliding friction; (b) melting due to regelation sliding; and (c) vertical straining of the ice. Therefore, high friction between particles and the bed will occur below large, heavy particles and/or where basal melting rates are high and ice is straining rapidly towards the bed. As already mentioned, this model assumes that frictional forces are independent of ice thickness and basal water pressure. Hallet (1979a) had to make several assumptions to be able to calculate the drag force exerted around particles by flowing ice, one of which was that the particles are spaced far apart. When this condition is satisfied, however, there is good agreement between the stresses predicted by this model and those measured in controlled laboratory experiments (Iverson, 1990).

4.5.4.3 SANDPAPER FRICTION MODEL

The sandpaper friction model was introduced by Schweizer and Iken (1992) for the case of debris-rich basal ice, where neighbouring particles are very close or in contact. Ice can no longer flow around the particles, and is simply the glue holding the mixture together. Taken as a whole, the debris-rich ice is deformable, and can adapt to the contours of the bed over which it flows. The basal ice layer is therefore everywhere in contact with the bed, but part of the

contact may consist of water-filled cavities between particles. The drag force at the base of the debris-rich ice is similar to that calculated by the Coulomb model, but is a function of water pressure *and* the area of the bed occupied by cavities between particles in the basal ice layer:

$$\tau_f = (p_i - sP_w).\tan \phi \qquad (4.18)$$

where s is the proportion of the bed occupied by cavities (varying between 0 and 1). The factor s is important because water-filled cavities tend to submerge small obstacles (asperities), smoothing the bed and reducing drag (Section 4.5.3). Equation (4.18) predicts values of basal drag somewhat lower that those derived from the Coulomb model, and yields more realistic values of basal sliding.

The 'sandpaper' friction model is thought to be the most appropriate where basal debris concentrations are greater than about 50 per cent by volume, and the Hallet model where concentrations are less than this (Schweizer and Iken, 1992; Willis, 1995). Whichever model is used, frictional drag has the effect of reducing the sliding velocity to values lower than those for clean ice at the same basal shear stress. This is not to say that glaciers with high basal debris concentrations will flow more slowly than clean-based glaciers. Rather, it means that where frictional drag is high, higher driving stresses are required to achieve given sliding velocities than in the case of low friction. Therefore, for two glaciers with equal balance velocities, the glacier with the higher basal drag would have to be thicker or steeper in order to create the required stresses to discharge the mass flux.

Schweizer and Iken (1992) incorporate basal friction into their sliding relation (equation (4.13)) by replacing the basal shear stress τ_b with the *effective value* τ'_b, given by

$$\tau'_b = \tau_b - \tau_f \qquad (4.19)$$

where τ_f is the frictional drag in equation (4.18). This means that when basal friction is high, the effective shear stress will be low, leading to low calculated ice velocities. Conversely, low basal friction will lead to high values of τ'_b and high sliding velocities. In the case of clean ice, τ_f disappears, and sliding velocities are simply those given in equation (4.13).

4.6 SPATIAL PATTERNS OF GLACIER MOTION

4.6.1 Variations between glaciers

Worldwide, glaciers flow at a very wide range of velocities, depending on the mechanisms of flow and the balance between driving and resisting forces. Cold-based glaciers, which are everywhere frozen to their beds, can flow only by internal deformation of the ice and the upper part of the bed, because sliding is negligible at subfreezing temperatures (Fig. 4.21; Boulton, 1979; Echelmeyer and Wang, 1987). The inverse relationship between ice creep rates and temperature (equation (4.5)) means that only low velocities are possible for cold-based glaciers. For example, the Meserve Glacier in the Dry Valleys of Antarctica has maximum velocities of only 2 m yr^{-1} (Chinn, 1988). Generally speaking, glaciers in cold regions have small balance velocities (Section 4.1), so that the low velocities associated with cold ice are sufficient to discharge the annual ice accumulation. However, where balance velocities are high, such as where ice from large accumulation areas is funnelled into narrow outlets, the ice will thicken, thus raising the basal temperature by reducing the conduction of geothermal heat towards the surface (Section 2.5.2). This may be sufficient to raise the bed of the glacier to the pressure melting point, thus allowing efficient sliding and increasing velocities to levels sufficient to discharge the annual balance. Steady-state conditions will occur when the physical processes of glacier motion are in equilibrium with the balance velocity

(Clarke, 1987b). This may not be possible, however, and the glacier may oscillate between rates of motion higher and lower than the balance velocity in a periodic cycle (Fowler and Johnson, 1995). This type of oscillatory behaviour is discussed in Section 4.8.

Where glacier beds are at the pressure melting point, efficient basal sliding or subsole deformation can occur, leading to very high glacier velocities in some cases. The world's fastest glacier is considered to be Jacobshavns Isbrae, a tidewater outlet of the Greenland ice sheet, which has a maximum observed flow rate equivalent to 8360 m yr^{-1} at midsummer near the calving front (Lingle et al., 1981; Echelmeyer and Harrison, 1990). Velocities in excess of 7000 m yr^{-1} have also been measured near the calving terminus of Glaciar San Rafael, an outlet of the North Patagonian ice cap, Chile (Fig. 4.22; Warren, 1993). The observed velocities on both Jacobshavns Isbrae and Glaciar San Rafael appear to be approximately equal to their balance velocities (Fig. 4.2; Clarke, 1987b). In the former case, high balance velocities are due to the funnelling of ice from a very large catchment, whereas in the latter they result from a combination of topographic funnelling and very high mass turnover rates (Yamada, 1988; Aniya, 1988; Kondo and Yamada, 1988; Warren, 1993).

Fig. 4.21 Cold-based glaciers in the Dry Valleys, Antarctica, among the most slowly moving glaciers in the world. (Photo provided by Richard Williams, USGS)

Fig. 4.22 San Rafael Glacier, North Patagonian icefield, Chile: one of the world's fastest glaciers, reflecting high mass turnover and efficient basal sliding. (Photo: Charles Warren)

Subglacial till deformation contributes towards high surface velocities on some glaciers and ice streams. Deforming layers occur beneath Ice Stream B, West Antarctica, and Columbia Glacier, Alaska, which achieve maximum velocities in excess of 800 m yr^{-1} and 4 km yr^{-1}, respectively (Alley et al., 1986, 1987a, 1987b; Whillans et al., 1987; Krimmel and Vaughn, 1987; Humphrey et al., 1993; Whillans and van der Veen, 1993). In both cases, however, the till appears to be too weak to support the calculated basal shear stresses, and therefore till deformation does not appear to be the only process controlling glacier motion. The velocity appears to be limited by the presence of rock knobs protruding through the till layer or some other form of 'sticky spots' on the bed (Kamb, 1991; Humphrey et al., 1993). It is not known whether basal motion is due to till deformation alone or some combination of deformation and sliding (cf. Iverson et al., 1995). Variations in till strain rates and the depth of the deforming layer mean that a wide range of glacier velocities can result from subglacial till deformation. Low till strain rates beneath Trapridge Glacier, Yukon, contribute to maximum surface velocities of only 27 m yr^{-1} (Clarke et al., 1984a; Clarke and Blake, 1991; Blake et al., 1992). Subsole deformation also occurs beneath Storglaciären, Sweden, contributing towards a surface velocity of around 14 m yr^{-1} (Iverson et al., 1995). These observations are important because they demonstrate that evidence for deforming layers at the beds of former glaciers and ice sheets is not necessarily evidence for fast glacier flow, as has been assumed by some authors (e.g. Clark, 1994).

Glaciers in most parts of the world have surface velocities intermediate between the very low values typical of thin, cold-based glaciers and the very high values associated with some ice streams and tidewater glaciers. For wet-based valley glaciers, surface velocities are typically a few tens of metres per year, attributable to a combination of basal sliding and ice creep (e.g. Raymond, 1971; Meier, 1974; Hambrey and Milnes, 1977). The wide range of glacier velocities has led to a classification of glaciers into *normal glaciers*, with velocities typically of the order of 10^1–10^2 m yr^{-1}, and *fast glaciers*, with velocities in the order of 10^2–10^3 m yr^{-1} (Budd, 1975; Clarke, 1987b). For normal glaciers, basal shear stresses are relatively high (40–120 kPa; Paterson, 1994), associated with relatively high bed strength. Velocity increases with the shear stress, as indicated by the sliding relations proposed by Schweizer and Iken (1992) (equation (4.13)). In contrast, for fast outlet glaciers and ice streams, basal shear stresses are generally very low (20 kPa for Ice Stream B), associated with low basal drag. The low bed strength of fast glaciers appears always to be due to the presence of pressurized water at the bed in some form of distributed drainage system such as linked cavities, porewater flow or braided canals (Clarke, 1987b; Kamb, 1991; Humphrey et al., 1993; Meier et al., 1995). It is not yet clear whether normal and fast glaciers form two distinct populations or are end-members of a continuum of glacier types. Certainly, the surging behaviour of some glaciers indicates two distinct modes of glacier flow, depending on the configuration and pressure of the basal water (Section 4.8). Variations in bed morphology and the strain response of subglacial rocks and sediments, however, may mean that for glaciers as a whole there is no clear dividing line between 'normal' and 'fast' populations.

4.6.2 Variations within glaciers

Where ice creep and subsole deformation contribute to glacier flow, velocities vary systematically with depth. Velocities are greatest at the surface, and are the cumulative effect of the strains in the underlying ice and sediment. In the case of ice creep, strain rates are largest near the base of the glacier, where the shear stress is at a maximum, so that the increase in velocity with height will be most rapid near the bed. Closer to the surface, velocity increases more slowly with height, because strain rates are small in shallower layers of the ice (Fig. 4.3). Paterson (1994) gives the velocity at the surface of a glacier as

$$u_s = u_b + [(2A/n + 1).(\rho_i.g.\sin a)^n h^{n+1}] \quad (4.20)$$

where u_s is surface velocity, u_b is the velocity at the base of the ice due to sliding and subsole deformation, A is a flow law parameter (equation (4.5)), n is the flow law exponent (equation (4.5)), ρ_i is ice density, g is gravitational acceleration, and h is the ice thickness.

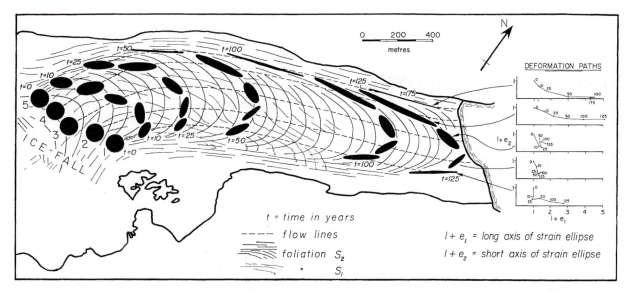

Fig. 4.23 Cumulative strain for Griesgletscher, Switzerland. Patterns of cumulative strain are indicated by the progressive elongation and rotation of ellipses derived from circles drawn at arbitrary points. Note that the highest cumulative strains are at the glacier margins. (From Hambrey et al., 1980. Reproduced by permission of the International Glaciological Society)

This relation is derived by integrating the strain rates given by the flow law for ice (equation (4.5)) with respect to height.

For subsole deformation, strain rates are greatest near the top of the deforming layer, where the effective normal stress is at a minimum. Velocities will therefore increase with height at an increasing rate (Fig. 4.12; Boulton and Hindmarsh, 1987; Alley, 1989).

On valley glaciers, drag at the valley sides will retard the motion of the glacier margins, so that velocities will be greatest near the centre-line. This pattern is illustrated in Fig. 4.23, which shows the displacement and cumulative strain of a set of originally circular regions on the surface of Griesgletscher, Switzerland (Hambrey et al., 1980). The cumulative strain (shown by the stretching of the strain ellipses) is greatest near the lateral margins, but velocities are greatest near the centre-line. The progressive bending of surface features such as ogive banding (Section 6.2.3) also reflects this pattern of surface velocities on valley glaciers. Taken together, vertical and horizontal changes in velocity in valley glaciers produce systematic velocity variations over glacier cross-sections. Figure 4.24 shows the distribution of longitudinal velocities in a cross-section of the Athabasca Glacier, Canada, measured using borehole data (Raymond, 1971). The physical mechanisms underlying such patterns are now well understood, and theoretical patterns match the observed pattern with a high degree of accuracy (Harbor, 1992a).

Velocities also vary along glacier flowlines. As noted in Section 4.1, mass balance discharges

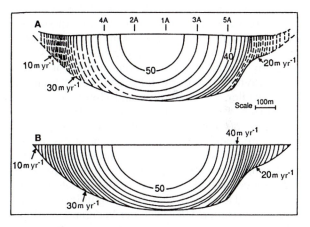

Fig. 4.24 Velocity in a transverse cross-section of Athabasca Glacier, Canada: (A) observed; (B) theoretical. (From Harbor, 1992a. Reproduced by permission of the International Glaciological Society)

increase from the source of a glacier to the equilibrium line, then decrease towards the snout. Thus, for a glacier with constant cross-section, mass balance velocities will follow the same pattern. However, longitudinal variations in glacier morphology will superimpose additional velocity variations on this simple pattern. Constrictions in a valley will tend to be associated with higher velocities, so that the discharge of ice can be maintained through a smaller cross-sectional area. Higher velocities are also often associated with glacier steepenings, such as icefalls, which tend to accelerate and thin the ice. Conversely, widenings of a valley profile tend to be associated with reduced glacier velocities.

4.7 TEMPORAL VARIATIONS IN GLACIER MOTION

The velocity of glaciers can vary on different timescales, ranging from hours to several years. Some of these variations are cyclic, such as daily or annual increases and decreases in velocity, but others are irregular. In this section we examine velocity variations on timescales of one year or less, and velocity changes due to shifts in glacier mass balance. Cyclic velocity variations on timescales of several years or more (glacier surge cycles) are discussed in Section 4.8.

4.7.1 Velocity variations and basal drainage

The single most important control on temporal variations in glacier motion is the distribution and pressure of water at the bed (Willis, 1995). Elevated water pressures serve to increase the rate of strain in subglacial deforming layers, owing to the reduction in frictional strength, and to increase basal sliding rates by increasing the spatial extent of water-filled cavities (Sections 4.4.2 and 4.5.3; Iken and Bindschadler, 1986; Boulton and Hindmarsh, 1987; Schweizer and Iken, 1992; Willis, 1995). Water pressures do not directly affect ice creep rates, which are mainly controlled by shear stress and ice temperature. Creep rates may, however, be indirectly influenced by water pressures, because water pressure-controlled variations in sliding rates on one part of a glacier will alter the longitudinal stresses elsewhere (Hutter, 1983; van der Veen and Whillans, 1989). Because most velocity fluctuations are dependent on variations in subsole deformation and basal sliding, cold-based glaciers which are frozen to their beds tend to flow at constant rates.

The supply of water due to basal melting does not vary significantly on short timescales on most glaciers, so fluctuations in basal water pressure and glacier motion are mainly due to variations in the supply of water from the surface and reorganizations ('switches') of the basal drainage system. Variations in water from the surface can follow daily or seasonal ablation cycles, or result from transient weather events such as prolonged clear-sky conditions or periods of rainfall. Meier *et al.* (1995) found that, on Columbia Glacier, Alaska, ablation rates and melt-water production also increased during periods when föhn-type winds supplied sensible heat to the glacier surface. Correlations between periods of high ablation rates and exceptionally high velocities have been observed on several glaciers, including White Glacier, Arctic Canada (Müller and Iken, 1973), Unteraargletscher, Switzerland (Iken *et al.*, 1983), Storglaciären, Sweden (Hooke *et al.*, 1983, 1989; Jansson, 1995), and Columbia Glacier (Meier *et al.*, 1995). Typically, short-term peaks in ablation and velocity are superimposed on a cycle of higher than average velocities during the ablation season (Fig. 4.25).

The supply of water to the bed can influence glacier velocities by increasing the amount of water temporarily stored at the bed in distributed basal drainage systems such as linked-cavity networks, braided canal systems and till porewater (Hodge, 1974; Iken *et al.*, 1983; Iken and Bindschadler, 1986; Kamb, 1987; Schweizer and Iken, 1992). In such systems, there is a positive correlation between discharge and water pressure, so that increased supply to the system results in elevated pressures (Section 3.4.4). Once water drains from the system faster than it is recharged, water pressures fall and sliding velocities are reduced. Evidence in support of this model

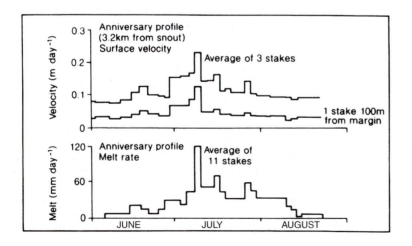

Fig. 4.25 Short-term velocity fluctuations on White Glacier, Axel Heiberg Island. The close association between velocity and melt rate suggests that surface meltwater reaching the bed exerts an important control on sliding rate. (From Willis, 1995, after Müller and Iken, 1973)

has been provided by Iken and Bindschadler (1986), who observed that periods of enhanced water pressure and sliding velocity on Findelengletscher, Switzerland, are accompanied by temporary uplift of the glacier surface, interpreted as the 'jacking' effect of pressurized water in cavities. The glacier surface subsides as the basal water pressure and sliding velocity return to lower levels, indicating closure of the cavities and removal of water from storage (Fig. 4.26). The effects of water storage and release have also been observed on other glaciers. For example, Naruse *et al.* (1992) found that on Glaciar Soler, an outlet of the Northern Patagonian icefield, velocity peaks occur about 7.5 hours before peak discharges of the glacier melt stream, and that velocities fall rapidly as water is drained from the bed.

Variations in discharge in Röthlisberger and Nye channels are not thought to influence glacier motion, because water is confined to a small proportion of the bed and transit times are small. The efficiency of channel systems in draining water that reaches the bed from the surface means that water pressures over much of the bed, and glacier sliding rates, can remain low even when discharges are high. Therefore, if the

drainage system 'switches' from a distributed to a channelized network during the course of a melt season, major drops in subglacial water pressures and sliding velocities can be expected. Such switches in the drainage system of Midtdalsbreen, Norway, do appear to be associated with a transition to lower overall sliding velocities (Willis *et al.*, 1990; Willis, 1995). Drainage reorganizations are also thought to play a major role in glacier surge cycles (Section 4.8).

An additional cause of velocity variations has been observed on Columbia Glacier, a large tidewater glacier in southern Alaska. Near the glacier margin, the velocity fluctuates in harmony with the tidal cycle, with the velocities being about 10 per cent higher at low tide than at high tide (Fig. 4.27; Walters and Dunlap, 1987; Krimmel and Vaughn, 1987). This is thought to be due to the greater back-pressure exerted by the deeper sea-water at high tide, presenting greater resistance to the forces driving glacier motion. Periods of increased velocity lasting for up to two days are superimposed on the tidal cycle, reflecting periods of rainfall and increased water supply to the bed (Krimmel and Vaughn, 1987). The tidal influence on velocity declines upglacier, and in the middle and upper reaches of the glacier, fluctuations in velocity are entirely due to variations in the water supply from the surface (Meier *et al.*, 1995; Kamb *et al.*, 1995).

Some glaciers do not exhibit significant velocity variations. Blue Glacier, Washington State, USA, shows very little increase in speed at the onset of the melt season, even though surface meltwater enters the glacier via moulins and is transported to the snout along the bed (Echelmeyer and Harrison, 1990). Engelhardt *et al.* (1978) found that basal sliding rates are very low during the summer, and the motion of the glacier appears to be almost entirely due to internal creep of the ice. During the summer months, the sliding velocity of Grinell Glacier, Montana, USA, remains remarkably constant at rates of 33 mm day^{-1} (12 m yr^{-1}) (R.S. Anderson *et al.*, 1982). The lack of significant variations in the basal motion of Blue and Grinell Glaciers may mean that basal water pressure never attains the separation pressure for their particular bed configurations, and that sliding rates remain independent of water pressure (Fig. 4.19). Harder to understand is the lack of seasonal velocity variations on Jacobshavns Isbrae, a fast-flowing outlet of the Greenland ice sheet, which terminates in a small floating ice shelf. Ice velocities on the ice shelf are very high (> 6000 m yr^{-1}), and gradually decline upglacier, with values of *c.* 3000 m yr^{-1} about 15 km from the grounding line and *c.* 1000 m yr^{-1} some 20 km further inland (Echelmeyer and Harrison, 1990). In all cases, seasonal variations in velocity are very small, despite the fact that significant

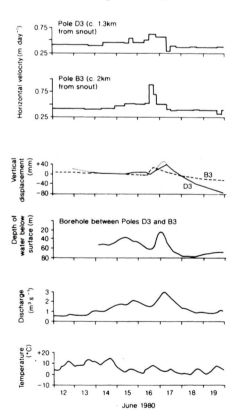

Fig. 4.26 Short-term velocity fluctuations on Findelen-gletscher, Switzerland. Note vertical displacement (uplift) and increased water level in borehole (increased water pressure) around the time of the velocity peak. (From Willis, 1995, after Iken and Bindschadler, 1986)

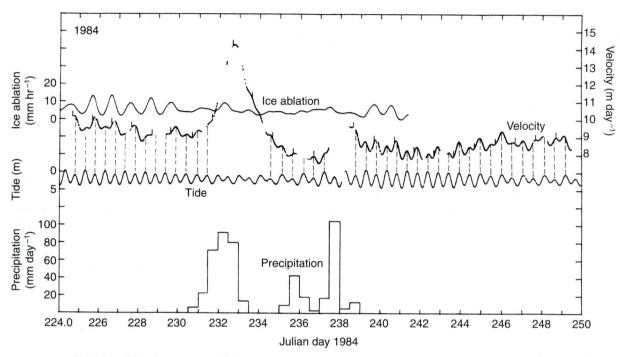

Fig. 4.27 Time series of velocity on Columbia Glacier, a tidewater glacier in Alaska. Note the close relationships between semi-diurnal velocity fluctuations and tidal cycles (shown inverted), and the velocity peaks during precipitation events. The breaks in the velocity record resulted from poor visibility. (From Krimmel and Vaughn, 1987. Reproduced by permission of the American Geophysical Union)

amounts of surface meltwater are produced during the ablation season and enter the glacier via moulins. Echelmeyer and Harrison (1990) suggested that this meltwater may not penetrate to the bed, but travels mainly in englacial conduits at higher levels in the ice. Motion of the glacier would therefore be unaffected by fluctuations in meltwater supply from the surface. Recent geophysical studies have shown that Jacobshavns Isbrae occupies a deep trough, and that basal shear stresses are in the region of 200–300 kPa (Clarke and Echelmeyer, 1996). Such high shear stresses imply that internal deformation of the ice, and not basal sliding, is the dominant mode of ice motion on this glacier.

4.7.2 Velocity variations and mass balance

Fluctuations in basal water pressure influence velocity by altering the resisting forces at the bed of a glacier. Velocity variations can also occur, over rather longer timescales, as a result of changes in the driving forces or shear stress. As we saw in Section 4.2, shear stresses in a glacier are primarily controlled by the thickness and surface slope of the ice (equation (4.4)), so shear stresses will increase or decrease if there is a change in the surface topography of the glacier, resulting from changes in mass balance. For

example, higher than average snowfall in the accumulation area will thicken the glacier and increase the surface gradient, leading to higher shear stresses. This results in increased velocities, which serve to discharge the excess mass towards the ablation area. Conversely, excessive ablation will thin the glacier, reducing shear stresses and velocities. Reduced velocities associated with glacier thinning have been observed by Hastenrath (1989) on Lewis Glacier, Mount Kenya, over the period 1974–86.

Increases in mass in glacier accumulation areas are transmitted downglacier in waves of increased velocity known as *kinematic waves*. The theory of kinematic waves on glaciers owes much to the work of Nye (1958, 1960, 1963, 1965b), and has been summarized by Paterson (1994). In essence, a zone of increased velocities will travel downglacier at a rate about three to five times faster than the ice velocity itself, depending on the relative contribution of ice creep and sliding to the motion of the glacier. Individual packets of ice travelling downglacier will first accelerate then decelerate as the wave passes. Kinematic waves are difficult to distinguish on real, non-surging glaciers, because the surface profile is constantly adjusting to changes in mass balance on several timescales. However, bulges of increased thickness, travelling downglacier faster than the ice itself, have been observed on several glaciers, including the Mer de Glace and Glacier des Bossons in the Mont Blanc massif, France, and Nisqually Glacier, USA (Lliboutry, 1958; Finsterwalder, 1959; Richardson, 1973).

4.8 SURGING GLACIERS

4.8.1 Overview

Some of the world's glaciers exhibit major periodic fluctuations in velocity over timescales ranging from a few years to several centuries, swinging between phases of rapid and slow flow. Such *surging glaciers* have attracted a great deal of attention because, like many instances of unusual or pathological behaviour, they offer a very instructive perspective on 'normality', and so can increase our understanding of the links between flow mechanisms, basal conditions and climatic inputs for glaciers in general.

Surging glaciers undergo characteristic changes in morphology and behaviour during a surge cycle (Fig. 4.28; Meier and Post, 1969; Raymond, 1987; Sharp,

1988a). The phase of rapid motion is termed the *surge* or *active phase,* and during it ice is rapidly transferred from the upper parts of the glacier (the *reservoir area*) towards the snout. The surge phase results in a rapid and dramatic advance of the glacier front, which, combined with thinning of the upper reservoir, produces a reduction in the overall gradient of the glacier. The period of slow flow between surges is the *quiescent phase*, during which ice builds up in the reservoir area and the snout stagnates and ablates *in situ*. The increase of mass in the upper part of the glacier and mass loss near the snout result in an increase in the overall glacier gradient, which continues until the next surge cycle is initiated. Maximum velocities during the active phase are typically 10 or more times the velocity during the quiescent phase (Raymond, 1987). For each surging glacier, the active and quiescent phases tend to be of relatively constant length, resulting in a uniformly periodic cycle, although there are large variations in cycle length between glaciers and between regions (Dowdeswell *et al.*, 1991). In the Svalbard archipelago, for example, the active phase of surging glaciers typically lasts for 4–10 years compared with only

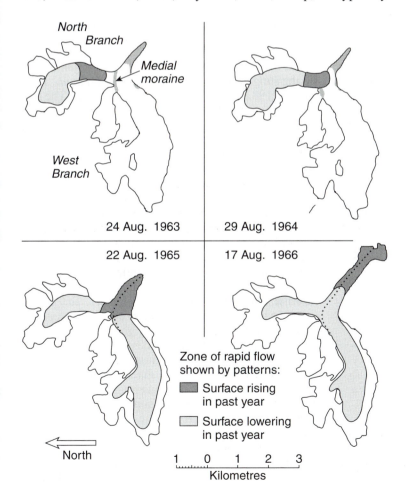

Fig. 4.28 Maps of Tyeen Glacier, Alaska, showing shifting zones of ice thickening and thinning associated with a surge cycle. (From Meier and Post, 1969. Reproduced by permission of the *Canadian Journal of Earth Sciences*)

1–3 years for surging glaciers in north-west North America, Iceland and the Pamirs (Fig. 4.29). However, although the active phase is longer in Svalbard, maximum ice velocities are comparatively low, ranging between 1.3 and 16 m day^{-1}, compared with velocities of 50 m day^{-1} measured on Variegated Glacier in Alaska. Thus, during a surge in Svalbard, mass is transferred over a longer period but at a lower rate than in North America. The quiescent phase is also relatively long for Svalbard glaciers (50–500 years) compared with other areas (20–40 years). The length of the surge cycle is unrelated to variations in glacier size (Post, 1969; Clarke *et al.*, 1986; Dowdeswell *et al.*, 1991).

Surging glaciers are not evenly distributed around the world's glaciated regions, but tend to cluster in relatively few areas. High concentrations of surging glaciers are found in Alaska, Yukon and British Columbia in north-western North America, Svalbard and Iceland in the North Atlantic region, and the Pamirs in western Asia (Meier and Post, 1969; Post, 1969; Thorarinsson, 1969; Dowdeswell *et al.*, 1991). Surging glaciers have also been recorded in Greenland, on Ellesmere Island and Axel Heiberg Island in Arctic Canada, in the Caucasus, Tien Shan and Karakoram Mountains in Asia, and in the Andes (Hattersley-Smith, 1969; Müller, 1969; Jeffries, 1984; Espizua, 1986; Weidick, 1988; Dowdeswell *et al.*, 1991). Both temperate and subpolar glaciers can exhibit surging behaviour, but polar glaciers, which are everywhere frozen to their beds, have never been

known to surge. There is also evidence that, within regions, the overall geometry of surge-type glaciers is significantly different from that of non-surging glaciers. Clarke *et al.* (1986) and Clarke (1991) found that, for glaciers in the Yukon, surge-type glaciers tend to be longer, wider and less steep than normal glaciers, and that glacier length shows the strongest individual correlation with surge tendency. Glacier length is also significantly correlated with surge behaviour for Svalbard glaciers (Hamilton and Dowdeswell, 1996). In addition, there is a tendency for surging glaciers to occur on certain rock types. In Svalbard, surging glaciers are most likely to occur on sedimentary rocks (Hamilton and Dowdeswell, 1996), whereas in the Alaska range many surging glaciers are found within the Denali Fault system but are absent from the hard, granitic rocks of the Coast Mountains (Post, 1969). The causal relationships between glacier length, bedrock geology and tendency to surge are unclear. However, glacier length and geology can be expected to influence the balance velocity and bed configuration respectively, two key variables that determine long-term glacier behaviour.

Glacier surges are not triggered by climatic fluctuations (Meier and Post, 1969), but instead result from oscillations in the internal workings of the glacier. It is clear that the rapid and large changes in glacier velocity that occur during surge cycles result from variations in basal sliding, which apparently reflect reorganizations in the subglacial drainage system (Kamb *et al.*, 1985; Kamb, 1987; Fowler, 1987b; Sharp, 1988a). The trigger for such reorganizations and the physics of glacier motion during the active and quiescent phases of surging glaciers have provided a major focus for research, and are discussed in Section 4.8.2.

The concentration of surging glaciers in certain regions, and systematic inter-regional variations in the length and duration of surge cycles, suggest that there is some link between surging behaviour and climatic inputs. An important step towards understanding the nature of this link was made by Budd (1975) as the result of numerical modelling experiments. He argued that surging and non-surging glaciers exhibit contrasting relationships with the local climate. In normal and continuously fast glaciers, the annual mass balance can be discharged by typical ice flow velocities. However, in surging glaciers, the climatic mass throughput is too great to be discharged by slow flow alone, but too small to sustain fast flow over the long term. Therefore, these glaciers build up slowly until fast flow is triggered, whereupon they drain out rapidly and exhaust the ice supply, and fast flow is halted. Once slow flow is re-established, ice builds up once more in the reservoir area, until the cycle begins again. The velocity of surging glaciers is thus constantly out of equilibrium with climate, either overshooting or undershooting the mass balance

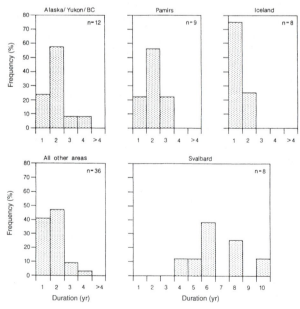

Fig. 4.29 Comparison of the duration of surge cycles in different parts of the world. (From Dowdeswell *et al.*, 1991. Reproduced by permission of the International Glaciological Society)

velocity as the glacier oscillates between fast and slow modes of flow (Fig. 4.2; Bindschadler, 1982; Sharp, 1988a). Budd (1975) went on to suggest that surging and non-surging glaciers may pass from one type to the other at times of changing climate, given sufficiently large changes in the mass balance.

According to Budd's model, the basal conditions and flow mechanisms of surging glaciers oscillate between two states, neither of which is sustainable in the long term. This idea has been taken further by Fowler (1987b), who showed that, for certain values of mass balance and bed roughness, the geometry, basal drainage system and velocity of modelled glaciers will undergo cyclic oscillations resembling surge cycles. The work of Budd and Fowler indicates that surging behaviour is linked with particular combinations of climate and basal conditions, although glaciological theory is not sufficiently advanced to allow exact values to be defined for real-world glaciers.

Surges leave distinctive structural imprints on glaciers. One of the most obvious is the presence of *looped medial moraines*, which are striking teardrop-shaped loops of debris on the surface of many surging glaciers (Gripp, 1929; Meier and Post, 1969; Croot, 1988c). Looped moraines record cyclic differences in velocity between a trunk glacier and its tributaries (Fig. 4.30). If tributary glaciers surge when the main trunk glacier is quiescent, they advance across the valley, forming loops which are carried downglacier by the next surge of the trunk. The intense shear characteristic of the margins of surging glaciers can form longitudinal *foliation* in the ice (Pfeffer, 1992; Section 6.2.4), whereas compression at the surge front causes folding, thrust faulting, crevassing, and thickening of basal debris sequences (Fig. 4.31; Sharp *et al.*, 1988; Lawson *et al.*, 1994; Section 5.4.4). Lawson (1996) used aerial photographs spanning the period 1947–83 to study the structural evolution of Variegated Glacier, Alaska, over two complete surge cycles. She found that surging creates distinctive crevasse patterns on the glacier surface. In the upper parts of the glacier, which were unaffected by surges, transverse crevasses record predominantly extending flow. In contrast, the middle zone of the glacier has a complex crevasse pattern, including longitudinal crevasses formed by compressive flow in advance of the surge velocity peak and transverse crevasses formed by extending flow behind the velocity peak. Finally, the lower zone is characterized by longitudinal crevasses. This zone lay downglacier of the final position of the velocity peak, and experienced only compressive flow.

4.8.2 Observations and mechanisms

Detailed time-series observations have been made on only a few surge-type glaciers, notably the Medvezhiy

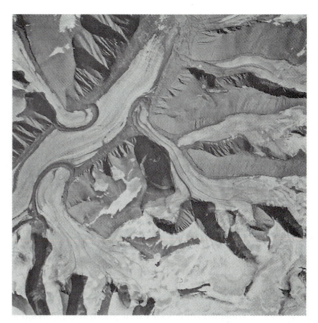

(a)

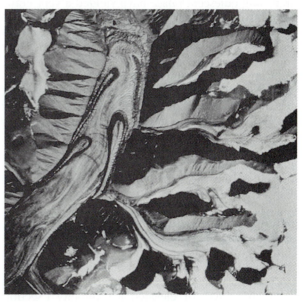

(b)

Fig. 4.30 Abrahamsbreen, Svalbard: (a) in 1969 before a surge; and (b) in 1990 after a surge. Note the formation of looped medial moraines on the trunk glacier by the transport of tributary lobes. (Aerial photos provided by Norsk Polarinstitutt)

Glacier in the Pamirs (Dolgushin and Osipova, 1975, 1978), Variegated, West Fork and Black Rapids glaciers, Alaska (Kamb *et al.*, 1985; Harrison *et al.*, 1994; Heinrichs *et al.*, 1996), and Trapridge Glacier, Yukon (Clarke *et al.*, 1984a; Clarke and Blake, 1991). Less detailed observations have also been

Fig. 4.31 The heavily crevassed snout of Eyjabakkajökull, Iceland, during the 1972 surge. (Photo: C.M. Clapperton)

made on several other glaciers (e.g. Meier and Post, 1969; Echelmeyer *et al.*, 1987).

The most complete record is for Variegated Glacier, where measurements of changes in the geometry, velocity, basal water pressure and meltwater discharge during the 1982/83 surge and the latter half of the preceding quiescent phase have yielded a very comprehensive picture of the evolution of the glacier and its basal drainage system (Kamb *et al.*, 1985; Raymond, 1987). During the quiescent phase, the reservoir area thickened and the receiving area thinned, resulting in a steepening of the glacier. At this time, surface velocities progressively increased, with maximum summer values rising from *c.* 0.2 m day^{-1} in 1973 to *c.* 1 m day^{-1} in 1981. These increases were largely due to rising ice creep rates resulting from increasing shear stresses, but a small contribution was also made by increased sliding rates. Throughout this period, measured velocities were much less than calculated balance velocities (Bindschadler *et al.*, 1977; Raymond and Harrison, 1988). Each June and July between 1978 and 1981, the thickening reservoir area underwent four to six waves of accelerated motion similar to surges, but on a smaller scale and termed *minisurges* (Kamb and Engelhardt, 1987). The minisurges propagated downglacier as kinematic waves of elevated ice velocity (Section 4.7.2), with the wave peaks passing downglacier at 250–400 m hr^{-1}, or *c.* 1000 times the ice velocity. As each wave passed through a point on the glacier, ice velocities increased abruptly (over a few hours) then decayed more slowly, over one day or so. The velocity waves were accompanied by waves of increased basal water pressure, with similar abrupt rises and slow falls. The minisurges were similar to short-term episodes of accelerated motion observed on non-surging glaciers (Section 4.7.1), and were attributed to temporary increases in sliding velocity caused by the growth of water-filled cavities at the bed associated with a pulse of meltwater under high pressure.

There was no apparent meteorological cause for increased meltwater, and Kamb and Engelhardt (1987) concluded that the water had been released from storage in reservoirs (probably subglacial). Hydrological aspects of the meltwater pulses have been discussed in detail by Humphrey *et al.* (1986).

The glacier surged in two phases, the first beginning early in 1982 and terminating in July of that year, and the second beginning in the winter of 1982/83 and lasting until early summer (Kamb *et al.*, 1985). The second phase was the more extensive, and affected almost the whole of the glacier. Each phase took the form of a kinematic wave of enhanced velocities propagating downglacier, the leading edge of which was a dramatic ice bulge or *surge front* (Raymond *et al.*, 1987). Peak velocities were in excess of 50 m day^{-1}, and occurred on the lower part of the glacier a short distance upglacier from the advancing surge front (Fig. 4.32). Behind the zone of peak velocities, the glacier was subject to large tensile stresses and was extensively crevassed. With the passage of the surge waves, the ice surface in the reservoir area fell, leaving blocks of ice stranded on the valley walls, marking the former maximum thickness. During the surge phases, the velocity fluctuated on hourly, daily and multi-day timescales. On several occasions, velocity peaks were followed by abrupt *slowdowns*, the last of which marked the termination of the surge. The slowdown events coincided with drops in subglacial water pressure, indicating that they reflect hydraulic controls on the sliding rate. Dye-tracing experiments conducted during the surge phase indicate low meltwater transit velocities (*c.* 0.02 m s^{-1}) and temporary water storage at multiple sites below the glacier, suggestive of an inefficient, distributed subglacial drainage system, probably a linked-cavity network (Sections 3.4.4 and 4.5.3; Kamb *et al.*, 1985; Kamb, 1987). Surge termination was accompanied by the rapid release of large amounts of stored water and a marked increase in transit velocities to *c.* 0.7 m s^{-1}, indicating a rapid transition from a distributed subglacial drainage network to an efficient conduit system.

The course of the whole surge cycle is inferred to be strongly controlled by the storage, distribution and pressure of water at the bed. The initiation of both surge phases during the winter led Raymond (1987) to argue that rapid sliding was triggered by rising water pressures at the bed, resulting from the closure of the conduit system that drained the glacier during summer. Conduit closure occurred by ice creep at a threshold value of ice overburden pressure below the thickening reservoir area. This is thought to have trapped subglacial water in a distributed system, leading to extensive cavity formation and accelerated sliding (Section 4.5.3; Kamb, 1987). The

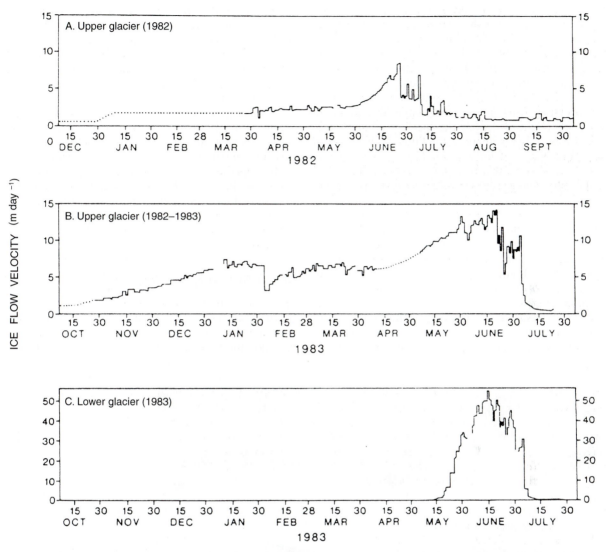

Fig. 4.32 Time series of velocity changes on the upper and lower Variegated Glacier during the 1982/83 surge. (From Kamb *et al.*, 1985. Reproduced by permission of the American Association for the Advancement of Science)

pressurized cavity system persisted until the following summer, when increased water discharges caused the re-establishment of efficient drainage, reducing subglacial water pressures and halting rapid sliding. The second surge phase repeated the pattern of the first, until rapid motion was again terminated by another release of stored water. Geometric changes in the glacier reservoir area prevented renewed surge activity the following winter, and the glacier entered the quiescent phase. The timing of surge initiation on other glaciers lends support to this model. The 1986/87 surge of Peters Glacier commenced in winter (Echelmeyer *et al.*, 1987), and the 1987/88 surge of West Fork Glacier began in late August, after the main Alaskan melt season (Harrison *et al.*, 1994).

Raymond's (1987) interpretation of the Variegated Glacier surge does not provide a general model of surge initiation and termination, however, because subpolar glaciers, which are partially frozen to their beds, also surge. No subpolar glacier has been monitored over a whole surge cycle, so that there are no data available on ice flow and drainage evolution during a surge. Nevertheless, extremely valuable observations have been made on Trapridge Glacier, which document the evolution of the glacier during its quiescent period (Clarke *et al.*, 1984a; Clarke and Blake, 1991). The margin of Trapridge Glacier is frozen to the bed, but further upglacier the basal ice is at the pressure melting point and resting on a bed of unfrozen, deformable sediment. Since 1969, when the glacier was first surveyed, an impressive wave-like bulge has formed on the glacier surface, forming an advancing lobe with a steepening frontal edge (Figs 4.33 and 4.34). The bulge first developed above the boundary between the cold-based ice at the margin and the wet-based ice upglacier, but has now overtaken the thermal boundary. Development of the bulge reflects a discontinuity in the flow of the glacier, which is slow in the frozen margin and more rapid where the bed is at the melting point. Much of the forward movement of the bulge is due to sub-sole deformation. Clarke *et al.* (1984a) and Clarke (1987c) proposed that a surge may be initiated by progressive changes to the water flow system over and through the porous subglacial sediment. According to this model, water is progressively retarded at the bed during the quiescent period, elevating subglacial porewater pressures and encouraging sediment dilatancy and rapid shear. The increased porosity of the dilatant till should lead to improved subglacial drainage, lowering strain rates and terminating the surge phase.

Several unanswered questions remain with this model. In particular, the *thermal dam* provided by the frozen marginal ice plays an unknown role in regulating surge behaviour. Surges may be initiated by the breakdown of the dam, or the glacier may overpass the dam by the development of extensive englacial thrust faults beneath the bulge. Evidence for the propagation of englacial thrusts has been documented by Clarke and Blake (1991), suggesting that the glacier may eventually surge by overriding its own margin. This mechanism may also have played a role in the 1973 surge of Medvezhiy Glacier (Dolgushin and Osipova, 1975). Trapridge Glacier is expected to surge within the next few years, and it is to be hoped that continued monitoring will allow the surge mechanism to be understood. Despite the great advances that have been made in recent years in our understanding of surges, a complete theory remains elusive, and awaits an expansion of the observational database.

4.8.3 Surging ice streams?

Some ice streams draining continental ice sheets have measured velocities that differ markedly from calculated balance velocities, suggesting the intriguing possibility that ice streams may undergo oscil-

Fig. 4.33 The wave-like bulge on Trapridge Glacier, Yukon. (Photo: Garry Clarke)

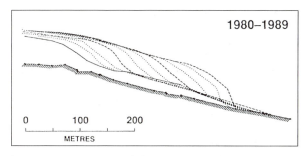

1980–1989

0   100   200
METRES

Fig. 4.34 Propagation of the wave-like bulge on Trapridge Glacier, 1980–89. The profile for 1980 is shown as a solid line; profiles for subsequent years are shown as dashed lines (1983, 1986, 1989) or dotted lines (all other years). (From Clarke and Blake, 1991. Reproduced by permission of the International Glaciological Society)

latory flow variations resembling surges. Ice Stream B in West Antarctica, for example, has measured velocities of *c.* 1000 m yr^{-1}, compared with a balance velocity of *c.* 400 m yr^{-1}, whereas neighbouring Ice Stream C has a balance velocity of *c.* 600 m yr^{-1} but measured velocities of only 5 m yr^{-1} (Fig. 4.2; Shabtaie and Bentley, 1987; Whillans *et al.*, 1987). Clearly, neither ice stream is in equilibrium with climate and catchment geometry: the discharge of Ice Stream B is unsustainable in the long term, whereas excess ice is accumulating in the catchment of Ice Stream C. Geophysical investigations of Ice Stream C reveal buried zones of marginal crevasses, similar to those found at the surface of active ice streams, suggesting that Ice Stream C was in a state of fast flow until about 250 years ago (Whillans *et al.*, 1987; Clarke, 1987b). It is possible that the two ice streams switch between fast and slow flow on a timescale of several

centuries. This switching may result from migration of the catchment boundaries, but this appears unlikely, and it is possible that the ice streams may be exhibiting some type of surge behaviour (Fastook, 1987; Clarke, 1987b).

It has also been suggested that parts of the large mid-latitude Pleistocene ice sheets may have surged. Recently, particular attention has been focused on the possible surging behaviour of the Hudson Strait ice stream of the last Laurentide ice sheet. Evidence from ocean cores indicates periodic advances of this ice stream and associated high discharges of icebergs into the Labrador Sea and North Atlantic (*Heinrich events*; Heinrich, 1988; Andrews and Tedesco, 1992; Bond *et al.*, 1992). Massive influxes of fresh water from melting icebergs may have exerted a strong influence on oceanic and atmospheric circulation in the North Atlantic region and further afield, possibly modulating patterns of global climate change (Bond *et al.*, 1992; Street-Perrot and Perrot, 1990). Instability of the ice stream has been attributed to rapid motion due to subsole deformation (Clark, 1994), and to periodic switching between periods of buildup and surge, called *binge–purge cycles* (MacAyeal, 1993a, b). Modelling studies indicate that the ice stream could have alternated between being cold-based during ice build-up and wet-based during purges, with the transition occuring at some critical ice thickness and gradient (MacAyeal, 1993b; Fowler and Johnson, 1995). While this model is compelling, recent evidence suggests that Heinrich events may be driven by climatic cycles unrelated to the internal dynamics of the Laurentide ice sheet (Bond *et al.*, 1993). Surging of the Laurentide and other Pleistocene ice sheets remains an intriguing, but unproven, possibility.

SUBGLACIAL PROCESSES

AN EPISODE IN THE HISTORY OF THE PASS OF LLANBERIS.

5.1 INTRODUCTION

The glacier bed is the least accessible, yet possibly the most important of all glacial environments. In Chapter 4, we saw how the glacier bed can be viewed as a shear zone, including the lower layers of the ice, the ice–bed interface, and deformable parts of the glacier substratum, and saw how the greater part of glacier motion occurs by deformation and slip in this zone. The subglacial shear zone is also of profound importance for glacial erosion, transport and deposition, for it is here that the glacier interacts directly with the landscape. Subglacial erosion encompasses the range of processes by which rock material passes between the immobile bed and the subglacial shear zone, and subglacial deposition encompasses the transfer of debris from the shear zone to the bed. Subglacial transport describes the mechanisms and effects of strain in the subglacial shear zone itself. Some of the most characteristic and spectacular glacial landforms are produced in the subglacial shear zone, including features such as ice-scoured rock surfaces, overdeepened troughs and drumlins.

A reasonable understanding of subglacial environments has emerged only very recently. Many years of painstaking study have been necessary to identify the main processes that occur at glacier beds, and to relate these processes to forms and sediments that survive in the geological record. Progress was slow because modern glacier beds, where processes of shear, erosion and deposition are currently active, are inaccessible or expensive and difficult to reach. Conversely, for the beds of now-vanished glaciers, where access is simple, the nature of the former subglacial processes may be very difficult to reconstruct (Fig. 5.1). In recent years, more rapid progress has been made by a creative interaction between observation of modern glacier beds, study of subglacial landforms and sediments, theoretical modelling, and experimentation in the field and laboratory. In particular, a great number of useful data have been collected from boreholes, tunnels, natural cavities and remote sensing techniques, which have produced a much more complete picture of subglacial conditions. These data have provided important input for glaciological theory, and have suggested new ways for interpreting glacial geological evidence. Many questions remain unanswered, however, and subglacial research remains an exciting and rapidly developing field of study.

In this chapter, we shall examine the processes of subglacial erosion and deposition, the ways in which sediment is transported in the subglacial shear zone, and the relationship between sediment dynamics and the formation and deformation of basal ice. One of the great recent breakthroughs in subglacial research has been the realization that many of the processes of erosion, strain and deposition at the glacier bed reflect the same fundamental set of forces (Boulton, 1974, 1975; Hallet, 1979a). Essentially, erosion, transport or deposition all depend on the balance between the imposed *shear stresses*, which tend to cause movement or failure of materials, and the *strength* of materials, which tends to resist movement or failure. Erosion and transport occur if the shear stresses are greater than the strength of bed materials, whereas material will be deposited or remain fixed to the bed if bed strength is greater than the applied stresses. This idea provides a powerful unifying framework for subglacial studies, and one which will be used throughout this chapter.

5.2 GLACIAL EROSION, TRANSPORT AND DEPOSITION

5.2.1 Basic concepts

In Section 4.4.2, we saw how the deformation of subglacial sediments depends on the balance between the applied shear stresses and the frictional and cohesive forces holding the sediment together. Exactly the same principle holds true for subglacial erosion and deposition, as illustrated in simple form in Fig. 5.2. Imagine a single particle embedded in the basal ice of a glacier, in contact with a rigid rock bed. The ice, which is sliding on the bed, exerts a shear stress on the particle. A frictional and cohesive force exists between the particle and the underlying rock, and tends to resist motion of the particle. In essence, if the shear stress is greater than the resisting force, then the particle will be dragged over the bed; and if the shear stress is less than the resisting force, then

Fig. 5.1 Subglacially eroded bedrock, Switzerland. Ancient glacier beds are easy to reach and study, but the processes responsible for their evolution can be understood only by studying modern glaciers, conducting laboratory experiments, and theoretical modelling. (Photo: N. Spedding)

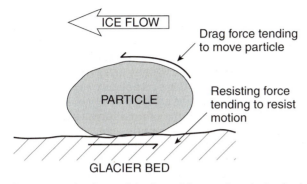

Fig. 5.2 A simple model of particles at the glacier bed, illustrating the balance of forces leading to subglacial erosion or deposition. If the drag force exerted by the ice on a particle exceeds the resisting forces, the particle will be set in motion; if the drag force is less than the resisting force, the particle will remain fixed to the bed or be deposited

the particle will be retarded against the bed and the ice will flow round or past the particle by some combination of plastic flow and regelation (Section 4.5.2). When shear stresses are high enough to set the particle in motion, *erosion* occurs: the particle has been detached from the bed and has entered glacial *transport*. The converse situation, where the particle comes to rest as a result of rising resisting force or falling shear stress, is *deposition*: the particle has left glacial transport and has become part of the substratum. These principles apply whether the 'particle' is an isolated rock fragment beneath a glacier, part of an assemblage of such particles, part of a subglacial deforming layer, part of a rock bed, such as a protuberance or a joint-bounded block, or even part of another particle. All components of the glacier bed and the subglacial shear zone can be seen as potentially movable particles which will undergo transport if the driving stresses can overcome the resisting forces, and will be deposited if they cannot. In the case of a rock bed, the frictional and cohesive resisting forces are generally much higher than for unconsolidated sediments, but they can still be analysed in much the same way.

5.2.2 Stress and frictional strength

5.2.2.1 INTRODUCTION

The basic model of erosion, deposition and transport illustrated in Fig. 5.2 is useful as an overall framework, but needs to be refined if it is to be of practical use in understanding subglacial processes. Most importantly, we need to be able to measure or calculate the stresses and resisting forces in any given situation.

Shear and *normal stresses* vary considerably over short distances on a rough glacier bed or around a particle. The magnitude of these stresses is very dependent on the shape of interacting surfaces, and pronounced *stress concentrations* develop around protuberances, crack tips and other sharp corners. Stress concentrations are areas where stresses are significantly higher than average, and play an important role in determining where breakage or deformation occurs. These spatial variations in stress can be expressed as *stress gradients*, and, other things being equal, failure and erosion are most likely to occur where stress gradients are high, because of the tendency for materials to move from areas where stresses are high to areas where they are low. Examples of stress concentrations and their influence on patterns of erosion are given in Section 5.3.

The *strength* of subglacial materials or contacts between particles also varies spatially and temporally. As described in Section 4.4.2, the strength of geological materials can be divided into *cohesion*, arising from chemical bonding and electrical forces, and *frictional strength*, arising from the interlocking of small protuberances (asperities) between surfaces. The cohesive strength of a particular material is usually regarded as a constant, although for unconsolidated sediments it may vary with the degree of packing (Clarke, 1987c). Frictional strength can be determined using the Coulomb equation (equation (4.6)), and is equal to the *effective normal force* pressing the material together multiplied by the *angle of internal friction*, a dimensionless index unique to each material. The frictional strength of subglacial materials is therefore highly variable, and can change dramatically as the stress conditions change. The cohesive strength and angle of internal friction for a range of geological materials are shown in Table 4.1.

Three models of subglacial friction, and their varying implications for basal sliding, were described in Section 4.5.4. These models – the *Coulomb, Hallet,* and *sandpaper* friction models (Schweizer and Iken, 1992) – also have differing implications for subglacial erosion, transport and deposition. Each is discussed in turn below.

5.2.2.2 COULOMB FRICTION MODEL

The Coulomb friction model assumes that the most important control on the strength of subglacial rock–rock contacts is the weight of the overlying ice, rock and water pressing the surfaces together, minus the water pressure at the contact. This relationship is described by the form of the Coulomb equation given in equations (4.6) and (4.8). The model was first employed in theoretical studies of subglacial erosion and deposition by G.S. Boulton (1974, 1975, 1979), who assumed that the interface between rock particles and the glacier bed is free of ice, and consists of *points of contact* and *cavities* (Fig. 5.3). Contact

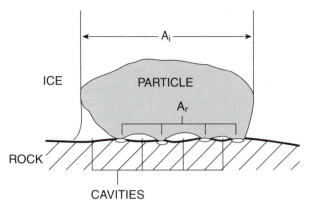

Fig. 5.3 Model of particle–bed contact according to Boulton (1974, 1975). The real area of contact of the particle (A_r) is much less than the apparent area of the particle (A_i). According to this model, air- or water-filled cavities can exist beneath the particle

between the particle and the bed occurs at protuberances or *asperities*, or via smaller particles trapped in the cavities. Owing to the presence of cavities beneath the particle, the real area of contact (A_r) is much smaller than the apparent area of contact (A_i). Water pressure in the cavities is assumed to vary independently of the ice overburden pressure. Boulton argued that, if the weight of the particle itself is ignored, then the frictional force between the particle and the bed (F_i) depends mainly on the weight of the overlying ice minus the water pressure in the cavities:

$$F_i = (\rho_i g h - P_w)A_i \tan \varphi \qquad (5.1)$$

where ρ_i is ice density, g is gravitational acceleration, h is ice thickness, P_w is water pressure beneath the particle, A_i is the apparent area of contact between the particle and the bed, and $\tan \varphi$ is the angle of internal friction.

If the water pressure is zero, the component in brackets is simply the weight of the ice column, whereas if the water pressure equals the ice overburden pressure, $F_i = 0$. Equation (5.1) gives the contact force over the whole apparent area of contact (A_i), but significant stress concentrations will occur at the real points of contact. The average normal stress at the contact points will be given by

$$\sigma_i = F_i/A_r \qquad (5.2)$$

where A_r is the real area of contact between the particle and the bed. The Coulomb model, therefore, predicts that very high contact forces can exist between rock surfaces beneath glaciers, particularly where ice is thick and subglacial water pressures are low (Boulton, 1974).

However, the model is probably valid only in special circumstances. Hallet (1979a), Drewry (1986)

and Iverson (1995) have argued that since ice behaves like a viscous fluid, it will deform completely around particles, pressing in on them with equal force in all directions. Therefore, the ice will buoy up the particle as well as press down on it, and the contact force will be completely independent of ice thickness. A useful analogy is a pebble resting at the bottom of a lake: the normal force exerted by the particle on the bed is simply due to the buoyant weight of the particle, and not the depth of the water. However, in some situations, Boulton's model does seem to be appropriate. Observations made in large cavities and tunnels show that small water- and air-filled cavities can exist around and beneath particles at the bed of glaciers (Fig. 5.4; Vivian and Bocquet, 1973; Boulton, 1974; Boulton *et al.*, 1979; Anderson *et al.*, 1982; Souchez and Lorrain, 1987). These small cavities form when ice in contact with the particle melts during the regelation sliding process and does not completely refreeze (Iverson, 1990, 1993; see Section 4.5.2.1). Water pressure in the cavities can be less than the ice overburden pressure if the water is able to drain away. Water drainage and the formation of low-pressure cavities are most likely to occur in the vicinity of an integrated subglacial drainage system, such as conduits or linked cavities (Sections 3.4.2 and 3.4.4). Schweizer and Iken (1992) have pointed out that the Coulomb friction model will probably also apply for rock fragments that have just been detached from bedrock, and have not yet been fully enveloped by the ice. Both this and the previous case are transient situations which occur before the ice surrounding a particle can find a new equilibrium.

The Coulomb friction model is also thought to give reasonable approximations of subglacial stresses for (a) the base of rigid slabs of debris-rich ice (Schweizer and Iken, 1992), and (b) in a subglacial deforming layer, where interstitial ice is absent (Boulton and Hindmarsh, 1987).

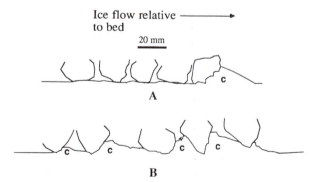

Fig. 5.4 Cavities around particles held between basal ice and the bed observed during laboratory experiments on the sliding of debris-rich ice. (From Iverson, 1993. Reprinted by permission of *Geology*)

5.2.2.3 HALLET FRICTION MODEL

In contrast to the Coulomb model, Hallet's (1979a, 1981) friction model regards basal friction as being independent of ice thickness and subglacial water pressure (Section 4.5.4.2). Instead, the contact force between a particle and the bed will be the sum of two components: (a) the buoyant weight of the particle; and (b) a drag force resulting from ice flow towards the bed. The *buoyant weight* is the weight of the rock particle minus the weight of the same volume of ice:

$$F_b = \tfrac{4}{3} \pi R_a^3 (\rho_r - \rho_i)g \qquad (5.3)$$

where F_b is buoyant weight, $\tfrac{4}{3} \pi R_a^3$ is the volume of the particle, ρ_r is density of the rock (*c.* 2700 kg m^{-3} for granite), ρ_i is the density of ice (*c.* 900 kg m^{-3}), and g is gravitational acceleration (9.81 m s^{-2}). Thus, for spherical granite particles,

$$\begin{aligned} F_b &= \tfrac{4}{3} \times 3.14 \times R_a^3 \times (2700 - 900) \times 9.81 \\ &= 7.396 \times 10^4 \, R_a^3 \end{aligned}$$

The cubic relationship between particle radius and the contact force means that the frictional strength due to buoyant weight is very much greater for large particles than for small ones.

Again assuming spherical particles, Hallet calculated the drag force between a particle and the bed due to ice flow towards the bed as

$$F_i = u_n \left[f 4\pi . \tan \varphi_i \, R^3 / (R_{ax}^2 + R_a^2) \right] \qquad (5.4)$$

where u_n is the ice velocity normal to the bed, f is a factor that modifies the drag force for the near-bed condition, φ_i is the angle of internal friction, R_a is the particle radius, and R_{ax} is the transition radius analogous to the controlling obstacle size in sliding theory (Section 4.5.2). Hallet suggests a representative value of 0.1 m.

Thus, high contact forces will arise when u_n is large; that is, when the ice is flowing rapidly towards the bed as a result of basal melting and vertical straining. Frictional forces will therefore tend to be highest for particles in contact with the upglacier sides of bumps on the bed, where u_n is large owing to pressure melting and enhanced creep.

The total frictional force F will be given by

$$F = F_b + F_i \qquad (5.5)$$

This contact force will all act across the real area of contact between the particle and the bed. If this is small, large stress concentrations will exist at the contact. High frictional drag will occur, therefore, below large, heavy particles, where basal melting rates are high and ice is straining rapidly towards the

bed, and where the real area of contact beneath a particle is small in relation to its radius.

This model and its modification by Shoemaker (1988) are applicable to the case of sparse basal debris, where ice flow around particles is not influenced by flow disturbances around its neighbours. For high debris concentrations, the sandpaper friction model is more appropriate (Schweizer and Iken, 1992).

5.2.2.4 SANDPAPER FRICTION MODEL

As described in Section 4.5.4.3, the sandpaper friction model is a modified form of the Coulomb model, in which basal drag is a function of the effective normal pressure and the area of the bed occupied by water-filled cavities between particles (equation (4.18)). The larger the area occupied by cavities, the smaller the average drag force on the bed. Considerable stress concentrations, however, will occur at the actual points of rock–rock contact between the debris-rich ice and the bed. The average magnitude of the normal stress at the contact areas is

$$\sigma_c = [(\rho_i gh - sP_w) \tan \varphi_i]/(1 - s) \qquad (5.6)$$

where $(1 - s)$ is the proportion of the bed occupied by rock–rock contacts. When $(1 - s)$ is small (i.e. when water-filled cavities cover much of the bed), the contact forces at particle–bed contacts will be high. The possibility that large stress concentrations can occur below debris-rich ice has potentially important implications for subglacial erosion and sedimentation.

5.2.2.5 SUMMARY

Schweizer and Iken (1992) argue that the Hallet model is the most applicable where basal debris is sparse, the sandpaper model is applicable where debris concentrations are high, and the Coulomb model applies in transient situations or where basal debris is rigid or ice-free. It is worth noting, however, that all three models are approximations of real conditions, and the use of three separate models to encompass variation in a single environment is probably unrealistic. In the future, it may be possible to formulate a single, unified model of subglacial friction. Such a model is likely to be complex, and would need to include particle size and shape, debris concentration, bed morphology, basal melting rates, water pressure and ice thickness. In general, however, the following points can be made regarding the friction below particles at the ice–bed interface:

1. Particle–bed friction increases with particle size.
2. Friction increases with basal melt rates, which cause the glacier to flow towards the bed, exerting drag on the particle.

3. Friction is greatest against the upglacier sides of bumps, where the ice is flowing towards the bed (Hallet friction) and where ice overburden pressures are highest (Coulomb and 'sandpaper' friction).

4. High friction is encouraged by the presence of low-pressure cavities beside or below particles, which serve to increase the effective overburden pressure below the particle.

5. If other factors are constant, friction will increase with debris concentration.

6. Friction is also affected by particle rolling, as a rolling particle will exert less drag on the glacier bed than one that is sliding. (An everyday example of this is what happens under the wheels of a car.)

In combination, these factors determine the friction between particles or between particles and the glacier bed. In turn, friction influences bed strength, so that variations in the factors controlling friction will determine the location of glacial erosion and deposition.

5.2.3 Melting and freezing

In Section 5.2.2 we saw how the frictional forces between rock particles, and the strength of subglacial materials, depend on particle size, debris concentration and whether rock particles are enveloped by ice, water or a mixture of the two. It is therefore useful to think of subglacial materials as three-component systems consisting of rock, ice and water in varying proportions, and material strength as a function of the particular rock–ice–water mixture. The proportions of ice and water will change during melting and freezing, and the proportion of rock will change by various processes such as the loss of meltwater from debris-rich ice, the formation of ice lenses, the addition of rock particles from the bed, and the removal of particles by meltwater and deposition on to the bed. Debris concentration can also be affected by flow patterns in the basal ice.

As noted in Section 2.5, melting and freezing can result from changes in temperature or changes in pressure. Heat for melting can be supplied from geothermal sources, sliding friction, flowing water or the atmosphere above the glacier, whereas heat can be lost to flowing water, the atmosphere, cold ice or a cold substratum. Pressure increases can cause melting by lowering the local melting point of ice to the ambient temperature. Conversely, reductions in pressure may elevate the local melting point above the ambient temperature, causing freezing. The relationship between pressure and temperature is important, because it means that melting and freezing can occur simultaneously in different parts of a glacier bed in

response to pressure variations. The effect of pressure on melting/freezing temperature as a glacier flows over bumps and other obstructions has important implications for glacial erosion (Robin, 1976), and is discussed in detail in Section 5.4.1. Melting and freezing can also occur simultaneously within straining debris-rich ice, as ice flows around individual particles by regelation (Iverson, 1993).

We now go on to describe the detailed processes of erosion, sediment transport and deposition in the subglacial environment.

5.3 THE EROSION OF HARD ROCK BEDS

Traditionally, subglacial erosion of hard rock beds is subdivided into two distinct processes: *abrasion,* or the grinding of fine-grained material, and *quarrying,* or the failure of larger blocks (Sugden and John, 1976; Drewry, 1986). The two processes have much in common, however, and both involve three basic stages: (a) *rock failure,* in which fragments are loosened from the bed; (b) *evacuation,* in which fragments are removed from their original position; and (c) *transport,* in which fragments are entrained by ice, water or in a subglacial deforming layer (Sugden and John, 1976; Röthlisberger and Iken, 1981). These stages can be recognized regardless of scale: they apply at the level of silt-sized particles abraded during the scoring of striations, or at the level of large blocks plucked from the bed. Evacuation and transport can also occur in the absence of failure if pre-existing loosened material exists at the bed. In this section, we examine the processes of rock failure and evacuation at different scales. Section 5.6 examines transport and erosion processes for deformable beds.

5.3.1 Mechanisms of rock failure

Rock failure can occur by breakage along pre-existing weaknesses or by the development of new fracture surfaces. Many rocks are seamed by weaknesses on many different scales, from large, readily visible joints and faults down to microscopic cracks and cleavage planes in mineral grains (Fig. 5.5). The size and density of such weaknesses are highly variable, and are major factors determining the strength of different rocks. For example, shale splits readily along closely spaced bedding planes, whereas joints in granite tend to be far apart. In some rocks, weaknesses can be continuous enough to completely isolate volumes of rock, such as joint-bounded blocks. When this is the case, rock failure is simply a matter of widening the existing fractures, a process known as *discontinuous rock-mass failure* (Addison, 1981).

(a)

(b)

Fig. 5.5 Joints on former glacier beds: (a) deep and widely spaced joints in abraded granite, Makinson Inlet, Ellesmere Island; (b) shallow and densely spaced joints on a limestone whaleback, Spitsbergen. (Photographs by D.J.A. Evans)

More usually, weaknesses are not continuous at first, but are extended as a result of imposed stresses until they join up, leading to failure.

The mechanisms of *crack growth* (or *crack propagation*) have been intensively studied by engineers seeking an understanding of how and why structures, such as bridges, skyscrapers and aircraft, fail in certain circumstances. A very readable engineer's account of crack growth and its implications is provided by Gordon (1976), and detailed studies of the mechanisms of rock fracture can be found in a volume edited by Atkinson (1987). Cracks grow as a result of *tensile stresses*, or stresses that tend to pull the walls of the crack apart. The presence of the crack causes pronounced stress concentrations around the crack tip, increasing the likelihood of crack growth and rock failure (Fig. 5.6). For two particles in contact, or a particle in contact with the glacier bed, tensile stresses are developed close to the contact where the relatively high normal stresses associated with the contact give way to the relatively low normal stresses in the surrounding area. In other

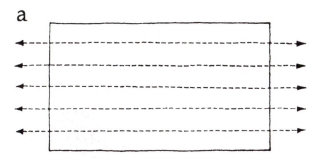

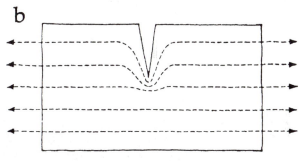

Fig. 5.6 Idealized stress patterns in a rectangular block under tensile stress (a) without, and (b) with a crack. The presence of the crack focuses stress trajectories (i.e. the paths along which stress is passed from molecule to molecule) around the crack tip, increasing the likelihood of failure there. (Modified from Gordon, 1976)

words, the tensile stresses are due to a *gradient* in the normal stress. This tensile stress causes the growth of favourably oriented cracks in the bedrock or particle, and may eventually lead to failure.

For any given stress, the likelihood of crack growth depends on the length of the crack, and longer cracks are more likely to grow than short ones. This is because crack growth releases stored strain energy from the surrounding rock (Section 4.2.2), which is then available to break apart more rock at the crack tip. The longer the crack, the more strain energy is released, and the easier it will be to make the crack grow. Engineers recognize that, for each material, there is a threshold crack size, dividing two very different types of behaviour. Above the threshold, the strain energy released by crack growth easily causes renewed growth, and the crack can grow in an explosive manner. In contrast, for short cracks, the small amount of strain energy released is insufficient to cause additional fracture, and crack growth is unlikely. The threshold length is called the *critical Griffith length* after the aircraft engineer A.A. Griffith, who laid the foundations of modern fracture mechanics in the 1920s. Cracks longer than the threshold length are *critical cracks*, and those below it are *subcritical cracks* (Fig. 5.7).

It is probable that much glacial erosion of rock beds is due to discontinuous rock mass failure and

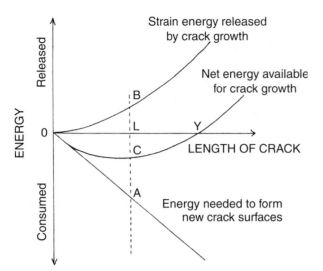

Fig. 5.7 Griffith model of crack growth. Line 0–A represents the energy requirement to grow cracks of length *L*; 0–B is the strain energy released by the progress of the crack; and 0–C is the net available energy in the system. For cracks shorter than length *Y*, more energy is needed for crack growth than is released; these subcritical cracks will not tend to grow. In contrast, for cracks longer than length *Y*, more energy is released by crack growth than is required for crack extension; these are critical cracks which can grow catastrophically. (Modified from Gordon, 1976)

the growth of critical cracks under applied shear stresses. As rock is stressed by overriding ice, critical cracks may grow rapidly, isolating volumes of rock and allowing erosion to take place. However, it is clear that in the long term, fracture of apparently intact rock can occur below glaciers. This is probably due to a phenomenon known as *subcritical crack growth*, where cracks below the critical Griffith length grow slowly, step by step, until they reach the threshold length, at which point rapid failure can follow. The main mechanism of subcritical crack growth is thought to be chemical weathering encouraged by the presence of water in the crack, but the extension of microscopic cracks may also be important, particularly if the material undergoes several cycles of stressing and stress release (Atkinson and Meredith, 1987).

5.3.2 Abrasion

Abrasion is generally regarded as 'the process whereby bedrock is scored by debris carried in the basal layers of the glacier' (Sugden and John, 1976, p. 152; see also Drewry, 1986; Iverson, 1990). In this book, we adopt a broader definition of abrasion, and use it to embrace the processes of *striation* (the scoring of bedrock) and *polishing*, in which the roughness of rock surfaces is reduced by small-scale brittle failure of protuberances.

5.3.2.1 STRIATION

Striae (singular: striation) are formed as asperities in rock particles are dragged over bedrock or clasts, scouring out thin grooves. As an asperity passes over a rock surface, there is a temporary stress concentration in the rock below the asperity, leading to the development of tensile stresses. If the tensile stress is sufficient to promote crack growth, brittle failure occurs below the asperity, and as the asperity moves on, a striation is left behind as a trail of damage. Striae are therefore the cumulative effect of small brittle failure events marking the passage of overriding particles (Fig. 5.8). Drewry (1986, p. 51) has described how the striation process is not continuous

> but comprises jerky steps. As grooving commences there is a build-up of elastic strain at the asperity tip. This is released giving rise to the impact of the asperity against the rock surface with subsequent production of rock chips. ... The sequence then recommences.

Evidence for this type of jerky abrading motion comes from studies of industrial rock cutting (Drewry, 1986), and from the micromorphology of natural striae. Although striae seem continuous to the naked eye, through the microscope they are seen to consist of numerous crescent-shaped fractures, each marking a discrete failure event (Fig. 5.9; Iverson, 1990, 1995).

The following factors are generally regarded as being the most important in controlling the efficiency of the striation process (e.g. Sugden and John, 1976; Drewry, 1986):

Relative hardness of rock surface and overriding clast

The striation process is most effective when the overriding clasts are much harder than the bed, because softer rocks are more likely to fail under imposed stresses than harder ones. This was clearly demonstrated by Boulton (1979) in a series of experiments conducted beneath Breidamerkurjökull, Iceland, in which plates of different types of rock were bolted on

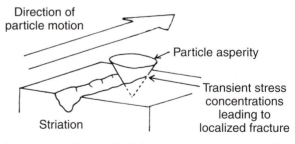

Fig. 5.8 Simple model of the striation process. (Modified from Drewry, 1986)

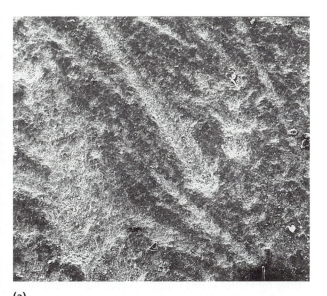

(a)

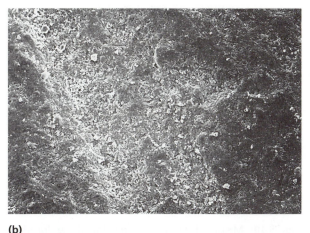

(b)

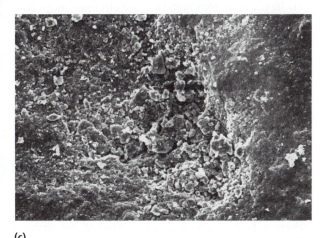

(c)

(d)

Fig. 5.9 Scanning electron micrographs of striae at a range of scales. Note the abundant evidence for brittle fracture within striae compared with the smooth appearance of the neck between striae. (Micrographs supplied by L.A. Owen)

to the glacier bed and left to be striated by the overriding ice (Fig. 5.10). Measured abrasion rates were found to be inversely proportional to bed hardness, and were several times higher for limestone (hardness = 180–210 kg mm^{-2}) than for basalt (hardness = 865–905 kg mm^{-2}).

Force pressing the clast against the bed

Other factors being equal, the tensile stresses around asperities in clasts being dragged over bedrock increase with the force pressing the clast towards the bed. The higher the contact force, the greater the energy that is available for the striation process. A familiar analogy is the use of sandpaper to wear down wood: the harder the sandpaper is pressed down, the more effective the sanding process for any given velocity. As described in Section 5.2.2, it is difficult to specify the contact forces between clasts held in ice and a glacier bed. According to the

Coulomb friction model, contact forces depend mainly on the weight of the overlying ice, minus the water pressure at the bed (Boulton, 1974, 1975). In contrast, the Hallet friction model states that contact forces depend on the buoyant weight of particles and the rate of ice flow towards the bed (Hallet, 1979a, 1981). Finally, the sandpaper friction model indicates that contact forces depend on the weight of the overlying ice and the area of the bed occupied by water-filled cavities (Schweizer and Iken, 1992). Although the predictions of these models differ in detail, in general, contact forces will be highest where rapid basal melting lowers large particles towards the bed, and where free drainage allows transient low basal water pressures.

Another factor influencing contact forces is the *rotation* of clasts held in the basal ice. If clasts rotate or roll along the bed, shear stresses transferred to the bed will be lower than for clasts held rigidly in the

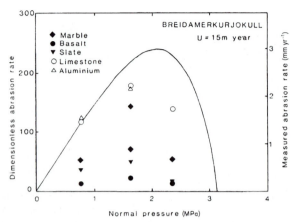

Fig. 5.10 Measured abrasion rates for three locations beneath Breidamerkurjökull, Iceland. Rates of abrasion were measured for platens of different materials bolted on to bedrock. Material hardnesses were as follows: marble = 450–510 kg mm^{-2}; basalt = 865–905 kg mm^{-2}; slate = 605–660 kg mm^{-2}; limestone = 180–215 kg mm^{-2}; aluminium = 50–60 kg mm^{-2}. The inverse relationship between hardness and abrasion rates is clearly apparent. Dimensionless abrasion rates predicted by Boulton (Coulomb friction) theory are also shown. The general pattern of abrasion rates first rising then falling with normal pressure is also apparent in the measured rates. (From Drewry, 1986, after Boulton, 1979)

ice, reducing the forces available for striating the underlying rock. The likelihood of fragment rotation is influenced by particle shape and by the pressure imposed by the overlying ice (Boulton, 1978; Iverson, 1990).

Velocity of clast relative to bed

The rate at which clasts are dragged over the bed is an important control on abrasion rates, because it governs the area of bedrock that is traversed by abrasive material in any given time. An important point is that the velocity of a basal clast relative to the bed is often less than the velocity of the ice, owing to the existence of drag forces between the clast and the bed, and the ability of the ice to flow past the clast by plastic flow and regelation. Clast velocity is therefore not a simple function of ice velocity, but also depends on particle–bed contact forces and debris concentration (Hallet, 1979a, 1981). When the contact forces between a particle and the bed are high, the particle tends to be retarded against the bed, reducing the rate at which it passes over the underlying rock. The limiting case occurs when contact forces are so high that the particle lodges against the bed, and can no longer abrade the underlying rock.

In some cases, reduction of particle velocity relative to the bed can actually *increase* the amount of abrasion, because it increases the amount of time

during which an asperity is pressed against one small region of the bed. This may encourage subcritical crack growth and failure of the bed, which might not have occurred if the asperity had passed quickly over the area.

Concentration of debris in ice at the abrading surface

If the velocity of clasts relative to the bed is constant, the amount of striation will increase with the number of particles in contact with the bed. However, clast velocity is strongly influenced by debris concentration, because the number of particle–bed contacts controls the amount of friction at the glacier bed. High friction tends to retard particles against the bed, reducing the amount of striation. Consequently, the efficiency of the striation process first increases with increasing debris concentration, then falls off. Modelling by Hallet (1981) suggests that maximum striation rates should occur at debris concentrations of 10–30 per cent, although in some circumstances considerable abrasion should be possible below debris-rich ice with 'sandpaper friction' (Section 5.2.2.4; Schweizer and Iken, 1992).

Removal of wear products

During the striation process, large numbers of small particles are created, and tend to accumulate around asperities in the striating clasts (Boulton, 1974; Iverson, 1990). If these wear products are not removed, they will tend to increase the effective area of contact between the clast and the bed, thus reducing clast–bed friction and the efficiency of the striating process. Fine wear products (<50 μm) can be removed by water flowing in thin films at the bed (Hallet, 1979b), whereas larger particles can be flushed out through channels or linked cavity systems.

Availability of basal debris

The striation process is most effective where particles are brought into contact with the bed by basal melting and ice deformation. Since ice flow towards the bed is greatest on the upglacier sides of bumps, debris availability, and hence the effectiveness of the striation process, will be highest in such areas (Hallet, 1979a, 1981; Iverson, 1995). Shoemaker (1988) has argued that when ice slides over a wavy bed, particles will tend to be elevated to the top of bumps, then transported at this high level. Therefore, those particles will contact the bed only near the crest of bumps located downglacier, limiting the effectiveness of the striation process (Fig. 5.11). Abrasion of the lower parts of bumps will therefore depend on the availability of debris freshly eroded from the bed, particularly the lee side of bumps immediately upglacier (Hallet, cited in Iverson, 1995).

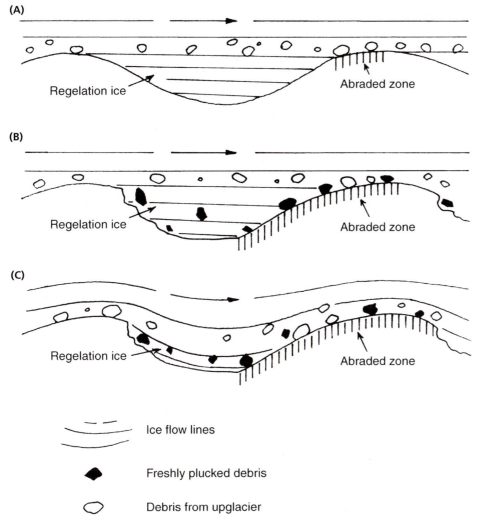

(A)

Regelation ice

Abraded zone

(B)

Regelation ice

Abraded zone

(C)

Regelation ice

Abraded zone

Ice flow lines

◆ Freshly plucked debris

◯ Debris from upglacier

Fig. 5.11 The effects of debris availability and ice flow mechanisms on abrasion. (A) Shoemaker (1988) model for sliding dominated by regelation. Debris elevated by upstream bumps stays at a high level, and so is able to abrade only bump crests. (B) Hallet model for sliding dominated by regelation. Freshly plucked debris is able to abrade all the stoss side of bumps. (C) Hallet model for sliding by enhanced deformation and regelation. Freshly plucked debris and debris from upglacier are able to abrade stoss sides, and abrasion can occur for a short distance downglacier of the bump crest. (Modified from Iverson, 1995)

The above factors have been incorporated into mathematical models of abrasion (e.g. Boulton, 1974; Hallet, 1979a, 1981; Drewry, 1986; Shoemaker, 1988; Iverson, 1995). The most widely accepted is that developed by Hallet, who regards abrasion rates as a function of bed hardness, debris concentration, particle size and particle velocity (both parallel to and towards the bed). Important experimental evidence in support of the Hallet model has been provided by Iverson (1990), who found that measured rates of abrasion were in reasonable agreement with those predicted by theory. It should be emphasized, however, that these results apply only to idealized conditions with sparse basal debris, and that theoretical understanding of the general striation process is still far from complete.

5.3.2.2 POLISHING

Most discussions of glacial abrasion assume that the abrasion of rock surfaces occurs as the sum of individual striating events, gradually wearing down and smoothing the surface. Careful examination of abraded rock surfaces, however, shows that areas of relatively smooth rock occur between striae. We suggest that these areas form by a distinct process of *polishing*, in which small protuberances are removed by overriding rock particles and ice. Small protuberances on a rock bed will tend to be the location of stress concentrations below overriding ice and basal debris (Section 5.2.2.1). The presence of these stress concentrations will locally increase the likelihood of bedrock failure by critical or subcritical crack growth, so that brittle failure is more likely to occur at protuberances than elsewhere. This will tend to remove the protuberances and smooth the bed. Stress concentrations will occur around bed protuberances even below flat, non-striating clast surfaces and masses of compressed rock flour, so that the breakage and removal of protuberances – or polishing – can be expected to be independent of the striation process, and may occur in its absence. Indeed, where the glacier bed and clasts in the basal ice are of equal hardness, polishing may be the dominant process of abrasion.

5.3.2.3 ABRASION BY CLEAN ICE V. ROCK PARTICLES

Until recently, it was thought that clean ice could not, by itself, cause bedrock failure, and that all subglacial abrasion was achieved by particles being dragged over the bed. However, Budd *et al.* (1979) conducted laboratory experiments that demonstrated that clean, sliding ice could erode rough granite slabs. Extrapolated erosion rates of up to 55 mm yr^{-1} were observed, maximum rates occurring at high normal and shear stresses and sliding velocities. Riley (1979) questioned the representativeness of these results, and suggested that they merely demonstrate the removal by ice of pre-existing loose material from the rock surfaces. He claimed that once loose material is removed, erosion rates will fall to more representative low values.

It is possible that the main mechanism of abrasion below clean ice is polishing, since all the drag below clean, wet-based ice is the form drag supported by protuberances, which will therefore be the locus of high stress concentrations (see Sections 4.5.2 and 5.3.2.2). This would also imply a reduction in abrasion rates with time as the roughness of the bed decreases, reducing the stress concentration effect. Whatever the effectiveness of ice as an erosive agent, it is agreed that subglacial erosion is much more rapid when rock fragments are present in the basal ice (Drewry, 1986). This is partly because rock fragments form hard, effective cutting tools, and partly due to the presence of asperities at the base of fragments, focusing the applied stresses on relatively small areas of the bed.

5.3.3 Quarrying

The fracture of large fragments (greater than *c.* 1 cm) from rock beds occurs by essentially the same process as abrasion. Temporary stress concentrations form below overriding ice or clasts, tending to enlarge and connect weaknesses in the rock and eventually leading to detachment of a fragment. The distribution of joints and other pre-existing cracks in rock can exert a strong influence on patterns of fracture, by providing the conditions for *discontinuous rock-mass failure* (Addison, 1981). Some authors have emphasized the role of pre-glacial weathering as a necessary precondition for large-scale erosion (see Sugden and John, 1976; Addison, 1981; Rastas and Seppälä, 1981; Lindstrom, 1988, and references therein). Crack growth may also occur by periglacial processes below the margins of glaciers, where heat

can be evacuated from the bed, especially during winter (e.g. R.S. Anderson *et al.*, 1982; Iverson, 1995).

Examination of fractures in former glacier beds and recent theoretical modelling, however, show that *stress patterns* are of primary importance in determining the location of failure surfaces by initiating the growth of suitably oriented microcracks. Failure of bedrock is most likely where there are large stress gradients, where tensile stresses tend to pull the rock apart (Section 5.2.2.1). Common situations where large stress gradients occur are (a) beneath large particles as they are dragged over the bed; and (b) near the lee side of obstacles or steps on the bed, particularly where cavities are present. Figure 5.12 shows modelled stress patterns associated with the passage of a particle over a flat glacier bed (Ficker *et al.*, 1980). In the case shown, a low-pressure cavity is present in the lee of the particle; this will occur if water melted during the regelation process is able to

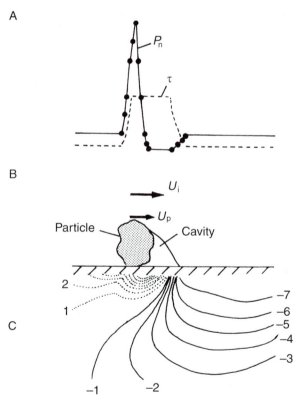

Fig. 5.12 Calculated stress patterns in a horizontal rock bed associated with a rock particle held in sliding ice. (A) Distribution of normal (P_n) and shear (τ) stresses at the rock surface. (B) The velocity of the particle (U_p) is less than that of the ice (U_i) owing to frictional retardation. As a result, a low-pressure cavity opens up in the lee of the particle. (C) Contoured values of the maximum stress in the rock. Positive values are compressive, negative values are tensile. (From Drewry, 1986, after Ficker *et al.*, 1980)

drain away freely (Sections 4.5.2.1 and 5.2.2.2). Maximum normal stresses (P_n) are developed below the point of contact, and maximum shear stresses (τ) occur below and for some distance in front of the particle. The pattern of maximum stresses within the bed shows compressive stresses below the particle and a curving zone of tensile stresses in front of and below the leading edge. The growth of critical microcracks in this zone of tensile stress may lead to the quarrying of curved rock fragments, leaving *crescentic fractures* and *chattermarks* like those found on subglacially eroded surfaces (Section 9.2.3). Subcritical cracks may grow slowly under repeated loading by overpassing particles, until the critical crack length is reached, leading to sudden crack growth and failure.

The stress distribution associated with ice flow over a bedrock obstruction was modelled by Morland and Boulton (1975) and Morland and Morris (1977). One case, where the average normal stress (P_n) is 100 kPa, is shown in Fig. 5.13. Local values of P_n follow a wave-like oscillation over the obstruction, and are highest halfway up the stoss side, where ice flow is towards the bed, and lowest halfway down the lee face, where ice flow is away from the bed. The highest shear stresses occur near the foot of the lee face, where stress may be high enough to initiate fracture of weak sedimentary rocks, such as siltstones, limestones and some sandstones. The modelled stresses are too low to account for the fracture of harder rocks such as schist, basalt and granite, and Morland and Boulton (1975) concluded that subglacial quarrying of such rocks must involve failure along pre-existing joints. More recent modelling, however, has shown that quarrying of intact hard rocks can occur if the presence of low-pressure cavities is taken into account.

The case of fracture near a step cavity is illustrated in Fig. 5.14 (Iverson, 1991). For a steady-state water-filled cavity, in which P_w is equal to the ice overburden pressure, bedrock stresses are at a maximum adjacent to the cavity. If water pressure drops, bedrock stresses increase dramatically while the ice adjusts towards a new equilibrium. Stress in the bedrock is generally vertically oriented and compressive, leading to the development of fractures that are also close to vertical, so that fragments will spall off approximately parallel to the back of the step. Conditions of rapidly fluctuating water pressure are probably common where supraglacial meltwater reaches the bed (Section 3.2.3.4), so that this mechanism may play a major role in subglacial quarrying.

Iverson (1991) also argued that water located in joints and microfractures plays an important role in the fracture of bedrock in the vicinity of subglacial cavities. He showed that water pressure in rock fractures cannot change as rapidly as can the water pressure in larger cavities, because there is a limit to the

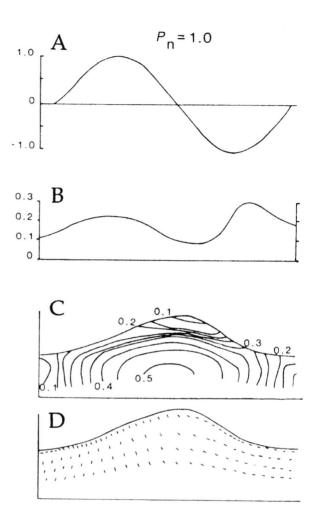

Fig. 5.13 Modelled stress distribution in a subglacial rock knob. (A) Distribution of normal stress at the interface; (B) maximum shear stress in surface rock layers; (C) maximum shear stress contours at depth (MPa); (D) orientation of principal stress axes. Ice flow from left to right. (From Drewry, 1986, after Morland and Boulton, 1975)

rate at which water can flow into or out of narrow fissures. This means that if the water pressure at the bed fluctuates, there will be a delay during which the water pressure in rock fractures will differ from that at the bed. This delay will be greatest for massive rocks with narrow fractures, and least for porous, well-jointed sedimentary rocks. If the water pressure at the ice–bed interface is falling, such as during waning discharge in a linked-cavity system, water pressure in rock fractures will remain relatively high while the system adjusts. This will tend to weaken the rock, because the water will support some of the confining pressure, reducing the rock-to-rock friction across the fracture (this effect is summarized in the Coulomb equation; Sections 4.4.2.2 and 5.2.2.2). At the same time, the reduced pressure in the adjacent cavity will mean that the proportion of the weight of

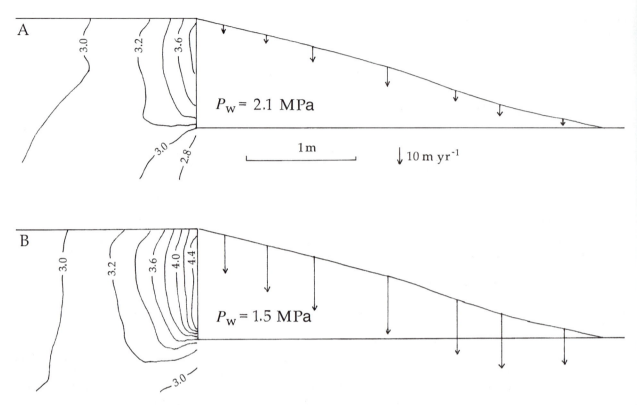

Fig. 5.14 Modelled principal stresses in bedrock upstream of a step cavity. (A) Steady-state case, where water pressure in the cavity (P_w) = 2.1 MPa. Principal stresses are at a maximum adjacent to the step. Downward-pointing arrows show the vertical component of ice flow in the cavity roof. (B) Stress pattern associated with a sudden drop of water pressure to P_w = 1.5 MPa. Note the dramatic increase in principal stresses and vertical ice velocities. (Modified from Iverson, 1991)

the glacier formerly supported by pressurized water will be shifted to the bedrock surface, increasing the local shear stress. The combination of reduced rock–rock friction, higher shear stresses and the presence of low-pressure cavities will create ideal conditions for rock fracture. In addition, high-pressure water in rock joints will also encourage long-term subcritical crack growth, further weakening the rock. Quarrying is even more likely to occur if falling cavity water pressures coincide with the passage of a large particle over the rock step.

Hallet (1996) modelled the controls on quarrying on a glacier bed composed of step cavities and rock ledges, and found that quarrying rates are sensitively dependent on basal sliding rates and the proportion of the bed occupied by cavities (Figs 5.14 and 5.15). According to this model, quarrying rates are small when there are few cavities at the bed, because average stresses on the area of ice–rock contact are relatively small. When cavities form on larger areas of the bed (i.e. when higher water pressures cause more extensive ice–bed separation), the basal stresses have to be supported on a smaller area of ice–rock contact, elevating the stress concentrations at these points and increasing the likelihood of crack growth and rock

fracture. For very extensive cavity formation, quarrying rates diminish, because the area affected by high loads is very small. There are few empirical data on subglacial quarrying rates which would allow testing of this model, but Hallet has argued that conditions favourable for rapid quarrying existed below Variegated Glacier during the 1982–83 surge, when effective pressures were small and an extensive system of linked cavities existed beneath the glacier (Section 4.8.2; Kamb et al., 1985; Kamb, 1987; Humphrey and Raymond, 1994). During the surge, the sediment flux in proglacial streams was very large, equivalent to basal erosion rates of about 0.5 m per 17-year surge cycle. Since erosion rates during quiescent phases are likely to have been small, it was concluded that quarrying rates may have been as high as 100 mm yr^{-1} during the surge. Such rates are at the upper limit of estimated subglacial erosion rates, indicating that conditions were optimal for quarrying (Section 5.3.6; Hallet et al., 1996).

5.3.4 Removal of fragments

Once rock fragments have been loosened from the glacier bed, they must be set in motion by ice or

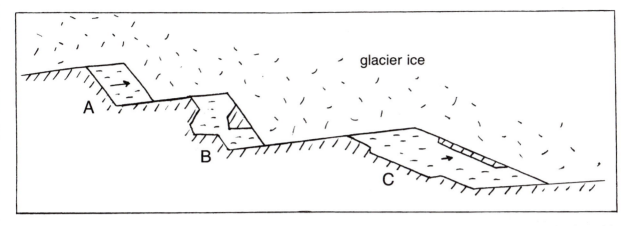

Fig. 5.15 Removal of rock fragments from a rock bed. (A) 'Active', high water-pressure cavity exerting a hydraulic jacking force. Fragment removal is unlikely because freezing will be inhibited. (B) 'Passive' low-pressure cavity, in which a rock fragment has been frozen on due to the Robin heat pump effect. (C) Combination of freezing and driving force at intermediate water pressures. (Redrawn from Röthlisberger and Iken, 1981)

flowing water before erosion is complete. A number of mechanisms have been proposed to account for how blocks and rock fragments can be removed from the bed by overriding ice. Robin (1976) argued that plucking can occur as the result of a *heat pump effect* in which basal ice locally freezes on to the bed as part of the regelation sliding process. As discussed in Section 4.5.2, glacier sliding over a rough bed involves large fluctuations in basal pressure in the vicinity of rock knobs. On the upglacier side of obstructions, where the glacier is flowing towards the bed, pressure is higher than average, causing spontaneous melting of the ice. Robin argued that melting occurs at the bed and at grain boundaries within the basal ice, from which water is expelled through veins. This process removes heat from the ice and lowers its temperature to the depressed pressure melting point. As the ice flows over the lee side of the obstruction, the pressure falls, but the temperature cannot rise to the new, elevated pressure melting point because the lost meltwater is not available to heat the interior of the basal ice. Therefore, cold patches will develop, causing the ice locally to freeze on to the bed. As a result, rock fragments can adhere to the ice and be removed from the bed.

The heat pump effect has been demonstrated experimentally by Röthlisberger and Iken (1981), who also considered the role of cavity formation on the plucking mechanism. We saw in Section 4.5.3 that the opening of basal cavities beneath a glacier is encouraged by locally high basal water pressures 'jacking' the glacier off the bed in lee-side zones. The increase of lee-side pressure associated with this process will tend to inhibit the formation of cold patches and the associated freezing on of rock debris. Therefore, Röthlisberger and Iken argued that the heat pump effect will tend to operate either in 'passive cavities' that are isolated from the main subglacial drainage network and experience only low water pressures, or during the closure of 'active cavities' as water pressures fall (Fig. 5.15). Note that conditions of falling water pressure, shown by Iverson (1991) to be conducive to rock fracture, will also be favourable for the freezing on of fragments. This emphasizes the importance of non-steady-state conditions for subglacial erosion.

The removal of rock fragments will also be encouraged if the friction between the fragment and the parent rock is reduced. One mechanism for friction reduction has been observed under Oksfjordjøkelen in Norway by Rea and Whalley (1994). The bed of this glacier is at the pressure melting point and consists of a series of abrupt, well-jointed rock steps aligned transverse to ice flow. Rea gained access to the glacier bed via a large cavity between two rock steps, and observed basal ice being squeezed under pressure into rock joints. The ice was very plastic, and was clearly under high stress. In one case, the ice had flowed into the joints around and beneath a large block on the bed, completely surrounding it (Fig. 5.16 (Plate 14)). The presence of this ice dramatically reduced the friction between the block and the bed, facilitating removal of the block by the overriding ice. Rea and Whalley suggested that such injection of plastic ice into rock joints may be a very important plucking mechanism, particularly where basal shear stresses are high and sliding rates low. It is probable that this process also requires low basal water pressures in order to be effective.

5.3.5 Erosion beneath cold ice

Sliding is negligible or non-existent below cold-based ice, owing to the high adhesive strength of ice and rock below the pressure melting point. Therefore, abrasion will not be an effective erosive mechanism beneath predominantly cold-based glaciers. Drewry (1986), however, has pointed out that clasts embedded in the base of cold-based ice will be subject to some drag force because the ice above the bed will be shearing by internal deformation. This drag force may be sufficient to cause rotation and slip of the clast over the bed, possibly resulting in a limited amount of abrasion. In comparison with predominantly wet-based glaciers, however, the amount of abrasion will be very small.

In contrast, plucking processes can be effective under certain conditions below cold ice. The main requirement is that the shear strength of at least part of the substratum should be low enough to allow brittle or ductile deformation under the glacially imposed stresses. This condition is most likely to be met for well-jointed rocks or unconsolidated sediments. Boulton (1979) has described glacial plucking of frozen dune sands below Wright Lower Glacier in the Dry Valleys region, East Antarctica, at temperatures of −15°C. The sands are partly cemented by ice but also contain ice-free pores and ice lenses derived from blown snow. The ice–sand interface is relatively smooth but with a few hummocks, reflecting the original dune morphology. In several places, blocks of sand have been displaced and incorporated within the ice, with varying amounts of disaggregation. There was evidence for both brittle and ductile failure of the sands. Blocks were found up to 2 m above the bed, and appear to have been incorporated within the ice by the differential ice flow around the flanks of hummocks. Experimental studies conducted below Urumqi Glacier 1, China, by Echelmeyer and Wang (1987), demonstrated ductile deformation of frozen subglacial sediments. The thickness of deforming sediment was 35 cm, and accounted for c. 60 per cent of the surface movement of the glacier. In contrast with the observations of Boulton (1979) and Echelmeyer and Wang (1987), England (1986) has demonstrated that a subpolar outlet glacier on Ellesmere Island, Arctic Canada, overrode an alluvial fan without causing any subglacial erosion.

Failure may possibly also occur in deeper, unfrozen parts of the substratum below cold-based glaciers. In this case, the base of the frozen material may act as a major décollement surface along which the overlying rocks, sediment and glacier ice will slide (Mathews and Mackay, 1960; Moran, 1971; Broster and Seaman, 1991). Moran (1971) and Moran et al. (1980) developed this idea to explain large thrust blocks of sediment that occur on the North American prairies. They postulated that, along the southern margin of the Laurentide ice sheet, the

ice was cold-based near the edge where the ice was thin, but was wet-based further upglacier. Meltwater produced in the wet-based zone would have flowed under pressure through subglacial aquifers towards the margin, where it would have been impeded by frozen and impermeable ground. This would have significantly raised porewater pressures beneath the frozen part of the substratum, decreasing frictional strength and increasing the likelihood of failure. Large blocks, frozen on to the glacier sole, could then be transported towards the margin. However, it is possible that, in the cases discussed by these authors, failure was proglacial in origin and that the blocks were subsequently overridden (see Sections 7.4.3 and 12.4.3).

5.3.6 Subglacial erosion rates

Subglacial erosion rates are very difficult to estimate, and few direct measurements have been made. Pioneering studies were conducted in the Swiss Alps between 1919 and 1925 by Alfred de Quervain and O. Lutschg, who drilled vertical holes into bedrock in front of advancing glacier snouts. When the ice retreated they were able to measure the amount of surface lowering, which ranged between 1.0 and 6.0 mm yr^{-1} for rocks ranging from vein quartz to gneiss (Embleton and King, 1975). More recently, direct measurements of abrasion were made below Breidamerkurjökull, Iceland, and Glacier d'Argentière, France, by Boulton (1974), who gained access to the glacier beds via artificial tunnels. Abrasion rates for rock platens fixed to the glacier bed ranged from 0.9 mm yr^{-1} for basalt below Breidamerkurjökull, to 36 mm yr^{-1} for marble below the rapidly sliding Glacier d'Argentière. Such measurements, however, are not representative of glacier beds in general, and cannot be used to infer long-term erosion rates. Furthermore, they do not take into account erosion by quarrying and chemical weathering.

An alternative approach is to estimate subglacial erosion rates from the sediment load of streams emerging from beneath glaciers. Bogen (1996) has shown that sediment yields show wide seasonal and inter-annual variations, which are partly due to switches in the subglacial drainage system. However, with care, sediment yields can be used to estimate long-term erosion rates (Hallet et al., 1996). Estimates range from 0.01 mm yr^{-1} for polar glaciers, to 1.0 mm yr^{-1} for small temperate glaciers in the Alps, to 10–100 mm yr^{-1} for large, fast-flowing glaciers in Alaska. Hallet et al. (1996) suggested that, in the absence of detailed knowledge of basal conditions, ice flux at the equilibrium line may be a useful indicator of erosion rates below temperate valley glaciers.

5.4 THE FORMATION AND DEFORMATION OF BASAL ICE

Ice close to the bed of a glacier (basal ice) tends to be distinctly different from the ice that makes up the rest of the glacier as the result of shear, melting, freezing and debris entrainment processes operating near the bed. Basal ice can have a very high debris content (>50 per cent by volume in some cases), and distinctive chemical and isotopic composition and structural characteristics (Hubbard and Sharp, 1989). Knowledge of the formation and characteristics of basal ice is important for several reasons. First, the concentration of debris in basal ice controls ice rheology and subglacial friction, both of which influence glacier motion (Sections 4.3.1 and 4.5.4). Second, the formation of basal ice is an important process for the removal of debris from the bed, in the form of solutes and solid particles. Third, the passage of debris-rich basal ice over a glacier bed can be a potent agent of subglacial erosion (Section 5.3). Fourth, basal ice can be an effective means of debris transport, and a consequent source of glacigenic sediment.

Basal ice is highly variable, and a range of processes contribute to its formation. The most important of these are Weertman regelation, net adfreezing, the entrainment of pre-existing ice, and ice deformation.

5.4.1 Weertman regelation

The refreezing of meltwater on the lee side of obstacles during regelation sliding will form thin layers of ice on the glacier sole. Because of its association with the sliding process, regelation ice is found below temperate glaciers, where the basal ice is at or very close to the pressure melting point throughout. Because of the nature of its formation, regelation ice is seldom more than a few centimetres thick, because ice formed in the lee of one obstacle is liable to be at least partially destroyed by melting against further obstacles downglacier (Hubbard and Sharp, 1989, 1993).

Regelation ice is commonly visibly laminated, consisting of closely spaced laminae of clear ice and layers of debris, usually fine-grained abrasion products (Fig. 5.17; Souchez and Lorrain, 1987; Hubbard and Sharp, 1993, 1995). Individual laminae are commonly 0.1–1 mm thick and form layers broadly parallel to the bed, although they may pinch out or merge laterally. Each lamina represents one refreezing event. The regelation process has also been invoked by some workers to account for another distinct type of ice, termed *clotted ice*, generally bubble-free ice containing dispersed, elongated 'clots' of

Fig. 5.17 Foliated regelation ice at the base of Slettmark-breen, Norway. The folding was probably produced by ice flow around obstacles on the bed. (Photo: D.I. Benn)

silt-sized debris (Sugden *et al.*, 1987; Souchez *et al.*, 1990; Knight *et al.*, 1994; Knight, 1994). According to Souchez and Lorrain (1991) and Knight (1994), clotted ice is probably equivalent to the *dispersed ice facies* described by Lawson (1979a). A regelation origin for clotted ice is uncertain, however, and alternative mechanisms have been proposed. Possibilities include the migration of fine particles along grain boundaries in the ice (Knight, 1987), the tectonic mixing of dirty and clean ice under high strains (Hubbard and Sharp, 1993), and the concentration of wind-blown debris during the formation of superimposed ice at the glacier surface, which is then transported to the bed by glacier flow (Koerner, 1989).

Hubbard and Sharp (1995) identified a distinct type of *clear ice* below temperate alpine glaciers. The ice is devoid of layering, but contains dispersed smears of debris and occasional clouds of deformed and flattened bubbles. They argued that this ice forms by internal melting and freezing and ice recrystallization during enhanced creep near the bed, as described by Lliboutry (1993).

The process of freezing results in isotopic fractionation, because heavy isotopes of oxygen and hydrogen (^{18}O and deuterium) preferentially bond into growing ice crystals (Section 2.4.2). During the initial stages of refreezing, therefore, regelation ice will be isotopically heavier than the remaining meltwater, but as freezing proceeds, each increment of ice will be isotopically lighter than the one before, owing to the depletion of heavy isotopes in the water. As a result, regelation ice often exhibits *compositional banding*, with each freezing cycle being represented by a downward gradation from isotopically heavy to isotopically light ice (Fig. 5.18; Lawson and Kulla, 1978; Souchez and Jouzel, 1984; Hubbard and Sharp, 1993). The bulk composition of regelation ice will be isotopically heavier than the parent meltwater

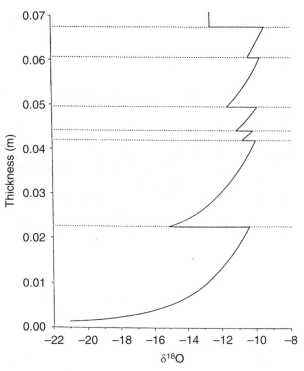

Fig. 5.18 Modelled isotopic variation in regelation ice, resulting from the preferential incorporation of heavy isotopes in the first ice to form in each cycle. (From Hubbard and Sharp, 1993. Reproduced by permission of the International Glaciological Society)

if some of the water is able to drain away before freezing is complete.

5.4.2 Net adfreezing

Basal ice forms by the regelation process when there is a rough balance between the amount of melting and refreezing. Where melting predominates over freezing, no basal ice will form, but when freezing predominates over melting, thick sequences of basal ice can develop. Such *net adfreezing* occurs where meltwater flows into cold areas of the bed from upglacier (Weertman, 1961c; Hubbard and Sharp, 1989; Hubbard, 1991). Cold areas can form for several reasons (Hubbard and Sharp, 1989). First, large-scale zones of freezing can occur where more heat is conducted away from the bed than can be provided by geothermal heat, sliding friction or incoming meltwater. Such zones may occur below thin ice close to glacier margins, on either a permanent or a seasonal basis. Second, cold patches on the bed may temporarily increase in size as the result of basal water pressure fluctuations (Robin, 1976; Hubbard and Sharp, 1989). For a wet-based glacier resting on a rough bed, the weight of the ice is supported by bedrock at high points of the bed and pressurized

water elsewhere. An increase in subglacial water pressure will result in a drop in the proportion of the weight of the glacier supported by bedrock, and a consequent drop in effective pressures at high points of the bed. This drop in pressure will depress the pressure melting point, causing basal freezing. A third cause of net adfreezing is the intrusion and circulation of cold air beneath a glacier. In winter, cold air can reach the bed via cavities, crevasses and tunnels, and can contribute to the formation of basal ice (R.S. Anderson *et al.*, 1982; Vivian and Bocquet, 1973).

The water required for the formation of basal ice can be supplied from several sources, including zones of net melting upglacier and surface meltwater carried to the bed down moulins and crevasses. Hubbard and Sharp (1995) described a distinctive type of basal ice formed by the freezing of standing water in bedrock hollows or cavities. This *interfacial ice* consists of alternating bands of ice crystals and debris-rich ice which undulate in parallel with the large-scale roughness of the bed.

The debris content of basal ice formed by net adfreezing is highly variable, depending on the conditions at the ice–bed interface. Very high debris concentrations can result from the downward migration of a freezing front into saturated sediments, so that the ice merely forms an interstitial cement between mineral grains (Boulton, 1970a; Harris and Bothamley, 1984). The descending glacier ice infiltrates the pore spaces between grains, replacing air or water with interstitial ice. In such cases, masses of pre-existing sediment frozen on to the base of the glacier can preserve delicate sedimentary structures such as cross-bedding and lamination. Adfreezing above a bedrock interface will usually result in basal ice with less debris, with the concentration depending on the proportions of water and ice that are available (Boulton, 1970a; Gow *et al.*, 1979; Hubbard and Sharp, 1989). As for regelation ice, ice formed by net adfreezing is generally isotopically heavier than meltwater derived from meteoric ice, because of fractionation and incomplete freezing.

Basal ice below subpolar glaciers is often distinctly stratified, with alternating debris-rich and debris-poor ice (Fig. 5.19 (Plate 15)). This type of basal ice has been termed the *stratified ice facies* by Lawson (1979a) and the *solid ice facies* by Knight (1987, 1994) and Hubbard and Sharp (1995). Stratification may reflect the passage of the glacier over different substrata (Gow *et al.*, 1979) or the effect of cyclic rejection of debris from the freezing front during slow freezing episodes (Hubbard, 1991), but may also develop as the result of complex overfolding during ice deformation (see below).

5.4.3 Entrainment of pre-existing ice

The stratified ice facies of subpolar and polar glacier margins has been explained by: (a) the overriding and entrainment of buried glacier ice; (b) the incorporation of aprons accumulating at the snout; and (c) net adfreezing (Goldthwait, 1960, 1961; Hubbard and Sharp, 1989). Any debris that is frozen on in an up-ice location will be elevated to higher englacial positions as it flows up over the basal ice of the frozen snout.

In regions characterized by permafrost, the retreat of glaciers containing high debris concentrations results in the preservation of considerable quantities of buried glacier ice. This arises through the release of debris on to the wasting glacier surface during snout thinning. As the basal ice in the snout zone is frozen to its permafrost bed, it is difficult to remove it from the landscape once it is covered by an overburden of debris that exceeds the active layer thickness (see Section 6.5). Buried glacier ice has been widely reported from Siberia (Astakhov and Isayeva, 1988), Banks Island and the Mackenzie Delta, Canada (Harry *et al.*, 1988; French and Harry, 1990), Greenland (Hooke, 1970) and the Canadian High Arctic (Hooke, 1973b; D.J.A. Evans, 1989b; D.J.A. Evans and England, 1993), where it documents former ice-marginal positions. In some cases, buried ice

is thought to have survived since the last glacial maximum. For example, buried glacier ice at Bluenose Lake, NWT, Canada, originally formed in a compressive frozen toe zone at the margins of an ice stream in Dolphin and Union Strait during the last glaciation (St Onge and McMartin, 1995). Sediment was elevated from the bed by submarginal thrusting, and melted out on the surface to form controlled hummocky moraine. Because deglaciation proceeded in a periglacial environment, the glacier ice that was covered by the supraglacial debris has been preserved to the present day.

Several glaciers on Ellesmere Island in the Canadian Arctic are presently readvancing over buried glacier ice and reincorporating it and its overburden (Fig. 5.20; D.J.A. Evans, 1989b). This reincorporation results in the reactivation of the buried glacier ice, and the former debris cover becomes an englacial debris band which attenuates during further glacier flow. In many cases the buried glacier ice on Ellesmere Island has remained continuous with the glacier snout and so forms a supraglacial 'ramp'. This ramp is subject to fluvial incision and the development of glacier karst (Section 6.5.4.2), which allows the repositioning of former surface debris within the buried ice. In this way, further prospective englacial debris bands are initiated prior to ramp reactivation.

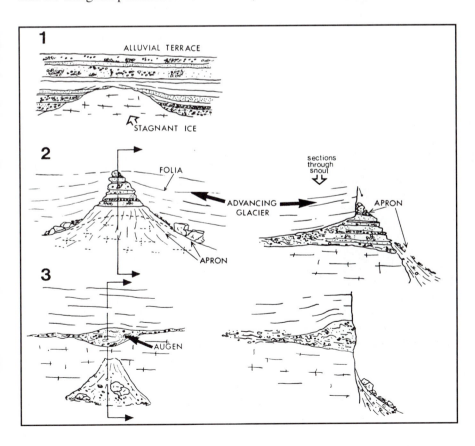

Fig. 5.20 An explanation of the origin of gravel pods in the cliffs of sub-polar glaciers. (1) Glacier recession results in the progradation of alluvium over stagnant ice (supraglacial debris concentrations may produce the same result). The buried ice does not melt because it is now well below the maximum depth of the active layer. (2) Glacier readvance and overriding of the ice-cored alluvium causes deformation and reworking into an apron. (3) Complete entrainment results in the isolation of an englacial gravel-rich clot, which marks the boundary between older, formerly stagnant ice and overriding glacier ice. (Modified from D.J.A. Evans, 1989b)

Advancing glaciers in cold regions often terminate in steep ice cliffs, below which frontal aprons of ice blocks and debris accumulate. Overridden and incorporated aprons are recognizable in the basal ice facies of many subpolar glaciers on Ellesmere Island in the Canadian High Arctic, where mosaics composed of individual ice blocks are still visible (Fig. 5.21; Shaw, 1977a; D.J.A. Evans, 1989b). A similar model involving the overriding of accumulations of wind-driven snow at glacier margins was proposed by Hooke (1973b) for the Barnes ice cap, Canada, and by Sharp (1984) for Skalafellsjökull, Iceland. The overriding of aprons by internally deforming glaciers leads to the incorporation and gradual attenuation of ice and sediment accumulations to produce complex basal ice layers of alternating debris- and ice-rich folia, which may exhibit well-developed folds and shear faults as a result of the differential movement induced either by (a) the varied rheological properties of the different ice–sediment mixtures, and/or (b) flow perturbations set up by irregularities

in the bed (Hudleston, 1976). Folding of basal debris-rich folia has been associated with differential movement between ice types by Hooke (1970, 1973b), a relationship that could be explained by the overriding and incorporation of pre-existing ice by a subpolar or polar glacier.

5.4.4 Ice deformation

The structure and thickness of basal ice layers can be strongly influenced by the deformation history of the ice. In zones of extending glacier flow, the ice is stretched and the basal ice layer will become thinner unless it is renewed by some other process such as freezing on. The effects of stretching can be observed where basal ice consists of layers with different flow responses to stress. Clean, coarse-grained ice tends to be quite ductile, and will stretch readily without breaking, whereas fine-grained or debris-rich ice tends to be stiffer and more susceptible to brittle deformation. When stretched, bands of fine or dirty ice will often break up, forming discontinuous lenses called *boudins* (Fig. 5.22; Hambrey and Milnes, 1975; Hubbard and Sharp, 1989; Hambrey, 1994). Boudins are familiar structures in metamorphic rocks, and are named after French black puddings, which they are said to resemble. Observation of boudinage and other deformation structures, therefore, can provide information on the relative competence of different components of basal ice layers (Knight, 1987; Hubbard and Sharp, 1989). The stretching of basal ice layers undergoing simple shear can also serve to rotate and lengthen lenses within the ice, creating new stratification (Hooke and Hudleston, 1978).

Compressive flow takes place where ice is decelerating. Common situations where this occurs are: (a) where ice flows against a topographic barrier; (b) the snouts of surging glaciers; or (c) where wet-based ice decelerates against cold-based ice downglacier. The latter case may occur on a seasonal basis, where a winter *cold wave* penetrates to the bed at the margin of a temperate glacier, or year-round at the margins of subpolar glaciers. Compressive flow will tend to shorten and thicken basal ice layers, and is thought to be responsible for many of the thick sequences of basal ice observed at glacier margins, particularly on sub-polar and surging glaciers. Thickening of basal ice sequences is commonly accompanied by folding and the development of thrust faults, often resulting in complex stratigraphies (Fig. 5.23; Hambrey and Müller, 1978; Sharp *et al.*, 1988; Knight, 1989). Impressive stratified sequences can be built up by the repetition of layers in recumbent folds, and by faulting. Indeed, where exposures of basal ice are limited, it may be difficult to distinguish stratification created by ice deformation from primary layering caused by cyclic adfreezing by using field

(a)

(b)

Fig. 5.21 (a) Frontal apron comprising dry calved ice blocks, Ellesmere Island. (b) Entrained ice blocks beneath foliated debris-rich ice, Eugenie Glacier, Ellesmere Island. (Photos: D.J.A. Evans)

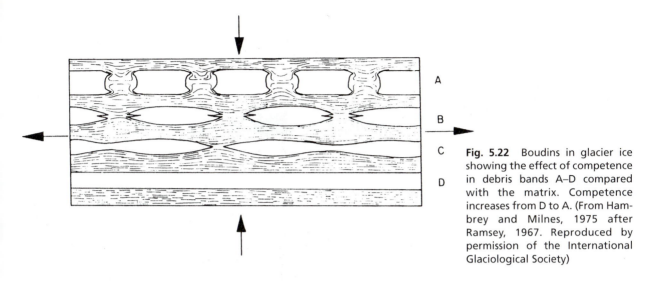

Fig. 5.22 Boudins in glacier ice showing the effect of competence in debris bands A–D compared with the matrix. Competence increases from D to A. (From Hambrey and Milnes, 1975 after Ramsey, 1967. Reproduced by permission of the International Glaciological Society)

Fig. 5.23 Folded and faulted basal debris-rich ice, Ellesmere Island. (Photo: D.J.A. Evans)

evidence alone. Careful studies of the isotopic composition of the ice may be required to determine its true origin. Layered ice sequences formed by complex folding at the margin of the Greenland ice sheet have been termed the *banded ice facies* by Knight (1989, 1994).

Thrusting has also been proposed as a possible mechanism for lifting debris from the glacier bed and incorporating it within the ice. Until recently, this process was questioned on theoretical grounds (Weertman, 1961c), but observations by Echelmeyer and Wang (1987) and Tison *et al.* (1993) have identified debris within shear planes rising from frozen glacier beds in China and Antarctica. It is believed that compressive flow near the glacier margins results in overthrusting of ice, which carries rock particles adhering to its base up over more slowly moving ice downglacier. According to Tison *et al.* (1993), this may be a major process for the formation of debris-rich ice below cold glaciers.

5.5 DEPOSITION FROM DEBRIS-RICH ICE

Subglacial deposition from debris-rich ice happens through the complex interaction of several factors, including the rate at which debris melts out from the base of the ice, the balance between frictional and driving forces acting on particles, debris grain size distribution, the configuration of the glacier bed, and the rate at which water can drain away from the site of deposition. We can recognize three basic processes of debris release: (a) *lodgement,* or deposition directly from sliding ice by frictional processes; (b) *melt-out,* or deposition due to the melting of stagnant or slowly moving ice; and (c) *deposition by gravity.* These three processes are not independent or mutually exclusive, and may operate together or in various sequences. In addition, sediment can be deposited from the base of subglacial deforming layers, as described in Section 5.6.2.

5.5.1 Lodgement

Lodgement is the plastering of glacial debris from the base of a sliding glacier on to a rigid or semi-rigid bed (Dreimanis, 1989). In effect, this occurs where the frictional drag between the debris and the bed is greater than the shear stress imposed by the moving ice, and is therefore sufficient to inhibit further movement of the debris (Boulton, 1975, 1982). The process of lodgement can occur for single particles or for masses of debris-rich basal ice.

Different situations where lodgement can occur are shown in Fig. 5.24. Where the glacier bed is rigid, friction between an overriding particle and the bed is mainly due to the interlocking of asperities beneath the particle and small roughness elements on the bed, and the resisting forces arise from the friction

(a) Frictional retardation against bed

(b) Prow of soft sediment or clast provides obstacle

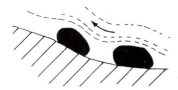

(c) Lodgement of debris-rich ice mass

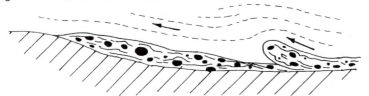

Fig. 5.24 The lodgement process. (a) Particles lodging by frictional retardation against a rigid bed. (b) Particles lodging against obstacles on a soft bed. Particles can plough up the bed, which then halts movement, or lodge against other particles. (c) Lodgement of debris-rich ice masses. The whole assemblage lodges against the bed, and melts out *in situ*. (Modified from Boulton, 1982)

between these essentially rigid elements (Section 5.2.2; Drewry, 1986). In contrast, where the glacier bed consists of deformable materials such as sand, gravel or till, particles protruding from overriding ice can plough up the bed, progressively consolidating it until it provides a large enough resistance to inhibit further movement. Particles may also be lodged against other particles already protruding from the bed, forming clusters (Boulton, 1975, 1982). Additionally, entire masses of debris-rich ice may lodge against the bed.

It will be remembered that the frictional drag between a particle and the bed is proportional to the force pressing the particle against the bed, and that different models have been proposed for calculating the contact forces (Section 5.2.2). According to Boulton (1974, 1975), the most important factor is the weight of the overlying ice minus the water pressure at the bed, whereas Hallet (1979a, 1981) contends that the relevant variables are the buoyant weight of the particle and the velocity of the ice towards the bed, and that ice thickness and water pressure are irrelevant. These two views make differing predictions about where lodgement is most likely to occur: Boulton's model predicts that lodgement will be encouraged below thick ice, but Hallet's model indicates that lodgement is most likely where basal melting rates are high, causing ice and debris to converge with the bed. Both the Boulton and Hallet models predict that lodgement will tend to occur on the upglacier (stoss) side of obstacles on the bed, but for different reasons. According to Boulton, lodgement should occur there because this is where pressure normal to the bed is highest (Fig. 5.25). The buildup of till on the upglacier side will mean that the obstruction will grow in that direction, forming a streamlined form parallel to glacier flow. According to the model, upglacier migration of the till covering

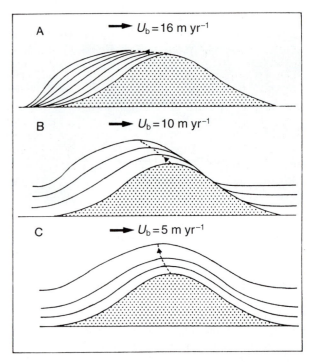

Fig. 5.25 Theoretical incremental patterns of lodgement around a rigid obstacle for different ice velocities at constant effective pressure. (After Boulton, 1982)

will be most rapid where ice velocities are high and effective overburden pressures are low. According to Hallet, however, preferential lodgement will occur on the stoss sides of obstacles because pressure melting (and, therefore, glacier flow towards the bed) is highest on the stoss side of obstructions. In general, Hallet's model of lodgement is probably the more accurate, because if lodgement were encouraged below thick ice, deep troughs and rock basins would tend not to form. It should be emphasized, however, that in some circumstances ice overburden and water pressure may well influence the lodgement process.

The processes of retardation and lodgement are intimately linked with glacier flow mechanisms, because lodged or slowly moving particles are, in effect, obstacles on the bed past which the glacier must flow (Boulton, 1975). For lodgement to occur, therefore, the frictional forces between particles and the bed must be greater than the resistance offered by the particle to glacier flow. As described in Section 4.5.2, there are two main mechanisms allowing ice to slide past obstacles on the bed: enhanced plastic deformation and regelation. Enhanced plastic deformation is thought to be most efficient around large obstacles, whereas sliding by pressure melting and regelation is most efficient for small obstacles. This led Boulton (1975) to propose that some sizes of particles will lodge more readily than others. He argued that small particles (< *c*. 1 cm across) will lodge readily because ice can easily flow past them by regelation sliding, and that large particles (> *c*. 10 cm across) will lodge readily because ice can flow past them by enhanced plastic deformation. Particles of intermediate size will tend to remain in transport, because for this size fraction the drag exerted by the overriding ice will be at a maximum, and therefore more likely to exceed the frictional resistance between the particle and the bed. This reasoning has been criticized by Hallet (1981), but some empirical support for the idea of selective size sorting has been provided by Humlum (1981) and Krüger (1994), who found that basal tills deposited below outlets from the ice cap Myrdalsjökull in Iceland were depleted in fine gravel material, relative to debris-rich basal ice in the glaciers. It was concluded that fine gravel was maintained in basal transport, in contrast to finer and coarser material, which was able to lodge against the bed as basal till.

It should be emphasized that lodgement is unlikely to result in the grain-by-grain accumulation of rigid deposits. Accumulating sediment below a wet-based, sliding glacier is likely to consist of some combination of immobile lodged particles, ploughing particles maintained in traction by the glacier sole, and shearing sediment masses (Boulton *et al.*, 1974; Boulton and Hindmarsh, 1987; Krüger, 1994). Sediment shearing may occur by either brittle or ductile deformation. In a study of basal tills at Slettmark-breen, Norway, Benn (1994a) found that large clasts tended to lodge on the bed, while the smaller clasts were more liable to remain in transport by shearing in a thin subglacial deforming layer. This effect may be partially due to the size sorting mechanisms proposed by Boulton, but probably also reflects the high drag force beneath large, heavy clasts (Hallet, 1979a), and the tendency for small particles to be carried along in the thin deforming layer, whereas larger clasts can penetrate the deforming layer and lodge against deeper, more rigid parts of the substratum.

Slabs of debris-rich ice will lodge against the bed if the frictional drag at the base of the slab exceeds the strength of the ice overlying the slab. Rheological differences between clean and debris-rich ice may encourage the development of a décollement plane immediately above the debris-rich ice. This plane can evolve into the new sliding base of the glacier, the ice below becoming part of the substratum. Once the slab has lodged, the interstitial ice will gradually melt out, releasing the debris, which will either become part of the immobile substratum or be remobilized as a subglacial deforming layer. The characteristics of sediments deposited by lodgement are described in Section 10.3.2.

5.5.2 Melt-out

Deposition by melt-out refers to direct sediment deposition through the melting of stagnant or very slowly moving debris-rich ice. Melt-out can occur in supraglacial or subglacial positions. Only subglacial melt-out is discussed here; supraglacial melt-out is described in Section 6.5.1. The heat required for subglacial melt-out can be supplied from geothermal sources, sensible heat from incoming meltwater, and, if the ice is thin, from the atmosphere. The amount of heat provided by these sources is generally not large, and typical rates of basal melt-out are in the range 5–12 mm yr^{-1} (Drewry, 1986; Paul and Eyles, 1990), although they may be much higher in volcanically active areas. Nobles and Weertman (1971) and Drewry (1986) have argued that melt-out rates are highest in hollows, owing to the convergence of lines of geothermal heat flow in these locations.

During the melt-out process, debris-rich ice undergoes *thaw consolidation*, or volume reduction consequent upon ice melt and meltwater drainage. The amount of consolidation depends on the original debris content of the ice, and will be large if the debris content is low. Where debris content is high, the volume of the deposited debris may be only slightly less than that of the parent ice, and quite delicate englacial structures may be preserved (see Section 10.3.5). The behaviour of released debris during thaw consolidation also depends on the balance between meltwater production and drainage away from the site (Paul and Eyles, 1990). If the meltwater produced during the melt-out process is free to drain away at a rate equal to or higher than the rate of production, debris will be deposited with little or no disturbance, other than that due to consolidation. However, where meltwater cannot drain away freely, porewater pressures will tend to rise during melt-out, decreasing the frictional strength of the deposited debris and increasing the likelihood of failure and remobilization. Paul and Eyles (1990) have argued that such conditions are very common where the debris content of basal ice is predominantly fine-grained. Remobilization of melted-out debris can also occur if the glacier bed is sloping, and, according to Paul and Eyles (1990), failure can occur on slopes as low as 8°.

5.5.3 Deposition by gravity

Basal debris may also be deposited by gravity into subglacial cavities, either in the lee of obstructions or below ice overhangs at the glacier margin (Fig. 5.26). The processes of debris release have been outlined by Peterson (1970), Boulton and Paul (1976) and Boulton (1982). If basal melting rates are high, debris may simply melt out from the roof of a cavity and fall to the floor. This process can be very important in ice-marginal cavities during the ablation season, as anyone who has crawled into one will be aware (Fig. 5.27). Additionally, slurries of saturated debris can flow into cavities from the ice–bed interface. Below thicker ice, clasts can be actively extruded from the basal ice into cavities by excess ice pressure (Peterson, 1970; Vivian, 1975; Boulton,

1982). A related phenomenon is the formation of *till curls*, or curved masses of debris-rich ice, which spall off from the bed of a glacier as it enters a cavity as a result of the different strain responses of debris-rich and clean ice to stress release (Peterson, 1970; Boulton, 1982). Debris from all cavity fill mechanisms depicted accumulates on the bottom of the cavity and may ultimately recouple with the glacier base, either frozen into the basal ice or as a subglacial deforming layer (Boulton, 1982).

5.6 DEFORMING BEDS: EROSION, TRANSPORT AND DEPOSITION

In Section 4.4 we saw how strain in a subglacial deforming layer occurs when stresses imposed by the overlying ice exceed the shear strength of the materials in the layer. *Erosion* associated with a subglacial deforming layer is simply the process by which material is detached from the underlying rigid substratum and set into motion by the applied shear stresses. *Deposition* from the deforming layer is the converse of this process, and occurs when all or part of the deforming layer comes to rest, either because its strength has increased or because the driving

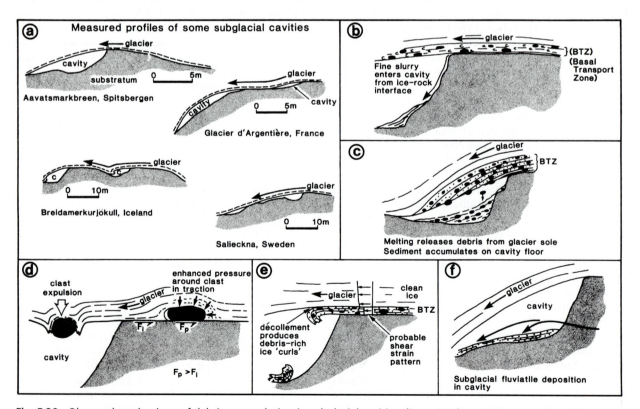

Fig. 5.26 Observed mechanisms of debris accumulation in subglacial cavities. (From Boulton, 1982. Reproduced by permission of Geo Books, Norwich)

Fig. 5.27 Debris melting out from the roof of a subglacial cavity. (Photo: D.I. Benn)

stresses have fallen. Much of the erosion and deposition associated with deformable beds can be viewed as the downward and upward movement, respectively, of the base of the deforming layer, occurring as conditions become more or less favourable for deformation (Fig. 5.28; Alley, 1989; Hart and Boulton, 1991; Hart, 1995a; Boulton, 1996). *Transport* occurs within the deforming layer itself. The rate of sediment transport depends on the thickness of the deforming layer, and the strain rate.

The deforming layer can *thicken* by: (a) downward migration of the base, which erodes deeper material incorporating it into the deforming layer; (b) the addition of material melting out from the overlying glacier; or (c) the arrival of more material from

upglacier than leaves; that is, a downglacier reduction in strain rates (Hart and Boulton, 1991; Hart, 1995a). In all cases, thickening is contingent on the local stress conditions being sufficient to maintain the deformation. Conversely, *thinning* of the deforming layer can occur by: (a) deposition at the base; (b) the freezing of material on to the overlying glacier; or (c) more material leaving the site than arrives from upglacier, or a downglacier increase in strain rate. The *strain rate* in a deforming layer is a function of shear stress, overburden pressures, water pressure and material properties (Section 4.4.2).

5.6.1 Erosion and transport

As noted above, erosion occurs by the downward movement of the base of a deforming layer. This mobilization of subsole materials has been termed *excavational tectonics* by Hart *et al.* (1990). It need not involve the disaggregation of all the newly eroded material, and intact fragments of unconsolidated sediments or weak rocks can be detached and rafted along within the deforming layer. Intact fragments can also be detached from their source because they overlie a weak failure surface, or *décollement surface*, such as a bedding plane or layer of weak sediment in the parent material. Detached fragments will not be disaggregated during transport if their shear strength exceeds the stresses that are causing the deformation. That is, the fragments must be significantly stronger than the deforming material for some reason, such as (a) dif-

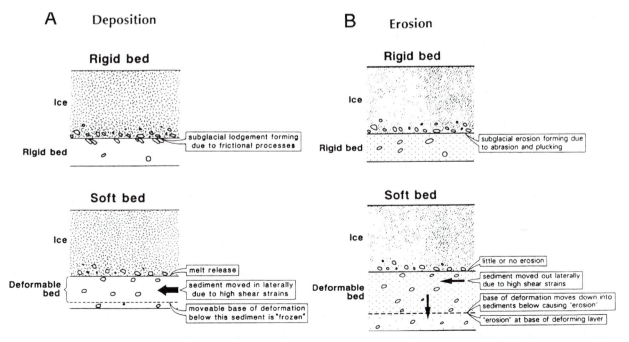

Fig. 5.28 Contrasting processes of (A) deposition and (B) erosion for rigid and soft glacier beds. (Modified from Hart, 1995a)

ferences in consolidation, (b) greater frictional or cohesive strength due to grain size differences, or (c) better drainage allowing pore water to escape. For unconsolidated sediments, sands, gravels and diamictons tend to be more resistant to deformation than silts and clays, owing to their higher coefficients of friction and/or higher permeabilities (Boulton and Jones, 1979; Boulton and Hindmarsh, 1987; Benn and Evans, 1996). Rafted fragments (or *rafts*) may vary greatly in size, from the centimetre scale (when they are termed *soft clasts, lenses, inclusions* or *boudins*) up to several kilometres (when they are termed *megablocks*) (Ruszczynska-Szenajch, 1987; Hart, 1995a).

In some cases, rafts and other fragile inclusions such as mollusc shells can be transported tens of kilometres in subglacial deforming layers without disaggregation, although they may be eroded at their edges. In other situations, rafts can suffer ductile deformation, but at lower strain rates than the surrounding deforming material, behaving like *augen* or *boudins* in metamorphic rocks (Hart and Boulton, 1991; Hart and Roberts, 1994; Hart, 1995a; Benn and Evans, 1996). Evidence from particle wear patterns shows that even hard rocks, such as basalt and gneiss, can undergo abrasion and fracture within a subglacial deforming layer (Benn, 1994a, 1995; Benn and Evans, 1996). The differential survival, wear and deformation of inclusions of widely differing strength – from shells to rock clasts – reflects the fact that inter-particle contact forces vary greatly in subglacial deforming layers. Where deformation occurs at very low shear stresses, contact forces may be low enough to permit the survival of very fragile rafts. In contrast, the brittle shear of stiff, stony till – such as in the B-horizon beneath Breidamerkurjökull, Iceland – requires relatively high shear stresses, and contact forces are high enough to cause particle abrasion and fracture (Boulton and Hindmarsh, 1987; Benn, 1995). Evidence for wear of rafts and rock clasts *in situ*, therefore, is a potentially important source of information on inter-particle contact forces and deforming layer strength.

Geological evidence suggests that a type of plucking process can occur beneath deforming sediments. In places where till overlies well-jointed rock, till has been observed penetrating joints and surrounding detached blocks, suggesting that the till has been squeezed under pressure around the blocks (Broster, 1991; Broster and Park, 1993). One possibility is that saturated till can act as a lubricant, reducing frictional drag to the point where blocks can be mobilized by the driving stresses.

It has also been suggested that abrasion of bedrock and other rigid surfaces can occur at the base of a deforming layer. Gjessing (1965) attributed meandering, striated grooves on rock surfaces to abrasion by saturated, fluid till, and MacClintock and Dreimanis (1964) argued that rock surfaces in the St Lawrence Valley had been abraded beneath deforming till below the Laurentide ice sheet. The horizontal, planed-off surfaces of some 'boulder pavements', or layers of boulders within till, may also be due to sub-till abrasion (Clark, 1991a).

5.6.2 Deposition

Deposition from a subglacial deforming layer will occur if the driving stresses fall or the shear strength of the layer increases so that the stresses are no longer high enough to sustain deformation. Shear stresses can change over time as the ice configuration changes during glacier advance and retreat, and will fall to zero below ice-divides or upon deglaciation. The most important factor causing short-term changes in shear strength is a change in porewater pressure. If the porewater pressure drops as the result of *dewatering*, the effective pressure forcing the mineral grains together will increase, causing the frictional strength of the sediment to rise. Dewatering may be temporary, reflecting short-term changes in the supply of meltwater to the bed, or may be due to longer-term changes in glacier configuration or drainage. Strength may also increase if the deforming layer freezes.

Deposition from a subglacial deforming layer will occur from the base up, because overburden pressures, and hence frictional strength, increase with depth, and the deepest sediments are therefore likely to come to rest first if shear stresses fall or strength rises. Deformation may, of course, cease first at higher levels if these parts of the layer are stronger than the underlying sediments, as in the case of sands overlying clays. In this case, however, the sands would not be considered to be *deposited*, since they would still be rafted along on the lower, deforming clays.

The deposition process may involve intermediate changes of state and style of deformation. For example, dewatering of dilatant, pervasively deforming till may result in collapse of the sediment and a transition to brittle deformation prior to deposition (Alley, 1989; Benn and Evans, 1996). Alternatively, dilatant till may come to rest with little or no change in its characteristics, as appears to be the case for the A-horizon at the margin of Breidamerkurjokull (Benn, 1995). Changes in state during deposition can exert a strong influence on the final characteristics of the deposit, particularly particle fabric and geotechnical properties. These characteristics are discussed in detail in Sections 10.3.3 and 10.3.4.

Another mechanism for deposition from a sub-glacial deforming layer has recently been proposed by Iverson *et al.* (1995), who made direct measurements of till deformation below boreholes through Storglaciären, Sweden. They showed that downglacier shear strain rates in the deforming layer *fell* at times of very high subglacial water pressure, even though the surface velocity of the glacier increased. This suggested that, when water pressures were high, the glacier was decoupling from its bed and sliding on an extensive water film at the ice–till interface. At such times, little or no till deformation occurred: the till had, in effect, been deposited. However, this deposition is likely to be temporary, because as water pressure falls and the glacier and the bed recouple, deformation will recommence.

Deposition from a subglacial deforming layer may be intimately associated with deposition by meltwater. Water flow in channels or canals at the ice–sediment interface, or flow within the sediment, can deposit lenses of sorted silt, sand and gravel, which may undergo subsequent transport and strain within the deforming layer (Sections 3.4.2, 3.4.5 and 3.4.6; Walder and Fowler, 1994; Clark and Walder, 1994; Benn and Evans, 1996). Such lenses may be hard to differentiate from rafts of the preglacial substratum that have become incorporated into the deforming layer, but their recognition is important because they provide important information on subglacial hydraulic conditions during deformation.

5.7 EFFECTS OF SUBGLACIAL TRANSPORT ON DEBRIS

5.7.1 Active transport

Frequent interactions between particles, and between particles and a rigid bed, result in their substantial modification during transport and deposition in the subglacial environment. Inter-particle stresses are often high enough to cause fracture and abrasion of particles, in much the same way as bedrock failure, described in Section 5.3. In view of its importance in modifying debris, Boulton (1978) introduced the term *active transport* for debris transport in the basal shear zone of glaciers, in contrast with *passive transport*, which refers to debris transport with little or no modification at higher levels within or on the glacier (Section 6.4.1). There are two main effects on debris characteristics as a result of active transport: (a) particle breakage reduces the size of particles and creates distinctive grain size distributions; and (b) the form or morphology of particles is changed as a result of breakage or wear.

5.7.2 Granulometry

Abrasion and breakage of particles during active transport produces grain size distributions that are very different from those resulting from periglacial mechanical weathering or transport by flowing water (Fig. 5.29). Distributions commonly exhibit a very

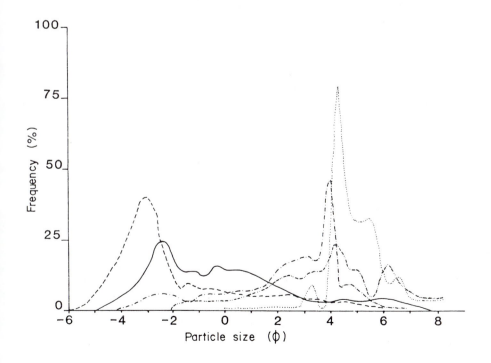

Fig. 5.29 Grain size distributions from debris-rich ice, Matanuska Glacier, Alaska. The curves with prominent peaks in the fine grain sizes (Ø 2–6) are typical of actively transported debris. For definition of the ø scale of particle size, see Appendix 2. (From Drewry, 1986, after Lawson, 1979a)

broad range of particle sizes, from clay or silt up to pebbles or larger, and are typically *bimodal* or *polymodal* (i.e. with two or more peaks in the grain size distribution). Such distributions reflect progressive particle size reduction or *comminution* as particles are fractured under applied stresses during shear. Boulton (1978) and Haldorsen (1981) attributed bi- and polymodal distributions to the effects of two processes: *crushing*, or the breakage of two interlocking grains, and *abrasion*, or the production of fine-grained fragments as two grains slide past each other. The products of crushing and abrasion were studied by Haldorsen (1981), who simulated subglacial particle wear using a *ball mill*, in which rock fragments and steel balls are tumbled in a drum. She found that crushing can produce a range of particle sizes from 0.016 to 2 mm, similar to one of the modes found in subglacial till samples from Norway (area A, Fig. 5.30). In contrast, experimental abrasion produced fragments in the range 0.002–0.063 mm, similar to the finest component of the till samples (area B, Fig. 5.30). Haldorsen interpreted coarser clast modes in the tills as residual rock fragments which could have been reduced further by subglacial crushing and abrasion, given sufficiently long transport distances (area C, Fig. 5.30). The identification of separate components in actively transported debris is facilitated by *Gaussian component analysis*, which statistically resolves polymodal distributions into a series of normal (bell-shaped or Gaussian) curves (Sheridan *et al.*, 1987). This technique was applied by Sharp *et al.* (1994) to debris from basal ice formed during the quiescent and surging phases of Variegated Glacier (Section 4.8.2), allowing the effects of different processes of wear and selective entrainment and deposition to be identified.

The progressive size reduction of subglacial debris during transport was studied by Dreimanis and Vagners (1971, 1972) by analysing the granulometry of tills with distance from known bedrock sources.

They found that coarse modes are well developed close to the debris source, and fine modes increase in importance with increasing transport distance, but there appeared to be a lower size limit beyond which no further comminution occurred, regardless of transport distance (Fig. 5.31). This limit was termed the *terminal grade*, and was interpreted as the particle size at which the available energy during transport is insufficient to cause further fracture. The existence of terminal grades can be partially explained by Griffith crack theory (Section 5.3.1), which shows that the energy required for crack growth and rock fracture is inversely proportional to crack length. The fracture of small particles can occur only by the growth of short cracks, and therefore requires greater energy input than the fracture of large particles by the growth of relatively long cracks (Bond, 1952; Drewry, 1986). Terminal grades are equivalent to the *limit of grindability* recognized by engineers, and appear to be related to the maximum stress concentrations that occur during shear. As would be expected, softer minerals, such as mica or feldspar, have smaller terminal grades than harder, resistant minerals such as quartz. Haldorsen (1983) found that the finer grades in experimentally crushed debris and basal tills are enriched in feldspar and sheet silicate minerals compared with coarser grades in which quartz is more abundant. However, 'terminal grades' should not be defined on the basis of sediment granulometry alone, as grain size modes may represent transient stages of comminution, which could have been further reduced given more time (Haldorsen, 1981; Drewry, 1986).

A new and interesting approach to interpreting particle size distributions has been taken by Hooke and Iverson (1995) and Iverson *et al.* (1996) using techniques developed by Sammis *et al.* (1987) for the study of tectonically crushed rocks. This method considers the *number* of grains per size fraction as shown on double logarithmic plots. Hooke and Iverson noted that, for subglacially sheared tills, the plot-

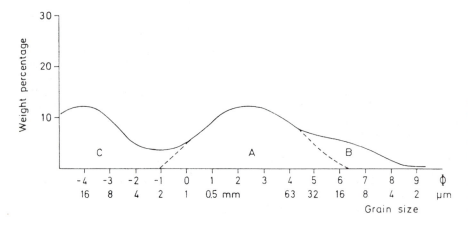

Fig. 5.30 Haldorsen's interpretation of the grain size distribution of 150 till samples. Area A: resistant crushed fraction; area B: component from abrasion; area C: residual clast mode. (From Haldorsen, 1981. Reproduced by permission of Scandinavian Academic Press)

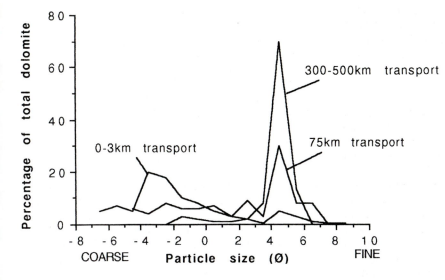

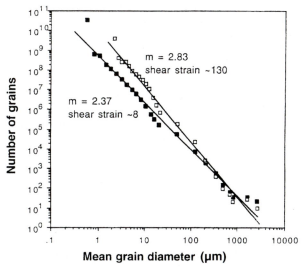

Fig. 5.31 Frequency distributions of dolomite in till samples from the Hamilton–Niagara region, Canada, showing the effects of progressive comminution during transport. (Redrawn from Dreimanis and Vagners, 1971)

Fig. 5.32 Plot of grain diameter v. number of grains for two experimentally comminuted sediment samples. The nearly linear plots suggest that the distributions have pseudo-fractal properties; i.e. they are self-similar at a range of scales; m is the fractal dimension of the samples (From Iverson et al., 1996. Reproduced by permission of the International Glaciological Society)

ted lines tend to be straight, indicating that the distributions are self-similar at all scales (i.e. they have *fractal* characteristics; Fig. 5.32). Sammis *et al.* (1987) argued that the slope gradients (*fractal dimensions*) of distributions produced entirely by grain fracture should be around 2.58, reflecting grain fracture near the contact between two similarly sized particles, where stress concentrations are highest. Progressive fracturing in these locations during shear will result in maximum cushioning for each particle and the minimization of stress concentrations. Calcu-

lated fractal dimensions for the subglacial samples were close to 3, leading Hooke and Iverson (1995) to propose a modified form of this mechanism for deforming subglacial tills, in which slippage and abrasion between grains occurs in addition to shear by intergranular fracture, producing an excess of fines (see also Section 4.4.2.5). Hubbard *et al.* (1996) pointed out that grain size distributions of actively transported debris are not truly fractal, because of the presence of distinct modes, although they found that the method was useful for inferring wear processes in modern alpine and Icelandic subglacial environments. The presence of modes in grain size distributions was not addressed by Hooke and Iverson (1995) and Iverson *et al.* (1996), but it could perhaps be explained by the influence of fracture distribution on fragment stability at different scales. It should be noted, however, that apparently fractal distributions also occur in non-glacial sediments (Hooke and Iverson, 1995), and may be produced by the mechanical mixing of distinct debris populations. Until such possibilities are studied in detail, therefore, the 'fractal' dimensions of subglacial debris should be interpreted with caution.

As noted in Section 5.6.1, not all debris in subglacial environments is produced by rock fracture; some may be entrained from pre-existing sediments beneath the glacier. As a result, grain size distributions may be inherited from other glacial or nonglacial processes. Subsequent shear of such sediment during active transport can modify the original distribution by particle comminution or mixing of the sediment with other debris.

5.7.3 Particle morphology

Active subglacial transport imparts very distinctive signatures on the *morphology*, or *form*, of sedimentary particles (Boulton, 1978; Benn and Ballantyne,

1994). In the following sections we will examine the form of *clasts*, or pebble to boulder-sized particles, and the morphology of *matrix particles* (clay, silt and sand), or *micromorphology*.

5.7.3.1 CLAST FORM

The effects of active subglacial transport and deposition on clast form have been studied by many researchers over the years (e.g. Holmes, 1960; Drake, 1972; Boulton, 1978; Krüger, 1984, 1994; Dowdeswell *et al.*, 1985; Benn and Ballantyne, 1994). It is useful to regard clast form as the sum of three characteristics at different scales, all of which are affected by active transport: (a) *shape*, or the relative dimensions of the long, intermediate and short axes of a clast; (b) *roundness*, or the degree of curvature of clast edges; and (c) *texture*, or the character of the clast surface (Griffiths, 1967; Barrett, 1980). To these three characteristics can be added *wear patterns*, or the distribution of different types of wear over the clast surface (Benn, 1994a, 1995).

Clast shape is dependent on both process and lithology. In mountain environments, actively trans-

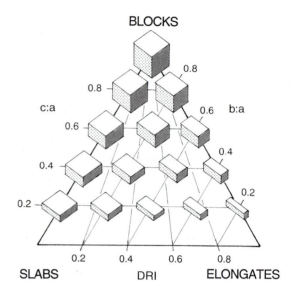

Fig. 5.33 The continuum of particle shape, showing variation between blocks, elongates, and slabs. The diagram is scaled using the ratios between the long (a), intermediate (b) and short (c) axes. DRI is the 'disk-rod index', equal to $(a-b)/(a-c)$. (From Benn and Ballantyne, 1993)

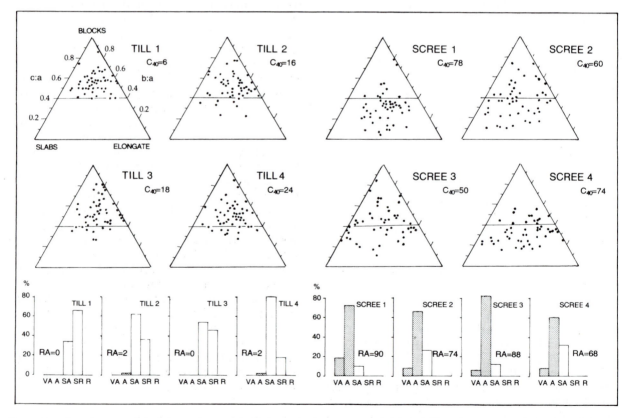

Fig. 5.34 Clast shape and roundness characteristics for actively transported debris (till) and scree near the margins of Storbreen, Norway. Roundness categories are as follows: VA, very angular; A, angular; SA, subangular; SR, subrounded; R, rounded. C_{40} is the percentage of clasts with $c{:}a$ ratios ≤ 0.4, and RA is the percentage of clasts in the VA and A categories. Note the low C_{40} and RA values for the till samples compared with the scree. (From Benn and Ballantyne, 1994. Reprinted by permission of Elsevier)

ported clasts of massive, coarse-grained rocks such as granite and gabbro tend to have compact, blocky shapes, in contrast with the more elongate and slabby shapes of periglacially weathered clasts of similar lithologies (Fig. 5.34; Ballantyne, 1982a; Vere and Benn, 1989; Benn and Ballantyne, 1994). This may reflect the initial shape characteristics of subglacially plucked clasts, or may be due to preferential breakage of clasts across their long axes during transport at the ice–bed interface (Drake, 1972; Benn, 1990). For fissile rocks such as shale and some limestones, however, actively transported clasts generally have more elongate shapes, indicating preferential fracture along bedding planes (Benn, 1990).

Roundness is affected in opposite ways by abrasion and fracture during active transport. Abrasion increases edge-rounding and creates polished facets, whereas fracture creates new, sharp edges and fresh faces. The balance between these two processes means that both angular and well-rounded forms tend to be rare in actively transported debris, and most clasts have intermediate roundness characteristics (i.e. subangular and subrounded; Fig. 5.34). Studies of clast-roundness evolution with distance from a known source have demonstrated that only short distances of active transport are necessary for

significant edge-rounding to occur (Drake, 1972; Humlum, 1985a).

Indices of shape and roundness can be plotted together on biaxial diagrams, allowing actively transported debris to be easily identified. Ballantyne (1982a) and Benn (1992b) found that the optimum shape and roundness indices for distinguishing actively and passively transported debris are based on clast *c:a* axial ratios and the percentage of angular and very angular clasts in a sample (Fig. 5.35a; see also Ballantyne, 1982; Benn and Ballantyne, 1994; Ballantyne and Benn, 1994a). These indices clearly highlight the blocky and edge-rounded character of actively transported clasts. Less powerful shape roundness diagrams were introduced by Boulton (1978) and have been widely used in studies of terrestrial and marine glacial environments (e.g. Domack *et al.*, 1980; Dowdeswell *et al.*, 1985; Hambrey, 1994; Fig. 5.35b). These diagrams show distinct between-sample differences in roundness, but the shape data have similar ranges in all samples. In these cases, therefore, plotting shape against roundness yields little more information than the use of roundness indices alone, because the shape index was insensitive to the actual variation in the data.

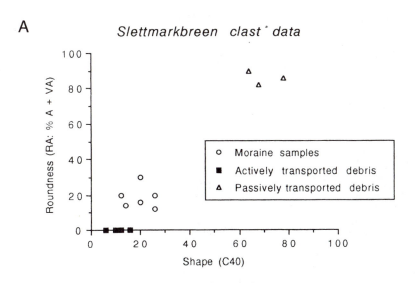

A — Slettmarkbreen clast data

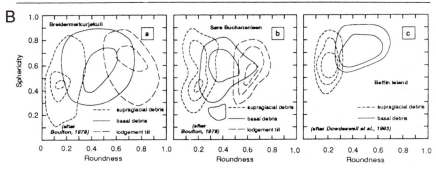

Fig. 5.35 The use of biaxial plots of shape and roundness to differentiate debris from different transport paths. (A) Actively and passively transported debris from Slettmarkbreen, Norway, plotted using C_{40} as a measure of shape and RA as a measure of roundness. Each point represents a sample of 50 clasts. Samples from a recessional moraine plot between the actively and passively transported samples, showing that the moraine is composed of debris from both transport paths. (B) Samples from Breidamerkurjökull, Søre Buchananisen and Baffin Island, plotted using Krumbein roundness and sphericity (shape) indices. The fields represent one sample each. Note that sphericity shows little variation between samples, owing to the insensitivity of the Krumbein index to the variation in the data. (From Hambrey, 1994, after Dowdeswell *et al.*, 1985. Reproduced by permission of the *Journal of Sedimentary Petrology*)

The *surface texture* of actively transported debris is also very distinctive. Interactions between particles and between particles and the bed result in polished and striated faces, particularly on fine-grained lithologies such as limestone, slate and basalt. Coarse-grained rocks such as granite tend not to bear clear striae, because the different mineral constituents offer varying resistance to erosion and strongly influence patterns of surface wear. Striae on clasts may be straight and parallel, indicating stability of the clast relative to the striating medium, or curved and exhibiting a wide variety of orientations, reflecting clast rotation and realignment during wear

(Fig. 5.36; Hicock, 1991; Benn, 1995). Straight striations aligned parallel to clast long axes are typical of the upper surfaces of clasts lodged beneath sliding ice (Sharp, 1982; Krüger, 1984; Benn, 1994a). Cross-cutting striae on individual clasts attest to clast rotation during their abrasion history. This may take place (a) in basal ice layers; (b) at the ice–substrate interface; or (c) in the deforming layer. Alternatively, a single clast with multiple striae alignments may have been subject to several phases of abrasion during different glacier events.

Humlum (1985a) studied the effects of subglacial transport on coarse debris by tracing changes in the

(a)

(b)

Fig. 5.36 Wear patterns on actively transported clasts from (a) Iceland and (b) Patagonia. (Photos: D.I. Benn)

size, shape and roundness of clasts with distance from a distinctive rhyolite rock knob on the foreland of Slettjökull, Iceland. He found a marked decrease in size and a slight increase in the proportion of equidimensional particles within the first 500 m of transport, but little change thereafter. These trends were attributed to selective transport, in which the largest particles tend to become lodged in the till beneath the glacier, whereas smaller, equidimensional particles tend to remain in transport until deposited when the glacier retreats. Humlum also found an increase in particle roundness with distance, indicating that abrasion caused some clast modification during transport.

An important aspect of clast morphology is the *asymmetry of wear patterns*. Clasts from fluvial or coastal environments are generally evenly worn on all parts of their surfaces, but this is often not the case for subglacially transported or deposited clasts, owing to the asymmetric distribution of stresses in the subglacial environment. Clasts lodged beneath sliding ice tend to develop asymmetric 'stoss-and-lee' forms owing to abrasion *in situ* of their upglacier (stoss) sides and plucking on their downglacier (lee) sides, similar to the erosion of a roche moutonnée in bedrock (Figs 5.37a, 5.38a; Boulton, 1978; Krüger, 1979; Sharp, 1982; Section 9.3.1). Krüger (1984) identified related 'double stoss-and-lee' forms, which he attributed to a two-stage process of clast wear at the ice–till interface (Figs 5.37b, 5.37c and 5.38b). Prior to deposition, the clast is ploughed through the underlying till, causing abrasion of the lower leading edge and fracture of the lower trailing edge. The clast then lodges against the bed and the stoss side is abraded while the lee side is fractured.

Sedimentological data indicate that clasts can also undergo asymmetric wear during shear in subglacial deforming layers. On the basis of detailed studies at Breidamerkurjökull, Benn (1995) argued that clasts rafted along within deforming till will be subject to stress fluctuations as the clasts override or are overridden by other clasts. The effects of stress fluctuations will be most pronounced at the upglacier and downglacier ends of clasts, owing to the presence of low-pressure zones in front of the upper leading face and behind the lower trailing face. If the clast cannot rotate out of the way (i.e. if the matrix is stiff and inter-clast contact forces are high), the passage of neighbouring clasts across the upper leading edge and below the lower trailing edge of the clast will set up steep stress gradients which may lead to fracture (Fig. 5.37d). Abrasion should also occur at the lower leading face and the upper trailing face. Flat, polished facets may be eroded on the upper and lower surfaces of clasts embedded within deforming sediment, particularly if the clast surface coincides with a shear plane (Fig. 5.37e, 5.37f).

5.7.3.2 MICROMORPHOLOGY

Matrix particles exhibit a range of shapes and surface textures, and several studies have sought diagnostic criteria to identify grains that have undergone active subglacial transport. Particular attention has focused on the surface texture of quartz grains as revealed through the scanning electron microscope (Whalley, 1996), because quartz is a durable, common mineral

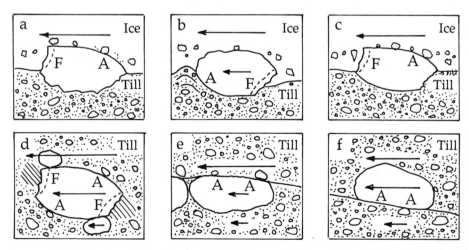

Fig. 5.37 Development of clast wear patterns beneath sliding ice and within a deforming layer, showing principal locations of abrasion (A) and fracture (F). Arrow lengths show relative velocities. (a) Lodged clast with stoss–lee form owing to stoss-side abrasion and lee-side fracture below sliding ice. (b), (c) Double stoss–lee morphology resulting from a two-stage process of ploughing (b) and lodgement (c). (d) Double stoss–lee clast resulting from a one-stage process within a deforming layer. Low-pressure zones are shaded. (e), (f) Flat, polished facets eroded on the upper and lower surfaces of clasts adjacent to a shear plane. (Modified from Krüger, 1984, and Benn and Evans, 1996)

(a) **(b)**

Fig. 5.38 (a) Stoss–lee clast on the foreland of Breidamerkurjökull, Iceland. (b) Double stoss–lee clast in a former subglacial deforming layer, Strait of Magellan, Patagonia. (Photos: D.I. Benn)

which exhibits fracture patterns that were thought to be relatively independent of mineral structure. However, results have been ambiguous, and the identification of distinct 'glacial' quartz textures remains elusive.

Krinsley and Doornkamp (1973) proposed that 'glacial' quartz grains can be recognized on the basis of their sphericity (high *c:a* ratios), angularity, and textures indicative of mechanical fracture, characteristics that were subsequently used to infer the presence of actively transported debris in the geological record (e.g. Blank and Margolis, 1975; Drewry, 1975). However, similar grains also occur in non-glacial environments characterized by high-energy transport or mechanical weathering (Whalley and Krinsley, 1974; N. Eyles, 1978a). N. Eyles (1978a) proposed that a more reliable diagnostic feature of subglacially transported quartz grains is the presence of *microblock texture*, consisting of a series of steps and angular corners on the grain surface. In turn,

Gomez *et al.* (1988) showed that microblock textures are not unique to subglacially transported grains, and concluded that quartz grain microtextures commonly reflect the history of crystallization, deformation and pre-glacial fracture experienced by the grain, rather than its glacial transport history.

An alternative view has been taken by Mahaney *et al.* (1988) and Mahaney (1990, 1995), who argue that glacially crushed quartz grains can be identified using a range of textures and fracture types. Indeed, these authors have used the relative abundance of certain microtextures to differentiate grains transported below mountain and continental glaciers. However, these conclusions are controversial (e.g. Clark, 1989), and require further rigorous testing. At present, the only definitive statement that can be made about quartz microtextures is that they are more ambiguous environmental indicators of active subglacial transport than the morphology of clast-sized particles.

CHAPTER
6

SUPRAGLACIAL AND ENGLACIAL ENVIRONMENTS

E, E, indicate the line of erosion.

6.1 INTRODUCTION

The supraglacial environment is the most familiar to mountain travellers, and gives glacierized landscapes much of their distinctive character (Fig. 6.1). Well-known topographical features such as medial moraines, crevasse fields and debris-covered glacier snouts are often, however, just the surface expression of deeper englacial structures that penetrate much of the body of the ice. Collectively, supraglacial and englacial structures contain a great deal of information on glacier formation, deformation and flow, and the transfer of debris through the glacial system. Furthermore, processes of debris movement and sorting on glacier surfaces exert a strong influence on landform evolution in many glacial environments. In this chapter, we examine the morphology and structure of glacier ice, processes and patterns of supraglacial and englacial debris transport, and the evolution of debris-covered glacier surfaces. Processes of snow and ice accumulation and ablation at the glacier surface are reviewed in Section 2.2, and supraglacial and englacial drainage are described in Section 3.3.

Fig. 6.1 Englacial conduit melting out from the surface of the Tasman Glacier, New Zealand. (Photo: M.P. Kirkbride)

6.2 ICE MORPHOLOGY AND STRUCTURE

Glacier ice is not uniform, but exhibits a wide variety of internal and superficial structures. In this section, we describe structures composed of glacier ice,

including crevasses, icefalls, ogive banding, and layering. Structures composed of rock debris form the subject of Section 6.3.

6.2.1 Crevasses

Crevasses are very prominent surface features on many types of ice mass, from small cirque glaciers to ice sheets. They form where ice is being pulled apart by tensile stresses that exceed the strength of the ice, and are usually oriented at right angles to the principal stress direction (Fig. 6.2; Nye, 1952; Meier, 1960; Hambrey, 1994; Paterson, 1994). Crevasses are therefore very useful indicators of the orientation of stresses in a glacier. Figure 6.3 illustrates three types of crevasse patterns which are common on valley glaciers. *Chevron crevasses* are linear fractures aligned obliquely upvalley from the glacier margins towards the centre-line, and form in response to the drag introduced by the valley walls (Fig. 6.3a). Because the ice is moving more rapidly at the centre of the glacier than at the margins, tensile stresses are set up that pull the ice downglacier from the margins towards the centre-line, tending to open up crevasses aligned at approximately 45° to the valley walls (Nye, 1952). Second, *transverse crevasses* form in valley glaciers subject to extending flow. Near the centre, the principal tensile stress is parallel to glacier flow, so that transverse crevasses open up at right angles to the centre-line and curve downstream, where the stress pattern is strongly influenced by the drag of the valley walls (Fig. 6.3b). Finally, *splaying crevasses* form where the glacier is subject to compressive flow which causes the glacier to expand laterally, modifying the stress pattern (Fig. 6.3c). In this case the crevasses are curved, and are roughly parallel to the flow direction towards the centre and bend outwards to meet the margins at angles of less than 45°. Similar crevasse patterns are also found near the

Fig. 6.2 Transverse crevasses on Myrdalsjökull, Iceland. (Photo: J. Wright)

(a)

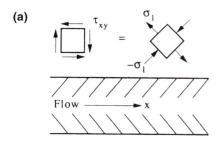

(b)

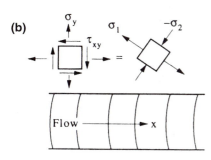

(c)

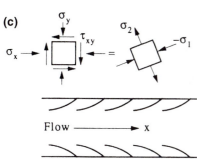

Fig. 6.3 Crevasse patterns in a valley glacier. The diagrams at the top show shear stresses and longitudinal and transverse normal stresses at the glacier surface near the upper margin (left panels) and the resultant principal stresses (right panels). (a) Chevron crevasses resulting from shear stress exerted by the valley walls. (b) Transverse crevasses resulting from a combination of shear stress and longitudinal tensile stress (extending flow). (c) Splaying crevasses due to a combination of shear stress and longitudinal compressive stress (compressive flow). (Modified from Nye, 1952)

margins of ice streams, where there are strong lateral gradients in ice velocity (Fig. 1.15).

A type of crevasse that is characteristic of cirque and valley glaciers is the *randkluft*, a fissure separating glacier ice from the rock wall above the head of the valley. These crevasses form partly because of preferential ablation adjacent to warm rock surfaces, but also reflect movement of the ice away from the rock wall (Mair and Kuhn, 1994). *Bergschrunds* are related features, consisting of deep, transverse crevasses near the heads of valley glaciers. It is often stated that bergschrunds separate immobile, cold-based ice at the head of the glacier from active, slid-

ing ice below, but Mair and Kuhn (1994) have shown that on Daunferner, a glacier in the Stubai Alps, Austria, ice is sliding both above and below a large bergschrund, which formed in a region where ice thickness and velocity were rapidly increasing. Comparative studies in other areas are few, but it seems likely that the patterns observed by Mair and Kuhn also occur on other glaciers. Randklufts and bergschrunds can present formidable barriers to climbers in alpine regions, and can form hidden hazards if concealed by drifted snow (Fig. 6.4).

Ice masses which are unconstrained by valley walls also exhibit distinctive crevasse patterns (Sugden and John, 1976). Piedmont lobes can have striking radial crevasse patterns related to the spreading of the ice (Fig. 6.5A), whereas on ice sheets and ice caps, transverse crevasses due to extending flow may extend straight across the ice for great distances (Fig. 6.2). Crevasse fields due to extending flow conditions are particularly common near the grounding lines of ice streams.

Once formed, crevasses will be carried by glacier flow into areas where stress conditions differ from those which formed them. Crevasses at the margins of valley glaciers and ice streams will tend to rotate because the part nearest the centre-line will have the highest velocity, and both splaying and transverse crevasses will tend to be straightened (Sugden and John, 1976). However, rotated crevasses will not remain open unless the direction of principal tensional stress also rotates by the same amount. As stress conditions change, crevasses will close up or be replaced by others with more suitable orientations. Crevasse patterns can change dramatically at times of changing stress regime, such as when flow is accelerating or decelerating through time. Figure 6.5 shows the crevasse patterns on Bodleybreen, a surging glacier on Svalbard, before and during a surge. In the pre-surge pattern, the snout area

Fig. 6.4 Crevasses on the upper Glacier du Géant, Mont Blanc. Bergschrunds are clearly visible curving below the steep ice slopes in the background. (Photo: D.I. Benn)

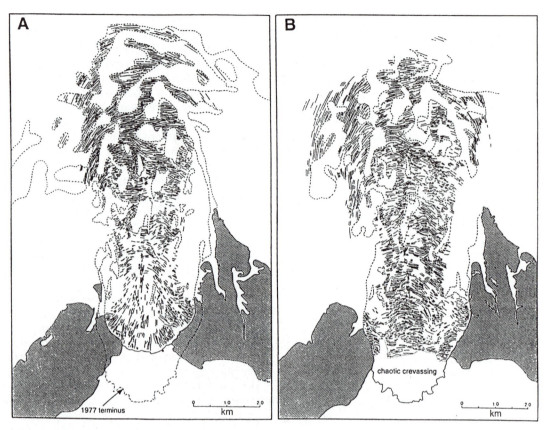

Fig. 6.5 Crevasse pattern maps of Bodleybreen (A) before and (B) during its surge. Note the radial crevasses on the pre-surge lobe and the widespread transverse crevasses opened up by extending flow during the surge. (From Hodgkins and Dowdeswell, 1994. Reproduced by permission of the International Glaciological Society)

displays radial crevasses indicative of lateral spreading, whereas during the surge the snout is traversed by numerous transverse crevasses reflecting large longitudinal tensile stresses (Hodgkins and Dowdeswell, 1994).

Upon closing, crevasses leave linear scars known as *crevasse traces* (Hambrey, 1975, 1994). These commonly take the form of prominent layers of blue ice formed by the freezing of meltwater in the crevasse prior to closure. Alternatively, thin layers of white, bubbly ice can mark the former position of snow-filled crevasses. In both cases, isolated rock particles or clusters of debris may be present within the crevasse trace, composed of material that had fallen into the original open crevasse. A more subtle type of trace forms during brittle fracture of the ice but without separation of the walls. If the rate of extension is not too great, an open crevasse does not form; instead, the stretching is taken up by the recrystallization of ice parallel to the fracture, creating a *tensional vein* analogous to those in rocks (Hambrey and Müller, 1978). Crevasse traces and tensional veins are usually initially close to vertical, but are rotated into other orientations by glacier flow.

6.2.2 Icefalls

Icefalls are steep zones on a glacier where ice flow is extremely rapid, sometimes exceeding 10 times that of the glacier elsewhere along its course. The acceleration of a glacier as it enters an icefall creates a zone of extreme extending flow where the ice thins and stretches under large tensile stresses, opening up large numbers of crevasses and breaking the ice surface into unstable ice blocks known as *séracs*. The deceleration of ice flow at the base of an icefall creates a zone of compressive flow, where crevasses are closed up and the ice thickens (Fig. 6.6).

Ice cover is not necessarily continuous through icefalls, and there are sometimes expanses of bare rock slabs or cliffs between the top of the icefall and the rest of the glacier below. Ice flow through the icefall takes the form of intermittent avalanches triggered by the collapse of séracs, piling up cones of broken ice blocks at the base. Such discontinuous icefalls are very common in high mountain environments, where accumulation areas can be separated from glacier snouts by rock cliffs many hundreds of metres high. Clearly, icefalls are extremely hazardous places to

Fig. 6.6 Hochstetter Icefall, Mount Cook, New Zealand. (Photo: M.P. Kirkbride)

be, and some, like the notorious Khumbu Icefall on Mount Everest, present serious obstacles to mountaineering expeditions and cost many lives.

6.2.3 Ogives

Ogives are repetitive arcuate bands or waves exposed on the surface of some valley glaciers, which form below some icefalls. They are convex downflow, with the amount of curvature increasing in each successive band as the result of the greater velocity of the central part of the glacier compared with the margins (Fig. 6.7). Two basic types of ogive have been recognized, but both have a similar origin, and intermediate types are common. *Banded ogives* are alternating bands of dark and light ice, and are the surface expression of three-dimensional spoon-shaped structures dipping upstream within the glacier. The dark bands consist of highly foliated, rather dirty ice, whereas the light bands tend to be composed of more uniform bubbly ice. *Wave* or *swell-and-swale ogives* are similar in planform, but consist of alternating ridges and troughs (Paterson, 1994).

Ogives are formed annually, each light–dark or ridge–trough pair representing one year of ice movement. They are therefore a useful means of determining ice velocities, as first suggested by the Scottish glaciologist James Forbes in the mid-nineteenth century (for this reason ogives are sometimes known as *Forbes bands*). The most likely mechanism for the formation of ogives was proposed by Nye (1958), who argued that they reflect seasonal variations in the passage of ice through icefalls. Because ice moves faster through an icefall than elsewhere, it is stretched and thinned as it accelerates into the upper part of the icefall. This thinning, plus the high concentration of crevasses in icefalls, means that ice passing through an icefall has a larger surface area than equivalent volumes of ice elsewhere on the glacier. As a result, elements flowing through an ice-

Fig. 6.7 Ogives on Saskatchewan Glacier, Canada. (Photo: D.J.A. Evans)

fall in summer will lose more ice by ablation than elements elsewhere, and will tend to collect more wind-blown dust and other superficial debris. At the base of the icefall, ice discharged during the summer will form troughs or dark bands as the ice is compressed once more. In contrast, ice passing though the icefall in winter will collect excess snow and become wave crests or light bands. The amplitude of wave ogives tends to decrease downglacier from a maximum of about 5 m near the base of the ice fall. This is probably due to differential melting, the troughs being protected from ablation by infills of drifted snow or thin covers of debris. Ogives will not form if ice takes more than six months to traverse the icefall (Waddington, 1986).

6.2.4 Layering and related structures

Englacial and supraglacial ice commonly exhibit various types of layering or foliation. In accumulation areas, the most common form of layering is *sedimentary stratification*, which reflects annual cycles of snow accumulation. The layers are generally parallel to the glacier surface and are best seen in the walls of crevasses (Fig. 2.1). Usually, the strata consist of relatively thick layers of coarse-grained white bubbly ice separated by thinner layers of bluish, dirty ice. The bubbly ice represents compacted winter snow, whereas the intervening blue layers consist of summer ablation surfaces and superimposed ice (Section 2.2). Dirt layers, formed when wind-blown dust is concentrated during summer melting, are often the most prominent feature of sedimentary layering. Successive layers are generally broadly parallel, although unconformities can occur, reflecting periods of irregular melting or wind-scouring on the glacier surface prior to renewed accumulation. Sedimentary layers can be a few millimetres to several metres thick, depending on the specific mass balance at that point on the glacier.

Ice breccia is a distinctive type of ice formed below intensively crevassed areas or ice cliffs, where fractured ice reconsolidates (Fig. 6.8). Voids between blocks can be infilled with refrozen meltwater or snow to create an aggregate seamed by randomly oriented veins.

Foliation is characteristic of deep, englacial ice in glacier accumulation areas and englacial and supraglacial ice in ablation areas. Individual layers consist of fine-grained white ice, coarse-grained clear (blue) ice or coarse-grained bubbly ice, with crystals in the fine layers usually less than 5 mm and those in the coarse layers around 1–15 cm (Allen *et al.*, 1960; Hambrey, 1994; Paterson, 1994). Most foliation develops from sedimentary stratification or crevasse traces which have undergone high strains and/or have been rotated by glacier flow (Hambrey, 1975, 1994; Hooke and Hudleston, 1978; Fig. 6.9).

Fig. 6.8 Ice breccia formed from a frontal apron of ice blocks, Ellesmere Island. (Photo: D.J.A. Evans)

Fig. 6.9 Foliation due to sedimentary stratification rotated by glacier flow, Flåajökull, Iceland. (Photo: D.I. Benn)

However, some foliation is thought to be an entirely metamorphic structure produced by alteration of the ice under high strains (Pfeffer, 1992). In detail, foliation can display structural variations related to the strain history of the ice. Isolated, tight fold hinges may be the only remaining remnants of large-scale folds that otherwise have been stretched and deformed beyond recognition. Boudinage can also occur (Section 5.4.4).

The orientation of foliation is determined by the original disposition of the primary or other structures from which the foliation is derived, and the subsequent sequence of deformation it has undergone during glacier flow. It is possible to recognize two main orientations, transverse and longitudinal, which may coexist with a cross-cutting relationship (Fig. 6.10; Hambrey, 1976a, 1994). *Transverse foliation* commonly consists of crevasse traces located downglacier from zones of transverse crevasses and icefalls. In valley glaciers, ice flow causes the progressive deformation of transverse foliation into a series of arcs similar in planform to ogives (Section 6.2.3). In three dimensions, the foliation dips downward towards the centre of the arc, the whole resembling a set of nested spoons (Allen *et al.*, 1960). Glaciers that pass through several icefalls may have several intersecting sets of arcuate foliation (Paterson, 1994). *Longitudinal foliation* is aligned parallel to glacier flow and originates by the rotation of sedimentary layering or crevasse traces (including transverse crevasses) during ice flow. It is most common at glacier margins or near the edges of flow units, where cumulative strains are highest, and it is normally steeply dipping. Where a broad accumulation basin feeds into a narrow tongue, longitudinal foliation may extend across the whole width of the ablation area (Hambrey and Müller, 1978).

6.3 SUPRAGLACIAL DEBRIS ENTRAINMENT

6.3.1 Sources

In Chapter 5, we considered the entrainment of debris at glacier beds. Debris can also enter glacial transport at the ice surface, and can be derived from a number of possible sources (Drewry, 1986; Hambrey, 1994; Kirkbride, 1995a): (a) mass movements from adjacent mountain slopes; (b) wind-blown dust; (c) volcanic eruptions; (d) salts and micro-organisms from sea-spray; (e) meteorites; and (f) pollutants from human sources.

Where supraglacial slopes are rare or absent, supraglacial debris entrainment is usually negligible, except in some volcanically active areas such as Iceland

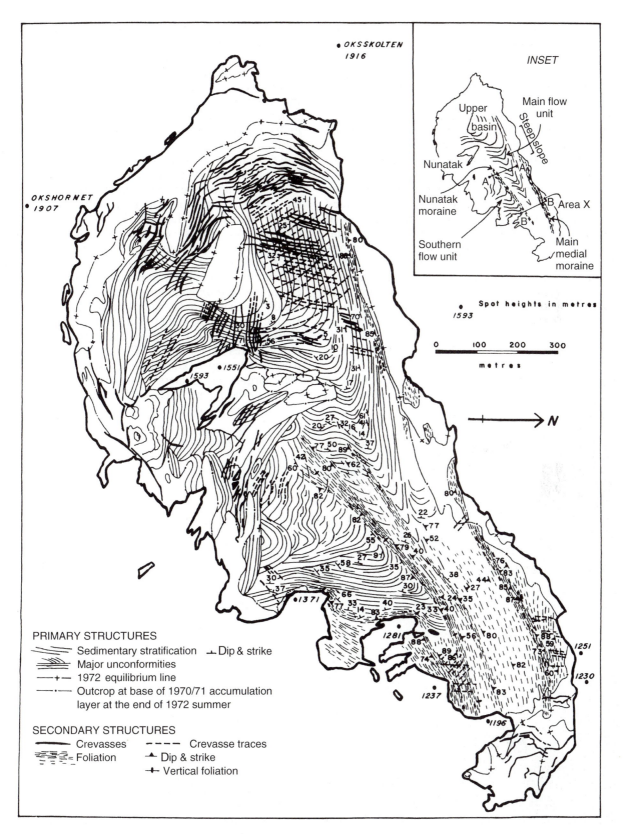

Fig. 6.10 Surface foliation and crevasse patterns on Charles Rabots Breen, Norway. (From Hambrey, 1976a. Reproduced by permission of the Geological Society of America)

where far-travelled ashfalls (tephra) can provide a major input. On ice sheet surfaces far from nunataks and volcanic sources, meteorites can constitute the most important source of large particles. Indeed, parts of the Antarctic ice sheet are valuable meteorite collecting grounds for scientists, because dark particles show up clearly against the otherwise clean ice surface. Recent finds in Antarctica include meteorite ALH 84001, which contains controversial evidence for ancient Martian micro-organisms (McKay *et al.*, 1996).

On valley and cirque glaciers, where ice surfaces are overlooked by valley walls, and where isolated peaks or *nunataks* protrude through ice sheet or ice cap surfaces, mass movements from slopes are by far the most important source of supraglacial debris (Figs 6.4 and 6.11). Many slope processes operate in mountain environments; they include rockfall, rock slides, snow and ice avalanching, debris flow and creep. In addition, debris can be transported on to glacier surfaces by stream flow.

6.3.1.1 ROCKFALLS AND ROCK SLIDES

Rock debris falling or avalanching from steep bedrock slopes is one of the most important sources of debris delivery to valley glaciers (Boulton and Eyles, 1979; Gordon and Birnie, 1986; Small, 1987a, b). Debris is released from rock slopes when pre-existing weaknesses are enlarged by mechanical weathering. In glacial environments, frost action is the principal mechanism, acting at the granular scale (microgelivation) or exploiting bedding planes or joints (macrogelivation) (Fig. 6.12; Tricart, 1956; Ballantyne and Harris, 1994). Two main processes are thought to be responsible for frost damage to rocks. First, water undergoes a 9 per cent expansion of volume upon freezing, which can force apart the sides of a water-filled pore or crack, causing crack extension (Bridgeman, 1912). To be most effective, this process requires a high degree of rock saturation and rapid freezing to temperatures around −5°C (McGreevy, 1981). Second, liquid water may migrate under capillary pressure towards freezing centres to feed expanding lenses of *segregation ice*, which gradually wedge apart the rock (Walder and Hallet, 1985, 1986; Hallet *et al.*, 1991). This process is likely to be most effective during sustained low temperatures of −4 to −15°C. Particularly effective frost weathering is known to occur in randklufts at the margins of glaciers and below snowpatches, owing to the microclimate and ready availability of

Fig. 6.11 Nunataks rising above an ice sheet surface, Queen Maud Land, Antarctica. (Photo: Helge Skappel, Norsk Polarinstitutt, Oslo)

Fig. 6.12 Frost-shattered rock (macrogelivation), 900 m above sea-level, southern Iceland. (Photo: A. Snowball)

water in these locations (Gardner, 1987; Ballantyne *et al.*, 1989). An alternative to frost weathering is *ordered water weathering* or *hydration weathering*, caused by the pressure exerted by the negatively charged ends of water molecules adsorbed on to clay surfaces (White, 1976).

Rockfalls associated with periglacial weathering tend to follow annual and diurnal temperature cycles. They are rare in cold weather and at night, when loose rock is cemented by ice, and are most common in warm conditions, particularly when cliffs receive direct sunlight. This fact is well known to alpinists, who aim to be out of range of active slopes before sunrise. The location of rockfalls is mainly controlled by rock slope form, aspect, lithology and geological structure. Rockfalls are most likely to occur on slopes underlain by well-jointed rocks, particularly if one of the joint sets dips out of the slope towards the glacier (Gardner, 1982; André, 1986; Gordon and Birnie, 1986).

Large-scale rock slope failures are less closely controlled by freezing cycles, although frost action can contribute to the weakening of slopes. Periods of heavy rainfall or rapid snow melt can promote failure by increasing the water pressure in joints and reducing the frictional strength of the rock, and events can also be triggered by seismic shocks in tectonically active areas (Shreve, 1966; Marangunic and Bull, 1968; Reid, 1969), or shocks associated with pulsed isostatic uplift during deglaciation (Sissons and Cornish, 1982a, b). Deglaciation also encourages large failures due to the exposure of oversteepened slopes, release of glacier-induced stresses, and changes to drainage conditions (Gardner, 1982; André, 1986).

Long-term rates of debris delivery to glaciers from rockfalls are poorly known. Measured and estimated rock wall retreat rates in present-day mountain environments range between 0.05 and 3.00 mm yr^{-1}

(Ballantyne and Harris, 1994). However, the amount of debris delivered on to glacier surfaces by individual rockfalls varies considerably, ranging from single particles (low-magnitude events) to huge rock avalanches (high-magnitude events). Low-magnitude events occur relatively frequently, providing a steady input of debris over long time periods, whereas high-magnitude events tend to be rare. The relative importance of low- and high-magnitude events in the long-term debris budget of valley glaciers is very poorly known, and is likely to be highly variable from place to place. Gordon *et al.* (1978) calculated that a single rock avalanche on to Lyell Glacier, South Georgia, in September 1975 supplied the same amount of debris as 93 years of 'normal' low-magnitude rockfalls on to the glacier. However, the recurrence interval of such events in this area is not known. A tentative magnitude–frequency curve for rockfall and rock slide events in part of the Rocky Mountains has been constructed by Gardner (1980), and suggests that over long timescales, falls of different magnitudes are of broadly equal importance. When they occur, however, large rockfalls and rock slides are major events that can profoundly influence local topography and glacier behaviour. A huge rock avalanche in December 1991 removed 20 m from the summit of Mount Cook (formerly 3764 m) in the New Zealand Alps, sending an estimated 14 million m^3 of rock and ice on to the Tasman Glacier below and obliterating the Hochstetter icefall *en route* (Fig. 6.13; Chinn *et al.*, 1992; Kirkbride and Sugden, 1992; Hambrey, 1994). By blanketing large areas of glacier surfaces with debris, large avalanches protect the ice from ablation, altering the mass balance. Several glaciers have been known to readvance after major rockfalls, including the Brenva Glacier in the Italian Alps, following a rockfall in 1922 (Porter and Orombelli, 1981), the Sherman Glacier, Alaska, after a rock slide triggered by the 1964 Alaskan earthquake (Shreve, 1966), and the Bualtar Glacier, Pakistan, after a major slope failure in 1986 (Hewitt, 1988; Gardner and Hewitt, 1990).

6.3.1.2 SNOW AND ICE AVALANCHING

Avalanching can be an extremely important mechanism for transporting debris on to glacier surfaces, especially in high mountain environments such as the Rockies, Alps and Himalaya, where large quantities of snow can accumulate on unstable slopes (Luckman, 1978; Owen and Derbyshire, 1989). The collapse of snow slopes or ice séracs can result in very large, destructive avalanches which can scour and pluck debris from underlying rock surfaces. Repeated avalanches from the same slope build out dirty avalanche cones at glacier margins, composed of a mixture of snow and crushed ice, ice blocks and

(a)

(b)

Fig. 6.13 (a) Small rock avalanche deposit on Storbreen. (Photo: D.I. Benn) (b) Large lobate rock avalanche deposit on the surface of the debris-mantled Tasman Glacier, resulting from the catastrophic slope failure of December 1991. (Photo: D.J.A. Evans)

(a)

(b)

Fig. 6.14 Snow avalanche cones, Miage Glacier, Italy. (Photos: D.I. Benn)

Fig. 6.15 Debris flow from sediment-covered valley sides on to a snow surface, Fåbergstollsbreen, Norway. (Photo: D.I. Benn)

rock. In high mountain environments such as the European Alps and the Himalaya, snow avalanching provides the single most important source of supraglacial debris (Figs 2.18 and 6.14).

6.3.1.3 DEBRIS FLOW AND WATER

Flowage of wet, unconsolidated sediments (debris flows) can deliver large quantities of sediment to the margins of some glaciers, especially where ice is overlooked by glacial deposits. This situation commonly occurs during deglaciation in mountainous terrain, when the oversteepened inner slopes of lateral moraines are exposed above ablating glaciers. Such slopes are generally composed of poorly sorted debris with a fine-grained matrix, and can degrade very rapidly, yielding numerous debris flows during times of snow melt or heavy rainfall (Fig. 6.15; Ballantyne and Benn, 1994b, 1996). The behaviour and

characteristics of debris flows are discussed in detail in Section 6.5. Flowing water from valley sides tends to be a relatively unimportant source of supraglacial debris, but can be locally significant in some glacier ablation areas (Evenson and Clinch, 1987).

6.3.2 Incorporation within glacier ice

Debris that accumulates on glacier surfaces can be incorporated into the body of the glacier in two main ways: (a) burial by snow and ice, and (b) by falling down crevasses or other holes in the glacier surface. Burial by snow and ice occurs mainly in glacier accumulation areas, where more snow falls each year than melts during the ablation season. In this case, debris on the glacier surface is progressively buried by snowfall and/or snow and ice avalanches, and is incorporated within the snowpack. With continued snow accumulation, the snow surrounding the debris is transformed into ice, and the debris is carried to progressively deeper levels within the glacier. Descent of the debris towards the glacier bed can also be encouraged by patterns of ice flow and basal melting (Fig. 6.16). A limited amount of debris burial can also occur at the lateral margins of glacier ablation zones, at the foot of avalanche cones (Fig. 6.14). Such cones represent localized areas of snow or ice accumulation, but in this case deep burial is unlikely, because the debris will be exhumed by melting once the surrounding snow is transported away from the active cone by glacier flow. Crevasses can swallow up debris in any part of the glacier, the only constraints being the presence of open crevasses and a source of debris. This mechanism can occur either where fresh rockfalls deliver debris into open fractures, such as lateral crevasses or the

junction of confluent glaciers, or where crevasses open up beneath a surface veneer of debris, such as in icefalls. The randkluft at the upper edge of a glacier can be a very important routeway for supraglacially entrained debris to pass into englacial and subglacial transport.

6.4 ENGLACIAL AND SUPRAGLACIAL TRANSPORT

6.4.1 Passive and active transport

The transport of rock debris in supraglacial and englacial positions has been termed *passive debris transport* by Boulton (1978) to distinguish it from *active transport* at the glacier bed (Section 5.7.1). During passive supraglacial and englacial transport, debris passes through the glacier system with little or no modification, for two main reasons. First, in supraglacial and englacial environments, debris concentrations are often low, and particles rarely come into contact with each other. Second, even where debris concentrations are high, such as in glacier ablation zones in high mountain environments, inter-particle contact forces are generally small, unlike those at glacier beds. However, debris modification can occur in supraglacial and englacial environments, by several processes:

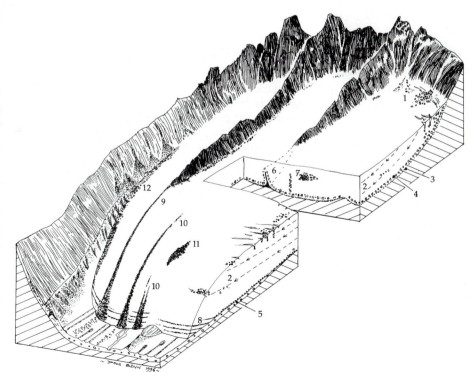

Fig. 6.16 Debris transport through a valley glacier. 1: Rockfall debris entrained supraglacially and buried by snow or ingested in crevasses; 2: ice flowlines transport debris away from the surface in the accumulation area and towards the surface in the ablation area; 3: basal tractive zone; 4: suspension zone; 5: basal till with limited deformation; 6: debris septum elevated from the bed below a confluence; 7: diffuse septum and cluster of rockfall debris; 8: debris elevated from the bed by marginal shearing; 9: ice–stream interaction medial moraine; 10: ablation-dominant medial moraines; 11: avalanche-type medial moraine; 12: supraglacial lateral moraine

1. Particle size reduction can occur on glacier surfaces by periglacial and chemical weathering processes, which may be accelerated by the abundant supply of meltwater.
2. Where supraglacial debris contains abundant large boulders, inter-particle stresses can be large enough to cause fracture and edge-rounding (Fig. 6.17).
3. Debris held within englacial shear zones can be abraded and fractured, as for debris in active transport (Section 5.7). For example, Hambrey (1976b) found that on Charles Rabots Breen, Norway, micaceous debris was ground to powder in a medial moraine across which there was significant shear between ice streams.
4. Sediment reworking and transport by supraglacial and englacial streams can result in significant debris modification on some glaciers, particularly where there is an abundant supply of debris and the internal and superficial drainage network is well developed (Kirkbride and Spedding, 1996).

Fig. 6.17 Crushing and particle wear at the contact between two large boulders on a debris-covered glacier surface. Glacier du Miage, Italian Alps. (Photo: D.I. Benn)

Fig. 6.18 Coarse, angular debris transported by snow avalanche, Lahul Himalaya, India. (Photo: D.I. Benn)

In general, however, supraglacial and englacial debris retains the characteristics of the parent material that was entrained by the glacier or elevated from the bed. If, as is often the case, the parent debris is periglacially weathered material that has fallen on to the glacier surface, it tends to be predominantly coarse-grained with a deficiency in the finer grades (fine sand and smaller). Individual particles are commonly angular or very angular, reflecting particle fracture during weathering and transport on to the glacier. Debris transported by snow avalanches tends to be particularly angular (Fig. 6.18). Particle shape is dependent upon lithology, especially the joint distribution in the parent bedrock, but in general elongate and slabby particles are common (Figs 5.34 and 5.35; Ballantyne, 1982a; Vere and Benn, 1989; Benn and Ballantyne, 1993, 1994). Of course, where supraglacial and englacial debris consists of reworked older deposits, it may differ greatly from rockfall material.

Debris may pass between passive and active transport in a variety of ways. Actively transported debris may be elevated from the bed by basal freezing or patterns of ice flow, and thus may enter passive englacial and supraglacial transport (Boulton, 1978). Conversely, passively transported debris may reach the glacier bed following basal melting, the opening of deep crevasses, and extensional flow. Furthermore, debris may pass between active and passive transport paths via englacial meltwater conduits (Kirkbride and Spedding, 1996). The types of pathways that debris can follow through the glacier system are described in the following sections.

6.4.2 Englacial transport paths

Most debris input to glaciers occurs at the bed or at the glacier margins below valley walls or nunataks (Sections 5.3 and 6.3). Consequently, most debris is concentrated close to the glacier margins or downstream from the confluence of flow units where marginal debris has been carried into mid-stream by glacier flow. Boulton and Eyles (1979) introduced the term *bed-parallel debris septum* (*septum* means a partition) to refer to the concentrated zone of debris close to the margins and bed of valley glaciers (Fig. 6.16). Broadly, the bed-parallel septum consists of two zones: the *basal tractive zone* in contact with the bed, and the *suspension zone* close to the bed but not in contact with it (Boulton, 1978; Boulton and Eyles, 1979; N. Eyles, 1983a; Small, 1987a). Debris transport in the basal tractive zone is described in detail in Sections 5.5, 5.6 and 5.7. Debris in the suspension zone generally does not undergo fracture or abrasion, and often consists of coarse, angular, valley-side debris (Boulton, 1978; Small, 1987a). However, the suspension zone may contain debris that has been

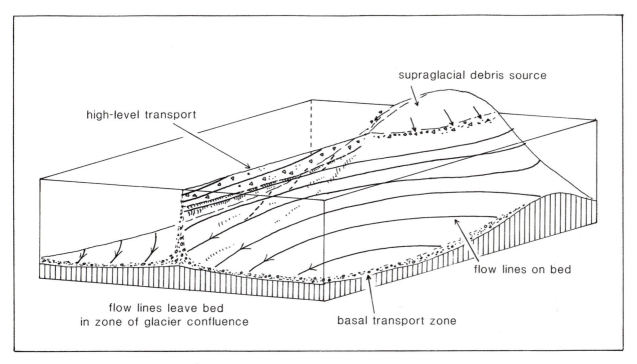

Fig. 6.19 Elevation of basal debris into high-level transport at a glacier confluence. (Modified from Boulton, 1978)

elevated from the bed by basal freezing or ice flow patterns, and it may therefore have shape characteristics and grain size distributions inherited from episodes of active transport (Section 5.7).

When ice flow units converge around bedrock obstructions or at valley confluences, flowlines can leave the bed, thus carrying debris into high-level transport midstream. Where this happens, two bed-parallel debris septa combine to form a *medial debris septum*, which can be vertical or inclined depending on the relative strength of the flow units (Fig. 6.19; N. Eyles and Rogerson, 1978a, b; Boulton and Eyles, 1979; Gomez and Small, 1985). In the case of converging flow around a basal obstruction or a glacier confluence in the accumulation area, the medial septum will not extend to the surface until exposed by melting in the ablation zone (Fig. 6.16). In contrast, if a glacier confluence is in the ablation zone, the resulting medial debris septum will usually extend from bed to surface.

Debris in medial septa is commonly concentrated into bands, which may be aligned parallel to glacier flow (longitudinal), or at a high angle to flow (transverse) (Small *et al.*, 1979; Gomez and Small, 1985; Small, 1987a). *Longitudinal bands* derived from elevated basal tractive zones consist of thin, concentrated bands of debris-rich ice. Such bands are sharply bounded and have debris characteristics typical of active transport (Fig. 6.20; Vere and Benn, 1989). Longitudinal bands can also originate as rock-

Fig. 6.20 Sharply bounded septum of elevated basal debris, Storbreen, Norway. (Photo: D.I. Benn)

falls on to the glacier surface, which subsequently become buried within the snowpack and rotate during glacier flow. The bands lie parallel to the longitudinal foliation in the ice, and in medial positions are usually vertical or steeply dipping. They tend to have more diffuse boundaries and lower debris concentrations than elevated basal debris, but dense clusters of coarse, angular particles are not uncommon (Small, 1987a; Vere and Benn, 1989). Individual bands measure a few metres to hundreds of metres long, and are commonly grouped together in subparallel chains, sometimes with several side by side or *en échelon* (Gomez and Small, 1985; Small,

1987a; Fig. 6.21a). *Transverse debris bands* can originate as rockfalls into crevasses, either in lateral positions, within icefalls, or at the confluence of ice streams. When newly formed, they are close to vertical, but are tilted by glacier flow to dip upglacier at progressively shallower angles. In some circumstances, transverse bands could be formed by the upthrusting of subglacial debris along shear planes, although in this case the debris should be clearly distinguishable from passively transported rockfall material in crevasse fills (Fig. 6.21b; Small, 1987a).

6.4.3 Supraglacial lateral moraines

Supraglacial lateral moraines are ice-cored accumulations of debris at the margins of valley glaciers (Boulton and Eyles, 1979). The debris may be entrained in the glacier accumulation area, then be transported by glacier flow to the ablation area where it melts out, or the debris may fall directly on to the ablation zone. In the latter case, supraglacial lateral

moraines can form by the coalescence of scree cones at the glacier margins (Figs 6.16 and 6.22). The reduction of ablation by a thick debris cover means that supraglacial lateral moraines commonly stand above the adjacent clean ice, which can result in sliding and slumping of debris down the moraine slopes on to the glacier surface or the valley side. Further consideration of the effect of debris on glacier ablation and debris reworking is given in Sections 6.4.4.2 and 6.5.

6.4.4 Medial moraines

6.4.4.1 CLASSIFICATION

Medial moraines are among the most striking features of valley glaciers, providing a graphic picture of the movement of both ice and debris (Fig. 6.23). The visible part of a medial moraine is the surface expression of a medial debris septum, which may be a shallow feature or extend to the base of the glacier

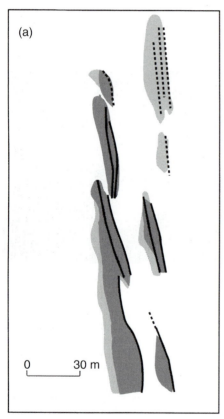

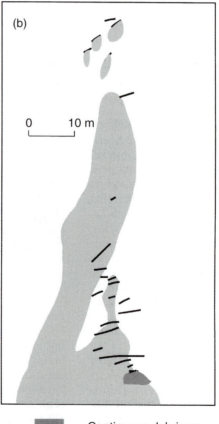

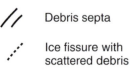

Debris septa

Ice fissure with scattered debris

Continuous debris on surface

Patchy debris on the surface

Fig. 6.21 Maps of debris septa. (a) Longitudinal septa exposed on the eastern medial moraine, Haut Glacier d'Arolla, Switzerland; (b) transverse septa, western medial moraine, Bas Glacier d'Arolla. (Redrawn from Small, 1987a)

Fig. 6.22 Supraglacial lateral moraine formed by the coalescence of scree cones at a glacier margin, Heimre Illabreen, Norway. (Photo: D.I. Benn)

(a)

(b)

Fig. 6.23 (a) Medial moraine on Shackleton Glacier, Antarctica (USGS). (b) Air photograph of part of the snout of Breidamerkurjökull, Iceland, in 1965, showing the close association between an englacial esker and a medial moraine. (Photograph by Landmaelingar Islands and University of Glasgow)

(Fig. 6.16; Section 6.4.2). However, the debris on the surface is generally more concentrated and laterally extensive than that within the glacier, owing to the effects of glacier ablation in redistributing the debris. A comprehensive classification of medial moraines was proposed by N. Eyles and Rogerson (1978a) on the basis of the relationship between debris supply and the morphological development of the moraine. Three broad types were recognized: (a) *ablation-dominant* (AD) moraines, which emerge at the surface as the result of the melt-out of englacial debris; (b) *ice-stream interaction* (ISI) moraines, which find immediate surface expression downstream from glacier confluences, often by the merging of two supraglacial lateral moraines; and (c) *avalanche-type* (AT) moraines, which are transient features formed by exceptional rockfall events on to a glacier surface. These three types are described in turn below.

6.4.4.2 ABLATION-DOMINANT (AD) MORAINES

Moraines of this type emerge in ablation areas by the melt-out of englacial debris septa formed further upglacier. N. Eyles and Rogerson (1978a) recognized three subtypes of ablation-dominant moraine, defined according to the mechanism of debris entrainment as follows.

Below firn-line type (AD1)

The debris in AD1 moraines is entrained in glacier ablation areas, in the lee of nunataks, where ice is heavily crevassed. Rockfall debris entering the crevasses comes to occupy shallow englacial positions, and is then exposed downglacier by ablation. As the englacial debris is released and accumulates on the surface, an ice-cored ridge develops because the debris insulates the underlying ice from ablation. As ridge relief increases downglacier, debris rolls, slides and flows down the flanks on to the adjacent

clean ice, increasing the width of the moraine. However, this process eventually ceases, because the debris does not extend the full depth of the glacier and the supply to the moraine is limited. Once the englacial debris supply is exhausted, the ridge form degrades, and the moraine becomes a low-relief veneer which is transported without further modification to the snout.

A prominent moraine of this type occurs on Austerdalsbreen, an outlet of the Jostedalsbre icefield in south-central Norway (N. Eyles and Rogerson, 1978a). The moraine forms at the junction of two icefalls, Odinsbreen and Thorsbreen, that drop 800 m from the snowfields above. Debris enters the glacier via numerous crevasses and progressively melts out downglacier, forming an upstanding ice-cored ridge. The ridge attains a maximum height of 12 m above the adjacent clean ice approximately 1200 m from the confluence, at which point the debris supply becomes exhausted. Thereafter, debris spreads laterally, the ridge declines in relief, and moraine width increases from 40 m to over 200 m in the terminal zone.

Above firn-line type (AD2)

The AD2 type of moraine is fed by debris in the accumulation area, either at the junction of flow units or at persistent rockfall sites such as gullies in the glacier back wall. During its transport through the glacier accumulation area, the debris becomes more deeply buried by annual snowfall, then melts out below the firn line as deep ice is exposed by ablation. Several variants of this moraine type can be recognized, according to the mechanism of debris entrainment. (a) Rock debris falls on to the glacier surface and is buried in the snowpack. The resulting medial moraine consists of longitudinal bands of coarse, angular debris, sometimes with several parallel bands which may be of different rock types. Examples from Storbreen and Søre Illåbreen, Norway, have been des-

cribed by Vere and Benn (1989) (Fig. 6.24b). (b) Rock debris falls into marginal crevasses, resulting in transverse debris bands that melt out to form beaded moraines. These may merge downglacier into a continuous cover of superficial debris. A well-studied example of this type occurs on the Glacier de Tsidjiore Nouve, Switzerland (Small et al., 1979; Small and Gomez, 1981; Small, 1987a). (c) Basal debris is elevated from the bed at a glacier confluence, forming a longitudinal septum rising upward from the bed but not immediately reaching the surface. At higher levels, englacial debris consists of rockfalls buried by snow or ingested in crevasses. As the resulting medial moraine melts out in the ablation zone, subglacial and rockfall debris become mixed together. This is the situation described as AD2 moraines by N. Eyles and Rogerson (1978a) using examples from Berendon Glacier, British Columbia, Canada (Fig. 6.24a). The quantity of englacial debris nourishing AD2 moraines is highly variable, leading to very different degrees of morphological development. If the amount of debris is small, the surface expression of the moraine consists of only a diffuse scatter of particles, but if a large amount of englacial debris is present, prominent upstanding ridges will develop (Small et al., 1979).

Subglacial rock knob type (AD3)

The debris forming the AD3 type of moraine is elevated from the bed in the lee of ice-covered rock knobs by converging ice flow. Basal debris-rich ice is rotated to form a vertical or steeply dipping longitudinal septum, which generally has a high debris

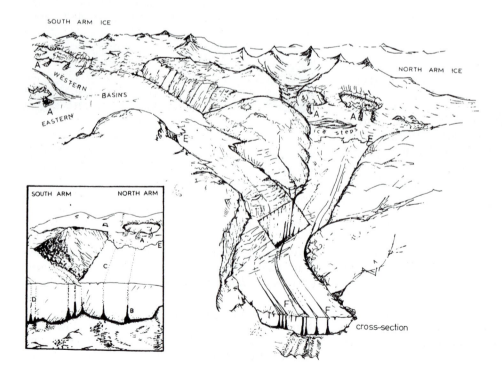

Fig. 6.24(a)

(b)

(c)

(d)

Fig. 6.24 Four medial moraines on Canadian and Norwegian glaciers. (a) Three-dimensional view of Berendon Glacier, British Columbia, showing the formation of an ice–stream interaction moraine at the junction of the North and South Arms (from N. Eyles and Rogerson, 1978b, reproduced by permission of the Geological Society of America); (b) Søre Illåbreen; (c) Storbreen; (d) Heimre Illåbreen. (Photos: D.I. Benn)

content, composed of abraded clasts in a fine-grained matrix. When the septum melts out from the ice, it forms a low, steep-sided ridge of poorly sorted till on the glacier surface, which gradually degrades on exposure to the weather (Vere and Benn, 1989). Debris accumulating on the adjacent ice retards ablation, increasing the relief of the moraine above the rest of the glacier. Debris then flows down the flanks of the moraine, broadening and thinning the cover of debris on the surface.

In an example of this type of moraine on Storbreen, Norway, the width of the debris septum decreases downglacier from the rock knob (Fig. 6.24c; Vere and Benn, 1989). The rock knob is now exposed by glacier retreat, revealing a massive, streamlined 'tail' of basal debris, tens of metres across at the base, immediately in the lee. The 'tail' progressively narrows downglacier, forming a 4 m wide septum 250 m from the rock knob, and is only 0.2 m wide at 280 m. Below this point, it narrows further to between 0.05 and 0.1 m. The narrowing of the septum results in a reduction of debris supply to the surface with distance downglacier, leading to a decline in the height of the moraine above the adjacent clean ice.

6.4.4.3 ICE–STREAM INTERACTION (ISI) MORAINES

The ISI type of moraine forms at the intersection of confluent valley glaciers below or close to the equilibrium line (Fig. 6.23a). The moraines find immediate surface expression at the confluence, commonly by the merging of supraglacial lateral moraines, although debris supply is usually supplemented downglacier by the melt-out of englacial debris, which may consist of passively transported rockfall material or elevated basal debris. As for ablation-dominant moraines, the morphological expression and evolution of ice–stream interaction moraines is influenced by the amount of available debris, large amounts of englacial debris encouraging the development of long-lived, high-relief moraine ridges. However, moraine morphology is also strongly influenced by the character of ice flow (N. Eyles and Rogerson, 1978b). Lateral compression between merging ice streams can counteract the tendency of moraine debris to spread out downglacier, particularly if the glacier is undergoing longitudinal extension (N. Eyles and Rogerson, 1978a; Smiraglia, 1989; Fig. 6.24a). Complex surface forms can also develop if the two ice streams have different velocities or discharges, causing one to shear up over the other (N. Eyles and Rogerson, 1978a, b).

6.4.4.4 AVALANCHE-TYPE (AT) MORAINES

The avalanche type of medial moraine is not a continuous feature from debris source to snout, but is an isolated, longitudinal concentration of debris passing downglacier (Fig. 6.24d). Avalanche-type moraines originate from large, low-frequency rockfall events, where the debris has subsequently been stretched out downglacier by ice flow. Clearly, a discrete medial moraine will not develop from rockfalls large enough to mask a considerable part of the glacier surface. N. Eyles and Rogerson (1978a) described a multi-ridged example almost 1 km long from the Berendon Glacier. Other examples, on the upper Tasman Glacier, New Zealand, are shown in Fig. 1.19.

6.5 DEBRIS-MANTLED GLACIER SNOUTS

Some glaciers have a continuous debris mantle covering the lower part of their ablation zones. Such glaciers are common in high-relief mountain environments, such as the Himalaya and the Southern Alps of New Zealand, where mass wasting processes deliver large volumes of debris to glacier surfaces in both the accumulation and the ablation zones (Fig. 6.25). Supraglacial debris entrained in the accumulation zone is buried in the snowpack or ingested in crevasses, and emerges in the ablation zone as debris septa (Section 6.3.2). Continuous debris mantles can also result from the melt-out of englacial debris that has been elevated from the bed along shear planes. In both cases, glacier ablation progressively concentrates debris on the surface, profoundly influencing patterns of ablation and ice flow. Uneven reworking and deposition of debris during glacier ablation is responsible for highly distinctive landform assemblages underlain by complex sediment successions. In this section, we describe the topographic evolution of ablating debris-mantled glacier ice, and the processes of debris reworking in the supraglacial environment. The sediment associations resulting from the ablation of debris-mantled glaciers are described in Sections 11.4, 12.3 and 12.6.2.

6.5.1 Ice ablation and debris accumulation

As we saw in Section 2.2.3.3, debris cover on a glacier surface profoundly influences surface ablation rates. Thin covers of sediment on the glacier (up to 5–10 mm) will promote melting, owing to their high thermal conductivity and lower albedo compared to bare ice. Once the debris concentration exceeds 5–10 mm it will insulate the underlying ice and retard its melt rate compared to surrounding bare ice surfaces (Østrem, 1959; Nakawo and Young, 1981, 1982). Below thick debris covers (> 1 m), the amount of melting is very small. The effects of debris cover

(a)

(b)

Fig. 6.25 Debris-mantled glacier snouts in the Himalaya. (a) Batal Glacier, India. Coarse, angular debris cover derived from extensive rockwalls. (b) Khumbu Glacier, Nepal. The debris-covered glacier surface (middle distance) is masked by snow cover. (Photos: D.I. Benn and B. Richards)

on ablation rates can be readily confirmed on debris-covered glacier ablation zones, where particles smaller than 1 cm thick are often found at the bottom of water-filled holes (*cryoconites*), whereas larger particles tend to be perched on top of pillars of protected ice (*glacier tables*) (Fig. 6.26). In extreme cases, the presence of very thick sediment accumulations may retard ablation to such an extent that vegetation colonizes the glacier surface. In most cases, this vegetation cover consists of grasses, sedges and alpine plants, but in temperate regions trees may become established, as on the surface of the Malaspina Glacier, Alaska (Fig. 6.27). In arid polar environments characterized by continuous permafrost, the surface sediment may eventually exceed the thickness of summer thawing and the buried glacier snout may never melt out, perhaps producing a rock glacier (Chinn and Dillon, 1987), or may proceed to ablate very slowly by the process of sublimation.

Fig. 6.26 Glacier table, where a large boulder has protected the underlying ice from ablation, Zongo Glacier, Cordillera Real, Bolivia. (Photo: D.I. Benn)

Fig. 6.27 Tree on the surface of the Malaspina Glacier, Alaska. (From R.P Sharp, 1988. Reproduced by permission of Cambridge University Press)

The primary process for the buildup of supraglacial debris mantles is the melt-out of debris from glacier ice, which adds new material to the base of the layer (Fig. 6.28). Rates of debris release depend on the debris concentration in the ice and the thickness of the debris overburden. As debris begins to melt out from glacier ice, debris accumulation rates will be high at first, because a thin debris cover promotes ablation by reducing the albedo of the surface

(Section 2.2.3). Once the debris cover thickens, debris accumulation rates slow down because the downward transfer of heat for ice melting is limited by the thermal conductivity and thickness of the debris layer. Drewry (1972) measured melt rates of clean and debris-covered ice and from these calculated the heat flow through both material types. Over a 26-day period a clean ice surface melted by 0.9 m (water equivalent), which equates to a mean heat flow of 130 W m^{-2}. This was considerably larger than the melt rates of debris-covered ice, which ranged from 0.21 to 0.57 m water equivalent and equated to a mean heat flow of 56–118 W m^{-2}. Drewry calculated a thermal conductivity (k_d) of 0.56 W m^{-1} K^{-1} for the debris cover at one of his experimental sites, which is much lower than a typical value of 2.10 W m^{-1} K^{-1} for the thermal conductivity of ice at 0°C (Drewry, 1972, 1986; Yen, 1981).

The effect of surface debris layer thickening on ablation rates for the underlying ice can be summarized in the following relationship:

$$A_{bb'} = k_d T_0 / (h_d L \rho_{i*}) \qquad (6.1)$$

where $A_{bb'}$ is the ablation rate for buried ice, k_d is the thermal conductivity of the debris layer, h_d is the debris layer thickness, ρ_{i*} is the density of ice with dispersed debris, L is the latent heat of fusion of ice, and T_0 is the temperature of the debris surface. This simply states that, if all other factors are held constant, ablation rates decline exponentially with increasing debris layer thickness.

Debris can also be released on to glacier surfaces by *sublimation*, or the direct transformation of ice to vapour, bypassing the intervening liquid phase (Section 2.2.3.1). This process is most effective on glacier snouts located in arid polar environments, such as parts of Antarctica, where the extreme cold renders melting ineffective. Once an arid polar glacier has developed a supraglacial debris cover which is thicker than the depth of seasonal melting (*active layer* depth), ablation by melt-out ceases, and further removal of ice can take place only by the process of sublimation (Bell, 1966; Shaw, 1988a). The sublimation process is an extremely slow one, and debris release rates have been calculated by Bell (1966) for the McMurdo Sound area, Antarctica. The time taken to remove interstitial ice from debris-rich glacier ice is given by

$$t = \rho_i \rho_a z^2 / 2\eta \, \Delta R \qquad (6.2)$$

where t is time, ρ_a is the density of air, ρ_i is the density of ice, z is till thickness, η is the viscosity of air, and ΔR is the difference in vapour density between the top of debris-rich ice and the atmosphere at the top of the debris layer. Calculations made by Bell (1966) indicate that it would take approximately 2000 years to release 1 m of till but 7000 years to

DEBRIS – CHARGED ICE   CLEAN ICE

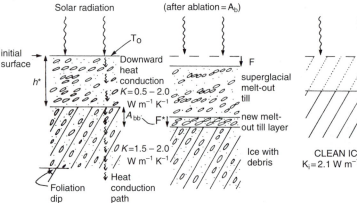

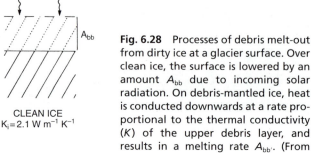

Fig. 6.28 Processes of debris melt-out from dirty ice at a glacier surface. Over clean ice, the surface is lowered by an amount A_{bb} due to incoming solar radiation. On debris-mantled ice, heat is conducted downwards at a rate proportional to the thermal conductivity (K) of the upper debris layer, and results in a melting rate $A_{bb'}$. (From Drewry, 1986)

release 2 m, showing that, as for supraglacial melt-out, debris release by sublimation becomes less effective as the overlying debris layer thickens.

The processes of debris layer thickening described above are often interrupted, however, by the lateral movement of debris under the influence of gravity or flowing water. Debris mobilization is an extremely important process on dirty glacier surfaces, and involves complex interactions between initial debris layer thicknesses, local ablation rates, and resulting changes in the ice surface topography, as described in the following section.

6.5.2 Debris cover and topographic development

6.5.2.1 DEBRIS DISTRIBUTION AND DIFFERENTIAL ABLATION

Continuous supraglacial debris mantles commonly exhibit great spatial variability in thickness, lithology and grain size, reflecting the distribution of debris sources and transport paths, and subsequent reworking on top of and within the glacier. The initial position of thicker debris accumulations reflects the distribution of englacial septa, which melt out of the ice, forming longitudinal or transverse concentrations of debris separated by areas of bare ice (Section 6.4.2). As ablation proceeds, continuing melt-out and lateral spreading of debris gradually result in complete coverage of the glacier surface, although spatial differences in debris thickness and composition commonly persist, at least during the initial stages of ablation. For example, longitudinal bands of granite and schist debris, reflecting variations in catchment lithology, can be traced the full length of the debris-mantled snout of the Khumbu Glacier (Fig. 6.29; Fushimi *et al.*, 1980). Debris variability on the glacier

surface is responsible for large differences in the thermal properties of the glacier over short distances, resulting in *differential ablation*. Local ablation rates are also influenced by debris lithology. Inoue and Yoshida (1980) found that on the Khumbu Glacier, ablation rates were greater below dark-coloured schistose debris than below light-coloured granitic debris, apparently because of differences in albedo (Fig. 6.30).

6.5.2.2 DIRT CONE DEVELOPMENT

Differential ablation removes more ice from some areas than others, producing upstanding dirt-covered mounds on the glacier surface. These mounds are commonly referred to as *dirt cones* even though the debris forms a relatively thin veneer and the mounds are predominantly composed of protected ice (Figs 6.31 and 6.32; Lewis, 1940a; Swithinbank, 1950; Drewry, 1972). Dirt cones are often elongate, because they represent either the release of continuous debris-rich septa or ice protected by the former bedload of supraglacial or englacial streams (Boulton, 1967; Hooke, 1970; Drewry, 1972; Knighton, 1973; Kirkbride and Spedding, 1996). The mounds will continue to grow (or rather, the troughs between will continue to deepen) until the gradient of the mound flanks is sufficient to cause sediment movement under gravity, exposing the ice core and allowing renewed melting. New mounds are then produced wherever the sediment comes to rest after falling or flowing from the old mound crest. Critical angles for such failure vary according to the grain size and water content of the sediment cover. For example, Drewry (1972) suggests a threshold grain size range of 0.2–0.6 mm for dirt cone inception, with optimum grain sizes lying between 0.6 and 5 mm. Below this threshold the low liquid limit of the finer sediment causes it to fail at low slope angles. In addition,

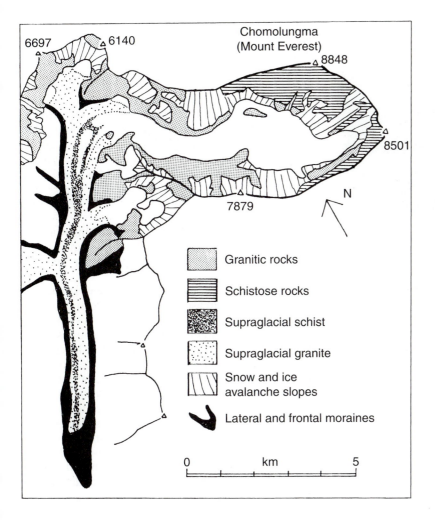

Fig. 6.29 Distribution of supraglacial debris types and catchment lithology, Khumbu Glacier, Nepal. (Modified from Fushimi *et al.*, 1980)

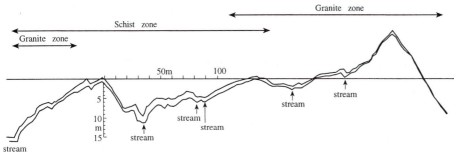

Fig. 6.30 Transverse profile of Khumbu Glacier, showing lowering of the surface profile between 27 July and 4 September 1978. Note that the largest amounts of surface lowering occur in the schist-covered zone. (Modified from Inoue and Yoshida, 1980)

coarser particles will tend to roll down ice slopes very easily, and even in situations where a thick coverage of coarse sediment exists, the increase in permeability allows air circulation and inhibits differential ablation.

Drewry (1972) developed a mathematical model of dirt cone evolution, in which growth is controlled by the *ablation differential*, or the difference between the ablation rate on the cone compared to the surrounding

ice. The growth is self-limiting, because increases in slope angles cause sediment failure, redistributing the debris and reducing the ablation differential. As sediment accumulations grow in thickness and extent on an ablating snout, they will ultimately reduce the ablation differential and eventually terminate the evolution of dirt cones. The model thus displays a negative feedback mechanism slows dirt cone growth and eventually initiates the waning phase.

Fig. 6.31 Dirt cones on stagnant glacier ice, Spitzbergen. (Photo: C.M. Clapperton)

6.5.3 Debris reworking

Debris accumulating on a glacier surface commonly undergoes several cycles of movement and remobilization before final deposition (Lawson, 1979a, 1981b, 1982; Paul and Eyles, 1990). The two most important processes of debris reworking are gravitational mass movement and transport by water.

6.5.3.1 GRAVITATIONAL REWORKING

Differential ablation leads to the production of relief on a glacier snout and the gradual steepening of slope angles on the flanks of dirt cones. Eventually, slope-steepening initiates failure and mass movement of the overlying debris by debris flow, sliding or other gravitational processes. Debris flows are particularly common, because of the abundance of meltwater, which saturates surface debris and lowers its strength (Fig. 6.33). Owing to this continuous change in the surface relief of the glacier and the high water contents or fluidity of supraglacial sediment assemblages, gravitational reworking occurs wherever sufficient debris occurs on sloping glacier surfaces. For example, Lawson (1979a) estimated that 95 per cent of the sediment

in the snout of the Matanuska Glacier, Alaska, undergoes resedimentation, predominantly by mass flowage on very low-angle slopes (1–7°).

The conditions for the mobilization of debris flows are very similar to those required for the movement of a subglacial deforming layer (Section 4.4.2). Flows are initiated when the shear stress exerted by the weight and surface slope of the debris layer exceeds the shear strength of the sediment. As for subglacial sediment, shear strength can be approximated by the Coulomb equation (Section 4.4.2.2):

$$\tau^* = c + N \tan \phi_i \qquad (6.3)$$

where c is cohesion, ϕ_i is the angle of internal friction, and N is effective normal stress (equal to the overburden pressure minus the porewater pressure: $p_i - P_w$).

Flowage will occur if: (a) sediment strength is reduced by an increase in porewater pressure; (b) the shear stress is increased by an increase in slope angle due to differential ablation; or (c) the shear stress is increased owing to an increase in debris layer thickness. This last case applies only where cohesion forms an important component of debris strength, because for cohesionless flows, strength increases faster than shear stress with increasing debris thickness.

The addition of sediment to a debris layer may be by mass movement from surrounding ice slopes (*allochthonous addition*) or from melt-out of underlying ice (*autochthonous addition*). Debris flows also serve to remove sediment from upstanding areas, exposing bare ice to renewed ablation (Fig. 6.34).

The reworking of supraglacial debris by mass flow processes has been studied by Boulton (1968, 1971) on Spitsbergen glaciers and by Lawson (1979a, 1981a, b, 1982) on the Matanuska Glacier, Alaska. Both researchers have emphasised that the behaviour, morphology and sedimentological characteristics of mass flows are strongly dependent on the water content. Lawson recognized four different types of flow, based upon the relationship between water content

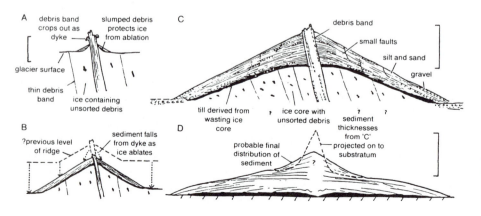

Fig. 6.32 The evolution of a dirt cone from the melt-out of an englacial debris band. Scale bars represent 3 m. (Modified from Boulton, 1967)

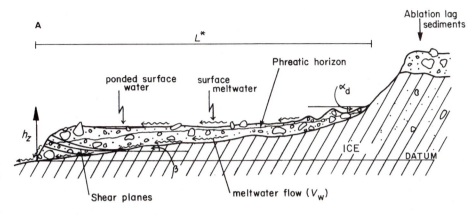

Fig. 6.33 Characteristics of supraglacial debris flows. (From Drewry, 1986)

(a)

(b)

Fig. 6.34 Removal and addition of sediment by debris flows. (a) Flow source area showing erosional scar. (b) Lobate snout of flow. (Photos: C.M. Clapperton)

and other factors such as the mean grain size, maximum flow thickness, shear strength, porosity and bulk density (Fig. 6.35 and Table 6.1). The success of this scheme is reflected in the fact that supraglacial mass movements are often referred to as 'Lawson flows'. The continuum ranges from stiff, high-strength and slow-moving flows (Type I) to fluidized, low-strength and rapidly evolving flows (Type IV).

Type I flows have a low water content (8–14 per cent by weight), and tend to behave as a rigid or semi-rigid 'plug' rafted along on a thin basal shear zone. Active flows form broad, non-channelized lobes, rather like miniature glaciers. *Type II flows* have a water content of *c.* 15–20 per cent, and consequently have lower shear strength. Active flows tend to occupy channels which are partly eroded by the flow and partly bounded by levées of coarse, bulldozed sediment, and run out into pronounced depositional lobes. As for Type I flows, a central, rigid plug zone is present, but shear occurs in a thicker basal and marginal zone. *Type III flows* have yet higher water content during flow (18–25 per cent), and still lower shear strength. Shear occurs throughout all or most of the sediment mass, with only poorly developed plug zones. Because of the low sediment strength, flows tend to be thin (generally < 0.5 m), rapidly moving and very erosive, cutting channels in pre-existing sediments. Flow is often turbulent, with sediment pulses sloshing irregularly down channels and feeding lobate terminal fans. *Type IV flows* have the highest water content (> 25 per cent) and very low strength. Active flows are rapidly moving slurries occupying narrow channels and feeding broad, thin terminal lobes, but because water is abundant during their formation, they are often reworked by stream flow. The sedimentological characteristics of debris flow deposits are discussed in Section 10.5.5.

6.5.3.2 MELTWATER REWORKING

Where debris-mantled glaciers have a low surface slope, meltwater streams meander tortuously over the surface and eventually incise deep channels into the snout, cross-cutting debris septa and reworking supraglacial and englacial sediment accumulations (Fig. 6.36). Reworking of sediment and debris-rich glacier ice by meltwater can take place on and within the glacier and at its margins, constituting an efficient grain-size sorting process by removing finer sediments from the snout area and depositing them in

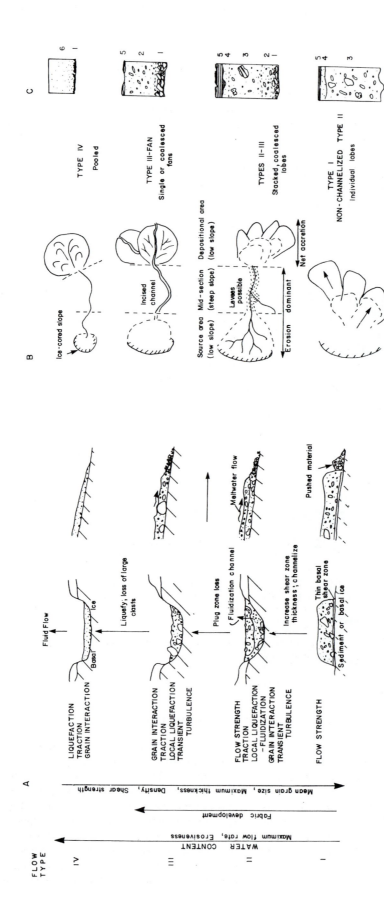

Fig. 6.35 Characteristics of debris flows and their relationship to water content. (A) Cross- and longitudinal profiles of active flows; (B) plan form; (C) sediment characteristics. Six typical sedimentary facies are recognized: (1) texturally homogeneous, increased gravel content of traction origin, massive to graded, pebble fabric weak to absent; (2) massive and texturally heterogeneous, absence of large grains (from settling), possible increase in silt and clay, weak pebble fabric; (3) massive and texturally distinctive, sometimes structured sediments, pebble fabrics absent, clasts commonly vertical; (4) massive and fine-grained (sand–clay) similar to matrix of facies 2, few coarse clasts due to settling during or after deposition; (5) stratified, diffusely laminated silts and sands (meltwater origin); (6) massive to partially or fully graded silty sand, fabric absent. Basal contacts may be conformable or unconformable, sharp, transitional or planar. (From Drewry, 1986, after Lawson, 1979a. Reprinted by permission of the Cold Regions Research and Engineering Laboratory)

Table 6.1 Attributes of active sediment flows

Attribute	Flow type			
	I	II	III	IV
Morphology	Lobate with marginal ridges; non-channelized	Lobate to channelized	Channelized	Channelized
Channel-wise profile	Body constant in thickness with planar surface; head stands above body; tail thins abruptly upslope	Body constant in thickness with ridged to planar surface; head stands above body (<type I); tail thins upslope	Mass thins from head to tail; irregular surface	Thin continuous 'stream'; planar surface
Thickness (m)	0.01–2 (0.5–1 typical)	0.01–1.4 (0.1–0.7 typical)	0.03–0.6	0.02–0.1
Bulk water content (wt %)*	~8–14	~14–19	~18–25	>25
Bulk wet density (g cm^{-3})	2.0–2.6	1.9–2.15	1.8–1.95	<1.8
Surface flow rates (cm s^{-1})	0.1–0.5	0.2–5	15–125	1–200
Typical length of flow (m)	10–300†	10–300†	100–400†	50–400†
Surface shear strength (kg cm^{-2})	0.4–1.5	0.6 or less	Not measurable	Not measurable
Approx. bulk mean grain size (mm)*	2–0.3	0.4–0.1	0.15–0.06	≤0.06
Flow character (laminar)	Shear in thin basal zone with override at head	Rafted plug with shear in lower and marginal zones	Discontinuous plug to shear throughout	Differential shear throughout
Grain support and transport	Gross strength	Gross strength in plug; traction, local liquefaction and fluidization, grain-dispersive pressures and reduced matrix strength in shear zone	Reduced strength, traction, grain-dispersive pressures; possibly liquefaction–fluidization, transient turbidity	Liquefaction; some traction; buoyancy (?)

Source: Lawson (1979a)
* Sample from central part of flow
† Maximum length of flow reflects boundary conditions of glacier terminus region

proglacial locations. In the absence of crevasse networks, stream channel positions on debris-mantled glacier snouts are often guided by debris concentrations and buried ice bodies, and therefore they are very effective in the redistribution and deposition of sediment at the margins of such obstacles. Sediment originating from the melt-out of debris septa can be carried away from the bases of dirt cones to be redeposited in crevasses and shallow ponds on the glacier surface. Deeper crevasses may accumulate thick sequences of stratified sediment in this way and may eventually develop into eskers, where streams excavate continuous tunnels through debris-rich ice.

Although such large-scale reworking by meltwater is often the most obvious, a considerable amount of sediment redistribution is undertaken by smaller flows. At the micro-scale, meltwater may move very fine particles between individual ice crystals by intergranular percolation. At the macro-scale, Lawson (1979a) reported sediment reworking on the

Fig. 6.36 Supraglacial fluvial deposits and meltwater streams cutting across debris septa, Kviarjökull, Iceland. (Photo: D.J.A. Evans)

Matanuska Glacier in two types of meltwater flow: (a) *sheet flows* 1–10 mm thick; and (b) *rill flows* 5–6 cm deep and up to 3 cm wide. Most material is moved in suspension, with minor amounts of coarse sand- and granule-sized particles being moved as a result of the low friction of the ice beds of the meltwater flows. Although the sedimentation rates of meltwater flows will vary widely according to debris concentration and distribution in the glacier, Lawson (1979a) recorded a range of 0.1–10 cm day^{-1} for meltwater flowing from ablating basal ice and backwasting ice-cored slopes on the Matanuska Glacier.

Crude sediment sorting takes place on ablating debris-rich ice by a combination of mass flow and fluvial processes. Using observations on the Matanuska Glacier, Lawson (1979a) described the influence of the slope angle of the ablating ice on sediment release and reworking. On high-angled slopes, the ablation process releases sediment which then slides, falls and rolls to the base of the slope. Silts and clays are carried off the slope in suspension in sheet and rill meltwater flows. If there is no surface drainage in the immediate vicinity of the ablating ice slope, all these sediments become reunited in localized hollows, where they become oversaturated by the continued melting of underlying ice in addition to inflows of surface meltwater. Effective drainage will result in the progressive coarsening of debris accumulating at the base of ablating debris-rich ice slopes. On low-angled slopes (< 20°), the debris released by ablation becomes rapidly coarsened as fines are transported away by meltwater flows, but this coarsening is inhibited once the coarse debris cover thickens. However, this negative feedback on low-angled slopes is observed only on a localized basis, owing to the large relief differences produced by differential ablation of the glacier surface. Excellent examples of the juxtaposition of

mass flow and fluvial processes on a stagnant glacier snout are provided by Boulton (1967, 1968) from Spitsbergen glaciers. During the early stages of ice wastage, fluvial sedimentary sequences are subject to the same continuous disturbance as other supraglacial debris and therefore are only crudely represented in the final depositional record. However, meltwater reworking during the advanced stages of deglaciation can result in large positive relief features with well-preserved internal structures. Where fluvial processes become dominant and large tracts of alluvium are deposited over shallow snouts, the proglacial area evolves into a kame and kettle topography (Section 11.4.3).

The margins of glaciers may be considerably modified by meltwater streams in situations where the valley sides or lateral moraines restrict lateral stream migration. Shallow snouts may eventually become heavily dissected and buried by alluvium in such situations. It has been suggested by D.J.A. Evans (1989b) that marginal streams are influential in the apron entrainment process by subpolar glaciers (cf. Maag, 1969) where they carry sediment across fluvially-cut ice steps into englacial positions; the sediments are then re-entrained by the closure of marginal ice tunnels or glacier overriding of aprons comprising dry-calved ice blocks and intervening gravel accumulations. This entrainment process, which may rework the same debris numerous times, is used to explain the debris-rich bands and elongate clots cutting across the folia in the snout cliffs and containing fluvially modified gravels.

The importance of subglacial and englacial drainage networks in eroding, transporting and depositing

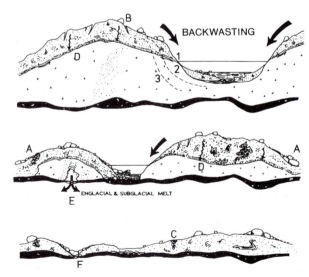

Fig. 6.37 Ablation of debris-covered glacier ice by backwasting, showing topographic inversion. (From N. Eyles, 1979. Reproduced by permission of the *Canadian Journal of Earth Sciences*)

sediment has been emphasized by Gustavson and Boothroyd (1987), who showed that around the stagnating margins of the Malaspina Glacier, Alaska, debris eroded from the bed by meltstreams is the single most important source of sediment. They argue that, in this case, the elevation of basal debris along shear planes within the ice is a comparatively minor mechanism for the supply of supraglacial debris. The importance of englacial meltstreams in the transport of debris in debris-rich ice has also been highlighted by Kirkbride and Spedding (1996).

6.5.4 Wastage of debris-mantled glaciers

The processes of differential ablation, debris reworking, and drainage development on debris-mantled glaciers result in very distinctive topographic evolution during deglaciation. The presence of a thick debris mantle also serves to decouple glacier response from climatic forcing, so that episodes of rapid ice wastage may lag behind climatic inputs by several decades or more, unlike clean glacier snouts, which respond rapidly to climatic warming.

6.5.4.1 TOPOGRAPHIC INVERSION

Dirt cone development and associated debris reworking serve to redistribute sediment on the glacier surface, changing the spatial pattern of differential ablation. Boulton (1971) showed how this leads to a sequence of *topographic inversion* or *reversal* on the glacier surface. As described in Section 6.5.2.2, topographic highs (dirt cones) are first produced where ice is protected by a debris concentration. As the slopes of the topographic high increase and debris flows to the surrounding low points on the glacier surface, the ice at the centre is exposed to the atmosphere and its ablation rates increase; the flowed sediment then acts to protect the ice in the surrounding depressions. Rapid ablation of the exposed ice cores of dirt cones and reduced ablation of debris-covered hollows eventually results in an inversion of the topography, with former hollows evolving into new dirt cones, and vice versa (Figs 6.37 and 6.38; see also N. Eyles, 1979; Kirkbride, 1993). Topographic inversion may take place numerous times before the supraglacial debris is finally deposited on the substrate, resulting in complex sediment assemblages consisting of faulted and folded fluvial, lacustrine and mass-movement deposits (Section 11.4).

Topographic inversion is an important process where debris cover is uneven enough to allow differential ablation. However, the development of very thick debris mantles reduces ablation rates over the entire glacier surface to such an extent that topographic inversion is inhibited, and a more subdued, low-relief

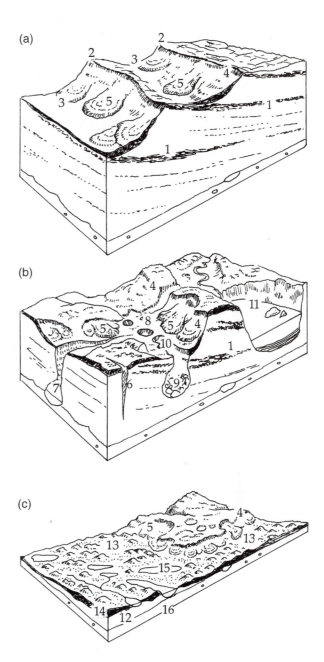

Fig. 6.38 Evolution of glacier karst on an ablating debris-mantled glacier. (a) Young stage; (b) mature stage; (c) old stage. 1: Debris bands in glacier ice; 2: ice-cored ridge; 3: depositional trough; 4: backwasting slope of exposed ice; 5: debris flow; 6: enlargement of crevasse by ice melt; 7: subglacial conduit; 8: sink holes; 9: collapsed tunnel roof; 10: enlargement of sinkhole by ice melt and collapse; 11: lake with backwasting margins; 12: dead ice; 13: ice-free hummocky terrain; 14: deposits of supraglacial environments; 15: lakes; 16: deposits of subglacial environments. (Redrawn from Krüger, 1994)

topography develops. This tends to be the case only on longer glaciers with very high debris content, such as the Khumbu Glacier (Iwata *et al.*, 1980).

6.5.4.2 GLACIER KARST

Where dirt cone development is inhibited, ablation tends to be focused where the debris cover is perforated by moulins, lake basins and other holes in the glacier surface. The slopes around such holes tend to be too steep to support debris, so that bare ice is exposed and ablation rates are high. Ablation proceeds by *backwasting*, or the parallel retreat of ice slopes around the hole or lake shore. This can be the most important component of ablation on the lower part of large debris-mantled glaciers such as the Khumbu and the Tasman (Inoue and Yoshida, 1980; Kirkbride, 1993). The importance of localized ablation in the vicinity of holes and fissures results in a distinctive sequence of glacier wastage similar to the development of karst features on limestone terrain, which led Clayton (1964) to introduce the term *glacier karst* to refer to the forms developed on wasting debris-mantled glaciers. Of course, on limestone terrain, fissures are enlarged by solution of calcium carbonate, whereas on glaciers they are enlarged by preferential melting of exposed ice. Clayton proposed a karst development cycle consisting of three stages: (a) *the young stage*, characterized by dirt cone growth; (b) *the mature stage*, in which extensive debris cover is broken only by moulins and crevasses, which develop into funnel-shaped sinkholes and finally into large water-filled depressions (*poljes* and *uvalas*); and (c) *the old stage*, in which only remnant, stagnant ice cores remain (Fig. 6.38). Each stage is associated with a thickening supraglacial debris cover ranging from less than a few centimetres to > 3 m thick. Very detailed descriptions of depositional and erosional processes, topographic development and sedimentology associated with each stage have been provided by Krüger (1994) for the margins of Kötlujökull, a debris-mantled outlet of the icecap Myrdalsjökull in Iceland.

Holes on debris-mantled glaciers are enlarged by backwasting and the collapse of the roofs of englacial and subglacial conduits, similar to the creation of large sinkholes on limestone when cave chambers collapse. Collapsed conduits are identifiable on glacier surfaces as chains of crater-like depressions filled with water. An example from the Tasman Glacier, New Zealand, is shown in Fig. 6.39 (Plate 16) and Fig. 6.40 (Kirkbride, 1993); other examples from Alaska have been described by Russell (1901), Tarr and Martin (1914) and Clayton (1964). Other karst-type features such as dry stream beds, tunnels and caves, sinking streams and springs were also identified on the Alaskan glaciers.

Glacier karst development on the Tasman Glacier over the past 40 years has progressed to the stage where sinkholes have been enlarged into supraglacial lakes (Fig. 6.40). Kirkbride (1993) has shown that the development of such lakes is related to the presence of a reverse slope beneath the glacier margin, such as the proximal edge of a sandur plain (outwash head) or a bedrock bar, which creates a high base level for englacial drainage (Fig. 6.41). Enlargement of moulins and the collapse of tunnel roofs perforate the debris mantle, leading to localized melting of the exposed ice walls. Initial growth of the lake basins

(a)

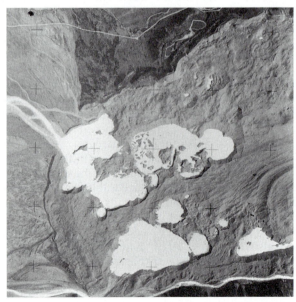

(b)

Fig. 6.40 Vertical aerial photographs of the Tasman Glacier, New Zealand, in (a) 1971 and (b) 1986. In the 1971 photograph, the chain of circular depressions probably records the collapse of an englacial conduit. By 1986 they had become enlarged into a series of lakes owing to preferential ablation around the margins of the depressions. (New Zealand Department of Scientific Research)

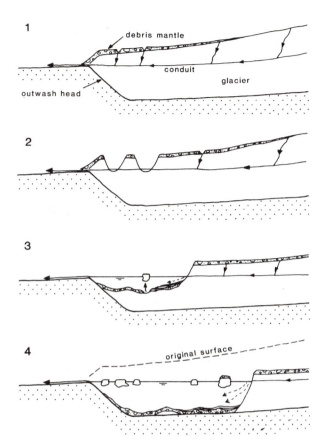

Fig. 6.41 Evolution of lake basins in glacier karst. (1) Slow melting under a continuous debris mantle; (2) collapse of conduit roofs exposes ice walls to rapid ablation; (3) growth and coalescence of lakes with potentially unstable ice floor; (4) disintegration of ice floor increases water depth and initiates rapid calving retreat. (From Kirkbride, 1993)

is by backwasting, but as the water deepens, the ice on the lake floor becomes increasingly unstable and blocks will calve off and float to the surface. Disintegration of the ice floor increases water depth, initiating rapid calving retreat of the lake margins. The total ablation associated with lake growth increases exponentially through time, because the area of bare ice increases as the lake perimeter enlarges, and because calving processes become more efficient as

the lake grows and deepens. Between 1973 and 1988 the average increase in lake diameter on the stagnating snout of the Tasman Glacier was $12 \, \mathrm{m \, yr^{-1}}$, translating into a growth in lake area of approximately $550,000 \, \mathrm{m^2}$ over the 15-year period (Chinn and Kirkbride, 1988; Kirkbride, 1993).

It is sometimes assumed that the presence of glacier karst implies that a glacier tongue is stagnant and no longer actively sliding or deforming internally. This has been shown to be untrue by Kirkbride (1995b), who showed that the Tasman Glacier is still moving within 300 m of its terminus position, even at advanced stages of karst development. Thinning of the glacier has resulted in a reduction in ice velocities, and a realignment of velocity vectors in response to the evolution of local ice surface topography. Similar patterns of ice flow were observed on the Khumbu Glacier by Fushimi (1977). Glacier stagnation can, however, occur over large areas in some cases. Low rates of ablation of debris-mantled ice mean that, during periods of negative mass balance and glacier recession, debris-covered glacier snouts can become cut off from more rapidly ablating ice further upvalley. The decrease in ablation with distance downvalley (due to thickening debris cover) reduces the overall gradient of the glacier snout, thus lowering rates of flow and creating a stagnant ice mass that wastes away *in situ*. This situation tends to occur at the snout of a surging glacier during its quiescent period, when rates of ice supply from the upper part of the glacier can no longer sustain flow in the lower part, and debris-covered areas of the snout become cut off from active parts of the glacier.

In summary, glacier karst is likely to develop where the following conditions are met: (a) debris cover insulates the ice and ablation is focused around holes in the glacier surface; (b) internal drainage networks are well developed so that tunnels can be enlarged by free-flowing meltwater; and (c) where ice is thin or englacial tunnels are close to the surface, so that conduits are not closed off by ice deformation at times of low flow. Where large conduits remain open, roof collapse is possible. In addition, glacier karst development can be maintained on actively moving glacier snouts if overall ablation rates are high.

CHAPTER
7

TERRESTRIAL ICE-MARGINAL ENVIRONMENTS

7.1 INTRODUCTION

So far we have presented information on processes associated directly with glacier ice, whether they take place in subglacial, englacial or supraglacial positions. In this chapter we present information on terrestrial processes associated with glacier margins. Ice-marginal environments are also often characterized by deep water where a variety of subaqueous processes are active, and these are covered separately in Chapter 8.

A visit to the snout of an actively receding temperate glacier reveals that a large number of processes contribute to the production of ice-marginal landform–sediment assemblages (Fig. 7.1). Indeed, the ice-marginal processes of erosion and redeposition that accompany ice recession from an area are so effective that it is surprising any subglacial and englacial features survive relatively intact. Examples of glacial landsystems presented in Chapter 12 demonstrate that deglaciated terrain often comprises a complex of primary glacigenic landform–sediment assemblages (drumlins, eskers, etc.) which are partially eroded and/or buried by ice-marginal forms and sediments (ice-contact fans, meltwater channels, pitted outwash, etc.), or even displaced by glacier readvance (push moraines, thrust-block moraines, etc.). We now review the processes associated with ice ablation, glacier re-advance and paraglacial reworking at the margins of glaciers. The more advanced (fluvial) stages of proglacial landform development are covered in Chapter 3.

7.2 GLACIER ADVANCE AND RETREAT

7.2.1 Advance and retreat cycles

The processes operating at glacier margins, and the types of sediments and landforms that are created, are sensitively dependent on whether the margin is advancing or retreating. Glacier margins advance or retreat in response to variations in ablation, which removes ice from the glacier, and the horizontal component of velocity, which carries ice forward. Boulton (1986) showed that a glacier margin will remain in the same position when

$$U_{tx} = a_b/\tan \alpha \qquad (7.1)$$

where U_{tx} is the horizontal velocity component, a_b is the ablation rate, α is the glacier surface slope, and ($a_b/\tan \alpha$) is the horizontal component of ablation.

Fig. 7.1 Margin of the Miage Glacier, Italy. Debris cover is redistributed by mass movement processes (left) and meltwater (centre). Overhangs in the exposure of clean ice (right) mark englacial thrust planes. (Photo: D.I. Benn)

Note that although the terminus *position* is stationary in this case, the ice itself is in motion but is removed from the leading edge at a rate equal to the velocity (Fig. 7.2). Terminus retreat occurs when $U_{tx} < a_b/\tan \alpha$, and advance occurs when $U_{tx} > a_b/\tan \alpha$.

Glacier margins commonly fluctuate over the course of a year as U_{tx} and a_b vary. In the winter, glacier sliding velocities tend to be low owing to the lack of meltwater at the bed (Sections 4.5.3 and 4.7.1). However, because ablation rates are generally negligibly small in winter compared with the summer, the relative magnitudes of U_{tx} and a_b over the year tend to produce a winter advance and a summer retreat. Winter readvances typically start late in the ablation season when melting at the margin no longer exceeds the forward motion of the glacier. The horizontal component of ablation is so low in late winter that the small winter flow velocity produces a small glacier advance. Although summer flow velocities are higher than those in winter, the high summer ablation rates force the glacier to undergo net retreat.

On longer timescales, glacier margins undergo advance and retreat in response to climatic change or internal instabilities such as surge cycles (Section 4.8). In essence, climatic influences on glacier terminus behaviour can be subdivided into factors causing

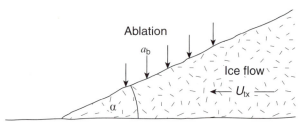

Fig. 7.2 Definition sketch showing the influence of velocity, ablation rate and ice surface slope on glacier advance and retreat

changes in ablation and those causing changes in accumulation. As discussed above, changes in ablation have a more or less instantaneous effect on the glacier front position by altering the balance between U_{tx} and a_b. Long-term increases in ablation rates will cause glacier recession, with the margin occupying more restricted positions each successive year. Conversely, long-term reductions in ablation will cause overall glacier advance. As we saw in Section 6.5.1, debris-mantled glacier snouts are rather insensitive to climatic change, owing to the damping effect of debris cover on ablation rates. Changes in accumulation also cause ice-front fluctuations by varying the amount of ice that must be discharged through the system, and causing changes in ice thickness and gradient which govern the magnitude of the driving stresses (Sections 4.1 and 4.7.2). In this case, however, advance or retreat of the margin will lag behind the climatic forcing because the signal must be transferred from the accumulation area to the snout by glacier flow. This time lag, or *response time*, is greatest for long or slowly moving glaciers, and least for small or rapidly moving glaciers (Johannesson *et al.*, 1989; Paterson, 1994). Glacier response times have been calculated using kinematic wave theory (e.g. Nye, 1960; Paterson, 1994), but can be more accurately determined using physically based flow models (e.g. van de Wal and Oerlemans, 1995).

7.2.2 Implications for geomorphic processes

Different geomorphic processes will dominate during periods of advance and retreat. During retreat, debris is shed from the ice and deposited on the newly deglaciated foreland by a range of processes, mainly gravitational and glacifluvial reworking. Overall rates of sediment accumulation will depend upon the debris content of the ice and the balance between the rates of retreat and debris delivery to the margin by glacier flow. Other factors being equal, rapid retreat will result in lower rates of debris release than slow retreat, because in the latter case more debris is transported towards the marginal zone by the forward movement of the glacier during each increment of glacier retreat. The most rapid sedimentation rates will occur when the ice margin is stationary, and all debris brought forward by the glacier is focused on a single site. The processes of ice-marginal deposition are reviewed in Section 7.3.

When glacier margins are advancing, debris can still be deposited on to the foreland by gravity and flowing water, but it will be overridden and/or relocated by the glacier as it advances over the site of deposition. The relocation of proglacial rocks and sediments by advancing glacier ice is a very important process, and one which produces some of the most impressive ice-marginal landforms. The processes of relocation are discussed in Section 7.4.

7.3 ICE-MARGINAL DEPOSITION

7.3.1 Processes and patterns of deposition

The processes of debris reworking and deposition in terrestrial ice-marginal environments are closely similar to those in supraglacial settings reviewed in Section 6.5.3. The most widespread are gravitational processes, such as debris flow, falling, rolling and sliding, and deposition from flowing water. The relative importance of gravitational and glacifluvial processes varies from glacier to glacier, according to the availability of debris and meltwater.

At stationary or slowly changing ice margins, gravitational processes deposit debris as ice-contact aprons of scree or debris flows. Where debris supply is low, such aprons form small ramparts along the ice edge, whereas larger amounts of debris build up substantial fans and cones (Fig. 7.3a). Debris supply is commonly variable along a glacier margin, and large debris cones may be deposited below re-entrants or crevasses in the ice margin, which funnel debris, or debris concentrations on the surface (N. Eyles, 1979). Rates of debris supply are highest around the margins of debris-mantled glaciers, where huge aprons of scree, or *lateral moraines*, can be built up around the entire glacier tongue (Fig. 7.3b; Section 11.3.3; Humlum, 1978; Boulton and Eyles, 1979; Small, 1983). Such large lateral moraines are common in high mountain environments such as the Himalaya, High Andes and the New Zealand Alps, where they can become sufficiently massive to dam back the glacier ice, impeding further advance. Therefore, periods of positive mass balance will cause the glacier to thicken, and if the rate of debris buildup at the margin keeps pace with the rate of ice thickening, the glacier can become perched high above the valley floor, hemmed in by its own moraines (Fig. 7.4a). Successive episodes of glacier expansion will tend to terminate at the same moraine, so that it may be built up on several occasions punctuated by non-active periods when vegetation, including trees, can colonize the outer slope (Fig. 7.3b). The damming effects of large lateral moraines are strikingly demonstrated when the barrier is breached for any reason (Fig. 7.4b). Lliboutry (1977) has argued that the 'crooked paths' followed by some debris-mantled Peruvian glaciers reflect the breaching of the moraine barrier by lake outburst floods at times of limited glacier extent. When the glacier expands once more, it is able to

(a)

(b)

Fig. 7.3 (a) Small rampart of ice-contact scree at the margin of Slettmarkbreen, Norway. The rampart is 0.5 m high. (b) Large lateral moraine at the margin of the debris-mantled Ghiacciaio del Miage, Italy, looking towards its accumulation area. Note how the glacier surface lies high above the valley floor, dammed by the lateral moraine. (Photos: D.I. Benn)

(a)

(b)

Fig. 7.4 (a) Debris-mantled glacier hemmed in by large lateral moraines, Huascaran–Chopicalqui massif, Cordillera Blanca, Peru. (b) Sequence of moraines marking the advance of a lobe of ice through a gap in a large lateral moraine, Ghiacciaio del Miage, Italy. (Photos: C.M. Clapperton and D.I. Benn)

advance through the breach, but is restrained by the remaining moraine along the rest of its margin (Fig. 7.5).

Substantial amounts of debris are reworked and deposited by *glacifluvial processes* at some ice margins, particularly the flanks of debris-mantled glaciers with well-developed supraglacial and englacial drainage networks (Gustavson and Boothroyd, 1987). Braided or meandering meltstreams can deposit sediment in troughs in the ice surface, burying the underlying ice and retarding ablation (Paul,

1983; D.E. Lawson, 1995). Topographic inversion and differential ablation can then result in the isolation of stagnant ice blocks, which may become completely buried by outwash deposits accumulating around the ice margin. Eventual melting of such buried ice blocks leaves water-filled *kettle holes* in the outwash surface (Fig. 7.6), whereas the melt-out of larger masses of ice buried below outwash produces a more irregular topography in which the areas between kettle holes are reduced to remnant upstanding ridges and mounds, which may follow the traces of former channels (*kame and kettle topography*, Section 11.4.3). Topographic development during the ablation of some ice margins is complicated by the emergence of sediment-choked subglacial and englacial conduits, forming upstanding ridges of sand and gravel or *eskers* (Section 11.2.9; Syverson *et al.*, 1994).

Characteristic aprons and ramps of debris form around the margins of some cold, polar and subpolar

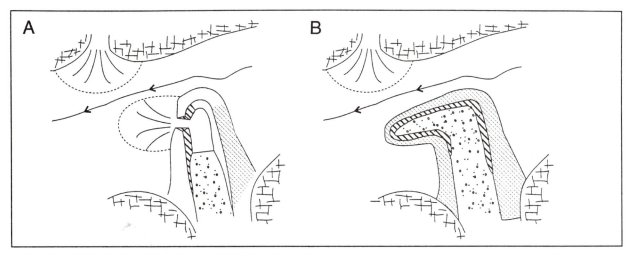

Fig. 7.5 Lliboutry's (1977) model of the development of a crooked debris-mantled glacier. (A) A large lateral moraine is breached by the outburst of a moraine-dammed lake at a time of limited glacier extent. (B) A subsequent glacier advance is funnelled through the breach but constrained elsewhere

Fig. 7.6 Kettle holes on a small outwash surface, Myrdal-sjökull, Iceland. (Photo: N. Spedding)

glaciers (Shaw, 1977a; D.J.A. Evans, 1989b; Fitzsimons, 1990). When advancing, such glaciers typically terminate in steep ice cliffs, from which ice blocks and debris topple and accumulate around the margin, a process known as *dry calving* (Figs 7.7, 7.8 and 7.9; Sections 2.2.3.1 and 5.4.3). During the brief summer ablation season, glacifluvial processes can also be important in such settings. Meltstreams deposit large amounts of sediment around the margins and at the foot of waterfalls that pour over the ice cliff. In addition, meltstreams can undercut the foot of the ice cliff, accelerating the dry calving process.

7.3.2 Ice-cored moraines

On many glaciers, debris cover is thickest very close to the margin. This may be due to the combined

Fig. 7.7 Debris apron accumulating at the margin of a cold, subpolar glacier: Eugenie Glacier, Ellesmere Island, Arctic Canada. (Photo: D.J.A. Evans)

effects of several processes, including the concentration of englacial debris at the surface by ablation, downslope movement of supraglacial debris towards the margin, the emergence of englacial meltstreams with high sediment loads, and the elevation of basal debris along submarginal shear planes (Fig. 7.10; Boulton, 1970a, 1977a; P.G. Johnson, 1971; Drozdowski, 1977; Krüger, 1994; Kirkbride and Spedding, 1996). During the ablation season, this debris cover retards melting of the underlying ice so that ablation at the margin is slower than that of cleaner

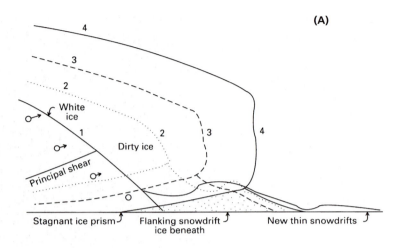

(A)

4

3

2

White ice

1

2 3 4

Dirty ice

Principal shear

Stagnant ice prism ⌐   Flanking snowdrift ⌐   New thin snowdrifts ⌐
                        ice beneath

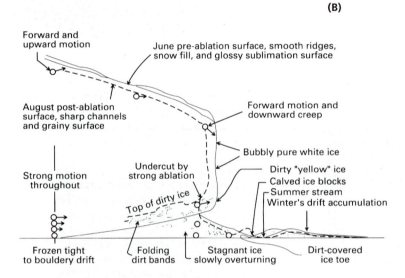

(B)

Forward and upward motion

June pre-ablation surface, smooth ridges, snow fill, and glossy sublimation surface

Forward motion and downward creep

August post-ablation surface, sharp channels and grainy surface

Bubbly pure white ice

Undercut by strong ablation

Dirty "yellow" ice
Calved ice blocks
Summer stream
Winter's drift accumulation

Strong motion throughout

Top of dirty ice

Frozen tight to bouldery drift

Folding dirt bands

Stagnant ice slowly overturning

Dirt-covered ice toe

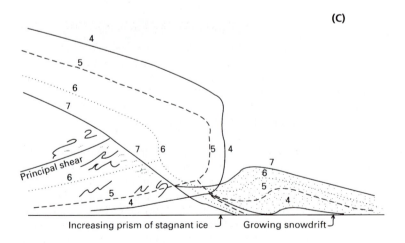

(C)

4

5

6

7

2

7 6 5 4

Principal shear

6

7

6

5

5

4

4

Increasing prism of stagnant ice ⌐   Growing snowdrift ⌐

Fig. 7.8 Apron formation at the margin of the Greenland ice sheet at Nunatarssuaq. The sequence shows: (A) ice cliff advance and apron (marked here as snowbank) overriding; (B) an ice cliff in relative equilibrium; and (C) ice cliff retreat and the production of a buried supraglacial ramp. Numbers show advance and retreat profiles. (Provided by Richard Williams, after Goldthwait, 1971)

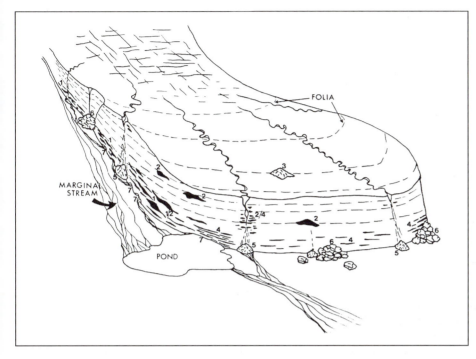

Fig. 7.9 Schematic diagram of the processes of debris accumulation around the margins of a subpolar glacier where there is no proglacial thrusting. 1: Debris bands cropping out at the surface; 2: pods or *augen* of debris; 3: supraglacial debris cone; 4: frontal debris bands; 5: waterfall debris piles; 6: apron of calved blocks; 7: thermoerosional niches cut by lateral streams and associated ice-marginal pond. (From Evans, 1989a)

Fig. 7.10 Debris bands uplifted towards the surface of a subpolar glacier, Ellesmere Island. (Photo: D.J.A. Evans)

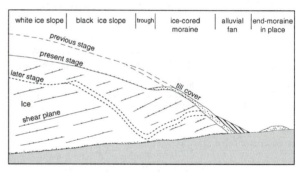

Fig. 7.11 Formation of an ice-cored moraine by differential ablation at a glacier margin. (Modified from Goldthwait, 1971)

stream networks may also develop glacier karst features in ice-cored moraines (Healy, 1975; Section 6.5.4.2).

7.4 RELOCATION OF PROGLACIAL MATERIAL BY GLACIER ICE

Relocation of previously deposited sediment by glacier ice is most important during periods of ice margin advance, during either the winter months or more prolonged phases of glacier expansion. At such times, proglacial material may be repositioned by direct glacigenic processes such as squeezing, pushing and glacitectonic disturbance. Relocation can also occur by squeezing of saturated material during periods of ice-margin standstill or retreat.

ice upglacier (Section 6.5.1). This can result in the isolation of upstanding debris-covered blocks of ice or snow known as *ice-cored moraines* at the ice margin (Fig. 7.11; Østrem, 1959, 1964; Østrem and Arnold, 1970; P.G. Johnson, 1971; Souchez, 1971; Rains and Shaw, 1981). Ice-cored moraines are analogous to supraglacial dirt cones (Section 6.5.2.2), and tend to be elongated parallel to the ice edge, reflecting the distribution of debris on the surface. The trough between ice-cored moraines and the main body of the glacier is commonly followed by meltstreams, thus strongly influencing patterns of deposition and erosion. Proglacial

7.4.1 Ice-marginal squeezing

During the ablation season, ice-marginal environments are often very wet, with an abundance of meltwater streams and standing water. As a result, ice-marginal sediments commonly have very high water contents and in many cases are easily liquefied and displaced by the processes of squeezing and pushing (Fig. 7.12). Pushing involves a forward movement by the glacier shoving the material from behind, whereas squeezing requires only static loading of water-saturated material by the ice mass. Squeezing occurs in response to the pressure gradient that exists between sediment overlain by glacier ice and unconfined sediment on the glacier foreland or below low-pressure cavities beneath the glacier. Sediment is therefore squeezed out from beneath the ice margin or infills cavities and basal crevasses (Fig. 7.13).

Fig. 7.12 Proglacial area of Breidamerkurjokull, showing an abundance of standing water. (Photo: D.I. Benn)

Fig. 7.13 Crevasse-squeeze ridge emerging on the surface of Flåajökull, Iceland. (Photo: D.J.A. Evans)

Squeezing of saturated debris into basal crevasses is particularly effective and widespread following glacier surges, when large numbers of crevasses open up and sediment is displaced vertically, often emerging on the glacier surface (Sharp, 1985a).

Extrusion of sediment from below glacier margins creates moraine ridges which mimic all the indentations of the ice margin (Price, 1970; Worsley, 1974). It has been suggested by Worsley (1974) that several independent squeeze ridges may be produced during one melt season owing to variability in the degree of saturation brought about by changing climatic conditions. In a study of moraines on Baffin Island, Andrews and Smithson (1966) suggested that the arrival of summer meltwater at the glacier sole is critical to the production of squeezed ice-marginal sediments. This meltwater may reach the ice base only during an advanced stage of the ablation season in such Arctic locations.

Squeezing often does not occur in isolation, and debris is commonly subject to relocation and reworking by several processes. A combination of squeezing and pushing has often been observed, because glaciers will maintain some forward motion during the early part of the ablation period when subglacial sediments are becoming saturated by meltwater (Sharp, 1984). Similar complexity is reported by Rogerson and Batterson (1982) from British Columbia, Canada. These authors document the squeezing of subglacial sediment into supraglacially dumped end moraines partially overlying thin glacier snouts.

7.4.2 Ice pushing

Ice pushing is simply the bulldozing of water-soaked ice-marginal sediment by an advancing glacier snout. It is distinct from glacitectonic thrusting, which is principally the result of large static loads beneath and immediately in front of ice margins (Section 7.4.3). Pushing affects sediment that accumulates at glacier margins by several processes including dumping, squeezing and glacifluvial deposition (Sharp, 1984). Sediment deposited during major glacier advances is usually overridden and enters the subglacial environment, so that ice-marginal pushing is generally most important during winter readvances by glaciers undergoing overall retreat. In such situations, a series of annual *push moraines* is produced (Section 11.3.2.1; Fig. 7.14 (Plate 17)). Hewitt (1967) has shown how push moraine formation at the margin of Biafo Glacier in the Karakoram Himalaya is part of an annual cycle of ice margin behaviour. During the summer ablation season, debris slumping off the ice surface builds up in aprons and fans of sediment along the margin. When ablation ceases in the winter, the glacier readvances and pushes and partially overrides the aprons. Forward movement of the glacier oversteepens the distal slopes of the aprons, which are then regraded by the downslope movement of debris. The lower, inner parts of the aprons are over-ridden by the glacier, forming a thickening wedge of till beneath the margin.

In some cases, bulldozing of proglacial sediments occurs in association with snowdrifts built up in the concavity in front of the ice margin (Birnie, 1977). The stresses imposed by the motion of the glacier can be transmitted through the snowbank, affecting proglacial sediments up to several metres in front of the glacier. Any moraines created by this process will mark not the position of the ice margin proper, but the former limit of the snowbank. Alternatively, the junction between the ice edge and the snowbank may buckle upwards, causing sediment to be pushed up between the two.

7.4.3 Proglacial glacitectonics

The large-scale tectonic disturbance and dislocation of unconsolidated sediments and weak rocks by glacier advances has been acknowledged for more than a century, and large-scale landforms ascribed to glacitectonic processes are widespread in formerly glaciated terrain. It is now widely acknowledged by glacial geologists and geomorphologists that glaciers affect the landscape through deformation as well as through erosion and deposition. Subglacial deformation is reviewed in Section 5.6. This section deals with the theory of ice-marginal (proglacial and submarginal) glacitectonic mechanisms and processes, whereas the structures and landforms created by glacitectonic processes are reviewed in Section 11.3.1.

7.4.3.1 THE MECHANICS OF GLACITECTONISM

Proglacial glacitectonics refers to the large-scale displacement of proglacial materials due to stresses imposed by glacier ice, and involves ductile or brittle deformation or a combination of the two (Fig. 7.15; Aber *et al.*, 1989; Hart and Boulton, 1991). Ductile deformation involves the production of large open folds in the sediments or rocks in front of an advancing glacier, which may develop into overfolds or begin to undergo internal thrusting owing to continued ice advance. In contrast, brittle deformation involves the thrusting of semi-coherent blocks along discrete planes of failure. In both cases the lower limit of deformation is usually marked by a basal failure plane or *plane of décollement*, which often coincides with a sedimentary discontinuity or bedding plane. Although all materials are capable of undergoing both ductile and brittle failure according to variations in the applied stress, temperature, strain

Fig. 7.15 Active proglacial thrusting, Axel Heiberg Island. (Photo: J. England)

rate and porewater pressure, glacitectonic disruption of frozen sediments appears to be dominated by brittle failure (e.g. Kalin, 1971; Klassen, 1982; D.J.A. Evans and England, 1991). The compression of proglacial materials by glacitectonic deformation is very similar to the processes which occur during continental collision and mountain-building (Croot, 1987; Aber, 1988a). Indeed, proglacial glacitectonic landforms can be regarded as scale models of mountain chains, and can be studied using well-founded principles of structural geology (see Twiss and Moores, 1992).

Proglacial glacitectonic deformation can be responsible for the formation of large thrust moraine complexes standing many tens of metres above the surrounding terrain (Section 11.3.1.2). The elevation of such large thrust masses depends upon a number of factors, the most important of which are low-strength proglacial sediments and high glacially imposed stresses. Sediment strength is dependent upon grain size and sorting, the existence of potential planes of failure, and porewater pressures. High porewater pressures result from subglacial drainage through inefficient distributed systems, such as porewater flow (Section 3.4), and can be further encouraged by the existence of proglacial permafrost, which acts to confine pressurized porewater in underlying unfrozen aquifers (Mathews and Mackay, 1960). The presence of weakened sediments, however, is not a sufficient condition for proglacial glacitectonic deformation, which also requires stresses large enough to elevate large masses of sediment above the glacier margin. It is clear that in most situations the shear stresses beneath glacier margins are too low to produce the observed deformation. The solution to this problem was developed by Rotnicki (1976), van der Wateren (1985) and Aber *et al.* (1989), and is known as the *gravity spreading model*. According to this model, proglacial sediment failure results from the translation of the *weight* of the glacier into lateral stresses, which push sediment wedges away from the load and then upward from their original position. The stress field producing a thrust block is therefore the product of the total weight of the ice mass and older thrust blocks.

The development of lateral stresses near a sloping ice margin is illustrated in Fig. 7.16. At any point below the glacier, the downward-oriented stress produced by the static weight of the ice column (the *normal stress* or *glaciostatic stress*) is calculated using equation (4.3):

$$\sigma_z = \rho_i gh$$

where ρ_i is the density of the ice, g is gravitational acceleration (9.81 m s^{-2}), and h is the ice thickness (m).

Since ρ_i is c. 900 kg m^{-3} and gravitational acceleration is 9.81 m s^{-2}, equation (4.3) reduces to:

$$\sigma_z = 8829h \qquad (7.2)$$

For $h = 100$ m, the normal stress would be c. 882,900 Pa or 882.9 kPa. Thus, for even moderate ice thicknesses, the glaciostatic stress far exceeds likely basal shear stresses, which usually lie in the region of 50–100 kPa.

Part of the glaciostatic stress is transferred to a horizontal stress, because of the tendency of subglacial materials to bulge outwards and press laterally against neighbouring particles in response to the load imposed by the overlying ice and sediment. The magnitude of the horizontal stress (σ_x) for any given glaciostatic stress (σ_z) depends on the material properties, and is calculated using *Poisson's ratio*, υ:

$$\sigma_x = (\upsilon\sigma_z)/1 - \upsilon \qquad (7.3a)$$

Since Poisson's ratio is close to 0.2 for many unconsolidated sediments, the horizontal component of the glaciostatic stress can be approximated as

$$\sigma_x = 2200h \qquad (7.3b)$$

Because the glaciostatic pressure is dictated by ice thickness, the horizontal component (σ_x) decreases in magnitude from the centre of the ice mass towards the ice margin, producing a lateral pressure gradient in the substatum (Fig. 7.16; Rotniki, 1976; van der Wateren, 1985; Aber *et al.*, 1989). This lateral pressure gradient between two points is given by:

$$\sigma_{x1} - \sigma_{x2} = 2200(h_1 - h_2) \qquad (7.4)$$

The horizontal stress differences are cumulative; that is, the stress difference over a given interval is passed on and added to the stress difference of the next, resulting in a maximum horizontal compressive stress beneath the margin (Fig. 7.16; Aber *et al.*, 1989). To this stress must be added the basal shear stress or *glaciodynamic stress*. Together, the glaciostatic and glaciodynamic stresses constitute the most important components of the total glacitectonic stress.

Failure will take place along a potential failure plane when the applied stress, in this case the total glacitectonic stress, equals or exceeds the *shearing resistance* (Hubbert and Rubey, 1959; Aber *et al.*, 1989). Failure conditions can be expressed using the familiar Coulomb equation (equations (4.6), (4.8) and (6.3)), and failure will occur when:

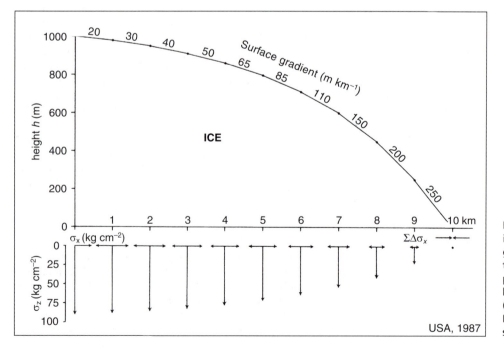

Fig. 7.16 The gradient in normal stress (σ_z) at a glacier margin, leading to a net horizontal compressive stress ($\Sigma\Delta\sigma_x$) below the ice edge. (From Aber et al., 1989. Reproduced by permission of Kluwer)

$$\sigma_{gt} \geq c + (p_i - P_w) \tan \varphi_i$$

where σ_{gt} is the total glacitectonic stress, p_i is the ice overburden pressure, P_w is the porewater pressure, c is cohesion, and $\tan \varphi_i$ is the angle of internal friction.

It can easily be seen that, where the cohesion is small, failure is most likely where P_w is large: as the porewater pressure approaches the overburden pressure, the stress required for the initiation of thrusting approaches zero. According to Aber *et al.* (1989), this situation arises either through the compaction of impermeable strata and the restriction of water escape routes, or through the transmission of water under pressure into a confined aquifer. In a homogeneous material, thrust faults normally develop at an angle of 30° to the horizontal, because their internal angle of friction lies around that value (Kulhawy, 1975). However, low-angle or subhorizontal thrust faults will exploit pre-existing weaknesses in the substratum such as bedding planes, lithological boundaries, incompetent strata and permafrost boundaries. Steeper thrust fault angles (> 30°) occurring in coarse-grained materials such as gravels attest to the existence of permafrost, because greater substrate rigidity is required to enable the development of such steep thrust fault angles. Movement of material along thrust planes due to gravity spreading results in the elevation of thrust blocks in front of the glacier margin (Fig. 7.17).

In addition to gravity spreading, two other mechanisms may contribute to the development of thrust systems (Aber *et al.*, 1989; Price and Cosgrove, 1990). First, *gravity gliding* or *gravity sliding* involves the deformation of sediment blocks as they move downslope under their own weight. This requires the uplifting of the ice-proximal zone by gravity spreading to produce a slope upon which blocks can slide. Second, *compression* or *push-from-behind* may occur by the direct shoving of sediment wedges by the forward movement of the glacier. In reality, gravity spreading, gravity gliding and compression can act in unison in proglacial settings.

The likelihood of glacitectonic disturbance is dictated by various factors, the most important of which are as follows (Fig. 7.18; Aber *et al.*, 1989):

1. *Slope of the proglacial area.* The presence of reverse slopes at the glacier margin, where ice advances against a topographic obstacle, increases the likelihood of proglacial glacitectonics (Bluemle and Clayton, 1984).
2. *Presence of weak layers in the substratum.* Failure is encouraged by weak layers which form potential décollement planes, and is most likely where such layers occur with a favourable orientation and at a suitable depth.

3. *The nature of ice–sediment contact in the proglacial area.* Where glaciers are partially buried by the proglacial sediments, any advance will act to drive a wedge of ice into the strata. This mechanism is thought to be responsible for the production of thrust blocks with sedimentary beds dipping away from rather than back towards the glacier snout in some Canadian Arctic thrust block moraines (D.J.A. Evans and England, 1991).
4. *Subglacial and proglacial drainage.* As noted above, failure is encouraged by high porewater pressures in proglacial and submarginal sediments and rocks. High water pressures can arise if impermeable sediments or permafrost occur at the glacier margin, both of which impede drainage and increase porewater pressures in underlying aquifers (Mathews and Mackay, 1960; Mackay and Mathews, 1964; Aber *et al.*, 1989). High proglacial and submarginal porewater pressures are also generated during glacier surges, when large amounts of water are discharged through distributed drainage systems (Section 4.8.2; Sharp, 1985b; Croot, 1988b; Mooers, 1990a). Surging also causes transient high water pressures by rapid loading by glacier ice. Although the presence of permafrost at glacier margins and glacier surging are conducive to proglacial tectonic disturbance, neither is a necessary condition, and thrusting can occur in other settings.

A distinct type of glacitectonic process has been described by Krüger (1993, 1994) on the basis of observations at the margin of Sléttjökull, Iceland. During winter, the penetration of a 'cold wave' from the surface causes a slab of basal till to be frozen on to the glacier sole. This slab is carried forward by the winter readvance of the glacier, whereupon it shears over the proglacial sediments, forming a small ridge (Fig. 7.19). The glacier melts back during the summer until the return of cold winter temperatures initiates another episode of freezing on and thrusting. Successive increments added to the proglacial ridge each year result in a moraine morphologically similar to, but genetically distinct from, conventional push moraines. A variation on this process has been described by Matthews *et al.* (1995) for the margin of a cirque glacier in Norway (Section 11.3.1.2).

7.4.3.2 GLACITECTONIC DEFORMATION

The shortening of proglacial sediments and rocks during glacitectonic deformation results in complex geological structures analogous in many ways to thrust fold belts at the collision points of drifting continents, and can be explained using the principles of structural geology (e.g. Bally *et al.*, 1966; Perry *et*

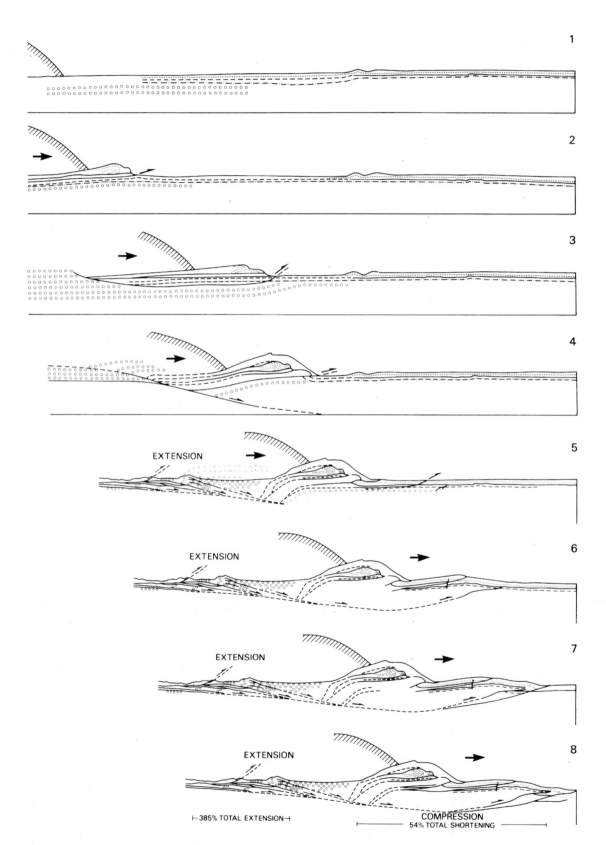

Fig. 7.17 Sequential development of glacitectonic thrusting, based on a thrust moraine in front of Eyjabakkajökull, Iceland. Note extension and excavation beneath the margin, and compression and ridge construction on the foreland. (From Croot, 1988b. Reproduced by permission of Balkema)

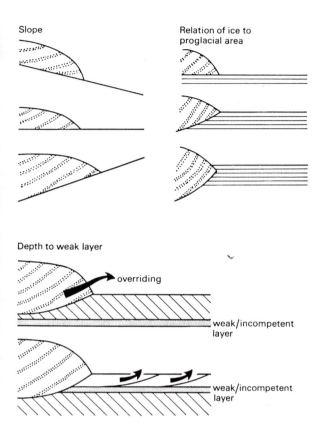

Slope

Relation of ice to proglacial area

Depth to weak layer

overriding

weak/incompetent layer

weak/incompetent layer

Fig. 7.18 Factors considered to be important in governing the style of glacitectonic deformation. (From Aber *et al.*, 1989. Reproduced by permission of Kluwer)

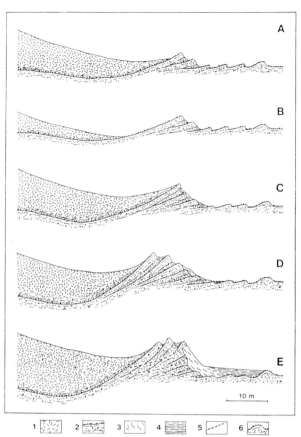

10 m

1   2   3   4   5   6

Fig. 7.19 Model for thrust moraine formation at the margin of Sléttjökull, Iceland. (1) Glacier; (2) basal till; (3) mass movement deposits; (4) outwash deposits; (5) thrust plane; (6) annual moraine. (From Krüger, 1993. Reproduced by permission of Scandinavian Academic Press)

al., 1984; Siddans, 1984; Aber, 1988a; Aber *et al.*, 1989; Pedersen, 1993).

Subhorizontal substrata affected by glacitectonic compression will often dislocate along low-angle failure planes called *thrust* or *overthrust faults* (low-angle reverse faults), which separate the overriding *hanging wall* and the overridden *footwall*. Such faults normally produce a 'staircase' pattern whereby the sole thrust climbs up through the strata, producing *flats* and *ramps*; flats are the subhorizontal parts of the thrust plane, whereas ramps are the parts that cut up through the strata at angles of around 30° (Fig. 7.20). Continued compression by glacier advance will result in the development of multiple thrust sequences. These may involve either *piggyback thrusting*, where new thrusts develop in the footwall, or *overstep thrusting*, where new thrusts develop in the hanging wall (i.e. they develop in sequence either forwards or backwards from the initial thrust; Figs 7.20c and 7.20d). Studies of glacitectonic structures by Rotnicki (1976) and van der Wateren (1985) suggest that piggyback thrusting is most common in proglacial deformation. More complex thrust masses are produced where *thrust slices* or *sheets* (horses) are imbricately stacked as a *duplex*. Duplexes there-

fore consist of a stack of horses bounded by a roof thrust and a floor thrust (Fig. 7.20e). Other common structures in glacitectonized sediments are *back thrusts*, which indicate a displacement in the opposite direction to the main thrust due to layer-parallel compression (Fig. 7.20f). Back thrusts will combine with frontal ramps to produce an uplifted hanging wall block called a *pop-up*, and may also truncate older thrust faults to produce *triangle zones* (Fig. 7.20g). Other types of faulting produced by compression in the proglacial zone are *anastomosing* and *conjugate* or *Riedel shear* patterns. These failures are characterized by two sets of similar faults which cross-cut each other at a consistent angle.

Ductile deformation during glacitectonic compression and shortening results in a wide range of fold types, depending on the magnitude and orientation of the applied stresses and the strain response of the sediments. Folds range from long-wavelength or *open* folds to *isoclinal* or *recumbent* (nappe) folds. Other fold types include chevron, sheath, kink-band,

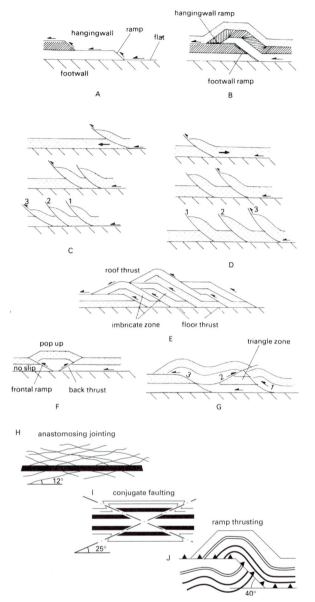

Fig. 7.20 Large-scale structures typically found in thrust zones: (A) shape of thrust surface with ramps and flats; (B) hangingwall geometry; (C) piggyback thrust sequence; (D) overthrust step sequence; (E) duplex-imbricate thrust slices contained between a floor thrust and a roof thrust; (F) pop-up structure formed by backthrusting; (G) triangle zone formed by backthrusting; (H) anastomosing jointing; (I) conjugate faulting; (J) ramp thrusting with distorted bedding in the footwall ramp. (Modified from Park, 1983, and Pedersen, 1993)

box and polyclinal folds (Fig. 7.21). Complexities may arise where different sedimentary layers respond differently to stress and produce disharmonic folds, in which the fold wavelengths of some layers are smaller than those of others. Studies of the three-dimensional form of glacitectonic folds reveal

that fold profiles typically vary along the fold axis (i.e. they are *non-cylindroidal*). Common examples of non-cylindroidal folds are *periclines*, *domes* and *basins* (Fig. 7.21). Domes are simply anticlinal structures which plunge at similar angles in all directions, whereas structural basins are produced where strata dip inwards in all directions. Periclines are elongated domes which are often aligned in an *en échelon* fashion as a result of compression. Where strata have undergone complex deformation histories, some folds may be superimposed on pre-existing folds. This may involve two separate phases of deformation, or may simply indicate shortening in more than one principal strain direction (Fig. 7.22; Ramsay, 1967).

Structures resulting from compressional deformation have been produced experimentally by Mulugeta and Koyi (1987), who subjected sand layers to 40 per cent shortening in a squeeze-box (Fig. 7.23). The resulting thrust mass was composed of three deformation domains which in order of formation are:

1. *Distal domain.* This consists of low-angle thrust faults demarcating thrust blocks with overturned drag folds, extension fractures and slumps. Note also the development of a sheath fold and thrust faults at the base of the outermost thrust block.
2. *Intermediate domain.* Thrust blocks are rotated into steeper positions, and thrust faults develop into concave-upward listric forms. Back-kinking also develops in the upper portions of the thrust blocks.
3. *Proximal domain.* The thrust blocks become vertically oriented and laterally compacted. The lateral compaction or shortening of the thrust mass results in the development of back thrusts and kink zones, which is in turn associated with underthrusting and thickening of the proximal zone.

These deformation domains can be recognized in proglacially thrust sediments (Humlum, 1985b; Eybergen, 1987; Croot, 1988b; Boulton *et al.*, 1989). Figure 7.17 illustrates the sequential formation of thrust slices outward from the margin of Eyjabakka-jökull, Iceland, and the progressive tilting of the oldest units (Croot, 1988b). Several examples of the internal structure of thrust moraines are described in Section 11.3.1.2.

Continued glacier advance over the top of thrust-block moraines results in the superimposition of subglacial deformation on proglacial tectonic structures (Section 4.4) and streamlining (see Sections 11.3.1.3 and 12.4.3). A two-phase model of glacitectonism was proposed by Aber (1982) in which the proglacial thrusting and stacking of substrata are followed by

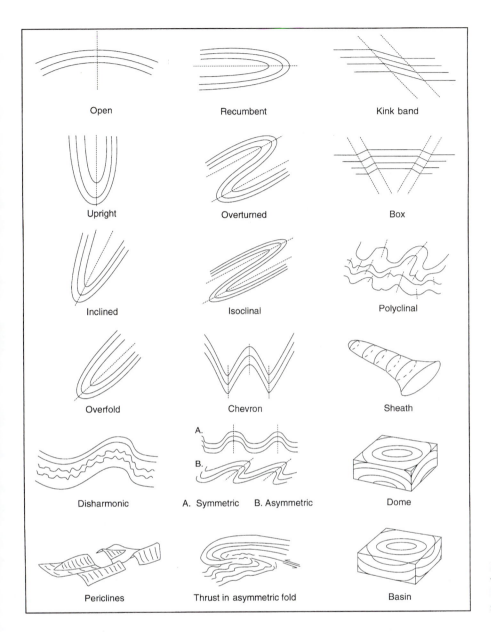

Fig. 7.21 Definitions of the various fold types that may be found in glacitectonized sediment sequences

glacial overriding. In this case, subglacial deformation takes over from proglacial deformation once the total energy consumed by the thrust mass equals the maximum amount that can be produced in the gradient stress field (van der Wateren, 1985), and the ice flows around the obstacle.

The expulsion of groundwater during glacitectonism is thought to be important in the development of some thrust systems. The migration of groundwater during thrusting is analogous to fluid and hydrocarbon migration into mid-continent regions during continental collisions (e.g. Oliver, 1986), when the construction of large thrust sheets results in the migration of fluids away from the disturbed strata. During glacitectonic thrusting, water is forced to flow along décollement planes, where it facilitates further displacement, and into aquifers. This water may be expelled at the surface in springs or even blow-outs if aquifers are pressurized (Fig. 7.24; Christiansen *et al.*, 1982; Bluemle, 1993). The patterns of groundwater flow beneath and in the marginal zones of ice sheets were modelled by Boulton *et al.* (1995) and Boulton and Caban (1995), who suggested that the aquifers beneath the Scandinavian ice sheet possessed a transmissivity capable of draining all the subglacial meltwater. Where confined aquifers become overpressurized owing to the presence of glacier ice, groundwater and liquefied sediment may burst through proglacial sediments, especially in areas of discontinuous permafrost, to produce large-

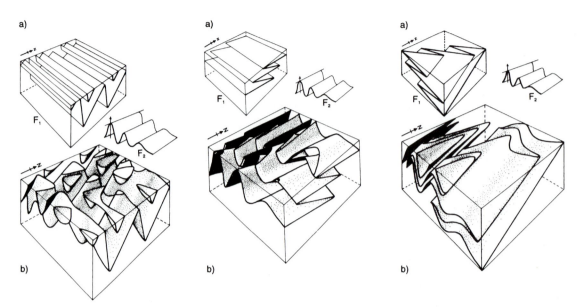

Fig. 7.22 Examples of superposition of two separate deformation events. Note that F_1 and F_2 in each case (a) refer to the fold patterns produced by the two separate deformation events. The superposition produces the complex patterns in each (b). (From Price and Cosgrove, 1990 after J. G. Ramsay. Reproduced by permission of Cambridge University Press)

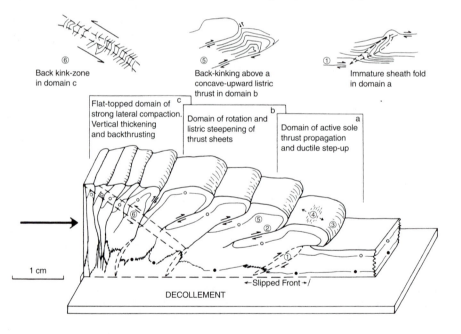

Fig. 7.23 An example of piggy-back-style thrusting, showing deformation of stratified sand after it was subjected to 40 per cent shortening in a squeeze box: (1) initiation of thrust; (2) thrust fault; (3) slump zone; (4) extension fractures; (5) back kink fold; (6) backthrust zone. (From Mulugeta and Koyi, 1987. Reprinted by permission of the Geological Society of America)

scale dewatering structures or *extrusion moraines* (Section 12.3.5).

7.5 GLACIERS AND PERMAFROST

Permafrost is defined as ground in which the temperature remains below 0°C for at least two consecutive years, and is widespread in Arctic Canada and Eurasia, and in some high mountain environ-

ments (Ballantyne and Harris, 1994). At depth, the ground remains permanently frozen, but seasonal melting occurs in a zone close to the surface known as the *active layer*. Permanently low temperatures at depth allow the survival of buried glacier and ground ice for long periods wherever the debris overburden exceeds the active layer thickness. This gives rise to distinctive forms in cold regions formerly occupied by glaciers, including *rock glaciers* and *buried glacier ice*. Both rock glaciers and buried ice, however, can be formed by

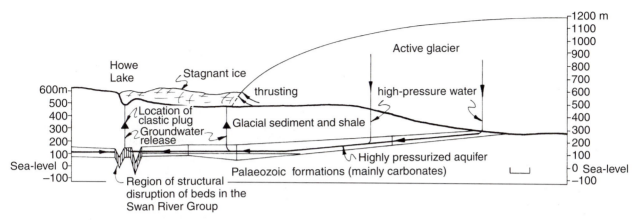

Fig. 7.24 A generalized and hypothetical cross-section through proglacial sediments during glacier advance, showing groundwater flow paths in subglacial and proglacial zones. This is the situation envisaged for the production of the Howe Lake blow-out feature in North Dakota. (From Bluemle, 1993. Reprinted by permission of the University of Regina)

non-glacial processes, and their interpretation is often somewhat controversial.

7.5.1 Rock glaciers

Rock glaciers are tongue-like or lobate masses of angular debris that resemble small glaciers, and which move downslope as a consequence of the deformation of internal ice or frozen sediments. They commonly have ridges, furrows and sometimes lobes on their surfaces, and have steep fronts down which surficial debris slides and tumbles, to be overridden by the advancing mass upslope (Fig. 7.25; Potter, 1972; Washburn, 1979; Ballantyne and Harris, 1994). Rock glaciers are among the most enigmatic forms produced in cold environments, and have inspired a wide range of conflicting views on their formation. Excellent reviews of rock glacier form, distribution, genesis and nomenclature are provided by Martin and Whalley (1987a), Whalley and Martin (1992) and Hamilton and Whalley (1995).

Several morphological and genetic classifications for rock glaciers have been proposed, and many conflicting descriptive terms are in use (e.g. Wahrhaftig and Cox, 1959; Outcalt and Benedict, 1965; Humlum, 1982; Ballantyne, 1984). Here we adopt a simple twofold genetic classification. First, *periglacial rock glaciers* (also known as talus-derived, talus-foot and protalus rock glaciers) originate by the deformation of the lower parts of talus slopes, forming a bulging lobe at the base (Fig. 7.26). They are therefore purely periglacial phenomena that do not involve the presence of glacier ice. Second, *glacial rock glaciers* form by the progressive burial and deformation of a core of glacier ice by a thick, bouldery debris mantle (Fig. 7.27; P.G. Johnson, 1980a, b; Barsch, 1987; Giardino and Vitek, 1988). Some researchers prefer to use the term 'rock glacier' exclusively to refer to periglacial phenomena, and

Fig. 7.25 Rock glacier at 4500 m in the Milang Valley, Lahul Himalaya, India. (Photo: D.I. Benn)

consider forms with cores of glacier ice to be debris-covered glaciers, and by definition not rock glaciers (e.g. Barsch, 1973, 1987; Haeberli, 1985). However, this disagreement is simply terminological, and the fact remains that similar deforming masses of rocky debris do evolve by both periglacial processes and the burial of glacier ice, and exposures of glacier ice have been reported from several rock glaciers (e.g. Lliboutry, 1953, 1955; Outcalt and Benedict, 1965; Potter, 1972; Benedict, 1973b; N. Eyles, 1978; Jackson and MacDonald, 1980; Whalley, 1983; Martin and Whalley, 1987b; D.J.A. Evans, 1993). Thus, rock glaciers may constitute prime examples of *equifinality*, wherein very similar features are produced by different processes (Fig. 7.28; Whalley and Martin, 1992).

Glacial rock glaciers commonly form in cirque basins where abundant debris is delivered to glacier surfaces by periglacial slope processes and the melt-out of englacial debris (Griffey and Whalley, 1979; Mayewski and Hassinger, 1980; Birnie and Thom, 1982; Whalley, 1983; Gordon and Birnie, 1986; P.G.

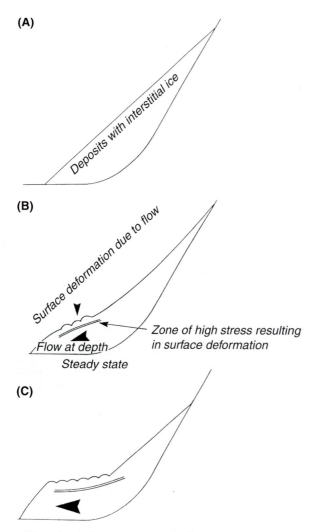

(A)

Deposits with interstitial ice

(B)

Surface deformation due to flow

Zone of high stress resulting in surface deformation

Flow at depth

Steady state

(C)

Flow due to ice content continuing

Fig. 7.26 Skin flow model of periglacial rock glacier formation. (From P.G. Johnson, 1984b. Reprinted by permission of the Association of American Geographers)

Johnson, 1987). The existence of permafrost is not essential (Martin and Whalley, 1987a). Debris will accumulate if it is delivered to the glacier surface more rapidly than it is advected away by glacier flow, so that rock glacier formation is favoured by low ice velocities and rapid debris accumulation rates. Such conditions are most likely for glaciers in an advanced stage of retreat, particularly those surrounded by high cirque headwalls. Eventually a point is reached where the stresses imposed by the weight and surface slope of the debris overburden are sufficient to cause deformation of the ice core, and the whole mass begins to creep downslope. It is sometimes assumed that the ice core is stagnant prior to renewed deformation, but there appears to be no reason why this must be the case.

Using the glacier ice flow law and assuming no basal sliding, Whalley and Martin (1992) showed that thin glacier snouts with low surface slope angles will flow very slowly (typically < 1.0 m yr^{-1}), and once ice thickness drops below approximately 30 m, no movement occurs at all (Fig. 7.29). Available data on rock glacier velocities appear to fit this theoretical rule and therefore satisfy a glacier ice core hypothesis in that wasting snouts are usually thin and possess low surface slopes. Anomalies occur wherever rock glaciers rest on steep slopes or where thicknesses exceed 30 m.

The sequence of events leading to the formation of glacial rock glaciers has been summarized by Whalley and Martin (1992) as follows:

1. A thin, wasting glacier snout is protected from melting by a rock debris cover.
2. The debris is derived from easily weathered bedrock cliffs which discharge talus directly on to the glacier surface.
3. The thin ice body flows at a slow rate which is predictable by glacier flow theory. Surface velocities are reduced over time, because the glacier ice thins as it flows and ablates.
4. The balance of debris supply to ice supply is critical to rock glacier formation.
5. Sliding rates are low or zero, even when ice is near the pressure melting point. If temperatures fall below 0°C, then flow rates decrease even further.

Finally, a variant of glacial rock glaciers forms by the deformation of ice-cored moraines in recently deglaciated terrain, where stagnant blocks of marginal glacier ice are buried by slope deposits before they can melt out (Section 7.3.2; Barsch, 1971). Ice-cored moraines develop into rock glaciers as the protected ice core begins to flow internally under the stresses imposed by the debris overburden (Messerli and Zurbuchen, 1968; Vere and Matthews, 1985). Such features have been referred to as *rock glacierized moraines* in the Canadian Arctic (Dyke *et al.*, 1982a; Dyke, 1990a; D.J.A. Evans, 1993), and may develop as a result of interstitial ice formation as well as by the flow of buried glacier ice.

7.5.2 Buried glacier ice

Massive subsurface ice bodies exist in permafrost regions, and have traditionally been explained by the growth *in situ* of ice lenses within the soil, as a result of *ice segregation* and *injection*. Essentially, segregation ice is formed by soilwater migration towards an advancing freezing front, creating lenses up to 10 m or more thick, and injection or intrusive ice is formed by the accumulation and freezing of pressurized water during permafrost aggradation (Mackay, 1972;

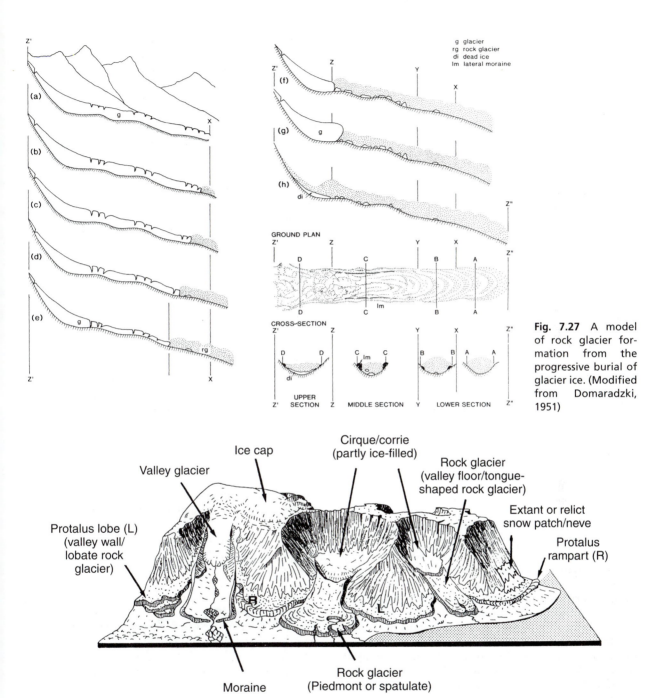

Fig. 7.27 A model of rock glacier formation from the progressive burial of glacier ice. (Modified from Domaradzki, 1951)

Fig. 7.28 The classification of rock glaciers according to morphology. (From Humlum, 1982. Reproduced by permission of the Norsk Geografisk Tidsskrift)

Ballantyne and Harris, 1994). The occurrence of massive ground ice bodies in formerly glaciated terrains and former ice-marginal areas has been used to suggest that they may be related either directly or indirectly to glaciation (Mackay, 1973; Mackay *et al.*, 1979; St Onge, 1994). For example, ground ice on Banks Island and the Mackenzie Delta, Canada, occurs near the outer limits of previous glaciations,

and this was regarded by Rampton (1974, 1988), Harry *et al.* (1988) and French and Harry (1990) as rather more than a coincidence. Both in this region of the western Canadian Arctic and in Siberia (Astakhov and Isayeva, 1988), there is a growing tendency to interpret some massive ground ice bodies as buried glacier ice, protected during deglaciation by a debris overburden which exceeds the active layer thickness.

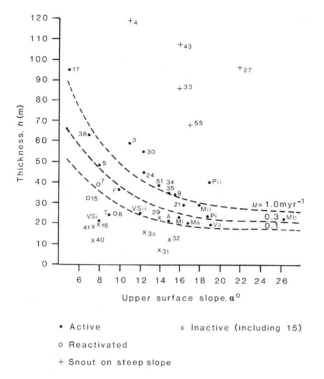

• Active   x Inactive (including 15)

○ Reactivated

+ Snout on steep slope

Fig. 7.29 Graph plotting rock glacier thickness (*h*) against upper surface slope (*α*). Dashed lines separate three surface velocities (*u*), the low-velocity band differentiating between inactive and active rock glaciers (note that those on steep slopes plot with anomalously high values). (From Whalley and Martin, 1992)

The very fact that some ground ice bodies occur in moraines or beneath areas of thick glacigenic sediment in regions of continuous permafrost suggests a strong possibility that they originated as glacier ice. This can be verified by detailed analysis of the ice bodies (French and Harry, 1990). The stratigraphy and crystallography of ground ice bodies provide a large amount of information on the origins of the ice. However, because glacier ice may have experienced numerous thawing and refreezing episodes during its formation and deformation, it may share many attributes with segregated or injection ice. This makes their differentiation difficult if not impossible in some locations. Diagnostic criteria for differentiating buried glacier ice and periglacial ground ice are outlined by French and Harry (1990) as: (a) ice–sediment contacts, which should be conformable in segregated ice but erosional in glacier ice sequences; (b) water/ice quality, which may change progressively with depth in segregated ice and trend downwards and upwards into the enclosing sediments, but would show no such trends in buried glacier ice; (c) air bubbles, which occur in trains extending downwards from the upper soil–ice con-

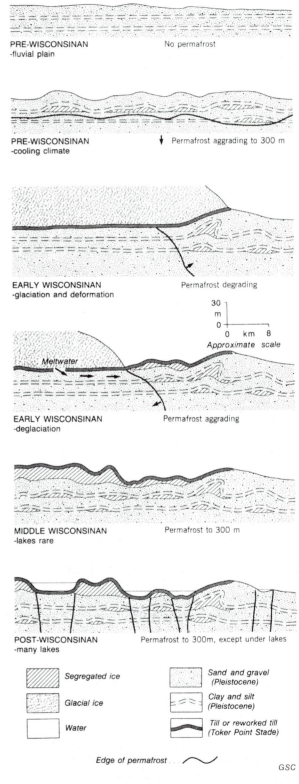

Fig. 7.30 Model of segregation ice development in association with glaciers, based on the Tuktoyaktuk Coastlands, NWT, Canada. (From Rampton, 1988. Reproduced by permission of the Geological Survey of Canada)

tact in segregation ice and indicate downward freezing – bubble type and distribution are far more variable in glacier ice; and (d) mineral inclusions, which in segregation ice are composed of soil from the same source of origin and in glacier ice are highly variable in composition. In addition, buried glacier ice may contain debris bands, striated pebbles, intensely folded folia and strong ice-crystal orientation indicative of ice that has undergone high cumulative strains.

A model which advocates ground ice growth through the indirect influence of nearby glaciers is that of Rampton (1974, 1988). He suggests that ground ice bodies are created during deglaciation when, owing to the hydraulic gradient produced by the glacier overburden pressure, large amounts of meltwater are driven beyond the glacier snout (Fig. 7.30). Permafrost degradation down to several hundred metres occurs beneath the ice sheet in Rampton's model, thus providing the aquifer for groundwater movement to the glacier margin. Deformation of other segregated ice bodies takes place at the same time as a result of glacier overriding, producing folded bands of mineral inclusions. French and Harry (1990) suggested a variant on this model, in which permafrost degradation takes place only in the uppermost layer, which then becomes saturated with meltwater under the glacier-imposed hydraulic gradient. Permafrost aggradation within the upper layers then produces segregation ice lenses which are juxtaposed with glacially deformed (pre-glacier advance) segregation ice and remnant buried glacier ice. The debris-rich basal glacier ice has high preservation potential in permafrost terrain because the insulating effect of supraglacial debris protects it from summer melting.

Buried glacier ice has been reported by D.J.A. Evans and England (1993) from terrain that was deglaciated some 10,000 years ago in the Canadian High Arctic. Remarkably old buried glacier ice has been reported by Sugden *et al.* (1995) from Antarctica, where radiometric dating techniques indicate an age of approximately 8 Myr!

7.6 PARAGLACIAL ACTIVITY

7.6.1 The paraglacial period

As glaciers retreat back from an area, newly deglaciated terrain is commonly subject to rapid change as fluvial, slope and aeolian systems relax towards non-glacial equilibrium states. The term *paraglacial* was introduced by Ryder (1971a, b) and later defined by Church and Ryder (1972) to encompass 'nonglacial processes that are directly conditioned by glaciation', characteristic of recently deglaciated environments. These processes, however, are not unique to such environments, and it is perhaps more useful to use the term 'paraglacial' to refer to the *period* of rapid environmental readjustment following glacier retreat (Church and Ryder, 1972; N. Eyles and Kocsis, 1988). Paraglacial activity is distinct from *periglacial processes*, which are characteristic of all cold, non-glacial environments, regardless of whether glacier ice is or was present in the catchment (Ballantyne and Harris, 1994).

The *paraglacial period* is characterized by high rates of sediment delivery from slopes and into fluvial and aeolian systems. This period of rapid response is triggered by the instability of unconsolidated glacigenic sediments (e.g. in lateral moraines and kame terraces) and oversteepened rock slopes once their support of glacier ice is removed. Sediment yields and denudation rates are highest immediately following deglaciation, then decline through time as sediment supply becomes exhausted and slopes relax towards more stable profiles (Fig. 7.31; Church and Slaymaker, 1989; Ballantyne and Benn, 1994b, 1996). The paraglacial period theoretically ends once sediment yields drop to rates typical of unglaciated catchments, although whether or not a landscape really fully adjusts following a glacial phase is difficult to ascertain. Certainly, delayed slope responses can occur many thousands of years after deglaciation (Ballantyne and Benn, 1996).

The concept of the paraglacial period was developed from sediment yields reconstructed from valley floor fans in North America, many of which were formed in the period following deglaciation in the late Pleistocene or early Holocene (Ryder, 1971a, b; Church and Ryder, 1972; Roed and Waslyk, 1973; Gardner, 1982; M.J. Clark, 1987; Beaudoin and King, 1994). Dating control on such sediments, however, is often poor, and rates of sedimentation and sediment yield are therefore subject to large errors. Recently, Ballantyne and Benn (1994b, 1996) and Ballantyne (1995) have studied landform development and sediment yields in two Norwegian valleys which have been deglaciated within the past two centuries, and demonstrated that very high rates of readjustment occurred immediately following the withdrawal of glacier ice. By comparing modern slope profiles with photographs taken earlier in the twentieth century, they calculated minimum average rates of surface lowering of some debris slopes of 50–100 mm yr^{-1}, with rates as high as 200 mm yr^{-1} in some locations (Fig. 7.32). Concomitant with slope erosion was the build up of debris cones on valley floors. By bracketing the period of fan growth by the age of trees growing on their surfaces and the age of moraines that they overlie, Ballantyne (1995) derived aggradation rates of 8–29 mm yr^{-1}, equiva-

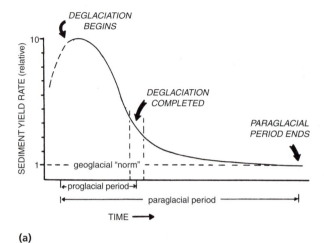

(a)

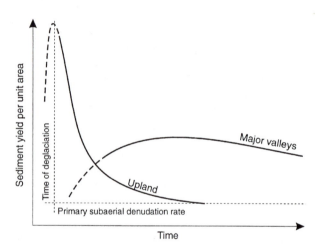

(b)

Fig. 7.31 Rates of sediment yield during the paraglacial period: (a) the paraglacial cycle of sedimentation (after Church and Ryder, 1972); (b) the paraglacial cycle of sedimentation, as modified by Church and Slaymaker (1989) to account for the effect of spatial scale on the temporal pattern of sediment yield. The time axis covers the postglacial period (< 15 kyr–10 kyr BP).

lent to average erosion rates of 37–93 mm yr^{-1}, in their catchment areas. In contrast with these high rates of sediment transfer, the cones studied by Ballantyne are now largely inactive, only 100–200 years after the withdrawal of glacier ice. These data are important because they demonstrate that, in these locations, the most active period of paraglacial resedimentation occurs within decades of deglaciation, and that the transition to lower sediment yields occurs within one or two centuries. It has yet to be established whether these timescales also apply to other recently deglaciated environments. In high mountain environments such as the Himalaya, where catchments are larger, and greater quantities of unconsolidated sediments are available, the period of

rapid paraglacial readjustment appears to be as much as several centuries or more (Fig. 7.33).

7.6.2 Gravitational reworking

Oversteepened and unstable debris slopes and rock slopes are common in recently deglaciated terrain, particularly in mountainous areas where glacial sediments and landforms are left perched high up on valley walls by receding glaciers. Such slopes undergo a period of high activity as they adjust towards nonglacial subaerial equilibrium. This is particularly evident on the ice-proximal sides of recently abandoned lateral moraines, where numerous slope failures can be observed (Fig. 7.34). The paraglacial readjustment of debris slopes commonly results in severe gullying and badland erosion, and reworking of glacigenic sediments into coalescent debris cones at the base of the valley walls (Ballantyne and Benn, 1994b, 1996).

The range of processes responsible for slope adjustment are common to very many environments, and can be usefully classified on the basis of water content and the type and rate of movement (Carson and Kirkby, 1972). In Fig. 7.35, common types of mass movement are differentiated on the basis of their velocity profiles, or the vertical distribution of strain within the slope. An alternative way of classifying mass movement types is presented in Fig. 7.36, which shows mass movements as a continuum determined by sediment concentration and velocity.

Rock avalanches, *rock slides* or *sturzstroms* are catastrophic failures of bedrock slopes resulting in the very rapid downslope transfer of debris. They are common in recently deglaciated high-relief valley systems where rock slopes have been weakened by pressure release or dilation once an ice load has been removed from the valley wall (e.g. Kieslinger, 1960; Bjerrum and Jorstad, 1968). Earthquake shocks, due to tectonic activity or isostatic uplift, are common triggers (e.g. Sissons and Cornish, 1982a, b). The retreat of valley walls and provision of debris to valley floors are critical to the erosion of glacial troughs over repeated periods of glaciation (Section 9.4.3). Rock avalanches may have long run-out distances, some boulders travelling across the valley floor and ascending the opposite valley wall. Various mechanisms have been suggested for the rapid transfer of materials over such long distances, each one emphasizing the tendency for the rock mass to flow rather than slide. This was explained by Kent (1966) as a fluidization process whereby air trapped within the rock debris acts as a fluid medium. Alternatively, Shreve (1968) proposed a cushion of air beneath the rock mass. These ideas have largely been superseded by those invoking numerous block collisions, whereby there is a

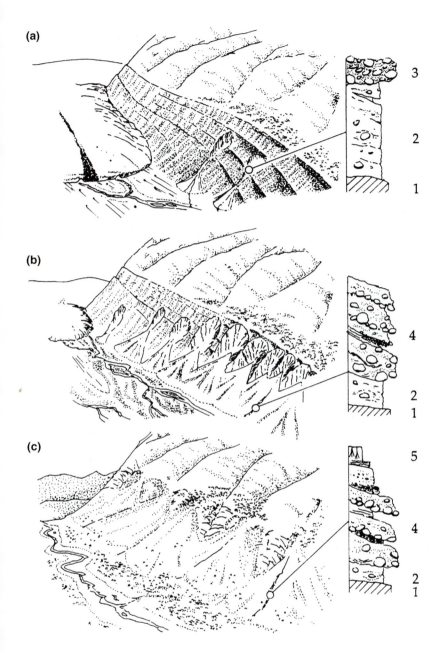

Fig. 7.32 Schematic block diagrams of the landforms and sediments associated with three stages in the paraglacial reworking of steep drift slopes in glaciated valleys. (a) Initial slopes exposed by glacier retreat, showing lateral moraines and the onset of gully incision; (b) advanced gully development and deposition of coalescing debris fans downslope; (c) exposed bedrock and stabilized, vegetated gullies and largely relict debris fans. By this stage, paraglacial reworking and slope adjustment have effectively ceased owing to diminution of debris supply. Facies key: 1: bedrock; 2: subaerial sediments relating to an earlier episode of paraglacial sedimentation; 3: ice-marginal deposits; 4: paraglacially reworked sediment (debris flows and intercalated slope-wash deposits); 5: soil horizons. (Modified from Ballantyne and Benn, 1996)

downslope transfer of kinetic energy and the mass acts like a cohesionless grain flow (Bagnold, 1954, 1956; Hsu, 1975). A review of the hazardous mass movements that may occur, and have historically occurred in recently deglaciated terrain, was presented by Eisbacher and Clague (1984).

Rockfall and *soilfall* are low-magnitude falls from rock slopes and debris slopes, respectively. They are particularly common in unstable, oversteepened slopes exposed by glacier retreat, and result in the accumulation of *talus* at the slope foot.

Snow avalanches from mountainsides can be effective agents of sediment transfer if failure takes place at the base of the snowpack or if the avalanche impacts upon snow-free debris slopes. Debris swept downslope accumulates in *avalanche boulder tongues*. These may take the form of *avalanche cones*, which are fan-shaped accumulations similar to debris flow cones, or, if the avalanche is strongly confined laterally, *roadbank tongues*, which are tongue-shaped embankments rising up to several metres above the level of the surrounding terrain (Rapp, 1959; Luckman, 1978; Ballantyne and Harris, 1994). Debris reworked by snow avalanches is often sharply angular, owing to high-energy particle collisions within the avalanche, and commonly forms precariously balanced accumulations where it has been let down on to the surface by snow melt.

Fig. 7.33 Large paraglacial fan derived from ice-marginal sediments, Milang Valley, Lahul Himalaya, India. (Photo: D.I. Benn)

Fig. 7.34 Rapidly degrading inner slopes of a lateral moraine exposed by glacier retreat, Pre de Bar Glacier, Italy. (Photo: D.I. Benn)

Slumps and *slides* are movements which take place along one or more shear planes, with varying amounts of internal deformation of the sliding mass. Failure is most likely to occur during periods of snow melt or heavy rainfall when pore- or cleftwater pressures are high within the slope. Slumps and slides may undergo partial or total disaggregation during downslope movement, and may evolve into flows.

Debris flows are particularly common in recently deglaciated environments, owing to the availability of unconsolidated sediments and abundant water supplied by ice and snow melt (Theakstone, 1982; Owen, 1991; Ballantyne and Benn, 1994b, 1996). They may feed substantial volumes of suspended sediment to proglacial streams via debris flow fans at the slope base. Individual debris flows commonly form elongate lobes with most sediment being deposited in a frontal deposit and marginal levées on either side of the debris flow channel. Successive flows build up conical debris fans. Because debris flow activity is often highest during spring snow melt, debris flows commonly accumulate on top of late-lying snow-

banks as *supranival flows*. On snow melt, the debris is let down on to the ground surface as an irregular hummocky deposit resembling some types of glacial moraines (Ballantyne and Benn, 1994b).

Further details of mass movement processes can be found in Carson and Kirkby (1972), Brunsden (1979), Brunsden and Prior (1984), Chorley *et al.* (1984) and Selby (1993).

7.6.3 Fluvial reworking

Fluvial systems are particularly sensitive to episodes of paraglacial activity, owing to the impact of large inputs of sediment (Church and Ryder, 1972). In upland catchments close to sediment sources, sediment yields can be expected to peak as deglaciation commences, but in more distal reaches, sediment throughputs should peak later, reflecting a time lag as the slug of sediment travels through the system. Thus in downstream reaches of large catchments, sediment loads will not begin to increase until after peak yields in proximal reaches (Church and Slaymaker, 1989; Fig. 7.31b). Furthermore, upland paraglacial cycles may be largely complete within a few centuries of deglaciation, whereas larger basins may still be responding to deglaciation for several millennia. Indeed, it is possible that the landscape may not have reached its non-glacial equilibrium before the onset of the next glaciation.

The role of the paraglacial cycle in the longer-term impact cycle of glaciation on river valleys has been assessed by Clague (1986), who argued that the early stage of glaciation brings about a peak in aggradation in the system, largely because of the damming of drainage basins (Fig. 7.37). Degradation occurs during glaciation as a result of subglacial erosion, but some aggradation may characterize the more advanced stages of glaciation, owing to till deposition. This aggradation then increases to its peak at the time of deglaciation, when large amounts of sediment are released in the freshly exposed terrain. The early part of the following non-glacial phase is then characterized by a short period of incision followed by a return to equilibrium conditions. The deglacial aggradation and degradation oscillation corresponds to the period of high sediment yields in the paraglacial cycle.

7.6.4 Aeolian reworking

On a regional scale, the extensive and often thick accumulations of *loess* (wind-blown silts and clays) and *cover sand* are invaluable archives of climatic and environmental changes during the Quaternary Period. Such material in mid-latitude North America and Europe travelled large distances from freshly deglaciated and periglacial terrain during phases of glaciation, and therefore has been considerably mod-

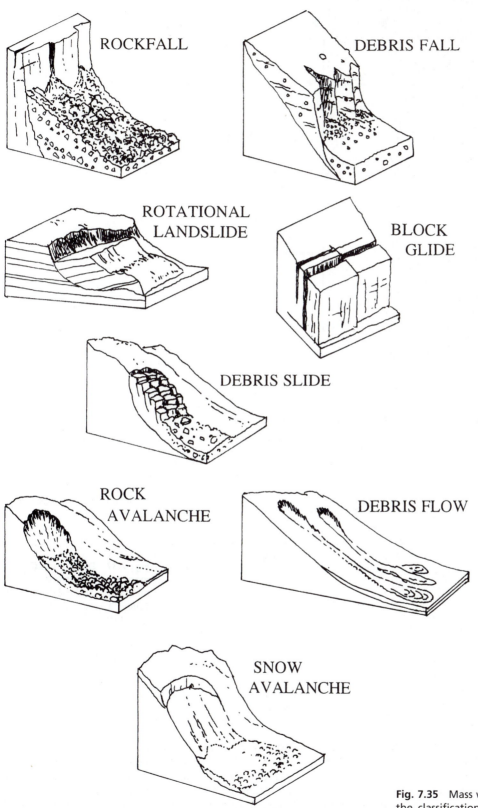

ROCKFALL

DEBRIS FALL

ROTATIONAL
LANDSLIDE

BLOCK
GLIDE

DEBRIS SLIDE

ROCK
AVALANCHE

DEBRIS FLOW

SNOW
AVALANCHE

Fig. 7.35 Mass wasting types according to the classification of Varnes (1958). (Modified from Selby, 1983)

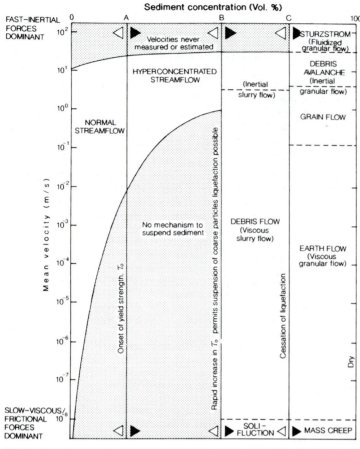

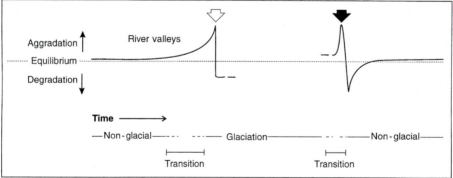

Fig. 7.36 Classification of mass movement types according to sediment concentration and velocity. (From Selby, 1993. Reproduced by permission of Oxford University Press)

Fig. 7.37 Diagrammatic summary of the patterns of sedimentation and erosion in river valleys based upon examples from British Columbia, Canada. This is depicted for a typical glacial cycle. The open arrow indicates the time of glacier overriding and the dark arrow indicates the time of deglaciation. (Modified from Clague, 1986)

ified by aeolian transport processes. The characteristics and palaeoenvironmental implications of loess are beyond the scope of this book, and the reader is referred to other texts for comprehensive coverage and references (e.g. Lowe and Walker, 1984; Williams *et al.*, 1993; Derbyshire, 1995). In this sec-

tion we concentrate specifically on aeolian processes active within the immediate proglacial zone.

It is the lack of stabilizing vegetation, the action of persistent glacier winds and large quantities of easily erodible sediment in proglacial areas that make them prime sites for aeolian erosion, transportation and deposition. In addition, the fluctuating discharges of proglacial streams allow for the intermittent drying and wind deflation of finer-grained alluvium. Wind erosion may affect any glacial landform–sediment assemblage, but vegetation colonization tends to stabilize most surfaces as the ice recedes. This implies that a zone of maximum potential aeolian activity is located very close to the ice margin. In an investigation into the textural changes of Icelandic till through time, Boulton and Dent (1974) demonstrated that silts and clays in freshly exposed subglacial tills are commonly transported by wind on to surfaces deglaciated more than 20 years previously where loess accumulates. This removal of fine-grained materials from the

tills leaves behind stone pavements, which may contain wind-faceted clasts (*ventifacts*) as a result of the sand-blasting effect of the strong localized winds. These features are common in Antarctica and are similar to the *desert pavements* produced in warm arid environments which protect underlying fine-grained materials from continued deflation.

This ice-proximal to distal zonation of aeolian processes is also reflected in the wind-blown sediments and forms of proglacial areas. The waning strength of glacier winds with distance from the ice margin is reflected in the grain size distributions of the aeolian deposits, which grade from medium to fine sands into silts and clays (loess) with distance

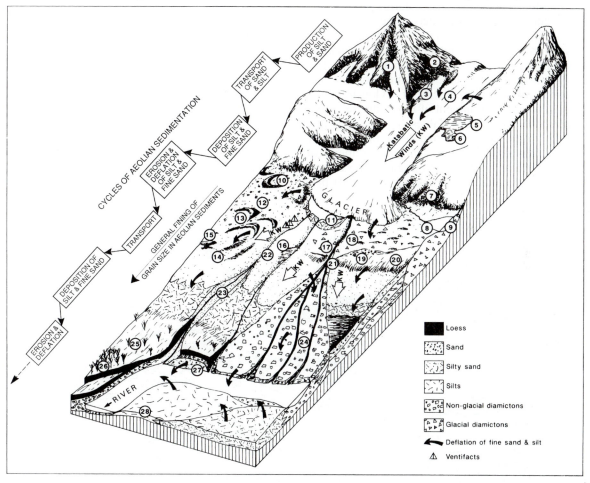

Fig. 7.38 The processes and sediment–landform associations in glaciaeolian environments: (1) silt and sand formation by weathering on high, steep valley slopes; (2) silts and sands produced and supplied to the glacier by rockfall and other mass movement processes; (3) silts deposited in a proglacial lake; (4) silts, sands and other rock debris falling into crevasses and incorporated in the glacier; (5) silts and sands washed into a small ice-marginal lake; (6) terraces comprising lacustrine silts and sands produced as lake dries up or drains; (7) fine sediments produced by overland flow and deposited at the base of the slope; (8) meltwater stream feeding a proglacial lake; (9) proglacial lake into which lacustrine silts and sands are deposited; (10) parabolic dune; (11) ice-contact lake; (12) and (13) barchan dunes; (14) longitudinal dunes; (15) deflation hollow; (16) deflation of lacustrine silts and sands in a dried-up proglacial lake; (17) rock-strewn surface; (18) hummocky moraine; (19) aeolian sand infilling depressions within till ridges; (20) end moraine; (21) meltwater stream dissecting end moraine; (22) cover sands; (23) floodplain sands and gravels; (24) alluvial fan; (25) river terraces capped by loess; (26) vegetated surface with formation of palaeosols; (27) fines being deflated from floodplain sediments; (28) loess hills. (From Derbyshire and Owen, 1996. Reprinted by permission of Butterworth-Heinemann)

(Fig. 7.38). The generally short travel distances are reflected in the surface morphology of individual grains under the scanning electron microscope. The grains have a frosted appearance due to physical impact chipping, but still exhibit physical signatures of their source of origin (e.g. fracture surfaces inherited from subglacial erosion and transport), unlike the further-travelled grains of the mid-latitude loess sheets (Krinsley and Doornkamp, 1973; Whalley and Krinsley, 1974).

The seminal work on the processes and deposits of aeolian geomorphology is that of Bagnold (1941), who identified three modes of sediment transport: (a) suspension; (b) saltation; and (c) surface creep. Suspension affects only the finest particles (silt and clay), whereas saltation affects sand, and creep affects sand and granules. The initial entrainment of a sediment particle is dependent upon various factors such as grain size and shape, sediment sorting and packing, surface roughness, moisture content and vegetation. A threshold velocity of 16 km hr^{-1} was suggested for the entrainment of loose, dry sediment by Ritter (1978), whereas values of approximately 30 km hr^{-1} have been reported for moist sand by Calkin and Rutford (1974). Recent measurements of summer katabatic winds at the margin of the Greenland ice sheet by van den Broeke *et al.* (1994) per-

sistently exceed the threshold velocity for dry sediment, and average wind speeds for the marginal areas of the Antarctic ice sheet are more than double the threshold velocity for moist sediment (Dudeney, 1987; Parish and Bromwich, 1989); it is therefore important not to underestimate the impact of aeolian processes in ice-marginal environments, with respect to both science and comfort!

Once entrained by the wind, a particle will behave according to its grain size, with silts and clays remaining in suspension by turbulence and sands resettling after a short travel distance. As sand grains strike the sediment surface they dislodge other particles, setting them in motion. This results in: (a) saltation, where the displaced grains become airborne within a zone less than 50 cm from the sediment surface; and (b) surface creep, where grains are pushed along the sediment surface by the saltating grains.

The deposition of wind-blown particles involves the three processes of wind ripple migration, grainfall and avalanching. Ripples form and migrate on a sediment surface as a response to bombardment by saltating grains. An upward limit on ripple growth is reached when the ripple crest reaches faster-flowing air, and grains are then removed easily. This helps to explain why coarser and therefore more stable grains are found on ripple crests and finer grains in the

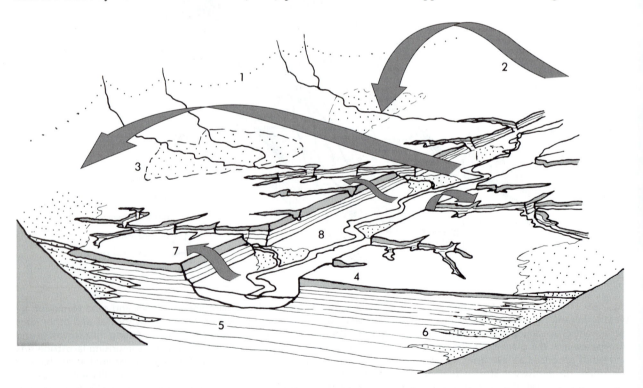

Fig. 7.39 A reconstruction of the environmental setting for loess deposition in the early Holocene (late paraglacial reworking of valley fills) based upon the South Thompson River valley, British Columbia, Canada. 1: Upper limit of loess deposition; 2: mountain and katabatic winds; 3: alluvial fans; 4: loess cap; 5: glacilacustrine silts; 6: debris flow deposits; 7: slope winds; 8: floodplain. (From Roberts and Cunningham, 1992. Reprinted by permission of John Wiley & Sons)

neighbouring troughs. Grainfall entails the fallout of grains from suspension or saltation and their accumulation in the lee of an obstacle. In proglacial areas this may involve clumps of vegetation or even individual boulders. Avalanching involves the mass movement of grains as slumps or flows where large quantities of sediment are being delivered to lee-side surfaces.

Because aeolian reworking is most effective when sediments are dry and unvegetated, it is during the early part of the paraglacial cycle that proglacial materials are most susceptible to wind deflation and transport. However, substantial wind-blown deposits are likely to be present only in the more arid of deglaciated catchments. These processes can be effective enough in some regions to produce the large and extensive aeolian forms usually associated with hot deserts, such as parabolic and barchan dunes (Webb and McKelvey, 1959; Calkin and Rutford,

1974; Koster, 1988; Koster and Dijkmans, 1988). Depositional patterns in such arid catchments provide important information on glaciaeolian processes. For example, Roberts and Cunningham (1992) document wedges of loess draping the lower slopes of the South Thompson River valley in British Columbia, Canada, which they interpret as the product of two components of wind flow (Fig. 7.39). The dominant downvalley (mountain and katabatic) winds and tangential secondary updraughts (slope winds) combine to carry silt from the thick glacilacustrine sediments deposited during deglaciation up the valley wall. The resulting wedge of loess feathers out upslope. The interaction of aeolian processes and landforms with glacial and proglacial processes is summarized in Fig. 7.38, which illustrates the numerous complex combinations of transportational and depositional histories that are involved in the glaciaeolian system.

CHAPTER
8

GLACILACUSTRINE AND GLACIMARINE ENVIRONMENTS

8.1 INTRODUCTION

Large parts of contemporary glacier margins calve into or are in contact with deep water. More than 90 per cent of the Antarctic ice sheet margin is presently in contact with the sea (Drewry *et al.*, 1982), and numerous glaciers elsewhere in the world terminate in the sea or lakes. The extent of glacier margins terminating in water was even more spectacular during Pleistocene glacial maxima, when continental-scale ice sheets in North America and Scandinavia descended to sea-level and contacted huge proglacial lakes (Section 1.7.2). Such water bodies act as sediment traps where thick successions can be preserved, providing valuable information on the former extent, behaviour and history of glaciers. Proglacial water bodies also profoundly influence glacier dynamics, acting as catalysts for glacier advance and retreat cycles which may be only weakly related to climatic forcing (Hughes, 1986, 1992; Warren, 1992).

Lake-water (*glacilacustrine* or *glaciolacustrine*) and sea-water (*glacimarine*, *glaciomarine* or *glacialmarine*) environments have much in common, and many important processes are common to both. There are also many important differences due, among other factors, to the difference in salinity between freshwater lakes and the sea. In this chapter, we discuss both glacilacustrine and glacimarine environments and processes, highlighting the similarities and differences where appropriate.

There is an important distinction between *glacier-contact* water bodies, in which glaciers extend into the water, and *non-glacier-contact* water bodies, which receive glacial meltwater and sediment input but are not in contact with glacier ice (Fig. 8.1; N.D. Smith and Ashley, 1985; Benn, 1989a; N. Eyles and C.H. Eyles, 1992). In the former case, interactions between water-body characteristics and glacier behaviour create very dynamic conditions, with rapid spatial and temporal changes in sediment transport and deposition rates, and potentially unstable glacier margins. In non-glacier-contact environments, however, the presence of glaciers in a catchment influences lakes and the sea to a lesser extent than where there is direct contact, and the glaciers themselves are little affected. Both glacier-contact and non-glacier-contact environments are *glacier-fed*, in that they receive important inputs of water and sediment from glaciers, in contrast with *glacial* lakes or marine basins, which owe their origin to glaciation but do not necessarily receive glacial inputs at present.

In this chapter, we begin by examining the behaviour of water-terminating glaciers, or the influence exerted by deep water on glacier dynamics in glacier-contact environments. We then go on to discuss the influence of glacial inputs on water-body characteristics in both glacilacustrine and glacimarine environments, and both glacier-contact and non-glacier-contact settings. We then review the range of depositional processes and environments characteristic of glacier-fed water bodies.

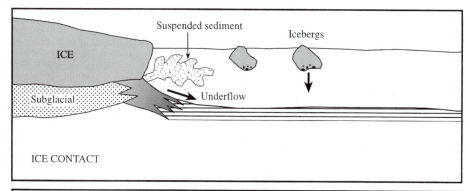

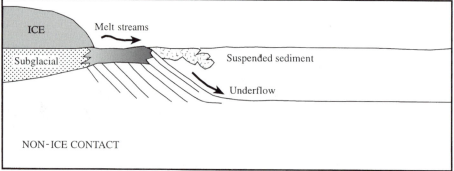

Fig. 8.1 Contrasting conditions in glacier-contact and non-glacier-contact environments. (Modified from N. Eyles and C.H. Eyles, 1992)

8.2 WATER-TERMINATING ICE MARGINS

The behaviour of glaciers that extend into deep water (*water-terminating glaciers*) is controlled by the fact that ice floats. This simple fact means that water-terminating glaciers are potentially unstable, prone to rapid breakup or advance depending on the interplay between water depth, glacier activity and climatic inputs. Given the extent of water-terminating glacier margins around Antarctica today, and around the great mid-latitude ice sheets in the past, it is clearly very important to understand the processes governing their behaviour as a key to predicting future glacier dynamics and interpreting the geologic record of climate change.

8.2.1 Buoyancy

Glacier ice has a density less than that of liquid water, because of the expansion of water upon freezing and the presence of trapped air bubbles (Section 2.2.2). Thus, a given volume of glacier ice is lighter than the same volume of water, so that ice will float provided the water is sufficiently deep. The depth of water required for flotation is given by the density contrast between ice and water:

$$h_{wf} = (\rho_i / \rho_w) h \qquad (8.1)$$

where h_{wf} is the water depth required for flotation, ρ_i is the density of ice (~ 900 kg m^{-3}), ρ_w is water density (~ 1000 kg m^{-3} for fresh water and ~ 1030 kg m^{-3} for sea-water), and h is the ice thickness. Thus

$$h_{wf} \simeq 0.9\, h \text{ (fresh water)}$$

$$\simeq 0.87\, h \text{ (sea-water)}$$

showing that glacier margins will float when in contact with water deeper than approximately nine-tenths of the ice thickness.

Glaciers entering water deeper than the critical depth will form *floating margins* or *ice shelves*, but only if the tensile strength of the ice is sufficient to withstand the pulling stresses at the margin. Temperate glacier snouts cannot form ice shelves because glacier ice at the pressure melting point has very low tensile strength, and readily breaks up by calving (Powell, 1984). Thus, temperate water-terminating snouts cannot survive in water deeper than the critical depth, and will calve back until grounded in shallower water. Glaciers with grounded marine termini are known as *tidewater glaciers* (Powell, 1984). Many of them occur in the fjords of Alaska, Chile

and Svalbard (Meier and Post, 1987; Dowdeswell, 1989; Warren *et al.*, 1995; Wiles *et al.*, 1995). Floating margins and ice shelves can form only in polar settings, such as the fjords of Arctic Canada and marine embayments around the coasts of Antarctica, where the ice is below the pressure melting point and has a high tensile strength.

The *grounding line* of a glacier is simply the zone at which a glacier-ice shelf starts to float or where a tidewater glacier ends as a vertical cliff (Hollin, 1962; Powell, 1984; Powell *et al.*, 1996). On the surface of a glacier-ice shelf, the position of the grounding line can be identified by the location of a pronounced decrease in slope profile, reflecting the transition to zero shear stress below the floating portion of the glacier.

8.2.2 Tidewater glaciers and other non-floating margins

Non-floating glacier margins, where the ice terminates at a cliff at the grounding line, are more common than ice shelves, and this also appears to have been the case during Quaternary glacial maxima. Tidewater glaciers occur in both open-coast and fjord locations and in both polar and temperate climates (Fig. 8.2). Therefore, although Antarctica is well known for its large ice shelves, tidewater glaciers actually comprise nearly 50 per cent of the coastline (Dubrovin, 1976).

Temperate water-terminating glaciers are commonly fast-flowing, and some, such as the San Rafael Glacier in Chile, are among the fastest glaciers on Earth (Meier and Post, 1987; Warren and Sudgen, 1993; Warren *et al.*, 1995). This tendency to fast flow is due to long-term interactions between climate, glacier activity and landscape evolution. Deep lake basins and fjords typically form in deeply dissected montane environments in maritime climatic settings, such as southern Alaska and Patagonia, where high

Fig. 8.2 Calving glacier margin, Hooke Glacier, Makinson Inlet, Ellesmere Island. (Photo: D.J.A. Evans)

snowfall and ablation drive the ice flux required for rapid erosion (Section 5.3; Hallet *et al.*, 1996). Such environments, therefore, have deep water bodies occupying overdeepened erosional basins, and the mass throughput necessary to sustain fast glacier flow. Furthermore, the presence of deep water at the glacier margins provides a high base-level for sub- and englacial drainage, encouraging high basal water pressures and rapid sliding (Section 4.5.3). The presence of fast, water-terminating glaciers in maritime mountain ranges is thus part of a self-reinforcing process of landscape change.

8.2.3 Ice shelves

Ice shelves consist of (a) the floating margins of outlet or valley glaciers (*glacier-ice shelves*); (b) floating ice sheets which are locally grounded and fed by surface accumulation and bottom freezing (*sea-ice ice shelves*); and (c) extensive floating shelves fed by a combination of glacier inflow, surface accumulation and bottom freezing (*composite ice shelves*; Section 1.4.3; Thomas, 1979b). An example of a small glacier-ice shelf in Arctic Canada is shown in Fig. 8.3, and part of the huge Ross ice shelf, a composite shelf receiving ice from the interior of Antarctica as well as surface snowfall, is shown in Fig. 8.4. In this section, we examine the flow, accumulation and melting of ice shelves. Calving processes are described in Section 8.2.4.

8.2.3.1 FLOW OF ICE SHELVES

Unlike in the case of grounded glaciers, there is no drag at the base of floating ice masses. As a result,

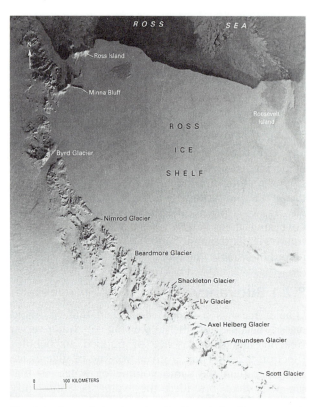

Fig. 8.4 Part of the Ross Ice Shelf and the Transantarctic Mountains, NOAA advanced very high resolution radiometer image, November 1980. (Image courtesy of Donald R. Wiesnet, National Oceanic and Atmospheric Administration)

Fig. 8.3 The floating and calving margin of Alfred Newton Glacier, Ellesmere Island, Arctic Canada. Note the development of radial crevasses beyond the grounding line, which cause the floating part of the snout to splay out as a series of ice tongues, and the heavy transverse crevassing, which reflects acceleration of the ice (down-draw) as it reaches the sea. (Photo: W. Blake Jr, Geological Survey of Canada, June 1982)

the basal shear stress is zero, a fact that has profound implications for the flow of ice shelves. For *unconfined ice shelves* without grounded margins, flow occurs entirely by *creep spreading* driven by the weight of the ice above the water-line (Weertman, 1957a; Thomas, 1973a, b; Sanderson, 1979; Paterson, 1994). With no basal shear stress, the ice shelf has no surface slope. Spreading causes thinning of the ice shelf, although this can be counteracted by surface accumulation, basal freezing or the influx of ice from glacier outlets, so that it is possible for a spreading shelf to maintain a constant thickness or even thicken. Unconfined ice shelves with zero or negative mass balances will always undergo thinning, and will eventually disappear.

Confined ice shelves are grounded at the sides by fjord margins, islands, promontories or high points on the bed. These grounded margins exert drag on the ice, increasing the resistance of the system to flow in much the same way as the margins of a valley glacier (Section 4.6.2). The existence of this drag allows the ice shelf to maintain a surface slope, which generates the stresses necessary to maintain fast flow (Fig. 8.5). Theoretical stresses and strain

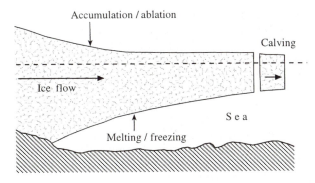

Fig. 8.5 Sketch of a confined ice shelf, showing inputs and outputs

rates in confined ice shelves have been calculated by van der Veen (1986) and summarized by Paterson (1994). This theory predicts that surface slopes will be steepest for narrow-confined ice shelves, and gentlest in wide ice shelves where drag at the margins is relatively unimportant, a prediction confirmed by field data (Fig. 8.6). High points on the bed, consisting of bedrock hills or shoals of sediment, are known as *pinning points*, and also exert an influence on ice shelf dynamics by locally increasing the basal drag and reducing losses by calving (Section 8.2.4).

8.2.3.2 ICE SHELF ACCUMULATION AND MELTING

Although calving at the ice shelf edge exerts the strongest influence on ice shelf stability (Section 8.2.4), the mass balance of many ice shelves is also critically dependent on the relative importance of

snowfall accumulation and ablation at the upper surface, and freezing and melting processes at the base (Fig. 8.5).

As we saw in Section 8.2.1, ice shelves form only in polar regions, where the ice is below the pressure melting point. In such regions, mean annual temperatures are often too low for significant melting to occur at the upper surface, so that ice shelf surfaces are often snowfall accumulation zones. Accumulation on ice shelves is also encouraged by their coastal location, close to oceanic moisture sources. In some areas, however, net ablation occurs at ice shelf surfaces.

Unlike in the case of grounded ice masses, basal melting and freezing processes contribute substantially to the mass balance of ice shelves. Conditions for melting or freezing are determined by the temperature of the ice and the water, and by the water salinity, and have been analysed mathematically by Doake (1976) and Drewry (1986). Increasing salinity reduces the local melting point of ice, thus increasing the likelihood of basal melting and suppressing basal freezing. Thus, freezing is favoured by factors that chill or reduce the salinity of the water beneath an ice shelf, such as the efflux of meltwater from basal drainage systems. The role of water pressure in determining thermal conditions and patterns of melting and freezing beneath an ice shelf have been investigated by Foldvik and Kvinge (1974, 1977), Doake (1976) and Robin (1979). Ice shelves generally thin down-flow, so that if water is forced to flow up an ice flowline it will descend because of the increase in ice thickness (Fig. 8.5). As the water descends, the increasing water pressure will depress the freezing point and so, in order to maintain equilibrium, the water will melt the overlying ice to lower its temperature. The reverse of this process will

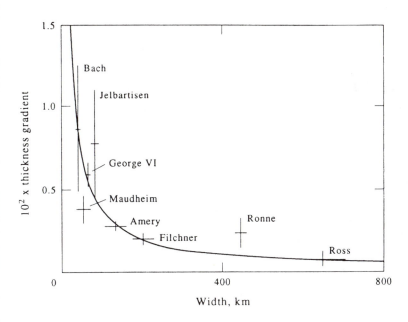

Fig. 8.6 Relation between the width of ice shelves and thickness gradient (change in thickness with distance downflow). (From Paterson, 1994. Reproduced by permission of Pergamon)

affect a water parcel that moves down a flowline and ascends to shallower depths.

Ice formed by basal freezing has characteristic crystallographic and isotopic signatures, which can be readily identified in ice cores (Souchez and Lorrain, 1991). For example, Site J9 on the Ross ice shelf revealed 6 m of frozen-on sea-ice at the base of a 416 m core (Clough and Hansen, 1979; Jacobs *et al.*, 1979; Zotikov *et al.*, 1979, 1980), and 158 m of sea-ice was found underlying 270 m of glacier ice in the Amery ice shelf by Morgan (1972).

Average bottom freezing rates for the base of Antarctic ice shelves are in the region of 30–50 mm yr^{-1}. These figures, however, mask considerable spatial variability, both between and within ice shelves. Rates of basal accretion by freezing-on have been calculated as 10–35 mm yr^{-1} for the Ross ice shelf and 300–600 mm yr^{-1} for the Amery ice shelf, where accumulation by basal freezing exceeds that attributable to snowfall (Fig. 8.7; Budd *et al.*, 1982). The Ross ice shelf can be subdivided into three zones based on patterns of basal melting and freezing (Souchez and Lorrain, 1991). *Zone 1* is an area of enhanced bottom freezing near to the grounding line, and results from the presence of fresh meltwater draining from beneath terrestrial ice streams. Localized basal melting also occurs in this zone in association with vigorous meltwater efflux points. *Zone 2* is an area of slow bottom freezing caused by effective upward heat transport through the ice shelf and the presence of cold isothermal water of low salinity. This gives way to an outer zone, *Zone 3*, where stronger circulation and greater heat exchange produces net basal melting.

Basal melting and freezing can change the character of ice shelves. For example, the Koettlitz Glacier, Antarctica, originates as a glacier-ice shelf and is transformed into a sea-ice ice shelf through a combination of surface ablation and bottom freezing (Gow and Epstein, 1972). Striking evidence of this transformation is provided by fish and other organic remains melting out from the ice shelf surface.

8.2.3.3 ICE SHELF STABILITY

In summary, ice shelves are potentially unstable features, and their formation and survival depends on several conditions being met (Powell, 1984):

1. Ice shelves form only in ice below the pressure melting point, which has high tensile strength. Accordingly, ice shelves are restricted to polar regions.
2. Ice shelf stability is favoured by net snow accumulation at the surface, and net freezing at the base. Conversely, net surface ablation and basal melting will result in ice shelf disintegration. Positive mass balance at the upper surface may counteract losses by basal melting, and vice versa.
3. Sheltered locations, such as embayments or fjords, and pinning points on the bed promote stability by sheltering the ice shelf from currents, waves and winds, and reducing calving. Locations can become more favourable to ice shelf growth by a fall in sea or lake level and/or the formation of sea ice beyond the ice shelf edge, protecting it from open water.
4. Stability is also encouraged by high ice discharges from terrestrial glacier feeders. Although thickening of feeding glaciers can initiate grounding-line advance, ice shelf expansion normally requires reduction of ablation by surface and basal melting, and particularly the reduction of calving at the distal edge.

8.2.4 Calving and icebergs

Calving is the production of icebergs by detachment of ice from a parent glacier terminating in water (Fig. 8.8). The calving zone of a tidewater glacier margin

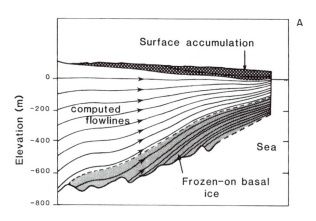

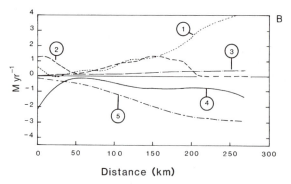

Fig. 8.7 (A) Longitudinal profile of the Amery ice shelf, Antarctica, showing ice flowlines, surface accumulation and basal freezing on. (B) Mass balance components for the Amery ice shelf: (1) surface accumulation rate; (2) basal ice accretion rate; (3) horizontal ice flux; (4) longitudinal strain thinning; (5) transverse strain thinning. (From Drewry, 1986, after Budd *et al.*, 1982)

(a)

(b)

(c)

Fig. 8.8 A calving event at the snout of Glaciar San Rafael, Chile. The calving front rises to approximately 45 m above the water line and the water depth is in excess of 250 m. In the final photo of the sequence a large wave has been produced by the collapse of the pinnacle in the second photo, and a new block of ice has risen up from below the water almost to the height of the snout cliff top. (Photos: C. R. Warren)

coincides with the grounding line, whereas the calving zone of an ice shelf may be up to hundreds of kilometres seawards of the grounding line (Swithinbank, 1955). Calving accounts for a large proportion of ablation from glaciers in contact with deep water: the total mass of ice lost by calving is estimated at 56 per cent and 77 per cent for the Greenland and Antarctic ice sheets, respectively (Jacobs *et al.*, 1992).

Calving occurs when the tensile stresses tending to pull a parcel of ice away from the margin exceed the strength of the ice. A model of the stresses acting on a floating glacier margin was developed by Reeh (1968), and considered the balance of forces across the floating ice face (Fig. 8.9). The main components of the force balance are the normal pressure exerted by the water on the ice (P_w), and the normal pressure exerted by the ice on the water (p_i). The normal pressure exerted by the water on a vertical, floating ice face at any given depth is given by

$$P_w = \rho_w g h_w \qquad (8.2)$$

where ρ_w is water density, g is gravitational acceleration and h_w is water depth.

The water pressure is thus zero at the water surface and rises linearly towards the ice base (Fig. 8.9B). Similarly, the outward cryostatic pressure exerted by the ice face towards the water column is given by

$$p_i = \rho_i g h \qquad (8.3)$$

where h is ice thickness, measured down from the surface.

Except for the ice base, where $p_i = P_w$, there is an imbalance in the stresses acting across the ice face, with a net horizontal hydrostatic force which reaches a maximum at the water line (Fig. 8.9C). In order to balance the forces, the floating ice front is distorted and warped downward at the snout. A potent combination of tensile and shear stresses is set up within the glacier, reaching a maximum at a distance from the ice front which is approximately equal to the ice thickness (Fig. 8.9D). This promotes fracturing at the ice surface, and calving occurs when fractures propagate to the ice base, producing icebergs with lengths comparable to the ice thickness. The widths of the icebergs will be dictated by the width of the ice shelf, provided it maintains uniformity across its margin. Reeh-type calving has been observed in the field by Holdsworth (1973a), and Reeh and Olesen (1986), but slight frontal upwarping, rather than the expected downwarping, occurs at the calving terminus of Jakobshavns Glacier in Greenland (Echelmeyer *et al.*, 1991; Warren, 1992).

Additional stresses are imposed by tides, changing lake levels, and storm waves during extreme weather events, each of which can facilitate crack propagation and calving (Holdsworth, 1974; Robin, 1979;

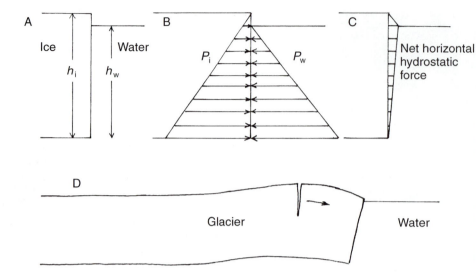

Fig. 8.9 The Reeh mechanism of iceberg calving. (A) Configuration of ice edge. (B) Magnitude of horizontal hydrostatic stresses exerted by the ice (p_i) and the water (P_w). (C) Net horizontal hydrostatic stress. (D) Resulting warping of the ice tongue and associated fracture propagation. (Adapted from Reeh, 1968, and Hughes, 1987)

Holdsworth and Glynn, 1981). Cyclical stresses caused by swell waves (Holdsworth, 1977; Holdsworth and Glynn, 1978), and collisions between icebergs and ice shelves (known as the 'big bang' theory of iceberg formation) can also weaken floating ice fronts (Swithinbank, 1969; Swithinbank *et al.*, 1977; Robin 1979).

Calving is also strongly dependent on weaknesses in the ice which locally reduce the tensile strength of floating glacier tongues. Radial and transverse crevasses are particularly important in controlling calving events (Fig. 8.3; Epprecht, 1987; Dowdeswell, 1989; Warren *et al.*, 1995), and accelerated calving can coincide with the arrival of crevasse fields at a glacier terminus (Powell, 1990). Calving by crevasse propagation is particularly likely when crevasses are full of water, because this will reduce the effective pressure at the crevasse walls and discourage closure by creep (Fastook and Schmidt, 1982).

Calving at grounded glacier margins has been examined by Iken (1977), Hughes (1989), Hughes and Nakagawa (1989), Funk (cited in Warren, 1992) and Kirkbride and Warren (1997). One of the most important factors is the gradual steepening of the ice front due to the vertical velocity gradient (Section 4.6.2), which increases the bending stresses at the ice surface and encourages downward crack propagation and toppling failure of spalled slabs of ice. As for ice shelf calving, the presence of crevasses exerts considerable control over the size and location of calving events. Ice cliff retreat is also accomplished by thermal erosion by warmer surface water, producing a deep notch along the base of the cliff near the water-line (Fig. 8.10). This notch undercuts the ice cliff, encouraging full-height slab calving. Thermal erosion near the water-line also produces a projecting

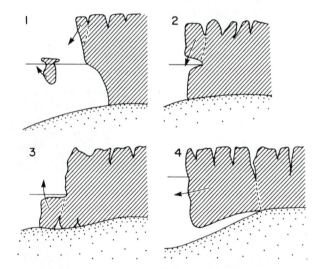

Fig. 8.10 Types of iceberg calving from a tidewater glacier terminus: (1) subaerial launch of a serac; (2) separation of a berg along a joint intersecting a thermoerosional notch; (3) subaqueous launch from an ice foot; (4) separation along a deeply incised crevasse. (From Syvitski *et al.*, 1987, after Lliboutry, 1965. Reproduced by permission of Springer)

ice foot below the water surface in front of the subaerial calving front. The density difference between ice and water renders this projection unstable, and blocks may bob up many tens of metres in front of the subaerial terminus (Warren, 1992; Warren *et al.*, 1995; Kirkbride and Warren, 1997).

The rate of retreat of an ice margin by calving is known as the *calving rate* or *calving velocity*. Despite the complexity of the calving process, there is a generally strong positive correlation between calving rates and water depth at the calving front, allowing rates of glacier retreat to be predicted from knowledge of subglacial topography (Fig. 8.11;

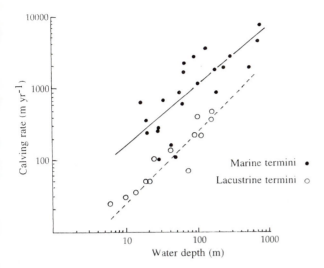

Fig. 8.11 Water depth and associated calving rates at the margins of a selection of tidewater glaciers (solid dots) and freshwater cliffs (open circles). (Data from Warren, 1992, and Warren *et al.*, 1995, compiled from various sources)

Brown *et al.*, 1982; Pelto and Warren, 1991; Warren, 1992). However, the relationships differ for freshwater and marine glacier margins, and for any given water depth, calving rates will tend to be greater for marine margins. The reasons for this difference in behaviour are not fully understood, but the greater buoyancy and enhanced melting of glacier ice in salt water, exposure to storms, and tidal flexure may all play a part (Warren, 1992). In a recent analysis of the calving behaviour of Columbia Glacier, Alaska, van der Veen (1996) questioned whether any simple relation exists between water depth and calving rate. He suggested that the calving rate is influenced by the effective pressure at the base of the ice, and that rapid calving will be initiated when the effective pressure approaches zero. This would imply that calving behaviour is a function of water depth, ice thickness and basal meltwater flux, because these quantities control the effective pressure at the base of tidewater glaciers. According to van der Veen, calving rates are controlled more by processes acting at the glacier bed (e.g. changes in the subglacial drainage system) than by what happens at the terminus. If true, this conclusion indicates that predictions of calving rates from water depth alone can never be more than first approximations.

The shape and size of icebergs released by calving events are related to glacier dynamics and morphology, particularly the distribution of crevasses. Dowdeswell (1989) found that for water-terminating ice masses around Svalbard, heavily crevassed surging glaciers release large quantities of small icebergs, *bergy bits* (< 10 m across) and *brash ice* (< 2 m across), whereas ice caps and large outlet glaciers

release large *tabular icebergs* (> 500 m across) and small, irregular icebergs. Valley glaciers in the same region typically calve irregular icebergs over 50 m across. Calving from ice shelves typically releases large tabular icebergs. Large tabular bergs form *ice islands,* which can be tracked over vast distances as they drift within pack ice without fragmenting for long periods of time (Koenig *et al.*, 1952; Rodahl, 1954; Hattersley-Smith, 1957a; Jeffries, 1987, 1992). Many Arctic and Antarctic ice shelves have undergone accelerated calving over the past century, probably owing to climatic warming causing, first, less widespread sea-ice (which formerly buttressed the ice shelf margins), and second, enhanced basal melting and surface ablation (D.J.A. Evans and England, 1992; Rott *et al.*, 1996). A good review of iceberg calving and fragmentation in Antarctica is provided by Kristensen (1983).

Worldwide, icebergs release vast amounts of fresh water into the oceans each year. Estimates of iceberg production around Antarctica range from 5×10^{14} to 12×10^{14} kg yr^{-1} (Mellor, 1967; Robe, 1980), and a further 2.15×10^{14} kg yr^{-1} is estimated to calve from the coasts around Greenland (Dowdeswell, 1987). Much smaller volumes of icebergs are derived from the ice shelves of Ellesmere Island and the glaciers of Zemlya Frantsa-Iosifa, Severnaya Zemlya, Svalbard, South Georgia, Alaska and Chile. Icebergs totalling 4200 km^2 in area calved from the Larsen ice shelf in the Antarctic Peninsula in a single month early in 1995 (Rott *et al.*, 1996). Given the size of such bergs, it is not surprising that some rather fanciful and very expensive schemes have been proposed for towing such icebergs to lower latitudes for use as a freshwater source!

8.2.5 Advance and retreat of calving ice margins

The relationship between calving rates and water depth means that glaciers terminating in deep water are potentially unstable and vulnerable to catastrophic retreat by rapid calving. Ice margin stability is therefore favoured by the presence of *pinning points* or constrictions in the enclosing channel or fjord (Fig. 8.12; Field, 1947; Mercer, 1961; Thomas, 1979b; Warren and Hulton, 1990; Greene, 1992). Such constrictions in the channel occupied by a glacier reduce losses by calving, and if the calving rate is less than the ice flux, the glacier will thicken and stabilize. However, if the glacier pulls back from a pinning point into deeper water, the calving rate may exceed the ice flux and rapid retreat will result. Because of the influence of topography on ice margin stability, calving glaciers exhibit non-linear responses to climatic forcing: a given climatic signal will trigger a large or small glacier response

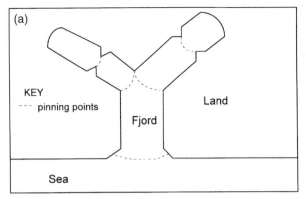

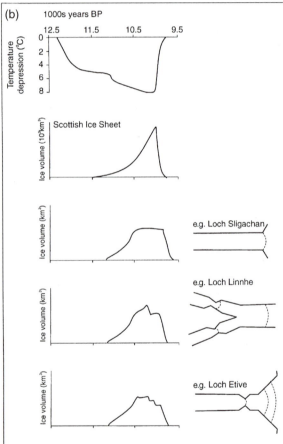

Fig. 8.12 **(a)** Pinning points located at changes in width of a fjord. Stable margin positions are favoured at narrowings. **(b)** Reconstructed retreat rates of Lateglacial fjord glaciers in Scotland, and their relationship to pinning points. (Modified from Greene, 1992, after Mercer, 1961)

depending on the position of the margin relative to local topography (Brown *et al.*, 1982; Mann, 1986; Clapperton *et al.*, 1989; Powell, 1991; Warren, 1992; Wiles *et al.*, 1995)

The links between climate change and the advance and retreat of fjord glaciers were first elucidated by Mercer (1961). If the glacier equilibrium line alti-

tude (ELA) falls, as a result of a climatic shift, the glacier margin will advance to a point where the fjord widens or deepens, thus increasing the calving rate to balance the ice flux. Conversely, a rise in ELA results in the retreat of the glacier snout to a point where the fjord narrows or becomes shallower, thus reducing calving rates. Glacier response, therefore, will depend on fjord configuration: glaciers occupying widening fjords should be relatively stable, because calving rates will increase if the glacier advances and decrease if the glacier retreats. On the other hand, glaciers occupying narrowing fjords will tend to be unstable; an advance of the glacier will reduce the calving rate, encouraging further advance, whereas retreat will increase the calving rate, leading to accelerating deglaciation.

Bottom topography has a similar influence on ice margin stability (Thomas, 1979b). Where the bed slopes away from the ice margin, a small increase in ice thickness will result in the advance of the grounding line into deeper water where calving rates are higher (Fig. 8.13a). On the other hand, where the bed slopes towards the ice margin, a small increase in ice thickness will initiate unstable advance into shallowing water, and a decrease in ice thickness will cause accelerating retreat into deeper water (Fig. 8.13b). This model has profound implications for the stability of marine-based ice sheets, such as the West Antarctic ice sheet, with beds largely lying well

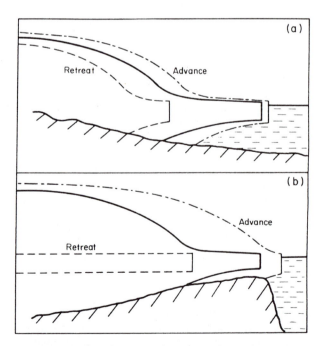

Fig. 8.13 The influence of bed configuration on the advance and retreat behaviour of water-terminating glaciers. (a) Bed sloping away from the ice, tending to limit the extent of advance and retreat; (b) bed sloping towards the ice, encouraging unstable advance and retreat

below sea-level. Such ice sheets are potentially vulnerable to increases in sea-level, which could accelerate calving and initiate catastrophic retreat (Hollin, 1962; Hughes, 1973, 1986; Mercer, 1978; Thomas and Bentley, 1978a; Sugden, 1988, 1993). Patterns of decay of marine-based ice sheets can be illustrated by deglaciation of the Hudson Bay sector of the Laurentide ice sheet. Deglaciation was slow until ice had retreated into Hudson Bay and away from the pinning points of Hudson Strait, whereupon it disintegrated rapidly, leaving only terrestrial remnants in Labrador, Keewatin and Baffin Island (Andrews and Falconer, 1969; Andrews and Ives, 1972; Paterson, 1972; Andrews and Peltier, 1976; Hughes, 1987).

Glaciers may advance into deep water by depositing thick sedimentary sequences at the grounding line, building up grounding-line fans or morainal banks which form shoals in front of the margin (Sections 8.5.2 and 8.5.3; Mayo, 1988; Powell, 1991; Hunter *et al.*, 1996). Given adequate sediment flux, the shoal can advance, conveyor-belt fashion, in front of the glacier by a combination of erosion on the subglacial side and deposition and debris reworking on the subaqueous side (Fig. 8.14). The presence of the shoal at the glacier margin reduces losses by calving, allowing the glacier to advance slowly through deep water behind the advancing shoal. However, excavation of sediment behind the shoal results in erosional overdeepening of a depression beneath the glacier (Wiles *et al.*, 1995; Björnsson, 1996), creating a situation where a small recession of the margin can initiate catastrophic calving retreat into deepening water. Thus, glacier retreat from grounding-line

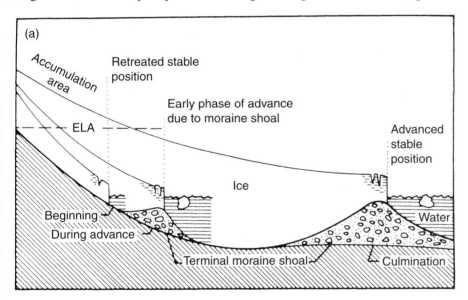

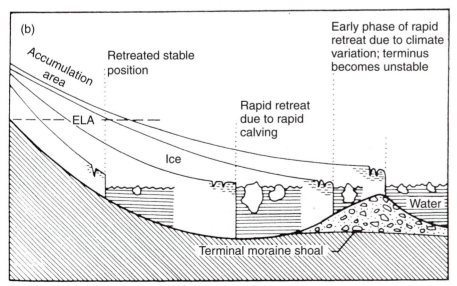

Fig. 8.14 Patterns of glacier advance and retreat in association with grounding-line shoals. (A) Slow advance behind a mobile shoal which protects the margin from calving losses. (B) Rapid retreat into deep water, leaving the shoal abandoned at the maximum position. (From Warren, 1992, after Mayo, unpublished)

shoals tends to be rapid, in contrast to advance by shoal migration, which is limited by sediment throughput rates and tends to be slow. The initial recession can be climatically triggered, or reflect sediment dynamics in the shoal. For example, rapid retreat of Muir Glacier, Alaska, appears to have been triggered by collapse of its morainal bank during the 1899 earthquake (Hunter *et al.*, 1996).

8.3 WATER-BODY CHARACTERISTICS

The input of meltwater and sediment from glaciers profoundly influences the characteristics of glacier-fed lakes and marine environments, including the vertical distribution of temperature and density and the orientation, strength and depth of currents. In turn, these water-body characteristics exert strong controls on processes and patterns of sedimentation. Although there are many similarities in the water characteristics of deep lakes and the sea, the influence of salinity and tidal movements in glacimarine environments results in very distinctive differences.

8.3.1 Lake stratification

Patterns of sediment transport and water flow in lakes are strongly dependent on the vertical temperature profile, or *thermal structure*, of the lake water. Particularly important is the relationship of water density to temperature. Pure water at atmospheric pressure has a maximum density of 1000 kg m^{-3} at a temperature of 3.98°C; above and below this temperature the density decreases (Fig. 8.15A). The rate of change in density increases as the temperature moves further away from 3.98°C, so that the density difference caused by a change from 19 to 20°C is 20 times greater than that caused by a change from 4 to 5°C. The influence of temperature on water density means that water bodies with different temperatures will have different buoyancies, and that variations in temperature with depth will subject a water body to vertical stacking according to the different densities produced. Because it is thermally induced, this stratification caused by density variation is referred to as *thermal stratification*.

Surface heating of a lake by solar radiation creates a warm upper layer overlying cold, denser water. The form of the temperature–depth curve, however, virtu-

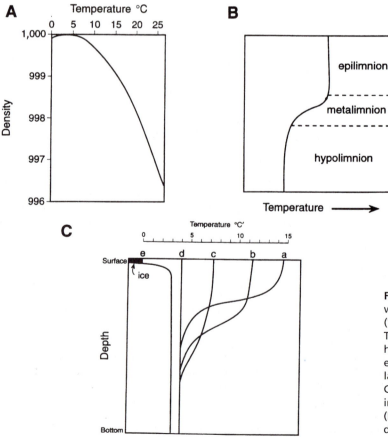

Fig. 8.15 Temperature characteristics of lake water. (A) Relationship between temperature (°C) and density (kg m^{-3}) for pure water. (B) Temperature profile resulting from surface heating and wind-mixing. (C) Hypothetical evolution of the thermal structure of a deep lake between (a) midsummer and (e) winter. Curve (d) shows isothermal conditions resulting from lake overturn in autumn and spring. (From N.D. Smith and Ashley, 1985. Reproduced by permission of SEPM)

ally never has the smooth exponential form that might be expected from Fig. 8.15A, because of vertical mixing processes caused by several factors. First, the water surface can be cooled by evaporation, back-radiation and sensible heat loss, setting up transient convection currents in the surface water layer according to the time of day and weather conditions. Second, winds will generate turbulence in the water column, and this initiates mixing and the downward transfer of heated surface waters.

A combination of surface heating and vertical mixing produces a characteristic three-layer water column (Fig. 8.15B; Smith and Ashley, 1985). The upper layer (the *epilimnion*) consists of warm, low-density water, which is generally isothermal owing to efficient vertical mixing. The lower layer, or *hypolimnion*, is made up of cold, dense and relatively undisturbed water. These two layers are separated by the *metalimnion* or *thermocline*, which is a zone of rapid temperature change. The depth of the thermocline is dictated by factors such as intensity and duration of solar radiation, the temperature and location of inflowing water, the presence of chemically stratified layers and, usually most importantly, wind strength. It has been shown by Ragotzkie (1978) that lakes with larger fetches have thicker midsummer epilimnions. Because of variability in wind activity, heating and cooling, the epilimnion may contain several smaller and short-lived thermoclines.

The thermal stratification depicted in Fig. 8.15B represents the midsummer situation when maximum stability is attained in the water column. The thermal structure of deep temperate lakes, however, changes through the year in response to variations in the energy balance (Fig. 8.15C). The midsummer curve (a) begins to change shape in late summer or early autumn when the lake surface waters begin to cool, deepening the thermocline (curves b and c). Sinking cool water sets up convection, progressively destabilizing and cooling the whole epilimnion. Curve d represents the situation in late autumn, when full-depth mixing has occurred and epilimnion temperatures have fallen to those of the hypolimnion. This process is known as *overturning*, and it produces an isothermal water body. Continued heat loss and mixing cool the lake to 4°C or less, but this rarely creates a reverse thermocline because of the very small density changes in the 0°–4°C temperature range (Fig. 8.15A), which are easily broken down by the wind. Ice formation (e) will occur once the surface layer reaches 0°C, but only if vertical mixing is suppressed during calm weather. Once this ice disappears in the following spring, the surface water will warm to hypolimnion temperatures and the lake will once again experience isothermal mixing followed by thermal stratification. This means that temperate lakes go through two overturning periods

in spring and autumn, and are thus referred to as being *dimictic*.

Monomictic lakes circulate only once per year, and can be subdivided into warm and cold varieties (Hutchinson, 1957). Warm monomictic lakes experience winter circulation above 4°C and have been reported from New Zealand, where the mild maritime climate prevents surface ice formation but allows the development of deep thermoclines (Irwin and Pickrill, 1982; Pickrill and Irwin, 1982). Cold monomictic lakes undergo summer circulation below 4°C, and this is thought to be the situation in some ice-contact lakes (Churski, 1973; Harris, 1976). Some lakes at high latitudes may be subject to different thermal conditions from year to year, because of the variability of duration and timing of the short ice-free season. For example, Hobbie (1961) reported a lake that varies from dimictic to cold monomictic from year to year. Some lakes may circulate at various times because of a lack of thermal stratification, and are referred to as *polymictic* (Gilbert and Church, 1983). Finally, very shallow lakes will not develop any thermal stratification because winds will continually mix the water column (Brewer, 1958).

The above examples are known collectively as *holomictic lakes*, which circulate completely to the bottom during at least part of the year. In contrast, *meromictic lakes* contain bottom water which is stabilized by dissolved material and which remains partly or wholly unmixed with overlying water, producing chemical stratification. As glacial meltwater generally has a low dissolved content, meromictic lakes are not well represented among glacier-fed lakes. However, meromictic lakes are common in basins recently isolated from the sea by glacio-isostatic rebound or still maintaining some contact with sea-water (Hattersley-Smith and Serson, 1964; Harris, 1976; Retelle, 1986b).

In glacier-contact lakes, the continuous supply of water at 0°C by channelled inflow, direct glacier-melting or melting icebergs will at the very least modulate the temperature. Quite often the cold water influx will inhibit or disrupt the formation of a thermal stratification of the water column, as outlined above. Churski (1973) has reported on the effects of subaqueous injections of cold water and ice-marginal melting into ice-contact lakes in Iceland (Fig. 8.16). These disruptions may grade downlake into thermally stratified water if the lake body is large enough.

Lake stratification also occurs as a result of variations in suspended sediment concentration, termed *sediment stratification* by N.D. Smith and Ashley (1985). Studies of Peyto Lake, Alberta, and Malaspina Lake, Alaska, have revealed gradual increases in suspended sediment with depth (Gustavson,

(a)

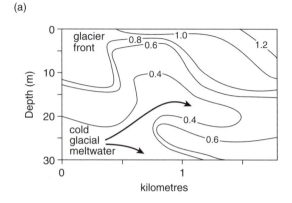

(b)

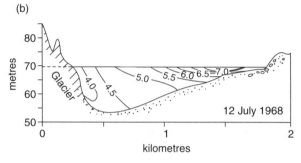

Fig. 8.16 Examples of the influence of glacier proximity on the thermal structure of glacier-contact lakes in Iceland. (a) Disruption of the thermal structure of Jökulsarlon by the injection of cold, dense meltwater below the lake surface. The glacier snout is 450 m to the left of the profile (after Harris, 1976). (b) Water-cooling adjacent to the margin of Skeiderarjökull (after Churski, 1973). (From N.D. Smith and Ashley, 1985. Reproduced by permission of SEPM)

1975b). The processes responsible for this density gradient are unclear, but are thought to relate to a combination of the upward diffusion of sediment from underflows and the downward settling of sediment from interflows, both of which are known to occur in Peyto and Malaspina Lakes.

8.3.2 Water inflow to glacier-fed lakes

In glacierized catchments, water will flow into lakes directly from glacier ice (via englacial and subglacial portals) and from both glacially fed and non-glacial rivers. In regions with marked seasonal temperature variations, maximum discharges of both water and sediment load occur during the summer melt season, with peak events coinciding with extreme weather or sudden releases of stored water (Sections 3.5 and 3.9.1).

As water enters a lake, turbulent eddies along the inflow margins cause it to mix with lake water, producing a two- or three-dimensional expansion of the flow. This is coupled with an exponential decrease in velocity of the inflow to produce an expanding and

decelerating *plume*. The shape and velocity patterns within the plume are governed by the discharge and velocity of the incoming water, channel shape, suspended sediment concentration and, most importantly, the density differences between the inflow and lake waters and the density stratification of the lake (Albertson *et al.*, 1950; Bates, 1953; Rao and Carstens, 1971; Bogen, 1983).

If the density of the diluted inflowing water is less than the density of the epilimnion, it will rise to the surface as an *overflow*. In contrast, if the incoming water is denser than the hypolimnion, it will flow along the bottom of the lake as an *underflow* (Fig. 8.17). Overflows and underflows were referred to as *hypopycnal* and *hyperpycnal* inflow, respectively, by Bates (1953), who described their shape as two-dimensional plane jets. *Interflows* occur where the density of the inflowing water is intermediate between the dense, cold lake bottom water and the low-density, warm lake surface water, and will move along the thermocline.

The larger the density differences between the inflow and lake water, the greater the energy required to mix them, so that turbulent diffusion is reduced. This usually results in the maintenance of a discrete density current through the lake. According to field measurements, underflow velocities vary greatly, and the flow may persist for hours or even days, especially in areas of high river discharges. Velocities ranging from a few centimetres per second to a few tens of centimetres per second are common (Gustavson, 1975b; Lambert *et al.*, 1976; Lambert and Hsu, 1979; Gilbert and Shaw, 1981; N.D. Smith *et al.*, 1982; Bogen, 1983), but values over 100 cm s^{-1} have been reported (Lambert, 1982; Weirich, 1984). Interflow velocities have been measured at 2 to 15 cm s^{-1} (Hamblin and Carmack, 1978; N.D. Smith, 1978; N.D. Smith and Ashley, 1985).

In unstratified or weakly stratified lakes where the inflowing water is of the same density as the lake water, a *homopycnal inflow* will result, in which the incoming plume quickly loses its identity by three-dimensional mixing (Fig. 8.17C). Homopycnal inflows are typical of shallow, freely circulating lakes which receive inflowing water of constant density, conditions which are not common in glacier-fed lakes. However, lakes fed by the lower stretches of glacier meltstreams, especially when further lakes act as sediment traps upstream, may receive homopycnal inflows. Quite often in these situations, more than 80 per cent of the sediment entering the lake leaves it again at its outlet because the inflowing water is of approximately the same temperature and the same suspended sediment concentration as the lake water, and so mixing occurs quickly and water stratification is inhibited (N.D. Smith *et al.*, 1982). As the sedimentation rates in such lakes are very low,

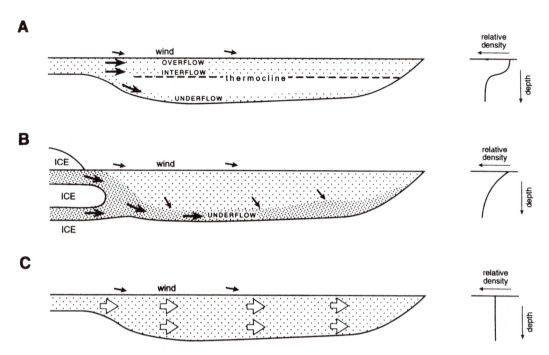

Fig. 8.17 Density stratification in glacier-fed lakes. (A) Thermal stratification, showing the positions of overflows, interflows and underflows. This situation is most common in non-glacier-contact lakes. (B) Sediment stratification, typical of glacier-contact lakes. (C) No stratification. (From N.D. Smith and Ashley, 1985. Reproduced by permission of SEPM)

the bottom sediment is commonly mixed by the action of organisms living on the lake bottom (*bioturbation*; Section 8.4.6.2).

The density of inflowing water is governed by temperature, dissolved solute content and suspended sediment concentrations. The suspended sediment concentration is the most important source of density variations in inflow water, because it varies over a wider range and changes over shorter periods of time than temperature, which varies over longer periods of days to months. Variability of inflow type and mixing patterns will relate to seasonal and weather-dependent stream discharges, stream temperatures, suspended sediment concentrations and seasonal changes in lake thermal stratification. In some lakes this may result in the production of different inflow types at different times of year (Carmack *et al.*, 1979; Pickrill and Irwin, 1982). Where discharge and density vary rapidly, overflows, interflows and underflows may coexist for short periods (Gilbert, 1973, 1975; Irwin and Pickrill, 1982; N.D. Smith *et al.*, 1982).

8.3.3 Water stratification and inflow in glacimarine environments

Fresh water is less dense than sea-water, so that meltwater entering glacimarine environments is relatively buoyant and will flow along the surface as an *overflow* or *meltwater plume*. The resulting water column has a two-layer density structure separated by the *pycnocline*, although this can be complicated by undercurrents and influxes of water from sub-basins separated by topographic barriers to produce multilayer circulation (Syvitski *et al.*, 1987). Underflows are rare in the glacimarine environment, because they require abnormally large suspended sediment concentrations to reduce buoyancy. Suspended sediment concentrations in subglacial streams are normally too low to compensate for the high densities of saline ocean water (Gilbert, 1983).

Water discharging into the sea from subglacial and englacial tunnels is subject to two sets of forces: (a) *buoyancy forces*, encouraging upwelling; and (b) *momentum forces*, tending to carry the water horizontally through the water column (Powell, 1990). When discharges are low, buoyancy forces are dominant and the meltwater rises rapidly to the surface as a *forced plume* (Fig. 8.18A), but at higher discharges the momentum forces dominate, and the efflux will issue as a subhorizontal *jet* before rising (Figs 8.18B, 8.18C). *Axisymmetric jets* are formed where the efflux travels through the free water column before rising, whereas *plane jets* travel along the sea floor before upwelling to form overflow or, more rarely, interflow plumes.

The behaviour of a turbulent jet once it is beyond the tunnel mouth depends largely upon the rate of mixing with the surrounding sea-water (Albertson *et al.*, 1950; List, 1982; Turner, 1986; Powell, 1990).

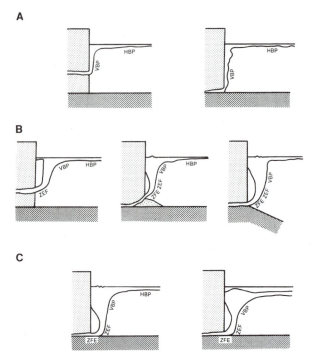

Fig. 8.18 Patterns of meltwater discharges from tidewater glaciers. (A) Forced plume dominated by buoyant forces. (B) Axisymmetric jets developing into axisymmetric plumes as momentum forces give way to buoyant forces. (C) Plane jets developing into axisymmetric jets and plumes. ZFE: zone of jet flow establishment; ZEF: zone of established jet flow; VBF: vertical buoyant plume; HBP: horizontal buoyant plume. (From Powell, 1990. Reprinted by permission of the Geological Society of London)

Three zones have been identified (Fig. 8.18). The *zone of flow establishment* (ZFE) occurs close to the tunnel mouth where the centre-line velocity equals the outlet velocity and the centre of the jet acts as a plug flow. Second, the *zone of established flow* (ZEF) begins at a distance of $6.2 \times D$ from a circular outlet (where D is the tunnel or pipe diameter), where the core velocity decelerates. In this zone, turbulent mixing occurs throughout the jet, and this can extend to beyond $200 \times D$ for axisymmetric jets and beyond $2000 \times D$ for plane jets. Clearly, bottom friction and suspended sediment concentration will cause these distances to vary. At the end of the ZEF the large amount of turbulence forces the residual axial velocity of the jet to decay to zero and, therefore, buoyancy forces take over. Third, the efflux rises in a *vertical buoyant plume* (VBP), and once at neutral buoyancy it becomes a *horizontal buoyant plume* (HBP). The roles of various jets and plumes in providing sediment to deep water bodies are discussed in Section 8.4.

General circulation in glacimarine environments is dominated by the seaward flow of surface meltwater

plumes from both glacier snouts and river mouths and compensatory landward flows of deeper saline water. The strength of this circulation is obviously linked to the intensity of meltwater inflow (Fig. 8.19). In Arctic fjords, surface plumes of anywhere between 5 and 31‰ salinity and temperatures of 2–4°C have been observed to overlie more saline, colder water (Lake and Walker, 1976; Schei *et al.*, 1979; Elverhoi *et al.*, 1983), but stratification due to subglacial meltwater efflux is uncommon around Antarctica.

Although surface plumes are readily visible near the efflux point, owing to their high suspended sediment concentration, they become less pronounced with distance because of mixing of the plume with the ambient water by molecular diffusion and turbulent entrainment (Drewry, 1986). Equal quantities of water are exchanged between layers, resulting in a reduction of the salinity of the ocean water. The amount of mixing and entrainment that takes place is a function of the balance between the *stabilizing forces* associated with density stratification and the *destabilizing forces* associated with the velocity gradient across the interface of the two water bodies. This is defined by the Richardson number (R_n):

$$R_n = (g/\rho_w)(d_{wc}\rho_w/d_z)/(d_{wc}U_w/d_z)^2 \qquad (8.4)$$

where g is gravitational acceleration, ρ_w is water density, U_w is water velocity, d_{wc} is the depth in the water column, (ρ_w/d_z) is the vertical density gradient and $(d_{wc}U_w/d_z)$ is the vertical velocity gradient.

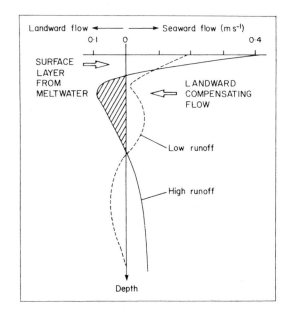

Fig. 8.19 A vertical velocity profile through the upper levels of a typical fjord. Meltwater is added to the surface layer, which flows seawards (left to right), and is compensated by a return flow at depth. Return flow occurs at deeper levels at times of low runoff. (From Drewry, 1986)

A Richardson number of < 0 indicates that the water column is unstable and mixing will occur, whereas $R_n > 0$ indicates stable conditions. Moreover, an R_n value > 0.25 would suggest that mixing due to shear instability is not significant (Dyer, 1973). In the absence of strong currents and circulation, the melting of glacier ice in contact with ocean water can result in dilution, producing ice-proximal brackish water (Matthews and Quinlan, 1975).

8.3.4 The role of surface ice

In glacilacustrine environments, the formation of surface ice in winter usually results in an abrupt change in surface temperatures in an otherwise stable temperature profile (Section 8.3.1; Fig. 8.15). This combines with reduced winter meltwater discharge and sedimentation rates, and negligible wind mixing of surface water, to produce a largely unstratified water column. In marine settings, sea-ice formation has a similar considerable impact on sea-water stratification but for slightly different reasons.

Sea-ice formation during the winter will eliminate wind mixing of surface waters and may act to dampen the effects of tides and currents (Gilbert, 1983). Furthermore, as sea-water is frozen, salt is rejected, and this produces a homogeneous water layer of high salinity below the ice. This high-salinity layer increases in thickness throughout the winter as more salt is released by ice growth and there is vertical mixing of deeper waters (Fig. 8.20; Syvitski *et al.*, 1987). At the same time, surface water temperatures are reduced, increasing the depth of the temperature maximum. In this way marine water

stratification caused by meltwater efflux (Section 8.3.3) essentially breaks down during the winter in areas where sea-ice forms. The thickness and duration of sea-ice obviously vary according to climate and local geography and oceanography, but generally they both increase with latitude. Some sheltered re-entrants, such as the fjords of the Canadian and Greenland High Arctic, are termed *frigid* in that they possess perennial sea-ice.

Water stratification in glacimarine environments is further complicated where glaciers or ice shelves block the mouths of fjords. Russell Fiord, Alaska, was transformed from an arm of the sea into a freshening ice-dammed lake by an advance of Hubbard Glacier in 1986, with profound implications for water-body characteristics and marine life (Krimmel and Trabant, 1992). On northern Ellesmere Island, Canada, the Ward Hunt ice shelf has dammed Disraeli Fiord, creating a 'lid' of fresh water 44 m deep on top of relict ocean water (Keys *et al.*, 1969; Keys, 1978). The salinity increases downward from 5 to 32‰ and the temperature drops from $-0.6°C$ to $-1.6°C$ at the pycnocline (Fig. 8.21).

8.3.5 Tides, currents and waves

Water column stratification and circulation in glacimarine environments and large glacier-fed lakes are affected by the actions of tides, currents and waves. Tidal currents and waves will also influence the stability and type of glacier front and the nature and extent of ice shelves (Sections 8.2.3 and 8.2.4). In general terms, tides and currents will enhance

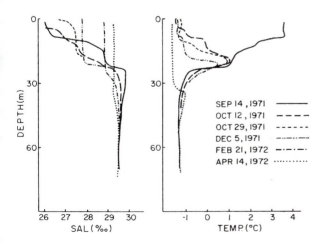

Fig. 8.20 Seasonal changes in water column temperatures and salinities in Cambridge Bay, Arctic Canada. (From Syvitski *et al.*, 1987, after Gade *et al.*, 1974. Reproduced by permission of Springer)

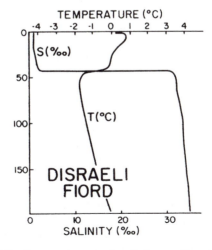

Fig. 8.21 Temperature and salinity profiles in Disraeli Fiord, Arctic Canada, due to the influence of the Ward Hunt ice shelf, which blocks the fjord mouth, encouraging the survival of a warm lid of fresh water. (From Syvitski *et al.*, 1987, after Keys, 1978. Reproduced by permission of Springer)

turnover in the water column and so minimize water stratification. With respect to sedimentation processes, they also influence the transport of suspended sediment and rework sediment on the ocean bottom (Sections 8.4.6.1 and 8.5.2). The impact of tidal forcing on currents and circulation in glacimarine situations is often more important during the winter, when the inputs of meltwater from glacial and fluvial systems are at their lowest.

Glacimarine environments are affected by tides in a number of ways. Below the surface, tides set up *barotropic currents* which are of more or less constant velocity with depth. The velocity may be affected by bathymetry in some marine re-entrants, such as fjords, where sills may constrict the flow and increase the velocity. However, tidal current velocity is predominantly related to tidal range. For example, tidal current velocities up to 0.31 m sec^{-1} have been measured in Glacier Bay, Alaska, which has a tidal range of 5 m (Matthews and Quinlan, 1975), compared with only 0.0027–0.0029 m sec^{-1} for Baffin Island fjords with a tidal range of 1.1 m (Gilbert, 1982).

Tides are also seen to affect the nature of surface plumes. Ebb tides, which reinforce flow away from the shore, will lead to plume velocity increases and decreasing plume thickness, whereas flood tides counteract seaward flow and will produce thicker but slower-moving surface layers. Water mixing is also enhanced by flood tides, which can hold surface plumes near to the terminus of tidewater glaciers. Vertically stacked turbid layers within the water column in Glacier Bay, Alaska, have been interpreted by Hoskin and Burrell (1972) as flocculated sediment layers resulting from each semi-diurnal tide (Section 8.4.3.3). Further variability in the behaviour of surface plumes is brought about by various interactions between periods of maximum stream discharge and ebb and flood tides. Tidal impacts on cyclic deposition of laminated sediment are reviewed in Section 10.6.2.

Surface currents are generated by wind shear. In glacial environments, katabatic winds blowing offshore will influence the flow direction and the longevity of overflow plumes, and will also determine the drift paths and drift speeds of icebergs. Larger-scale oceanic currents such as long-shelf currents, upwellings, geostrophic currents and those triggered by subaqueous slope failures are also important in glacimarine settings, and are a function of land mass positioning, global ocean currents and submarine topography. Subaqueous slope failures are temporally discontinuous and relatively unpredictable, but they are capable of producing considerable impacts through current activity and wave action, and very large failures can initiate devastating tidal waves or *tsunamis* (e.g. Dawson *et al.*, 1988).

8.3.6 The Coriolis effect

Near-surface inflow plumes discharging into water bodies are also affected by the Earth's rotation. Because of the curvature of the Earth, surface currents will be deflected in order to conserve angular momentum, an effect which is most pronounced close to the poles. This *Coriolis effect* causes plumes to be deflected to the right in the northern hemisphere and the left in the southern hemisphere, similarly to the deflection of atmospheric wind systems (H.D. Johnson and Baldwin, 1986; Barry and Chorley, 1992). Plumes will be deflected until they impinge upon a shoreline, along which they then flow (Hamblin and Carmack, 1978). Transmissivity profiles from Lake Louise, Alberta, which provide a measure of the amount of light transmitted through the water column according to the suspended sediment concentration, provide evidence of such a deflected interflow (Fig. 8.22; N.D. Smith and Ashley, 1985). Such inflow patterns have considerable effects on lake bottom sedimentation patterns in a cross-lake direction, which are superimposed on the normal proximal–distal downlake patterns.

8.3.7 Bathymetry

The bottom topography of both marine and lacustrine basins may have considerable impact on water circulation and associated sedimentation patterns. Common topographic elements are submerged bedrock hills or depositional features such as moraines or morainal banks, subaqueous channels, and enclosed basins created by erosional overdeepening or uneven sedimentation. Local topography creates sub-basins within larger depositional lake or marine basins, and may have distinct circulation and deposition patterns (Fig. 8.23). The transfer of sediment from one sub-basin to another will be impeded to a greater or lesser extent depending on the depth of the topographic barrier relative to dominant current flow (e.g. underflows, interflows or overflow plumes). Bottom topography has the greatest influence on the flow direction of underflows, which are deflected away from topographic high points and steered towards the deepest parts of basins (N.D. Smith *et al.*, 1982).

Water stratification in fjords can be modified by the presence of deep, stagnant bottom water behind bedrock or sediment sills at the seaward end. The salinity of this water will be related to the magnitude of the deep-water renewal and the strength and mixing depths of freshwater plumes. Where a shallow sill restricts circulation, fjord bottom waters are isolated from the open sea, and deep saline water only rarely penetrates the fjord, thus producing *euxinic* conditions. For a thorough review of fjord circulation patterns, see Syvitski *et al.* (1987). Where shal-

% Transmissivity

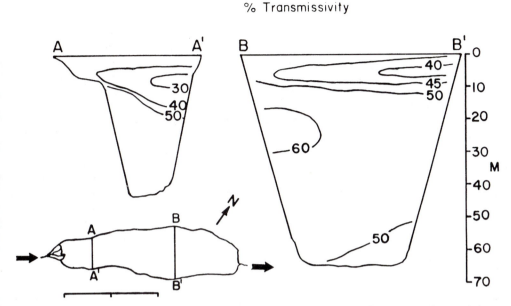

Fig. 8.22 Transmissivity profiles of two cross-sections of Lake Louise, Alberta, Canada, on 12 August 1975, showing an interflow deflected to the right-hand shore by the Coriolis effect. (From N.D. Smith, 1985. Reprinted by permission of SEPM)

low basins have been glacioisostatically lifted above present sea-level, the sill will help to isolate sea-water within the depression as a terrestrial lake is produced by uplift. This results in stratified lakes wherein saline bottom water can be preserved (Hattersley-Smith and Serson, 1964; Hattersley-Smith *et al.*, 1970; Retelle, 1986b).

8.4 DEPOSITIONAL PROCESSES

8.4.1 Sediment inputs

Sediment can enter glacimarine and glacilacustrine environments from several sources (Fig. 8.24; Powell, 1991; Powell and Domack, 1995). (a) *Supraglacial, englacial and subglacial debris* can be released directly into the water in ice-contact environments, from either glacier margins or calved icebergs. (b) *Unfrozen subglacial sediment* can emerge from beneath glaciers at the grounding line, either as a result of localized squeeze processes or as part of large-scale subsole deformation. (c) *Glacier meltstreams* can carry sediment directly into lakes or the sea from englacial and subglacial portals or supraglacial channels. Alternatively, meltwater can pass through proglacial channels before entering standing water. (d) Sediment can also be delivered from non-glacial sources by rivers and slope processes. In this case, the sediment load may consist of paraglacially reworked glacigenic sediment (Section 7.6).

The relative contribution of sediment from these sources varies from basin to basin, depending on the lithology, topography and tectonic setting of the catchment, glacier activity and morphology, and the climatic regime. Glacier meltstreams are generally by far the most important sediment source at temperate glacier margins (Powell and Domack, 1995; Hunter *et al.*, 1996). In such settings, thick, debris-rich basal ice is rare, and basal sediment is effectively flushed out by subglacial streams, so that direct sediment input from debris-rich ice or icebergs tends to be of minor importance. In contrast, ice shelves may have thick sequences of basal debris-rich ice, much of which can directly enter the water following basal melting. Glaciers originating in mountain terrain commonly have extensive covers of supraglacial debris, which can enter lakes and the sea by gravitational or glacifluvial processes.

The overall magnitude of sediment supply is also highly variable. Calculated sedimentation rates for proximal glacimarine environments range from $30–100 \, mm \, yr^{-1}$ in Svalbard (Elverhoi *et al.*, 1983) to $2000–9000 \, mm \, yr^{-1}$ for Alaskan fjords (Powell, 1983a). Extreme values of $13,000 \, mm \, yr^{-1}$ were reported by Cowan and Powell (1991) for an ice-proximal basin in front of McBride Glacier, Alaska.

8.4.2 Deposition from jets

Where subglacial or englacial streams discharge directly into the lacustrine or marine environment, large quantities of fluvial bedload and suspended

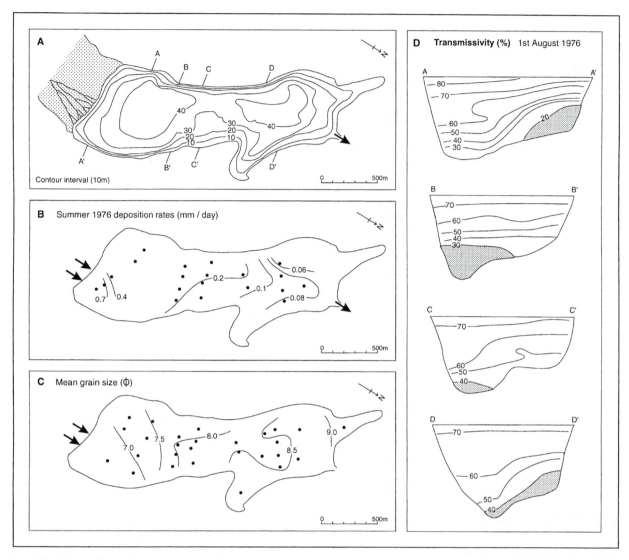

Fig. 8.23 The effects of bottom topography on sedimentation in the underflow-dominated Peyto Lake, Alberta, Canada (modified from N.D. Smith *et al.*, 1982): (A) bathymetric map showing two sub-basins and cross-profiles A–A′ to D–D′; (B) summer sedimentation rates measured at points marked by dots; (C) mean grain size distribution showing correspondence with bathymetry; (D) transmissivity profiles along the sections A–A′ to D–D′ showing that the lowest values (highest suspended sediment concentrations = stipple) are controlled by bathymetry and positions of inlet and outlet

load can be transported and deposited by jets (Section 8.3.3; Powell and Molnia, 1989; Powell and Domack, 1995). The highest sediment concentrations in jets occur close to the efflux point in the zone of flow establishment (ZFE). Bedload and suspended sediment loads can be very high, and sediment can be transported out on to the lake or sea floor as *hyperconcentrated flows* or as *bedload traction carpets* similar to those in turbid subaerial streams (Section 3.8.1). The coarsest sediment is deposited rapidly close to the efflux point as flow decelerates, and the finer material is transported further by the turbid jet and plume, which maintain some momentum (Pow-

ell, 1990; T.G. Stewart, 1991). In glacimarine environments, fine suspended sediment is commonly uplifted from the bed in buoyant plumes, promoting more rapid deposition of the coarser material. Sediment deposited from jets forms sheets and channelized lenses of gravel and sand, which typically display rapid downstream fining. The distance over which such fining takes place relates to the magnitude of the meltwater discharge from the tunnel, being greater during the melt season. Flow can also be erosive, scouring out channels which are later filled by renewed sediment accumulation (Rust and Romanelli, 1975).

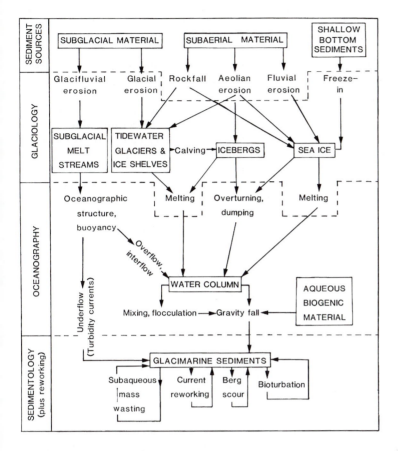

Fig. 8.24 Model of the glacimarine sedimentary system, illustrating the sediment sources, glaciological and oceanographic constraints, and subaqueous processes. Boxes represent sediment stores and unboxed terms represent processes. (From Dowdeswell, 1987)

8.4.3 Settling from suspension

Suspended sediment in overflows, interflows and underflows can be transported for considerable distances in lakes and the sea, and then gradually settles out through the water column. The processes of transport and suspension settling are essentially similar in glacilacustrine and glacimarine settings, although the salt content of sea-water and tidal effects introduce some important differences.

8.4.3.1 SEDIMENT DISPERSION IN OVERFLOWS AND INTERFLOWS

In glacilacustrine settings, overflows occur only when the inflowing water is less dense than the lake water, therefore overflows cannot carry a large amount of suspended sediment. This means that inflowing meltstreams with high suspended sediment concentrations will generally produce underflows rather than overflows. Furthermore, overflows will tend not to be produced in thermally stratified lakes but will form during the late spring and winter when incoming water is more likely to contain small suspended sediment loads and be more buoyant than the lake water. Because they occur at the lake surface, overflows are susceptible to wind stress, and because of the effects

of the wind are often very difficult to differentiate from shallow interflows.

Overflows are much more common in glacimarine environments, as a result of the high density of sea-water (c. 1030 kg m^{-3}) compared with fresh water (c. 1000 kg m^{-3}). This means that incoming water can have suspended sediment concentrations as high as 30 kg m^{-3} and still remain buoyant enough to form overflows. Water upwelling from submerged englacial and subglacial conduits can be observed at the termini of many tidewater glaciers in the form of turbid sediment plumes, surface water disturbances and currents (Fig. 8.25 (Plate 18); Cowan and Powell, 1990; Powell, 1990; Powell and Domack, 1995).

Interflows are common in thermally stratified lakes, and may dominate the fine-sediment transport in distal lakes which receive only light sediment loads (N.D. Smith, 1978). Interflows will also alternate with the more dominant underflows in lakes receiving high sediment loads (Gilbert, 1975; Gustavson, 1975b; Wright and Nydegger, 1980). When viewed from the air, the outer edge of a turbid overflow plume can be seen to possess a very sharp contact with the lake water into which it is discharging. This marks the position of the 'plunge line', where the plume sinks to reach lake water of similar density. Between the inflow point and the plunge line the plume spreads out two-dimensionally and decelerates as it mixes with the lake water. The turbulence in the plume entrains lake water as the plume sinks. This increases the volume of the plume and sets up compensatory surface counter-currents to replace the water lost to the sinking plume. The plunge line essentially represents the convergence point of the lakeward-moving plume and the shoreward-moving counter-currents (Fig. 8.26). Once the interflow plume reaches the depth at which its density equals that of the surrounding water (usually around the level of the metalimnion), it spreads out horizontally. Clay-sized material can remain suspended in the interflow, owing to upward-directed currents, but the coarser fractions will drop out progressively in a downlake direction.

High density underflows are an important means of dispersing suspended sediment in both glacilacustrine and glacimarine environments, and are discussed in Section 8.4.4.4.

8.4.3.2 SUSPENSION SETTLING

In still water, suspended particles will fall down through the water column with a velocity that reflects the buoyant weight of the particle and the drag forces resisting motion. The buoyant weight and drag forces are influenced by several factors, particularly the size, density and shape of the particle, and the density and viscosity of the water. For spherical particles, the settling velocity is given by Stokes' law:

$$v_f = (g/18\eta)(\rho_p - \rho_w)D^2 \qquad (8.5)$$

where v_f is the velocity of fall, g is gravitational acceleration, η is water viscosity, ρ_p is particle density, ρ_w is water density, and D is grain diameter.

This shows that fall velocity increases with particle size and the density difference between the particle and the water, and decreases with the water viscosity. Fall velocities are large for large particles, and very small for small particles, as is strikingly demonstrated by considering the settling of quartz spheres through salt water (Allen, 1985). Terminal settling

Fig. 8.26 Turbid inflow plumes in Peyto Lake, Alberta, Canada, showing sharp contacts between lake water and lakeward edges of plumes (plunge points). (From N.D. Smith and Ashley, 1985. Reprinted by permission of SEPM)

velocities are 0.00979 m sec^{-1} for a 100 μm particle (fine sand), 0.0000979 m sec^{-1} for a 10 μm particle (silt), and only 0.000000979 m sec^{-1} for a 1 μm clay particle. At these rates, settling times through a 100 m column of salt water are approximately 3 hours (fine sand), 11 days (silt) and 3 years (clay). Since particles will remain in suspension if vertical motions in the water equal or exceed the settling velocity, it is easy to see that fine particles can be transported for considerable distances in well-mixed waters.

The density term in Stokes' law shows that particles will settle more rapidly in fresh water than in salt water. Additionally, particles with high density, such as iron-rich minerals, will settle out more rapidly than low-density particles of the same size. Settling velocities for non-spherical particles will differ from those calculated by Stokes' law. Platy particles, such as shell fragments, mica flakes and single clay particles, settle more slowly, owing to their high surface area relative to their volume (Allen, 1984, 1985).

8.4.3.3 FLOCCULATION AND PELLETIZATION

The behaviour of suspended clay particles in sea-water is strongly influenced by the process of *flocculation*, or the bunching of clay particles into aggregates called *floccules* or *flocs*. Flocculation takes place when the repulsive charge that normally exists between clay particles is reduced by the electrolytic action of salt water, allowing particles to stick together if they collide (Lerman, 1979; Drewry, 1986). A salinity of only 3–4‰ will promote flocculation (Kranck, 1973). The larger floccules sink faster than isolated clay particles, causing them to be deposited alongside fine silts. Floccule sizes increase with water depth and salinity, and Syvitski and Murray (1981) have observed floccules of 10–70 μm in a glacier-fed fjord in British Columbia.

Aggregation of clay particles also occurs by the production of *faecal pellets* by zooplankton (Syvitski *et al.*, 1987). This pelletization can produce agglomerations up to 100 μm across (Syvitski and Murray, 1981), and faecal pellets comprise up to 18 per cent of the suspended particulate matter in the waters just offshore of the Ross ice shelf, Antarctica (Carter *et al.*, 1981). Similar processes operate in glacilacustrine environments, and N.D. Smith and Syvitski (1982) have reported faecal pellets produced by sediment-ingesting zooplankton in Bow Lake, Canada. These pellets promote the deposition of fine silts and clays which would otherwise remain in suspension until they were flushed out of the lake system.

A final process of particle aggregation is *agglomeration*, or the attachment of mineral grains to organic detritus by cohesion (Syvitski *et al.*, 1987). This process is unimportant in ice-proximal environments where suspended sediment concentrations are high and organic content is low, but may be important for the removal of fine clay particles from the water column in distal settings.

8.4.4 Gravitational processes

Gravitational mass movements are very important agents of sediment transfer and deposition in many glacilacustrine and glacimarine environments, particularly where unconsolidated debris rests on steep, potentially unstable slopes. As for terrestrial slope failures, several distinct types of subaqueous mass movement can be recognized (Stow, 1986; Nemec, 1990; Prior and Bornhold, 1990). Here, we describe three basic types: (a) *sediment gravity flows* (including cohesive and cohesionless debris flows and turbidity currents); (b) *debris falls*; and (c) *slumps and slides* (Fig. 8.27).

8.4.4.1 SEDIMENT GRAVITY FLOWS

Sediment gravity flows or *mass flows* are sediment–water mixtures that flow downslope under the force

of gravity. Subaqueous flows can consist of debris introduced directly into standing water from ice margins or meltstreams, or material remobilized by subaqueous slope failures. As for subaerial flows, movement occurs when the stresses imposed by the weight and surface gradient of the material exceed its yield strength (Section 6.5.3.1). High stresses and flow mobilization may be triggered by high deposition rates, slope oversteepening by ice-push or erosion, iceberg calving, wave impacts or even earthquake shocks. In glacilacustrine and glacimarine settings, mass flows result in the rapid, strongly episodic transfer of large amounts of debris, which can build up thick sediment successions. Flows can also be erosive, scouring out pre-existing sediment and depositing it further downslope.

Subaqueous mass flows can exhibit a range of flow behaviour depending on their grain size distribution, density and strength (Middleton and Hampton, 1973; Enos, 1977; Nemec and Steel, 1984; Postma, 1986; Nemec, 1990). An important distinction can be made between *debris flows*, which are high-concentration plastic slurries, and *turbidity currents* (or *fluidal flows*), which are rapidly moving turbulent underflows. These two types can be further subdivided according to the dominant flow mechanism, although it should be recognized that one flow type can evolve into another as a result of changes in water content and strength during flow (Hampton, 1972; Prior and Bornhold, 1989; Nemec, 1990). Here, we describe mass flows under three headings: (a) cohesive debris flows; (b) cohesionless debris flows; and (c) turbidity currents (Postma, 1986). The distinction between *cohesive* and *cohesionless* flows is important, because the presence or absence of matrix cohesion strongly influences flow behaviour and the character of the resulting deposits (Nemec and Steel, 1984; Postma, 1986; Nemec, 1990).

8.4.4.2 COHESIVE DEBRIS FLOWS

Cohesive subaqueous debris flows contain some clay matrix material which, when mixed with water, acts as a fluid with cohesive strength (A.M. Johnson, 1970; Middleton and Hampton, 1973). Large clasts, boulders and blocks of pre-existing sediment can be rafted along at high levels of the flow, partly because of the cohesive strength of the matrix and partly because particles enveloped in a dense slurry have low buoyant weight (Rodine and Johnson, 1976). The upper part of the flow commonly consists of a semi-rigid 'plug' riding passively on a thin basal shear zone, as for Type I subaerial flows (Section 6.5.3.1; Lawson, 1981b, 1982). Unlike subaerial flows, however, the uppermost parts of subaqueous cohesive flows can become diluted during flow, owing to turbulent mixing with the overlying water column.

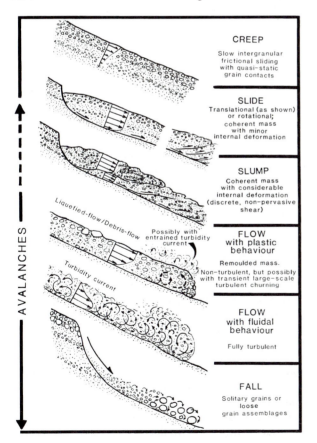

Fig. 8.27 Range of gravity-driven sediment transport processes on subaqueous slopes. The velocity profiles are schematic and not to scale. (From Nemec, 1990. Reproduced by permission of Blackwell)

This can cause transformation of part of the flow into a low-density, turbulent underflow, altering the characteristics of the final deposit (Hampton, 1972).

The velocities of cohesive debris flows increase non-linearly with shear stress. Increasing shear stresses also cause thickening of the basal shear zone at the expense of the plug zone, and for high stresses (or low-strength flows) the plug zone may disappear, as for Type IV subaerial flows. Deposition occurs once the shear stress falls below the shear strength again and the basal shear zone thickness shrinks to zero, a process known as *freezing*. This has nothing to do with the transformation of water to ice, so when flow behaviour in glacial environments is being described, the term 'freezing' should be used with caution.

8.4.4.3 COHESIONLESS DEBRIS FLOWS

Cohesionless debris flows are high-concentration, non-turbulent flows in which cohesion contributes little or nothing to sediment strength. They tend to occur in sandy and gravelly materials (Nemec and Steel, 1984; Postma, 1986; Nemec, 1990). Grains are supported during transport by one or more mechanisms, including: (a) *dispersive pressure*, or the upward component of the pressure exerted by neighbouring shearing grains; (b) *buoyancy effects* due to the density of the surrounding matrix, which makes grains lighter than they would be in water; and (c) *liquefaction and fluidization*, or the upward movement of pore fluid during dewatering (Middleton and Hampton, 1973; Lowe, 1976b; Nemec, 1990). Various subtypes of cohesionless debris flows have been defined on the basis of the dominant particle support mechanism during transport, including *grain flows*, *fluidized flows* and *liquefied flows*, although these mechanisms rarely work in isolation (Middleton and Hampton, 1973; Lowe, 1979; Nemec and Steel, 1984). Cohesionless debris flows can originate by the failure of cohesionless materials on subaqueous slopes, or by the gravitational transport of sediment delivered to subaqueous slopes by stream flow or dumping from glacier margins.

The grains in cohesionless flows move by sliding, rolling, bouncing and rattling off one another. The upward dispersive pressure exerted during movement causes expansion or dilation of the flowing mass and tends to force larger particles to the surface. This process can be illustrated by shaking a bowl of sugar to bring the larger particle aggregates (produced by those who insist on dipping wet spoons into the bowl) to the top. At the same time, finer particles are filtered to the base of the flow through the gaps between larger particles, a process known as *kinetic sieving*. The overall effect of these vertical movements is to produce *inverse grading* or upward coarsening of the flow.

Liquefied flows are triggered by elevated porewater pressures associated with transient shocks, such as iceberg calving events, earthquakes, and sudden loading by overpassing sediment. These shocks will cause the grains to be suspended in fluid for a short period during which flow is possible. As grains begin to settle out and regain contact with each other after the initial shock, they displace porewater upwards. This means that a liquefied flow will 'freeze' from the bottom upward. Commonly, the upwards flow of water is not uniform and is concentrated in pipes, producing sand volcanoes on the surface and ball and pillow structures within the sediments. The temporary suspension or upward movement of grains by escaping fluid is called *fluidization* or *dewatering*, and is not synonymous with liquefaction, which refers to the downward settling of grains through the fluid (Lowe, 1976b).

The morphology of cohesionless flows generally consists of an upper erosive channel and a lower depositional lobe (Prior and Bornhold, 1990). On steep slopes, cohesionless flows may grade into debris falls (Section 8.4.4.5) and exhibit downslope coarsening or fall sorting, owing to the tendency for the larger particles with most momentum to travel furthest.

8.4.4.4 TURBIDITY CURRENTS (UNDERFLOWS)

In turbidity currents (also known as *underflows* or *fluidal mass flows*), particles are kept aloft within the flow body by turbulent suspension (Middleton, 1966a, b, c). The suspended particles render the flow denser than the surrounding water, causing it to move downslope along the bed. Turbidity currents are subdivided into head and tail regions, the head being 1.5 to 2 times the thickness of the tail (Fig. 8.28). The turbulent head moves downslope, mixing with the ambient water, and requires constant transfer of denser fluid from the tail to maintain momentum. This can be sustained for hours where the turbidity current is fed by dense underflowing meltstreams, but most flows eventually dissipate once the influx of suspended sediment from the initiating disturbance is exhausted.

Sediment is transported by turbidity currents either as *suspended load* or as *bedload* swept along by the overpassing density current. As flows decelerate, particles settle out of suspension and become part of the bedload or the static bed. The higher settling velocities of larger particles (Section 8.4.3.2) mean that they will be deposited first, while finer material remains in suspension, resulting in *normal grading* or upward fining. Turbidity currents can also be erosive, scouring bottom sediments to form subaqueous channels, often bounded by levées (Brodie and Irwin, 1970; Lambert, 1982; Carlson *et al.*, 1989, 1992; Prior and Bornhold, 1990).

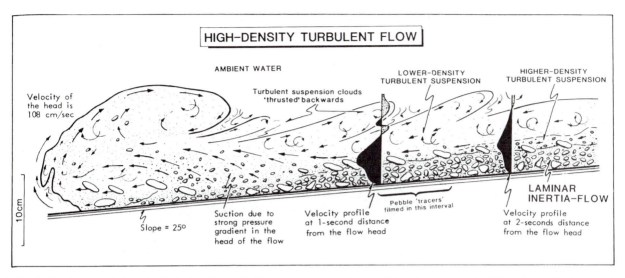

8.28 Main characteristics of a high-density turbidity current, drawn from a laboratory-generated flow. (From Postma *et al.*, 1988b. Reproduced by permission of Elsevier)

Turbidity currents can be initiated in at least three different ways (Nemec, 1990):

1. They may form directly from sediment-laden meltwater discharging from meltstreams or submerged conduit mouths, in the form of hyperpycnal underflows (Section 8.3.2). Such underflows are important agents for transporting and depositing sediment in glacier-fed lakes, but are less common in glacimarine environments, because suspended sediment concentrations in meltwater inflows are not usually large enough to overcome buoyancy effects (Section 8.3.3). In glacier-fed lakes, the sediment-laden inflowing water is often considerably denser than the lake water, and quasi-continuous underflows occur as a result. Such underflows are generated during maximum melt periods, when the thermal stratification of the lake water can be totally overwhelmed.
2. Turbidity currents may evolve less directly, by intense sediment fallout from suspension plumes, which can blanket large areas of subaqueous slopes with dense, mobile suspensions (Syvitski and Farrow, 1989). This situation is most common in proximal glacimarine environments where there is abundant material in high-level suspension.
3. They may evolve from the progressive downflow dilution of debris flows, resulting from turbulent mixing with the overlying water column (Hampton, 1972). Such sediment failures are usually associated with high sedimentation rates at subaqueous ice-proximal or ice-contact depocentres such as fans and deltas.

Flow in case 3 will tend to occur in short-lived surges, whereas cases 1 and 2 range from brief surges to more or less continuous 'streams' flowing for hours or days (Prior *et al.*, 1987; Nemec, 1990).

8.4.4.5 DEBRIS FALLS

Wherever subaqueous slopes are very steep, sediment may be subject to rapid downslope movement by *debris fall*, similar to rockfall in terrestrial environments. The term *grain fall* is also used in the literature, although Nemec (1990) has suggested that it be avoided because of its use in reference to suspension sedimentation. Debris may fall as single particles or as masses of strongly dispersed particles; each particle is driven in a downslope direction by its own momentum and will bounce, slide or roll when impacting the bottom. Therefore, debris fall sediments will display a downslope coarsening similar to the fall sorting that occurs on terrestrial scree slopes (Fig. 8.29). In addition, a fining-upward or normally graded sequence may be produced by a debris fall, owing to the tendency for larger particles to be deposited first and then to be overrun by the finer-grained tail of the avalanche.

8.4.4.6 SLUMPS AND SLIDES

As for subaerial environments, sediment on subaqueous slopes can fail along internal shear planes, and undergo downslope transport as *slumps* or *slides* (Stow, 1986; Prior and Bornhold, 1988; Nemec, 1990). Slide sheets undergo variable amounts of internal deformation, and may disaggregate during transport and evolve into debris flows. They are identifiable on images of present-day sea and lake floors from transverse compressional ridges on the slide

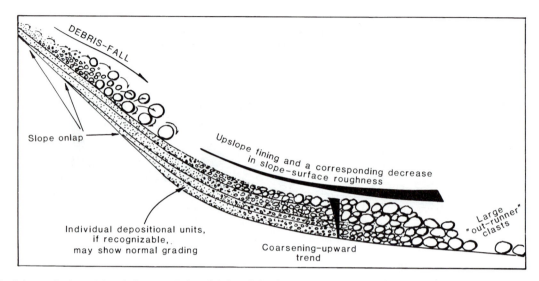

Fig. 8.29 Schematic depositional characteristics of debris falls. (From Nemec, 1990. Reproduced by permission of Blackwell)

surface, which may mark the surface trace of dipping shear faults within the slope.

8.4.5 Melt-out from ice shelves and icebergs

Debris ranging in size from clay and silt up to boulders can be introduced directly to ice-contact glacilacustrine and glacimarine environments by melt of debris-rich ice beneath ice shelves or dumping from icebergs. In the case of iceberg dumping, debris may be transported hundreds or even thousands of kilometres from the ice margin prior to deposition, thus providing a very efficient form of debris dispersion.

8.4.5.1 ICE RAFTING

Debris can be rafted out into lakes and the sea by floating ice, to be released into the underlying water column as the ice melts. In addition to icebergs derived from glacier and ice shelf margins, sea- and lake ice can also transport considerable quantities of debris. Clastic material in ocean sediment cores is often interpreted as an indicator of expanded glaciers, but other forms of rafting should be considered (Clark and Hanson, 1983; Sejrup *et al.*, 1984; Reimnitz and Kempema, 1988; Gilbert, 1990).

Bergs calved from glaciers may contain englacial or subglacial debris bands or spreads of supraglacial debris (Powell and Domack, 1995). Abundant debris is typical of bergs derived from valley and outlet glaciers in high-relief terrain, surging glaciers and some subpolar glaciers. In contrast, bergs from ice shelf margins generally have low debris concentrations, because much of the basal debris may have

been lost by basal melting prior to calving (Drewry and Cooper, 1981).

Sea- and lake ice rafting can be subdivided into active and passive types (Fig. 8.30). *Active sea/lake ice rafting* involves either: (a) freezing of bottom sediment directly on to thickening surface ice (Lauriol and Gray, 1980); (b) the incorporation of suspended sediment directly into forming sea- or lake ice (Campbell and Collin, 1958; Ackley, 1982; Barnes *et al.*, 1982; Dionne, 1984; Reimnitz and Kempema, 1987, 1988); or (c) formation of anchor ice on the sea or lake floor (Dayton *et al.*, 1969; Reimnitz *et al.*, 1987; Reimnitz and Kempema, 1988). In contrast, *passive sea/lake ice rafting* involves the deposition of sediment on to existing ice surfaces by: (a) rockfalls

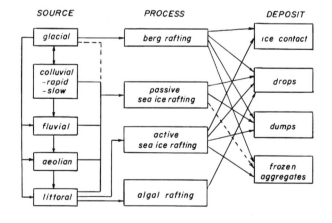

Fig. 8.30 Debris sources, transport processes and deposits involved in rafting in the glacimarine environment (excluding extraterrestrial and volcanic sources). (From Gilbert, 1990. Reprinted by permission of the Geological Society of London)

and avalanches from overlying slopes (Luckman, 1975); (b) fluvial deposition on to shore-ice surfaces before they break up in the spring (Keys, 1978); (c) aeolian deposition (McKenna *et al.*, 1986, 1986); and (d) littoral deposition, whereby material is moved up through cracks in the shore-ice during high tide (Dionne, 1984; Gilbert, 1990).

One other unusual form of rafting in the marine environment involves the lifting of clasts by marine algae, popularly known as seaweed. Gilbert (1984, 1990) has demonstrated that once the alga *Fucus vesiculosus* has grown to three times the weight of the clast to which it is attached, it will float the clast. This is because the alga possesses gas-filled vesicles in its thalli, producing buoyancy. A large number of algae are also found frozen into sea-ice and anchor-ice in littoral areas. Although the relative importance of algal rafting is difficult to assess, it is responsible for the introduction of individual dropstones to fine-grained sediments far from their source area, and this is normally regarded as indicative of iceberg or at least sea-ice rafting. The problem of equifinality here is exacerbated by the fact that the subpolar zone of ice rafting, at least in the northern hemisphere, overlaps a large part of the zone of algal rafting (Emery, 1963).

Drifting ice gradually melts as it is carried by ocean and lake currents, releasing its sediment load into the underlying sediment column (Fig. 8.31). Debris release from individual bergs is typically episodic, reflecting fragmentation and overturning events triggered by changes in the shape and centre of gravity of the melting berg. As icebergs overturn, any debris that has accumulated on the surface is suddenly dumped into the water. The amount of fragmentation and overturning decreases with distance from the coast as icebergs slowly attain equilibrium positions with low centres of gravity (Bellar *et al.*, 1964; Ovenshine, 1970; Orheim, 1980). A typical debris release history for a tabular iceberg is shown in Fig. 8.32 (Drewry and Cooper, 1981; Dowdeswell, 1987). Most of the sediment is released by the berg within the first 200 km of the ice margin, and the berg is essentially clean after 400 km. This model is based upon observations of icebergs at varying distances from the ice margin, as well as laboratory and theoretical analyses of iceberg melting rates in water at 0°C. Melting rates vary between $0.1 \, \text{m yr}^{-1}$ and $100 \, \text{m yr}^{-1}$, depending on debris content (Martin and Kauffman, 1977; Gade, 1979; Budd *et al.*, 1980; Russell-Head, 1980; Drewry, 1986). Calculations of deposition rates from suspension and ice rafting are presented in Table 8.1.

The most prominent impact of ice rafting is the introduction of coarse material to more distal locations where deposition is dominated by suspension sedimentation from overflow and interflow plumes. Debris can be released and deposited in four different ways (Fig. 8.31; Thomas and Connell, 1985; Gilbert, 1990). (a) *Iceberg drop* involves the deposition of a single particle, usually after release from the berg by melting. (b) *Iceberg dumps* refer to the simultaneous deposition of a number of particles in a single event, usually associated with the overturn of an iceberg and

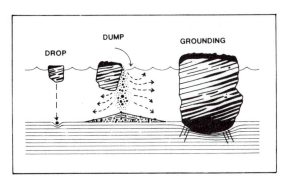

Fig. 8.31 Dropping and dumping of debris from icebergs, and sediment deformation by grounded bergs. (From Thomas and Connell, 1985. Reproduced by permission of SEPM)

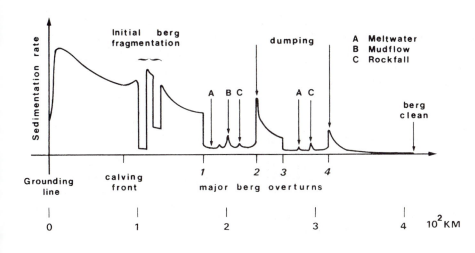

Fig. 8.32 Hypothetical debris-release history of a drifting and melting iceberg showing the effects of calving, fragmentation and overturn. (From Drewry, 1986)

Table 8.1 Qualitative estimates of variations in the distribution and quantity of entrained sediment in different types of ice mass

	Position of debris transport		
	Basal	Englacial	Supraglacial
Ice shelves			
from ice sheet	■■	■	■
via mountains	■■■	■■	■
Outlet glaciers			
ice stream	■■■■	■	■
via mountains	■■■■	■■	■■
Ice cliffs			
ice sheet edge	■■	■	■
near mountains	■■■	■■	■■

Source: Drewry (1986)
Sediment:
    substantial   ■■■■
    moderate   ■■■
    small   ■■
    trace   ■

the tipping of surface material into the water. Alternatively, concentrations of sediment from former englacial tunnels or crevasse fills can be released during melt-out and/or iceberg overturn. (c) *Frozen aggregates* are sediments held together by interstitial ice which are released when ice rafts are broken apart. (d) *Ice-contact* rafted sediments are deposited by an iceberg that is grounded on the sea floor.

Dowdeswell and Murray (1990) have suggested that five factors govern the amount of sediment deposited by iceberg rafting where the parent ice mass is a tidewater glacier: (a) the distribution and concentration of sediment in the parent glacier; (b) the rate of iceberg calving; (c) the rate of melting at the iceberg base; (d) the temperature of the ocean water; and (e) the velocity, stability and drift track of the iceberg. Some of these factors may counteract each other. For example, where sea-ice acts as a buttress to a calving glacier, it will reduce the rate of calving and therefore reduce the iceberg sedimentation rate. This will be counteracted by the restriction placed upon the drift speed and distance by the sea-ice buttress, which will increase ice-proximal sedimentation rates. Where icebergs are derived from ice shelves, undermelt and bottom freezing must be taken into account. Before calving takes place, undermelt may remove basal debris, whereas bottom freezing may protect it (Section 8.2.3.2).

8.4.5.2 ICEBERG SCOURING

As grounded icebergs are dragged over the lake or sea bed by currents, they plough curved, flat-bottomed troughs or furrows (Fig. 8.33). Large examples can be incised up to 20 m deep and 250 m wide, and stretch for several kilometres, but they are generally much smaller. The trough bottoms often exhibit a micro-topography of ridges and grooves with a relief of up to 30 cm. These are thought to be the product of dragging of bed material which has been mechanically incorporated into the iceberg keel. Sand volcanoes on top of this micro-topography attest to the liquefaction and dewatering of bed material during and immediately after scouring (Hodgson *et al.*, 1988). On either side of the main furrows, linear scour *berms* of displaced blocky material can reach heights of 6 m. The chaotic criss-crossing of iceberg furrows reflects the varied drift patterns of icebergs on the lake or ocean surface (Fig. 8.34). These can be

Fig. 8.33 Iceberg plough mark on the floor of a former lake, Eyjabakkajökull, Iceland. (Photo: C.M. Clapperton)

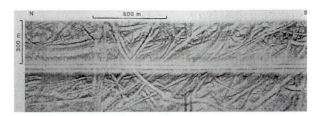

Fig. 8.34 Side-scan sonar image of iceberg plough marks on the Barents Shelf. (From Solheim, 1991. Reproduced by permission of the Norsk Polarinstitutt)

affected by winds as well as dominant current patterns. The deformation structures created by iceberg ploughing are described in Section 10.6.4.

8.4.6 Reworking by currents and biological activity

8.4.6.1 CURRENT REWORKING

Lake and sea currents are generated by a number of mechanisms including wind shear, tides, storm waves, long-shelf currents, upwellings, geostrophic currents, glacier and iceberg melting, and subaqueous sediment gravity flows. Generally, currents that reach the bottom of a water body tend to winnow the fine-grained materials and leave coarse lag deposits, termed a *palimpsest lag* by Powell (1984). They may also erode pre-existing material to produce erosional contacts or cross-cutting relationships between sedimentary units. Particularly erosive sediment gravity flows may be capable of scouring small-scale bedforms or flute marks similar to those produced in fluvial environments (Pettijohn and Potter, 1964; Dzulynski and Walton, 1965; Allen, 1971a).

8.4.6.2 BIOTURBATION

Disturbance of sediment by bottom-dwelling marine organisms is referred to as *bioturbation*. The wide range of structures produced by bioturbation are usually called *trace fossils*, and where they can be recognized and assigned to species are often termed *lebensspuren*. A classification scheme suggested by Seilacher (1953) includes five different types of lebensspuren in soft bottom sediments: (a) *resting traces*, where creatures rest in one place for a period of time; (b) *crawling traces*, made by animals making their way across the bed (Fig. 8.35); (c) *browsing traces* made by animals searching for food on the sediment surface; (d) *feeding structures*, which include burrows and similar structures created by organisms moving through the sediment for food; and (e) *dwelling structures*, where organisms create semi-permanent burrows. The various forms of trace

Fig. 8.35 A trace fossil from Holocene glacimarine silts in Clements Markham Inlet, north-east Ellesmere Island, Arctic Canada. The markings are on a bedding plane that has been opened like a book (left surface is imprint, right surface is cast). The central marking was made by the side-to-side motion of a fish's tail and the outer imprints were made by the pushing movement of its fins. A further unidentified trail crosses the fish trail from top to bottom. (Photo: T.G. Stewart)

fossils are illustrated in most sedimentology texts (e.g. Crimes and Harper, 1970; Reading, 1986; Reinick and Singh, 1980; Allen, 1982a; Leeder, 1982).

Because of the high sedimentation rates in proximal glacilacustrine and glacimarine environments, bioturbation is not usually intensive. Bioturbation can, however, result in the partial or complete destruction of primary depositional structures in more distal positions where sedimentation rates are low (Gilbert, 1982).

8.5 DEPOSITIONAL ENVIRONMENTS

There is a very wide range of glacilacustrine and glacimarine depositional environments, ranging from the margins of small ice-marginal lakes in glaciated valleys to the floors of the deep oceans. For the purposes of description, however, it is useful to group them into five general types: (a) grounding-line fans; (b) morainal banks; (c) grounding-line wedges beneath ice shelves; (d) deltas; and (e) distal environments.

8.5.1 Grounding-line fans

Grounding-line fans (also known as *subaqueous outwash fans*; Rust and Romanelli, 1975) are fan-shaped depocentres formed where subglacial and englacial streams emerge at subaqueous conduit mouths (Fig. 8.36; Powell, 1990; Powell and

Domack, 1995). Coarse bedload is deposited close to the tunnel mouth as the efflux jet decelerates, building up steep, unstable slopes at the glacier margin. Slope failures and renewed sediment discharge from the tunnel mouth feed mass flows that transport sediment radially away from the margin. Flows are likely to be cohesionless, owing to the removal of fine-grained matrix material in suspension. As flows travel over the fan surface, turbulent mixing with the overlying water results in their progressive dilution, and they may evolve into turbidity currents.

In glacimarine environments, buoyant forces cause sediment-laden meltwater to detach from the bed and rise as turbid plumes (Section 8.3.3). Upward velocities can be high enough to transport particles as coarse as medium sand to the surface (Cowan *et al.*, 1988; Cowan and Powell, 1990). The suspended sediment spreads laterally until it is able to settle through the water column. On the basis of detailed observations at tidewater margins in Alaska, Cowan and Powell (1990) suggested that suspension settling of fine particles (clay floccules and silt) is most likely during ebb tides, when internal turbulence within plumes is at a minimum, whereas only the coarsest particles rain out during flood tides. The resulting sedimentation cycles produce rhythmically laminated sediment packages known as *cyclopsams* and *cyclopels*, described in detail in Section 10.6.2. Very high sediment concentrations may increase plume density sufficiently to produce marine underflows (Powell, 1990).

Powell (1990) has considered the effects of varying water and sediment discharge on fan evolution

in glacimarine environments (Fig. 8.37). For low discharges, efflux jets have low momentum and so rise immediately to the surface. Deposition on the fan is dominated by mass flows fed by bedload dumped at the conduit mouth, supplemented by rain-out from the plume. At moderate water discharges and high sediment discharges, efflux jets can travel across the fan surface before rising, resulting in a zone of traction deposition at the fan apex flanked by a zone of mass flow activity. At high discharges, jet momentum is high enough to delay buoyant rising of the suspended sediment plume, and deposition on the fan is dominated by tractive currents. Barchanoid dunes and mass flows may occur at the point of detachment.

In glacilacustrine environments, sediment-laden meltwater is generally denser than the surrounding lake water, and will tend to hug the lake floor as underflows (Sections 8.3.2 and 8.4.4.4). Deposition on grounding-line fans is therefore likely to be dominated by mass flows, with comparatively minor inputs from high-level suspended sediment.

Oscillations of the glacier margin introduce further complications to the development of grounding-line fans (Fig. 8.38). Glacier advance may produce glacitectonic thrusts and folds in the outwash, and may even lead to till deposition and morainal bank formation (Section 8.5.2; G.W. Smith, 1982; Powell, 1983a). In contrast, ice margin retreat can separate the efflux jet from the fan top, focusing deposition in the backslope zone. A combination of rapid deposition and backslope reworking can produce very

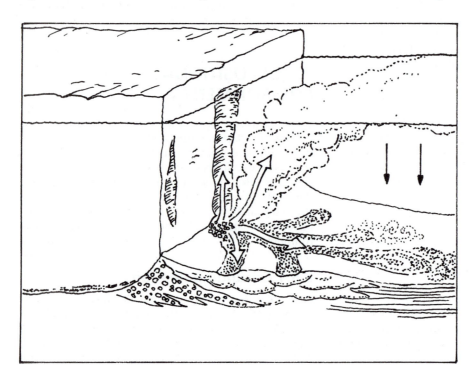

Fig. 8.36 Processes of sediment transport and deposition at a marine grounding-line fan. (Adapted from Powell and Domack, 1995)

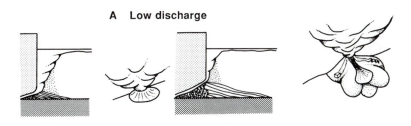

A   Low discharge

B   Moderate water discharge, high sediment discharge

C   High discharge

Fig. 8.37 The effects of discharge variations on grounding-line fan development. For explanation see text. (From Powell, 1990. Reprinted by permission of the Geological Society of London)

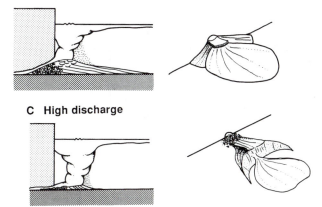

A.   Backflow effects and backslope deposition

B.   Underflows

Fig. 8.38 The effects of glacier fluctuations and high suspended sediment concentration on grounding-line fan development. A: Glacitectonized fan produced by glacier advance during deposition. Left-hand panel: glacier advance deforms the fan top, and horizontal jet flow causes a backflow near the fan apex. Right-hand panel: glacier retreats slightly and jet discharges on to the fan backslope, which fills with coarse sediment. B: Underflows produced when high suspended sediment load falls back on to the fan surface. (From Powell, 1990. Reprinted by permission of the Geological Society of London)

complex sedimentation patterns (Powell, 1990). Prolonged ice margin stability may allow fans to build up to water level, forming ice-contact deltas (Section 8.5.4).

8.5.2 Morainal banks

Morainal banks are elongate masses of sediment formed along glacier gounding lines by a variety of processes similar to those operating at terrestrial ice margins (Sections 7.3 and 7.4). Primary sediment delivery to morainal banks can be from several sources (Fig. 8.39; Powell, 1981a; Powell and Molnia, 1989; Powell and Domack, 1995). (a) Supraglacial debris can slump or fall into the water down the terminal cliff. This process is particularly effective during calving events, when it is known as *calve dumping* (Powell, 1990). (b) Englacial and subglacial debris bands melt out below the water line, releasing debris directly into the water. (c) Debris is dispersed over a wider area by iceberg melt-out and overturn. (d) Unfrozen sediment can emerge from beneath the glacier margin, either as the result of squeezing or as the output from subglacial deforming layers (Andrews, 1963; N.D. Smith, 1990; Benn, 1996a). (e) As for grounding-line fans, considerable quantities of sediment can be delivered by meltstreams. Indeed, deposition on some morainal banks is dominated by subaqueous meltwater discharge, either because subglacial drainage is by sheet flow and emerges simultaneously along large parts of the margin, or because conduit mouths migrate rapidly back and forth along the grounding line (G.W. Smith, 1982; Powell, 1990).

In addition, sediment can be deposited on the proximal side of morainal banks by subglacial processes such as lodgement and/or deposition from

Fig. 8.39 Processes of sediment transport and deposition at a morainal bank. (Adapted from Powell and Domack, 1995)

a subglacial deforming layer (Holdsworth, 1973b; Barnett and Holdsworth, 1974; Benn, 1996a). At oscillating glacier margins, morainal banks can also be built up or modified by ice-push or thrusting (Powell, 1981a; Boulton, 1986).

Sediment accumulating in morainal banks is subject to gravitational reworking, and sediment can be transported away from the margin by debris flows, slumping, rockfall and other processes. Debris flows may be cohesive or cohesionless, depending on the availability of fine-grained matrix in the parent material.

Given the large variety of depositional and deformational processes active at glacier grounding lines, morainal banks are best regarded not as distinct depositional systems, but as part of a continuum ranging from isolated grounding-line fans and meltwater-dominated morainal banks to frontal dump, squeeze and push moraines (Powell and Domack, 1995). The sedimentology, form, structure and genesis of morainal banks are discussed in detail in Section 11.6.2.1.

8.5.3 Deposition below ice shelves

As noted in Section 8.2.1, ice shelves can form only in polar settings, where ice is below the pressure melting point. Thus, meltwater-related processes tend to be of much less importance at ice shelf grounding lines than at temperate grounding-line cliffs. Instead, sediment transport and deposition are dominated by a combination of gravitational

processes and ice-marginal tectonics (Powell and Domack, 1995).

Not surprisingly, modern ice-shelf grounding lines are difficult and expensive to reach, and few direct observations have been made. Powell *et al.* (1996) conducted a pioneering study beneath the floating tongue of Mackay Glacier, Antarctica, using a robotic submersible equipped with a camera. Their observations showed that the most widespread depositional process beneath the floating margin is rain-out of debris released by melting of basal debris-rich ice (Fig. 8.40). This produces a thin and patchy drape of sediment which overlies fluted subglacial till deposited when grounded ice extended further seaward. At the grounding line itself, ridges of rubble up to several metres high are produced by dumping and ice-push, and soft till emerging from beneath the glacier is redeposited in low-angle debris flows.

Apart from these observations, understanding of sedimentary processes beneath ice shelves is based on hypothetical modelling and interpretation of seismic profiles and ancient sedimentary successions. Some of this work is considered in Section 11.6.3.

8.5.4 Deltaic environments

Deltas are sediment masses built out into standing water by fluvially transported debris delivered to the shore (Fig. 8.41). They form by a combination of fluvial *aggradation* above water level, and *progradation* on the delta front, where bedload is rapidly deposited when the incoming stream decelerates on

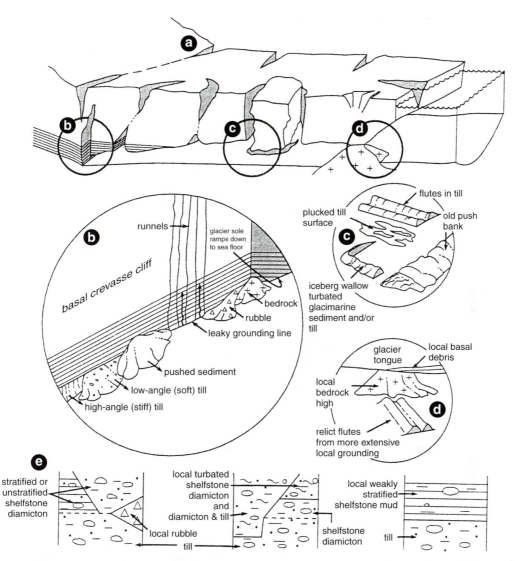

Fig. 8.40 Conceptual model of important sedimentary processes beneath a glacier ice shelf, based on observations made by a robotic submersible beneath the Mackay Glacier Tongue, Antarctica. (a) General setting; (b) grounding line; (c) sea floor beyond grounding line; (d) pinning point; (e) idealized facies sequences. Not to scale. (From Powell *et al.*, 1996. Reproduced by permission of the International Glaciological Society)

contact with standing water. Fine-grained suspended sediment is carried beyond the delta by underflows, interflows and overflows (depending on plume buoyancy), to be deposited in the deeper parts of the lake or marine basin. Delta-front profiles usually display steep upper slopes giving way to easier-angled slopes towards the base of the delta, reflecting proximal-to-distal decreases in sedimentation rates and grain size (Bogen, 1983).

In glacierized catchments, deltas can form in three main settings: (a) *Ice-contact deltas* are fed directly by supraglacial, englacial or subglacial streams, and the ice edge and delta front are separated by a relatively short sandur surface. Ice-contact deltas com-

monly evolve from grounding-line fans as sediment accumulates at the glacier margin. The emergence of a delta at the margin of Riggs Glacier, Glacier Bay, Alaska, is shown in Fig. 8.42. (b) *Glacier-fed deltas* are formed by proglacial rivers and may be some considerable distance from, and at lower altitude than, the glacier margin. (c) *Non-glacial deltas* receive no input from glacier melt, but may still have high sediment input from paraglacial reworking of relict glacigenic deposits (Section 7.6).

Delta surfaces in glacierized catchments are, in effect, sandur plains, and exhibit a similar range of channel and inter-channel forms (Section 3.9). The frontal edge of the delta, or delta shoreline, can take

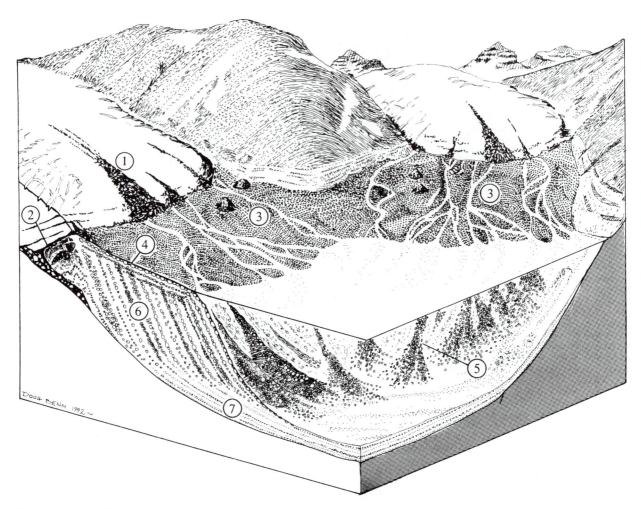

Fig. 8.41 Block diagram of Gilbert-type deltas. 1: Supraglacial debris; 2: subglacial debris; 3: braided sandur surface; 4: topsets; 5: delta front; 6: foresets; 7: bottomsets. (From Benn, 1992a)

Fig. 8.42 Aerial photographs of Riggs Glacier, Glacier Bay, Alaska, showing the growth of a grounding-line fan to a delta. (A) 21 June 1978, when there was 60 m of water at the tidewater cliff; (B) 24 August 1988, when a delta had aggraded from a grounding-line fan and prograded so that its surface lay above high tide. This provides a growth rate of approximately 106 m^3 yr^{-1}. (From Powell, 1990. Reprinted by permission of the Geological Society of London)

a variety of forms depending on the rate of channel mouth progradation, the stability of the distributary channels on the delta surface, and the intensity of wave action (Wright and Coleman, 1973; Wright, 1977; Syvitski *et al.*, 1987). Irregular delta fronts will be produced where channels are stable and prograde rapidly, thus focusing deposition in narrow zones along the delta front. Channel stability is affected mainly by the ratio of bedload to suspended load carried by the stream, and channels will be relatively stable where the amount of bedload is low. A fan-shaped or multi-lobate delta front will be produced by shifting distributary channels. Wave action will tend to smooth out irregularities in delta shorelines, and consistent longshore currents will deflect sediment downcurrent, forming shore-parallel bars.

In areas protected from longshore currents, smooth shorelines tend to be produced in deep basins where gravel bed rivers with unstable channels are prograding, whereas irregular shorelines are associated with stable distributaries carrying fine-grained sediments into shallow basins. Therefore, smooth shorelines tend to characterize ice-contact and proximal glacier-fed deltas, which receive abundant coarse bedload (Gilbert, 1975; Pickrill and Irwin, 1983; Syvitski *et al.*, 1987; England, 1992). Irregular shorelines are more characteristic of glacier-fed deltas in more distal environments (N.D. Smith, 1975; Syvitski *et al.*, 1987).

Once sediment arrives at the delta front, it can be deposited in various ways. (a) Coarse bedload avalanches down the delta front in intermittent cohesionless debris flows, building out prograding *foreset beds* (Nemec, 1990; Section 11.6.5). (b) The suspended load of the river (sand, silt and clay) is transported further out into the water by underflows, interflows or overflows, depending on sediment concentration and the buoyancy of the inflowing water. Deposition by continuous suspension fallout results in marked proximal-to-distal fining, with proximal areas receiving the sands and finer fractions being deposited in more distal positions. Whenever distributaries switch position, sediments deposited by suspension fallout will tend to blanket older coarse-grained sediments below abandoned channel mouths. (c) In glacilacustrine environments, underflows can scour the delta front, redistributing sediment further out towards the basin floor. (d) Actively prograding delta fronts are subject to episodic slope failure because sediment delivery to their upper slopes results in oversteepening. Tension and rotational failure at the delta top produces pull-aparts and slump terraces, whereas compression at the delta base produces pressure ridges and contorted beds (Fulton and Pullen, 1969; Pickrill and Irwin, 1983; Schwab and Lee, 1983, 1988; Schwab *et al.*, 1987;

Postma *et al.*, 1988b). The preservation of contorted beds within these features suggests that they probably fail very slowly (Gilbert, 1975; N.D. Smith, 1978). However, more rapid subaqueous failure can produce debris flows and turbidity currents.

Active sedimentary processes and associated morphological evolution of delta fronts have been studied by Prior and Bornhold (1989, 1990) and Carlson *et al.* (1992), using a combination of side-scan sonar imaging, seismic profiling, coring and grab sampling. This work shows that subaqueous delta surface forms can extend several kilometres from the river mouth (Kostaschuk and McCann, 1983; Prior and Bornhold, 1988, 1990). For example, Bornhold and Prior (1990) reported that the provision of sediment by annual snow melt floods, jökulhlaups and glacier outbursts to the Noeick River delta in British Columbia, Canada, has resulted in the quasi-continuous feeding of underflows to the delta front. This has produced a proximal–distal sequence of delta surface forms characterized by sand and gravel swales, channels, chutes with crescentic flutes, sand splays and bedforms.

8.5.5 Distal environments

8.5.5.1 GLACILACUSTRINE ENVIRONMENTS

Distal glacilacustrine environments tend to have low sedimentation rates, mainly from underflows, interflows and overflows and, in glacier-contact lakes, ice rafting. Underflows are most active during the summer ablation season, when they are generated by flood events in glacial and non-glacial streams and slope failures in the lake basin. Cold, sediment-laden water sinks down the sides of the lake basin, and can travel considerable distances below the lake. Underflows extending 5 to 6 km from the source stream have been reported from Lillooet Lake, British Columbia, by Church and Gilbert (1975). Sediment dispersal is strongly guided by topography, which focuses underflows into low points on the lake floor (Ashley, 1995). Deposition from overflows and interflows tends to be strongly seasonal, governed by lake stratification and the presence of surface ice. Suspension settling is generally most rapid following the breakdown of stratification and lake overturn in the autumn. Settling of the clay fraction is also encouraged during winter, when the formation of surface ice suppresses wind shear and surface mixing. The dominance of underflows in the summer and suspension settling from overflows and interflows in the autumn and winter gives rise to a characteristic annual depositional cycle in many distal glacilacustrine environments. This cycle produces the classic *varve* couplets, consisting of variable but generally silty summer layers and blanket-like clay-rich winter layers. The characteristics of varves and related sediments are discussed in Section 10.6.1.

In general, sedimentation rates and grain size both decline in a downlake direction (Gilbert, 1975; N.D. Smith, 1978; N.D. Smith *et al.*, 1982). This pattern, however, may be overprinted by other factors, such as patterns of circulation in the lake and subaqueous or subaerial slope failures. In glacier-contact lakes, sediment is transported into distal positions by drifting icebergs. Patterns of sediment accumulation depend on the rate of iceberg melting and their drifting velocity. In some lakes, accumulation of ice-rafted debris is greatest in the most distal part, near the lake outflow point, where bergs are carried by surface currents (McManus and Duck, 1988). Lake bathymetry is also important, because bergs will become grounded in water shallower than the critical depth required for flotation, so that sills may act as barriers to the transfer of ice-rafted debris.

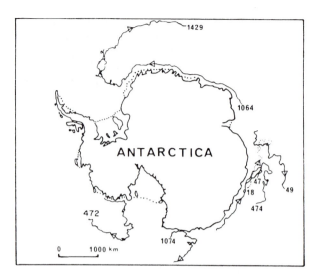

Fig. 8.43 Observed iceberg trajectories around Antarctica. (From Drewry, 1986, after Tchernia and Jeanin, 1983)

8.5.5.2 CONTINENTAL SHELVES

The supply of suspended sediment to present-day continental shelves is highly variable. In the waters around Antarctica, suspended sediment concentrations are low, because of the limited discharge from glacial meltstreams from the predominantly cold-based ice sheet. In contrast, continental shelves adjacent to humid temperate glacierized regions, such as the northern Gulf of Alaska, receive substantial amounts of suspended sediment carried offshore by glacial meltwater (Carlson *et al.*, 1990).

The importance of ice-rafted debris varies considerably, according to the number and debris content of icebergs and their drift tracks. Around Antarctica, icebergs are steered by easterly winds on the inner shelf, then by westerlies as they drift northwards (Fig. 8.43; Tchernia and Jeanin, 1983; Drewry, 1986). Sedimentation rates, however, are low, because of generally low debris content. In the northern hemisphere, most icebergs are trapped in fjords and do not contribute substantially to sediment budgets on many parts of the continental shelves (Hambrey, 1994). There are some exceptions to this, such as Baffin Bay, which receives sediment rafted by icebergs from outlet glaciers in east and west Greenland and small glaciers in the Canadian Arctic. Iceberg rafting was very important in the North Atlantic during several episodes in the last glaciation (Heinrich events), when massive calving from expanded ice sheets transported debris to around 40° N. Analysis of ocean cores shows that sedimentation rates decreased towards the east and south, away from the main source of the icebergs in Hudson Strait (Dowdeswell *et al.*, 1994).

Vast areas of the high-latitude continental shelves are constantly scoured by the keels of drifting icebergs. Icebergs presently scour the ocean bottom in water depths of 235 m in the Arctic and sub-Arctic (Hodgson *et al.*, 1988), and apparently to the remarkable depth of 500 m in Antarctica (Barnes and Lien, 1988). Relict Pleistocene scours are reported at present-day water depths of 700 m (Josenhans and Woodworth-Lynas, 1988), which, even when an arbitrary 100 m is taken off for the lower full-glacial sea-level, document the passage of icebergs of considerable proportions. The impact of iceberg scouring is readily apparent in side-scan sonar images of the sea bed and in aerial photographs of former lake or sea beds (Fig. 8.34).

Reworking by debris flows, turbidity currents and other gravitational processes can be important on glacially influenced shelves, owing to the low strength of rapidly deposited fine-grained sediment on the sea floor (N. Eyles and C.H. Eyles, 1992). Slope instability can be triggered by seismic shocks and stresses imposed by grounding icebergs. Sediment is also redistributed and sorted by deep ocean currents.

Biological productivity is high in polar seas in areas where suspended sediment concentrations are low and sea-ice breaks up during the summer, owing to upwelling of nutrient-rich bottom water (Hambrey, 1994). In a zone several hundred kilometres wide around Antarctica, the principal source of sediment is the skeletons of *diatoms*, which rain to the sea floor, forming thick deposits of *siliceous ooze* (Domack, 1988; Dunbar *et al.*, 1989; Hambrey, 1994). Areas of high diatom productivity tend to be less widespread in the Arctic, largely because of the extent of perennial sea-ice. Other components of polar fauna, such as foraminifera, radiolaria and molluscs, also supply sediment to the sea floor, but rarely account for significant amounts of the total. Such fauna, however, are sensitive indicators of water salinity and temperature, and their fossils in sea floor sediment provide valuable information concerning past environmental change (Lowe and Walker, 1984; Bradley, 1985; Andrews *et al.*, 1996).

PART
2
GLACIATION

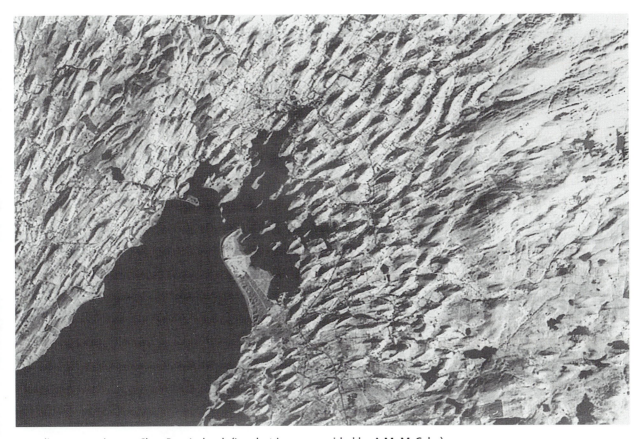

Drumlins converging on Clew Bay, Ireland. (Landsat image provided by A.M. McCabe)

.12 (Plate 1)
at false-colour
osite image of
ökull, Iceland. The
orium line is
ted by the grey to
colour change on
e cap surface. Note
all outlet glaciers of
bsidiary ice dome of
to the south, which
tween the large
glaciers
merkurjökull and
rarjökull.
rarsandur also
out vividly to the
n left of the image.
e produced by
naelingar Islands)

(a)

.13 (Plate 2) The
land ice sheet. Local ice
nd ice domes are shown
en and ice-free areas are
n in grey. (a) Flow lines of
reenland ice sheet,
ding a clear impression of
ajor ice divides; (b) the
e topography of the
land ice sheet, defining
najor domes; (c) the
ated altitude of the
tion limit for Greenland;
e distribution of
oitation in Greenland and
ent Ellesmere Island
$^{-2}$ yr^{-1}). (Provided by
d Williams, USGS)

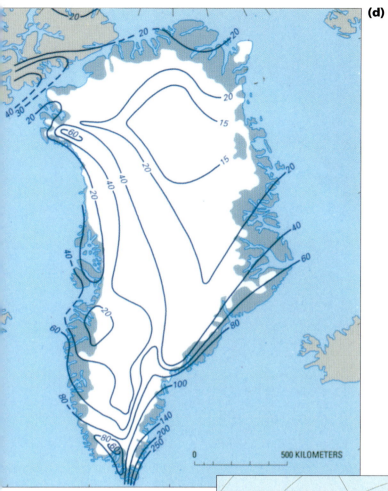

(d)

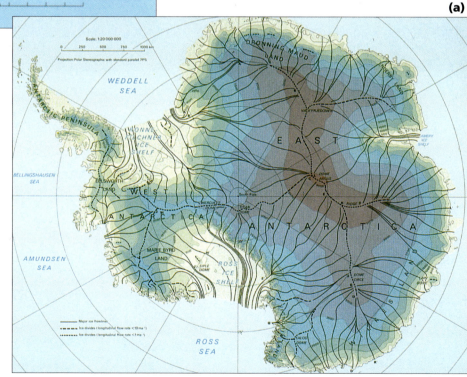

(a)

1.16 (Plate 3) (a) Surface
graphy, ice divides and
ice flow lines of the
ctic ice sheet; (b) section
gh the Antarctic ice sheet
s bedrock topography from
e Entrance, West Antarctica,
vocoresses Bay, East
ctica. (From Drewry, 1983.
nted by permission of the
Polar Research Institute)

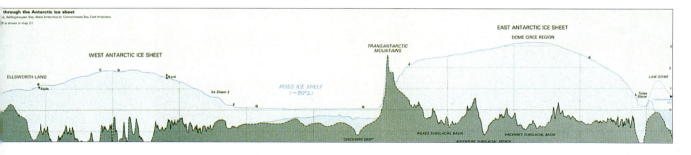

Fig. 1.18 (Plate 4) Salyu 6 photograph of the Southern Patagonian icefield of Chile and Argentina. (From Short and Blair 1986, Geomorphology from Space, NASA)

Fig. 1.23 (Plate 5) The piedmont lobe of the Malaspina Glacier, Alaska showing the spectacular fold structures produced by ice flow into the lowlands. (Photo: Michael Hambrey)

(b)

(d)

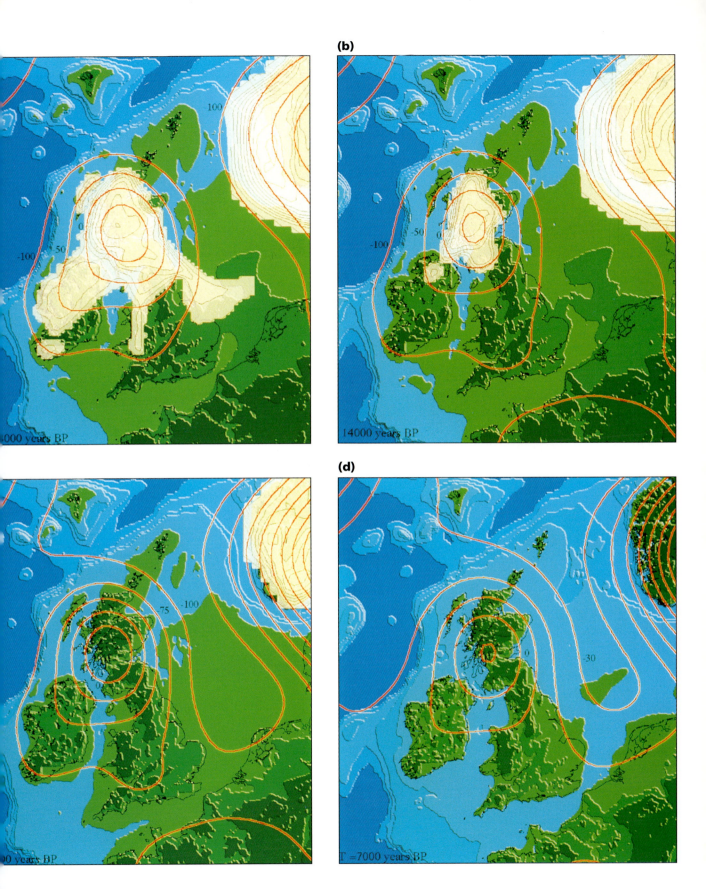

.45 (Plate 6) Theoretical isobases and ice sheet limits for the British Isles and adjacent north-west Europe for time
since the last glacial maximum, according to a model produced by Lambeck (1995). Ages of each time slice are in the
m left corner of each map. Deglaciation starts at 18 kyr BP. Palaeowater depth contours are at 50, 100, 150 and 200 m.
se contours are 50 m for 18 and 14 kyr BP, 25 m for 12 kyr BP, and 10 m for 7 kyr BP. (Maps kindly provided by Kurt
eck)

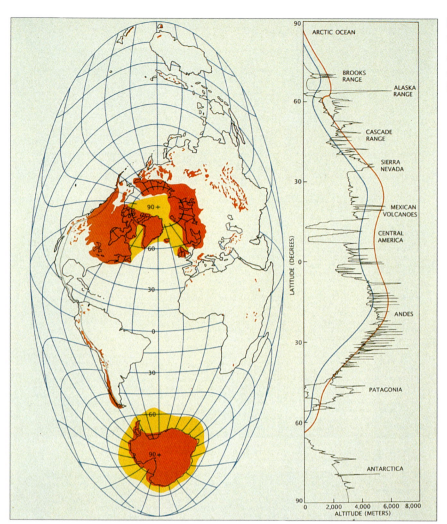

Fig. 1.47 (Plate 7) Topographic profile from north to south through the Americas showing glacier equilibrium-line altitudes for the present (red) and the last glacial maximum (blue) (From Broecker and Denton, 1990a)

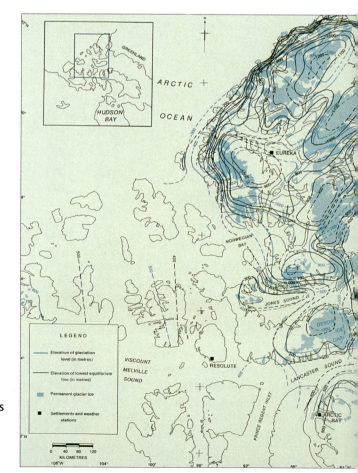

Fig. 1.52 (Plate 8) Glaciation levels and equilibrium-line altitudes in the Queen Elizabeth Islands, Arctic Canada, as reconstructed by Miller *et al.* (1975). (Reproduced by permission of INSTAAR, University of Colorado)

(b)

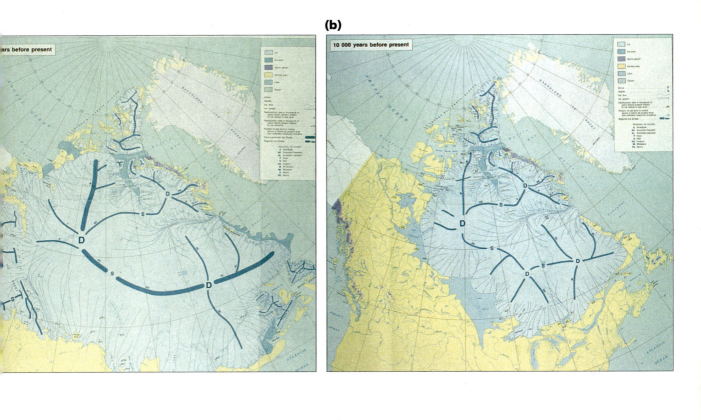

(d)

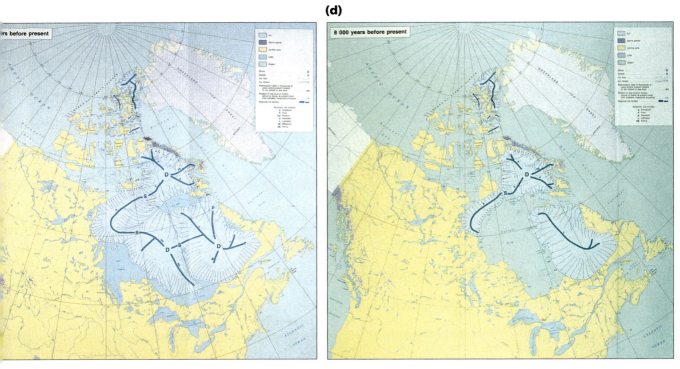

.60 (Plate 9) Palaeogeographic reconstructions of the Laurentide ice sheet at various times during the last glacial (From Dyke and Prest, 1987. Reproduced by permission of the Geological Survey of Canada)

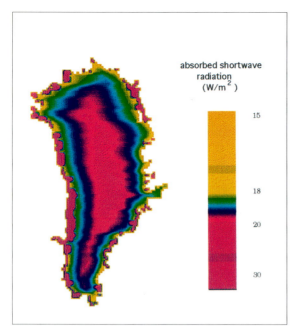

Fig. 2.9 (Plate 10) The absorbed short-wave radiation over the Greenland ice sheet as calculated by an energy balance model. Minimum values occur in the vicinity of the equilibrium-line where albedo is high and cloud cover relatively common. In the interior, clear sky conditions are frequent, counteracting the high albedo of the snow surface. (From van de Wal and Oerlemans, 1994. (Reproduced by permission of Elsevier)

Fig. 2.24 (Plate 11) Landsat false-colour image of the southern central part of Spitsbergen showing the transient snowline on the glaciers (the line where blue ice changes to white snow cover). The image was taken on 18 July 1976 and was provided by Richard Williams, USGS

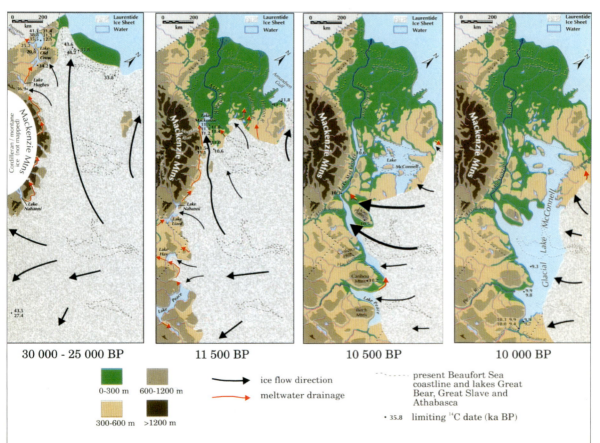

30 000 - 25 000 BP   11 500 BP   10 500 BP   10 000 BP

0-300 m   600-1200 m

300-600 m   >1200 m

→ ice flow direction

→ meltwater drainage

- - - - present Beaufort Sea coastline and lakes Great Bear, Great Slave and Athabasca

• 35.8   limiting ¹⁴C date (ka BP)

Fig. 3.7 (Pla Palaeogeogra maps of the west margin Laurentide ic sheet from th glacial maxim to 10 kyr BP showing the extent of Gla Lake McConr during the ea stages of deglaciation. (From Lemme al., 1994b)

Fig. 3.9 (Plate 13) Historical evolution of proglacial lakes in front of Breidamerkurjökull, Iceland, based on air photographs and ground surveys. (University of Glasgow, based on Howarth and Price, 1969; Price and Howarth, 1970; Price, 1982)

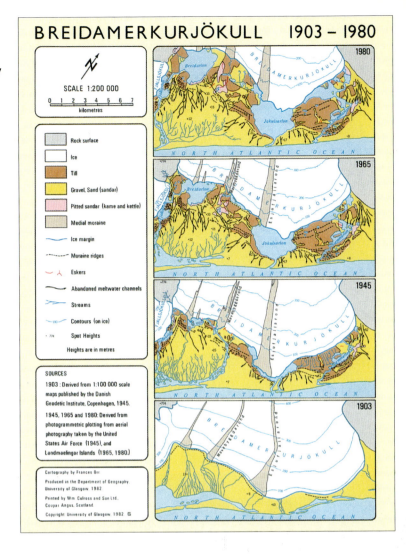

BREIDAMERKURJÖKULL 1903 – 1980

SCALE 1:200 000

0 1 2 3 4 5 6 7
kilometres

	Rock surface
	Ice
	Till
	Gravel, Sand (sandar)
	Pitted sandar (kame and kettle)
	Medial moraine
	Ice margin
	Moraine ridges
	Eskers
	Abandoned meltwater channels
	Streams
	Contours (on ice)
	Spot Heights
	Heights are in metres

SOURCES

1903 : Derived from 1:100 000 scale maps published by the Danish Geodetic Institute, Copenhagen, 1945.

1945, 1965 and 1980: Derived from photogrammetric plotting from aerial photography taken by the United States Air Force (1945), and Landmaelingar Islands (1965, 1980.)

Cartography by Frances Orr
Produced in the Department of Geography, University of Glasgow. 1982
Printed by Wm Culross and Son Ltd. Coupar Angus, Scotland
Copyright University of Glasgow. 1982

5.16 (Plate 14) Glacier ice being injected into oints in bedrock below Oksfjordjökelen, Arctic vay. (Photograph by B.R. Rea)

Fig. 5.19 (Plate 15) Stratified basal ice exposed at the margin of the subpolar Eugenie Glacier, Ellesmere Island. (Photo: D.J.A. Evans)

Fig. 6.39 (Plate 16) Supraglacial lakes on the Tasman Glacier. (Photo: D.J.A. Evans)

Fig. 7.14 (Plate 17) A push moraine being formed at the margin of Fjallsjökull, Iceland. (Photo D.J.A. Evans)

Fig. 8.25 (Plate 18) Turbid overflow upwelling at a glacier ice shelf margin, Leffert Glacier, Ellesmere Island. (Photo: D.J.A. Evans)

9.40 (Plate 19) Troughs
ed by deep erosion by
t glaciers. (a) Yosemite Valley,
rnia.(Photo: Landform
s.) (b) Large lake occupying
erdeepened trough,
devatn, Norway. (Photo:
enn.) (c) Sognefjord,
ay. (Photo: D.J.A. Evans.)
ndsat image of part of the
of west Greenland, with
reenland ice sheet visible
top right. The longest
in this view is Søndre
nfjord. Note the dendritic
rn of the fjords and the
erous lakes indicative of
scour on the highlands.
dsat image 21654-14185
-3-79, Canadian Centre for
te Sensing)

(a)

(b)

(c)

(d)

(e)

Fig. 10.10 (Plate 20) Glacitectonites: (a) penetrative glacitectonite derived from glacilacrustrine sediments, Loch Quoich, Scotland; (b) detail of the laminated diamicton within the penetrative glacitectonite of Loch Quoich; (c) large-scale penetrative glacitectonite in glacilacustrine sediments at Lake Pukaki, New Zealand; (d) vertical sequence from non-penetrative glacitectonite to penetrative glacitectonite and deformation till carapace in glacilacustrine delta deposits, Drymen, Scotland (photos: D.J.A. Evans); (e) banded tills, Slip Inn, Leicestershire (photo: D.I. Benn)

10.25 (Plate 21) Hyperconcentrated flow
~~·~~sits comprising poorly to moderately sorted
~~els~~ and stratified diamictons capping planar and
~~-~~bedded gravels and sands. These sediments were
~~·~~sited during the catastrophic release of ice-
~~·~~ned lake waters in a tributary valley to Dobbin
~~·~~Ellesmere Island. (Photo: D.J.A. Evans)

11.14 (Plate 22) Landsat false-colour satellite
~~·~~e of part of north-western England showing
~~·~~rumlin swarms of the area. Note how the
~~·~~er ice responsible for the streamlining of the
~~·~~in was guided by the major valley systems.
~~·~~ecambe Bay is to the left and the Yorkshire
~~·~~s are clearly picked out to the top right. (Image
~~·~~ded by the British Geological Survey)

Fig. 11.73 (Plate 23) Eastward view across the
outwash plain north of Lake Pukaki, New Zealand,
showing a staircase of lateral moraines and kame
terraces dating back over the last 60 kyr. The
terminal moraine of the Birch Hill advance
(12 kyr–8 kyr BP) is visible in the foreground
(Photo: D.J. A. Evans)

(b)

12.8 (Plate 24) Example of moraine deposition by a subpolar outlet glacier fed by a plateau ice cap and carrying a
~~·~~se debris load, near Dobbin Bay, south-east Ellesmere Island, Arctic Canada: (a) prominent fresh trimline (moraine)
~~·~~rding the retreat of small piedmont lobe; (b) ground photograph of the trimline, showing it to be composed of a thin
~~·~~le veneer with scattered boulders (person is standing at the former ice margin). (Photos: D.J.A. Evans)

Fig. 12.29 (Plate 26) False-colour Landsat image of gla landforms on the west side of the Strait of Magellan, southern Chile. Image processed by D.R. Payne

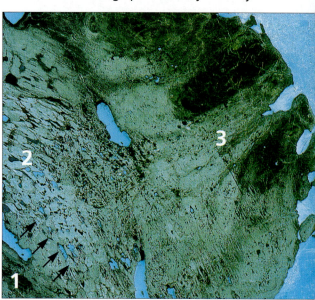

Fig. 12.11 (Plate 25) Landsat image of southern Wollaston Peninsula showing ground moraine (I), hummocky moraine (II), end/lateral and shear moraines (III), and streamlined forms (IV). The Colville moraine is marked C, CI, CII. White arrow in Dolphin and Union Strait indicates north and the figure references are to Sharpe's (1988b) aerial photographs

Fig. 12.41 (Plate 28) Landsat image of south-eastern Prince of Wales Island in the central Canadian Arctic, showing three superimposed drumlin fields (1–3 in order of decreasing age). The lateral shear moraine is marked by arrows. (Image provided by the Geological Survey of Canada and geomorphology by Dyke and Morris, 1988)

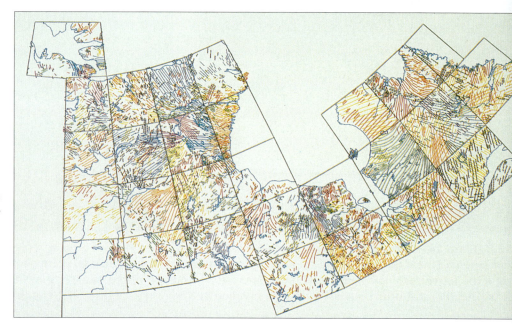

Fig. 12.36 (Plate 27) The pattern of cross-cutting glacier flow lineations observed on Landsat images of the area covered by the centre of the Laurentide ice sheet. Each NTS quadrant has its own colour coding of flow sets, but continuity between NTS quadrants can be traced. (From Boulton and Clark, 1990b. Reproduced by permission of the Royal Society of Edinburgh)

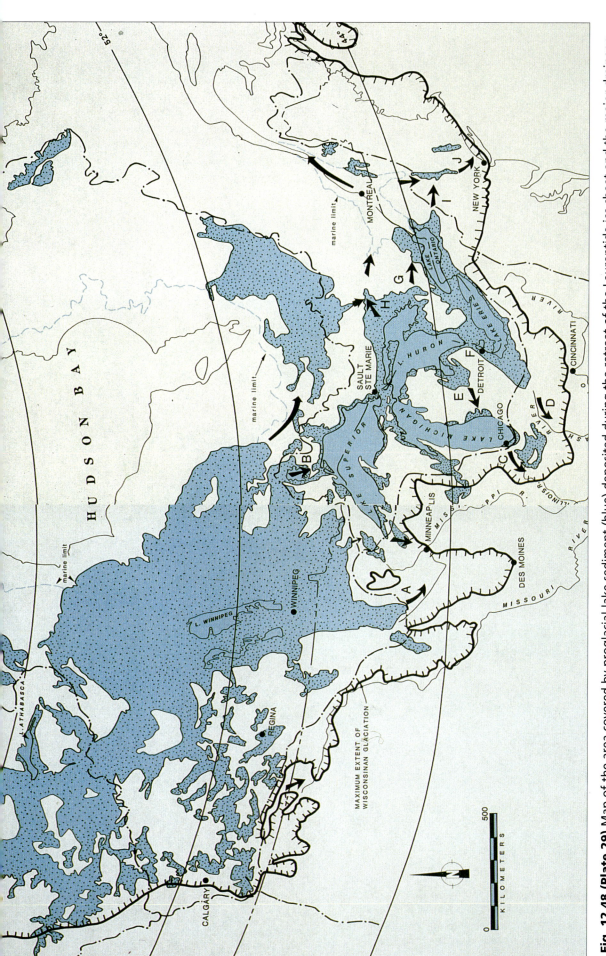

Fig. 12.48 (Plate 29) Map of the area covered by proglacial lake sediment (blue) deposited during the retreat of the Laurentide ice sheet and the major drainage pathways (spillways) draining the lakes into the Mississippi River and Atlantic Ocean (arrows). The pathways are: A = Minnesota River Valley, B = Eastern Agassiz outlets, C = Chicago outlet, D = Wabash River Valley, E = Grand River Valley, F = Port Huron outlet, G = Kirkfield (Fenelon Falls) outlet, H = North Bay outlet, I = Mohawk Valley, J = Hudson Valley. Lake sediments in the St. Lawrence Valley and Hudson Bay are covered by later marine sediments and are therefore not marked. (From Teller, 1987. Reproduced by permission of the Geological Society of America)

Fig. 12.53 (Plate 30) A subhorizontal lateral moraine deposited on Knud Peninsula at the margin of an outlet glacier of the Prince of Wales Icefield, which occupied Hayes Fiord, south-eastern Ellesmere Island, Arctic Canada during the last glaciation. At approximately 120 m above sea-level (sea-ice visible bottom right), this moraine attests to the fact that the glacier was close to flotation at this point. Small lakes were dammed in the side valleys and on the cols on the plateaux between fjords. (Photo: D.J.A. Evans)

Fig. 12.77 (Plate 31) Rock glaciers on Snaefell, Iceland. (Photo: D.J.A. Evans)

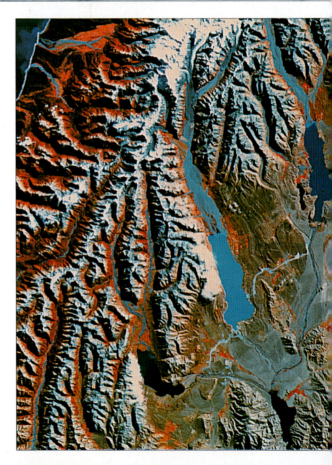

Fig. 12.80 (Plate 32) False-colour satellite image of glaciated valleys in the Mackenzie drainage basin, South Island, New Zealand. Nested latero-terminal moraines are very prominent around the margins of Lake Pukaki, near the centre of the image. (Landcare Research NZ Ltd)

CHAPTER
9

EROSIONAL FORMS AND LANDSCAPES

ROCHE MOUTONNÉE WITH BLOCS PERCHÉS, NEAR PONT-Y-GROMLECH.

9.1 INTRODUCTION

Erosion in glacial environments results in some of the most familiar and characteristic forms and land-forms, such as striated rock surfaces, roches mouton-nées, cirques and troughs, and creates some of the most spectacular landscapes on Earth. The presence of glacier ice in a catchment also strongly influences the evolution of erosional forms beyond the ice margin, such as proglacial meltwater channels. Erosional forms exist on a wide variety of scales, reflecting the operation of glacial processes over different amounts of time. In this chapter, we discuss the characteristics and origin of erosional forms at four scales:

1. *Small-scale forms.* These are superficial erosional marks such as striae and friction cracks, which commonly record single rock failure events.
2. *Intermediate-scale forms.* These include bed-forms, depressions or channels which, although locally impressive, are small compared with the ice flow unit responsible for their formation.
3. *Large-scale forms.* These are erosional forms which are comparable in scale with the associated glacier or ice stream, such as cirques and troughs.
4. *Landscapes of glacial erosion.* At the largest scale, erosional landscapes are distinctive assemblages of landforms that record long-term, regional patterns of glaciation.

Some relationships between these forms, and characteristic dimensions, are shown in Table 9.1. Note that the dimensions of forms at different 'scales' may overlap. The important point to bear in mind, however, is that small-scale erosional marks are commonly superimposed on intermediate-scale bedforms and channels, which in turn are superimposed on large-scale ice-eroded channels or surfaces, which combine to form erosional landscapes, in a spatial hierarchy of forms.

At all scales, it is useful to bear in mind the sets of variables that control or influence processes and patterns of erosion, and the form, size and distribution of erosional forms. Sugden and John (1976) grouped these variables into four categories:

1. *Glaciological variables* encompass the characteristics of the ice, particularly conditions at the bed. They include basal shear and normal stresses, subglacial water pressures and drainage system configuration, flow direction and basal velocity, thermal regime, and the amount of debris held in the basal ice. Basal thermal regime is particularly important, because sliding and significant erosion can occur only where the basal ice is at the pressure melting point (Section 5.3). Cold-based ice tends to protect the underlying substratum, although locally impressive quarrying can occur below cold-based ice margins. Significant debris entrainment is possible in zones of net freezing, but not in zones of net melting.
2. *Substratum characteristics* incorporate the physical characteristics of the bed, and include the structure, lithology, joint distribution and degree of weathering of hard rock beds, and the thickness and composition of unconsolidated sediments. The erosion of hard rock beds is strongly influenced by lithology and the degree of preglacial weathering. Thus, glacial erosion is thought to have been particularly effective in the early stages of the present ice age, owing to the widespread availability of deep-weathered regolith (saprolites) formed during the Tertiary period. After the removal of large areas of this regolith during early glaciations, the erosional capability of later glaciations was reduced because of the occurrence at the ground surface of unweathered bedrock. Another important factor is the permeability of the substratum, which influences the efficiency of drainage at the bed.
3. *Topographic variables* encompass the morphology of the glacier bed at a wide range of scales, from small-scale roughness elements up to the relative relief of the whole glacierized catchment. At the smallest scale, topographic variables influence local patterns of glacier flow and determine the location of stress concentrations at the bed, whereas at the largest scale, relief influences the location of glacier masses, and their morphology, dynamics and efficiency as agents of erosion.
4. *Temporal variables* include the duration of glaciation and changes in any of the above variables over time.

In combination, these variables influence the modification of glacier beds in such a way that they become more efficient pathways for the evacuation of glacier ice or meltwater. This framework focuses attention on the links between process and form, and encourages recognition of equilibrium forms (Sugden and John, 1976).

9.2 SMALL-SCALE EROSIONAL FORMS

A very wide range of small-scale erosional forms has been recognized, and a variety of descriptive terms are used in the literature (e.g. Laverdiere *et al.*, 1979, 1985). In this book, we have adopted the most widely used terminology, and describe small-scale forms under four headings: (a) striae; (b) rat tails; (c) chattermarks, gouges and fractures; and (d) P-forms. The general characteristics of many small-scale forms are depicted in Fig. 9.1.

Table 9.1 Size classification of glacial erosional forms

Process	Relief type	Relief shape	Micro m^{-2} (1 cm)	m^{-1}	m^0 (1 m)	m^1	m^2 (100 m)	m^3	m^4 (10 km)	m^5	m^6 (1000 km)	Macro m^7
Areal ice flow	Eminence	Streamlined				Whaleback	Rock drumlin — Crag and tail		Streamlined-spur		Landscape of areal scouring	→
	Eminence	Part-streamlined				Roche moutonnée – Flyggberg						
	Depression	Streamlined	Striae	P-form			Groove					
	Depression	Part-streamlined					Rock basin					
Linear flow in rock channel	Depression	Streamlined							Trough		Landscape of ice sheet linear erosion	
Interaction of glacial and periglacial	Depression							Alpine trough — Cirque			Valley glacier landscape	
	Eminence						Residual summit or horn				Nunatak landscape	

Source: after Sugden and John (1976)

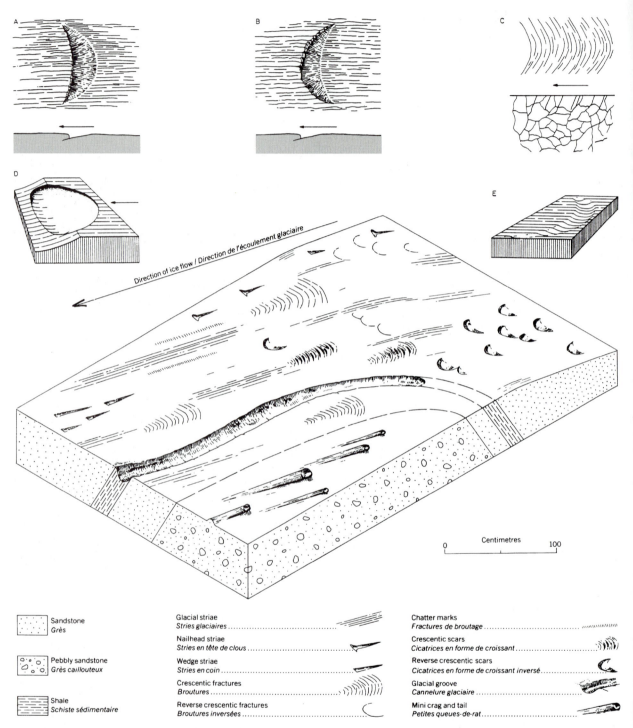

Fig. 9.1 Small-scale forms of glacial erosion. (A) lunate fracture (plan and section); (B) crescentic gouge (plan and section); (C) crescentic fractures (plan and section); (D) conchoidal fracture; (E) sichelwanne. Main diagram: striae, fractures, grooves and associated features. (After Prest, 1983, and Embleton and King, 1975)

9.2.1 Striae

Striae (singular: *striation*) are scratches incised into bedrock or clast surfaces, and have long been recognized as evidence for scoring by particles embedded in glacier ice (e.g. Agassiz, 1838; Forbes, 1843; Chamberlin, 1888; Gilbert, 1903, 1906a). They are direct results of subglacial abrasion and, according to experimental studies and field observations, can be eroded rapidly if optimum conditions are satisfied (Section 5.3.2; Lister *et al.*, 1968; Veyret, 1971; Hope *et al.*, 1972; Boulton and Vivian, 1973; Boulton, 1974; Atkinson, 1984; Drewry, 1986; Iverson, 1990).

The morphology and distribution of striae reveal much about their mode of formation. Close inspection of striae reveals that their edges and bases are rough and composed of numerous arcuate fracture surfaces, attesting to their excavation by indentation fracture beneath asperities in overpassing clasts (Section 5.3.2.1; Fig. 5.9; Lawn and Wilshaw, 1975; Iverson, 1990). Indentation theory suggests that the width and depth of striae depend upon the shape of the asperity, and that they increase with the load pressing the particle against the bed (Drewry, 1986). Striae may widen downglacier, owing to the progressive blunting of the indenting asperity as it is dragged across the bed. The widening may be gradual (*wedge striations*) or abrupt (*nail-head striations*), depending on the rate of asperity blunting relative to the velocity of the abrading clast (Fig. 9.1). Rarely, striae become narrower downglacier. The reasons for this are unclear, although one possibility is that they reflect reductions in the normal force pressing the indenting asperity against the bed. Striae also commonly terminate abruptly at a deep, blunt end next to the narrow up-ice end of another striation. This *en échelon* pattern is thought to relate to the rotation or 'flip-out' of the striating clast, which lifts one asperity clear of the substrate but brings a new, sharper asperity in contact along an adjacent flowline (Edelman, 1951). Although individual striae may maintain a straight line for more than a metre, they sometimes deviate from the mean flow direction if the striator clast has been rotated while still in contact with the substrate.

The orientations of striae may vary considerably on a single rock outcrop. While striae on the flat upper surface of an outcrop are commonly parallel with each other and deviate only slightly from the average ice flow direction, striae on uneven surfaces generally deviate markedly, owing to the irregularities in the basal flow of the glacier (Fig. 9.2; Virkkala, 1960).

Bedrock outcrops often display two or more sets of striae, recording separate ice flow events (Fig. 9.2). The sequence of striating events cannot always be determined, except where shallow striae cut across deeper striae produced by older ice flows. Cross-cutting relationships may result from shifts in ice divides or dispersal centres during a single glaciation, or from separate glacial episodes. Many glacial geomorphologists have employed multiple striae directions in regional reconstructions of former glacier flow. For example, multiple striae were used in conjunction with radiocarbon-dated till stratigraphies by Veillette (1986) to reconstruct former ice

flow directions on the Ontario–Quebec border, Canada. He concluded that an early Wisconsinan west-south-west flow was cross-cut by two late Wisconsinan flows aligned towards the south-south-west and south-south-east, the later striae sets documenting changes in flow direction during the same glacial phase. Similarly well-preserved records of cross-cutting striae in Scandinavia allow shifts in ice

(a)

(b)

(c)

Fig. 9.2

(d)

Fig. 9.2 Striae: (a) cross-cutting striae on moulded bedrock, Chandra Tal, Lahul Himalaya; (b) variable striation directions reflecting complex ice flow around a bedrock bump, St Jonsfjorden, Spitsbergen; (c) multiple striation directions on moulded surface and lee-side cavity face, St Jonsfjorden, Spitsbergen. Note three ice flow directions on the upper surface and transverse ice flow on the lee-side face (photos: D.J.A. Evans); (d) nail-head striae. (Photo provided by C. Laverdiere (Laverdiere *et al.*, 1985))

dispersal centres during the Weichselian and Younger Dryas to be reconstructed (Anundsen, 1990; Kleman, 1990). A complex pattern of multiple ice-flow directions relating to the late Wisconsinan and older glaciations of New Brunswick, Canada, was documented by Rampton *et al.* (1984) in an exhaustive study of till provenance, glacial landforms and striae patterns. They found evidence for several pre-late Wisconsinan and six late Wisconsinan phases of ice flow, during which a complex pattern of striae directions was etched on to the rock outcrops of the province. Similar evidence is employed by Stea (1994) in reconstructing the complex ice-flow history of nearby Nova Scotia.

Two separate sets of cross-cutting striae are documented by Gray and Lowe (1982), Gregory (1986) and Sharp *et al.* (1989a) in the Llyn Llydaw cirque complex, north Wales. The first set records former ice flow from south-west to north-east at the base of the last (Dimlington Stadial) ice sheet, whereas the overlying set were cut by a smaller Loch Lomond Stadial glacier flowing from west to east (Fig. 9.3). The preservation of earlier striae attests to the lower sliding velocity and lower erosional capacity of the Loch Lomond Stadial cirque glacier. Sharp *et al.* (1989a) have used these cross-cutting striae in an attempt to quantify the amount of erosion by the Loch Lomond Stadial glacier. They argued that the almost total absence of striae of < 0.4 mm width in the Dimlington Stadial set suggests that they have been erased by the Loch Lomond Stadial ice, placing

a limiting value on the amount of erosion that occurred during that event.

9.2.2 Rat tails

Rat tails are small residual longitudinal ridges extending down-ice from resistant rock knobs or nodules (Figs 9.1 and 9.4). They are essentially small-scale equivalents of crag and tails (Section 9.3.3), created by the removal of less resistant material to either side. In some examples, elongate scalloped troughs curve round the up-ice side and flanks of the rat tail in an elongate sickle-shaped depression. Most researchers regard rat tails as the product of differential abrasion of bedrock surfaces, and the presence of lateral troughs as evidence for small-scale streaming of ice around the obstacle.

9.2.3 Chattermarks, gouges and fractures

Fracture marks or cracks in bedrock record the removal of rock flakes by subglacial quarrying (Fig. 9.1; Sections 5.3.3 and 5.3.4). These are variably known as chattermarks, crescentic gouges or crescentic fractures (*sichelbrüche* or *parabelrisse*), conchoidal fractures (*muschelbrüche*) and lunate fractures (Laverdiere and Bernard, 1970). *Chattermarks* are usually only a few centimetres wide, and often occur at the base of shallow grooves. Their open or concave sides face down-ice, and thus may provide a sense of former ice flow direction. They commonly occur as a series of closely spaced fractures nested one inside the other, resulting from repeated fracture events beneath a single overpassing clast, possibly in association with 'stick-slip' motion (Figs 9.1 and 9.5; Chamberlin, 1888; Gilbert, 1906a; Lahee, 1912; Ljungner, 1930; Harris, 1943; Okko, 1950; Wintges and Heuberger, 1980).

Larger arcuate gouges and fractures may measure from a few centimetres to more than a metre across. *Crescentic gouges* and *lunate fractures* are similar except that their horns are aligned in opposite directions: the horns of lunate gouges point down-ice, whereas those of crescentic gouges point up-ice (Fig. 9.5; Harris, 1943; Dreimanis, 1953; MacClintock, 1953). Both may be present together in the same rock outcrop. *Crescentic gouges* and *lunate fractures* are bounded by two fracture planes: one that dips in a down-ice direction (principal fracture) and one that is vertical and constitutes the down-ice termination of the principal fracture (Figs 9.1A and 9.1B). The principal fracture plane may be gently or steeply dipping, but it always dips down-ice, providing a useful method of reconstructing former glacier flow direction (Harris, 1943; Dreimanis, 1953). An additional type of gouge, referred to as a *conchoidal frac-*

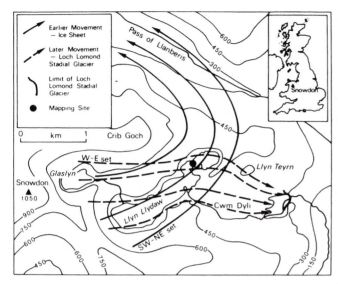

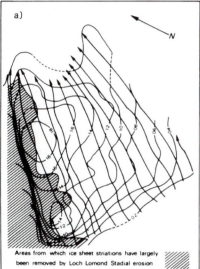

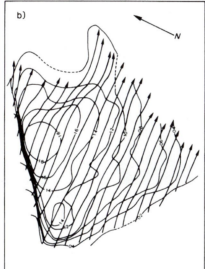

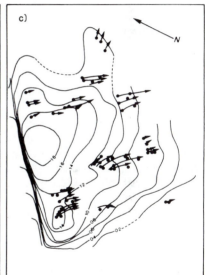

Fig. 9.3 Cross-cutting striae on a bedrock outcrop in the Snowdon massif, Wales. Upper panel: location map. Lower panel: (a) pattern of striae produced by ice sheet flow; (b) pattern of striae produced by a local valley glacier during the Loch Lomond (Younger Dryas) Stadial; (c) flow pattern inferred from striae in lee-side cavities. (From Sharp *et al.*, 1989a. Reproduced by permission of Longman)

ture, was identified by Ljungner (1930). Conchoidal fractures are produced when the fracture plane is concave upward, the final form telling us very little about former ice flow direction.

Crescentic fractures are arcuate fracture planes which may dip either up-ice or down-ice and may have their horns pointing upglacier or downglacier (Harris, 1943; Dreimanis, 1953). Fractures with up-ice-pointing horns commonly occur in isolation, whereas those with down-ice-pointing horns generally occur in a series, the width of individual fractures decreasing in a down-ice direction (Fig. 9.5; Prest, 1983). Series of crescentic fractures therefore provide a clear sense of former ice flow direction.

9.2.4 P-forms

R. Dahl (1965) introduced the term *P-forms* or *plastically moulded forms* to refer to smoothed depres-

Fig. 9.4 Rat tails. Ice flow from left to right (Photo provided by C. Laverdiere (Laverdiere *et al.*, 1985))

(a)

(b)

(c)

Fig. 9.5 Cracks and gouges. (a) Crescentic fractures and striations. Ice flow is from left to right (Photo provided by C. Laverdiere (Laverdiere *et al.*, 1985)). (b) Crescentic gouges, Nigardsbreen, Norway. Ice flow is from right to left. (c) Chattermarks in grooves across striations, St Jonsfjorden, Spitsbergen. Ice flow is from left to right. (Photos: D.J.A. Evans)

sions eroded into bedrock. The name reflects Dahl's conclusion that they are formed by abrasion below plastically deforming ice, although this conclusion has been questioned by Kor *et al.* (1991), who introduced the alternative non-generic term *S-form* or *sculpted form*. In this book, we adopt the original, widely accepted terminology, while recognizing that P-forms might originate by a variety of glacial and fluvial erosional processes.

P-forms exhibit a wide variety of shapes and sizes, and may be classified into three broad types according to whether they are parallel or transverse to ice flow, or non-directional (Fig. 9.6).

1. *Transverse forms* are aligned at right angles to ice flow and include the following types. (a) *Muschelbrüche* (singular: *muschelbruch*) are shaped like mussel shells, with sharp convex-upflow rims and indistinct downflow margins (Ljungner, 1930). Note that muschelbrüche have also been interpreted as scars left by the removal of quarried rock fragments (Section 9.2.3). (b) *Sichelwannen* (singular: *sichelwanne*) are sickle-shaped depressions, often containing striae, with horns pointing downglacier (Fig. 9.7). (c) *Comma forms* are similar to sichelwannen, but with one horn missing or less well developed than the other. (d) *Transverse troughs* are elongate hollows with a relatively steep, planar upflow slope and a gentler downflow slope which is commonly scalloped.
2. *Longitudinal forms* are aligned parallel to ice flow, and include the following. (e) *Spindle flutes* are spindle-shaped depressions with a pointed end oriented upflow and broadening downflow to a distinct, rounded termination (closed spindles) or a smooth, gradual ramp (open spindles). (f) *Cavettos* are curvilinear, undercut channels eroded into steep or vertical rock faces. They may contain striae and crescentic gouges. (g) *Furrows* are elongate, flow-parallel grooves which may be straight, curved or winding in planform (Figs 9.7b and 9.7c). They may have smaller P-forms or striae on their floor and walls.
3. *Non-directional forms* are of two main kinds. (h) *Undulating surfaces* are low-amplitude undulations found on the lee sides of bedrock humps. (i) *Bowls* and *potholes* are near-circular depressions a few centimetres to several metres deep (Fig. 9.7d).

P-forms vary widely in size. For example, H.T.U. Smith (1948) reported grooves in the Mackenzie Valley, north-west Canada, which measure 12 km long, 30 m deep and 100 m wide. However, individual P-forms are usually much smaller than this, although collectively they may cover very large areas (Fig. 9.8).

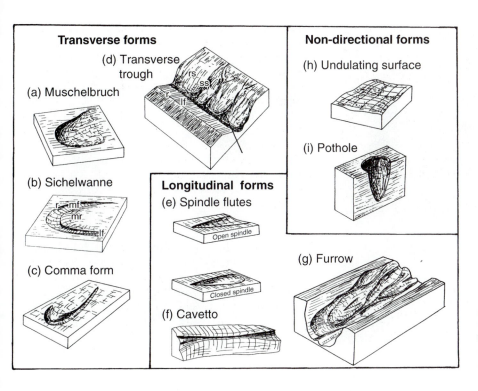

Transverse forms

(d) Transverse trough

(a) Muschelbruch

(b) Sichelwanne

(c) Comma form

Longitudinal forms

(e) Spindle flutes

Open spindle

Closed spindle

(f) Cavetto

(g) Furrow

Non-directional forms

(h) Undulating surface

(i) Pothole

Fig. 9.6 Classification of P-forms. (From Kor *et al.*, 1991. Reproduced by permission of the *Canadian Journal of Earth Sciences*)

Four media have been invoked to explain the erosion of P-forms (Gjessing, 1965; Gray, 1981, 1982b): (a) *debris-rich basal ice*; (b) *saturated till* flowing between the ice base and bedrock; (c) *subglacial meltwater* under high pressure; and (d) *ice–water mixtures*. Gjessing (1965, 1967a) concluded that most forms reflect erosion by saturated till, which he envisaged moving as a viscous liquid driven by pressure differences at the glacier bed (Section 4.4). Subsequent studies have tended to stress the importance of either debris-rich ice or flowing water, media with strikingly different viscosities and flow behaviour.

Several early research papers on P-forms (e.g. Ljungner, 1930; Hjulstrom, 1935; Ebers, 1961; R. Dahl, 1965) concluded that they are the product of

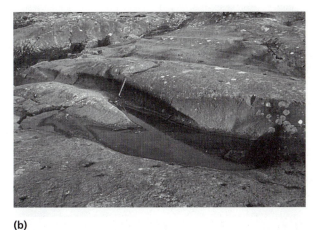

(b)

Fig. 9.7 (a)

(c)

(d)

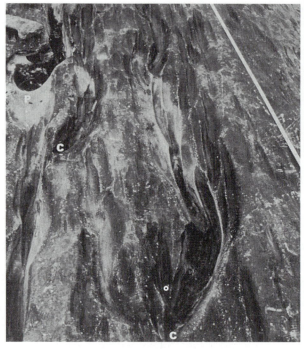

(g)

(e)

(f)

Fig. 9.7 Examples of P-forms. (a) Sichelwanne at Loch na Keal, Isle of Mull, Scotland; (b) curved channel at Loch na Keal (Photos: Landform Slides); (c) groove with chattermarks, Maringouin Cape, Bas Saint-Laurent, Canada (Photo: C. Laverdiere); (d) large pothole, Finland (Jari Väätäinen, from Andersen and Borns, 1994. Reproduced by permission of Scandinavian University Press); (e) large bedrock flutes, with lateral furrows accentuated, in the median strip of Highway 401 at Wilton Creek, Ontario; (f) small-scale flutes from Wilton Creek, Ontario; (g) flutes and scallops and their association with trough terminations at convergences (C) and a pothole (P) at Wilton Creek. (From Shaw, 1988. Reproduced by permission of the *Canadian Journal of Earth Sciences* and Elsevier)

subglacial meltwater erosion, and this was an especially attractive interpretation wherever P-form assemblages include potholes (e.g. Gjessing, 1967a; Holtedahl, 1967). Potholes on former glacier beds are morphologically similar to those scoured out by turbulent subaerial streams, and have detailed features strongly suggestive of fluvial flow. For example, some potholes have rising, spiralling grooves incised on their walls (Brogger and Reusch, 1874; Faegri, 1952; Kor *et al.*, 1991). Early researchers suggested that subglacial potholes marked the position of *plunge pools* at the lower end of moulins carrying meltwater from the glacier surface (e.g. Brogger and Reusch, 1874; Upham, 1900; Gilbert, 1906b; Fuller, 1925; Marr, 1926; Faegri, 1952). However, the impermanence of moulins and the unlikely scenario of water descending from the glacier surface to the bed in one unbroken fall cast doubt

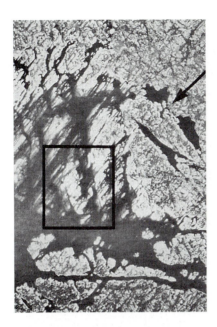

(a)

(b)

Fig. 9.8 Aerial photographs of the Henvey Inlet area, Ontario, showing widespread P-forms. (a) General view showing furrows and intervening rock drumlins. Ice flow was towards the south-west (arrow), cutting across the NW–SE-trending rock structure. Arrow length equivalent to 0.5 km. (b) Detail of boxed area showing a group of sichel-wannen (s), furrows (f) and rock drumlins (d). (From Kor *et al.*, 1991. Ontario Ministry of Natural Resources air photograph 9-4529-36-192)

on this as a general explanation, prompting the development of other ideas. In a review of pothole erosion, Alexander (1932) proposed that subglacial potholes were *eddy-holes* cut in the bed of subglacial conduits, on the basis of the presence of spiral grooves suggestive of vortex flow. Several later studies pro-

posed that potholes were initiated by cavitation erosion where turbulent subglacial or englacial streams struck bedrock at an angle (Section 3.6.2; Hjulstrom, 1935; Streiff-Becker, 1951; Higgins, 1957; R. Dahl, 1965).

Recently, some workers have argued that all P-forms are eroded by subglacial meltwater, on the grounds of their morphological similarity to scours eroded by fluids (Bernard, 1971a, b, 1972; Sollid, 1975; Allen, 1982a; Shaw and Kvill, 1983; Shaw and Sharpe, 1987). For example, Allen (1971a) drew attention to the fact that sichelwannen closely resemble bedforms produced by fluvial erosion of soft bedrock. The possibility that such forms could be eroded by flowing water was demonstrated experimentally by Hjulstrom (1935), who reported the production of sichelwannen in a metal plate by the process of cavitation in turbulent water flowing around a bolt. Shaw (1988c, 1994), Sharpe and Shaw (1989) and Kor *et al.* (1991) argued that P-forms are eroded by vortices in turbulent subglacial meltwater streams which impinge upon the rock bed over short distances (see Bernal and Roshko, 1986). Sichelwannen and comma forms were explained as the products of erosion by horseshoe vortices set up in subglacial meltwater by bedrock obstacles in the boundary layer (Fig. 9.9). The fluvial hypothesis is based almost entirely on a form analogy between P-forms and non-glacial forms scoured by fluid flows, and the similarity between such forms and scours eroded around obstacles by wind and water is indeed very striking. However, the fluvial hypothesis cannot explain the striated surfaces of many P-forms, or the fact that longitudinal P-forms are commonly parallel to adjacent striae over long distances.

The presence of striae in P-forms is often cited as evidence of their erosion by glacier ice (Carney, 1910; H.T.U. Smith, 1948; Goldthwait, 1979), but opponents of the ice erosion hypothesis always allude to the fact that striae occur only on parts of many P-forms and often do not occur at all, suggesting that glacial erosion merely ornamented the P-forms after they were cut by meltwater (Chamberlin, 1888; Ebers, 1961; Sollid, 1975; Shaw, 1988c, 1994). It has also been argued that many P-forms are too tortuous to have been eroded by basal ice. However, Boulton (1974) observed P-forms in contact with debris-rich basal ice beneath Breidamerkur-jökull and Glacier d'Argentière. Boulton (1974, 1979) argued that P-forms are eroded where basal debris concentrations are higher than average, owing to differential ice flow around bedrock protuberances. Ice flows more rapidly around the sides of obstacles than over the tops, concentrating debris-rich ice in the troughs between obstacles, providing conditions favourable for the erosion of cavettos and grooves on the flanks of roches moutonnées (Fig.

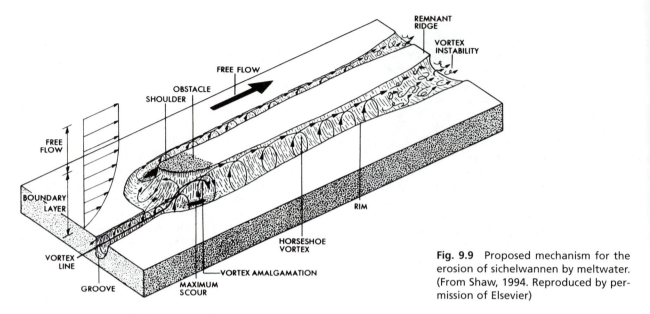

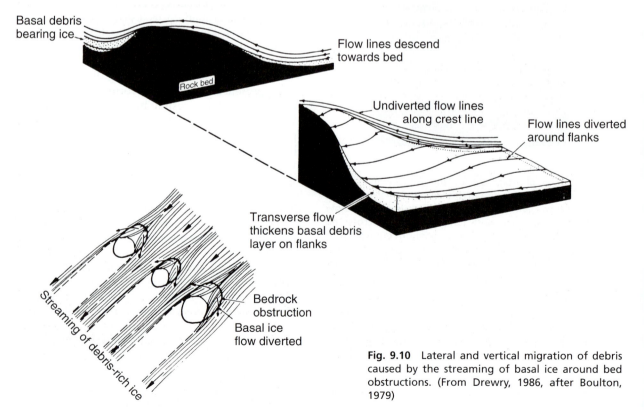

Fig. 9.9 Proposed mechanism for the erosion of sichelwannen by meltwater. (From Shaw, 1994. Reproduced by permission of Elsevier)

9.10). In addition, Boulton (1974) observed that sichelwannen in subglacial positions occur near points of cavity closure, which he considered to reflect enhanced basal ice pressure against the bed and streaming of debris-rich ice around bedrock obstructions. In Boulton's view, it is the enhanced plastic flow of basal ice around obstructions and its separation into debris-rich and debris-poor streams

that is responsible for the differential abrasion of bedrock substrates into most P-forms. He accepts that potholes and other fluvial forms may coexist, but they relate to phases of fluvial erosion at the bed and should not be considered as glacially abraded P-forms merely because they are juxtaposed; potholes within other P-forms are explained by Boulton (1974) as the products of alternating fluvial and

Fig. 9.10 Lateral and vertical migration of debris caused by the streaming of basal ice around bed obstructions. (From Drewry, 1986, after Boulton, 1979)

glacial erosion which may take place on a yearly basis or over longer timescales. Furthermore, where ice streams over the top of a pothole it may undergo pressure fluctuations, and therefore P-forms may emanate from potholes.

If glacier ice is responsible for the erosion of P-forms, then abrasion forms such as striae should occur not just on the walls of individual forms but also in the tightest corners. It is sometimes claimed that ice is unable to stream in tight folds and in different directions over small distances, which represents a major flaw in the ice abrasion hypothesis. However, observations made by Rea and Whalley (1994) in a large cavity beneath an outlet of Oksfjordjøkelen showed that ice can be squeezed through narrow openings and can turn acute corners into cavities, resulting in a wide range of striae alignments over small distances. Furthermore, Demorest (1938) documented striae cut at 90° to the main glacier flow direction, owing to localized ice deformation into transverse trenches. Striae cut vertically into the lee-side faces of rock steps in the proglacial area of Oksfjordjøkelen attest to the abrasion of widening joint systems by small ice streams moving in a direction totally different from that of the main glacier flow (Rea, 1994; Rea and Whalley, 1994). An example of striae conforming to a P-form on a whaleback is shown in Fig. 9.11.

A hybrid origin for at least some P-forms was proposed by Boulton (1974), who grouped them into: (a) glacially abraded forms such as cavettos, sichelwannen and troughs; (b) fluvially eroded forms such as potholes; and (c) intermediate forms, which are less easy to classify but have been affected by fluvial and glacial processes. Boulton's approach accepts the role of both subglacial fluvial erosion and glacial abrasion but assigns different P-forms to those processes. The fact that some of the forms described by Shaw (1988c, 1994) carry a strong fluvial signa-

ture is not in doubt, but those who advocate fluvial erosion exclusively for all P-forms are in danger of rejecting some of the rare and invaluable subglacial observations available to glacial research. Because the P-forms of the Isle of Mull, Scotland, contain fluvial and glacial signatures, Gray (1992) proposed a two-phase evolution regardless of the specific P-form type: fluvial erosion by corrasion and cavitation followed by glacial abrasion. Such interpretations are prompted by the fact that striae commonly only partially cover the surfaces of P-forms, possibly recording partial ornamentation of fluvial forms by glacier ice. However, the distribution of striae in grooves is perhaps better explained by the fact that the glacially eroded grooves acted as drainage routeways during deglaciation and underwent cosmetic changes by limited meltwater erosion.

9.3 INTERMEDIATE-SCALE EROSIONAL FORMS

Intermediate-scale erosional forms reflect the interaction between geology, topography, and patterns of ice and water flow. For this reason, they can yield important insights into former glacial conditions, particularly when morphological studies are combined with modern glaciological theory. In this section, we consider roches moutonnées, whalebacks and rock drumlins, crag and tails, and channels.

9.3.1 Roches moutonnées

Roches moutonnées (singular: *roche moutonnée*) are asymmetric bedrock bumps or hills with abraded up-ice or stoss faces and quarried down-ice lee faces (Fig. 9.12). The name was first introduced by de Saussure (1786), based on a fancied resemblance to

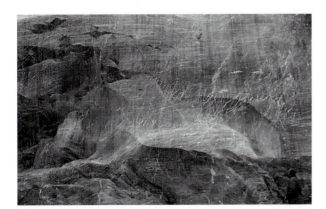

Fig. 9.11 Striae conforming to P-form on the side of a whaleback, St Jonsfjorden, Spitsbergen. Ice flow from right to left. (Photo: D.J.A. Evans)

Fig. 9.12 Roche moutonnée on the Isle of Skye, Scotland. Ice flow from right to left. Note the quarried lee face and the abraded stoss face. (Photo: D.I. Benn)

the wavy wigs of that period, which were called moutonnées after the mutton fat used to hold them in place. Roches moutonnées range in size from less than 1 m to several hundreds of metres across. Some researchers have regarded roches moutonnées as one part of an erosional continuum, ranging from small *asymmetrical rocks* to major *bedrock steps* (Lewis, 1947; Rudberg, 1973; Laverdiere *et al.*, 1985), although the term 'roche moutonnée' usefully describes asymmetric erosional forms at all scales. Examples of large roches moutonnées are the asymmetrical granite hills of New England (up to 1.3 km long and 50 m high; Jahns, 1943) and Deeside, northeast Scotland (up to 150 m high; Sugden *et al.*, 1992; Fig. 9.13). Very large asymmetrical hills, known as *flyggbergs,* occur in parts of Sweden and are up to 3 km long and 350 m high (Rudberg, 1954). Large asymmetric hills may have smaller roches moutonnées superimposed on their surfaces (Glasser and Warren, 1990).

The distribution of small-scale erosional forms on the surfaces of roches moutonnées in part of Finland has been studied by Rastas and Seppälä (1980). Striae are widespread on the stoss sides, except for steep upglacier-facing surfaces, and striae and friction cracks occur together on gently sloping stoss-side surfaces (Fig. 9.14). Polished facets are confined to the flanks and gently sloping surfaces of the lee sides, and plucked surfaces are found on steep downglacier-facing parts.

The large-scale form and surface morphology of roches moutonnées reflect the distribution of stresses in bedrock humps beneath sliding ice, and the associated processes and patterns of rock failure. On the stoss side of humps, normal stresses are higher than average and particles held in basal ice are brought towards the bed, increasing the effectiveness of abrasion (Section 5.3.2; Boulton, 1974; Hallet, 1979a). In contrast, normal pressures on the lee side are lower than average, encouraging the formation of cavities and suppressing abrasion. Instead, the presence of cavities on the lee side of bumps promotes crack propagation and fracture in the bedrock just upstream from the cavity, particularly under conditions of fluctuating water pressures (Sections 5.3.3 and 5.3.4; Iverson, 1991; Sugden *et al.*, 1992; Hallet, 1996). The preferential action of abrasion and quarrying on the stoss and lee sides, respectively, of bedrock bumps leads to the evolution of the classic roche moutonnée form. Study of preglacial joint structures in large roches moutonnées in New England by Jahns (1943) indicated that 33 m of rock had been removed by quarrying from the leesides compared with a maximum of only 4 m on their abraded stoss ends. In this case, therefore, quarrying was a far more effective process of subglacial erosion than abrasion. Recent work suggests that quarrying is generally the

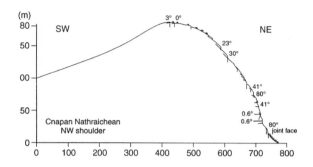

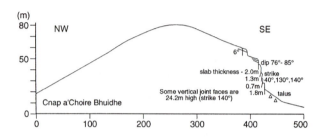

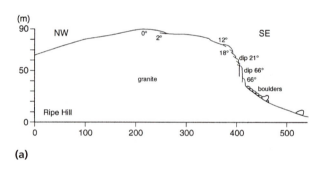

(a)

(b)

Fig. 9.13 Large roches moutonnées on Deeside, Scotland. (a) Measured cross-profiles. (Modified from Sugden *et al.*, 1992) (b) Giant roche moutonnée. Note trees for scale. (Photo: N. Glasser)

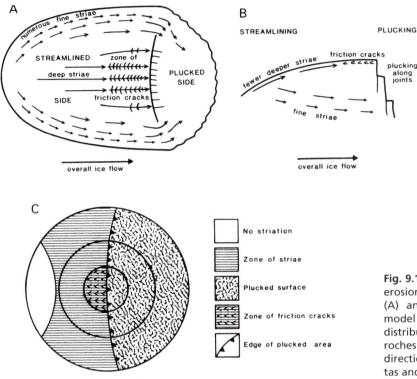

Fig. 9.14 Idealized distribution of small-scale erosional forms on roches moutonnées in plan (A) and profile (B) view. C: Stereographic model (upper-hemisphere projection) of the distribution of small-scale erosional features on roches moutonnées. Arrow indicates ice flow direction. (From Chorley *et al.*, 1984, after Rastas and Seppälä, 1981. Reproduced by permission of Methuen)

more important process (Boulton, 1979; Drewry, 1986; Iverson, 1995), thus explaining why roches moutonnées are such widespread forms in glaciated hard-rock terrain.

The important role of cavities in the quarrying process suggests that roches moutonnées will tend to form where low-pressure cavities exist at the glacier bed. Such cavities are most likely below thin ice where the average ice overburden pressure p_i is low, and in areas where subglacial water pressures undergo large fluctuations due to variations in the supply of meltwater from the glacier surface (Sections 4.5.3 and 5.3.3; Iverson, 1991; Sugden *et al.*, 1992). It may be expected, therefore, that roche moutonnée formation will be encouraged below thin, temperate valley glaciers and near the melting margins of ice sheets. Empirical support for this idea has been provided by Sugden *et al.* (1992), who noted that boulder trains originating from the leesides of roches moutonnées in north-east Scotland extend only a few hundred metres downflow. Such short transport distances indicate that the trains relate to a pulse of erosion at the end of the last glacial cycle, when ice was thin and meltwater was abundant.

The morphology of roches moutonnées is also strongly influenced by bedrock structure (Matthes, 1930; Demorest, 1937; Zumberge, 1955; Gordon, 1981; Rastas and Seppälä, 1981; Laitakari and Aro, 1985; Glasser and Warren, 1990; Sugden *et al.*,

1992). Quarrying and plucking are encouraged by the presence of favourably oriented joint systems and other pre-existing fracture surfaces (Fig. 9.15). In some situations, limited plucking can even occur on stoss-side surfaces. The influence of jointing on the evolution of a granite roche moutonnée was reconstructed by Sugden *et al.* (1992) and is shown in Fig. 9.16. The initial stages of quarrying exploit sheet joints developed parallel to the original ground surface, yielding arcuate slabs. As erosion progresses, deeper vertical and horizontal tectonic joints are exposed, leading to the creation of a stepped lee-side profile.

Some researchers have suggested that roches moutonnées are essentially remnants of preglacial weathering which have been only slightly modified by glacial erosion (Leiviska, 1907; Davis, 1909; Sahlstrom, 1914; Demorest, 1939; Mattsson, 1960, 1962; Lindmar-Bergstrom, 1988; Lindstrom, 1988). According to this view, roches moutonnées are immature glacial erosional forms which would be completely removed by lee-side cliff retreat under prolonged glaciation. Lindstrom (1988) regarded this 'weathering hypothesis' as an alternative to the 'classical glacial theory', but the two models are in fact complementary. Some roches moutonnées do appear to be slightly modified preglacial hills (Jahns, 1943; Rudberg, 1954; Sugden and John, 1976; Lindstrom, 1988), but in many areas, such as the floors of cirques

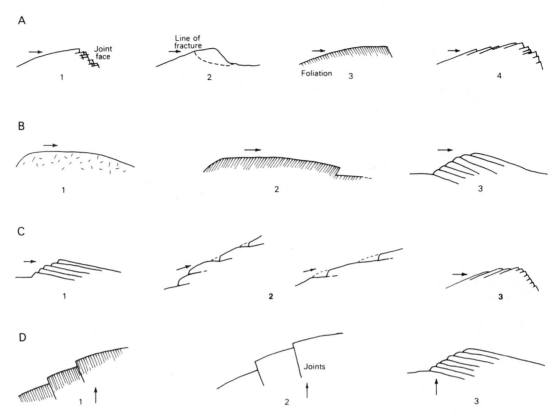

Fig. 9.15 Schematic relationships between bedrock structure and glacial erosional forms. (A) Quarried lee slopes associated with 1: jointing; 2: massive rocks; 3: steep down-ice dipping foliation; 4: up-ice dipping joints. (B) Patterns of abrasion associated with 1: massive rocks; 2: steep up-ice-dipping foliation; 3: down-ice-dipping joints. (C) Quarried stoss slopes associated with 1: down-ice-dipping joints; 2: up-ice-dipping joints with layers truncated on up-ice sides; 3: up-ice-dipping joints with layers truncated on down-ice sides. (D) Quarried lateral slopes (transverse cross-sections) associated with three structural patterns. (From Gordon, 1981. Reproduced by permission of Scandinavian University Press)

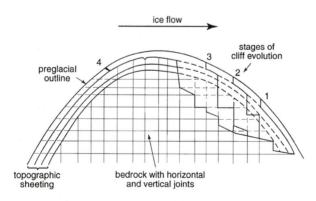

Fig. 9.16 Model of evolutionary stages in the formation of roches moutonnées based upon examples in eastern Scotland: stage 1 involves the removal of lee-side surface slabs and the exploitation of topographic sheeting; stage 2 involves the up-ice migration of vertical faces with a limitation on the depth of crack propagation; stage 3 involves continued back-wearing and successive cycles of deepening; stage 4 involves the production of the typical roche moutonnée profile with a staircase cliff and an abraded upper surface. (Modified from Sugden et al., 1992)

and troughs, they are clearly entirely due to differential glacial erosion. Between these end-members there is likely to be a continuum of forms with varying degrees of inherited topography.

9.3.2 Whalebacks and rock drumlins

Whalebacks and *rock drumlins* are elongate, smoothed bedrock bumps which lack the quarried lee faces characteristic of roches moutonnées (Linton, 1963; Sugden and John, 1976; I.S. Evans, 1996). Whalebacks are approximately symmetrical, looking rather like the backs of whales breaking the ocean surface (Fig. 9.17), whereas rock drumlins (also known as *tadpole rocks*; Dionne, 1987) are asymmetrical, with steeper stoss faces and gently tapering lee sides. Both are commonly ornamented with abundant striae, friction cracks and P-forms. Whalebacks up to 1 km long have been described by I.S. Evans (1996) from the Coast Mountains of British Columbia, Canada.

The absence of quarried lee faces on whalebacks and rock drumlins is thought to imply that low-pres-

(a)

(b)

Fig. 9.17 Erosional bedforms on the foreland of Nigards-breen, an outlet glacier of the ice cap Jostedalsbre, Norway. (a) Whaleback (Photo: D.I. Benn); (b) rock drumlin. (Photo: D.J.A. Evans)

sure cavities did not exist at the glacier bed during their formation (I.S. Evans, 1996). Cavity formation is suppressed below thick ice where average ice overburden pressures are high, and in such situations abrasion may take place over most of the bed, creating smoothed, symmetrical whalebacks. According to Evans, whalebacks can form below ice a few hundred metres thick, but are best developed where ice was 1–2 km thick. Asymmetrical rock drumlins are thought to occur where abrasion is focused on the stoss side, and both abrasion and plucking are suppressed on the lee side. Such conditions may exist where either: (a) normal stresses over the lee side of bumps are low enough to inhibit abrasion, but too high to allow cavity formation; or (b) cavities form, but are not subject to the fluctuations in water pressure that encourage quarrying. The latter condition is most likely to occur where surface meltwater is prevented from reaching the bed, such as beneath ice streams in polar areas which have surface layers of cold ice (R. le B. Hooke, cited in I.S. Evans, 1996).

Whalebacks and rock drumlins are widespread on the floors of troughs in the Coast Mountains of British Columbia, which channelled ice streams draining the Cordilleran ice sheet during the last glaciation (I.S. Evans, 1996). Such forms may be characteristic of erosion beneath thick, fast-flowing ice streams and outlet glaciers.

Several other ideas have been proposed to explain the formation of whalebacks and rock drumlins. These include: (a) the survival of preglacial bedrock hills (Lindstrom, 1988); (b) the remodelling of roches moutonnées and the removal of quarried faces due to changing ice flow directions (Veillette, 1986; Anundsen, 1990); and (c) bedrock structure which is unfavourable to the development of plucked lee faces (Gordon, 1981). Subglacial fluvial erosion was also considered by Kor *et al.* (1991), because rock drumlins are often ornamented with what they interpret as fluvially cut P-forms. The occurrence of P-forms, however, may simply reflect efficient abrasion due to high overburden pressures, high sliding velocities, and large areas of intimate ice–bed contact beneath thick ice (Sections 5.3.2 and 9.2.4).

9.3.3 Crag and tails

Erosional crag and tails are elongate, streamlined hills consisting of a resistant bedrock crag at the up-ice end, and a tapering tail of less resistant rock extending down-ice. They are produced by the streaming of ice around the obstacle, and the protection of the 'tail' from erosion. The classic example of a crag and tail is Edinburgh Castle and the Royal Mile (Fig. 9.18; Sissons, 1967, 1971; D.J.A. Evans and Hansom, 1996). In this case, the 'crag' is a volcanic plug of hard, resistant basalt and the 'tail' is composed of less resistant ash and sedimentary strata. In this part of Scotland, soft strata have been preferentially removed over large areas, leaving the more resistant volcanic plugs and their associated residual tails standing prominently on the urban skyline. The great length of the Royal Mile 'tail' indicates a long 'pressure shadow' in the lee of the crag, possibly reflecting high sliding velocities below this part of the last ice sheet. Erosional tails may also be cut in pre-existing Quaternary sediments. However, good exposures are necessary to differentiate such features from large flutings and cavity fills deposited in the lee of basal obstructions (Sections 11.2.3 and 11.2.7).

9.3.4 Channels

Channel forms, cut by glacial meltwater, are a very distinctive feature of many glacial landscapes. Indeed, in polar environments, where glacier ice carries a relatively small debris load and is commonly frozen

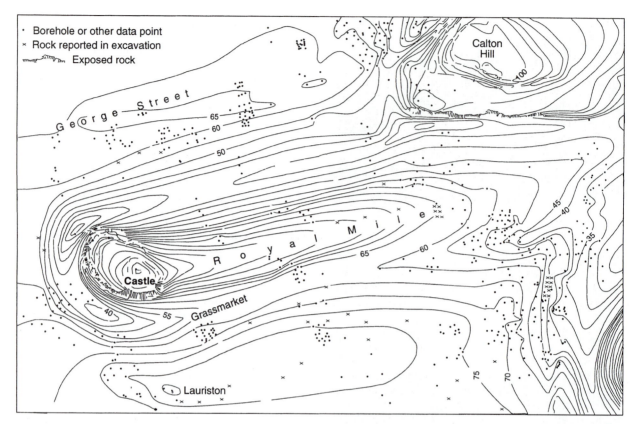

Fig. 9.18 Crag and tail: map of bedrock surface around Edinburgh Castle and the Royal Mile. (From D.J.A. Evans and Hansom, 1996, modified from Sissons, 1971)

to its bed, the erosional impact of meltwater is often the only imprint left by past glaciations. In this section, we describe channels formed in subglacial, ice-marginal and proglacial environments, and discuss their formation in terms of the theory of meltwater flow and the erosional processes detailed in Chapter 3.

9.3.4.1 NYE CHANNELS

Nye channels are erosional features cut in bedrock and consolidated sediments by subglacial drainage. Most extend for a few tens to a few thousands of metres, and are up to a few tens of metres wide. (The very largest features, however, form major valley systems up to 100 km in length known as *tunnel valleys*. The origin of these large, complex forms is discussed in Section 9.3.4.2.) Channels can occur as isolated features or as part of complex, many-branched systems extending over large areas. Such systems can take the form of *dendritic networks*, similar to subaerial channel patterns, recording efficient subglacial drainage along discrete conduits (Section 3.4.2), or *anastomosing systems* in which multiple channels split and rejoin. Anastomosing systems may represent former distributed drainage in linked-cavity networks (Section 3.4.4), but may

also be time-transgressive patterns resulting from channel migration and switching. Thus, careful work is needed in the interpretation of former subglacial channel networks to determine whether all parts of the system were occupied contemporaneously (e.g. Sissons, 1963; Booth and Hallet, 1993).

Water-filled subglacial channels follow paths dictated by the hydraulic gradient, which is a function of the ice surface slope, subglacial topography and glacier sliding velocity (Section 3.4.2). For equilibrium conditions, the hydraulic gradient is primarily controlled by the ice surface slope, so that meltwater channels will generally tend to parallel glacier flowlines. For this reason, subglacial meltwater channels are commonly known as *ice-directed channels*. The characteristic undulatory long profiles of such channels as they cross topographic barriers is one of the clearest criteria for distinguishing subglacial from subaerial channels (Mannerfelt, 1945; Sissons, 1958a, 1960a, b, 1961a, 1963; Derbyshire, 1962; Bowen and Gregory, 1965).

Many of the characteristics of ice-directed channels can be illustrated by superb examples cut on the northern slope of the Cairngorm Mountains, Scotland (Fig. 9.19). The channels trend obliquely across the hillside, guided by the former ice surface, which

Fig. 9.19 Large meltwater channel cut into bedrock, Cairngorm, Scotland. (Photo: C.M. Clapperton)

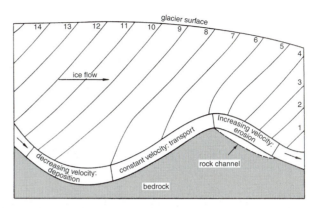

Fig. 9.20 Schematic diagram showing the erosion of subglacial meltwater channels across a divide. Equipotential contours are numbered. (From Sugden and John, 1976)

in this case declined towards the north-east (Sugden and John, 1976; Gordon, 1993). Most channels are discontinuous, and channels commonly begin at the crests of transverse drainage divides or spurs, a pattern that reflects variations in the behaviour of water flowing over a wavy bed (Shreve, 1972; Sugden and John, 1976; Booth and Hallet, 1993). Figure 9.20 shows how the spacing of equipotential contours varies over bedrock bumps, and influences patterns of meltwater erosion and deposition. Equipotential lines are most widely spaced on the upstream sides of bumps, and most closely spaced in their lee. Consequently, the erosive power of subglacial meltwater, which is a function of discharge and hydraulic gradient, will be greatest on the lee side of obstructions on the bed, encouraging channel incision. Conversely, limited erosion or even deposition will occur on the upstream side of obstructions, where the hydraulic gradient is low, explaining the alternation between channels and eskers along some drainage pathways (e.g. Sissons, 1961b, 1963).

At a local scale, the paths taken by subglacial channels are influenced by the bed topography as well as the ice surface slope. In particular, channels will tend to avoid high points on the bed, which are areas of high hydraulic potential, and commonly cross topographic barriers at low points or cols (Mannerfelt, 1945, 1960; Sissons, 1958a; Gjessing, 1960; Derbyshire, 1961, 1962; Embleton, 1964a; Booth and Hallet, 1993). The influence of both ice thickness and basal topography on subglacial drainage has been examined by Booth and Hallet (1993), who compared channel networks cut below the Lateglacial Puget Lobe of the Cordilleran ice sheet with theoretical patterns predicted by Shreve's (1972) theory (Section 3.2.3). Subglacial channels are particularly well developed in this area, and are typically 10–100 m deep, 50–150 m wide, and up to 8 km long. Furthermore, the former ice surface can be reconstructed with considerable confidence

because of the abundant iceflow indicators and well-constrained ice limits, allowing basal equipotential contours to be drawn. There is generally remarkably good agreement between observed and predicted patterns, with many of the major channel systems coinciding with or closely similar to reconstructed flow paths (Fig. 9.21). Some degree of uncertainty is inevitable in such reconstructions because: (a) the present distribution of channels may not represent the former drainage system at a single point in time, but is probably time-transgressive; and (b) the true basal topography is commonly partially obscured by later deposits. The success of the reconstruction despite these problems strongly supports the idea that drainage below the Lateglacial Puget lobe was through a dendritic system of water-filled Nye channels directed by ice pressure and local topography. Such reconstructions, therefore, add considerably to our understanding of past glacier dynamics.

Below valley glaciers, high relative relief exerts an important control on the location of subglacial channels, which commonly follow valley axes. Focused meltwater erosion can produce deep, narrow *slot gorges*, such as the 2 m wide, 30 m deep gorge at Berekvam, Norway, described by Holtedahl (1967), or the 10–20 m wide, 60 m deep Corrieshalloch Gorge in Northern Scotland. However, it is often difficult to determine to what extent pre-existing fluvial gorges were influential in guiding subglacial meltwater, and how effective deglacial and postglacial subaerial streams have been in deepening such forms (e.g. de Martonne, 1957; Gjessing, 1965, 1966).

Subglacial meltwater flow is also strongly influenced by local topography if channels are only partially filled with water for much of the time (Section 3.2.3.4). In this case, water pressure is atmospheric, and hydraulic potential is governed solely by elevation. Such conditions are most likely

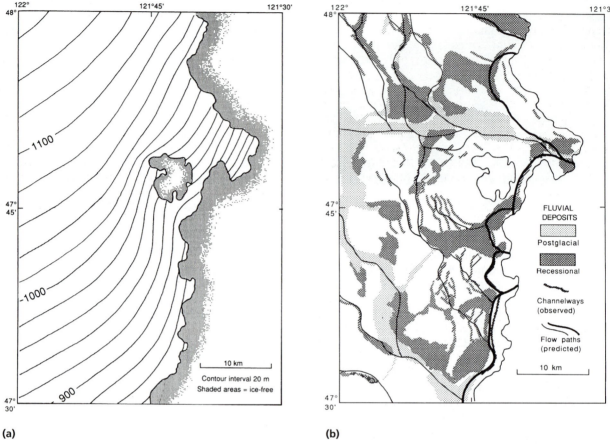

Fig. 9.21 Subglacial drainage of the eastern part of the late Pleistocene Puget lobe of the Cordilleran ice sheet. (a) Reconstructed ice surface; (b) comparison of observed channelways and flow paths predicted from a numerical model. (From Booth and Hallet, 1993. Reproduced by permission of the Geological Society of America)

close to glacier margins where ice is thin and tunnel creep-closure rates are low, and where most meltwater is derived from the surface and is thus subject to large diurnal fluctuations. Fossil channel systems may exhibit increasing conformity with local topography with decreasing elevation, with predominantly ice-directed channels on the upper slopes giving way to downslope-oriented channels near the valley floors. Such networks are commonly time-transgressive systems consisting of higher, older elements controlled by the ice surface slope and lower, more recent slope-directed elements cut below thinning ice in the later stages of deglaciation (e.g. Sissons, 1958b, 1961b; Clapperton, 1971b; Clapperton and Sugden, 1972; Young, 1978; Gordon, 1993).

Anastomosing systems of Nye channels record several generations of channel incision or former linked-cavity drainage networks (Section 3.4.4). Very clear examples of former linked-cavity systems have been exposed on limestone bedrock by recent retreat of the Glacier de Tsanfleuron, Switzerland (Sharp *et al.*, 1989b), Castleguard Glacier, Alberta,

Canada (Hallet and Anderson, 1980), and Blackfoot Glacier, Montana, USA (Walder and Hallet, 1979). Such systems can be subdivided into five geomorphic units, each associated with a different set of subglacial processes: (a) areas of intimate ice–bedrock contact; (b) Nye channels incised into bedrock; (c) lee-side cavities on the down-ice side of bedrock obstacles; (d) surface depressions filled with calcite precipitates; and (e) karst sinkholes (Fig. 9.22). *Areas of intimate ice–bedrock contact* are striated, but also display solutional hollows and calcite precipitates formed during regelation sliding (Section 4.5.2; Hallet, 1976a; Sharp *et al.*, 1989b). The *Nye channels* are aligned largely parallel to former glacier flow, and form an anastomosing rather than an arborescent drainage pattern. Many channels have blind terminations, although it is possible to detect evidence of limited water flow from one channel to another or from channels into cavities. Therefore, the Nye channels act as links between lee-side cavities, and together the channels and cavities act as interconnected drainage routeways. Three types of *lee-side cavity* were recognized by

Sharp *et al.* (1989b): (a) large cavities aligned oblique to the former ice flow direction; (b) elongate channels aligned parallel to former ice flow and continuous with Nye channels; and (c) small cavities. The large cavities were apparently fully integrated into the subglacial drainage system, whereas the small cavities were poorly connected. *Surface depressions* or hollows that are almost totally filled by precipitates are located at the heads and margins of Nye channels. They are differentiated from other channels and cavities by their shallowness and lack of elongation, and are thought to document cavity closure and meltwater freezing during periods of low discharge. *Karst sinkholes* occur in association with Nye channels and elongate cavities, indicating

a connection between the subglacial and subterranean drainage systems. Sinkholes occur at both the upstream and downstream ends of Nye channels, suggesting that they feed meltwater into as well as receiving meltwater from the channels; a sinkhole may provide meltwater during periods when the subterranean system is full of water or is blocked by glacier ice.

Very impressive fossil anastomosing and dendritic channel networks have been described by Sugden *et al.* (1991) and Denton *et al.* (1993) from the Upper Wright Valley, Antarctica (Fig. 9.23). Individual channels are up to 50 m deep, and in several places are punctuated by huge potholes tens of metres deep and across. Sugden *et al.* (1991) concluded that the

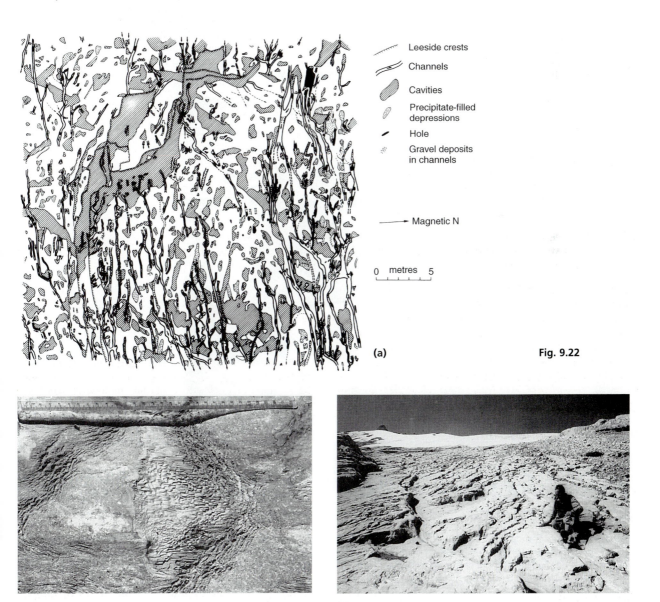

Leeside crests

Channels

Cavities

Precipitate-filled depressions

Hole

Gravel deposits in channels

Magnetic N

0 metres 5

(a)

(b)

(c)

Fig. 9.22

(d)   **(e)**

Fig. 9.22 Former linked-cavity drainage system exposed on the foreland of Glacier de Tsanfleuron, Switzerland. (a) Map of part of the proglacial area, showing the distribution of erosional features. White areas represent areas of former intimate glacier–bed contact. (From Sharp *et al.*, 1989b. Reproduced by permission of John Wiley & Sons) (b) Detail of an area of intimate glacier–bed contact, showing striated bedrock, solution furrows and subglacially precipitated calcite. Ice flow was from left to right. (c) Network of anastomosing Nye channels. (d) Former lee-side cavity, showing water-eroded surfaces. (e) Precipitate-filled depression at the margin of a Nye channel. (Photos: © M.J. Sharp)

Fig. 9.23 The Labyrinth, a spectacular series of subglacial meltwater channels exposed in front of the Upper Wright Glacier, Dry Valleys, Antarctica. The channels stand at the head of the trough of Wright Valley, and were cut when the glacier was more extensive and wet-based. (US Navy Photograph TMA 2448, 244. Photo kindly supplied by D.E. Sugden)

channels were cut below expanded tongues of the East Antarctic ice sheet, probably by the catastrophic release of water stored in subglacial or surficial lakes.

9.3.4.2 TUNNEL VALLEYS

Tunnel valleys (also known as *rinnentaler* or *tunneldale*; Grube, 1979, 1983; Hinsch, 1979; Kuster and

Meyer, 1979) are large, overdeepened channels cut into bedrock or sediment, which can reach > 100 km long and 4 km wide (Figs 9.24 and 9.25; Ó Cofaigh, 1996). They can occur in isolation or as parts of dendritic or anastomosing patterns extending over very large areas, and have been recognized in many areas formerly covered by Pleistocene ice sheets, including North America, north Germany, Denmark, Poland, the floor of the North Sea, Britain and Ireland (Ehlers and Linke, 1989; N. Eyles and McCabe, 1989b; Wingfield, 1989, 1990; Brennand and Shaw, 1994). They share many characteristics with Nye channels, including undulatory bed-long profiles, overdeepened basins along their floors, and hanging tributary valleys (e.g. Bowen and Gregory, 1965; Linke, 1983; Ehlers and Linke, 1989; Ó Cofaigh, 1996). Individual tunnel valleys usually have wide, relatively flat bottoms and steep sides,

and the numerous troughs that occur along their lengths may be occupied by lakes (*rinnenseen*; Woldstedt, 1926, 1954). These characteristics provide clear evidence that they are excavated by subglacial meltwater flowing under hydrostatic pressure, as was recognized by early researchers (e.g. Ussing, 1903, 1907; Werth, 1907; Madsen, 1921; Koch, 1924). Further evidence for a subglacial origin is the tendency for tunnel valleys to terminate abruptly at major moraines, where they may grade into large subaerial ice-contact fans (e.g. Milthers, 1948; Hansen and Nielsen, 1960; Patterson, 1994). The surfaces of these fans may lie up to 100 m above the tunnel valley bottom, reflecting deposition from pressurized meltwater emerging from beneath the ice.

Tunnel valleys may be completely infilled by thick sedimentary successions, including glacigenic, glacifluvial, glacilacustrine, glacimarine and non-glacial

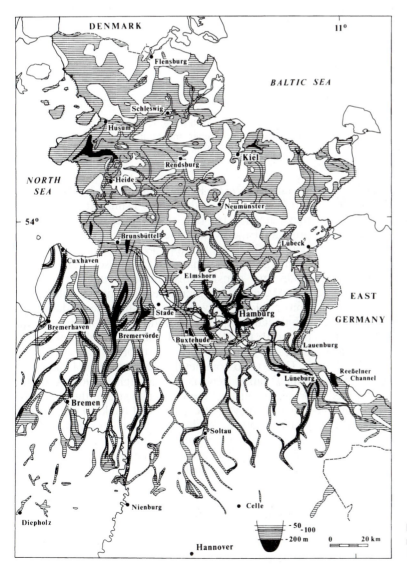

Fig. 9.24 Elsterian buried tunnel valleys, north-west Germany. (From Ehlers *et al.*, 1984. Reproduced by permission of Elsevier)

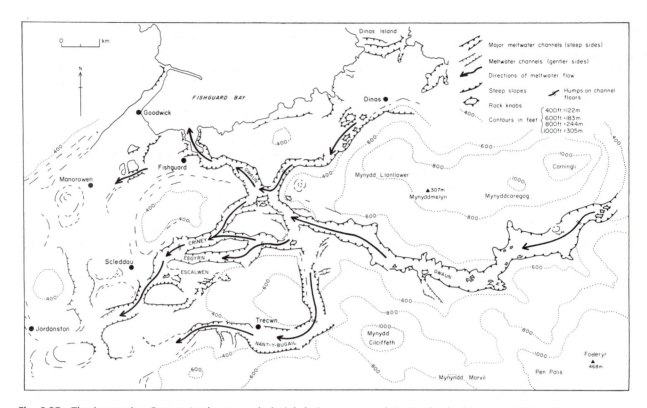

Fig. 9.25 The impressive Gwaun–Jordanston subglacial drainage network in Pembrokeshire, west Wales. The steep-sided and flat-floored channels are cut in bedrock, are up to 45 m deep and have a combined maximum length of 20 km. (From Sugden and John, 1976)

deposits, and may not have any clear topographic expression on the surface (Ó Cofaigh, 1996). In such cases, detailed studies of borehole logs, geophysical data and sedimentary exposures are necessary to determine their morphology and extent. For example, buried tunnel valleys incised into chalk bedrock up to 75 m below present sea-level in East Anglia were discovered only after the analysis of borehole evidence of anomalous thicknesses of Anglian glacigenic sediments (Woodland, 1970). Some care needs to be taken wherever old subaerial drainage networks have been re-excavated or partially cross-cut by later subglacial tunnel valleys. For example, early Pleistocene (preglacial) river valleys lie buried beneath later glacial sediments on the southern Alberta prairies in Canada, and the occurrence of chains of elongate depressions demarcates the positions of subglacial channels, which are often centred over the old buried valleys (Stalker, 1961; Tokarsky, 1986; D.J.A. Evans and Campbell, 1995). In such situations, the long profiles of the buried valley bottoms reveal that they are normal fluvial forms, whereas the superimposed chains of depressions on the prairie surface indicate that subglacial meltwater excavated some of the glacial sediments filling the valley.

Hansen (1971), Krüger (1983) and Lykke-Andersen (1986) suggest that some Danish tunnel valleys may in fact be old river valleys excavated by proglacial meltwater or even glacier ice (cf. Woldstedt, 1926, 1952; Gripp, 1975). Clearly, ice sheet submarginal and marginal processes can modify landforms considerably through one or several glacial cycles, and therefore many tunnel valley sections could be polygenetic. Despite the tendency for subglacial streams to reoccupy their former courses, the north German tunnel valleys largely cut across the early Pleistocene river networks.

There is no completely satisfactory explanation of tunnel valley genesis (Ó Cofaigh, 1996). A major problem concerns the very large size of the channels, which, if they ever experienced bankfull conditions, would imply water discharges far in excess of those that could be maintained by steady-state basal melting. Two main theories have been advanced to overcome this difficulty: (a) tunnel valleys could result from progressive excavation of sediment by normal discharges in conjunction with subglacial sediment deformation; or (b) they could be excavated by extremely large, transient discharges associated with catastrophic lake drainage events. These ideas are discussed in turn below.

Drainage over subglacial deforming sediment

Tunnel valley genesis has been explained by Shoemaker (1986a) and Boulton and Hindmarsh (1987) as the result of steady-state meltwater drainage over subglacial deforming sediment. They argued that where subglacial discharges cannot be accommodated by Darcian groundwater flow, piping failure will initiate subglacial drainage conduits which will help to maintain glacier stability (Section 3.4.5). According to this model, deforming sediment will tend to creep into the conduit, whereupon it is flushed out by meltwater. Continued sediment excavation by this process is argued to result in the formation of a valley system much larger than the active conduit (Fig. 9.26). The final stage of this process occurs when discharges fall, and the conduit fills with sediment and contracts by ice creep.

There are two main problems with this model (Ó Cofaigh, 1996). First, there is little field evidence in support of the idea that sediment deformation occurs

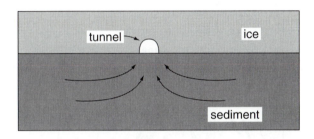

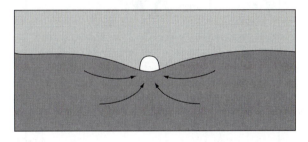

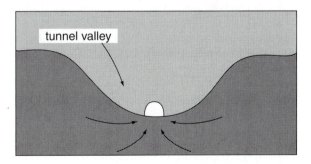

Fig. 9.26 The evolution of a sediment-floored tunnel valley according to Boulton and Hindmarsh (1987). Sediment creeps into the tunnel, and is then carried away by subglacial meltwater. (Provided by C. Patterson, after Boulton and Hindmarsh, 1987)

in conjunction with tunnel valley excavation (Barnett, 1990), although many tunnel valleys do occur in areas where there are extensive deformation tills (Mooers, 1989a; Patterson, 1994). Second, the excavation of deep channels by sediment deformation into conduits is at odds with recent ideas of subglacial hydrology, which indicate that sediment deformation should be suppressed in the vicinity of conduits (Sections 3.4.5 and 3.4.6; Alley, 1992b; Hubbard et al., 1995). Furthermore, Walder and Fowler (1994) argue that excess drainage over deformable beds should take the form of broad, shallow anastomosing canal systems rather than major conduits. It should be emphasized, however, that little is known about drainage conditions on soft glacier beds, and that much more research is necessary before a complete theory is available.

Catastrophic meltwater floods

Several researchers have invoked catastrophic drainage events to account for the excavation of tunnel valleys. Extensive tunnel valley systems in northern Germany (Fig. 9.24) are believed by Ehlers (1981), Ehlers et al. (1984) and Ehlers and Linke (1989) to have been cut by jökulhlaup lake drainage below the margins of the Scandinavian ice sheet. There is evidence for several generations of erosion, and Ehlers and Linke (1989) envisage a complex excavation history involving outburst floods and intervening periods of resculpturing and widening by glacier ice.

Catastrophic floods have also been implicated by Wingfield (1990) to account for extensive networks of elongate, blind-ended depressions more than 100 m deep on the floor of the North Sea in the marginal zone of the former Scandinavian and British ice sheets, where seismic profiling has enabled the identification of three generations of incision (Fig. 9.27; Stoker et al., 1985; Cameron et al., 1987; Wingfield, 1989). Although there is controversy surrounding the interpretation of the North Sea incisions (see review in Ehlers and Wingfield, 1991), their similarity to the tunnel valleys of northern Europe has prompted a genetic comparison with those terrestrial examples. Wingfield (1990) proposed that the incisions were excavated by rapid headward erosion of channels during jökulhlaups from large subglacial lakes. Features similar to those identified on the North Sea floor were reported by Boyd et al. (1988) from the Scotian shelf, Atlantic Canada, and were interpreted as subglacial tunnel valleys cut beneath either a pre- or early Wisconsinan ice sheet extending to the shelf edge.

Extensive tunnel valley networks in south-central Ontario, Canada, have been interpreted as the result of exceptionally large catastrophic lake drainage beneath the Laurentide ice sheet by Brennand and

Shaw (1994). These networks consist of multiple, large anastomosing channels with undulating long profiles and deep, overdeepened basins, and commonly contain eskers running along their floors (Fig. 9.28). The case for catastrophic lake drainage in this region is highly controversial, and is based on radical interpretations of many landform assemblages, including drumlins, flutings and bedrock erosional marks as the products of huge subglacial sheet floods (Section 11.2.6; Shaw and Kvill, 1984; Shaw, 1989, 1994; Shaw et al., 1989). The inferred drainage events are too large and extensive to have been supplied by subglacial meltwater, and Shoemaker (1992b) has suggested that they originated by reversed subglacial drainage from proglacial lakes supplemented by supraglacial melting. There is no independent field evidence for this scenario, and there are strong reasons for regarding it as physically implausible (Walder, 1994; Ó Cofaigh, 1996). There is therefore a need for critical testing of the outburst flood hypothesis, employing a wide range of sedimentological and geomorphological evidence from wide areas.

A radial network of tunnel valleys formed beneath the Superior lobe of the southern Laurentide ice sheet in Minnesota, USA (Fig. 9.29), has been interpreted as resulting from either (a) simultaneous incision by catastrophic discharges (Wright, 1973), or (b) multigenerational incision during steady-state, stable drainage during ice sheet retreat (Mooers, 1989a; Patterson, 1994). Like their north European counterparts, the Minnesota tunnel valleys terminate at sand and gravel fans, which are often up to 85 m higher than the valley floors and form part of end moraine complexes, documenting the deposition of outwash by subglacial meltwater escaping from the ice margin under pressure. The distribution of the fan complexes and other glacial landforms in the area lends strong support for their formation in several stages during ice retreat, but the mechanisms of tunnel valley excavation remain uncertain (Ó Cofaigh, 1996). Resolution of the debate concerning the origin of these and other tunnel valley complexes is clearly important, and has wide-reaching implications for the nature of drainage below large ice sheets and the relative contribution of catastrophic and steady-state conditions to landscape evolution and environmental change.

9.3.4.3 ICE-MARGINAL (LATERAL) CHANNELS

Water draining along the margins or in submarginal zones of glaciers can be responsible for considerable incision of sediment and bedrock, producing *lateral channels* that mark former ice margin positions (Fig. 9.30). Such features are particularly well developed at the margins of subpolar glaciers where meltwater

cannot penetrate to the frozen bed of the glacier. Following deglaciation, lateral meltwater channels are left perched on valley sides, and are usually distinguishable from subglacial and subaerial channels by their planform and distribution. Lateral channels may terminate abruptly where the meltwater drained down englacial or subglacial tunnels. Marginal and submarginal channels can form nested *inset sequences,* which have been used to reconstruct glacier recession patterns. Ice-marginal channels have been differentiated from submarginal channels by Embleton (1964b) on the basis of their gradients; typical

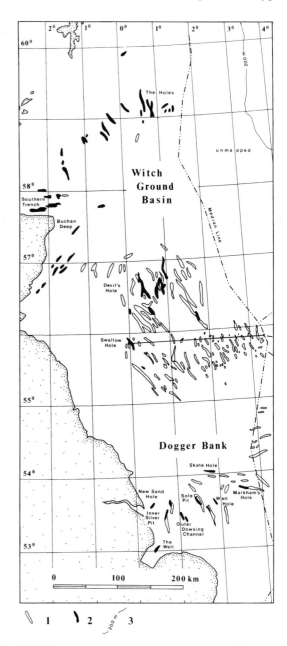

Fig. 9.27 (a)

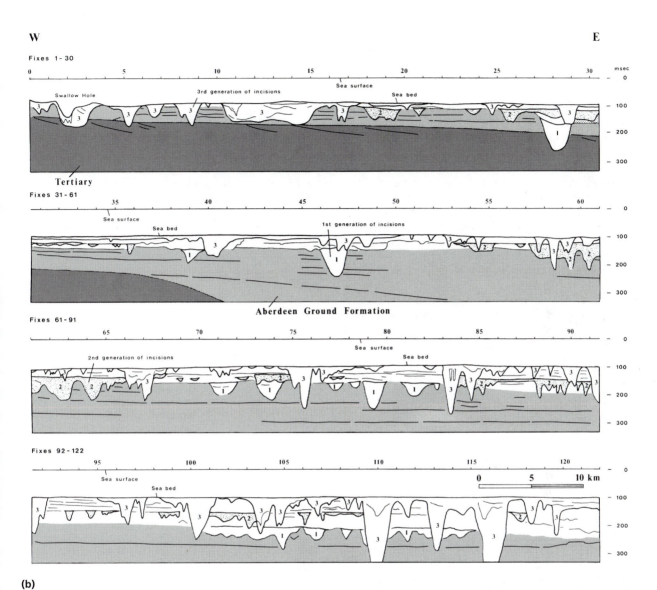

(b)

Fig. 9.27 (a) Major subglacial incisions in the British sector of the North Sea. 1: Major incisions; 2: enclosed deeps; 3: 200 m depth contour. (b) Interpretations of west–east seismic profiles, showing three generations of incisions. (Modified from Ehlers and Wingfield, 1991)

gradients are 1:300 for marginal channels, but as great as 1:45 to 1:25 for submarginal channels.

The evolution of lateral meltwater channels at the margins of subpolar glaciers on Axel Heiberg Island in the Canadian Arctic has been studied by Maag (1969). Because of the occurrence of permafrost, meltwater will often excavate into the glacier rather than into bedrock or frozen sediment and therefore will drain submarginally; in successive summer seasons streams may reuse old channels even though during the winter they may fill up with snow or partially close owing to ice creep (Fig. 9.31). In

addition, meltwater routes may alternate between marginal and submarginal positions, and may drain alternately over rock/sediment and glacier ice (D.J.A. Evans, 1989b). Furthermore, the simultaneous erosion of parallel marginal and submarginal channels was documented by Schytt (1956) for the Moltke Glacier, Greenland. Consequently, lateral channels associated with subpolar glaciers can be discontinuous and often document submarginal as well as ice-marginal drainage. Maag (1969) also stressed the importance of (a) valley-side slope angle in the production of lateral channels, steep slopes being eroded into benches rather than channels, and (b) the nature of the substrate. Incision rates as high as 1 m per 24 hours, due to the combined effect of fluvial and thermal erosion, were recorded in frozen gravels on Axel Heiberg Island. Incision rates and channel depths in bedrock will obviously vary according to lithology and discharge magnitudes during channel occupancy.

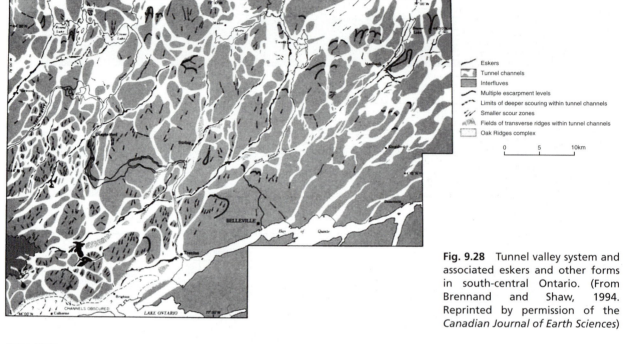

Eskers
Tunnel channels
Interfluves
Multiple escarpment levels
Limits of deeper scouring within tunnel channels
Smaller scour zones
Fields of transverse ridges within tunnel channels
Oak Ridges complex

0   5   10km

Fig. 9.28 Tunnel valley system and associated eskers and other forms in south-central Ontario. (From Brennand and Shaw, 1994. Reprinted by permission of the *Canadian Journal of Earth Sciences*)

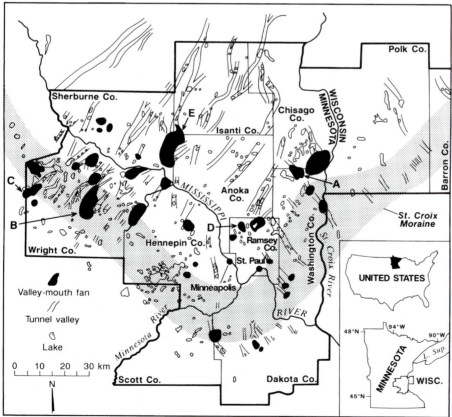

Valley-mouth fan
Tunnel valley
Lake

0 10 20 30 km
N

Fig. 9.29 (a)

(b)

Fig. 9.29 Tunnel valleys and associated forms of the Late Wisconsinan Superior Lobe. (a) Tunnel valleys, valley mouth fans and the St Croix moraine (shaded) in east-central Minnesota and Wisconsin, USA. A: Comfort Lake tunnel valley and fan; B: Buffalo Lake tunnel valley and fan; C: Goose Lake tunnel valley and fan; D: Arsenal fan and Lino Lakes tunnel valley system; E: Elk River fan complex. (b) Aerial photograph of Comfort Lake fan and associated tunnel valleys partially delineated by elongate lakes. (From Patterson, 1994. Aerial photograph by Mark Hurd Aerial Surveys)

Lateral meltwater channels cut along the cold-based northern margins of the Laurentide ice sheet have been mapped by Dyke (1983, 1993b) and Dyke *et al.* (1992), who used them to reconstruct regional patterns of ice recession (Fig. 9.30). Extensive flights of inset meltwater channels with very shallow gradients are evident on the upper slopes of the fjords and intervening plateaux of south-east Ellesmere Island in the Canadian Arctic, documenting meltwater drainage along the margins of low-gradient fjord glaciers. Retreat patterns of subpolar glaciers have been reconstructed for large areas of Ellesmere Island using prominent lateral meltwater channels in conjunction with other ice-marginal accumulations (Bednarski, 1986; Lemmen, 1989; England, 1990; D.J.A. Evans, 1990b). Lateral channels have also been mapped by Brown (1993) in north-east Scotland, where they are inferred to have formed during the retreat of cold-based lobes of the last (Devensian) ice sheet.

Although lateral meltwater channels are best developed along cold-based glacier margins, they have been reported from regions where glaciers were apparently temperate. For example, in the Canadian Cordillera, lateral meltwater channels demarcate the receding margins of valley glaciers debouching from the mountain icefield (Fulton, 1967; Tipper, 1971; Dyke, 1990b). These are explained by Dyke (1993b) as the product of either: (a) high subglacial water pressures deflecting some surface water along the glacier margins; (b) spring snow melt on surrounding slopes and meltwater drainage along the glacier margin while the bed was still frozen after the penetration of the winter cold wave; or (c) cold-based conditions in the outer marginal zone of otherwise temperate glaciers. Where numerous, closely spaced lateral channels have been cut in a slope in association with the retreat of an ice margin, it has been suggested that each channel may record the drainage of one melt season (e.g. Mannerfelt, 1945, 1949; Sissons, 1958b). Although caution is warranted with such interpretations, especially in situations where meltwater may have originated from ice-contact lakes, Dyke (1993) proposed that the spectacular lateral meltwater channels of the central Canadian Arctic display a very fine temporal resolution, perhaps of 'melt event' scale.

The diversion of large rivers by advancing glaciers can result in the excavation of large-scale channels along the glacier margins. Such channels, known as *urstromtaler*, mark successive marginal positions of the Scandinavian ice sheet in Poland, Germany and the Netherlands (Woldstedt, 1955; Neef, 1970; Woldstedt and Duphorn, 1974; ter Wee, 1983; Ark-

hipov *et al.*, 1995), with continuations on the German Bight sector of the North Sea (Figge, 1983). The urstromtaler mark the positions of the main European rivers after they were diverted from their normal south–north drainage direction by the advancing ice

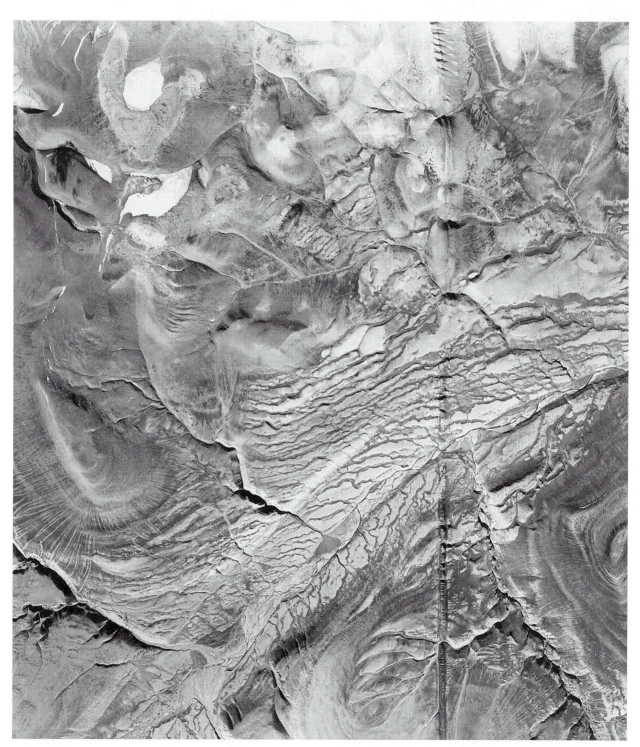

Fig. 9.30(a)

(b)

Fig. 9.30 Lateral meltwater channels cut at the margins of predominantly cold-based glaciers at the end of the last glaciation in the Canadian Arctic. (a) Eastern Borden Peninsula, Baffin Island. (b) The head of Flagler Bay, Ellesmere Island. (Aerial photos: Department of Energy, Mines and Resources, Canada)

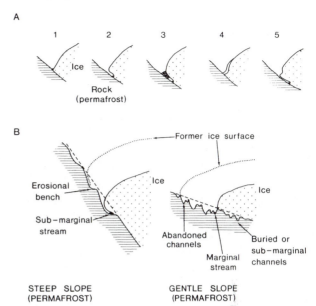

Fig. 9.31 Meltwater channel development at the margins of subpolar glaciers based on examples from Axel Heiberg Island, arctic Canada. (A) The development of a marginal/submarginal channel. 1: marginal drainage at the beginning of the melt season; 2: marginal stream undercuts the ice edge; 3: collapse of the ice margin and submarginal drainage; 4: beginning of the subsequent melt season – snow covers previous year's depression and the ice cave survives; 5: drainage reuses the former channel. (B) The effects of slope on the morphology of ice-marginal meltwater channels. (From Embleton and King, 1975, after Maag, 1969)

sheet to flow in a westerly direction along the ice margin.

9.3.4.4 PROGLACIAL CHANNELS AND FLOOD TRACKS

Channels and gorges cut by proglacial streams can achieve impressive dimensions, owing to the erosive power of high discharges and sediment loads during peak flows. In areas formerly occupied by glacier ice, such channels can originate as subglacial Nye channels and undergo subsequent subaerial enlargement after deglaciation, and the relative contribution of subglacial and subaerial erosion can be unclear (Section 9.3.4.1).

The very largest proglacial channels, however, are cut by glacial outburst floods or jökulhlaups (Section 3.5). The immense erosive capacity of such floods is perhaps most clearly demonstrated by the *Channeled Scablands* of Washington State, USA (Baker, 1981; Baker *et al.*, 1987), which are the deeply dissected remnants of the formerly extensive loess and basalt deposits of the Columbia Plateau. This landscape was first attributed to erosion by a large jökulhlaup, referred to as the Spokane Flood, by J. Harlen Bretz

in a series of controversial publications (e.g. 1923a, b, 1927, 1969; Bretz *et al.*, 1956). It was not until the 1960s that Bretz's ideas of a proglacial flood were finally accepted, largely because of the absence of an obvious source for a flood large enough to accomplish such massive erosion. Comprehensive coverage of the controversy surrounding the acceptance of Bretz's flood hypothesis has been provided by Gould (1980), Baker (1981) and Baker *et al.* (1987), who regard the Spokane Flood as one of the most valuable of what William Morris Davis endearingly entitled 'outrageous geological hypotheses', which challenge accepted thinking and offer a new paradigm.

The Channeled Scablands are now known to have been created by catastrophic drainage of Glacial Lake Missoula, a large ice-dammed lake ponded up by the margin of the late Pleistocene Cordilleran ice sheet, and for this reason the Washington floods are sometimes referred to as the Lake Missoula floods (Fig. 9.32). Glacial Lake Missoula in Montana had a maximum volume of approximately $2500 \, km^3$, and on several occasions drained catastrophically through the ice dam, releasing up to $2184 \, km^3$ of water (comparable to the volume of Lake Ontario;

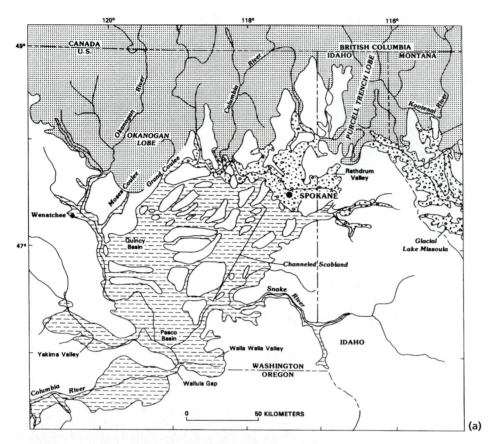

(a)

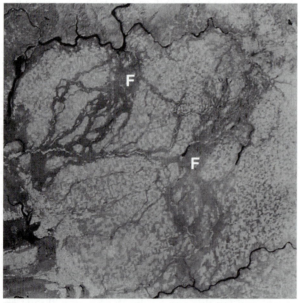

(b)

Fig. 9.32 (a) The course of the Lake Missoula floods (horizontal dash pattern) and the location of the Channeled Scabland of Washington. The Cordilleran ice sheet (dot pattern) and ice-dammed lakes (coarse stipple pattern) are also shown. (From Baker *et al.*, 1987. Reproduced by permission of the Geological Society of America) (b) Landsat image of the Columbia Basin showing the flood tracks (F)

1985; Waitt and Thorson, 1983; Baker and Bunker, 1985). The awesome erosional capacity of these floods is manifest in the Channeled Scablands, which comprise 'a great anastomosis of flood channels and recessional gorges replete with a bizarre assemblage of erosional and depositional landforms such as rock basins, giant cataract alcoves, large residual "islands", great bars of gravel, and giant current dunes' covering an area of approximately 40,000 km^2 (Baker *et al.*, 1987; Figs 9.32 and 9.33). The escaping waters stripped a cover of loess up to 60 m thick and plucked out basalt blocks from channel walls and floors, in many cases leaving isolated mesas or tablelands, known locally as *scabs*, between channels. In confined channels, water depths attained 100–200 m, reaching almost 300 m in the narrow eastern entrance to the Columbia River gorge. Some channels were enlarged by the headward retreat of major waterfalls; one of the largest, Dry Falls, was over 5 km wide and

Clarke *et al.*, 1984b; Clarke, 1986). The floodwaters scoured across the northern part of the Columbia Plain towards the Pasco Basin and then down the Columbia River valley towards the Pacific Ocean (Pardee, 1942; Baker, 1973a, b; Waitt, 1980, 1984,

Fig. 9.33 Various features of the Channeled Scabland of Washington/Idaho, USA. (a) West Potholes cataract. The water spilled from a 3 km wide lip, over 100 m cliffs and into 40 m deep plunge basins. Boulders 30 m in diameter were carried from the cataract and deposited in huge bars downstream. (b) Dry Falls cataract complex. (c) Butte-and-basin scabland development in Lenore Canyon. (d) Giant current ripples near Spirit Lake, Idaho. The gravel ripples are spaced 85 m apart and are 4 m high on average. (From Baker *et al.*, 1987. Reprinted by permission of the Geological Society of America)

120–130 m high (Fig. 9.33b). Baker (1973a, 1981) and Baker *et al.* (1987) concluded that the greater part of this erosion was achieved in one or a few exceptionally large floods, perhaps over a period of hours, although Waitt (1980, 1984, 1985) has argued that there were up to 40 flood events. Various calculations of the flood discharges from Glacial Lake Missoula have been attempted, and range from 2.7 to $21 \times 10^6 \, \text{m}^3 \, \text{sec}^{-1}$, astounding figures equivalent to between 2 and 20 times the mean flow of all the world's rivers into the oceans (Baker, 1973a; Clarke *et al.*, 1984b; Clarke, 1986; Baker *et al.*, 1987; O'Connor and Baker, 1992).

The erosional landforms of the Channeled Scablands have been grouped into three categories by Baker *et al.* (1987):

1. *Scabland erosion complexes* are formed by the fluvial incision of bedrock. Experiments show that

forms such as longitudinal grooves, potholes and transverse erosional ripples eventually become incised by inner channels, which migrate upstream by knickpoint recession. This has been modelled for the Channeled Scablands by Baker *et al.*, (1987), who inferred a five-stage evolutionary sequence. During phases I and II the loessic mantle was streamlined and incised. Phase III was marked by plucking of the underlying well-jointed basalt lava flow by turbulent vortices or *kolks*. Pothole enlargement and coalescence during phase IV led to the development of butte and basin topography, and phase V was characterized by the growth of inner channels and cataract recession.

2. *Streamlined erosional residuals* are isolated loess hills with approximately lemniscate planforms, interpreted as erosional islands which have been streamlined to reduce the drag or resistance to flowing water (Chorley, 1959; Baker and Kochel,

1978; Baker, 1979; Komar, 1983, 1984). Lemniscate forms are defined and discussed in Section 11.2.4.1.

3. *Scour marks* are produced wherever obstacles exist within scabland channels. Such obstacles will initiate a horseshoe-shaped vortex and scour hole at their upflow boundaries and a wake vortex in the downstream zone of flow. Analogous small-scale forms can be observed in wind-blown sand or snow in the vicinity of fence posts or other obstacles. Scour marks may be associated with lee-side depositional forms known as *pendant bars*, depending on flow conditions.

Giant depositional features also occur in the Channeled Scablands, including *bars* tens of metres high on the floor of channels, and *giant current ripples* between 1 and 15 m high and spaced between 20 and 200 m apart (Bretz *et al.*, 1956; Baker *et al.*, 1987). When viewed from above, the crest forms of the gravel waves are similar to those of normal sand ripples or dunes but are composed of cross-bedded gravel and boulders (Fig. 9.33d; see Sections 3.8.3, 10.4.3 and 10.4.4). The large size and coarse composition of such bedforms make the immense size of the Missoula floods all the more impressive.

Since the acceptance of the Missoula flood hypothesis, proglacial flood tracks have been identified in a number of locations. Cataclysmic flood tracks of a size comparable with those produced by the Missoula floods have been recognized at the margins of proglacial lakes in Siberia by Rudoy *et al.* (1989), Rudoy (1990), Baker *et al.* (1993) and Rudoy and Baker (1993). Similarly, Fraser *et al.* (1983) have identified flood tracks in the unglaciated terrain of the Wabash Valley of Indiana. Large channels identified on the floor of the North Sea basin and the English Channel have been interpreted as the erosional products of large meltwater discharges either beneath or in front of the retreating Scandinavian/British ice sheet (N.D. Smith, 1985; Wingfield, 1990; Ehlers and Wingfield, 1991; Section 9.3.4.2). Other channel systems and streamlined residuals thought to have been cut by jökulhlaups have been described from the Great Plains of North America (Kehew, 1982; Kehew and Clayton, 1983; Kehew and Lord, 1986, 1987, 1989), Sweden (Elfström and Rossbacher, 1985) and Norway (Longva and Thoresen, 1991).

Spillways are distinctive types of proglacial channels, produced where water decants from ice-dammed or proglacial lakes over cols or low points on the watershed. Spillways can be cut over a long period of time by regulated overflow water, although catastrophic jökulhlaup-like release of water down spillway courses can produce scabland topography (Kehew, 1982; Kehew and Lord, 1986; Bryan *et al.*, 1987). An impressive example of a spillway is at Newtondale on the North York Moors, England; it is thought to have been cut by water overflowing from an ice-dammed lake on the north side of the Moors (Fig. 9.34). The spillway carried the water through the unglaciated highlands of the moors into Glacial Lake Pickering, which was located to the south of the high terrain and dammed by a glacier lobe on the east Yorkshire coast. An extensive network of deglacial spillways occurs throughout the Canadian provinces of Alberta, Saskatchewan, Manitoba and Ontario and the US states of North Dakota, Minnesota, Michigan and Wisconsin, documenting overflow from proglacial lakes during the retreat of the Laurentide ice sheet (Taylor, 1960; St Onge, 1972; Christiansen, 1979; Teller and Clayton, 1983; Teller, 1985; Kehew and Lord, 1986, 1987, 1989; Teller and Thorleifson, 1987; Teller and Mahnic, 1988). Quite often the release of water from one lake caused receiving lakes downstream rapidly to incise their outlets, thus producing a chain reaction of catastrophic drainages (Kehew and Lord, 1989). Some narrow bedrock canyons, such as the Ouimet Canyon in Ontario (100 m wide and 100 m deep; Kor and Teller, 1986), and much larger channels such as the Souris Spillway of Saskatchewan and North Dakota (1 km wide and 45 m deep; Kehew, 1982), were excavated over short periods of time during catastrophic drainage.

The characteristics of spillways produced by catastrophic glacial lake outbursts in the North American Great Plains have been studied by Kehew (1982), Kehew and Clayton (1983), Kehew and Lord (1986, 1987), Bryan *et al.* (1987) and D.J.A. Evans (1991a). Figure 9.35a shows a fourfold evolutionary sequence of spillway development proposed by Kehew and

Fig. 9.34 Oblique aerial photograph of Newtondale, North Yorkshire, England. This feature is thought to be a proglacial lake spillway cut across the unglaciated uplands of the North York Moors by meltwater draining from a glacier margin impinging on the northern flanks of the high land. Note the North York Moors railway line in the valley bottom. (Aerial photograph BA30 of the Cambridge University Collection of Aerial Photography)

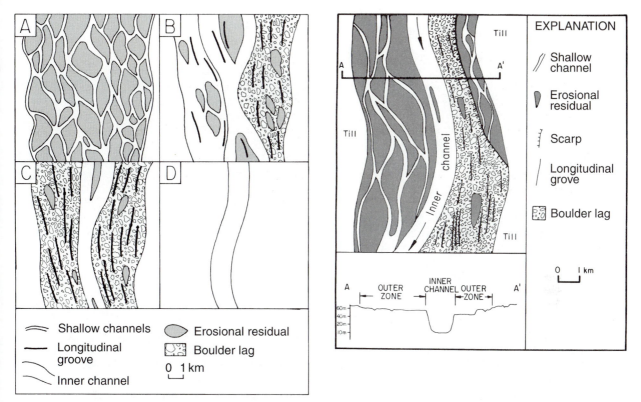

Fig. 9.35a Schematic diagram of four possible stages in spillway development: A: shallow anastomosing channels with irregular interchannel residuals and no inner channel; B: fewer, more streamlined residuals with an anastomosing channel pattern still evident in places and with a poorly developed inner channel; C: well-developed inner channel, lemniscate residuals and longitudinal grooves; D: larger inner channel with terrace remnants, few residuals and no outer zone. (From Kehew and Lord, 1986. Reproduced by permission of the Geological Society of America)

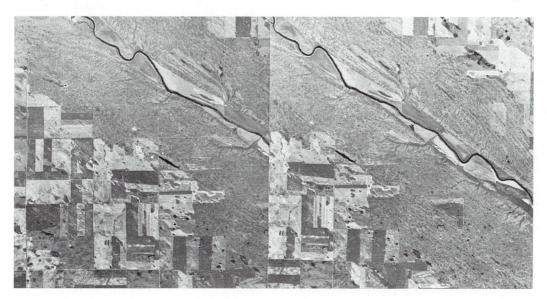

Fig. 9.35b Aerial photograph stereopair of the Souris spillway near Hitchcock, Saskatchewan, showing the outer zone of boulder lag, not in use for crop production. (After Kehew and Lord, 1989. Reproduced by permission of the Canadian Association of Geographers)

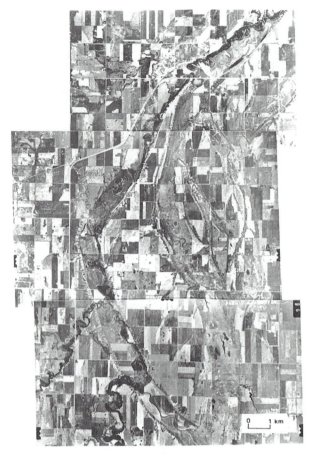

(c)(i)

(c)(ii)

Fig. 9.35c Characteristics of the inner channels of spillways on the North American Great Plains: (i) aerial photograph mosaic of erosional residuals in the Souris spillway between Lake Souris and Lake Hind, Manitoba (aerial photographs A24965-214, A24966-09 and A24966-40, Manitoba Department of Natural Resources); (ii) streamlined erosional hill in the Souris spillway (Photo: Alan Kehew). The hill is 1.6 km long, 0.5 km wide and 20 m high

(d)(i)

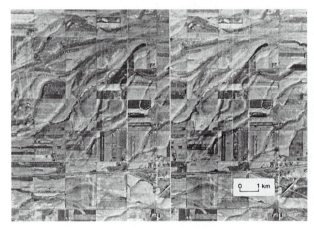

(d)(ii)

Fig. 9.35d Characteristics of the outer channels of spillways on the North American Great Plains. (i) The Souris spillway near Hitchcock, Saskatchewan. The spillway outer zone (10 km wide) extends from the foreground to the cultivated land in the distance, contains conspicuous longitudinal grooves and is covered by boulder lag deposits. The inner channel in the centre of the photo is 1 km wide and 35 m deep. (Reproduced by permission of the Geological Society of America) (ii) Stereopair of erosional anastomosing channels near Minot, North Dakota. The channel incisions have isolated a large number of erosional residuals. (From Kehew and Lord, 1986; Aero Service Corporation photographs 340-2471 and 2472)

Clayton (1983) and Kehew and Lord (1986, 1987). (A) In the early stages of a flood, no channel exists to carry the outburst discharge, and a system of anastomosing channels is eroded. (B) As discharges rise, a central channel is cut, leaving erosional remnants of the earlier anastomosing channel system in an outer zone. Scouring of the outer zone produces longitudinal grooves and streamlined erosional hills, and boulder lags represent particles too large to be carried along by the flow. (C) Channel incision and scouring of the outer zone continue. (D) Flow gradually becomes increasingly focused in the inner channel, causing deep incision. The final form of inner channels of flood spillways is characterized by: (1) a

(a)

trench-like shape; (2) uniform width and side slopes; (3) regular meander bends; (4) occasional bifurcation to form parallel or anastomosed channels separated by linear ridges or streamlined erosional residuals; and (5) occasional isolated and often streamlined erosional residuals. Such channels can be 1–3 km wide and 25–100 m deep. Deposits formed by catastrophic floods are described in Section 11.5.1.3.

Icebergs carried in jökulhlaups can exert a considerable impact on the landscape by scouring out grooves and depressions. Longva and Thoresen (1991) have identified a variety of iceberg scours and gravity craters formed during a jökulhlaup from Glacial Lake Nedre Glamsjo in Norway (Fig. 9.36). Peak discharges were responsible for the erosion of anastomosing channels and intervening streamlined mounds with a maximum relief of 2 m, together with iceberg scour marks up to 1 m deep and 2 km long. As current velocities gradually decreased, the icebergs were arrested while still scouring, and where they finally came to rest they produced *gravity craters*, consisting of shallow semicircular depressions up to 30 m wide. Some icebergs were refloated after making one gravity crater, and thus produced *crater chains*; this may have been achieved by a fast-drifting iceberg with a rocking motion, by reflotation after berg breakup, or by short pulses of increased

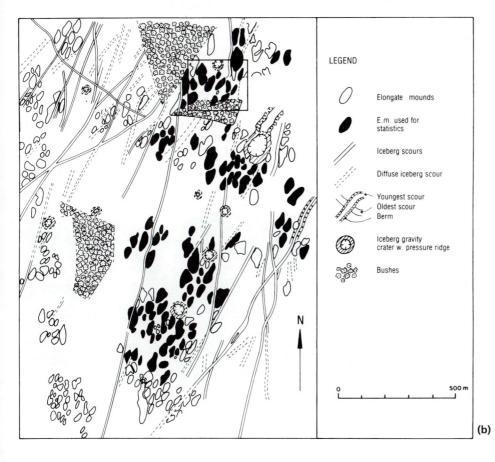

(b)   Fig. 9.36

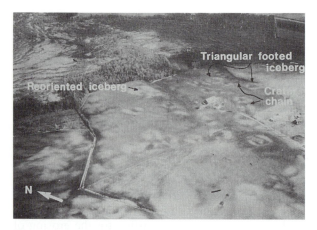

(c)

Fig. 9.36 Iceberg scours and gravity craters in flood tracks from Glacial Lake Nedre Glamsjo, Norway. (a) Aerial photograph of an area near the junction of the Vorma and Glomma rivers, showing elongate mounds (light patches) and iceberg grounding and scour features (dark lines). (Aerial photograph by Fjellanger Widerøe A/S) (b) Geomorphological interpretation of the area shown in (a). (Modified from Longva and Thoreson, 1991) (c) Iceberg gravity craters and crater chains (Photo: T.H. Bargel)

discharge. Some icebergs appear to have been reorientated without being refloated, producing a close cluster of gravity craters. Similar features on the former bottom of Glacial Lake Agassiz have been interpreted by Mollard (1983) in a variety of ways, including iceberg settling structures to groundwater extrusion craters. However, the association of the Norwegian craters with an obvious flood track appears to support their origin by iceberg impact. Kettle holes created by the melt-out of buried ice blocks are described in Section 11.4.6.

Smaller-scale features produced by catastrophic flooding by glacial meltwater include scratches or 'striations' on bedrock surfaces. Although normally attributed to direct glacial abrasion, striations have been associated with jökulhlaup-type floods by McCarroll *et al.* (1989). Such scratches are distinguishable from glacially abraded striations because they have low length:width ratios and a tapering shape, indicating a striking blow rather than steadily applied pressure. Furthermore, they possess numerous cross-cutting directions which should not be misinterpreted as the products of directional changes in glacier flow.

9.4 LARGE-SCALE EROSIONAL LANDFORMS

Large-scale erosional landforms constitute some of the most impressive geomorphological features on the Earth's surface, providing striking testimony to the immense erosive potential of glacier ice. In this section, we discuss the characteristics and genesis of basins and overdeepenings in bedrock and unconsolidated sediments, troughs and fjords, cirques, and strandflats.

9.4.1 Rock basins and overdeepenings

Basins eroded into bedrock by glaciers occur at a wide variety of scales, from small hollows between roches moutonnées to large overdeepenings occupying the full width of cirques and troughs. Large overdeepenings are common in deglaciated valleys, where they are commonly occupied by lakes (Fig. 9.37), and have been identified beneath many existing glaciers in the course of echo-sounding studies. Four overdeepenings have been discovered beneath Storglaciären, a small glacier in northern Sweden (Hooke, 1991). In addition, numerous rock basins can occur together in ice-scoured *knock and lochan* topography (Linton, 1963; Rea and Evans, 1996), which is related to areal scour and reviewed in Section 9.5.1.

The formation, size and shape of rock basins are controlled by glaciological variables, such as the thermal regime and stress conditions at the glacier bed, and substratum characteristics, particularly bedrock structure and lithology. The main processes of rock basin erosion are quarrying and abrasion (Sections 5.3.2 and 5.3.3), although some authors have argued that direct erosion by meltwater is important in some cases (e.g. Kor *et al.*, 1991). Effective quarrying and abrasion can occur only where the glacier bed is predominantly wet-based, such as the beds of temperate glaciers and the base of fast-flowing ice streams. Additionally, quarrying is most effective where there are large water-pressure variations at the bed, which encourage the large transient stress gradients required for rock failure (Iverson,

Fig. 9.37 Rock basins seperated by a rock bar (riegel) in a cirque in Kananaskis Country in the Canadian Rockies. (Photo: D.J.A. Evans)

1991; Hallet, 1996). The role of water pressure variations on the erosion of large overdeepenings below valley glaciers has been examined by Hooke (1991) using data from Storglaciären. He argued that quarrying will be focused on the lee side of major steps on the bed, because these locations will favour the formation of surface crevasses, which allow surface meltwater to reach the bed (Fig. 9.38). Fluctuating water pressures will encourage rapid erosion downglacier from rock steps, overdeepening the bed profile. In turn, the evolving glacier bed topography favours continued surface crevassing, so that overdeepenings of the glacier bed will be amplified over time by a positive feedback. A similar process may operate on a small scale, whereby cavities become enlarged by differential erosion.

Bedrock structure and lithology control bed resistance to quarrying and abrasion, and thus exert a strong influence on the location and morphology of rock basins (Sugden and John, 1976; Gordon, 1981). In addition, rock weaknesses may have been partially excavated by non-glacial processes prior to glacier advance. Preglacial weathering can isolate resistant bedrock protuberances and weaken well-jointed rock in the intervening areas, preparing the way for effective subglacial erosion (Rudberg, 1973). In a study of rock basins on the Canadian Shield, Brochu (1954) demonstrated that some rock types (e.g. schist) are more susceptible to rock basin excavation than others (e.g. granite and gabbro), because of their different fracture densities. Similar lithological controls on rock basin development are emphasized by I.S. Evans (1994) on the basis of studies of cirques in the English Lake District and the Cayoosh Range of British Columbia, Canada.

Shallow basins are commonly elongated, because: (a) they are guided by lithological changes or joint/fault systems in exposed strata (Zumberge, 1952; Linton, 1963; Nougier, 1972); (b) glacier ice excavates preferentially along those zones of weakness lying subparallel to the ice flow direction (Virkkala, 1952); and (c) preglacial fluvial relief may be preferentially overdeepened (Zumberge, 1955; Rudberg, 1954, 1973). Very large overdeepenings occurring at

the junctions of different rock types have been attributed to glacial erosion by Davis (1920) and Tricart and Cailleux (1962). For example, the depressions hosting Great Bear and Great Slave lakes were regarded by White (1972) as the products of glacial overdeepening at the junction between the resistant rock of the Canadian Shield and the less resistant Devonian and Cretaceous rocks to the west.

9.4.2 Basins in soft sediments

Erosional basins also occur in areas underlain by soft sediments, and are thought to result from three main processes:

1. Where the sediments form a rigid bed, erosion may occur by *plucking*, as for rock beds. Other factors being equal, erosion rates should be higher than for hard rock beds, so that the location and form of basins may simply reflect the distribution of pre-existing sediments.
2. *Glacitectonic thrusting* near glacier margins can excavate large basins from submarginal positions, the eroded material being repositioned beyond the ice margin as ice-thrust masses (Section 7.4.3; Croot, 1988b; Aber *et al.*, 1989). Basins formed in this way commonly occur immediately up-ice of thrust-block moraines marking major ice margin positions. Thrust masses and basins occurring together are known as *hill–hole pairs*, and in some cases the dimensions of the thrust masses correspond closely to those of the basin. Hill–hole pairs are reviewed in detail in Section 11.3.1.1. A conceptual model of glacitectonic disturbance and basin overdeepening based on structural studies of thrust moraines in Germany and Spitsbergen is shown in Fig. 9.39. The lateral margins of basins formed by thrusting are commonly straight, and parallel to the former ice flow direction, reflecting the location of *tear faults* at the boundaries of the thust mass. Where ice-marginal thrust belts are overridden by continued glacier advance, excavational basins and thrust masses may appear smoothed or even streamlined.

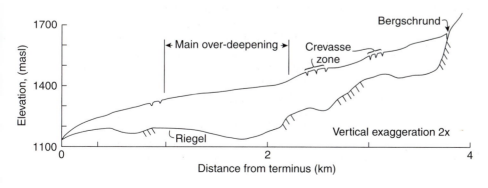

Fig. 9.38 Longitudinal section through Storglaciären, Sweden, showing the location of surface crevasse zones, the bedrock profile, and inferred zones of enhanced quarrying. (Modified from Hooke, 1991)

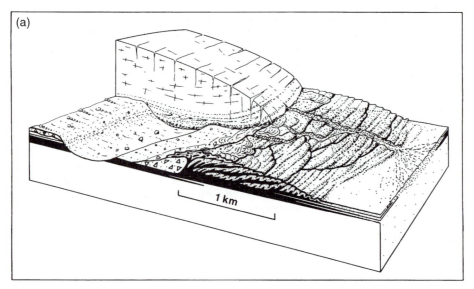

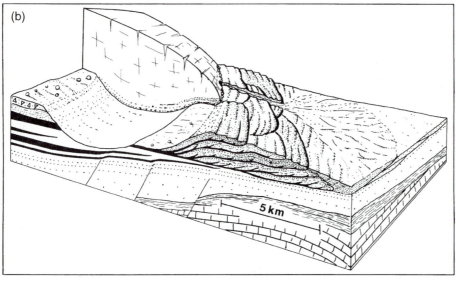

Fig. 9.39 Basin excavation associated with thrust block moraine formation at the margins of (a) Holmstrom-breen, Spitsbergen (modern) and (b) Dammer Berge, Germany (ancient). (From van der Wateren, 1995. Reprinted by permission of Butterworth-Heinemann)

3. On a *deformable bed*, erosion may occur by the downward migration of the base of the deforming layer and the net advection of deforming sediment away from the site, termed *excavational deformation* by Hart *et al.* (1990), Hart and Boulton (1991) and Hart (1995a) (Section 5.6.1). Boulton (1996a) has argued that erosion by this process should be most rapid just upglacier from glacier equilibrium lines, where the mass balance velocity is rising downglacier and glacier flow is extensional (see Section 4.1). According to this view, therefore, erosional basins should occur beneath the lower part of the glacier accumulation area. However, several other variables will complicate this simple picture, including the influence of sediment properties and water pressures on deformation rates, and the influence of topography on glacier thicknesses and velocities. Furthermore,

subglacial deformation may preferentially enlarge pre-existing basins produced by ice-marginal tectonics and other processes.

9.4.3 Troughs and fjords

The most spectacular manifestations of glacial erosion are troughs and fjords carved by ice flow through major rock channels (Fig. 9.40 (Plate 19)). The largest troughs on Earth, the Thiel and Lambert Troughs, are presently occupied by outlet glaciers of the Antarctic ice sheet and are approximately 1000 km long, >50 km wide and up to 3.4 km deep (Morgan and Budd, 1975; Drewry, 1983). Very long and deep open fjords (>100 km long and up to 3 km deep from mountain crest to sea bottom) occur along the coasts of Greenland, Norway and the Canadian Arctic islands of Ellesmere and Axel Heiberg, the longest in each region being Nordvestfjord/Scoresby Sund (300 km), Sognefjord (220 km) and Greely Fiord/Nansen Sound (400 km), respectively. Fjords are well developed also along the glaciated coasts of British Columbia in Canada, southern Chile, Greenland, Iceland, Spitsbergen, the south-west corner of New Zealand's South Island, and western Scotland, reflecting the present and past distribution of low-altitude outlet glaciers (Fig. 9.41; Syvitski *et al.*, 1987). A summary of fjord characteristics is provided in Table 9.2.

9.4.3.1 TROUGH CROSS-PROFILES

The cross-profiles of troughs and fjords are often referred to as U-shaped, although in reality they tend to be asymmetric, with one steep and one gentler slope (e.g. Rudberg, 1973), or approximately parabolic (Svensson, 1959; Graf, 1970). The form of many troughs and fjords can be approximated by the formula for a parabola:

$$V_d = aw^b \tag{9.1}$$

where w is the valley half-width, V_d is valley depth, and a and b are constants.

For many troughs, the exponent b is approximately equal to 2. This formula produces a mathematical parabola or an endless curve, therefore a complete description of trough form also requires the relative dimensions to be defined. Graf (1970) suggested using a *form ratio*:

Table 9.2 Typical characteristics of major fjord coastlines

Fjord district	Fjord stage*	Tidal range†	River discharge‡	Climate	Sedimentation rate§
Greenland	1, 2	Low	Medium to high	Subarctic to Arctic	Medium to high
Alaska	1, 2, 3, 4	High	Low to high	Subarctic maritime	Medium to high
British Columbia	3, 4	High	Medium to high	Temperate maritime	Medium to high
Canadian Maritimes	4, 5	Low to medium	Low to high	Subarctic to temperate maritime	Low
Canadian arctic archipelago	1, 2, 3, 4	Low to high	Low to medium	Arctic desert to maritime	Low to medium
Norwegian mainland	3, 4	Low	Low to medium	Subarctic to temperate maritime	Low
Svalbard	2, 3	Low	Low	Arctic island	Medium
New Zealand	4, 5	Medium	Low to medium	Temperate maritime	Low to medium
Chile	2, 3, 4	Low	Low to high	Temperate to subarctic maritime	Medium to high
Scotland	4, 5	Low to high	Low	Temperate maritime	Low

Source: Syvitski *et al.* (1987)
* Stage 1 = glacier-filled; stage 2 = retreating tidewater glaciers; stage 3 = hinterland glaciers; stage 4 = completely deglaciated; stage 5 = fjords infilled
† low = <2 m mean range; medium = 2–4 m mean range; high = >4 m mean range
‡ low = <50 m^3 s^{-1} mean annual discharge; medium = 50–200 m^3 s^{-1}; high = >200 m^3 s^{-1}
§ low = <1 mm yr^{-1} averaged over entire fjord basin; medium = 1–10 mm yr^{-1}; high = >10 mm yr^{-1}

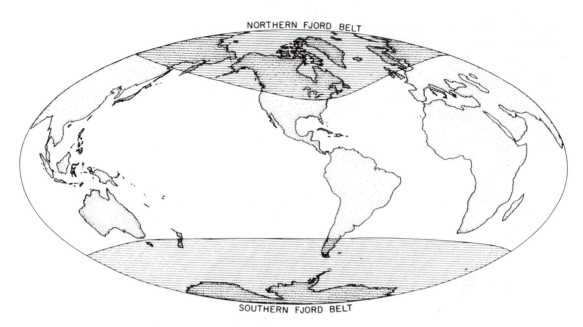

Fig. 9.41 The worldwide distribution of fjords. (From Syvitski *et al.*, 1987. Reprinted by permission of Springer-Verlag)

$$\mathrm{Fr} = V_{\mathrm{d}}/\mathrm{WI} \qquad\qquad (9.2)$$

where WI is the valley top width. Graf obtained form ratios of 0.242 to 0.445 for troughs in the Beartooth Mountains of Montana and Wyoming.

However, trough cross-profiles may depart significantly from this ideal form. For example, Sognefjord exhibits a major break in slope at present sea-level, with relatively gentle slopes above the water-line and steep slopes below sea-level. Nesje and Whillans (1994) have argued that the upper slopes have been modified by subaerial weathering and denudation, whereas the lower slopes preserve their glacial form, owing to the support provided by the water.

The evolution of trough cross-profiles has been modelled by Harbor *et al.* (1988) and Harbor (1992b) using theoretical velocity distributions in a valley glacier cross-section (Fig. 9.42). For a V-shaped valley, modelled basal velocities are highest part-way up the valley sides, and lowest below the glacier margins and centre-line. If it is assumed that erosion rates are proportional to the sliding velocity, the most rapid erosion will occur on the valley sides, causing broadening and steepening of the valley. Eventually, an equilibrium profile is attained, which continues to deepen over time. The effects of pulsed erosion over multiple glacier expansion and contraction cycles have been modelled by Harbor (1992b). Variations in glacier thickness over time strongly influence patterns of erosion, producing a series of breaks in slope and overhangs on the trough walls above the downcutting glacier (Fig. 9.42b). Such slopes are likely to be unstable and prone to collapse, thus modifying the

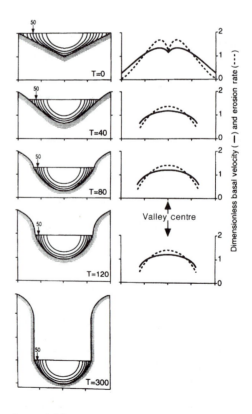

Fig. 9.42 (a)

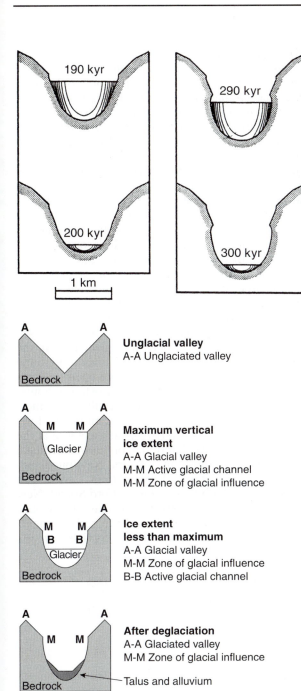

Fig. 9.42 Evolution of trough cross-profiles as modelled by Harbor *et al.* (1988) and Harbor (1992b). (a) Evolving valley cross-sections at different time steps ($T = 0$–300) and corresponding basal velocities and erosion rates. (b) The development of irregular cross-profiles due to glacier expansion and contraction on 100 kyr cycles. (c) Schematic evolution of troughs by erosion in the active glacial channel (B–B) and subaerial slope modification in the zone of glacial influence (M–M). At the maximum vertical ice extent, the active glacial channel and the zone of glacial influence coincide. (From Harbor, 1992b. Reprinted by permission of the Geological Society of America)

form of the trough by paraglacial reworking (Section 7.6). For this reason, oversteepened, ice-free slopes in troughs are termed the *zone of glacial influence*, in which slope evolution is strongly influenced by the contemporary or former presence of glacier ice (Fig. 9.42c).

Parabolic cross-profiles may develop with the aid of pressure release mechanisms in the underlying bedrock. It has long been recognized that freshly exposed rock will undergo expansion (*dilation* or *dilatation*) due to removal of the confining pressure (Bain, 1931), and engineers involved in road and rail construction are well aware of the problems of pressure release in freshly cut rock faces. Rock dilation results in the development of fractures parallel to the ground surface known as dilation joints or sheeting, particularly on massive rocks where there are few pre-existing discontinuities along which strain may be accommodated (e.g. Jahns, 1943; Lewis, 1954; Battey, 1960; Linton, 1963; Sugden and John, 1976). Dilation critically weakens rock masses, facilitating subsequent subglacial erosion (Section 5.3.1). The most obvious time for dilation to take place is after deglaciation, when the ice overburden is removed and freshly eroded rock surfaces are exposed, although it is probably also important beneath active ice once glacial erosion has removed rock slabs that have been fully released along dilation joints during a previous non-glacial period. Thus, pressure release may serve to accelerate subglacial erosion in favourable locations.

Erosion rates in troughs that carry major outlet glaciers are likely to be greater than those below tributary glaciers, resulting in *hanging valleys* perched above the main trunk floor. Hanging valleys may exhibit little or no evidence of glacial erosion and may retain pre-glacial profiles, particularly in areas where the ground adjacent to troughs was occupied by thin, cold-based ice (Sugden, 1974).

9.4.3.2 TROUGH LONG PROFILES

Troughs have been classified by Linton (1963) as: (a) *alpine*, cut by valley glaciers emanating from high ground; (b) *Icelandic*, with a closed *trough head* at the upper end, having been cut by ice spilling over from a plateau ice cap or ice sheet; (c) *composite* – essentially *through troughs* or *through valleys* open at both ends and cut beneath an ice sheet; and (d) *intrusive* or *inverse*, cut against the regional slope where ice impinges upon the downstream ends of highland valleys (e.g. New York Finger Lakes; Clayton, 1965; Mullins and Hinchey, 1989).

Trough heads represent important process thresholds in the landscape, and may form either beneath ice sheets at the upper end of ice streams where erosion rates suddenly increase, or at the head of valley glaciers. Souchez (1966, 1967b) has suggested that trough heads retreat upvalley in much the same way as a nick point on a stream profile, whereas Nesje and Whillans (1994) emphasize the role of cyclic subaerial denudation and debris removal by glacier ice. In both cases, the trough head marks the upper position of effective subglacial erosion. Its location may be influenced by several factors, including: (a) a pre-existing valley step; (b) increases in ice discharge due to the convergence of two or more tributary glaciers; or (c) lithological or structural variation in the bedrock leading to a downglacier increase in erodibility.

In contrast with river valleys, troughs and fjords commonly have *overdeepenings* on their floors (Section 9.4.1). Overdeepened basins commonly form where ice discharges are relatively high, such as at the junctions of tributary valleys, or at narrowings in the valley profile (Hattersley-Smith, 1969). Overdeepened sections terminate at a *sill* or *threshold*, commonly located where ice flow becomes less constrained and velocities decrease. This is clearly illustrated by the relationship between topography and basin depth along Inugsuin Fjord on Baffin Island, Canada, which has been overdeepened by glacier ice converging on it from the surrounding mountains (Fig. 9.43; Løken and Hodgson, 1971).

The area of deepest erosion in troughs and fjords marks the long-term average position of maximum ice discharge, which may broadly coincide with the equilibrium line on the ice surface. In coastal mountains, the shallowing of fjords and the occurrence of thresholds may be associated with the increased buoyancy and eventual flotation of glacier ice in the marine environment (Crary, 1966), in addition to the flow divergence induced by the more open topography (Shoemaker, 1986b). The glacioisostatic rebound that accompanies deglaciation can result in the emergence of fjord thresholds to produce numerous low-lying bedrock islands (*skerries*), such as those at the mouths of the Norwegian fjords, or land-locked lake basins, such as those in western Newfoundland, Canada (Berger *et al.*, 1992). Some fjords have simple, single basin forms (e.g. Milford Sound, New Zealand; Bruun *et al.*, 1955), whereas others have several basins and multiple sills (e.g. Hardangerfjord, Norway; Holtedahl, 1975; Syvitski *et al.*,

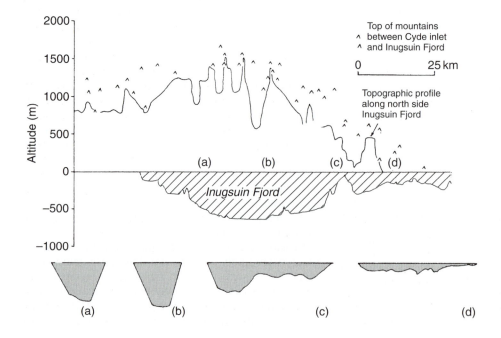

Fig. 9.43 Relationship between the depth of Inugsuin Fjord, Baffin Island, and surrounding topography. The fjord is deep and narrow in the vicinity of high mountains but wide and shallow towards the coast. (From Sugden and John, 1976, after Løken and Hodgson, 1971)

1987). These contrasting forms give rise to profoundly different postglacial sedimentation patterns.

Another aspect of trough/fjord form critical to understanding glacial erosion patterns is the stepped long profile, which comprises basins and intervening *riegels,* or transverse rock bars. Irregularities in the glacier bed may arise from several factors, the most important of which are: (a) spatial variations in bedrock lithology and structure (Matthes, 1930; King, 1959); (b) changing glacier discharge as a result of ice confluence or diffluence (Veyret, 1955; Holtedahl, 1967; Glasser, 1995); and (c) pre-existing aspects of the relief (Klimaszewski, 1964; Bakker, 1965).

9.4.3.3 THE INFLUENCE OF GEOLOGY AND TOPOGRAPHY

Several researchers have argued that the location and planform of troughs and fjords reflects pre-existing topography or geological structures. The sinuous or meandering courses that some troughs and fjords display are often interpreted as preglacial legacies, where glacier ice has excavated pre-existing fluvial valleys (e.g. Bretz, 1935; Ahlmann, 1941; Gjessing, 1966; Holtedahl, 1967; Funder, 1972; Augustinus, 1992). Such interpretations are supported where traces of largely unmodified fluvial valleys occur in the same landscape (e.g. Linton, 1963; Clayton, 1965; Sugden, 1968), or fjord systems display dendritic patterns such as in the Canadian Arctic and Fiordland, New Zealand (Pelletier, 1966; Sugden, 1978; Augustinus, 1992). The juxtaposition of glacially over-deepened and unmodified valleys is a product of differential modification according to valley alignment and former ice flow direction. Those valleys oriented parallel to the ice flow are most likely to carry large volumes of ice and experience the most intense glacial erosion. The overdeepening of valleys aligned parallel to ice flow is thought to be a product of the greater ice thicknesses over such sites and the increased frictional heat produced by preferential ice flow downvalley (Haefeli, 1968). Basal ice in these locations is, therefore, prone to pressure melting and basal sliding.

Open-ended or *through troughs,* especially those which have breached pre-existing watersheds, are more difficult to explain, but probably result from the streaming of ice down preglacial fluvial valleys and the headward erosion of trough heads (Dury, 1953; Linton and Moisley, 1960; Gjessing, 1966). Such a change in the basal topography over time can result in the capture of glacier ice and a radical change in basal flow directions. Alternatively, glacier ice may exploit regional zones of structural weakness in the bedrock such as fault systems (Bretz, 1935; Funder, 1972) and therefore bear no resemblance to preglacial drainage patterns; troughs excavated along such zones are entirely glacial in origin.

The importance of geologic structures and pre-existing structural forms (e.g. grabens) to the funnelling of glacier ice has been stressed by numerous researchers. The close affinity between major ice streams and geologic features is certainly obvious at some locations; for example, Lambert Glacier in Antarctica occupies the Lambert Graben, a major structural feature. Just as the sinuosity of troughs and fjords has been linked to pre-existing fluvial valleys, so trough and fjord alignments have been linked to bedrock lineaments such as faults and intrusions. The rectilinear pattern of some fjord systems has been linked to intersecting lines of fracture at a regional scale, and Gregory (1913, 1927) argued for a predominantly tectonic origin for fjords. Despite the claims of Nicholson (1963) that most fjords in Norway cross-cut major fault systems, careful mapping of structural lineaments and intrusions by, for example, Randall (1961) for northern Norway, Holtedahl (1967) and Nesje and Whillans (1994) for southern Norway, and England (1987) for Ellesmere Island in the Canadian Arctic, has revealed a close relationship between the orientation of such bedrock features and fjords or valleys. Furthermore, the remarkably straight and parallel cliffs of Somerset Island, Devon Island, and the north-west of Baffin Island in the Canadian Arctic, which form virtually unbroken coastlines in places up to 200 km long, have been cited as evidence for large-scale block faulting by Kerr (1980) and England (1987). This interpretation for the inter-island channels of the central Canadian Arctic is supported by the occurrence of remnants of formerly continuous Tertiary river valleys on the islands (horsts) and on the floors of the channels (grabens; Bornhold *et al.,* 1976; Dyke *et al.,* 1992). Rift valleys in non-glaciated regions provide clear examples of tectonically controlled landforms, and it is not unreasonable to suggest that such landforms exist in glaciated regions, especially where geological evidence provides support for a tectonic origin (Kerr, 1980). However, it is most likely that a continuum of landforms exists, ranging from tectonically controlled grabens through glacially modified river and fault systems to entirely glacially eroded troughs and fjords. The fjords and troughs of northern Ellesmere Island contain several contrasting elements, such as straight walls, rectilinearity and sinuous reaches, indicating that a combination of factors, some perhaps more dominant than others, is critical to the glacial excavation of such landscapes (Fig. 9.44).

An argument in favour of the tectonic origin of some inter-island channels, such as those of the central Canadian Arctic archipelago, is the lack of an integrated upland glacier source area, like those surrounding the fjord coastlines of Norway, British

Fig. 9.44 Satellite image of north-western Ellesmere Island, Arctic Canada, showing remarkably long, straight fjord walls and both rectilinear and sinuous fjord patterns. (Image provided by Martin Jeffries; numbers refer to sites in Jeffries *et al.*, 1992)

Columbia, Chile and New Zealand, and the absence of a nearby continental interior that could supply large volumes of ice. In addition, some of the largest channels, such as Lancaster Sound, are oriented transverse to the former ice flow of the Laurentide ice sheet. In contrast, the troughs and fjords presently occupied by the outlet glaciers of the Greenland and Antarctic ice sheets lie parallel to ice flow and are much more likely to have been eroded by the ice draining through them.

9.4.4 Cirques

9.4.4.1 CIRQUE MORPHOLOGY

Cirques (also known as *corries or coires* in Scotland, and *cwms* in Wales) are among the most characteristic forms of glacial erosion in mountainous terrain (Fig. 9.45). Evans and Cox (1974) defined a cirque as

> a hollow, open downstream but bounded upstream by the crest of a steep slope (headwall), which is arcuate in plan around a more gently sloping floor. It is 'glacial' if the floor has been affected by glacial erosion while part of the headwall has developed subaerially, and a drainage divide was located sufficiently close to the top of the headwall for little or none of the ice that fashioned the cirque to have flowed in from outside.

A typical cirque has a flat-floored or overdeepened basin connected to a steep backwall by a concave slope. This profile is a product of subglacial erosion on the cirque floor and lower backwall, and subaerial frost action on the upper backwall, the exact form being controlled by bedrock structure (McCabe, 1939; Battey, 1960; Haynes, 1968). Most descriptions of cirques emphasize their steep headwalls, their flat or overdeepened rock floors and their simple arcuate shapes (e.g. Lewis, 1938; Sugden, 1969).

However, they are rarely simple features; the headwall may be poorly developed or missing, they may have rectilinear planforms, or cirques can occur nested one within the other (e.g. Derbyshire, 1968; Haynes, 1968; Gordon, 1977; Bennett, 1990). The following types can be defined:

1. *simple cirques*, which are distinct, independent features;
2. *compound cirques*, in which the upper part consists of two subsidiary cirques of approximately equal size;
3. *cirque complexes*, in which the upper part consists of more than two subsidiary sidewall or headwall cirques;
4. *staircase cirques*, where two or more cirques occur one above the other; and
5. *cirque troughs*, where the cirque marks the upper end of a trough (Gordon, 1977; Bennett, 1990).

Some cirques may have attributes of more than one category, for example, compound or staircase cirques forming trough heads. A wide range of cirque sizes exists, ranging from valley-side niches less than 50 m from backwall to lip, up to cirques with floors several kilometres long and forming *alpine valley heads* (Gordon, 1977; Haynes, 1995).

Attempts to provide general classifications of cirque form have employed a variety of dimensions (e.g. Manley, 1959; Andrews, 1965; I.S. Evans, 1969; Andrews and Dugdale, 1971; Gordon, 1977). Some commonly used measurement parameters are: *length* from headwall to threshold; *width* between the sidewalls; *depth* from headwall crest to floor; *threshold elevation*; *area*; *volume*; and *aspect* (Gordon, 1977; Bennett, 1990). The long profiles of cirques have been generalized mathematically. Lewis (1960), for example, fitted cirque profiles to arcs of circles, and Haynes (1968) used logarithmic curves of the form:

Fig. 9.45 (a)

(b)

Fig. 9.45 Typical cirque forms. (a) A cirque presently occupied by a small glacier below the aptly named Cirque Mountain, Torngat Mountains, Labrador, Canada. Several Neoglacial moraines can be seen in the foreground. (Photo: D.J.A. Evans) (b) Map of the cirques surrounding Cader Idris, north Wales. The lakes Llyn Cau and Llyn y Gadair occupy rock basins, but are also partially moraine-dammed. (From Embleton and King, 1975)

$$y = k(1 - x)e^{-x} \qquad (9.3)$$

where y is altitude, x is distance along the profile, and k is a constant describing the concavity of the long profile.

Where $k = 2$, the basin is deep with a steep headwall, and where $k = 0.5$, the cirque is shallow with a flat or outward-sloping floor. Thus high k values correspond to well-developed, overdeepened cirques (Fig. 9.46).

Several attempts have been made to reconstruct sequences of cirque evolution using *ergodic models*, in which a continuum of forms in space is used as a substitute for evolutionary development through time. This approach is common in geomorphology, where long-term processes cannot be measured, and is based on the idea that in areas with evolving landforms there will be examples at various stages of development, which can be arranged in an hypothetical developmental sequence (e.g. Schumm and Lichty, 1965). In general, as cirque size increases, the degree of enclosure of the planform and long profile increases (Gordon, 1977; I.S. Evans and Cox, 1995), indicating that cirques develop from hollows by progressive retreat of the backwall and downcutting of the floor (Fig. 9.47). It has been argued that this evolution reflects the progressive growth of the occupying snow and ice body, from a snowpatch in a *nivation hollow*, through a larger snowpatch in a *nivation cirque*, to a glacier in a true cirque (e.g. Matthes, 1900; Wright, 1914; Russell, 1933; Boye,

1952; Watson, 1966; I.S. Evans and Cox, 1974; Vilborg, 1977). However, this view is probably highly simplistic, and cirques are likely to evolve from a variety of pre-existing hollows in several episodes of glacier occupancy (I.S. Evans and Cox, 1995).

9.4.4.2 PROCESSES OF CIRQUE FORMATION

The enlargement of mountainside hollows can occur by snowpatch erosion or *nivation*, which embraces a range of weathering and transport processes accelerated by late-lying or permanent snow (Thorn, 1979;

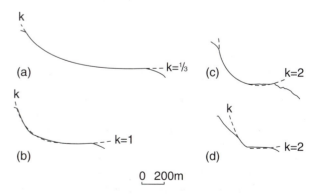

Fig. 9.46 *k*-curves fitted to cirque long profiles in Scotland. Eighty-one per cent of the sampled long profiles closely resemble *k*-curves, although some, such as (d), do not. (From Haynes, 1968. Reprinted by permission of Scandinavian University Press)

A PLAN

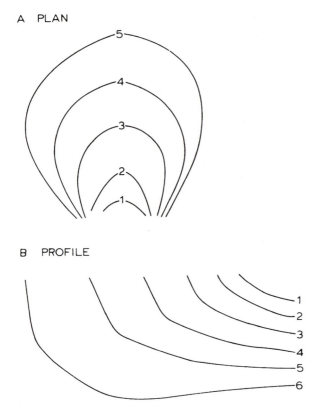

B PROFILE

Fig. 9.47 Suggested model of cirque development by the progressive enlargement and deepening of a hollow. (From Gordon, 1977. Reprinted by permission of Scandinavian University Press)

Ballantyne and Harris, 1994). Nivation processes include intensive freeze–thaw activity, enhanced chemical weathering, slopewash, debris transport by snow creep or slip, and solifluction, which are encouraged by repeated freeze–thaw cycles and/or abundant meltwater during the summer months (Ballantyne, 1978; Thorn, 1979; Thorn and Hall, 1980). The relative importance of these processes, and their variation with climate and snowpatch thickness, are very poorly known. Mountainside hollows can also originate by fluvial erosion or large-scale slope failures. True cirque formation occurs once hollows are occupied by glacier ice; that is, where the ice is sufficiently thick and steep to generate the stresses required for internal creep deformation and basal sliding. Ballantyne and Benn (1994a) argue that the threshold length between snow patches and glaciers is approximately 30–70 m from backwall to toe.

Glaciers enlarge cirques by subglacial quarrying and abrasion, which erode the floor, and subaerial slope retreat, which supplies debris to the glacier surface. Effective subglacial erosion and cirque downcutting require that the ice is at least partially wet-based, and overdeepened rock basins mark sites

where subglacial erosion was particularly effective. Possible scenarios include: (a) cirque occupancy by small, subpolar glaciers, which are wet-based where the ice is thickest, but cold-based in the thinner marginal zone (Richardson and Holmlund, 1996); (b) occupancy by temperate glaciers, with overdeepening occurring in areas where sliding velocities were highest, in the region of the glacier equilibrium line; and (c) enhanced erosion due to the availability of surface meltwater via bergschrunds or crevasse fields (Hooke, 1991).

Backwall retreat occurs mainly by a combination of mechanical weathering and mass movements (Section 6.3). Debris delivered to the glacier surface is transported away by glacier flow. The least-known aspect of cirque erosion is what happens in the transition zone between subglacial and subaerial process domains, at the upper margin of the glacier. It has been argued that mechanical weathering may be particularly effective in the *randkluft* (the crevasse between a cirque glacier and the backwall, Section 6.2.1), because of favourable temperature cycles and abundant meltwater. However, data collected in randklufts demonstrate that temperatures are commonly stable at or below freezing and that there is often very little evidence of frost shattering (Battle and Lewis, 1951; Thompson and Bonnlander, 1956; Battle, 1960; Gardner, 1987). Measurements taken by Gardner (1987) in the randkluft of the Boundary Glacier in the Canadian Rockies indicate that considerable frost shattering takes place on the lower backwall rather than in the randkluft, where summer temperatures do not rise above freezing. During the ablation season the randkluft migrates downslope and thus extends the rock surface area exposed to freeze–thaw weathering. Over longer periods of glacier thickening and thinning, large areas of cirque backwalls can be exposed to mechanical breakdown in this way. Furthermore, during the ablation season, subglacial quarrying may be facilitated downglacier of the randkluft or bergschrund, owing to large diurnal variations in the supply of meltwater from the surface (Section 9.4.1; Hooke, 1991; Iverson, 1991).

The definition of cirques cited in Section 9.4.4.1 excludes those features fed by ice from outside the cirque basin. However, this is difficult to demonstrate in many situations, such as where cirques are incised into the margins of plateaux. Ice is likely to form first on the high plateau at the onset of glacial conditions and spill over cirque headwalls, so modifying their form (McDougall, 1995). Thus, in such settings, long-term cirque evolution over several glacial cycles may reflect varying conditions, a possibility that presents problems for ergodic models.

Some calculations of cirque erosion rates and development cycles have been attempted in a variety of settings. The erosion rates of cirques cut into the sides of radiometrically dated volcanoes in Antarctica have been calculated by Andrews and LeMasurier (1973) as $360-460\,mm\,kyr^{-1}$. For these cirques, rates of backwall erosion were far greater $(5800\,mm\,kyr^{-1})$ than sidewall erosion rates $(800\,mm\,kyr^{-1})$, indicating that these cirques will tend to become more elongate over time. This pattern is not true of all cirques, however, as relative backwall and sidewall erosion rates will be strongly dependent on geological and microclimatic factors (Bennett, 1990; Bennett and Glasser, 1996). Larsen and Mangerud (1981) calculated erosion rates of $500-600\,mm\,kyr^{-1}$ for a cirque glacier in western Norway, indicating that up to 125 kyr would be required to erode the cirque. Somewhat lower erosion rates of $60\,mm\,kyr^{-1}$ were calculated by Andrews and Dugdale (1971) for cirques on eastern Baffin Island, Canada, by assuming that the cirque was occupied by ice for 1 million years. Anderson (1978) calculated erosion rates of $8-76\,mm\,kyr^{-1}$ for cirque glaciers on Baffin Island, indicating that 2–14 million years was required for cirque production. Studies of sediment concentration in meltstreams indicate present-day erosion rates of c. $200-600\,mm\,kyr^{-1}$ for cirque and valley glaciers in Norway (Schneider and Bronge, 1996; Østrem et al., cited by Richardson and Holmlund, 1996).

9.4.5 Strandflats

Named after the prominent examples of the Norwegian coast, *strandflats* are extensive, undulating rock platforms located close to sea-level around the coasts of high-latitude land masses including Norway, Greenland, Spitsbergen, Iceland, Scotland, Ireland and Antarctica. Where partially submerged, they appear as a zone of low rocky islands or skerries, known as *skjaergard* in Norway (Reusch, 1894; Nansen, 1904, 1922; O. Holtedahl, 1929; Strom, 1948; H. Holtedahl, 1960). Strandflats are up to 50 km wide, cut across geological structures, and usually end abruptly at an inland cliffline or break of slope (Fig. 9.48). In some instances they may extend short distances up fjords, but they are generally associated with the shallow water depths of fjord mouths. When viewed parallel to the coast, strandflats are remarkably horizontal, but they may possess very small seaward slopes, in part owing to postglacial glacioisostatic displacement. In some locations several strandflats may occur at various heights above and/or below present sea-level.

Strandflat formation has been explained by four main mechanisms, acting in combination: (a) frost action, combined with active sea-ice rafting; (b) marine erosion; (c) subaerial erosion; and (d) subglacial erosion. *Frost action* is considered to be an effective erosive agent on periglacial shores, owing

to a potent combination of frequent cold temperatures and abundant water (Matthews et al., 1986; Dawson et al., 1987). Furthermore, frost-shattered debris can be readily removed by sea-ice which freezes on to the near-shore sea floor in winter (the *icefoot*), then breaks up and floats away in spring and summer (Dionne, 1973). Together, frost shattering and ice rafting are thought to be capable of rapid coastal erosion (Nansen, 1904, 1922; Sissons, 1974b; Dawson, 1980, 1982; Larsen and Holtedahl, 1985). *Marine erosion* includes quarrying and abrasion by large, high-energy storm waves (Chorley et al., 1984). Some strandflats exhibit striated and smoothed bedrock outcrops, providing evidence for *glacial erosion*.

Abrupt inner margins and clifflines are consistent with marine erosion, although strandflats tend to be very wide, with small or no seaward slopes, unlike marine platforms, which have prominent seaward slopes. It is difficult to reconcile the large width and low slope of strandflats with a purely marine origin, because the energy of large waves is dissipated as they travel through shallow water, severely reducing their erosive capability at the shoreward margin of a wide platform. These problems prompted Ahlmann (1919) and Nansen (1922) to suggest that preglacial subaerial denudation produced a peneplain, which was later modified by wave action during the Quaternary. An alternative scenario was envisaged by O. Holtedahl (1929), Dahl (1947) and H. Holtedahl (1959) in which strandflats are formed by a two-stage glacial process involving (a) the coalescence of cirque and valley floors by glacial headward erosion (Fig. 9.49), followed by (b) ice sheet erosion. Because many strandflats are wider and best-developed on wave-exposed coastlines, Strom (1948) suggested that marine abrasion was followed by glacier modification of the inner margins. A worldwide review of strandflats by Guilcher et al. (1986) highlighted their probable polygenetic nature, but stressed the impact of glacial erosion in the evolution of the more extensively developed and wider forms. Guilcher et al. provided evidence for modern-day frost action and coastal erosion of the strandflats of Spitsbergen and the South Shetland Islands, Antarctica, but suggested that the Spitsbergen examples are so wide that they must have been initiated either by glacial erosion by piedmont lobes or as Tertiary erosion surfaces. Similarly, Guilcher et al. invoked a preglacial planation surface as a precursor for the small strandflats of western Ireland.

It appears from the literature that enough evidence is available to support a combined frost action and marine genesis for narrow strandflats with seaward tilts, which should therefore be termed *marine platforms* (e.g. John and Sugden, 1971, 1975; Curl, 1980; Hansom, 1983b). The wider, often horizontal

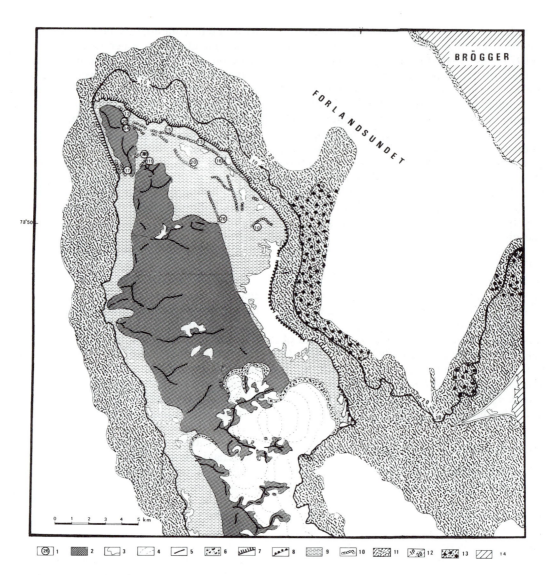

Fig. 9.48 Strandflats and associated features on northern Prince Charles Island, Spitsbergen. 1: Altitude of coastal features in metres; 2: mountain; 3: lagoon; 4: glacier; 5: mountain ridge; 6: moraine; 7: cliff; 8: coastal pebble ridge; 9: emerged strandflat; 10: raised pebble ridge; 11: submarine strandflat; 12: submarine contour, in metres; 13: submarine strandflat with pebble ridges; 14: emerged area not shown in detail. (Modified from Guilcher *et al.*, 1986)

platforms of glaciated coasts, which commonly merge with glaciated cirque, valley and fjord floors and which contain direct evidence of glacial erosion, are more likely to have been at least partially shaped by glaciation. The fact that the strandflats of Norway merge inland with the preglacial (or *paleic*) surface lends strong support to the theory that they are glacially modified erosion surfaces (*etchplains* and *peneplains*) which have escaped deep ice erosion because of the reduction in erosive capacity at the ice sheet margins. One intriguing possibility is that strandflat erosion may occur by planation beneath partially floating glacier tongues near marine ice sheet margins. However, the abrupt inner margins of

many strandflats, which are marked by bedrock cliffs extending over long distances and are transverse to former glacier flow, are difficult to explain by glacial erosion, and therefore probably document at least a partial marine influence in strandflat genesis. Porter (1989) has argued that strandflats formed during 'average' glacial conditions prevailing throughout most of the Quaternary (Section 1.7.2.7), when relative sea-level along many glaciated coasts was similar to that of today, and coastal zones were subject to repeated alternations between periglacial, marine and subglacial erosive processes during glacier advance and retreat cycles.

(a)

(b)

Fig. 9.49 (a) A strandflat linked to the floors of deep troughs at the margin of the Eggum massif, Vestvagoy, northern Norway. Beaches on the strandflat have been constructed from reworked moraines. (Photo provided by A. Guilcher) (b) A narrow strandflat near Cape Sabine, Ellesmere Island, Arctic Canada. (Photo: D.J.A. Evans)

9.5 LANDSCAPES OF GLACIAL EROSION

Sugden (1974) and Sugden and John (1976) defined distinctive *landscapes of glacial erosion*, consisting of regional associations of erosional landforms. Their classification is based on both morphology and processes of erosion, and therefore sheds light on the links between glacier dynamics and landscape evolution. Five landscapes of glacial erosion were recognized: (a) *landscapes of areal scouring*, which everywhere bear signs of glacial erosion; (b) *landscapes of selective linear erosion*, in which erosion is confined to troughs between unmodified plateau remnants; (c) *landscapes of little or no glacial erosion*, which are essentially unmodified preglacial landscapes that have survived one or more periods of

glacial occupancy; (d) *alpine landscapes*, consisting of dendritic networks of troughs separated by ridges; and (e) *cirque landscapes*, in which essentially discrete cirques are set in a hill or mountain massif. These landscapes reflect the long-term influences of pre-existing topography, glacier and ice sheet morphology, basal thermal regime and mass flux, and develop cumulatively over multiple glacial cycles. They are therefore more likely to relate to 'average' glacial coverage during the Quaternary than to 'instantaneous' ice configurations at glacial maxima (Porter, 1989).

9.5.1 Areal scouring

Landscapes of areal scouring are extensive tracts of subglacially eroded bedrock, consisting of rock knobs, roches moutonnées and overdeepened rock basins (Figs 9.50 and 9.51). In reference to this characteristic morphology, Linton (1963) referred to such terrain as *knock and lochain* topography, from the Scots Gaelic words *cnoc*, meaning 'knoll', and *lochain*, meaning 'small lake'. The location of high and low points in areally scoured terrain reflects bedrock lithology and structure. Hollows and rock basins occur where joint density is high or where less resistant rocks crop out, whereas knolls are underlain by more resistant rocks. However, relief amplitude is generally low, and is typically less than 100 m. Spectacular examples of areally scoured landscapes occur on the Canadian Shield, in west Greenland and in north-west Scotland (Linton, 1962, 1963; Sugden, 1974; Gordon, 1981; Rea and Evans, 1996).

The widespread evidence for abrasion and plucking in landscapes of areal scouring indicates that they develop below wet-based ice, in situations where flow is laterally extensive and not focused into narrow channels. The link between scoured landscapes and thermal conditions was used by Andrews *et al.* (1985a, b), who identified the location of former ice streams on Baffin Island by mapping zones of high lake density. A cautious approach should be adopted when attempting reconstructions of former ice masses from erosional landscapes, however, because of the distinct possibility that the present landscape is a palimpsest produced during several glacial stages.

The amount of erosion represented by landscapes of areal scouring is very difficult to determine, and has been the subject of much debate. White (1972) suggested that the Canadian Shield has been subject to deep erosion by successive Laurentide ice sheets, with the greatest amount of erosion located beneath the ice sheet centre in the vicinity of Hudson Bay. This idea has been criticized by Gravenor (1975), Sugden (1976, 1978) and Higgs (1978), who argued that the Hudson Bay basin is a very ancient feature which contains Palaeozoic sedimentary rocks, and

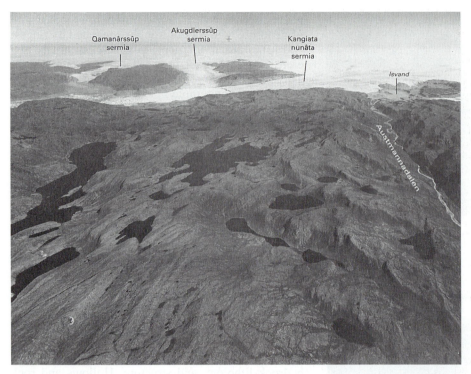

Fig. 9.50 View across are-ally scoured terrain towards the southern margin of the Greenland ice sheet. To the right of this view areal scour has only partially modified an older cirque terrain. A landscape of selective linear erosion can be seen near the ice sheet margin. Aust-mannadalen is the valley descended by Fridtjof Nansen after his first cross-ing of the ice sheet in 1888. (Photograph by Danish National Survey and Cadas-tre; provided by Richard Williams, USGS)

that the Canadian Shield must have been eroded to something like its present level by Palaeozoic times. Thus, glacial erosion may have been responsible for only relatively minor changes to the landscape. Bell and Laine (1985) estimated the amount of Pleis-tocene glacial erosion by the Laurentide ice sheets from the volume of glacial-age sediment on land and below the adjacent seas. There are many uncertain-ties with this approach, but they concluded that the amount of glacial erosion during the past 3 million years probably lay in the range 120–183 m, averaged over the whole area of glacial occupancy. Since part of the area occupied by the ice sheets experienced net deposition, the average amount of erosion in the Shield area was possibly even higher. These esti-mates thus provide indirect evidence for deep erosion of the Shield, although part of the glacial erosion could have consisted of the stripping off of weaker sedimentary strata. Interestingly, the classic land-scapes of areal scouring in north-west Scotland form part of a land surface that can be traced beneath Late Precambrian sedimentary rocks (McCulloch, 1819; A.D. Stewart, 1991), showing that in this area glacial erosion has exhumed and modified an extremely ancient erosion surface.

9.5.2 Selective linear erosion

Landscapes of selective linear erosion are character-ized by deep troughs separated by essentially unmodified plateau surfaces (Fig. 9.40 (Plate 18)); Sugden, 1968, 1974). Striae or polished surfaces may

cover the trough sides right up to the cliff top, show-ing that the troughs were once completely filled with erosive ice, whereas periglacial forms such as block-fields and tors may be preserved on the plateaux a few tens of metres from the trough edge. In planform, the troughs may form a dendritic network or a more complex interconnected pattern. Landscapes of selective linear erosion are well represented in east and north Greenland, the Allegheny plateau area of North America, Scandinavia, the Cairngorms of Scotland, and some Arctic islands (Sugden, 1968, 1974; Sugden and John, 1976).

It is generally agreed that such landscapes develop beneath ice sheets, with the troughs mark-ing former ice streams and the intervening plateaux marking areas of slowly moving or cold-based ice. The presence of troughs is thought to reflect positive feedbacks between subglacial topography, ice veloc-ity, basal temperature and erosion rates. Where ice sheets occupy irregular topography, basal melting will be encouraged within valleys or troughs, where the ice is thickest. Basal sliding provides another source of heat at the glacier bed, promoting further basal melting and increased sliding rates. As a result, erosion rates will be highest within valleys, deepening and enlarging them, and promoting more basal melting, fast sliding and erosion. Thus, pre-existing valleys in a landscape will tend to be exploited by ice flow and transformed into troughs. In contrast, high points beneath the ice sheet will tend to be occupied by thin, cold-based ice which does not erode but protects the pre-existing land sur-

Fig. 9.51 Landscape of areal scour near Loch Laxford, north-west Scotland. (RCAHMS aerial photograph, Crown Copyright)

face. Pre-existing valley systems will also encourage faster than average ice flow by collecting ice from large catchment areas, thus increasing the mass balance velocity (Section 4.1).

Sugden and John (1976) have noted that in Greenland, Scandinavia and Scotland, landscapes of selective linear erosion are most common in areas removed from the maritime fringes of ice sheets. This association probably reflects the relatively low mass turnover in areas removed from oceanic moisture sources, where rapid, erosive flow can occur only in favourable locations.

It is inherently difficult to derive quantitative assessments of the amount of glacial erosion represented by troughs, because of the unknown depth of preglacial valleys and an incomplete understanding of the tectonic contribution to relief production. By using the reconstructed *paleic* or preglacial surface in the Sognefjord drainage basin and subtracting the present-day topography of the fjord, Nesje *et al.* (1992) calculated that $7610 \, \text{km}^3$ of material has been removed during successive Quaternary glaciations (Fig. 9.52). Depending upon the amount of Quaternary time over which glaciations have been eroding Sognefjord, this yields a range of values from 102 to $330 \, \text{cm} \, \text{kyr}^{-1}$ for ice stream erosion rates. The reconstruction of the paleic surface allows us to view the nature of the preglacial relief and provides evidence for the channelling of ice and the concentration of glacial erosion by pre-existing topography. Nesje and Whillans (1994) proposed the following evolutionary development of Sognefjord as a representative example of the Norwegian fjords (Fig. 9.53):

- *Phase 1*, deep chemical weathering and erosion during the Mesozoic and early Tertiary to produce the paleic surface just above sea-level;

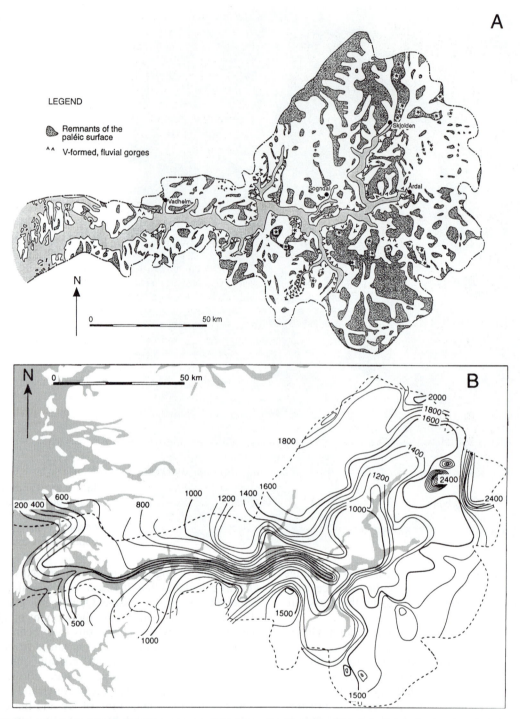

Fig. 9.52 The erosion of Sognefjord, west-central Norway. (A) The Sognefjord drainage basin, showing remnants of the paleic surface, V-shaped valleys and gorges (V-symbol) and Tertiary soil remnants (stars). (B) Reconstruction of the paleic surface based on interpolation of paleic surface remnants; elevations in metres. The dashed line marks the edge of the modern drainage basin. (From Nesje and Whillans, 1994. Reprinted by permission of Elsevier)

- *Phase 2*, a preglacial phase of Tertiary uplift, which was greater towards the west coast, and predominantly fluvial incision of valleys guided by rock fracture patterns;

- *Phase 3a*, interglacials and interstadials when subaerial processes erode the fractured rock, especially at valley heads, and fill valley floors with the debris (the paleic surface not greatly affected);

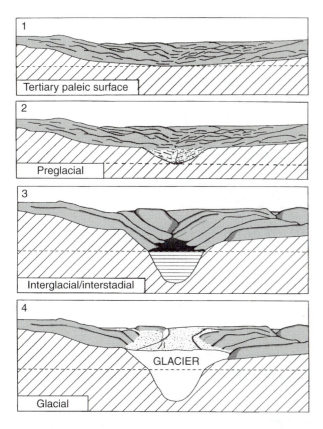

Fig. 9.53 Schematic reconstruction of the main phases of development of Sognefjord, Norway. The dashed line represents sea-level. (From Nesje and Whillans, 1994. Reprinted by permission of Elsevier)

- *Phase 3b*, repeated glaciation throughout the Quaternary resulting in overdeepening of pre-existing valleys and a uniform but very restricted net erosion of the paleic surface.

This reconstruction emphasizes the role of slope mass wasting during interglacials and interstadials, on the basis of the observation that the largest accumulations of coarse debris today occur in fans at valley heads and in deltas, both located below bedrock fracture zones.

The dimensions of troughs appear to be scaled to the amount of ice which discharged through them, in the same way as river channels are adjusted to bankfull discharges. A relationship between trough size and ice discharge was first noticed by Penck (1905), who proposed the *law of adjusted cross-sections*. This relationship has been tested by Haynes (1972) for Greenland, Roberts and Rood (1984) for British Columbia, and Augustinus (1992) for New Zealand, by correlating parameters such as trough width, depth, length and cross-sectional area with the ice catchment area. The first four parameters ideally should be measured relative to preglacial erosion sur-

faces, if they are identifiable, between the troughs being studied. The assessments undertaken by Haynes (1972), Roberts and Rood (1984) and Augustinus (1992) confirm that relationships exist between trough size parameters and glacier contributing area, which is the equivalent of the drainage basin in fluvial systems. Roberts and Rood (1984) proposed that trough length is proportional to glacier contributing area following a power function similar to that derived by Hack (1957) for fluvial systems:

$$L_f = a\mathrm{GB}^b \qquad (9.4)$$

where L_f is trough length, GB is glacier contributing area, and a and b are constants.

There are significant correlations between various trough parameters and glacier contributing area, indicating that the erosion of troughs and fjords increases in direct proportion to the volume of ice discharged through them (Fig. 9.54). Moreover, comparisons made by Augustinus (1992) show that fjords in British Columbia are 2.5 times deeper and 2.4 times longer than New Zealand fjords for the same glacier contributing areas, indicating more intense glacial erosion on the British Columbia coast. This is probably due to several factors, the most important being (a) water depths immediately offshore, which are shallower in British Columbia and therefore less capable of floating outlet glaciers, and (b) palaeoglacial flowlines, which are considerably longer and fed by an inland ice sheet in British Columbia. In addition, the relative contribution of preglacial fluvial and tectonic processes in both areas cannot be fully assessed, although Ryder (1981) suggested that the British Columbia fjords are too deep to be explained by glacial erosion alone, and a strong structural control is evident in the rectilinear pattern of fjord development.

Problems arise where troughs or fjords have been accessed by glacier ice flowing from ice sheet interiors and overtopping watersheds, such as occurred in the through troughs and fjords of the Torngat Mountains of Labrador (Ives, 1978; Clark, 1991b) and eastern Baffin Island during Laurentide ice sheet glaciations (Dowdeswell and Andrews, 1985). Such patterns of ice flow will result in disproportionate trough sizes relative to local drainage areas. Haynes (1972) suggested that trough size–drainage area assessments can therefore be used to differentiate those features eroded by local as opposed to regional ice. However, even within local ice sheet centres, ice streams may cross from one drainage basin to another. The amount of watershed breaching by an ice mass has been quantified by Haynes (1977) and Augustinus (1992) using the interconnectivity of troughs (Fig. 9.55). Connectivity is high at former ice dispersal centres where glacier ice was capable of

(a)

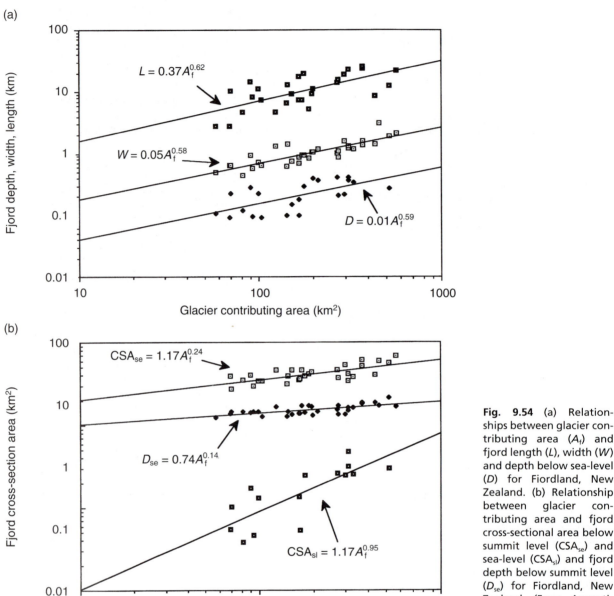

(b)

Fig. 9.54 (a) Relationships between glacier contributing area (A_f) and fjord length (L), width (W) and depth below sea-level (D) for Fiordland, New Zealand. (b) Relationship between glacier contributing area and fjord cross-sectional area below summit level (CSA_{se}) and sea-level (CSA_{sl}) and fjord depth below summit level (D_{se}) for Fiordland, New Zealand. (From Augustinus, 1992. Reprinted by permission of Elsevier)

overtopping and incising through preglacial watersheds, and low in peripheral areas.

9.5.3 Landscapes of little or no glacial erosion

Numerous examples exist of preglacial landforms, sediments and weathering horizons that have survived intact despite occupation of the area by glacier ice, perhaps on numerous occasions. Such landscapes of little or no glacial erosion owe their preservation to protection by cold-based ice caps or sectors of ice sheets (Sugden, 1974, 1978; Whillans, 1978; Roald-

set *et al.*, 1982; England, 1986, 1987; Kleman, 1992, 1994; Dyke, 1993b; Kleman and Bergstrom, 1994).

For example, land surfaces of probable Tertiary age are preserved within the limits of late Wisconsinan ice caps in the central Canadian Arctic archipelago (Dyke *et al.*, 1992; Dyke, 1993b; Fig. 9.56). The land surfaces display dendritic fluvial valleys, tors, blockfields, patterned ground and gelifluction terraces, features characteristic of subaerial weathering and erosion in a periglacial climate. Similar rock types which are known to have been ice-scoured during the last glaciation bear little or no weathered regolith, indicating that the weathered landscapes

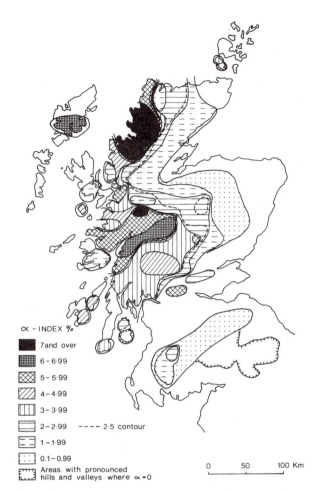

∝ - INDEX %

■	7 and over
▤	6 - 6·99
▨	5 - 5·99
▨	4 - 4·99
▥	3 - 3·99
☰	2 - 2·99 - - - - 2·5 contour
☷	1 - 1·99
⋰	0·1 - 0·99
⌁	Areas with pronounced hills and valleys where ∝ = 0

0   50   100 Km

Fig. 9.55 Valley connectivity indices for Scotland, showing areas of high trough connectivity for the former ice dispersal centres of the western Highlands. (From Haynes, 1977. Reproduced by permission of Scandinavian University Press)

represent tens of thousands of years of preglacial subaerial modification. In such areas, the only signs of glacial modification may be marginal meltwater channels cut during deglaciation (Sections 9.3.4.3 and 12.3.2). Palaeo-surfaces, often containing Tertiary weathering residues, exist in numerous highland terrains which are dissected by troughs and fjords (Ambrose, 1964; Bird, 1967; Sugden and John, 1976; Nesje and Whillans, 1994). In such areas, selective linear erosion marks the position of former wet-based ice streams, whereas the intervening remnant preglacial landscapes record the location of cold-based ice. Gellatly *et al.* (1988) and Rea *et al.* (1996) have described blockfields melting out from beneath cold-based plateau ice caps in the mountains of Arctic Norway. The plateaux contain no evidence for glacial modification, but rise above deeply dissected, ice-moulded terrain. The presence of protective, cold-based ice caps on high plateaux has

(a)

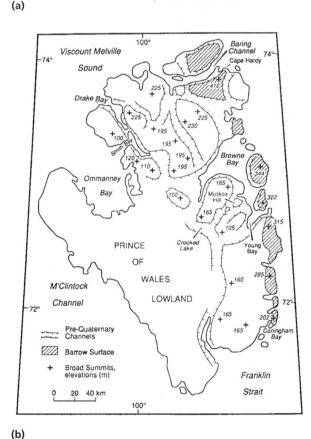

(b)

Fig. 9.56 Landscape of restricted glacial erosion: (a) Landsat image of Prince of Wales Island, Canadian central Arctic archipelago, showing fragments of Tertiary valley systems; (b) the pre-Quaternary physiographic elements of Prince of Wales Island, which have survived presumably numerous ice sheet glaciations. (From Dyke *et al.*, 1992; reprinted by permission of the Geological Survey of Canada)

important implications for glacier reconstructions, because blockfields on high surfaces could be mistaken for evidence of former nunataks (see E. Dahl, 1965; Section 12.6.1.2).

At lower latitudes, deep-weathered bedrock or saprolites exist in some areas that have been covered by ice sheets on a regular basis throughout the Pleistocene (e.g. Roaldset *et al.*, 1982; Bouchard, 1985; Hall, 1985, 1986; Peulvast, 1985; Hall and Sugden, 1987; Godard, 1961, 1989), attesting to erosional selectivity in glacial systems. On the basis of studies in Scotland, Glasser (1995) argued that erosional selectivity reflects preferential ice streaming along preglacial valleys, and correspondingly minor erosion of interfluves. He has shown that the most intense erosion occurs where the topography favours convergence of flow, implying that preglacial relief is critical to patterns of erosion by glacier ice during multiple glaciations (cf. Klimaszewski, 1964).

9.5.4 Alpine landscapes

Alpine landscapes are among the most instantly recognizable landscapes of glacial erosion. They form in areas of repeated valley glaciation, and comprise networks of troughs and cirques deeply incised into mountain massifs. During glacial occupancy, the interfluves are commonly ice free, and are typically reduced to narrow *aretes* and sharp rock peaks or *horns* (Fig. 9.57). An ergodic model of the evolution of an alpine landscape is shown in Fig. 9.58. *Stage A* represents a fluvial landscape, with rounded interfluves, sinuous valley floors, interlocking spurs and tributaries graded to the main valley floor. In *Stage B*, small cirque and niche glaciers occupy favourable locations, and accomplish limited erosion. *Stage C* represents occupation of the landscape by confluent valley glaciers, and *Stage D* represents the mature alpine landscape, consisting of troughs, hanging valleys, truncated spurs and narrow interfluves. At times of maximum glacier cover, ice may spill from one catchment to another over low points on the watershed, creating glacial breaches (Fig. 9.59).

Alpine landscapes are most spectacularly developed in high-relief, tectonically active mountain environments, such as the Himalaya, the North American Cordillera, the Andes and the New Zealand Alps. In such areas, relative relief from the valley floors to the summits can be in excess of 3000 m, attesting to very high erosion rates linked to rapid uplift, slope instability, and high throughput of water and ice. Relative relief is generally lower for tectonically stable alpine environments, and is least for old, relict mountain terrains such as the Highlands of Scotland and the mountains of western Scandinavia. The relationship between relative relief and tectonic activity suggests that alpine environ-

ments may undergo a long-term evolutionary sequence mirroring that of the mountain belt in which they occur. Glaciation will be initiated during mountain uplift, when the mountain mass intersects the regional snow line. Glacial incision and alpine landscape development will reach a maximum when rapidly uplifting, high-altitude terrain extends high above the snow line. Once uplift ceases, relative relief is reduced by erosion at an exponentially decreasing rate. Thus, mountain belts that have been tectonically active in the recent geological past (such as the European Alps) still have high-relief alpine terrain, whereas ancient mountain belts (such as the Scottish Highlands) exhibit much more subdued alpine landscapes. This simple evolutionary sequence will, of course, be interrupted by global and regional climatic change, inundation of the massif by ice sheets, and other factors, but serves as a useful basic framework for understanding alpine landscapes.

An assessment of the long-term development of an alpine landscape has been made by Kirkbride and Matthews (1997) for the Ben Ohau Range, South Island, New Zealand. This north–south-trending mountain range trends obliquely to the regional north-west to south-east rise in glacier equilibrium-line altitudes, which resulted in a marked gradient in past glacier distribution along the range. As a consequence, the south end of the range is a non-glacial landscape of fluvial incision, whereas the north end is a classic alpine glacial landscape of troughs and narrow, pinnacled intervening ridges. By combining detailed morphometric analyses of these landscapes with the reconstructed climatic and tectonic history of the region, Kirkbride and Matthews were able to demonstrate that valley forms develop recognizable glacial profiles after *c.* 70,000 years of glacier occupancy, cirques become established after 200,000 years of occupancy, and fully developed troughs require at least 320,000 years of occupancy. It was not possible to determine absolute rates of erosion, owing to uncertainty surrounding the amount of fluvial incision of the valleys, and spatial variations in glacier mass throughput and erosiveness.

9.5.5 Cirque landscapes

Cirque landscapes are characterized by independent cirques incised into upland terrain. The appearance of the landscape is strongly influenced by cirque density, and may range from isolated cirque basins cut into the edges of plateaux to densely packed cirques separated by narrow aretes (Fig. 9.60).

Because of their assumed association with small glaciers, cirques have been used as indicators of long-term average palaeoclimatic conditions. Two principal cirque attributes have been employed: *orientation* and *altitude*. Numerous studies have

Fig. 9.57 Aerial photograph of a low-relief alpine landscape occupied by a complex of valley and cirque glaciers, Wootton Peninsula, north-west Ellesmere Island, Arctic Canada. (Aerial photograph provided by the Department of Energy, Mines and Resources, Canada)

recognized the tendency for the cirques within a mountain range to be preferentially oriented in certain directions. In the mid-latitudes of the northern hemisphere, cirques predominantly face between north and east (e.g. Seddon, 1957; Schytt, 1959; King and Gage, 1961; Andrews, 1965; Temple, 1965; Sissons, 1967; Unwin, 1973; I.S. Evans, 1977; Vilborg, 1977; Bennett, 1990). In contrast, in the south-

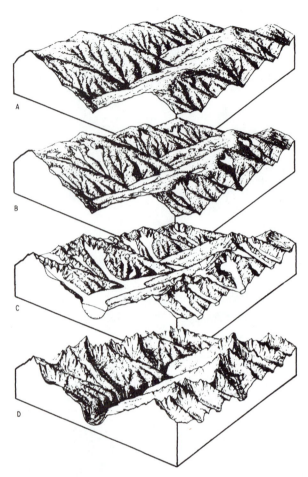

Fig. 9.58 Ergodic model of the development of an alpine landscape from a fluvial precursor. (From Flint, 1971. Reproduced by permission of John Wiley & Sons)

Fig. 9.59 Deep glacial breach, Cuernos del Paine, Chile. The breach is cut through a granite batholith (pale) and its shale capping (dark). (Photo: C.M. Clapperton)

ern hemisphere mid-latitudes, orientations are predominantly between north-east and south-east. (e.g. Galloway, 1963; Derbyshire, 1964; Clapperton, 1971a). At the higher latitudes of the Antarctic Pen-

Fig. 9.60 Cirque landscape, verging on alpine landscape: Torngat Mountains, Labrador, Canada. (Aerial photograph provided by the Department of Energy, Mines and Resources, Canada)

insula, the predominant orientations shift to north and south, with secondary peaks of north-west and south-west (Haynes, 1995). Cirque orientations for specific regions can be portrayed effectively in *cumulative vector diagrams* (Fig. 9.61). Successive legs, proportional to the number of cirques with particular orientations, are drawn on the diagram, beginning with the orientation approximately opposite to the vector mean. The straight line joining the first and last legs is the *resultant vector*, and can be used to summarize the distribution. A measure of cirque preferred orientation or *cirque asymmetry* can be calculated by expressing the length of the resultant vector as a percentage of total length of all the vectors, a quantity known as *vector strength*. I.S. Evans (1977) proposed a set of asymmetry classes based on vector strength (Table 9.3).

Such diagrams and statistics give an immediate impression of the preferred orientation of cirques in any particular region. Strong to extreme asymmetry may reflect structural and topographic controls, but such factors are unlikely to explain marked cirque asymmetry over large regions unless combined with climatic variables. The tendency for cirques in mid-latitudes to face eastwards reflects the accumulation of snow and glacier growth in the lee-side hollows of upland terrains located in the westerly wind belts (e.g. Enquist, 1917; Gloyne, 1964; King, 1974; Graf, 1976; Benn, 1989b). The poleward aspects of many cirques (north- and south-facing in the northern and southern hemispheres respectively) result from reduced ablation and preferential glacier survival in such locations, where direct solar insolation is lowest

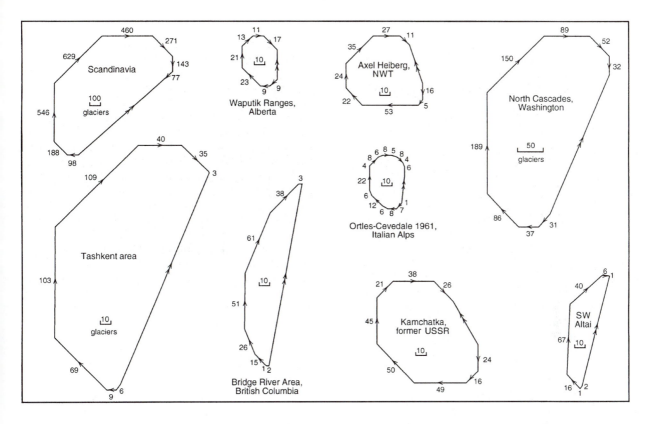

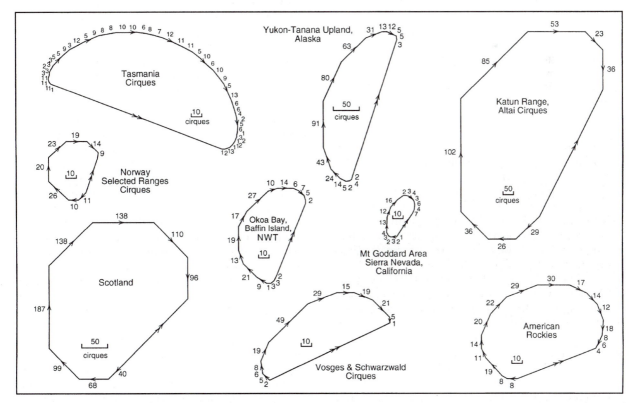

Fig. 9.61 Cumulative vector diagrams of glacier aspects (upper) and cirque orientations (lower). All glacier vector diagrams are at the same scale except for the Cascades (half scale) and Scandinavia (one-tenth scale). Cirque vector diagrams are at the same scale except for Scotland and Yukon-Tanana (half scale) and Katun (one-fifth scale). North is towards the top of the page. Resultant vectors are indicated by double arrows. (Modified from I.S. Evans, 1977)

Table 9.3 Definition of cirque orientation asymmetry categories

Vector strength	Cirque asymmetry
100%	Extreme
80%	Strong
60%	Marked
40%	Weak
20%	Symmetric

(Wendler and Ishikawa, 1974). Low cirque asymmetry may indicate that several generations of cirques, each relating to different palaeoclimates, may exist in the same area (e.g. Clark, 1991b). For example, Haynes (1995) has argued that in the Antarctic Peninsula the largest alpine valley heads with predominantly lee-side and poleward aspects (south-east and south) date to early marginal glacial conditions, whereas the smaller cirques, which face predominantly windward (north-west), formed during more recent, less marginal, glacial conditions. Therefore, weak and windward vectors may arise in locations where glaciation is less marginal and cirques are well developed on the windward sides of mountain terrains where the highest direct precipitation occurs. In some regions strong windward vectors may result from this precipitation control (e.g. Andrews and Dugdale, 1971; Aniya and Welch, 1981). The tendency for cirque asymmetry to be influenced by total glacier ice cover of a region (intensity of glacierization) was termed the *law of decreasing glacial asymmetry with increasing glacier cover* by I.S. Evans (1977). The greater asymmetry in cirque orientation is thus conceivably a measure of the marginality of a region to glacierization over the long glacial history of the Earth and may explain patterns of greater asymmetry in mid- and low-latitude upland terrains.

Regional trends in *cirque elevation* have also been interpreted in terms of palaeoclimate. For example, cirque floor altitudes in Scotland and north Wales rise from south-west to north-east, suggesting that there was a similar altitudinal distribution of cirque glaciers at various times during the Pleistocene (Sissons, 1967; Robinson *et al.*, 1971; Unwin, 1973; Bennett, 1990). This pattern most probably reflects former precipitation gradients, with conditions being most favourable for glacier growth near oceanic moisture sources (Section 1.6.2). Similar patterns have been noted elsewhere in the world, including Scandinavia (Ljunger, 1948), the western USA (Meier, 1960) and the Mount Everest region of Nepal (Müller, 1980). Palaeoclimatic reconstructions based on cirque floor altitudes must, however, be treated with caution. At best, they can only offer information on average conditions during the Pleistocene, as cirques must

have been initiated and enlarged on several occasions under varying climates. Additionally, cirque floor altitudes may have been influenced by regional uplift or subsidence since their formation, particularly in tectonically active regions such as New Zealand, the Himalaya and the American Cordillera (Holmes, 1993). Finally, the altitude of a cirque basin does not necessarily bear a close correspondence to the equilibrium-line altitude of the glacier which formed it. For example, Richardson and Holmlund (1996) found that in northern Sweden, the cirque occupied by Passglaciären probably underwent most rapid erosion on occasions when it was occupied by the head of an extended valley glacier, rather than a small cirque glacier as at present. Thus in this case, the cirque floor is probably a perched rock basin eroded below the upper reaches of a large valley glacier, and at times of more restricted cirque glaciation it is an essentially fossil form bearing no clear relationship to the prevailing climatic conditions.

9.5.6 Continent-scale patterns of erosion

An important attempt to interpret the large-scale distribution of landscapes of glacial erosion in terms of ice sheet thermal regime was made by Sugden (1977, 1978), who considered patterns of erosion beneath the Laurentide ice sheet. First, he reconstructed the surface morphology and basal temperature of the ice sheet using a simple glaciological model. The resulting thermal pattern consisted of a wet-based inner area surrounded by an outer zone of cold-based ice, which in turn was followed by a zone of wet-based ice in some areas. This pattern was then compared with the distribution of landscapes of areal scouring, selective linear erosion, and little or no glacial erosion determined from satellite images. The intensity of glacial erosion was estimated from the density of lake basins. Sudgen found a broad correspondence between the reconstructed thermal zones and the observed patterns of erosion. Landscapes of areal scour coincide with inferred wet-based areas, whereas landscapes of little or no glacial erosion occur where the ice was calculated to be cold-based. The zone of most intense erosion (identified from areas with high lake density) coincided with the modelled transition between upglacier wet-based ice and downglacier cold-based ice, conditions conducive to widespread plucking and debris entrainment (Fig. 9.62). The closeness of fit between reconstructed thermal regime and glacial landscapes is remarkable, especially considering the simplicity of the glaciological model and the fact that the mapped erosional landforms are extremely unlikely to have formed synchronously during the ice sheet maximum.

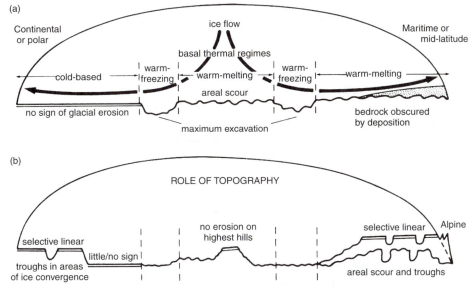

(a)

Continental
or polar

ice flow

basal thermal regimes

Maritime or
mid-latitude

cold-based | warm-freezing | warm-melting | warm-freezing | warm-melting

areal scour

no sign of glacial erosion

maximum excavation

bedrock obscured
by deposition

(b)

ROLE OF TOPOGRAPHY

selective linear

no erosion on
highest hills

selective linear

Alpine

troughs in areas
of ice convergence

little/no sign

areal scour and troughs

Fig. 9.62 A model of erosion by ice sheets. The left-hand side represents polar or continental conditions and the right-hand side mid-latitude or maritime conditions. (a) Thermal regimes and idealized erosional effects. (b) Schematic effect of topography, showing how uplands experience minimal erosion or selective linear erosion. Massifs protruding above the ice sheet surface near the margin are sculpted into alpine landscapes. In cold-based zones, troughs will form in areas of flow convergence. (From Chorley *et al.*, 1984. Reproduced by permission of Methuen)

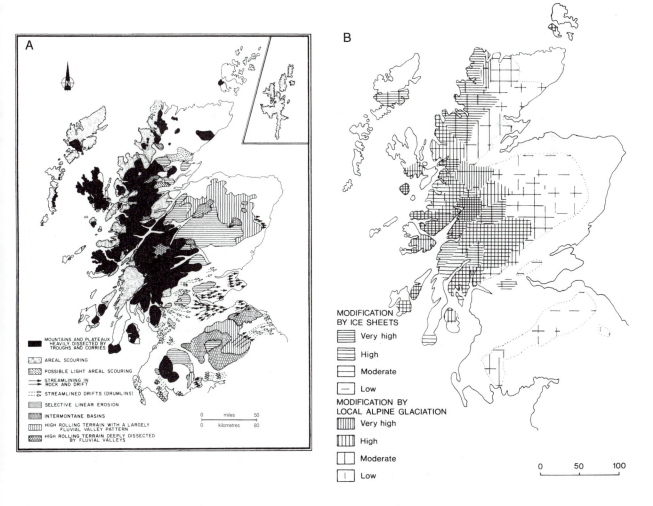

A

MOUNTAINS AND PLATEAUX
HEAVILY DISSECTED BY
TROUGHS AND CORRIES

AREAL SCOURING

POSSIBLE LIGHT AREAL SCOURING

STREAMLINING IN
ROCK AND DRIFT

STREAMLINED DRIFTS (DRUMLINS)

SELECTIVE LINEAR EROSION

INTERMONTANE BASINS

HIGH ROLLING TERRAIN WITH A LARGELY
FLUVIAL VALLEY PATTERN

HIGH ROLLING TERRAIN DEEPLY DISSECTED
BY FLUVIAL VALLEYS

0   miles   50
0   kilometres   80

B

MODIFICATION
BY ICE SHEETS

Very high
High
Moderate
Low

MODIFICATION BY
LOCAL ALPINE GLACIATION

Very high
High
Moderate
Low

0   50   100

Fig. 9.63

C

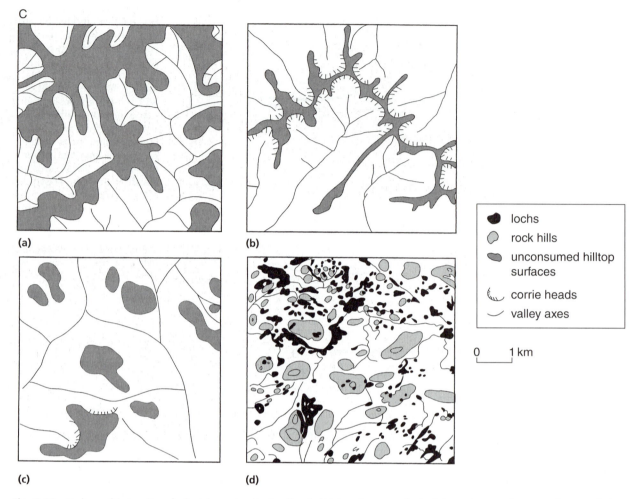

(a)   (b)

(c)   (d)

lochs

rock hills

unconsumed hilltop
surfaces

corrie heads

valley axes

0   1 km

Fig. 9.63 Style and intensity of glacial erosion in Scotland. (A) Landscapes of glacial erosion, and areas of streamlining. (From Haynes, 1983. Reprinted by permission of the author) (B) Intensity of landscape modification by ice sheets and alpine glaciers. The intensity of ice sheet glaciation was measured by a valley connectivity index, and the intensity of mountain glaciation is reflected in the proportion of valleys which terminate in cirques. (From Haynes, 1977. Reprinted by permission of Scandinavian University Press) (C) Idealized examples of the different types of glacial erosion: (a) proglacial landscape preserved; (b) mountain glaciation; (c) interconnected valleys of ice sheet glaciation; (d) areal scouring. (From Haynes, 1983. Reproduced by permission of the author)

Indeed, basal thermal zones must have migrated during glacier expansion and retreat, and may have also shifted in association with climate change, resulting in palimpsest erosional landscapes (Holm-lund and Naslund, 1994; Haynes, 1995). It is probably significant that a similar modelling exercise conducted by Gordon (1979) for Pleistocene ice sheets in Scotland revealed a generally poor fit between reconstructed thermal regimes and patterns of erosion. Further complications are introduced by topography and substrate geology. High massifs will experience minimal erosion, selective linear erosion, or areal scour and trough formation, depending on their position relative to the 0° isotherm at the bed and the degree of flow convergence. Massifs protruding above the ice sheet surface close to the mar-

gin will develop into alpine landscapes, which will be more rugged in maritime than in continental environments. It should be noted, however, that alpine landscapes can also develop during local glaciations, at times when the large ice sheets are more restricted.

An important factor influencing large-scale patterns of glacial erosion is former glacier mass balance, particularly the availability of precipitation (Sugden and John, 1976). Other things being equal, higher precipitation totals will increase the mass turnover, and therefore the erosive capability, of the glacier system (Sections 2.3.3 and 4.1.1; Andrews, 1972). For example, the amount of glacial erosion in Scotland is greater in the west than the east, owing to the greater proximity of the western Highlands to oceanic

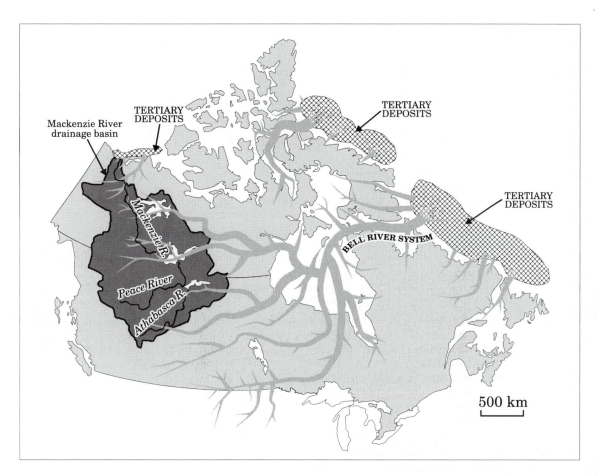

Fig. 9.64 Reconstruction of preglacial Bell River System, Canada, and the glacial diversion of the headwaters to form the modern Mackenzie River drainage basin. (From Duk-Rodkin and Hughes, 1994. Figure provided by A. Duk-Rodkin)

moisture sources (Fig. 9.63; Linton, 1963; Haynes, 1983; Hall and Sugden, 1987). In the eastern part of the country, successive ice sheets probably had low mass turnover, allowing the survival of preglacial landscape elements. The zonation of such landscapes has provided estimates of the intensity of subglacial erosion associated with ice sheets or phases of alpine glaciation (Clayton and Linton, 1964; Clayton, 1974; Sugden, 1974, 1978; Haynes, 1977, 1983).

One of the most spectacular effects of glacial erosion is the diversion of major river systems. Many of the river systems of northern Europe were div-

erted from their preglacial courses by the ice sheet advances during the middle Pleistocene Saalian glaciation (Gibbard, 1988; Arkhipov *et al.*, 1995; Section 9.3.4.3). For example, the river Thames, England, was diverted southwards from its preglacial course by the advance of the Anglian ice sheet (Rose, 1983; Gibbard, 1985). Figure 9.64 shows the reconstructed Tertiary drainage system of the Canadian Shield, and the present drainage basin of the Mackenzie River resulting from glacial diversion of the original west-to-east flow direction of the precursor Bell River (Duk-Rodkin and Hughes, 1994).

CHAPTER
10

SEDIMENT FACIES

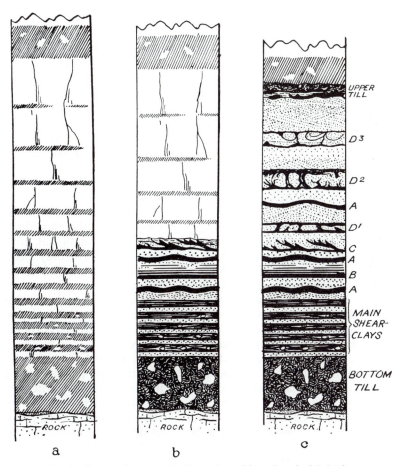

Fig. 1.—Progressive stages in the undermelting of englacial detritus. (a) Glacier section with "bottom", "banded" and, at the top, "overriding" dirts; (b) undermelt in progress, and (c) nearly completed.

10.1 INTRODUCTION

Sediment sequences in glacial environments can be bewildering in their variety and complexity, and reflect the ever-changing character of their processes of deposition. Active depositional centres shift position on a daily, seasonal and random basis, and as glaciers advance and retreat. Debris is subject to deformation, reworking and resedimentation by active and stagnant ice, flowing water, gravity and wind, and can undergo several cycles of reworking before final deposition and preservation in the geological record. In consequence, glacigenic and associated sediments exhibit enormous variability, and include sediments found in many other sedimentary environments as well as several unique to glacial settings. Additionally, many glacial sedimentary processes are difficult or impossible to observe directly, owing to the inaccessible or highly dangerous environments in which they operate.

Understanding the complex range of glacigenic sediments, and the landforms and landscapes that they underlie, has therefore provided glacial geologists with major problems, and a seemingly endless source of controversy and argument. Many solutions have been proposed to the problems of description and classification, as described in a detailed review by Dreimanis (1989). In recent years, powerful systematic approaches to the study of glacial sediments have been developed, as part of a wider movement in sedimentology as a whole (Boulton, 1972b; Walker, 1976, 1992a; Dreimanis, 1989). These approaches recognize that glacial depositional systems exhibit order on many different levels, either as a series of steps in time, or on a range of scales in space.

The time dimension is important for understanding the origin of deposits, because different sediment properties develop at different points of the erosional, transport and depositional history of the debris as it travels through the glacial system. The sequence of steps can be regarded as a type of *debris cascade system* (Chorley *et al.*, 1984), which can be used as a basis for sediment classification (Boulton and Deynoux, 1981; Dreimanis, 1982b, 1989). This topic is discussed in Section 10.2.1, and the evolution of glacial landscapes on long time-scales is considered in Chapter 12. The spatial dimension allows the position of deposition of a sediment to be established with respect to adjacent sediments, the wider environment, and the landscape as a whole. Study of the spatial arrangements of sediments at a hierarchy of scales allows the development of *facies models*, or summaries of how individual deposits are nested together to form depositional systems (Walker, 1976, 1992a). The idea of facies models is introduced in Section 10.2.2 and developed throughout Chapters 10, 11 and 12.

Sections 10.3–10.7 describe the range of deposits that commonly occur in glacial and proglacial environments, and their origin and diagnostic properties.

10.2 SEDIMENT CLASSIFICATION AND DESCRIPTION

10.2.1 Debris cascades and sediment classification

Sedimentary deposits can be viewed as outcomes of a series of processes extending back in time. In Fig. 10.1, this debris cascade system is broken down into three stages: debris source; transport path; and depositional process. The *debris source* is the primary input to the system, and may be subglacial (e.g. plucked and abraded bedrock, or overridden sediments) or extraglacial (e.g. rock walls and pre-existing deposits). The debris is then carried along one or more *transport paths*, including active or passive glacial transport, rivers and streams, subaerial or subaqueous slopes, iceberg rafting, and the air. Debris may pass between transport paths many times prior to deposition. *Depositional processes* refer to the mechanisms that lay down the final deposit, and include glacial, fluvial, gravitational and aeolian processes. Of these, only glacial depositional processes are unique to glacial environments; processes from the other categories can operate in non-glacial environments as well as on glaciers. For example, fluvial processes can deposit sediment in subglacial tunnels, on proglacial braid plains or in unglaciated valleys.

Figure 10.1 provides a background for the genetic classification of sediments found in glacial environments. The main criterion for the classification is depositional process, but it recognizes that earlier stages of debris history also influence the properties of deposits. Examples of properties acquired during each of the three stages are as follows:

1. The *debris source* controls the lithology of the particles in a sediment, but can also influence particle morphology and grain size distribution (Gomez *et al.*, 1988; Benn and Ballantyne, 1994).
2. The *transport path* determines the wear processes experienced by particles as they move through the system, so exerts a strong influence on particle morphology (Boulton, 1978; Dowdeswell *et al.*, 1985; Ballantyne and Benn, 1994). Grain size distributions are also modified during transport as the result of progressive wear and the preferential transport of particular size grades by water, wind and gravitational processes (Church and Gilbert,

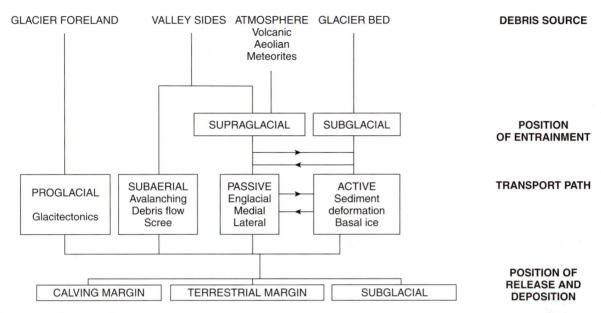

| GLACIER FORELAND | VALLEY SIDES | ATMOSPHERE | GLACIER BED | DEBRIS SOURCE |

Fig. 10.1 Debris cascade system

1975; Humlum, 1985a; Werritty, 1992). Particle lithology can also be influenced through contrasts in the durability of rock types under different types of transport (Slatt and Eyles, 1981). For glacially and gravitationally deformed sediment, aspects of the strain history can be reflected in sediment fabric and structure (Banham, 1977; Hart and Roberts, 1994; Benn, 1994b; Benn and Evans, 1996).

3. *Depositional processes* generally exert the strongest influence on sediment properties, and can control the geometry and extent of beds, sedimentary structures (e.g. lamination, cross-bedding and grading), grain size distributions, particle morphology, fabric, geotechnical properties (e.g. porosity, shear strength and permeability), and others. Sediment properties are also influenced by post-depositional processes, such as diagenesis, faulting and folding, frost-heaving, and winnowing and leaching of fine material by wind and water.

It is important to realize that sediment characteristics are little affected by the *environment* of deposition. The properties of debris flow deposits, for example, depend on the grain size distribution of the parent debris and the water content during flow and deposition, and do not vary systematically between supraglacial, subglacial and non-glacial environments. The same can be said for other sediment types including fluvial sediments. This has important implications for sediment classification. Many classification schemes have been proposed for glacigenic deposits in which deposits are named according to

the process *and* position of deposition, which has resulted in a profusion of named types of 'till' and related deposits (see Dreimanis, 1989; Brodzikowski and van Loon, 1991). In practice, however, such schemes may be unworkable because, for deposits of vanished glaciers, the environment of deposition is not usually obvious from the deposit itself and must be inferred from careful analysis of the surrounding sediments and landforms, and even then it may be ambiguous. A more logical approach is to classify sediments by the *processes of formation*, as revealed through their physical properties (Lawson, 1981a, b, 1989; N. Eyles *et al.*, 1983).

In classifying sediments deposited in glacial environments, a distinction is sometimes made between *primary deposits*, laid down by uniquely glacial agencies, and *secondary deposits*, which have undergone some form of reworking by non-glacial processes (Lawson, 1981b; Dreimanis, 1989). According to this distinction, primary deposits include tills deposited from a subglacial deforming layer, or by lodgement or melt-out, whereas secondary deposits are composed of glacial debris that has been remobilized and deposited by gravitational flowage, stream flow or other agencies. The boundary between primary and secondary deposits, however, is arguable, and it is not always clear at what point sediments lose their uniquely glacial character and become some other kind of deposit. This difficulty is reflected in the debate over the definition of the word *till*. This debate kept the Till Work Group of the International Union for Quaternary Research (INQUA) occupied for more than 15 years in the 1970s and 1980s, and is still not resolved to everyone's agreement. Most members of the Till Work Group accepted a broad definition of till as 'sediment that has been transported and deposited by or from glacier ice, with little or no sorting by water' (Dreimanis, 1989). This definition would embrace all primary deposits and some secondary deposits, particularly those formed at or near glacier margins by gravitational reworking processes. However, many glacial geologists prefer the more restrictive definition proposed by Lawson (1981b): 'Till is . . . sediment deposited directly by glacier ice, and has not undergone subsequent disaggregation and resedimentation.' According to this definition, only sediments deposited by primary, uniquely glacial processes can be classified genetically as tills, and mass movement deposits from glacial settings are grouped with similar deposits from other environments, rather than with tills. Lawson's definition is adopted in this book, because it emphasizes the importance of depositional processes, rather than position of deposition, in sediment classification. Approaches to defining the position and environment of deposition of a sediment are discussed in the following section.

10.2.2 Facies and facies models

10.2.2.1 SPATIAL HIERARCHIES

The position of deposition and environmental significance of a sediment can be understood only with reference to the surrounding sediments and rocks. In modern environments, sediments are rarely deposited in isolation, but are laid down as part of *assemblages* that reflect the range of processes active in that environment. Such assemblages can be recognized at a wide range of scales, from the very small to that of a whole continent or ocean basin. For example, cross-bedded sands may be part of an assemblage of sand and gravel infilling a fluvial channel; in turn, the channel fill could be part of an assemblage of channel and bar deposits in a braided river system; and the braided river system part of a yet larger assemblage of deposits laid down during a glaciation. The environmental context of a sediment can therefore be defined at different levels of a spatial hierarchy, beginning with the immediate locality and panning out to wider and wider horizons. At each successive level, the controls on the sedimentary system become larger in scale and longer-lasting in effect. For our example of fluvial sands, at the local level the main controls on deposition are the shape of the immediate river bed and the short-term flow conditions as determined by rainfall, glacier melt and release of stored water. At the largest scale, the formation, location and extent of a braided river system is controlled by global factors such as long-term climatic cycles, relative sea-level fluctuations and tectonics.

This hierarchical approach to sedimentology, therefore, is a powerful means of describing how sediments, landforms and landscapes fit together, and determining how organization in the landscape reflects the organization of depositional processes and external controls in the environment. It forms the basis of *facies modelling*, or constructing descriptive and predictive models of relationships between different deposits (Walker, 1976, 1990, 1992a; Reading, 1986).

In the following chapters, we will discuss glacial sediments and associated landforms at four levels of organization, based on those defined by Walker (1992a; Fig. 10.2). From the smallest to the largest scale, these are: (a) *facies*, or individual deposits; (b) *sediment–landform associations*; (c) *depositional systems*, or *landsystems*; and (d) *glacial systems tracts*, or large-scale linkages of depositional systems. The boundaries between these levels are to some extent arbitrary, and differ slightly from those recognized by other authors. Those defined below have been tailored to the particular problems of description of glacial sediments, landforms and environments.

10.2.2.2 FACIES

The fundamental unit of this spatial hierarchy is the individual sedimentary deposit or *facies* (from the

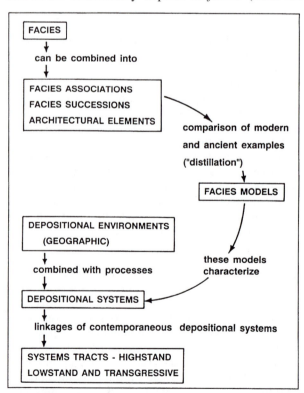

Fig. 10.2 Hierarchical sediment classification. (From Walker, 1992a. Reproduced by permission of the Geological Association of Canada)

Latin word meaning 'aspect' or 'appearance of' something). In this context the word 'facies' refers to a body of sediment with a distinctive combination of properties that distinguish it from neighbouring sediments (Reading, 1986; Walker, 1992a). Facies are distinct sediment types deposited by single processes or groups of processes acting in close association. When we wish to refer only to the objective, physical characteristics of a deposit, with no reference to depositional processes, the term *lithofacies* is used. Examples of lithofacies are *cross-bedded sand* and *silt/clay laminae*. *Genetic facies* are named sedimentary units which state or imply a specific mode of formation. The genetic facies equivalents of the previous lithofacies examples might be *dune-bedded sands* and *varves*, which imply an origin by bedform migration and annual lacustrine sedimentation cycles, respectively. Note that genetic facies carry varying amounts of environmental information, because some can occur in an extremely wide variety of environments whereas others are formed in a very limited number of settings. For example, the presence of lodgement till carries more information about the former distribution of glacier ice than a debris flow deposit, because the former requires uniquely glacial processes for its deposition.

10.2.2.3 SEDIMENT–LANDFORM ASSOCIATIONS

The next level in the hierarchy is the *sediment–landform association* or *facies association*, defined as an assemblage of facies that are genetically related (Walker, 1992a). Facies associations consist of sediments that were laid down beside each other or in an unbroken vertical succession. An unbroken *vertical* transition between two facies reflects a *lateral* shift in depositional processes, such as where fluvial bars migrate downstream and bury older channel deposits, or where glacier retreat causes ice-marginal deposits to be superimposed on subglacial sediments. The use of vertical facies associations to infer the spatial patterns of depositional processes is known as *Walther's law of facies* (Walther, 1894; Blatt *et al.*, 1980; Reading, 1986).

The boundaries of facies associations are marked by *surfaces of erosion or non-deposition* or *bounding discontinuities,* which represent breaks in deposition of varying duration. Similar, more extensive discontinuities are also used to define depositional systems and systems tracts at higher levels in the hierarchy (Van Wagoner *et al.*, 1988; Walker, 1990, 1992a). Sediment–landform associations are facies associations that have surface expression in a landform which is genetically related to the underlying facies. A related concept is the *architectural element*, which is a facies association with characteristic three-dimensional geometry, which may be buried by younger deposits (Miall, 1985a; Walker, 1992a). An example of a sediment–landform association is a streamlined ridge of deformed sediments aligned parallel to glacier flow, characteristic of a fluted till surface; and an example of an architectural element is a buried river channel filled with cross-bedded sands and gravels. In glacial environments, sediment–landform associations can also be defined on the basis of distinctive deformation structures, such as thrust-block moraines, which are composed of dislocated wedges of sediment separated by low-angle reverse faults.

Facies associations and sediment–landform associations place facies within a *context*, indicating the local stratigraphic, structural and genetic relationship between closely associated facies. They therefore imply particular depositional environments or subenvironments, such as a proximal reach of a braided river or an advancing ice margin, in a way that lithofacies or genetic facies seldom can. Subglacial, ice-marginal, glacifluvial, glacilacustrine and glacimarine sediment–landform associations are discussed in Chapter 11.

10.2.2.4 DEPOSITIONAL SYSTEMS, LANDSYSTEMS AND SYSTEMS TRACTS

Facies associations and sediment–landform associations can be grouped together into *depositional systems* or *landsystems*. These are large-scale groupings of facies deposited in the same overall environment, such as the infill of a major proglacial lake or the assemblage of sediments and landforms deposited along the margins of a retreating glacier. Glacial landsystems also include erosional forms that were created broadly contemporaneously with sediments in the same environment. Examples include *subglacial landsystems*, consisting of areas of ice-scoured bedrock, subglacial tills, fluted moraines and eskers; and *glaciated valley landsystems*, consisting of troughs, rockwalls, lateral moraines and fluvial forms. Summary models of glacial depositional systems and landsystems have been provided by Boulton and Paul (1976), Boulton and N. Eyles (1979), N. Eyles (1983c), Powell (1984), C.H. Eyles *et al.* (1985) and N. Eyles and C.H. Eyles (1992). In Chapter 12, we describe glacial landsystems under four headings: (a) subglacial; (b) ice-marginal and supraglacial; (c) subaqueous; and (d) glaciated valleys.

In recent years, petroleum geologists have developed the idea of *sequence stratigraphy* as a means of predicting the location of hydrocarbon reservoirs (Brown and Fisher, 1977; Van Wagoner *et al.*, 1988; Posamentier *et al.*, 1988; Galloway, 1989; Walker, 1992a). Sequence stratigraphy links depositional systems into genetically related *systems tracts,* and provides a powerful means of analysing the complex relationships between facies in a depositional basin, and unravelling the history of environmental changes contained in the geological record. The approach has been applied to glacial systems by some researchers (e.g. Boulton, 1990; N. Eyles and C.H. Eyles, 1992; Martini and Brookfield, 1995), although its potential is still largely unrealized. The application of sequence stratigraphy to glacial studies is discussed in Section 12.2.2.

The study of the large-scale distribution of depositional systems and landsystems is one of the most exciting branches of glacial geology, because it is at this scale that the impact of glaciation is revealed at its most impressive. Large-scale studies allow long-term cycles of landscape change to be reconstructed, and the links between glacial cycles and global oceanographic and climatic change to be identified.

10.2.3 Sediment description

From the foregoing, it is clear that a comprehensive description of a sediment should include details of: (a) its properties, allowing aspects of its erosional, transport and depositional history to be reconstructed; (b) its geometry; and (c) its position with respect to the adjacent sediments and the land surface at a range of scales.

Sediment properties provide the basis for defining and describing lithofacies. The choice of properties to be described must be guided by the object of the investigation, because different properties convey different types of information (Section 10.2.1). Properties commonly described in glacial process studies are grain size distributions, structures, fabric, particle form, matrix properties and density. Some of the techniques for measuring lithofacies properties are briefly described in this book; additional information can be found in comprehensive textbooks by Bradley (1985), Goudie (1990) and Gale and Hoare (1991). A particularly important characteristic is sediment texture, defined in terms of the dominant grain size and the grain size distribution. The definitions of clay, silt, sand and gravel particles are shown in Appendix 2. Poorly sorted sediments are termed *diamictons* (Table 10.1) and their lithified equivalents are termed *diamictites* (Flint *et al*., 1960a, b; Harland *et al*., 1966). Some researchers use the overall term *diamict* to cover both diamictons and diamictites (e.g. N. Eyles *et al*., 1983).

Sedimentologists have devised various shorthands or *lithofacies codes* for recording lithofacies properties in the field. These codes consist of letters denoting sediment lithology, plus additional letters for internal structures or inclusions. A widely used scheme for the deposits of braided rivers was introduced by Miall (1977, 1978) and was later extended by N. Eyles *et al*. (1983) to include codes for diamictons (Fig. 10.3). The diamicton code is less satisfactory than that for fluvial sediments because it omits several important diagnostic properties, such as density, particle morphology and fabric. Reliance on lithofacies coding alone yields insufficient information for correct interpretation, so the codes should be used only as a supplement to other observations (Dreimanis, 1984b; Karrow, 1984b; Kemmis and Hallberg, 1984; Hambrey, 1994). Other coding schemes have been devised for specific purposes. Ghibaudo (1992) proposed a useful scheme for describing subaqueous mass flow deposits, incorporating code letters for textural variation and a range of internal structures (Fig. 10.4).

The geometry and position of lithofacies are usually portrayed graphically in vertical profiles, two-dimensional logs, maps or some combination of the three. *Vertical profiles* are measured vertical sections on which bed thickness and lithology are shown, alongside additional information such as structural data and the nature of contacts between lithofacies (e.g. gradational, erosional, deformed). Modal grain size of lithofacies is shown on vertical profiles using the width of the bed, providing a rapid impression of

Table 10.1 Classification of glacial deposits and associated facies used in this book

Primary glacigenic deposits (tills)	*Glacifluvial deposits*	*Gravitational mass-movement deposits*	*Deposits from suspension settling and iceberg activity*
Lodgement till	Plane-bed deposits	Scree	Cyclopels
Glacitectonite	Ripple cross-laminated facies	Debris-fall deposits	Cyclopsams
Deformation till	Cross-bedded facies (dunes and	Gelifluction deposits	Varves
Comminution till	antidunes)	Slide and slump deposits	Dropstone mud
Melt-out till	Gravel sheets	Debris-flow deposits	Dropstone diamicton
Sublimation till	Silt and mud drapes	Turbidites	Undermelt diamicton
	Hyperconcentrated flow		Iceberg contact deposits
	deposits		Ice-keel turbate

Code	Description
Diamictons	Very poorly sorted admixture of wide grain size range
Dmm	Matrix-supported, massive.
Dcm	Clast-supported, massive.
Dcs	Clast-supported, stratified.
Dms	Matrix-supported, stratified.
Dml	Matrix-supported, laminated.
_ _ _ (c)	Evidence of current reworking.
_ _ _ (r)	Evidence of resedimentation.
_ _ _ (s)	Sheared.
_ _ _ (p)	Includes clast pavement(s).
Boulders	Particles > 256 mm (b-axis)
Bms	Matrix-supported, massive.
Bmg	Matrix-supported, graded.
Bcm	Clast-supported, massive.
Bcg	Clast-supported, graded.
Bfo	Deltaic foresets.
BL	Boulder lag or pavement.
Gravels	Particles of 8–256 mm
Gms	Matrix-supported, massive.
Gm	Clast-supported, massive.
Gsi	Matrix-supported, imbricated.
Gmi	Clast-supported, massive (imbricated).
Gfo	Deltaic foresets.
Gh	Horizontally bedded.
Gt	Trough cross-bedded.
Gp	Planar cross-bedded.
Gfu	Upward-fining (normal grading).
Gcu	Upward-coarsening (inverse grading).
Go	Openwork gravels.
Gd	Deformed bedding.
Glg	Palimpsest (marine) or bedload lag.
Granules	Particles of 2–8 mm
GRcl	Massive with clay laminae.
GRch	Massive and infilling channels.
GRh	Horizontally bedded.
GRm	Massive and homogeneous.
GRmb	Massive and pseudo-bedded.
GRmc	Massive with isolated outsize clasts.
GRmi	Massive with isolated, imbricated clasts.
GRmp	Massive with pebble stringers.
GRo	Open-work structure.
GRruc	Repeating upward-coarsening cycles.
GRruf	Repeating upward-fining cycles.
GRt	Trough cross-bedded.
GRcu	Upward coarsening.
GRfu	Upward fining.
GRp	Cross-bedded.
GRfo	Deltaic foresets.
Sands	Particles of 0.063–2 mm
St	Medium to very coarse and trough cross-bedded.
Sp	Medium to very coarse and planar cross-bedded.
Sr(A)	Ripple cross-laminated (type A).
Sr(B)	Ripple cross-laminated (type B).
Sr(S)	Ripple cross-laminated (type S).
Scr	Climbing ripples.
Ssr	Starved ripples.
Sh	Very fine to very coarse and horizontally/plane bedded or low angle cross-laminated.
Sl	Horizontal and draped lamination.
Sfo	Deltaic foresets.

Code	Description
Sfl	Flasar bedded
Se	Erosional scours with intraclasts and crudely cross-bedded.
Su	Fine to coarse with broad shallow scours and cross-stratification.
Sm	Massive.
Sc	Steeply dipping planar cross-bedding (non-deltaic foresets).
Sd	Deformed bedding.
Suc	Upward-coarsening.
Suf	Upward-fining.
Srg	Graded cross-laminations.
SB	Bouma sequence.
Scps	Cyclopsams.
_ _ _ (d)	With dropstones.
_ _ _ (w)	With dewatering structures.
Silts & Clays	Particles of <0.063 mm
Fl	Fine lamination often with minor fine sand and very small ripples.
Flv	Fine lamination with rhythmites or varves.
Fm	Massive.
Frg	Graded and climbing ripple cross-laminations.
Fcpl	Cyclopels.
Fp	Intraclast or lens.
_ _ _ (d)	With dropstones.
_ _ _ (w)	With dewatering structures.

SYMBOLS

Diamict
stratified
sheared
jointed

Gravel
Sand
Fines
Laminations
with intraclasts
with dropstones
with loading structures

Contacts
Erosional
Conformable
Loaded
Interbedded

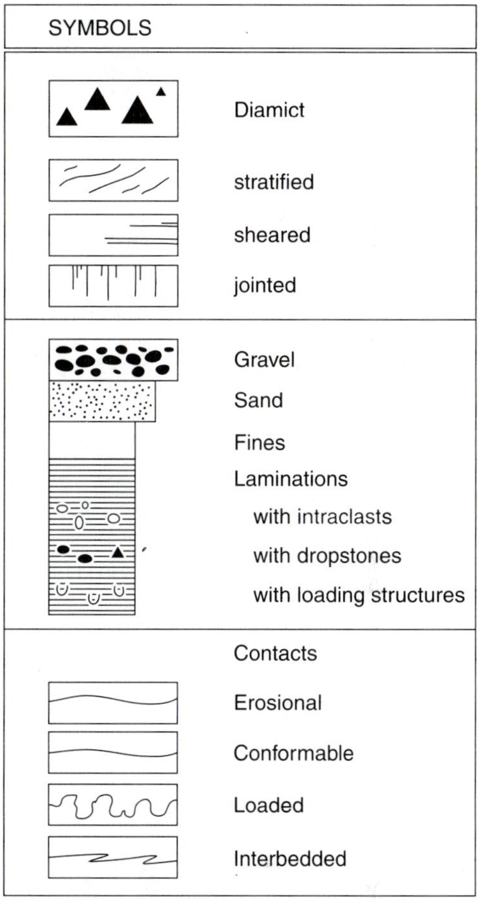

Fig. 10.3 Lithofacies coding scheme modified from Miall (1978) and N. Eyles *et al.* (1983), and including modifications from other sources

(a)

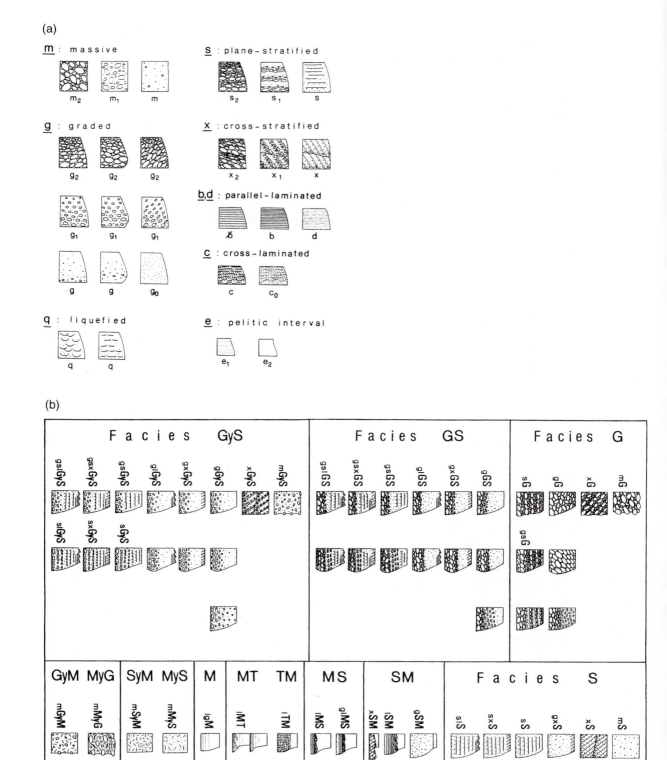

Fig. 10.4

(C)

Facies G (gravel beds)
mG :   massive gravel
xG :   cross-stratified gravel
sG :   plane-stratified gravel
gG :   graded gravel
gsG :   graded to plane-stratified gravel

Facies GS (gravel–sand couplets)
gGS :   graded gravel–sand couplet
gxGS : graded to cross-stratified gravel–sand couplet
gsGS : graded to plane-stratified gravel–sand couplet
glGS : graded to laminated gravel–sand couplet
gsxGS : graded to plane-stratified to cross-stratified gravel–sand couplet
gslGS : graded to plane-stratified to laminated gravel–sand couplet

Facies GyS (gravelly sand beds)
mGyS :   massive gravelly sand
xGyS :   cross-stratified gravelly sand
gGyS :   graded gravelly sand
gxGyS : graded to cross-stratified gravelly sand
gsGyS : graded to plane-stratified gravelly sand
glGyS : graded to laminated gravelly sand
gsxGyS : graded to plane-stratified to cross-stratified gravelly sand
gslGyS : graded to plane-stratified to laminated gravelly sand
sGyS :   plane-stratified gravelly sand
sxGyS : plane-stratified to cross-stratified gravelly sand
slGyS : plane-stratified to laminated gravelly sand

Facies S (sand beds)
mS :   massive sand
xS :   cross-stratified sand
gS :   graded sand
gxS :   graded to cross-stratified sand
gsS :   graded to plane-stratified sand
glS :   graded to laminated sand
gsxS : graded to plane-stratified to cross-stratified sand
gslS : graded to plane-stratified to laminated sand
sS :   plane-stratified sand
sxS :   plane-stratified to cross-stratified sand
slS :   plane-stratified to laminated sand

Facies SM (sand–mud couplets)
gSM :   graded sand–mud couplet
glSM : graded to laminated sand–mud couplet
lSM :   laminated sand–mud couplet
xSM :   cross-stratified sand–mud couplet
tgSM : thin-bedded, graded sand–mud couplet
trSM : thin-bedded, rippled sand–mud couplet

Facies MS (mud–sand couplets)
glMS : graded to laminated mud–sand couplet
lMS :   laminated mud–sand couplet

Facies TM (silt–mud couplets)
lTM :   laminated silt–mud couplet
gTM :   graded silt–mud couplet

Facies MT (mud–silt couplets)
lMT :   laminated mud–silt couplet
gMT :   graded mud–silt couplet

Facies M (mud beds)
lgM :   laminated to graded mud
gM :   graded mud

Facies MyS (muddy sand beds)
mMyS : massive muddy sand
gMyS : graded muddy sands

Facies SyM (sandy mud beds)
mSyM : massive sandy mud
gSyM : graded sandy mud

Facies MyG (muddy gravel beds)
mMyG : massive muddy gravel
gMyG : graded muddy gravel

Facies GyM (gravelly mud beds)
mGyM : massive gravelly mud
gGyM : graded gravelly mud

Fig. 10.4 Facies classification of subaqueous flow deposits proposed by Ghibaudo (1992): (a) letter codes for internal structures; (b) facies and subfacies classification; (c) facies codes and descriptions. (From Ghibaudo, 1992. Reproduced by permission of Blackwell)

the overall character of a succession (Fig. 10.5). Vertical profiles are ideal for portraying the lithology of cores, and are also often used for recording outcrop data. However, vertical profiles give very limited information on the geometry of facies and facies associations, and the relative importance and extent of bounding surfaces, so where outcrop data are available, *two-dimensional logs* are preferable. Two-dimensional logs vary from sketches to very detailed measured drawings, and are a good method for showing lithofacies geometry and the larger-scale geometry (*architecture*) of associations. Lithological data can also be shown. Two-dimensional logs can be constructed relatively quickly by drawing on to overlays on photographs. Several examples of two-dimensional logs are given in Chapter 11. Ideally, facies geometry and architecture should be logged in three dimensions, but data of this quality are very rarely available. *Landform maps* have been important tools for glacial geomorphologists for many years, and when combined with two-dimensional logs can give a very good summary of the three-dimensional form of sediment assemblages.

In the following sections, we describe the principal genetic facies commonly found in glacial, proglacial, glacilacustrine and glacimarine environments, using the classification shown in Table 10.1. Shoreface and tidal sediments are not described, and interested readers are referred to Elliott (1986a), Walker and Plint (1992) and Dalrymple (1992) for good recent reviews. A genetic approach is taken for two main reasons. First, it emphasizes the links between facies characteristics and the processes of formation,

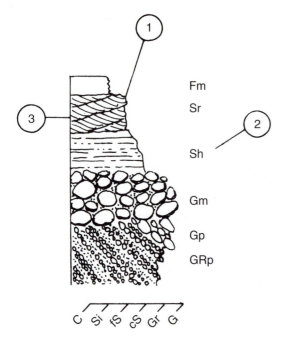

Fig. 10.5 Conventions used in vertical profile logs. (1) Log width indicates modal particle size, as depicted on the lower scale, where C = clay, Si = silt, fS = fine sand, cS = coarse sand, Gr = granules and G = gravel; (2) facies codes; (3) graphical symbols indicate sediment structures (e.g. ripples, planar bedding, etc.)

providing continuity with Part One of this book. Second, genetic facies may have a range of lithofacies characteristics depending on variations in the source material and the relative importance of various depositional processes. A purely descriptive approach to discussing glacigenic and related sediments would therefore be over-complex and repetitive.

For each genetic facies covered in this chapter, we provide a definition, a brief account of the processes of deposition, and typical distinguishing characteristics, such as grain size distribution, particle morphology, fabric, internal structures, and facies dimensions and geometry.

10.3 PRIMARY GLACIGENIC DEPOSITS (TILL)

10.3.1 Overview

Primary glacigenic deposition, as defined in Section 10.2.1, can occur supraglacially by the melt-out and sublimation of debris-rich ice, or subglacially by some combination of lodgement, deposition from a deforming layer, and melt-out. As we saw in Sections 5.5 and 5.6, subglacial deposition probably usually

involves complex interactions between these basic processes. As a result, subglacial tills are enormously variable, and defy attempts to classify them into distinct, exclusive categories. A useful approach is to regard tills deposited by each of the three primary processes as end-members of a continuum, and tills deposited by more than one process as occupying intermediate positions between these end-member types (Dreimanis, 1989). This approach has been refined by Hicock (1990, 1993), who portrayed the continuum of glacial depositional processes first as a triangular prism, incorporating gravitational flowage (Fig. 10.6), then as an octahedron incorporating gravitational flowage and the activities and effects of water. The evolution of particular tills can then be shown as curves showing the changing influence of the end-member processes through time (*spaghetti diagrams*).

In the following sections we describe genetic till facies associated with each end-member depositional process, then consider the characteristics of intermediate tills. Sediment deposited from subglacial deforming layers is described under two headings: *glacitectonites*, in which some structural characteristics of the parent materials survive, and *deformation tills*, where they do not.

10.3.2 Lodgement till

Lodgement till can be defined as 'sediment deposited by plastering of glacial debris from a sliding glacier sole by pressure melting or other mechanical processes' (Dreimanis, 1989). Lodgement tills are commonly overconsolidated due to dewatering under ice overburden pressure. They usually have high bulk density and penetration resistance, and in some cases have the appearance of concrete. They may be mas-

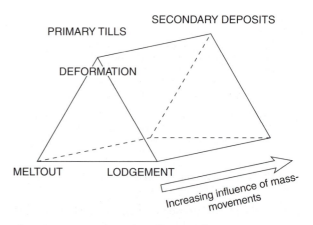

Fig. 10.6 Triangular prism diagram showing a continuum of subglacial till forming processes. (From Benn and Evans, 1996, after Hicock, 1990)

sive in appearance, but more usually are traversed by numerous subhorizontal joints, giving the till a fissile structure (Fig. 10.7). These joints are sometimes polished and striated, indicating that some are former shear planes (Boulton *et al.*, 1974; Benn, 1994a), although it has also been suggested that joint development is related to dewatering processes (Muller, 1983). Most lodgement tills are diamictons with grain size distributions that reflect debris crushing and abrasion during subglacial shear (Section 5.7.2). Typical distributions are bimodal or multimodal, with distinct modes in the silt (rock flour) and gravel (clast) ranges. The precise positions of these modes are strongly influenced by the lithology and mineralogy of the parent materials, particularly their susceptibility to crushing and abrasion processes (Haldorsen, 1981, 1983).

Pebble and boulder *a*-axis fabrics have consistent strong preferred orientations parallel to ice flow, and *a–b* planes tend to have a gentle upglacier imbrication (Fig. 10.8; Dowdeswell and Sharp, 1986; Benn, 1994a; Hart, 1994a; Krüger, 1994). *a*-axis fabric shape is characterized by low isotropy and moderate to high elongation. These characteristics reflect strongly constrained shear at the ice–till interface: if particles are bridged between ice and till, movement of the ice will tend to rotate the particle until its *a*-axis and *a–b* plane are parallel to the plane of shear, although an upglacier imbrication will develop, because of the tendency for till matrix to plough up in front of the particle. Additionally, the *a*-axes will rotate into parallelism with the direction of shear, because of the drag imposed by the surrounding ice and till matrix (Lindsay, 1970; Benn, 1994a, 1995; Benn and Evans, 1996). Matrix particles can also show strong preferred orientations parallel to shear, particularly in the vicinity of micro-shears and other fault planes (Owen and Derbyshire, 1989; Owen, 1994).

One of the most distinctive characteristics of lodgement tills is the abundance of asymmetric *stoss-and-lee* clasts, resembling miniature roches moutonnées, with smoothed upglacier (stoss) sides and fractured downglacier (lee) sides, formed after clasts have become lodged and continue to be overridden by active ice (Figs 5.36–5.38; Boulton, 1978; Krüger, 1979; Sharp, 1982; Benn, 1994a). The plucked lee faces often show a very strong downglacier preferred orientation. *Double stoss-and-lee* clasts have also been described from lodgement tills by Krüger (1984, 1994).

10.3.3 Glacitectonite

The term *glacitectonite* refers to subglacially sheared rocks and sediments, and was introduced by Banham (1977), who made the analogy with tectonite, a metamorphic rock with structures and fabrics imprinted by its tectonic history. Banham used the term to embrace sheared materials that retain some primary structures, such as lamination or cross-bedding (*exodiamict glacitectonite*), and diamictons, in which all primary structures have been destroyed by shear (*endiamict glacitectonite*). In this book 'glacitectonite', is used more restrictively, and is defined as rock or sediment that has been deformed by subglacial shearing but retains some of the structural characteristics of the parent material, which may consist of igneous, metamorphic or sedimentary rock, or unlithified sediments. It therefore is used in the same way as 'exodiamict glacitectonite' as defined by Banham (1977). Diamictic products of subglacial shear are referred to as *deformation till* in this book (Section 10.3.4). Some researchers object to using 'glacitectonite' on the grounds that the suffix '-ite' should be applied only to lithified sediments or other rocks (e.g. Elson, 1989). This view, however, is

Fig. 10.7 Lodgement till traversed by subhorizontal joints, Isle of Skye, Scotland. Note the abraded clasts. (Photo: D.I. Benn)

Fig. 10.8 Lodged boulders are commonly aligned with their long axes parallel to the direction of glacier flow, as for these boulders on the foreland of Slettmarkbreen, Norway. (Photo: D.I. Benn)

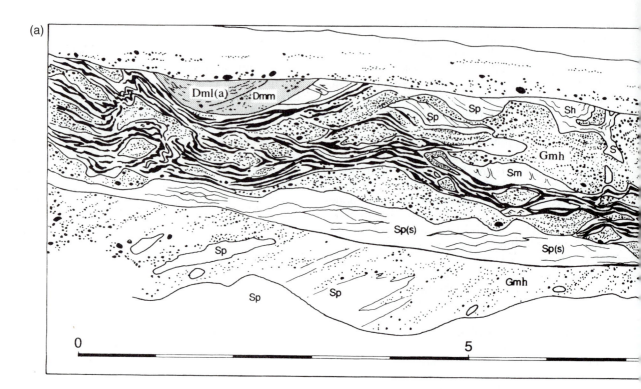

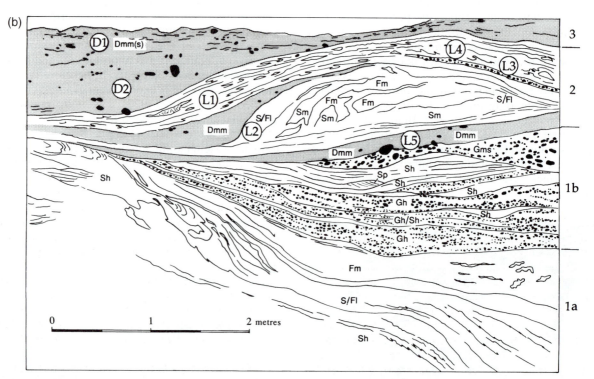

Fig. 10.9 Glacitectonites from Drymen, Scotland. (a) Glacitectonite comprising sand and gravel pods separated by attenuated silt and clay beds. Note the erosional contact with underlying sand and gravel foreset beds, which have been cannibalized to produce the glacitectonite. (b) Vertical sequence comprising non-penetrative glacitectonite of folded sands and silts (unit 1a) and sands and gravels (unit 1b), overlain by penetrative glacitectonite of silts, sands and diamicton (unit 2), and capped by a deformation till (unit 3). Circled codes represent positions of fabrics presented by Benn and Evans (1996). (Reproduced by permission of Elsevier)

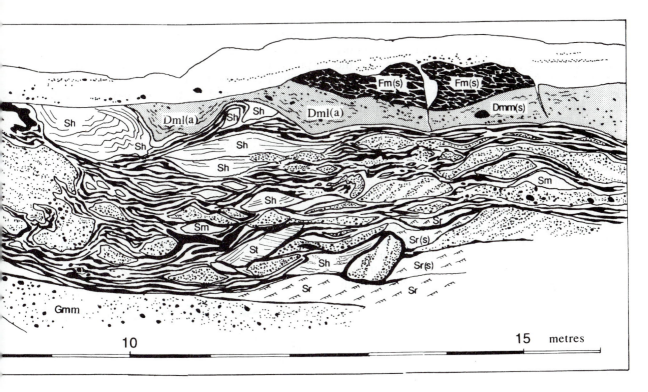

incorrect. There is no such restriction on the use of '-ite', and the words 'diatomite' and 'turbidite', for example, are used by geologists and Quaternary scientists to refer to lithified and unlithified materials alike. Furthermore, no viable alternative word for this type of deposit has been proposed (Pedersen, 1989; Dreimanis, 1989; Benn and Evans, 1996).

The characteristics of glacitectonites partly reflect the nature of the parent material and partly their history of strain in a subglacial deforming layer. They can display evidence for brittle or ductile deformation, or a combination of the two. Some glacitectonites are breccias, broken up and displaced along fault planes (brittle deformation), whereas others are folded by pervasive ductile deformation (Fig. 10.9). The most complex deformation structures tend to occur in inhomogeneous materials with widely varying strengths, such as interbedded gravels, sands, silts and clays. In such cases, the fine-grained units can be highly deformed, with complex elongated folds, whereas coarse-grained units (e.g. sands or gravels) commonly form relatively undeformed pods or lenses (e.g. Berthelsen, 1979; Hart and Roberts, 1994; Benn and Evans, 1996). This reflects variations in strain response to stress, primarily due to contrasts in sediment frictional strength and permeability (Section 5.6). Pods of stiff sediment commonly have streamlined forms, recording erosion and deformation of their edges during shear within more rapidly deforming material, and are analogous

to *augen* and *boudins* in metamorphic rocks (Fig. 10.9). Glacitectonites that have undergone high shear strains sometimes exhibit a distinctive subhorizontal banding (sometimes called *banded* or *laminated tills*). Individual bands are commonly lithologically distinct, and represent different rock and sediment types that have been highly attenuated but not mixed during strain. Figure 10.10e (Plate 20) shows a middle Pleistocene example from the English Midlands in which bands of pure chalk fragments alternate with bands derived from limestones and mudrocks. Bands tend to be bent around pebbles and other stiff inclusions, resulting from deflection of the deforming material. Banded glacitectonites with scattered pebbles can be mistaken for laminated glacilacustrine and glacimarine deposits with dropstones, but can be distinguished by the presence of lithological banding, asymmetric folds around pebbles, and attenuated folds among the banding (Hart and Roberts, 1993). Banded glacitectonites can also be mistaken for other types of deposit, such as debris flows (e.g. N. Eyles, 1994) and melt-out tills (e.g. Shaw, 1982).

Pebble fabrics in glacitectonites can reflect the primary fabric of the parent material or the subglacial strain history, depending on position. In thrust planes or zones of intense shear, pebble *a*-axes are commonly closely parallel to the plane of shear (low isotropy) and show strong preferred orientations parallel to glacier flow (high elongation), reflecting strongly constrained shear in narrow zones bounded by stiffer material (Fig. 10.11).

Vertical sequences in glacitectonites commonly display increasing deformation upsection, as the result of patterns of strain in the subglacial deforming layer (Banham, 1977; Pedersen, 1989; Benn and Evans, 1996; Sections 4.4.2 and 5.6). An idealized full sequence might consist (from the base up) of: (a) undisturbed bedrock or sediment; (b) glacitectonite with mildly distorted primary structures; (c) glacitectonite with widespread, penetrative shear structures; and (d) deformation till. This sequence reflects the fact that overburden pressures, and hence sediment frictional strength, decrease upwards, so that the most intense deformation and the highest cumulative strains will be found at the top of the sequence (Boulton and Hindmarsh, 1987; Alley, 1989; Hart *et al.*, 1990). It should be emphasized, however, that vertical patterns of deformation will also vary with sediment grain size, consolidation and other properties (Fig. 10.12).

10.3.4 Deformation till

Deformation till can be defined as rock or sediment that has been disaggregated and completely or largely homogenized by shearing in a subglacial deforming layer. Many synonyms have been proposed, including *deformed till*, *soft-bed till*, *deforming bed till*, *shear till* and *endiamict glacitectonite*, but all post-date 'deformation till', coined by Elson (1961, 1989). Elson's original definition would also include glacitectonite (Section 10.3.3), but we believe it is useful to have separate terms. Deformation tills may consist entirely of local materials mobilized by glacially applied stresses, but they can also contain far-travelled lithologies which have either undergone long-distance transport by deformation or melted out from glacier ice prior to incorporation in the deforming layer.

The destruction of pre-existing structures under very high cumulative strains makes deformation tills problematical to identify (Boulton, 1987; Hart *et al.*, 1990). Particular disagreement has arisen over the interpretation of massive tills deposited by the Laurentide ice sheet, and some researchers have argued that they provide evidence for widespread subsole deformation (Boulton and Jones, 1979; Alley, 1991; Hicock and Dreimanis, 1992a), while others contend that they may be melt-out tills (Clayton *et al.*, 1989; Mickelson *et al.*, 1992; Ronnert and Mickelson, 1992). Deformation tills are best identified on the basis of a range of characteristics, none of which may be uniquely diagnostic in itself.

Perhaps the clearest evidence for deformation is the presence of *deformed inclusions*, such as streaked-out or folded pods of sand or soft rock.

Inclusions can behave as passive markers, highlighting patterns of strain in the surrounding till, or may behave as relatively rigid regions, and exhibit pressure shadow effects and boudinage (Berthelsen, 1979; Hart and Roberts, 1994; Benn and Evans, 1996). Deformed inclusions may be particularly

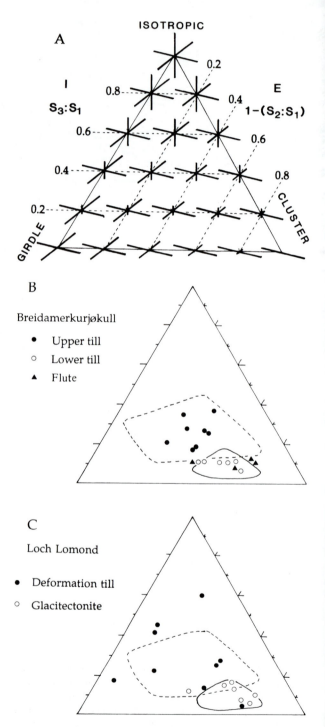

Fig. 10.11

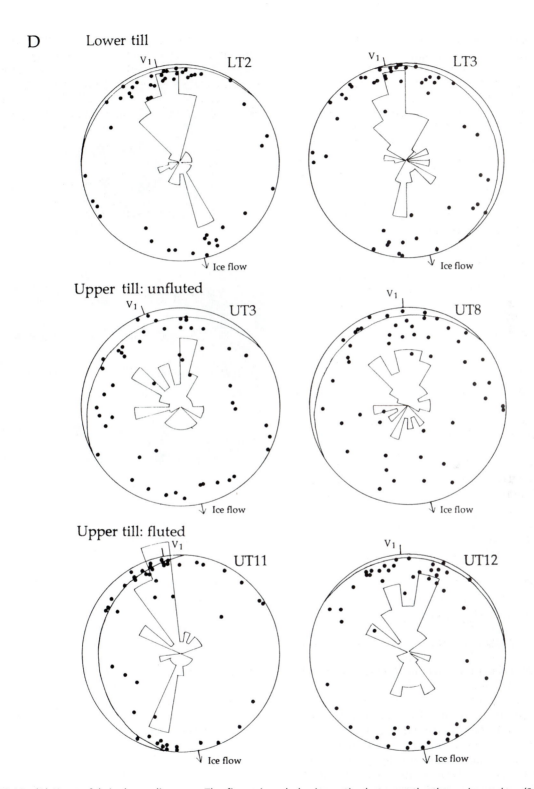

Fig. 10.11 (A) Key to fabric shape diagrams. The figure is scaled using ratios between the three eigenvalues (S_1, S_2, S_3) calculated from fabric data. All fabrics can be represented as points on the continuum between isotropic, girdle and cluster types. (From Benn, 1994a) (B) Fabric shapes for deformation till from Breidamerkurjökull, Iceland. (C) Fabric shapes for deformation till and glacitectonite from Loch Lomond, Scotland. Also shown are envelopes for fabric data from Skalafellsjökull: upper till (dashed line) and lower till (solid line). (D) Stereoplots and rose diagrams of deformation till fabrics, Breidamerkurjökull. (From Benn, 1995)

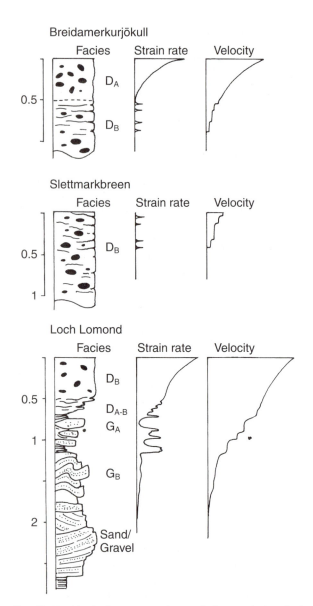

Fig. 10.12 Vertical patterns of cumulative strain in glacitectonites and deformation tills in Iceland, Norway and Scotland. D$_A$, ductile deformation till; D$_B$, brittle deformation till; G$_A$, penetrative glacitectonite; G$_B$, non-penetrative glacitectonite. (From Benn and Evans, 1996. Reproduced by permission of Elsevier)

abundant near the base of till units, where cumulative strains will be lower and the material less homogenized, and there may be a gradational contact with underlying glacitectonite (Hart, 1995a; Benn and Evans, 1996). In some cases, it is possible to trace inclusions back into glacitectonite or undisturbed parent materials (e.g. Benn, 1996a). However, there is often a sharp, non-gradational boundary between till units and underlying glacitectonite, reflecting either vertical rheological 'jumps' in the former subglacial deforming layer

(a)

(b)

Fig. 10.13 (a) Deformation till at Breidamerkurjökull, Iceland, showing the massive, dilatant A-horizon (above the penknife) and the fissile, denser B-horizon (below the knife). (Photo: A. Snowball) (b) Boulder pavement at the base of deformation till, Tierra del Fuego, Chile. (Photo: D.I. Benn)

(Alley, 1989), or two unrelated episodes of subglacial deposition.

The density and structure of deformation tills vary with their strain history. This can be illustrated by the two till horizons below Breidamerkurjökull, Iceland, that were studied by Boulton et al. (1974), Boulton and Hindmarsh (1987) and Benn (1995) (Section 4.4; Fig. 10.13). The upper (A) horizon of the till, which deformed by continuous ductile flow while in a dilatant state, has low shear strength and a porous, bubbly texture, and is apparently structurally massive. In contrast, the lower (B) horizon deformed much more slowly, apparently by brittle or brittle-ductile shear along discontinuous shear planes, and has higher shear strength, is compact, and has a fissile, platy structure. Void ratios for the upper till are 0.485–0.650 (mean = 0.585), markedly higher than those for the lower till (0.395–0.500, mean = 0.450). Ronnert and Mickelson (1992) have warned that high void ratios do not necessarily mean that till has deformed in a dilatant state; it can also arise from the melt-out of debris-rich ice.

The grain size distribution of deformation tills can vary widely. Some distributions simply reflect mixing of the parent materials, with little or no modification by crushing or abrasion (Elson, 1989), whereas others show clear evidence for grain fracture and comminution during shear (e.g. Boulton *et al.*, 1974; Hooke and Iverson, 1995; Benn, 1995, 1996a). The amount of grain size modification depends on the relative importance of dilation and grain fracture in the shearing process, which in turn reflects till strength and porewater pressures (Sections 4.4.2.4 and 4.4.2.5). Dilatant, low-strength tills can deform with little or no grain-crushing because grains can climb over each other during shear. In contrast, stiff, high-strength tills may deform mainly by the fracture of interlocking grains, leading to distinctive grain size distributions (Hooke and Iverson, 1995). Individual deformation till units are expected to have rather uniform grain size distributions over large areas, as a result of thorough mixing of materials under high cumulative strains (Alley, 1991).

Clast and matrix fabrics also vary with till strain history. Clast *a*-axis fabrics from dilatant tills are highly variable, and tend to exhibit a wide range of dip values (high isotropy) and weak to moderate preferred orientations (low elongation) (Dowdeswell and Sharp, 1986; Benn, 1994b, 1995; Hart, 1994a). The range of fabric shapes from modern and inferred ancient deformation tills is very similar to that from debris flow deposits, because the processes of particle orientation are similar in dilatant tills and shearing debris flows (Fig. 10.11). The preferred orientations of deformation till fabrics, however, will tend to be fairly spatially consistent, in contrast to the high spatial variability of debris flow fabric maxima. Fabric maxima in dilatant deformation tills generally parallel the direction of shear, although transverse maxima also occur (Hicock and Dreimanis, 1992a; Hart, 1994a; Benn, 1995; Benn and Evans, 1996). It should be remembered that the direction of shear in a subglacial deforming layer is governed by local stress gradients, which may diverge markedly from the flow of the overlying ice (Benn, 1994a). The variability of fabric shapes associated with dilatant tills probably reflects the low matrix strength and the consequent freedom of particles to rotate in any direction during shear. Particle *a*-axes can realign rapidly in response to transient strains, and may roll or tumble through the deforming matrix. On the foreland of Breidamerkurjökull, fabric shape displays pronounced variations within short distances in the vicinity of lodged boulders, reflecting local variations in former strain conditions (J.V. Kidd, unpublished data). In contrast, fabrics in fluted moraines on the foreland tend to exhibit consistently low isotropy and high elongation parallel to the flute axis, reflecting the formation of the flutes by sediment deformation into and along ice-walled grooves in the base of the glacier, where strains are likely to have been uniform and extensional (Boulton, 1976; Benn, 1995; Benn and Evans, 1996).

Fabrics from tills that have undergone brittle deformation, such as the lower till from Breidamerkurjökull, are similar to lodgement till fabrics, with low isotropy and moderate to high elongation parallel to the direction of shear (Fig. 10.11; Hart, 1994a; Benn, 1994a, 1995; Benn and Evans, 1996). This is because particle rotation will be strongly constrained by the stiff till matrix, and orientation mechanisms will consequently be similar to those that operate at the ice–till interface.

Hart (1994a, 1995a) has argued that there is an inverse relationship between fabric 'strength' (i.e. S1 values) and deforming layer thickness. Data from the upper till horizon at Breidamerkurjökull, however, show that this relationship is not consistent, and fabric organization can vary widely within a single deforming layer of relatively uniform thickness (Benn, 1995, Benn and Evans, 1996). We would emphasize, therefore, the value of interpreting fabric characteristics in terms of strain *type*, rather than deforming layer thickness.

Deformation tills exhibit a wide variety of microstructures and fabrics, depending on granulometry and strain history (Owen and Derbyshire, 1988; Menzies and Maltman, 1992; van der Meer, 1993; Owen, 1994). Faulting or folding may be easily recognizable where they affect sorted sediments, such as laminated silts and clays (e.g. van der Meer, 1993), but former deformation of diamictons can also be identified from more subtle features. Structures indicative of small-scale brittle deformation include discrete microshears (particularly in clay-rich sediments), brecciation, boudins or augen, and crushed quartz grains (Owen and Derbyshire, 1988; Menzies and Maltman, 1992; van der Meer, 1993). Pervasive deformation is recorded by various features indicating rotation of particles or particle aggregates, such as the rounded sediment 'pebbles' described by van der Meer (1993). Van der Meer has suggested that there may be an upward transition between microstructures recording brittle deformation and those typical of pervasive deformation, reflecting vertical patterns of till deformation (Boulton and Hindmarsh, 1987; Alley, 1989). This raises the interesting possibility that macro- and microstructures and fabrics can be used in conjunction to reconstruct till strain history.

Distinctive particle shapes may evolve beneath and within deforming subsole sediments, provided shear stresses and inter-particle contact forces are high enough to cause wear (Figs. 5.36–5.38; Section 5.7.3). Double stoss-and-lee forms reflect abrasion at the lower leading and upper trailing edges, and fracture at the lower trailing and upper leading edges as till deforms around clasts (Benn, 1996; Benn and Evans, 1995). Clasts can also develop flat, polished facets on their upper and lower surfaces, parallel to the plane of shear. The development of double stoss-and-lee forms and facets implies that the clasts have a quasi-stable attitude in the deforming till. Where lodged clasts are overridden by deforming till, the upper surfaces can become polished and striated, with little or no wear occurring in other parts of the clasts (MacClintock and Dreimanis, 1964; Clark, 1991a). In weak, dilatant tills deforming under low shear stresses, inter-particle contact forces are liable to be low, and particle shapes in the resulting deposit will be inherited from earlier episodes of erosion, transport and deposition. In such cases, polished facets and stoss-and-lee forms will bear little or no relation to the direction of shear (Benn, 1995; Benn and Evans, 1996).

Hicock (1991), Clark (1991a) and Boulton (1996a) have argued that some *boulder pavements* – horizons of clasts one or more particles thick – at the base of massive till units may be evidence for former subglacial deforming layers. Clark (1991a) suggested that large clasts in a weak deforming layer will sink down through the till to a level where the matrix is strong enough to support their weight, explaining the concentration of large clasts at a single level (Fig. 10.13b). This proposed mechanism is analogous to the well-known tendency for large clasts in debris flows to sink to the bottom (Section 10.5.5). Boulton (1996a) has argued that sub-till boulder pavements are formed during excavational deformation at the base of a deforming layer. According to this model, downward movement of the base of a deforming layer into an older till unit will result in the preferential mobilization of fine material and the accumulation of large particles which resist entrainment. The Boulton model therefore predicts that the boulders in sub-till pavements will be lithologically similar to those in the underlying sediments, whereas Clark's model predicts that they should be similar to those remaining in the overlying till. Clast pavements can, however, form by other mechanisms, and are not uniquely diagnostic of deformation tills (Section 10.7). Careful study of clast fabric in the pavements, and the relationship between the pavement and the overlying till, may be necessary to establish the origin (Hicock, 1991).

Comminution till is a distinctive type of till formed by subglacial deformation, first recognized by Elson (1961, 1989). These tills have very high density and shear strength, and are composed of crushed and powdered local bedrock with a grain size distribution similar to that produced in industrial crushing and grinding. They are thought to form during progressive crushing and abrasion in a subglacial deforming layer, leading to a reduction of void space and an increase in frictional strength. The till is deposited when its strength exceeds the driving stress. Elson illustrated this idea with the analogy of using abrasives to grind glass lenses and mirrors. As finer and finer abrasives are used in the grinding process, more and more water needs to be added to maintain void space and lubricate the working surface, owing to the increasing surface area of the abrasive and powdered glass. If insufficient water is added, the tool and mirror can suddenly stick together, and can be difficult to separate without breakage. If this type of process occurs beneath glaciers, it is most likely to be under well-drained conditions where basal shear stresses are high, such as under steep, slow-moving glaciers with surface profiles governed by the flow of ice rather than weak, deforming beds. Warren (1987) has described overconsolidated diamictons from Southern Ireland, which he interpreted as subaerial mass movement deposits that have been restructured and compacted by overriding ice. Warren proposed the new term *diagenetic till* to describe this type of deposit, but its mode of formation seems to be closely similar to that of comminution till.

10.3.5 Melt-out till

Melt-out till is sediment released by melting of stagnant or slowly moving debris-rich glacier ice, and directly deposited without subsequent transport or deformation. It can be deposited as (a) *subglacial melt-out till* by bottom melting of debris-rich ice at the base of a stagnant glacier or a stagnant zone beneath an active glacier; or (b) *supraglacial melt-out till* by top melting of glacier ice, often beneath a cover of debris (Boulton, 1971; Dreimanis, 1989). There have been very few studies of tills in modern environments where till genesis by melt-out can be clearly demonstrated, and many of the proposed characteristics of melt-out tills have been inferred from the geological record (e.g. Haldorsen and Shaw, 1982). Interpretations of ancient tills are often controversial, however, and diagnostic criteria based on geological evidence alone are not accepted by all researchers. The most detailed study of modern melt-out tills is by Lawson (1979a, b), who compared the characteristics of glacier ice and associated deposits at the margin of Matanuska Glacier, Alaska. Important observations have also been made in Svalbard by Boulton (1970a, b, 1971) and at the margin of Burroughs Glacier, Alaska, by Mickelson (1973, 1986), Ronnert and Mickelson (1992), and Ham and Mickelson (1994).

The properties of melt-out tills are derived primarily from the ice source, but are usually modified to some extent by depositional and postdepositional processes. Modification will generally be greatest where the parent ice has a low debris content, and where water produced during melting cannot drain freely away. Poor drainage conditions lead to elevated porewater pressures in the till, increasing the likelihood of deformation and reworking (Boulton, 1971; Paul and Eyles, 1990). Preservation of primary characteristics is also dependent on the local depositional slope, and failure and remobilization of the till can occur on slopes as low as 3° if drainage is poor (Paul and Eyles, 1990).

Melt-out tills may inherit foliation from the parent ice, including discontinuous layers and lenses, textural and compositional banding, and flow structures (Lawson, 1979a, b). The dip of the foliation will be less than that in the ice source, because of the volume reduction and compaction consequent on ice melting and dewatering (Fig. 10.14; Boulton, 1971). Possible Pleistocene examples of foliated melt-out tills have been described by Shaw (1979, 1982, 1983b), although these examples appear to be very similar to glacitectonites and deformation tills, and the foliation could have originated in subglacial deforming layers rather than in debris-rich ice. Indeed, it may be very difficult in some cases to distinguish between these two origins without very detailed study. Some melt-out tills are structureless and massive, and lack any visible foliation (Lawson, 1979a).

The granulometry and particle morphology of melt-out tills are inherited from the parent ice, and may be characteristic of either active or passive transport. Melt-out tills can be texturally identical to the parent debris-rich ice, but can show a relative depletion of fine material, owing to leaching by escaping porewater (Lawson, 1979a). Skins of silt and clay on the upper surfaces of clasts provide evidence for this process (Lundqvist, 1989b). According to Haldorsen (1981, 1983), melt-out till is generally coarser and contains more angular material than lodgement till in the same area. Subglacial melt-out tills may contain high percentages of edge-rounded and striated clasts indicative of wear during shear, but should not display consistently oriented stoss-and-lee clasts, which develop during lodgement and ploughing beneath moving ice.

The porosity of freshly deposited till is variable. Lawson (1979a) found that supraglacial melt-out till on Matanuska Glacier was loose and had high porosity, whereas subglacial melt-out till was much more compact. In contrast, Ronnert and Mickelson (1992) described fresh subglacial melt-out till with high porosity. The main controls on porosity are the debris concentration in the parent ice, effective overburden pressures and drainage conditions.

Pebble *a*-axis fabrics will reflect the original englacial fabrics, but with some overprinting during dewatering and consolidation (Boulton, 1970b; Law-

son, 1979a, b). Englacial fabrics often have strong preferred orientations parallel to the direction of shear, because of the rotation of clasts by the surrounding deforming ice (Fig. 10.15; Glen *et al.*, 1957; Lawson, 1979a; Ham and Mickelson, 1994). Preferred orientations of pebbles in ice are commonly parallel to ice flow directions, but may be transverse in zones of compressive flow (Boulton, 1970a). Melt-out till fabrics collected by Lawson (1979a, b) faithfully reflect former ice flow directions, unlike those of glacigenic debris flow deposits in the same area (Fig. 10.16). The melt-out till fabrics, however, display a reduction in dip values and an increase in dispersion relative to englacial fabrics. Compaction during the melt-out process reduces the range of dip values and 'flattens' the fabric, reducing its isotropy, and clast interactions during settling weaken the preferred orientation, reducing the elongation (Fig. 10.15; Lawson, 1979a, b; Benn, 1994b). Fabric elongation, however, remains higher than that typically associated with lodgement tills or 'brittle' deformation tills.

Lawson's data-set has been widely used as a reference for the interpretation of ancient tills. However, the data-set is small and from a single location, so it would be unwise to assume that all melt-out tills will share the same fabric characteristics.

10.3.6 Intermediate varieties of subglacial till

The properties of many subglacial tills suggest that they have complex histories, involving various combinations and cycles of lodgement, deformation and melt-out. Examples include till complexes formed by all three processes at Bradtville, Ontario (Dreimanis *et al.*, 1986; Hicock, 1990); and hybrid lodgement and deformation tills described by Hicock and Dreimanis (1992b) and Benn (1994a). The difficulty of reconstructing till history is illustrated by Ham and Mickelson (1994), who described subglacial tills exposed by recent retreat of Burroughs Glacier, Alaska. The tills are known to have been frozen into the glacier sole, and to have subsequently melted out. Some characteristics of the till, however, indicate that it is not a simple melt-out till. First, slickensided shear surfaces are present throughout, particularly in the uppermost 0.5 m, which has a pronounced subhorizontal fissile structure, possibly indicative of subsole deformation. Second, stoss-and-lee boulders and lee-side flutes are common at the upper surface of the till, indicating modification of the till below sliding ice. Third, fabric shape shows a broad scatter of isotropy and elongation values, overlapping with those associated with debris-rich ice, lodgement tills and deformation tills. These characteristics suggest that the till may have undergone several cycles of erosion, transport and deposition beneath the glacier, involving repeated melt-out and freezing.

Benn (1994a, 1995) and Benn and Evans (1996) have described tills from the forelands of Slettmarkbreen, Norway, and Breidamerkurjökull, Iceland, which apparently formed by a combination of subsole deformation and lodgement. At both sites, large boulders on the surface of the till commonly exhibit stoss–lee forms with a consistent orientation of plucked lee faces in the downglacier direction, indicative of lodgement at the ice–till interface. However, the frequency of stoss–lee clasts and of downglacier-facing lee faces decreases

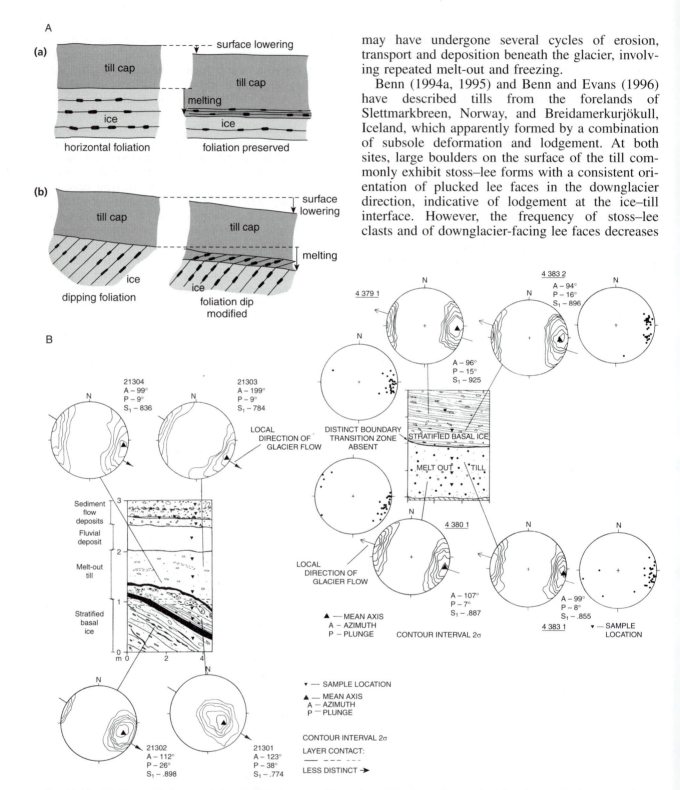

Fig. 10.14 Melt-out till characteristics. (A) The preservation and modification of supraglacial melt-out tills showing (a) horizontal ice surface and folia and (b) dipping ice surface and folia and the resulting modification of fabric. (From Sugden and John, 1976, after Boulton, 1971) (B) Stratigraphic and fabric characteristics of melt-out tills from Matanuska Glacier, Alaska. Left is surface melt-out till, which displays preservation of pebble orientations from the glacier ice but with increased scatter and decreased dip angle. Right is basal melt-out till, which displays virtually identical pebble fabrics to the glacier ice. (Modified from Lawson, 1979b)

with clast size, and many of the smaller clasts show evidence of remobilization in a thin subglacial deforming layer following deposition (Fig. 10.17). The tendency for large boulders to lodge while finer material forms a mobile bed is thought to reflect two main factors: (a) large boulders will tend to lodge most readily because the drag between clasts and the bed is directly proportional to their buoyant weight (Hallet, 1979a); and (b) larger clasts will be more likely to 'bridge' across the deforming layer and lodge against the rigid substrate, while smaller material can be carried along within the till. The tills from Slettmarkbreen and Breidamerkurjökull can therefore be regarded as hybrid lodgement and deformation tills, the relative importance of lodgement and deformation varying with clast size.

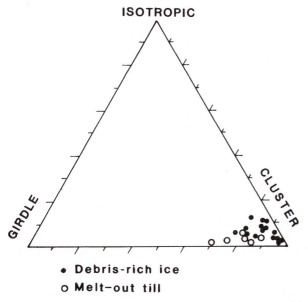

Fig. 10.15   Melt-out till fabrics. (Data from Lawson, 1979b)

10.3.7 Sublimation till

Sublimation till is defined as sediment released by sublimation of glacier ice, and directly deposited without subsequent transport. It is therefore closely related to melt-out till, except that ice is lost by sublimation, or the direct transformation of ice to vapour, rather than melting (Bell, 1966; Shaw, 1977a, 1989). Sublimation without associated melting is a form of freeze-drying, and requires very cold, arid conditions, such as those currently experienced in Antarctica. The formation of sublimation till requires contact with the atmosphere (possibly through pore spaces), but may be derived from supraglacial, englacial or subglacial debris sequences.

The process of sublimation is extremely slow and gentle, and very delicate englacial structures can be inherited from the parent ice, including foliation, augen, laminae bent round clasts, and attenuated folds (Fig. 10.18; Shaw, 1977a; Lundqvist, 1989b). In studies of modern sublimation tills in the Dry Valleys region, Antarctica, Shaw (1977a, c) described massive facies grading upwards into foliated facies, a transition that he attributed to strain patterns in the basal layers of the ice. The slow loss of interstitial ice means that sublimation till can be extremely loosely packed, and liable to collapse at the slightest touch (Lundqvist, 1989b). If wetted, it tends to flow readily. Such tills will therefore have very low preservation potential.

Grain size, particle morphology and fabrics in sublimation tills will be closely similar to those in the parent ice. Leaching of fines should not occur during deposition, reflected by the absence of silt skins on the upper surfaces of clasts (Lundqvist, 1989b). Fabrics should faithfully preserve original englacial fabrics, although minor modification may occur during settling. A rose diagram presented by Lundqvist (1989b) showed strong preferred clast orientations parallel to the former ice flow direction.

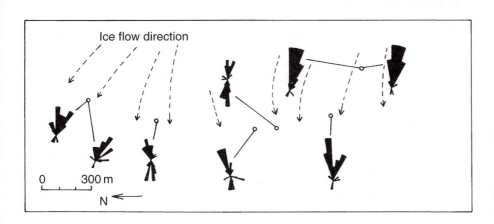

Fig. 10.16 The spatial pattern of pebble fabrics in melt-out tills on the foreland of Matanuska Glacier, Alaska. Fabric maxima show close parallelism with former glacier flow lines. (Redrawn from Lawson, 1979b)

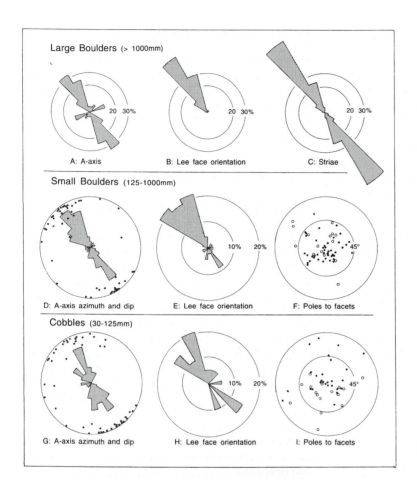

Fig. 10.17 Fabric data from the tills in the foreland of Slettmarkbreen, Norway. For the pole to facet data (F, I), facets on the upper surfaces of clasts (solid circles) are plotted on upper hemisphere equal-area nets, while facets on the lower surfaces (open circles) are plotted on standard lower hemisphere equal-area nets. (From Benn, 1994a. Reprinted by permission of Blackwell)

Fig. 10.18 Debris-rich glacier ice passing vertically upwards into sublimation till, Vega Island, Antarctic Peninsula. (Photo: S. Rubulis)

10.3.8 Tills and magnetism

As in any other sediment, the individual silt-sized grains that make up till may be aligned with the Earth's magnetic field as they are deposited. This *palaeomagnetism* is assessed in many Quaternary stratigraphic studies and is discussed in detail by Stupavsky and Gravenor (1984), Thompson and Oldfield (1986) and Oldfield (1991). Subaqueous sediments, such as those that are produced in glacilacustrine and glacimarine settings, are deposited in low-stress environments, and so their tendency to align themselves with Earth's magnetic field is unhindered. This results in high degrees of *natural remanent magnetism*. In contrast, magnetic particles in subglacial deposits are more likely to be aligned parallel to the direction of shear, and commonly show close agreement with clast fabric orientations (Fuller, 1964; Owens, 1974; Boulton, 1976).

The bulk magnetic susceptibility of tills, or the ability of a sample to become magnetized, also has been employed to differentiate them in chronostratigraphic sequences and to determine ice flow directions (Gravenor and Stupavsky, 1974; Puranen, 1977; Day and Morris, 1982; Chernicof, 1983). Because the susceptibility of a till is related to the concentration of magnetic minerals, its bedrock source can often be pinpointed on the basis of procedures similar to those of till geochemistry. For example, susceptibility will drop off in a down-ice direction from a source outcrop such as a lead–zinc or sulphide ore body. The use of mineral magnetic assemblages of glacial diamictons is a potentially powerful tool for differentiating multiple depositional and palaeo-ice flow events, provided sample grain size is carefully considered. Magnetic parameters can be strongly related to particle size, and their variability in a bulk sample will therefore reflect grain size rather than sediment provenance (Oldfield *et al.*, 1985). The mineral magnetic assemblages of the tills of the Isle of Man have been analysed by Walden *et al.* (1987, 1992), who have been able to differentiate the provenance of these sediments and in addition isolate provenance differences between stratigraphically equivalent till units on the east and west coasts of the island. Factor analysis was employed by Walden *et al.* (1992) to demonstrate the discrete clustering of mineral magnetic parameters and the small intra-unit variation compared with inter-unit variability (Fig. 10.19). The Isle of Man study and a similar study undertaken on the tills of the Midland Valley of Scotland (Walden *et al.*, 1995) reveal a close agreement between remanent magnetism and palaeo-ice flow directions previously reconstructed by clast

fabric analysis, till petrology and ice flow lineation mapping.

The techniques involved in the mineral magnetic analyses of glacial sediments, particularly tills, are described by Walden *et al.* (1992). References to general applications of mineral magnetics are found therein and are further reviewed in Thompson and Oldfield (1986).

10.4 FLUVIAL DEPOSITS

10.4.1 Terminology and classification

Sediments deposited from flowing water are common in many glacial environments because of the seasonal abundance of meltwater. Fluvial facies can be deposited in subglacial and englacial conduits, in supraglacial and proglacial streams, and near subaqueous meltwater portals, and are therefore important consituents of many subglacial, ice-marginal, proglacial and subaqueous sediment–landform associations. Individual lithofacies reflect local sediment supply and water flow conditions rather than the position of deposition, so that some facies may be deposited in all the above environment types. In each environment, however, facies occur in different combinations, forming distinctive vertical successions and architectural elements that reflect larger-scale channel patterns and their change through time.

In fluvial systems, it is possible to recognize a hierarchy of depositional units, representing the storage of sediment at different scales (Jackson, 1975; Ashley, 1990; Miall, 1992). At the smallest scale, sediment is deposited in *bedforms* such as ripples and dunes on channel floors. These bedforms are repetitive, mobile structures that migrate downcurrent in response to the erosion of sediment from their upstream faces (where shear stresses are highest) and the deposition of sediment on their lee faces (where shear stresses are low). The form, wavelength and height (amplitude) of bedforms reflect specific combinations of sediment grain size and flow velocity, as shown in phase diagrams based on flume experiments (Fig. 3.37; Section 3.8.3). The smallest bedforms (ripples) reflect flow conditions very close to the bed, in the viscous sublayer, and are classed as *microforms*, whereas the larger bedforms (dunes) reflect conditions in the outer part of the boundary layer of the flow, and are termed *mesoforms* (Jackson, 1975; Allen, 1982a, 1985). Microforms can be superposed on the upstream faces of mesoforms. In turn, bedforms occur within larger depositional units such as longitudinal and bank-attached bars (*macroforms*), and channel fills. These units respond to large-scale flow patterns and long-term fluvial regime, and tend to be active only during peak discharges. At yet

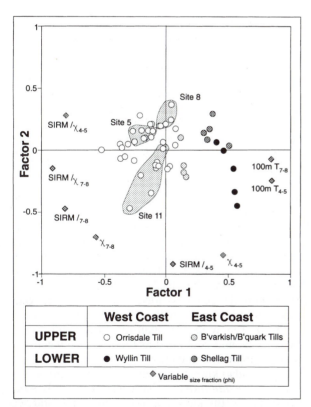

Fig. 10.19 Results of factor analysis to show the discrete clustering of mineral magnetic parameters for the tills of the Isle of Man. Variables are: SIRM = saturation isothermal remanent magnetization; χ = initial low-field mass-specific magnetic susceptibility; 100m T = backfield ratio (4–5 and 7–8 = phi particle size). Orrisdale till matrix samples from individual sites plot as discrete clusters (e.g. samples from sites 5, 8 and 11 are contained in shaded areas), illustrating that intra-unit variation is less within sites than between sites. (Modified from Walden *et al.*, 1992)

larger scales, macroforms and channel fills are nested within channel systems, valley fills and basin fills (Miall, 1991, 1992).

Fluvial sediments are characterized by stratification and other sedimentary structures at several scales, which record the migration and accretion of bedforms and larger depositional units (Allen, 1983; Miall, 1985a, 1991, 1992). In the following sections, we describe genetic facies formed by the accretion and migration of bedforms below flowing water. We also consider facies formed by settling of suspended particles from portions of flows that have come to a halt, and facies deposited from hyperconcentrated flows, or sediment–water mixtures that are transitional between fluvial flows and debris flows. The ways in which genetic facies combine in larger units (architectural elements) representing channel fills and bar forms are discussed in Section 11.5.1.2.

10.4.2 Plane bed deposits

Net deposition of sediment on flat, sandy stream beds produces beds of horizontally bedded or laminated sand, with the laminae recording minor fluctuations in flow velocity or sediment supply. Lamination can be strong or faint, depending on depositional conditions (Fig. 10.20). Grain size tends to be very fine to coarse sand (0.0625–1.00 mm). For units deposited under upper flow regime conditions (Section 3.8.3), bedding planes display thin, linear grooves and ridges aligned parallel to the former flow direction, called *parting lineation*. This lineation forms in response to instabilities in the boundary layer of the water flow (Allen, 1982a, 1985), and is often observed in sandstones that split along bedding planes. It is much harder to detect in unlithified Quaternary sediments.

Sedimentation on plane beds under the lower flow regime also results in horizontally bedded and laminated sands. In this case the sand tends to be coarse

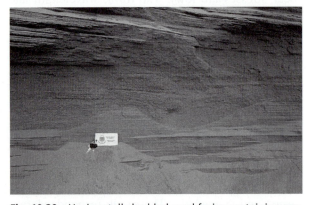

Fig. 10.20 Horizontally bedded sand facies containing erosional scours, Drumheller, Alberta, Canada. (Photo: D.J.A. Evans)

to very coarse (0.8–2.00 mm), and parting lineation is absent.

10.4.3 Ripple cross-laminated facies

Sand facies deposited by the migration and vertical accretion of current ripples (Section 3.8.3) display distinctive internal structures called *ripple cross-lamination*, the form of which depends on the balance between downcurrent migration and suspension sedimentation during ripple formation (Jopling and Walker, 1968; Allen, 1970, 1985; Ashley *et al.*, 1982). Three basic types of cross-lamination can be recognized (Jopling and Walker, 1968; Allen, 1973).

Type A develops when more sediment is eroded from the stoss sides during ripple migration than is added from suspension. Successive lee-side positions of the ripples are preserved as cross-laminae that dip down-current, but the stoss positions are not preserved, because of net erosion. Instead, individual sets of lee-side laminae are separated by diachronous erosion surfaces that dip upcurrent (Fig. 10.21). The angle of climb of these surfaces is variable, and reflects the net amount of deposition from suspension averaged over the entire bedform. There are two limiting cases for Type A cross-lamination, based on the angle of climb. In the first case, the angle of climb is zero relative to the base of the bedform, recording ripple development by migration only, with no addition of sediment from suspension. The ripples form isolated sets sometimes known as *starved ripples*. In the second case, the angle of climb is identical to the angle of the original stoss faces, recording an exact balance between deposition from suspension and erosion on the stoss face of the bedform. This type is also known as *critical cross-lamination* (Allen, 1985).

Type B cross-lamination displays successive stoss face positions as well as the lee face positions, and records net deposition from suspension over the whole bedform (Fig. 10.21). Net deposition is still highest on the lee faces, however, owing to the redistribution of sediment by the water current. The angle of climb is now greater than the slope angle of the stoss faces, and is defined as the angle between successive ripple crest positions.

Type S, also known as *sinusoidal cross-lamination* or *draped lamination*, is characterized by only small variations in thickness between stoss-side and lee-side laminae, and a very steep angle of climb (60°–90°). Ripple profiles are only weakly asymmetric or symmetrical. This type of cross-lamination records small or no ripple migration (i.e. weak or zero current flow) and a dominance of suspension sedimentation.

Vertical gradations between two or more of these ripple types are often observed in sandy fluvial and glacilacustrine sediments, recording temporal changes in current flow and suspended sedimentation rates. Upward transitions from A–B–S, A–B and B–S are characterized by an upward increase in the angle of climb paralleled by a decrease in mean sediment size, and are formed during waning flows and an increase in deposition from suspension. Upward transitions from S–B–A, S–B or B–A have a decreasing angle of climb and an increase in mean sediment size, and record rising flows and a decrease in the importance of suspension sedimentation. Excellent examples of different types of transition are illustrated by Jopling and Walker (1968) and Ashley *et al.* (1982).

10.4.4 Dunes

Sand and gravel facies produced by migration and vertical accretion of dunes (Section 3.8.3) have a cross-bedded internal structure, resembling large-scale ripple cross-lamination. Two basic types are recognized, although more detailed classification schemes are used in detailed sedimentological studies (Allen, 1963, 1982a).

Planar cross-bedded gravels and sands originate by the migration of transverse, two-dimensional dunes with approximately planar lee surfaces. The thickness of sets gives a minimum height for the original bedform, and generally ranges between 0.25 m and 4 m or more for gravels and from 0.05 to at least 5 m for sands (typically less than 1 m; Miall, 1977). The thickness of lee-side beds may be as much as 0.4 m in larger gravel sets, recording a considerable amount of sediment transport over the original dune surface. Planar cross-bedded gravels and sands can also form by the growth of small deltas from bars into deep channels, but this is thought to be rare.

Trough cross-bedded gravels and sands form by the migration of three-dimensional dunes or two-dimensional dunes with scalloped lee faces (Miall, 1977; Ashley, 1990). The cross-beds have curved surfaces, concave up and concave downstream, reflecting deposition in lee-side hollows. The maximum dip of cross-beds may reach 30°. Sets typically range from 0.2 m to 3 m in thickness for gravels and 0.05 m to 0.6 m for sands (Miall, 1977). Trough cross-bedding may occur in solitary scoops eroded into other facies, or form cosets (multiple sets) of cross-cutting units called *festoon cross-bedding*.

(a)

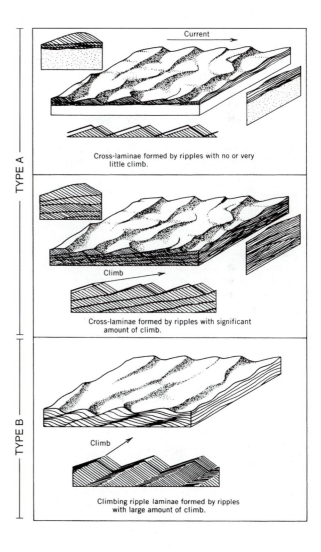

(b)

Fig. 10.21 Rippled sand facies: (a) illustrating climbing relationships, Drumheller, Alberta, Canada (Photo: D.J.A. Evans); (b) the production of cross-laminated rippled sands and their relationship with rates of bedform climbing. (From Fritz and Moore, 1988. Reproduced by permission of John Wiley & Sons)

Planar and trough cross-beds can be cross-cut by low-angle erosion surfaces called *reactivation surfaces* (Fig. 10.22; Boothroyd and Ashley, 1975; Collinson, 1986). These surfaces, which may be planar, curved or irregular, record reworking of dune slipfaces, often during lower stages of flow. Renewed cross-bed deposition above the reactivation surface indicates a re-establishment of downcurrent dune migration.

Measurement of the direction of dip of cross-beds in planar and trough cross-bedded gravels and sands is a useful indicator of palaeoflows. It should be remembered, however, that in braided or meandering rivers, bedforms migrate at a variety of angles to the main trend of the river, and that reconstructions based on a few measurements may give a misleading view of former river flow directions. Large samples typically show a broad scatter of palaeoflow direc-

tions, the centre of which gives a good indication of the general direction of flow.

10.4.5 Antidunes

Sediment facies deposited by the migration of antidunes can form *foresets* or *backsets*, depending on the direction of bedform migration (Allen, 1982a, 1985; Rust and Gibling, 1990). Backsets typically dip upstream at a shallow angle, and provide the clearest diagnostic feature of antidunes. Additionally, downstream-dipping surfaces may have smaller bedforms superimposed upon them, a situation which cannot occur on dunes.

10.4.6 Scour and minor channel fills

Scours are overdeepened parts of river beds related to disturbances of the current around dunes or other obstructions. They are elongate parallel to flow, typically asymmetric transverse to flow, and may be up to 0.45 m deep and 3 m wide (Miall, 1977). The subsequent infilling of scours with sand, gravel or pebbly sands produces cross-bedded facies similar in some respects to trough cross-beds formed by dune migration. However, scour fills show more internal variation, and sometimes display minor sedimentary structures such as ripple cross-lamination and parting lineation on bedding planes, indicating that the infill

(a)

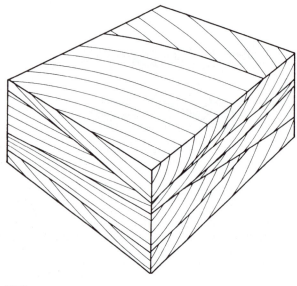

(b)(i)

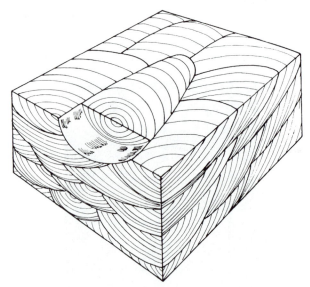

(b)(ii)

Fig. 10.22 Sand and gravel cross-bedding: (a) a complex sequence of cross-bedded sands and gravels in glacifluvial outwash, Alberta, Canada (Photo: D.J.A. Evans); (b) block diagrams showing (i) planar and (ii) trough cross-bedding. Bold lines denote reactivation surfaces. (From Reinick and Singh, 1980. Reproduced by permission of Springer-Verlag)

does not consist of lee-side slipfaces. Scour erosion and infill may be quite widely separated in time, unlike the development of erosional and depositional surfaces in festoon cross-bedding.

10.4.7 Gravel sheets

Gravel sheets form by clast-by-clast accretion on low-relief parts of river beds, such as longitudinal and bank-attached bars (Hein and Walker, 1977; Miall, 1977; Collinson, 1986). Sediment movement and deposition are strongly episodic, occurring during floods and subsequent waning flows; between floods, gravel sheets are commonly exposed above water level. Typical dimensions for individual depositional units range up to 1 m in thickness and tens to hundreds of metres in lateral extent.

Gravel sheets consist of massive to crudely bedded clast-supported gravels (Fig. 10.23; Miall, 1977). The clast-supported framework reflects grain-by-grain deposition from bedload. During this process, fine gravel particles can become trapped in the spaces between large clasts, resulting in a poorly sorted deposit, although crude horizontal stratification can develop during fluctuating flows. Clusters of large clasts sometimes occur, as a result of the trapping of clasts upstream of obstructions, in a process similar to the development of clast clusters in lodgement tills (Brayshaw, 1984). Clast accumulation in gravel sheets can result in fining-up units as the deposit builds to shallower water levels, but coarsening-up units can also occur as the result of downstream bar migration (Miall, 1985a). Gravel sheets can be openwork, but fine matrix commonly occupies the spaces between clasts, owing to filtering of sand or silt into the framework following deposition. Clast fabrics typically show upstream imbrication of *a–b* planes, and a preferred orientation of *a*-axes transverse to flow (Rust, 1972a, 1975; Boothroyd and Ashley, 1975; Hein, 1984). This fabric represents the most stable alignment of pebbles beneath flowing water, maximizing the strength of the bed and minimizing the stress imposed by the flow, and is widely regarded as a good palaeocurrent indicator.

10.4.8 Silt and mud drapes

During waning flows, pools of stagnant water are commonly left in abandoned channels, allowing fine suspended sediment to settle out without being entrained as bedload. This sediment forms drapes of mud, silt and sometimes very fine sand that blanket the underlying deposits (Fig. 10.24; Miall, 1977). Drapes will tend to be thickest in the middle of the pool, where the water is deepest, forming thin, concave-up lenses of variable lateral extent. Typical thicknesses range from a few millimetres to a few centimetres. Beds are commonly massive, but can be laminated if sediment delivery occurred in pulses. Fluvial mud drapes must not be confused with glacimarine mud drapes discussed in Section 10.6.

10.4.9 Hyperconcentrated flow deposits

The deposits of hyperconcentrated stream flows (Section 3.8.1) form sheets or channelized lenses, commonly tens of centimetres to a few metres thick and several metres to tens of metres across. Particularly extensive deposits, hundreds or thousands of metres across, can be formed by catastrophic outburst floods, or jökulhlaups, caused by failure of moraine-dammed lakes (Section 3.5; Lliboutry *et al.*, 1977; Lord and Kehew, 1987; Maizels, 1989a, b). In general, deposits consist of matrix-supported gravels

Fig. 10.23 Gravel sheets in Late Devensian glacial outwash, Lackford, south-east England. Note the intact mammoth tusk protruding from the gravels to the right of the trowel, attesting to short travel distances. (Photo: D.J.A. Evans)

Fig. 10.24 Sandy and silty mud drapes in a vertical sequence of gravel sheets, Alberta, Canada. (Photo: D.J.A. Evans)

or diamictons, reflecting mass transport and rapid deposition. Units may be entirely massive or display internal structures such as crude stratification and basal inverse- or inverse-to-normal grading (Fig. 10.25 (Plate 21); Maizels, 1989b; Todd, 1989). Stratification can consist of crude horizontal bedding or cross-bedding, and probably records pulsed deposition from surges in the flow (Maizels, 1989b). There is commonly an upward increase in textures, structures and sediments indicative of fluvial processes, such as better sorting, well-developed bedding and cross-bedding, and winnowed gravels. These characteristics are thought to result from the transition from hyperconcentrated to normal stream flow during falling discharges, whereupon formerly suspended sediment is reworked as bedload (Maizels, 1989; Todd, 1989).

Pebble fabrics in the massive parts of hyperconcentrated flow deposits tend to show preferred *a*-axis orientations parallel to flow and upflow *a*-axis imbrication (Todd, 1989), characteristics typical of sheared, concentrated sediment–water mixtures (Hein, 1982; Massari, 1984; Benn, 1995). No fabric shape data have been published.

10.5 GRAVITATIONAL MASS-MOVEMENT DEPOSITS

10.5.1 Overview

The abundance of steep, unstable slopes in many glacial environments makes gravitational mass movements very important agents of sediment reworking and deposition. Gravitational processes operate in subglacial cavities, on the surface and around the margins of glaciers, and on proglacial and subaqueous debris slopes. Deposits formed by gravitational mass wasting processes are usually classified according to the geometry of the moving mass, the velocity and mechanism of movement, and water content during transport and deposition (Middleton and Hampton, 1973; Selby, 1982; Postma, 1986). Five basic types of deposit are described in the following sections: (a) *fall deposits*, composed of particles that have fallen, rolled, bounced or slid down steep slopes; (b) *gelifluction deposits*, formed by the slow downslope movement of seasonally frozen debris; (c) *slump and slide deposits*, formed by the sliding of intact blocks of sediment or rock, with varying amounts of internal deformation; (d) *debris flow deposits*, resulting from the flowage of concentrated sediment-water mixtures; and (e) *turbulent flow deposits* or *turbidites*, deposited from turbulent underflows below standing water.

It should be emphasized that, in nature, mass movements commonly undergo transformations from one type to another during transport. For example, slumps can evolve into debris flows if the component debris becomes liquefied and disaggregated; and subaqueous debris flows can partially or completely transform into turbulent underflows as a result of sediment mixing with the overlying water column (Sections 7.6.2 and 8.4.4). As a result of transformations during movement, mass movement deposits may exhibit lateral and vertical changes in structure, grain size, sorting and other properties, which can be used to reconstruct the sequence of events during transport and deposition. Another important point is that debris can undergo several cycles of movement and redeposition by gravitational and other processes. This is particularly true in ice-contact environments where the topography is subject to major changes due to ice melt and advance, and for steep subaqueous slopes which may undergo alternating phases of oversteepening and failure (Sections 6.5 and 8.5). Such multiple reworking episodes can form complex successions that require careful study for their correct interpretation; some examples are described in Chapter 11.

10.5.2 Fall deposits

Fall deposits form by the accumulation of debris at the foot of slopes, following transport by falling, rolling, sliding and bouncing down the slope. Fall deposits can be predominantly coarse-grained or diamictic, and can accumulate by numerous low-magnitude events or single high-magnitude events. In this book, diamictic deposits are termed *debris fall deposits*, coarse fall deposits constitute *rockfall deposits* or *talus*, and very coarse deposits from single high-magnitude rock slope failures are *rock avalanche deposits*. Fall deposits can be deposited in subaerial and subaqueous environments, and to a much more limited extent in subglacial cavities. The characteristics of fall deposits tend to differ in each of these three types of environment, and are discussed in turn below.

10.5.2.1 SUBAERIAL FALL DEPOSITS

The extent and geometry of subaerial fall deposits are very variable, depending on debris supply and the position of deposition. At the base of ice and snow slopes, deposits tend to accumulate in cones and wedge-shaped bodies, which may be only tens of centimetres thick where debris supply is limited. Where debris supply is high, talus deposits below ice and snow slopes can be several metres thick and coalesce into ramparts tens or hundreds of metres across (Lawson, 1981b, 1989; Ballantyne, 1987; D.J.A. Evans, 1989b; Ballantyne and Harris, 1994). Warren (1988) proposed the new term *protalus till* for talus that accumulates at the snout of a glacier after falling from a supraglacial source area and travelling over the ice by gravitational processes. This lies outside the definition of 'till' adopted in this book, and is not used here. Extensive fall deposits can also form below debris slopes, such as steep-sided moraines, recently deglaciated slopes and unglaciated hillsides, and take the form of cone-shaped or sheet-like bodies, depending on the nature of the source area (Ballantyne and Harris, 1994). *Talus* is a loose, clast-dominated deposit, with varying amounts of void space and interstitial matrix (Fig. 10.26). In *openwork* deposits, matrix material is completely absent. Deposits may be massive and structureless, but crude stratification and vertical size-sorting may be present. Particle morphology is commonly typical of periglacially weathered or passively transported debris, with high percentages of angular and slabby or elongate clasts, although this is not exclusively the case, and talus derived from glacier ice or pre-existing sediment can display a variety of clast morphologies. Fabrics can be isotropic, particularly in small talus bodies (Lawson, 1979b, 1989), but more commonly have some degree of preferred orientation, parallel or transverse to the depositional slope. Strong downslope preferred orientations develop where clasts slide down the depositional surface prior to deposition, or where postdepositional shearing of the deposit occurs, whereas more isotropic fabrics arise where particles interlock with a rough talus surface at varying angles (Fig. 10.27; Benn, 1994b). Talus and avalanche deposits often display marked *fall sorting* or downslope coarsening, owing to the greater momentum of large particles and their consequent tendency to travel furthest. *Debris fall deposits* have been described by Lawson (1979a, 1981a, 1989), who called them 'slope colluvium' and 'ice-slope colluvium'. The deposits are typically chaotic, consisting of unsorted and texturally diverse materials,

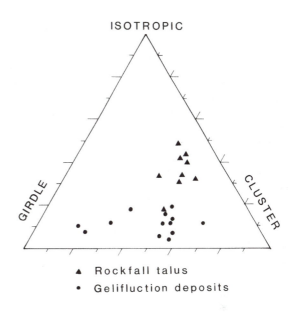

Fig. 10.27 Slope deposit fabric shapes. (Data provided by C.K. Ballantyne)

sometimes enclosing intact blocks of pre-existing sediment. The action of running water between fall events can form interbeds and lenses of sorted sediments and winnowed lag horizons.

10.5.2.2 SUBAQUEOUS FALL DEPOSITS

Fall deposits can show up very clearly on side-scan sonar images of present-day subaqueous slopes as lobate, fall-sorted spreads of coarse debris (Prior and Bornhold, 1990; Nemec, 1990). They can be difficult to identify in section in ancient deposits, mainly because debris falls tend to become enveloped by later fine-grained sediments, causing them to resemble diamictic debris flows. Diagnostic criteria might be isolated clasts overlain by layered drapes of fine sediment, indicating that the fine drape was deposited after the clasts, or evidence that large clasts have ploughed through the underlying sediment before coming to a halt.

10.5.2.3 FALL DEPOSITS IN SUBGLACIAL CAVITIES

Glacial debris dropping from the roof of a subglacial cavity can form diamictic fall deposits. They are most likely to be preserved on the lee side of roches moutonnées, where they form wedge-shaped masses thinning rapidly in the downglacier direction. The deposits can be massive and structureless, or exhibit crude stratification. Lenses of water-sorted sediments may be present (Hillefors, 1973).

Fig. 10.26 A talus slope in the Torngat Mountains, northern Labrador, Canada. (Photo: D.J.A. Evans)

10.5.3 Gelifluction deposits

Gelifluction is the slow downslope flow of sediment associated with the seasonal thawing of ground ice (Harris, 1981; Ballantyne and Harris, 1994), and is an important process of subaerial sediment reworking in periglacial environments. Flow occurs by intergranular creep and slip along planes of weakness formed by the melting of lenses of segregation ice. Velocities decrease with depth and are typically a few centimetres per year at the surface. Displacement can extend down as much as 3 m, but depths of 0.5 m are more common (Ballantyne and Harris, 1994). *Gelifluction deposits*, or *gelifluctates*, form extensive sheets composed of numerous downslope-oriented lobes. They are typically matrix-supported diamictons, although they can be clast-supported in places (Fig. 10.28). Granulometry and particle form are inherited from the parent material. Structurally, gelifluctates can appear massive but a crude stratification is often apparent, arising from slope-parallel shear surfaces, often marked by thin stringers of clay or silt. Fabric data from lobate gelifluction deposits on Ben Wyvis, Scotland, are shown in Fig. 10.27 (Ballantyne, 1981; Benn, 1994b). Fabric shape is closely similar to that from lodgement tills and 'brittle' deformation tills, probably because the fabrics developed under similar conditions of closely constrained shear. In the gelifluctates, however, fabric maxima are oriented downslope and bear no relation to regional ice movement. The strength of downslope preferred orientation is highly variable, probably reflecting varying styles of strain, vertical velocity gradients, and amounts of downslope movement.

10.5.4 Slide and slump deposits

Slides and slump deposits form sheet-like masses *c.* 10 cm to >100 m thick. In slides, the internal beds are mainly undisturbed, but may exhibit compressional or tensional deformation near the toe and head zones, respectively (Sections 7.6.2 and 8.4.4.6). Slumps display more extensive evidence of internal deformation, consisting of overfolds, box folds, imbrication of soft clasts, and branching thrust faults (Allen, 1982a; Stow, 1986). In large slumps and slides, the style of internal deformation can be very similar to that caused by proglacial glacitectonics (Section 7.4), and correct interpretation will require consideration of the surrounding facies, palaeoslope and surface topography. Slumps commonly generate debris flows, in the form of either local, small-scale flows, or large flows resulting from the large-scale transformation of the whole mass. Where subaqueous mass movement disrupts sediment but does not result in its long-distance transport and disaggregation, it produces what is known as gravity-induced

(a)

(b)

Fig. 10.28 Gelifluction deposits. (a) Active solifluction terraces, 1000 m above sea-level, Scottish Highlands. (b) Section through gelifluction deposits, Isle of Skye, Scotland. The sandy layers are shear surfaces that may mark the position of former ice lenses. (Photos: D.I. Benn)

soft sediment deformation (N. Eyles and Clark, 1985). This can lead to the production of sediment mélange consisting of flow structures and attenuation features such as boudins similar to those found in glacitectonite (Visser *et al.*, 1984).

10.5.5 Debris flow deposits

Debris flow deposits are formed by the gravitational flowage of sediment–water mixtures. Flowage can take place in both subaerial and subaqueous environments (Sections 6.5.3.1 and 8.4.4). In many texts, the term *flow till* is used to refer to subaerial sediment flows deposited in direct association with glacier ice or from freshly deposited till (e.g. Hartshorn, 1957; Sugden and John, 1976; Dreimanis, 1989; Hambrey, 1994). Similarly, the terms *subaquatic flow till* and *submarine flow till* (Evenson *et al.*, 1977; Hicock *et al.*, 1981) have been employed for subaqueous flow deposits. However, because processes and products of debris flow can be identical in glacial and non-glacial environments, it is preferable to employ terms that emphasize mechanisms of flow transport and deposition, rather than the origin of the sediment or the location of the flow (Lawson, 1979a, 1981a, 1989; Lowe, 1979; Gravenor *et al.*, 1984; Nemec and Steel, 1984; Postma, 1986; Nemec, 1990). In this book, the general term *debris flow deposit* is used in preference to 'flow till' with its environmental implications. The prefix *glacigenic* may be added where direct association with glacier ice can be demonstrated (Lawson, 1979a). Subaerial and subaqueous debris flow deposits commonly have distinct characteristics, owing to the role of the water column in modifying subaqueous flows (Fig. 10.29). They are therefore described separately below. A number of good sedimentology books are available in which further treatment of sediment gravity flows is provided (e.g. Reading, 1978; Reinick and Singh, 1980; Allen, 1982a, 1985; Leeder, 1982; Collinson and Thompson, 1989).

10.5.5.1 SUBAERIAL DEBRIS FLOW DEPOSITS

Subaerial sediment flow deposits are usually, but not always, diamictons with a fine-grained matrix (clay, silt or sand) and variable clast content. Bulk grain size distributions are essentially those of the parent debris, because of the unimportance of sorting and winnowing during transport. Because inter-particle contact forces are low during flowage, particle morphology is inherited with little or no modification from the source material, except for fragments of very weak rocks such as shale. Particle morphology may be characteristic of actively or passively transported glacigenic debris, water-worn sediment, or some mixture, depending on the source.

Deposits form lobate sheets or channelized lenses, generally a few centimetres to 2 m or more thick, and a few square metres to hundreds or even thousands of square metres in area. Sequences of flows can build up to over 10 m thick and cover areas of several tens of thousands of square metres (Lawson, 1979a). In section, individual flows form tabular or lens-shaped units, and if flows are channelized, units can have concave-up, erosional bases and flat tops (Fig. 10.30). Where successive flows are similar in texture, individual units may be difficult to distinguish, but boundaries are commonly marked by basal concentrations of clasts, upper washed horizons, interbeds of silt, sand or gravel, or more subtle bedding

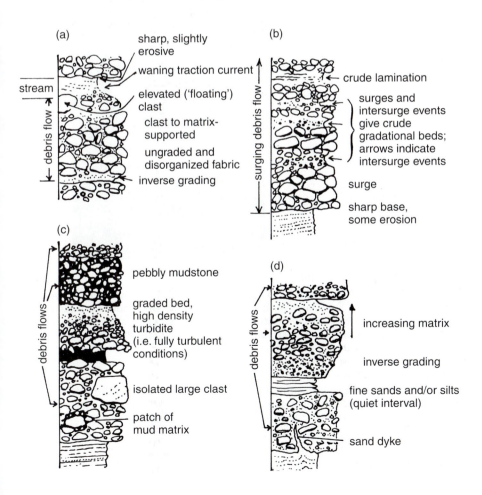

Fig. 10.29 Examples of different structures resulting from debris flows in (a), (b) subaerial and (c), (d) subaqueous settings. (Redrawn from Nemec, 1990)

(a)

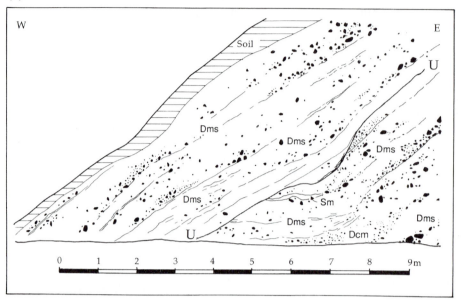

(b)

Fig. 10.30 Subaerial debris flow deposits, Isle of Skye, Scotland. (a) Section through a moraine, showing stacked debris flows. Note the basal clast clusters, planar and channelized contacts, and unconformity 'U'. (b) Photograph of part of the section. (Photo: D.I. Benn)

structures. At first acquaintance, sediment flow deposits can appear totally chaotic, but close examination often reveals a surprising amount of internal organization which can be used to reconstruct the processes of deposition in some detail.

The internal structure of sediment flow deposits is strongly related to the water content during flow and deposition. Four basic types of flow deposit, based on the flow types defined by Lawson (1979a, 1982), can be identified (Section 6.5.3.1). *Type I flow deposits* generally consist of a clast-rich, poorly sorted deposit incorporating blocks of pre-existing sediments, representing mass transport in a semi-rigid plug (Fig. 10.30; Rodine and Johnson, 1976; Shultz, 1984). The only visible structures may be zones of short, curving shear planes near the base, and thin silt or sand horizons near the top, reflecting post-depositional washing by water. Concentrations of large clasts may be present at the margins of the flow.

Type II flow deposits typically have concave-up lower surfaces in transverse section, with a massive, poorly sorted interior underlain and flanked by more organized, sheared sediment (Lawson, 1979a; Shultz, 1984; Cas and Landis, 1987). The sheared horizons tend to show distinct size sorting, with concentrations of clasts at the base owing to sinking of coarse particles through saturated, liquefied matrix, and their subsequent transport as bedload (Fig. 10.30). The sediment immediately above such clast concentrations can be depleted in coarse material relative to the rest of the deposit, and may be inversely graded. Curving shear surfaces and streaked-out smears of silty clay may be present. Washed horizons

commonly occur at the flow top, as the result of surface water flow and ponding following deposition.

Type III flow deposits resemble the lower, sheared part of Type II flows, with pronounced basal clast concentrations. There are often marked lateral and longitudinal variations in the thickness of deposits and basal clast concentrations, reflecting discharge fluctuations or *surges* in the parent flows (Fig. 10.29). Pulsed flow can also result in a crudely stratified deposit, with interbedded diamicton, clast concentrations and washed horizons. Individual particles can be thicker than the whole deposit, projecting above the flow surface.

Type IV flow deposits consist mainly of silt and sand, sometimes with thin basal layers of coarse sand or fine gravel, because flows with high water content cannot support clasts owing to their very low matrix strength. Individual depositional units are thin (0.02–0.1 m), but can build up thick sequences of thin laminae.

Pebble fabrics in sediment flow deposits are very variable, and display a wide range of isotropy and elongation values (Fig. 10.31; Lawson, 1979a, b; Mills, 1991; Benn, 1994b). Fabric shape and preferred orientations are dependent on patterns of strain within the parent flow. Fabrics in non-deforming plug zones can be isotropic, but well-organized fabrics can develop in zones of intense shearing as particles are rotated by the deforming matrix (Lawson, 1979a). Particle *a*-axes and *a–b* planes can be aligned parallel to the frontal and lateral margins of flows, where flow is compressive, whereas *a*-axis fabric maxima will tend to be parallel to flow below the central parts, where flow is extending (Fig. 10.32; Boulton, 1971; Owen, 1991). In all cases, preferred orientations will show no systematic relationship to regional ice flow directions (Lawson, 1979a, b).

10.5.5.2 SUBAQUEOUS DEBRIS FLOW DEPOSITS

For subaqueous debris flows, an important distinction is made between *cohesive* and *cohesionless* flows, because the presence or absence of matrix cohesion exerts an important control on flow behaviour and the character of the resulting deposit (Section 8.4.4; Nemec and Steel, 1984; Postma, 1986; Nemec, 1990). A similar distinction can also be made for subaerial flows, but flows that behave as cohesionless materials are unimportant in terrestrial glacial environments.

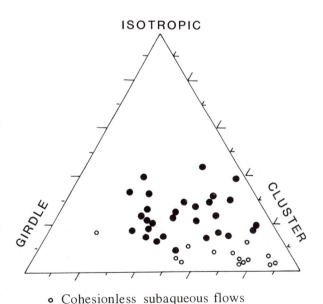

Fig. 10.31 Fabric shape of subaerial and cohesionless subaqueous debris flow deposits. (Subaerial fabric data from Lawson, 1979a, and Mills, 1991. Subaqueous fabric data kindly provided by F. Massari)

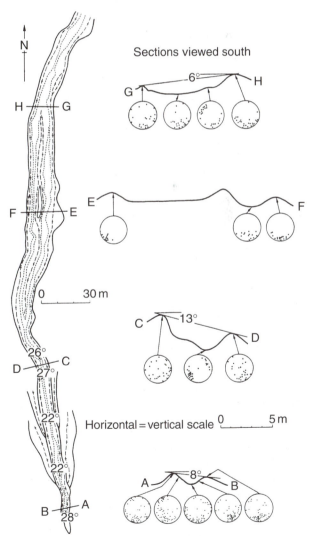

Fig. 10.32 Fabric patterns in a subaerial debris flow, near the Batura Glaciers, Pakistan. (From Owen, 1991. Reproduced by permission of the *Zeitschrift für Geomorphologie*)

Cohesive debris flow deposits

Deposits of cohesive debris flows form sheet-like or lobate beds, with planar or slightly scoured bases. Individual units are typically decimetres to several metres thick, and a few metres to hundreds of metres in lateral extent. Very extensive debris flow deposits, thousands of metres long, have been identified on glacially influenced submarine slopes (Wright and Anderson, 1982; C.H. Eyles *et al.*, 1992; King, 1993). Such large-scale debris flows are often produced by the downslope disaggregation of subaqueous slumps and slides, and can be recognized by the presence of blocks of the substratum in the deposits (Kurtz and Anderson, 1979; N. Eyles, 1987).

Former plug zones in cohesive debris flow deposits are typically composed of massive, matrix-supported diamicton or muddy sands and gravels. Very large clasts and intact or remoulded soft-sediment rafts may be present (C.H. Eyles, 1987; N. Eyles, 1987; Ghibaudo, 1992). The upper parts of flows may pass vertically into normally graded and stratified gravels, sands and silts, deposited from turbulent flows derived from the debris flow (Fig. 10.29; Walker, 1975; Krause and Oldershaw, 1979). The relative thickness of the debris flow and turbulent flow components of such two-layer deposits is variable, and depends on the extent to which flow modification occurred. The deposits of turbulent flows (turbidites) are described in detail in Section 10.5.6.

Basal sheared zones of cohesive debris flow deposits can be identified by the presence of attenuated wisps of clay or silt, shear planes and normal or inverse grading (Fig. 10.33). Normal grading reflects the sinking of large particles through weak, shearing sediment, whereas inverse grading is generally attributed to some combination of upward clast migration due to shearing and a sieving effect (these mechanisms are explained under *cohesionless debris flows* below). Another possible mechanism for the development of inverse grading was proposed by Broster and Hicock (1985), who argued that normally graded beds on the upper surfaces of cohesive flows (formed by turbulent mixing) could be rolled over the nose of the flow, conveyor belt fashion, to be redeposited upside down beneath the flow as an apparently inversely graded layer.

Cohesionless debris flows

Cohesionless debris flow deposits typically form dipping sheet-like or lobate masses centimetres to metres thick. Depositional dips are variable, but are usually in the range 10°–37°, although they can be as low as 3° for liquefied or fluidized flows (Section 8.4.4.3; Middleton and Hampton, 1973). Bases are commonly erosive, and can form deep, channelized scours (chutes) aligned downslope (Walker, 1975; Rust, 1977; Postma, 1984b; N. Eyles *et al.*, 1987; Prior and Bornhold, 1989,

(a)

(b)

(c)

Fig. 10.33 Subaqueous mass flow deposits: (a) normally graded flow unit capped by laminated sand and silt, Loch Quoich, northern Scotland; (b) thin lenses of diamicton with silt partings, Loch Quoich; (c) subaqueous graded mass flow deposits in the foresets of a glacilacustrine delta, Drymen, Scotland (Photos: D.I. Benn and D.J.A. Evans)

1990; Carlson *et al.*, 1992; Lønne, 1993). Incorporation of underlying material into eroding flows can result in gradational 'welded' basal contacts.

Deposits typically consist of gravel, sand or pebbly sand, and may be well or poorly sorted, depending on the source material (Figs 10.29 and 10.33). Flows fed by subaerial streams tend to be well sorted, reflecting the grain-size characteristics of fluvial bedload, whereas flows fed directly by subglacial conduits or derived by reworking of heterogeneous sediments can contain a wide range of grain sizes (Thomas, 1984a; Mastalerz, 1990; Lønne, 1993). Torrential subaerial streams can also introduce poorly sorted material to subaqueous slopes (Martini, 1990). Ice-proximal flow deposits can contain an abundance of boulder-sized material. Units may be massive or exhibit inverse or inverse-to-normal grading. *Inverse grading* is particularly common at unit bases, and is thought to result from two main processes. First, *dispersive pressure* generated by colliding grains tends to be greatest around large grains, which preferentially migrate upwards towards zones of lower strain rate (Bagnold, 1956). Second, a *kinetic sieving* mechanism occurs, in which small particles fall down through the spaces between large ones, displacing the large grains upward (Middleton, 1970; Middleton and Hampton, 1973). Shake a box of popcorn, a bowl of sugar or a large packet of crisps to see this effect in action. *Normal grading* is typically only weakly developed, because strong normal grading implies that grains were able to move freely enough relative to one another to allow differential gravitational settling, conditions usually associated with cohesive or turbulent flows. Particle fabrics in gravelly cohesionless debris flows commonly show preferred *a*-axis orientations parallel to flow, and upflow *a*-axis imbrication, resulting from particle rotation in a shearing medium (Fig. 10.31; Rust, 1977; Hein, 1982; Massari, 1984). Imbrication angles are variable, and can increase or decrease upward depending on patterns of strain in the flow during final deposition (Walker, 1977; Massari, 1984).

Some sandy flow deposits exhibit distinctive *dewatering structures*, indicating formerly liquefied or fluidized flow (Middleton and Hampton, 1973; Lowe, 1976b). Dewatering structures record the disruption and upward displacement of sediment by escaping pore fluid, and include subvertical *pipes, swirled lamination* and *dish structures*. The latter form groups of short, concave-up laminae, 4–50 cm wide and 1–2 cm deep, which are sometimes tightly curved, and form by the breakage and deformation of horizontal structures by upwelling fluid.

10.5.6 Turbidites

Turbidites are the deposits of fully turbulent sediment-gravity flows, or *turbidity currents*, beneath standing water (Section 8.4.4.4; Walker, 1992b). They are probably responsible for the majority of laminated and graded beds in glacier-contact and glacier-fed lakes (Shaw, 1977b; Shaw and Archer, 1978, 1979; Shaw *et al.*, 1978), and are also important in periglacial and glacimarine settings (C.H. Eyles *et al.*, 1985; Prior and Bornhold, 1989; Brodzikowski and van Loon, 1991). Turbidites are particularly characteristic of the mid- and lower zones of deltas and subaqueous fans (Section 8.5).

Deposition from turbidity currents happens as the flow slows down or becomes less turbulent, by a combination of settling from suspension and reworking as bedload beneath higher, still-moving parts of the flow (Walker, 1992b). These processes give rise to a general proximal-to-distal fining of turbidite beds, reflecting the rapid deposition of coarse material and the transport of finer material into deeper parts of the basin. They also produce a characteristic graded vertical sequence within the bed, known as the *Bouma sequence* (Bouma, 1962; Walker, 1992b). The full Bouma sequence is, from the top down (Fig. 10.34):

- massive clay or silt (E);
- interlaminated silt and/or clay (D);
- ripple cross-laminated sand and silt (C);
- planar laminated sand (B); and
- massive or normally graded sand or gravel (A).

Division A is the product of rapid deposition from suspension of the coarsest material with little or no subsequent reworking. Parallel lamination (Division B) forms as sand is transported and deposited as bedload in the upper flow regime (Section 10.4.2), driven by the flow above. The sand is derived either from the overlying current or from reworking of Division A. Continued deceleration of the flow results in a transition from the upper to the lower flow regime, and a consequent change in bedforms developed at the base of the flow. Ripple cross-lamination reflects the downflow migration of ripples during the reworking of Division B and/or the influx of new sediment from suspension. Climbing ripples (Section 10.4.3) record net deposition of sediment, and *convolute lamination* can develop by the deformation of laminae during loss of excess porewater. Division D forms by the settling of suspended fines during the last stages of the flow, with lamination reflecting pulses in flow velocity. Division E represents the gradual settling of residual suspended load in quiet water conditions.

Many variations on this basic sequence can exist. Parts of the sequence can be missing or repeated, depending on flow conditions. High-concentration flows produce thick A and B units with subordinate or missing C to E units, whereas low-concentration flows can produce base cut-out sequences consisting

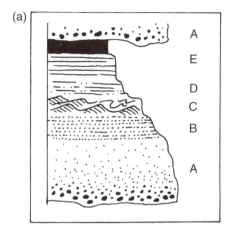

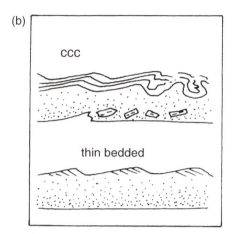

(c)

(d)

Fig. 10.34 Subaqueous underflow deposits (turbidites). (a) Classic 'Bouma' turbidite sequence. Letter codes are explained in the text. (b) Examples of 'base cut-out' distal turbidites. Unit 'ccc' consists of climbing ripples, convolute lamination and rip-up clasts. The 'thin bedded' unit consists of graded sands and starved ripples (redrawn from Walker, 1992b). (c) Normally graded, ripple cross-laminated and convolute laminated sands from glacilacustrine bottom sets, Achnasheen, Scotland. (d) Laminated and convolute laminated sands and silts, with gravelly lens in glacimarine bottom sets, Sognefjord, Norway. (Photos: D.I. Benn)

of Divisions C to E only. Repeated divisions can form by pulsed flows or flows reversing their direction of travel after rebounding from topographic highs (Haughton, 1994). 'Floating' clasts, much larger than the surrounding material, can occur in Divisions A and B. These are transported in the flow by gliding along boundaries between regions with contrasting density, in much the same way as a water-skier skims along the boundary between water and air (Postma *et al.*, 1988b). Outsized clasts are deposited at high levels if the surrounding flow rapidly immobilizes or 'freezes'.

10.6 DEPOSITS FROM SUSPENSION SETTLING AND ICEBERG ACTIVITY

In glacilacustrine and glacimarine environments, sediment can be carried at high levels in the water column in overflow and interflow plumes of fine suspended sediment, as ice-rafted debris, and by vegetation such as seaweed (Sections 8.4.3 and 8.4.5; Powell, 1990; Gilbert, 1990). Suspended sediments and debris released from icebergs, ice shelves and floating vegetation settle through the water column to be deposited. Gravitational settling also delivers biogenic material, such as plant and animal microfossils, to lake and sea floors, forming fossiliferous muds and biogenic oozes. Also included in this section are deposits and sedimentary structures formed by grounding of icebergs.

10.6.1 Varves and other glacilacustrine overflow/interflow deposits

Suspended sediment carried into lakes as high-level overflows and interflows spreads laterally and gradually settles out through the water column with velocities proportional to particle size (Section 8.4.3; Drewry, 1986). Fine particles may also be deposited following ingestion by zooplankton and egestion as faecal pellets (N.D. Smith and Syvitski, 1982), or the death of organisms with mineral skeletons or phytoliths (Lowe and Walker, 1984; Bradley, 1985). The resulting sediments form blanket-like drapes which typically display proximal to distal fining and thinning (N.D. Smith *et al.*, 1982; Drewry, 1986). Deposits will also tend to be thicker towards the right or left shore in the northern and southern hemispheres, respectively, owing to Coriolis deflection of the circulating plumes (Section 8.3.6; N.D. Smith and Ashley, 1985).

The deposits will tend to exhibit different structures below shallow and deep water, respectively. Below shallow water (generally above the thermocline), overflow and interflow deposits will tend to form massive muds and silts, owing to sediment mixing by bioturbation, wave disturbance and wind-generated currents (Sturm and Matter, 1978; N.D. Smith and Ashley, 1985). Below deeper water, the deposits typically form laminated fining-up sequences of silt and clay (Figs 10.35 and 10.36). The laminae reflect fluctuations in the grain size and quantity of incoming sediment, as a result of daily, meteorological or annual water and sediment discharge cycles (Church and Gilbert, 1975). Short-term cycles produce thin, normally graded laminae with sharp basal contacts and gradual fining-up within each unit. In contrast, annual cycles can produce distinct silt–clay couplets

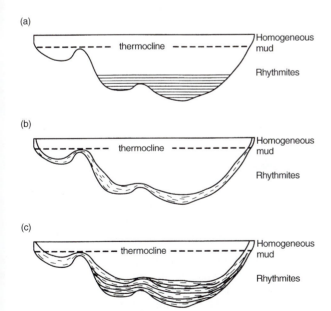

Fig. 10.35 Spatial variations in lake sediment characteristics under different dispersal mechanisms: (a) overflow–interflow; (b) underflow; (c) overflow–interflow and underflow. (Modified from Smith and Ashley, 1985)

known as *varves*, with sharp contacts between the coarse and fine components, as a result of marked differences in sediment supply between summer and winter (DeGeer, 1912; Ashley, 1975; N.D. Smith, 1978). The coarse component is deposited from overflow and interflow plumes generated during the ablation season, and the fine component records gradual settling of the finest material during winter when there is little or no incoming water to the lake (Fig. 10.37). Winter sedimentation commonly takes place below a cover of surface ice. Thin layers of coarser material can occur within winter clay layers in varves, deposited during winter storms or as a result of sediment reworking on the lake floor (Shaw *et al.*, 1978). Varves show a proximal to distal decrease in the thickness of the silt component relative to the clay, and also an overall reduction in mean grain size (Ashley, 1975). They also display marked year-to-year variations in thickness, which is valuable in establishing correlations between different parts of a basin (N.D. Smith, 1981).

Laminated glacilacustrine sediments can be deposited by the combined action of overflows, interflows and underflows (turbidity currents). It is relatively straightforward to distinguish turbidites and overflow/interflow deposits because the latter are regularly and thinly bedded, and do not contain current structures such as ripple cross-lamination (cf. Section 10.5.6). The relative thicknesses of the underflow and overflow/interflow components of composite sequences vary with position with respect to sediment sources. In locations close to sediment influx points, turbidites tend to predominate, with thin clay layers representing short-term pauses in sedimentation (N.D. Smith and Ashley, 1985). In contrast, sedimentation in distal locations tends to be dominated by settling from overflows and interflows, with only minor input from turbidity currents (Shaw *et al.*, 1978). The underflow and overflow/interflow components in composite laminated deposits represent very different rates of sedimentation. Turbidity currents in glacial lakes may last only a few minutes but deposit several centimetres of sediment; whereas quiet-water settling of clays may deposit only a few millimetres or less over many months. Thus laminated sediments may represent depositional cycles on many different timescales. The term 'varve' should be used only where a seasonal depositional cycle can be clearly demonstrated (N. Eyles and C.H. Eyles, 1992).

10.6.2 Cyclopels and cyclopsams

Cyclopels and *cyclopsams* (the latter with a silent 'p') are rhythmically bedded sediments deposited from overflow and interflow plumes in glacimarine environments (Section 8.3.3). The terms were intro-

CLASSICAL VARVES

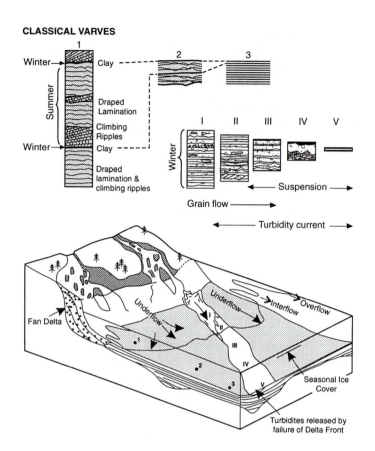

Fig. 10.36 Schematic model to show the internal characteristics of laminated sediments (varves v. other rhythmites) and their positions of deposition. (Modified from N. Eyles and Miall, 1984)

Rhythmite type	Surge deposit	Varve	
Sediment dispersal mechanisms	Slump-generated Surge currents	Suspension settling	Winter
		Overflow-interflows Underflows Surge currents	Summer
Time for depositing each rhythmite	Minutes	1 year	

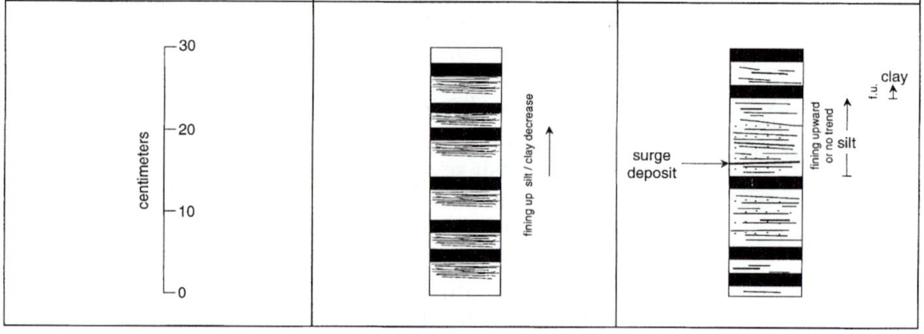

Fig. 10.37 A comparison of surge deposits and varves. (Modified from N.D. Smith and Ashley, 1985)

duced by Mackiewicz *et al.* (1984) using the Greek root *cycl* to refer to their cyclic nature and the Greek words *pel* and *psam* to denote muddy and sandy units, respectively. Overflow and interflow plumes of suspended sediment tend to be more important in glacimarine environments than glacial lakes, owing to the buoyancy of fresh meltwater in dense, saline sea-water. Turbid meltwater plumes emerging from

efflux points below grounded ice can rise to the surface or to intermediate depths, and can be transported for many tens of kilometres from the ice front (Syvitski *et al.*, 1987). The buoyancy of freshwater plumes in sea-water also means that coarser grains, up to sand size, can be transported in suspension.

Deposition from turbid plumes typically produces couplets of silt and mud (cyclopels) and sand and mud (cyclopsams). Each couplet has a sharp lower contact, recording the sudden onset of deposition of the coarsest material, and is normally graded, recording the gradual settling of progressively finer material (Fig. 10.38; Mackiewicz *et al.*, 1984; Cowan and Powell, 1990). Sedimentation rates can be very high: as much as 15.4 cm of sand in 19 hours near the margins of temperate tidewater glaciers (Powell and Molnia, 1989). Cyclopels and cyclopsams were first

recognized and defined in cores (Mackiewicz *et al.*, 1984), and have been identified in a variety of glacimarine settings (Powell and Molnia, 1989; Cowan and Powell, 1990).

The grading in cyclopels and cyclopsams records variations in sediment supply and settling rates, controlled by fluctuating meltwater stream discharge, tides and wind shear (Cowan *et al.*, 1988; T.G. Stewart, 1991). Tidal influence was clearly demonstrated

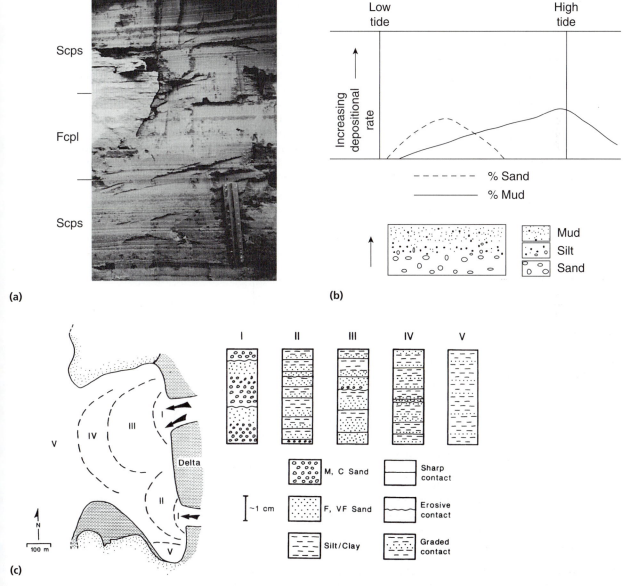

Fig. 10.38 Cyclopels and cyclopsams. (a) Cyclopels and cyclopsams in a vertical sequence in glacimarine sediments in Clements Markham Inlet, Ellesmere Island. (Photo: T.G. Stewart) (b) The relationship between cyclopel/cyclopsam deposition and tides. (From Cowan and Powell, 1990. Reproduced by permission of the Geological Society of London) (c) The distribution of sequences of sedimentary structures in Riggs embayment, Glacier Bay, Alaska. The pattern of sedimentation is directly related to distance from stream mouths, Zone I being dominated by proximal downslope processes and Zones II–V being characterized by more distal suspension settling. (From Philips *et al.*, 1991. Reprinted by permission of the Geological Society of America)

by Cowan and Powell (1990) in a study of sedimentation in McBride Inlet, Alaska, near the tidewater snout of the McBride Glacier. Sediment traps positioned on the sea floor collected the largest amounts of sand within the two hours following each low tide, and the least amount around one hour after high tide, when silt and mud deposition was more important. This pattern reflects the dynamics of sediment movement during the tidal cycle. When the tide is rising, water velocities are high, maintaining coarse material at high levels. This material gradually sinks towards the bottom during high and ebb tide, finally settling out in the slack water around low tide. This cycle happens against a background of continuous settling of fine silt and clay particles. Thus, each day, two laminae will be produced (one for each tidal cycle), plus additional laminae related to fluctuations in sediment and meltwater influx, wind patterns, and sediment reworking on the sea floor (Cowan *et al.*, 1988).

Cyclopsams are generally deposited within 1 km of the point where the meltwater plume enters the sea. Cyclopels may also be deposited in proximal environments during periods of low discharge, but are more common between one and several kilometres of the efflux point (Mackiewicz *et al.*, 1984; Powell and Molnia, 1989). Both cyclopsams and cyclopels can be interbedded with turbidites, other mass flow deposits and ice-rafted debris, particularly in proximal environments (Domack, 1984; Powell and Molnia, 1989).

10.6.3 Ice-rafted debris and undermelt deposits

Material rafted into marine or lacustrine environments by floating ice, in the form of icebergs or ice shelves, ranges in size from clay to boulders (Ovenshine, 1970; Powell, 1981; Molnia, 1983; Gilbert, 1990). Collectively, such material is known as *ice-rafted debris* (IRD), and can be very prominent in glacimarine and glacilacustrine depositional sequences. Isolated clasts dropped on to a lake or sea bed from floating ice are termed *dropstones*, and may occur in both massive and laminated sediments. They also occur as outsized clasts in current-bedded bottom sediments, but can be confused with isolated particles that have rolled or bounced out beyond coarse mass flows in more proximal environments. Dropstones can also be transported by floating vegetation, particularly genera of seaweed such as bladderwrack that have gas-filled sacs (Section 8.4.5.1; Gilbert, 1990). The existence of dropstones in a sediment does not, therefore, necessarily imply the former presence of icebergs. Where the surrounding sediments are laminated, dropstones can be identified by the deformation or penetration of underlying laminae, as described by Thomas and Connell (1985; Fig. 10.39).

Sustained rain-out from below ice shelves or multiple icebergs can produce thick deposits of mud, pebbly mud or diamicton. Many terms have been proposed for deposits formed by rain-out below ice shelves, including *undermelt till* (Dreimanis and Lundqvist, 1984; Parkin and Hicock, 1989; King *et al.*, 1991), *subaquatic melt-out till* (Dreimanis, 1979; Boulton, 1972b), *subaqueous basal till* (Link and Gostin, 1981), *grounding-line till* (Powell, 1984) and *dropped para-till.* In the latter term, the Greek prefix 'para-' indicates that the material is closely related to till, but by strict definition is not a till. However, because deposition by rain-out below ice shelves is by settling under gravity, we prefer to avoid the term 'till', and recommend *undermelt diamicton*, as proposed by Gravenor *et al.* (1984). Where deposition was from icebergs, the terms *dropstone diamicton* and *dropstone mud* should be employed, depending on their grain size characteristics. A cut-off point of 10 per cent clasts per unit area is a convenient division between dropstone diamictons and muds (Powell, 1984). In practice, it may not be possible, using internal characteristics alone, to distinguish under-

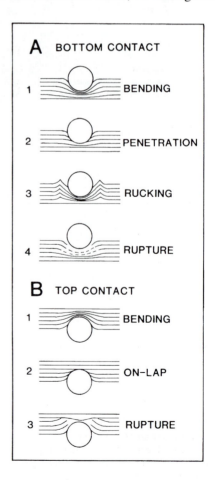

(a)

Fig. 10.39

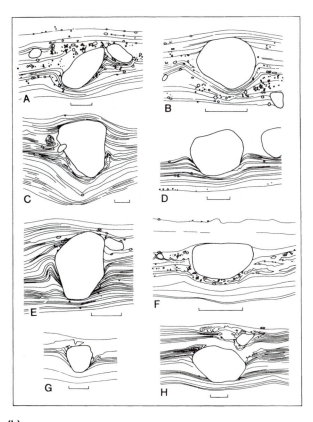

(b)

Fig. 10.39 Dropstone structures: (a) the nature of bottom and top bedding contacts; (b) structures observed in glacilacustrine sediments in Scotland. (From Thomas and Connell, 1985. Reproduced by permission of SEPM)

melt diamicton from dropstone diamicton deposited from multiple icebergs.

Dropstone diamictons and muds form extensive, blanket-like sheets draped over the pre-existing topography, and can be massive or weakly stratified (Fig. 10.40). Stratification can be particularly pronounced if ice rafting occurs alongside other depositional processes, such as the settling of suspended plumes, bottom tractive current activity or sediment gravity flow (Powell, 1984). Stratification will tend to be destroyed by bioturbation, resulting from the activities of bottom-dwelling invertebrates. Other signs of biological activity in dropstone muds can include trace and body fossils, including shells and microfossils. Sequences of dropstone diamictons and muds may be only a few centimetres thick below small glacier-contact lakes (Benn, 1989a), but can build up to be tens of metres thick on glacially-influenced coasts.

The grain size distributions of dropstone diamictons and dropstone muds are highly variable, depending on the character of the parent debris and the degree of fall sorting, water-column winnowing and bottom winnowing that occurs prior to deposi-

tion (e.g. Anderson *et al.*, 1980). Clast-rich dropstone diamictons are typically associated with proximal environments, where they form by debris melt-out below floating glacier ice or frequent icebergs (Hambrey, 1994). If fine-grained sediment is winnowed out in the water column, the deposited sediment can be a clast-supported diamicton or gravel (Powell, 1984). Dropstone muds are matrix-supported diamictons with strongly bimodal particle size distributions,

(a)

(b)

(c)

Fig. 10.40

(d)

Fig. 10.40 Dropstone muds, diamictons, and iceberg dump structures: (a) stratified diamictons and dropstone muds, Loch Quoich, Scotland; (b) dropstone in glacimarine silts and clays, illustrating bending in lower contact and on-lap of upper contact, Phillips Inlet, Ellesmere Island; (c) iceberg dump structure in glacilacustrine subaqueous sands, Drymen, Scotland (Photo: D.J.A. Evans); (d) iceberg dump structure in glacilacustrine sediments, Achnasheen, Scotland. (Photo: D.I. Benn)

reflecting dominant suspension sedimentation and minor quantities of dropstones.

Particle morphology is inherited from the parent glacial debris. Characteristics typical of actively transported debris (subrounded and subangular clasts, faceting and striations, and crushed quartz grains) are common, indicating the importance of basal debris (Domack *et al.*, 1980), but on a local scale passively transported debris may predominate, reflecting supraglacial debris sources or debris eroded beneath cold ice. No systematic differences in particle morphology are evident between ice-proximal and ice-distal environments (Hambrey, 1994).

Dropstone *a*-axis fabrics will be influenced by the character of the bottom sediment. Where bottom sediments are relatively stiff, but allow penetration by clasts, they will tend to preserve the high dip angles of clasts that fall vertically through the water column and hit the bottom nose first, producing a weak cluster fabric with a vertical preferred orientation. Conversely, where bottom sediments are either very soft or compacted, vertically impacting clasts will tend to fall sideways because either the sediment cannot hold the clast upright or the clast cannot penetrate the sediment, resulting in a girdle fabric (Domack and Lawson, 1985). Preferred orientations can also develop in response to bottom current activity and reorientation by mass movements. More isotropic fabrics can reflect a combination of these processes and the varying influence of particle shape.

The rate of deposition of ice-rafted debris is a function of the debris content of the ice (debris concentration and debris layer thickness) and the fre-

quency of iceberg passage, which reflects the calving rate and distance from the glacier margin (Dowdeswell and Dowdeswell, 1989; Dowdeswell and Murray, 1990).

Iceberg dump mounds are formed where icebergs roll over and rapidly dump large quantities of debris in a single location (Thomas and Summers, 1982; Thomas and Connell, 1985; Dowdeswell *et al.*, 1994). These are planar-based mounds of coarse materials which can occur in isolation, offlap each other, or be stacked vertically and interstratified with finer-grained sediments (Fig. 10.41). They are typically symmetrical cones with a maximum height of 2 m and a width to height ratio of 6:1. Asymmetrical cones are produced by multiple dumping events as a result of repeated overturning of, or basal melt-out from, the same iceberg. Mound bases can be horizontal or inclined, depending on the form of the substratum on which they lie. Clasts at the base of the mounds will commonly deform underlying fine-grained sedi-

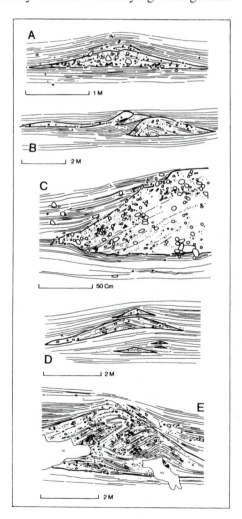

(a)

Fig. 10.41

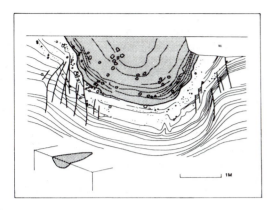

(b)

Fig. 10.41 Iceberg dump and grounding structures observed in glacilacustrine sediments in Scotland: (a) includes single symmetric structure (A), stacked asymmetric, laterally overlapping structure (B), detailed marginal contact (C), vertically stacked, symmetric structure (D) and compound, stacked asymmetric structure (E); (b) an iceberg grounding structure with three-dimensional view. (From Thomas and Connell, 1985. Reproduced by permission of SEPM)

ments. The mounds themselves consist of poorly sorted gravels with bedding dipping outwards from the cone centre, or comprise a central core of clast-rich diamicton grading upwards and outwards to poorly sorted gravel. These coarse sediments can be interstratified with fine sediments towards the top flanks. Iceberg dump events can introduce a wide range of sediment sizes with different settling velocities to the water column, and turbid underflows can transport suspended sediments laterally over a wide area, forming graded turbidites (Elverhoi and Roaldset, 1983; Gilbert, 1990). The open texture of some diamictons in iceberg dump mounds led Thomas and Connell (1985) to suggest that rafted blocks of debris-rich ice had dropped, then melted out on the bottom. A related process is the rafting of frozen pre-existing sediments by sea-ice (Gilbert, 1990). Thomas and Connell (1985) proposed the term *iceberg dump till* for the sediments in iceberg dump mounds, but we would include these sediments in the category of iceberg *dropstone diamictons* in line with Powell (1984), because they are not tills in the strict sense. The dump structures described above do not display any evidence for faulting, scour or other deformation associated with iceberg contact, and so are not deposited directly by ice.

10.6.4 Iceberg grounding structures and sediments

Diamictons and clusters of clasts can be deposited directly from icebergs without intervening gravitational transport if bergs lodge on the sea- or lake bed for long periods and debris melts out *in situ*

(Lavrushin, 1968; Fecht and Tallman, 1977). Diamicton produced by this process has been termed *iceberg till* (Dreimanis, 1979; Powell, 1984) on the grounds that the sediment is deposited directly from glacier ice and is therefore a till. However, although glacier ice is involved, it is no longer part of a glacier, and this usage is not adopted in this book. Instead, we use the term *iceberg contact deposits*.

Iceberg plough and scour marks have been reported widely from both ancient and modern lake and ocean settings (Section 8.4.5.2; Belderson *et al.*, 1973; Belderson and Wilson, 1973; Harrison and Jolleymore, 1974; Syvitski *et al.*, 1983; Woodworth-Lynas and Guigné, 1990; Dowdeswell *et al.*, 1993). Icebergs can also disturb bottom sediments at tidewater margins, if bergs fall through the water column and strike the bottom before floating away (Powell, 1981a). The frequency of iceberg scouring is so intense in some locations that all primary depositional stuctures can be destroyed, producing massive, structureless diamicton from pre-existing sediments (Vorren *et al.*, 1983; Dowdeswell *et al.*, 1994). Fine bottom sediment can also be resuspended during this process, and the remaining sediment can be depleted of fines depending on currents (Marienfeld, 1992). Massive diamictons produced by ice keel scouring are termed *ice-keel turbate* by Woodworth-Lynas and Guigné, 1990), and can be very difficult to differentiate from non-stratified dropstone diamictons, as well as some subglacial tills, unless some identifiable deformation structures are preserved.

Distinctive faulting structures can be created by single iceberg grounding or ploughing events (Thomas and Connell, 1985; Woodworth-Lynas and Guigné, 1990). An idealized sequence of events was presented by Woodworth-Lynas and Guigné (1990), as shown in Fig. 10.42. As the iceberg ploughs forward, slices of sediment are thrust forward and to the sides along low-angle thrust faults. The fault surfaces are highly polished and slickensided, indicating the direction of thrusting. The outward movement of the thrust slices is accompanied by downward movement of a central wedge of sediment, which is typically folded and seamed with microfaults. Downward normal faulting also occurs at the outside margins of the ridges ('berms') of thrust sediment.

10.6.5 Fossiliferous deposits and biogenic oozes

Biological activity can be very important in glacimarine and some glacilacustrine environments, and body or trace fossils can constitute an important component of some facies. Where input of clastic particles is high, such as in proximal environments, fossils may make up only a small volume of the sediment, but in other areas, fossils can make up the bulk

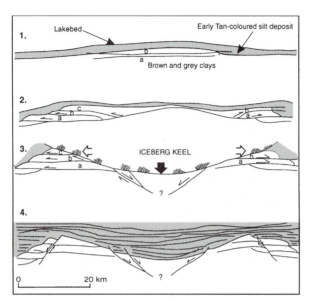

Fig. 10.42 A model to show the progressive development of fault structures in lake or marine sediments during scouring by a drifting iceberg. (1) Sediment is compacted in front of the scouring iceberg keel and as a result the sediment surface bulges and develops horizontal faults; (2) as scouring continues, the dominant motion of the displaced sediment is radially away from the scour axis, and berms are produced by the horizontal displacement of clay slabs; (3) as the iceberg keel passes over the site, the final emplacement of stacked clay slabs takes place at the scour margins. Vertical loading by the keel causes shallow foundation failure, and shallow faults develop beneath the scour trough; (4) as the keel moves past the site, the stress release causes high-angle normal faulting of the clay slabs on the outer berm margins. Sediment then fills the iceberg scour trough, and faults may be reactivated by the resultant overburden pressure. (Modified from Woodworth-Lynas and Guigné, 1990)

of deposits. Sediments made up mainly or entirely from microfossils are termed 'biogenic oozes'. The range of species present in fossiliferous sediments is an important source of palaeoenvironmental data, such as water depth, temperatures, turbidity and salinity (Lowe and Walker, 1984; Bradley, 1985; Andrews *et al.*, 1996).

10.7 WINNOWING STRUCTURES (LAGS, COQUINAS AND BOULDER PAVEMENTS)

Currents flowing in both marine and lacustrine environments can rework bottom materials by transporting away fine-grained sediments leaving coarse *lags*, especially during periods of non-deposition or reduced sediment influx to the deep water (Emery, 1968; Marlowe, 1968; Stanley, 1968; Edwards,

1975). These have been termed *palimpsest lags* by Powell (1984) because each lag surface in a vertical sedimentary succession represents an incomplete record of former depositional processes. Powell (1984) went on to suggest that glacier readvance over a palimpsest lag would produce a striated clast pavement similar to those produced in subglacial situations (Section 10.3.4).

Coquinas are a type of lag deposit consisting of shell-rich bands within water-lain muds and sands. Examples have been described from the early Pleistocene Yakataga Formation, Alaska, which contain whole and broken bivalves, gastropods, brachiopods, worm tubes and barnacles in a matrix of sand or mud, and form extensive planar or concave-up features (Bergen and O'Neil, 1979; Lagoe *et al.*, 1989; N. Eyles and Lagoe, 1990). The fauna represented in the coquinas prefer shallow-water environments with little or no sediment accumulation, suggesting that the coquinas formed when increased current energy prevented mud deposition. Coquinas within channel fills probably represent shell-rich lags produced by channelized currents, which scoured and winnowed shell-bearing sediments.

Striated *clast* or *boulder pavements* which may have originated partly by current winnowing have been described by C.H. Eyles (1988b) and N. Eyles and Lagoe (1990) from the Yakataga Formation, and by McCabe and Haynes (1996) from late Pleistocene sediment sequences in Ireland. In the Alaskan example, the clasts making up the pavements do not have the attributes normally associated with subglacial lodgement or deformation, such as stoss and lee forms and consistent *a*-axis fabrics (Sections 10.3.2 and 10.3.4), although their upper surfaces do bear consistently oriented striae, unlike striae formed by sea- or shore-ice (Hansom, 1983a; Dionne, 1985). The clast pavements were interpreted by C.H. Eyles (1988) as the products of winnowing of diamictons by storm waves and currents, followed by glacial abrasion of the coarse lag by later ice advance. Because there is a lack of glacitectonic disturbance of surrounding sediments, N. Eyles and Lagoe (1990) suggested that the ice margin was partially floating and touched down only locally on high points on the sea bed. The Irish boulder pavements described by McCabe and Haynes (1996) occur within water-lain muds and sands. The boulders are closely packed in a mosaic-like pattern which is generally one boulder thick. Most boulders have striated surfaces, and about 50 per cent bear two or more sets of striae, with no clear dominant striation direction. The boulders are thought to have been delivered by ice rafting, and organized into a pavement by winnowing in the near-shore zone. The inconsistent direction of the striae suggests that they were incised by freely floating ice floes.

CHAPTER
◇ 11 ◇

SEDIMENT–LANDFORM ASSOCIATIONS

MORAINES AND ROCHES MOUTONNÉES AT THE MOUTH OF CWM-GLAS.

11.1 INTRODUCTION

Glacial depositional landforms and facies associations exhibit a sometimes bewildering variety, reflecting a very wide range of processes, topographic settings and glacial environments. Many landforms and associations can be observed forming in modern glacial environments, and in consequence these types have proved relatively easy to classify and interpret. Other landforms and associations, however, are known only from long-deglaciated terrains, such as drained lake or shelf sea floors or the former beds of the great mid-latitude ice sheets, and may be of a scale, morphology and structure that have never been observed in active depositional settings. As a result, genetic models for such associations must be based on interpretation of the geomorphological and geological record. Interpretation of ancient sediments and landforms should be based on detailed, systematic observations, and guided, wherever possible, by appropriate modern analogues and sound theory. Unfortunately, there has always been a tendency in glacial geology for some researchers to indulge in unconstrained speculation, and many weird and wonderful ideas have appeared in the literature to account for certain landforms. In this chapter, we concentrate on those interpretations which are well supported by observations or theory, and generally mention less plausible or outmoded ideas only in passing. This approach has allowed us to illustrate the broad range of glacigenic sediment–landform associations and their origins without becoming too distracted by the many curiosities and antiques that clutter the literature.

The classification of glacial sediment–landform associations inevitably presents difficulties, and many approaches have been tried in the past (e.g. Embleton and King, 1975; Sugden and John, 1976; Brodzikowski and Van Loon, 1991). We have chosen to classify them according to depositional environment, reflecting the structure of Part One of this book. Sediment–landform associations are described below under five general headings: (a) subglacial associations; (b) ice-marginal moraines; (c) supraglacial associations; (d) proglacial associations; and (e) glacilacustrine and glacimarine associations. There are some problems with this approach, because some glacial depositional landforms are shaped in more than one environment. Cupola hills (Section 11.3.1.3), for example, are ice-marginal glacitectonic moraines that have been modified by subglacial processes; and ice-walled lake plains (Section 11.4.4) are supraglacial glacilacustrine associations. We have placed such awkward cases alongside the associations with which they have the closest genetic affinity.

11.2 SUBGLACIAL ASSOCIATIONS

Subglacial associations are among the most enigmatic products of glaciation, and a reasonable understanding has begun to emerge only in recent years as the result of a considerable research effort. This section begins with a discussion of subglacial facies associations that lack surface expression in landforms, either because of erosion or because of burial under younger sediments. Such associations may constitute the only record of glaciation in early Quaternary or older strata, and their correct interpretation is very important for reconstructions of former glacier behaviour. We go on to describe subglacial bedforms, such as drumlins, flutings and Rogen moraine, which are distinctive sediment–landform associations produced beneath active glaciers. Lee-side cavity fills and crevasse-fill ridges are then described, followed by a discussion of eskers. The section concludes with consideration of subglacial volcanic forms.

11.2.1 Subglacial facies associations

Sedimentation and tectonic processes in subglacial environments can result in complex, highly variable associations of diamicton and sorted sediments. Many facies models have been proposed in the literature, which explain such associations in terms of shifting subglacial environments and processes. Such models, however, are of varying usefulness, and may be strongly influenced by ruling hypotheses and untested assumptions about till genesis and depositional environments. Furthermore, the quantity and quality of the sedimentological data employed in the construction of such models are very variable. As a result, any one subglacial facies association can be interpreted in very different ways by different researchers. Several good examples of subglacial facies models can be found in recent volumes edited by van der Meer (1987), Dardis and McCabe (1994) and Warren and Croot (1994).

An informative example of changing interpretations of a glacigenic facies association is recent work on the Catfish Creek Formation, exposed along the northern shore of Lake Erie, Canada. The Catfish Creek Formation is an extensive association of interstratified diamictons, sands, gravels and laminated muds laid down during the late Wisconsinan glaciation (Fig. 11.1; Dreimanis *et al.*, 1986). The diamicton units are variably laminated and massive, and some units exhibit isoclinal fold structures or augen-shaped sand and silt inclusions. Sand interbeds are common, interfingering with diamicton units. The Catfish Creek Formation was interpreted by Evenson *et al.* (1977) as an ice-contact subaqueous association consisting of subaqueous debris flows (Section 10.5.5.2)

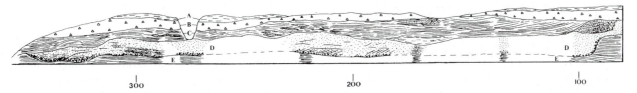

Fig. 11.1 The stratigraphic context of the Catfish Creek till, Ontario, Canada. A = Lake Maunee silt, B = Port Stanley till, C = stratified Catfish Creek till, D = sand and gravel, E = laminated clays and silts. (Modified from Gibbard, 1980)

and water-sorted sediments. Folding and augen structures within diamicton units were regarded as the result of remobilization of pre-existing sediments by sub-aqueous flowage. This model was challenged by Gibbard (1980), who reinterpreted the association as the result of rain-out beneath a floating ice shelf, combined with limited postdepositional flowage and intermittent meltwater discharge. Subsequently, Dreimanis *et al.* (1986) and Hicock (1992, 1993) conducted detailed analyses of some exposures, including systematic studies of clast morphology and fabric, and concluded that the Catfish Creek Formation records complex switching between subsole deformation, lodgement, flowage, and deposition by meltwater in subglacial cavities. Under this interpretation, isoclinal folds and soft sediment inclusions in laminated diamicton units resulted from subsole deformation of glacilacustrine sediments. As discussed in Section 10.3.3, such features are characteristic of some glacitectonites. The changing interpretations of the Catfish Creek Formation reflect changing fashions in glacial geology, and emphasize the need for rigorous analyses using a wide range of descriptive criteria when attempting to establish the origin of complex sediment successions (Hicock, 1992; Hart and Roberts, 1994; Benn and Evans, 1996).

Several scenarios have been proposed to account for vertical and lateral facies variability in subglacial sediment associations. These include: (a) rheologic superposition of different till facies; (b) intermittent subglacial drainage during till deposition; (c) lobal interactions between adjacent ice flow units; and (d) alternations between subglacial and ice-marginal, glacilacustrine or glacimarine environments at oscillating ice margins. Sedimentary successions resulting from cases (c) and (d) do not comprise single facies associations, but represent two or more superimposed associations formed in different environments or events. As such, they are best discussed in Section 12.4.2. Successions formed by rheologic superposition and alternate till deposition and subglacial drainage are described in turn below.

11.2.1.1 RHEOLOGIC SUPERPOSITION

Rheologic superposition can be defined as the successive formation of subglacial tills with different deformational histories (Hicock, 1992; Hicock and Dreimanis, 1992b; Hicock and Fuller, 1995), and has been inferred from till successions that exhibit systematic vertical changes in fabric characteristics and structures. For example, switches between ductile and brittle deformation at a glacier bed may result in partial overprinting of existing till characteristics with new fabrics and structures (Sections 5.6 and 10.3.4). This can take the form of the collapse and brittle shear of a ductile A-horizon, producing boudinage and fracturing of soft-sediment deformation structures (Hicock and Dreimanis, 1992a; Hicock, 1992), or the dilation and ductile deformation of hitherto stiff tills, resulting in the rotation of striated, faceted particles away from their original orientations (e.g. Hicock, 1991, 1992; Benn, 1994a, 1995; Benn and Evans, 1996; Fig. 11.2). Preservation of a vertical succession of till types, recording a full history of deformation at the site, clearly requires net accretion of till, or *constructional deformation* (Hart and Boulton, 1991; Hart, 1995a).

Another example of rheologic superposition is till formed by basal melt-out of debris-rich ice followed by remobilization in a deforming layer. In this case, structures and textures formed during the shear of debris-rich ice are overprinted during subglacial shear. Shaw (1987) described facies associations in the Edmonton area, Canada, which he interpreted as interbedded melt-out tills, deformation tills and subglacial channel fills. However, great care must be taken when interpreting such complex successions, because many till characteristics previously assumed to represent inherited englacial structures are now known to be characteristic of glacitectonites (Sections 10.3.3 and 10.3.5).

11.2.1.2 INTERMITTENT SUBGLACIAL DRAINAGE

Several researchers have described till successions that contain interbeds and lenses of sorted sediments (e.g. N. Eyles *et al.*, 1982b; Åmark, 1986; Brown *et al.*, 1987; Shaw, 1987; D.J.A. Evans *et al.*, 1995; Benn and Evans, 1996). Sorted interbeds commonly take the form of broad lenses of silt, sand or gravel with concave-up lower contacts and nearly planar upper contacts. Lenses may exhibit well-preserved internal structures such as plane bedding and ripple cross-stratification (Sections 10.4.2 and 10.4.3), but varying degrees of tectonic folding and attenuation are common at the contacts with overlying tills (Fig. 11.3; N. Eyles *et al.*, 1982b; Åmark, 1986; D.J.A. Evans *et al.*, 1995). Clark and Walder (1994) have argued that such successions represent deformation tills and the infills of former braided canal systems developed at the ice–till interface (Section 3.4.6; Walder and Fowler, 1994). According to this model, the till units are formed by subsole deformation when the glacier is coupled to the bed, and the canal systems are formed when discharges are too high to be evacuated through the bed, and bed separation occurs over large areas. Clearly, the survival of canal fills in a deformational environment requires that they experience only low cumulative strains, either because (a) deformation persists for a short period following canal formation; (b) strain persists but strain rates are very low; or (c) net accretion of deformation till occurs and the base of the deforming layer migrates upward, so deeper parts of the deforming layer become immobile (i.e. there is constructional deformation; Hart and Boulton, 1991).

Figure 11.3 shows interbedded tills, sands and silts at Skipsea, Yorkshire, northern England, described by D.J.A. Evans *et al.* (1995) and Benn and Evans

(1996). The tills are variably massive to laminated, and contain clear deformational structures such as isolated folds and smeared-out lenses of chalk and unconsolidated sediments. The sorted lenses exhibit both primary depositional structures and tectonic deformation, and commonly grade upward into complexly folded and attenuated interbeds of silt, sand and diamicton. From the base up, the Skipsea sequence was interpreted as: (a) a lower deformation till; (b) subglacial canal fills representing meltwater flow at the ice–till interface; and (c) glacitectonized sorted sediments and deformation till resulting from incorporation of part of the channel fill sequence into a new deforming layer, together with till advected into the site. This interpretation implies constructional deformation at the site, and upward migration of the base of the deforming layer during sediment accretion.

Isolated channel fills, possibly representing Nye channels on a former glacier bed, have also been reported from subglacial till successions. An example from late Pleistocene sediments exposed on the coast of Tierra del Fuego, southern Chile, is shown in Fig. 11.4. The channel is incised downward into compact till, interpreted as a hybrid lodgement–brittle deformation till (Section 10.3.6), and is bounded above by a horizontal erosion sur-

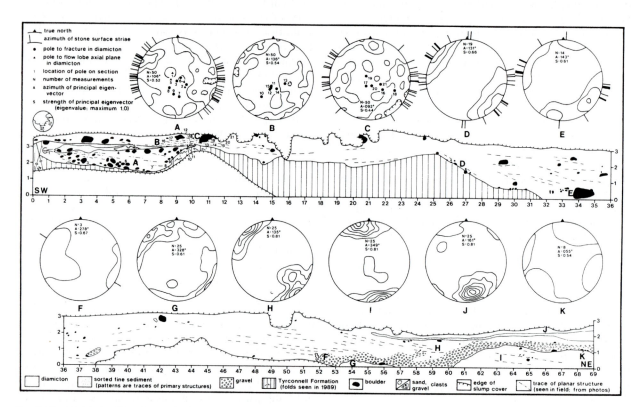

Fig. 11.2 Multiple till and stratified sediment sequence near Bradtville, Ontario, Canada, interpreted as the product of lobal interactions and rheological superposition. (From Hicock, 1992. Reproduced by permission of Scandinavian University Press)

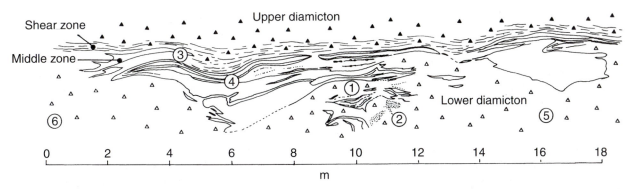

Fig. 11.3 Stratigraphic section including two diamictons or tills (triangle pattern) separated by sand and gravel interbeds (no pattern) at Skipsea, east Yorkshire, England. The sands and gravels are former subglacial canal fills. Note the glacitectonic deformation of the sands and gravels and their partial ingestion into a shear zone in the upper diamicton. (From Benn and Evans, 1996. Reproduced by permission of Elsevier)

Fig. 11.4 Subglacial channel fill, Tierra del Fuego. C.M. Clapperton for scale. (Photo: D.I. Benn)

face. The association of the channel and compact till is compatible with the hydraulic theory of Walder and Fowler (1994), which predicts that drainage over rigid or slowly deforming sediments is likely to be via dendritic conduits, in contrast with the braided canals postulated for weak, deformable substrata.

Study of former subglacial drainage networks from subglacial facies associations is still very new, and much remains to be learned. Clearly, careful attention will need to be paid to establishing whether lenses of sorted sediments result from subglacial meltwater activity or were rafted in from upglacier within a deforming layer. Additionally, techniques need to be developed to determine the relationship between channel formation and the contemporary state of the bed (i.e. whether it was rigid or deforming). With careful research design, however, it should be possible to test whether current models of subglacial drainage are plausible, and to derive approximate palaeodischarges from grain size and bedform data.

11.2.2 Subglacial bedforms

Subglacial bedforms are longitudinal or transverse accumulations of sediment formed below active ice (Rose, 1987b; Menzies and Rose, 1989). Longitudinal forms are streamlined features aligned parallel to ice flow, and can be divided into *drumlins*, *flutings* and *megaflutings*; the most important transverse bedforms are *Rogen moraines*. Bedforms occur in fields which are positioned in relation to ice divides and ice streams at various times during active glaciation. This means that they are related not just to substrate morphology, local stress variations and sediment supply, but also to ice flow and sediment deformation histories. Longitudinal forms such as drumlins and flutings are clearly very similar to whalebacks and rock drumlins (Section 9.3.2), suggesting that some common factors underlie their formation.

The distinction between drumlins, flutings and megaflutings is based upon the length and elongation ratio of the bedform. The elongation ratio is defined as:

$$E = l/w \qquad (11.1)$$

where E is elongation ratio, l is maximum bedform length and w is maximum bedform width.

Elongation ratios and lengths of some drumlins, flutings and megaflutings are shown in Fig. 11.5 (Rose, 1987b). According to Rose, drumlins are large forms (> 100 m long axis) with elongation ratios up to about 7:1; flutings are less than 100 m long with elongation ratios in the range 2:1 to 60:1 or even more; and megaflutings are elongate forms with long axes greater than 100 m. Drumlinoid forms longer than *c.* 1000 m are termed streamlined hills, and very large elongate forms many tens of kilometres long, hundreds of metres wide and more than 25 m high have been recognized on satellite images and termed mega-scale glacial lineations by Clark (1993).

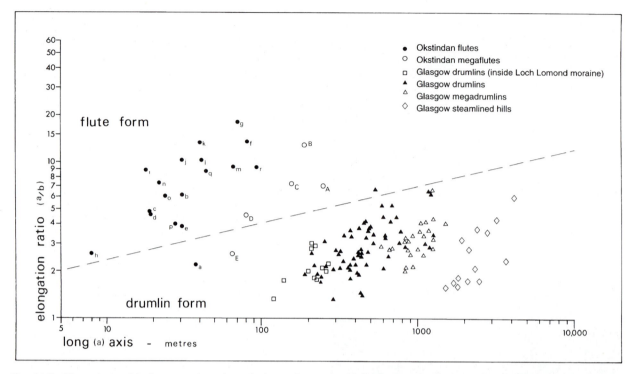

Fig. 11.5 The relationship between length and elongation ratio of flutings, megaflutings, drumlins, megadrumlins and streamlined hills from the Glasgow area, Scotland and Okstindan, Norway. The dashed line represents a possible quantitative differentiation of flutings and drumlins. (From Rose, 1987b. Reprinted by permission of Balkema)

However, the dimensions of 'drumlins' and 'flutings' described in the literature commonly differ from those defined in Fig. 11.5. In recognition of this, Rose has stressed that drumlins, flutings and megaflutes form a continuum and any lines drawn between them can only ever be arbitrary.

In this section we consider the morphology and composition of flutings, megaflutings, drumlins and Rogen moraine, and discuss modern theories for their origin.

11.2.3 Flutings

Flutings, also known as *flutes* or *fluted moraines*, are elongate streamlined ridges of sediment aligned parallel to former glacier flow (Fig. 11.6; Boulton, 1976; Rose, 1987b; Gordon *et al.*, 1992). They are generally a few tens of centimetres to a few metres high and wide, and occur in groupings of subparallel ridges on many modern glacier forelands, and some older glacial landscapes (e.g. Sissons, 1967; Gordon *et al.*, 1992; Bennett, 1995). Interesting early observations of flutings were made by Gilbert (1904a), Grant and Higgins (1913), Tarr and Martin (1914), Ray (1935), Dyson (1952), Nichols and Miller (1952), Hoppe and Schytt (1953) and Schytt (1963). Flutings have low preservation potential because they are readily degraded by wind and water, and

consequently are much more common on modern forelands than on older terrain (Boulton and Dent, 1974). Flutings commonly begin downglacier from lodged boulders or clast clusters, although this is not always the case. They are generally straight or very gently curving, but may swerve abruptly around boulders before resuming a straight course (Figs 11.6 and 11.7; Gordon *et al.*, 1992). They tend to consist mainly of subglacial till, although cores of deformed, pre-existing sediments have been reported (Paul and Evans, 1974; Boulton, 1976, 1987; Rose, 1989b; Eklund and Hart, 1996).

Several models of fluting formation have been proposed (Gordon *et al.*, 1992). The most widely accepted model regards flutings as the product of subglacial sediment deformation in the lee of obstructions on the bed (Dyson, 1952; Boulton, 1976; Benn, 1994a; Eklund and Hart, 1996). According to this interpretation, flutes are initiated when weak, saturated sediment is squeezed under pressure into small lee-side cavities behind obstructions. The sediment flows more readily in response to stress than the ice, so the cavity becomes filled up and is prevented from closing. The sediment-filled cavity can then evolve into a fluting by one of two mechanisms. (a) As sediment flows into the lee-side cavity, the drop in pressure may cause it to freeze and adhere to the basal ice. The frozen sediment is then carried

(a)

(b)

Fig. 11.6 Flutings on modern glacier forelands in Iceland. (a) Fresh flutings, Breidamerkurjökull. Note deflection of the nearest flute around the boulder. Ice flow from right to left. (Photo: D.I. Benn) (b) Degraded flute, Skalafellsjökull, showing its relationship to a lodged, faceted boulder. Note the striae and the perched erratic on the up-ice side of the boulder. Ice flow away from viewer. (Photo: D.J.A. Evans)

forward by the ice, and new material is added by deformation into the cavity at the upglacier end (Schytt, 1962, cited in Eklund and Hart, 1996). (b) The sediment within the cavity can remain unfrozen, and new sediment is added at the downglacier end as the cavity is carried forward by ice flow (Fig. 11.8; Boulton, 1976; Morris and Morland, 1976; Åmark, 1980; Benn, 1994a). Flutings formed by the second mechanism can become frozen if they propagate into areas where the base of the glacier is below the pressure melting point. Flutings composed of frozen sediment have been observed at the margins of modern glaciers (Boulton, 1976; Hart, 1995b).

Evidence for sediment deformation in flutings is provided by folds and faults in sorted sediments, magnetic fabrics and clast fabrics (Paul and Evans, 1974; Boulton, 1976; Benn, 1994a; Eklund and Hart,

1996). At several sites, converging *herringbone fabrics* have been reported, in which particles on the flanks of flutings tend to be oriented obliquely downglacier towards the fluting axis. Benn (1994a) showed that in a fluting on the foreland of Slettmarkbreen, Norway, herringbone fabrics do not conform to the ice flow direction recorded by striae on lodged boulders within the flute, showing that they record patterns of strain within deforming till rather than shear at the ice–bed interface (Fig. 11.9). In contrast, herringbone fabrics in flutings at Austre Okstindbreen, Norway, were interpreted by Rose (1989b) as a record of former patterns of ice flow over the bed (Fig. 11.10). According to this interpretation, ice flow was convergent at the proximal end of the fluting and divergent at the distal end, owing to flow disturbances in the basal ice set up by boulders, bedrock knobs and boulder clusters at the ice–bed interface. No boulder striation data are available for these flutes, but striations on boulders in other flutes on the foreland indicate parallel ice flow over the features, as for the Slettmarkbreen example (Rose, 1992). The Okstindbreen fabrics, therefore, may also record oblique strain patterns within till, reflecting stress gradients set up in the vicinity of boulders. An interesting discussion of fabric patterns in flutings and their significance has been provided by Eklund and Hart (1996).

Benn and Evans (1996) have argued that there may be two contrasting types of fluting: (a) *tapering flutings*, which become lower and narrower with distance downglacier; and (b) *parallel-sided flutings*, which maintain constant cross-profiles for considerable distances. Tapering flutings occur on the forelands of Slettmarkbreen and Okstindbreen, Norway (Figs 11.9 and 11.10), whereas examples of parallel-sided flutings occur in front of Breidamerkurjökull and Skalafellsjökull in Iceland. In contrast with the herringbone fabric patterns in tapering flutings, fabrics from parallel-sided flutes may be either parallel to the flute axis or converge only slightly downglacier (Boulton, 1976; Benn, 1995).

Benn and Evans (1996) have suggested that the contrasting forms of parallel-sided and tapering flutes may reflect differences in till rheology, or strain response to stress. Parallel-sided flutes at Breidamerkurjökull are composed of weak, dilatant deformation till which can deform much more readily than ice (Boulton and Hindmarsh, 1987; Benn, 1995). Therefore, pressure differences within grooves in the basal ice are equilibrated by till flow rather than ice creep, so that groove closure is prevented and the fluting can propagate downglacier all the way to the margin. In contrast, the till in the Norwegian tapering flutes was probably stiff and non-dilatant, and there is likely to have been a lower contrast between the viscosity of the ice and till

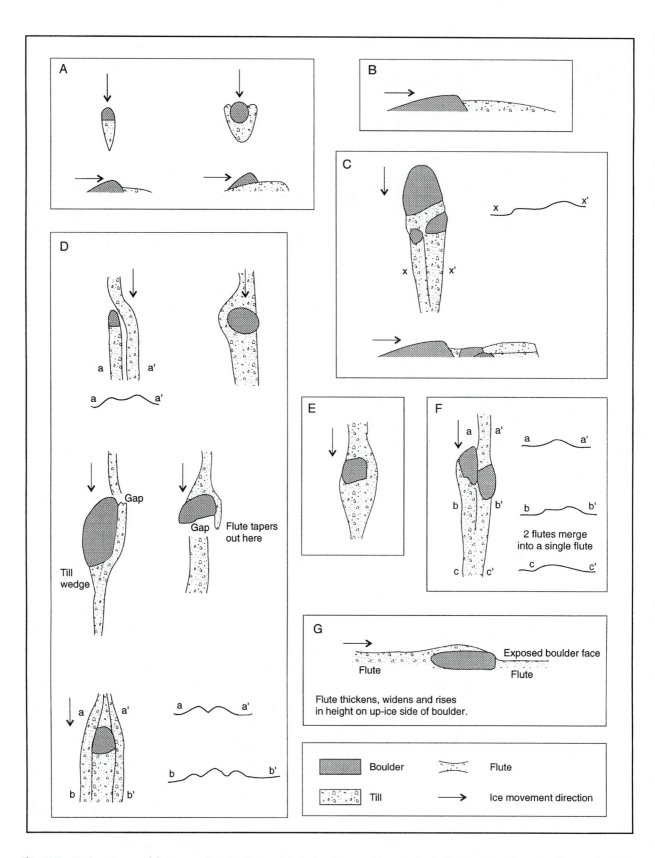

Fig. 11.7 Various types of fluting and their relationship to boulders on the glacier bed. (Modified from Gordon *et al.*, 1992)

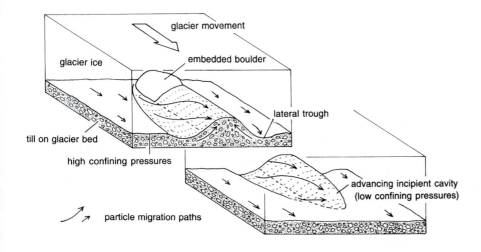

Fig. 11.8 A model of fluting formation by the subsole deformation of sediment into an advancing incipient cavity. (From Benn, 1994a. Reprinted by permission of Blackwell)

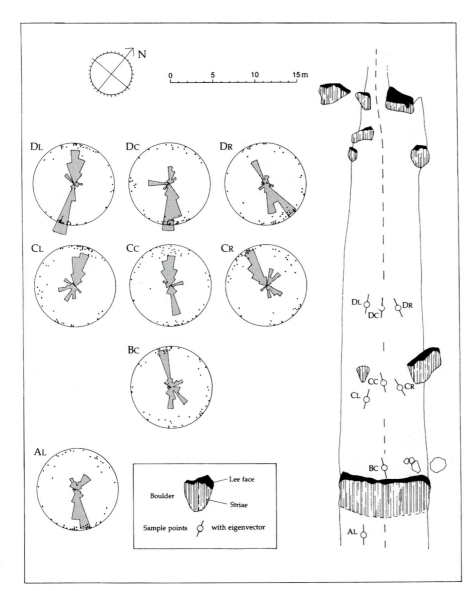

Fig. 11.9 'Herringbone' pattern of clast fabrics in a tapering flute on the foreland of Slettmarkbreen, Norway. (From Benn, 1994a. Reprinted by permission of Blackwell)

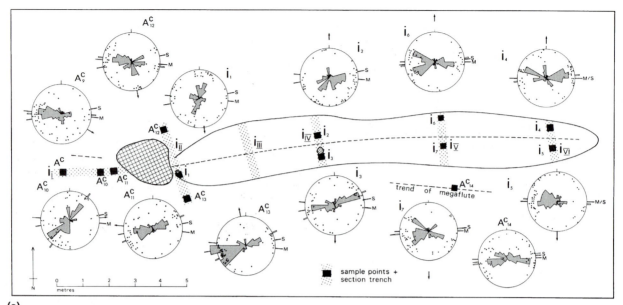

(a)

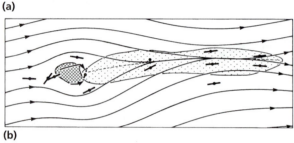

(b)

Fig. 11.10 (a) The clast fabrics of a fluting superimposed on a megafluting at Austre Okstindbreen, Norway; (b) reconstructed ice flow patterns of the superimposed fluting based upon clast fabrics. (From Rose, 1989b. Reprinted by permission of Elsevier)

(Benn, 1994a, b). Therefore, grooves in the basal ice may have been able to contract by ice deformation, leading to downglacier decay of the flute form. According to this view, the length of tapering flutings is a function of the relative strain rates of ice and till, whereas the length of parallel-sided flutings is the product of the ice velocity and flow duration. Fluting length, however, is ultimately limited by the position of the ice margin relative to the point of initiation.

Flutings have been observed frozen into the base of glaciers, leading Gordon *et al.* (1992) to suggest that they may originate by the melt-out of longitudinal thickenings of basal debris-rich ice. This implies that the characteristics of these flutes are inherited from strain patterns in the basal ice layers rather than from the ice–bed interface (Rose, 1989b) or subsole deformation (Benn, 1994a). However, there is the possibility that the flutings described by Gordon *et al.* (1992) were produced by the deformation of unfrozen sediment beneath thicker temperate ice and then frozen to the glacier sole during ice thinning. Annual freezing and thawing cycles have been invoked in a variant of the subsole deformation model by Shaw (1985).

The flutings studied by Rose (1989b) at Austre Okstindbreen are composed of a distinctive till unit, allowing the rates of sediment transfer involved in fluting formation to be estimated. Rates range from $0.0077 \, \mathrm{m^3 \, m^{-2} \, yr^{-1}}$ to $0.38 \, \mathrm{m^3 \, m^{-2} \, yr^{-1}}$ (cubic metres per square metre of bed area per year). Application of these figures to larger drumlins and flutings at other locations reveals that subglacial bedforms may be created within a period of between 4 and 400 years irrespective of their size, thus indicating rapid development. It should be noted, however, that these sediment transfer rates are for one glacier at one time period, and may not be typical.

An erosional origin has been suggested for some flutings. For example, the 'fluted ground moraine' described by Prest (1968) consists of ridges with crests at the same height as the surrounding till surface, giving the impression that the flutings are a product of the erosion of furrows rather than the construction of ridges. Evidence that 'erosional fluting' occurs on a small scale is provided by *sole marks* below the bases of some subglacial tills and on some recently deglaciated surfaces (Westgate, 1968; Ehlers and Stephan, 1979; Shaw, 1982; Hart, 1995b). These sole marks appear to have been ploughed by clasts protruding from the glacier sole, which were dragged over the substratum, and clasts tend to occur at their downglacier ends, in contrast with constructional flutings, which commonly have clasts at their upglacier ends. It is unlikely that this mechanism could account for any larger flutings exposed on modern glacier forelands. Certain crag and tails (Section 9.3.3) can also be regarded as isolated flutings formed by differential erosion around protecting obstacles. Flutings on the foreland of the southern Norwegian glacier Blåbreen, which contain cores of pre-existing stratified outwash downflow of stoss boulders embedded in the outwash, may possibly be explained in this way.

11.2.4 Drumlins, megaflutings and mega-scale lineations

Drumlins and mega-scale flutings are among the most enigmatic of glacial landforms, and over the past hundred years a very large body of literature and a great many theories of their formation have been published (e.g. Upham, 1892, 1894; Tarr, 1894; Alden, 1905; Fairchild, 1929; Czechowna, 1953; Menzies, 1984; Menzies and Rose, 1987, 1989). In fact, Sugden and John (1976) have gone so far as to state that 'there are almost as many theories of drumlin formation as there are drumlins'. The problem of determining the origin of large-scale streamlined forms stems from the lack of observations of analogous modern environments, so that researchers must base their arguments on the morphology, internal composition and distribution of features that formed below long-vanished glaciers, where little is known of former subglacial conditions. Recent advances have been made by combining detailed sedimentological studies of drumlin composition with modern glaciological theory, but many important questions remain. Menzies (1979) provided a very thorough review of the literature up to the late 1970s. Here we concentrate on the more recent literature, focusing on the most plausible theories of formation.

11.2.4.1 MORPHOLOGY

The term *drumlin* is derived from the Gaelic word *druim* meaning a rounded hill (Close, 1867). A useful general definition of drumlin form was given by Menzies (1979):

> typically smooth, oval-shaped hills or hillocks of glacial drift resembling in morphology an inverted spoon or an egg half-buried along its long axis. Generally the steep, blunter end points in the up-ice direction and the gentler sloping, pointed end faces in the down-ice direction, these two ends being respectively known as the stoss and lee sides.

Drumlin long axes are oriented parallel to the direction of ice flow, with higher and wider stoss ends which taper down to a pointed lee end. Some typical drumlin shapes are shown in Fig. 11.11. Long, narrow drumlins have been termed *spindle* forms, and broader, often asymmetrical drumlins are known as *parabolic* forms (Shaw, 1983a; Shaw *et al.*, 1989). More complex forms also exist, such as *transverse*

asymmetrical drumlins or *superimposed drumlins*, consisting of longitudinal ridges superimposed on transverse or oblique hills (Fig. 11.11b; Rose and Letzer, 1977; Rose, 1987b). Compound spindle drumlins with multiple crests and asymmetrical spindle drumlins with overdeepened hollows along one margin have been observed in Chile by Clapperton (1989) (Fig. 11.12).

With increasing elongation ratios, drumlins grade into *megaflutings* or *megaflutes*. Using Landsat satellite imagery, Clark (1993) has identified particularly large *mega-scale glacial lineations*. These lineations are not easily recognizable on aerial photographs until they are compared with linear features identifiable by a distinct grain on satellite images (Fig. 11.13a). The coverage of individual aerial photographs is not large enough to view the mega-scale lineations, but they often show up as a series of aligned drift mounds (Fig. 11.13b). The lengths of mega-scale glacial lineations range from 8–70 km, widths from 200–1300 m and spacings from 300 m to 5 km. One such lineation identified by Clark in the James Bay lowlands, Canada, is thought to be the largest subaerially exposed ice flow landform in the world at 70 km long.

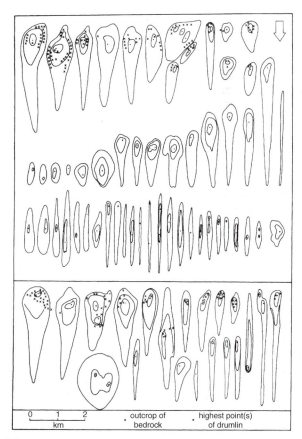

```
0    1    2          . outcrop of      . highest point(s)
    km                 bedrock           of drumlin
```

(a)

Fig. 11.11

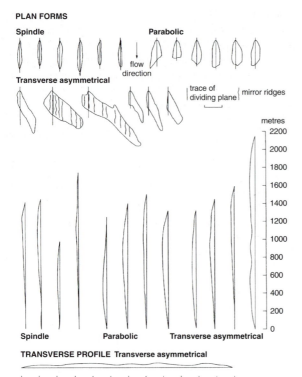

PLAN FORMS

Spindle   Parabolic

flow direction

Transverse asymmetrical

trace of dividing plane   mirror ridges

metres

Spindle   Parabolic   Transverse asymmetrical

TRANSVERSE PROFILE Transverse asymmetrical

0   200   400   600   800 1000 1200 1400 1600 1800 2000 2200 2400   metres
(drawn without vertical exaggeration)

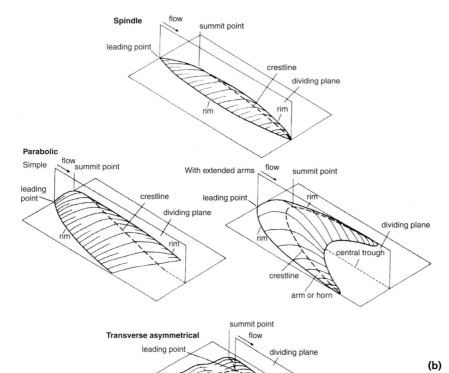

Spindle
flow
summit point
leading point
crestline
dividing plane
rim   rim

Parabolic
Simple
flow summit point
leading point
crestline
dividing plane
rim   rim

With extended arms
flow summit point
leading point
rim
rim
dividing plane
central trough
crestline
arm or horn

Transverse asymmetrical
summit point
flow
leading point
dividing plane
rim

(b)

Fig. 11.11 Typical drumlin and fluting shapes in (a) central Finland (from Gluckert, 1973) and (b) Canada. (From Shaw, 1983a. Reproduced by permission of the International Glaciological Society)

Fig. 11.12 Asymmetrical drumlins, Strait of Magellan, southern Chile. Ice flow away from viewer. (Photo: D.I. Benn)

Numerous indices have been devised to describe drumlin morphology (e.g. Mitchell, 1994). Chorley (1959) and Reed *et al.* (1962) described drumlin shapes as *lemniscate loops*, defined by the relation:

$$r = a^2 \cos k \qquad (11.2)$$

where r is the form of the loop, a is the long axis length, and k is a dimensionless constant defining loop elongation.

The k value is calculated by:

$$k = a^2/4A \qquad (11.3)$$

where A is the drumlin area.

The various shapes of streamlined forms were regarded by Chorley (1959) as those which presented the minimum disturbance to the medium moving around them. He suggested that large k values, which are associated with greater elongation, indicate a small resistance to flow. They may therefore develop where shear stresses are low and basal ice velocities are high.

A slightly different approach was taken by Reed *et al.* (1962), who used the *rose curve* to approximate the outline of drumlins:

$$R = a \cos k \qquad (11.4)$$

A comparison of the lemniscate loop and rose curve by Doornkamp and King (1971) showed that rose curves provide a much better fit to observed drumlin shapes, because they are generally narrower than lemniscate loops. Further mathematical formulae for idealized drumlin shapes are discussed by Smalley and Warburton (1994), together with some interestingly 'off the wall' ideas of drumlin formation.

11.2.4.2 DISTRIBUTION

Although assessments of individual drumlin shapes are important (I.S. Evans, 1987; Mills, 1980, 1987;

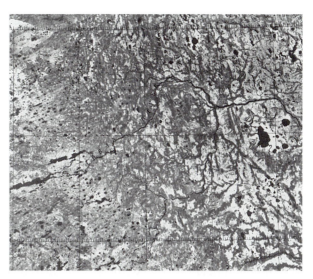

(a)

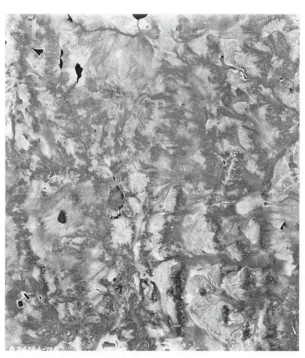

(b)

Fig. 11.13 Mega-scale glacial lineations in the James Bay lowlands, Canada, viewed at satellite image and aerial photograph scales (after Clark, 1993). (a) Landsat image displaying strong NNW–SSE-aligned megalineations in the right-hand part, and ENE–WSW-aligned Rogen moraine on the left. (b) Aerial photograph of the central part of the megalineations in the Landsat image showing that the lineations are not easy to detect at this scale, probably due to postglacial dissection. (Aerial photograph from the Department of Energy, Mines and Resources, Canada)

Coude, 1989), the analysis of drumlin distributions and the relationships between drumlins and other landforms tell us more about the evolution of stream-lined shapes and thus the subglacial processes that formed them (Trenhaile, 1971, 1975; Menzies, 1979; Patterson and Hooke, 1996).

The distribution patterns of drumlins have been analysed at several scales. Drumlins tend to be concentrated in *fields*, often numbering several thousand individuals (Vernon, 1966; Shilts *et al.*, 1987; Aylsworth and Shilts, 1989a, b; Haavisto-Hyvarinen *et al.*, 1989). Within such fields, drumlins may occur in close association with rock drumlins (Section 9.3.2) and Rogen moraine (Section 11.2.5; Fig. 11.13). Drumlin fields commonly form broad bands, aligned either transverse or parallel to former glacier flow directions. Transverse fields may reflect zones of drumlin formation behind former ice margins, whereas longitudinal drumlin fields may mark the position of fast ice streams or glacier lobes in former ice sheets (Wright, 1957, 1962; Rose and Letzer, 1977; Jones, 1982; Piotrowski and Smalley, 1987; Rose, 1987b; Dyke and Morris, 1988; Goldstein, 1989; Heikkinen and Tikkanen, 1989; Patterson and Hooke, 1996). In the British Isles, drumlins are most common in lowland areas or in large valley systems which acted as conduits for the evacuation of ice from the last ice sheets (e.g. Hollingworth, 1931; Clapperton, 1970; Smalley and Warburton, 1994; Mitchell, 1994; Fig. 11.14 (Plate 22)). This is not to say that drumlins are restricted to lowland terrains, and several researchers have stressed that drumlin fields cannot be correlated with particular topographic settings or substrate lithologies (cf. Gravenor and Meneley, 1958; Aronow, 1959; Reed *et al.*, 1962; Hill, 1973; Patterson and Hooke, 1996). Drumlins occurring in areas of greater relief are often developed on the lee side of spurs, where they are known as *drumlinoid drift tails* (Embleton and King, 1975; Mitchell, 1991).

Drumlin distribution within a swarm may appear to be regular, but is more likely to be random (Smalley and Unwin, 1968). However, non-random distributions have also been noted in real-world examples by Vernon (1966), Hill (1973), Jauhiainen (1975) and Mills (1980). The occurrence of bands of drumlins lying abreast of one another within drumlin fields has been used by Smalley and Piotrowski (1987) and Smalley and Warburton (1994) to suggest that they reflect a banding of bed properties. The influence of pre-existing variations in small-scale topography and substrate properties on drumlin distribution has been stressed by Boulton (1987) and is discussed in Section 11.2.4.4. Drumlins are found on a wide variety of substrata. In a review of the literature, Patterson and Hooke (1996) found that unconsolidated sediments make up 34 per cent of drumlin substrata, 18

per cent being till and 16 per cent being stratified sediments. The remaining 66 per cent are rock, divided between shales and slates ($\frac{1}{3}$), crystalline rocks ($\frac{1}{3}$), carbonates ($\frac{1}{4}$), and sandstones, conglomerates and basalts. Menzies (1979) has pointed out that factors such as joint and fracture systems, depth of weathering, lithology and mineralogy need to be assessed when making generalizations about substrates and drumlin occurrence (Jahns, 1943; Virkkala, 1963; Feininger, 1971; Trainer, 1973).

Patterson and Hooke (1996) have argued that drumlins form in areas where basal shear stresses are low and porewater pressures are high. They found that, in situations where it is possible to reconstruct the ice-surface form for particular drumlin-forming events, reconstructed basal shear stresses fall in the range 11–35 kPa. Interestingly, this range is closely similar to that for modern glaciers known to be underlain by deforming substrata. Patterson and Hooke also argued that drumlins commonly occur upglacier from ice-thrust landforms thought to have formed at frozen glacier margins. They argued that the presence of a frozen toe would have blocked subglacial drainage, leading to elevated porewater pressures. As they point out, high porewater pressures are compatible with low bed strength and low shear stresses. It should be noted, however, that neither former shear stresses nor basal water pressures are known for the vast majority of drumlin fields, and that generalizations are probably premature.

Several studies have assessed the morphometry of drumlin fields, including characteristics such as drumlin spacing, density and distribution, and their variation within and between fields (Jewtuchowicz, 1956; Gravenor and Meneley, 1958; Reed *et al.*, 1962; Heidenreich, 1964; Vernon, 1966; Baranowski, 1969; Barnett and Finke, 1971; Doornkamp and King, 1971; Trenhaile, 1971, 1975; Hill, 1973; Crozier, 1975). *Drumlin spacing*, defined as the perpendicular distance between adjacent drumlins (Reed *et al.*, 1962), may exhibit normal or multimodal distributions for individual fields, and is highly variable between fields. Such inter-field variability indicates that the glacial processes responsible for drumlin formation differ according to location (Menzies, 1979). *Drumlin density*, the number of drumlins per unit area, varies from 19.3 km^{-2} in Appleby, England, to 1.8 km^{-2} in Nova Scotia, Canada (Gravenor, 1974; Menzies, 1979), but great variability may occur within individual fields (Reed *et al.*, 1962; Vernon, 1966; Doornkamp and King, 1971; Hill, 1973; Jauhiainen, 1975). In New York State, for example, drumlin densities range from 3.37 to 8.39 km^{-2} (Miller, 1972). Drumlin densities may decrease or increase in a downglacier direction (e.g. Fairchild, 1929; Hollingworth, 1931; Charlesworth, 1957; Smalley and Unwin, 1968; Hill, 1973), although Menzies (1979) suggested that measurements of drumlin density are not instructive unless they are put into the context of the local topography.

It cannot be assumed that drumlins within a swarm were synchronously deposited, or that their forms were in equilibrium with the most recent glacial event. Evidence for time-transgressive (diachronous) deposition of drumlins within individual drumlin fields in Ireland has been presented by Synge and Stephens (1960), Vernon (1966) and Hill (1971). More recently, Mooers (1989b) has argued that drumlins in central Minnesota, USA, were formed incrementally during retreat of the Superior lobe of the Laurentide ice sheet, with active drumlin formation occurring only in a 20–30 km wide zone below the margin. Similarly, Patterson and Hooke (1996) found that drumlins commonly occur in transverse belts behind former ice-margin positions during glacier retreat. It is now generally accepted that overprinting of one set of drumlins on top of another can result from ice-divide migration, which is a response to changes in the overall shape of the ice sheet. Previous forms may be completely obliterated, or they may survive as remnant hills or buried till stratigraphies (Stea and Brown, 1989; Section 12.4). The remoulding of pre-existing forms and sediments is referred to as *drumlinization*.

11.2.4.3 COMPOSITION

The composition of drumlins and megaflutings is extremely varied. Some have rock cores mantled by a superficial *carapace* of till, which surrounds the core like an eggshell (e.g. Hart, 1995b), but most are composed entirely of unconsolidated sediments. Many researchers have reported drumlins with cores of sorted sediments covered by a till carapace (Lemke, 1958; Aronow, 1959; Krüger and Thomsen, 1984; Boulton, 1987; Boyce and Eyles, 1991; Bluemle et al., 1993). In some cases, the sediment core clearly consists of overridden pre-existing materials (e.g. Krüger and Thomsen, 1984; Boyce and Eyles, 1991), but in others the age relationships between sorted sediments and drumlinization are not so obvious. Some researchers have suggested that sorted sediments within drumlins were deposited contemporaneously with drumlin formation (e.g. Dardis and McCabe, 1983, 1987; Dardis, 1985, 1987; Hanvey, 1987, 1989; Dardis et al., 1984; McCabe and Dardis, 1989; Shaw et al., 1989; Shaw, 1993). Sediment cores may be undeformed, or display complex deformation structures such as overfolds and thrusts (e.g. Slater, 1929; Jewtuchowicz, 1956; Krüger and Thomsen, 1984; Stanford and Mickelson, 1985; Boulton, 1987; Wysota, 1994). Complexly deformed sorted sediments have been described from megaflutings in North Dakota, USA,

by Bluemle et al. (1993). Drumlins and megaflutings may also consist entirely or predominantly of till (e.g. Gravenor and Meneley, 1958; Newman and Mickelson, 1994; Nenonen, 1994).

Till fabric studies commonly show systematic variations in the preferred orientation of clasts in individual drumlins, revealing former patterns of ice flow or sediment deformation. Fabrics from surface tills tend to be parallel with the former ice flow direction on the top of drumlins, but parallel to the contours on the flanks, suggesting divergent ice or till flow at the stoss ends and convergent flow at the lee ends or tails of the obstacles (e.g. Andrews and King, 1968; Savage, 1968; Krüger and Thomsen, 1984). Within till cores, fabrics may bear no relationship to drumlin form, and appear to record till accretion prior to drumlinization (e.g. Gravenor and Meneley, 1958; Grant, 1975; Stea, 1980; Stea and Brown, 1989; Newman and Mickelson, 1994). Fabric preferred orientations can vary upsection within drumlin cores, probably as a result of changing ice flow directions during till accretion (Hill, 1971; Pessl, 1971; Stea and Brown, 1989).

11.2.4.4 PROCESSES OF FORMATION

Many theories of drumlin and megafluting formation have been proposed over the past century, and as research on subglacial processes has advanced, so older theories have been cannibalized for spares, and the most feasible aspects have been built into new theories. A thorough review of the history of ideas on drumlin formation is beyond the scope of this book, but good reviews and bibliographies of the earlier literature have been provided by Menzies (1979, 1984).

In essence, drumlins and megaflutings can be explained as the result of either (a) erosion of the intervening hollows, (b) accretion of sediment in hills, or (c) some combination of the two. *Erosion* of material from hollows and swales could produce drumlins from the remnants of formerly more extensive pre-existing sediments. Early proponents of *drumlin accretion* argued that drumlins may be built up by the successive additions of till as a series of concentric shells (e.g. Fairchild, 1929; Flint, 1947). This model can explain only those drumlins consisting of concentric layers of till, or rock-cored drumlins with a till carapace; it cannot account for the occurrence of sorted sediments within drumlins. Furthermore, it does not address the question of why drumlins should begin to grow in some areas but not others. Many other mechanisms of sediment accretion have been proposed, including differential sediment dilation (Smalley and Unwin, 1968), frost heaving (Baranowski, 1969), and deposition below hypothetical helicoidal flow cells in basal ice (Shaw and Freschauf, 1973). Modern research has shown that such models are physically implausible, and they find little support today.

The most widely accepted model of drumlin formation is based on recent ideas of sediment erosion and redistribution within subglacial deforming layers (Boulton, 1987; Sections 4.4 and 5.6). According to this model, regions within a deforming layer that are stronger or stiffer than average will remain static or deform slowly, in contrast with the intervening weaker areas, which will undergo higher strain rates. The evolution of an inhomogeneous bed under this process is illustrated in Fig. 11.15. In this example, the initial bed consists of gravel bodies within finer-grained sediments. The gravel bodies are permeable and well drained, leading to low porewater pressures and high sediment strength, whereas the finer-grained intervening areas are less well drained and consequently weaker. There is therefore a spatially variable bed response to glacially imposed stresses: the gravel bodies form rigid or slowly deforming cores, whereas the finer-grained material is more likely to undergo pervasive deformation, forming far-travelled sheaths of highly attenuated glacitectonite or deformation till around the stiffer cores. Streamlining of the residual gravel bodies forms drumlins, which are essentially equivalent to boudins in metamorphic rocks. Note how the flow lines around the drumlins come to resemble the till fabric patterns found in drumlin carapaces (Section 11.2.4.3).

Over time, some cores may become *derooted* and mobile, although still moving more slowly than the surrounding weak sediment (Fig. 11.16g, h). When cores have been transported from their original positions, the initial distribution of deformable and non-deformable materials on the substrate may be impossible to reconstruct. Derooted cores are also

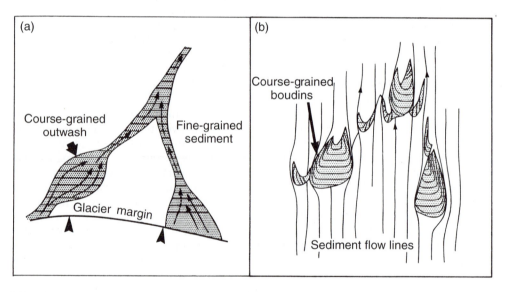

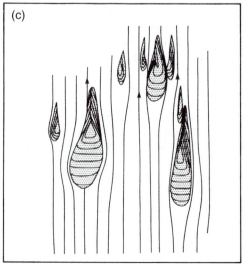

Fig. 11.15 Hypothetical example of drumlin formation as a result of the contrasting rheological properties of sediment overridden by a glacier: (a) initial distribution of coarse-grained proglacial outwash; (b), (c) the progressive development of drumlins from coarse-grained sediment masses, with rapid deformation of the intervening weaker sediments. (From Boulton, 1987. Reproduced by permission of Balkema)

characterized by highly deformed stratigraphies, whereas stable cores reveal very little deformation.

Boulton's (1987) theory appears to explain the data available on drumlins with cores of rock, till and stratified sediment. Cores of rock and consolidated till can form stiff, resistant regions around which sheaths of deforming sediment can flow. Well-drained stratified sediment will also form resistant cores and may reveal very few signs of disturbance despite having been overridden by glacier ice (Fig. 11.16i). Modern Icelandic examples of drumlins formed from pre-existing glacifluvial sediments have been described by Krüger and Thomsen (1984) and Krüger (1994). The drumlins formed from an inter-stream area of coarse bar deposits which acted as a subglacial obstacle. Signs of disturbance in stratified cores are commonly best developed in the tail of the drumlin (Evenson, 1971; Whittecar and Mickelson, 1979; Stanford and Mickelson, 1985; Stephan, 1987) where, Boulton (1987) suggests, strong convergence should dominate during drumlin formation. That stratified sediments can resist very high stresses during glacial transport and can retain much of their original structure is illustrated by the presence of undeformed boudins in glacitectonites (Hart and Roberts, 1994; Hart, 1995a, 1996; Benn and Evans, 1996). General descriptions of drumlin sediment, as reviewed by Karrow (1981), suggest that drumlins tend to form in sandy rather than clay-rich tills. Boulton (1987) considered this to be compatible with his theory because the coarser-grained sediments are better drained and more likely to form resistant cores. Resistant cores composed of a wide range of sediment types as well as coarse-grained materials were reported from the Peterborough drumlin field, Canada, by Boyce and Eyles (1991), who suggested that hollows between the cores were eroded by excavational deformation (Section 5.6.1).

Boulton (1987) used the drumlins of northern Saskatchewan to illustrate his theory (Fig. 11.17). Several of these drumlins appear to be connected and to form traces of shear folds on a horizontal plane. Boulton suggested that the barchanoid bodies represent competent, stiff sediment cores and the limbs represent attenuated deformable sediment. Interestingly, Shaw (1989) cited this same drumlin field as evidence for a completely different theory of drumlin formation (Section 11.2.6).

According to Clark (1993), *mega-scale glacial lineations* form by differential patterns of subglacial sediment deformation set up by variations in bed characteristics, and their great length reflects rapid ice flow and/or long periods of time for development. Using late Quaternary chronologies for the James Bay lowlands, which contain mega-scale glacial lineations produced by the former Laurentide ice sheet, Clark (1993) calculated typical maximum estimates of the periods of time available for the production of these large streamlined landforms and concluded that they are produced by fast glacier flow (400–1600 m yr^{-1}), at velocities typical of modern ice streams. Former ice streaming has been invoked by D.J.A. Evans (1996) to explain megaflutings more than 50 km long in southern Alberta, Canada. The production of mega-scale lineations over long periods of time (low velocity/high duration) rather than during fast flow phases (high velocity/low duration) is becoming less acceptable as an explanation as more evidence for dynamic and unstable ice sheets is uncovered (cf. Dyke et al., 1982b; Andrews et al., 1983; Dyke and Prest, 1987; Boulton and Clark, 1990a, b).

Stratified sediments in numerous drumlins in Ireland have been interpreted as subglacial glacifluvial deposits, rather than pre-existing sediment cores (Dardis and McCabe, 1983, 1987; Dardis et al., 1984; Dardis, 1985, 1987; Hanvey, 1987, 1989; McCabe and Dardis, 1989; Dardis and Hanvey, 1994). These authors argue that downglacier-dipping, sorted sediments exposed near the lee side of drumlins are contemporaneous with drumlin streamlining, and were deposited as *lee-side stratification sequences* in water-filled cavities. Thus, according to this model, the drumlins behaved rather like roches moutonnées with lee-side cavities, in which water-sorted sediments were deposited (Fig. 11.18; Section 11.2.7). Subglacial shearing is thought to modify drumlin form during and following lee-side deposition, deforming the upper stratified beds, forming a till carapace, and then producing a streamlined form. The evolution of the sediment sequences in the Galway Bay drumlins in Ireland is explained by McCabe and Dardis (1989) as their being the products of subaqueous glacimarine sedimentation followed by drumlinization. Water throughflow and escape structures in the drumlin sediments are explained by McCabe and Dardis (1994) as products of the release of glacitectonically induced high porewater pressures, which were elevated during glacier overriding of drumlin stoss sides. This model of drumlin modification by subglacial meltwater has some similarities with controversial ideas developed by John Shaw and co-workers in Canada, who attribute drumlins and megaflutings (and many other types of glacial landforms) to catastrophic subglacial megafloods. The megaflood hypothesis is considered in Section 11.2.6.

11.2.5 Rogen moraine

The term *Rogen moraine* was introduced by Lundqvist (1969) to describe the distinctive morainic landscape around Lake Rogen in Sweden (Fig. 11.19;

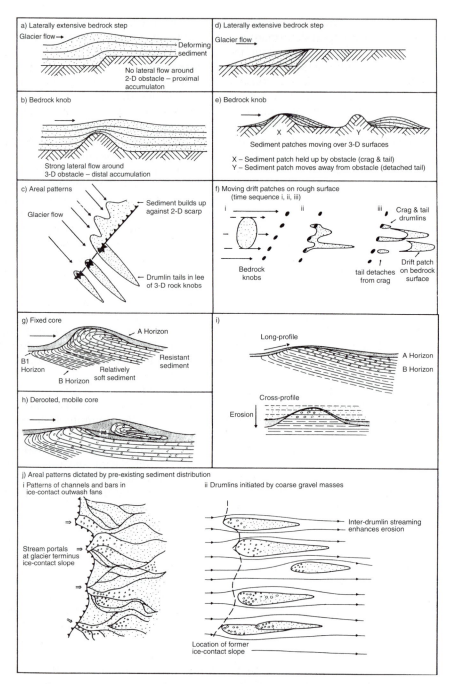

Fig. 11.16 Summary schematic diagram showing drumlin-forming processes, internal structures and distribution patterns according to the deforming bed model. (a), (d) Streamlined nose builds up on the proximal side of a laterally extensive bedrock step. (b), (e) Streamlined tail builds up on the distal side of a bedrock knob, owing to enhanced flow around the flanks. The drumlin forms a standing wave when the subglacial sediment is thick, although sediment does move through the drumlin. When subglacial sediment (drift) patches move over the rock substrate, the drumlin can move past the retarding knob. (c) Drumlin noses build up on the proximal side of a step of uniform height. The locations of noses reflect subglacial sediment 'streams'. Where gaps occur in the scarp, sediment streams over and flows into low-pressure points in the lee of the inter-gap knobs. (f) Subglacial sediment patches moving over bedrock are retarded by crags to produce crag and tails and detached tails. (g), (h) Drumlins initiated by stiff sediment obstacles that form deforming cores and which may be fixed or mobile. (i) Drumlins initiated by stiff sediment obstacles which form rigid, fixed cores. Weaker sediment to either side is eroded by subsole deformation. (j) An example of how drumlin distribution patterns can reflect original sedimentary inhomogeneities (e.g. coarse gravels at ice-contact outwash fan apices and gravel bars). Such drumlins can become derooted and mobile. (From Boulton, 1987. Reproduced by permission of Balkema)

Lundqvist, 1989a). It is applied to fields of coalescent crescentic ridges up to 30 m high and up to 100 m wide lying transverse to former ice flow. The arcuate forms are aligned with their outer limbs bent downglacier. In North America these features are usually referred to as *ribbed moraine* (Lee, 1959; Hughes, 1964; Cowan, 1968; Aylsworth and Shilts, 1989a, b). Fields of Rogen moraine in Scandinavia (Lundqvist, 1969) and in Labrador/Quebec and Newfoundland, Canada (Bouchard, 1989), lie in elongate depressions, but similar linear fields of Rogen moraine in Keewatin, Canada, appear to be independent of topography, although local relief is very slight (Aylsworth and Shilts, 1989a). Individual Rogen ridges have asymmetric profiles with shallow up-ice and steep down-ice flanks, giving the impression of a fish-scale texture in aerial photographs (Shilts, 1977; Shilts *et al.*, 1987; Aylsworth and Shilts, 1989a).

An important characteristic of Rogen moraines is their gradual transition into, or association with, drumlins or megaflutings (Figs 11.13 and 11.20). Rogen moraines in topographic basins commonly pass into drumlins on topographic highs (Markgren and Lassila, 1980; Lundqvist, 1969, 1989a), although Lundqvist (1989a) and Aylsworth and Shilts (1989a) pointed out that both landforms can also occur in association on flat ground. Aylsworth and Shilts (1989) further point out that Rogen moraine does not grade into drumlins in Keewatin but lies alongside them with an abrupt boundary. The close association between Rogen moraine and drumlins, together with zones of transition between the two forms, suggests that they may be deposited contemporaneously by similar subglacial processes.

Rogen moraines are composed of a variety of sediment facies, including massive and laminated diamicton, gravels, sands and silts, which commonly show evidence of deformation. In Sweden, Rogen moraines contain characteristic associations of laminated tills with attenuated interbeds of sorted sediments known as *Kalix till* (Beskow, 1935; Lundqvist, 1969) and *Sveg till* (Shaw, 1979; Dreimanis, 1989; Lundqvist, 1989c). Laminae are wrapped around large clasts and augen-like inclusions, and the whole mass is deformed by folds and thrust planes that record downglacier overturning and transport (Fig. 11.21; Shaw, 1979; Bouchard, 1989). Similar facies associations in Rogen moraine in Newfoundland have been described by Fisher and Shaw (1992), although other associations also occur in this area, including disturbed masses of crudely bedded gravels and stratified stony diamicton.

Various theories of Rogen moraine formation have been proposed, and were summarized by Lundqvist (1989a). Most of these theories are either implausible or cannot explain all the characteristics of Rogen

moraine, and here we concentrate on two, involving formation by: (a) deformation of debris-rich ice; and (b) deformation of pre-existing bed materials. The *debris-rich ice model* proposes that Rogen moraine is composed of thrust slices of basal ice, stacked up by compressive flow beneath the glacier (Fig. 11.22). Such thrusting is thought to occur in topographic hollows where ice flows against obstacles on the bed, or at the junction between warm-based, sliding ice located upglacier and cold-based ice downglacier (Sugden and John, 1976; Shaw, 1979; Sollid and Sorbel, 1984; Bouchard, 1989). Shaw (1979) argued that this model is supported by the characteristics and structure of the Sveg tills, which he interpreted as basal melt-out tills derived from folded and dislocated debris-rich ice. According to this model, therefore, the distinctive Rogen moraine morphology is inherited from stacked slices of basal ice, and the internal structures and sedimentology preserve the former englacial deformation structures. Under this model, the tendency for Rogen moraine to occur in topographic lows, with a transition to drumlins on topographic highs, has been interpreted as the result of a transition from compressive to extending flow at the glacier bed.

The *bed deformation model* was proposed by Boulton (1987), who interpreted Rogen moraines as part of a continuum of streamlined forms resulting from subsole sediment deformation. According to Boulton, Rogen moraines represent early stages of the drumlinization of transverse ridges of sediment on the glacier bed, and they can develop from drumlins following a change in glacier flow direction (Fig. 11.23). The deformation of weak bed materials around transverse ridges results in the preferential downglacier transport of the extremities of the features, producing the characteristic concave-downglacier planform. Different rates of sediment transport within the ridge cause fragmentation of the original ridge, creating numerous short crescentic ridges. The shape of Rogen moraines is thus regarded as analogous to that of barchan dunes. Boulton argued that continued attenuation of Rogen moraines will result in barchanoid drumlins, then ellipsoidal drumlins, then flutings, as the bed adjusts to the new ice flow conditions. Thus, Rogen moraines and flutings are regarded as end-members of a continuum of bedforms, representing an evolutionary time series. Spatial transitions between moraine types can therefore be explained in terms of variations in the cumulative strain experienced by the bed materials, reflecting variations in basal ice velocity and bed strength.

Boulton (1987) did not consider the internal composition and structure of Rogen moraines, but the published evidence does appear to be consistent with his model. The Sveg tills (Fig. 11.21) appear to

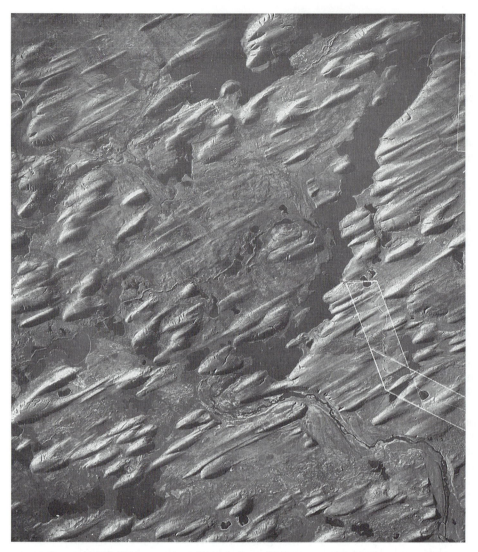

(a)

Fig. 11.17 The Livingstone Lake drumlins, Saskatchewan. (a) Aerial photograph of parabolic, transverse asymmetrical and spindle drumlins, northern Saskatchewan (ice flow from bottom left, which is from the north-east); (b) sketch of the drumlins and eskers from the aerial photograph (from Boulton, 1987). The shapes are thought to represent fold traces and the major drumlins represent stable sediment masses around which other material has deformed to produce attenuated fold limbs. A transverse lineation occurs between I6 and B9, documenting a possible earlier ice flow directional indicator (fluting) which has been partially remoulded. (c) Aerial photograph of transverse asymmetrical drumlins, northern Saskatchewan. (Aerial photographs by the Department of Energy, Mines and Resources, Canada. Reprinted by permission of Balkema)

be very similar to glacitectonites formed by the shear and attenuation of pre-existing sediments (Section 10.3.3; Hart and Roberts, 1994; Benn and Evans, 1996), and so may record strain within the bed rather than within debris-rich ice. Clearly, re-examination of the sedimentology of Rogen moraine will be required to resolve the problem of their origin.

The idea that Rogen moraine can be formed by subglacial megaflood events is considered in the following section.

11.2.6 Megaflood hypothesis for subglacial bedform genesis

Drumlins, Rogen moraines and flutings, together with tunnel valleys and many bedrock erosional forms, have been radically reinterpreted by John Shaw and co-workers, who emphasize the role of subglacial meltwater erosion and deposition. According to this model, drumlins and Rogen moraines represent the infillings of giant scours cut upwards into basal ice by large subglacial floods (Fig. 11.24; Shaw, 1983a; Shaw and Kvill, 1984). In support of

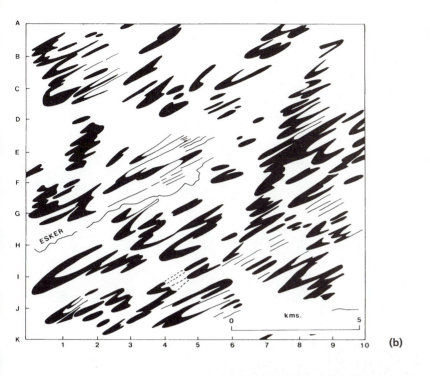

ESKER

kms.

0   5

(b)

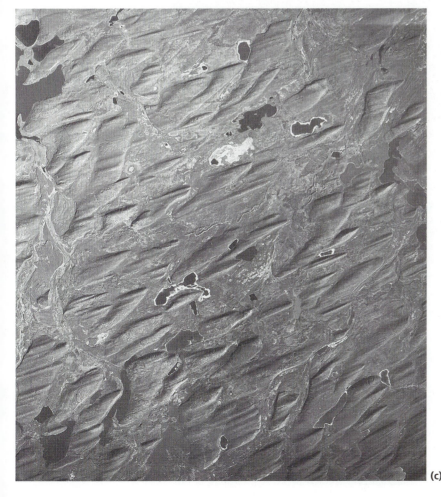

(c)   **Fig. 11.17**

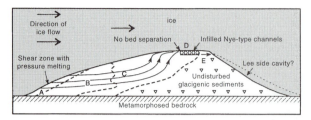

Fig. 11.18 A depositional model for water throughflow in the Kanrawer drumlin, Ireland. Zone A is a glacitectonic shear zone, Zone B is dominated by glacitectonic shunting, diamict disaggregation and tortuous water throughflow, Zone C is characterized by water throughflow along bed discontinuities and vertical escape routes, Zone D represents points on the drumlin surface where the water expelled from the drumlin excavates channels along the soft bed of the ice–substrate interface, and Zone E is where water escape routes form interconnecting passages to lee-side cavities. (Modified from McCabe and Dardis, 1994)

(a)

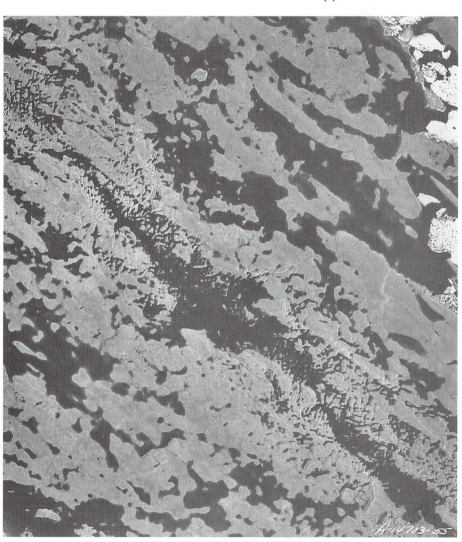

Fig. 11.19 (b)

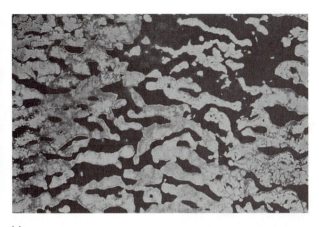

(c)

Fig. 11.19 Rogen moraine. (a) Type area around Lake Rogen, Sweden. (Photo: Jan Lundqvist, 1963) (b) Rogen moraine near Rankin Inlet, NWT, Canada, giving the land surface a fish scale appearance. (c) Rogen moraine in NWT, Canada. (Photographs provided by the Department of Energy, Mines and Resources, Canada)

this idea, they cited the striking similarity of form of subglacial bedforms and scour marks made at the base of turbulent underflows, arguing that these forms share a common origin (Fig. 11.25). Furthermore, Shaw (1983a), Shaw and Kvill (1984) and Sharpe (1987) suggested that the waning stage of the flood events fills the eroded cavity with stratified sediments, apparently explaining the occurrence of undisturbed stratified sediments whose bedding conforms to the surface shape of the landform. The integrity of the stratified sediments and the lack of glacitectonic structures was interpreted as evidence that the material has not been subject to high stresses in the subglacial environment. This interpretation of stratified sediments in subglacial bedforms clearly conflicts with that of Boulton (1987), who viewed them as cores of pre-existing material. As shown in Section 11.2.4.4, however, well-drained stratified cores need not undergo significant glacitectonic

(a)
Fig. 11.20

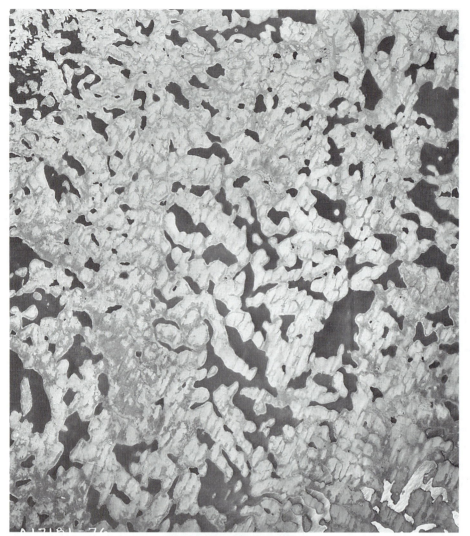

(b)

Fig. 11.20 Transitional subglacial bedforms produced beneath the Keewatin sector of the Laurentide ice sheet. (a) Flutings transitional to Rogen moraine, Dubawnt Lake (ice flow from right); (b) drumlinized Rogen moraine (ice flow from top). (From Aylsworth and Shilts, 1989a. Aerial photographs provided by the Department of Energy, Mines and Resources, Canada)

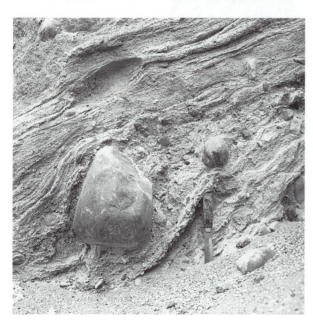

Fig. 11.21 Laminated diamicton (Sveg till) exposed in Rogen moraine near Sveg, Sweden. (Photo: J. Lundqvist)

Stage 1 Formation of initial folds in compressive zone

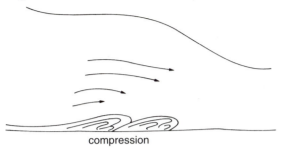

compression

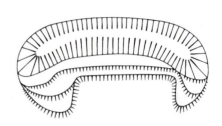

(see Shilts 1977, fig 5)

Stage 2 Development of major folds and thrusts

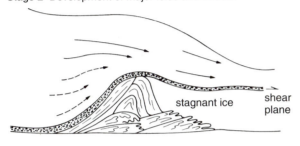

stagnant ice

shear plane

Stage 3 Stagnation and undermelt

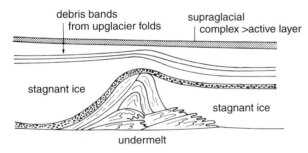

debris bands
from upglacier folds

supraglacial
complex >active layer

stagnant ice

stagnant ice

undermelt

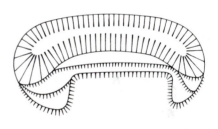

Stage 4 (with change in scale) Development of large thicknesses of fluvial deposits and slumping from
the exposed ridge

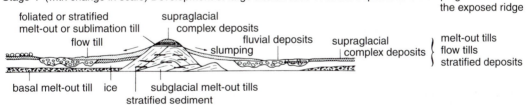

foliated or stratified
melt-out or sublimation till

flow till

supraglacial
complex deposits

fluvial deposits

slumping

supraglacial
complex deposits

melt-out tills
flow tills
stratified deposits

basal melt-out till   ice

subglacial melt-out tills

stratified sediment

Stage 5 Final landform and sediment complex

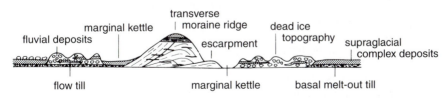

fluvial deposits

marginal kettle

transverse
moraine ridge

escarpment

dead ice
topography

supraglacial
complex deposits

flow till

marginal kettle

basal melt-out till

(a)
Fig. 11.22

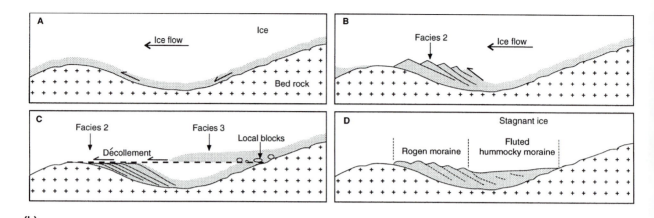

(b)

Fig. 11.22 The sequential evolution of Rogen moraine and associated deposits: (a) from Shaw (1979) (Reproduced by permission of Scandinavian University Press); (b) modified from Bouchard (1989)

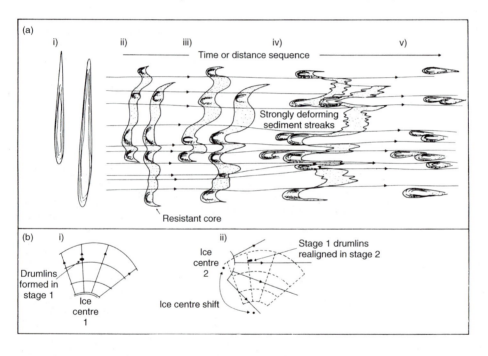

Fig. 11.23 A schematic reconstruction of the progressive transformation of flutings to Rogen moraine and drumlins by subglacial deformation. (a) i) Original flutings produced by earlier ice flow direction; ii–iii) Rogen moraine stage; iv–v) drumlin stage. This can be a time sequence or a distance sequence. (b) Explanation of the change in ice flow direction in (a) by a shift in the ice dispersal centre. (From Boulton, 1987. Reproduced by permission of Balkema)

deformation if they are strong enough to withstand the stresses applied by overriding ice.

Shaw *et al.* (1989) and Shaw (1993) have expanded the megaflood hypothesis by suggesting that it can explain the form and internal structure of till-cored drumlins and flutings. In this variation, the subglacial meltwater excavates the substrate *between* the bedforms, so that the drumlins or flutings are erosional remnants of pre-existing sediments. In this case, a form analogy is made between drumlin morphology and the lemniscate loops and streamlined forms of the fluvially eroded loessic hills of the Missoula Scablands (Section 9.3.4.4). Thus, the megaflood hypothesis attempts to explain the diverse characteristics of subglacial bedforms by different mechanisms. If bedforms contain sorted sediments,

they are interpreted as scour infillings; if they do not, they are interpreted as erosional remnants which survived the flood. It is therefore an example of an unfalsifiable hypothesis: no matter what the evidence, a way can be found to explain it in terms of the flood. As such, it is not a testable scientific theory, but a body of ideas that could conceivably have produced drumlins, Rogen moraine and other landforms.

Nevertheless, much, if not all, of the evidence cited in favour of the megaflood hypothesis can be satisfactorily explained in terms of Boulton's bed deformation model. For example, the Athabasca giant flutings in Alberta, Canada, were used as evidence for the fluvial erosional hypothesis, because they contain interbedded and disturbed tills and strat-

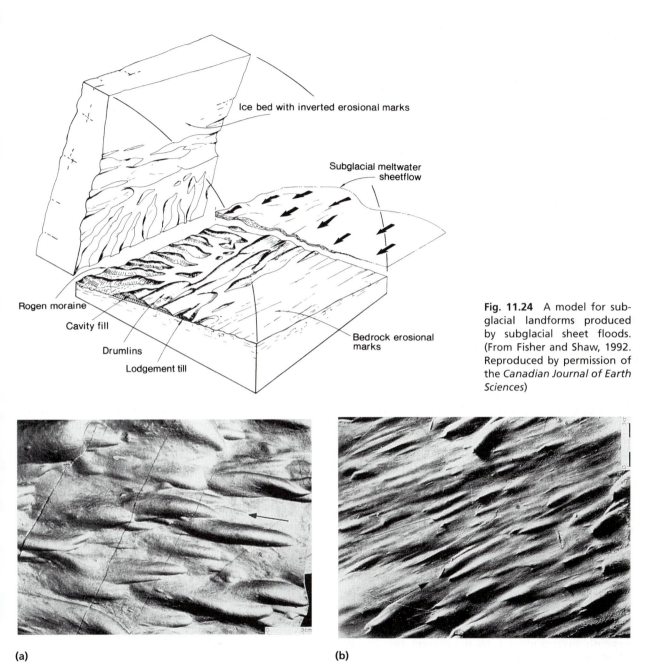

Fig. 11.24 A model for subglacial landforms produced by subglacial sheet floods. (From Fisher and Shaw, 1992. Reproduced by permission of the *Canadian Journal of Earth Sciences*)

(a)

(b)

Fig. 11.25 Streamlined erosional marks associated with turbidites: (a) narrow, parabolic and spindle flute casts; (b) longitudinal obstacle scour moulds cut behind small tool marks. The arrow indicates flow direction. (From Dzulinski and Walton, 1965)

ified sediments. However, the sediments have all the characteristics of glacitectonite (Section 10.3.3), and therefore are just as likely to have been produced by subsole deformation. Clearly, form analogy is not always a satisfactory criterion upon which to base genetic interpretations. Streamlined and rippled forms are produced in a wide range of environments wherever two media shear past one another, and can be observed in cloud formations, river beds, wind-blown sand dunes and snow, at the base of turbidity currents and other mass movements, and on glacier beds. The use of sole marks beneath turbidites as analogues for drumlins involves a major leap of faith, given the huge difference in the scale of the features. Much remains to be learned about the dynamics of shear in the subglacial environment, but it is clear that the coupled shearing of basal ice and glacier substrata, and slip at the ice–bed interface, can produce a variety of streamlined, transverse and non-aligned forms. Processes of ice creep, basal sliding and bed deformation are now well known, so that the bed deformation theory of drumlin and Rogen moraine formation is based upon well-founded physical principles. In contrast, the megaflood hypothesis relies upon form analogies and largely untested numerical modelling (Shoemaker, 1992a, b, 1994; Walder, 1994), and must also identify a source for the very large subglacial discharges required to create the large drumlins and Rogen moraine.

The megaflood hypothesis requires vast amounts of water to produce the extensive drumlin fields in places such as Livingston Lake, northern Saskatchewan, studied by Shaw (1983a) and Shaw and Kvill (1984). Thus, possible reservoirs need to be identified and confirmed by independent evidence. Shoemaker (1991, 1992a, b) has proposed a possible meltwater source at the centre of the Laurentide ice sheet, but this has raised several problems, and the idea is not taken seriously by most researchers (Section 9.3.4.4). However, it may not be wise to dismiss out of hand the idea that certain subglacial bedforms may have been created by large, transient drainage events. Subglacial drainage of large ice-marginal, supraglacial and subglacial lakes is well known from modern environments, and there is strong evidence that in some cases water is carried in distributed systems rather than conduits (Section 3.5.2). It is also clear that some Pleistocene jökulhlaup events were of a much higher magnitude than any observed in historical times (Sections 1.7.2.4 and 9.3.4.4). The megaflood hypothesis, however, must be rigorously tested by critical analysis of the morphology, internal structure and sedimentology of drumlins and Rogen moraine, and the consideration of all possibilities for their explanation. Successful resolution of the deforming bed/megaflood debate is clearly within the capabilities of glacial geologists.

11.2.7 Lee-side cavity fills

In Section 11.2.3, we saw how flutings can form beneath modern glaciers by the deformation of saturated till into cavities on the lee side of boulders and other obstructions. Sediments are also deposited in lee-side cavities on hard rock beds, but by different processes. On hard beds, cavities cannot close by bed deformation, and their size is determined by the velocity of the ice, the effective normal pressure, and the shape of the obstruction (Sections 3.2.4.1 and 3.4.4; Boulton, 1982). Such cavities can fill with sediment as debris falls from the roof or is washed in by meltwater (Fig. 5.26; Sections 5.5.3 and 10.5.2.3). The sediment assemblages produced in the lee of bedrock obstacles have been termed *lee-side tills* by Hillefors (1973). The accumulation of sediments in lee-side cavities tends to flatten out irregularities on the bed, producing landforms such as depositional

crag and tails (Fig. 11.26; Theakstone, 1967; Vivian, 1975; Boulton, 1979, 1982).

Sedimentological evidence for ancient lee-side tills in the Canadian Rocky Mountains has been provided by Levson and Rutter (1986, 1989a, b), who described massive diamictons with rare, steeply dipping sand and gravel lenses located on the lee side of bed obstructions. The clasts within the diamicton are striated and possess a strong fabric with a dominant downvalley plunge of a-axes, and the sand lenses also dip consistently in a downvalley direction. The sediments are truncated by an overlying lodgement and melt-out till sequence, suggesting that the glacier eventually contacted the lee-side till once the cavity was full.

11.2.8 Crevasse-fill ridges

The term *crevasse filling* was first used by Flint (1928) for sediments filling all types of supraglacial depression, and the idea that ridges could be formed by flowage or squeezing of subglacial till into basal crevasses was introduced by Gripp (1929). Basal crevasse-fill ridges have been reported from a number of recently deglaciated glacier forelands (Hoppe, 1952a; Okko, 1955; Mackay, 1960; Galon, 1973; Mickelson and Berkson, 1974; Haselton, 1979; Clarke et al., 1984a). Examples from Bruarjökull, a surging glacier in Iceland, are shown in Fig. 11.27. The most thorough study of crevasse-fill ridges was by Sharp (1985a), who examined recently exposed examples from the foreland of Eyjabakkajökull, an Icelandic surging glacier. The features are 1–2 m high till ridges which overlie fluted basal till. The ridges could be traced into the glacier snout, where they continued as crevasse traces, and often emerged on the wasting glacier surface as debris dykes. The ridges are arranged in a pattern that mimics the radial and transverse crevasse patterns on the glacier snout. The ridges have three different types of junction with

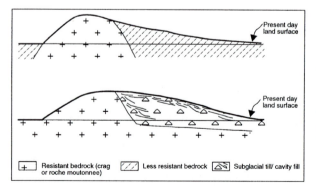

Fig. 11.26 Crag and tail forms: (a) erosional form in bedrock; (b) cavity fill form in lee of rock knob. Ice flow is from left to right

Fig. 11.27 Crevasse-squeeze ridges at the margin of Bruar-jökull, Iceland. These diamicton ridges were squeezed up into basal crevasses during the final stages of a surge of the glacier in 1963. Note how the ridges can be traced into the glacier snout, where they continue as crevasse traces. Although many of these forms appear to be random hummocks, they appear as linear forms on aerial photographs. (Photo: D.J.A. Evans)

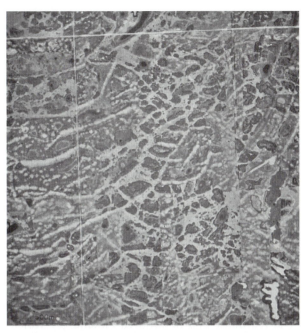

Fig. 11.28 Aerial photograph of drift ridges in Alberta, Canada, showing intersecting pattern. (Aerial photograph provided by Energy, Mines and Resources, Canada)

flutings on the till surface: (a) flutings exhibit strike-slip displacement along the line of ridges; (b) flutings pass undisturbed through ridges; and (c) flute crest-lines rise to intersect ridge crests. These relationships indicate that the ridges were formed after or simultaneously with the flutings. In the first case, flutings appear to have been displaced along the line of crevasses while still enveloped by basal ice; in the second, the ridge appears to have been superimposed on the fluting; and in the third, both fluting and ridge were apparently squeezed up into a crevasse. The till making up the ridges is massive, but clasts tend to dip at steep angles, suggesting that the surrounding matrix has been squeezed upward. Sharp (1985a) argued that the ridges formed at the termination of the 1972 surge of Eyjabakkajökull, when the heavily crevassed glacier snout stagnated and sank into its bed material. Basal till was squeezed up into basal crevasses because of the local differences in overburden pressure between open crevasses at atmospheric pressure and areas lying beneath glacier ice.

Since the work of Hoppe (1952a, 1957), several researchers have interpreted certain Pleistocene landforms as crevasse-fill ridges. However, for ancient landforms a crevasse-squeeze origin is seldom certain, and other interpretations are often possible. Gravenor and Kupsch (1959) described *linear disintegration ridges*, consisting of linear ridges oriented parallel, transverse and oblique to ice flow, from parts of western Canada. The ridges make intersecting 'waffle' or 'box' patterns reminiscent of crevasse patterns, leading to the interpretation of some as supraglacial and subglacial crevasse-fill ridges (Fig. 11.28). However, debris lodgement in zones of shear-

ing or thrusting, where active ice flows against stagnant ice near the margin, has also been suggested (Gwynne, 1942; Gravenor and Kupsch, 1959; Prest, 1983). In addition, some closely spaced ridges oriented parallel to outer terminal moraines were interpreted as annual push moraines (Section 11.3.2.1). Basal crevasse-fill ridges should be exposed only where the former supraglacial debris cover was very thin, otherwise the low-amplitude ridges would be buried during ice wastage. At some locations, possible crevasse-fill ridges do disappear beneath supraglacially derived landforms, especially where they lie adjacent to end moraines (Gravenor and Kupsch, 1959; Stewart *et al.*, 1988).

Kleman (1988) has described a series of till ridges from the southern Norwegian–Swedish mountains, which appear to be well-preserved crevasse-squeeze ridges. The ridges trace out rectilinear zigzag patterns, but in places resemble subglacial drainage networks in planform, and Kleman concluded that they were formed by squeezing of basal till into subglacial crevasses and meltwater conduits. The excellent preservation of the landforms was attributed to a transition to cold-based ice during deglaciation.

A crevasse-fill origin has been proposed by Stewart *et al.* (1988) for some Pleistocene corrugated ground moraine in Iowa, which consists of curvilinear 1–2 m high ridges separated by 1–2 m deep troughs or swales. These ridges are concave upglacier and are parallel to the outer terminal Bemis moraine and mimic the former retreating ice margin.

They were interpreted as push moraines by Gwynne (1942, 1951) and Foster and Palmquist (1969). The sediments within the Iowa ridges are distinctly different from those observed in Iceland by Sharp (1985a), in that large amounts of stratified sediment occur in some ridges. Furthermore, there is no evidence of sediment squeezing up into basal crevasses. A crevasse-fill origin was nonetheless preferred by Stewart *et al.* (1988), who envisaged thickening of basal till in some crevasses and glacifluvial deposition followed by melt-out till deposition in others.

During the 1950s and 1960s, several researchers argued that subglacial squeezing could account for large landforms in areas affected by continental ice sheets (e.g. Hoppe, 1952a; Stalker, 1960). Features referred to as *moraine plateaux* or *plains plateaux* in western Canada were interpreted by Stalker (1960) as the products of squeezing of subglacial till into basal ice cavities. Similar forms were ascribed to subglacial deformation both in northern Sweden (by Hoppe, 1952a, 1957, 1959), where they are called *Veiki moraines*, and in northern Finland (by Kujansuu, 1967, who names them *Pulju moraines*). These features are differentiated from kame plateaux (Section 11.4.3) by their composition of fine-grained tills. They were thought to document the advanced stages of ice stagnation when subglacial cavities were being excavated and basal tills were becoming saturated by the excess meltwater. The squeezing of subglacial material into cavities was also suggested by Hoppe (1948, 1959) to account for *Kalixpinnmo hills* in Norrbotten, Sweden, which consisted of elongate ridges of fine sand and silt oriented parallel to the former glacier margin. Hoppe argued that the ridges are subglacial accumulations of fluviglacial sediment which have been deformed by glacier flow.

The size and density of the landforms described by Hoppe and Stalker imply that, if they are subglacial ice-squeezed forms, the base of the ice must have been covered by large scallops, cavities and crevasses in the final stages of deglaciation. Although such a network of large cavities has been postulated by Shaw (1983a), Shaw and Kvill (1984) and Shaw *et al.* (1989), there is little evidence that these forms are characteristic of stagnating glaciers.

More recent research on the Veiki moraines of Sweden (Lagerback, 1988) and the hummocky terrain in Canada attributed by Stalker to basal squeezing (Tsui *et al.*, 1989) indicates that many areas are more likely to be overridden glacitectonized moraine fields, and therefore relate to active glaciation rather than stagnation (Section 11.3.1.3). This interpretation is especially attractive wherever hummocks are composed of folded and thrust-faulted bedrock. Unless additional evidence is forthcoming, a squeeze origin for large morainic landforms remains conjectural.

Squeezing of saturated sediment into transverse basal crevasses has also been proposed as a mode of origin for some De Geer moraines (Andrews, 1963; Mickelson and Berkson, 1974; Lundqvist, 1989c; Zilliacus, 1989). This idea is discussed in Section 11.6.2.2.

11.2.9 Eskers

The term *esker* is an English rendering of the Irish word *eiscir* meaning 'ridge', and refers to elongate, sinuous ridges of glacifluvial sand and gravel (Fig. 11.29; Warren and Ashley, 1994). They are known as *osar* (singular: *ose*) in Scandinavia. Eskers are the infillings of ice-walled river channels, and may record deposition in subglacial, englacial or supraglacial drainage networks. A general genetic classification of eskers proposed by Warren and Ashley (1994) recognizes the following basic types (Fig. 11.30): (a) *tunnel fills*, formed in englacial and subglacial conduits, and exposed by ice ablation; (b) *ice channel fills*, deposited in subaerial ice-walled channels; (c) *segmented tunnel fills*, formed during pulsed glacier retreat; and (d) *beaded eskers*, consisting of successive subaqueous fans deposited during pulsed retreat of a water-terminating glacier.

11.2.9.1 PLANFORM AND DISTRIBUTION

The planform of eskers is extremely variable, and can take the form of single continuous ridges of uniform cross-sectional profile, single ridges of variable height and width, single low ridges linking numerous mounds or beads, or complex braided systems of confluent and diffluent esker ridges (Auton, 1992; Fig. 11.31). Most eskers are aligned subparallel to the direction of former glacier flow, reflecting former

Fig. 11.29 An esker system in Lake Rorstromssjon, Angermandland, Sweden, illustrating bifurcations in the middle distance and the association between the esker and an elongate bedrock depression. (Photo: Erling Lindstrom)

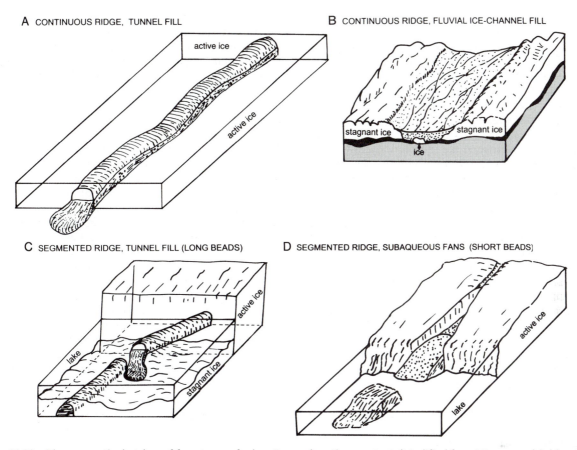

A CONTINUOUS RIDGE, TUNNEL FILL

B CONTINUOUS RIDGE, FLUVIAL ICE-CHANNEL FILL

C SEGMENTED RIDGE, TUNNEL FILL (LONG BEADS)

D SEGMENTED RIDGE, SUBAQUEOUS FANS (SHORT BEADS)

Fig. 11.30 Diagrammatic sketches of four types of esker. For explanation see text. (Modified from Warren and Ashley, 1994)

meltwater flow towards the ice margin. However, some eskers are aligned transverse to former glacier flow, a situation most common for former ice-walled channels guided by ice-cored moraines (Mannerfelt, 1945; Bird, 1967; Johansson, 1994).

Subglacial tunnel fills are the infillings of former Röthlisberger channels, and record the routeing of water at the glacier bed. Where water flow is in pressurized conduits, the hydraulic gradient and direction of flow are controlled by ice surface topography and, to a lesser extent, the form of the bed (Section 3.2.3). Therefore, eskers formed in these conditions can have up-and-down long profiles, with some sections climbing over topographic obstacles (Shreve, 1972). Studies of the distribution of Pleistocene eskers in Maine (Shreve, 1985a, b) and recent eskers exposed by the retreat of Burroughs Glacier, Alaska (Syverson *et al.*, 1994), have demonstrated good agreement between observed esker paths and calculated drainage pathways based on the hydraulic theory of Shreve (1972). Where tunnels are at atmospheric pressure, water flow is not influenced by the ice surface gradient, but follows the local slope just like subaerial streams, so that eskers deposited in such tunnels are aligned directly downslope. These were

Fig. 11.31 The Carstairs esker system, Lanarkshire, Scotland, showing both single ridges and braiding pattern of esker development. (Cambridge University Aerial Photograph Collection)

termed *valley eskers* by Bird (1967), and are typical of thin glacier margins where tunnel creep closure rates are low (Johansson, 1994; Syverson *et al.*, 1994). They are usually shorter and straighter than normal eskers, and are commonly associated with kame terraces or lateral drainage channels at their upper ends, and with kame and kettle topography at their lower ends (Sections 11.4.3 and 11.4.5; Sugden and John, 1976).

Shreve (1985a) related the morphology of tunnel-fill eskers to hydraulic conditions beneath the glacier, particularly the rates of melting and freezing of the tunnel walls. He argued that *multiple-crested eskers*, consisting of several subparallel branches, form where the active channel has a tendency to migrate laterally rather than incise itself into the overlying ice. Such conditions are thought to be typical of gently ascending reaches, where tunnel melting rates are low because of the downstream increase in elevation head (Section 3.2.3). Single-thread, *broad-crested eskers* are thought to be associated with zones of net freezing of the walls in steeply ascending reaches, where tunnels are likely to be wide, low and stable. Finally, *sharp-crested eskers* were attributed to zones of net melting, where the tunnel can melt its way into the overlying ice as the floor fills with sediment. However, sharp-crested eskers may also form by the lowering of englacial tunnel fills on to the land surface (Price, 1973).

The upstream or downstream ends of some subglacial tunnel-fill eskers are connected to Nye channels eroded into bedrock or sediment, possibly recording synchronous patterns of deposition and erosion at the glacier bed (Section 9.3.4.1). It is unlikely, however, that esker and channel systems extending over hundreds of kilometres were formed by a single tunnel system of that length. Rather, they were probably formed in segments during ice sheet retreat, when the marginal zone of ablation and its associated channel system migrated towards the former ice sheet centre. Clear evidence for temporal changes is provided by eskers lying within Nye channels, indicating channel bed erosion followed by deposition and tunnel infilling, probably as a result of falling discharges (e.g. Booth and Hallet, 1993). The presence of eskers within Nye channels implies that the subglacial conduit was a relatively stable feature of the glacier bed.

Tunnel-fill eskers in North Dakota have been associated with glacitectonically disturbed groundwater flow by Bluemle and Clayton (1984). Near the town of Anamoose an esker starts at the margins of the depression in a hill–hole pair (Section 11.3.1.1). The depression has been excavated in a buried valley filled with sand and gravel, which acted as a closed aquifer prior to glacitectonism. The elevated groundwater pressures produced by glacier overriding of aquifers are critical to most cases of glacitectonic thrusting (Section 7.4.3; Moran, 1971). The formation of the esker in this situation is explained by Bluemle and Clayton (1984) using the vivid analogy of popping a cork from a champagne bottle. The pressure in the bottle (the aquifer) was released when the cork (the glacitectonized hill) was removed by thrusting. This allowed groundwater to escape from the aquifer, thus explaining the initiation of an esker at the margins of the excavated hole.

Supraglacial and *englacial eskers* were first described by Crosby (1902) and Tanner (1932, 1934, 1937). Examples in the process of formation, and recently deposited ice-walled channel fills, have been described from many modern glacial environments (e.g. Lewis, 1949; Meier, 1951; Stokes, 1958; Petrie and Price, 1966; Price, 1966, 1969, 1973; Welch and Howarth, 1968; Gustavson and Boothroyd, 1987; Kirkbride and Spedding, 1996). In Alaska and Iceland, surveys have demonstrated that the melt-out of glacier ice has gradually lowered eskers on to the substrate (Price, 1973). The draping of such glacifluvial landforms on to high-relief substrates provides an alternative explanation for eskers that climb over uphill sections of the former glacier bed. Supraglacially and englacially derived eskers will contain heavily disturbed if not structureless sediments after deposition, in contrast with the generally well-preserved stratified sediments of subglacial tunnel fills (e.g. Brennand, 1994).

11.2.9.2 COMPOSITION

Eskers are composed of a wide variety of facies, ranging from sorted silts, sands, gravels and boulders to matrix-supported gravels. These materials are usually not far-travelled, most particles being no further than 15 km from their outcrop source (Hellaakoski, 1931; Trefethen and Trefethen, 1945; Lee, 1965). Individual beds may be ripple cross-laminated, horizontally bedded, cross-bedded or massive, depending on flow conditions during deposition (Section 10.4). Small-scale cross-bedded units in eskers represent the prograding avalanche faces of migrating bars (Shaw, 1972; Bannerjee and McDonald, 1975), whereas larger cross-bedded units are likely to be deltaic foresets deposited in subglacial or proglacial pools. Although the sedimentary structures observed in tunnel-fill eskers are similar to those found in open-channel fluvial deposits, there are some important distinctions relating to tunnel hydraulics. Bannerjee and McDonald (1975) pointed out that standing waves cannot form in full-pipe (conduit) flows, and that antidune bedforms cannot be deposited in such settings (Sections 3.8.3 and 10.4.5). Therefore, antidunes and occasional backset bedding are taken to indicate flow in former open ice-walled channels rather than tunnel deposition. It should be remembered, however, that subglacial conduits may not always be water-filled (Section 3.2.3.4). More recently, Brennand (1994) has identified very large antidunes in the eskers of south-central Ontario, Canada, which she explains as closed-conduit sediments because of the striking esker ridge continuity, low variability in palaeocurrent direction and up-and-down esker long profiles. According to Brennand, these large bedforms were deposited when hyperconcentrated floodwaters flowed from constricted to expanded reaches and therefore were transformed from a supercritical to a subcritical state. Saunderson (1977) argued that matrix-supported gravels within eskers document sliding-bed transport of hyperconcentrated flows, and that such conditions may be unique to full pipe flows. More recent work has shown that matrix-supported gravels can be deposited in subaerial and subaqueous flows in a wide range of environments (e.g. Maizels, 1989b; Todd, 1989), and cannot therefore be used as diagnostic criteria for pipe flows.

The central core of eskers can display cyclic sequences of gravel and sand (Bannerjee and McDonald, 1975; Ringrose, 1982), the base of each cyclic sequence often being marked by a scoured erosional surface (Shaw, 1972). These sequences display a fining upwards from cross-bedded and horizontally bedded, imbricated gravels to horizontally bedded, plane-bedded and trough cross-bedded sands (cf. Lobanov, 1967). Such fining-upward sequences with erosive bases record fluctuations in discharge and sediment availability. The depositional cycles or rhythmicity in esker sediments have been interpreted as annual by Bannerjee and McDonald (1975) and seasonal by Brennand (1994). Within the large cyclic sequences, cosets of climbing ripple drift or cross-lamination are believed to document anything between hours and tens of hours of deposition (Allen, 1971b). This corresponds well to the typical discharge fluctuations of glacial meltwater streams.

Other depositional trends in esker sediments include coarsening-upward sequences and fining outwards from the core (Shaw, 1972; Saunderson, 1975). The fining outwards usually involves a gradation from cross-bedded sand and gravel near the core to horizontally bedded and cross-laminated sand with laminated silt interbeds towards the margins (Fig. 11.32). This is thought to represent sidewall melting and esker widening, an interpretation supported by the occurrence of faulting in the marginal zone. Saunderson (1975) suggested that only the gravel core is the esker proper; sediment draping of esker cores and esker–deltaic deposition are discussed more fully below.

In all situations, some slumping will occur at the margins of an esker ridge once the ice walls are

Fig. 11.32 A coarse-grained gravel and sand esker core with marginal drapes of ripple-bedded and cross-laminated sands and silts with minor gravel beds, Rooskagh esker, Ireland. (Photo: D.J.A. Evans)

removed by melting. The degree of internal disturbance by folding and faulting relates to the thickness of underlying ice or the magnitude of side slumping, and the most heavily disturbed esker sediments will be produced wherever the tunnel was located englacially or supraglacially (Price, 1973). The sediment sequences within eskers commonly form anticlinal structures or arched bedding when viewed in cross-section. Anticlines are interpreted as the product of slumping at the esker sides after removal of the supporting ice walls, and arched bedding as the result of simultaneous deposition on the crest ridge and flanks of the esker (Grano, 1958; Shreve, 1972). As tunnels collapse or change shape, or streams change position or size, one depositional sequence may be truncated and partially infilled or overlain by another (Terwindt and Augustinus, 1985).

11.2.9.3 SEGMENTED AND BEADED ESKERS

Segmented and beaded eskers are generally interpreted as time-transgressive landform assemblages deposited close to glacier margins during deglaciation. The distal ends of segments may mark the positions of ice-marginal portals during temporary glacier stillstands. Larger beads represent substantial depocentres at the efflux points of subglacial tunnels, either in subaerial or subaqueous environments (e.g. Bannerjee and McDonald, 1975; Rust and Romanelli, 1975; Thomas, 1984a, b; Warren and Ashley, 1994). In addition, Gorrell and Shaw (1991) and Brennand (1994) have described complex associations of esker ridges, beads and fans in Ontario, Canada, which they attributed to deposition in conduits, lateral water-filled cavities beneath the ice, and grounding-line fans beneath a floating glacier tongue (Fig. 11.33). Wherever subglacial meltwater exits from the glacier into water, a subaqueous fan or ice-contact delta will be produced, depending on water depth (Section 11.6.2). For example, Sharpe (1988a) has described glacimarine subaqueous fans on the floor of the former Champlain Sea Basin and on Victoria Island, Canada, which emanate from eskers and record the rapid deposition of sediment beyond the confines of an ice-walled conduit. Exposures in these landforms show that esker cores continue beneath the fan forms and are overlain or draped by turbidites, sediment flows and rhythmites (Fig. 11.34). This partial burying of esker forms by subaqueous fan sediments probably attests to a combination of high sediment loads and slow glacier retreat rates. Detailed sedimentological descriptions of Pleistocene fan deltas emanating from eskers have been given by Martini (1990). The distal parts of the fan deltas record deposition from cohesionless debris flows and turbulent underflows in standing water, but the proximal and upper parts are very variable,

depending on hydraulic conditions and relative water level.

Gorrell and Shaw (1991) have described in detail the sedimentology of a complex assemblage of esker ridges, beads and fans in Ontario, Canada. The northern (proximal) part of the complex consists of a single esker ridge, *c.* 8 m high and 70–139 m wide. This passes distally into a ridge and bead complex, in which lateral beads are connected to the main esker ridge by minor ridges (Fig. 11.33a). The distal part of the complex consists of a series of overlapping fans which emanate from anastomosing esker ridges. The fans were deposited in standing water, either the Champlain Sea or a proglacial ice-dammed lake. The main esker ridge has a core of gravels, which record deposition from bedload and hyperconcentrated flows in a subglacial conduit. The lateral beads in the central part of the complex are composed of stacked successions of plane-bedded, cross-bedded and mas-

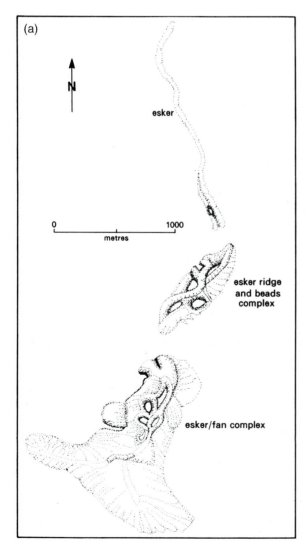

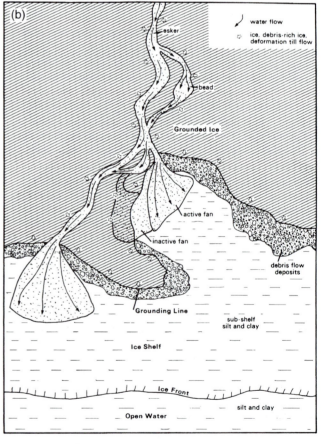

Fig. 11.33

(c)

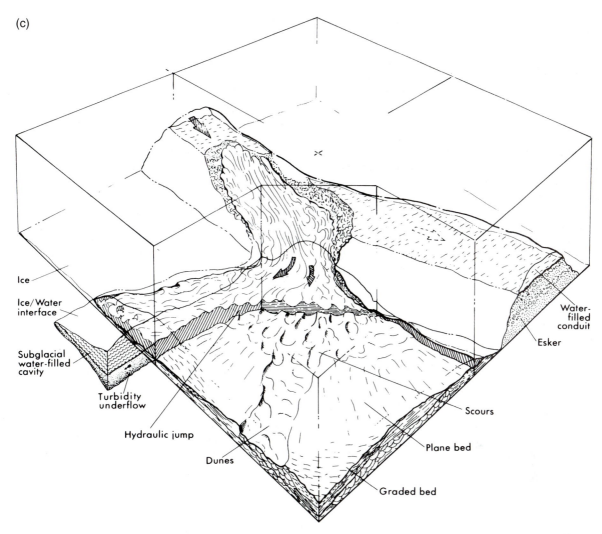

Fig. 11.33 The esker, bead and fan complex at Lanark, Ontario, Canada: (a) morphology of the esker, bead and fan complex; (b) reconstruction of the depositional environment in which esker, bead and fan complexes may be produced; (c) reconstruction of the depositional environment in which lateral fans are deposited. (From Gorrell and Shaw, 1991. Reprinted by permission of Elsevier)

sive sand, which display a pronounced proximal-to-distal fining away from the main ridge. In the esker–fan complex, a number of secondary ridges diverge from the main esker ridge. Two of these ridges contain stacked sand beds which display multiple fining-upwards sequences, each of which represents a distinct pulse of sediment. Gorrell and Shaw interpreted the lateral beads and secondary ridges as the infills of subglacial cavities and distributary channels which 'soaked up' water and sediment at times of high water pressure in the main conduit. The cavities and distributaries are thought to have been 'pinched off' from the main conduit at times of low flow. The opening and closing of the connections was likened to the action of a valve which allowed water to escape laterally when water pressures in the main

conduit exceeded the local ice overburden pressure. Each pulse of sediment is thought to represent an opening event, when water debouched from the main conduit. According to Gorrell and Shaw, this implies that the whole glacier snout was delicately poised close to the flotation point. However, this is not necessarily so. The 'valves' may be analogous to orifices in linked-cavity networks, which open up at times of high discharge and water pressure, and close as discharge falls (Section 3.4.4). Orifices open when the basal water pressure locally exceeds ice overburden (i.e. where effective pressure falls to zero), but overburden pressures can remain high over most of the bed. The esker complex may represent a drainage system which locally switched between single-thread channelized flow and distributed flow, but was not

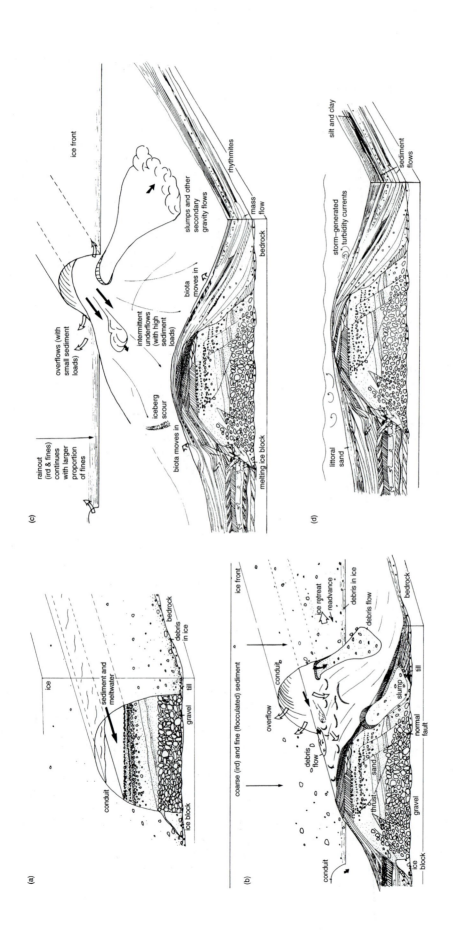

Fig. 11.34 A model to explain the deposition of an esker and subaqueous glacimarine fan in the late Quaternary Champlain Sea near Ottawa and Montreal, Canada: (a) tunnel fill (esker) sedimentation; (b) ice-marginal retreat and removal of support for tunnel fill sediment causes normal faulting and mass flows; (c) further ice retreat and the draping of the esker/fan core with fine-grained sediments; (d) marine regression and littoral reworking of esker/fan sediments. (Modified from Sharpe, 1988a)

necessarily associated with widespread near-flotation and very low average effective pressures. The map in Fig. 11.33b shows the ice margin at the time of esker formation as a floating terminus located some distance away from the efflux point. However, Gorrell and Shaw presented no firm evidence that the margin was floating, and the ice may have terminated at the grounding line.

Despite these criticisms, Gorrell and Shaw's study is an excellent example of detailed sedimentological analysis, and raises many interesting questions about esker genesis. Further discussion of the valve mechanism and additional sedimentological data from esker systems in Ottawa have been given by Brennand (1994).

The relationship between eskers and ice-contact fans of glacifluvial sediment in a terrestrial setting in Skåne, Sweden, has been described by Hebrand and Åmark (1989). In this area, eskers are composed of interlocking fans, each consisting of ridge, hummock and terrace components, composed of downstream-fining sequences of gravel and sand (Fig. 11.35). Each sequence documents deposition in a restricted, debris-covered and largely stagnant outer margin of the ice sheet. The progressive recession of the active ice and the development of new stagnant ice led to the accretion of new fans in an up-ice direction. The downstream transition from ridges to hummocks and then to terrace remnants within each individual fan records deposition first in tunnels and then as a series of unrestricted distributaries on the downwasting ice surface.

11.2.10 Subglacial volcanic forms (tuyas or Stapis)

Tuyas are flat-topped volcanoes built during subglacial eruptions (Mathews, 1947; van Bemmelen and Rutten, 1955; Jones, 1969; Björnsson, 1975). The initiation of a tuya is shown in Fig. 11.36a, which shows how lava erupts directly into a subglacial water reservoir. As the reservoir expands it is filled with pillow lavas, hyalotuffs, hyaloclastites, conglomerates and compound lavas to form a steep-sided and flat-topped rock mass whose size and shape reflect the size of the reservoir. Eruption may cease before the tuya and associated cavity reach the surface of the glacier, or the tuya may develop into a Surtseyan or subaqueous-type volcano occupying a glacial lake (Fig. 11.36b–d). Smellie and Skilling (1994) reported tuyas of this type from Antarctica. Smellie and Skilling also reported evidence that volcanically heated meltwater and lavas exploited and enlarged subglacial tunnels. As a result, the volcanic sediments possess a stratigraphic architecture similar to that of eskers (Fig. 11.37b).

11.3 ICE-MARGINAL MORAINES

The present and former margins of glaciers and ice sheets are demarcated by a wide variety of depositional features produced by the complex interactions of numerous glacigenic and paraglacial processes. In this section, we describe ice-marginal moraines under four headings: (a) glacitectonic landforms; (b) push and squeeze moraines; (c) dump moraines; and (d) latero-frontal fans and ramps. These moraine types are formed by the deposition of sediment around the edge of active glacier snouts, or by glacially induced stresses. It must be emphasized that the four moraine types discussed in this section are not distinct, exclusive categories, and that in nature composite moraines are not uncommon. Moraines that originate supraglacially or englacially and are lowered on to the ground surface by glacier ablation are described in Section 11.4, and ice-marginal moraines deposited in subaqueous settings are discussed in Section 11.6.

Some general terms are applied to many moraine types regardless of genesis. The outermost moraine ridge formed at the limit of a glacier advance is known as a *terminal moraine*. Younger moraines nested within a terminal moraine are termed *recessional moraines*, because they are formed during overall glacier recession even though they may have been deposited or pushed up during minor readvances or standstills. Terminal and recessional moraines may be subdivided into *frontal* and *lateral* components, or described as *latero-frontal moraines*. Where it can be demonstrated that recessional moraines formed on a yearly basis, they are known as *annual moraines*.

11.3.1 Proglacial glacitectonic landforms

The idea that glaciers can deform bedrock and Quaternary sediment into large thrust moraines is almost as old as glacial theory itself, and many early researchers identified glacitectonic landforms in a wide range of locations including Sweden, Denmark, Germany, the New England offshore islands, Minnesota, Poland, Alberta and Spitsbergen (Torell, 1872, 1873; Erdmann, 1873; Johnstrup, 1874; Merrill, 1886a, b; Sardeson, 1905, 1906; Hopkins, 1923; Lewinski and Rozycki, 1929; Gripp, 1929). The most prolific early researcher on glacitectonic landforms was G. Slater, who published papers on features in England (Slater, 1927a, b), Denmark (Slater, 1927c, d), Canada (Slater, 1927e), the USA (Slater, 1929) and the Isle of Man (Slater, 1931). During the middle part of the twentieth century, considerable interest was focused on glacitectonic landforms in Denmark

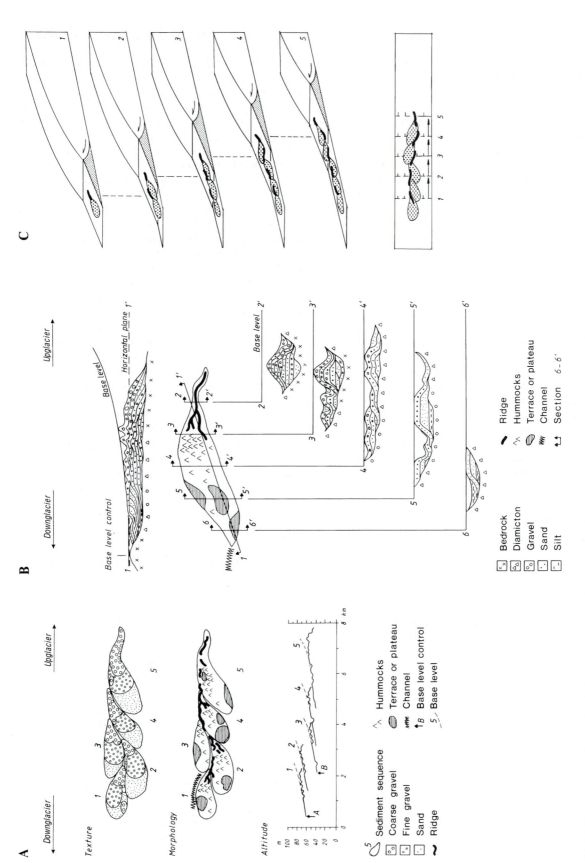

Fig. 11.35 Conceptual models of internal sediment sequences and evolution of the eskers of Skåne, Sweden: (A) the morphology, grain size distribution and arrangement of sediment sequences; (B) the internal composition of sediment sequences; (c) the formation of eskers in a zone of stagnant debris-covered ice (shaded) fringing active receding ice. (From Hebrand and Åmark, 1989. Reproduced by permission of Scandinavian University Press)

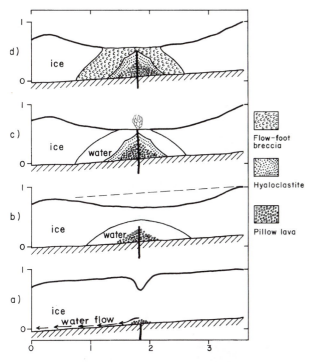

Fig. 11.36 The sequential development (from bottom to top) of a tuya. (From Bjornsson, 1975. Reprinted by permission of Jökull)

(Jessen, 1931; Gry, 1940), the Netherlands (de Jong, 1952; Maarleveld, 1953; Rutten, 1960), Poland (Jahn, 1956; Kozarski, 1959; Dylik, 1961; Galon, 1961), and Canada (Byers, 1959; Mackay, 1959; Kupsch, 1962; Christiansen, 1961) and since that time research on all aspects of glacitectonics has blossomed into a thriving subdiscipline of glacial geology and geomorphology.

Excellent coverage of the history of glacitectonic study is available in Aber *et al.* (1989). Numerous examples of modern studies together with compre-

hensive bibliographies on the subject of glacitectonics are available in van der Meer (1987), Croot (1988a), Aber (1993a) and Warren and Croot (1994). These volumes constitute the reports from workshops and conferences organized by the INQUA Work Group on Glacial Tectonics, which was established in 1982 as a recognition of the significance of glacitectonic processes and products.

In this book, glacitectonic sediment–landform associations are classified using the scheme proposed by Aber *et al.* (1989). Four basic types are recognized: (a) hill–hole pairs; (b) composite ridges and thrust-block moraines; (c) cupola hills; and (d) megablocks and rafts (Table 11.1). Glacitectonic moraine ridges are defined as those in which glacitectonized pretectonic and syntectonic sediments constitute >25 per cent of the unit area of the moraine. Other types of moraine, such as push moraines, may contain small amounts of glacitectonized sediment (Section 11.3.2.1). Glacitectonic landforms may be composed of pre-Quaternary bedrock, pre-existing Quaternary sediments or contemporaneous sediment (Moran, 1971; Andrews, 1980), and where they have been overridden by glacier ice after construction they may be covered by a carapace of glacitectonite or till (Section 10.3.3). More prolonged and active glacier overriding will result in the streamlining and moulding of glacitectonic features to produce subglacial bedforms (Section 11.2.4). The principles of proglacial glacitectonics are explained in Section 7.4.3.

11.3.1.1 HILL–HOLE PAIRS

A hill–hole pair was defined by Bluemle and Clayton (1984) as 'a discrete hill of ice-thrust material, often slightly crumpled, situated a short distance downglacier from a depression of similar size and shape'. Ice-thrust hills may be found at distances of

Table 11.1 Characteristics of glacitectonic landforms

Landform	Height (m)	Area (km²)	Primary material	Primary morphology
Large composite ridge	100 to 200	20 to >100	Bedrock	Subparallel ridge and valley system, arcuate in plan
Hill–hole pair	20 to 200	<1 to >100	Variable	Ridged hill associated with source depression
Small composite ridge	20 to <100	1 to >100	Quaternary strata	Subparallel ridge and valley system, arcuate in plan
Cupola hill	20 to >100	1 to 100	Variable	Smoothed dome to elongate drumlin with till cover
Megablock/raft	0 to <30	<1 to 1000	Bedrock	Often concealed, flat buttes or irregular hills

Source: Aber *et al.* (1989)

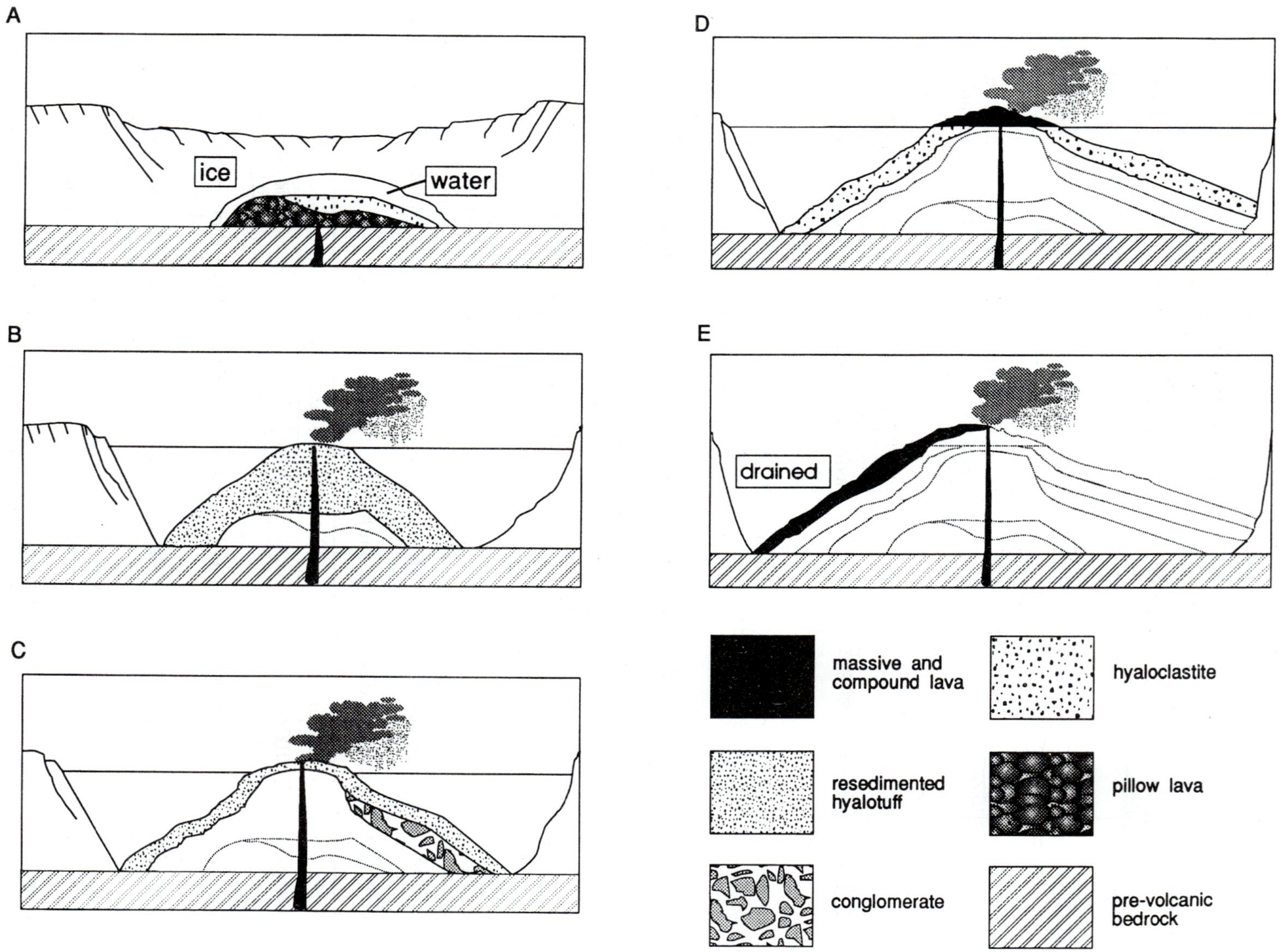

A

ice

water

B

C

D

E

drained

hyaloclastite

pillow lava

pre-volcanic
bedrock

massive and
compound lava

resedimented
hyalotuff

conglomerate

Fig. 11.37(a)

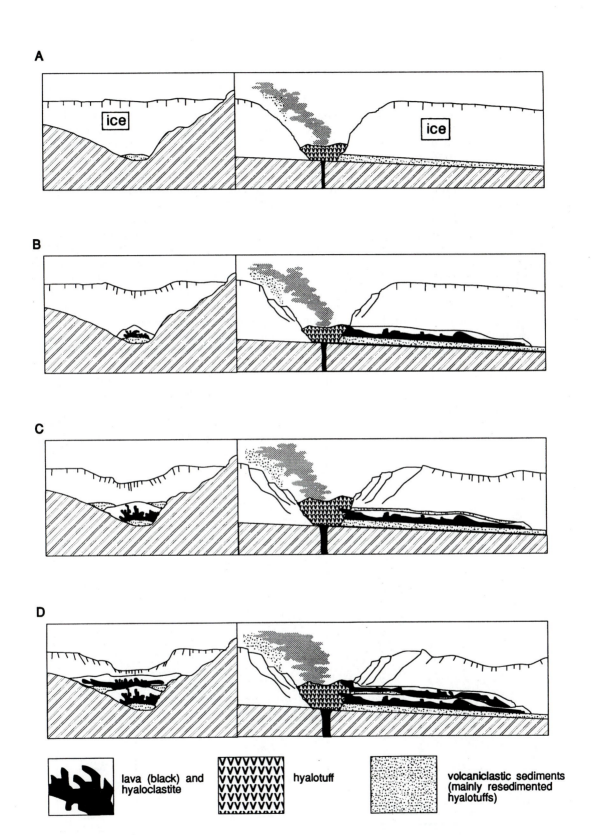

(b)

Fig. 11.37 Alternative interpretations of subglacial volcanic landform (tuya) production: (a) cone building stage A–D is followed by a cone-extending or tuya stage E; (b) flushing of volcanically heated meltwater and vitric tephra in a downvalley direction beneath the capping glacier (A and C) is followed by lava effusion (B and D). Note that the left and right sides of the figure are cross- and down-valley sections respectively. (From Smellie & Skilling, 1994. Reproduced by permission of Elsevier)

lava (black) and hyaloclastite

hyalotuff

volcaniclastic sediments (mainly resedimented hyalotuffs)

A

ice   ice

B

C

D

up to 5 km from source depressions, but are usually closer (Bluemle, 1970; Carlson and Freers, 1975; Clayton *et al.*, 1980; Bluemle and Clayton, 1984). However, source depressions can become infilled with younger sediment, and are not always visible as surface features. Conversely, glacitectonic depressions are sometimes found without associated hills (Ruszczynska-Szenajch, 1976, 1978; Andriashek and Fenton, 1988), perhaps because positive relief features have been destroyed by subglacial erosion.

The typical morphology of a hill–hole pair is as follows (Aber *et al.*, 1989): (a) the hill has an arcuate or crescentic planform, which is concave upglacier; (b) the surface of the hill is traversed by a series of transverse, subparallel ridges and depressions; (c) the hill has an asymmetric cross-profile, with the highest point and steeper slopes on the convex or downglacier side; and (d) the topographic depression is approximately the same shape and area as the hill, located on the concave or upglacier side. These morphologic features may be less obvious where prolonged subglacial modification has taken place after formation.

Many of these features can be identified on the Wolf Lake hill–hole pair, Alberta, Canada (Fig. 11.38). The parallel-sided margins of Wolf Lake have been interpreted as strike-slip tear faults, marking the lateral boundaries of the thrust mass (Andriashek and Fenton, 1988). Thrusting is thought to have occurred in association with a small sublobe of the ice margin,

with the basin excavated beneath the margin, and the hill thrust up in a proglacial position (cf. Stephan, 1985). The Wolf Lake example and other hill–hole pairs in the North American Great Plains do not appear to have been greatly modified by subglacial processes, and are thought to have been formed in a 2–3 km wide frozen-bed zone of the Laurentide ice sheet during standstill or minor readvances that punctuated overall retreat (Moran *et al.*, 1980).

Another excellent example of a hill-hole pair, Herschel Island in Mackenzie Bay, Canada, is one of the largest so far recognized on the Earth's surface, with an area $> 100 \text{ km}^2$ (Mackay, 1959; Rampton, 1982). Composed of preglacial Pleistocene sediments thrust up from a now submerged source depression, Herschel Island is arcuate in shape with an upglacier concave face and an asymmetrical profile. Trellised stream patterns on the island record postglacial fluvial erosion along fracture zones (Aber *et al.*, 1989). Sea cliffs reveal internal structures such as low-angle thrust faults, synclines and anticlines, tilted beds which occasionally approach subvertical angles, overturned folds, repeated strata, slickensides, and segregated ice lenses which originally formed horizontally but now parallel the deformed enclosing strata (Mackay, 1959; Mackay and Stager, 1966; Mackay *et al.*, 1972). The morphology and internal structures of Herschel Island indicate deformation from the south-east, which is consistent with other

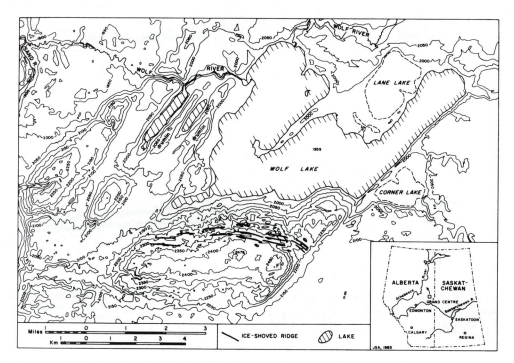

Fig. 11.38 A topographic map of the Wolf Lake hill–hole pair and associated drumlins, Alberta, Canada. Elevations in feet, contour intervals at 50 ft, perennial lakes shown by diagonal shading and ephemeral lakes shown by dashed outlines. (From Aber *et al.*, 1989. Reprinted by permission of Kluwer)

glacial geomorphological evidence of a glacier lobe moving along the Yukon coastal plain from that direction during the last (Buckland) glaciation (Rampton, 1982). The enclosed ice lenses indicate that the Herschel Island sediments were permafrozen before being glacitectonized.

Landforms in Alberta, Canada, which may be subglacially modified hill–hole pairs have been described by Stalker (1973a), and termed *murdlins* because they resemble inverted drumlins. Murdlins consist of elongate arcuate ridges with their open ends facing upglacier and central troughs terminating at mounds on the downglacier loops. The lateral ridges and downglacier mound of each murdlin are derived from the central trough. Stalker (1973a) envisaged erosion of the trough by narrow tongues of active ice pushing through belts of dead ice, but since he provided no evidence for this interpretation, murdlin genesis is still open for further discussion. It is interesting to note that the forms of murdlins are similar to the fold traces reconstructed from the northern Saskatchewan drumlin field by Boulton (1987) (Section 11.2.4.4), suggesting that they may belong to the subglacial bedform continuum. This does not necessarily preclude a glacitectonic origin for such features, because the cores of streamlined terrain may contain intact pre-streamlining structures. Thus, hill–hole pairs could be the initiating points for certain streamlined subglacial bedforms.

The recognition of hill–hole pairs has proved critical to safe and efficient coal-mining operations in Alberta, Canada. Glacitectonic thrusting has removed coal from near-surface seams to produce depressions of up to 1 km^2 in area, which constitute gaps in formerly continuous strata. Problems encountered in open-pit mining include large-scale failures on the benches used for positioning heavy mining machinery. These failures are common wherever the benches are cut in the glacitectonically transported materials, which are heavily fractured and cut by numerous shear planes (Fenton *et al.*, 1986; Aber *et al.*, 1989).

11.3.1.2 COMPOSITE RIDGES AND THRUST-BLOCK MORAINES

The most common and distinctive types of glacitectonic landforms are *composite ridges* and *thrust-block moraines*. These forms have been termed *push moraines* by some researchers (e.g. Boulton, 1986; Etzelmüller *et al.*, 1996), but we reserve this term for moraines formed by bulldozing, or push-from-behind, rather than glacitectonic processes, which excavate and elevate proglacial materials (Section 11.3.2.1). Composite ridges are composed of multiple slices of up-thrust and contorted bedrock and/or

unconsolidated sediments, which are commonly interlayered with and overlain by glacigenic and glacifluvial sediment (Prest, 1983; Aber *et al.*, 1989). They are subdivided by Aber *et al.* (1989) into large (> 100 m of relief) and small (< 100 m of relief) composite ridges. This size differentiation is essentially arbitrary, and the processes involved in the production of both large and small composite ridges are the same. However, Aber *et al.* (1989) suggested that large ridges usually contain a considerable volume of pre-Quaternary bedrock, whereas small ridges are composed mostly of unconsolidated Quaternary strata. The term *thrust-block moraine* was introduced by D.J.A. Evans and England (1991) to describe composite ridges comprising proglacially thrust masses of permafrozen blocks of Quaternary sediments at the margins of Arctic subpolar glaciers. Composite ridges are associated with proglacial or submarginal glacitectonics, and they mark the positions of glacier standstill or readvances.

In planform, composite ridges comprise arcuate suites of subparallel ridges and intervening depressions, arranged in a concave upglacier pattern and conforming to the general shape of the glacier margin that produced them (Fig. 11.39). Individual ridge

Fig. 11.39 Aerial photo of composite ridges at the margin of Rabotsbreen, Svalbard. Note the concentric transverse ridge crests, and deep meltwater channels incised through the distal slopes. (Norsk Polarinstittut Photo No. 4W4/300)

crests correspond to the crests of internal folds, the upturned ends or edges of thrust blocks or resistant upturned beds. The depth of disturbance associated with composite ridges is generally a few tens of metres, and may be up to 200 m for the very largest features (Kupsch, 1962; Aber *et al.*, 1989). The internal structures of large composite ridges are very similar to the products of thin-skinned tectonics in thrust and folded mountain belts like the Canadian Rockies, even though composite ridges are at least one order of magnitude smaller than true mountain ranges (Aber *et al.*, 1989). The material in composite ridges commonly forms imbricately stacked sheets, with individual thrust slices showing varying degrees of internal deformation. Overturned and rootless folds of bedrock and Quaternary sediments in the composite ridges at Flade Klint, Denmark, have been thrust up to 100 m from their probable source depression (Gry, 1940).

Probably the most spectacular exposures through a large composite ridge are located at Møns Klint in Denmark. Several ice advances during the last glaciation have piled up numerous chalk scales with intervening Quaternary sediment to produce a composite ridge characterized by > 150 m of structural relief (Fig. 11.40; Johnstrup, 1874; Haarsted, 1956; Aber, 1985a). The structures within the cliffs at Møns Klint include imbricately thrust anticlines and individually folded and stacked chalk floes, which can be traced as a series of ridges on the rugged highland of Høje Møn.

Very extensive composite ridges occur in the area of the Missouri Coteau, a hilly upland belt stretching from North Dakota, USA, to southern Alberta and Saskatchewan, Canada (Slater, 1927e; Fraser *et al.*, 1935; Byers, 1959; Kupsch, 1962). The boundary of the Missouri Coteau in Canada is demarcated in places by a prominent north-east-facing escarpment

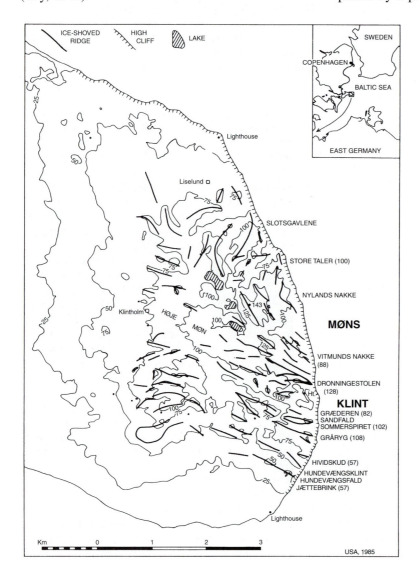

(a)

Fig. 11.40

(b)

Fig. 11.40 The composite ridge of Møns Klint, Denmark. (a) Topographic map of eastern Møn showing the ice-shoved ridges, lake basins and chalk cliff names and heights in metres (modified from Aber, 1985). Former ice flow direction was from the northeast. (b) Aerial photograph of Møns Klint and Høje Møn, showing how the chalk masses form high cliffs and continue inland as sharp-crested ridges. The more rugged area of ridges is covered in trees. Chalk masses are visible in the fields as light tones. The scale bar represents 0.5 km. (Aerial photograph reproduced by permission of the Geodotisk Institut, Denmark)

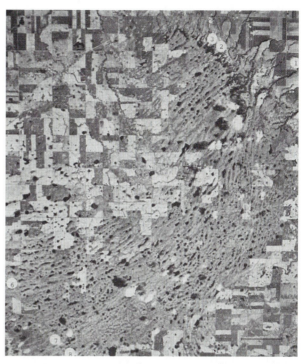

Fig. 11.41 Aerial photograph of the eastern and northern Dirt Hills composite ridges, Saskatchewan, Canada, showing the arcuate sub-parallel ridges and depressions arranged in a concave upglacier pattern. The scale bar represents 2 km. (Aerial photograph reproduced by permission of the Department of Energy, Mines and Resources, Canada)

marking the boundary between the Saskatchewan Plain and the Alberta Plain, which is underlain by more resistant rocks. During the overall retreat of the Laurentide ice sheet, readvances by ice lobes penetrated embayments in the escarpment, and led to the folding and thrusting of bedrock scales in an imbricate pattern to produce large composite ridges such as the Dirt Hills and Cactus Hills (Fig. 11.41). The Dirt Hills and Cactus Hills rise up to 150 m above the Missouri Coteau and cover an area of approximately 1000 km^2 (Christiansen, 1961; Parizek, 1964; Aber, 1993b). Impressive amounts of horizontal and vertical displacement are recorded by these composite ridges; some formerly horizontal Cretaceous strata are now tilted vertically and are located 200 m above their normal stratigraphic position (Christiansen, 1971a; Christiansen and Whitaker, 1976; Aber, 1993b). As in the case of Møns Klint, the trends of the individual ridges on the Dirt and Cactus Hills trace the underlying glacitectonic structures in the displaced bedrock. Although the bedrock is thought

to have been in a thawed state during thrusting (Aber, 1993b), the basal décollement for the Dirt and Cactus Hills was in mudstones which may remain unfrozen even in permafrost (Mathews and Mackay, 1960; Mackay and Mathews, 1964). The interruption of aquifer drainage by an advancing glacier lobe is thought to have been responsible for the production of Prophets Mountains, North Dakota, another large composite ridge located on the Missouri Coteau (Bluemle, 1981; Bluemle and Clayton, 1984). Natural groundwater flow is towards the east, but during ice advance from that direction the flow was reversed, leading to increased porewater pressures which facilitated glacitectonic disturbance (Section 7.4.3). Other examples of large composite ridges are the Bride Moraine on the Isle of Man (Thomas, 1984c), the Cromer Ridge in Norfolk, England (Hart, 1990), the Weichselian Main Stationary Line in Denmark (Pedersen *et al.*, 1988), the Dammer Berge in north Germany (van der Wateren, 1987; van Gijssel, 1987) and the North Antrim end moraine (Shaw and Carter, 1980).

There are numerous examples of small composite ridges, both in front of modern glacier snouts and in the ancient landform record. Spectacular composite ridges composed of outwash and glacimarine sediments form continuous moraines around the snouts of many Spitsbergen glaciers (Fig. 11.39), and have been the subject of research for the past hundred years (Garwood and Gregory, 1898; Lamplugh, 1911; Gripp and Todtmann, 1925; Gripp, 1929; Croot, 1988b). In Spitsbergen and Iceland, composite ridges are strongly associated with surging glaciers, suggesting that conditions during surges are particularly conducive to proglacial tectonics (Sharp, 1985b; Croot, 1987, 1988b, c; Drozdowski, 1987). The most important factors underlying this association are (a) rapid advance of the snout; (b) extreme compressive deformation at the margin; and (c) elevated porewater pressures associated with the release of meltwater. The role of porewater pressures is highlighted by the general absence of composite ridges where proglacial areas slope steeply away from the

glacier, and water is able to drain away freely (Croot, 1988c).

The internal structures of recent composite ridges at Eyjabakkajökull, an Icelandic surging glacier, have been studied by Croot (1988b) (Fig. 11.42). The principal subglacial structures are low-angle thrust faults dipping downglacier. Transport of thrust slices over these faults resulted in overall extension. Proglacial structures record compression, and include stacked thrust sheets composed of outwash, tephra, turf, organic soil and peat beds, which form asymmetric ridges. In the more distal parts of the moraine, smaller asymmetric ridges are the surface expression of anticlines or overturned anticlines.

Thrust-block moraines produced by proglacial thrusting of blocks of frozen outwash and raised marine sediments have been documented in the

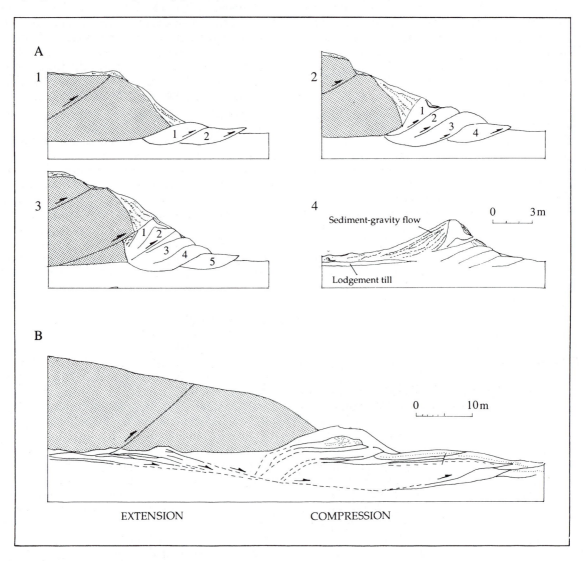

Fig. 11.42 Examples of thrust moraine formation from Iceland: (A) Hofdabrekkujökull (after Humlum, 1985b); (B) Eyjabakkajökull, after Croot, 1988b)

Canadian Arctic (Kalin, 1971; Klassen, 1982; Evans, 1989a; D.J.A. Evans and England, 1991). In most examples, the bedding within individual blocks and block surfaces dip gently towards the glacier, suggesting that the moraines are composed of either imbricately stacked scales or partially rotated deep-seated blocks (Kalin, 1971). In some cases, however, bedding may dip away from the glacier snout, suggesting deep-seated wedging of the advancing snout beneath proglacial sediments (D.J.A. Evans and England, 1991). The Canadian Arctic thrust-block moraines occur in areas where permafrost is often in excess of 700 m thick, but they are located in valley bottoms either below the marine limit or below former proglacial lake shorelines relating to the last glaciation. This suggests that permafrost may have been partially degraded below deep water bodies prior to the Holocene advances responsible for thrusting. Therefore, block failure could have been initiated along an unfrozen *talik* between newly aggraded postglacial surface permafrost and the degraded top of the full-glacial permafrost. A similar scenario may also apply to the Herschel Island hill–hole pair, where sea-level had dropped prior to the Buckland glacial advance, allowing permafrost aggradation to an unknown but perhaps shallow depth (Hill *et al.*, 1985). Etzelmüller *et al.* (1996) have noted that thrust-block moraines in Svalbard, which is within the region of continuous permafrost, also occur only below the marine limit. They argued that the presence of salt plays an important role in maintaining liquid water in permafrozen glacimarine sediments, weakening them sufficiently to allow large-scale thrusting. Thrust-block moraines occur at the margins of cold-based glaciers in the Dry Valleys area, Antarctica, where they tend to be composed of

lacustrine sediments. Fitzsimons (1996) has argued that thrusting takes place when cold glaciers override weak, unfrozen lake-floor sediments. Unfrozen areas occur below either saline lakes or wet-based lakes which have a permanent ice cover but do not undergo full-depth freezing. As the glacier enters the lake, sediment blocks become frozen on to the base of the glacier, but can be thrust forward because their deeper layers remain unfrozen. In summary, therefore, the formation of thrust-block moraines in permafrost regions is thought to occur only within areas where unfrozen or only partially frozen sediments occur.

Ancient examples of small composite ridges include the Brandon Hills in Manitoba, Canada (Aber, 1988b; Aber *et al.*, 1989), and the Utrecht Ridge and associated hills and basins in the central Netherlands (Fig. 11.43; van der Wateren, 1981, 1985; van den Berg and Beets, 1987; de Gans *et al.*, 1987). Both examples are composed predominantly of glacifluvial sands and gravels. On the basis of structural evidence from the Utrecht Ridge, van der Wateren (1985) estimated that the imbricately stacked and gently folded thrust blocks have been pushed up at least 100 m from the basal décollement. The thrust blocks strike parallel to the ridge crest and dip at 35–40° towards the former ice margin. The composite ridges of the central Netherlands have probably been excavated from the glacial basins that are now buried beneath the post-Saalian sediment cover, but as the basins were also partially excavated by subglacial meltwater they do not represent classic hill–hole pairs (van den Berg and Beets, 1987; de Gans *et al.*, 1987). Small composite ridges were constructed at the margins of outlet glaciers in Scotland during the Younger Dryas Stadial (Section 1.7.2.4).

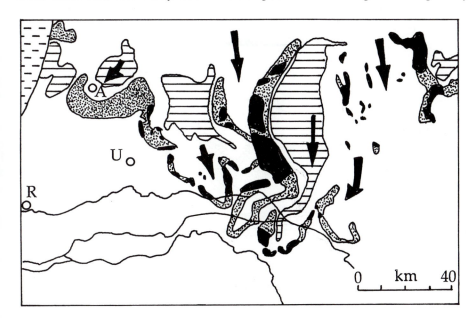

Fig. **11.43** Composite glacitectonic ridges and excavated basins in the central Netherlands. Surface ridges are shown in black, buried ridges in stipple, and basins in horizontal hatching. Arrows represent former ice flow directions. The ridges form part of an extensive Saalian thrust moraine system stretching from the Netherlands across northern Germany. A: Amsterdam; U: Utrecht; R: Rotterdam. (Modified from Aber *et al.*, 1989, after van den Berg and Beets, 1987)

Proglacially thrust blocks of raised marine sediments demarcate the former margins of glacier lobes at the Lake of Menteith in the Forth valley (Price, 1983; D.E. Smith, 1993), which may also consitute a hill–hole pair, and at Loch Don on the Isle of Mull (Benn and Evans, 1993).

The influence of ice-marginal fan deposits on the development of thrust moraines has been considered by Boulton (1986). Figure 11.44 shows that the deposition of outwash fans at a glacier margin encourages proglacial glacitectonic deformation by providing a wedge of sediment in front of the snout. Glacifluvial incision will modify the form of the feature (cf. Fig. 11.39) and syntectonic deposition will supply new material for subsequent deformation. Recent thrust ridges at Höfdabrekkujökull, Iceland, have originated in this way, with a minor amount of excavation of the foreland at the ice margin (Krüger, 1985; Humlum, 1985b; Fig. 11.42). Thrusting is initiated adjacent to the ice margin, and propagated on to the foreland as successive thrust slices. The form of the moraine is substantially modified by debris flows during and following formation. Pleistocene

examples of syntectonic outwash deposition have been described by van der Wateren (1987).

A distinctive type of thrust moraine has been described by Krüger (1993, 1994), on the basis of annual observations at the margin of Sléttjökull, Iceland. The moraine forms by incremental stacking of sediment slabs in a three-phase process of: (a) early winter freeze-on of sediment below the glacier margin; (b) winter advance of the glacier margin with its frozen-on sediment slab; (c) summer melt-out of the sediment slab. If the glacier margin reaches approximately the same position in successive years, a moraine ridge will be constructed consisting of imbricate slabs of melt-out till interbedded with debris flow deposits.

A variation on Krüger's model has been proposed by Matthews *et al.* (1995) to account for moraine formation at Styggedalsbreen, Norway (Fig. 11.45). According to this *double-layer annual melt-out model*, two stratigraphic layers are added to embryonic moraines at the end of each seasonal cycle, rather than the single till unit in Krüger's model. One layer results from the melt-out of frozen-on till at the

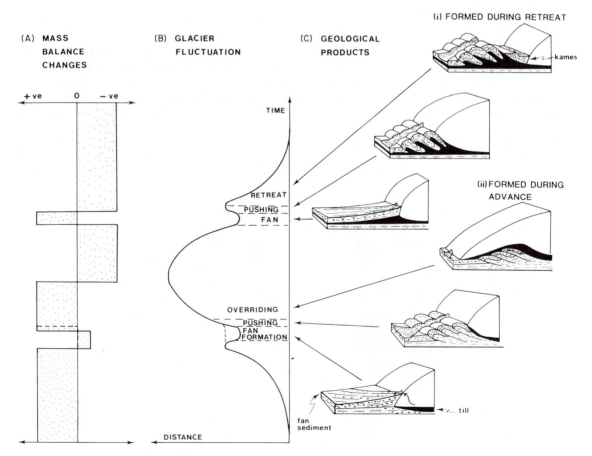

Fig. 11.44 Relationships between mass balance changes, glacier fluctuations and the outwash fans and thrust moraines formed as a result of ice-marginal fluctuations. (From Boulton, 1986. Reproduced by permission of Blackwell)

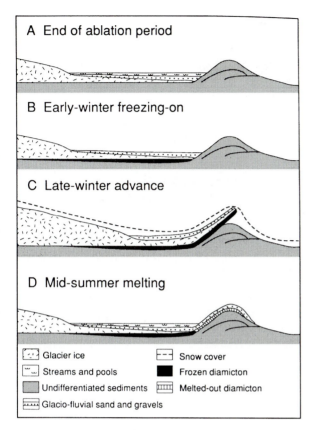

A End of ablation period

B Early-winter freezing-on

C Late-winter advance

D Mid-summer melting

Glacier ice

Streams and pools

Undifferentiated sediments

Glacio-fluvial sand and gravels

Snow cover

Frozen diamicton

Melted-out diamicton

Fig. 11.45 A 'double-layer annual melt-out' model for moraine formation based upon the moraine of Styggedals-breen, Norway: (A) late summer; (B) early winter; (C) late winter; (D) midsummer. Ridge development is incremental as a result of annual cycles at the glacier snout. (From Matthews *et al.*, 1995, Reproduced by permission of Scandinavian University Press)

ice base, and the other from the deposition of glacifluvial sands and gravels as a supraglacial unit. During the winter, the thin glacier snout advances and undergoes up-arching over the embryonic moraine. The till and glacifluvial sedimentary units are then lowered on to the moraine during summer melting of the thin snout. Matthews *et al.* (1995) claim that this process involves no pushing, but similar snout up-arching at Flåajökull, Iceland, is associated with sediment deformation, bulldozing and fluting construction, presumably owing to the fact that the ice margin continues to advance during the early summer when sediments on moraine proximal slopes have thawed.

11.3.1.3 CUPOLA HILLS

Cupola hill is the English translation of the Danish term *kuppelbakke*, which was introduced by Smed (1962) to describe irregular hills with the general characteristics of glacitectonic landforms but lacking

a hill–hole pair relationship and/or the transverse ridged morphology of composite ridges (Clayton *et al.*, 1980; Bluemle and Clayton, 1984). The characteristics of cupola hills can be summarized as follows: (a) they have a dome-like morphology varying from near-circular to oval shapes with lengths of 1–15 km and heights of 20–100 m; (b) internally, they are composed of detached and deformed floes of Quaternary sediments, older strata or bedrock; and (c) they have a carapace of till which truncates the underlying structures. These characteristics indicate that cupola hills are subglacially overridden hill–hole pairs or composite ridges, and represent early stages in the development of drumlins or megaflutings (Section 11.2.4.4). In cases where the subglacial modification has been slight, transverse ridge morphology may be partially preserved and the cupola form will tend to lie transverse to former ice flow (cf. Fernlund, 1988; Pedersen *et al.*, 1988). With increasing subglacial modification, ridge forms tend to be replaced by more smoothed, elongated forms.

Examples of cupola hills composed entirely of Quaternary sediments include the island of Ven in Sweden, and Ristinge Klint, Denmark (Aber *et al.*, 1989). Ven is composed of gentle synclines disturbed by numerous thrust faults and overturned folds. The glacitectonic disturbance responsible for these structures has produced two major transverse ridges, the surface of which is muted by a cover of discordant till (Adrielsson, 1984). In contrast, the topography of Ristinge Klint is subdued and drumlinized, similar to cupola hills on the nearby island of Aero (Smed, 1962). Cliff sections reveal that the area of Ristinge is underlain by more than 30 imbricately stacked scales of Quaternary sediments, each scale being up to 20 m thick and containing the same stratigraphic sequence of Saalian, Eemian and Weichselian strata (Fig. 11.46; Madsen, 1916; Ehlers, 1978; Sjorring *et al.*, 1982; Sjorring, 1983). The internal stratification of each scale is little disturbed except for prominent drag folds associated with the thrust faults between scales. Thrusting of the scales is thought to have occurred at the ice margin during glacier advance and deposition of the upper, discordant tills; surface streamlining took place during subsequent ice overriding.

Excellent examples of cupola hills occur in the late Wisconsinan end-moraine systems that make up the offshore islands of New England, USA. The islands consist of glacitectonically disturbed floes or thrust scales of Cretaceous, Tertiary and Quaternary strata (Schafer and Hartshorn, 1965; Sirkin, 1980; Oldale and O'Hara, 1984). Sections at Gay Head Cliff, Martha's Vineyard, expose a series of imbricates scales, each 20–30 m thick and dipping upglacier, interrupted by overturned folds, low-angle thrust faults and occasional downdropped blocks or grabens. In places, the strata have been uplifted by over 160 m.

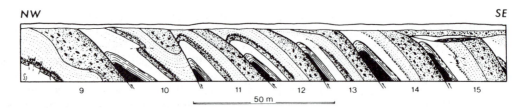

NW   SE

9   10   11   12   13   14   15

50 m

Fig. 11.46 The internal structures of a cupola hill, Ristinge Klint, Denmark. This shows stacked scales of Quaternary strata (9–15), each separated by a thrust fault and associated drag folds. (From Sjorring *et al.*, 1982. Reprinted by permission of the Danish Geological Survey)

These impressive tectonic structures were recognized as being of glacial origin as long ago as the late nineteenth century (Merrill, 1886b; Hollick, 1894; Woodworth, 1897; Upham, 1899), although Shaler (1888, 1898) held the view that they were due to orogenic processes. The imbricate structures are truncated by a thin and discordant till cover which was deposited when the Martha's Vineyard composite ridge was partially overrun by glacier ice (Kaye, 1964b). This overriding produced the cupola hill of Gay Head, which lies transverse to the former ice flow direction and possesses gently sloping sides and numerous irregular surface knobs and depressions.

Other impressive examples of cupola hills have been described from the southern Barents Shelf (Saettem, 1994), and the island of Møn, Denmark (Aber *et al.*, 1989). At the latter locality, structural evidence indicates that there were two phases of proglacial glacitectonic disturbance.

Viewed from the air, cupola hills may appear aligned in arcuate or transverse belts which conform to the former margins of glaciers. Where such belts occur they represent the positions of former large composite ridges that have been modified by subglacial moulding. A good example of an approximately 20 km wide belt of rounded oval-shaped ridges occurs on the southern Alberta prairie around the city of Brooks (Fig. 11.47). These features lie on the crest of a wide ridge transverse to the former ice flow direction. Exposures produced by badlands erosion on the northern flank of the cupola belt reveal shallow anticlinal folds relating to glacitectonic disturbance (D.J.A. Evans and Campbell, 1992). These cupola hills delimit marginal lobes of the Laurentide ice sheet, which probably underwent compressive flow where they advanced against higher topography. This appears to have initiated glacitectonic disturbance and composite ridge construction, which was later subglacially moulded as ice overrode the site (D.J.A. Evans, 1996).

11.3.1.4 MEGABLOCKS AND RAFTS

Megablocks and rafts are dislocated slabs of rock and unconsolidated strata which have been transported

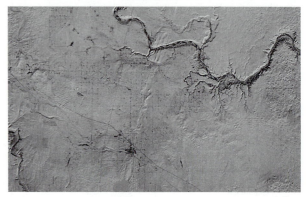

(a)

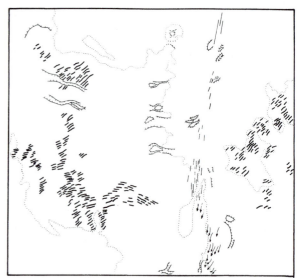

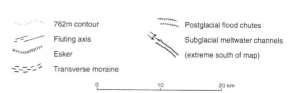

.... 762m contour   ~~~~~~ Postglacial flood chutes

Fluting axis   Subglacial meltwater channels

Esker   (extreme south of map)

Transverse moraine

0   10   20 km

(b)

Fig. 11.47 An arcuate belt of oval-shaped ridges (cupola hills) in the vicinity of Brooks, southern Alberta, Canada: (a) enhanced Landsat image of the cupola hills; (b) interpretive sketch map of the cupola hills and associated glacial features. (From D.J.A. Evans, 1996)

from their original position by glacier action (Jahn, 1950; Dellwig and Baldwin, 1965; Ruszcynska-Szenajch, 1976, 1987; Stalker, 1973b, 1976a; Sauer, 1978). Traditionally, such rafts were assumed to have been frozen on to the base of cold-based ice sheets, although it is now recognized that failure may occur along a basal décollement in a subglacial deforming layer (Sections 5.3.5 and 5.6.1). Megablocks may originate from uprooted cupola hills, although it is often difficult if not impossible to demonstrate detachment from the parent substrate unless there is exceptionally good exposure. Where they are sub-horizontal, megablocks and rafts may form flat-topped buttes or plateaux and can be mistaken for bedrock outliers. Similarly, buried megablocks can be mistaken for bedrock if borehole logs or exposures are not available.

Although megablocks and rafts contain coherent masses of bedrock or sediment, they are also commonly traversed by shear zones, faults and brecciated zones, and may be folded (Aber, 1985a). A limestone megablock near Topeka, Kansas, is quite heavily fractured and contains some rotated blocks (Dellwig and Baldwin, 1965). The block measures $50\,\text{m} \times 150\,\text{m}$ and is only 1–2 m thick. It is separated from underlying striated limestone by a 30 cm zone of brecciated shale (the shale is normally 12 m thick) and glacial sediment. This particular block has been transported less than 1 km by horizontal sliding, but some far-travelled megablocks have been documented. Rafts of Jurassic clay > 20 m thick at Łuków, Poland, are derived from Lithuania, more than 300 km to the north-east, and are thought to have been transported in a frozen state (Jahn, 1950; Ruszczynska-Szenajch, 1976). A similar transport distance has been calculated for the Cooking Lake megablock near Edmonton, Alberta, Canada, which is 10 m thick and covers an area of approximately $10\,\text{km}^2$ (Stalker, 1976).

A good example of the subsurface expression of a megablock occurs in the vicinity of the Qu'Appelle Valley, Saskatchewan (Fig. 11.48). This block of

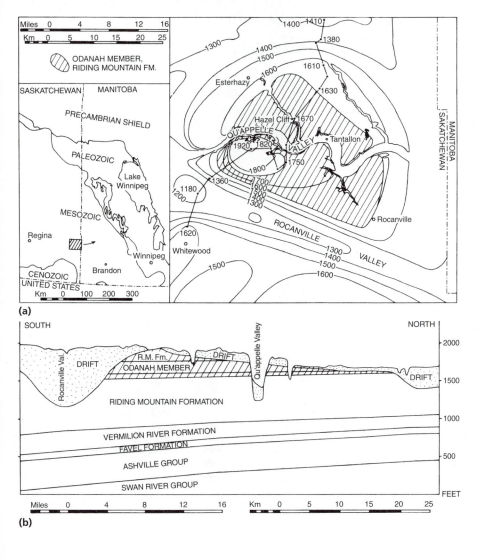

(a)

(b)

Fig. 11.48 The megablock of the Qu'Appelle Valley, Saskatchewan, Canada. (a) Bedrock contour map of Esterhazy and vicinity. The 1600 ft contour defines the position of the megablock of the Cretaceous Odanah Member and Riding Mountain Formation (shaded). Black dots show test wells along the line of section used in Fig. 11.48b, and elevations of the bedrock surface in the test wells are shown in feet. (b) Subsurface cross-section of the megablock with a vertical exaggeration of 42×. (From Aber *et al.*, 1989, after Christiansen, 1971b, and Sauer, 1978. Reprinted by permission of Kluwer)

Cretaceous shale is approximately $1000\,km^2$ in area with a maximum thickness of $100\,m$, and was discovered by borehole logging and bedrock contour mapping by Christiansen (1971b) and Sauer (1978). A well drilled at the western end of the megablock penetrated $80\,m$ of brecciated and deformed bedrock of the Riding Mountain Formation and then $2\,m$ of till. A large number of smaller megablocks composed of Cretaceous strata and sandwiched between Quaternary sediments have been described from southern Alberta, Canada, by Stalker (1973b, 1976a).

Steeply dipping rafts are more likely to have some surface expression, although they may be buried and difficult to identify without extensive borehole data. A series of detached chalk rafts occur in till overlying limestone in quarry exposures at Kvarnby in Sweden, suggesting that they were removed from the Baltic Sea bed and transported more than $25\,km$ before being imbricately stacked at angles of up to $80°$ (Ringberg, 1980, 1983; Ringberg *et al.*, 1984). These subsurface rafts lie beneath a structurally controlled low-amplitude hummocky topography which would be difficult to differentiate from a cupola hill without exposures. Similarly, chalk rafts in the Anglian till of Hertfordshire, England, have been identified in borehole records and as pale-toned patches in darker, freshly seeded fields on aerial photographs (Hopson, 1995). Unlike the Swedish examples, the Hertfordshire chalk rafts have been displaced only a short distance (mostly <1 km) from their origin by the southerly-flowing Anglian ice sheet.

11.3.2 Push and squeeze moraines

11.3.2.1 PUSH MORAINES

Push moraines are small moraine ridges, usually less than $10\,m$ in height, produced by minor (often annual) glacier advances (Fig. 11.49; Section 7.4.2; Worsley, 1974; Boulton and Eyles, 1979; Rabassa *et al.*, 1979; Rogerson and Batterson, 1982; Sharp, 1984). They can be produced at either subaqueous or terrestrial ice margins, but we describe subaqueous push moraines separately in Section 11.6.2.2. Morphological terms that incorporate possible push moraines include *washboard moraines* (Mawdsley, 1936; Gwynne, 1942; Elson, 1957; Thorarinsson, 1967; Nielson, 1970), *cross-valley moraines* (Andrews, 1963; Andrews and Smithson, 1966) and *corrugated ground moraine* (Prest, 1968, 1983; Elson, 1969). The term 'push moraine' has also been employed by some researchers to refer to large-scale glacitectonized moraine ridges (e.g. Rutten, 1960; Kalin, 1971; Johnson, 1972; Boulton, 1986; Goldthwait, 1989; Etzelmüller *et al.*, 1996). The definition adopted in this book excludes such glacitectonized

moraines, although it should be noted that some moraines are formed by a combination of proglacial glacitectonics and ice-marginal bulldozing, and small-scale glacitectonic structures may be present in annual push moraines. A working definition is that glacitectonic moraines contain at least 25 per cent glacitectonic structures which retain enough sedimentary bedding to enable identification of primary depositional process (Section 11.3.1; Humlum, 1985b; Krüger, 1985).

Push moraines vary widely in composition, and can consist of subglacial till, mass movement deposits, water-sorted sediments or large boulders, depending on the nature of the sediment on the glacier foreland. Where glacier margins are debris-covered, push moraines can consist mainly of supraglacial material dumped on to the forefield dur-

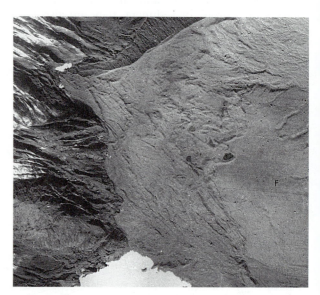

(a)

(b)
Fig. 11.49

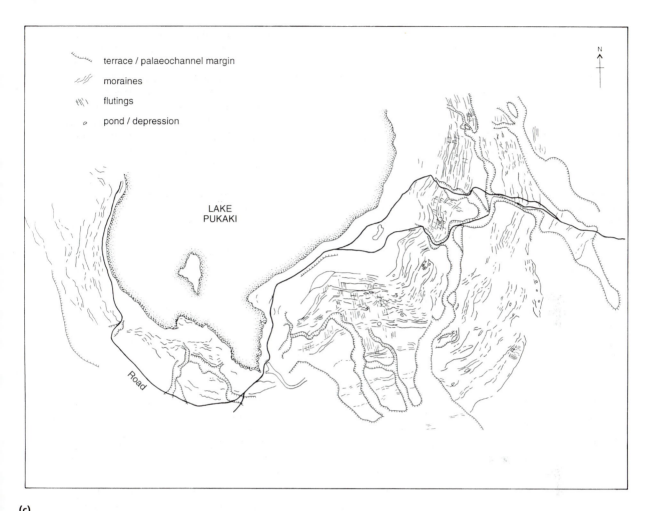

(c)

Fig. 11.49 Examples of push moraines with saw-tooth patterns. (a) Push moraines at the margin of Fjallsjökull, Iceland. Note the crevassed glacier snout, which is responsible for the pecten and thus the saw-tooth pattern of moraine ridge development. Small flutings and boulders (F) are visible on the flat terrain on the distal edge of the outer moraines. The distance across this view is approximately 200 m. (Aerial photograph (1965) Reproduced by permission of Landmaelingar Islands) (b) The cross-valley moraines of the Isortoq River, Baffin Island. Note the saw-tooth nature of the moraine ridge morphologies. (Photo: Jack Ives) (c) Map of nested end moraines with saw-tooth morphology and associated flutings at the southern shore of Lake Pukaki, New Zealand

ing the summer and pushed up during winter (Hewitt, 1967). Push moraines commonly have asymmetric cross-profiles, with gentle proximal and steep distal slopes. The ice-proximal slopes of fresh push moraines are often covered with small flutings, recording glacier overriding and deformation. As the glacier retreats during the summer, sediment melting out from the ice base and material at the crests of the moraines tend to feed subaerial debris flows, and so delicate surface forms like the flutings get buried or destroyed.

Clast *a*-axis fabrics in push moraines on the foreland of Skalafellsjökull, Iceland, have been studied by Sharp (1984), and summarized in a general model (Fig. 11.50). Beneath the ridge, undisturbed lodge-ment till (or deformation till) has strong preferred clast orientations and low upglacier dips (1), contrasting with weaker fabrics and steeper upglacier dips of the proximal disturbed till (2) and the downglacier dips of the distal disturbed till (3). Surface sediments are characterized by weak fabrics of subaerial sediment gravity flows (5), weak ridge-parallel orientations of ice slope colluvium dropped on to the ridge crest from the glacier (6), and size-sorting of distal face avalanche debris (7). These fabric patterns help to illustrate the internal structure of push moraines, which consists of an asymmetric fold with an axial plane dipping upglacier. This structure is a reflection of the overall cross-sectional morphology of the moraines.

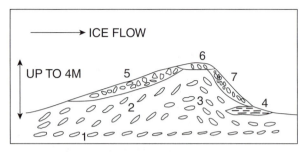

Fig. 11.50 Idealized reconstruction of the internal fabrics of an annual moraine ridge based upon examples at Skalafellsjökull, Iceland: 1) undisturbed lodgement till – strong fabric, low up-glacier dip; 2) disturbed lodgement till – weaker fabric with steeper up-glacier dip; 3) disturbed lodgement till – down-glacier dipping fabric; 4) laminated sands and silts overriden during ridge-forming advance; 5) sediment mass flow deposits – no preferred orientation, slope-conformable dip; 6) ice slope colluvium – weak preferred orientation parallel to ridge crest; 7) avalanched debris with downslope size sorting. (From Sharp, 1984. Reprinted by permission of the International Glaciological Society)

Fig. 11.51 Aerial photograph of the saw-tooth moraines on the glacier foreland of Bødalsbreen, southern Norway. (Photo: Fjellanger Widerøe)

Push moraines are broadly arcuate in planform, but in detail are often irregular and winding, reflecting the morphology of the glacier snout. Where the glacier snout is indented by radial crevasses, push moraines have striking *saw-tooth* planforms, with downvalley-pointing teeth and upvalley-pointing notches (Figs 11.49 and 11.51; Matthews *et al.*, 1979). Annual push moraines in front of Skalafellsjökull and Fjallsjökull, Iceland, have saw-tooth forms where the glaciers are heavily crevassed and curvilinear forms where the glaciers are less crevassed (Price, 1970; Sharp, 1984). At Bødalsbreen, Norway, Little Ice Age saw-tooth moraines have undulating crest lines, with high points at the notches and low points at the teeth. The teeth have asymmetric cross-profiles with steeper distal slopes, but the notches tend to be more symmetrical. The greater heights of the notches were explained by Matthews *et al.* (1979) as the product of accumulation of bulldozed debris in the recesses formed by radial crevasses in the glacier margin, whereas the lower heights of the teeth reflect the spreading of debris around advancing projections of ice between crevasses. Micro-scale saw-tooth forms also occur on the shallow proximal slopes of some saw-tooth moraines, recording minor oscillations of the glacier snout. Owing to the debris flow activity that characterizes freshly deposited moraines, such delicate forms do not have a good preservation potential.

The maximum height of the Bødalsbreen saw-tooth moraines is 9 m, although much larger Pleistocene saw-tooth moraines have been recorded (e.g. Andrews and Smithson, 1966). In New Zealand, saw-tooth moraines with notch crests up to 30 m high occur within last (Otiran) glaciation latero-frontal moraine complexes in the vicinity of Whataroa. It is unlikely, however, that such large features originated solely by pushing. Smaller features of a similar age are well preserved among the push moraines of the southern end of Lake Pukaki, on the eastern flank of the New Zealand Alps.

The planform of push moraines also depends on glacier activity. It is common to find younger moraines overlying or completely consuming parts of older moraines wherever summer retreat is small and/or some winter readvances are more substantial than usual (Fig. 11.49; Sharp, 1984; Boulton, 1986). Where glacier snouts are quasi-stable or have reached a position where they are in equilibrium with their environment, large terminal moraines may accumulate by annual accretion. Similarly, the absence of moraines relating to individual years within a field of annual moraines can be explained by unusually large winter advances or longer-term readvances. These can result in ice advance that exceeds several years of annual retreat, and so winter push moraines can be destroyed or incorporated into a larger moraine ridge (Rogerson and Batterson, 1982; Boulton, 1986; Krüger, 1993, 1994).

Four types of push moraine ridge were observed in front of Skalafellsjökull, Iceland, by Sharp (1984) (Fig. 11.52). *Type A moraines* are the most common, and possess fluted proximal slopes of deformation till recording glacier overriding. *Type B moraines* contain a type A moraine core but also have an ice core and a surface veneer of resedimented debris on the proximal slope. This results from the burying of thin ice margins by sediment gravity flows. *Type C moraines* are also superimposed on type A cores and are formed wherever debris bands crop out on the ice

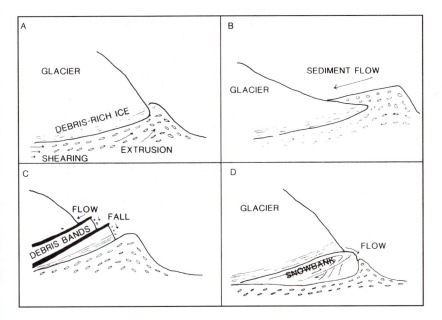

Fig. 11.52 Annual moraine ridge formation based upon examples at Skalafellsjökull, Iceland: (A) simple ridges comprising deformed lodgement till; (B) ridges with ice core incorporated by the flow of ridge-top sediments back over the glacier margin; (C) ridge with an ice core isolated beneath thick englacial debris bands which are exposed by backwasting of overlying ice; (D) ridge formed at the distal end of a marginal snowbank which has been pushed forward by the glacier. The flow of debris from the glacier has incorporated the snowbank into the ridge. (From Sharp, 1984. Reprinted by permission of the International Glaciological Society)

surface. As clean ice melts back, the debris bands insulate underlying ice to produce an ice core on the proximal slope of the moraine. Debris flows are then common as the ice slowly melts out. *Type D moraines* are produced wherever marginal snowbanks are pushed forward and overridden by the glacier (cf. Østrem, 1963, 1964; Birnie, 1977). The snowbank may be incorporated into the moraine where debris from the glacier falls over it and is deposited on the pushed ridge. The snowbank will later melt out and initiate debris flows.

Sharp (1984) argued that sediment delivery to the margin of Skalafellsjökull was by subglacial deformation, as observed beneath nearby Breidamerkurjökull (Sections 4.4 and 5.6; Boulton and Jones, 1979; Boulton and Hindmarsh, 1987). It has been suggested by Boulton (1979) that the till deformation results in a net discharge of sediment towards the glacier margin and the excavation of a submarginal depression parallel to the ice front. Therefore, the depths of depressions between recessional push moraines should vary according to the retreat rate. Sharp (1984) suggested that the material eroded from the depressions by subglacial deformation would accumulate at the glacier margin and may even be extruded from beneath the ice, as a result of the ice-marginal pressure gradient. This would explain the upglacier dip of the clast *a*-axes on the proximal flanks of push moraines. It was thought most likely that extrusion would occur at some time during the winter, presumably before or after any ice-marginal freezing and when ablation is at its lowest. A glacier readvance at the same time results in the bulldozing of the extruded sediments and the production of the asymmetric fold.

11.3.2.2 SQUEEZE MORAINES

The formation of minor moraine ridges by squeezing of saturated sediment from beneath ice margins has been observed in front of several glaciers (Section 7.4.1; Price, 1970; Worsley, 1974). They are particularly common along the margins of the large southern outlet glaciers of Vatnajökull, Iceland, such as Fjallsjökull, Flåajökull, Skalafellsjökull, Breidamerkurjökull and Heinabergsjökull, many parts of which are underlain by poorly drained, fine-grained till (Price, 1970; Benn, 1995). Newly formed squeeze moraines have steep or vertical sides and are rarely more than 1 m high. Price (1973) reported steeply dipping or vertical pebble fabrics in fresh squeeze moraines in Iceland. Squeeze moraines tend to degrade rapidly, and are commonly subject to reworking by ice push, so that unmodified squeeze moraines are probably rare in the geomorphological record. Large, possible sublacustrine squeeze moraines on Baffin Island have been described by Andrews and Smithson (1966) (Section 11.6.2.2).

11.3.3 Dump moraines and ice-marginal aprons

Material accumulating at the surface of a glacier through the melt-out of debris-rich folia is ultimately subject to remobilization by mass flowage, fall or fluvial transport. Where such material exists at the glacier margin, its remobilization may result in its being dumped on to the adjacent terrain during ice recession (Section 7.3). Depending on the rate of debris accumulation and glacier activity, a variety of sediment–landform associations will be created. For retreating glaciers, dumping of supraglacial material on to the former subglacial surface slowly emerging from beneath the ice produces a thin veneer of coarse-grained diamicton, or perhaps only sporadic collections of boulders (N. Eyles, 1979). *Dump moraines* will form where the ice margin remains stationary during debris accumulation (Boulton and Eyles, 1979), although they will be bulldozed into push moraines if the glacier undergoes a subsequent readvance (Section 11.3.2.1). Dump moraine size is related to supraglacial debris volume and the length of the standstill period. Small dump moraines will form where glaciers remain stationary during the winter but retreat during the summer, each moraine marking one winter's increment of debris accumulation. The largest and most spectacular dump moraines, however, form at the margins of debris-mantled valley glaciers which occupy similar positions for considerable periods.

The dumping of large quantities of debris around glacier margins builds up *latero-frontal dump moraines* (Fig. 11.53; Boulton and Eyles, 1979; Small, 1983, 1987b). Supraglacial debris slides, rolls, flows and falls from the glacier margins and is deposited as ice-contact scree (Section 7.3.1; Fig. 11.54). In mountain environments, the debris may be predominantly passively transported rockfall material, consisting of coarse, angular clasts with little matrix. However, actively transported debris can also crop out near valley glacier margins as a result of the elevation of basal debris septa, introducing matrix-rich debris to some dump moraines (Section 6.4.1; Small, 1983, 1987b). On some latero-frontal moraines, there is a tendency for clast roundness to increase down-moraine, towards the glacier snout position (Matthews and Petch, 1982). Benn and Ballantyne (1994) showed that, on the Little Ice Age terminal moraine of Storbreen, Norway, this increase in clast roundness is due to an increasing proportion of actively transported debris in the moraines closer to the former glacier centre-line (Fig. 11.55). In lateral

positions, debris in the moraine is dominated by passively transported rockfall material.

The accumulation of dumped material at the margin of a thickening glacier results in a wedge-shaped moraine with crude internal bedding and clast fabrics dipping away from the glacier at angles of between 10° and 40° (Fig. 11.54a, part A; Boulton and Eyles, 1979; Small, 1983, 1987b). Coarse, bouldery layers within the moraine may be derived from glacially transported rock avalanche material (Humlum, 1978). The deposition of lateral dump moraines against valley sides creates swales between the moraine and the valley wall, features rather inaccurately referred to as *ablation valleys* in the older literature (e.g. Oestreich, 1906; Mason, 1930). Because their formation has little to do with ablation, the alternative term *valley-side depression* is more appropriate (Hewitt, 1993). During glacier thinning, dump moraines are abandoned and their inner faces are subject to collapse and reworking (Ballantyne and Benn, 1994b). A new inset dump moraine may be formed within the older one if the glacier restabilizes at a new position, but such moraines will be unstable if they are deposited on top of dead ice masses (Fig. 11.54a, parts B, C). Alternatively, renewed thickening of the glacier may dump material on top of the old dump moraine, which may become completely buried. In this case, periods of non-deposition may be recorded by palaeosols or even buried trees. Such buried organic material provides a very valuable source of palaeoclimatic data in alpine environments (Röthlisberger *et al.*, 1980).

Where dump moraines are deposited on steep valley sides, debris derived directly from the glacier is interdigitated with sediment from the valley walls (Fig. 11.56). For example, lateral moraines recently abandoned by the receding Tasman Glacier, New Zealand, contain interdigitated beds of, first, stratified debris flows and fall-sorted debris avalanche material with beds dipping away from the glacier margin; second, subhorizontally bedded alluvium deposited as kame terraces along the ice margin; and third, debris flows and fall-sorted screes from valley-side paraglacial fans. Removal of lateral support by glacier recession results in rotational failure of the inner slopes of the moraines. In many cases, such failures clearly extend up into the screes which were deposited against the lateral moraines.

Lateral dump moraines on steep valley sides have poor preservation potential, owing to rapid paraglacial reworking of their ice-proximal faces once the support of glacier ice is removed. On the other hand, dump moraines deposited beyond the confines of cirques or steep valleys are more likely to be preserved, because paraglacial activity is generally less significant after glacier recession. Such moraines can act as major barriers to subsequent

Fig. 11.53 Large lateral moraines in the Koa Rong massif, Lahul Himalaya, India. (Photo: D.I. Benn)

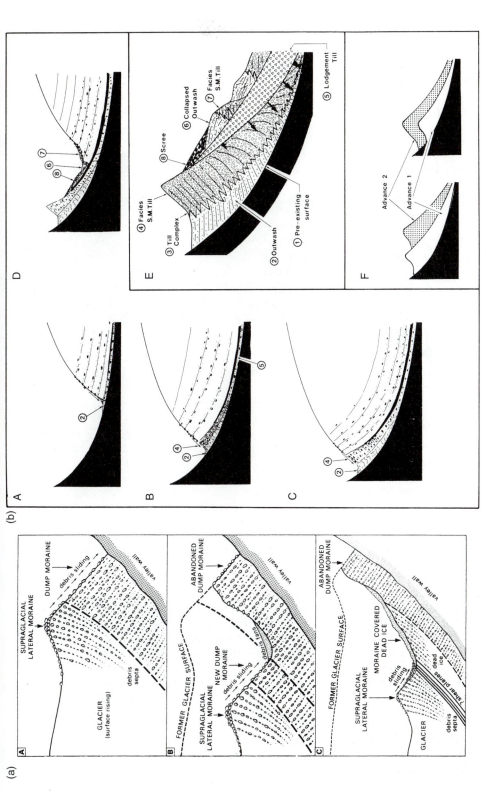

Fig. 11.54 Schematic reconstruction of the evolutionary stages of lateral moraines. (a) Based on the moraines of Glacier de Tsidjiore Nouve, Switzerland. (A) Formation of lateral dump moraine by sliding of debris from the crest of a supraglacial lateral moraine on a rising glacier surface. (B) Formation of abandoned dump moraine by glacier recession and lowering glacier surface, followed by a limited readvance and formation of new dump moraine. (C) Formation of abandoned dump moraine and debris-covered dead ice mass by a rapid lowering of the glacier surface. Note the possible formation of shear planes between the dead and flowing ice. (From Small, 1983. Reproduced by permission of the International Glaciological Society) (b) the sequential development of lateral moraines by the process of sediment dumping from supraglacial positions, where A–D represent simple moraine development during a single glacier advance and retreat cycle. The numbers refer to the features labelled in E which depicts the sedimentology of the moraine produced in D. The arrows in the S.M. till (supraglacial morainic till) indicate a zone of glacitectonic disturbance produced during ice overriding. Note that lake sediments may also be deposited between the moraine and the ice margin even though they are not included in this model. In F the relationships between two separate landform–sediment suites are shown where, on the left, advance 1 is more extensive than advance 2 and, on the right, advance 1 is less extensive than advance 2. (From Boulton and Eyles, 1979. Reproduced by permission of Balkema)

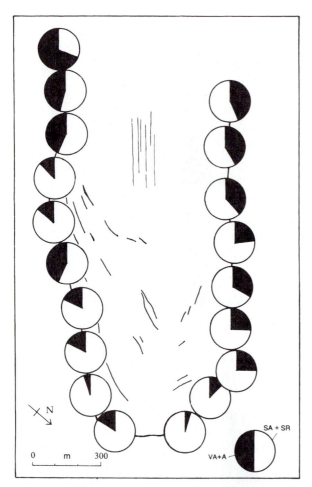

Fig. 11.55 Storbreen clast roundness data. (From Benn and Ballantyne, 1994. VA and A denote very angular and angular clasts (passively transported debris), and SA and SR denote sub-angular and sub-rounded clasts (actively transported debris). Reproduced by permission of Elsevier)

glacier advances, and may be constructed over several glacial cycles. For example, the 100 m high lateral moraines of Kviarjökull, Iceland, are the product of incremental accumulation of Little Ice Age morainic debris over the top of an older latero-frontal moraine (Fig. 11.57; Thorarinsson, 1956a). The older moraine was responsible for restricting the flow of Kviarjökull, forcing it to remain as a linear outlet glacier, unlike other outlet glaciers of the Vatnajokull ice cap which descended on to coastal lowlands as piedmont lobes during the Little Ice Age.

Distinctive dump moraines known as *frontal* or *ice-marginal aprons* form around the margins of some arctic and polar glaciers (Shaw, 1977c; D.J.A. Evans, 1989b; Fitzsimons, 1997). These moraines form where debris and ice blocks accumulate at the base of steep, terminal ice cliffs by a combination of dry calving and limited melting. When forming, the aprons have steep ice-contact slopes and gentler ice-

distal slopes determined by the properties of the debris. However, they will be subject to considerable modification following deglaciation, owing to the ablation of incorporated ice blocks.

11.3.4 Latero-frontal fans and ramps

Deposition by debris flows and glacifluvial processes around stationary glacier margins can result in *lat-*

(a)

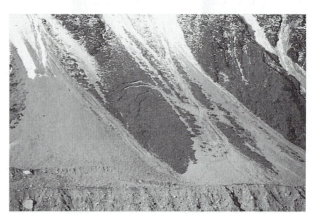

(b)

(c)
Fig. 11.56

(d)

Fig. 11.56 The lateral moraines, paraglacial alluvium and slope deposits of glacier margins in tectonically active mountain environments. (a) Subhorizontally bedded alluvium exposed by the rotational failure of lateral moraines at the receding margin of the Tasman Glacier, New Zealand. The more recent lateral moraines which occur in the foreground and to the left of this view comprise interbedded supraglacial sediment dipping towards the valley side and reworked alluvium dipping towards the glacier. Older lateral moraines can be seen in the middle distance, where they have remained in contact with the alluvium. Because the alluvium was deposited between lateral moraine and the valley wall it is strictly paraglacial valley fill rather than a kame terrace. (b) The steep ice-contact slope of the left lateral moraine of the Tasman Glacier and rotational failure scars in partially vegetated screes. The rotational failure is a response to the removal of lateral support by the Tasman Glacier during its recent recession. (c) Section through a lateral moraine and debris flow fan at the margin of the Batal Glacier, Lahul Himalaya. Note the stratification and change in bedding angle in the vicinity of the debris flow fan, attesting to the interdigitation of glacial and non-glacial sediments in lateral moraines. (d) Latero-frontal moraines of the Hooker Glacier, New Zealand, showing the burying of older moraine ridges by paraglacial fans fed by avalanches and debris flows. (Photos: D.J.A. Evans)

Fig. 11.57 The large latero-terminal moraine complex of Kviarjökull, Iceland. The moraine exceeds 100 m in height in places and was produced as a result of the large debris volume provided by the precipitous rock walls that confine the glacier in its middle reaches. (Aerial photograph, 1980, provided by Landmaelingar Islands)

Fig. 11.58 Large debris-flow dominated ice-marginal fans, Batal Glacier, Lahul Himalaya. (Photo: D.J.A. Evans)

ero-frontal fans and ramps, consisting of coalescent debris fans which descend from the glacier snout (Fig. 11.58; Owen and Derbyshire, 1989; Kuhle, 1990). The outer slopes of such forms have much shallower gradients than latero-frontal dump moraines built up by rockfall and related processes. The inner slopes are steep, consisting of the former ice-contact face, so that the whole landform has a pronounced asymmetric cross-profile. In some situations where moraines are absent, fans and ramps provide the only evidence of former ice-marginal positions. Fans deposited predominantly by glacifluvial processes grade distally into sandar, and are described in Section 11.5.1.1.

Large debris-flow dominated ice-contact fans occur at the margins of many debris-mantled Karakoram valley glaciers (Jones *et al.*, 1983; Owen, 1991; Owen and Derbyshire, 1993). The fan morphology develops over a considerable period, sometimes involving several generations of deposition with intervening periods of abandonment. During fan accumulation, active depocentres shift position, reflecting changes in the morphology of the ice margin and fan surface. As the fan progrades, surface angles tend to become progressively reduced. Meltwater exiting from englacial positions or draining the glacier surface locally excavates the ice-contact fan

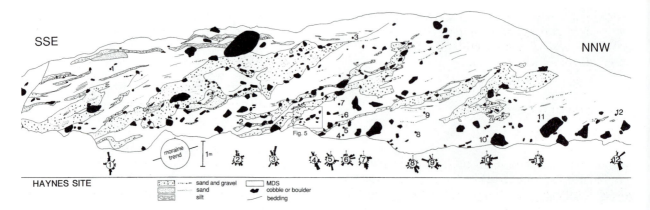

Fig. 11.59 Log of a cross-section through a Pleistocene dump moraine, Durango, Colorado, USA. Rose diagrams are of pebble fabrics from matrix-supported diamicton units (MDS), interpreted as debris flow deposits. (From Johnson and Gillam, 1995. Reprinted by permission of the Geological Society of America)

or ramp faces to produce inset meltwater fans. The Karakoram ice-contact fans can reach heights of hundreds of metres, attesting to very high rates of debris transfer in this high-altitude, tectonically active region. Because debris flow fans and ramps are deposited partly over glacier ice, they may be characterized by pitted or degraded surfaces after ice melt-out. When ice-contact fans are abandoned by glacier retreat, the inner faces are susceptible to collapse and may provide material for lateral moraine development during ice recession.

Like lateral dump moraines, large fans and ramps may prove to be insurmountable obstacles for later glacier advances, and can act to constrain or channel the ice (see Section 7.3.1). An example of this occurs on the foreland of the Batal Glacier in the Lahul Himalaya (Fig. 11.58; Owen *et al.*, 1996a). Early glacier advances led to the production of piedmont lobes, but later events were constrained by debris flow ramps and were consequently less extensive.

The sedimentology of the fans and ramps is dominated by massive, often bouldery diamictons with beds up to tens of metres thick and dipping away from the glacier snout. Good examples of the internal structure of dump moraines have been provided by Johnson and Gillam (1995), who conducted a detailed study of late Pleistocene moraines at Durango, Colorado (Fig. 11.59). Sections in the moraines expose interbedded debris flows and water-sorted sediments dipping away from the former glacier margin, comprising a series of ice-marginal fans. Limited exposures of melt-out till in low-level, ice-proximal positions show that, in places, sediment was deposited on top of debris-rich ice.

Well-exposed examples of coalescent ice-contact fans up to several hundred metres high occur along the west coast of South Island, New Zealand (Fig. 11.60). Coastal sections through the fans expose

Fig. 11.60 Stratified diamictons and crudely bedded gravels deposited by avalanches, debris flows and slopewash in a Pleistocene lateral debris fan, Wanganui River, South Island, New Zealand. (Photo: D.J.A. Evans)

crude to well-developed interbeds of clast-supported and matrix-supported diamictons, poorly sorted gravels and boulder to cobble-sized rubble units, recording deposition from debris flows and other mass movements. Occasional pockets of glacilacustrine sediment, which have in some instances been glacitectonized, attest to the occurrence of short-lived proglacial lakes ponded between the ice margin and the fans during ice retreat phases. The dip directions of the beds within the fans vary from one exposure to another, indicating deposition from a series of depocentres located at the ice margin and partially reworked during glacial advance and retreat cycles.

Sedimentologically, ice-contact fans and ramps are very similar to paraglacial slope deposits (Section 11.5.2), but can be recognized by the presence of a former ice-contact slope.

11.4 SUPRAGLACIAL ASSOCIATIONS

This section describes sediment–landform associations that originate englacially or supraglacially, then evolve as they are progressively lowered on to the substratum by glacier ablation. This section begins with medial moraines, which are supraglacial associations resulting from longitudinal debris concentrations in the parent ice. The other associations discussed in this section are hummocky moraine, controlled moraine, kame and kettle topography, ice-walled lake plains, kame terraces and pitted sandar. Many of these associations result from topographic inversion and/or glacier karst development on debris-mantled glaciers (Section 6.5.4), and are commonly referred to as *ice-stagnation topography*. However, as N. Eyles (1983b) and Kirkbride (1993) have shown, ice stagnation may occur only during a late stage in the development of such forms, and then only in a relatively narrow marginal zone, so that the presence of widespread supraglacial associations need not imply widespread glacier stagnation. Hummocky moraine, kames and ice-walled lake plains are related features, which reflect deposition predominantly by mass movement, glacifluvial and glacilacustrine processes respectively. In supraglacial environments, however, these processes commonly operate in close proximity, and hybrid sediment–landform associations may frequently be observed.

11.4.1 Medial moraines

Medial moraines are prominent features on the surfaces of many glaciers (Section 6.4.4), although they do not have good preservation potential because they contain relatively little debris and are subject to intense reworking during glacier melt. They sometimes can be traced from modern glacier surfaces on to proglacial surfaces, where they may be recognizable as diffuse spreads of debris with distinctive lithological composition or particle morphologies. However, such moraines commonly have no marked surface expression, and may be very difficult to recognize once the glacier foreland becomes vegetated. Recognition of ancient medial moraines is easier where they consist of large boulders which may be traced back to their source outcrops (Fig. 11.61). An unusually clear example of an ancient medial moraine emanating from mountainous topography

occurs on the Isle of Jura, Scotland (Dawson, 1979). The moraine is known locally as Scriob na Caillich, or the Witch's Slide, and consists of angular quartzite blocks arranged in a series of parallel ridges which extend 3.5 km north-westward from the slopes of Beinn an Oir (Fig. 11.62). The moraine was produced by the last (late Devensian) ice sheet, when Beinn an Oir stood above the ice surface as a nunatak.

In certain situations *ice stream interaction*-type medial moraines have a high preservation potential. Very large ice stream interaction medial moraines, >100 m high in some areas, have been produced by the coalescence of ancient piedmont glaciers with well-developed lateral moraines on the west coast of New Zealand (Fig. 11.63). Because these features accumulated essentially as lateral moraines, they contain sediments characteristic of ice-marginal deposition (Section 11.3.3). However, the medial moraine cores commonly display glacitectonic structures, and in places, carapaces of subglacial till record glacier coalescence and overriding. The interesting corollary is that such overridden interlobate features may well lie at the core of some megaflutings.

Large medial moraines, referred to as *interlobate moraines*, occur on the Canadian Shield and document the coalescence of the margins of different sectors of the former Laurentide ice sheet. The Harricana interlobate moraine, formerly separating the Hudson Bay and Labrador sectors of the receding ice sheet, stretches for 1000 km southwards from James Bay to Lake Simcoe, Ontario (Wilson, 1938; Hardy, 1977, 1982; Vielette, 1986, 1988) and com-

Fig. 11.61 Bouldery medial moraine near Glen Torridon, Scottish Highlands. The moraine is composed entirely of passively transported angular quartzite blocks. (Photo: D.I. Benn)

(A)

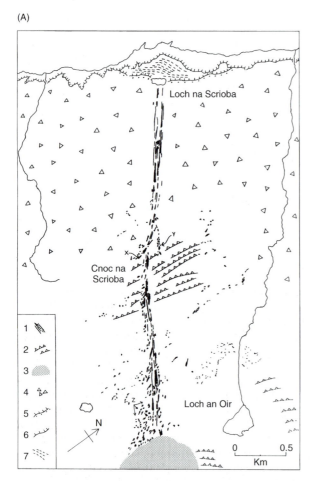

(B)

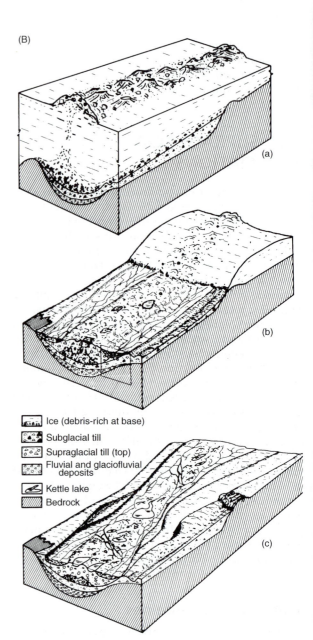

Fig. 11.62 Examples of ancient medial moraines. (A) The Witch's Slide, a fossil medial moraine on the Isle of Jura, Scotland. 1: bouldery ridges; 2: quartzite bedrock escarpments; 3: talus-covered slopes of Beinn an Oir; 4: till; 5: Lateglacial marine limit; 6: raised cliffs; 7: raised beach ridges. The land surface slopes gradually from *c.* 400 m in the south-east to sea-level in the north-west. (Modified from Dawson, 1979) (B) The sequential development of a medial moraine in the centre of the Athasbaca Valley near Jasper, Alberta, Canada. (a) A well-developed medial moraine forms a substantial debris cover between the Athabasca and the Miette valley glaciers. (b) Bedrock benches on the valley sides become free of ice first, leaving a well-preserved fluted topography with little supraglacially derived material. Debris, derived from the medial moraine and by melt-out of the medial debris septum, accumulates on the ice surface and eventually develops into a linear, ice-cored moraine. Ice-marginal debris flow and fluvial deposits develop into kame terraces along the sides of the decaying, debris-covered ice mass. (c) During the final melt-out phase, a chain of kettle lakes forms in the area of the thickest supraglacial debris cover, and large gullies, faults and other collapse structures develop in the kame terrace deposits, owing to ice block decay. Morainal debris is also reworked by braided streams to produce a dissected and subdued series of kettle holes. (From Levson and Rutter, 1989b. Reprinted by permission of the *Canadian Journal of Earth Sciences*)

prises a series of ridges up to 10 km wide and 100 m high. The Burntwoodknife interlobate moraine, formerly separating the Hudson Bay and Keewatin sectors of the receding Laurentide ice sheet, covers a distance in excess of 500 km on the west side of Hudson Bay and is composed of broad ridges of glaciofluvial sediment and till. Individual segments are up to 12 km long, 4 km wide and 60 m high (Fig. 11.64; Dredge *et al.*, 1986; Klassen, 1986; Dyke and Dredge, 1989). The large quantities of sands and gravels comprising these moraines were deposited in proglacial lacustrine environments during the final stages of the Laurentide ice sheet, attesting to the large-scale reworking of medial moraine debris dur-

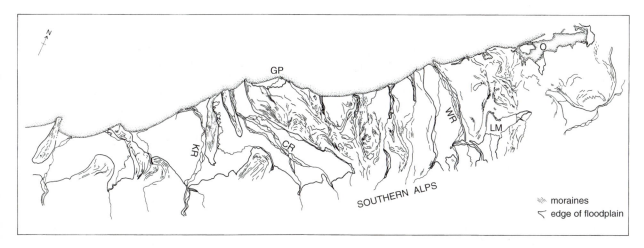

(a)

ing glacier recession. In fact these features are regarded more as glacifluvial than supraglacial in origin (Dredge and Cowan, 1989; Vincent, 1989).

Similarly, continuous esker systems may demarcate the former positions of medial moraines in glaciers or ice sheets. The deposition of a continuous esker ridge is clearly related to sediment supply, and this may be assured wherever englacial meltwater streams coincide with medial moraine positions. An example of glacifluvial reworking of a medial moraine in Breidamerkurjökull, Iceland, is documented by Price (1969, 1973). A low-relief kame and kettle topography centred on the bottom of the Athabasca Valley near Jasper in Alberta, Canada, is interpreted as a former supraglacial medial moraine by Levson and Rutter (1989b). This feature has also been locally reworked by glacial meltwater and is closely associated with glacifluvial landforms such as kame terraces (Fig. 11.62b).

11.4.2 Hummocky moraine and controlled moraine

The term *hummocky moraine* has been employed in a wide range of senses. As a purely descriptive term, it has been applied to moundy, irregular morainic topography exhibiting varying degrees of order, ranging from entirely chaotic assemblages of mounds to suites of nested transverse ridges (Sissons, 1967; Benn, 1992b; Attig and Clayton, 1993; Bennett and Boulton, 1993a). Genetically, it has been used to encompass landforms with a variety of origins, including deposition in association with active and stagnant ice, and glacitectonic deformation (e.g. Hoppe, 1952a; Gravenor and Kupsch, 1959; Stalker, 1960; Aario, 1977; Bennett and Boulton, 1993a). Many authors, however, have employed the term 'hummocky moraine' in a more restrictive sense, to refer to moraines deposited during the melt-out of debris-mantled glaciers (e.g. Harker, 1901; R.P.

(b)

Fig. 11.63 (a) the impressive latero-frontal and ice stream interaction medial moraines of the New Zealand west coast, where large quantities of extraglacial debris accumulated at the lateral margins of glaciers debouching from the Southern Alps during the last glaciation. O = Okarito Lagoon, LM = Lake Mapourika, WR = Waiho River, GP = Gillespie Point, CR = Cook River, KR = Karangarua River. (b) Large lateral and medial moraines on the Payachatas massif, western Bolivia. The moraines demarcate a series of ice tongues that fringed the northern volcano during the late Pleistocene. The southern volcano was created by more recent volcanic activity, and hence has no moraines of that age. (Image processed by D. Payne)

Sharp, 1949; Hoppe, 1952a, 1959; Winters, 1961; M.J. Sharp, 1985b), and it is in this sense that the term is used in this book. This genetic definition should be applied only to suites of moraines whose origin has been established by detailed mapping and sedimentological analyses, because generalized descriptions of surface morphology alone are insufficient to determine moraine genesis. For example, many tracts of so-called hummocky moraine in the Highlands of Scotland, originally believed to be chaotic ice stagnation topography (e.g. Harker, 1901; Sissons, 1967, 1974c, 1979a), have subsequently been recognized as nested push and dump moraines, or drumlins draped by supraglacial and ice-marginal deposits (Bennett and Glasser, 1991; Benn, 1992; Benn *et al.*, 1992; Bennett, 1994; Bennett and Boulton, 1993a, b; see Section 12.6.1.3).

Hummocky moraine deposited from debris-mantled ice can appear chaotic or linear in pattern, depending on the distribution of debris in the parent glacier and the patterns of debris redistribution and reworking during ice wastage. Occurrences with pronounced transverse linear elements are termed *controlled moraines*.

Hummocky moraine is the end product of *topographic inversion* cycles during the ablation of debris-mantled ice (Section 6.5.4.1). During ice ablation, debris is transferred away from topographic highs on the glacier surface by mass movements and meltwater, exposing ice cores to renewed melting and creating new depressions in former high points. Further debris reworking and topographic development is achieved by meltwater streams meandering between dirt cones (Section 6.5.2), the collapse of englacial tunnels, and the enlargement of supraglacial lake basins (Section 6.5.4.2). Several cycles of debris reworking and topographic inversion may occur before the debris is finally deposited in a series of irregular mounds and ridges (Fig. 11.65).

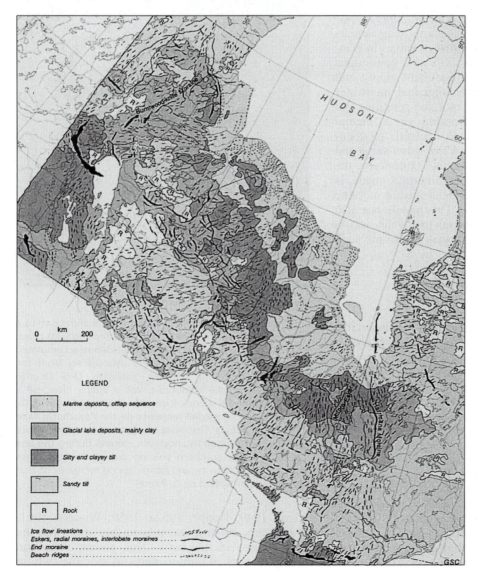

(a)

Fig. 11.64

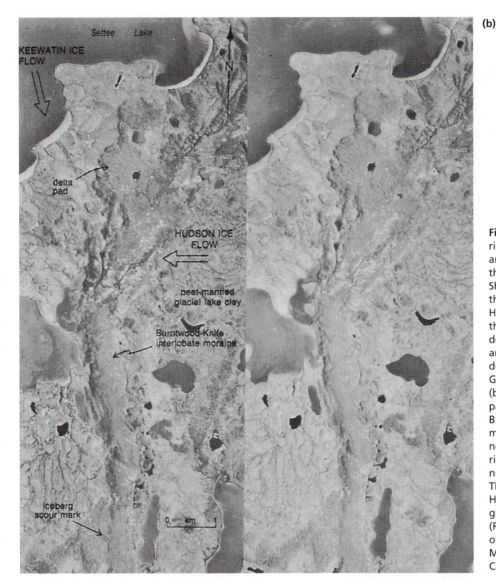

(b)

Fig. 11.64 (a) Surface materials, major moraines, eskers and ice flow indicators on the south-west Canadian Shield. Note the positions of the Burntwoodknife and Harricana moraines, marking the locations of interlobate deposition. (From Dredge and Cowan, 1989. Reproduced by permission of the Geological Survey of Canada) (b) Aerial photograph stereo-pair of a segment of the Burntwoodknife interlobate moraine near Settee Lake, northern Manitoba. The ridge is characterized by a number of attached deltas. The flow directions of the Hudson Bay and Keewatin glacier ice are indicated. (Reproduced by permission of the Department of Energy, Mines and Resources, Canada)

The distribution of *controlled moraines* is inherited from debris concentrations or septa within the parent glacier (Section 6.4.2). Transverse debris concentrations close to glacier margins may result from the thrusting of debris along shear planes (Goldthwait, 1951; Bishop, 1957; Souchez, 1967a), the emergence of basal debris frozen on to the glacier sole (Boulton, 1970a), or reworked ice-marginal aprons (Hooke, 1970; Shaw, 1977a; Hudleston, 1976; D.J.A. Evans, 1989b; Section 7.3.2). Such transverse septa control the pattern of differential ablation on the glacier surface, producing linear ice-cored ridges which eventually become separated from the snout to form *ice-cored moraines* (Section 7.3.2; Østrem, 1963, 1964). The morphological expression of englacial structures is always most striking while the moraines retain ice cores. However, uneven sediment redistribution during ice-core

wastage means that the final deposits tend to consist of discontinuous transverse ridges with intervening hummocks, preserving only a weak impression of the former englacial structure. Reworking by supraglacial meltwater systems also acts to destroy large tracts of controlled moraine before they can be deposited on the ground surface (Fig. 11.66). Detailed descriptions of these processes on modern Svalbard glaciers have been presented by Boulton (1967, 1972b) and Drozdowski (1977). Patterns of debris reworking and deposition can also be controlled by crevasses or other lines of weakness within the ice, which can guide the final stages of glacier disintegration (Gravenor and Kupsch, 1959). Moraines formed by the melt-out of englacial debris concentrations were termed *shear moraines* by Bishop (1957) and Souchez (1967a), and Thule–Baffin moraine by Weertman (1961c). However, the

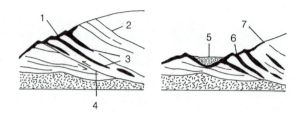

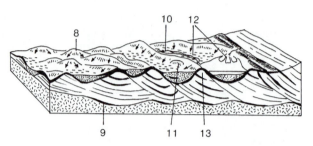

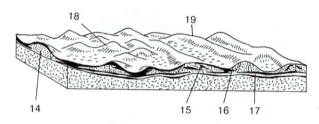

Fig. 11.65 Chaotic hummocky moraine, Sandfellsjökull, Iceland. (Photo: D.J.A. Evans)

broader term 'controlled moraine' is preferable because it encompasses the wide range of initial conditions that can result in the inheritance of debris concentrations by moraine form.

Another model of controlled moraine development has been proposed by Rains and Shaw (1981), on the basis of field studies in Antarctica. According to this model, uneven distribution of debris on glacier surfaces is determined by ice topography, rather than englacial debris concentrations (Fig. 11.67). Debris accumulates in *cusps* melted on to the glacier surface, and between thrust blocks of ice near the margin, and is then lowered on to the ground surface as hummocks of coarse-grained debris inset behind linear moraine ridges.

The sedimentology of hummocky moraine is complex, reflecting multiple cycles of redeposition during its formation. Typical facies associations consist of interbedded debris flows and other mass movement deposits, laminated lacustrine sediments, and glacifluvial sands and gravels, in varying degrees of disturbance (N. Eyles, 1979; Paul, 1983. The constituent debris may be actively or passively transported, depending on the position of the moraine relative to debris transport paths in the parent glacier. In some cases, debris from different sources and transport paths may remain segregated within the moraine, giving rise to compositional stratification (Benn, 1992b). An exposure through a conical hummock in a field of chaotic hummocky moraine on the Isle of Skye, Scotland, is shown in Fig. 11.68. The exposure reveals diamicton lenses interbedded with water-sorted gravels and sands, all displaying faults, slumps and other collapse structures, and is interpreted as a series of debris flow deposits and glacifluvial sediments which have undergone syndepositional deformation and reworking relating to the meltout of buried ice. Other well-exposed examples of complex sediment successions in hummocky moraine are on the Lleyn Peninsula in

Fig. 11.66 Model of controlled moraine development from transverse debris concentrations in a glacier snout. 1: debris released at glacier surface; 2: foliation; 3: debris band; 4: shear fault; 5: outwash collecting between stagnant ice ridges; 6: supraglacial debris accumulating from the melt-out of debris bands; 7: relatively clean ice surface resulting from mass flowage of melt-out debris; 8: small mass flows on the surface of stagnant ice hummocks; 9: lodgement till; 10: supraglacial channels reworking mass flows from stagnant ice; 11: large supraglacial mass flow; 12: delta accumulation in supraglacial ponds; 13: elongate ice ridge protected from melting by the accumulation of supraglacial sediment as debris bands melt out; 14: absence of collapse structures in sediment deposited directly on to bedrock; 15: collapse structures; 16: interdigitating mass flow diamictons and glacifluvial sands and gravels; 17: subglacial melt-out tills; 18: hummocks that may possess some linear trends; 19: controlled moraine ridge. (Modified from Boulton, 1972b)

north Wales (Boulton, 1977b; McCarroll and Harris, 1992), where interbedded debris flow diamictons, sands and gravels overlie subglacial sediment associations.

In certain situations, debris-rich basal ice may melt out contemporaneously with supraglacial sedimentation, resulting in a two-layer stratigraphy consisting of a lower assemblage of meltout tills and an upper assemblage of debris flows, glacifluvial sands and gravels, and glacilacustrine silts. An excellent example of this stratigraphic succession was described by

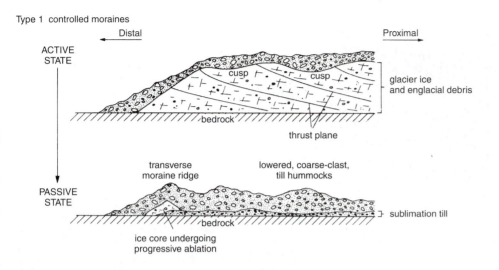

Type 1 controlled moraines

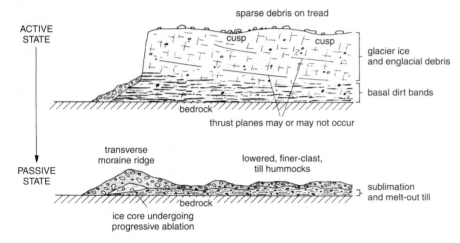

Type 2 controlled moraines

Fig. 11.67 The evolution of two types of controlled moraine based upon examples from Antarctica. Type 1 moraines are produced where there is abundant debris on the ice surface, and Type 2 moraines involve a sparse supraglacial debris cover. (From Rains and Shaw, 1981. Reproduced by permission of the International Glaciological Society)

Johnson *et al.* (1995) from the hummocky moraine of western Wisconsin, where supraglacial sediment is thought to have been lowered vertically on to a basal melt-out sequence. Variations in the thickness of basal debris concentrations and thus the melt-out till are considered to be important to the production of the final hummocky forms in western Wisconsin, rather than the effects of supraglacial topographic reversal alone.

11.4.3 Kame and kettle topography

Kames are steep-sided, variously shaped mounds composed chiefly of sand and gravel, formed by supraglacial or ice-contact glacifluvial deposition (Holmes, 1947). The term has a Scottish origin and is derived from the word *cam* or *kaim* meaning 'crooked and winding or steep-sided mound', and originally encompassed eskers (Geikie, 1894). In the glacial geomorphological literature, however, the term 'kame' is employed for discontinuous or ter-

race-like features, in contrast with more continuous esker ridges. In practice, it is not possible to draw a firm line between kames and eskers, and many transitional forms exist.

The term 'kame' has been used in several, sometimes confusing, senses in the literature. It is commonly used as a noun but has been used as an adjective, as in *kame plateau, kame terrace, kame delta* and *kame moraine*. Many of these terms are of limited usefulness, and duplicate more accurate terminology. Kame deltas, for example, are simply ice-contact deltas (Section 11.6.5.2). In this section, we discuss kame and kettle topography, comprising assemblages of mounds and hollows. Kame terraces, or ice-contact valley-side terraces, are described in Section 11.4.5.

Kame and kettle topography forms tracts of mounds and ridges (kames) and intervening hollows (kettles or kettle holes). The kettle holes may be filled with water, and represent areas of subsidence caused by the melting of buried ice, whereas the

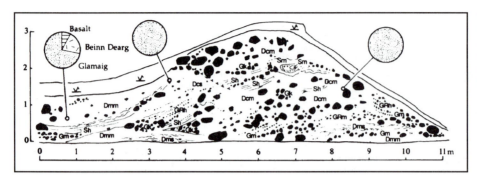

Fig. 11.68 An exposure through hummocky moraine on the Isle of Skye, Scotland. Pie diagrams show proportions of basalt and Beinn Dearg and Glamaig granites, and suggest compositional stratification. (From Benn, 1992b)

kames are upstanding masses of glacifluvial deposits (Fig. 11.69). Kettle holes also occur on otherwise continuous outwash surfaces, known as *pitted outwash* (Section 11.4.6). The difference between kame and kettle topography and pitted outwash reflects the relative importance of glacifluvial deposition and the melt-out of buried ice in the form of the final landform assemblage. Kame and kettle topography may appear completely chaotic, but some degree of lineation or pattern is not uncommon (Paul, 1983). Linear patterns provide evidence for *controlled deposition*, reflecting former englacial and supraglacial debris concentrations, crevasse patterns and/or drainage systems.

Kame and kettle topography forms where large quantities of debris are reworked by supraglacial and englacial drainage systems during the final stages of glacier wastage (Fig. 11.70; Cook, 1946a, b; Clayton, 1964). Kames are commonly found in association with hummocky moraine (Section 11.4.2), from which they can be differentiated by their morphology and composition. The presence of flat or gently sloping plateau surfaces, representing remnant glacifluvial terraces, is particularly diagnostic (Christiansen, 1956; Hartshorn, 1957; Gravenor and Kupsch, 1959; Winters, 1961; Clayton, 1964; Marcussen, 1977). Additionally, kames tend to have smoother slopes than hummocky moraine, because of their less variable internal composition. Exposures in kames reveal bedded sand and gravel facies with variable internal geometry depending on conditions during deposition. Interbeds of lacustrine and mass flow deposits may also be present (Huddart, 1983). There is usually some degree of internal folding and faulting due to the removal of supporting ice, particularly towards kame margins. The greatest amounts of disturbance will be associated with supraglacial or englacial deposits let down on to the substratum (Boulton, 1972b; Karczewski, 1974; Huddart, 1983; Hambrey, 1984; Brodzikowski and van Loon, 1991). The most common structures are steep normal faults bounding down-faulted masses (McDonald and Shilts, 1975).

Fig. 11.69 Kame and kettle topography on the foreland of Sandfellsjokull, Iceland. (Photo: D.J.A. Evans)

The presence of glacifluvial sediments can often be inferred even where good exposures are lacking. One useful criterion is the presence of numerous rabbit burrows in mounds, indicating the occurrence of easily excavated sand. The morphology of surface clasts is also a useful indicator of water-worn material (King and Buckley, 1968; Sugden and John, 1976). Kames represent proximal glacifluvial deposits, and clasts tend to be less well rounded than in more distal outwash deposits, but better-rounded than in subglacial facies. It should be noted, however, that clasts may have been subject to complex transport and depositional histories, and care must be taken not to attach too much significance to roundness statistics, unless large samples are taken.

Elongate kames may be produced by the filling of surface crevasses by glacifluvial sediment (Flint, 1928; Holmes, 1947; Johnson, 1975). If such sediments do not penetrate close to the bed, any resulting landforms will have poor preservation potential, owing to severe disturbance during glacier ablation. An example of a series of possible supraglacial crevasse-fill ridges occurs on north-west Ellesmere Island, Canada (Fig. 11.71; D.J.A. Evans, 1990a). The ridges were deposited along the coalescence zone of two outlet glaciers during the last glaciation and now

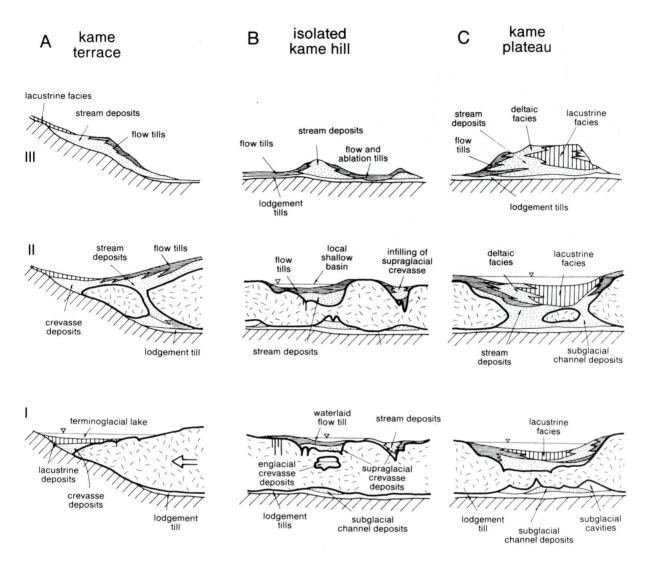

Fig. 11.70 Schematic models for the production of various types of kame: (A) kame terraces; (B) isolated kame hills or mounds; (C) kame plateaux. (From Brodzikowski and van Loon, 1991. Reproduced by permission of Elsevier)

trend obliquely across the contours of a valley side. They probably represent the fluvially reworked lower portion of a medial moraine in the ablation zone, because they are composed of sand and gravel hummocks and, although heavily disturbed by melt-out, contain pockets of well-sorted sediments.

11.4.4 Ice-walled lake plains

Ice-walled lake plains are irregularly shaped, flat-topped mounds which represent the infillings of former ice-walled lakes (Section 6.5.4.2), which are left as positive relief features when the surrounding ice melts (Clayton and Cherry, 1967; Parizek, 1969). Many ice-walled lake plains have raised rims around the central plateau which are typically composed of glacifluvial or mass movement deposits. These rims record late-stage deposition from the ice walls into the infilled lake basin, or subsidence of the central area due to melting of a central ice core. In many cases, however, the central parts of the mounds show little evidence of subsidence or collapse, indicating that deposition, at least in its final stages, took place on firm ground. Late Pleistocene examples in western Wisconsin are shown in Fig. 11.72 (Johnson *et al.*, 1995). The plains range in area from 1 to 13 km^2, with a typical area of around 2 km^2, and have similar surface altitudes. They commonly have a rim standing 3–10 m above the centre of the plain, partly composed of diamicton that flowed from the ice slopes surrounding the lakes. The interiors of the plains consist of up to 20 m of glacilacustrine silts, whereas stream- and wave-sorted sand and gravel commonly occurs round the margins.

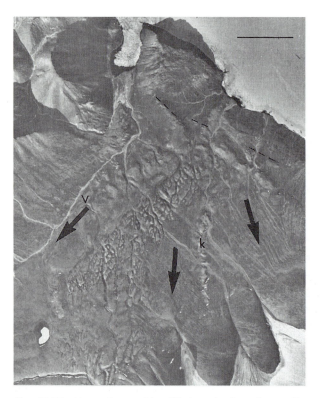

Fig. 11.71 Linear kame ridge (k) deposited at the confluence of two glacier snouts, Phillips Inlet, north-west Ellesmere Island, Arctic Canada. Note the lateral meltwater channels cut at retreating glacier margins (direction of retreat shown by arrows; v = main valley). Scale bar represents 1 km. (Aerial photograph reproduced by permission of the Department of Energy, Mines and Resources, Canada)

11.4.5 Kame terraces

Kame terraces are gently sloping depositional terraces perched on valley sides, which are deposited by meltwater streams flowing between glacier margins and the adjacent valley wall (Fig. 11.73 (Plate 23); Mannerfelt, 1945, 1949; Flint, 1957; Sissons, 1958b). Flights of kame terraces may occur on valley walls, documenting periodic reduction of the glacier surface during terrace formation (Sissons, 1958b, c). During deglaciation, the ice-contact faces of kame terraces are very unstable and are thus prone to collapse through debris flows and landslides. Additionally, melt-out of buried ice within kame terraces often produces a kettled or pitted surface. In a study of a kame terrace in Alaska, McKenzie (1969) discovered that only 4 m of sand and gravel was overlying approximately 35 m of glacier ice. In such situations the terrace form is not likely to survive final melt-out of buried ice.

Although composed predominantly of fluvial sands and gravels, kame terraces may also contain lacustrine sediments which collect in ephemeral ponds, and debris flow diamictons derived directly from the glacier surface (e.g. Huddart and Lister, 1981; de Jong and Rappol, 1983; Levson and Rutter, 1989b). Coarsening-upward sequences have been reported from some kame terraces, which can be interpreted as indicating an increase in stream discharge concomitant with glacier retreat (John, 1972; Sugden and John, 1976; Levson and Rutter, 1989b). Kame terraces can sometimes be confused with formerly continuous outwash surfaces that have been reduced to valley-side terraces by postglacial incision (Section 11.5.1.1). However, outwash terrace remnants on opposite valley sides should have similar gradients and altitudes, whereas paired kame terraces may have different gradients, reflecting differences in ice-margin morphology (Gray, 1975).

A number of features identified as ancient kame terraces may in fact be slope deposits which accumulated between lateral moraines and the valley side, but can easily be differentiated from true kame terraces if exposures are available. An example of a terrace deposited on top of a lateral moraine at the margin of the Tasman Glacier, New Zealand, is shown in Fig. 11.56a.

11.4.6 Pitted sandar

Pitted sandar, also referred to as *kettled sandar* or *kettled outwash plains*, are sandar which are cratered by hollows left by the melt-out of isolated buried blocks of glacier ice (Maizels, 1977). The ice blocks may originate in two main ways: (a) as remnants of a glacier snout, detached from the rest of the glacier by differential ablation (Rich, 1943; Price, 1969; Gustavson and Boothroyd, 1987); and (b) as icebergs transported on to the sandur surface by floodwaters, particularly jökulhlaups (Churski, 1973; Galon, 1973; Klimek, 1973; Maizels, 1992). In either case, deposition of glacifluvial sediments may partially or completely bury the ice blocks, leaving hollows when the ice melts. Although case (b) refers to proglacial environments, pitted sandar formed by the melt-out of transported ice blocks are included in this section because of their close genetic similarity to supraglacial pitted sandar.

Patterns of deposition on supraglacial sandar are strongly controlled by patterns of ablation of the underlying ice. For example, the late Pleistocene Orrisdale Outwash Member of the Isle of Man has been interpreted as a supraglacial sandur in which patterns of sedimentation were controlled by transverse ridges of dead ice, perhaps reflecting englacial debris concentrations (Fig. 11.74; Thomas *et al.*, 1985). A series of marginal sandar were built up between the dead-ice ridges until melt-out and fluvial

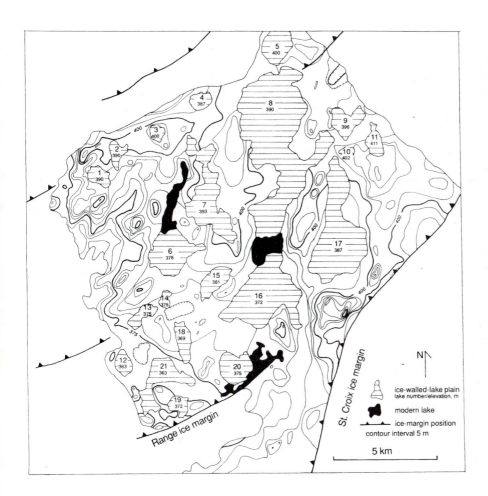

Fig. 11.72 Map of ice-walled lake plains among hummocky moraine in western Wisconsin, USA. (From Johnson *et al.*, 1995. Reproduced by permission of Scandinavian University Press)

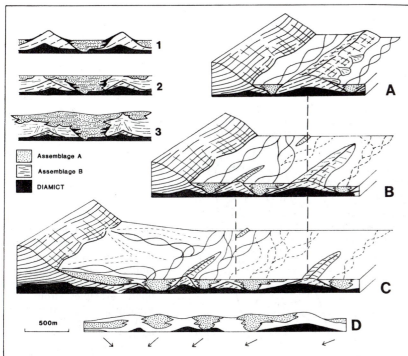

Fig. 11.74 Depositional sequence produced in an ice-marginal supraglacial sandur based upon interpretations of the Orrisdale Outwash Member on the Isle of Man. Fluvial sedimentation is guided by the changing supraglacial topography of melting ice. Stages A–C show sequential evolution of the sandur system to produce the final facies assemblage in D. The inset 1–3 shows the stages in the development of a marginal sandur and traces the progressive melt-out of stagnant ice ridges and the widening of the sedimentation system. Assemblage A comprises typical gravel and sand glacifluvial outwash and Assemblage B comprises predominantly overbank fines. (From Thomas *et al.*, 1985. Reprinted by permission of John Wiley & Sons)

aggradation caused outwash widening and sandur coalescence. A particular type of pitted outwash sometimes occurs in mountainous terrain where large segments of wasting glacier lobes become isolated in valley bottoms and then buried by outwash (Fleisher, 1986). Ice melt-out produces large depressions termed *dead-ice sinks*. Numerous closely spaced sinks can coalesce to form a single elongate depression with a complex bottom topography, referred to as a *dead-ice moat* (Fig. 11.75).

Features resulting from the melt-out of ice blocks transported by jökulhlaups on Myrdalssandur, Iceland, have been studied in detail by Maizels (1992). Ice melt produces *boulder ring structures*, between 3 m and 40 m in diameter, consisting of near-circular, boulder-rich rims surrounding a central depression (Fig. 11.76). The rims are highest on the downstream

flanks of the structures, and gaps may occur in the rim on the upstream flanks. Four principal types of ring structure were observed, which reflect the amount of sediment held within the ice block (Fig. 11.77). Type 1 is a *normal kettle hole* produced by the collapse of outwash materials surrounding a clean ice block; Type 2 is a *rimmed kettle*, which possesses a diamicton drape in the central depression and low-amplitude bouldery diamicton rims; Type 3 is a *crater kettle*, which possesses a thick diamicton lining and thick boulder-covered diamicton rims; and Type 4 is a *till-fill kettle*, produced by the melt-out of a heavily sediment-laden ice block and therefore constituting a mound of diamicton centred over a depression in the outwash.

The morphology and sedimentology of the boulder ring structures were linked to the sediment con-

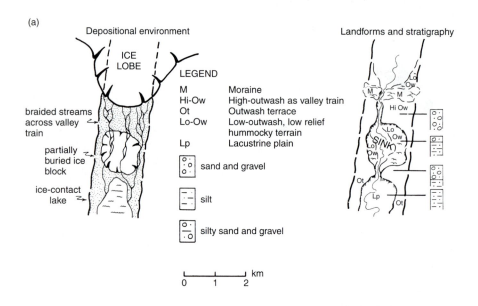

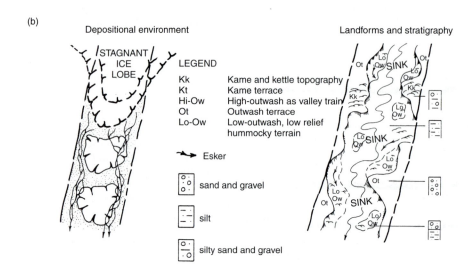

Fig. 11.75 The development of dead-ice sinks (a) and moats (b) with associated landforms and stratigraphies in deglaciated valleys based upon examples from the Appalachian Plateau of New York. (From Fleisher, 1986. Reprinted by permission of the Geological Society of America)

Fig. 11.76 Ring structures indicative of iceberg melt-out after a jökulhlaup on a distal residual bar surface of Myrdalssandur, Iceland. Ring diameters are 20–25 m and the view is looking upstream. (Photo: Judith Maizels)

centration of ice blocks with the aid of laboratory experiments. Maizels (1992) introduced an *index of rim morphology* (RMI) which approximates the relative volumes of rim ridges and hollows in ring structures, and so is a volumetric measure of sediment concentration of the ice block relative to its diameter:

$$\text{RMI} = \frac{2wh}{DH} \times 100 \qquad (11.5)$$

where w is the width of the rim ridge, h is the height of the rim ridge, D is the diameter of the structure, and H is the depth of the hollow. The dimensions of the ring structures vary with both the sediment concentration and the depth of burial of the ice block; rim height and width increase in direct relation to sediment concentration and inversely with depth of burial. This relationship is demonstrated by the relation:

$$\text{RMI} = 23.62D_b - 1.02bC \qquad (11.6)$$

where D_b is the depth of burial relative to ice block height, C is the sediment concentration by volume, and b is an empirically determined constant.

Using these formulae, Maizels (1992) calculated that, for ice blocks buried at depths of $0.3 \times H_i$ (where H_i is the height of the ice block), the formation of Types 2, 3 and 4 ring structures requires sediment concentrations of 5–20 per cent, 20–60 per cent and >60 per cent respectively.

Kettle holes may become completely infilled with sediments if sandur aggradation continues during the melting of buried ice blocks. In this case, no sur-face expression remains, but buried kettle holes are preserved in the sedimentary record as downfaulted blocks bounded by steep normal faults (Fig. 11.77: Type 1). The amount of displacement of such faults will decrease upward if downfaulting occurred during sediment aggradation. This is because the deeper layers which were deposited immediately above the ice block will experience the greatest amount of subsidence, whereas shallower layers, deposited after some of the block has already melted, will experience less. The uppermost layers, deposited after complete melt-out of the block, will be undeformed.

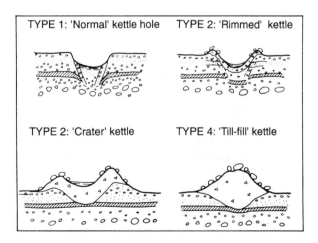

Fig. 11.77 Schematic reconstructions of the four types of ring structure produced by the melting *in situ* of ice blocks on sandur surfaces. Types 1 to 4 represent a progressive increase in sediment concentration of the parent ice block. (Modified from Maizels, 1992)

11.5 PROGLACIAL ASSOCIATIONS

This section discusses sediment–landform associations deposited in terrestrial proglacial environments by glacifluvial and slope processes.

11.5.1 Sandar and valley trains

Most proglacial rivers carry large amounts of suspended sediment and bedload, and this is characteristically deposited in extensive, gently sloping outwash plains known by the Icelandic term *sandar* (singular: *sandur*). Narrow tracts of outwash hemmed in by valley sides in mountainous terrain are termed *valley trains*. The streams responsible for sandar and valley trains are typically braided, owing to a combination of steep gradients, abundant bedload, cohesionless bank and bed material, and fluctuating discharges (Section 3.9.2; Fahnestock, 1963; Cheetham, 1979; Miall, 1992). The migration of braided channels and intervening bars during outwash accumulation results in very distinctive facies associations, careful study of which allows the former hydrological environment to be reconstructed in some detail. Some sandar are periodically inundated by jökulhlaups, or glacier outburst floods (Section 3.5), and exhibit facies associations which differ in many respects from those deposited by 'normal' meltstreams fed by glacier ablation cycles.

This section examines the morphology of sandar and valley trains, goes on to describe the sedimentology of 'normal' and jökulhlaup-influenced sandar, and concludes by briefly examining methods for reconstructing palaeodischarges from ancient glacifluvial deposits.

11.5.1.1 MORPHOLOGY

The proximal ends of sandar and valley trains may be located at: (a) former glacier terminus positions; (b) the outlets of periodically draining ice-marginal lakes; or (c) some distance below former ice margins where stream velocities fall, such as where narrow, confined gorges open out into broad valley floors. Where sandar commence at a former ice margin, feeder channels **may** occupy gaps in moraines. Alternatively, moraines may be absent, and the upper edge of the sandur may simply be marked by an abrupt upglacier-facing scarp known as an *ice-contact slope* which records the location of the ice when the sandur was being deposited. Such ice-contact slopes are also known as *outwash heads*, and may be the only evidence for former glacier limits in cases where ice-marginal deposition was dominated by glacifluvial processes.

Outwash heads sometimes form *ice-contact fans*, consisting of cone-shaped sediment accumulations radiating out from former feeder points and grading down on to a lower sandur surface. The geometry of such fans reflects the routeing of meltwater, which can be strongly constrained by the ice-marginal topography. Two contrasting examples, from the respective forelands of Fjallsjökull and Breidamerkurjökull, Iceland, were described by Boulton (1986). The north-eastern margin of Fjallsjökull is banked against a hillside, forcing meltwater to drain parallel to the glacier snout and between abandoned moraines, resulting in narrow, linear fans. In contrast, the southern margin of Breidamerkurjökull is unconfined by topography, and therefore meltwater from the glacier produced a series of coalescent outwash fans, which formed a ramp sloping down from the margin towards the coast. In its proximal zone this outwash was deposited over the thinning ice margin, thereby producing ice stagnation topography upon melt-out of buried ice. The characteristics of outwash deposited over buried ice are described in Sections 11.4.3 and 11.4.6.

A characteristic feature of many sandur surfaces is the presence of *terraces* at higher levels than the active channel. Indeed, in the case of ancient sandur plains, the entire surface may take the form of terraces perched high above the modern river channel. Flights of several terraces may be present. Depositional terraces record episodes of sediment aggradation (sandur formation) followed by incision. Incision may be focused in the central part of the sandur, leaving *paired terraces* on both sides of the valley, but if incision is accompanied by migration of the active channel towards one valley side, a single *unpaired terrace* may be preserved on the opposite side (Fig. 11.78). The switch from aggradation to incision may result from external forcing, such as changes in climate or base level, or from internal changes within the system, such as local sediment dynamics (Dawson and Gardiner, 1987). In glaciated catchments, the most common causes of terrace formation are fluctuations in sediment supply or fluvial discharge. For example, sandur aggradation will be encouraged during the paraglacial period in the early stages of deglaciation, when large quantities of unconsolidated sediment can enter the fluvial system (Sections 7.6 and 11.5.2.1). When the supply of sediment begins to be exhausted, incision will occur if rivers maintain high capacity for sediment transport; this will be the case particularly if abundant meltwater is still available from receding glaciers. Major terrace systems in many glaciated regions are known to have formed in full-glacial conditions, when large amounts of sediment were introduced into lowland catchments. In Britain, the Main Terrace in the Severn basin can be traced southward from ice-contact deposits at the limit of the last (Devensian) glaciation (Shotton, 1977; Dawson and Bryant, 1987; Dawson

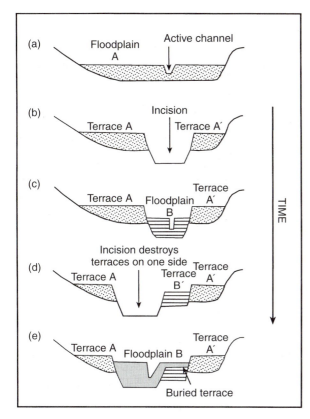

Fig. 11.78 Glaciofluvial terrace formation. (Modified from Lowe and Walker, 1984)

and Gardiner, 1987), whereas some high fluvial terraces in the Thames basin have been correlated with the earlier Anglian glaciation (Gibbard, 1985, 1988).

Incision in one part of a sandur will release sediment from storage, sending a pulse of sediment into lower parts of the system, which may initiate local aggradation. Thus terrace formation may occur at different times in proximal and distal parts of a catchment, depending on local sediment dynamics. The tendency for fluvial catchments to exhibit diachronous patterns of aggradation and incision has been termed *complex response* by Schumm (1977) and Patton and Schumm (1981), who warn against too simplistic an interpretation of terrace sequences in terms of climatic change. Further discussion of river terrace formation can be found in Dury (1970), Lowe and Walker (1984) and Dawson and Gardiner (1987).

11.5.1.2 BRAIDED RIVER FACIES ASSOCIATIONS

In braided river environments, the aggradation and migration of bedforms and larger-scale dunes, bars and channels produce a wide range of facies associations. Such associations can be usefully viewed as sediment packages at a range of scales (Allen, 1983;

Miall, 1985a, 1992; Bristow, 1996). At the smallest scale, sediment packages comprise individual *laminae* or *beds* laid down from single pulses of sediment (Section 10.4; McKee and Weir, 1953; Allen, 1963, 1982a). At larger scales, stratification is defined by *bounding surfaces*, which represent former surfaces of erosion or non-deposition lying between depositional units. Several orders of bounding surfaces can be recognized, with progressively wider lateral extent and environmental significance (Fig. 11.79; Allen, 1982a, 1983; Miall, 1985a, 1992). *Sets* of strata deposited during the migration of single bedforms are bounded above and below by *first-order surfaces*; and *cosets*, or groups of sets deposited during migration of assemblages of bedforms, are bounded by *second-order surfaces*. Successively higher-order surfaces (up to eighth order according to Miall, 1992) define progressively larger depositional units. According to Miall (1985a, 1992), fourth and fifth order surfaces define *architectural elements*, representing large-scale components of the fluvial landscape, such as longitudinal bars and channel fills. Architectural elements combine to form depositional systems representative of different fluvial styles. Miall (1985a, 1992) defined eight basic *architectural elements* characteristic of fluvial deposits, which can then be used to reconstruct fluvial depositional styles from sediment exposures (Fig. 11.80). The eight elements are:

1. *Channels* (CH). This element represents the infills of channels. Typical facies include ripple cross-laminated sands and cross-bedded sands and gravels, recording the downstream migration of bedforms within the channel. With care, transverse sections can be used to reconstruct former channel dimensions, geometry and patterns of migration (Bristow, 1996).

2. *Downstream accretion macroforms* (DA). This element comprises cosets of downstream-dipping cross-beds resting on a flat or channelized base, representing the downstream migration of bar fronts. DA elements are particularly characteristic of braided streams. They commonly have a complex internal geometry consisting of multiple cosets bounded by a hierarchy of erosion surfaces, which represent the migration of superimposed bedforms (e.g. ripples superimposed on dunes superimposed on bars).

3. *Lateral accretion macroforms* (LA). These forms consist of sets of cross-beds that dip transversely or obliquely to the main channel trend, recording the lateral migration of bank-attached or mid-channel bars. LA macroforms also develop on the insides of channel bends, where deposition occurs, while erosion takes place on the opposite bank. Facies may be predominantly gravelly or

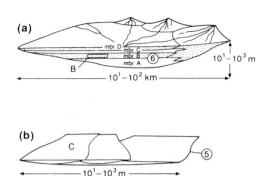

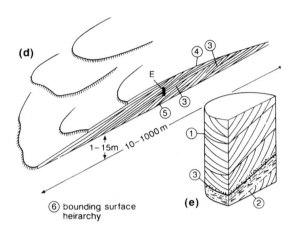

(a)

mbr D mbr C
mbr B
mbr A
B

$10^1 - 10^3$ m

$10^1 - 10^2$ km

(b)

C

$10^1 - 10^3$ m

(d)

E

1–15m 10–1000 m

⑥ bounding surface
heirarchy

(e)

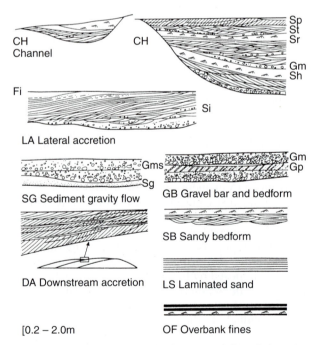

(c)

sand flat
CH
D
lateral cr
point bar
LA
DA

10–3000 m

Fig. 11.79 Scales of depositional events in a fluvial system, showing a hierarchy of bounding surfaces. Circled numbers represent the ranks of the bounding surfaces, from (1) first-order surfaces marking the migration of bedforms to (6) sixth-order surfaces between major depositional cycles. In diagram (c) the two-letter codes refer to the architectural elements defined in the text and Figure 11.80. (From Miall, 1992. Reproduced by permission of the Geological Association of Canada)

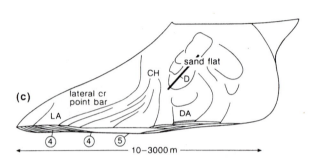

CH
Channel

CH

Sp
St
Sr

Gm
Sh

Fi

Si

LA Lateral accretion

Gms

Sg

SG Sediment gravity flow

Gm
Gp

GB Gravel bar and bedform

DA Downstream accretion

SB Sandy bedform

LS Laminated sand

[0.2 – 2.0m

OF Overbank fines

Fig. 11.80 Eight architectural elements of fluvial deposition. (From Miall, 1992. Reproduced by permission of the Geological Association of Canada)

sandy. The vertical thickness of LA and DA macroforms reflects the original bar top to channel height, minus any subsequent erosion, and is typically less than a few metres.

4. *Gravel bars and bedforms* (GB). Low-relief gravel surfaces form a major element of many sandar, and are represented in sections as extensive tabular sheets of massive gravel and sets of trough and planar cross-bedded gravel (Sections 10.4.4 and 10.4.7). This element commonly occurs on top of LA or DA macroforms, recording the migration of bar tops over former channel margin positions.

5. *Sediment gravity flow* (SG). This element consists of channelized or tabular lenses of poorly sorted gravel or diamicton, deposited by debris flows or related processes such as hyperconcentrated flows (Sections 10.4.9 and 10.5.5). Such elements are common on the proximal reaches of some sandar and along the margins of some valley trains, where they represent mass flows from glacier margins or steep valley sides.

6. *Sandy bedforms* (SB). This element includes lenses and sheets of horizontally bedded, ripple cross-laminated and cross-bedded sand, and records deposition on bar surfaces and in chutes and minor channels. This element commonly occurs on top of gravel bars and bedforms, where

fining-upward sequences may provide evidence for deposition during waning flow.

7. *Laminated sand sheets* (LS). Laminated sand sheets are extensive tabular units of horizontally bedded sand, deposited by ephemeral sheet floods. They are uncommon in proximal sandar, but occur in some distal successions (e.g. Landvik and Mangerud, 1985).

8. *Overbank fines* (OF). This element consists of blanket-like drapes of massive and laminated silt, and forms a major component of the floodplain deposits of meandering non-glacial rivers. However, OF elements occur locally within sandur

deposits, particularly in distal environments where they infill abandoned channels.

These architectural elements can be recognized in many sandur and valley train deposits, although not all researchers employ the terminology proposed by Miall (1985a, 1992). For example, Dawson (1985), Dawson and Bryant (1987) and Dawson and Gardiner (1987) have described extensive exposures in the Main Severn Terrace near the southern margin of the last British ice sheet, and recognized three main architectural elements (Fig. 11.81). Association A is equivalent to Miall's GB element, and consists of lat-

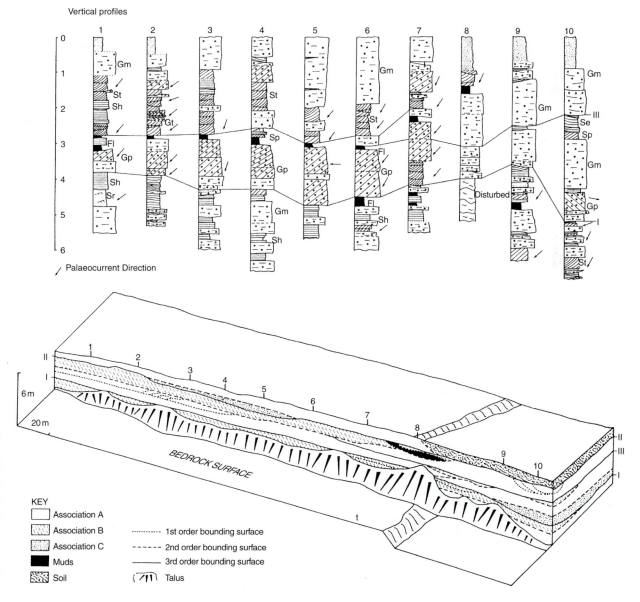

Fig. 11.81 Vertical profiles and block diagram showing the facies architecture in late Devensian outwash gravels at Holt Heath, Worcestershire, England. (From Dawson and Bryant, 1987. Reproduced by permission of SEPM)

erally extensive sheets of massive and cross-bedded gravel which were interpreted as bar cores. Association B is composed of discontinuous, thin sheets of massive and horizontally laminated sand, with subordinate amounts of gravel, isolated cross-bedded or ripple cross-laminated sand, and thin mud drapes. The association was interpreted as bar top deposits laid down by shallow, waning flows, and is apparently equivalent to Miall's SB element. Association C consists of cosets of cross-stratified sand and gravel occupying broad, shallow channels. This association was interpreted as the infills of channels by migrating bedforms, and is equivalent to element CH. The geometry of these elements in the exposures, and their relationship to one another, records sediment aggradation in a dynamic braided river environment, in which channels and bars constantly shifted position.

Another good example of architectural analysis of braided river deposits is a study of last glacial (Würm) outwash in northern Switzerland by Siegenthaler and Huggenberger (1993). Three elements were recognized. (a) The first was extensive tabular sheets of poorly sorted gravel with a matrix of silt and sand. These occur at high levels in the deposit as single beds or sets of beds up to several metres in thickness, and were interpreted as former 'traction carpets' deposited from hyperconcentrated floods (Section 10.4.9; Todd, 1989). This is probably equivalent to Miall's sediment gravity flow (SG) element. (b) The second element was horizontally bedded, well-sorted gravels. This element comprises sheets a few metres to tens of metres wide and 1–2 dm thick. Sheets may grade laterally into cross-bedded gravels or abut inclined surfaces of older gravel. This association is similar to Miall's GB element, and was interpreted as the deposits of migrating gravel sheets or incipient mid-channel bars within wide, shallow channel reaches (Section 3.9.3; Ferguson, 1993). (c) The third type of deposit was large-scale trough fills. These range from a few metres to more than 100 m wide and 0.5–6 m deep, and are filled with cross-beds of sand and gravel. Cross-bedding is generally sub-parallel to the trough margins, giving the fills an onion-like appearance. This association is interpreted as the infills of deep, migrating scour pools, and is a special case of Miall's CH element. The pool and mid-channel gravel sheets (associations (c) and (b)) were interpreted as the products of moderate-magnitude floods, and the poorly sorted gravel sheets (association (a)) as the deposits of high-magnitude floods. Facies and structures of low-magnitude flows were not recorded, indicating that differential sediment preservation can play an important role in the final character of sandur deposits.

Rigorous three- or two-dimensional analysis of glacifluvial sediments is very time-consuming and requires extensive exposures. As a result, many studies have utilized one-dimensional *vertical profile analysis* to record and interpret glacifluvial successions, using Walther's law to infer spatial patterns of deposition from conformable facies associations (Sections 10.2.2.3 and 10.2.3). Repetitive vertical transitions between facies can be identified quantitatively by *Markov chain analysis*, which determines the probability that particular sequences will occur (Fraser, 1982; Landvik and Mangerud, 1985; Thomas *et al.*, 1985; Dawson and Bryant, 1987). Sequences with a high probability of occurrence are interpreted as recurrent associations representing common depositional cycles. This approach has its limitations, as it ignores the type of bounding surface between units and therefore tends to group together facies which are not genetically related. Furthermore, it obscures the true geometrical relations between facies by a statistical averaging procedure. Nevertheless, it allows the identification of repetitive facies sequences which record common depositional events such as major floods, the migration of bars, and the infilling and abandonment of channels.

The range of variation in braided river deposits was illustrated by Miall (1977, 1978), who defined several type examples of braided river successions based on published descriptions of well-known fluvial sequences (e.g. Williams and Rust, 1969; Rust, 1972b, 1978; Boothroyd and Ashley, 1975). The type examples, or facies models, which are most representative of glacifluvial sequences are named the Trollheim-type, Scott-type, Donjek-type, and Platte-type assemblages, named after the North American rivers where they were first recognized (Table 11.2). Miall (1977, 1978) represented the models using vertical profiles, which can be equated with three-dimensional architectural element diagrams presented by Miall (1985a), as shown in Figs 11.82 and 11.83. The models do not represent distinct and separate styles of deposition, but fixed points on a continuum of variation. Nevertheless, they are useful points of reference which exemplify contrasting styles of sedimentation in a range of sandur settings.

The *Trollheim-type* vertical facies succession is dominated by massive, clast-supported gravels and matrix-supported gravels, representing braid bars and debris flow deposits, respectively. This succession is equivalent to stacked architectural elements GB and SG, and is typical of proximal settings where debris supply is large relative to water discharge.

The *Scott-type* succession is named after the Scott outwash fan in Alaska studied by Boothroyd and Ashley (1975), and is typical of fluvially dominated proximal sandar. The dominant sediments are stacked massive to cross-bedded gravel units (GB), which record the aggradation and migration of longi-

Table 11.2 Facies assemblages associated with vertical facies models of Miall (1977, 1978)

Model type	Environment	Main facies	Minor facies
Trollheim type	Proximal rivers (predominantly alluvial fans) subject to debris flows	Gms, Gm	St, Sp, Fl, Fm
Scott type	Proximal rivers (including alluvial fans) with stream flows	Gm	Gp, Gt, Sp, St, Sr, Fl, Fm
Donjek type	Distal gravelly rivers (cyclic deposits)	Gm, Gt, St	Gp, Sh, Sr, Sp, Fl, Fm
South Saskatchewan type	Sandy braided rivers (cyclic deposits)	St	Sp, Se, Sr, Sh, Ss, Sl, Gm, Fl, Fm
Platte type	Sandy braided rivers (virtually non-cyclic)	St, Sp	Sh, Sr, Ss, Gm, Fl, Fm
Bijou Creek type	Ephemeral or perennial rivers subject to flash floods	Sh, Sl	Sp, Sr

tudinal bars. Channel scours underlie gravel units but are often difficult to identify, owing to the similarity of grain sizes between units. Thin lenses of sand, representing deposition in abandoned channels or bar-edge sand wedges, are interbedded with the gravel units. Fining-upward gravel to sand units can also be observed, and document deposition through the waning stages of high flow.

The *Donjek-type* is typical of intermediate reaches of sandar, where average particle sizes are smaller than in proximal environments. Intermediate reaches also tend to exhibit several topographic levels, with a marked differentiation into active channels and inter-fluves which are inundated only during exceptional flood events. The succession consists of: (a) massive gravels (GB); (b) cross-bedded, horizontally bedded and ripple cross-laminated sands (SB); and (c) drapes of fine-grained sediments (OF) in repeated fining-upward sequences. These sequences document the aggradation and then abandonment of channels (deposition of gravel followed by sands and then fines), reflecting frequent channel and bar migration (Bluck, 1979). Similar fining-upward sequences have been identified by Markov chain analysis of late Palaeozoic glacifluvial deposits in India (Casshyap and Tewari, 1982). An important component of this

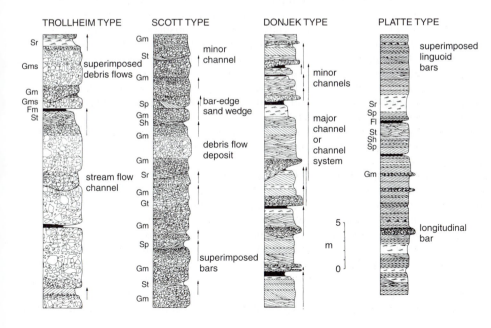

Fig. 11.82 Four vertical facies type associations representative of different sandur environments. (From Miall, 1978. Reprinted by permission of the Canadian Society of Petroleum Geologists)

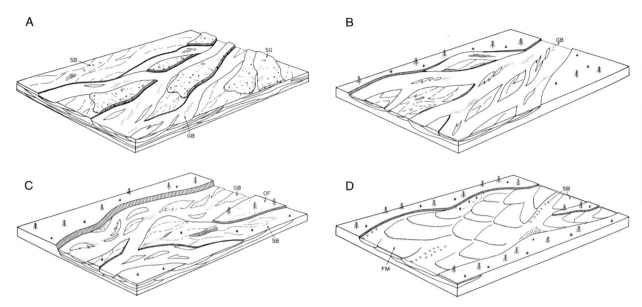

Fig. 11.83 Block diagrams showing the braiding patterns associated with the type fluvial facies successions shown in Fig. 11.82. (A) Proximal sandur with braid bars and debris flow deposits, equivalent to the Trollheim-type vertical facies succession. (B) Fluvially dominated proximal sandur, equivalent to the Scott-type succession. (C) Multi-level intermediate sandur, equivalent to the Donjek-type succession. (D) Distal sandur, equivalent to the Platte-type succession. (From Miall, 1985. Reproduced by permission of Elsevier)

vertical sequence is the occurrence of diamicton clasts at the base of the massive gravel facies, interpreted as ice-proximal bar deposits. Diamicton clasts are often regarded as indicative of rapid aggradation of jökulhlaup deposits (Section 11.5.1.3). In contrast to the Scott type, which contains >90 per cent gravel, the Donjek type contains anywhere between 10 and 90 per cent gravel, and therefore contains a wider range of recognizable bedforms.

The *Platte-type* vertical facies succession is regarded as typical of distal reaches of sandar, although it is named after the non-glacial Platte River of Colorado and Nebraska (Smith, 1971; Blodgett and Stanley, 1980; Miall, 1983b). Runoff is spread between numerous shallow distributaries, and predominantly sandy bedload is transported in migrating linguoid bars, dunes and other bedforms. The resulting facies association is dominated by sandy bedforms with abundant cross-bedding (SB). Minor gravel lenses and overbank fines may also be present.

Miall (1983b) also adopted a fifth sandur model, applicable to the extreme distal reaches of glacially influenced rivers where the sediment source is predominantly wind-blown material (loess) and the resultant deposits are dominated by silts. The model was proposed by Rust (1978), on the basis of studies of the Slims River in the Yukon, which is characterized by low-relief channels and bars. The resulting *Slims-type* association consists of massive, laminated and ripple cross-laminated sandy silts deposited in drapes and migrating bedforms. According to Miall

(1983b), no Quaternary or ancient successions of this type have been described.

The *lithology*, *morphology* and *grain size distribution* of particles in glacifluvial sediments provide important sources of information on sediment provenance and transport. *Clast lithological analysis*, or statistical analysis of the rock types of river gravels or finer particles, has often been employed to correlate widely separated terrace fragments or sediment exposures (Bridgland, 1986). For example, Rose (1987a, 1989c) used particle lithology to correlate scattered exposures of Middle Pleistocene gravels and sands in the English Midlands and East Anglia, indicating the former existence of a major west- to east-flowing drainage system. The appearance of extra-basinal, erratic material in fluvial sediments can be used to determine when glaciers first appeared in a catchment, providing a powerful method of palaeogeographic reconstruction (e.g. Shotton, 1977; Green et al., 1980; Gibbard, 1988).

On many sandar, particle morphology and *grain size* show systematic downstream variations as a result of selective transport of particles and progressive abrasive wear (cf. Werritty, 1992). For example, Wightman (1986) demonstrated that glacifluvial gravels in front of two Norwegian glaciers display a downstream reduction in maximum grain size and an increase in particle roundness (Fig. 11.84). Such trends may be blurred by several factors, including local variations in depositional environment, and the

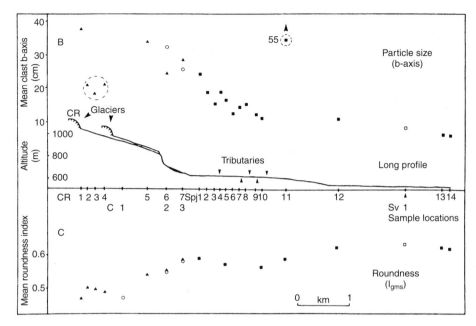

Fig. 11.84 Downstream variation in grain size (top panel) and roundness (lower panel) in front of Charles Rabotsbreen (CR) and Corneliussenbreen (C), Norway. (From Wightman, 1986. Reproduced by permission of the Quaternary Research Association)

introduction of older sediment by bank erosion or tributary streams.

11.5.1.3 JÖKULHLAUP-DOMINATED FACIES ASSOCIATIONS

Sandar that are periodically inundated by glacier outburst floods or jökulhlaups (Section 3.5) exhibit very distinctive facies successions, which can be used to reconstruct the type of flood event and flow characteristics. The most exhaustive work on jökulhlaup deposits is that of Maizels (1989a, b, 1993), who compared the sedimentology and morphology of jökulhlaup-dominated and 'normal' sandar in southern Iceland.

Maizels recognized three types of facies succession on Icelandic sandar, controlled by the runoff and sediment supply regimes of each glacifluvial system. *Type I* or *non-jökulhlaup sandar* occur where runoff is ablation-related and seasonal, and have the characteristics described in Section 11.5.1.2. *Type II* or *limno-glacial jökulhlaup sandar* develop where catastrophic floods are triggered by ice-dammed lake drainage, such as drainage of Lake Graenalon at the margin of Skeidararjökull. *Type III* or *volcano-glacial jökulhlaup sandar* occur where catastrohpic runoff is triggered by subglacial volcanic eruptions, such as the 1918 jökulhlaup on Myrdalssandur following an eruption of Katla below Myrdalsjökull (Jonsson, 1982; Tómasson, 1996). Maizels identified several vertical facies associations, which she was able to relate to different flow regimes and sediment supplies. The associations formed by jökulhlaups were labelled Types A–C, with subtypes, and are illustrated in Fig. 11.85.

Limno-glacial jökulhlaup sandar are dominated, at least in their proximal zones, by coarsening-upward gravel units interpreted as flood surge deposits (profile C5, Fig. 11.85). The gravels are clast-supported and dominantly subrounded. These deposits are overlain by a fining-upward sequence of cobble and pebble gravels, then horizontally laminated pumice silts, sands and fine gravel, interpreted as a post-surge waning flow sequence. The whole inverse-to-normally graded succession records rising then falling discharges over a single jökulhlaup event. Rounded till balls have been observed in such deposits, indicating that local tills had been eroded by the flood, but deposition rates had been too rapid for complete disaggregation of the eroded blocks (Fraser *et al.*, 1983).

Volcano-glacial jökulhlaup sandar are dominated by lithofacies profile types A and B, but profile types C and other facies associations may occur as minor components on a localized scale (Fig. 11.85). *Type A* profiles consist of massive to crudely bedded granules of tephra and pumice, deposited by hyperconcentrated flows. The deposits may include clasts, stringers or inclusions of other materials that were easily incorporated into the flow. *Type B* profiles, especially B3, are the most widespread facies association on volcano-glacial jökulhlaup sandar. They comprise massive granule gravels capped and sometimes underlain by horizontally bedded or trough cross-bedded granule gravels. Profile Type B3 is characterized by four distinct units. Unit 1 consists of coarse-grained, crudely bedded, clast-supported gravels or pumice granules, interpreted as pre-surge sediments derived from pre-existing sandur deposits. Unit 2 is composed of massive to crudely bedded pumice granules, interpreted as the products of hyperconcentrated flows associated with the main flood surge. Unit 3 comprises trough cross-bedded pumice granules which are separated from Unit 2 by an erosional contact, and Unit 4 comprises horizontally bedded pumice granules and sands. Units 3 and

4 are interpreted as the products of post-surge conditions during which deposition was by increasingly shallow fluid flows.

The variability in vertical facies sequences from non-jökulhlaup and jökulhlaup sandar is summarized in Fig. 11.86, which relates the different runoff and

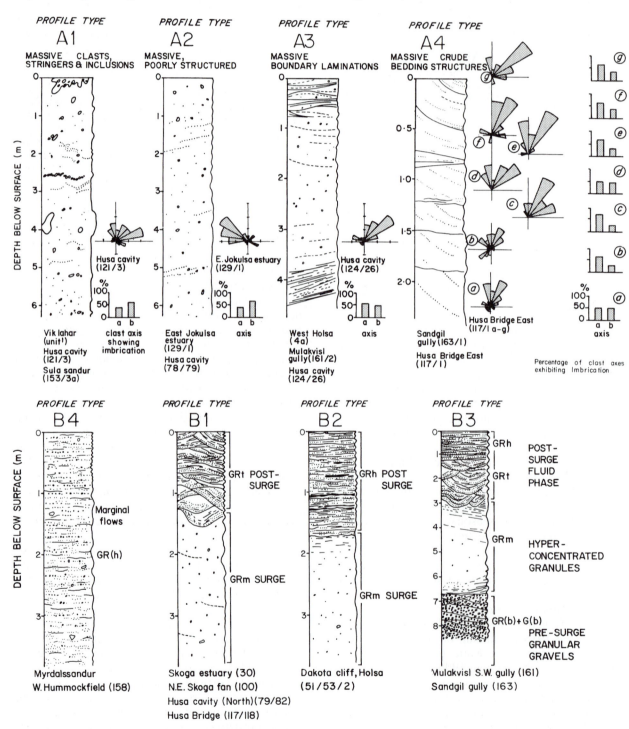

Fig. 11.85

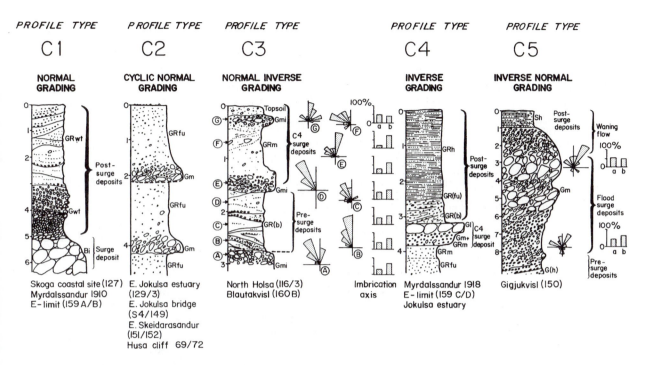

Fig. 11.85 Vertical lithofacies profiles (Types A, B and C) from the jökulhlaup sandar of Iceland. Facies codes explained in Fig. 10.3. (From Maizels, 1993. Reproduced by permission of Elsevier)

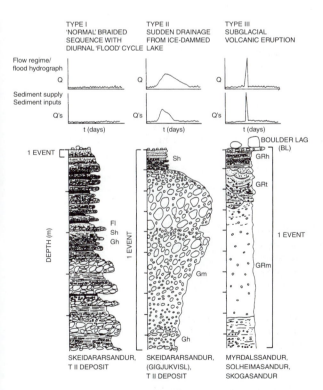

Fig. 11.86 Representative vertical lithofacies profiles for Icelandic sandar subject to different runoff regimes, with associated hydrographs and sediment supply characteristics. (From Maizels, 1993. Reproduced by permission of Elsevier)

sediment supply regimes of the three sandur types (I–III) to characteristic vertical profile models. The large difference in thickness of sedimentary units related to one discharge event between jökulhlaup and non-jökulhlaup sandar is very striking. The distinctive inverse-to-normal grading of the limno-glacial jökulhlaup sandur profile reflects the rising and falling flood stage. In contrast, the volcano-glacial jökulhlaup sandur profile has relatively constant grain size distribution throughout, reflecting the predominance of easily disaggregated volcanic eruption products. As a result, rising and falling flood stages, and changing sediment transport mechanisms, are represented by differences in sediment structure.

Figure 11.87 shows how facies associations deposited by jökulhlaups vary systematically with flow type and grain size distribution of the available sediment. A continuum of flow types is shown along the horizontal axis, ranging from high-concentration, cohesive debris flows to low-concentration, turbulent fluid flows. Sediment grain size distribution is represented on the vertical axis, which shows the relative abundance of a coarse gravel mode (c) and a finer-grained pumice fragment mode (f). This continuum diagram simply demonstrates that deposits will increase in coarseness as the source materials get coarser, so that the granular sediments of volcano-glacial sandar plot towards the top of the diagram

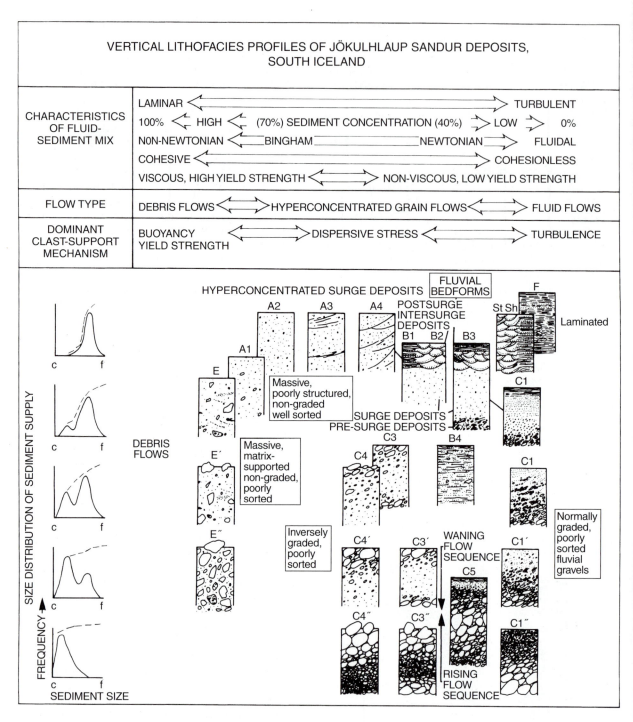

Fig. 11.87 Typical vertical lithofacies profiles of Icelandic sandar deposits classified according to the characteristics of the fluid–sediment mix and relative availability of sediment (horizontal axis) and size distribution of the source sediments (vertical axis). On the sediment size curves, c = coarse-grained and f = fine-grained materials. (From Maizels, 1993. Reproduced by permission of Elsevier)

and the gravelly sediments of the limno-glacial and non-jökulhlaup sandar plot towards the base. Furthermore, the profiles display increasing organization towards the right of the diagram, with decreasing sediment concentration. At lower sediment concentrations, particles are more able to move relative to one another, allowing the development of inverse or normal grading and other sedimentary structures.

Sediments deposited in association with jökulh-laup spillways on the North American Great Plains have been studied by Lord and Kehew (1987) (see Section 9.3.4.4). The outburst floods evacuated through the spillways were highly erosive, so deposits are rarely preserved. However, sediment eroded during spillway incision was deposited in wider reaches of channels in the form of *gravel bars*. Unlike bars in normal proglacial outwash, the spill-way bars are composed of homogeneous masses of massive, matrix-supported and very poorly sorted pebbly cobble gravels. Boulders up to 3 m in diame-ter may also occur in the bar sediments. Lord and Kehew estimated that the outburst flows were hyper-concentrated, with sediment–water concentrations being as high as 40 per cent by weight. The conclu-sion that the flows were hyperconcentrated is similar to the findings of Maizels (1989a, b, 1993) in her studies of the deposits of the 1918 jökulhlaup from Katla. However, recent reconstructions of the Katla flood indicate that overall sediment concentrations were low, and that flow was fluidal rather than hyper-concentrated (Tómasson, 1996). One possible reason for this anomaly is that the Katla flood may have comprised an upper fluidal flow and a lower hyper-concentrated traction carpet (cf. Todd, 1989), and that the deposits represent only the lower flow com-ponent.

11.5.1.4 PALAEODISCHARGES AND VELOCITIES

Several researchers have attempted to reconstruct palaeodischarges and velocities from sediment sequences (e.g. Cheetham, 1976, 1980; Church, 1978; Maizels, 1983; Dawson and Gardiner, 1987). Such reconstructions are important, because it is of great interest to understand how long-term hydrolog-ical changes in a catchment influence landscape development and sediment yields.

Palaeohydrological methods employ empirical relationships between flow velocities and channel-form variables derived for modern rivers. By mea-suring variables that can be observed in ancient deposits, such as channel morphology and slope, sediment grain size and bed roughness, the methods then work backwards to derive likely values of veloc-ities when the palaeochannel was bankfull. Dis-charges are then calculated by multiplying velocity by channel cross-sectional area. There are several potential sources of error in this approach, and they include the following (Dawson and Gardiner, 1987):

1. Palaeochannel width and depth may be difficult to reconstruct from exposures, for several reasons. First, the upper part of the channel may have been truncated by subsequent erosion. Second, the available exposure may not provide a true trans-verse section, distorting the cross-profile. Third, it may be difficult to identify contemporaneous sur-faces on opposite channel banks, owing to over-printing during channel migration and infilling.
2. Channel slope can be hard to measure or estimate.
3. Bed friction is a very important factor, but is diffi-cult to parameterize from sedimentary data.
4. Suspended sediment concentration exerts a strong influence on flow viscosity and velocity, but is unknown for ancient rivers.
5. For braided river systems, total discharge is dis-tributed through several channels, but it may be impossible to determine from sedimentological evidence just how many channels were occupied simultaneously. One way of avoiding this diffi-culty is to identify the main channel and assume that it carried a fixed proportion of the total dis-charge. Mosley (1983) found that in the braided Rakaia River, New Zealand, the main channel car-ried between 47 and 93 per cent of the flow, with an average of 71 per cent. The very large range of these figures underlines the uncertainty of the method.

According to Dawson and Gardiner (1987), the errors introduced by these factors can be consider-able: up to three orders of magnitude, with the major contribution being provided by uncertainties about channel form. Nevertheless, such methods do enable researchers to place bounds on the likely behaviour of former rivers during periods of sandur aggrada-tion.

11.5.2 Paraglacial associations

Paraglacial activity, or the rapid readjustment of glaciated landscapes to non-glacial conditions fol-lowing deglaciation (Section 7.6), results in distinc-tive sediment–landform associations on slopes and valley floors.

11.5.2.1 PARAGLACIAL FANS AND TERRACES

Once glaciers have retreated from a drainage basin in part or whole, unstable sediments are made available to rivers, resulting in the rapid aggradation of thick valley fills and alluvial fans at the mouths of tributary valleys or gullies (Fig. 11.88; Jackson et al., 1982). Detailed descriptions of early Holocene fans and ter-raced valley fills in the mountainous terrain of British Columbia have been provided by Ryder (1971a, b), Church and Ryder (1972), Clague (1986), N. Eyles and Kocsis (1988) and others. The valley fills are deeply incised, and modern river channels may lie over 200 m below the fan surfaces, reflecting reduced sediment supplies at the end of the paraglacial period. The occurrence of Mazama Ash, dating to

6.6 kyr BP, at shallow depths on fans throughout south-central British Columbia indicates that most fan deposition took place during the paraglacial period of 10–6 kyr BP. Rivers now largely transport only the material that is being made available by current denudation processes and so are in equilibrium with their normal weathering environment.

Owen (1989) studied paraglacial terraces in the Karakoram Mountains of Pakistan, and recognized six main types. (a) *Morainic terraces* include ice-contact sediment produced by subglacial, proglacial and slope processes, which have been subsquently terraced by paraglacial river activity. (b) *Glacifluvial terraces* comprise dissected ice-contact fans and outwash plains or valley trains. (c) *Fluvial terraces* are produced by the dissection of more distal river sediments many tens of kilometres away from the nearest glacier snout. (d) *Debris terraces* are produced by the dissection of mass movement deposits which originate from rockfalls, debris flows, landslides and rock glacier movement. The stratigraphy of such forms is dominated by crudely bedded matrix- and clast-supported diamictons dipping down towards the valley floor. (e) *Lacustrine terraces* are produced by the dissection of lake sediments following the drainage of ice-dammed or landslide-dammed lakes. (f) *Fan terraces* are the dissected remnants of polygenetic forms produced during the early stages of deglaciation, when slopes are at their most unstable, by the processes of debris flows. Clast lithological analysis reveals that they are composed predominantly of resedimented till.

Debris fans can also develop below hanging glaciers in steep mountain terrain, where debris avalanching from the glacier front accumulates in fans on benches or valley floors. Such *glacial avalanche-fed fans* may be very difficult to distinguish from other types of fan, although they may be expected to display some internal deformation due to the melt-out of buried ice blocks.

All the terrace forms identified by Owen (1989) are subject to considerable modification during the paraglacial period. Indeed, any valley that is approaching the more stabilized phase of its paraglacial activity may be subject to renewed incision if deglaciation in neighbouring catchments acts as a trigger.

11.5.2.2 PARAGLACIAL SLOPE DEPOSITS

Paraglacial slope deposits form wedge- or cone-shaped slope-foot accumulations, the detailed morphology and sedimentology of which reflect the character of the source sediment and the processes of reworking and deposition. *Talus deposits* may consist of unmodified rockfall material (Section 10.5.2.1), or may be modified by other slope processes such as

(a)

(b)

Fig. 11.88 Paraglacial landform examples. (a) The incised valley fill of the Fraser Valley near Clinton, British Columbia, Canada. The Fraser River incised the valley fill during the early Holocene, after the end of the last glaciation. (Aerial photograph BC1087–46 by the Province of British Columbia) (b) Large valley-side debris fan, Lahul Himalaya, India. (Photo: D.I. Benn)

snow avalanching, debris flow or internal creep deformation. Unmodified rockfall talus is characterized by a straight upper slope of 30–40° and a basal concavity, whereas modified talus generally has a more pronounced concavity and is broken up into surface gullies and depositional lobes (Ballantyne and Harris, 1994). Internally, talus deposits may consist of openwork gravels or diamictons, and commonly exhibit crude slope-parallel bedding.

Thick sequences of diamictons can result from successive debris flows, shallow landsliding, and slower mass movements such as gelifluction and soil creep (Ballantyne and Harris, 1994). In lowland Britain, such deposits are commonly referred to as *head*, a term introduced by De la Beche (1839). Head deposits can form in non-glaciated periglacial environments, but a paraglacial origin can be recognized from the presence of reworked glacigenic material, such as erratics and striated clasts. Examples of paraglacially reworked till deposits have been described from the south Wales coalfield (Harris and Wright, 1980; Wright and Harris, 1980; Wright, 1983, 1991), the 'blue head' at Morfa Bychan, west Wales (Vincent, 1976; Harrison, 1991; Ballantyne and Harris, 1994), and the Cheviots of northern England (Harrison, 1993). In each of these examples, the period of most active till remobilization and slope deposition is thought to have been immediately following deglaciation, when freshly exposed glacigenic sediments were unvegetated and unstable. Glacigenic sediments reworked by debris flows are particularly common in glaciated mountain environments, where a potent combination of abundant debris and steep slopes contributes to high debris mobility. Examples from the Karakoram Mountains have been described by Owen (1991), and from Fåbergstolsdalen, Norway, by Ballantyne and Benn (1994b). Such accumulations can be distinguished from non-glacial slope deposits by the presence of erratics, and from in-situ tills by slope-parallel bedding and downslope-oriented clast fabrics.

Large-scale rock failure (sturztroms) in steep bedrock cliffs has also been documented in freshly deglaciated terrain and has been explained in some situations by the pressure release occasioned by the removal of glacier ice from valley sides. Extensive boulder fields lying in valley bottoms and at the base of bedrock cliffs record former catastrophic slope failures (e.g. Porter and Orombelli, 1980; Dawson *et al.*, 1986; Owen *et al.*, 1995). Such deposits may have well-defined lobate fronts and arcuate transverse ridges on their surfaces, reflecting flow conditions just prior to final deposition.

11.6 GLACILACUSTRINE AND GLACIMARINE ASSOCIATIONS

Glacilacustrine and glacimarine depositional processes and environments have become much better known in recent years, mainly by a combination of well-designed process studies near modern ice margins and detailed sedimentological studies of ancient deposits (Sections 8.4 and 8.5). This section reviews the range of sediment–landform associations deposited in standing water, including ice-marginal forms, such as grounding-line fans, morainal banks and subaqueous moraines, and proglacial forms such as deltas.

11.6.1 Grounding-line fans

Grounding-line fans (also known as *subaqueous fans*) are masses of sediment that build up where subglacial meltwater exits from tunnels directly into deep water, and sediment load is deposited rapidly as a result of the sudden drop in stream velocity (Section 8.5.1; Rust and Romanelli, 1975; Cheel and Rust, 1982; C.H. Eyles *et al.*, 1985). Grounding-line fans differ from deltas (Section 11.6.5) in that the coarse sediment enters at the base of the water column instead of at the top, although grounding-line fans can build up to the water surface to form deltas if a glacier remains quasi-stationary for a period of time and sedimentation rates are high enough (Powell, 1990). Because of their association with subglacial tunnels, grounding-line fans are often found at points along esker systems, where they mark former glacier margin positions; in such cases they may be referred to as *esker beads* (Section 11.2.9.3; Bannerjee and McDonald, 1975; Visser *et al.*, 1987). Subaqueous fans formed in glacimarine and glacilacustrine environments may be very similar, and may be indistinguishable by internal characteristics alone. However, in glacimarine environments, the buoyancy of inflowing waters tends to produce plumes which carry fine-grained sediments away from the fan, unless sediment concentrations are very high (Powell, 1990). The removal of fine-grained sediments in plumes means that glacimarine fans tend to be predominantly coarse-grained with a general lack of underflow deposits (turbidites). The positioning and general shapes of the subaqueous or grounding-line fans produced in association with various tunnel jet discharges and plumes are discussed in Section 8.5.1.

Internally, grounding-line fans exhibit crude to well-developed *foreset bedding* dipping down from the fan apex parallel to the fan surface, generally at angles of 10–30° (Fig. 11.89). Near the fan apex, beds commonly consist of coarse-grained, poorly sorted, massive to normally graded gravels, deposited from high-density cohesionless debris flows and gravelly traction carpets (Sections 8.4.2, 8.4.4.3 and 10.5.5.2). Units may have channelized, erosive bases, recording scour and fill on the fan surface. Because of the dynamic nature of grounding-line fans and the oversteepening of fan surfaces by rapid sedimentation rates, sediment remobilization by gravity flows is common, producing stratified diamicton units occupying scoured channels (McCabe *et al.*, 1984). Coarse sediments at the fan apex grade distally into better-sorted, finer-grained facies, resulting from episodic turbidity currents and cohesionless debris flows triggered by influxes of sediment-laden meltwater or slope failures on the fan surface (Cheel and Rust, 1982; Postma *et al.*, 1983).

Fig. 11.89 Foreset bedding comprising coarse, poorly sorted gravels and sands, Younger Dryas grounding-line fan, Achnasheen, Scotland. (Photo: J. Merritt)

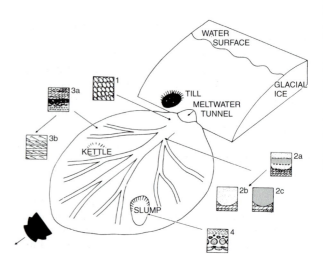

Fig. 11.90 A depositional model for a single subaqueous outwash fan fed directly by a subglacial meltwater tunnel: (1) proximal gravel facies; (2) channel facies; (3) interchannel facies; (4) slump facies. The palaeocurrent distribution from all facies of the fan is shown at the fan front. (From Cheel and Rust, 1982. Reprinted by permission of Geobooks)

Various types of cross-bedding and cross-lamination may be present, recording the downflow migration of dunes and ripples at the base of turbid underflows (Sections 10.4.3 and 10.4.4; Cheel and Rust, 1982; Shaw, 1975). Proximal-to-distal fining reflects the drop-off of flow velocities with distance from the tunnel mouth, whereas multiple vertical sequences of upward-fining units record waning discharges in individual flow events. Towards the fan margin, high suspension sedimentation rates produce climbing ripple drift and massive sands. Ball and pillow and dish structures are common, owing to dewatering as sediment is compacted by fan aggradation (Cheel and Rust, 1982). Ice-rafted debris up to boulder size may also occur in fan deposits.

The overall geometry of grounding-line fans is often complex, with cross-cutting foreset beds recording migration of the efflux point and avulsion of distributary channels on the fan surface. Distally, bottom currents transporting sediment to the fan margin may behave like meandering rivers, causing the depositional environment to switch repeatedly between channels and overbank or channel overspill zones. The resulting facies associations typically consist of fine-grained laminated sediments incised by channelized lenses of coarser facies.

A schematic model of facies distribution on glacilacustrine grounding-line fans was developed by Cheel and Rust (1982), on the basis of Pleistocene examples from Canada (Fig. 11.90). At the fan apex are *proximal gravels* (1), with upflow imbrication recording deposition from jets. Clast-supported diamictons may also be deposited from hyperconcentrated flows in such locations (N. Eyles and McCabe, 1989a, b). Further downfan is a system of steep-sided, bifurcating channels filled with channel facies (2). Three types of *channel facies* were recognized by Cheel and Rust (1982). *Type A channels* are filled with horizontally stratified sand, with discon-

tinuous beds of imbricate gravel along channel bases and between sand units. *Type B channels* are filled with massive sands, and *Type C channels* are filled with subhorizontally laminated sands. The deeper Type A channels are the more proximal, and contain sediments documenting erosive currents carrying gravel bedload. This gravel was excavated from the sediment surrounding the channels, which explains its discontinuous nature. Some inversely graded sands occur in all channel types, recording deposition by grain flows, derived either from the channel walls or directly from the tunnel mouth. Between channels, two types of *interchannel facies* (3) are found. *Type A* consists of proximal cross-stratified and massive to normally graded sands and fine gravels, whereas *Type B* consists of distal climbing ripple drift with occasional silt drapes. The massive to normally graded sands in the proximal facies were deposited by sediment gravity flows during periods of high sediment discharge, whereas the silt drapes of the distal facies record periods of suspension sedimentation when discharges were low.

Cheel and Rust (1982) also identified a slump facies, resulting from slope failures. This facies consists of deformed sediments occupying broad channels, and exhibits evidence of dewatering such as ball and pillow and dish structures. Deformation structures occurring in abrupt lateral contact with undisturbed strata, whose beds are bent sharply upwards at the contact owing to fluidization, were interpreted as the product of melt-out of buried ice blocks. Similarly, McDonald and Shilts (1975) have reported fault systems within subaqueous fans relating to the melt-out of buried ice blocks.

Faulting and folding in grounding-line fan deposits may result from gravitational collapse, ice thrusting, or the grounding of icebergs (Fig. 11.91). N. Eyles and McCabe (1989b) reported syndepositional folding within subaqueous gravels deposited at glacimarine tunnel mouths in eastern Ireland. In these situations, the increasing load of thickening gravel bodies causes sagging into underlying underconsolidated sediments, producing increasing dip angles from the top to the base of the gravel beds (Fig. 11.92). Similarly, inverted gravel flame structures are produced by diapiric intrusion into underlying materials possessing high porewater pressures. Thrust faults can be initiated in subaqueous fans wherever the glacier snout readvances after or during deposition (Powell, 1990). Deformation of sediments by thrusting, iceberg grounding and gravitational reworking may be so pervasive that primary structures can be difficult to distinguish, and in some cases proximal fan deposits have a completely churned diamictic appearance with only isolated, deformed remnants of sorted facies (e.g. Benn, 1996a).

Facies successions in grounding-line fans relating to ice retreat have been recognized by T.G. Stewart (1991) for a subpolar tidewater glacier setting in the Canadian Arctic (Fig. 11.93). Five lithofacies associations were recognized, documenting a proximal-to-distal sequence. *LA I* consists of interbedded diamictons, sands and silts, interpreted as proximal subaqueous debris flows and ice-rafted material with suspension-derived cyclopels. *LA II* comprises normally and inverse graded matrix to clast-supported sandy gravel with some planar cross-stratification, interpreted as the products of high-velocity turbulent, hyperconcentrated and debris flows on a channelized fan surface. *LA III* occurs mostly in channels, and consists of massive, normal and inverse-graded sands with water escape structures, sand intraclasts or rip-ups. This lithofacies is interpreted as high-density turbidity current and mass flow deposits in subaqueous fan channels. Occasional Bouma sequences record intermittent low-density turbidity currents. *LA IV* comprises partial Bouma sequences, climbing ripple drift, starved ripples and draped lamination, and is interpreted as the product of turbidity current sedimentation in overbank and interchannel areas. *LA V* consists of interlaminated cyclopsams and cyclopels representing suspension sedimentation from turbid plumes. Vertical transitions from LA I to LA V document progressively more distal deposition on the fan surface, reflecting glacier snout retreat.

The palaeoclimatic significance of grounding-line fans is that they represent standstill positions of retreating glaciers terminating in deep water. The large sediment accumulations indicate either high discharges in subglacial tunnel systems; and/or that meltwater carries abundant sediment derived from subglacial drainage basins. Because subaqueous fans are fed by subglacial tunnels, they may be expected to be absent where glacier margins are cold-based. However, D.J.A. Evans (1990a) and T.G. Stewart (1991) have reported subaqueous fans relating to retreating subpolar glaciers in the Canadian Arctic,

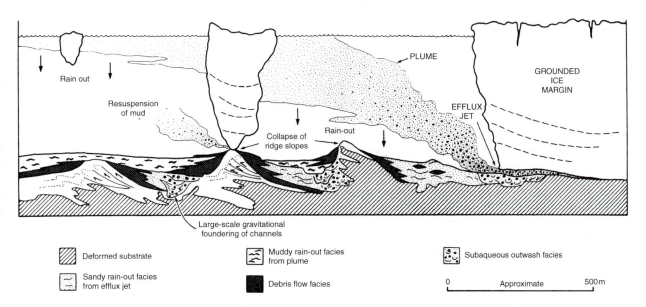

Fig. 11.91 Schematic reconstruction of the depositional and deformational processes associated with a retreating tidewater glacier building grounding-line fan complexes. (From Eyles and McCabe 1989a. Reprinted by permission of Pergamon)

where debris-rich basal ice was excavated by sub-glacial meltwater debouching into deep water. These isolated fans were interpreted as the product of a transient switch to temperate snout conditions as a response to early Holocene climatic warming (cf. D.J.A. Evans and England, 1993).

In glacimarine and glacilacustrine basins with pronounced relief, facies sequences similar to those in grounding-line fans can also form in distal locations far removed from the ice front, where subaqueous valleys and swales funnel sediment gravity flows and turbidity currents. Channelized underflows are able to erode the substratum, forming cut-and-fill sequences and depositional fans (Miall, 1985b; Walker, 1992b). Such sequences have been described from Early Proterozoic glacimarine deposits in Ontario (Miall, 1985b), and distal parts of the Yakataga Formation in Alaska (C.H. Eyles and Lagoe, 1990; Section 12.5.3).

11.6.2 Subaqueous moraines

Subaqueous moraines are transverse moraines deposited at or close to the grounding lines of water-

(a)

(b)

(c)

Fig. 11.92 Characteristic sedimentary structures produced by rapid sedimentation in ice-proximal subaqueous fans. (a) Gravel layers oversteepened by syndepositional foundering at the base of a section in proximal glacimarine sediments infilling a subglacial tunnel valley at Killiney, eastern Ireland. (b), (c) Gravel-filled dykes at the base of crudely graded gravels in proximal glacimarine sediments infilling a subglacial tunnel valley near Bray, eastern Ireland. The dykes were loaded and injected into underlying saturated diamictons in response to the rapid emplacement of gravels. (From Eyles and McCabe, 1989b. Reprinted by permission of Blackwell Scientific)

terminating glaciers. Many alternative terms have been employed, including *morainal banks* (Booth, 1986; Powell, 1990), *sublacustrine moraines* (Drewes *et al.*, 1961; Holdsworth, 1973b; Barnett and Holdsworth, 1974), *cross-valley moraines* (Andrews, 1963; Andrews and Smithson, 1966) and *De Geer moraines* (Hoppe, 1959; Sugden and John, 1976; Larsen *et al.*, 1991). These terms have been used in overlapping senses, and there is no universally agreed set of definitions. In this book, the terms 'morainal bank' and 'composite grounding-line fans' are used for large subaqueous moraine ridges formed mainly by deposition at glacier grounding lines, and the term 'De Geer moraine' for smaller, narrow, sharp-crested ridges which commonly occur in fields of closely spaced subparallel moraines. The non-genetic term *washboard moraine* has sometimes been employed by some researchers for closely spaced suites of sublacustrine moraines (e.g. Mawd-

SUBAQUATIC OUTWASH FAN

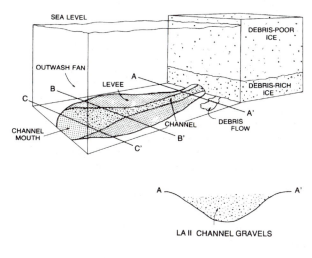

LA II CHANNEL GRAVELS

LA III AMALGAMATED & GRADED SAND

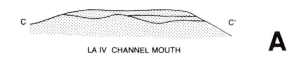

LA IV CHANNEL MOUTH

A

TURBID PLUME DEPOSITION

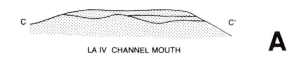

B

Fig. 11.93 The depositional processes and lithofacies associated with the production of subaqueous fans at the margins of subpolar glaciers: (A) morphology of the subaquatic outwash with cross-sections through proximal, intermediate and distal sediments; (B) reconstruction of turbid plume deposition fed by subglacial meltwater discharges. (From Stewart, 1991. Reprinted by permission of the Geological Society of America)

sley, 1936; Lawrence and Elson, 1953; Elson, 1957; Nielson, 1970). The occurrence of large numbers of closely spaced subaqueous moraines in some areas prompted De Geer (1897) to interpret them as *annual moraines*, a usage which has been adopted more recently by Liestøl (1976) and Boulton (1986) for

subaqueous push moraines formed in shallow marine environments. An annual origin may be impossible to prove, however, and it is preferable to use the term 'De Geer moraine' for continuous and closely spaced subaqueous moraine ridges which may or may not be annual.

In planform, individual subaqueous moraines may be linear or have sinuous crestlines (Fig. 11.94). In some cases they are slightly concave upglacier, apparently reflecting the form of calving bays in the ice front (e.g. Sissons, 1982; Benn, 1989a, 1996a). The morphology and sedimentology of subaqueous moraines are highly variable, and a large number of depositional models have been proposed. The more important models are discussed in the following sections.

11.6.2.1 MORAINAL BANKS AND COALESCENT SUBAQUEOUS FANS

Wherever glacier grounding lines remain quasi-stable for long periods of time and/or sediment fluxes to the glacier margin are high, subaqueous moraines will form by the accumulation of sediment along the glacier margin. Debris may be delivered from the ice surface, the melt-out of debris-rich ice and icebergs, the release of unfrozen till from a subglacial deforming layer, or reworking of older deposits (Sections 8.4.1 and 8.5.2). This debris can be reworked in conveyor-belt fashion by erosion off the subglacial face and deposition on the marine slope (Powell, 1984). Subaqueous moraines that extend up into shallow water can also be exposed to wave energy, leading to winnowed and sorted gravels and sands in their upper parts (Powell, 1984).

Some subaqueous moraines consist mainly of *coalescent grounding-line fans*, deposited at the same

Fig. 11.94 Sinuous, low-amplitude morainal bank deposited at the grounding line of a subpolar glacier at Cape Armstrong, Ellesmere Island, Arctic Canada. This moraine marks the maximum of the last glaciation, and its small volume attests to the small debris turnover of subpolar glaciers advancing over bedrock. (Photo: D.J.A. Evans)

time at the efflux points of adjacent meltwater conduits. The Salpausselkä moraines of Finland are spectacular examples, consisting of transverse chains of ice-contact fans and deltas deposited during standstill of the southern margins of the Scandinavian ice sheet where it contacted deep water (Sauramo, 1929; Okko, 1962; Virkkala, 1963; Gluckert, 1977, 1986; Fyfe, 1990; Section 12.5.2). Coalescent grounding-line fans also demarcate former glacier margins in the fjords of Norway, where they have been referred to as *Tronder moraines* by Sollid and Reite (1983). Large-scale moraines formed of partially glacitectonized grounding-line fans occur in eastern Maine, USA (Ashley *et al.*, 1991). The fans are arranged laterally along former grounding-line positions and are associated with tunnel mouth and esker deposits. The whole depositional complex covers an area of $250\,km^2$ over a $40\,km$ wide zone and contains a volume of approximately 5 billion m^3 of sediment. The complex marks a series of standstill positions of an actively retreating interlobate ice margin, deposited over the 1000-year period from 13.5 kyr to 12.5 kyr BP.

Some subaqueous moraines are formed mainly by *dumping* and other gravitational processes at the foot of terminal ice cliffs. This is especially the case where subpolar glaciers with thick basal debris sequences contact the sea, releasing large amounts of debris in environments where there is only limited subglacial meltwater (Powell, 1984). At many glacier margins, however, gravitational deposits are interbedded with subaqueous outwash, particularly at the margins of temperate glaciers where meltwater is abundant. Pleistocene examples of interbedded gravitational and outwash deposits have been reported from the east-central coast of Ireland by McCabe *et al.* (1984, 1987), McCabe (1985, 1986) and N. Eyles and McCabe (1989b). The sediments include channelized subaqueous outwash and debris flow facies interdigitating with ice-rafted dropstone muds, and with some evidence for glacitectonic deformation. They underlie arcuate belts of hummocky topography up to several kilometres in width, and are interpreted as glacimarine sediment deposited during a temporary halt in glacier retreat from the Irish Sea basin.

Exposures in a subaqueous moraine deposited near the margin of a Lateglacial ice-dammed lake at Achnasheen, Scotland, were described by Benn (1996a). The ice-proximal portion of the moraine consists of deformation tills and glacitectonites derived from pre-existing glacilacustrine deposits, whereas the ice-distal portion is underlain by subaqueous outwash interbedded with cohesive to cohesionless debris flow deposits (Fig. 11.95). The debris flows grade laterally into the deformation till, indicating that sediment was delivered from the glacier to the lake from the subglacial deforming layer, as well as from meltwater systems.

The size of moraines built up by outwash and gravitational processes is ultimately limited by debris supply from the glacier. The limit of the last glaciation in part of northern Ellesmere Island, Canada, is marked by subaqueous moraines less than 1 m high (D.J.A. Evans, 1990a, b). These occur where former glaciers advanced from plateaux on to lowlands covered by very patchy sediment veneers. Supraglacial and subglacial debris loads were thus so insignificant that only low-amplitude diamicton ridges were deposited over bedrock at the grounding line (Fig. 11.94). In contrast, large, extensive subaqueous moraines can be produced over short periods of time where debris is abundant, provided that glacier stability is encouraged by glaciodynamic or topographic factors.

The role of bathymetry in the production of subaqueous moraines has been assessed by Crossen (1991) and G.W. Smith (1982) for the Gulf of Maine. Small, closely spaced moraines (De Geer moraines) were deposited in deeper water where rapid ice retreat allowed time for the deposition only of small ridges at the ice margin. Where calving rates were reduced in shallower water, more substantial subaqueous fans developed. If such fans aggrade to sea-level, ice-contact Gilbert-type deltas form (Section 11.6.5; Powell, 1990). When ice retreats into very shallow water or on to land, melting replaces calving and the meltwater produced feeds sediment to the terrestrial portions of Gilbert-type outwash deltas.

11.6.2.2 DE GEER MORAINES

De Geer moraines are named after the pioneering Swedish geologist G. De Geer (Hoppe, 1959). They commonly occur in fields of closely spaced ridges in association with subaqueous sediments, and mark the intermittent retreat of water-terminating glaciers (Fig. 11.96). Many models have been proposed to account for their formation.

Some researchers conclude that De Geer moraines mark former ice-margin positions, where sediment is deposited or pushed up during brief standstill or minor readvances. Boulton (1986) has reported sequences of closely spaced subaqueous moraines on the sea bed in front of Alpha Glacier, Cambridge Fiord, east Baffin Island, which appear to document annual winter pushing (Fig. 11.97). These moraines are characterized by steeper distal than proximal slopes, lobate forms with acute interlobe angles pointing upglacier, and truncated sections due to partial overriding by later moraines. All of these characteristics are typical of terrestrial push moraines, as described in Section 11.3.2.1. The formation of sub-

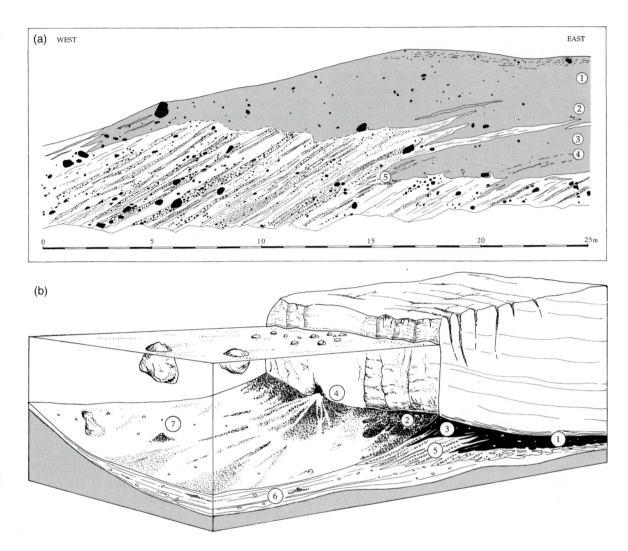

Fig. 11.95 Lateglacial morainal bank deposited in an ice-dammed lake, Achnasheen, Scotland. (a) Exposure through the morainal bank; (b) reconstruction of the depositional environment. Numbers in (a) refer to fabric samples and in (b) are: 1, subglacial till; 2, mass flows; 3, mass flow diamictons; 4, subglacial meltwater portal; 5, subaqueous outwash; 6, laminated lake sediments; 7, iceberg dump mound. (From Benn, 1996. Reproduced by permission of Scandinavian University Press)

aqueous moraines with steep ice-distal slopes and gentle ice-proximal slopes has also been linked to the calving of floating glacier snouts, perhaps on a yearly basis (Aartolahti, 1972; Holdsworth, 1973b). According to this model, the gentle ice-proximal slopes are produced by glacier overriding prior to the calving of a section of the snout (Fig. 11.98a). On the basis of observations at Generator Lake, Baffin Island, Holdsworth (1973a, b) linked the calving events to annual lake-level changes, although major calving events may also be triggered by annual ice-marginal thinning by summer ablation. Moraine formation by squeezing of saturated sediment out from beneath the ice margin was proposed by Andrews and Smithson (1966) (Fig. 11.98b).

Some researchers have argued that De Geer moraines are formed in basal crevasses a short distance behind glacier margins in similar ways to terrestrial crevasse fills (e.g. Elson, 1957; Hoppe, 1957; Stromberg, 1965; Zilliacus, 1989). This interpretation is influenced by the patterns made by some fields of subaqueous moraines, which are similar to those of crevasses in glacier lobes (Lundqvist, 1989). According to this model, transverse crevasses are formed in areas of extending flow and then widened during partial flotation of the ice margin (Zilliacus, 1976, 1981, 1987a, b, 1989; Lundqvist, 1989c). Moraines are thought to form when the water level drops and the glacier is lowered on to the substratum, which is squeezed up into the basal

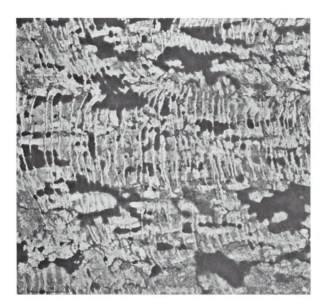

Fig. 11.96 De Geer moraines on the lowlands surrounding Hudson Bay, Canada. (Aerial photograph provided by Energy, Mines and Resources, Canada)

crevasses (Fig. 11.99). Tabular icebergs would later break off along the transverse crevasses, thus ensuring that the moraines were not later overridden by the glacier. This interpretation of subaqueous moraines has been used by Lundqvist (1989) to support reconstructions of massive-scale downdraw and crevassing of portions of the Scandinavian ice sheet in response to rising sea-levels. Zilliacus (1989) has suggested that such moraines document surging behaviour along glacier margins (cf. Solheim, 1986), providing a direct comparison with the terrestrial crevasse-fill ridges reported by Sharp (1985a).

Features referred to as 'lift-off moraines' have been identified in seismostratigraphic investigations of the Scotian and mid-Norwegian shelves by King and Fader (1986) and King *et al*. (1991). The production of lift-off moraines is thought to be related to fracture patterns in the base of a retreating glacier snout, involving a genesis similar to that proposed by Lundqvist (1989). It is uncertain whether or not lift-off moraines are produced individually by single events or simultaneously as groups (King *et al*., 1991). According to Beaudry and Prichonnet (1991), subaqueous moraines in the Chapais area, Quebec, were formed as glacifluvial infills in basal crevasses. From the published sedimentological data, however, a subglacial origin is not proven, and it is equally likely that the moraines formed along the grounding line as coalescent subaqueous fans.

11.6.3 Grounding-line wedges: 'till deltas' and 'till tongues'

The term *grounding-line wedge* was introduced by Powell and Domack (1995) to refer to dipping diamicton beds overlain by horizontal sheets of diamicton, mainly subglacial till. The term was intended to replace the earlier term *till delta* (Alley *et al*., 1989), to avoid defining all of the sediment as till, and to avoid any association with sea-level that could be implied from the word 'delta'.

Grounding-line wedges are thought to form near the grounding lines of glaciers underlain by deforming till. Till emerging from beneath the glacier is redistributed by subaqueous debris flows, producing diamicton beds that dip away from the margin (Fig. 11.100). Successive debris flows were inferred to build out a prograding wedge of sediment beyond the grounding line. Subsequent grounding-line advance over these deposits causes them to be disconformably overlain by deformation till. Using seismic data, Alley *et al*. (1987c, 1989) have identified possible modern examples of grounding-line wedges near the junction between Ice Stream B and the Ross ice shelf, Antarctica. In this case, the grounding-line wedge is thought to be a sediment sink for deforming till emerging from beneath the fast-flowing ice stream at the point where the ice begins to float. The evidence for till deformation below Ice Stream B is reviewed in Section 4.4.1. Other possible grounding-line wedges have been identified by Boulton (1990) on the basis of geophysical studies of the sea floor around Svalbard. Major wedges of sediment consisting of outward-dipping diamicton beds overlain by subglacial till occur near the mouths of major troughs, where subglacial sediment discharge was highest. Boulton (1990) referred to such accumulations as *trough-mouth fans*, and suggested that they may be characteristic depositional systems at the tidewater or floating termini of fast-flowing ice streams and outlet glaciers.

Exposures of a possible Pleistocene grounding-line wedge on the shores of Lake Erie, Canada, have been described by Dreimanis (1987). The sediments were deposited near the grounding line of the Lake Erie lobe of the Laurentide ice sheet, which terminated in a deep proglacial lake, and were interpreted as subaqueous debris-flow tongues fed by 'dilated lodgement till' emerging from beneath the glacier. Additional sedimentological evidence for subaqueous flows fed by emergent deformation till has been provided by Benn (1996a) (Fig. 11.95; Section 11.6.2.1).

Powell and Domack (1995) argued that the *till tongues* reported from the Scotian and Norwegian continental shelves by King and Fader (1986) and King *et al*. (1987, 1991) also represent ancient

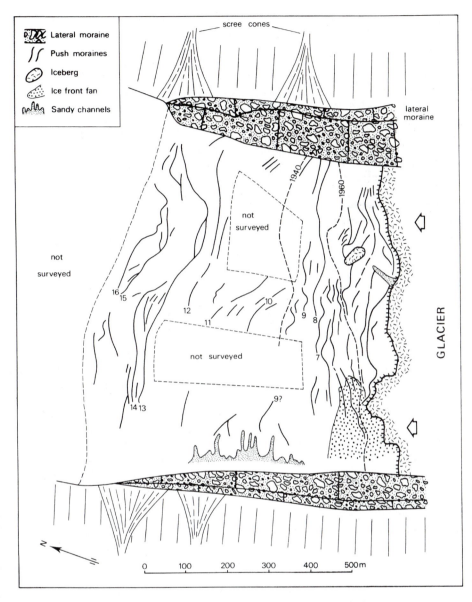

Fig. 11.97 Annual De Geer moraines (subaqueous push moraines) on the sea bed in front of Alpha Glacier, Cambridge Fiord, Baffin Island. (From Boulton, 1986. Reprinted by permission of Blackwell Scientific)

grounding-line wedges (Fig. 11.101). These 'till tongues' were identified from high-resolution seismic records, and consist of tongues of massive diamicton interdigitating with stratified glacimarine deposits, including rhythmites and dropstone diamictons. Similar depositional systems have been reported from the Hebrides Slope by Stoker (1990), where dipping beds of probable debris flows occur at the shelf edge, just beyond the outer limit of subglacial tills deposited by the Scottish ice sheet.

King *et al.* (1991) interpreted 'till tongues' by a slightly different model from that proposed for grounding-line wedges by Alley *et al.* (1989) and Powell and Domack (1995). According to this model, 'till tongues' are formed by three major processes operating in the zone where a glacier snout begins to

float. (a) Subglacial deposition behind the grounding line was assumed to be by lodgement and melt-out, rather than subglacial deformation (Sections 10.3.2 and 10.3.5). (b) Downglacier from the grounding line, debris melting out from the base of the ice shelf is deposited after falling through no more than a few metres of water-forming *undermelt diamictons* (Section 10.6.3; Gravenor *et al.*, 1984). (c) Sediment is redeposited beyond the grounding line by sediment gravity flows. The flows were assumed to be fed by melt-out of debris-rich ice and failure of pre-existing deposits. The lateral continuity of individual till tongues over distances of >300 km suggests that they could not be entirely the product of mass flows (King *et al.*, 1991), but mass flows could extend the downslope boundaries of predominantly subglacially

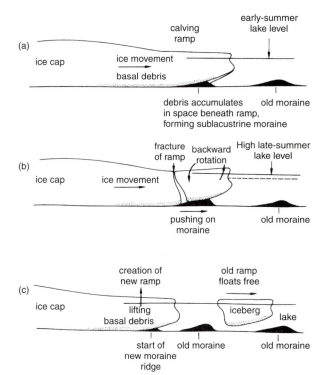

(A)

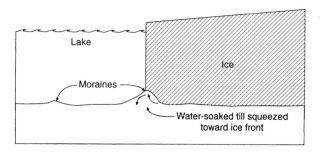

(B)

Fig. 11.98 Models for the formation of De Geer moraines at glacier grounding lines: (A) moraine formation by accumulation of sediment and glacitectonics at the grounding line, followed by calving (from Sugden and John 1976, after Holdsworth, 1973); (B) ice-marginal squeezing (modified from Andrews and Smithson, 1966)

deposited till tongues. These downslope-feathered boundaries of ancient till tongues (*till tongue tips*) are located in present-day water depths of >700 m and so were deposited in minimum depths of >575–600 m, if 100–125 m is subtracted for full glacial conditions.

The structure of grounding-line wedges will reflect patterns of grounding-line advance and retreat. Glacier advance will result in progressive on-lap of subglacial facies over the top of old mass flow deposits

(Figs 11.100 and 11.101), whereas retreat will result in the deposition of undermelt diamictons and mass flows on top of till sheets. Oscillation of the grounding line will therefore be recorded by interdigitation of subglacial and subaqueous deposits. Furthermore, distinctive stratigraphies should result from episodes of sea-level rise and fall (Boulton, 1990; Section 12.5.1).

Grounding-line wedges are thought to be characteristic of settings where large amounts of debris are available, but meltwater discharge is negligible. In consequence, grounding-line wedges may be formed mainly by polar and subpolar glaciers flowing on deforming beds, or with thick basal ice sequences (Alley *et al.*, 1989; King *et al.*, 1991; Powell and Domack, 1995). The glaciers may or may not have floating termini. Where meltwater is abundant, such as at the tidewater margins of temperate glaciers, deposition is more likely to produce morainal banks or grounding-line fans (Sections 11.6.1 and 11.6.2; Powell and Domack, 1995).

11.6.4 Ice shelf moraines

Ice shelf moraines are formed where debris is deposited or relocated around the margins of ice shelves or floating glacier tongues. The crestlines of such moraines tend to be horizontal or have very low gradients, reflecting the low gradient of ice shelves, which have zero basal shear stresses (Section 8.2.3). The gradients of ancient ice shelf moraines, however, may be steeper, as a result of postglacial isostatic tilting, so their gradients should be corrected with reference to the sea-level record (Section 1.5).

Only a small amount of detailed work has been undertaken on modern and ancient ice shelf moraines. A modern example formed at the margin of the George VI Sound ice shelf, Antarctica, has been described by Sugden and Clapperton (1981). The ice shelf is glacier-fed and flows from the Antarctic Peninsula across George VI Sound to Alexander Island, decreasing in thickness from 400–500 m to 100–300 m. Near the coast of Alexander Island, complex pressure ridges form on the ice shelf as it grounds on the island or meets local outlet glaciers (Fig. 11.102). A horizontal moraine ridge has been formed over a distance of 120 km along the promontories of Alexander Island, where basal debris is brought to the glacier surface by ablation and thrusting. Sediments within the moraine include: (a) subglacial debris entrained on Alexander Island and the Antarctic Peninsula; (b) glacifluvial and lacustrine sediments from ice-marginal streams and ponds; and (c) marine sediments and fauna derived from freezing-on at the grounding line on Alexander Island.

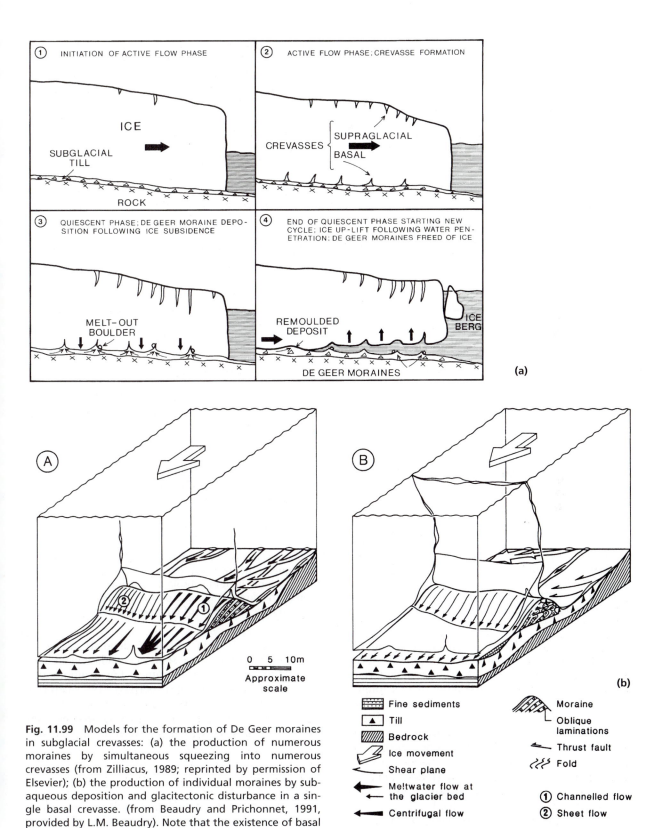

Fig. 11.99 Models for the formation of De Geer moraines in subglacial crevasses: (a) the production of numerous moraines by simultaneous squeezing into numerous crevasses (from Zilliacus, 1989; reprinted by permission of Elsevier); (b) the production of individual moraines by subaqueous deposition and glacitectonic disturbance in a single basal crevasse. (from Beaudry and Prichonnet, 1991, provided by L.M. Beaudry). Note that the existence of basal crevasses is hypothetical, and conflicts with glaciological theory.

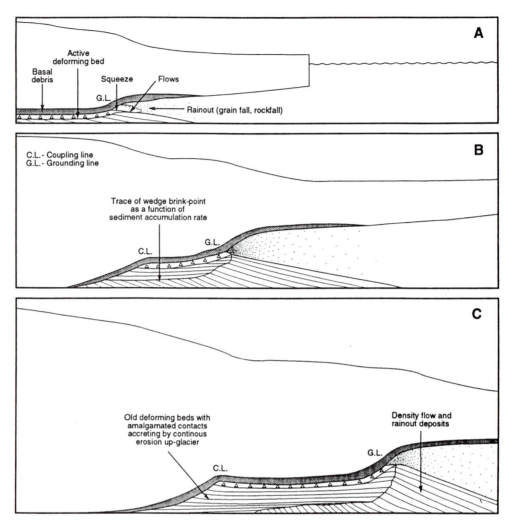

Fig. 11.100 Depositional model for the formation of grounding-line wedges at the margin of a polar ice stream. (A) Till emerging from a subglacial deforming layer feeds prograding subaqueous debris flows. (B) Deformation till accumulates behind the grounding line, overlapping older debris flow deposits. The inner limit of subsole deformation is known as the coupling line. (C) Till erosion above the coupling line and migration of the grounding line produce the overall geometry of the grounding-line wedge. (From Powell and Domack, 1995. Reproduced by permission of Butterworth-Heinemann)

England *et al.* (1978) have described ancient ice shelf moraines from north-east Ellesmere Island, Arctic Canada (Fig. 11.103). The horizontal moraines mark the former margins of floating glacier tongues, and contain marine shell fragments thought to have been entrained at the grounding line or transferred to the surface of the ice shelf by bottom freezing of sea-water. Another ancient example, produced at the north-western margin of the Laurentide ice sheet in Viscount Melville Sound, was described by Hodgson and Vincent (1984). The moraines consist of extensive linear belts of till (Winter Harbour Till), between the altitudes of 60 and 150 m above present sea-level, along the shores of Byam Martin, Melville, Banks and Victoria islands and extending eastwards to north Baffin Island (Dyke *et al.*, 1992; Klassen, 1993; Fig. 11.103).

The accumulation of slope deposits along the grounded margins of ice shelves can also result in the formation of ice shelf moraines. For example, considerable amounts of debris derived from steep fjord walls occur along the margins of the northern Ellesmere Island ice shelves. Such debris on Arctic ice islands has been used to determine the ice shelf source of the islands (Jeffries, 1992).

11.6.5 Deltas

Deltas are masses of sediment built out into standing water by subaerial streams, by a combination of fluvial processes above the water-line and some combination of gravitational mass movement and suspension settling below water-level. They are

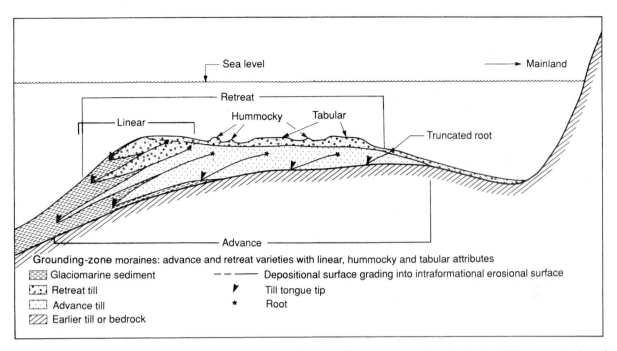

Fig. 11.101 Till tongue stratigraphies and associated moraines: idealized reconstruction of a typical till tongue stratigraphy produced during a complete glacial cycle, based on examples from the mid-Norwegian shelf. (From King *et al.*, 1991. Reprinted by permission of the Geological Society of America)

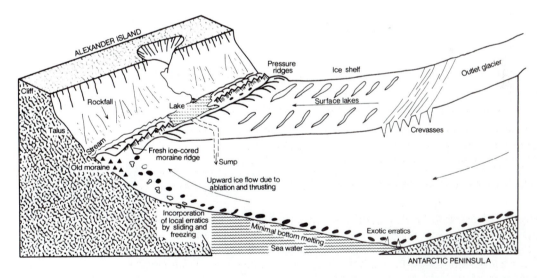

Fig. 11.102 A conceptual model of the formation of the George VI Sound ice shelf and ice shelf moraine, Antarctica. (From Sugden and Clapperton, 1981. Reproduced by permission of the International Glaciological Society)

major features on sea coasts and lake shorelines in all climatic zones, although here we concentrate on those deltas produced in glacierized basins. Deltas may form in association with glaciers in two main settings: (a) *glacier-fed deltas* receive sediment from glacial meltwater streams following transport across an intervening land surface; and (b) *ice-contact deltas* are built out directly from glacier mar-

gins, after evolving from grounding-line fans or other subaqueous depositional systems. The sedimentology of deltas in these two settings is reviewed in the following sections. Further information on deltas can be found in texts by Wright (1978), Coleman (1982), Miall (1984), Elliott (1986b), Nemec and Steel (1988), and Collela and Prior (1990).

11.6.5.1 GLACIER-FED DELTAS

The morphology and sedimentology of deltas are strongly influenced by the gradient of the feeder river and water depth, leading to a range of delta types (Fig. 11.104; Postma, 1990). The most common in glacierized catchments are Types 2–6, consisting of shallow-water *Hjulström-type deltas* with gently

(a)

sloping fronts, and intermediate- to deep-water *Gilbert-type deltas* and *subaqueous debris cones* with steep fronts. Deltas intermediate in character between Hjulström-type and Gilbert-type deltas have been termed *Salisbury-type deltas* (Salisbury, 1896; Church and Gilbert, 1975). The delta types depicted in Fig. 11.104 represent steady-state forms which reflect specific environmental conditions. In glacierized landscapes, however, depositional environments do not remain stable for long periods of time, and so deltas may evolve from one type to another in response to changing sea-levels, proximity of glacier ice, sediment availability and river discharges.

Hjulström-type deltas are deposited in shallow waters at the distal end of sandur plains, and were first identified in Iceland by F. Hjulström (1952). They are characterized by three physiographic zones, consisting of (a) a subaerial delta plain; (b) a gently sloping delta front dominated by coarse-load deposition and commonly influenced by waves; and (c) a very gently sloping prodelta dominated by hemipelagic sedimentation.

Gilbert-type deltas are named after the pioneering American geologist G.K. Gilbert, and are the most

(b)

Fig. 11.103 Examples of ice shelf moraines. (a) Ice shelf moraine on north-east Ellesmere Island (from England *et al.*, 1978). (b) Viscount Melville Sound ice shelf moraine (Winter Harbour till, dashed line) and the limits of the Bolduc till (dotted line), perhaps from an earlier ice shelf, at Winter Harbour, Melville Island (from Hodgson and Vincent, 1984). (Aerial photographs provided by the Department of Energy, Mines and Resources, Canada)

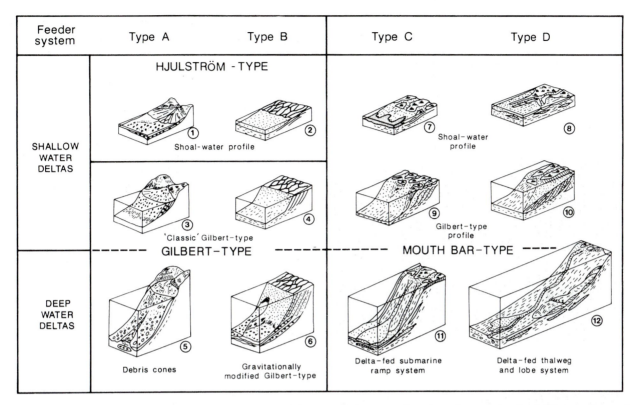

Fig. 11.104 Classification scheme for deltas, showing delta morphology as a function of water depth and the gradient of the feeder river system. Feeder system types are: A: very steep; B: steep; C: moderate; and D: low-gradient. (From Postma, 1990. Reprinted by permission of Blackwell Scientific)

common in glaciated basins. They have three main components: (a) topsets, consisting of fluvial sediments deposited on the subaerial delta surface; (b) foresets, deposited by gravitational processes on the delta foreslope; and (c) bottomsets, deposited by underflows on gentle slopes at the foot of the delta front (Section 8.5.4). *Topsets* are essentially sandur deposits, and incorporate the range of facies described in Sections 10.4 and 11.5.1.2. *Delta foresets* may be composed of a variety of gravel and sand facies, depending on sediment supply and depositional mechanism (Fig. 11.105). In general, sorting is better than in grounding-line fan foresets, owing to the effects of subaerial fluvial transport on grain size distribution. Remobilization of debris on the delta foreslope can, however, result in remixing of sediment, producing poorly sorted beds. Common facies include massive to inversely graded openwork gravels, sandy gravels, gravelly sands, and sands deposited by cohesionless debris flows; fall-sorted gravels deposited by debris falls; and normally graded gravels and sands deposited by high-density turbulent underflows (Postma and Cruickshank, 1988; Prior and Bornhold, 1988, 1989, 1990; Nemec, 1990; Martini, 1990). Because of the high-energy conditions on upper delta foreslopes, fine-grained

sands and silts are rare and clays are absent. Delta foresets commonly form repeated fining-upward units documenting fluctuations in sediment delivery on a variety of timescales. Such fluctuations may result from diurnal or weather-related changes in fluvial discharge, or switches in sediment influx positions on the delta surface (Gustavson *et al.*, 1975; Cohen, 1979; Clemmensen and Houmark-Nielsen, 1981; Thomas, 1984a; Bornhold and Prior, 1990; Mastalerz, 1990). Cross-bedding, recording the downflow migration of dunes, is rare, although examples of *backsets* have been reported (Clemmensen and Houmark-Nielsen, 1981; Postma *et al.*, 1983; Postma, 1984b; Postma and Roep, 1985; Collela *et al.*, 1987; Postma and Cruickshank, 1988; Martini, 1990). These are sets of upflow-dipping cross-strata which dip in the opposite direction as compared with the foreset beds with which they are embedded (Fig. 11.106). Nemec (1990) suggested that backsets could be the product of hydraulic jumps where turbulent flows encounter obstructions on the delta slope. This is a process similar to the production of antidunes (Sections 3.8.3 and 10.4.5).

Erosive events on upper delta foreslopes result in downslope-oriented *chutes* and *channels* up to 200 m wide and 20 m deep. These are cut by turbidity currents and erosive debris flows originating at bowl-shaped slump scars or stream channel mouths (Hoskin and Burrell, 1972; Prior *et al.*, 1981; Syvitski and Farrow, 1983; Prior and Bornhold, 1988, 1989, 1990; Weirich, 1989; Bornhold and Prior, 1990). These chutes and channels become infilled with turbidites, debris flows and suspension deposits (Collinson, 1970; Massari, 1984; Postma, 1984a, b;

(a)

(b)

Fig. 11.105 Delta foreset bedding: (a) sand and gravel foreset beds overlain by topset gravels, near Sandane, southern Norway; (b) details of sand and gravel foreset beds in a glacilacustrine delta, Drymen, Scotland. Note the grading within individual gravel beds and the soft sediment clast above the hammer. (Photos: D.J.A. Evans)

Postma and Cruickshank, 1988; Nemec, 1990). In section, former chutes and channels are recognizable as cut-and-fill sequences within the foreset bedding, or by units with *welded contacts* at their bases recording partial incorporation of the underlying sediment.

The foresets of Gilbert-type deltas become easier-angled and finer-grained downslope. On lower delta foreslopes, typical facies are sands and silts deposited by turbid underflows triggered by the collapse of upper delta foresets or fresh sediment influx events (Jopling and Walker, 1968; Gustavson *et al.*, 1975; Shaw, 1975; Cohen, 1979; Clemmensen and Houmark-Nielsen, 1981; Thomas, 1984a, b). Multiple fining-upward sequences, a few centimetres to a few tens of centimetres thick, are common, typically consisting of massive or planar laminated sands grading upward into ripple cross-laminated sands and silt drapes, recording progressively waning flow during individual underflows (Ashley *et al.*, 1982). Partial or repeated sequences can result from fluctuations and surges within flows. In three dimensions, such sequences form overlapping lobes of sediment emanating from the shifting influx points of distributary delta surface channels (N.D. Smith and Ashley, 1985). The bottomsets of Gilbert-type deltas consist of distal turbidites and fine-grained silts and clays deposited from high-level suspension (Sections 10.5.6 and 10.6.1).

The fronts of deep-water Gilbert-type deltas may be subject to frequent slope failures and gravitational reworking. In such *gravitationally modified Gilbert-type deltas* (Type 6, Fig. 11.104), coarse, poorly sorted material is transferred to the delta foot by slumps and cohesionless debris flows (Massari, 1984; Postma, 1984a, 1990; Postma and Roep, 1985; Postma *et al.*, 1988a). Typical structures include compressional slump folds; low-angle, rising and conjugate shear faults; sediment-filled tension cracks; shear zones beneath sediment slides; and distorted beds disturbed by water escaping from underlying strata (Jones and Preston, 1987; Nemec *et al.*, 1988; Nemec, 1990).

Subaqueous debris cones or *underwater conical deltas* lack a subaerial delta plain component and

(a)

Fig. 11.106

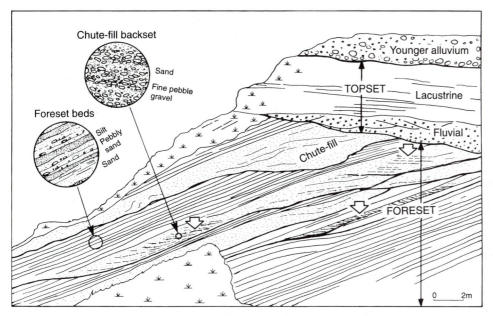

Fig. 11.106 Examples of backset bedding in the foresets of Gilbert-type deltas. (a) Unit of pebbly sand (arrowed) with up-slope-dipping cross-strata (backsets) on an outwash delta foreslope in the Netherlands. The delta prograded towards the left. (b) Backsets (arrowed) in a Gilbert-type delta in Calabria, Italy. The delta prograded to the left. (Modified from Nemec, 1990)

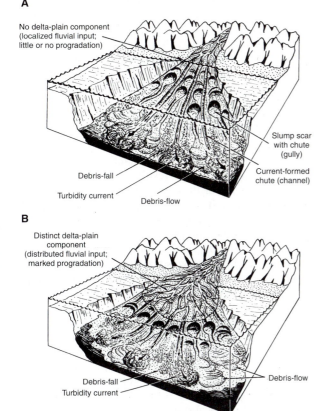

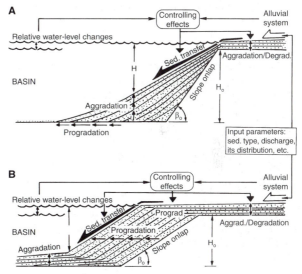

Fig. 11.107 Profiles of (A) an underwater conical delta, and (B) a Gilbert-type delta. (From Nemec, 1990. Reprinted by permission of Blackwell Scientific)

form in deep water (Type 5, Fig. 11.104, and Fig. 11.107; Postma, 1990; Nemec, 1990). Continued sediment delivery to an underwater conical delta often eventually leads to the shallowing of water depths until foresets are able to prograde at water-

level, allowing the development of a delta plain. Thus, conical underwater deltas may act as precursors to Gilbert-type deltas. The development of underwater conical and Gilbert-type deltas under non-steady-state conditions in fjords in British

Columbia, Canada, has been studied by Prior and Bornhold (1990) using side-scan sonar imagery of modern examples (Fig. 11.108). Prior and Bornhold argued that delta morphology changes through time in response to changes in, first, sediment supply, and second, subaqueous relief and fan gradients. In British Columbia, sediment supply has declined during the Holocene, owing to glacier recession and the stabilization of deglaciated slopes in the drainage basins, which essentially starves the offshore delta slopes. Second, *fan gradients* have gradually been reduced during postglacial time as the deltas have aggraded and prograded into formerly deep water, reducing the gravitational stresses on the delta surfaces considerably. Taken together, reduced sediment supplies and gravitational stresses have conspired to replace high-energy sediment dispersal with low-energy processes. According to this model, changing processes of sediment dispersal result in a fourfold facies succession consisting of (1) coarse-grained avalanche deposits with slopes in excess of 20°; (2) cohesionless debris flows (inertia flows) on slopes averaging 12°; (3) turbidites; and (4) reworked sediments resulting from gravitational instability on the delta front (Fig. 11.109). During this succession, the delta front is increasingly starved of coarser material, and slope instability increases in importance as the foreslope experiences loading by distributary mouth bar growth and the passage of turbid underflows (Fig. 11.110).

11.6.5.2 ICE-CONTACT DELTAS

Ice-contact deltas (sometimes referred to as *kame deltas*; Price, 1973; Sharp, 1982; Feenstra and Frazer, 1988) may be deposited in ice-marginal or supraglacial lakes, or in the sea. They form by processes similar to those that produce glacier-fed deltas, and exhibit similar facies successions, although there are some important differences in morphology and facies architecture. Ice-contact deltas are disconnected from subaerial feeder basins, and form isolated plateaux terminating in steep bluffs or *ice-contact slopes* at their ice-proximal ends. They may be perched on topographic highs, where the only plausible source of sediment and water is a former glacier margin. Pitting or kettle holes are common on ice-proximal surfaces, recording the melt-out of buried glacier ice (Fig. 11.111). In glacimarine settings, ice-contact deltas are particularly useful in reconstructing glacial and relative sea-level histories.

The sedimentology and internal structure of ice-contact deltas reflect the close proximity of glacier ice. Debris flow diamictons may be widespread both on the delta surface and in ice-proximal foresets, reflecting direct debris input from the glacier surface

(Shaw and Archer, 1979; Shaw, 1988b; Lønne, 1993; Benn and Evans, 1993). The majority of ice-contact deltas originate as subaqueous grounding-line fans, which then aggrade to water level (Powell, 1990), so that their ice-proximal cores will consist of the sedimentary sequences described in Section 11.6.1 (e.g. Thomas, 1984a; Lønne, 1993; Benn and Evans, 1996). The ice-distal topsets, foresets and bottomsets of ice-contact deltas, however, may be indistinguishable from those of glacier-fed deltas (Section 11.6.5.1), although palaeocurrent indicators may be more varied than in other delta sequences, owing to changing influx points at the glacier margin. Figure 11.112 shows a longitudinal section through an ice-contact delta deposited at the margin of a piedmont glacier that occupied Loch Lomond, Scotland, during the Younger Dryas. At the central and ice-proximal (right-hand) parts of the section, the deposits consist of cross-cutting sandy and gravelly subaqueous flows, interpreted as subaqueous fans. These deposits pass distally into gravel and sand foreset beds, which are overlain by horizontal topsets. At the ice-proximal end, the sequence has undergone extensive subglacial deformation, whereby the fan deposits have been remoulded into glacitectonite and deformation till (Sections 10.3.3 and 10.3.4), whereas the distal parts of the delta are traversed by low-angled thrust faults. The section records the evolution of subaqueous fans into an ice-contact delta, which was subsequently partially overridden and deformed by a minor readvance of the Loch Lomond glacier.

The proximal parts of ice-contact deltas may also undergo slumping, as a result of the melt-out of buried glacier ice or the removal of support from the ice-contact face. Typical structures include slump folds and normal faults, and it can sometimes be demonstrated that slumping occurred syndepositionally; that is, delta deposition took place while underlying and adjacent ice was wasting. For supraglacial deltas, disturbance may be so extensive that the original morphology of the delta is almost totally destroyed.

Relationships between developmental sequences in ice-contact deltas and ice retreat rates have been considered by Shaw (1977b). At relatively stable glacier termini, normal Gilbert-type deltas can develop because of the low ice-retreat to sedimentation ratio (Fig. 11.113A). If the terminus then retreats a short distance to another stable position, as often happens in glacilacustrine and glacimarine settings, the first delta will subside and be buried by a second deltaic sequence (Fig. 11.113B). Subsidence in each delta occurs only on the ice-contact face. If the glacier margin retreats rapidly, the increased ice-retreat to sedimentation ratio means that delta formation is replaced by deposition of grounding-line fans and

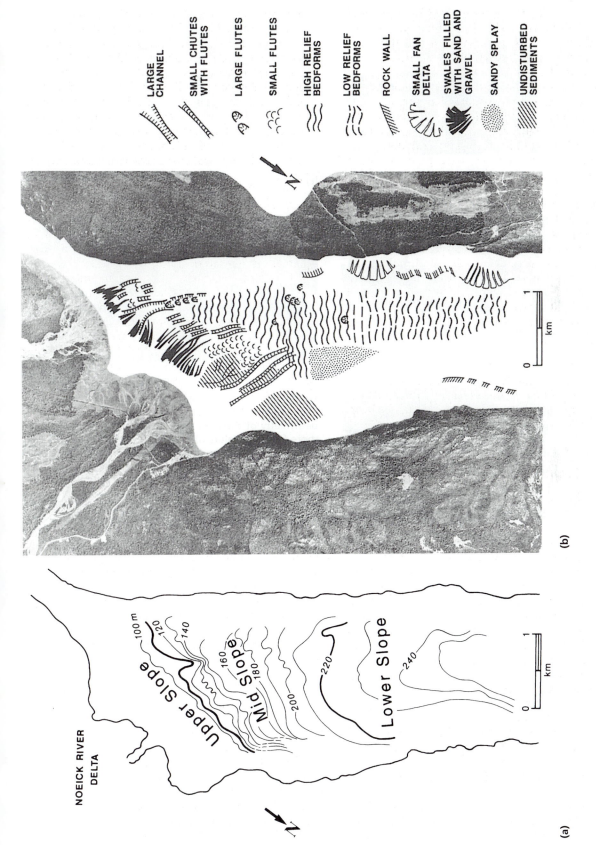

Fig. 11.108

(c) i

(c) ii

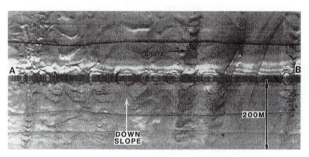

(c) iii

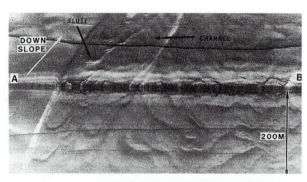

(c) iv

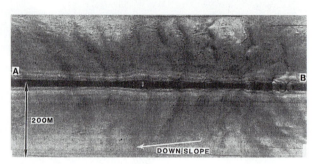

(c) v

Fig. 11.108 The morphology of the subaqueous extension of the Noeick River Delta, British Columbia, Canada: (a) the bathymetry of the fjord bottom (South Bentinck Arm) offshore from the delta; (b) the distribution of fjord floor features; (c) side-scan sonographs of sea floor features: i – upper slope ridges and swales, ii – downslope convergent chutes on the upper fan, iii – prominent chute containing flute trains on the upper fan, iv – major channel containing broadly spaced flute marks, v – transverse, long-wavelength bedforms (antidunes?) on the mid-slope. (From Bornhold and Prior, 1990. Reprinted by permission of Blackwell Scientific)

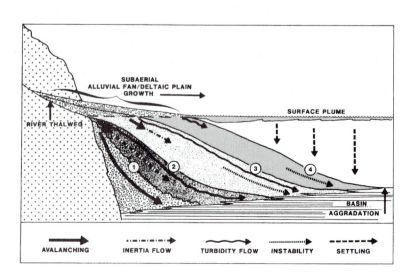

Fig. 11.109 Sequential evolution of glacier-fed deltas based on examples in British Columbia fjords. The sequence begins with debris avalanching and then progresses to inertia flows, to turbidity flows and finally to slope instability. (From Prior and Bornhold, 1990. Reprinted by permission of Blackwell Scientific)

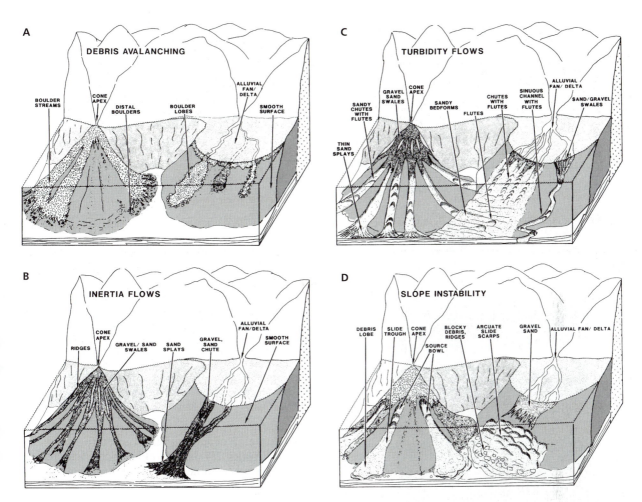

Fig. 11.110 Schematic morphology of some subaqueous mass movements. A: Debris avalanching (coarse-grained cohesionless debris flow and fall); B: inertia flows (sandy cohesionless debris flow); C: turbidity flows (underflows); D: slope instability (slumps, slides, and cohesive debris flows). (From Prior and Bornhold, 1990. Reprinted by permission of Blackwell Scientific)

distal rhythmites, including varves (Fig. 11.113B: c and d).

Changing water depth during delta formation can also be recorded in internal structures. Falling water level results in delta incision and the formation of inset deltas graded to the new base level (e.g. Sissons, 1982), whereas rising water level results in 'stacked' delta sequences one above the other (Fig. 11.114; Thomas, 1984a). Mastalerz (1990) has interpreted Pleistocene deltaic sequences near Neple, eastern Poland, in terms of an annual cycle of water level changes. According to his interpretation, the cycle consists of the following stages (Fig. 11.115). (a) *Summer*. Water levels are high because glacier ablation rates are at a maximum, and deposition is by normal foreset progradation. (b) *Winter*. Ablation ceases, water levels fall, and the ground surface freezes. Deltaic deposition ceases. (c) *Spring*. The ground thaws, but water levels remain low, and large-

Fig. 11.111 Ice-contact delta on the Knud Peninsula, Ellesmere Island. The delta (bottom right) is connected to a contemporaneous subhorizontal lateral moraine which extends into the distance along the fjord wall. Both features were deposited at the margins of a retreating fjord glacier at the end of the last glaciation. (Photo: D.J.A. Evans)

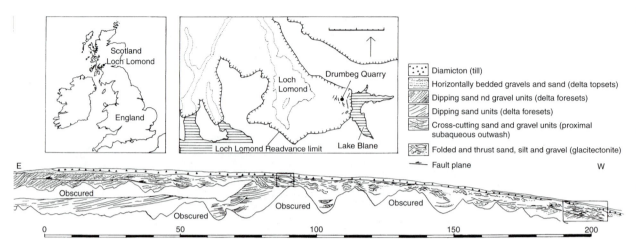

(a)

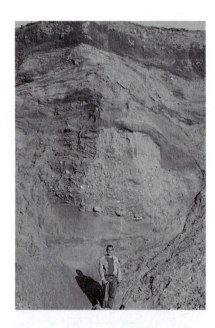

(b)

Fig. 11.112 (a) Section through an ice-contact delta at Drumbeg Quarry, near Drymen, Loch Lomond, Scotland (from Benn and Evans 1996), with boxes locating examples of glacitectonites at two locations in the exposure (left box = Fig. 10.9b and right box = Fig. 10.9a); (b) subaqueous gravelly mass flow units cross-cutting subaqueous cross-bedded sands at the centre of the section; glacitectonite and deformation till cap the sequence; (c) delta foreset beds overlying the cross-bedded sands at the east end of the section. Glacitectonic disturbance is evident as low-angle shear planes. (Photos: D.J.A. Evans)

scale slumping occurs on the exposed foreslope. (d) *Summer*. Water levels rise once more, and foreset progradation is resumed. It is not known how typical such sequences are of ice-contact deltas.

11.6.6 Deepwater sedimentary successions controlled by submarine channels

In Section 10.6, mud drapes were described as materials that blanket the bottom topography of deep water masses. Basins with pronounced relief will tend to produce local thickening of various subaqueous sediments. Similarly, features such as submarine channels will funnel sediment gravity flows and turbidity currents. Any topography produced by erosion into unconsolidated bottom sediments by powerful traction currents will also lead to gravity reworking and channelling of subsequent currents (Miall, 1985a). This will result in the production of channel fills composed of various sedimentological sequences. Many such features occur in ice-proximal and ice-contact fans, and these are discussed above. Channel fills of more distal locations are discussed here.

Density underflows and turbidites fed from adjacent ice margins may erode or infill channels on relatively shallow shelf slopes. Such a depositional

(c)

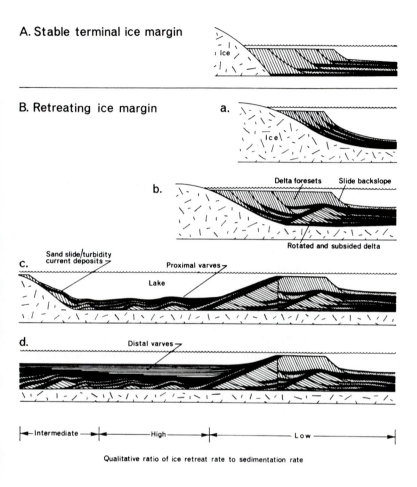

A. Stable terminal ice margin

B. Retreating ice margin

a.

b. Delta foresets   Slide backslope

Rotated and subsided delta

c. Sand slide/turbidity current deposits   Proximal varves

Lake

d. Distal varves

|←— Intermediate —→|←——— High ———→|←——————— Low ———————→|

Qualitative ratio of ice retreat rate to sedimentation rate

Fig. 11.113 Reconstructions of depositional sequences in alpine glacial lakes during (A) glacier occupancy and ice-contact delta production; and (B) glacier retreat and ice-contact/supraglacial deposition of deltaic and distal sediments. Note delta collapse and the qualitative ratio of ice retreat to sedimentation rate. (From Shaw, 1977. Reprinted by permission of Scandinavian University Press)

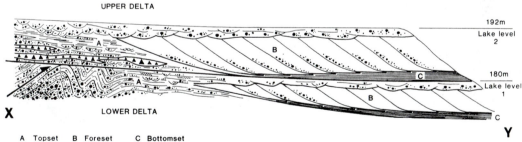

UPPER DELTA

192m
Lake level 2

180m
Lake level 1

X   LOWER DELTA

Y

A Topset   B Foreset   C Bottomset

Fig. 11.114 Schematic representation of deltaic sediments at Rhosesmor, Wales, resulting from a rise in water level from 180 m to 192 m. (From Thomas, 1984a. Reproduced by permission of John Wiley & Sons)

environment is presented later in Fig. 12.64 for the early Pleistocene Yakataga Formation in Alaska. This model depicts an ice-maximum stage when variably graded gravels and sands infill channelized scours, while palimpsest lags are striated on topographic highs. During intermediate and ice-minimum stages these channel fills are blanketed first by dropstone diamictons and muds and then muds from turbid plumes. The general lack of ice-rafted debris on areas of the Antarctic shelf (e.g. Domack, 1982), probably because of the loss of debris by floating ice margins and icebergs within a short distance seaward of the

grounding line, has led C.H. Eyles *et al.* (1985) to suggest that downslope resedimentation by gravity flows is the major contributor of sediment to offshore canyon systems (see Section 12.5.3 on the ice shelf zone of Powell's (1984) model).

Subglacial tunnel valleys infilled with glacimarine sediments after deglaciation have been reported by N. Eyles and McCabe (1989b) from the east coast of Ireland where ice retreat occurred in contact with deep water (Fig. 11.116). The subglacial tunnels, once cut by meltwater and then abandoned in deep water by retreating ice, acted as depocentres for sed-

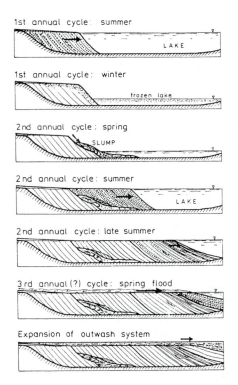

Fig. 11.115 Model of seasonally controlled delta development, based on sediments exposed near Neple, Poland. (From Mastalerz, 1990. Reproduced by permission of the International Association of Sedimentologists)

iment gravity flows composed of weakly graded to massive gravels and massive diamictons. Coarse-grained sediment was directed to the offshore channels during deglaciation by high meltwater discharges moving along the subglacial portions of the channels. The gravity flow facies were then draped by mud from suspended sediment plumes and ice-rafted debris. The tunnel valleys reported by Eyles and McCabe are tributaries to a larger valley system which continues for more than 100 km offshore and which is infilled with glacimarine sediment deposited during the deglaciation of the Irish Sea basin (Whittington, 1977). The distance travelled along the tributary channels by individual gravel-rich sediment gravity flows is unknown, but proximal to distal and upward fining relating to ice-marginal retreat would presumably characterize the stacked glacimarine facies in the channels.

11.6.7 Lake floor patterns

A variety of patterns have been identified on the plains of former glacial lakes in North America by Mollard (1983). These are referred to as reticulate, polygonal, cellular, orbicular, vermicular, doughnut-like and brain-like micro-relief, with reference to their general appearance on aerial photographs (Fig. 11.117), and are often associated with iceberg scour marks. The patterns are best-developed in deposits

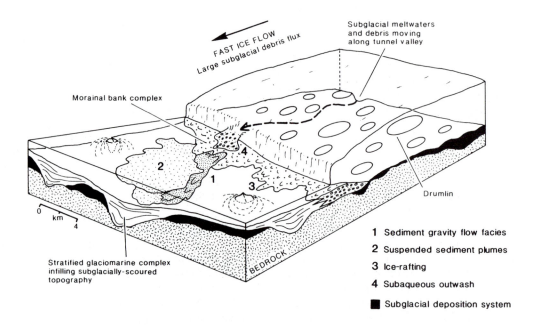

Fig. 11.116 Idealized depositional setting for the margins of the Irish Sea basin according to Eyles and McCabe (1989b). This shows the relationship between drumlins produced by fast ice flow and the construction of morainal banks by sediment being pumped to a tidewater glacier margin. Fine-grained sediment is carried by plumes and deposited as mud drapes over the subglacial landsystem in more distal locations. (Reprinted by permission of Blackwell Scientific)

containing large quantities of montmorillonite clays, which are highly plastic and have a strong tendency to swell and shrink. Sections in the micro-relief forms often exhibit folds, convolutions and diapirs, which indicate considerable disturbance in what were once subhorizontally bedded lake sediments. Mollard (1983) summarized eight different hypotheses which have been provided as explanations for the striking patterns observed on former lake plains:

1. Slow melt-out of debris-rich or debris-covered glacier ice buried beneath lake sediments. This may explain doughnut-like mounds which are similar to higher-relief hummocky moraine forms such as prairie mounds. The underconsolidated tills produced by the melt-out process become loaded by the accumulating lake clays, thus initiating the doughnut-like patterns.
2. Settling of glacier or lake ice fragments into soft lake muds as a consequence of lake drainage. The horizontal and vertical squeezing of mobile sediments by grounded icebergs to produce craters is common in flood tracks (Section 9.3.4.4). However, it is unlikely that the dense patterns illustrated in Figure 11.117b could be produced by icebergs.
3. Formation and melting of lenses and wedges of ground ice. Reticulate and cellular patterned ground forms are found at the margins of many lakes in periglacial environments, indicating intermittent inundation and exposure of bottom sediments by fluctuating water levels, and sediment modification by wave action (e.g. Dionne, 1978). However, individual cells on glacial lake plains are up to four times larger than terrestrial permafrost patterned ground forms.
4. Differential shrinkage of montmorillonite lacustrine clays with veneers or inclusions of clayey silts. After lake drainage and the gradual lowering of the water table, areas composed of expansive montmorillonite clays lose far more moisture and thus develop into topographic hollows. Some of

(a)

(b)

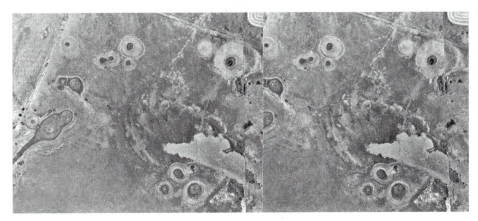

(c)

Fig. 11.117 Typical lake floor patterns on the Canadian prairies: (a) iceberg scour marks on the former floor of Glacial Lake Agassiz; (b) orbicular patterns near Maryfield, Saskatchewan (from Mollard, 1983); (c) stereopair of raised rims around piping orifices, near Rosetown, Saskatchewan (average 100 m across). (Aerial photographs reproduced by permission of Energy, Mines and Resources, Canada)

the resulting criss-crossing patterns are similar to those produced by iceberg scouring except that the relief of the furrows is inverted and they are now positive rather than negative relief features. This may be a product of the differential shrinkage that has taken place in lake bottom clays (in which furrows were scoured during lake drainage) and more silty sediments (deposited in the furrows during the late stages of lake sedimentation) since glacial lake drainage.

5. Differential consolidation of near-surface materials due to desiccation. This is well known in expansive clays, where the patterns so produced are referred to as *gilgai* (Gustavson, 1975a).

6. Groundwater piping from sublacustrine aquifers. During the period following glacial lake drainage, doughnut-like mounds and orbicular patterns can be produced in areas where artesian groundwater migrates upwards to the surface along vertical pipes. At the surface, the escape of the groundwater produces small craters or piping orifices.

7. Earthquake shock waves. Any ground-shaking generated by earthquake shock waves would induce disturbance in sediments with high moisture contents. After glacial lake drainage a desic-cated crust gradually forms over supersaturated and underconsolidated muds. Vigorous ground-shaking may result in the foundering of the higher-density crust in the underlying muds and the creation of convolutions and diapirs. The highly mobile muds may also burst through the surface crust as small eruptions a few metres in height, explaining the cellular patterns on some lake plains.

8. Wave action and drift-ice erosion along a retreating shoreline. Repeated scouring and pitting of bottom muds by drift ice may be responsible for the patterns observed on some ancient lake plains. In addition, pits may be produced by the lifting of mud and marsh vegetation which has frozen to the underside of lake or sea-ice (Dionne, 1968).

Despite their having been recognized over 40 years ago (Horberg, 1951; Mollard and Liang, 1951; Nikiforoff, 1952; Colton, 1958), research on the patterned ground of former lake plains is still in its infancy. Not only do the details of each of the eight major hypotheses need to be verified, but also these miscellaneous lake floor patterns must be differentiated from glacigenic and terrestrial periglacial features.

CHAPTER 12

LANDSYSTEMS AND MEGAGEOMORPHOLOGY

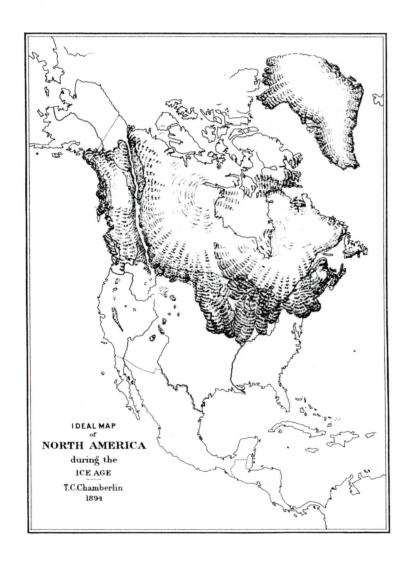

IDEAL MAP
of
NORTH AMERICA
during the
ICE AGE

T.C.Chamberlin
1894

12.1 INTRODUCTION

The study of large-scale patterns of glacial sediments and landforms and their significance is one of the most fascinating and controversial branches of glacial geology. When viewed at the scale of an entire glaciated basin, the character and distribution of glacial sediments and landforms can reveal shifting patterns and processes of glaciation over complete glacial cycles, and provide a valuable complement to short-term studies of modern glacial environments. However, reconstructions of former glacial environments are only as good as the observations and assumptions that go into them, and it is at the largest scale that errors in interpretation become most magnified. For example, if a certain diamicton unit is interpreted as a basal till, the implications for former glacier extent and dynamics are very different from those that arise if it is interpreted as a distal glacimarine deposit. Furthermore, the interpretation of ancient landforms relies on analogues with modern examples, and an inappropriate choice of analogue can lead to potentially misleading conclusions. This problem is becoming less serious as the range of modern landsystems and their controls become better known, although much remains to be learned.

Despite the difficulties raised by the interpretation of ancient landscapes, painstaking work over the past decades has revealed a very detailed picture of ancient ice masses and their evolution. This is particularly the case for the great Pleistocene ice sheets of North America and Europe, where the geological record documents complex but coherent patterns of environmental change over much of the last glacial cycle. Furthermore, the ice sheet record can be linked with regional and global climatic and oceanographic changes, and an integrated theory of glacial–interglacial cycles is emerging (Broeker and Denton, 1990a, b; Hughes, 1992).

In this chapter we review basin-scale distributions of glacial landforms and sediment packages, focusing on what such features reveal about long-term patterns of environmental change. This provides an opportunity to synthesize the material presented in previous chapters, and to consider the evidence for former climates, hydrology, glacier dynamics, patterns of erosion and deposition, and sea-level change over complete glacial cycles in a variety of settings. Although we focus on the glacial geology of the Pleistocene, the period for which the evidence is most complete and accessible, it should not be forgotten that the Earth's glacial record extends into the much more distant past, and that an understanding of glacial systems can reveal a great deal about the geography of ancient continents and seas. In the following sections, we review glacial landsystems under four broad headings: (a) ice-marginal and supraglacial; (b) subglacial; (c) subaquatic; and (d) glaciated valley landsystems. We conclude by reviewing evidence for past glaciations of Mars, where glacial geomorphological studies meet their most challenging frontier.

12.2 APPROACHES TO LARGE-SCALE INVESTIGATIONS

Investigations of glacial geology at regional or even continental scales have a long history (e.g. Geikie, 1877, 1894; Upham, 1895; Tyrrell, 1898a, b; Flint, 1943). In recent times, such studies have emphasized the links between sediment–landform associations and genetic processes, providing a framework for using geological evidence to reconstruct long-term patterns of environmental change. Two main approaches are discussed here: (a) the study of glacial landsystems and process–form models; and (b) sequence stratigraphy. Other, more specialized approaches are discussed in later sections of the chapter.

12.2.1 Landsystems and process–form models

The concept of *landsystems* was first introduced in Australia by geomorphologists and agricultural land surveyors in an attempt to classify land capability based upon relationships between geology, surface relief and climate (Christian, 1957; Mabbutt, 1968; Ollier, 1977). Landsystems were defined as areas of common terrain attributes, different from those of adjacent areas, in which recurring patterns of topography, soils and vegetation reflect the underlying geology, past erosional and depositional processes, and climate (Christian and Stewart, 1952; Stewart and Perry, 1953). The identification of landsystems therefore constitutes a holistic approach to terrain evaluation, wherein the geomorphology and subsurface materials that characterize a landscape are genetically related to the processes involved in their development. According to this approach, the landsystem is the upper level in a hierarchy of terrain classification (Lawrance, 1972). At the lowest level are *land elements*, or individual landforms; at the intermediate level are *land facets* or groups of land elements; and *landsystems* are composites of linked land facets. Glacial examples of land elements, facets and landsystems might be drumlins, drumlin fields, and the whole assemblage of forms representing the former glacier bed. Terrain evaluation is usually restricted to the uppermost few metres of the ground (Mitchell, 1973), but N. Eyles (1983c) has stressed that glacial landsystems must also include all genetically related subsurface sediments. The development of sediment–landsystem models allows subsurface conditions to be predicted from surface morphology, an ability that has profound implications for resource assessment and civil engineering works (N. Eyles, 1983c).

Type examples of *glacial landsystems* were introduced by Boulton and Paul (1976), and developed further by Boulton and N. Eyles (1979) and N. Eyles (1983c). Three glacial landsystems were defined by Eyles (1983a, c): (a) the *subglacial landsystem,* formed at the glacier bed; (b) the *supraglacial landsystem*, which is draped over the former glacier bed during ice retreat; and (c) the *glaciated valley landsystem*, which is characteristic of mountain areas. These three categories provided powerful unifying concepts when they were first introduced, but some limitations have since become apparent. First, Eyles's system is not inclusive, and omits landsystems formed in proglacial and glacimarine environments, and other important glacial settings. Second, the landsystem approach does not deal explicitly with spatial patterns of erosion and sedimentation, or their variation during glacial advance–retreat cycles.

The landsystem approach can be placed in a dynamic context in *process–form models*. Such models were first applied to glacial landscapes by Clayton and Moran (1974), who used spatial patterns of glacial landforms and sediments in North Dakota to infer changes in glacial dynamics during ice margin advance and retreat. Process–form models emphasize the interrelationships of specific landform–sediment associations at both local and regional scales, and their genetic significance in terms of migrating zones of deposition and erosion. Process–form models can be applied at a wide range of scales, from the marginal zone of a fluctuating valley glacier to reconstructions of the great Pleistocene ice sheets (e.g. Sugden, 1977; Dyke *et al.*, 1982b; Sharp, 1985b; Dyke and Prest, 1987; Boulton and Clark, 1990b; Mooers, 1990b).

Glacial landscapes commonly consist of superimposed depositional systems of different ages, resting upon older erosional surfaces. Such landscapes have been compared to a *palimpsest,* or a parchment which was reused by early monastic scholars when writing materials were in short supply (Chorley *et al.*, 1984). Older text was erased to make way for other writing, but the erasing was not totally effective, and the older inscriptions tended still to be partially visible. Thus, a palimpsest provides a vivid metaphor for landscapes in which the products of ancient processes are partially visible below more recent forms. Each new generation of the landscape inherits the topography, substrate type, drainage characteristics and other factors established by the previous generation, so that older landscape components act as important constraints on subsequent landscape evolution. In glaciated terrains, large-scale erosional forms are generally the cumulative effect of multiple glacial cycles (Section 9.5), whereas the sedimentary record typically records only the most recent glaciation. This is particularly true of continental interiors, where widespread erosion and sediment reworking during glacier advances ensure that older glacigenic sequences rarely survive within younger glacial limits. Glacimarine sequences have greater preservation potential, particularly in subsiding basins such as the central graben of the North Sea, where > 1 km of Quaternary sediments are preserved. On tectonically stable shelves, glacimarine sequences are more likely to be removed by glacial or proglacial erosion, and rarely exceed 100 m in thickness (Section 12.5; N. Eyles and C.H. Eyles, 1992; N. Eyles, 1993).

Even when applied at a very general level, the landsystem approach can reveal striking patterns of glacial erosion and deposition. For example, Fig. 12.1 shows large-scale glacial geomorphological zones in North America and Britain (N. Eyles and Dearman, 1981; N. Eyles *et al.*, 1983). These maps give a clear impression that glacial landscapes exhibit systematic spatial organization, and hint at important underlying climatic, geological, topographical and glaciological controls. In later parts of this chapter, we examine the organization and genetic significance of glacial landscapes in detail, and review the emerging picture of the long-term impact of glaciation.

12.2.2 Sequence stratigraphy

A new and powerful approach to the study of glacial depositional systems has arisen from the concepts of *sequence stratigraphy*, which deals with large-scale depositional cycles in the rock record. Originally developed by petroleum geologists from the Exxon Production Research Company in Houston, Texas, as a tool for locating potential hydrocarbon reservoirs, sequence stratigraphy seeks to identify large-scale *sequences* of conformable, genetically related strata bounded by unconformities or other major discontinuities (Brown and Fisher, 1977; Van Wagoner *et al.*, 1988; Posamentier *et al.*, 1988; Galloway, 1989; Walker, 1992a; Martini and Brookfield, 1995). Within such sequences, strata may be grouped into *systems tracts*, defined as linkages of contemporaneous depositional systems (Fig. 12.2). The original emphasis was on shallow marine environments, and systems tracts were originally defined in terms of major erosional and depositional episodes associated

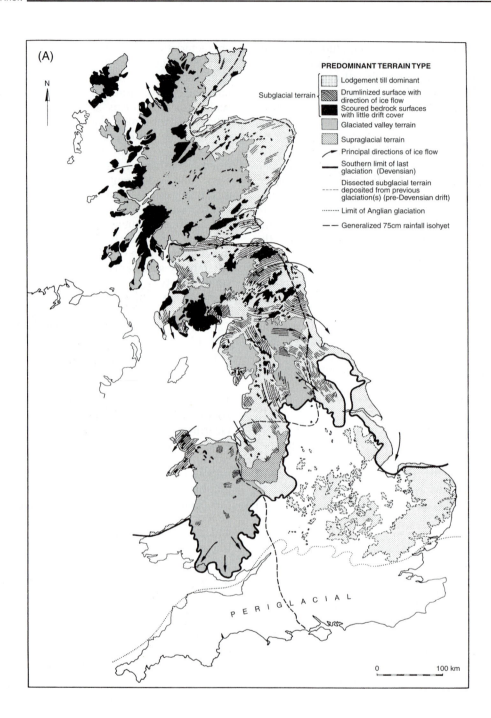

(A)

PREDOMINANT TERRAIN TYPE

Subglacial terrain

	Lodgement till dominant
	Drumlinized surface with direction of ice flow
	Scoured bedrock surfaces with little drift cover
	Glaciated valley terrain
	Supraglacial terrain

Principal directions of ice flow

Southern limit of last glaciation (Devensian)

Dissected subglacial terrain deposited from previous glaciation(s) (pre-Devensian drift)

Limit of Anglian glaciation

Generalized 75cm rainfall isohyet

PERIGLACIAL

0   100 km

Fig. 12.1

with global cycles of sea-level change. Three types of systems tract were recognized, formed respectively during periods of marine transgression, highstand and regression. Thoughtful reviews of the concepts of sequence stratigraphy and its methodological problems have been provided by Walker (1990, 1992a).

Some principles of sequence stratigraphy have been applied to glacial depositional systems by Boulton (1990), N. Eyles and C.H. Eyles (1992) and Mar-

tini and Brookfield (1995). It has been found necessary, however, to modify the approach before it can be applied to the glacial case. First, patterns of local sea-level change in glaciated areas are complicated by the isostatic depression of the crust by ice loading (Section 1.5.1.2), and variation in crustal loading across a glaciated basin can cause one part to experience a fall in relative sea-level while sea-levels are rising in another. The grouping of contemporaneous depositional systems into systems tracts on the basis

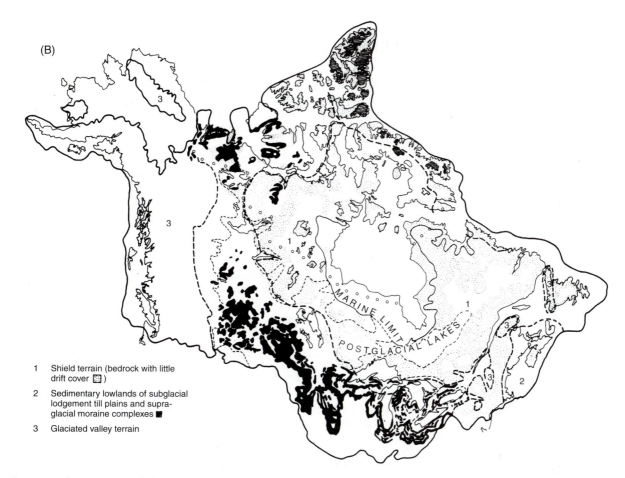

Fig. 12.1 The coverage of glacial landsystems in: (A) Britain (From N. Eyles and Dearman, 1981. Reproduced by permission of the International Association of Engineering Geologists); and (B) North America (modified from N. Eyles *et al.*, 1983). Present ice cover is shown only for the Arctic. Southern limit of continuous permafrost shown by open circles. Maximum outermost limit of glaciation shown by thick black line.

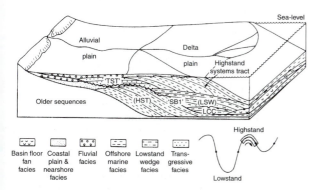

Fig. 12.2 Hypothetical systems tract development during sea-level cycles. Highstand systems tracts (HST) are formed during periods of high sea-level; transgressive systems tracts (TST) are formed during rising sea-level; and lowstand wedges (LSW) are formed at times of low sea-level. Sequence boundary 1 (SB1) is an erosion surface separating two *sequences*. LCC = Levéed channel complex. (From Walker, 1992a. Reproduced by permission of the Geological Association of Canada)

of regional marine erosion surfaces is thus impractical (N. Eyles and C.H. Eyles, 1992). Second, glacial depositional systems located on land, in inland lakes or below grounded glacier ice are generally controlled by factors other than global sea-level change. The logical basis for defining *glacial systems tracts* is the *glacier advance and retreat cycle*. Within the maximum glacier limits, depositional systems deposited during advance and retreat will be separated by a time-transgressive subglacial erosion surface or a wedge of subglacially deposited or deformed sediment (Section 12.4.1; Berthelsen, 1978; Boulton, 1996a, b). Beyond the ice limit, complete depositional sequences may be preserved in the form of distal–proximal–distal successions, recording the migration of depositional systems in response to glacier advance and retreat (Section 12.5.1; Boulton, 1990; N. Eyles and C.H. Eyles, 1992). Throughout this chapter, we emphasize the role of the glacial cycle in controlling the large-scale sedimentary architecture of glaciated basins.

12.3 ICE-MARGINAL AND SUPRAGLACIAL LANDSYSTEMS

This section reviews landsystems produced at the margins of ice sheets and ice caps, where supraglacially entrained debris is rare or absent. In such situations, debris for the construction of ice-marginal landforms must be derived from the glacier foreland or the bed, and elevated to higher levels by either proglacial or englacial glacitectonic processes, or pressurized meltwater flowing in englacial conduits. The processes and patterns of debris entrainment, elevation and deposition at glacier margins are strongly influenced by glacier thermal regime and ice margin activity, allowing some of the climatic and dynamic characteristics of former ice masses to be reconstructed from geological evidence.

This section examines ice-marginal and supraglacial landsystems under four headings: (a) temperate glacier margins; (b) subpolar margins; (c) polar-continental margins; and (d) surging glaciers. Temperate margins are mainly wet-based but may experience seasonal freezing near the snout during cold conditions in the winter; subpolar margins are characterized by cold-based conditions near the snout and wet-based conditions further upglacier; and polar-continental margins are cold-based with insignificant production of meltwater at the surface. Surging glaciers may be of temperate or subpolar type. In reality, there is an unbroken continuum of variation between temperate and polar-continental margins, but it is important to emphasize the distinctiveness of ice-marginal processes and products in different climatic settings. At temperate glacier margins, melting occurs at the glacier surface and the bed, and meltstreams are important agents of sediment transport and deposition. Landsystems produced in such settings record alternating glacifluvial and gravitational processes, locally modified by glacitectonic deformation. In addition, extensive proglacial rivers may heavily rework glacigenic landforms following deposition. Subpolar margins are commonly affected by glacitectonic processes, owing to strong compressive flow as wet-based ice decelerates against cold-based ice near the snout. These processes are capable of elevating large amounts of debris above the glacier bed and foreland. Meltwater is available on the glacier surface but not the bed, and can have an important impact on ice-marginal landsystems. Polar-continental margins may be subject to tectonic processes similar to those undergone by subpolar margins, but meltwater is of negligible importance in these arid settings, so ice-marginal sediment accumulations are less prone to reworking after ice recession.

Ice-marginal and supraglacial landsystems can be locally impressive, forming high-relief, irregular terrain in otherwise flat landscapes. When viewed at a regional scale, however, such landsystems occupy a relatively small proportion of the area occupied by former ice sheets, and commonly contain 'windows' through which underlying subglacial landsystems are visible. Thus, in regions affected by ice sheet glaciation, the subglacial landsystem is commonly the most prominent in the geomorphological record (Section 12.4). The most impressive ice-marginal and supraglacial landsystems are those produced by valley glaciers in mountainous terrains where considerable quantities of debris are delivered to glacier surfaces (Section 12.6). We discuss the landsystems formed at temperate, subpolar, polar-continental and surging glacier margins in turn below, focusing on modern examples. We then examine the applicability of these landsystems to Wisconsinan ice-marginal and supraglacial landforms deposited at the southern margin of the Laurentide ice sheet, as a case study of the continental-scale distribution of Pleistocene landforms.

12.3.1 Temperate glacier margins

Temperate glacier margins can be defined as those which are mainly wet-based for at least part of the year, and generally occur in areas of discontinuous or no permafrost. A narrow frozen zone develops below the margin of many temperate glaciers during the winter, owing to the penetration of a *cold wave* from the atmosphere through the thin ice at the edge of the glacier (Boulton, 1977a; Harris and Bothamley, 1984; Krüger, 1994). Additionally, small and discontinuous areas of net freezing can exist below some temperate margins in areas of discontinuous permafrost (e.g. Lawson, 1979a). At present, broad temperate glacier lobes are found mainly in southern Alaska, Iceland and Patagonia. Mountain glaciers in many mid- and low-latitude areas also have temperate margins, but the landsystems they create are generally strongly influenced by topography and supraglacial debris supply (Section 12.6).

Beneath temperate glaciers, sequences of debris-rich basal ice are typically thin or absent (Section 5.4.1; Hubbard and Sharp, 1989, 1993). Although some englacial thrusting and folding can occur in association with seasonally frozen ice at the margin, the amount of debris elevated from the bed to glacier surfaces is generally small (Boulton, 1977a). Therefore, in situations where supraglacial or glacifluvial debris sources are of minor importance, deposition at temperate glacier margins typically produces small dump moraines, or push and squeeze moraines derived from sediment exposed on the glacier foreland (Fig. 12.3; Sections 7.3, 7.4, 11.3.2 and 11.3.3). Imbricate thrust moraines are most commonly associated with subpolar and surging glaciers, but also occur at some temperate ice margins. Modern examples have been described from Iceland, where thrusting mainly affects ice-contact fans built up against the ice front but can also relocate slabs of subglacial till (Krüger, 1985, 1987, 1993, 1994; Humlum, 1985b; Boulton, 1986). The sole thrusts of these imbricate moraines are at shallow depth (1–2 m), and may be associated with the base of a seasonally frozen layer (Krüger, 1994). In contrast, the sole thrusts of tectonized ridges in front of surging Icelandic glaciers may be several tens of metres below the surface (Croot, 1988b; Section 12.3.4).

Glaciers without extensive debris cover respond rapidly to seasonal temperature fluctuations, so glacier margins tend to oscillate on an annual basis. As a result, suites of recessional moraines are commonly formed during deglaciation, except where glacier velocities are low and/or ablation continues throughout the year (Sections 7.2 and 11.3; Sharp, 1984; Boulton, 1986; Krüger, 1987, 1994). The areas between recessional moraines represent periods of rapid glacier retreat, and commonly exhibit well-preserved subglacial landforms (Figs 12.3, 12.4). A complete sequence of annual moraines dating from the period 1965–83 (and probably up to the present) exists on the foreland of Breidamerkurjökull, Iceland (Fig. 12.5). The relationships between annual temperature cycles, ice margin fluctuations and moraine formation at this site were studied by Boulton (1986), who showed that the recessional moraines are formed at the culmination of winter readvances superimposed on overall glacier retreat. Well-preserved sequences of recessional moraines deposited by ancient glaciers may also have an annual origin,

Fig. 12.3 Aerial photograph of the foreland of Heinabergsjökull (H), Iceland. Annual push moraines (p) record readvances that interrupted glacier recession, while other parts of the foreland have been reworked by meltstreams. Flutings (f) are also visible in the inner zone of push moraines in front of Heinabergsjökull. All these features were produced during the retreat of the glaciers from their Little Ice Age maximum limits (m). Scale bar represents 500 m. (Aerial photograph (1989) reproduced by permission of Landmaelingar Islands)

but this is difficult to test given the accuracy of currently available dating methods.

Meltwater can have a profound influence on the landsystems formed at temperate glacier snouts. Where subglacial and englacial drainage systems are well-developed, sands and gravels can accumulate within conduits, and meltstreams emerging at the glacier surface can deposit considerable quanti-

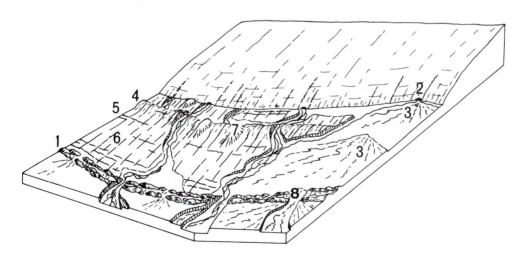

Fig. 12.4 Model of ice-marginal, subglacial and proglacial landsystems on the foreland of Myrdalsjökull, Iceland. 1: outermost push moraine dating to the Little Ice Age; 2: ice cave and stream outlet; 3: outwash fan produced at various ice-marginal positions; 4: old push moraine overridden by the glacier during its advance to the Little Ice Age limit; 5: annual (push) moraines produced during recession from the Little Ice Age limit; 6: small flutings; 7: drumlin (old moraines drumlinized); 8: proglacial outwash deposited when ice was at Little Ice Age maximum. (Modified from Krüger, 1987)

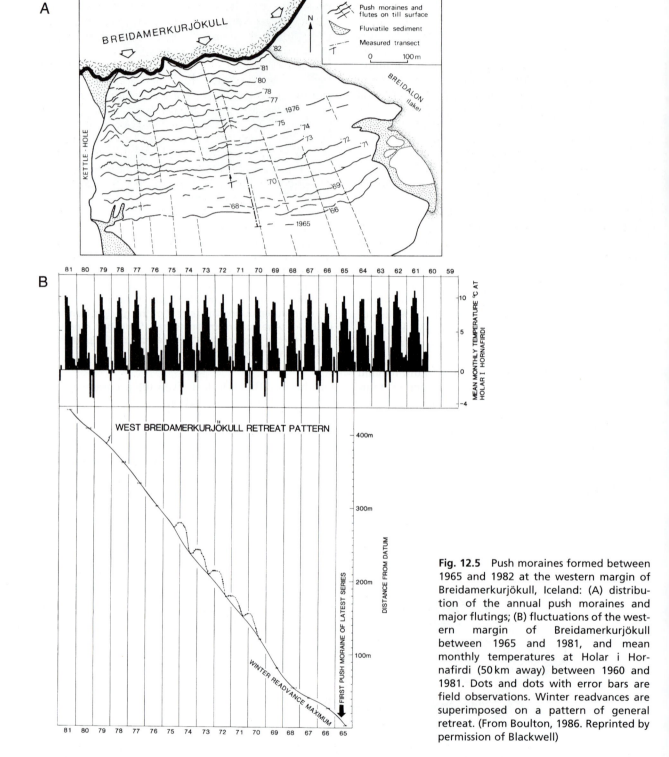

Fig. 12.5 Push moraines formed between 1965 and 1982 at the western margin of Breidamerkurjökull, Iceland: (A) distribution of the annual push moraines and major flutings; (B) fluctuations of the western margin of Breidamerkurjökull between 1965 and 1981, and mean monthly temperatures at Holar i Hornafirdi (50 km away) between 1960 and 1981. Dots and dots with error bars are field observations. Winter readvances are superimposed on a pattern of general retreat. (From Boulton, 1986. Reprinted by permission of Blackwell)

ties of glacifluvial sediment in supraglacial positions. Indeed, on some glaciers, glacifluvial sediment constititutes the major part of the englacial and supraglacial sediment load close to the margins (Gustavson and Boothroyd, 1987; Kirkbride and Spedding, 1996). During deglaciation, uneven ablation of sediment-covered ice can lead to the development of glacier karst (Section 6.5.4.2; Clayton, 1964). Sediment deposited within conduits and in supraglacial outwash fans and lakes produces a complex assemblage of landforms including esker systems, kame ridges and plateaux, and pitted outwash (Sections 11.2.9 and 11.4; Fig. 12.6a). The surface topography of such assemblages is commonly highly irregular,

consisting of discontinuous ridges, mounds and meandering erosional channels, with a relative relief of up to several tens of metres. Modern examples have been described from the Malaspina Glacier, southern Alaska, by Gustavson and Boothroyd (1987). Landforms at different topographic levels and locations record changing depositional environments during glacier downwasting and recession (Young, 1975, 1978; Gustavson and Boothroyd, 1987; Gray, 1991; Johnson and Menzies, 1996). Distinctive spatial associations of glacifluvial landforms deposited during ice wastage have been termed *morphosequences* (Jahns, 1941; Koteff, 1974; Koteff and Pessl, 1981). In such sequences, proglacial outwash passes upvalley into kame and kettle topography, which is inset within ice-marginal kame terraces and moraines. Kame and kettle topography is locally refashioned into suites of river terraces by proglacial and postglacial streams.

(a)

(b)

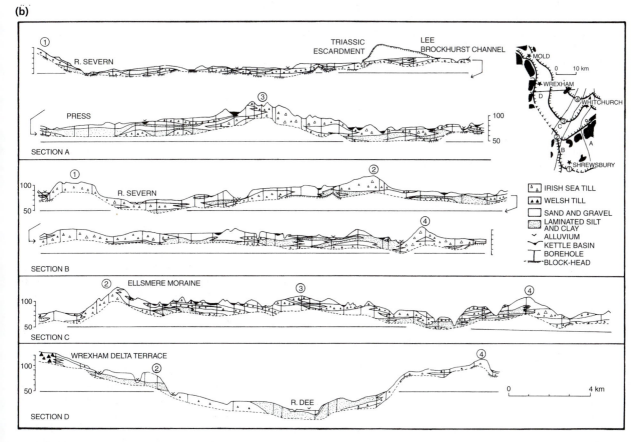

Fig. 12.6 (a) Kame and kettle landscape: Bruarjökull, Iceland. Outwash has prograded over a low-gradient glacier snout, and progressive melt-out of buried ice has led to the development of kettle holes and linear depressions and channels. (Photo: D.J.A. Evans) (b) Simplified cross-sections through the complex stratigraphic successions associated with ice recession from the coalescence zone of the Irish Sea and Welsh ice sheets at the close of the last glaciation. The interbedded sediments represent former ice-marginal sandar, ice-contact alluvial fan ramps, proglacial sandar, and ice-contact and proglacial lakes. Supraglacial deposition on a receding ice margin was interrupted by minor standstills and ice front oscillations. The inset map shows the locations of rock outcrops (shaded) and cross-sections A–D. (From Thomas, 1989. Reprinted by permission of John Wiley & Sons)

An excellent example of the stratigraphic complexity produced during ice margin retreat was reported by Thomas (1989) from the Cheshire–Shropshire lowland, England, in the former coalescence zone of the Devensian Irish Sea and Welsh ice sheets. A thick supraglacial debris cover was influential in the formation of prominent moraine ridges, the tills of which interdigitate with the glacifluvial and glacilacustrine sediments of ice-contact fans, ribbon sandar and ice-contact and proglacial lakes (Fig. 12.6b).

In temperate environments, subglacial, ice-marginal and supraglacial landsystems are commonly substantially modified by proglacial glacifluvial deposition and erosion during and following deglaciation. Landforms such as drumlins, eskers and moraines may be heavily reworked by meltwater and/or buried by outwash sediments so that they are left as remnants or inliers within sandur plains (Fig. 12.3). Even large ice-marginal moraines can be dissected or partially destroyed by meltstreams directly after emplacement (Price, 1969; Price and Howarth, 1970; Krüger, 1987, 1994; Thompson, 1988). In some cases, virtually all the depositional legacy of a temperate ice lobe takes the form of glacifluvial sediments.

12.3.2 Subpolar glacier margins

Subpolar margins are characterized by an outer cold-based zone and an upglacier wet-based zone. At present, such glacier margins are found mainly in areas of continuous permafrost north of the Arctic Circle, in Greenland, the islands of the Canadian Arctic, and the Arctic Ocean archipelagos of Svalbard, Novaya Zemlya and Severnaya Zemlya. However, the high-altitude ice caps and glaciers in the mountains of central Asia (e.g. the Kunlun and Tien Shan; Huang Maohuan, 1992) also lie in permafrozen regions, and have many similarities with those at high latitudes. Subpolar glaciers also occur in the Antarctic Peninsula. Regions of continuous permafrost have low mean annual temperatures and generally low precipitation. For example, at sea-level in Svalbard, mean annual temperature (MAT) is c. −6°C and precipitation is c. 200–400 mm yr^{-1}. Permafrost is c. 100 m deep in coastal areas, thickening to more than 400 m in the central mountains (Liestøl, 1977). At Resolute in the Canadian Arctic, MAT is −16°C and precipitation is 860 mm yr^{-1}, and permafrost in the region is up to 600 m deep.

The thermal regime and geomorphic impact of glaciers in permafrost regions are determined by ice thickness and activity (Dyke, 1993b). Ice caps growing on thick permafrost must initially be entirely cold-based, but wet-based zones can develop in response to subglacial heating from two sources: (a)

geothermal heat; and (b) *strain heating* produced by glacier flow (Section 2.5.2). According to Dyke, thawing of the bed is most likely where glacier velocities are highest; that is, in the vicinity of the glacier equilibrium line and in areas of flow convergence (Section 4.1.1). Thus, wet-based ice will be most extensive below glaciers with relatively high mass throughput and/or strong converging flow. In the upper parts of glacier accumulation areas, cold-based zones are likely to persist, because of low ice velocities and downward advection of cold firn. Cold-based ice occurs beneath the glacier margins where ice is thin, velocities are low, and proglacial permafrost extends below the glacier. Thus in permafrost regions, small low-activity glaciers are likely to be cold-based throughout.

The presence of wet-based ice upglacier from a frozen margin in subpolar glaciers has a profound impact on processes of debris entrainment, transport and deposition. First, erosion below wet-based, sliding ice provides a source of debris which can be transported towards the margin in regelation ice or a subglacial deforming layer. Erosion rates may be small where sliding velocities are low, but can be significant below outlet glaciers (Dyke, 1993b; Hallet et al., 1996; Section 5.3.6). As this debris is transported into the frozen zone at the glacier margin, sequences of debris-rich basal ice develop as a result of net adfreezing (Section 5.4.2; Weertman, 1961c; Hubbard and Sharp, 1989; Hubbard, 1991). These debris sequences are commonly thickened and elevated towards the glacier surface by strong compressive deformation resulting from ice deceleration at the frozen margin (Section 5.4.4; Goldthwait, 1951; Boulton, 1972a; Echelmeyer and Wang, 1987; Knight, 1989, 1994; Tison et al., 1993; Hambrey and Huddart, 1995; Etzelmüller et al., 1996). Thick basal debris sequences also form where glaciers override buried ice dating from earlier glacial events (Section 5.4.3; Hooke, 1970, 1973b; D.J.A. Evans, 1989b; D.J.A. Evans and England, 1992). The long survival of buried ice in permafrost environments means that debris-rich ice may be recycled many times over several glacial cycles as glaciers advance and retreat.

Compressive stresses at the margin are transmitted to the glacier foreland, and where thick accumulations of unconsolidated sediments are present, composite ridges and thrust-block moraines are constructed by proglacial tectonics (Section 11.3.1.2; Kalin, 1971; D.J.A. Evans, 1989a, b, 1990a; D.J.A. Evans and England, 1991; Hambrey and Huddart, 1995; Etzelmüller et al., 1996). Thrusting is most effective in wide valley bottoms close to sea-level where fine-grained glacimarine and lacustrine sediments crop out, although thrusting may also affect frozen gravels (Kalin, 1971; D.J.A. Evans and England, 1991; Lehmann, 1992; Hambrey and Huddart, 1995). Proglacial thrusting is commonly accompanied by englacial thrusting and the elevation of basal debris to the ice surface, which is then reworked by water and gravitational processes during deglaciation (Boulton, 1972b; Hambrey and Huddart, 1995; Etzelmüller *et al.*, 1996). The combination of proglacial and englacial thrusting, therefore, creates a landsystem in which outer thrust-block moraines pass upvalley into ice-cored chaotic hummocky moraine, controlled moraine or kames (Figs 12.7 and 12.10). The generally low discharges of meltstreams in Arctic environments mean that such landsystems are only locally reworked into outwash deposits, and tend to be well preserved.

Where proglacial thrusting does not occur, glaciers commonly terminate in steep ice cliffs along which elevated basal debris is exposed (Fig. 7.10). Ice margin positions are recorded by *frontal aprons* built up from fallen or inwashed debris (Sections 7.3.1 and 11.3.3; D.J.A. Evans, 1989b). Following ice margin retreat, belts of hummocky and controlled moraine are deposited upvalley from frontal aprons, as the result of melt-out and reworking of englacial debris. If the debris cover is thicker than the depth of summer melting (the *active layer thickness*), ice cores may survive indefinitely within frontal aprons and associated hummocky and controlled moraine.

In cases where bedrock is close to the surface or glacier activity is low, moraines formed at the margins of Arctic-type glaciers may amount to no more than a line of boulders, attesting to very low turnover of debris (Fig. 12.8 (Plate 24)). Recessional moraines are uncommon, owing to the scarcity of debris. Basal tills are rare and usually occur only as a veneer <50 cm thick (D.J.A. Evans, 1990a). The amount of available debris reaches a minimum where glaciers are predominantly or entirely cold-based. Former glacier margins in such areas are most often marked by *lateral meltwater channels* cut into bedrock (Section 9.3.4.3; D.J.A. Evans, 1990b; Dyke, 1993b). These lateral channels form because surface meltwater is prevented from penetrating the glacier by low ice temperatures, so that it must drain over the surface and around the margin of the glacier. Dyke (1993b) described spectacular examples of lateral channels from the central Canadian Arctic, where they commonly form nested sets that clearly delineate successive positions of retreating ice lobes (Fig. 12.9). Where best-developed, lateral channels are very closely spaced and record retreat with a very fine temporal resolution, and in some cases possibly resolve individual (annual) melt events. Lateral channels may be the only evidence left by predominantly or wholly cold-based glaciers and ice caps, such as the late Wisconsinan ice caps on Somerset and Cornwallis Islands in the central Canadian Arctic. In these areas, channels are incised into otherwise unmodified periglacial land surfaces consisting of felsenmeer, tors and gelifluction terraces (Dyke, 1993b). On south-east Ellesmere Island, where glacier activity was higher, owing to moisture advected from the open water of northern Baffin Bay, tills and moraines are more widespread, but lateral meltwater channels incised into bedrock remain important components of the glacial landscape.

Where cold glacier tongues descend to the sea, as in parts of the Canadian Arctic archipelago, glacier ice shelves can form *ice shelf moraines* along the margins of fjords and inter-island channels (Sections 1.4.3.1 and 11.6.4; England *et al.*, 1978; Hodgson and Vincent, 1984; Dyke, 1987; Dyke and Prest, 1987; Hodgson, 1994). In fjord landscapes, ice shelf moraines tend to be composed predominantly of coarse, angular debris derived from local bedrock slopes, whereas at the margins of inter-island channels they commonly contain considerable quantities of reworked marine and glacimarine sediment. Ice shelf moraines have been reported from the Antarctic Peninsula by Sugden and Clapperton (1981), but polar examples are otherwise rare in the glacial literature (Section 11.6.4).

The regional distribution of glacial landforms on Victoria Island, Arctic Canada, has been used by Sharpe (1988b) to reconstruct phases of activity at the northern margins of the Laurentide ice sheet. The distributions of four *landform sets* and their relationships to topography along one transect on Wollaston Peninsula are depicted on Fig. 12.10 and Fig. 12.11 (Plate 25). *Landform set I* consists of 'ground moraine', which comprises a thin, low-relief glacial sediment cover with small recessional moraines, meltwater channels and ice stagnation topography. It is restricted to the high plateau surface of the peninsula. *Landform set II* consists of hummocky moraine and occurs largely as a broad belt around the peninsula, associated with the major bedrock scarps at the peninsula margins. *Landform set III* incorporates lateral moraines and deformed (shear) moraines including the complex Colville end moraines. These were deposited along the southern escarpment of the peninsula and include glacimarine sediments where the ice margin contacted the sea. The shear moraine is interpreted as the product of deposition in a shear zone at the margin. *Landform set IV* consists of streamlined landforms, and includes drumlins and flutings of the lowland to the south of the peninsula. The drumlins trend parallel to the southern escarpment of the plateau, whereas the flutings document ice flow towards the escarpment and the shear moraine. Sharpe used these landform sets to suggest a complex series of glaciological conditions brought about by changing ice thicknesses and localized topographic controls. Landform set I was produced by the retreat of thin, active ice undergoing extending flow,

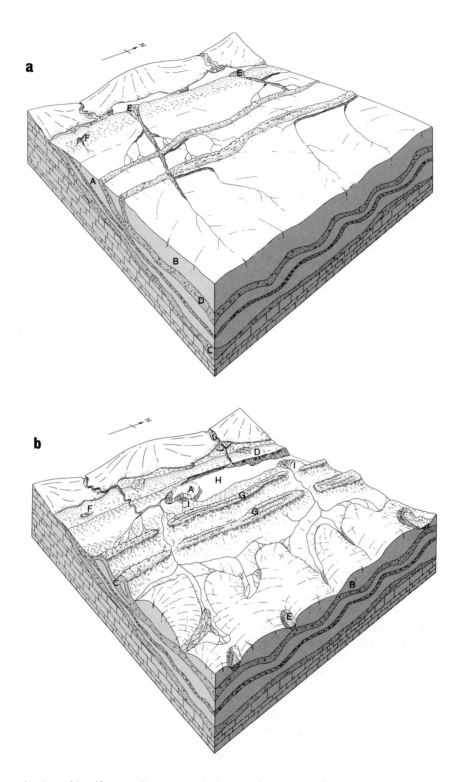

Fig. 12.7 The production of landform–sediment associations at the margin of the Laurentide ice sheet in the vicinity of Bluenose Lake, NWT, Canada, at the end of the last glaciation. (a) Ice-maximum phase. A, inactive ice; B, active ice; C, D, thick englacial debris band producing transverse controlled moraine; E, outwash; F, moulin and kame. (b) Glacial Lake Bluenose phase. Controlled moraine ridges and ice-marginal kame and lacustrine features are produced as ice undergoes recession; large tracts of buried glacier ice still exist in this region. A, ice-contact delta; B, stagnant ice; C, thick englacial debris band; D, kame terrace; E, isolated supraglacial kame; F, isolated kame; G, controlled transverse moraine; H, Glacial Lake Bluenose; I, glacilacustrine deltas. (Modified from St Onge and McMartin, 1995)

whereas the thicker sediment cover in landform set II records compressional flow followed by stagnation. Landform set III records ice-marginal sedimentation by compressional flow against the escarpment and possible shearing during a glacier surge, or at least a reactivation of fast glacier flow, in the lowlands to the south of Wollaston Peninsula, where landform set IV was deposited.

12.3.3 Polar-continental glacier margins

Distinctive glacial landsystems are formed around the margins of glaciers in the polar deserts fringing parts of the Antarctic ice sheet, such as the Vestfold Hills in East Antarctica and the Dry Valleys in the Transantarctic Mountains. These areas are cold and windy, and have a moisture deficit because sublimation exceeds precipitation. For example, in the Wright Valley in the Dry Valleys region, MAT is $-19.8°C$, and precipitation is only $10\,mm\,yr^{-1}$ (Denton *et al.*, 1993). Because of the low temperature and humidity, 90 per cent of glacier ablation is by sublimation and only 10 per cent by melting (Chinn,

1980). Local glaciers are entirely cold-based, and are nearly free of debris, although some of the larger terrestrial outlet tongues of the Antarctic ice sheet are partly wet-based with cold-based margins, and have stacked basal debris sequences similar to those of Arctic-type glaciers (Shaw, 1977a, c; Lundqvist, 1989b; Denton *et al.*, 1993).

The extreme aridity and limited importance of melting in this environment have important implications for glacial deposition and landsystem development. Where thick basal ice sequences crop out, marginal aprons accumulate by gravitational processes, but with very limited reworking by meltwater. Within the marginal aprons, belts of hummocky and controlled moraines can develop from stacked and thrust englacial debris sequences. The topography of controlled moraines is inherited from englacial structures, such as debris bands and folds in debris-rich basal ice, and commonly includes ridges transverse to glacier flow (Shaw, 1977a, c; Rains and Shaw, 1981). Rates of ablation are extremely slow, owing to the predominance of sublimation (Section 6.5.1), and features that appear to be terminal and hummocky moraines in fact commonly constitute a thin veneer of debris overlying buried ice (Rains and

Fig. 12.9 Lateral meltwater channels cut along the margins of retreating fjord glaciers on Knud Peninsula, Ellesmere Island. (Aerial photograph provided by the Department of Energy, Mines and Resources, Canada)

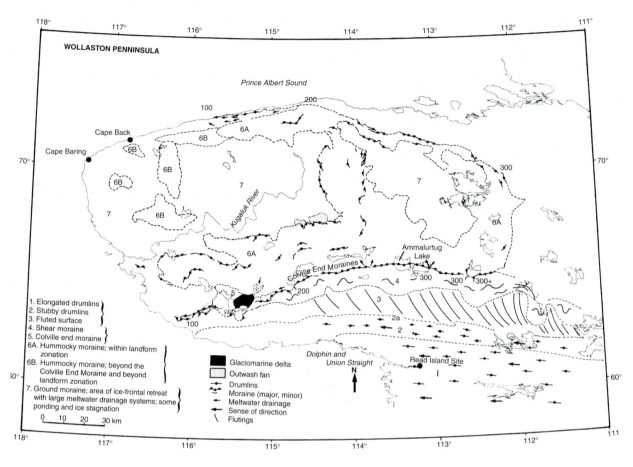

(a)

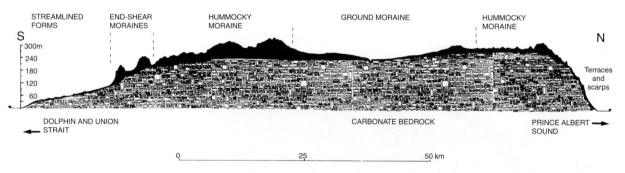

(b)

Fig. 12.10 The glacial landforms of Wollaston Peninsula, Victoria Island, Arctic Canada; (a) the distribution of landform sets and subsets for the whole of Wollaston Peninsula; (b) cross-section through Wollaston Peninsula showing the elevation control on glacial landform zonation. Note that the hummocky moraine and end/shear moraines are draped adjacent to or over a major bedrock scarp. (From Sharpe, 1988b. Reprinted by permission of the *Canadian Journal of Earth Sciences*)

Selby, 1972; Pickard, 1983; Lundqvist, 1989b; Fitzsimons, 1990). Lundqvist (1989b) has described impressive ice-cored moraines from the Vestfold Hills. The moraines are sharp-crested and up to 25 m high, but are composed mainly of clean ice with a thin debris cover nowhere more than 2 m thick. Aprons and hummocky and controlled moraines are mainly composed of debris flow and fall deposits and sublimation tills (Sections 10.5.2.1, 10.5.5 and 10.3.7).

Although meltwater discharges are small in Antarctica, ephemeral proglacial streams are capable of undercutting ice-cored moraines and accelerating the ablation process (Rains and Selby, 1972; Pickard, 1983). Where such streams dissect sublimating ice-cored ridges, they deposit small quantities of stratified sands and gravels which interdigitate with mass flow diamictons liberated from the buried ice. Entirely cold-based glaciers without stacked basal debris sequences leave very little imprint on the landscape, and end moraines commonly consist of little more than low ridges (Fig. 12.12). Unlike in Arctic environments, marginal meltwater channels are rare.

Thrust-block moraines occur in front of some glaciers in the Dry Valleys region (Fitzsimons, 1996, 1997). These moraines are up to a few tens of metres high, and consist of upglacier-dipping blocks of gravels, sands and occasional organic muds. The formation of these moraines appears to require special conditions, because most proglacial sediment is deeply frozen and ice-cemented, and has a high shear strength. The thrust-block moraines studied by Fitzsimons are adjacent to ice-marginal lakes, and he argued that they formed by the proglacial thrusting of unfrozen lake bottom sediments. Lake floors in the Dry Valleys can remain unfrozen below surface ice cover, and Fitzsimons argued that when glaciers advanced into wet-based lakes, ice and debris were frozen on to the glacier sole, and thrust upward at the margin (Section 11.3.1.2). The presence of the lakes, therefore, appears to have created rare conditions conducive to thrusting in this extremely cold, arid environment.

The extreme aridity of polar desert environments means that glacial landforms undergo very little modification after deglaciation. Strong katabatic winds winnow out finer material and shape surface clasts into faceted forms or *ventifacts*, but otherwise

there is little paraglacial activity. Denton *et al.* (1993) have presented convincing evidence that the Dry Valleys region is essentially a fossil landscape in which Pliocene and even Miocene landforms are preserved virtually intact. For example, sequences of small recessional moraines near the margin of the Taylor Glacier record minor glacier fluctuations over the past 2.2 million years, and some surface till units in the area pre-date volcanic ashes deposited 4.4 million years ago (Marchant *et al.*, 1993b). The preservation of such forms shows that the Dry Valleys have experienced a cold, arid climate since at least the late Miocene, and indicates that the Antarctic ice sheet is very unlikely to have melted during the Pliocene (Denton *et al.*, 1993; Section 1.7.1).

12.3.4 Surging glaciers

Glacier surge cycles (Section 4.8) produce very distinctive suites of landforms and sediments, owing to a potent combination of erosional, deformational and depositional processes (Sharp, 1985b, 1988b). The debris content of the margins of surging glaciers is typically very high, for several reasons. First, rapid sliding and the widespread development of cavities at the glacier bed is conducive to high rates of abrasion and quarrying (Clapperton, 1975; Humphrey and Raymond, 1994; Sharp *et al.*, 1994; Hallet *et al.*, 1996). Data from the 1982/83 surge of Variegated Glacier, Alaska, indicate that subglacial erosion rates during surges are among the highest in the world (Sections 5.3.3 and 5.3.6). In contrast, erosion rates during quiescent phases are likely to be very low, because of the low sliding rates. Second, severe compressional deformation at the advancing snouts of surging glaciers results in extensive thrust faulting and folding, which thickens and elevates basal ice sequences (Section 5.4.4; Clapperton, 1975; Sharp *et al.*, 1988, 1994). Near the snout, basal debris is commonly elevated to the glacier surface, forming a zone of debris-mantled ice which expands upglacier when the glacier ablates. Third, the wastage of debris-mantled ice during the quiescent period of a surge cycle produces extensive areas of ice-cored moraines, which may be overridden and reincorporated into the glacier during the next surge (Raymond *et al.*, 1987). In this case, the supraglacial debris mantle produced by one surge forms part of the englacial debris load of the next.

Surging glaciers also tend to be associated with widespread subglacial and proglacial glacitectonic deformation. Where the glacier bed is composed of till or other unconsolidated sediment, high porewater pressures during surges encourage subsole deformation, transporting weak, dilatant sediment towards the snout (Clarke *et al.*, 1984a; Clarke, 1987c; Richards, 1988). Near the margin, stresses

Fig. 12.12 Moraine sequence formed at the margin of the Taylor Glacier, Dry Valleys, Antarctica. (Photo courtesy of D.E. Sugden)

transmitted to proglacial sediments commonly result in extensive tectonic thrusting, particularly if the substratum has been weakened by high porewater pressures (Croot, 1988b, c). In addition, large discharges of meltwater and sediment associated with glacier surges are responsible for major changes in deposition rates in proglacial lakes and sandar (Humphrey *et al.*, 1986; N.D. Smith, 1990; Humphrey and Raymond, 1994).

The spatial distribution of sediments and landforms produced during a glacier surge cycle has been studied by Sharp (1985b) at Eyjabakkajökull, Iceland. Five landscape elements were identified (Fig. 12.13):

1. The maximum limit of the surge is delimited by *composite thrust-block moraines* formed by glacitectonic deformation of the glacier foreland (see Section 11.3.1.2).
2. *Excavational basins* occur within the surge limit reflecting submarginal erosion, possibly associated with thrust moraine formation.
3. *Fluted moraines and crevasse-fill ridges,* composed of deformation till, are exposed over large areas of the former glacier bed. The flutings record deformation of the bed during the surge phase (Section 11.2.3), whereas the crevasse-fill ridges were produced at the termination of the surge, when the weak till beneath the glacier was compacted and forced under pressure into basal crevasses (Section 11.2.8).
4. Areas of *chaotic hummocky moraine* blanket the ice-proximal flanks of thrust-block moraines and parts of the former glacier bed. The hummocky moraine is formed by the reworking of supraglacial debris following the surge, when the glacier snout stagnates. This supraglacial debris is derived from englacial debris bands and medial moraines. Controlled moraine formation may reflect englacial debris distribution or the infillings of supraglacial crevasses with debris flows or glacifluvial sediments.
5. *Sandur surfaces* blanket parts of all the above elements, and occupy fluvially eroded gaps in thrust-block moraines and hummocky moraines.

These five elements record deposition and deformation during a rapid glacier advance–retreat cycle (Fig. 12.14). During the advance (surge) phase, rapid sliding and subsole deformation of a well-lubricated bed produces fluted tills, and glacitectonic processes close to the snout result in excavation below the margin and proglacial thrust moraine formation. At the termination of the surge, crevasse-fill ridges formed at the bed, then hummocky moraine and outwash were deposited during glacier recession in the quiescent phase. Sharp (1985b) suggested that this assem-

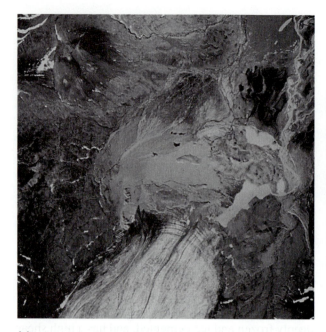

(a)

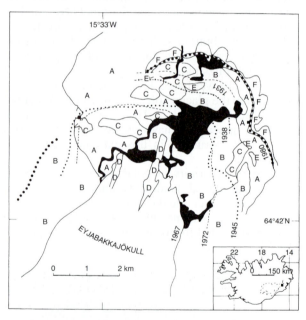

(b)

Fig. 12.13 Landform associations formed during the 1890 surge of Eyjabakkajökull, Iceland: (a) aerial photograph (1993) of the snout and foreland of Eyjabakkajökull showing concertina esker, crevasse-fill ridges, thrust-block moraine, excavational basin, and flutings produced during surges of the glacier margin (aerial photograph provided by Landmaelingar Islands); (b) geomorphological map showing former glacier margins and surge-related landform assemblages. A, sandur; B, fluted till with crevasse-squeeze ridges; C, chaotic hummocks; D, medial moraine; E, thrust ridge; F, lobate thrust ridge. (From Sharp, 1985b. Reproduced by permission of the University of Washington)

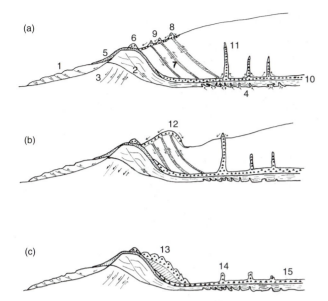

Fig. 12.14 Schematic model for the evolution of the sediment–landform associations at the margins of Eyjabakka-jökull: (a) immediately after a surge; (b) early quiescent phase; (c) late quiescent phase. 1: Lobate complexes of ridges formed by gravity spreading of peats and silts; 2: ridges of glacitectonically thrust peats and silts; 3: normal faults in sands and gravels in the core of the main ridge; 4: water escape structures; 5: outwash gravels; 6: ridge of resedimented diamictons; 7: englacial debris bands raised from the bed along thrust faults; 8: debris dykes on the glacier surface; 9: sediment gravity flows; 10: subglacial lodgement till; 11: crevasse-fill ridges; 12: formation of upraised marginal rim of debris-covered ice by differential ablation; 13: chaotic hummocky ice-cored topography; 14: crevasse-fill ridges on fluted lodgement till surface; 15: fluted lodgement till surface. (From Sharp, 1985b. Reprinted by permission of University of Washington)

blage of landforms may assist in the recognition of palaeosurges in the geologic record. Croot (1988c) argued that composite thrust ridges in Svalbard are exclusively developed in front of surging glaciers, although not all surging glaciers produce such ridges, because other preconditions are not satisfied. He suggested that, if such an association were to be shown for other areas, the presence of composite ridge systems could be used as a key indicator of former surging glaciers. However, it is now known that composite thrust ridges are formed in front of many non-surging glaciers, including some in Svalbard (Etzelmüller et al., 1996). Thrust ridges and associated areas of hummocky moraine form where deformable proglacial materials are subjected to large compressional stresses, and although these conditions are met in front of many surging glaciers, they are also commonplace in subpolar and polar-continental environments (Sections 12.3.2 and 12.3.3; Kalin, 1971; Klassen, 1982; D.J.A. Evans, 1989a; D.J.A. Evans

and England, 1991; Lehmann, 1992; Fitzsimons, 1996).

Sharp (1985a) considered that crevasse-fill ridges may be particularly diagnostic of former surges. Basal crevasse-squeeze ridges are certainly indicative of wet-based conditions close to the margin, which usefully allows the landforms left by surging glaciers to be differentiated from those of subpolar glaciers. Unfortunately, crevasse-squeeze ridges have low preservation potential, and are difficult to identify in ancient landsystems (Section 11.2.8). Another type of landform associated with surging glaciers has been described by Knudsen (1995), who identified *concertina eskers* on the foreland of the surging glacier Bruarjökull in Iceland (Fig. 12.15). These landforms originated as ordinary eskers, but have undergone shortening and crumpling in the downglacier direction as the result of compressive deformation of the glacier snout during a surge. Concertina eskers may provide strong evidence for former surges, particularly if they occur within the type of landsystem described by Sharp (1985b). To date, however, none have been recognized in Pleistocene glacial landscapes.

The geomorphological imprints of recent glacier surges have been identified off the coast of Svalbard from side-scan sonographs (Solheim and Pfirman, 1985; Solheim, 1986, 1991). Within the glacier limits, submerged landforms exhibit a zonation similar to that described by Sharp (1985b) for terrestrial environments. An outer push or thrust moraine lies distal to numerous subglacial crevasse-fill ridges. Beyond the outer moraine, iceberg furrows crisscross the sea floor, recording the passage of icebergs calved from the surge front.

12.3.5 The southern margin of the Laurentide ice sheet

Ice-marginal and supraglacial landforms deposited by the Pleistocene ice sheets can yield very detailed information on ice dynamics over a glacial cycle. A large number of glacial chronologies have been proposed for the ice sheets, and recent references can be found in Ehlers (1983b), Ruddiman and Wright (1987), Fulton (1989), Ehlers et al. (1991), and Clark and Lea (1992). In this section, we describe some of the landsystems associated with the maximum position and retreat of the southern margin of the Wisconsinan Laurentide ice sheet, focusing on the possible implications for former ice dynamics and basal thermal regime.

In the Prairie areas of Alberta, Saskatchewan and North Dakota, ice-marginal and supraglacial landsystems commonly consist of large glacitectonic thrust moraine systems passing proximally into wide belts of hummocky moraine (Fig. 12.16; Clayton and

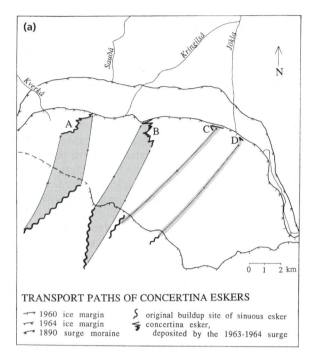

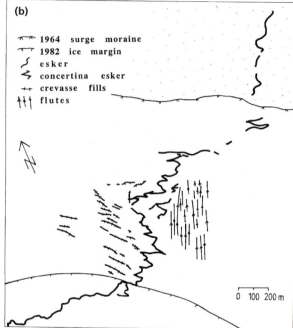

(c)

Fig. 12.15 Concertina eskers, Bruarjökull, Iceland. (a) Transport paths of formerly sinuous eskers to their present positions as concertina eskers by the 1963/64 surge. (b) Concertina esker A and associated crevasse-fill ridges and flutings. (From Knudsen, 1995. Reprinted by permission of Elsevier) (c) Aerial photograph (1993) of part of the snout and foreland of Bruarjökull showing concertina eskers, crevasse-fill ridges, which can be traced on to the ice margin, small push moraines, and flutings produced during surging. (Aerial photograph provided by Landmaelingar Islands)

Moran, 1974; Moran *et al.*, 1980; Rampton, 1982; Clayton *et al.*, 1985). The hummocky moraine has been attributed to the melt-out and reworking of englacial debris sequences which were elevated from the glacier bed by compressive deformation and/or net adfreezing near the margin (Section 5.4). This interpretation is compatible with the evidence for proglacial thrusting, and suggests that this part of the ice sheet margin was characterized by glacitectonic activity during readvances or standstills, and localized stagnation of debris-mantled ice during retreat. Some mounds and hummocks may be *extrusion moraines*, formed by the upwelling of pressurized meltwater from confined aquifers (Fig. 12.16; Boulton and Caban, 1995).

The presence of thrust blocks and hummocky moraines on the prairies has been interpreted as evidence that this part of the Laurentide ice sheet had subpolar margins, consisting of cold-based outer zones and a warm-based interior (Section 12.3.2; Boulton, 1972a; Moran *et al.*, 1980; Rampton, 1982). Frozen glacier margins are likely to have developed in this area, as a result of low air temperatures and widespread permafrost in the northern prairies at the last glacial maximum. Some belts of hummocky moraine occur where the ice sheets advanced against topographic obstacles such as escarpments, where compressive flow is likely to have encouraged transportation of debris to the ice

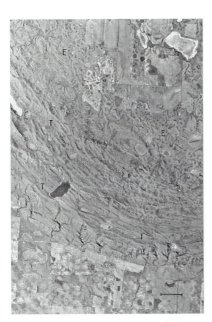

Fig. 12.16 The thrust block moraine (T) of the Dirt Hills, Saskatchewan, Canada. This view shows that numerous rimmed depressions (E; possible extrusion moraines) occur on the ice-proximal flanks of the thrust blocks. Scale bar represents 1 km. (Aerial photograph reproduced by permission of the Department of Energy, Mines and Resources, Canada)

surface (Straw, 1968; Worsley, 1969; N. Eyles and Menzies, 1983; Paul, 1983; Sharpe, 1988b; Fig. 12.17). In western Canada, for example, part of the southern margin of the Laurentide ice sheet is marked by a belt of hummocky moraine up to several hundred kilometres wide, and Paul (1983) suggested that this moraine belt was initiated by compressive flow against the Missouri Coteau and Manitoba and Missouri escarpments. Once a wide, stagnating ice sheet margin is produced by flow against the escarpments, it forms an obstacle to ice advance, initiating compressive flow even further upglacier; debris is then transported to the glacier surface over a wider area and the width of the stagnating ice margin increases, forming hummocky moraine in time-transgressive increments.

Mooers (1990b) has argued that the basal thermal regime of the Laurentide ice sheet margins may have varied over comparatively short distances. His conclusions were based upon studies of the former Rainy and Superior lobes in Minnesota, USA, which both advanced to the St Croix moraine during the last (Wisconsinan) glaciation but were responsible for producing different landform–sediment associations (Fig. 12.18). The portion of the St Croix moraine deposited by the Rainy Lobe comprises large proglacial outwash plains, glacially thrust masses, and chaotic hummocky moraine and kames, but lacks

streamlined (drumlin) forms and features associated with subglacial meltwater. Recessional margins are marked by similar landform–sediment associations but thrust masses are more common. In contrast, the Superior Lobe was responsible for depositing closely spaced recessional ridges with sparse hummocky moraine and kames. In addition, drumlins and subglacial drainage features are abundant, and in places the recessional positions are marked by the termination of eskers and tunnel valleys as proglacial outwash fans. On the basis of this evidence, Mooers suggested that the two lobes had different subglacial thermal regimes during their retreat. He argued that both lobes had frozen outer zones when the ice sheet stood at its maximum position, resulting in the elevation of basal debris and hummocky moraine formation. In the case of the Rainy Lobe, the frozen zone appears to have persisted during deglaciation, producing wide hummocky moraine belts and ubiquitous thrust masses, and discouraging the formation of drumlins and subglacial meltwater features. Mooers argued that meltwater could not penetrate to the glacier bed, but drained from the ice surface to produce outwash fans at the ice margin. In contrast, the margin of the Superior Lobe appears to have switched from a cold-based to a temperate condition during retreat, allowing surface meltwater to reach the bed and forming eskers and tunnel valleys. The presence of drumlins and the limited distribution of hummocky moraine also appear to indicate wet-based conditions.

The association of proglacially thrust ridges and hummocky moraine is characteristic of many modern surging glaciers (Section 12.3.4), and some researchers have suggested that surges may have affected the southern margin of the Laurentide ice sheet in the Prairie and Great Lakes regions (e.g. Wright, 1973; Clayton et al., 1985; P.U. Clark, 1994). Two main lines of evidence support this interpretation. First, where the ice sheet was underlain by fine-grained sedimentary rocks and sediments, the ice margin was divided into numerous low-gradient ice lobes (Fig. 12.19; Clayton et al., 1985; Fisher et al., 1985). Low ice margin gradients are characteristic of glaciers with weak, well-lubricated beds, conditions which are typical of surging glaciers but not subpolar margins. Interestingly, during the later stages of deglaciation when the ice sheet had pulled back to areas underlain by hard rocks or sandy sediment, the ice margin was less broken up into lobes, which suggests that possible surging behaviour was linked to the presence of deformable sediment (P.U. Clark, 1994). Second, detailed radiocarbon dating evidence indicates that individual lobes, such as the Lake Michigan, Des Moines, Huron–Erie and Erie–Ontario lobes, underwent rapid advance and retreat cycles, perhaps indicative of surging behaviour. The possi-

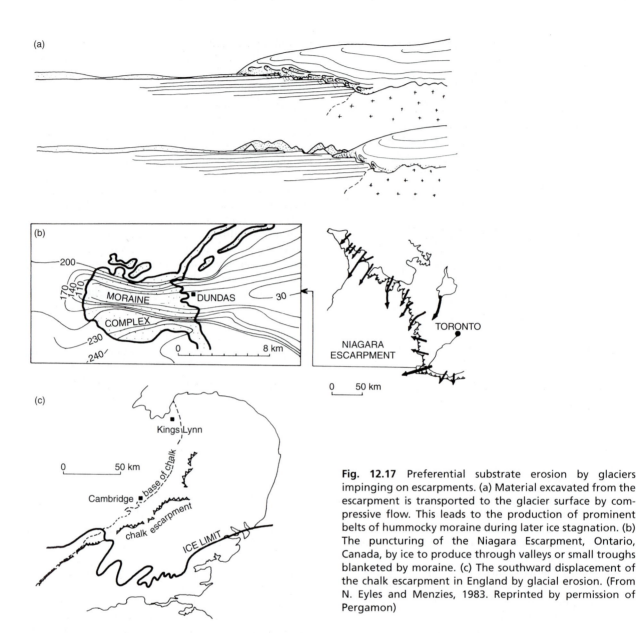

Fig. 12.17 Preferential substrate erosion by glaciers impinging on escarpments. (a) Material excavated from the escarpment is transported to the glacier surface by compressive flow. This leads to the production of prominent belts of hummocky moraine during later ice stagnation. (b) The puncturing of the Niagara Escarpment, Ontario, Canada, by ice to produce through valleys or small troughs blanketed by moraine. (c) The southward displacement of the chalk escarpment in England by glacial erosion. (From N. Eyles and Menzies, 1983. Reprinted by permission of Pergamon)

bility that parts of the Laurentide ice sheet were prone to surging has important implications for ice sheet stability and global climate change, and remains a very important topic for future research (MacAyeal, 1993b; P.U. Clark, 1994; Marshall *et al.*, 1996).

The south-eastern margin of the Wisconsinan Laurentide ice sheet is marked by a series of end moraines on Long Island, Cape Cod and surrounding islands. These are predominantly proglacial glacitectonic ridges composed of imbricate, thrust slices of glacilacustrine and glacifluvial sediment (Mills and Wells, 1974; Sirkin, 1982; Oldale, 1982; Oldale and O'Hara, 1984). Several generations of thrust ridges are present, recording oscillation of the ice sheet margin. It is unclear whether permafrost existed in

the area during the formation of the moraines, although Oldale and O'Hara (1984) believed that the thrusting of coherent masses of loose sand and gravel would have been facilitated if the ground was frozen. Recession of the ice sheet produced abundant ice-contact glacifluvial and glacilacustrine landforms, including kame and kettle topography, eskers and pitted sandar (Goldsmith, 1982; Oldale, 1982; Gustavson and Boothroyd, 1987). According to Koteff (1974) and Koteff and Pessl (1981), this sediment was elevated to the former glacier surface along englacial shear planes, whereupon it was redeposited by glacifluvial processes in a glacial karst environment. The elevation of basal debris by shearing was referred to as the 'dirt machine', and is characteristic of subpolar margins (Section 12.3.2). More

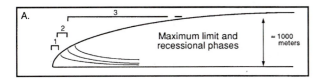

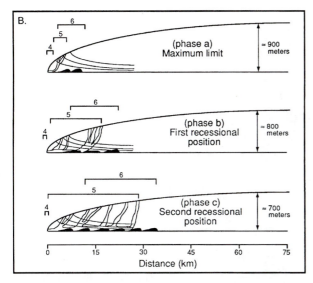

Fig. 12.18 Conceptual models of spatial and temporal variations in glacial landform development based upon evidence from central Minnesota and documenting glaciation by the Rainy and Superior lobes of the southern Laurentide ice sheet. (A) Conditions varied little during ice recession. Zone 1 was an area of debris-covered ice, because of thrusting of subglacial sediments. Zone 2 was an area of basal freezing-on of sediment and Zone 3 was the thawed bed area. (B) Suites of landforms suggest pronounced variations in landform development during ice recession. Zone 4 was an ice-marginal area with a sediment cover of variable thickness. A frozen toe was present at the maximum limit but absent during recession. Zone 5 was an area of tunnel valley and esker formation, while drumlins were forming in Zone 6. (From Mooers, 1990b. Reprinted by permission of the Geological Society of America)

recently, Gustavson and Boothroyd (1987) have argued that the temperate margin of the Malaspina Glacier, southern Alaska, provides a better analogue for Pleistocene depositional conditions in New England. Pressurized water flowing in englacial conduits is capable of transporting large quantities of sediment, and forms extensive supraglacial debris mantles on some temperate glaciers (Section 12.3.1). According to Gustavson and Boothroyd, this mechanism provides a more plausible explanation for the supraglacial landsystems of New England than the englacial shearing proposed in the 'dirt machine' hypothesis, strongly suggesting that the south-east margins of the Wisconsinan Laurentide ice sheet were temperate during the early stages of deglaciation. Climatic conditions are certainly likely to have

been milder in this area than in the Prairies, because of the proximity of the Atlantic Ocean, which would have provided a source of heat to the atmosphere in winter.

The distribution and character of ice-marginal, submarginal and supraglacial landsystems in North America also reflect temporal changes in ice dynamics during deglaciation. As noted above, moraine systems on the Prairies and in the Great Lakes region record highly dynamic ice margins which deposited and reworked large quantities of sediment during multiple advance and retreat cycles. When the ice sheet retreated northwards on to the hard substrate of the Canadian Shield, it continued to retreat actively, and deposited overlapping concentric moraines and short, discontinuous eskers (Shilts *et al.*, 1987; Fig. 12.20: Zone B). In many places, these moraines were deposited in proglacial lakes ponded against the ice margin (e.g. Glacial Lake Agassiz; Section 1.7.2.5; Teller, 1987). In the later stages of deglaciation, recessional moraines were deposited further apart, and extensive, well-integrated esker networks indicate regional downwasting (Fig. 12.20: Zone A). The transition from active retreat to more rapid deglaciation probably reflects climatic change, but may be due in part to destabilization of the ice margins as they retreated into deepening proglacial lakes (Section 12.5.4).

12.4 ICE SHEET BEDS

The supraglacial and ice-marginal landsystems discussed in the foregoing section provide a record of environmental change during deglaciation, but for a direct record of ice sheet buildup and occupancy we must look to former glacier beds. Large parts of the areas occupied by the former Pleistocene ice sheets are, in effect, fossil glacier beds, providing a window on to subglacial processes and environments which are only poorly known from modern examples. Recently, large-scale studies of former glacier beds have shown that the great ice sheets were highly dynamic entities with complex histories of advance and retreat, ice-divide migration and internal dynamics, and that they shifted their configuration to adapt to climatic change, the position and depth of proglacial water bodies, and basal conditions (e.g. Dyke and Prest, 1987; Boulton and Clark, 1990a, b; C.D. Clark, 1993, 1994). In this section, we discuss some of the techniques used to unravel the geologic record of former glacier beds, then examine the types of subglacial landsystem associated with hard rock and deformable beds. We then go on to examine regional distributions of subglacial landforms and associated sediment dis-

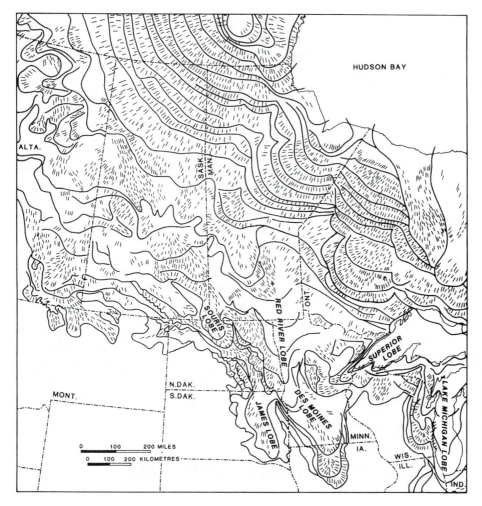

Fig. 12.19 Recessional ice-margin positions of the Laurentide ice sheet in the Prairie region, North America, reconstructed from landform evidence. (From Clayton *et al.*, 1985. Reproduced by permission of Scandinavian University Press)

persal patterns, and their implications for ice sheet dynamics.

12.4.1 Kineto-stratigraphy

Large-scale patterns of ice flow and subglacial conditions may be preserved in the geologic record in the form of distinct till units, glacitectonic deformations, subglacial bedforms and surfaces of erosion or non-deposition. Assemblages of subglacial deposits and landforms created during the advance and retreat of an ice sheet are typically exceedingly complex, and often defy analysis by the traditional methods of sedimentary stratigraphy. In areas with soft substrata, major glacitectonic dislocations, responsible for features such as megablocks and rafts (Section 11.3.1.4), can provide considerable problems for stratigraphers and mineral prospectors (Christiansen and Whitaker, 1976; Sauer, 1978; Aber *et al.*, 1989). Furthermore, the products of several glacial advances may be superimposed, and glacitectonic dislocation and subglacial deformation often rework all or part of any pre-existing glacial

and non-glacial strata. Because of partial erosion or patchy deposition, tills relating to any one glaciation can be discontinuous, so that individual events are recognizable only through the type, style and orientation of glacitectonic structures. As a result, new methods of analysis have been developed by glacial geologists to help them unravel the ice sheet record.

Before we discuss such methods, it is useful to consider theoretical patterns of subglacial deposition and erosion created during an ice sheet advance and retreat cycle. A helpful overview has been provided by Boulton (1996a), who modelled debris transport below a simple ice sheet by assuming that debris transport occurs entirely by subsole deformation at rates scaled to the balance velocity of the ice sheet. This is clearly highly simplified, and does not deal explicitly with other important subglacial processes such as the erosion of hard substrata, the entrainment of debris into basal ice layers, and deposition by lodgement and melt-out. However, the model does provide interesting insights into large-scale time-transgressive patterns of erosion and deposition, which can act as a basis for examining more complex cases.

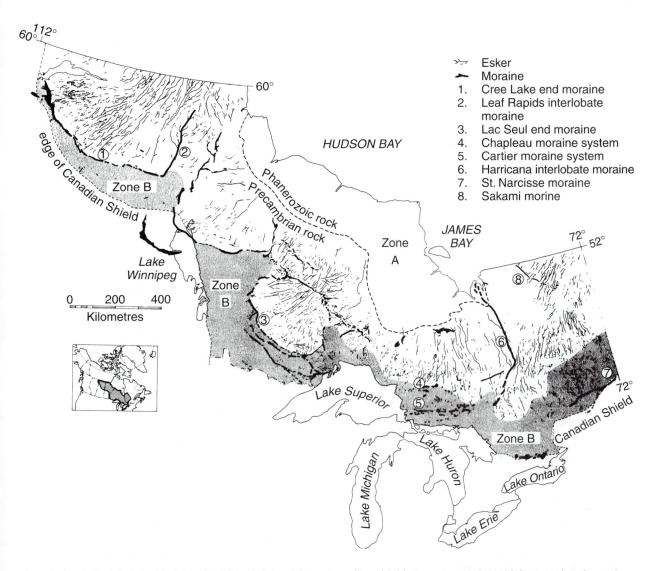

Fig. 12.20 Glacial geomorphological zones on the southern Canadian Shield. Zone A contains widely spaced end moraines and well-integrated esker systems produced by regional downwasting, and Zone B contains overlapping concentric moraines and short, discontinuous eskers produced during ice-marginal surging into proglacial lakes. Note the highlighting of individual glacier lobes by interlobate and terminal moraines. (Reprinted by permission of the Geological Society of America)

Results of a typical model run are shown in a time–space diagram in Fig. 12.21. The ice sheet is initiated at a nucleation point at time (*t*) = 0, and expands to the north and south, reaching its maximum 12,000 years after inception. It then contracts into the area where it first developed, finally vanishing at *t* = 24,000 years. The model predicts that there will be an outer zone of deposition near the ice sheet margin and an inner zone of erosion, although there will be no erosion below the ice divide, where the balance velocity is zero. The history of deposition and erosion at any given site reflects the migration of the depositional and erosional zones during ice sheet advance and retreat. At each site, deposition will

occur during ice sheet advance. The resulting till will be removed from most sites by the subsequent erosion phase, although it will be preserved near the ice sheet limit. As the ice sheet retreats, the zone of deposition passes over each site in turn, depositing a retreat-phase till. After final deglaciation, the overall pattern of deposition and erosion comprises four major zones:

- *Zone 1*. The ice-divide zone, with slight erosion owing to ice-divide migration and a thin till deposited during the retreat phase (sites E and F).
- *Zone 2*. A zone of strong erosion, in which the advance-phase till and part of the preglacial sub-

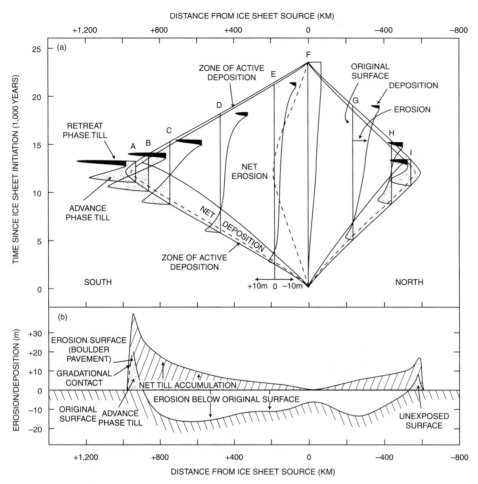

Fig. 12.21 (a) Time–distance diagram showing modelled shifting patterns of subglacial erosion and deposition associated with the growth and decay of an ice sheet. Graphs A–I represent deposition and erosion at selected locations. (b) Resultant pattern of deposition and erosion at the end of the glacial cycle. (From Boulton, 1996a. Reprinted by permission of the International Glaciological Society)

stratum are removed. The erosion surface is capped by the retreat-phase till (sites B, C, D, G and H).

- *Zone 3.* A zone where both advance and retreat-phase tills are preserved, but with an intervening erosion surface. Some of the advance-phase till has been removed, but erosion was insufficient to bite down into the preglacial surface (sites A and I).

- *Zone 4.* A zone of continuous till deposition and no erosion. This zone occurs only very close to the ice sheet limit.

More complex stratigraphies will result from oscillations of the ice margin and the migration of different basal thermal regimes (Boulton, 1996a, b). Below real ice sheets, patterns of erosion and deposition are also strongly influenced by lateral variations in flow conditions, such as the presence of ice streams or variations in substrate geology and drainage. Nevertheless, Fig. 12.21 does highlight the fact that subglacial till units are time-transgressive deposits, and do not represent a 'snapshot' of conditions at an instant in time.

Berthelsen (1973, 1978) developed a powerful geological approach to the study of time-transgressive subglacial sequences. The approach is known as *kineto-stratigraphy*, and focuses on the *structural* record of subglacial environments in much the same way as event stratigraphy is used by structural geologists to study the evolution of orogenic belts. This approach subdivides subglacial sequences into *kineto-stratigraphic units*, defined as sedimentary units 'deposited by an ice sheet or stream possessing a characteristic pattern and direction of movement' (Berthelsen, 1973). Individual units are therefore grouped according to their directional elements, such as fabrics, folds and faults (Fig. 12.22). The lower limit of a kineto-stratigraphic unit is defined as the lower boundary of the sediments deposited by the ice flow event, or the lower limit of associated penetrative and intense subglacial deformation. Thus, kineto-stratigraphic units may consist of basal tills and underlying penetrative glacitectonite (Sections 10.3.3 and 10.3.4; Banham, 1977; Berthelsen, 1978; Pedersen, 1989; Benn and Evans, 1996). Ice flow events may also cause non-penetrative deformation of older strata; such disturbance is termed *extra-*

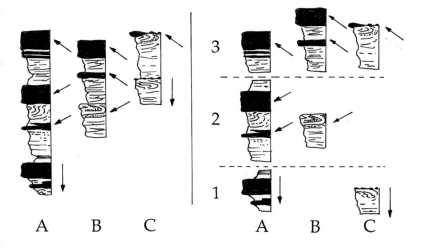

Fig. 12.22 Hypothetical kineto-stratigraphic units at three sites (A, B and C). At site A, three units are preserved, recording ice flow from the north (unit 1), north-east (unit 2) and south-east (unit 3). At site B, unit 1 is missing; unit 2 has been erosionally truncated and is represented by penetrative deformation structures which are unrelated to the overlying unit 3. At site C, unit 2 is missing and there is a major erosional hiatus at the top of unit 1. In some cases, it is possible to infer 'missing' kineto-stratigraphic units from surviving extra-domainal deformation structures. (Redrawn from Berthelsen, 1978)

domainal deformation, and does not form part of the overlying kineto-stratigraphic unit. Deformation of sediment within kineto-stratigraphic units is referred to as *domainal deformation*.

Multiple kineto-stratigraphic units, defined on the basis of their directional elements, could be formed within a single glacier advance–retreat cycle. For example, two units could be formed by the advance and retreat of an ice sheet with an intervening period of erosion (Fig. 12.21; Boulton, 1996a, b). Alternatively, multiple units could be formed as a result of changes in ice flow caused by ice-divide migration or internal reorganization of flow. Using the terminology introduced in Section 12.2.2, the sediments formed below an ice sheet during a whole glacial cycle comprise a *subglacial systems tract*, whereas the component kineto-stratigraphic units record dynamic events within the cycle.

Kineto-stratigraphy and similar approaches and methods have been adopted by many researchers (e.g. Aber, 1979; N. Eyles *et al.*, 1982b; Houmark-Nielsen, 1987, 1988; Albino and Dreimanis, 1988; Hicock, 1992; Pedersen, 1993; D.J.A. Evans, 1994; Hicock and Fuller, 1995). When combined with the study of contemporaneous erosional surfaces, bedforms and extra-domainal deformation, kineto-stratigraphy and related methods provide very powerful tools for unravelling the dynamic history of ice sheets.

12.4.2 Subglacial landsystems

12.4.2.1 HARD BEDS

At the scale of continental ice sheets, areas characterized by 'hard' and 'soft' beds tend to occur in geographically distinct areas. Former hard beds exist in *shield terrain*, or ancient continental crust, where the predominant rock types are complexly structured

igneous and metamorphic rocks, mostly of Precambrian age (Eyles, 1983c). Examples of shield areas are the Canadian and Scandinavian Shields and the Lewisian basement of north-west Scotland. Subglacial deposits in such areas tend to be thin and patchy, overlying landscapes of areal scour or the floors of troughs in *bedrock–drift complexes* (N. Eyles, 1983c; Fig. 12.23; see Sections 9.5.1 and 9.5.2). Sediment cover is thickest between bedrock highs, and is commonly streamlined at a wide range of scales, forming flutings, megaflutings, mega-scale lineations and rock-cored drumlins (Sections 11.2.3 and 11.2.4). Lee-side cavity fills may occur on the down-ice side of bed protrusions (Section 11.2.7; Hillefors, 1973). Cavity fills are most likely to form during glacier retreat, when ice is thin, overburden pressures are low and abundant meltwater is available, leading to the opening of water-filled cavities at the bed.

Subglacially streamlined forms such as megaflutings and drumlins commonly record the last (deglacial) ice flow directions, although earlier ice flow directions may be recorded in some areas where older lineations were not obliterated by deglacial flow. Rogen moraine occurs almost exclusively in shield terrains (Bouchard, 1989). This association has been attributed by Bouchard to the irregular bedrock topography in such areas, which, he argued, would encourage local shearing and thickening of debris-rich basal ice. Alternatively, Rogen moraine in shield areas may represent older streamlined forms which have been partially modified following a shift in the ice divide position (Section 11.2.5; Boulton, 1987; Boulton and Clark, 1990a, b; Clark, 1994). Cross-cutting lineations provide valuable evidence of ice sheet dynamics, and are discussed in detail in Section 12.4.4.

Long esker networks are also common in areas of hard substrata, where they record late stages of

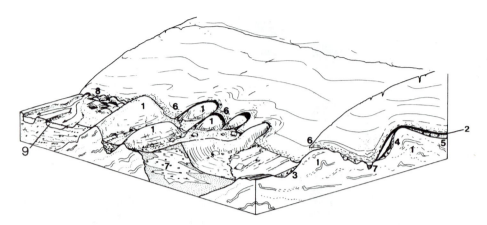

Fig. 12.23 The subglacial landsystem in an area of hard substrata: 1, abraded and streamlined rock knobs; 2, basal debris; 3, lodgement till on low-relief rock surface; 4, lee-side cavity fill; 5, basal melt-out till; 6, debris melting out at ice surface and dumped by gravity on the freshly exposed subglacial surface; 7, subglacial esker with gravel core; 8, hummocky or kettled outwash surface produced by the melt-out of ice buried by outwash fans; 9, the proglacial stream carries subglacial abrasion products. (From N. Eyles, 1983a. Reproduced by permission of Elsevier)

drainage below ablating ice sheets (Shilts *et al.*, 1987; Aylsworth and Shilts, 1989a, b). Clark and Walder (1994) have argued that eskers are characteristic of hard-rock terrains because channelized conduit flow is possible at the ice–bed interface (Sections 3.4.2 and 3.4.6; Walder and Fowler, 1994). They suggested that eskers should be rare in areas underlain by deformable substrata, where drainage is expected to consist of a distributed canal system or porewater flow. Clark and Walder presented evidence in support of this view by comparing the regional distribution of eskers and hard substrata in North America and parts of Europe. At a continental scale, eskers are more widespread on hard substrata (Fig. 12.24), although in detail the picture is much more complex. Eskers do occur on soft substrata, in some cases superimposed on drumlins and other bedforms, and it is possible that on both hard and soft beds subglacial drainage switched between channelized and distributed systems as ice thickness and gradient and meltwater supply changed over the course of a glacial cycle. However, the work of Clark and Walder suggests interesting new directions for the study of ice sheet drainage and its relationship with glacier bed conditions, and deserves to be followed up in detailed local and regional studies.

12.4.2.2 SOFT BEDS

Soft glacier substrata occur in areas underlain by weak sedimentary rocks, or where accumulations of Quaternary sediments mask hard lithologies. Expansive sedimentary bedrock lowlands exist at the margins of the North American and Scandinavian shields, which provided an outer zone of potentially deformable sediment below the great Pleistocene ice

sheets. In such areas, the subglacial landsystem typically consists of fluted and drumlinized terrain underlain by varying thicknesses of subglacial sediments. Late-stage drainage events may be recorded by eskers and/or meltwater channels (Fig. 12.25; N. Eyles, 1983c; N. Eyles and Menzies, 1983; N. Eyles and C.H. Eyles, 1992). Subglacial sediments may comprise thin till units perched above older sedimentary successions, as at Breidamerkurjökull, Iceland, where approximately 2 m of fluted deformation till overlies several metres of proglacial gravels and sands. In many areas occupied by Pleistocene ice sheets, however, the subglacial landsystem may be underlain by tens of metres of vertically stacked subglacial tills and intervening stratified sediments. Till units may be laterally continuous for many hundreds of kilometres, with remarkably uniform sedimentological characteristics and lithological composition (e.g. Perrin *et al.*, 1979; Kemmis, 1981; Holm, 1981). Widespread, uniform till units have been attributed to efficient debris mixing within basal ice or during repeated entrainment and deposition during regelation sliding (Kemmis, 1981; Rappol and Stoltenberg, 1985). More recently, however, Boulton (1987) and Alley (1991) have argued convincingly that such till units represent sediment homogenization within subglacial deforming layers (see also Section 10.3.4; Hicock and Dreimanis, 1992a; Hart and Roberts, 1994; Benn and Evans, 1996). Boulton (1996b) has shown that extensive till units in the Netherlands are very unlikely to have formed by the melt-out of debris-rich ice because: (a) till thicknesses would have required basal ice thicknesses of up to 100–200 m, far in excess of anything observed in modern glaciers; (b) such thicknesses of debris-rich ice would require extremely high freezing rates

Fig. 12.24 The regional distribution of eskers in North America, showing the limit of the Wisconsinan ice sheet (heavy line) and the boundary between Canadian Shield rocks and younger sedimentary rocks (lighter line). Note the preferential occurrence of eskers on shield rocks. (From Clark and Walder, 1994. Reproduced by permission of the Geological Society of America)

at the glacier bed, which is contrary to other evidence (Boulton *et al.*, 1995); and (c) if thick basal ice sequences were present, their exposure at the ice margin would produce huge dump moraines, which are not found in the Netherlands. At present, many researchers favour the view that extensive, massive tills deposited in 'soft-bed' regions are deformation tills, although this interpretation remains to be tested by detailed sedimentological analysis in many areas.

In places, subglacial successions extend down to rock head, which may exhibit a range of erosional forms from small-scale striae to large roches moutonnées (Fig. 12.25; N. Eyles *et al.*, 1982b; N. Eyles, 1983c; N. Eyles and Menzies, 1983). Blocks from the bedrock surface may be observed in various stages of removal, the earliest stage sometimes being

represented by blocks surrounded by till squeezed into the bounding joints (Broster, 1991; Broster and Park, 1993). The efficiency of plucking appears to be dictated by bedrock type and the intensity of its weathering prior to ice overriding. For example, the high concentration of limestone blocks in the Dummer Moraine, Ontario, Canada, was regarded by Shulmeister (1989) as the product of efficient plucking of weathered karst terrain.

Till units may be separated by gradational or erosional contacts, clast pavements, or intervening beds of sorted sediments. In deformation tills, clast and boulder pavements may represent large particles that sank to the base of weak dilatant till (Clark and Hansel, 1989; Clark, 1991a), or the lower limit of excavational deformation beneath a deforming layer

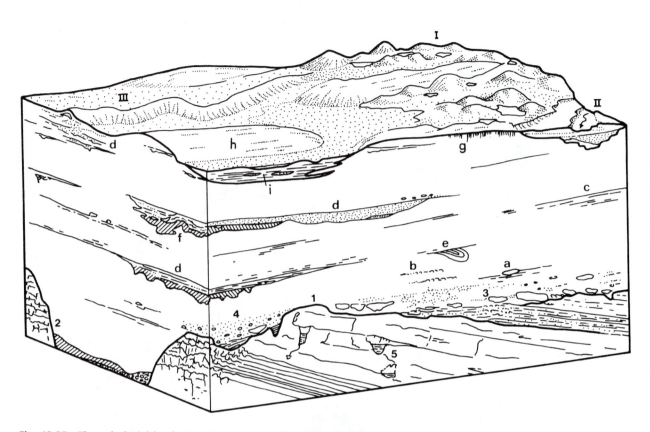

Fig. 12.25 The subglacial landsystem in an area with a low-relief limestone terrain and where multiple stacked till units have been deposited during a single glaciation. This subglacial landsystem is draped/reworked by: (I) hummocky kame and kettle topography; (II) outwash cut into the till surface and comprising stratified sands and gravels; (III) esker deposited during ice wastage and therefore not truncated by subglacial tills like other channel fills in the subglacial landsystem. The base of the subglacial landsystem is characterized by: (1) striated rock head; (2) buried channel/valley with a fill of subglacial sands and gravels and till; (3) glacitectonized rock head, with rock rafts and boulder pavements; (4) lowermost till, comprising local lithologies, which thickens in the lee sides of rock protuberances as lee-side cavity fills; (5) cold-water karst. The sediments of the subglacial landsystem are characterized by: (a) predominantly preferentially aligned, faceted clasts; (b) crude shear lamination produced by the smearing of soft lithologies (deformation till/glacitectonite); (c) slickensided bedding planes (fissility) produced by glacitectonic shear; (d) stratified gravels, sands and clays deposited in subglacial cavities, pipes or canals and truncated by overlying tills (the base of each till unit may be fluted) – they constitute lenses which are elongated in the direction of ice flow and typically internally disturbed by folding and faulting due to post-depositional deformation by glacier/till overriding; (e) folded and sheared-off channel fill; (f) diapiric intrusion of till squeezed up into subglacial cavity; (g) vertical joints produced by postdepositional pedogenic processes; (h) drumlinized surface of upper till sheet; (i) inter-drumlin depressions filled with postglacial solifluction debris and peat. The horizontal scale may range from 10 m to 10 km and the vertical scale may be 10 cm to 100 m high. (Modified from N. Eyles, 1983a)

(Boulton, 1996a, b; Section 10.3.4). Inter- and intra-till lenses of sorted sediments may represent: (a) the infills of former braided canal systems; (b) rafted remnants of the glacier substratum; (c) water-lain components of melt-out till sequences; or (d) proglacial or interglacial deposition between glacier advances. Stratified intra-till lenses containing primary depositional structures, flat upper surfaces and concave-up bases are thought to be particularly diagnostic of *in situ* canal fills (Section 11.2.1.2; Clark and Walder, 1994; D.J.A. Evans *et al.*, 1995; Benn and Evans, 1996). The detailed investigation of sorted lenses within basal tills remains an important objective for future research, which could provide much-needed information on former subglacial drainage systems. Inter-till lenses of sorted sediments attributable to episodes of subaerial or subaqueous deposition between periods of glacier occupancy have been recognized by many researchers (e.g. W.H. Johnson and Hansel, 1990). The correct identification of such proglacial sediments is important because they contain information on glacial history, former glacier dynamics and climate change, and because misinterpretation of proglacial components as the infills of subglacial channels or lakes leads to erroneous conclusions about the former subglacial environment.

In many localities, stacked till units record flow interactions between ice lobes (e.g. N. Eyles *et al.*, 1982b; Hicock, 1992; Hicock and Fuller, 1995). Near the junction between two adjacent ice flow units, changes in relative ice discharges can cause gradual or abrupt shifts in the ice flow direction, as one flow unit becomes dominant over the other. As a result, debris from different source areas can be transported to the site, producing marked *compositional stratification* in which the lithological composition of till units varies systematically upsection (Broster, 1986). Such compositional changes are commonly paralleled by shifts in the preferred orientation of particle fabric modes, allowing sequences of ice flow changes to be reconstructed in some detail (Section 12.4.1; Berthelsen, 1973, 1978; N. Eyles *et al.*, 1982b). For example, kineto-stratigraphic analysis of till successions combined with mapping of ice-marginal sediment–landform associations has been used to reconstruct the history of the Late Weichselian ice sheet in Denmark (Fig. 12.26; Houmark-Nielsen, 1988). In this region, the earliest evidence for Weichselian glaciation consists of a basal till containing erratics from the Scandinavian mountains, and deformation of pre-existing strata towards the south, recording ice advance from the north (the Norwegian Advance). Subsequently, deposition of a complex suite of subglacial, ice-marginal and proglacial sediments and widespread glacitectonic deformation was associated with ice flow from the east and north-east (the Main Weichselian Advance). Within the limits of this advance, progressively less extensive till units and fragmentary moraine belts mark readvances that punctuated ice retreat. Finally, a new generation of tills and glacitectonic structures record renewed ice advance, this time from the east and south-east (the Young Baltic Advance).

Vertical compositional changes can also occur in the absence of changes in flow direction. Boulton (1996b) has shown that till composition reflects the duration of glaciation, particularly the length of time available to transport debris from distant locations. Theory suggests that till composition should be dominated by local lithologies near the base and shift to predominantly far-travelled lithologies at the top, reflecting the passage of successive waves of debris from further and further upglacier. Boulton (1996b) applied his sediment transport model to till sequences in Illinois described by W.H. Johnson and Hansel (1990), and showed that the observed sequence could have been created during a single glacial cycle during which progressively more distant debris was delivered to the marginal zone (Fig. 12.27). Of course, this does not prove that the sequence originated in this way, but Boulton's work is an important step in applying glaciological theory to the interpretation of complex stratigraphies.

12.4.3 Regional distributions of subglacial landforms

Subglacial sediments and landforms exposed by the retreat of the great Pleistocene ice sheets commonly exhibit systematic spatial patterns, in which different types of landforms occur in distinct zones aligned parallel or transverse to former ice flow. As we have seen in Chapters 5, 10 and 11, several factors influence the formation of subglacial sediments and landforms, including strength and hydraulic conductivity of the bed, basal thermal regime and the availability of meltwater, ice velocity, shear stress and effective overburden pressure. Thus, landform zonation can be viewed as a mosaic produced in a variety of sub-environments beneath the parent ice sheet, either at a single moment in time or as a time-transgressive palimpsest resulting from ice advance or retreat. Elongate, flow-parallel zones may mark the former position of flow units such as ice streams and marginal lobes, or meltwater drainage pathways (Section 11.2.4.2), whereas transverse zones may record downflow variations in basal conditions and their migration through time.

Ice sheet advance over an area can be recorded by distinctive landform–sediment zones, particularly in regions underlain by weak rocks or unconsolidated sediments. For example, in the northern Great Plains of North America, arcuate belts of hummocky ice-thrust terrain and excavational basins commonly lie at the distal edges of drumlin fields. Some of the thrust belts have a subdued, streamlined appearance, and contain cupola hills and drumlins with low length:width ratios (Fig. 12.28). The excavational basins and thrust ridges document glacitectonic disturbance near the ice margin, probably encouraged by cold-based conditions below the snout (Sections 11.3.1 and 12.3.5; Clayton and Moran, 1974; Bluemle and Clayton, 1984; Bluemle *et al.*, 1993), and the streamlined terrain and drumlin belts record the subsequent advance of wet-based ice over the former marginal zone. In regions where subglacial deformation was intense and prolonged, glacitectonized thrust masses may lie buried, their upper surfaces intensely sheared and incorporated within the subglacial deforming layer. The amount of subglacial modification commonly decreases down-ice, and landform assemblages exhibit a continuum from drumlins with high length:width ratios, through drumlins with low length:width ratios and cupola hills, to largely unmodified thrust masses. Such landform associations apparently represent progressively less effective or short-lived subglacial moulding towards the ice limit.

Another example of landform zonation near a former ice limit is illustrated in Fig. 12.29 (Plate 26), which shows part of the western side of the Strait of

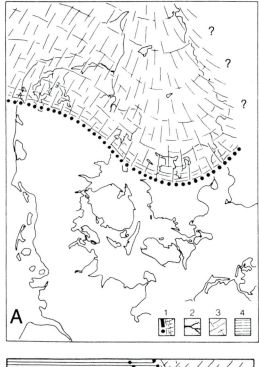

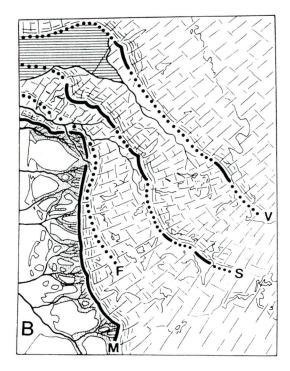

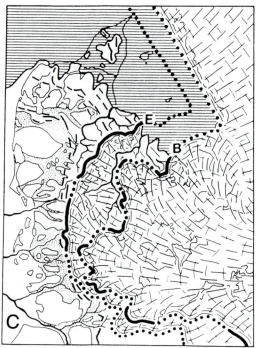

Fig. 12.26 Reconstructions of successive stages of the Weichselian ice sheet in Denmark. (A) Norwegian Advance; (B) Main Weichselian Advance; (C) Young Baltic Advance. 1: Major ice-marginal features; 2: sandur deposits and associated drainage systems; 3: ice flowlines; 4: late-glacial Younger Yoldia Sea. (From Houmark-Nielsen, 1988. Reproduced by permission of Balkema)

Magellan in southernmost South America. The main features of the landscape record the expansion of ice lobes to the north from high source areas near the Pacific coast during the last glaciation (Clapperton 1989; Porter *et al.*, 1992; Clapperton, *et al.*, 1996b). The ice terminated at a belt of composite thrust ridges banked against high ground to the north. The outermost ridges are sharp-crested and unmodified,

but towards the south there is a wide zone in which large numbers of meltwater channels wind between upstanding sediment masses with streamlined, drumlinized surfaces. Upglacier of this zone is a series of overdeepened basins, then a broad belt of spindle-shaped drumlins. Fieldwork in this area suggests that this assemblage of landform zones resulted from the advance of an ice lobe with a cold-based margin and

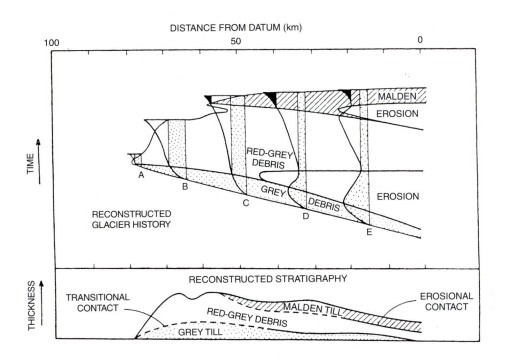

Fig. 12.27 Schematic diagram showing how the three-part till stratigraphy observed by W.H. Johnson and Hansel (1990) could have originated during a single glacial cycle. (From Boulton, 1996b. Reproduced by permission of the International Glaciological Society)

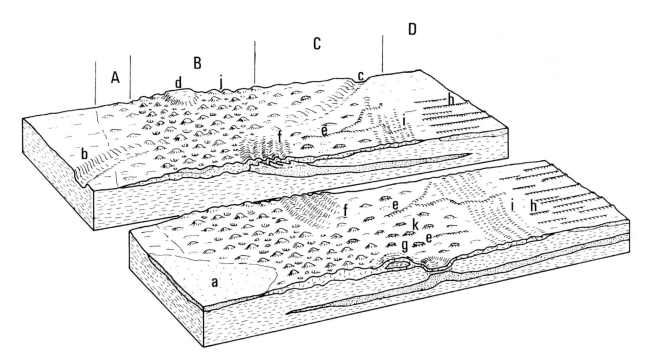

Fig. 12.28 Idealized zonation of glacial landforms and sediments on the prairies of North Dakota. A: Proglacial suite; B: supraglacial suite; C: transitional (submarginal) suite; D: subglacial suite. a: Proglacial lake; b: proglacial meltwater channel; c: subglacial meltwater channel or tunnel valley; d: ice-walled lake plain; e: esker; f: transverse thrust moraines cupola hills; g: prairie mound; h: flutings; i: transverse recessional moraines; j: hummocky moraine; k: isolated kames. (Modified from Clayton and Moran, 1974)

a wet-based interior. Thrusting and excavational tectonics near the advancing ice margin was followed by streamlining below wet-based ice as the glacier advanced, and the spindle drumlins were formed in the area of longest ice occupancy. The meltwater channels are late-stage features documenting submarginal and proglacial drainage during deglaciation.

Multiple ice advances introduce considerable complexity to subglacial landform–sediment sequences. Van der Wateren (1995) proposed a model of glacitectonic overprinting associated with ice margin oscillations, based upon the identification of five structural styles typical of proglacial and subglacial environments (Fig. 12.30). *Style A* represents the undeformed foreland, composed of outwash relating to the glacial advance; *Styles B, C* and *D* represent varying degrees of proglacial glacitectonic deformation; and *Style E* records subglacial deformation. Style E is further subdivided into compressive, or constructional, deformation (Ec), and extensional, excavational deformation (Ee). Constructional deformation is recorded by accretionary deformation tills and glacitectonites, and excavational deformation by areas of overdeepening, erosion and substrate streamlining (see also Sections 5.6, 7.4.3, 9.4.2, 10.3, 11.2.1; Hart *et al.*, 1990; Hart and Boulton, 1991). Oscillations of the ice margin will produce sediment successions in which different structural styles are superimposed. A single advance of the ice margin results in the overprinting of styles A to D by style E, whereas ice margin retreat then readvance will result in the overprinting of style E by styles B, C and D (Fig. 12.31). Such overprinting is implied in the deformation model of drumlin formation (Boulton, 1987; D.J.A. Evans, 1996), in which drumlins or flutings containing partially disturbed sand and gravel cores are interpreted as streamlined remnants of pre-existing sediment (i.e. E overprints A). The

recognition of structural styles is therefore a powerful complement to kineto-stratigraphic analysis, and allows complex ice margin fluctuations to be deduced from sediments and landforms. Van der Wateren's structural model is similar in many ways to the process–form model of continental ice sheet margins developed by Clayton and Moran (1974).

Boyce and Eyles (1991) reported down-ice changes in drumlin morphometry and sedimentology in the Peterborough drumlin field south of the Dummer moraine, Ontario, which they attributed to variations in substratum characteristics and duration of glaciation (Fig. 12.32). In the northern part of the drumlin field, spindle drumlins with length : width ratios of >6 lie directly on bedrock, in a landsystem typical of hard substrata. These grade into less elongate drumlins towards the south, where ice overrode thick accumulations of Late Wisconsinan outwash. South of the Oak Ridges moraine, the drumlins have length : width ratios of <3 and are composed of deformation till unconformably overlying thick sequences of older, largely undisturbed sediments. In the northern rock-floored area, eskers record the drainage of meltwater in discrete conduits, whereas eskers are rare or absent in the south, indicating that meltwater evacuation may have taken place through the deforming substrate. Boyce and Eyles (1991) considered the down-ice variations in drumlin morphometry to be a function of decreasing duration of subglacial deformation towards the ice margin. In the northern part of the area, pre-existing sediment is thought to have been stripped off by subglacial erosion beneath the ice sheet interior (see Figs 12.1, 12.20; Boulton, 1996a).

Aylsworth and Shilts (1989a, b) have mapped the regional distribution of glacial landforms on the Canadian Shield beneath the former Keewatin sector of the Laurentide ice sheet (Fig. 12.33). In this area,

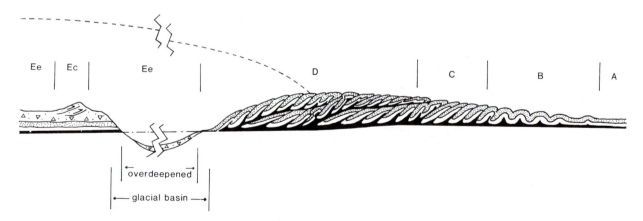

Fig. 12.30 Glacitectonic structural styles A–E: (A) undeformed foreland; (B) Jura-style proglacial folding; (C) low-angle thrust structures; (D) nappe zone; and (E) subglacial zone, comprising compressional (Ec) and erosional (Ee) subzones. (From van der Wateren, 1995. Reproduced by permission of Butterworth-Heinemann)

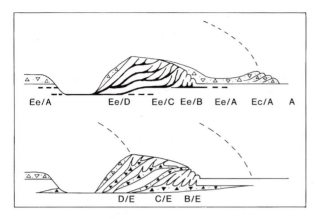

Fig. 12.31 Overprinting of structural styles: (upper) advance sequence, where style E overprints styles A–D and the triangle pattern depicts till relating to the ice advance; (lower) readvance sequence, where styles B–D overprint style E and the black triangle pattern depicts till relating to older advance. (From van der Wateren, 1995. Reproduced by permission of Butterworth-Heinemann)

glacial landforms are arranged in four broadly concentric zones around the Keewatin ice divide. *Zone 1* is located in the former ice divide area, and is characterized by low-relief hummocky moraine deposited during final wastage of the ice sheet. *Zone 2* consists of Rogen moraine (ribbed moraine), drumlins and eskers, and grades distally into *Zone 3*, which is characterized by drumlins and eskers. The drumlins and eskers of Zone 3 decrease in size and frequency towards *Zone 4*, which comprises bedrock surfaces with sparse glacial sediment cover. Although the Keewatin ice divide underwent changes in position throughout the last glaciation, the landform zones provide a clear impression of spatial variations in depositional processes, and their changes through time. The radial pattern of eskers throughout the region clearly records meltwater flow away from the ice divide towards the retreating ice margins. Many eskers form integrated dendritic networks with tributaries as high as fourth order. The preservation of such systems suggests that the ice was relatively inactive during the final stages of deglaciation, and that at the time of esker formation much of the ice sheet surface may have been below the equilibrium line altitude. The Rogen moraine in Zone 2 is arranged in a horseshoe-shaped zone 200–250 km wide around the region of the ice divide, and is rare elsewhere. Individual fields form trains oriented parallel to ice flow, and are commonly composed of coarse, bouldery material. Aylsworth and Shilts suggested that the Rogen moraine was formed in areas where bed materials and ice dynamics favoured subglacial thrusting of coherent blocks of entrained debris. An alternative possibility is that the Rogen moraines were originally streamlined bedforms and

were subsequently partially modified by ice flowing from a different direction when the ice divide shifted position (Section 11.2.4.4; Boulton, 1987; Clark, 1993, 1994). The drumlins in Zones 2 and 3 generally record radial flow towards the margins, although Clark (1993, 1994) recognized superimposed lineations in this area which apparently record realignment of flow during deglaciation. Some drumlin fields form elongate trains oriented parallel to regional ice flow, and contain high concentrations of far-travelled debris. The boundaries of these trains of drumlinized exotic debris are very sharp, and concentrations of far-travelled components drop from percentages of several tens to zero within a few kilometres at the lateral margins. Aylsworth and Shilts suggested that these trains were formed beneath ice streams or fast-flowing sectors of the ice sheet. Drumlins are less common in the region of the divide itself, where flow velocities are likely to have been very low. The scarcity of depositional landforms in Zone 4 probably reflects the lack of sufficient unconsolidated sediment in this area of the shield. Large-scale patterns of drift lineations, and their implications for ice sheet dynamics, form the subject of the following section.

12.4.4 Cross-cutting lineations and ice sheet dynamics

Regional mapping of flow-parallel and transverse landforms can reveal large-scale patterns of ice sheet flow and their changes through time, yielding valuable information on ice sheet evolution and dynamics. Reconstructed flow patterns for ice sheet maxima demonstrate that ice sheets were not simple domes from which ice spread radially, but rather comprised multiple accumulation and dispersal centres (Tyrrell, 1898a, b, 1913; Shilts *et al.*, 1979; Shilts, 1980; Dyke *et al.*, 1982b; Boulton *et al.*, 1985; Punkari, 1995). Furthermore, it is possible to reconstruct the migration of ice divides and the switching of ice streams during a glacial cycle by mapping cross-cutting and superimposed subglacial bedforms, tracing indicator erratics, and undertaking kinetostratigraphic analysis of exposures (Fig. 12.34; Sections 11.2.1.1 and 11.2.4.4). In other words, the study of palimpsest subglacial landsystems on a regional scale allows us to view the evolution of the last great ice sheets through time (Fig. 1.60 (Plate 9); Dyke and Prest, 1987; Boulton and Clark, 1990b). Such studies represent the integration of vast amounts of work conducted at the local scale, and are a fitting tribute to many decades of painstaking observations by field geologists.

Superimposed drumlins and megaflutings documenting changing ice flow directions have been reported by numerous researchers (e.g. Fairchild,

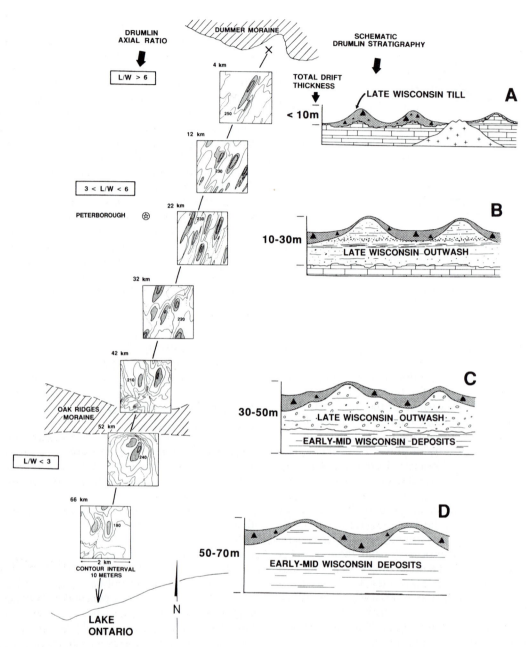

Fig. 12.32 The variability in drumlin stratigraphy and morphometry (expressed as length : width ratios, *L/W*) along a 70 km flowline through the Peterborough drumlin field, Ontario, Canada. (From Boyce and Eyles, 1991. Reproduced by permission of the Geological Society of America)

1929; Hare, 1959; Rose and Letzer, 1977; Mollard, 1984; Riley, 1987; Rose, 1987b, 1989b; Mitchell, 1991, 1994). Multiple-limbed drumlins exhibiting evidence of more than one ice flow direction have been used by Stea and Brown (1989) to reconstruct changes in ice divide positions in Nova Scotia, Canada, and complex lineation patterns, thought to relate to different flow directions at various stages in the life cycle of continental ice sheets, have been identified by Synge and Stephens (1960), Vernon (1966),

Prest *et al.* (1968), Hill (1971), Shilts (1980), Dyke *et al.* (1982b), Boulton *et al.* (1985), D.J.A. Evans (1985), and Dyke and Morris (1988). In addition, several researchers have argued that transitions from one subglacial bedform to another record remoulding by changing ice flows in response to migrating ice divides (Boulton, 1987; Lundqvist, 1989a; Boulton and Clark, 1990a, b; C.D. Clark, 1993, 1994).

Analysis of Landsat satellite images and aerial photographs of the area occupied by the Wisconsinan

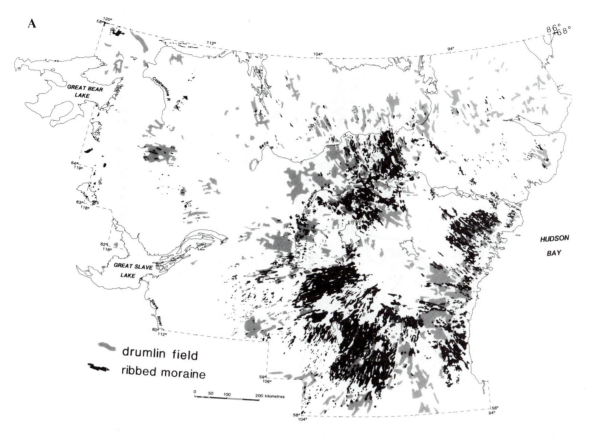

A

GREAT BEAR
LAKE

GREAT SLAVE
LAKE

HUDSON
BAY

drumlin field
ribbed moraine

0   50   100   200 kilometres

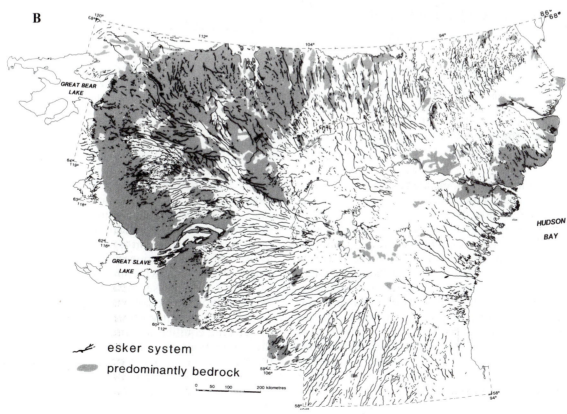

B

GREAT BEAR
LAKE

GREAT SLAVE
LAKE

HUDSON
BAY

esker system
predominantly bedrock

0   50   100   200 kilometres

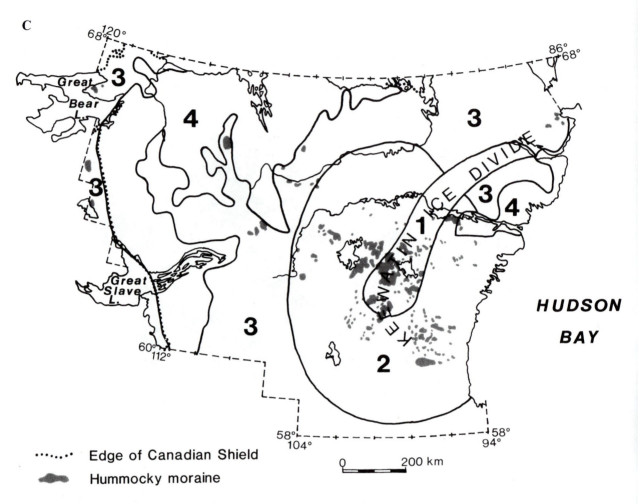

Fig. 12.33 The distribution of glacial landforms in Keewatin, Canada: (A) the distribution of drumlins and ribbed (Rogen) moraine; (B) the distribution of eskers relative to areas predominantly of bedrock; (C) concentric zones around the area of the former Keewatin ice divide based upon landform assemblages. The distribution of hummocky moraine is also illustrated. Zone 1 = devoid of eskers and subglacial lineations, Zone 2 = Rogen moraine, drumlins and eskers, Zone 3 = drumlins and eskers, and Zone 4 = bedrock with minimal drift. (From Aylsworth and Shilts, 1989a, b. Reprinted by permission of Elsevier (A and B) and the Geological Survey of Canada (C))

Laurentide ice sheet has enabled Boulton and Clark (1990a, b) and Clark (1993, 1994) to reconstruct changing glacial flow directions. Numerous *ice flow–landform assemblages* were identified, composed of drumlins, megaflutings and mega-scale lineations, which commonly are superimposed on or cross-cut each other (Fig. 12.35; Fig. 12.36 (Plate 27)). Clark (1993, 1994) identified two relative age indicators which enable the sequence of ice flows to be reconstructed: (a) *simple superimposition*, where one set of streamlined lineations is superimposed over another set with a different orientation; and (b) *pre-existing lineation deformation*, where deformation during a more recent ice flow phase alters the form or continuity of pre-existing lineations. According to Clark, the degree of modification of earlier lineations

forms a continuum, ranging from: (a) a situation where there is no modification of the pre-existing lineation; to (b) superimposition of smaller forms on the surface of older lineations; to (c) substantial breaching or deformation of the pre-existing lineation; to (d) total reorganization of sediment into a new orientation (Fig. 12.37A). This model of bedform modification has been used to explain the formation of Rogen moraine (Section 11.2.5; Boulton, 1987). As Clark (1993) explained, each streamlined lineation is in dynamic equilibrium with glacier flow, and if the flow direction changes, lineations will become obstructions in an otherwise planar substrate. This means that they are very susceptible to modification and thus become sources of sediment for renewed subglacial debris transport.

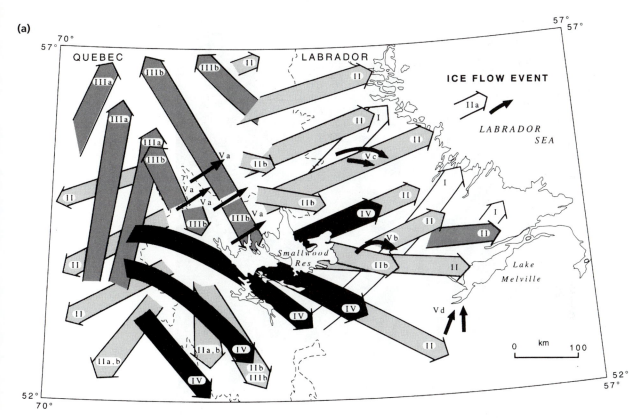

(a)

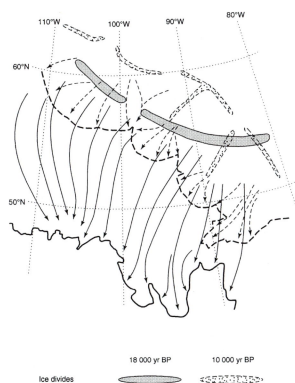

(b)

Fig. 12.34 (a) Summary of ice flow events in central Labrador/Quebec, Canada, based upon striae, glacially streamlined landforms and till composition. (From Klassen and Thompson, 1993. Reprinted by permission of the Geological Survey of Canada) (b) Relationships between ice divide shifts, ice margin positions and glacier flow directions in the south-western sector of the Laurentide ice sheet between 18 kyr and 10 kyr BP. Note that parallel flow lines are replaced by later radial flow patterns. (Modified from Dyke and Prest, 1987)

A number of factors are thought to influence the degree of bedform modification consequent on changes in the ice flow direction, including ice flow velocity, sediment properties, ice thickness and the duration of individual flow phases. The influence of flow velocity is illustrated in Fig. 12.37B, which shows the theoretical velocity distribution along a flowline of a simple ice sheet (Boulton, 1987; Clark, 1993, 1994). The balance velocity is at a minimum below the ice divide, and increases downflow to reach a maximum in the vicinity of the equilibrium line, from where it decreases outwards towards the margin. If ice flow velocity is the dominant factor controlling the degree of bedform modification when flow changes direction, the greatest amount of modification should be expected close to the margin. Thus, bedforms in this area are likely to record only the most recent flow direction, whereas cross-cutting

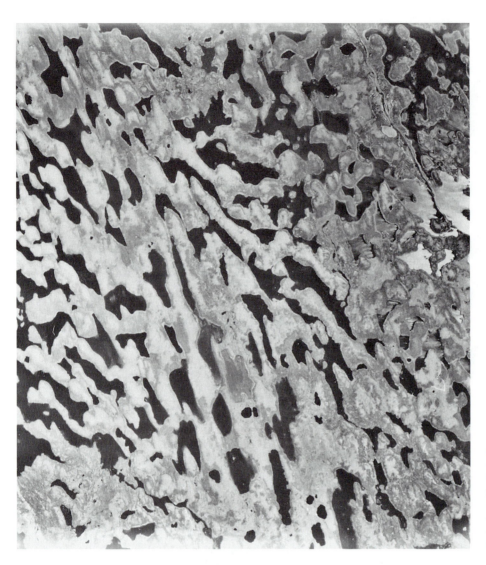

Fig. 12.35 Aerial photograph of part of the Boyd Lake area, NWT, Canada, showing cross-cutting drift lineations. (Aerial photograph from the Department of Energy, Mines and Resources, Canada)

lineations should be expected to be best-preserved further upglacier. Such spatial relationships do in fact exist and have been identified by Boulton and Clark (1990a, b) in Canada (Fig. 12.38). Larger numbers of preserved lineations document less intense modification, some of which are located in the former areas of ice divides such as Keewatin, west of Hudson Bay. This model also explains why drumlins are commonly formed in belts upglacier from recessional moraines, and record the ice flow directions that prevailed during deglaciation (Section 11.2.4.2; Mooers, 1989b; Patterson and Hooke, 1996). Differences in the degree of bedform modification could also result from lateral variations in ice-flow velocity, such as those near the margins of ice streams.

Cross-cutting lineations also provide evidence for the nature of ice flow changes. If the changes in ice flow direction were gradual, and there was a continuous adjustment to changing flow, we should expect

to see either a whole series of intermediate orientations, or a single final orientation. However, only a small number of discrete ice flow directions are observed in the landform record in any one area, which led Clark (1993) to suggest two possible explanations: (a) punctuated ice flow shifts, in which periods of relative stability and lineation formation are separated by brief episodes during which ice flow direction changes rapidly; or (b) an on–off mechanism of lineation generation, which involves a steady change in the ice flow direction but with an on–off lineation production mechanism that records snapshots of the changing flow regime. Clark (1993) favoured the punctuated ice-flow model because it best explains the continent-wide coherent sets of lineations reported by Boulton and Clark (1990a, b). However, the 'on–off mechanism' may be applicable where basal conditions were near the threshold for bed erosion, deposition or deformation.

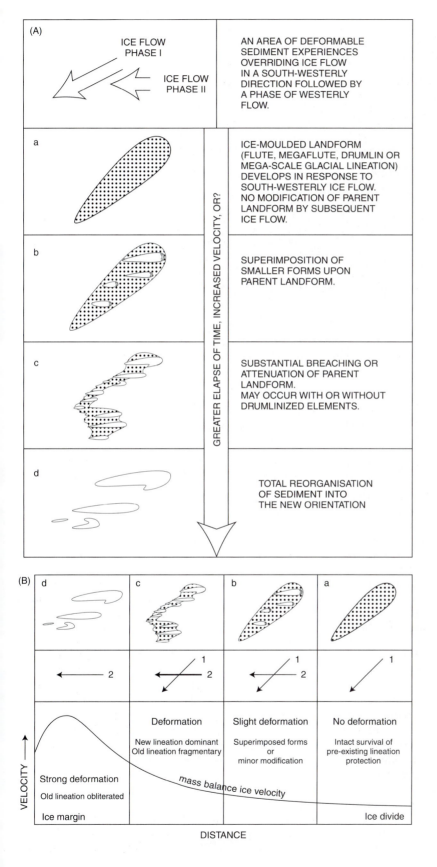

Fig. 12.37 (A) Theoretical continuum of subglacial modification of a lineation when overridden by ice flowing from a different direction. (B) The theoretical continuum of subglacial modification transferred to a transect from ice divide to margin. Note that this assumes that the ice flow velocity is the main determinant in the degree of subglacial modification. (From Clark, 1993. Reproduced by permission of John Wiley and Sons)

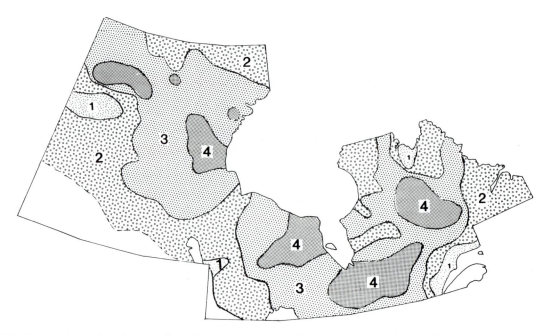

Fig. 12.38 The numbers of glacier flow lineations per 2 per cent of mapped area for the area formerly covered by the centre of the Laurentide ice sheet. These are regarded by Boulton and Clark (1990b) as generalized indices of glacial erosion. Large numbers of lineations indicate less intense erosion and thus mark the positions of former ice divides. (Modified from Boulton and Clark, 1990b)

Boulton and Clark (1990b) have compiled maps of successive ice flow directions for the Canadian Shield based on the distribution of cross-cutting lineations (Fig. 12.36 (Plate 26)). The sequence of ice flow events in Hudson Bay is compatible with till fabric data and the pattern and sequence of erratic dispersal. Changes in ice flow through time are linked to changes in the positions of ice divides and therefore changes in the shape of the various sectors of the ice sheet. If shifts in ice dispersal are not taken into account when interpreting ice flow lineation maps, then 'bogus' ice divides will exist in the ice sheet reconstruction. This is illustrated in Fig. 12.39, which depicts all the ice flow lineations associated with three successive ice dispersal centres. If only the strongest lineations are used in reconstructing the ice divide then its position will be bogus. Similarly, bogus divides will result from incorrect interpretation of lineations resulting from a shift in a principal ice divide and its associated subsidiary divides.

Obviously, large numbers of early ice flow indicators are obliterated by later ice flows, and this constitutes a major problem in producing a reconstruction of multiple flow events. If an ice divide does not shift significantly during a glacial cycle then older forms may be preserved beneath it, because the amount of erosion is limited by low ice velocities.

A very clear example of shifting flow conditions reconstructed from drift lineations, regional till composition and erratic dispersal patterns is illustrated in Fig. 12.40, which shows the inferred ice flow history

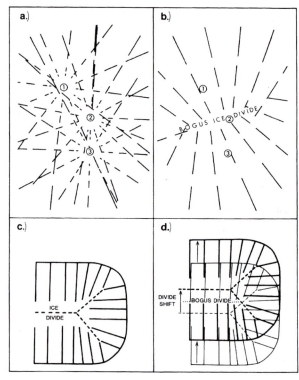

Fig. 12.39 Glacier flow lineations in relation to ice divide shifts. In (a) and (b) the ice centre shifts through positions 1–3, remaining stable at each position for a considerable period. Note how bogus ice divides may be interpreted from the geomorphological record. (From Boulton and Clark, 1990. Reprinted by permission of the Royal Society of Edinburgh)

of Prince of Wales Island in Arctic Canada, associated with the shifting of the M'Clintock ice divide (Dyke, 1983, 1984b; Dyke and Morris, 1988; Dyke *et al.*, 1992; Hodgson, 1994). The presence of bedforms was used as evidence for warm-based conditions, whereas the preservation of bedforms relating to older flow directions was taken to indicate cold-based ice. Furthermore, downstream transitions from drumlins to Rogen (ribbed) moraine were interpreted as the product of shifts from a wet, sliding bed to a freezing zone. The edges of individual ice streams dating to different flow phases are very clear on satellite imagery where bedforms cross-cut each other (Fig. 12.41 (Plate 28)), and the till plumes produced by each flow are lithologically distinct. The western edge of the phase 2 ice stream ends abruptly as a *lateral shear moraine*, consisting of a streamlined ridge of till, 68 km long and up to 1 km wide, which was formed by lateral shearing at the boundary between warm-based and cold-based ice.

The final stages of ice retreat on Prince of Wales Island were characterized by the stabilization of the ice divide towards the south of the island and the draw-down of ice into calving bays as a result of glacioisostatically higher sea-levels. This initiated the production of smaller fields of drumlins and flutings in topographic depressions. These final stages of deglaciation on Prince of Wales Island are similar to those envisaged by N. Eyles and McCabe (1989a) for the land surrounding the Irish Sea basin at the end of the last glaciation, when prominent drumlin swarms were produced by ice streaming down major valley systems in Scotland, northern England and Northern Ireland in response to rising sea-levels (Section 12.5.4).

Ice directional indicators recording various former ice movement directions during the same glaciation have been mapped at regional scales also in Scandinavia and northern Europe (Nordkalott Project 1986a, b; Donner, 1995). Several studies have shown that the Scandinavian ice sheet initially accumulated as a single elongate dome over the western mountains, then the major central divide migrated to the east over the continental interior (e.g. Ljungner, 1948; Vorren, 1977; Lundqvist, 1986; Ringberg, 1988). This explains the early north–south ice flow and later east–west ice flow over Denmark and northern Germany (Fig. 12.26; Ehlers, 1983a, 1990; Ehlers *et al.*, 1984; Houmark-Nielsen, 1988). During deglaciation, the ice divide migrated back towards the western highlands. However, more complex ice flow patterns involving subsidiary domes have been suggested in order to explain complex cross-cutting ice flow indicators (Lagerlund, 1983; Anundsen, 1990). The deglaciation of Scandinavia occurred over large areas in contact with deep lakes or marine waters, and complex moraine systems were deposited in arcuate

belts which demarcate the lobate pattern of the ice sheet. The form of marginal ice lobes can be reconstructed from the subglacial legacy, which consists of overlapping suites of streamlined landforms, each suite comprising drumlins and megaflutings fanning out to an arcuate end moraine (Section 12.4.3; Gluckert, 1973, 1974; Lundqvist, 1977, 1988; Haavisto-Hyvarinen *et al.*, 1989; Heikkinen and Tikkanen, 1989; Donner, 1995).

12.4.5 Sediment dispersal patterns

The occurrence of boulders in areas far removed from their bedrock source has long been recognized as firm evidence for glacial transportation (Agassiz, 1838b; Geikie, 1877, 1894). For example, fragments of microgranite originating from the small island of Ailsa Craig off the west coast of Scotland occur in the tills surrounding the Irish Sea basin, documenting glacier ice flow southwards from Scotland towards Ireland, Wales and west-central England (Fig. 12.42). Various exotic rock fragments originating from Scandinavia occur in the tills of eastern England (Ehlers and Gibbard, 1991), northern Germany, Denmark and the Netherlands (Flint, 1971; Overweel, 1977). Other erratic trains have an immediately obvious provenance where they emanate directly from their source outcrop, such as the Assynt sandstone boulder trains of north-west Scotland, which extend down-ice from mountains of Torridonian sandstone (Fig. 12.43; Lawson, 1990, 1995; Lawson and Ballantyne, 1995). Some of the largest glacially transported blocks are the quartzites of the *Foothills erratics train*, which lie on the Cretaceous bedrock prairie of southern Alberta more than 375 km from their source in Jasper National Park (Fig. 12.44; Stalker, 1956, 1976a; Mountjoy, 1958; Tharin, 1969). The largest block is estimated to weigh approximately 16,000 tons (Trenhaile, 1990).

Erratics and erratic trains can be used to reconstruct former transportation pathways, although it should be remembered that debris transport histories may involve several glacial cycles during which the ice divides and ice flow vectors may shift dramatically. Nonetheless, the distribution of erratics and fine-grained components of till provides glacial geologists with a powerful tool for reconstructing the patterns and history of ice dispersal in studies of ice sheet dynamics (Clark, 1987; Boulton, 1996a). Additionally, the study of dispersal patterns has great practical applications for mineral prospecting in glaciated terrains (e.g. Jones, 1973; Shilts, 1982a; DiLabio and Coker, 1989; Kujansuu and Saarnisto, 1990).

Indicator erratics are those for which a definite source area is known, and *indicator fans* are regional fields over which the erratics from particular source

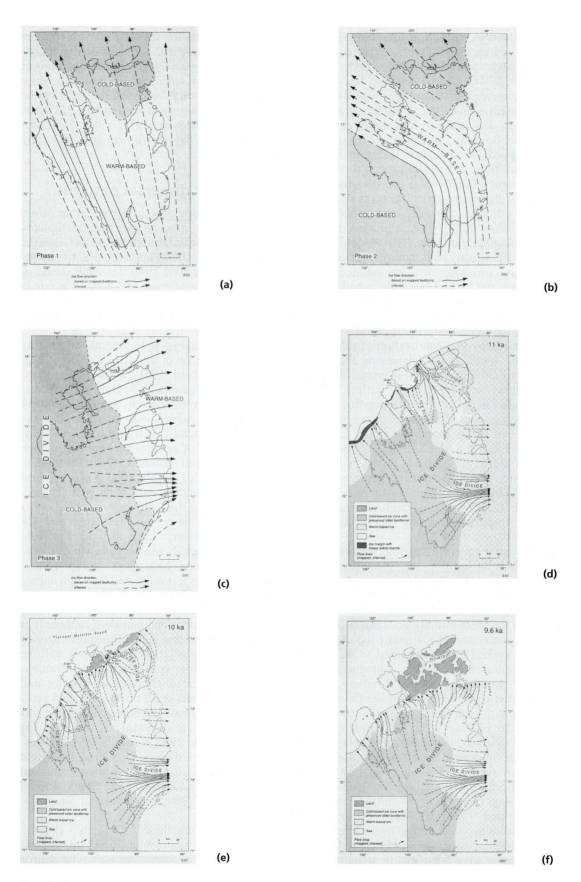

Fig. 12.40

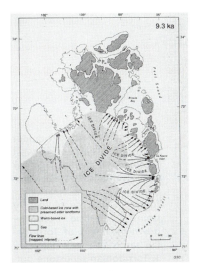

Fig. 12.40 Changing ice flow directions and basal thermal regime during the last glaciation of Prince of Wales Island, Arctic Canada. (a–c) Shifting flow conditions during the last glacial maximum. (d–g) Flow conditions and associated ice margins for 11, 10, 9.6 and 9.3 kyr BP, showing the migration of ice divides and ice streams, the distribution of cold- and warm-based ice and the extent of the deglacial high sea-level. The solid flowlines on these maps represent glacial bedforms that are still visible even though later flows from different directions overrode them. (From Dyke *et al.*, 1992. Reprinted by permission of the Geological Survey of Canada)

rocks have been dispersed by glacier flow (Fig. 12.45; Flint, 1971). Indicator fans encompass the range of transport directions produced by shifting ice divides and dispersal centres, and can be identified using a range of sediment sizes from erratic blocks down to the fine-grained matrix of tills. The concentrations of indicator erratics vary systematically along ice flowlines. Within indicator outcrops, concentrations increase rapidly downglacier, reflecting the addition of new material from the glacier bed, but concentrations drop off rapidly down-ice of the outcrop margin (Fig. 12.46; Hellaakoski, 1930; Gillberg, 1965, 1967; Virkkala, 1969; Linden, 1975; Perttunen, 1977). This simple picture may be complicated by shifting patterns of erosion and sediment accumulation during glacier advance and retreat, and in some cases the peak in erratic concentration may be displaced down-ice from the boundary of the source outcrop (Boulton, 1996a). The up-ice and down-ice limits of an indicator plume are known as the *head* and *tail*, respectively. Debris dispersal by modern glaciers has been studied by DiLabio and Shilts (1979), who demonstrated that head and tail zones are identifiable in the lateral moraines of glaciers crossing from one bedrock type to another.

The transport distances of the majority of indicator erratics are relatively short. A good index of transport distance is the *half-distance value*, or the distance

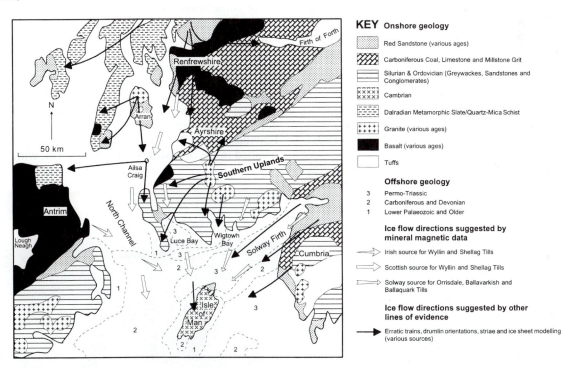

Fig. 12.42 The geology of the Irish Sea basin and surrounding areas together with various ice flow directions based upon erratic distributions, drumlins, striae, ice sheet modelling and mineral magnetic data. (Modified from Walden *et al.*, 1992)

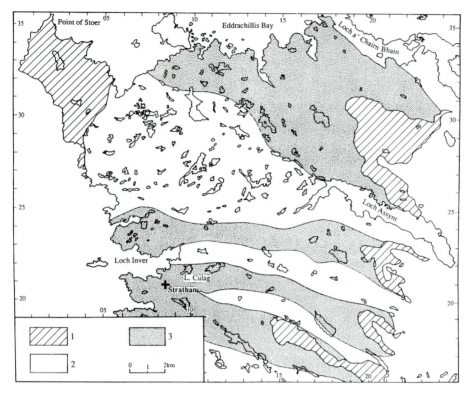

Fig. 12.43 The boulder trains of Assynt, north-west Scotland, indicating glacier flow from east to west. Torridonian sandstone erratics are found in discrete plumes or trains emanating from the down-ice flanks of Torridonian sandstone rock outcrops: 1, Torridonian sandstone outcrops; 2, areas devoid of Torridonian sandstone erratic boulders; 3, areas where Torridonian sandstone erratic boulders have been found. (From T.J. Lawson, 1995. Reprinted by permission of the Quaternary Research Association)

from the nearest possible source to the point at which the frequency of indicator erratics is half of its maximum. The half-distance value varies with the material resistance to abrasion and crushing and the depth to the source rock beneath superficial sediment (Bouchard and Salonen, 1990). Salonen (1986, 1987) found that the average half-distances for particles in tills in Finland are only 5 km. Very far-travelled erratics (up to 1200 km in northern Europe) may have been transported during several glacial cycles, and part of their journey may have been by iceberg rafting (Marcussen, 1973; Spjeldnaes, 1973; Overweel, 1977).

Dyke and Morris (1988) suggested a twofold classification scheme for erratic trains based upon examples in Canada (Fig. 12.47). (a) The *Dubawnt type* is named after the Dubawnt Sandstone in central Keewatin, which is a relatively restricted source outcrop from which material was dispersed as a plume over the surrounding bedrock by ice moving at a uniform rate over an entire region. Another example of this type of dispersal train is the 40 km long plume located downflow from the Strange Lake alkalic intrusion in northern Labrador, Canada (Batterson, 1989). (b) The *Boothia type*, based on examples on the Boothia Peninsula, is characterized by debris plumes that extend from small parts of large source areas by zones of more rapid ice flow (ice streams) within the ice sheet.

Vertical changes in till composition at a particular site may result from: (a) the complete erosion or burial of some source outcrops; (b) a shift in ice flow direction so that material is transported from different source areas; or (c) time-dependent patterns of erratic transport along a flowline (Section 12.4.1). Stea *et al.* (1989) and Turner and Stea (1990) have linked till characteristics to ice divide migrations in Nova Scotia during the last glaciation. Early shifts in regional ice divides led to the influx of far-travelled debris to the area, forming the Hartlen Till and Lawrencetown Till, and later local ice dispersal resulted in the deposition of the Stony Till, with local

Fig. 12.44(a)

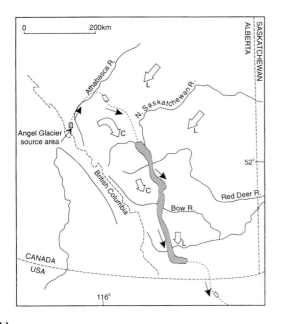

(b)

Fig. 12.44 The Foothills Erratics Train, Alberta, Canada. (a) The Okotoks erratic near Calgary, southern Alberta. This huge quartzite boulder is part of the Foothills Erratics Train, which provides an impression of glacial transport paths over a distance of more than 375 km. (Photo: D.J.A. Evans) (b) The transport path of the Foothills Erratics Train from Jasper in the Rocky Mountains to Montana, USA. (Modified from Klassen, 1989)

lithological signatures. The identification of palaeo-ice lobes from till lithology has been attempted in a number of settings, and considerable success has been achieved in determining the source areas of individual till sheets in, for example, the Great Lakes region of North America (Gwyn and Dreimanis, 1979), the Canadian Prairies (Shetsen, 1987), eastern England (Madgett and Catt, 1978), central Quebec and Labrador, Canada (Klassen and Thompson, 1989, 1993), and Finland (Hirvas and Nenonen, 1985). Dispersal patterns beneath former ice divides tend to be complex, and subglacial material can be transported in several different directions as the result of ice divide migration. The resulting dispersal pattern consists of multiple plumes emanating from a source outcrop, termed an *amoeboid pattern* by Shilts (1993).

Indicator fans or *dispersal fans* provide important clues to the location of mineral outcrops or ore bodies in glaciated terrain, where elongate plumes of mineral-enriched tills extend down-ice from source outcrops (Shilts, 1976, 1993; DiLabio, 1981, 1990a, b). The mineral content of the plumes is known as a mineral *float*. Ore enrichment in tills can be assessed by pebble counts and/or geochemical and minerological analyses of the fine-grained matrix. Because ore bodies are often hidden by till, the geochemical or mineralogical characteristics of till samples, col-

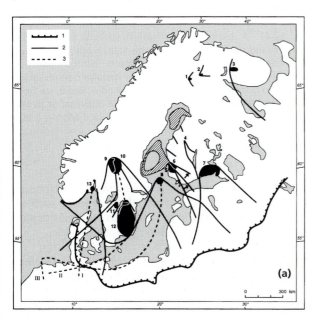

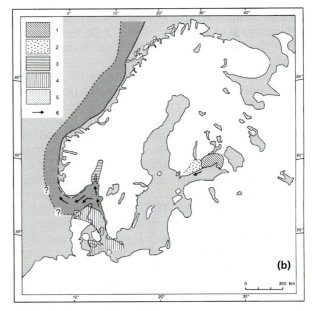

Fig. 12.45 (a) Glacial erratic indicator fans produced by the Scandinavian ice sheet based upon various sources. Key: 1, outer limit of Weichselian glaciation; 2, lateral limit of fan; 3, path of dispersal. Source areas: 1, Jatulian sandstone and conglomerate; 2, Nattanen granite; 3, Umptek and Lujarv-Urt nepheline syenite; 4, Lappajarvi impactite; 5, Vehmaa and Laitila rapakivi granite; 6, Jotnian sandstone (Satakunta); 7, Viipuri rapakivi granite; 8, Aland rapakivi granite; 9, Jotnian (Dala) sandstone; 10, Dala porphyries; 11, Cambro-Silurian limestone; 12, Smaland granite; 13, Oslo rhomb porphyries and larvikite. (b) The dispersal pattern of ice-rafted material along the western Scandinavian coast: 1, Viipuri rapakivi granite source area; 2, area of dispersal for Viipuri rapakivi granite; 3, Oslo rhomb porphyries; 4, flint; 5, area of dispersal for rhomb porphyries and flint; 6, major currents. (From Donner, 1995. Reproduced by permission of Cambridge University Press)

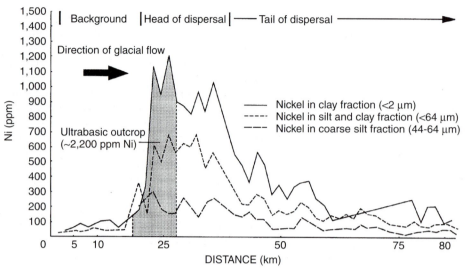

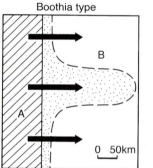
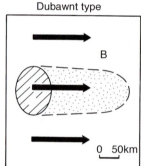

Fig. 12.46 Dispersal curves for nickel concentrations in fine and coarse fractions of tills downglacier from an ophiolitic complex at Thetford Mines, Quebec, Canada. Nickel is particularly rich in the clay fraction because of the dominance of nickel-rich serpentine, which is preferentially reduced to clay sizes. (From Shilts, 1993. Reprinted by permission of the *Canadian Journal of Earth Sciences*)

lected on a regular grid and then contoured, often reveal plume-shaped areas of metal enrichment or dispersal trains. The largest dispersal trains can be hundreds of kilometres long but derived from very small ore bodies; they are hundreds to thousands of times larger than their source area, thus representing a much larger exploration target for prospectors than the ore body itself (DiLabio, 1990a). Some dispersal trains can be identified visually on the basis of their colour, such as the carbonate till plumes that stretch across Prince of Wales Island, Somerset Island and Boothia Peninsula in the central Canadian Arctic archipelago (Dyke, 1983, 1984b; Dyke and Morris, 1988; Dyke *et al.*, 1992) and central Baffin Island (Tippett, 1985).

Variations in nickel concentrations in till in part of Quebec, Canada, are shown in Fig. 12.46 (Rencz and Shilts, 1980; Shilts, 1993). The dispersal curves show only background levels of nickel up-ice from the source ultrabasic bedrock outcrop, and head and tail zones down-ice from the outcrop. The concentrations of nickel vary with grain size as a result of differential abrasion and crushing during subglacial transport (see Section 5.7.2; Dreimanis and Vagners, 1971, 1972; Haldorsen, 1983). Each mineral has a different mode at which its chemical signature is strongest, and this must be taken into account when analysing till samples for their geochemical signatures (Shilts, 1971, 1975). Such *chemical partitioning* is discussed further by DiLabio (1982), Nikkarinen *et al.* (1984), Shilts (1984a, 1991) and Shilts and Wyatt (1989). Anomalous geochemical signals can be detected in till samples because of weathering and pedogenesis, the depth of which is dictated by such factors as the water table and grain size (e.g. Shilts, 1975, 1976, 1984a; Rencz and Shilts, 1980; Peuraniemi, 1984; Shilts and Kettles, 1990). Therefore, till samples for geochemical analysis and indicator

Fig. 12.47 A twofold classification scheme for dispersal trains based upon the shape of the source outcrop and ice flow characteristics. The Boothia type is formed under ice streams and the Dubawnt type under sustained regional ice flow. Two different rock types are marked A and B, ice flow direction is indicated by arrows, and the dispersal of debris from rock type A is stippled. (From Dyke and Morris, 1988. reproduced by permission of the Canadian Association of Geographers)

tracing must be taken from unweathered horizons in order to avoid post-depositional alterations, such as the destruction of labile ore minerals and their depletion in till samples from above the water table.

Changes in the lithological composition of glacifluvial materials have also been used to determine transport patterns in glacial depositional systems (Shilts, 1984b; Ryder, 1995). Studies on the lithology of materials in Scandinavian eskers have revealed that transport distances from source areas vary according to the comminution rate of the different rock types. In eskers the increase in clasts of a particular rock type over its outcrop area begins further downflow from the up-ice margin of the outcrop than in tills. In addition, the concentration of rocks from such sources is far higher in esker sediments than in tills (Hellaakoski, 1930; Lillieskold, 1990). In glacifluvial systems, rocks that disintegrate easily to form smaller grain sizes will travel further (Virkkala, 1958; Gillberg, 1968). Multiple high-concentration peaks of garnet in glacifluvial materials associated with the Salpausselkä end moraines, Scandinavia, have been explained by Perttunen (1989) as the result of ice-marginal standstills.

12.5 SUBAQUATIC DEPOSITIONAL SYSTEMS

Subaquatic depositional systems form a very important component of the record left by former ice sheets, for several reasons. First, large parts of the Pleistocene ice sheets terminated in the sea or large proglacial lakes, and large volumes of sediment were deposited in standing water at and beyond the ice margins (Fig. 12.48 (Plate 29)). The present Antarctic ice sheet has predominantly marine margins, and exerts a strong influence on sedimentation on the surrounding continental shelves. Second, subaquatic sediments tend to have high preservation potential, and commonly preserve a more complete record of glaciation than terrestrial depositional systems. The continental shelves off northern Europe, North America and Antarctica are draped by thick sequences of glacimarine sediments, and much of the evidence for pre-Cenozoic glaciations consists of subaquatic facies (N. Eyles, 1993). Third, study of subaquatic depositional systems yields important insights into the dynamics of former ice sheets, including the patterns of deglaciation, the location of the grounding lines of former ice streams, and episodes of rapid calving retreat and ice rafting (e.g. Heinrich events).

In this section, we review the main controls on the large-scale stratigraphic architecture of subaqueous deposits, then discuss the depositional record in fjords, on continental shelves, and in former large proglacial lakes and epicontinental seas. Subaqueous depositional systems in glaciated valleys are discussed in Section 12.6.

12.5.1 Stratigraphic architecture

In recent years, many researchers have studied the large-scale form or *stratigraphic architecture* of subaquatic depositional systems as a key to understanding environmental change in glacially influenced basins (e.g. Boulton, 1990; N. Eyles and C.H. Eyles, 1992; Hambrey, 1994). A large body of information is available on subaqueous depositional systems for hundreds of formerly glaciated sites, and many depositional models have been proposed (e.g. Drewry and Cooper, 1981; Nelson, 1981; Mode *et al.*, 1983;

Molnia, 1983b; Gravenor *et al.*, 1984; C.H. Eyles *et al.*, 1985; Benn, 1989a; Fyfe, 1990; Anderson *et al.*, 1991; Brodzikowski and van Loon, 1991; Ashley, 1995; Powell and Domack, 1995). It has become clear that, although the lateral and vertical relationships between sediment–landform associations such as morainal banks, grounding-line wedges and distal drapes are very varied, they reflect the interaction of relatively few groups of variables. The most important are (a) the topography and tectonic setting of the basin; (b) the extent, configuration and dynamics of glacier ice and floating ice and the position of associated depocentres; and (c) changes in water depth due to eustatic and isostatic cycles and the life-cycle of proglacial lakes (Powell, 1984, 1990, 1991; Boulton, 1990; N. Eyles and C.H. Eyles, 1992; Martini and Brookfield, 1995; Powell and Domack, 1995). The influence of tectonic setting on the stratigraphic architecture of glacially influenced continental shelves is discussed in Section 12.5.3. The influence of glacial variables and that of water depth are examined in turn below.

In Sections 8.5 and 11.6 we saw that subaqueous depositional systems are strongly influenced by the proximity of grounded glacier ice. Accordingly, it is useful to subdivide such depositional systems according to their position of formation relative to the ice front. Powell (1984) defined four depositional zones, each of which is associated with distinctive sets of processes and stratigraphic architecture. These are: (a) subglacial; (b) ice-proximal; (c) ice shelf; and (d) ice-distal. The zones were originally defined for glacimarine environments, but are also applicable to large proglacial lakes.

The *subglacial zone* lies behind the glacier grounding line, and is reviewed in detail in Section 12.4. Facies deposited in this zone form downglacier-thickening sheets of basal till and associated lenses of water-sorted material. Sediment dynamics in the subglacial zone exert an important control on the delivery of sediment to the subaqueous environment, either by meltwater, by subglacial till deformation, or within sequences of basal ice.

The *ice-proximal zone* encompasses subaqueous environments adjacent to the glacier grounding line. The thickness of sediment packages deposited in the ice-proximal zone is related to the glacier snout retreat rate: slow retreat or quasi-stable conditions will encourage the accumulation of large volumes of sediment, whereas during rapid calving retreat a large proportion of debris will be rafted away, restricting the amount of ice-proximal deposition. Sediment associations include push and thrust moraines, morainal banks, deltas, De Geer moraines, and grounding-line fans (Sections 11.3 and 11.6). The type of association depends on a number of factors, including ice velocity and calving rate, sediment supply, input of meltwater from the ice, and water depth and salinity. For example, at the tidewater margins of temperate glaciers, large amounts of coarse debris are deposited in grounding-line fans, and fine-grained sediments are carried away in turbid plumes to form cyclopsams and cyclopels (Section 8.5). In contrast, grounding lines below ice shelves in polar environments are more commonly associated with grounding-line wedges composed of mass flow deposits, because of the limited availability of subglacial meltwater (Sections 8.2.3 and 8.5.3). Sediment gravity flows are common in the ice-proximal zone, because of high sedimentation rates, ice push, iceberg calving, storm waves, tidally induced pressures below ice shelves, and morainal bank collapse. Subaqueous outwash and sediment flows both commonly form prograding beds dipping away from the glacier margin, known as *clinoforms*. The upper parts of such clinoforms may be erosionally truncated during glacier advance, forming a subhorizontal erosion surface overlain by basal till.

The *ice shelf zone* occurs in polar climates where ice below the pressure melting point floats and produces glacier and sea-ice shelves (Section 8.2.3). Deposition in this zone is dominated by undermelt and rain-out, forming drapes of dropstone diamicton and mud (Section 10.6.3). Patterns of deposition vary with the debris content of the ice and the location of basal melting and freezing zones beneath the ice shelf, but in general, deposits are thickest near the grounding line and thin distally. Below some ice shelves, basal freeze-on of sea-water allows significant amounts of basal debris to be transported to the ice shelf edge, forming dropstone deposits in more distal locations. The geometry of undermelt deposits is to a large extent determined by the local topography. In areas with significant relief, reworking by mass movements relocates sediment and forms thick infills in topographic basins.

The *ice-distal zone* is dominated by ice-rafted debris and muds formed by the settling of suspended mineral grains and microfaunal remains (Section 10.6). C.H. Eyles and N. Eyles (1983) and C.H. Eyles *et al.* (1985) argued that sedimentary sequences in distal glacilacustrine and glacimarine environments are determined by the relative importance of rain-out, current reworking and gravitational resedimentation (Fig. 12.49). In glacimarine environments, deposition and reworking by these three processes can produce sequences of massive and stratified sediment tens to hundreds of metres thick, and spanning several glacial cycles. The long-term preservation potential of such sequences is high, and they are well represented in the ancient geological record (N. Eyles, 1993). Iceberg scouring is common in high-latitude distal glacimarine environments, churning the

bottom sediments and producing massive, structure-less ice-keel turbates (Section 10.6.4; Vorren *et al.*, 1983; Woodworth-Lynas and Guigné, 1990; Dowdeswell *et al.*, 1994).

Depositional systems deposited in each of these zones can be superimposed, recording glacial advance and retreat cycles. This is illustrated in Figs 12.50 and 12.51, which show changing patterns of sedimentation associated with the growth and decay of a water-terminating ice sheet (Boulton, 1990; N. Eyles and C.H. Eyles, 1992). During ice sheet advance, the ice-proximal zone migrates over the former ice-distal zone, depositing ice-contact facies associations on top of distal sediments. Such associations tend to be eroded, transported or redeposited in the subglacial zone once glacier ice advances over the site. The ice sheet buildup phase is therefore typically represented by a sheet of subglacial till, and underlying glacimarine deposits are likely to survive only close to the margin. As the ice sheet retreats, the ice-proximal and distal zones migrate in turn over the former subglacial zone, commonly blanketing subglacial forms with a fining-upward drape. More complex successions result from ice-marginal oscillations, and where deposition is also influenced by sediment from non-glacierized catchments or glacial meltstreams.

Water depth exerts a strong control on the distribution and character of subaqueous depositional systems, owing to its influence on ice margin stability, the space available for sedimentation, and base levels for subaerial erosion (N. Eyles and C.H. Eyles, 1992; Martini and Brookfield, 1995). In glacimarine environments, patterns of sedimentation are influenced by isostatic depression and rebound during glacial cycles, which may be in phase or out of phase with global eustatic sea-level changes. The influence of sea-level change and sedimentation rates on depositional architecture in glacimarine environments has been considered by Bednarski (1988) (Table 12.1), using concepts similar to those of sequence stratigraphy (Section 12.2.2; Curray, 1964). Figure 12.52 shows a hypothetical glacimarine depositional sequence produced during a sea-level transgression–regression cycle. The cycle begins during deglaciation, when receding ice allows marine waters to flood the basin, forming an erosional transgression surface (VII). A sequence of marine deposits is then laid down on top of the erosion surface, interfingering with terrestrial glacigenic sediments at the landward end (VI and V). The marine limit marks the highest level reached by the sea. Subsequent regression can result in erosion, which truncates glacimarine sediments deposited during transgression and highstand (I). Alternatively, glacimarine sediment can be reworked into littoral features such as beaches, storm ridges or tidal flats (II, III and IV). Such shoreface

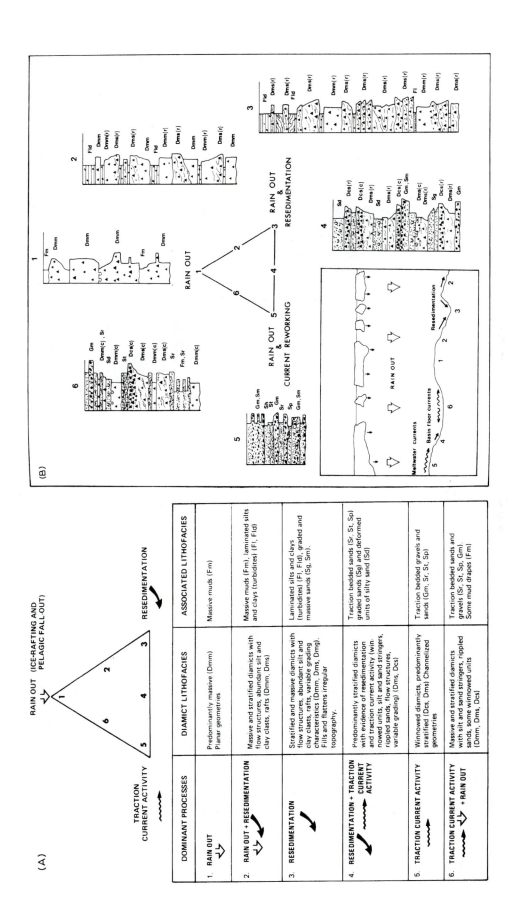

Fig. 12.49 Model of distal subaqueous sedimentation as a continuum between three end-member processes: rain-out (from suspended and ice-rafted sediment), traction current activity and gravitational resedimentation. (From C.H. Eyles *et al.*, 1985. Reproduced by permission of Elsevier)

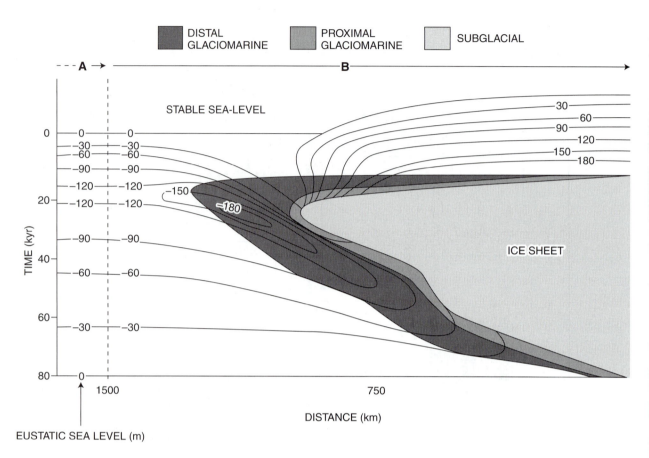

Fig. 12.50 Time–distance diagram showing the migration of subglacial and glacimarine environments during a glacial cycle. The isolines show changes in sea-level due to glacioeustasy and glacioisostasy. (From N. Eyles and C.H. Eyles, 1992, after Boulton, 1990. Reproduced by permission of the Geological Association of Canada)

reworking is very common around isostatically uplifted coasts, limiting the amount of glacimarine sediment exposed at the surface (Benn and Dawson, 1987). In glacilacustrine environments, the opening and closing of outlets by ice or sediment can cause multiple rapid changes in water level, resulting in complex depositional systems separated by erosion surfaces (Martini and Brookfield, 1995).

Water level changes have the largest effect on patterns of sedimentation in the shallower parts of a basin. In contrast, changing water levels may have little effect in deeper areas, where facies variations depend more simply on the proximity of glacier ice. In the following sections, we discuss the stratigraphic architecture of sediments deposited in fjords, on continental shelves and in proglacial lakes, with emphasis on the controlling environmental variables.

12.5.2 Fjords

Glacial stratigraphic architecture in fjords is strongly influenced by topography. Overdeepened basins act as sediment sinks, whereas bedrock sills provide pin-ning points where ice-proximal depositional systems can accumulate. Furthermore, the presence of steep flanking slopes encourages gravitational reworking of sediment and debris input from ice-free terrain. Fjords can be subdivided into high- and low-relief types. High-relief fjords have lateral slopes which are too steep for the accumulation of sediment, and tend to be characterized by precipitous rock slopes flanking a sediment-floored trough. In contrast, low-relief fjords contain more extensive low-gradient surfaces on which sediment can accumulate, encouraging the preservation of a wide variety of sediment–landform associations.

The sedimentary infill of fjord basins can be usefully divided into three broad units deposited during glacier advance, maximum and retreat, respectively:

1. *Advance phase.* Within the glacier limit, the advance stage of a glacial cycle is typically represented by a basal till unit or an erosion surface. Till thickness generally increases downfjord towards the glacier limit, reflecting erosion and downglacier transport of pre-existing sediment

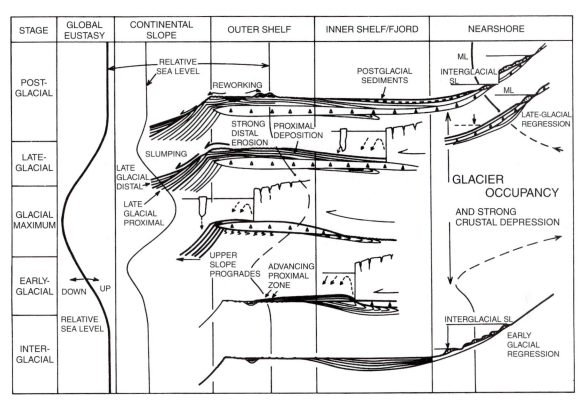

Fig. 12.51 A model of glacimarine architecture in space and time, showing sediments produced on different parts of a continental margin throughout a glacial cycle. Relative sea-levels are also depicted for each zone. (From Boulton, 1990. Reprinted by permission of the Geological Society of London)

and bedrock fragments (Boulton, 1990). Beyond the glacier limit, extensive blankets of mud and diamicton may record distal glacimarine deposition during the advance phase, but are typically buried by younger glacimarine sediments.

2. *Maximum phase.* Glacier limits within fjords are commonly located at pinning points where reduced calving rates encourage ice margin stability (Section 8.2.5; Mercer, 1961; Warren and Hulton, 1990; Greene, 1992). Typical sediment–landform associations include push or thrust moraines, grounding-line fans, morainal banks and deltas (Section 11.6). In low-relief fjord basins, lateral moraines may be preserved on the valley sides, but in high-relief settings such moraines are generally rapidly destroyed by paraglacial reworking after glacier retreat. Lateral moraines with exceptionally low gradients indicate the former extent of floating ice shelves (Fig. 12.53 (Plate 30)); Section 11.6.4). Beyond the ice limit, blanket-like drapes of fine-grained sediments and diamictic muds record suspension sedimentation from turbid overflows and dumping from icebergs (Sections 10.6.2 and 10.6.3; Powell and Molnia, 1989). The relative amount of suspended sediment and ice-rafted debris varies with topographic set-

ting, calving rates, and the availability of debris and meltwater. The highest rates of suspension sedimentation are associated with temperate fjord glaciers, where discharges of turbid meltwater are high during the ablation season and thick sequences of cyclopels and cyclopsams can accumulate rapidly (Section 10.6.2; Cowan and Powell, 1990).

3. *Retreat phase.* During glacier retreat, depositional zones migrate upfjord, and progressively more distal facies are laid down on top of older units. Pauses in glacier retreat may be marked by push or thrust moraines, grounding-line fans or morainal banks (Boulton, 1986). Substantial ice-marginal accumulations such as large morainal banks and deltas tend to be associated with topographic pinning points. At trough margins, glacimarine facies may be overlain by subaqueous fans fed by subaerial streams on the valley sides (Section 11.6.5.1). Fan aggradation is thought to be most rapid in the paraglacial period immediately after deglaciation, when large amounts of readily entrained glacigenic sediment are available.

Depositional systems associated with the retreat phase can be illustrated by a set of models developed

Table 12.1 Classification of transgressions and regressions that occur during a glacial cycle in a fjord

Classification*	Glacial cycle	Conditions	Results
—	Glacial	Ice occupying fjord	Glacigenic deposition
VII	Deglaciation	Submergence >> sedimentation	Overstep marine over glacial deposits
VI†‡		Transgression > sedimentation	Thin veneer of littoral sand overlain by discontinuous marine sediments
	Decreasing ice load		
V†‡ (approaching marine limit)		Transgression ≥ sediment supply	Marine onlap
VIII† (of short duration)	Uplift = eustatic sea-level rise	Geographically stable coast	Marine limit reached
I	Sea-level rise < uplift	Emergence >> sedimentation	No beaches, dissection of uplifted marine sediments
II†		Emergence > sedimentation	Mostly wave cut beaches, regressive strandline
↑ isostatic zone I			
III	Decreasing rebound		
IV† ↓		Emergence ≥ sedimentation	Marine offlap, regressive beaches, delta building
VIII ↑		Net erosion from wave action, local subsidence from compaction	
isostatic zone I/II			
V ↓	Collapse of the forebulge	Transgression	Marine onlap

Source: Bednarski (1988)
* From Curray (1964) and explained in Figure 12.52
† Shorelines may be preserved
‡ Areas beyond the glacial limit

by Powell (1981a) for Alaskan fjord glaciers (Fig. 12.54). *Facies association I* results from rapid calving retreat in deep water. Large amounts of sediment are transported by icebergs, producing widespread dropstone muds and diamictons. Small morainal banks are formed during brief glacier standstills or readvances. *Facies association II* is characteristic of slowly retreating or stationary glacier margins in shallow water. These conditions encourage the deposition of large grounding-line fans or morainal banks. *Facies association III* is formed by glaciers terminating in very shallow water. In such situations, the glacier margin melts more rapidly than it calves, resulting in a gently sloping front. Deposition from meltstreams or mass flows produces ice-contact subaqueous fans that pass distally into turbidites. *Facies association IV* forms at fjord margins that receive sediment from a land-based glacier margin. Glacier-fed deltas are produced where glacial meltstreams enter the fjord, and deltaic bottomsets, foresets and topsets prograde over older facies. These four associations can develop in sequence during glacier retreat into shallowing water, or alternate if the retreating margin migrates through a series of basins (Fig. 12.55).

Boulton (1990) drew attention to the influence of fjord relief on the geometry of sedimentary infills

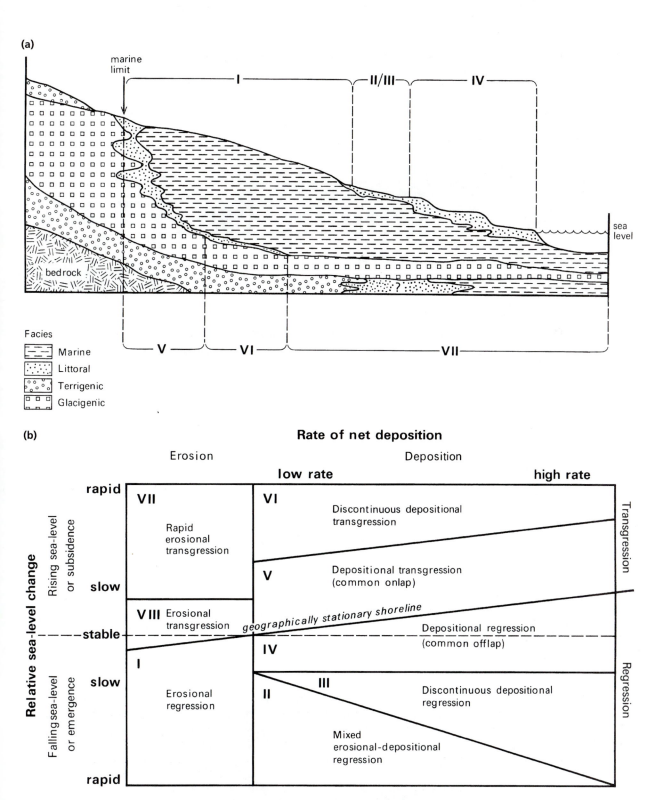

Fig. 12.52 (a) A reconstruction of the depositional sequence in an Arctic fjord during full-glacial conditions, based upon Clements Markham Inlet, Arctic Canada. The stratigraphies numbered I–VII document the depositional sequence produced during a glacial–deglacial cycle at various locations. (b) Classification of transgressions and regressions based upon the work of Curray (1964). (From Bednarski, 1988. Reprinted by permission of Les Presses de l'Université de Montréal)

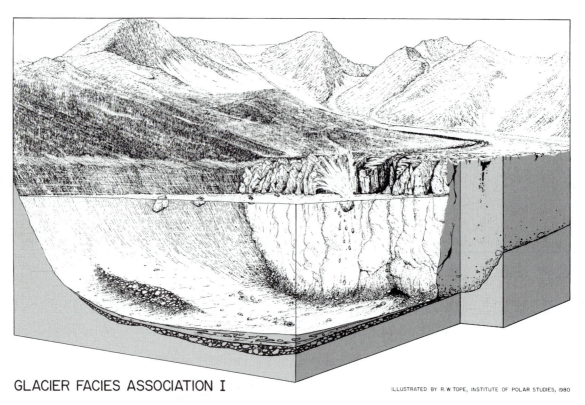

GLACIER FACIES ASSOCIATION I

ILLUSTRATED BY R.W.TOPE, INSTITUTE OF POLAR STUDIES, 1980

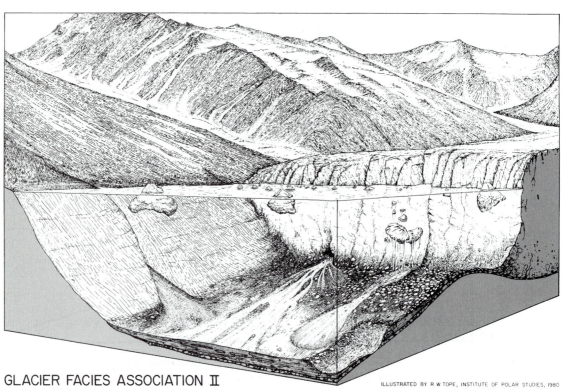

GLACIER FACIES ASSOCIATION II

ILLUSTRATED BY R.W.TOPE, INSTITUTE OF POLAR STUDIES, 1980

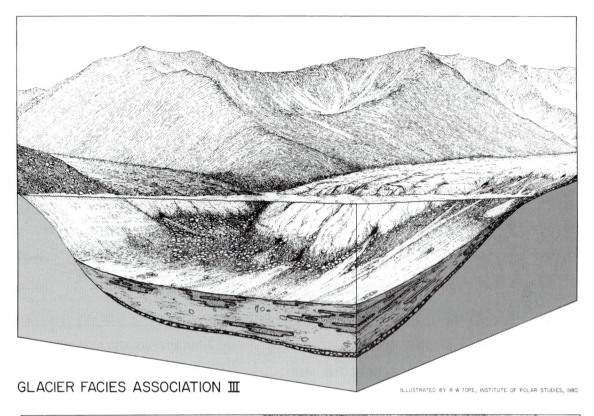

GLACIER FACIES ASSOCIATION III

ILLUSTRATED BY R.W.TOPE, INSTITUTE OF POLAR STUDIES, 1980

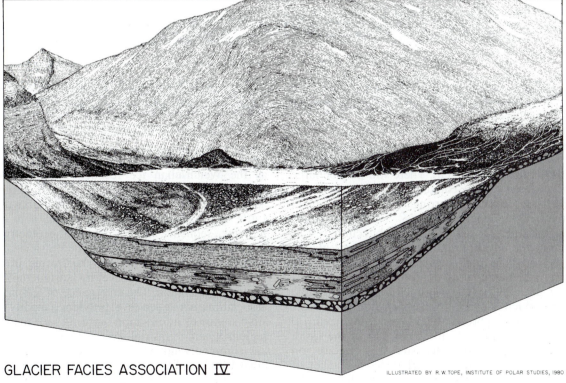

GLACIER FACIES ASSOCIATION IV

ILLUSTRATED BY R.W.TOPE, INSTITUTE OF POLAR STUDIES, 1980

Fig. 12.54 Depositional models for retreating tidewater glaciers based on Alaskan examples. I: rapidly retreating glacier in deep water; II: slowly retreating glacier in shallow water; III: slowly retreating glacier in very shallow water; IV: terrestrial glacier margin supplying sediment to a delta. (Drawings by R.W. Tope, from Powell, 1981a. Reproduced by permission of the International Glaciological Society)

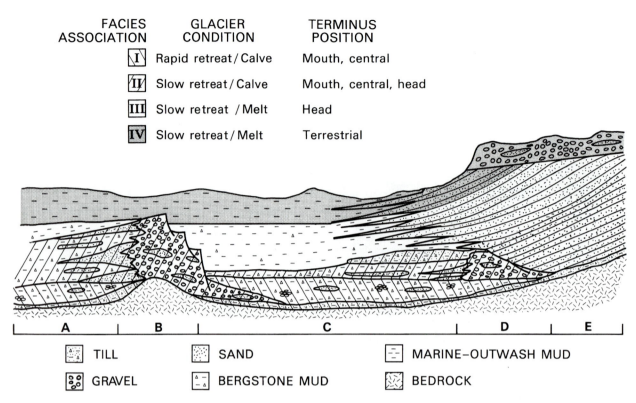

FACIES ASSOCIATION | GLACIER CONDITION | TERMINUS POSITION

I	Rapid retreat / Calve	Mouth, central
II	Slow retreat / Calve	Mouth, central, head
III	Slow retreat / Melt	Head
IV	Slow retreat / Melt	Terrestrial

TILL   SAND   MARINE–OUTWASH MUD

GRAVEL   BERGSTONE MUD   BEDROCK

Fig. 12.55 Hypothetical section through glacimarine sediment stratigraphy associated with retreating glacier snout. The depositional environments of Facies Associations I–IV are depicted in Fig. 12.54. At interval A the ice front terminated in deep water and retreated rapidly by calving. At interval B calving continued but recession was slowed by a channel constriction. At interval C the ice once again calved rapidly in deep water. At interval D the ice had reached the fjord head and recession slowed; calving was replaced by surface melting. At interval E the ice front became terrestrial and an outwash delta prograded out over all previous facies. (From Powell 1981a. Reprinted by permission of the International Glaciological Society)

(Fig. 12.56). In high-relief fjords, slumping from steep margins relocates sediment on the trough floor, producing flat-lying *infill sequences* such as that in Cambridge Fiord, Baffin Island. In low-relief fjord environments, sediment cover is more widely distributed, forming an extensive *draped sequence* (e.g. Kongsfjorden, Spitsbergen). There still may be large local differences in sediment thickness, with the thickest sequences occurring close to sediment sources and in topographic lows.

An example of stratigraphic architecture resulting from the advance and retreat of a fjord glacier is given in Fig. 12.57, which shows the results of geophysical studies in Krossfjorden, Svalbard (Sexton *et al.*, 1992). The fjord consists of multiple basins, which clearly influence patterns of sedimentation. Three broad zones were recognized: (a) *Outer fjord complex*. In the outer part of the fjord, hummocky glacigenic deposits lie directly on bedrock, and in turn are overlain by a blanket of massive and laminated sediment. The hummocky deposits were laid down during the late Weichselian glacial maximum, and the overlying sediments were deposited during subsequent retreat. (b)

Fan complex. This occurs in an elongated upper basin separated from the outer fjord complex by a bedrock sill. The lowermost sedimentary unit consists of a massive diamicton sheet, possibly basal till, which is overlain by laminated silt and sand. Much of the laminated component was deposited as the Weichselian ice receded upfjord. (c) *Ice-proximal unit*. This comprises a series of sub-aquatic moraines formed during and following the Little Ice Age maximum. The distal part of the moraine sequence has a fan-like geometry, and oversteps the older 'fan complex'.

Glacimarine sediments and landforms in fjords may be raised above sea-level by isostatic uplift following deglaciation. The emergence of ice-proximal associations above sea-level provides a readily accessible source of information on former ice margins in regions where terrestrial ice-marginal accumulations are commonly not well preserved. Emergent morainal banks and grounding-line fans occur in most Norwegian fjords, where they are referred to as *Tronder moraines* (Sollid and Reite, 1983). Larsen *et al.* (1991) mapped the distribution of emergent De Geer moraines in the Møre area of western Norway,

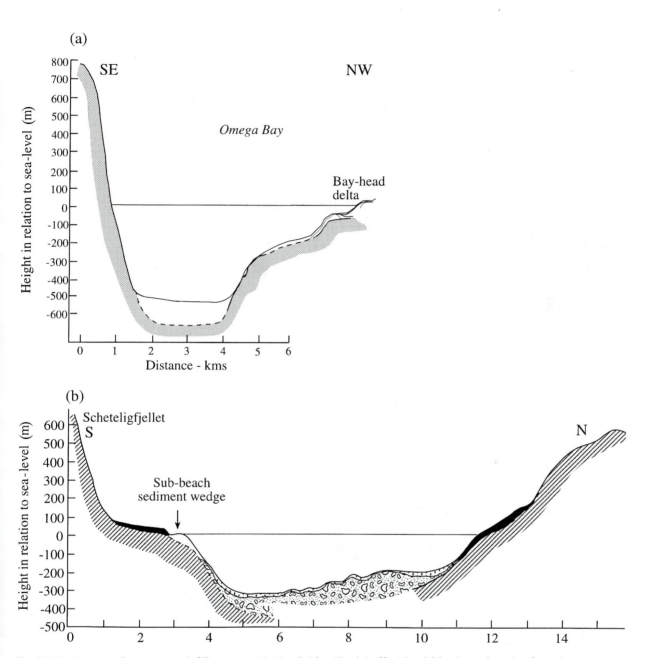

Fig. 12.56 Contrasts between an 'infill sequence' in Cambridge Fiord, Baffin Island (a), where slumping from the steep margins produces a flat-bottomed, slump and turbidite infill sequence, and a 'draped sequence' in Kongsfjorden, Spitsbergen (b), where the fjord slopes are much lower; till (coarse pattern) overlies bedrock and is in turn draped by marine sediments. (Modified from Boulton, 1990)

and showed that they record a very detailed picture of deglaciation. Glacier retreat rates varied between fjords, as a consequence of differences in topography and ice discharge.

Emergent glacimarine sediments and landforms are subject to modification by waves, currents and icebergs in the near-shore zone (Hambrey, 1994). Raised beaches and deltas are formed where there is abundant sediment supply, and spectacular flights of

terraces occur along the margins of some isostatically raised fjord coastlines. The altitude of raised shorelines provides important data on the history of sea-level change and glacier fluctuations (Section 1.5). In fjords, the marine limit commonly declines towards the ice accumulation centre, recording the occupancy of inner basins by glacier ice during isostatic uplift of deglaciated outer coasts (e.g. Sissons and Dawson, 1981).

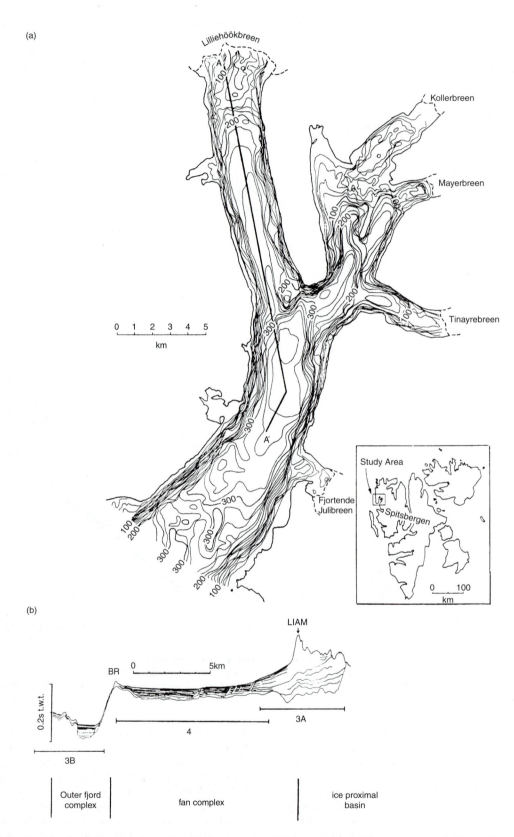

Fig. 12.57 Fjord basin architecture from Krossfjorden, Svalbad. The acoustic profile (b) is locked on the bathymetric map (a). LIAM = Little Ice Age moraine; BR = bedrock ridge. (From Sexton *et al.*, 1992. Reproduced by permission of Elsevier)

The lowest-relief fjord landsystems are those of inter-island channels. Some of the best-preserved glacial landforms associated with inter-island channels are those of Lancaster Sound and Bylot Island in the Canadian Arctic archipelago. The outer margins of ice streams at the north-eastern margin of the Laurentide ice sheet are clearly demarcated on Bylot Island by the Eclipse moraines (Klassen and Fisher, 1988; Klassen, 1993). These moraines show that Bylot Island was surrounded by ice streams in its neighbouring channels and that the local mountain glaciers were overwhelmed by Laurentide ice at that time; the strength of the regional ice was such that it back-filled bays and valleys along the Bylot Island coast (Fig. 12.58). A similar situation arose during the last glaciation of south-eastern Ellesmere Island, where outlet glaciers from the Prince of Wales ice-field advanced into local fjords and floated to form calving ice shelves. The former margins of these glaciers are marked by subhorizontal lateral moraines, such as those in Hayes Fiord and Flagler Bay, which terminate in ice-contact glacimarine deltas. These fjord glaciers and their moraines also dammed small lakes in the low cols on the plateaux separating the fjords, resulting in the deposition of ice-contact glacilacustrine deltas.

12.5.3 Continental shelves

As described in Section 1.2.1, glaciation of shelf areas may arise through (a) the advance of terrestrial margins in a seaward direction; (b) the thickening and grounding of sea-ice encouraged by lower sea-level; or (c) a combination of (a) and (b). The impact of glaciation on a continental shelf may be *direct* where the shelf has hosted a grounded glacier mass, or *indirect* where it merely received distal glacimarine sediments during a glacial cycle. Patterns of sedimentation are dependent on a number of interrelated factors, including the extent and configuration of glacier ice, basin history and tectonic setting, bathymetry, oceanography, glacial thermal regime, and sediment supply (e.g. Powell, 1984; King *et al.*, 1987; C.H. Eyles and Lagoe, 1990; Hambrey *et al.*, 1992; N. Eyles, 1993).

The influence of ice margin configuration and thermal regime on patterns of sedimentation on continental shelves was considered by Powell (1984), who defined eight glacimarine regimes, based on whether: (a) the ice source is a valley glacier or ice sheet; (b) ice at the grounding line is cold- or wet-based, or whether net freezing or melting occurs; and (c) the glacier front is a tidewater cliff or ice shelf (Fig. 12.59). The regimes can be summarized as follows:

Regimes 1 and *3* occur in polar environments where glaciers and ice sheets, respectively, with melting/freezing bases form ice shelves. Sea-ice traps tabular icebergs close to the glacier margin, limiting the transport of ice-rafted debris. The glaciers in regime 1 tend to carry more englacial debris, resulting in greater amounts of rain-out, whereas the larger drainage basins involved in regime 3 encourage greater subglacial stream discharges and glacifluvial sediment transport.

Regimes 2 and *4* are typical of polar glaciers and ice sheets, respectively, which are frozen to their beds and form ice shelves. Tabular icebergs are again trapped near to the glacier ice margin by sea-ice. Overall, debris is less abundant than in regimes 1 and 3, deposition rates are low, and sedimentation is interrupted by erosional episodes resulting in palimpsest lags (Section 10.7; Powell *et al.*, 1996).

Regime 5 occurs where a valley or outlet glacier with a melting base ends as a tidewater front. These glaciers transport large amounts of debris to the marine environment. Sea-ice does not restrict iceberg drift, and so ice-rafted debris is more widely distributed. Morainal banks are more susceptible to disturbance by icebergs and waves than in ice shelf regimes and therefore contain less diamicton and more gravel. In addition, subglacial discharges produce subaqueous fans. *Regime 7* represents an ice sheet with a melting/freezing base and a tidewater front. Deposits are similar to those in regime 5 except that less high-level englacial debris is available.

Regimes 6 and *8* refer to cold-based glaciers and ice sheets, respectively, with tidewater fronts. These contribute far less sediment to the marine environment than the other regimes. Lithofacies in regime 6 are similar to those of regime 2, although more sediment is released close to the grounding line. Regime 8 contributes almost no sediment to the marine environment, and palimpsest lags are common.

Typical lithofacies associations formed in regimes 1–8 are shown in Fig. 12.60 and Table 12.2.

Recently, it has become apparent that the tectonic setting exerts a strong influence on the style and preservation of glacigenic sediments on continental shelves (N. Eyles, 1993; Hambrey, 1994). The tectonic setting of continental shelves can be subdivided into (a) passive margins, (b) convergent margins, where oceanic crust is subducted below continental crust, and (c) rift basins, where the crust is under tension and the basin floor subsides. We discuss examples of glacimarine sequences in each type of setting below, with reference to ice mass configuration, thermal regime and other controlling variables.

A well-studied example of a glacially influenced *passive continental margin* is the eastern Canadian shelf, extending from the Scotian Shelf off Nova Scotia to Baffin Bay (Figs 12.61 and 12.62; Piper and Normark, 1989; Josenhans and Fader, 1989). The physiography of the shelf records repeated episodes

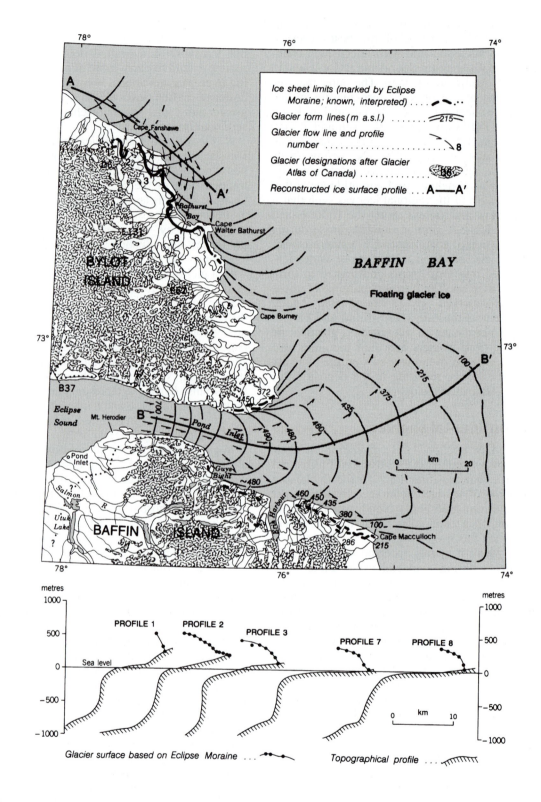

(a)

Glacier surface based on Eclipse Moraine ... •—•—•   Topographical profile ... ⊤⊤⊤⊤⊤

Fig. 12.58

(b)

Fig. 12.58 (a) The extent of the Eclipse moraine on Bylot and Baffin Islands, Arctic Canada, and the reconstructed glacier surface profiles based upon the Eclipse moraine on valley walls (profiles 1–3 and 7–8) and drawn along the centre-lines of the two glacier lobes (A-A′ and B-B′). (From Klassen, 1993. Reprinted by permission of the *Canadian Journal of Earth Sciences*) (b) Aerial photograph of the Cape Fanshawe area, Bylot Island, showing the extent of valley backfilling by foreign (Laurentide ice sheet) glacier ice. The limits of the Eclipse moraine and drift are marked by the broken line. (Aerial photograph reproduced by permission of the Department of Energy, Mines and Resources, Canada)

of erosion under both glacial and non-glacial conditions, including Tertiary fluvial incision, moderate glacial scouring, and intense selective linear erosion. Ice-rafted debris first appears off eastern Canada in sediments dating to the late Pliocene (*c.* 3 Myr BP), but despite this long record of glaciation the quantity of Quaternary sediment is meagre, amounting to only a few tens of metres in many areas and thickening to *c.* 200 m in the offshore extensions of fjord basins. The limited amount of sediment reflects glacial, marine and subaerial erosion of the shelf, encouraged by low glacial sea-levels, which has severely limited the survival of sediment over multiple glacial cycles. It is thought that large amounts of sediment have been evacuated from the shelf by grounded glaciers on several occasions, causing sediment to accumulate on the continental slope by a process of progradation or 'outbuilding' (N. Eyles, 1993). According

to Josenhans and Fader (1989), the uppermost glacial–postglacial sequence on the eastern Canadian Shelf thins towards the north. Off Nova Scotia and Newfoundland, former terminus positions of the Laurentide ice sheet are marked by large submerged moraines and thick diamicton sequences which interfinger with distal glacimarine deposits (Fig. 12.62). Fields of lift-off moraines are common within basins inside the ice sheet limits. Further north, off the coast of Labrador, sediment cover is thinner, and till tongues (grounding-line wedges) are not evident. This contrast is thought to result from differences in glacier thermal regime and ice dynamics: in the northern part of the shelf, deposition rates may have been limited by low sediment supply owing to colder conditions, and rapid ice sheet retreat. Much of the continental shelf off eastern Canada is traversed by iceberg scour marks (Section

Table 12.2 Lithofacies occurring in glacimarine regimes 1–7 of Powell (1984)

Facies	Regimes							
	1	2	3	4	5	6	7	8
Subglacial till	Gc	Gu-r	Gc	Gr	Gc	Gu-r	Gc	Gr
Dropstone diamicton	Sc		Sc-u		BPc		BPc	
Stratified dropstone diamicton	Sc		Su		BPc-u			
Dropstone mud	Sc, BDc	Su, Bu			BPu, BDc	BPr	Bc	
Stratified dropstone mud			Su-r		BDu-r		BPc	
Grounding line sediment/ morainal bank	Pc	Pu-r	Pc	Pr	Pc	Pu-r	Pc	Pr
Subaqueous outwash fan	Pu		Pu		Pc		Pc	
Interlaminated sand/silt/clay	Pu		Pu-r		BPc-u			
Subaqueous sediment gravity flow deposits	Pc, Sc, Bc	Pu, BPr*	Pc, Sc, Bc	BPr*	Pc, Bc	BPr*	Pc, Bc	BPr*
Interstratified gravity flow deposits and dropstone diamicton	Pc*		Pc*					
Interstratified gravity flow deposits and dropstone mud					Pc*			
Interstratified gravity flow deposits and dropstone diamicton and mud							BPc*	
Palimpsest lag		Bc-u	Bc-u	Sc, Bc				Bc
Biogenic bank and/or debris	BDu	Br	BDu		BPc-u		BDr	
Deep-sea biogenic mud (ooze)			BDu	BDu-r		BDu-r		BDr
Dropstone ooze	BDc	BDr						
Interstratified dropstone mud and ooze					BDc		BDc	

Sedimentation zones: G, subglacial; P, ice-proximal; S, ice shelf; B, iceberg; BP, iceberg-proximal; BD, iceberg-distal
Frequency of occurrence: c, common (large, thick); u, uncommon (small, thin); r, rare
* = facies occur when the grounding line is at the edge of the continental shelf

10.6.4). Iceberg scours can be attributed to different glacial stages, providing an important source of environmental information. For example, modern iceberg scours are abundant off Labrador, but only relict scours occur off Nova Scotia, recording a more southerly extension of iceberg drifting under full-glacial conditions.

Another example of a glacially influenced passive continental margin is the Barents Shelf between Spitsbergen and mainland Scandinavia. Geophysical studies in this area have revealed striking evidence for a Weichselian marine-based ice sheet which either formed on the shelf or flowed southwards from Spitsbergen (Vorren *et al.*, 1989, 1990; Solheim *et*

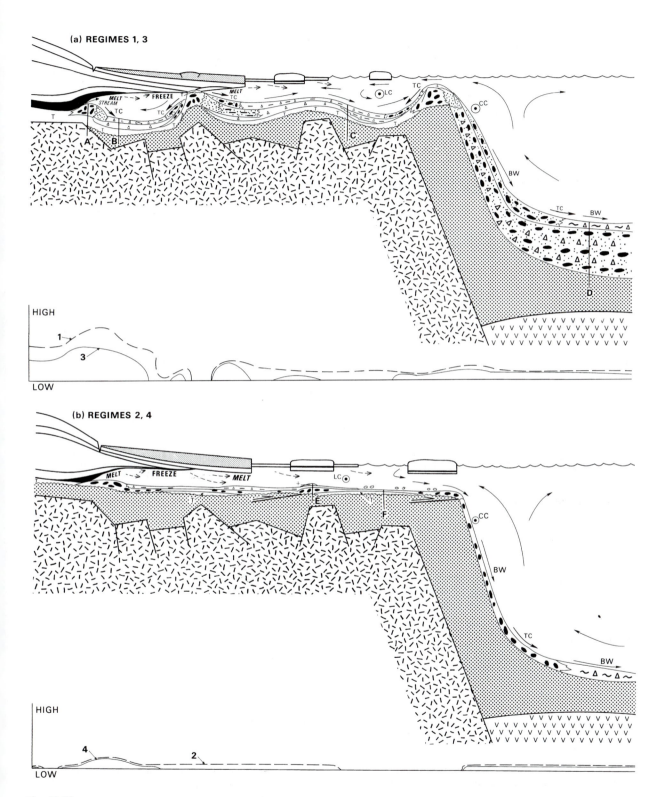

Fig. 12.59

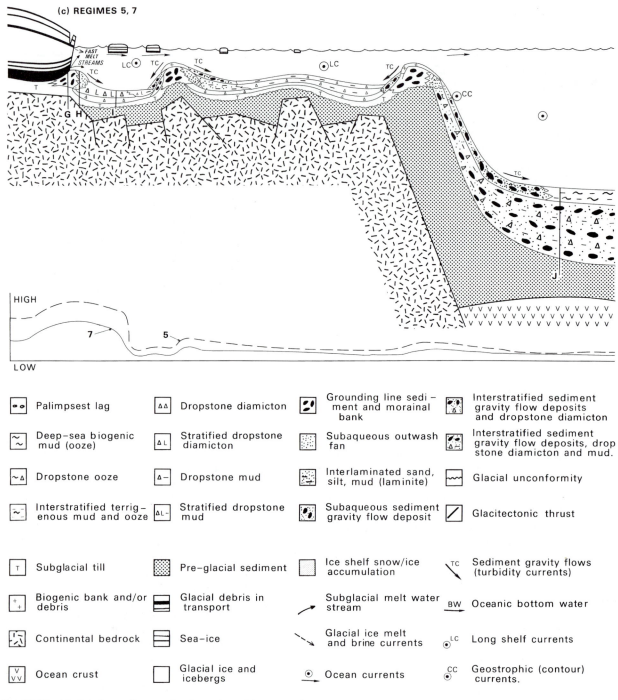

(c) REGIMES 5, 7

⊙⊙ Palimpsest lag	ΔΔ Dropstone diamicton	Grounding line sediment and morainal bank	Interstratified sediment gravity flow deposits and dropstone diamicton
∼ Deep-sea biogenic mud (ooze)	ΔL Stratified dropstone diamicton	Subaqueous outwash fan	Interstratified sediment gravity flow deposits, dropstone diamicton and mud.
∼Δ Dropstone ooze	Δ– Dropstone mud	Interlaminated sand, silt, mud (laminite)	Glacial unconformity
∼ˍ Interstratified terrigenous mud and ooze	ΔL– Stratified dropstone mud	Subaqueous sediment gravity flow deposit	Glacitectonic thrust
T Subglacial till	Pre-glacial sediment	Ice shelf snow/ice accumulation	TC Sediment gravity flows (turbidity currents)
+ + Biogenic bank and/or debris	Glacial debris in transport	Subglacial melt water stream	BW Oceanic bottom water
Continental bedrock	Sea-ice	Glacial ice melt and brine currents	⊙LC Long shelf currents
V VV Ocean crust	Glacial ice and icebergs	⊙ Ocean currents	CC Geostrophic (contour) currents.

Fig. 12.59 Models for sedimentation in various tidewater glacier and ice shelf settings: (a) regimes 1 (outlet glacier) and 3 (ice sheet) with melting/freezing base ending as an ice shelf or floating glacier tongue; (b) regimes 2 (valley or outlet glacier) and 4 (ice sheet) with frozen base ending as an ice shelf or floating glacier tongue; (c) regimes 5 (valley or outlet glacier) and 7 (ice sheet) with a melting/freezing base ending as a tidewater front; (d) regimes 6 (valley or outlet glacier) and 8 (ice sheet) with frozen base ending as a tidewater front. Lines A–J show the location of the vertical profiles in Fig. 12.60. Curves show relative rates of sedimentation for each regime. (From Powell, 1984. Reprinted by permission of Elsevier)

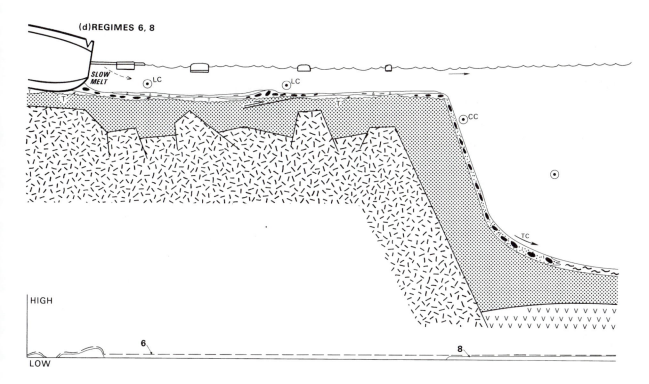

Fig. 12.59(d)

al., 1990). Fields of flutings have been identified by side-scan sonar on the north-central Barents Shelf in up to 340 m of water, recording the movement of grounded, wet-based ice over weak, unconsolidated sediment (Solheim *et al.*, 1990). In many places, the flutings are cut across by De Geer moraines and iceberg plough marks, but otherwise they have undergone little modification, suggesting that deglaciation was rapid, possibly by ice sheet destabilization and catastrophic calving retreat. The stratigraphic architecture of the glacigenic succession on the Barents shelf is complex (Vorren *et al.*, 1989, 1990). In places, stratified clinoforms are overlain by extensive horizontal diamicton sheets, possibly recording the progradation of grounding-line wedges and subsequent overriding by grounded ice (Section 11.6.3). Elsewhere on the shelf, there is widespread evidence for subglacial erosion, including a vast areal scour zone through which glacier ice was funnelled to the western shelf edge. At its maximum, the ice sheet extended to the shelf break, where large amounts of proximal glacimarine sediment were deposited by sliding and other gravitational processes. The thickest sediment sequences are located on the shelf slope immediately below local troughs which acted as efficient pathways for the transport of debris.

A 5 km thick sequence of glacimarine sediment deposited at a *convergent plate margin* is preserved in the Yakataga Formation in the Gulf of Alaska. The Pacific Plate is being subducted northwards below the North American continental plate, producing a structurally complex forearc basin (Fig. 12.63). Temperate glaciers flowing from the rapidly uplifting coastal mountains have delivered large volumes of fine-grained sediments to the shelf (Powell and Molnia, 1989; C.H. Eyles and Lagoe, 1990; N. Eyles, 1993; Powell and Domack, 1995). The Yakataga Formation records the rapid infilling of the forearc basin on the landward side of an *accretionary wedge* (C.H. N. Eyles *et al.*, 1991; Eyles, 1993). Basin infilling has resulted in a shallowing-upward succession from deep-water debris flows and turbidites to glacially influenced shallow marine facies deposited in a 'ponded' basin. Rapid uplift has raised parts of the Yakataga Formation above sea-level, and impressive sections can be examined on the south coast of Alaska and adjacent offshore islands, most notably Middleton Island, where exceptionally well-exposed successions have been dated to 1.8–0.7 Myr BP (Plafker and Addicott, 1976; C.H. Eyles and Lagoe, 1990). The succession on Middleton Island consists mainly of massive and stratified diamicton units with associated boulder pavements and coquinas. Channel fills of gravelly sand are common at the base of the succession, and extensive blankets of mud occur at the top. C.H. Eyles and Lagoe (1990) argued

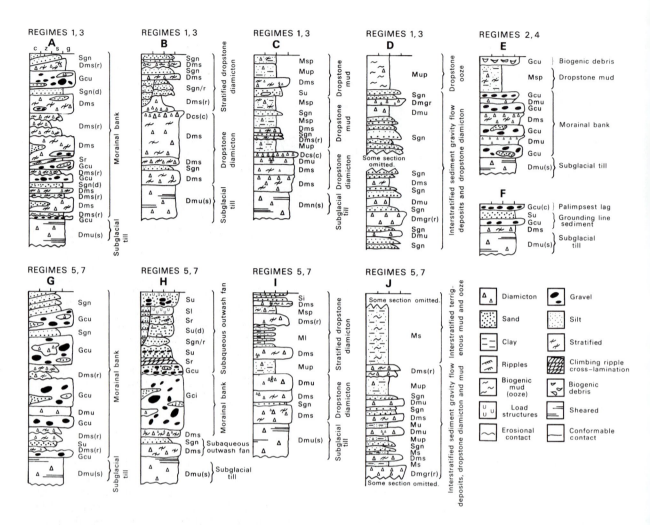

Fig. 12.60 Hypothetical facies sequences for sections located in Figs 12.59a–d. D, diamicton; G, gravel; S, sand; M, mud; u, unstratified; gn, normally graded; gr, reversely graded; i, imbricate; and p, pebbly. (From Powell, 1984. Reproduced by permission of Elsevier)

that the sediments record changing depositional environments during full-glacial to interglacial conditions, and proposed a threefold interpretive sequence (Fig. 12.64):

1. *Full-glacial stage.* Striated boulder pavements indicate that grounded or partially grounded ice extended to the shelf edge, although apparently no basal tills are exposed on Middleton Island. Beyond the ice front, channel-fill associations were deposited on the continental slope, possibly fed by meltwater debouching from the glacier.
2. *Intermediate ice cover.* Extensive muddy diamicton units were deposited on the shelf, recording deposition from icebergs and the settling of suspended sediment. The abundance of ice-rafted debris suggests that many coastal glaciers reached

tidewater, but did not extend far on to the shelf. At present, large quantities of suspended sediment are carried into the gulf by glacial meltstreams, and accumulation rates may have been even higher when ice cover was more extensive. Sedimentation rates were variable, however, and coquinas located on topographic highs indicate that such areas were starved of sediment.

3. *Ice minima.* Extensive mud drapes with few dropstones record deposition during interglacial conditions, when sediment was delivered mainly by turbid rivers discharging into the gulf. The restricted amount of ice-rafted debris in these sediments indicates that few glaciers had tidewater margins. Such conditions are similar to those of today and may have been typical of interglacial periods in this area.

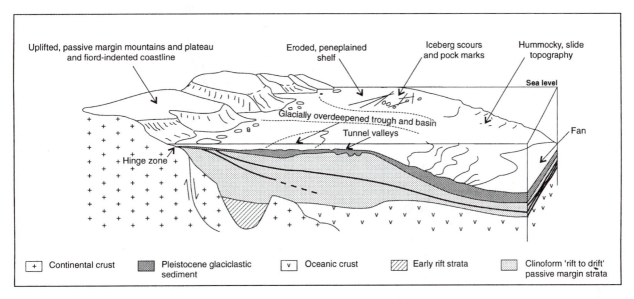

Fig. 12.61 Model of a glaciated passive plate margin based upon the eastern Canadian and north-west European continental shelves. In the absence of marked subsidence, the accumulation of thick Pleistocene glacigenic sediments is restricted. (Modified from N. Eyles, 1993)

Subsiding rift basins contain the thickest sequences of glacimarine sediments. For example, the Ross and Weddell Sea basins off Antarctica form parts of a single rift system that contains a record of glaciation extending back to the Early Oligocene (*c.* 36 Myr BP). A drill core in the McMurdo Sound, Ross Sea, reached the base of a 700 m thick glacial section which contains a rich record of glacial fluctuations and environmental change (Barrett, 1989). The lower part of the succession is dominated by turbidites and thin diamictites, and represents the infill of a deep basin adjacent to the uplifting Transantarctic Mountains, whereas the upper part is predominantly diamictic and records repeated advances and recessions of grounded ice across the continental shelf. N. Eyles (1993) has made the interesting suggestion that the presence of fast-flowing ice streams draining into the Ross Embayment (including the intensively studied Ice Stream B) reflects structural controls in this rifted basin. Another example of a glacially influenced subsiding basin is the North Sea, where the Pleistocene sequence reaches a maximum thickness of *c.* 920 m in the central graben (N. Eyles, 1993; Hambrey, 1994).

Much of the Earth's pre-Cenozoic glacial record has been interpreted as glacimarine, and regional studies of facies variation have been used to reconstruct the palaeogeography of ancient glaciations. A thorough and readable summary of the literature has been provided by N. Eyles (1993).

12.5.4 Large proglacial lakes and epicontinental seas

During deglaciation of the Pleistocene ice sheets, ice margins were commonly in contact with large water bodies, either proglacial lakes ponded between the ice and topographic barriers, or arms of the sea which flooded isostatically depressed and glacially overdeepened parts of continental interiors (Fig. 12.48 (Plate 28)); Section 1.7.2.5; Teller, 1987; Dawson, 1992). The extent and evolution of many of these water bodies have been reconstructed from the sedimentary record, in combination with shoreline evidence, pollen and microfaunal records, and dating programmes (e.g. Hyvärinen and Eronen, 1979; Eronen, 1983; Bjorck and Digerfeldt, 1986; Teller, 1987, 1995; Lundqvist, 1989c). This research has shown that the presence of large proglacial water bodies encouraged rapid ice-margin retreat by calving, and may have triggered some episodes of ice advance or surges by encouraging low effective pressures below the glacier margin (Lundqvist, 1989c; Fyfe, 1990). It is therefore important to understand the sedimentary record of proglacial lakes and epicontinental seas in order to distinguish climatic and non-climatic controls on ice sheet behaviour.

Depositional systems in such settings are similar to those of continental shelves (Ashley, 1995), although there are some important differences:

1. Fresh lake water and brackish shallow marine conditions reduce the importance of suspended-sediment transport in overflows and interflows, so

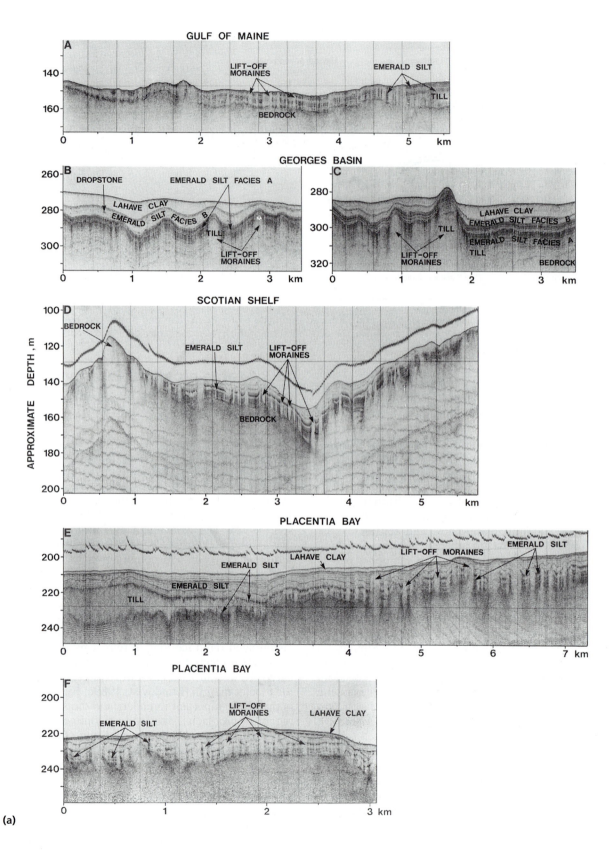

Fig. 12.62

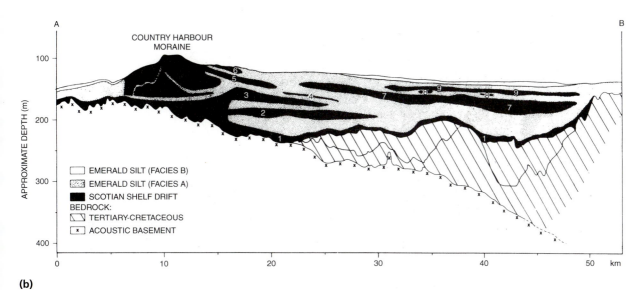

(b)

Fig. 12.62 (a) Huntec DTS profiles of the Scotian Shelf, eastern Canada, showing lift-off (De Geer) moraines overlain by distal glacimarine silts. (from King and Fader, 1986) (b) Interpretive diagram of seismic profiles from part of the Scotian Shelf, showing a major moraine complex and till tongues interbedded with glacimarine silts. (From Josenhans and Fader, 1989. Reprinted by permission of the Geological Survey of Canada)

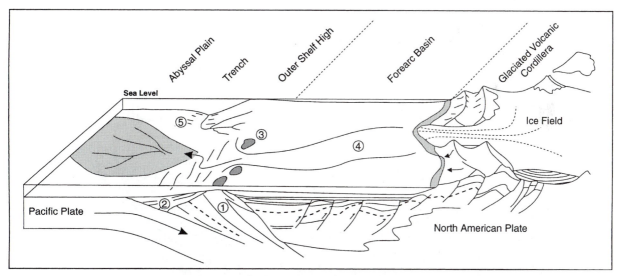

Fig. 12.63 Model of a glaciated active plate margin based upon the Gulf of Alaska basin. Note the tectonic disturbance of a thick Pleistocene stratigraphy, forming an accretionary wedge and an infilled forearc basin. 1, accretionary wedge; 2, slope basins in areas of compressional ridging; 3, outer shelf high above accretionary wedge; 4, shallow marine trough crossing shelf; 5, dendritic gully on outer slope.(Modified from N. Eyles, 1993)

that underflow deposits are better represented in ice-proximal and distal systems (Section 8.3).

2. In lakes, the seasonal alternation between deposition by underflows in summer and settling of clay particles in winter produces annual varves (Section 10.6.1). Varves form extensive blankets in parts of North America, Europe and Scandinavia, and record annual cycles of sedimentation in proglacial lakes during deglaciation (De Geer, 1912; Sauramo, 1923; Ashley, 1975; Brodzi-

kowski and van Loon, 1991). When studied in combination with moraine sequences and biostratigraphic data, varves can yield very detailed chronologies, allowing deglaciation histories to be constructed at a resolution not often possible for terrestrial settings.

3. Proglacial water bodies are prone to sudden changes in water level, as a result of the opening and closing of outlets by glacier margin fluctuations and periodic jökulhlaups and recharge events

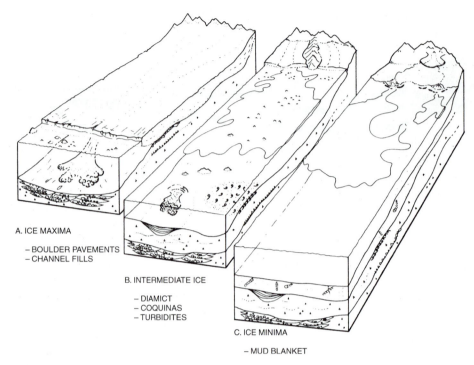

A. ICE MAXIMA

– BOULDER PAVEMENTS
– CHANNEL FILLS

B. INTERMEDIATE ICE

– DIAMICT
– COQUINAS
– TURBIDITES

C. ICE MINIMA

– MUD BLANKET

Fig. 12.64 Model of glacimarine sedimentation on the continental shelf, Gulf of Alaska, based upon the stratigraphic succession on Middleton Island: (A) maximum ice cover; (B) intermediate ice cover; (C) minimal ice cover (similar to the present day). (From C.H. Eyles and Lagoe, 1990. Reproduced by permission of the Geological Society of London)

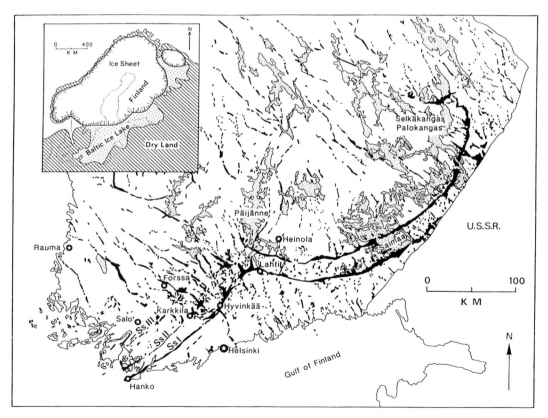

Fig. 12.65 The Salpausselkä I, II and III moraines of southern Finland and associated eskers. Inset: reconstructed palaeogeography at the time of their deposition. (From Fyfe, 1990, after Eronen, 1983. Reprinted by permission of Scandinavian University Press)

(Section 3.5). Cyclic water level changes can be recorded by discrete sediment packages bounded above and below by erosion surfaces (Section 12.5.1; Fig. 12.2), and can add considerable complexity to stratigraphic sequences associated with glacier fluctuations (Mastalerz, 1990; Martini and Brookfield, 1995).

4. Proglacial lakes and epicontinental seas can receive substantial amounts of sediment from recently deglaciated land-surfaces surrounding the basin as well as from the glacier margin (Martini and Brookfield, 1995). In lakes and arms of the sea, even ice-distal parts of the basin can receive coarse river-borne sediment. Sediment sequences at basin margins can consist of interfingering assemblages derived directly from the ice and from the deglaciated area, and may contain a combined record of water level changes and ice margin fluctuations.

In southern Scandinavia and North America, some major moraine systems were deposited at the margins of proglacial lakes and epicontinental seas during pauses in the retreat or readvances of the last ice sheets (e.g. Falconer *et al.*, 1965; Fogelberg, 1970; Hillaire-Marcel *et al.*, 1981; Aylsworth and Shilts, 1989b; Fyfe, 1990; Brandal and Heder, 1991). Some of the moraine systems record climatically controlled oscillations of the ice margin. For example, the Salpausselkä moraines in southern Finland delimit the margins of a large ice lobe that advanced into the deep waters of the Baltic Ice Lake during the Younger Dryas (Fig. 12.65; Lundqvist, 1989c; Fyfe, 1990; Fastook and Holmlund, 1994; Section 1.7.2.4). The Salpausselkä moraines are composed almost entirely of ice-contact glacifluvial and glacilacustrine sediments deposited in a variety of water depths. Fyfe (1990) showed that the depositional systems making up the outermost moraine (Salpausselkä I) show systematic relationships with water depth and the inferred hydrology of the ice sheet margin. Where the ice terminated in shallow water, the moraine consists of discrete grounding-line fans or deltas connected to large esker networks, recording sediment delivery via conduits (R-channels, Section 3.4.2). Sedimentation was focused around channel exits, allowing deposits to build up to water level in places. In contrast, where the ice terminated in deeper water (e.g. 40–100 m, south-west of Hyvinkää, Fig. 12.65), the moraine forms a narrow, nearly continuous ridge, and feeder esker systems are absent. Sediment was delivered to the ice margin at many points by a distributed subglacial drainage system, possibly a braided canal network. Fyfe (1990) argued that distributed drainage systems developed below the ice lobe where submarginal water pressures were increased by deep water at the terminus. Spatial differences in subglacial drainage may have influenced ice dynamics, and

caused local variations in ice margin behaviour which were unrelated to climatic forcing (Lundqvist, 1989c; Sharp, 1992; Fastook and Holmlund, 1994).

Some moraine systems deposited during pauses in the retreat of the Pleistocene ice sheets appear to have been controlled by spatial and temporal changes in water level rather than climatic events. For example, moraines such as the Lac Daigle–Manitou–Matamek moraine, the Roulier moraine, and the 600 km long St Narcisse moraine in eastern Canada are located on topographic highs where the retreating Wisconsinan ice margin became temporarily anchored (Hillaire-Marcel and Occhietti, 1980). Other moraines apparently formed when ice margins stabilized following a drop in water level. For example, the 500 km long Sakami moraine in Quebec documents subaqueous deposition in response to changing water depths along the western margin of the receding Labrador sector of the Laurentide ice sheet (Fig. 12.66; Fig. 1.60 (Plate 9)). As the Hudson Bay and Labrador sectors became separated at approximately 8.1 kyr BP, they began calving into proglacial Lake Ojibway (Dyke and Prest, 1987; Vincent, 1989). Water depths were suddenly reduced when, at approximately 8 kyr BP, the Hudson Bay sector receded enough to allow Lake Ojibway to drain northwards into the Tyrrell Sea, which was the precursor to the present Hudson Bay. The new Tyrrell Sea depths were some 200 m shallower than those of Lake Ojibway, and the sudden drop in water level apparently caused a halt in the retreat of the western margin of the Labrador sector, allowing the Sakami moraine to build up while the ice sheet profile readjusted to the new situation. Hillaire-Marcel *et al.* (1981) proposed the term *re-equilibration moraines* for moraines formed in such circumstances. Extensive closely spaced, smaller moraines occur on both sides of the Sakami moraine, documenting active retreat at rates of 173–239 m yr^{-1} in Lake Ojibway and then at 200 m yr^{-1} in the Tyrrell Sea (Mawdsley, 1936; Norman, 1938; Shaw, 1944; Ignatius, 1958; Vincent, 1977, 1989).

Spatial and temporal relationships between depositional systems show that some Pleistocene ice lobes underwent rapid retreat by calving into deep lakes or arms of the sea. As we saw in Section 8.2.5, catastrophic retreat is likely where thinning ice occupies an overdeepened basin, and the ice margin retreats into deepening water. Evidence for rapid ice sheet retreat and marine flooding in the Irish Sea basin has been presented by N. Eyles and McCabe (1989a, 1991). Complex assemblages of glacigenic sediments and landforms occur on the floor of the Irish Sea and on the adjacent land surfaces, formed during the advance and retreat of the Late Devensian British ice sheet. According to Eyles and McCabe, the glacial stratigraphy of the basin can be resolved into *sub-*

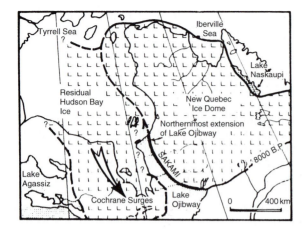

(a)

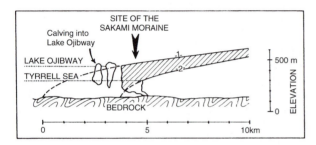

(b)

Fig. 12.66 (a) Location of the Sakami moraine and contemporaneous ice margins; (b) schematic ice margin behaviour during the formation of the Sakami moraine. (From Hillaire-Marcel *et al.*, 1981. Reproduced by permission of the Geological Society of America)

glacial and *glacimarine depositional systems* and used to reconstruct shifting environments during deglaciation (Fig. 12.67). The subglacial landsystem, consisting mainly of drumlins and other streamlined landforms, is linked stratigraphically in a down-ice direction to morainal bank complexes, which in turn grade distally into glacimarine sediments overlying

earlier subglacial sediments in the Irish Sea (C.H. Eyles and N. Eyles, 1984a; McCabe *et al.*, 1987; N. Eyles and McCabe, 1989b, 1991; McCabe and Haynes, 1996). The stratigraphic relationships between the terrestrial and marine depositional systems can be used to establish an *event stratigraphy* for the basin, describing the sequence of deglaciation and sea-level change (Fig. 12.68). N. Eyles and McCabe (1989a) concluded that high relative sea-levels caused by isostatic subsidence of the basin helped to trigger rapid calving retreat of the ice margin. In turn, this initiated draw-down and fast streaming of glacier ice in the lowlands surrounding the northern margins of the basin, resulting in subglacial streamlining of sediment into extensive drumlin fields (Fig. 12.69 (Plate 30)). The stabilization of ice margins at approximately the present-day coastline of the Irish Sea basin resulted in the deposition of morainal banks at the outer edges of the drumlin fields. This interpretation of the glacial geology of the Irish Sea basin has been criticized, and questions remain regarding the significance of some critical sites (e.g. Thomas and Dackombe, 1985; Dackombe and Thomas, 1991; Huddart, 1991, 1994; McCarroll and Harris, 1992; Walden, 1994). The event stratigraphy proposed by N. Eyles and McCabe (1989a, 1991) may need to be revised in certain respects, but it remains a powerful framework for understanding the processes and patterns of environmental change at the termination of the last glaciation (McCabe, 1996).

12.6 GLACIATED VALLEY LANDSYSTEMS

Valley glaciers produce very distinctive assemblages of sediments and landforms, which were referred to as the *glaciated valley landsystem* by Boulton and Eyles (1979) and N. Eyles (1983d). In fact, the glaciated valley landsystem may contain ice-marginal, supraglacial, subglacial, proglacial and subaquatic landsystems in close proximity to or superimposed

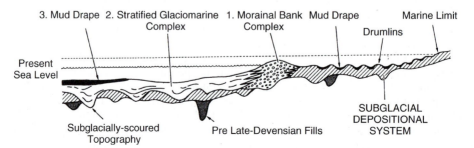

Fig. 12.67 The generalized stratigraphic relationship between the subglacial landsystem and glacimarine sediments of the continental shelf landsystem based upon deposits in the central Irish Sea basin according to N. Eyles and McCabe (1989a). (Reproduced by permission of Elsevier)

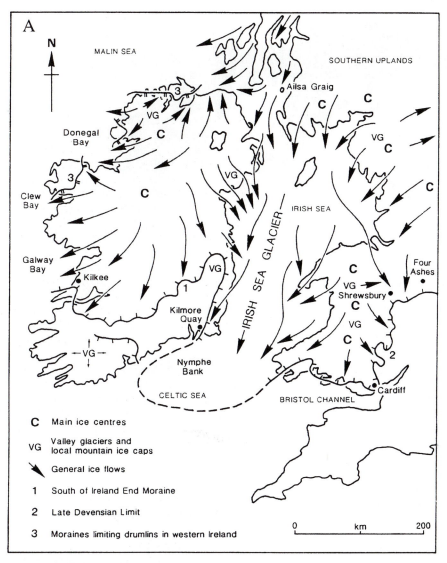

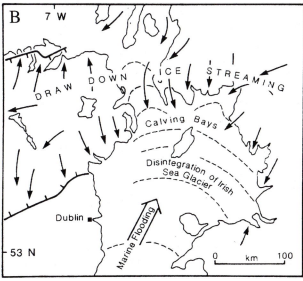

Fig. 12.68 Irish Sea palaeogeography during the late Devensian glaciation, showing (A) ice flow and dispersal centres, and (B) the final stage of disintegration by calving. (From N. Eyles and McCabe, 1989a. Reproduced by permission of Elsevier)

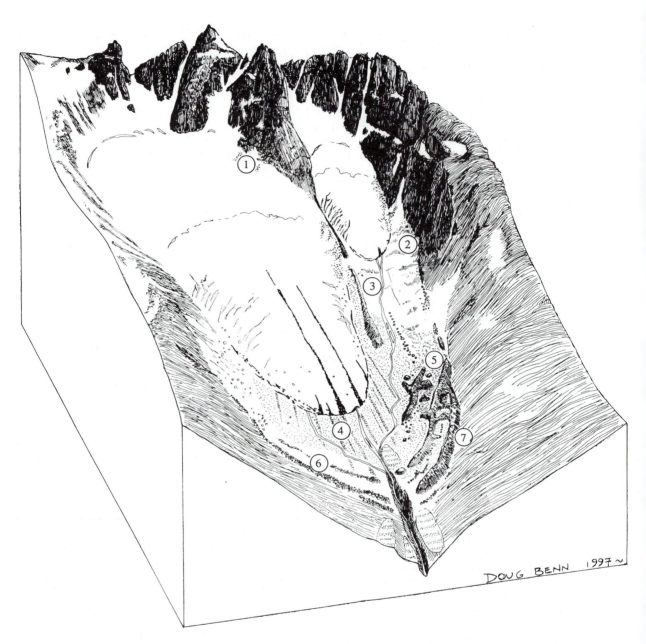

Fig. 12.69 Block diagram of a valley glacier in 'low-relief' mountain terrain. 1: Supraglacially entrained debris. 2: Periglacial trimline above ice-scoured bedrock. 3: Medial moraine. 4: Fluted till surface. 5: Paraglacial reworking of glacigenic deposits. 6 and 7: Lateral moraines, showing within-valley asymmetry

on one another, recording the migration of very different depositional environments. Thus, many of the depositional models described in Sections 12.3–12.5 may be applied to glaciated valleys, with appropriate modifications to account for differences in scale and topographic setting. Glaciated valley settings are, however, unique, owing to the importance of valley sides as debris sources and topographic confinement of depositional basins. The influence of these factors varies from valley to valley, and it is useful to subdivide glaciated valley landsystems into low-relief and high-relief types. Low-relief settings include valleys in ancient mountain belts where the vertical distance between valley floor and summit ridges is generally less than 1000 m (e.g. the glaciated valleys of Scotland, Norway and Labrador). High-relief settings are characterized by extensive steep valley sides rising thousands of metres above the valley floor. Such environments are commonly associated with young or tectonically active fold mountains such as the European and New Zealand Alps, the High Andes and the Himalaya, where relative relief may be 3000 m or more. It should be emphasized, however, that the high- and low-relief classification is applied at the scale of individual valleys, and that both types may occur within a single mountain massif (e.g. Owen and Derbyshire, 1989, 1993).

12.6.1 Low-relief mountain environments

12.6.1.1 SUBGLACIAL SEDIMENTS AND LANDFORMS

Valley glaciers in low-relief mountain terrain carry relatively small amounts of supraglacial debris, so that subglacial landsystems commonly escape burial during deglaciation and may be very well exposed at the surface (Fig. 12.69 (page 606); N. Eyles, 1979). Indeed, the beds of some former valley glaciers are so clearly preserved that they provide splendid opportunities to study subglacial systems at several scales, from that of the bed roughness up to the entire glacier system (e.g. Sharp *et al.*, 1989a; Benn, 1991; Hubbard and Sharp, 1993). Figure 12.70 shows the bed of a former cirque glacier on the Isle of Skye, Scotland. Broadly, the bed can be divided into three zones which occur in sequence from the head of the glacier to the snout:

1. *Erosional zone.* The upper part of the cirque floor is characterized by extensive areas of ice-moulded bedrock, recording net erosion of the bed. Striae, roches moutonnées and an overdeepened rock basin document abrasion and quarrying of the bed by sliding, debris-charged basal ice.
2. *Intermediate zone.* The erosional zone passes downvalley into an intermediate zone where there is evidence for both erosion and deposition. Till deposits are thin and discontinuous, and are restricted to the swales between roches moutonnées. This distribution may reflect debris streaming between obstructions (Boulton, 1982). Lee-side cavity fills occur on the downglacier flanks of rock knobs.
3. *Depositional zone.* This zone is in the lower part of the cirque, and is characterized by continuous till cover. The till is overconsolidated and has a fissile structure, and is interpreted as lodgement till or high-strength, brittle deformation till (Sections 10.3.2 and 10.3.4). On aerial photographs, the till has a faintly lineated surface, although flutings are not easily recognizable on the ground.

These three zones bring to mind the patterns of subglacial erosion and deposition associated with continental ice sheets, albeit on a much smaller scale (Section 12.4), and record downglacier transport of debris towards the margin. Tills deposited on the beds of former valley glaciers have been studied by Levson and Rutter (1989a, b), Rose (1989b), Benn (1994a), and others.

In situations where glaciers extended from the confines of valleys on to fringing lowlands, the depositional zone can resemble the soft-bed subglacial landsystem described in Section 12.4.2.2, with drumlinized surfaces underlain by deformation tills and glacitectonites (Rose, 1981, 1987b; Rose and Letzer, 1977; Thorp, 1991; Benn and Evans, 1996). Examples of this type of landsystem occur around the margins of the Scottish Highlands, where piedmont glaciers flowed into sediment-floored lowlands during the Younger Dryas (Ballantyne, 1989; Thorp, 1991; Benn, 1992b; Benn and Evans, 1996).

12.6.1.2 TRIMLINES AND WEATHERING ZONES

The upper limits of glacier occupancy in glaciated valleys can be preserved as erosional 'tidemarks', or *trimlines,* on valley sides (Fig. 12.69 (page 606); Flint, 1971). In recently deglaciated terrain, trimlines are often instantly recognizable from differences in vegetation cover either side of the former glacier limit. Below the limit, all pre-advance vegetation has been stripped off, leaving bare rock, glacigenic deposits and sparse pioneer vegetation; whereas above the limit, diverse plant communities or even forests record a much longer period of vegetational development. With the passing of centuries, such *vegetation trimlines* become harder to recognize, as plant communities within glacier limits approach the local climax vegetation. As a result, vegetation trimlines are not preserved in landscapes that have been deglaciated since the early to mid-Holocene. In older glacial landscapes, *periglacial trimlines* may be preserved where frost-shattered terrain on upper slopes gives way abruptly to ice-scoured bedrock on lower ground (Fig. 12.71). Such trimlines owe their origin to the removal of pre-existing regolith by a glacier advance and/or periglacial weathering of the upper slopes during the period of glacier occupancy. Several criteria may be used to identify periglacial trimlines, including: (a) the distribution of landforms such as roches moutonnées, and periglacial blockfields and tors (Thorp, 1981, 1986; Porter and

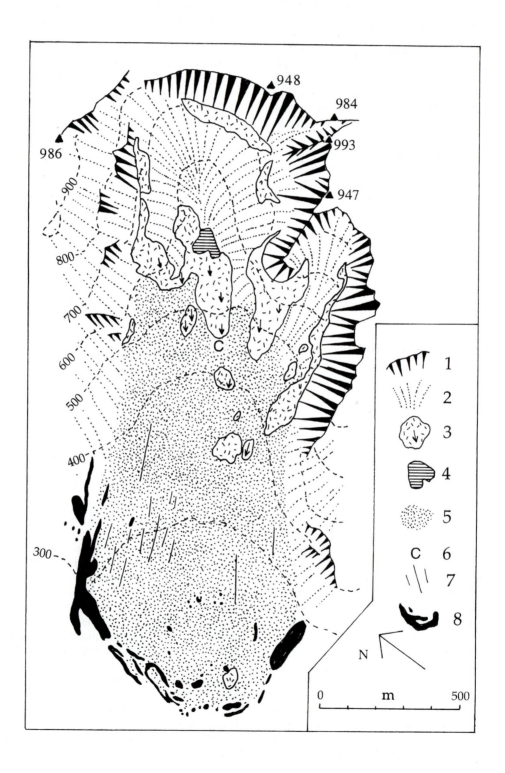

(a)

Fig. 12.70

(b)

Fig. 12.70 (a) Geomorphological map of Coire Lagan, Isle of Skye, Scotland, showing the distribution of ice-scoured bedrock and basal till formed during the Loch Lomond Readvance (Younger Dryas). In the upper part of the valley, large areas of bedrock are blanketed by postglacial scree: 1, frost-shattered rock walls; 2, scree; 3, ice-scoured rock with striae; 4, rock basin lake; 5, basal till; 6, lee-side cavity fill; 7, flutings; 8, moraine ridges and mounds. Contours in metres. (b) View of the upper part of Coire Lagan showing rock peaks and ice-scoured bedrock. (Photo: C.K. Ballantyne)

Orombelli, 1982; Ballantyne, 1990b); (b) the depth of rock joints, which tend to be deeper above a trimline (Ballantyne, 1982b); (c) an abrupt downslope limit to the distribution of long-term weathering products, such as the clay mineral gibbsite (Ballantyne, 1994; Ballantyne and McCarroll, 1995); and (d) small-scale weathering characteristics, such as rock texture and intact hardness (Ballantyne and McCarroll, 1995). The clarity of trimlines varies with rock type and structure, and the degree of postglacial weathering and slope activity. The clearest trimlines are to be found on spurs composed of resistant crystalline rocks, where they may be pinpointed to within 10 m or less (Thorp, 1981; Benn, 1989b). Former nunataks may also be delimited by ice-marginal moraines (Fig. 12.72).

Other explanations have been proposed to account for downslope weathering limits on mountain slopes. (a) Weathering limits could be explained by former altitudinal climatic gradients, with the most intense weathering occurring on the upper slopes, where climatic conditions were most severe. Such weathering gradients, however, should be gradual, whereas trimlines are generally abrupt. Additionally, trimlines typically have marked downvalley gradients that record the former glacier surface, while altitudinal weathering zones should be subhorizontal over large distances. (b) Some transitions between periglacially weathered and ice-scoured terrain may represent the junction between cold-based and wet-based ice

below an ice sheet. In this case, the preservation of periglacially weathered terrain on high ground can be attributed to protection below thin, cold-based ice, whereas glacial erosion on lower ground could have occurred below thicker, wet-based ice (Section 9.5.6). Such 'thermal trimlines' should be elevated on the upglacier sides of high ground and depressed on the downglacier sides, owing to the effect of longitudinal stresses on the pressure melting point of ice. In contrast, periglacial trimlines have smooth regional gradients consistent with theoretical ice sheet profiles (Ballantyne and McCarroll, 1995).

Taken together with lateral moraines and other ice-marginal landforms, trimlines allow the dimensions of former valley glaciers to be reconstructed with considerable accuracy, yielding important input for glaciological models or palaeoclimatic studies (Porter and Orombelli, 1982; Duk-Rodkin *et al.*, 1986; Ballantyne, 1989; Thorp, 1991). Good examples of periglacial trimlines have been described by Thorp (1981, 1986) and Ballantyne (1989), who used the distribution of periglacial and glacial landforms to determine the limits of Loch Lomond (Younger Dryas) Stadial glaciers in Scotland. More recently, Ballantyne (1990b, 1994) and Ballantyne and McCarroll (1995) have shown that there is an older, higher set of trimlines in Scotland, marking the upper limit of the last (Late Devensian) ice sheet. Thus, three *weathering zones* can be recognized. (a) Above the upper trimlines, blockfields and gibbsitic soils represent the weathered surfaces of nunataks that stood above the ice surface. Weathering has proceeded at least since the penultimate glaciation. (b) At lower altitudes, but outside the Loch Lomond Readvance limits, is a zone of 'frost-fretted' terrain, in which rock surfaces display the gross forms of glacial erosion but have been subject to minor joint widening and disaggregation by periglacial weathering since ice sheet deglaciation. (c) Within the limits of the Loch Lomond Readvance, rock surfaces have been subject to weathering only in the Holocene, and glacial features are well preserved.

The recognition of weathering zones is an important method of correlating glacial limits where there is a lack of datable material. For example, multiple glaciations of the Torngat Mountains and fjords of northern Labrador, Canada, have been differentiated on the basis of weathering zones and glacial geomorphological evidence (Ives, 1978; Evans and Rogerson, 1986; Clark, 1988). During successive glaciations in this region, trunk glaciers occupied progressively lower positions on fjord and trough walls, and ice-marginal accumulations deposited by these glaciers are more intensely weathered with age (Boyer and Pheasant, 1974; Dyke, 1979; Fig. 12.73). The degree of weathering of lateral moraines has also been used by Burbank and Kang (1991) to establish

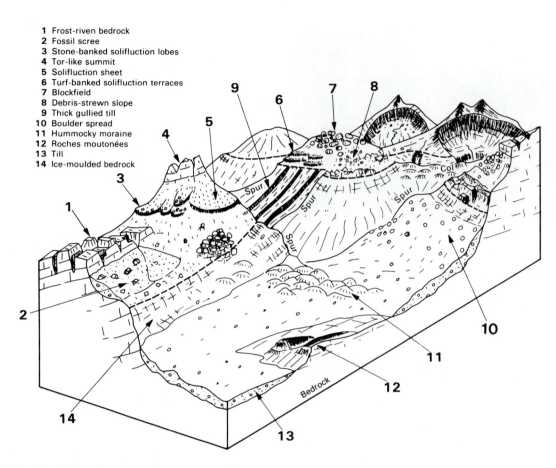

1 Frost-riven bedrock
2 Fossil scree
3 Stone-banked solifluction lobes
4 Tor-like summit
5 Solifluction sheet
6 Turf-banked solifluction terraces
7 Blockfield
8 Debris-strewn slope
9 Thick gullied till
10 Boulder spread
11 Hummocky moraine
12 Roches moutonées
13 Till
14 Ice-moulded bedrock

Fig. 12.71 Schematic diagram of periglacial trimlines and drift limits, based upon Scottish examples. The dashed line represents the upper limit of the most recent glaciation in this glaciated valley landsystem and separates the ice-scoured bedrock and drift at lower elevations from the higher-altitude frost-shattered periglacial terrain. (From Thorp, 1986. Reproduced by permission of Norwegian University Press)

a relative glacial chronology for the Rongbuk Valley to the north of Mount Everest.

12.6.1.3 ICE-MARGINAL MORAINES AND RELATED LANDFORMS

The margins of valley glaciers in low-relief mountain areas are commonly delimited by latero-frontal dump and push moraines (Fig. 12.70a). In the Scottish Highlands, such moraines are typically 2–5 m high, although they may be a few tens of metres high where abundant debris was available (Sissons, 1967, 1977a; Ballantyne, 1989; Benn, 1990, 1992b). Little Ice Age moraines in the Norwegian mountains have a similar range of sizes (Matthews and Petch, 1982). Moraine characteristics are strongly influenced by catchment lithology, and in valleys underlain by resistant, crystalline rocks, lateral moraines may be little more than ridges of boulders. In contrast, thrust moraines may occur where glaciers come into contact with thick, unconsolidated sediments such as

glacimarine clays and silts (e.g. Gray and Brooks, 1972; Benn and Evans, 1993). Meltwater deposition at some glacier margins is recorded by ice-marginal ramps and fans.

Lateral moraines may be much larger on one valley side than the other, a phenomenon known as *within-valley asymmetry* (Fig. 12.74; Matthews and Petch, 1982). At least four causes of within-valley asymmetry have been identified. (a) Lateral moraines may be best-developed on valley sides where there are extensive rock walls, which supply larger amounts of debris to that side of the glacier (Matthews and Petch, 1982; Benn, 1989b). Debris may be delivered to the glacier directly via rockfalls or indirectly by the subglacial incorporation of pre-glacial screes. Thus, in this case, within-valley asymmetry records differential slope retreat of the valley sidewalls. The orientation of rockwalls may reflect climatic factors such as the amount of direct insolation or patterns of snow-blow and accumulation. (b) Where lateral moraines are formed by pushing of pre-

Fig. 12.72 Aerial photograph stereopair of moraines and drift limit isolating a nunatak in the Glenlyon area, Yukon Territory, Canada. (From Duk-Rodkin *et al.*, 1986. Reproduced by permission of the Geological Survey of Canada)

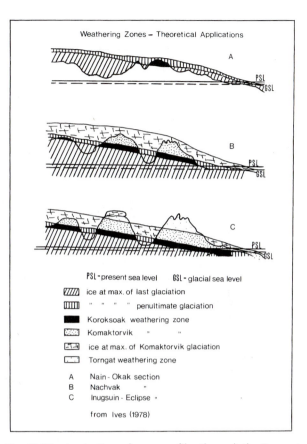

Fig. 12.73 A selection of cross-profiles through the Torngat Mountains/eastern seaboard of Labrador and Baffin Island, Canada, illustrating the inset weathering zones associated with multiple glaciations of various ages. Note that the oldest (Torngat) weathering zone represents nunataks that lay above the upper margins of the oldest (Komaktorvik) glaciation. (Modified from Ives, 1978)

existing materials, within-valley asymmetry could result from differences in the thickness and type of sediment on the foreland (Matthews and Petch, 1982). (c) Cross-valley differences in lithology or structure can influence rates of debris supply, to either the surface or the bed of the glacier (Benn, 1990). (d) Within-valley asymmetry may occur due to differences in glacier dynamics on either side of a valley (Bennett and Boulton, 1993a). For example, if one side of a glacier maintains a stable position for several seasons, a large moraine can be deposited, but if the other margin is in overall retreat, the glacier may leave a series of smaller moraines. In this case, the amount of debris deposited on each valley side may be the same, but an impression of asymmetry arises because deposition on one side is focused on a smaller area. Cross-valley differences in glacier behaviour may be due to climatic factors, such as the amount of shading.

According to N. Eyles (1979, 1983b), the majority of the debris transported by valley glaciers is derived from mass wasting of valley walls. However, in many low-relief glaciated valleys, subglacially entrained debris may be of equal or greater importance. For example, Benn (1992b, 1993) used clast lithological analysis to show that Lateglacial moraines in many glaciated valleys on the Isle of Skye, Scotland, are composed predominantly of debris eroded from the bed (Fig. 12.75). In these cases, subglacial debris entrainment appears to have been most effective

below the glacier margins and in areas underlain by well-jointed rocks.

Glacier retreat in low-relief mountain environments may be recorded by recessional moraines superimposed on subglacial sediments and landforms. The size and spacing of such moraines reflect the availability of debris and the frequency and duration of standstills or readvances of the margin (N. Eyles, 1979). Rapid retreat will result in a thin, patchy veneer laid down on the subglacial surface, except where debris cover was high, such as along medial moraines. The most prominent fossil medial moraines generally extend downvalley from spurs between confluent valleys (Section 11.4.1).

In some areas, the lower slopes and floors of valleys are choked by complex assemblages of mounds and ridges. In Scotland, such assemblages were classified as 'hummocky moraine' by J.B. Sissons and co-workers (e.g. Sissons, 1967, 1974c, 1977b), and interpreted as chaotic products of glacier stag-

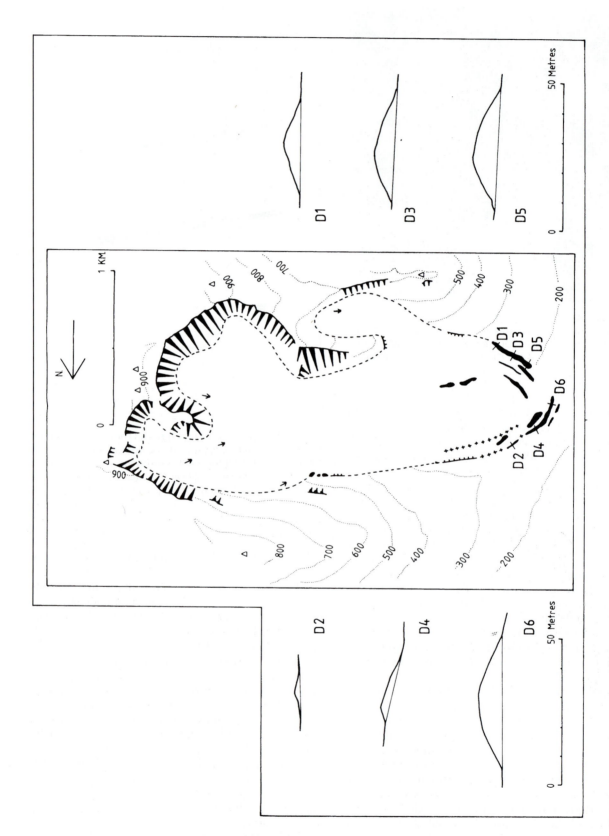

Fig. 12.74 Map of Coire a'Ghreadaidh, Isle of Skye, Scotland, showing within-valley asymmetry of lateral moraines. (From Benn, 1989b. Reproduced by permission of John Wiley & Sons)

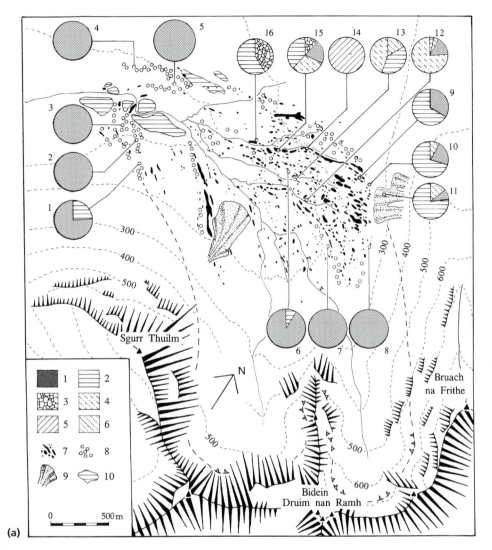

Fig. 12.75 Evidence for subglacial erosion of debris during the Loch Lomond Stadial, Coire na Creiche, Isle of Skye, Scotland. (a) Glacial landforms and debris lithology. The pie-charts show the lithology of large surface boulders (samples 1–8) and clasts within moraine ridges (samples 9–16). Key: 1: Gabbro; 2: Basalt; 3: Acid breccia; 4: Rhyolite; 5: Feldsparphyric dolerite; 6: Trachyte; 7: Moraine ridges and mounds; 8: Glacially transported boulders; 9: Debris fans; 10: Outwash terraces. The cirque backwalls are composed mainly of gabbro and basalt, and the other lithologies crop out on the cirque floor. (b) Reconstruction of the Coire na Creiche glacier during deglaciation. A: glacier limit; B: gabbro boulders delivered from cirque backwalls; C: accumulation of scree at glacier margins; D: moraines composed of debris eroded from the bed. (From Benn, 1992b)

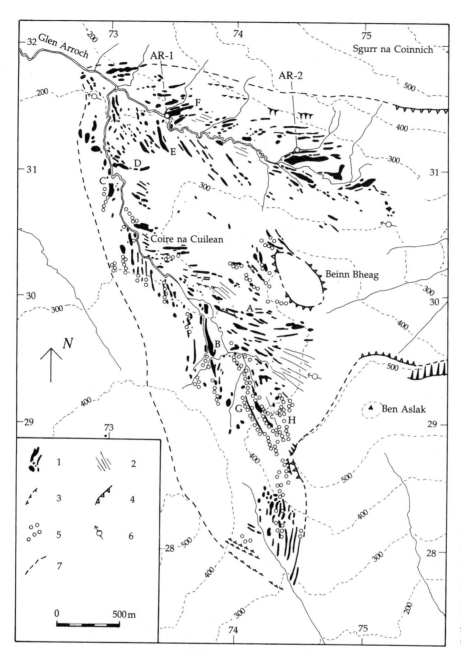

Fig. 12.76 Recessional moraines, Glen Arroch, Isle of Skye, Scotland. 1, moraine ridges and mounds; 2, fluted moraines; 3, bench; 4, trimline; 5, boulders; 6, striae; 7, glacier limit. (From Benn, 1992b)

nation. Recently, however, much of this so-called 'hummocky moraine' has been shown to consist of closely spaced recessional moraines (Fig. 12.76; Benn, 1990, 1992b, 1993, 1996b; Bennett and Glasser, 1991; Bennett and Boulton, 1993a, b; Bennett, 1994). The moraines commonly form converging cross-valley pairs of sharp-crested ridges and chains of hummocks up to 12 m high but mostly only 2–3 m. The ridges have an undulating long profile, which is responsible for the hummocky appearance when viewed from the ground. The sedimentology, cross-profiles and planforms of the moraines indicate that they are dump and push moraines, with varying amounts of larger-scale proglacial tectonics. This evidence shows that the moraines were formed during oscillations of the ice margins during active retreat, rather than widespread stagnation, and underlines the importance of making detailed observations at a range of scales when assessing the environmental significance of glacial landforms (Benn *et al.*, 1992; Bennett and Boulton, 1993b; Bennett, 1994; Benn, 1996b). Local and regional patterns of moraine ridges have been used to reconstruct the deglacial history of the

Isle of Skye and the northern Scottish Highlands during and following the Younger Dryas Stade (Section 1.7.2.4). The reinterpretation of Scottish 'hummocky moraine' as the products of active glacier retreat essentially marks a return to the ideas of early researchers such as T.F. Jamieson (1874) and J.K. Charlesworth (1956). A good review of the debate surrounding the origin of Scottish moraines has been provided by Bennett (1994).

Assemblages of chaotic hummocky moraine may occur in pockets within fields of recessional moraines, possibly as a result of the isolation and stagnation of sediment-covered ice remnants (Benn, 1992b). The occurrence of chaotic hummocky

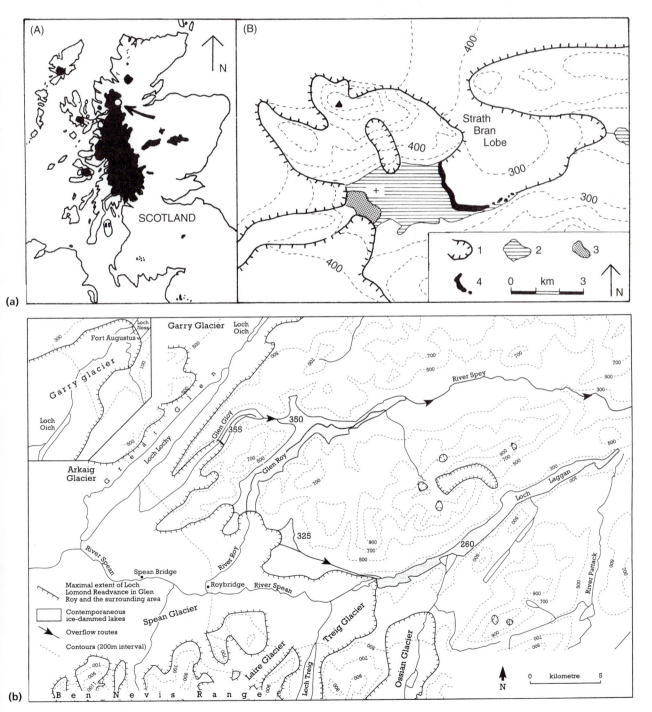

Fig. 12.78

(c)

Fig. 12.78 Loch Lomond (Younger Dryas) Stadial ice-dammed lakes in Scotland. (a) Reconstruction of the Achnasheen ice-dammed lake, showing the location of deltas and subaqueous moraines: 1, glacier margin; 2, lake; 3, delta; 4, moraine (From Benn, 1996a); (b) reconstruction of the Glen Roy lake (from Sissons, 1979d); (c) the Parallel Roads of Glen Roy: fossil lake shorelines. (Photo: D.I. Benn)

moraine is not necessarily indicative of widespread glacier stagnation, as was once thought, and many examples of hummocky moraine have been observed at active ice margins (e.g. Okko, 1955; Vikingsson, 1978; Kaldal, 1978; N. Eyles, 1979, 1983b). Large areas of hummocky moraine can be produced by *incremental marginal stagnation*, whereby chaotic moraines are deposited in a 10–500 m wide marginal zone of debris-mantled ice which retreats upvalley during deglaciation (Charlesworth, 1957; N. Eyles, 1983b, d). Transverse lineations may be constructed subsequently within the chaotic hummocky moraine by thrusting by more active ice upglacier. Assemblages indicative of more general stagnation do occur on some flat-lying valley floors, although they are rare (Benn, 1992b). Such assemblages may consist of chaotic hummocky moraine inset within 'staircases' of subhorizontal kame terraces.

12.6.1.4 PROGLACIAL LANDFORM ASSEMBLAGES

Following glacier retreat, ice-marginal and supraglacial landforms are subject to paraglacial reworking (Section 7.6), but not to such a large degree as in high-relief mountain environments. Paraglacial reworking is particularly intense at the foot of gullies which focus meltwater and snow avalanches, and in such locations glacigenic deposits are commonly reworked into debris fans and cones (Ballantyne, 1995; Ballantyne and Benn, 1996). Where slopes are steep, moraines may be completely destroyed, leaving a mantle of debris-flow diamictons (Ballantyne and Benn, 1996). Benn (1990) found that on the Isle

of Skye, the maximum slope angle on which Lateglacial moraines are preserved is *c.* 20°, and sedimentological evidence indicates that on steeper slopes, debris reworking occurred shortly after deglaciation. Rock glaciers have been described from many low-relief mountain environments (e.g. N. Eyles, 1978b; Giardino *et al.*, 1987; Ballantyne and Harris, 1994). Rock glaciers form where debris supply is high relative to snowfall, so in low-relief mountain ranges where debris supply is relatively low, rock glaciers tend to occur in the more arid areas (Fig. 12.77 (Plate 31)). This appears to be the case in Scotland, where the highest density of Lateglacial rock glaciers is in the Cairngorm Mountains in the drier, eastern part of the Highlands (Ballantyne and Harris, 1994).

Glacifluvial deposits are commonly well preserved in low to moderate-relief glaciated valleys. The focusing of meltwater flow by valley sides results in the erosion of gorges or the deposition of ribbon-like valley trains along valley axes, where subglacial or ice-marginal landsystems tend to be buried or reworked. Maizels (1995) has mapped the valley train in front of the Glacier des Bossons in the French Alps, where glacifluvial activity has almost completely reworked any ice-marginal deposits. River terraces are striking components of many glaciated valley landscapes (Section 11.5.1.1). Staircases of terraces occur along the floors of many glaciated valleys, and the highest members may show signs of ice-marginal deposition (*kame terraces*; Section 11.4.5). Younger, lower terraces record deposition and erosion by proglacial meltwater and, in deglaciated basins, postglacial streams. In some instances, terraces can be correlated on the basis of their weathering characteristics and soil development (e.g. Robertson-Rintoul, 1986).

Ice-dammed lakes are common in many mountainous environments, as a result of the blocking of side or trunk valleys by expanded glacier tongues (Section 3.2.4.3). Good examples of landform and sediment assemblages formed in Pleistocene ice-dammed lakes are found in Glen Roy and near Achnasheen in the Highlands of Scotland (Fig. 12.78; Sissons, 1978, 1979d, 1981; Benn, 1989a, 1996A). In Glen Roy, water level was controlled by cols on the watershed, and the lake waters rose and fell as successive cols were blocked by glacier advance or exposed by retreat. The former water levels are recorded by very prominent shorelines known locally as the 'Parallel Roads', which remain strikingly clear after approximately 10,000 years of weathering and erosion. At Achnasheen, lake level was controlled by an ice dam, and was consequently probably less stable. This appears to be reflected in the shorelines, which are much more numerous and less distinct than those at

Glen Roy (Benn, 1989a). At both locations, assemblages of morainal banks, deltas and distal glacilacustrine sediments preserve a record of glacier fluctuations and shifting depocentres during the lifespan of the lakes. Moraine-dammed lakes tend to be relatively unimportant components of low-relief glaciated valley environments, because of the typically small volumes of sediment in end and recessional moraines.

12.6.2 High-relief mountain environments

In high-relief mountain environments, the floors of glaciated valleys are commonly filled by very large volumes of sediment, particularly in tectonically active ranges such as the Southern Alps of New Zealand, the Himalaya and the Andes, where sediment yields are high (Fig. 12.79 (Plate 32); Fig.

12.80; Owen and Derbyshire, 1989, 1993; Clapperton, 1993; Owen, 1994; Kirkbride, 1995a).

Valley glaciers in high-relief settings typically have extensive covers of supraglacial debris, and sedimentation around glacier margins forms large latero-frontal dump moraines and ice-contact fans and ramps (Fig. 12.79 (Plate 32)); Humlum, 1978; Boulton and Eyles, 1979; Small, 1983; Owen and Derbyshire, 1989, 1993; Owen, 1994). Such landforms can constitute major barriers to glacier flow, so that repeated glacier advances may terminate at the same location and contribute to moraine-building, resulting in very large landforms which can exceed 100 m in height (Section 11.3.3). A depositional model for the margins of debris-covered glaciers has been presented by Owen and Derbyshire (1988, 1993), who termed it the *Ghulkin-type association* after the glacier of that name in the Karakoram Mountains (Fig. 12.80). The ice-contact sediments

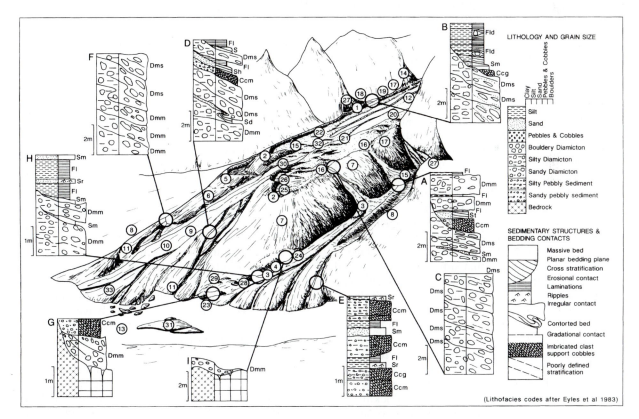

Fig. 12.80 Landforms and sediments produced in glaciated valley landsystems of high relief based upon the Ghulkin Glacier, Karakoram Mountains. Note the dominance of diamictons produced by mass movement processes: (1) truncated scree; (2) termino-lateral dump moraine; (3) laterally drained outwash channel; (4) glacifluvial outwash fan; (5) slide moraine; (6) slide-debris flow cones; (7) slide-modified lateral moraine; (8) abandoned lateral outwash fan; (9) meltwater channel; (10) meltwater fan; (11) abandoned meltwater fan; (12) bare ice; (13) trunk valley river; (14) debris flow; (15) flow slide; (16) gullied lateral moraine; (17) lateral moraine; (18) ablation valley lake; (19) ablation valley; (20) supraglacial lake; (21) supraglacial stream; (22) ice-contact terrace; (23) lateral lodgement till; (24) roche moutonnée; (25) fluted moraine; (26) diffluence col; (27) high-level till remnant; (28) diffluence col lake; (29) fines washed from supraglacial debris; (30) ice-cored moraines; (31) river alluvium; (32) supraglacial debris; (33) dead ice. (From Owen and Derbyshire, 1993. Reproduced by permission of Routledge)

consist of stacked debris flows dipping away from the glacier margin (Sections 11.3.3 and 11.3.4), and following glacier retreat these are dissected by melt-streams and overlain by glacifluvial fans and valley trains. The inner faces of lateral moraines are typically oversteepened, and are thus highly susceptible to paraglacial reworking. Indeed, where such moraines are overlooked by steep valley sides, they may not survive for very long after deglaciation, and the only record of glacier occupancy may be in the form of glacial erratics in debris fans (Owen, 1991; Owen et al., 1995).

Debris-mantled glaciers can be expected to have relatively muted responses to climatic change, because large latero-frontal moraines limit the amount of glacier advance, and extensive supraglacial debris cover inhibits ablation of the underlying ice (Sections 2.2.3.3 and 6.5.2). As described in Section 12.6.1.3, the retreat of valley glaciers in low-relief mountain environments may be punctuated by numerous oscillations of the ice margin, which are recorded by minor dump and push moraines. In contrast, debris-mantled glaciers are very unlikely to exhibit such minor oscillations, and will tend to remain at the limits imposed by latero-frontal moraines until advance or retreat is triggered by significant climatic change. Thus, the landform record of a retreating debris-mantled glacier might consist of major moraine complexes, representing periods of stability, separated by extensive tracts of hummocky moraine deposited during episodes of ice-margin wastage and stagnation.

The presence of thick supraglacial and ice-marginal sediments means that subglacial landsystems are rarely exposed in high-relief environments. The existence of deeply incised glacial troughs and the presence of actively transported debris in latero-frontal moraines indicate that subglacial erosion and transport must be effective, although it is likely to be volumetrically less significant than supraglacial debris entrainment and transport in most basins (e.g. Small, 1987a, b; Gardner and Jones, 1993).

Temporary lakes are common features of high-relief mountain environments, and may be formed in three main situations (Owen, 1989; Owen and Derbyshire, 1993): (a) where moraines are deposited by a glacier emanating from a side-valley, blocking the drainage of the trunk valley; (b) behind lateral-frontal moraines or within moraine complexes; or (c) where drainage is blocked by rockfalls, landslides and debris flow fans. Although they may be short-lived, such temporary lakes can have profound effects on the landscape. They may be infilled rapidly by silts and other sediments, forming extensive terraces stretching upvalley from the dam. For example, during the winter of 1857/58, a landslide at Serat Pungurh in the Hunza Valley in the Karakoram

Mountains dammed back a lake approximately 10 km long. The dam was breached about six months later, releasing a large flood down the Indus, causing much loss of life (Mason, 1929). During this six-month period, up to 10 m of silt accumulated in the lake basin, and it now forms the lowest of three terraces in the valley (Owen and Derbyshire, 1993). These high deposition rates are typical of the Karakoram region. Jökulhlaups released from temporary lakes also have high geomorphological impacts, eroding and reworking large volumes of sediment (Sections 3.5, 9.3.4.4 and 11.5.1.3). Flood tracks may be preserved in the landscape in the form of channels and bouldery debris fans extending downvalley from the moraine breach (Clague and Evans, 1994; Coxon et al., 1996).

Valley fills including extensive supraglacial and proglacial lake deposits have been described from the mountains of British Columbia, Canada (Shaw, 1977b; Shaw and Archer, 1979; N. Eyles et al., 1987; Ryder et al., 1991). Supraglacial lakes were common during the wastage of the Cordilleran ice sheet in British Columbia, where stagnant glacier ice was isolated in deep valley bottoms and broke up into lake basins (see Section 6.5.4.2). Clague and Evans (1994a) have argued that environmental conditions were similar to those occurring today in the St Elias mountains near the Canadian–Alaskan border, where major debris-mantled glacier tongues have thinned dramatically during the past few hundred years and stagnated in situ. Sediment sequences deposited in former supraglacial lakes in British Columbia record deposition in multiple basins, and are dominated by subaquatic mass flow deposits. The sediments are affected by numerous deformation structures caused by the melt-out of buried ice. In some areas, valley-side delta and kame terraces are preserved, recording former lake levels or positions of the ice surface (Fig. 12.81).

Following ice retreat, glacial sediments and landforms are commonly subject to extensive reworking by glacifluvial and mass movement processes. As rivers adjust to new base levels and sediment loads, they incise existing sediment piles and redeposit debris in terraced valley trains (Fig. 12.82). Paraglacial processes are particularly effective, owing to the abundance of steep slopes, the action of high-energy processes such as snow avalanching and ablation-triggered floods, and the widespread availability of unconsolidated sediment and unstable bedrock (e.g. Eisbacher and Clague, 1984; Owen et al., 1995). Trimlines are rarely well preserved, but may be visible on spurs and ridges. A model for deglaciated valleys in the Karakoram was presented by Owen (1989) and shows that mass movement features, lacustrine and fluvial sediment accumulations, and river terraces may be much more widespread

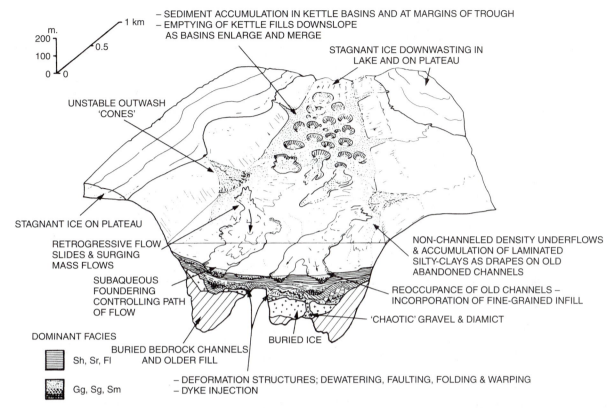

Fig. 12.81 Model for sedimentation in a supraglacial lake system in an area of moderate to high relief, based on the stratigraphy of the Fraser River valley, British Columbia, Canada. (From N. Eyles *et al.*, 1987. Reproduced by permission of Blackwell)

Fig. 12.82 A valley train or proglacial outwash tract emanating from a breach in the most recent terminal moraine of the Tasman Glacier, New Zealand, and fed directly by proglacial/supraglacial lake waters. Note the incision and abandonment of older outwash surfaces. (Photo: D.J.A. Evans)

than glacigenic landforms (Fig. 12.83). Former ice margins are preserved only as intermittent trimlines or drift limits where slopes were relatively stable. The supply of extraglacial debris from surrounding slopes may be so abundant that receding glacier ice becomes buried, resulting in the evolution of a

glacial rock glacier. In addition, the high production rates of rock debris in high-relief terrains can facilitate the development of permafrost rock glaciers on steep scree slopes (Section 7.5.1).

Paraglacial reworking of glacial landforms and sediments is less effective where glaciers advanced out from high-relief mountainous regions to the foothills, and in such settings the preservation potential of the substantial ice-marginal landforms is greater. For example, the classic latero-frontal moraines of the Pinedale Glaciation in the USA were deposited in the foothills of the Rocky Mountains, where paraglacial processes were responsible for merely cosmetic changes (e.g. Fig. 12.84). Similarly, some of the most impressive latero-frontal moraines on Earth occur around the fringes of the New Zealand Alps, where glaciers were supplied with large quantities of debris by steep, tectonically active slopes. In the long valley systems emanating from the eastern side of the Southern Alps divide, numerous inset lateral moraines and kame terraces form continuous ice-marginal deposits which stretch over tens of kilometres. For example, kame terraces and lateral moraines skirt the margins of Lake Pukaki and partially back-fill tributary valleys (Fig. 12.79 (Plate 32)). At the southern end of the lake a natural dam is composed of linear hummocky moraine with super-

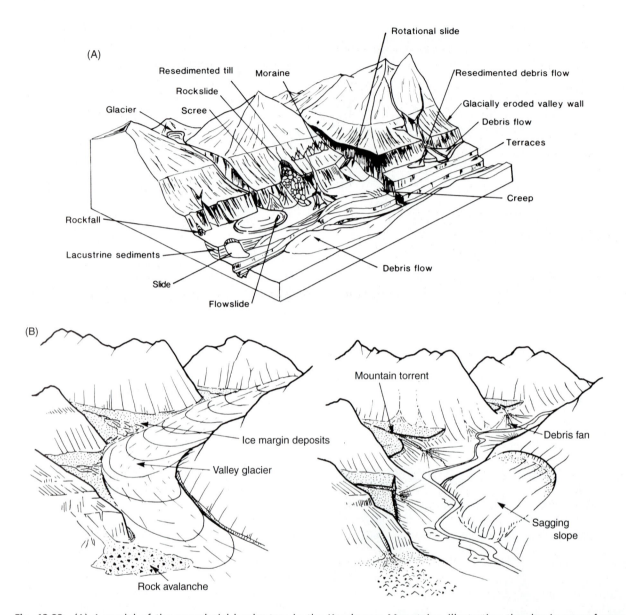

Fig. 12.83 (A) A model of the paraglacial landsystem in the Karakoram Mountains, illustrating the dominance of mass movement processes and fluvial incision of valley fills immediately after deglaciation. (From Owen, 1991. Reprinted by permission of E. Schweizerbartische/Gebruder Bornträger Verlagsbuchhandlung) (B) A sketch showing the influence of glaciation on paraglacial activity in a high relief landscape. During glacial conditions (left) ice-contact fluvial sediments accumulate in tributary valleys. After deglaciation (right) the fluvial sediments are incised and cannibalized to produce debris fans, and slopes may fail or sag. (From Eisbacher and Clague, 1984. Reprinted by permission of the Geological Survey of Canada)

imposed saw-tooth push moraines. Extensive and thick glacilacustrine sediments on the ice-proximal side of this end moraine complex have been glacitectonized in places, indicating that the ice margin oscillated in a moraine-dammed lake. The Tasman, Hooker and Mueller glaciers, which coalesced to produce the glaciated valley landsystem in the Lake Pukaki basin, have now receded to their own steep-sided valleys, where paraglacial processes are heavily reworking any valley-side glacigenic accumulations.

12.7 EXTRATERRESTRIAL GLACIATION: THE FINAL FRONTIER

It is appropriate to end this book by briefly reviewing one of the newest and most exotic fields of glacial study: the nature of glaciation on other planets. Our ever-improving technological capabilities allow us to view extraterrestrial geomorphology at increasingly

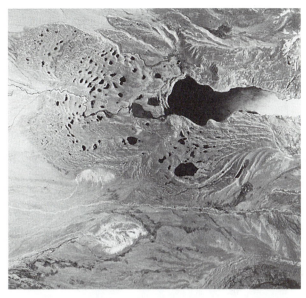

(a)

(b)

Fig. 12.84 Latero-frontal moraine systems. (a) Large latero-frontal moraines deposited by piedmont glacier tongue, Green Creek, Sierra Nevada, USA. Numbers 1–3 indicate lateral moraines of different phases of glaciation, and EM is the end moraine of lateral moraine 3. (From R.P. Sharp, 1988. Reprinted by permission of Cambridge University Press) (b) the latero-frontal moraines emanating from the Cordillera Apolobamba on the Peru–Bolivia border. Note the fluted terrain and stream-lined frontal moraines produced by glacier overriding, and the saw-tooth pattern visible in the outermost frontal moraines.

greater levels of accuracy, particularly on one of our nearest neighbours, Mars. The picture that is emerging from Mars suggests a long and complex geological history, possibly including multiple glaciations.

It has long been known that permanent ice caps exist in both polar regions of Mars (Sharp, 1974). However, they are quite unlike Earth's ice caps in composition and behaviour. Despite being up to 1000 km wide, the Martian ice caps are only tens of metres thick, and are composed of perennial water ice covered on a seasonal basis by solid carbon dioxide (CO_2 or *dry ice*) precipitated from the thin Martian atmosphere (Sharp, 1974). Over the winter pole, blizzards of frozen CO_2 deposit a few metres of dry ice on the ice cap at temperatures as low as $-123°C$, while at the opposite pole the ice cap gives up CO_2 by sublimation, although temperatures never rise high enough to melt water ice. Thus, the ice caps advance and recede each year, but by processes different from those on Earth. The extreme cold and aridity of the Martian polar environment means that the ice caps do not undergo significant sliding or internal deformation, and so are incapable of making a geomorphological impact on the planet's surface.

The geomorphology of Mars, however, indicates that environmental conditions were very different in the past. Widespread fluvial landforms on the planet's surface indicate that the Martian atmosphere was once denser, and liquid water was abundant, and possible glacial landforms have been recognized in many areas (Carr, 1986; Squyres, 1989; Kargel and Strom, 1992a, b, 1996; Parker *et al.*, 1993). Furthermore, the areas immediately beyond the current ice caps have an uncharacteristically low density of meteorite impact craters, suggesting that these areas were protected or modified by expanded ice cover. This evidence raises the intriguing possibility that in the ancient past, Mars was affected by glacial cycles similar to those of Earth.

Images obtained by orbiting spacecraft, such as the Viking probe, show that the surface of Mars is complex and varied, displaying many landforms similar to those on Earth and others which remain enigmatic. Many researchers have described surface features which are closely similar to glacial landforms found on Earth (Squyres and Carr, 1986; Kargel and Strom, 1990, 1991, 1992a; Ruff and Greeley, 1990; N. Johnson *et al.*, 1991; Kargel *et al.*, 1991; Metzger, 1991; Strom *et al.*, 1991). Such landforms include possible moraines, kame and kettle topography, sinuous esker ridges, tunnel valleys, sandar, glacilacustrine basins, streamlined grooves and megaflutings, rock glaciers, cirques, aretes and horns (Figs 12.85 and 12.86). However, these forms are open to other interpretations. For example, the sinuous ridges

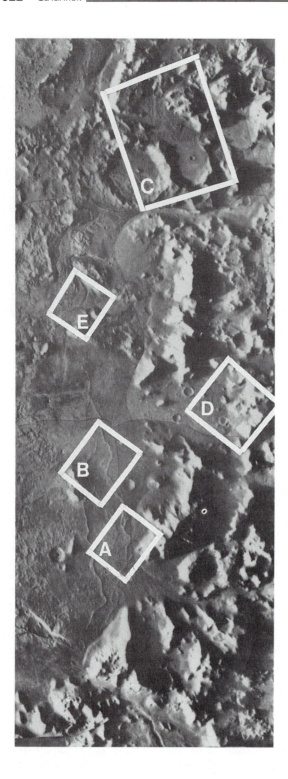

Fig. 12.85 Landforms on the Martian surface interpreted as glacially modified features. The location is the Argyre basin and the scene length is 650 km with the illumination from the left. Boxed areas are shown in detail in Fig. 12.86. (Photograph taken by the Viking Orbiter and provided by J.S. Kargel)

interpreted as eskers by some researchers have been regarded as lava tubes, exhumed dikes, linear sand dunes or spits and bars by others.

One line of evidence lends strong support to the idea that glacial landforms exist on the surface of Mars. In some areas, landforms are arranged in regional associations very similar to those formed by the advance and retreat of ice caps or mountain glaciers on Earth. In other words, the similarity between some Martian landforms and glacial features on Earth applies at all observable scales. Kargel and Strom (1992a) have mapped landsystems which appear to record the advance and retreat of a large ice cap centred on a volcanic upland (Fig. 12.87). Their reconstruction suggests that ice flowed down from the high ground into a nearby basin, forming an arcuate band of terminal moraines and ice stagnation topography at its limit. Part of the ice margin appears to have terminated in a large proglacial lake. Within the glacier limits, possible drumlin fields and grooved terrain suggest that large areas of the ice cap were wet-based.

In some areas, assemblages of closely spaced ridges form *thumbprint* or *mottled terrain*, which has been interpreted as suites of recessional moraines deposited by dynamically active ice during deglaciation (Carr *et al.*, 1980; Scott and Underwood, 1991). Kargel and Strom (1992a) drew attention to the common association between thumbprint terrain and possible esker systems, tunnel valleys and glacilacustrine basins, and suggested that it may consist of subaqueous De Geer moraines. The spatial zonation of these landforms is strikingly similar to that left by the retreat of the Laurentide ice sheet in the North American prairies (Sections 12.3.5 and 12.4.3).

Many parts of Mars exhibit spectacular dendritic channels and canyons which must have been cut by surface stream systems. In addition, large *outflow channels* and *streamlined residuals* have been identified by Sharp and Malin (1975), and compared to the Channeled Scablands of Washington by Baker and Kochel (1978), Baker (1979) and Baker *et al.* (1987). The close resemblance of the Martian channel systems to landscapes eroded by large catastrophic floods raises the interesting problem of where floodwaters could have originated. The evidence that glaciers and proglacial lakes may once have existed on Mars suggests that some floods could have been jökulhlaups. However, it has also been suggested that water could have burst out from highly pressurized subsurface aquifers following geothermal melting of permafrost (Carr, 1981; Howard, 1981).

Obviously, the very inaccessibility of the Martian surface makes it impossible to propose anything other than interpretations based upon form analogy.

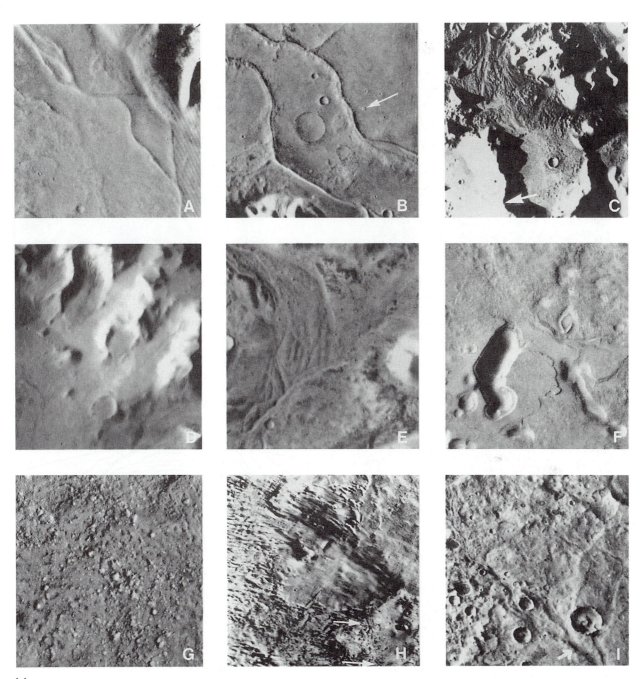

(a)
Fig. 12.86

This entails a necessary and not altogether safe assumption, that the terrestrial forms and processes that we know so well are suitable analogues for those of other planets. On a planet with an atmosphere and climate so unlike our own, it is possible that these features cannot be explained by any terrestrial analogues, and we should remember that we are only just beginning to discover the secrets of forms, processes and environments outside our own Earth–atmosphere system.

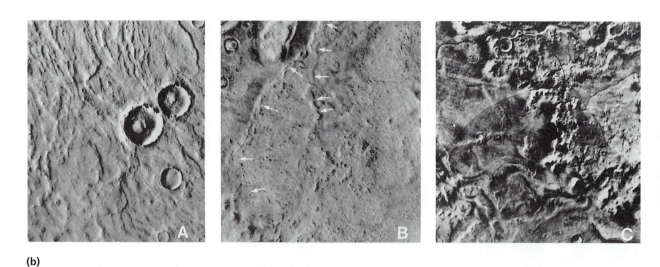

(b)

Fig. 12.86 Individual landforms thought to be indicative of former glaciation of the Martian surface. (a) Argyre basin forms: (A) anastomosing sinuous ridges; (B) sinuous ridges with double-ridged, rounded and sharp-crested sections. Arrowed dome may be a pingo; (C) sharp-crested semicircular mountain ridge (arrow) thought to be a crater modified into a cirque; (D) lobate rock glaciers flowing from cirques; (E) braided fluvial erosional and/or depositional features; (F) complex terrain formed by ice disintegration, glacilacustrine deposition and cliff erosion by wave action; (G) pitted terrain interpreted as kame and kettle topography; (H) sculptured terrain eroded by ice and/or meltwater. Arrows indicate segmented esker; (I) quasi-dendritic esker (arrow top left) and tunnel valley system (arrow lower right), traceable to braided complex in (E). (b) Hella area forms: (A) lineated forms possibly produced by glacial scouring; (B) cuspate ridges (arrows) interpreted as terminal moraines; (C) possible glacilacustrine plains (left and bottom), wave-cut cliffs (prominent diagonal structure), and changes in channel morphology as they enter the former lake area. (Photographs taken by Viking Orbiter and provided by J.S. Kargel)

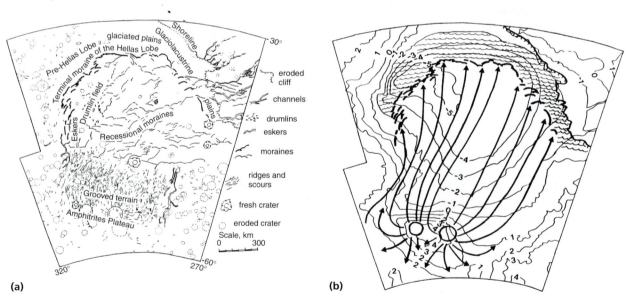

Fig. 12.87 Martian ice cap reconstruction: (a) glacial geomorphology map of the Hellas area of the Martian surface; (b) former direction of ice flow (arrows) and position of a possible proglacial lake, based upon glacial geomorphology. Contours in kilometres. (From Kargel and Strom, 1992a. Reproduced by permission of the Geological Society of America)

APPENDIX

1

NOTATIONS USED IN FORMULAE

α	glacier surface slope
$\Delta p/d$	porewater pressure gradient
ΔR	the difference in vapour density between the top of debris-rich ice and the overlying atmosphere
$\Delta^{18}O$	the variation in the proportion of the oxygen isotopes ^{16}O and ^{18}O
η	fluid viscosity
$\varnothing$	hydraulic potential
φ_i	angle of internal friction
$\varnothing_0$	a constant in the formula describing hydraulic potential of a conduit
$\varnothing_e$	potential due to elevation in the formula describing hydraulic potential of a conduit
λ	bump wavelength on the glacier bed
ρ_a	the density of air
ρ_i	the density of ice ($c.$ 900 kg m^{-3})
$\rho_i{}^*$	density of ice with dispersed debris
ρ_p	the density of a particle
ρ_r	the density of rock ($c.$ 2700 kg m^{-3} for granite)
ρ_w	the density of water (1000 kg m^{-3} for fresh water and 1030 kg m^{-3} for sea-water)
ρ_w/dz	vertical density gradient in a column of water
σ	normal or glaciostatic stress in a glacier
σ_{gt}	total glacitectonic stress
σ_i	average normal stress at the contact point of a particle and the bed
σ_x	horizontal stress in a glacier
τ	shear stress
τ_b	basal shear stress in a glacier
τ'_b	effective basal shear stress
τ_f	basal friction or frictional drag in a glacier
τ^*	the total shear strength of a material (used in the Coulomb equation)
A	cross-sectional area at right angles to water flow in a channel; also used in Glen's flow law as a constant relating to ice temperature; drumlin area
A_b	area of ablation on a glacier
$A_{bb'}$	the ablation rate for buried ice
A_c	area of accumulation on a glacier
A_i	the apparent area of contact between a particle and the bed
A_r	the real area of contact between a particle and the bed
$A(x)$	glacier cross-sectional area
a	a constant dependent on material properties in the flow law in equation 4.9; surface slope of ice; long-axis length; miscellaneous constants
a_b	ablation rate
B_n	net balance (glacier mass balance)
b	a constant dependent on material properties used in the flow law in equation (4.9); miscellaneous constants
b_{nb}	mass balance gradient in the ablation area of a glacier
b_{nc}	mass balance gradient in the accumulation area of a glacier
BR	balance ratio (ratio between accumulation and ablation gradients)

b_x	specific net balance of point upglacier from point x (used in ice discharge equation)
C	sediment concentration by volume
C_{en}	solute concentration in englacial component of meltwater stream flow
C_{sub}	solute concentration in subglacial component of meltwater stream flow
C_t	solute concentration in meltwater stream flow
c	cohesion
D	the average intermediate axis of a particle or grain diameter (in metres)
D_b	depth of burial relative to ice block height
$\bar{d}_b$	average net annual ablation in the ablation area of a glacier
$\bar{d}_c$	average net annual accumulation in the accumulation area of a glacier
$d_{wc}U_w/dz$	vertical velocity gradient in a column of water
d_{wc}	depth in the water column
E	evaporation; elongation ratio
E_s	strain rate
F	the total frictional force
F_b	buoyant weight
F_i	the frictional force between a particle and the bed
Fr	form ratio
f	a factor that modifies the drag force for the near-bed condition
GB	glacier contributing area
g	gravitational acceleration (9.81 m s^{-2})
grad φ	gradient in hydraulic potential
grad H	ice surface gradient
grad z	gradient of englacial passage
H	altitude of glacier ice surface; depth of a hollow
H_i	height of an ice block
h	glacier ice thickness (m); height of rim ridge
h_d	debris layer thickness
h_w	water depth
h_{wf}	water depth required for the flotation of ice
K	a constant dependent on material properties used in the flow law in equation (4.9)
k	hydraulic conductivity; thermal conductivity; miscellaneous constants
k_d	thermal conductivity of a debris layer
L	latent heat of fusion of ice
L_f	trough length
l	maximum bedform length
N	effective normal stress (the difference between water pressure and ice overburden pressure)
n	Manning's n roughness coefficient; used also as unrelated constants in Glen's flow law, usually representing 3, and in the basal sliding velocity equation (4.13)
P	precipitation
P_c	critical water pressure (relating to glacier bed sliding)
P_s	separation pressure
p	the perimeter of a channel that is under water (used also as an empirically determined constant in the basal sliding equation, equation (4.14)
p_i	ice overburden pressure
p_m	the pressure change due to melting or contraction of conduit walls
P_w	water pressure
Q_e	latent heat transfer
Q_{en}	englacial meltwater discharge component (used in meltstream solute content equation)
Q_h	sensible heat transfer
Q_l	net long-wave radiation flux
Q_m	energy available to melt ice
Q_{max}	peak flood discharge (m^3 s^{-1})
Q_{mean}	mean discharge (m^3 s^{-1})
Q_s	net short-wave radiation flux
Q_{sub}	subglacial meltwater discharge component (used in meltstream solute content equation)
Q_t	total discharge in glacial meltwater stream (used in meltstream solute content equation)

$Q(x)$	discharge of ice through a cross-section at a distance x from the highest point on a glacier
q	an empirically determined constant used in the basal sliding equation, equation (4.14)
R	runoff; form of a drumlin, as given by the rose curve
R_a	particle radius
R_{ax}	transition radius analogous to the controlling obstacle size in sliding theory (0.1 m)
R_n	Richardson number (calculated to determine the stability of a water column)
RMI	index of rim morphology
r	hydraulic radius of a channel; form of a lemniscate loop
S	channel bed slope
s	the proportion of a glacier bed occupied by cavities
$\tan \varphi_i$	coefficient of friction
T_i	sum of the mean daily temperature
T_0	temperature of debris surface
t	time
U_{tx}	horizontal velocity component of glacier flow
U_w	velocity of water entering a lake or ocean
u_b	basal velocity of a glacier
u_n	glacier ice velocity normal to the bed
u_s	glacier surface velocity
V_d	valley depth
V_0	total volume of water drained from an ice-dammed lake (m^3)
v	flood peak velocity
v_f	the fall velocity of a particle
$v(x)$	average ice velocity through cross-section x on the glacier
WI	valley top width
w	valley half-width; maximum bedform width; width of a rim ridge
w_x	glacier width
z	elevation; till thickness
z_b	area-weighted mean altitude of the ablation area of a glacier
z_c	area-weighted mean altitude of the accumulation area of a glacier

APPENDIX

2

GRAIN SIZE SCALES

Phi (Ø) scale	Size (mm)	British sieve sizes	US sieve no.	Wentworth scale
				boulder
−8	256			
−6	64	63 mm		cobble
−4	16	20 mm		
−2	4	6.3 mm	5	pebble
−1.75	3.36	3.35 mm	6	
−1.5	2.83		7	granule
−1.25	2.38		8	
−1	2	2.00 mm	10	
−0.75	1.68		12	
−0.5	1.41		14	very coarse
−0.25	1.19	1.18 mm	16	sand
0	1		18	
0.25	0.84		20	
0.5	0.71		25	coarse sand
0.75	0.59	600 µm	30	
1	0.50		35	
		425 µm		
1.25	0.42		40	
1.5	0.35		45	medium sand
1.75	0.30	300 µm	50	
2	0.25	212 µm	60	
2.25	0.210		70	
2.5	0.177		80	fine sand
2.75	0.149	150 µm	100	
3	0.125		120	
3.25	0.105		140	
3.5	0.088		170	very fine
3.75	0.074		200	sand
4	0.0625	63 µm	230	
4.25	0.053		270	
4.5	0.044		325	coarse silt
4.75	0.037		400	
5	0.031			
6	0.0156			medium silt
7	0.0078			fine silt
8	0.0039			very fine silt
9	0.0020			
10	0.00098			
11	0.00049			
12	0.00024			clay
13	0.00012			
14	0.00006			

APPENDIX

2

GRAIN SIZE SCALES

Phi (Ø) scale	Size (mm)	British sieve sizes	US sieve no.	Wentworth scale
				boulder
−8	256			
−6	64	63 mm		cobble
−4	16	20 mm		
−2	4	6.3 mm	5	pebble
−1.75	3.36	3.35 mm	6	
−1.5	2.83		7	granule
−1.25	2.38		8	
−1	2	2.00 mm	10	
−0.75	1.68		12	
−0.5	1.41		14	very coarse
−0.25	1.19	1.18 mm	16	sand
0	1		18	
0.25	0.84		20	
0.5	0.71		25	coarse sand
0.75	0.59	600 µm	30	
1	0.50		35	
		425 µm		
1.25	0.42		40	
1.5	0.35		45	medium sand
1.75	0.30	300 µm	50	
2	0.25	212 µm	60	
2.25	0.210		70	
2.5	0.177		80	fine sand
2.75	0.149	150 µm	100	
3	0.125		120	
3.25	0.105		140	
3.5	0.088		170	very fine
3.75	0.074		200	sand
4	0.0625	63 µm	230	
4.25	0.053		270	
4.5	0.044		325	coarse silt
4.75	0.037		400	
5	0.031			
6	0.0156			medium silt
7	0.0078			fine silt
8	0.0039			very fine silt
9	0.0020			
10	0.00098			
11	0.00049			
12	0.00024			clay
13	0.00012			
14	0.00006			

REFERENCES

Aario, R. 1972. Association of bed forms and paleocurrent patterns in an esker delta, Haapajarvi, Finland. *Ann. Acad. Scientarium Fennicae*, series A, part III, 111.

Aario, R. 1977. Classification and terminology of moraine landforms in Finland. *Boreas* 6, 87–100.

Aartolahti, T. 1972. On deglaciation in southern and western Finland. *Fennia* 114, 1–84.

Aber, J.S. 1979. Kineto-stratigraphy at Hvideklint, Møn, Denmark and its regional significance. *Bulletin of the Geological Society of Denmark* 28, 81–93.

Aber, J.S. 1982. Model for glaciotectonism. *Bulletin of the Geological Society of Denmark* 30, 79–90.

Aber, J.S. 1985a. The character of glaciotectonism. *Geologie en Mijnbouw* 64, 389–395.

Aber, J.S. 1985b. Definition and model for Kansan glaciation. *Ter-Qua Symposium Series* 1, 53–60.

Aber, J.S. 1988a. *Structural Geology Exercises with Glaciotectonic Examples*. Hunter Textbooks, Winston-Salem.

Aber, J.S. 1988b. Spectrum of constructional glaciotectonic landforms. In Goldthwait, R.P. and Matisch, C.L. (eds), *Genetic Classification of Glacigenic Deposits*. Balkema, Rotterdam, 281–292.

Aber, J.S. (ed.) 1993a. *Glaciotectonics and Mapping Glacial Deposits*. Canadian Plains Research Centre, University of Regina.

Aber, J.S. 1993b. Geomorphic and structural genesis of the Dirt Hills and Cactus Hills, southern Saskatchewan. In Aber, J.S. (ed.), *Glaciotectonics and Mapping Glacial Deposits*. Canadian Plains Research Centre, University of Regina, 9–35.

Aber, J.S., Croot, D.G. and Fenton, M.M. 1989. *Glaciotectonic Landforms and Structures*. Kluwer, Dordrecht.

Ackley, S.F. 1982. Ice scavenging and nucleation: two mechanisms for incorporation of an alga into newly forming sea ice. *EOS Transactions of the American Geophysical Union* 63, 54.

Addison, K. 1981. The contribution of discontinuous rock-mass failure to glacier erosion. *Annals of Glaciology* 2, 3–10.

Adie, R.J. 1975. Permo-Carboniferous glaciation of the southern hemisphere. In Wright, A.E. and Moseley, F. (eds), *Ice Ages: Ancient and Modern*. Seel House Press, Liverpool, 287–300.

Adrielsson, L. 1984. Weichselian lithostratigraphy and glacial environments in the Ven–Gumslov area, southern Sweden. Lundqua Thesis 14, Lund University, Sweden.

Agassiz, L. 1838a. On the polished and striated surfaces of the rocks which form the beds of glaciers in the Alps. *Proceedings of the Geological Society of London* 3, 321–322.

Agassiz, L. 1838b. On glaciers and the evidence of their having once existed in Scotland, Ireland and England. *Proceedings of the Geological Society of London* 3, 327–332.

Agassiz, L. 1847. *Nouvelles études et experiences sur les glaciers actuels, leur structure, leur progression et leur action physique sur le sol*. Paris and Neuchâtel.

Ageta, Y. and Higuchi, K. 1984. Estimation of mass balance components of a summer-accumulation type glacier in the Nepal Himalaya. *Geografiska Annaler* 66A, 249–255.

Ahlmann, H.W. 1919. Geomorphological studies in Norway. *Geografiska Annaler* 1, 3–148 and 193–252.

Ahlmann, H.W. 1924. Le niveau de glaciation comme fonction de l'accumulation d'humidité sous forme solide. *Geografiska Annaler* 6, 224–272.

Ahlmann, H.W. 1935. The Fourteenth of July Glacier. *Geografiska Annaler* 17, 167–218.

Ahlmann, H.W. 1941. The main morphological features of north-east Greenland. *Geografiska Annaler* 23, 148–182.

Ahlmann, H.W. 1946. The Froya Glacier in 1939–40. *Geografiska Annaler* 28, 239–257.

Ahlmann, H.W. 1948. Glaciological research on the North Atlantic coasts. *Royal Geographical Society Research Series* 1.

Ahlmann, H.W. and Thorarinsson, S. 1938. Vatnajökull: scientific results of the Swedish–Icelandic investigations 1936–38, Chap. V – The ablation. *Geografiska Annaler* 20, 171–233.

Aitkenhead, N. 1960. Observations on the drainage of a glacier-dammed lake in Norway. *Journal of Glaciology* 3, 607–609.

Albertson, M.L., Dai, Y.B., Jensen, R.A. and Rouse, H. 1950. Diffusion of submerged jets. *Transactions of the American Society of Civil Engineers* 115, 639–664.

Albino, K. and Dreimanis, A. 1988. A time-transgressive kinetostratigraphic sequence spanning 180° in a single section at Bradtville, Ontario, Canada. In Croot, D.G. (ed.), *Glaciotectonics. Forms and Processes*. Balkema, Rotterdam, 11–20.

Alden, W.C. 1905. The drumlins of southeastern Wisconsin. *US Geological Survey Bulletin* 273.

Alexander, H.S. 1932. Pothole erosion. *Journal of Geology* 40, 305–337.

Allen, C.R., Kamb, W.B., Meier, M.F. and Sharp, R.P. 1960. Structure of the Lower Blue Glacier, Washington. *Journal of Geology* 68, 601–625.

Allen, J.R.L. 1963. The classification of cross-stratified units, with notes on their origin. *Sedimentology* 2, 93–114.

Allen, J.R.L. 1965. Sedimentation in the lee of small underwater sand waves, an experimental study. *Journal of Geology* 73, 95–116.

Allen, J.R.L. 1970. A quantitative model of climbing ripples and their cross-laminated deposits. *Sedimentology* 14, 5–26.

Allen, J.R.L. 1971a. Transverse erosional marks of mud and rock: their physical basis and geological significance. *Sedimentary Geology* 5, 167–185.

Allen, J.R.L. 1971b. A theoretical and experimental study of climbing-ripple cross-lamination, with a field application to the Uppsala esker. *Geografiska Annaler* 53A, 157–187.

Allen, J.R.L. 1973. A classification of climbing-ripple cross-lamination. *Journal of the Geological Society of London* 129, 537–541.

Allen, J.R.L. 1982a. *Sedimentary Structures*. Elsevier, Amsterdam.

Allen, J.R.L. 1982b. Late Pleistocene (Devensian) glaciofluvial outwash at Banc-y-Warren, near Cardigan (west Wales). *Geological Journal* 17, 31–47.

Allen, J.R.L. 1984. Experiments on the settling, overturning and entrainment of bivalve shells and related models. *Sedimentology* 31, 227–250.

Allen, J.R.L. 1985. *Principles of Physical Sedimentology*. Allen & Unwin, London.

Allen, J.R.L. 1993. Studies in fluviatile sedimentation: bars, bar complexes and sandstone sheets (low-sinuosity braided streams) in the Brownstones (L. Devonian), Welsh Borders. *Sedimentary Geology* 33, 237–293.

Allen, P. 1975. Ordovician glacials of the central Sahara. In Wright, A.E. and Moseley, F. (eds), *Ice Ages: Ancient and Modern*. Seel House Press, Liverpool, 275–286.

Alley, N.F. 1973. Glacial stratigraphy and the limits of the Rocky Mountain and Laurentide ice sheets in southwestern Alberta, Canada. *Canadian Petroleum Geology Bulletin* 21, 153–177.

Alley, R.B. 1983. In search of ice stream sticky spots. *Journal of Glaciology* 39, 447–454.

Alley, R.B. 1989. Water-pressure coupling of sliding and bed deformation. *Journal of Glaciology* 35, 108–139.

Alley, R.B. 1991. Deforming bed origin for southern Laurentide till sheets? *Journal of Glaciology* 37, 67–76.

Alley, R.B. 1992a. Flow-law hypotheses for ice-sheet modeling. *Journal of Glaciology* 38, 245–256.

Alley, R.B. 1992b. How can low-pressure channels and deforming tills coexist subglacially? *Journal of Glaciology* 38, 200–207.

Alley, R.B., Blankenship, D.D., Bentley, C.R. and Rooney, S.T. 1986. Deformation of till beneath Ice Stream B, West Antarctica. *Nature* 322, 57–59.

Alley, R.B., Blankenship, D.D., Bentley, C.R. and Rooney, S.T. 1987a. Till beneath Ice Stream B: 3. Till deformation: evidence and implications. *Journal of Geophysical Research* 92, 8921–8929.

Alley, R.B., Blankenship, D.D., Bentley, C.R. and Rooney, S.T. 1987b. Till beneath Ice Stream B: 4. A coupled ice-till flow model. *Journal of Geophysical Research* 92, 8931–8940.

Alley, R.B., Blankenship, D.D., Rooney, S.T. and Bentley, C.R. 1987c. Continuous till deformation beneath ice sheets. In Waddington, E.D. and Walder, J.S. (eds), *The Physical Basis of Ice Sheet Modelling*. IAHS Publication 170, 81–91.

Alley, R.B., Blankenship, D.D., Rooney, S.T. and Bentley, C.R. 1989. Sedimentation beneath ice shelves: the view from Ice Stream B. *Marine Geology* 85, 101–120.

Allison, I. and Kruss, P. 1977. Estimation of recent climatic change in Irian Jaya by numerical modelling of its tropical glaciers. *Arctic and Alpine Research* 9, 49–60.

Allison, I. and Peterson, J.A. 1989. Glaciers of Irian Jaya, Indonesia. In Williams, R.S. and Ferrigno, J.G. (eds), *Satellite Image Atlas of the World*. USGS Professional Paper 1386-H, 1–20.

Alt, B.T. 1978. Synoptic climate controls of mass balance variations on Devon Island ice cap. *Arctic and Alpine Research* 10, 61–80.

Åmark, M. 1980. Glacial flutes at Isfallsglaciären, Tarfala, Swedish Lapland. *Geologiska Foreningens i Stockholm Forhandlingar* 102, 251–259.

Åmark, M. 1986. Glacial tectonics and deposition of stratified drift during formation of tills beneath an active glacier: examples from Skåne, southern Sweden. *Boreas* 15, 155–171.

Ambach, W. 1955. Uber die Strahlungsdurchlassigkeit des Gletschereises. *Sitzungsberichte der Osterreichischen Akademie der Wissenschaften. Math-naturwiss. Klasse*, Abt. 2, Bd. 164, Ht. 10, 483–494.

Ambach, W. 1985. Characteristics of the heat balance of the Greenland Ice Sheet for modelling. *Journal of Glaciology* 31, 3–12.

Ambrose, J. W. 1964. Exhumed paleoplains of the Precambrian shield of North America. *Journal of Science* 262, 817–857.

Andersen, B.G. 1965. The Quaternary of Norway. In Rankama, K. (ed.), *The Quaternary 1*. Interscience, New York, 91–138.

Andersen, B.G. 1979. The deglaciation of Norway 15,000–10,000 BP. *Boreas* 8, 79–87.

Andersen, B.G. 1980. The deglaciation of Norway after 10,000 BP. *Boreas* 9, 211–216.

Andersen, B.G. 1981. Late Weichselian ice sheets in Eurasia and Greenland. In Denton, G.H. and Hughes, T.J. (eds), *The Last Great Ice Sheets*. Wiley, New York, 3–178.

Anderson, B.G. and Borns, H.W. 1994. *The Ice Age World*. Scandinavian University Press, Oslo.

Andersen, J.L. and Sollid, J.L. 1971. Glacial chronology and glacial geomorphology in the marginal zones of the glaciers Midtdalsbreen and Nigardsbreen, south Norway. *Norsk Geografisk Tidsskrift* 25, 1–38.

Anderson, J.B., Brake, C.F., Domack, E.W., Myers, N. and Wright, R. 1983. Development of a polar glacial-marine sedimentation model from Antarctic Quaternary deposits and glaciological information. In Molnia, B.F. (ed.), *Glacial-Marine Sedimentation*. Plenum Press, New York, 233–264.

Anderson, J.B., Kennedy, D.S., Smith, M.J. and Domack, E.W. 1991. Sedimentary facies associated with Antarctica's floating ice masses. In Anderson, J.B. and Ashley, G.M. (eds), *Glacial Marine Sedimentation: Paleoclimatic Significance*. Geological Society of America Special Paper 261, 1–25.

Anderson, J.B., Kurtz, D.D., Domack, E.W. and Balshaw, K.M. 1980. Antarctic glacial marine sediments. *Journal of Geology* 88, 399–414.

Anderson, J.B., Kurtz, D.D. & Weaver, F.M. 1979. Sedimentation on the Antarctic continental slope. In Doyle,

L.J. and Pilkey, O.H. (eds), *Geology of Continental Slopes*. SEPM Special Publication 27, 265–283.

Anderson, J.B., Kurtz, D.D. and Weaver, F.M. 1982. Sedimentation on the West Antarctic continental margin. In Craddock, C. (ed.), *Antarctic Geoscience*. University of Wisconsin, Madison, 1003–1012.

Anderson, L.W. 1978. Cirque glacier erosion rates and characteristics of Neoglacial tills, Pangnirtung Fiord area, Baffin Island, NWT, Canada. *Arctic and Alpine Research* 10, 749–760.

Anderson, R.S., Hallet, B., Walder, J. and Aubry, B.F. 1982. Observations in a cavity beneath Grinell Glacier. *Earth Surface Processes and Landforms* 7, 63–70.

André, M.F. 1985. Lichenométrie et vitesses d'évolution des versants arctiques pendant l'Holocène (Région de la Baie du Roi, Spitsberg, 79° N). *Revue de Géomorphologie Dynamique* 1985, 49–72.

André, M.F. 1986. Dating slope deposits and estimating rates of rockwall retreat in northwest Spitsbergen by lichenometry. *Geografiska Annaler* 68A, 65–75.

Andrews, D.E. 1980. Glacially thrust bedrock: an indication of late Wisconsin climate in western New York State. *Geology* 8, 97–101.

Andrews, J.T. 1963. The cross-valley moraines of north-central Baffin Island, NWT, a descriptive analysis. *Geographical Bulletin* 19, 49–77.

Andrews, J.T. 1965. The corries of the northern Nain–Okak section of Labrador. *Geographical Bulletin* 7, 129–136.

Andrews, J.T. 1970. *A Geomorphological Study of Postglacial Uplift with Particular Reference to Arctic Canada*. Institute of British Geographers, Special Publication 2.

Andrews, J.T. 1971. Quantitative analysis of the factors controlling the distribution of corrie glaciers in Okoa Bay, east Baffin Island (with particular reference to global radiation). In Morisawa, M. (ed.), *Quantitative Geomorphology: Some Aspects and Applications*. New York State University Press, 223–241.

Andrews, J.T. 1972. Glacier power, mass balances, velocities and erosion potential. *Zeitschrift für Geomorphologie* Suppl. Bd. 13, 1–17.

Andrews, J.T. (ed.) 1974. *Glacial Isostasy*. Dowden, Hutchinson & Ross, Stroudsburg.

Andrews, J.T. 1975. *Glacial Systems: An Approach to Glaciers and their Environments*. Duxbury Press, North Scituate, Mass.

Andrews, J.T. 1978. Sea level history of Arctic coasts during the upper Quaternary: dating sedimentary sequences and history. *Progress in Physical Geography* 2, 375–407.

Andrews, J.T. 1982. Comment on 'New evidence from beneath the western North Atlantic for the depth of glacial erosion in Greenland and North America'. *Quaternary Research* 17, 123–124.

Andrews, J.T. 1987a. The Late Wisconsin glaciation and deglaciation of the Laurentide Ice Sheet. In Ruddiman, W.F. and Wright, H.E. (eds), *North America and Adjacent Oceans During the Last Deglaciation*. Geological Society of America, *The Geology of North America*, Vol. K–3, 13–37.

Andrews, J.T. 1987b. Glaciation and sea level: a case study. In Devoy, R.J.N. (ed.), *Sea Surface Studies: A Global View*. Croom Helm, London, 95–126.

Andrews, J.T. 1989. Quaternary geology of the northeastern Canadian shield. In Fulton, R.J. (ed.), *Quaternary Geology of Canada and Greenland*. Geological Society of Canada, Geology of Canada, No.1, 276–302.

Andrews, J.T., Austin, W.E.N., Bergsten, H. and Jennings, A.E. (eds) 1996. *Late Quaternary Palaeoceanography of the North Atlantic Margins*. Geological Society Special Publication 111.

Andrews, J.T. and Barnett, D.M. 1979. Holocene (Neoglacial) moraine and proglacial lake chronology, Barnes Ice Cap, Canada. *Boreas* 8, 341–358.

Andrews, J.T. and Barry, R.G. 1978. Glacial inception and disintegration during the last glaciation. *Annual Review of Earth and Planetary Sciences* 6, 205–228.

Andrews, J.T., Barry, R.G. and Draper, L. 1970. An inventory of the present and past glacierization of Home Bay and Okoa Bay, east Baffin Island, NWT, Canada, and some climatic and palaeoclimatic considerations. *Journal of Glaciology* 9, 337–362.

Andrews, J.T. and Boulton, G.S. (1971/1972) Englacial debris in glaciers: correspondence and replies. *Journal of Glaciology* 10, 410–411; 11, 155–156.

Andrews, J.T., Clark, P. and Stravers, J.A. 1985a. The patterns of glacial erosion across the eastern Canadian Arctic. In Andrews, J.T. (ed.), *Quaternary Environments: Eastern Canadian Arctic, Baffin Bay and West Greenland*. Allen & Unwin, London, 69–92.

Andrews, J.T. and Dugdale, R.E. 1971. Quaternary history of northern Cumberland Peninsula, Baffin Island, NWT; Part V: Factors affecting corrie glacierization in Okoa Bay. *Quaternary Research* 1, 532–551.

Andrews, J.T. and Falconer, G. 1969. Late glacial and postglacial history and emergence of the Ottawa Islands, Hudson Bay, NWT: evidence on the deglaciation of Hudson Bay. *Canadian Journal of Earth Sciences* 6, 1263–1276.

Andrews, J.T. and Ives, J.D. 1972. Late and postglacial events (<10,000 BP) in the eastern Canadian arctic with particular reference to the Cockburn Moraines and breakup of the Laurentide ice sheet. In Vasari, Y., Hyvarinen, H. and Hicks, S. (eds), *Climatic Changes in Arctic Areas during the Last 10,000 Years*. Univ. Ouluensis Acta, Ser. A, Sci. Natur. No. 3, Geol. No. 1, 149–174.

Andrews, J.T. and King, C.A.M. 1968. Comparative till fabrics and till fabric variability in a till sheet and a drumlin: a small scale study. *Proceedings of the Yorkshire Geological Society* 36, 435–461.

Andrews, J.T. and LeMasurier, W.E. 1973. Rates of Quaternary glacial erosion and corrie formation, Marie Byrd Land, Antarctica. *Geology*, October, 75–80.

Andrews, J.T. and Matsch, C.L. 1983. *Glacial Marine Sediments and Sedimentation: An Annotated Bibliography*. Geo Abstracts, Norwich.

Andrews, J.T. and Miller, G.H. 1972. Quaternary history of Northern Cumberland Peninsula, Baffin Island, NWT, Canada. Part IV: Maps of the present glaciation limits and lowest equilibrium line altitude for north and south Baffin Island. *Arctic and Alpine Research* 4, 45–59.

Andrews, J.T. and Miller, G.H. 1985. Holocene sea level variations within Frobisher Bay. In Andrews, J.T. (ed.), *Quaternary Environments: Eastern Canadian Arctic, Baffin Bay, and Western Greenland*. Allen & Unwin, London, 585–607.

Andrews, J.T. and Peltier, W.R. 1976. Collapse of the Hudson Bay ice center and glacio-isostatic rebound. *Geology* 4, 73–75.

Andrews, J.T., Shilts, W.W. and Miller, G.H. 1983. Multiple deglaciations of the Hudson Bay lowlands, Canada, since deposition of the Missinaibi (last interglacial?) Formation. *Quaternary Research* 19, 18–37.

Andrews, J.T., Shilts, W.W. and Miller, G.H. 1984. Reply to A.S. Dyke's discussion. *Quaternary Research* 22, 253–258.

Andrews, J.T. and Smithson, B.B. 1966. Till fabrics of the cross-valley moraines of north-central Baffin Island, NWT, Canada. *Bulletin of the Geological Society of America* 77, 271–290.

Andrews, J.T., Stravers, J.A. and Miller, G.H. 1985b. Patterns of glacial erosion and deposition around Cumberland Sound, Frobisher Bay and Hudson Strait, and the location of ice streams in the eastern Canadian Arctic. In Woldenburg M. (ed.), *Models in Geomorphology*. Allen & Unwin, London, 93–117.

Andrews, J.T. and Tedesco, K. 1992. Detrital carbonate-rich sediments, northwestern Labrador Sea: implications for ice-sheet dynamics and iceberg-rafting (Heinrich) events in the North Atlantic. *Geology* 20, 1087–1090.

Andrews, J.T. and Webber, P.J. 1964. A lichenometrical study of the northwestern margin of the Barnes Ice Cap: a geomorphological technique. *Geographical Bulletin* 22, 80–104.

Andrews, J.T. and Webber, P.J. 1969. Lichenometry to evaluate changes in glacial ice mass budgets: as illustrated from north central Baffin Island, NWT. *Arctic and Alpine Research* 1, 181–194.

Andriashek, L.D. and Fenton, M.M. 1989. Quaternary stratigraphy and surficial geology, Sand River Map Sheet 73L, Alberta. Terrain Sciences Department, Alberta Research Council, Bulletin 57.

Aniya, M. 1988. Glacier inventory for the Northern Patagonia Icefield, Chile, and variations between 1944/45 to 1985/86. *Arctic and Alpine Research* 20, 179–187.

Aniya, M. and Welch, R. 1981. Morphometric analysis of Antarctic cirques from photogrammetric measurements. *Geografiska Annaler* 63A, 41–54.

Anundsen, K. 1990. Evidence of ice movement over southwest Norway indicating an ice dome over the coastal district of west Norway. *Quaternary Science Reviews* 9, 99–116.

Arkhipov, S.A., Bespaly, V.G., Faustova, M.A., Glushkova, O.Y., Isayeva, L.L. and Velichko, A.A. 1986a. Ice-sheet reconstructions. *Quaternary Science Reviews* 5, 475–483.

Arkhipov, S.A., Ehlers, J., Johnson, R.G. and Wright, H.E. 1995. Glacial drainage towards the Mediterranean during the Middle and Late Pleistocene. *Boreas* 24, 196–206.

Arkhipov, S.A., Isayeva, L.L., Bespaly, V.G. and Glushkova, O.Y. 1986b. Glaciation of Siberia and northeast USSR. *Quaternary Science Reviews* 5, 463–474.

Armentrout, J.M. 1983. Glacial lithofacies of the Neogene Yakataga Formation, Robinson Mountains, southern Alaska Coast Range, Alaska. In Molnia, B.F. (ed.), *Glacial-Marine Sedimentation*. Plenum Press, New York, 629–666.

Armstrong, J.E. 1981. Post Vashon Wisconsin glaciation, Fraser Lowland, British Columbia. *Geological Survey of Canada Bulletin* 322.

Armstrong, T.E., Roberts, B. and Swithinbank, C.W.M. 1973. *Illustrated Glossary of Snow and Ice*. Scott Polar Research Institute, Cambridge.

Armstrong, T.E., Roberts, B. and Swithinbank, C.W.M. 1977. Proposed new terms and definitions for ice features. *Polar Record* 18, 501–502.

Arnborg, L. 1955. Ice-marginal lakes at Hoffellsjokull. *Geografiska Annaler* 37, 202–228.

Aronow, S. 1959. Drumlins and related streamline features in the Warwick–Tokio area, North Dakota. *American Journal of Science* 257, 191–203.

Ashley, G.M. 1975. Rhythmic sedimentation in glacial Lake Hitchcock, Massachusetts-Connecticut. In Jopling, A.V. and MacDonald, B.C. (eds), *Glaciofluvial and Glaciolacustrine Sedimentation*. SEPM Special Publication 23, 304–320.

Ashley, G.M. 1990. Classification of large-scale subaqueous bedforms: a new look at an old problem. *Journal of Sedimentary Petrology* 60, 160–172.

Ashley, G.M. 1995. Glaciolacustrine environments. In Menzies, J. (ed.), *Glacial Environments, Vol. 1: Modern Glacial Environments: Processes, Dynamics and Sediments*. Butterworth-Heinemann, Oxford, 417–444.

Ashley, G.M., Boothroyd, J.C. and Borns, H.W. 1991. Sedimentology of late Pleistocene (Laurentide) deglacial-phase deposits, eastern Maine: an example of a temperate marine grounded ice-sheet margin. In Anderson, J.B. and Ashley, G.M. (eds), *Glacial Marine Sedimentation: Paleoclimatic Significance*. Geological Society of America Special Paper 261, 107–125.

Ashley, G.M., Southard, J.B. and Boothroyd, J.C. 1982. Deposition of climbing-ripple beds: a flume simulation. *Sedimentology* 29, 67–79.

Ashmore, P.E. 1991. How do gravel-bed rivers braid? *Canadian Journal of Earth Sciences* 28, 326–341.

Astakhov, V.I. and Isayeva, L.L. 1988. The 'ice-hill': an example of 'retarded deglaciation' in Siberia. *Quaternary Science Reviews* 7, 29–40.

Ashworth, P.J. and Ferguson, R.I. 1986. Interrelationships of channel processes, changes and sediments in a proglacial river. *Geografiska Annaler* 68A, 261–371.

Atkinson, B.K. 1984. Subcritical crack growth in geological materials. *Journal of Geophysical Research* 89B, 4077–4114.

Atkinson, B.K. (ed.) 1987. *Fracture Mechanics of Rock*. Academic Press, London.

Atkinson, B.K. and Meredith, P.G. 1987. The theory of subcritical crack growth with applications to minerals and rocks. In B.K. Atkinson (ed.), *Fracture Mechanics of Rock*. Academic Press, London, 111–166.

Atkinson, T.C., Briffa, K.R. and Coope, G.R. 1987. Seasonal temperatures in Britain during the past 22,000 years, reconstructed using beetle remains. *Nature* 325, 587–592.

Attig, J.W. and Clayton, L. 1993. Stratigraphy and origin of an area of hummocky glacial topography, northern Wisconsin, USA. *Quaternary International* 18, 61–67.

Aubert, D. 1980. Les stades de retrait des glaciers du Haut-Valais. *Bulletin Murithienne* 97, 101–169.

Augustinus, P.C. 1992. Outlet glacier trough size–drainage area relationships, Fiordland, New Zealand. *Geomorphology* 4, 347–361.

Auton, C. 1992. Scottish landform examples – 6: The Flemington eskers. *Scottish Geographical Magazine* 108, 190–196.

Aylsworth, J.M. and Shilts, W.W. 1989a. Bedforms of the Keewatin Ice Sheet, Canada. *Sedimentary Geology* 62, 407–428.

Aylsworth, J.M. and Shilts, W.W. 1989b. Glacial features around the Keewatin Ice Divide, Districts of Mackenzie and Keewatin. Geological Survey of Canada, Paper 88–24.

Bagnold, R.A. 1941. *The Physics of Blown Sand and Desert Dunes*. Methuen, London.

Bagnold, R.A. 1954. Experiments on gravity-free dispersion of large solid spheres in a Newtonian fluid under shear. *Proceedings of the Royal Society of London* A225, 49–63.

Bagnold, R.A. 1956. The flow of cohesionless grains in fluids. *Royal Society of London Philosophical Transactions*, Series A 249, 235–297.

Bai, Z. and Yu, X. 1985. Energy exchange and its influence factors on mountain glaciers in west China. *Annals of Glaciology* 6, 154–157.

Bain, G.W. 1931. Spontaneous rock expansion. *Journal of Geology* 39, 715–735.

Baker, V.R. 1973a. Paleohydrology and sedimentology of Lake Missoula flooding in eastern Washington. Geological Society of America, Special Paper 144.

Baker, V.R. 1973b. Erosional forms and processes for the catastrophic Pleistocene Missoula floods in eastern Washington. In Morisawa, M. (ed.), *Fluvial Geomorphology*. Allen & Unwin, London, 123–148.

Baker, V.R. 1978. Large-scale erosional and depositional features of the Channeled Scablands. In Baker, V.R. and Nummendal, D. (eds), *The Channeled Scablands*. NASA, 81–115.

Baker, V.R. 1979. Erosional processes in channelized water flows on Mars. *Journal of Geophysical Research* 84, 7985–7993.

Baker, V.R. 1981. *Catastrophic Flooding: The Origin of the Channeled Scabland*. Dowden, Hutchinson & Ross, Stroudsburg, Penn.

Baker, V.R., Benito, G. and Rudoy, A.N. 1993. Paleohydrology of late Pleistocene superflooding, Altay Mountains, Siberia. *Science* 259, 348–350.

Baker, V.R. and Bunker, R.C. 1985. Cataclysmic late Pleistocene flooding from glacial Lake Missoula, a review. *Quaternary Science Reviews* 4, 1–41.

Baker, V.R., Greeley, R., Komar, P.D., Swanson, D.A. and Waitt, R.B. 1987. Columbia and Snake River Plains. In Graf, W.L. (ed.), *Geomorphic Systems of North America*. Geological Society of America, Centennial Special Volume 2, 403–468.

Baker, V.R. and Kochel, R.C. 1978. Morphometry of streamlined forms in terrestrial and Martian channels. National Aeronautics and Space Administration, *Proceedings of the 9th Lunar and Planetary Science Conference*, 3193–3203.

Bakker, J.P. 1965. A forgotten factor in the interpretation of glacial stairways. *Zeitschrift für Geomorphologie* 9, 18–34.

Balise, M.J. and Raymond, C.F. 1985. Transfer of basal sliding variations to the surface of a linearly viscous glacier. *Journal of Glaciology* 31, 308–318.

Ballantyne, C.K. 1978. The hydrologic significance of nivation features in permafrost areas. *Geografiska Annaler* 60A, 51–54.

Ballantyne, C.K. 1981. Periglacial landforms and environments in the Northern Highlands of Scotland. Unpublished PhD thesis, University of Edinburgh. 2 vols.

Ballantyne, C.K. 1982a. Aggregate clast form characteristics of deposits near the margins of four glaciers in the Jotumheimen Massif, Norway. *Norsk Geografisk Tidsskrift* 36, 103–113.

Ballantyne, C.K. 1982b. Depths of open joints and the limits of former glaciers. *Scottish Journal of Geology* 18, 250–252.

Ballantyne, C.K. 1984. The late Devensian periglaciation of upland Scotland. *Quaternary Science Reviews* 3, 311–343.

Ballantyne, C.K. 1987. Some observations of the morphology and sedimentology of two active protalus ramparts, Lyngen, northern Norway. *Arctic and Alpine Research* 19, 167–174.

Ballantyne, C.K. 1989. The Loch Lomond Readvance on the Isle of Skye, Scotland: glacier reconstruction and palaeoclimatic implications. *Journal of Quaternary Science* 4, 95–108.

Ballantyne, C.K. 1990a. The Holocene glacial history of Lyngshalvoya, northern Norway: chronology and climatic implications. *Boreas* 19, 93–117.

Ballantyne, C.K. 1990b. The late Quaternary glacial history of the Trotternish escarpment, Isle of Skye, Scotland, and its implications for ice sheet reconstruction. *Proceedings of the Geologists' Association* 101, 171–186.

Ballantyne, C.K. 1994. Gibbsitic soils on former nunataks: implications for ice sheet reconstruction. *Journal of Quaternary Science* 9, 73–80.

Ballantyne, C.K. 1995. Paraglacial debris-cone formation on recently-deglaciated terrain. *The Holocene* 5, 25–33.

Ballantyne, C.K. 1996. Periglacial trimlines in the Scottish Highlands. *Quaternary International* 38/39, 119–136.

Ballantyne, C.K. and Benn, D.I. 1994a. Glaciological constraints on protalus rampart development. *Permafrost and Periglacial Processes* 5, 145–153.

Ballantyne, C.K. and Benn, D.I. 1994b. Paraglacial slope adjustment and resedimentation following recent glacier retreat, Fåbergstølsdalen, Norway. *Arctic and Alpine Research* 26, 255–269.

Ballantyne, C.K. and Benn, D.I. 1996. Paraglacial slope adjustment during recent deglaciation: implications for slope evolution in formerly glaciated terrain. In Brooks, S. and Anderson, M.G. (eds), *Advances in Hillslope Processes*, Wiley, Chichester, 1173–1195.

Ballantyne C.K., Black, N.M. and Finlay, D.P. 1989. Enhanced boulder weathering under late-lying snow patches. *Earth Surface Processes and Landforms* 14, 745–750.

Ballantyne, C.K. and Harris, C. 1994. *The Periglaciation of Great Britain*. Cambridge University Press, Cambridge.

Ballantyne, C.K. and McCarroll, D. 1995. The vertical dimensions of Late Devensian glaciation on the mountains of Harris and southeast Lewis, Outer Hebrides, Scotland. *Journal of Quaternary Science* 10, 211–223.

Balling, N. 1980. The land uplift in Fennoscandia, gravity field anomalies and isostasy. In Morner, N.A. (ed.), *Earth Rheology, Isostasy and Eustasy*. Wiley, Chichester, 297–322.

Bally, A.W., Gordy, P.L. and Stewart, G.A. 1966. Structure, seismic data and orogenic evolution of southern Canadian Rocky Mountains. *Bulletin of the Canadian Petroleum Geologists* 14, 337–381.

Balson, P.S. and Jeffery, D.H. 1991. The glacial sequence of the southern North Sea. In Ehlers, J., Gibbard, P.L. and Rose J. (eds), *Glacial Deposits in Great Britain and Ireland*. Balkema, Rotterdam, 245–253.

Banham, P.H. 1977. Glacitectonites in till stratigraphy. *Boreas* 6, 101–105.

Bannacef, A., Beuf, S., Biju-Duval, B., DeCharpal, O., Gariel, O. and Rognon, P. 1971. Example of cratonic sedimentation: Lower Palaeozoic of Algeria Sahara. *American Association of Petroleum Geologists Bulletin* 55, 2225–2245.

Bannerjee, I. and McDonald, B.C. 1975. Nature of esker sedimentation. In Jopling, A.V. and McDonald, B.C. (eds), *Glaciofluvial and Glaciolacustrine Sedimentation*. SEPM Special Publication 23, 132–154.

Baranowski, S. 1969. Some remarks on the origin of drumlins. *Geographia Polonica* 17, 197–208.

Bard, E., Hamelin, B. and Fairbanks, G. 1990. U–Th ages obtained by mass spectrometry in corals from Barbados: sea level during the past 130,000 years. *Nature* 346, 456–458.

Barnes, H.L. 1956. Cavitation as a geological agent. *American Journal of Science* 254, 493–505.

Barnes, P.W. and Lien, R. 1988. Icebergs rework shelf sediments up to 500 m off Antarctica. *Geology* 16, 1130–1133.

Barnes, P.W., Reimnitz, E. and Fox, D. 1982. Ice rafting of fine grained sediment, a sorting and transport mechanism, Beaufort Sea, Alaska. *Journal of Sedimentary Geology* 52, 493–502.

Barnett, D.M. and Holdsworth, G. 1974. Origin, morphology and chronology of sublacustrine moraines, Generator Lake, Baffin Island, NWT, Canada. *Canadian Journal of Earth Sciences* 11, 380–408.

Barnett, F.M. and Finke, P.G. 1971. *Morphometry of Landforms: Drumlins*. US Army Natick Lab., Earth Science Lab., Technical Report ES–63.

Barrett, P.J. 1980. The shape of rock particles: a critical review. *Sedimentology* 27, 291–303.

Barrett, P.J. (ed.) 1989. Antarctic Cenozoic glacial history from the CIROS–1 drillhole, McMurdo Sound. DSIR Bulletin 245, Wellington, New Zealand.

Barrett, P.J., Adams, C.J., McIntosh, W.C., Swisher, C.C. and Wilson, G.S. 1992. Geochronological evidence supporting Antarctic deglaciation three million years ago. *Nature* 359, 816–818.

Barry, R.G. 1985. The cryosphere and climate change. In MacCracken, M.C. and Luther, F.M. (eds), *Detecting the Climatic Effects of Increasing Carbon Dioxide*. US Department of Energy, Washington DC, 111–148.

Barry, R.G. 1992. *Mountain Weather and Climate*, 2nd edition. Routledge, London.

Barry, R.G., Andrews, J.T. and Mahaffy, M.A. 1975. Continental ice sheets: conditions for growth. *Science* 190, 979–981.

Barry, R.G. and Chorley, R.J. 1992. *Atmosphere, Weather and Climate*, 6th edition. Routledge, London.

Barsch, D. 1971. Rock glaciers and ice-cored moraines. *Geografiska Annaler* 53A, 203–206.

Barsch, D. 1973. Refraktionsseismische Bestimmung der Obergrenze des gefrorenen Schuttkorpers in verschiedenen Blockgleschern Graubundens, Schweizer Alpen. *Zeitschrift für Gletscherkunde und Glazialgeologie* 9, 143–167.

Barsch, D. 1987. The problem of the ice-cored rock glacier. In Giardino, J.R., Shroder, J.F. and Vitek, J.D. (eds), *Rock Glaciers*. Allen & Unwin, London, 45–53.

Barsch, D. 1988. Rock glaciers. In Clark, M.J. (ed.), *Advances in Periglacial Geomorphology*. Wiley, Chichester, 69–90.

Bass, D.W. 1980. Stability of icebergs. *Annals of Glaciology* 1, 43–47.

Bates, C.C. 1953. Rational theory of delta formation. *American Association of Petroleum Geologists Bulletin* 37, 2119–2162.

Batterson, M.J. 1989. Glacial dispersal from the Strange Lake alkalic complex, northern Labrador. In DiLabio, R.N.W. and Coker, W.B. (eds), *Drift Prospecting*. Geological Survey of Canada, Paper 89–20, 31–40.

Battey, M.H. 1960. Geological factors in the development of Veslgjuv-botn and Veslskautbreen. In Lewis, W.V. (ed.), *Norwegian Cirque Glaciers*. Royal Geographical Society Research Series 4, 5–10.

Battle, W.R.B. 1960. Temperature observations in bergschrunds and their relationship to frost shattering. In Lewis, W.V. (ed.), *Norwegian Cirque Glaciers*. Royal Geographical Society Research Series 4, 83–95.

Battle, W.R.B. and Lewis, W.V. 1951. Temperature observations in bergschrunds and their relationship to cirque erosion. *Journal of Geology* 59, 537–545.

Beaudoin, A.B. and King, R.H. 1994. Holocene palaeoenvironmental record preserved in a paraglacial alluvial fan, Sunwapta Pass, Jasper National Park, Alberta, Canada. *Catena* 22, 227–248.

Beaudry, L.M. and Prichonnet, G. 1991. Late glacial De Geer moraines with glaciofluvial sediment in the Chapais area, Quebec, Canada. *Boreas* 20, 377–394.

Beck, R.A., Burbank, D.W., Sercombe, W.J., Olson, T.L. and Khan, A.M. 1995. Organic carbon exhumation and global warming during the early Himalayan collision. *Geology* 23, 387–390.

Bednarski, J. 1986. Late Quaternary glacial and sea level events, Clements Markham Inlet, northern Ellesmere Island, arctic Canada. *Canadian Journal of Earth Sciences* 23, 1343–1355.

Bednarski, J. 1988. The geomorphology of glaciomarine sediments in a high arctic fiord. *Géographie Physique et Quaternaire* 42, 65–74.

Behrendt, J.C. and Cooper, A. 1991. Evidence of rapid Cenozoic uplift of the shoulder escarpment of the Cenozoic West Antarctic rift system and a speculation on possible climate forcing. *Geology* 19, 315–319.

Belderson, R.H., Kenyon, N.H. and Wilson, J.B. 1973. Iceberg plough marks in the northeast Atlantic. *Palaeogeography, Palaeoclimatology, Palaeoecology* 13, 215–224.

Belderson, R.H. and Wilson, J.B. 1973. Iceberg plough marks in the vicinity of the Norwegian Trough. *Norsk Geologisk Tiddskrift* 35, 323–328.

Bell, M. and Laine, E.P. 1985. Erosion of the Laurentide region of North America by glacial and glaciofluvial processes. *Quaternary Research* 23, 154–174.

Bell, R.A.I. 1966. A seismic reconnaissance in the McMurdo Sound region, Antarctica. *Journal of Glaciology* 6, 209–221.

Bellar, P., Tourenq, J. and Vernhet, S. 1964. Un échantillon de moraine interne du glacier de l'Astrolabe (Terra Adélie). *Revue de Géographie Physique et de Géologie Dynamique* 6, 115–121.

van Bemmelen, R.W. and Rutten, M.G. 1955. *Tablemountains of Northern Iceland*. Leiden.

Benedict, J.B. 1973a. Chronology of cirque glaciation, Colorado Front Range. *Quaternary Research* 3, 584–599.

Benedict, J.B. 1973b. Origin of rock glaciers. *Journal of Glaciology* 12, 520–522.

Benn, D.I. 1989a. Controls on sedimentation in a Late Devensian ice-dammed lake, Achnasheen, Scotland. *Boreas* 18, 31–42.

Benn, D.I. 1989b. Debris transport by Loch Lomond Readvance glaciers in northern Scotland, basin form and the within-valley asymmetry of lateral moraines. *Journal of Quaternary Science* 4, 243–254.

Benn, D.I. 1990. Scottish Lateglacial moraines: debris supply, genesis and significance. Unpublished PhD thesis, University of St Andrews.

Benn, D.I. 1991. Glacial landforms and sediments on Skye. In Ballantyne, C.K., Benn, D.I., Lowe, J.J. and Walker, M.J.C. (eds), *The Quaternary of the Isle of Skye, Field Guide*. Quaternary Research Association, Cambridge, 35–67.

Benn, D.I. 1992a. The Achnasheen Terraces. *Scottish Geographical Magazine* 108, 128–131.

Benn, D.I. 1992b. The genesis and significance of 'hummocky moraine': evidence from the Isle of Skye, Scotland. *Quaternary Science Reviews* 11, 781–799.

Benn, D.I. 1993. Scottish landform examples – 9: Moraines in Coire na Creiche, Isle of Skye. *Scottish Geographical Magazine* 109, 187–191.

Benn, D.I. 1994a. Fluted moraine formation and till genesis below a temperate glacier: Slettmarkbreen, Jotunheimen, Norway. *Sedimentology* 41, 279–292.

Benn, D.I. 1994b. Fabric shape and the interpretation of sedimentary fabric data. *Journal of Sedimentary Research* A64, 910–915.

Benn, D.I. 1995. Fabric signature of subglacial till deformation, Breidamerkurjökull, Iceland. *Sedimentology* 42, 735–747.

Benn, D.I. 1996a. Subglacial and subaqueous processes near a glacier grounding line: sedimentological evidence from a former ice-dammed lake, Achnasheen, Scotland. *Boreas* 25, 23–36.

Benn, D.I. 1996b. Glacier fluctuations in western Scotland. *Quaternary International* 38, 137–147.

Benn, D.I. and Ballantyne, C.K. 1993. The description and representation of particle shape. *Earth Surface Processes and Landforms* 18, 665–672.

Benn, D.I. and Ballantyne, C.K. 1994. Reconstructing the transport history of glacigenic sediments: a new approach based on the co-variance of clast shape indices. *Sedimentary Geology* 91, 215–227.

Benn, D.I. and Dawson, A.G. 1987. A Devensian glaciomarine sequence in Western Islay, Inner Hebrides. *Scottish Journal of Geology* 23, 175–187.

Benn, D.I. and Evans, D.J.A. 1993. Glaciomarine deltaic deposition and ice-marginal tectonics: the 'Loch Don Sand Moraine', Isle of Mull, Scotland. *Journal of Quaternary Science* 8, 279–291.

Benn, D.I. and Evans, D.J.A. 1996. The interpretation and classification of subglacially-deformed materials. *Quaternary Science Reviews* 15, 23–52.

Benn, D.I. and Gemmell, A.M.D. 1997. Calculating equilibrium line altitudes of former glaciers: a new computer spreadsheet. *Glacial Geology and Geomorphology*. In press.

Benn, D.I., Lowe, J.J. and Walker, M.J.C. 1992. Glacier response to climatic change during the Loch Lomond Stadial and early Flandrian: geomorphological and palynological evidence from the Isle of Skye, Scotland. *Journal of Quaternary Science* 7, 125–144.

Benn, D.I. and Owen, L.A. 1997. The role of the South Asian monsoon and mid-latitude cooling in Himalayan glacial cycles: review and speculative discussion. *Journal of the Geological Society*, in press.

Bennett, M.R. 1990. The cwms of Snowdonia: a morphometric analysis. Queen Mary and Westfield College Research Papers in Geography 2.

Bennett, M.R. 1994. Morphological evidence as a guide to deglaciation following the Loch Lomond Readvance: a review of research approaches and models. *Scottish Geographical Magazine* 110, 24–32.

Bennett, M.R. 1995. The morphology of glacially fluted terrain: examples from the Northwest Highlands of Scotland. *Proceedings of the Geologists' Association* 106, 27–38.

Bennett, M.R. and Boulton, G.S. 1993a. A reinterpretation of Scottish 'hummocky moraine' and its significance for the deglaciation of the Scottish highlands during the Younger Dryas or Loch Lomond Stadial. *Geological Magazine* 130, 301–318.

Bennett, M.R. and Boulton, G.S. 1993b. The deglaciation of the Younger Dryas or Loch Lomond Stadial ice-field in the Northern Highlands, Scotland. *Journal of Quaternary Science* 8, 133–146.

Bennett, M.R. and Glasser, M.F. 1991. The glacial landforms of Glen Geusachan, Cairngorms: a reinterpretation. *Scottish Geographical Magazine* 107, 116–123.

Bennett, M.R. and Glasser, N.F. 1996. *Glacial Geology: Ice Sheets and Landforms*. Wiley, Chichester.

Benson, C.S. 1961. Stratigraphic studies in the snow and firn of the Greenland Ice Sheet. *Folia Geographica Danica* 9, 13–37.

Bentley, C.R. 1984. Some aspects of the cryosphere and its role in climatic change. In Hansen, J.E. and Takahashi, T. (eds), *Climate Processes and Climate Sensitivity*. American Geophysical Union, Washington DC, 207–220.

Bentley, C.R. 1987. Antarctic ice streams: a review. *Journal of Geophysical Research* 92, 8843–8858.

Benxing, Z. 1989. Controversy regarding the existence of a large ice sheet on the Qinghai-Xizang (Tibetan) plateau during the Quaternary Period. *Quaternary Research* 32, 121–123.

van den Berg, M.W. and Beets, D.J. 1987. Saalian glacial deposits and morphology in the Netherlands. In van der Meer, J.J.M. (ed.), *Tills and Glaciotectonics*. Balkema, Rotterdam, 235–251.

Bergen, F.W. and O'Neil, P. 1979. Distribution of Holocene foraminifera in the Gulf of Alaska. *Journal of Paleontology* 53, 1267–1292.

Berger, A. 1988. Milankovitch theory and climate. *Reviews of Geophysics* 26, 624–657.

Berger, A.R., Bouchard, A., Brookes, I.A., Grant, D.R., Hay, S.G. and Stevens, R.K. 1992. Geology, topography and vegetation, Gros Morne National Park, Newfoundland. Geological Survey of Canada, Miscellaneous Report 54.

Bergsten, F. 1954. The land uplift in Sweden from the evidence of old water marks. *Geografiska Annaler* 36, 81–111.

Bernal, L.P. and Roshko, A. 1986. Streamwise vortex structures in plane mixing layers. *Journal of Fluid Mechanics* 170, 499–525.

Bernard, C. 1971a. Les marques sous-glaciaires d'aspect plastique sur la roche en place (p-forms): observations sur la bordure du Bouclier canadien et examen de la question – 1. *Revue de Géographie de Montréal* 25, 111–127.

Bernard, C. 1971b. Les marques sous-glaciaires d'aspect plastique sur la roche en place (p-forms): leur rapport avec l'environnement et avec certaines marques de corrasion – 2. *Revue de Géographie de Montréal* 25, 265–279.

Bernard, C. 1972. Les marques sous-glaciaires d'aspect plastique sur la roche en place (p-forms): interpretation génétique – 3. *Revue de Géographie de Montréal* 26, 177–191.

Berthelsen, A. 1973. Weichselian ice advances and drift successions in Denmark. *Bulletin of the Geological Institute University of Uppsala*, 5, 21–29.

Berthelsen, A. 1978. The methodology of kineto-stratigraphy as applied to glacial geology. *Bulletin of the Geological Society of Denmark* 27, 25–38.

Berthelsen, A. 1979. Recumbent folds and boudinage structures formed by subglacial shear: an example of gravity tectonics. *Geologie en Mijnbouw* 58, 253–260.

Berthelsen, A., Konradi, P. and Petersen, K.S. 1977. Kvartaere lagfolger og strukturer i Vestmons klinter. *Dansk Geologisk Forening, Arsskrift for 1976*, 93–99.

Beskow, G. 1935. Praktiska och kvartargeologiska resultat av grusinventeringen i Norrbottens lan. Foredragsreferat. *Geologiska Foreningens i Stockholm, Forhandlingar* 57, 120–123.

Beverage, J.P. and Culbertson, J.K. 1964. Hyperconcentrations of suspended sediment. *Journal of the Hydraulics Division, American Society of Civil Engineers* 90, 117–126.

Bickerton, R.W. and Matthews, J.A. 1992. On the accuracy of lichenometric dates: an assessment based on the 'Little Ice Age' moraine sequence of Nigardsbreen, southern Norway. *The Holocene* 2, 227–237.

Bickerton, R.W. and Matthews, J.A. 1993. 'Little Ice Age' variations of outlet glaciers from the Jostedalsbreen ice-cap, southern Norway: a regional lichenometric-dating study of ice-marginal moraine sequences and their climatic significance. *Journal of Quaternary Science* 8, 45–66.

Biju-Duval, B., Deynoux, M. and Rognon, P. 1981. Late Ordovician tillites of the Central Sahara. In Hambrey, M.J. and Harland, W.B. (eds), *Earth's Pre-Pleistocene Glacial Record*. Cambridge University Press, Cambridge, 99–107.

Bindschadler, R. 1982. A numerical model of temperate glacier flow applied to the quiescent phase of a surge-type glacier. *Journal of Glaciology* 28, 238–265.

Bindschadler, R. 1984. Jakobshavns Glacier drainage basin: a balance assessment. *Journal of Geophysical Research* 89, 2066–2072.

Bindschadler, R. 1993. Siple Coast Project research of Crary Ice Rise and the mouths of Ice Streams B and C, West Antarctica: review and new perspectives. *Journal of Glaciology* 39, 538–552.

Bindschadler, R., Harrison, W.D., Raymond, C.F. and Crosson, R. 1977. Geometry and dynamics of a surge-type glacier. *Journal of Glaciology* 18, 181–194.

Bird, J.B. 1967. *The Physiography of Arctic Canada*. Johns Hopkins University Press, Baltimore.

Birnie, R.V. 1977. A snow-bank push mechanism for the formation of some 'annual' moraine ridges. *Journal of Glaciology* 18, 77–85.

Birnie, R.V. and Thom, G. 1982. Preliminary observations on two rock glaciers in South Georgia, Falkland Islands Dependencies. *Journal of Glaciology* 28, 377–386.

Birot, P. 1968. *The Cycle of Erosion in Different Climates*. Batsford, London.

Bishop, B.C. 1957. Shear moraines in the Thule area, northwest Greenland. US Snow, Ice and Permafrost Research Establishment, Research Report 17.

Bishop, J.F. and Walton, J.L.W. 1981. Bottom melting under George VI Ice Shelf, Antarctica. *Journal of Glaciology* 27, 429–447.

Bjerrum, L. and Jorstad, F. 1968. Stability of rock slopes in Norway. Norwegian Geotechnical Institute Publication 79, 1–11.

Bjorck, S. and Digerfeldt, G. 1986. Late Weichselian–early Holocene shore displacement west of Mt Billingen, within the Swedish end moraine zone. *Boreas* 15, 1–18.

Bjorlykke, K. 1969. Contribution to discussion 'Late Precambrian glaciation in Scotland'. *Proceedings of the Geological Society of London* 1657, 177–198.

Bjorlykke, K., Englund, J.O. and Kirkhusmo, L.A. 1967. Latest Precambrian and Eocambrian stratigraphy of Norway. *Norges Geologiske Undersoegelse* 251, 5–17.

Björnsson, H. 1974. Explanation of jökulhlaups from Grimsvötn, Vatnajökull, Iceland. *Jökull* 24, 1–26.

Björnsson, H. 1975. Subglacial water reservoirs, jökulhlaups and volcanic eruptions. *Jökull* 25, 1–12.

Björnsson, H. 1976. Marginal and supraglacial lakes in Iceland. *Jökull* 26, 40–50.

Björnsson, H. 1981. Radio-echo sounding maps of Storglaciären, Isfallsglaciären and Rabots Glaciar, northern Sweden. *Geografiska Annaler* 63A, 225–231.

Björnsson, H. 1992. Jökulhlaups in Iceland: prediction, characteristics and simulation. *Annals of Glaciology* 16, 95–106.

Björnsson, H. 1996. Scales and rates of glacial sediment removal: a 20 km long, 300 m deep trench created beneath Breidamerkurjökull during the Little Ice Age. *Annals of Glaciology* 22, 141–146.

Black, R.F. 1973. Cryomorphic processes and micro-relief features, Victoria Land, Antarctica. In Fahey, B.D. and Thomson, R.D. (eds), *Research in Polar and Alpine*

Geomorphology. Proceedings of 3rd Guelph Symposium on Geomorphology, 11–24.

Blake, E., Clarke, G.K.C. and Gerin, M.C. 1992. Tools for examining subglacial bed deformation. *Journal of Glaciology* 38, 388–396.

Blake, W. 1970. Studies of glacial history in arctic Canada. I: pumice, radiocarbon dates, and differential postglacial uplift in the eastern Queen Elizabeth Islands. *Canadian Journal of Earth Sciences* 7, 634–664.

Blake, W. 1975. Radiocarbon age determinations and postglacial emergence at Cape Storm, southern Ellesmere Island, arctic Canada. *Geografiska Annaler* 57A, 1–71.

Blake, W. 1977. Glacial sculpture along the east-central coast of Ellesmere Island, Arctic Archipelago. Geological Survey of Canada, Paper 77–1C.

Blake, W. 1981. Neoglacial fluctuations of glaciers, southeastern Ellesmere Island, Canadian arctic archipelago. *Geografiska Annaler* 63A, 201–218.

Blake, W. 1992a. Shell-bearing till along Smith Sound, Ellesmere Island–Greenland: age and significance. *Sveriges Geologiska Undersokning* Ca81, 51–58.

Blake, W. 1992b. Holocene emergence at Cape Herschel, east-central Ellesmere Island, Arctic Canada: implications for ice sheet configuration. *Canadian Journal of Earth Sciences* 29, 1958–1980.

Blake, W. 1993. Holocene emergence along the Ellesmere Island coasts of northernmost Baffin Bay. *Norsk Geologisk Tidsskrift* 73, 147–160.

Blank, R.G. and Margolis, S.V. 1975. Pliocene climatic and glacial history of Antarctica as revealed by south east Indian Ocean deep-sea cores. *Geological Society of America Bulletin* 86, 1058–1066.

Blankenship, D.D., Bell, R.E., Hodge, S.M., Brozena, J.M., Behrendt, J.C. and Finn, C.A. 1993. Active volcanism beneath the West Antarctic ice sheet and implications for ice-sheet stability. *Nature* 361, 526–529.

Blankenship, D.D. and Bentley, C.R. 1987. The crystalline fabric of polar ice sheets inferred from seismic anisotropy. In Waddington, E.E. and Walder, J.S. (eds) *The Physical Basis of Ice Sheet Modelling*. IAHS Publication 170, 17–28.

Blankenship, D.D., Bentley, C.R., Rooney, S.T. and Alley, R.B. 1986. Seismic measurements reveal a saturated porous layer beneath an active Antarctic ice stream. *Nature* 322, 54–57.

Blankenship, D.D., Bentley, C.R., Rooney, S.T. and Alley, R.B. 1987. Till beneath Ice Stream B. I. Properties derived from seismic travel times. *Journal of Geophysical Research* 92, 8903–8911.

Blatt, H., Middleton, G.V. and Murray, R.C. 1980. *Origin of Sedimentary Rocks*, 2nd edition. Prentice-Hall, Englewood Cliffs, NJ.

Blodgett, R.A. and Stanley, K.O. 1980. Stratification, bedforms and discharge relations of the Platte braided river system, Nebraska. *Journal of Sedimentary Petrology* 50, 139–148.

Bloom, A.L. 1967. Pleistocene shorelines: a new test of isostasy. *Geological Society of America Bulletin* 78, 1477–1494.

Bloom, A.L. 1970. Holocene submergence in Micronesia as the standard for eustatic sea-level changes. *Quaternaria* 12, 145–154.

Bloom, A.L., Broecker, W.S., Chappell, J., Mathews, R.K. and Mesolella, K.J. 1974. Quaternary sea level fluctuations on a tectonic coast: new ^{230}Th/^{234}U dates from the Huon Peninsula, New Guinea. *Quaternary Research* 4, 185–205.

Bluck, B.J. 1974. Structural and directional properties of some valley sandur deposits in southern Iceland. *Sedimentology* 21, 533–554.

Bluck, B.J. 1979. Structure of coarse grained braided stream alluvium. *Transactions of the Royal Society of Edinburgh* 70, 181–221.

Bluemle, J.P. 1970. Anomalous hills and associated depressions in central North Dakota. *Geological Society of America, Abstracts with Programs* 2, 325–326.

Bluemle, J.P. 1981. Geology of Sheridan County, North Dakota. *North Dakota Geological Survey, Bulletin* 75, part 1.

Bluemle, J.P. 1993. Hydrodynamic blowouts in North Dakota. In Aber, J.S. (ed.), *Glaciotectonics and Mapping Glacial Deposits*. Canadian Plains Research Center, University of Regina, 259–266.

Bluemle, J.P. and Clayton, L. 1984. Large-scale glacial thrusting and related processes in North Dakota. *Boreas* 13, 279–299.

Bluemle, J.P., Lord, M.L. and Hunke, N.T. 1993. Exceptionally long, narrow drumlins formed in subglacial cavities, North Dakota. *Boreas* 22, 15–24.

Bobrowsky, P. and Rutter, N.W. 1992. The Quaternary geologic history of the Canadian Rocky Mountains. *Géographie Physique et Quaternaire* 46, 5–50.

Bockheim, J.G., Wilson, S.C., Denton, G.H., Andersen, B.G. and Stuiver, M. 1989. Late Quaternary ice surface fluctuations of Hatherton Glacier, Transantarctic Mountains. *Quaternary Research* 31, 229–254.

Bogen, J. 1983. Morphology and sedimentology of deltas in fjord and fjord valley lakes. *Sedimentary Geology* 36, 245–267.

Bogen, J. 1996. Erosion rates and sediment yields of glaciers. *Annals of Glaciology* 22, 48–52.

Bond, F.C. 1952. Third theory of comminution. *Transactions of the American Institute of Mechanical Engineering* 193, 484.

Bond, G. and 13 others. 1992. Evidence for massive discharges of icebergs into the North Atlantic Ocean during the last glacial period. *Nature* 360, 245–249.

Bond, G., Broeker, W., Johnsen, S., McManus, J., Labeyrie, L., Jouzel, J. and Bonani, G. 1993. Correlations between climate records from North Atlantic sediments and Greenland ice. *Nature* 365, 143–147.

Bond, G. and Lotti, R. 1995. Iceberg discharges into the North Atlantic on millennial timescales during the last glaciation. *Science* 267, 1005–1010.

Booth, D.B. 1986. Mass balance and sliding velocity of the Puget lobe of the Cordilleran ice sheet during the last glaciation. *Quaternary Research* 25, 269–280.

Booth, D.B. and Hallet, 1993. Channel networks carved by subglacial water: observations and reconstruction in the eastern Puget Lowland of Washington. *Geological Society of America Bulletin* 105, 671–682.

Boothroyd, J.C. and Ashley, G.M. 1975. Processes, bar morphology, and sedimentary structures on braided outwash fans, northeastern Gulf of Alaska. In Jopling, A.V. and McDonald, B.C. (eds), *Glaciofluvial and Glaciola-*

custrine Sedimentation. SEPM Special Publication 23, 193–222.

Boothroyd, J.C. and Nummendal, D. 1978. Proglacial braided outwash: a model for humid alluvial fan deposits. In Miall, A.D. (ed.), *Fluvial Sedimentology*. Canadian Society of Petroleum Geologists, Memoir 5, 641–668.

Bornhold, B.D., Finlayson, N.M. and Monahan, D. 1976. Submerged drainage patterns in Barrow Strait, Canadian Arctic. *Canadian Journal of Earth Sciences* 13, 305–311.

Bornhold, B.D. and Prior, D.B. 1990. Morphology and sedimentary processes on the subaqueous Noeick River delta, British Columbia, Canada. In Colella, A. and Prior, D.B. (eds), *Coarse-Grained Deltas*. International Association of Sedimentologists, Special Publication 10. Blackwell, Oxford, 169–181.

Bouchard, M.A. 1985. Weathering and weathering residuals on the Canadian Shield. *Fennia* 163, 327–332.

Bouchard, M.A. 1989. Subglacial landforms and deposits in central and northern Quebec, Canada, with emphasis on Rogen moraines. *Sedimentary Geology* 62, 293–308.

Bouchard, M.A. and Salonen, V.P. 1990. Boulder transport in shield areas. In Kujansuu, R. and Saarnisto, M. (eds), *Glacier Indicator Tracing*. Balkema, Rotterdam, 87–107.

Boulton, G.S. 1967. The development of a complex supraglacial moraine at the margin of Sorbreen, Ny Friesland, Vestspitsbergen. *Journal of Glaciology* 6, 717–736.

Boulton, G.S. 1968. Flow tills and related deposits on some Vestspitsbergen glaciers. *Journal of Glaciology* 7, 391–412.

Boulton, G.S. 1970a. On the origin and transport of englacial debris in Svalbard glaciers. *Journal of Glaciology* 9, 213–229.

Boulton, G.S. 1970b. On the deposition of subglacial and melt-out tills at the margins of certain Svalbard glaciers. *Journal of Glaciology* 9, 231–245.

Boulton, G.S. 1971. Till genesis and fabric in Svalbard, Spitsbergen. In Goldthwait, R.P. (ed.), *Till: A Symposium*. Ohio State University Press, Columbus, 41–72.

Boulton, G.S. 1972a. The role of the thermal regime in glacial sedimentation. In Price, R.J. and Sugden, D.E. (eds), *Polar Geomorphology*. Institute of British Geographers, Special Publication 4, 1–19.

Boulton, G.S. 1972b. Modern arctic glaciers as depositional models for former ice sheets. *Journal of the Geological Society* 128, 361–393.

Boulton, G.S. 1974. Processes and patterns of subglacial erosion. In Coates, D.R. (ed.), *Glacial Geomorphology*. State University of New York, Binghamton, 41–87.

Boulton, G.S. 1975. Processes and patterns of subglacial sedimentation: a theoretical approach. In Wright, A.E. and Moseley, F. (eds), *Ice Ages: Ancient and Modern*. Seel House Press, Liverpool, 7–42.

Boulton, G.S. 1976. The origin of glacially-fluted surfaces: observations and theory. *Journal of Glaciology* 17, 287–309.

Boulton, G.S. 1977a. *Guide to Breidamerkurjökull, Iceland*. INQUA Commission on Lithology and Genesis of Quaternary Deposits.

Boulton, G.S. 1977b. A multiple till sequence formed by a Late Devensian Welsh ice-cap: Glanllynnau, Gwynedd. *Cambria* 4, 10–31.

Boulton, G.S. 1978. Boulder shapes and grain-size distributions of debris as indicators of transport paths through a glacier and till genesis. *Sedimentology* 25, 773–799.

Boulton, G.S. 1979. Processes of glacier erosion on different substrata. *Journal of Glaciology* 23, 15–38.

Boulton, G.S. 1982. Subglacial processes and the development of glacial bedforms. In Davidson-Arnott, R., Nickling, W. and Fahey, B.D. (eds), *Research in Glacial, Glacio-fluvial, and Glacio-lacustrine Systems*. Geo Books, Norwich, 1–31.

Boulton, G.S. 1983. Debris and isotopic sequences in basal layers of polar ice sheets. In Robin, G. de Q. (ed.), *The Climatic Record in Polar Ice Sheets*. Cambridge University Press, Cambridge, 83–88.

Boulton, G.S. 1986. Push moraines and glacier contact fans in marine and terrestrial environments. *Sedimentology* 33, 677–698.

Boulton, G.S. 1987. A theory of drumlin formation by subglacial sediment deformation. In Menzies, J. and Rose, J. (eds), *Drumlin Symposium*. Balkema, Rotterdam, 25–80.

Boulton, G.S. 1990. Sedimentary and sea level changes during glacial cycles and their control on glacimarine facies architecture. In Dowdeswell, J.A. and Scourse, J.D. (eds), *Glacimarine Environments: Processes and Sediments*. Geological Society Special Publication 53, 15–52.

Boulton, G.S. 1996a. Theory of glacial erosion, transport and deposition as a consequence of subglacial sediment deformation. *Journal of Glaciology* 42, 43–62.

Boulton, G.S. 1996b. The origin of till sequences by subglacial sediment deformation beneath mid-latitude ice sheets. *Annals of Glaciology* 22, 75–84.

Boulton, G.S. and Caban, P.E. 1995. Groundwater flow beneath ice sheets: Part II – Its impact on glacier tectonic structures and moraine formation. *Quaternary Science Reviews* 14, 563–587.

Boulton, G.S., Caban, P.E. and van Gijssel, K. 1995. Groundwater flow beneath ice sheets: Part I – Large scale patterns. *Quaternary Science Reviews* 14, 545–562.

Boulton, G.S. and Clark, C.D. 1990a. A highly mobile Laurentide Ice Sheet revealed by satellite images of glacial lineations. *Nature* 346, 813–817.

Boulton, G.S. and Clark, C.D. 1990b. The Laurentide Ice Sheet through the last glacial cycle: drift lineations as a key to the dynamic behaviour of former ice sheets. *Transactions of the Royal Society of Edinburgh, Earth Sciences* 81, 327–347.

Boulton, G.S. and Dent, D.L. 1974. The nature and rates of post-depositional changes in recently deposited till from south-east Iceland. *Geografiska Annaler* 56A, 121–134.

Boulton, G.S., Dent, D.L. and Morris, E.M. 1974. Subglacial shearing and crushing, and the role of water pressures in tills from south-east Iceland. *Geografiska Annaler*, 56A, 135–145.

Boulton, G.S. and Deynoux, M. 1981. Sedimentation in glacial environments and the identification of tills and tillites in ancient sedimentary sequences. *Precambrian Research* 15, 397–420.

Boulton, G.S. and Eyles, N. 1979. Sedimentation by valley glaciers: a model and genetic classification. In Schluchter, C. (ed.), *Moraines and Varves*. Balkema, Rotterdam, 11–23.

Boulton, G.S. and Hindmarsh, R.C.A. 1987. Sediment deformation beneath glaciers: rheology and sedimentological consequences. *Journal of Geophysical Research* 92, B9, 9059–9082.

Boulton, G.S. and Jones, A.S. 1979. Stability of temperate ice caps and ice sheets resting on beds of deformable sediment. *Journal of Glaciology* 24, 29–43.

Boulton, G.S., Jones, A.S., Clayton, K.M. and Kenning, M.J. 1977. A British ice sheet model and patterns of glacial erosion and deposition in Britain. In Shotton, F.W. (ed.), *British Quaternary Studies: Recent Advances*. Oxford University Press, Oxford, 231–246.

Boulton, G.S., van der Meer, J.J.M., Ruegg, G.H.J., Beets, D.J., Riezebos, P.A., Castel, I.I.M., Hart, J.K., Quinn, I., Thornton, M. and van der Wateren, D.F.M. 1989. Preliminary report on the Glacitecs 84 expedition to Spitsbergen. Internal Report of the Fysisch-Geografisch en Bodemkundig Laboratorium, University of Amsterdam.

Boulton, G.S., Morris, E.M., Armstrong, A.A. and Thomas, A. 1979. Direct measurement of stress at the base of a glacier. *Journal of Glaciology* 22, 3–24.

Boulton, G.S. and Paul, M.A. 1976. The influence of genetic processes on some geotechnical properties of tills. *Journal of Engineering Geology* 9, 159–194.

Boulton, G.S., Smith, G.D., Jones, A.S. and Newsome, J. 1985. Glacial geology and glaciology of the last mid-latitude ice sheets. *Journal of the Geological Society of London* 142, 447–474.

Boulton, G.S. and Vivian, R. 1973. Underneath the glaciers. *Geographical Magazine* 45, 311–319.

Bouma, A.H. 1962. *Sedimentology of Some Flysch Deposits*. Elsevier, Amsterdam.

Bowen, D.Q. and Gregory, K.J. 1965. A glacial drainage system near Fishguard, Pembrokeshire. *Proceedings of the Geologists' Association* 76, 275–282.

Bowman, I. 1920. *The Andes of Southern Peru*. American Geographical Society Special Publication 1.

Boyce, J.I. and Eyles, N. 1991. Drumlins carved by deforming till streams below the Laurentide ice sheet. *Geology* 19, 787–790.

Boyd, D., Scott, D.B. and Douma, M. 1988. Glacial tunnel valleys and Quaternary history of the outer Scotian Shelf. *Nature* 333, 61–64.

Boye, M. 1952. Névés et erosion glaciaire. *Revue de Géomorphologie Dynamique* 3, 20–36. English translation 'Névés, firn and glacial erosion' in Evans, D.J.A. (ed.), *Cold Climate Landforms*. Wiley, Chichester, 205–220.

Boyer, S.J. and Pheasant, D.R. 1974. Delimitation of weathering zones in the field area of eastern Baffin Island. *Geological Society of America Bulletin* 85, 805–810.

Bradley, R.S. 1985. *Quaternary Palaeoclimatology*. London, Chapman & Hall.

Bradley, R.S. 1990. Holocene paleoclimatology of the Queen Elizabeth Islands, Canadian high arctic. *Quaternary Science Reviews* 9, 365–384.

Bradley, R.S. and Serrez, M.C. 1987. Mass balance of two high arctic plateau ice caps. *Journal of Glaciology* 33, 123–128.

Braithwaite, R.J. 1981. On glacier energy balance, ablation and air temperature. *Journal of Glaciology* 27, 381–391.

Braithwaite, R.J. 1985. Calculation of degree-days for glacier–climate research. *Zeitschrift für Gletscherkunde und Glazialgeologie* 20, 1–8.

Braithwaite, R.J. and Olesen, O.B. 1985. Ice ablation in West Greenland in relation to air temperature and global radiation. *Zeitschrift für Gletscherkunde und Glazialgeologie* 20, 155–168.

Braithwaite, R.J. and Olesen, O.B. 1989. Calculation of glacier ablation from air temperature, west Greenland. In Oerlemans, J. (ed.), *Glacier Fluctuations and Climatic Change*. Kluwer, Dordrecht, 219–233.

Braithwaite, R.J. and Olesen, O.B. 1990a. Response of the energy balance on the margin of the Greenland ice sheet to temperature changes. *Journal of Glaciology* 36, 217–221.

Braithwaite, R.J. and Olesen, O.B. 1990b. A simple energy balance model to calculate ice ablation at the margin of the Greenland ice sheet. *Journal of Glaciology* 36, 222–228.

Brandal, M.K. and Heder, E. 1991. Stratigraphy and sedimentation of a terminal moraine deposited in a marine environment: two examples from the Ra-ridge in Østfold, southeast Norway. *Norsk Geologisk Tidsskrift* 71, 3–14.

Brayshaw, A.C. 1984. Characteristics and origin of cluster bedforms in coarse-grained alluvial channels. In Koster, E.H. and Steel, R.J. (eds), *Sedimentology of Gravels and Conglomerates*. Canadian Society of Petroleum Geologists Memoir 10, 77–85.

Brazel, A.J., Chambers, F.B. and Kalkstein, L.S. 1992. Summer energy balance on West Gulkana Glacier, Alaska, and linkages to a temporal synoptic index. *Zeitschrift für Geomorphologie* 86, 15–34.

Brazier, V., Owens, I.F., Soons, J.M. and Sturman, A.P. 1992. Report on the Franz Josef Glacier. *Zeitschrift für Geomorphologie*, Supp. 86, 35–49.

Brennan, T.A. 1994. Macroforms, large bedforms and rhythmic sedimentary sequences in subglacial eskers, south-central Ontario: implications for esker genesis and meltwater regime. *Sedimentary Geology* 91, 9–55.

Brennand, T.A. and Shaw, J. 1994. Tunnel channels and associated landforms, south-central Ontario: their implications for ice-sheet hydrology. *Canadian Journal of Earth Sciences* 31, 505–522.

Bretz, J.H. 1923a. The Channeled Scabland of the Columbia Plateau. *Journal of Geology* 31, 617–649.

Bretz, J.H. 1923b. Glacial drainage on the Columbia Plateau. *Geological Society of America Bulletin* 34, 573–608.

Bretz, J.H. 1927. Channeled Scabland and the Spokane Flood. *Journal of the Washington Academy of Science* 17, 200–211.

Bretz, J.H. 1935. Physiographic studies in east Greenland. In Boyd, L.A. (ed.), *The Fjord Region of East Greenland*. American Geographical Society Special Publication 18, 161–266.

Bretz, J.H. 1969. The Lake Missoula floods and the Channeled Scabland. *Journal of Geology* 77, 505–543.

Bretz, J.H., Smith, H.T.U. and Neff, G.E. 1956. Channeled Scabland of Washington: new data and interpretations. *Geological Society of America Bulletin* 67, 957–1049.

Brewer, M.C. 1958. The thermal regime of an arctic lake. *American Geophysical Union Transactions* 39, 278–284.

Bridge, J.S. 1993. The interaction between channel geometry, water flow, sediment transport and deposition in braided rivers. In Best, J.L. and Bristow, C.S. (eds), *Braided Rivers*. Geological Society of London Special Publication 75, 13–71.

Bridgeman, P.W. 1912. Water, in the liquid and five solid forms, under pressure. *American Academy of Arts and Sciences, Proceedings* 47, 441–558.

Bridgland, D.R. (ed.) 1986. *Clast Lithological Analysis*. Quaternary Research Association, Cambridge.

Briggs, D. and Smithson P. 1985. *Fundamentals of Physical Geography*. Routledge, London.

Bristow, C. 1996. Reconstructing fluvial channel morphology from sedimentary sequences. In Carling, P.A. and Dawson, M.R. (eds), *Advances in Fluvial Dynamics and Stratigraphy*. Wiley, Chichester, 351–371.

Broccoli, A.J. and Manabe, S. 1987a. The effects of the Laurentide Ice Sheet on North American climate during the last glacial maximum. *Géographie Physique et Quaternaire* 41, 291–299.

Broccoli, A.J. and Manabe, S. 1987b. The contributions of continental ice, atmospheric CO_2, and land albedo to the climate of the last glacial maximum. *Climate Dynamics* 1, 87–99.

Brochu, M. 1954. Lacs d'érosion différentialle glaciaire sur le Bouclier canadien. *Revue de Géomorphologie Dynamique* 6, 274–279.

Brodie, J.W. and Irwin, J. 1970. Morphology and sedimentation in Lake Wakatipu, New Zealand. *New Zealand Journal of Marine and Freshwater Research* 4, 479–496.

Brodzikowski, K. and van Loon, A.J. 1991. *Glacigenic Sediments*. Elsevier, Amsterdam.

van den Broeke, M.R., Duynkerke, P.G. and Oerlemans, J. 1994. The observed katabatic flow at the edge of the Greenland ice sheet during GIMEX–91. *Global and Planetary Change* 9, 3–15.

Broecker, W.S. and Denton, G.H. 1990a. What drives glacial cycles? *Scientific American*, January, 43–50.

Broecker, W.S. and Denton, G.H. 1990b. The role of ocean-atmosphere reorganisations in glacial cycles. *Quaternary Science Reviews* 9, 305–343.

Broecker, W.S., Kennett, J.P., Teller, J.T., Trumbore, S., Bonani, G. and Wolfli, W. 1989. Routing of meltwater from the Laurentide Ice Sheet during the Younger Dryas cold episode. *Nature* 341, 318–321.

Brogger, W.C. and Reusch, H.M. 1874. Giant kettles at Christiania. *Quarterly Journal of the Geological Society of London* 30, 750–771.

Broster, B.E. 1985. Multiple flow and support mechanisms and the development of inverse grading in a subaquatic glacigenic debris flow. *Sedimentology* 32, 645–647.

Broster, B.E. 1986. Till variability and compositional stratification: examples from the Port Huron Lobe. *Canadian Journal of Earth Sciences* 23, 1823–1841.

Broster, B.E. 1991. Glacitectonic deformation in sediment and bedrock, Hat Creek, British Columbia. *Géographie Physique et Quaternaire* 45, 5–20.

Broster, B.E. and Hicock, S.R. 1985. Multiple flow and support mechanisms and the development of inverse grading in a subaquatic glacigenic debris flow. *Sedimentology* 32, 645–657.

Broster, B.E. and Park, A.F. 1993. Dynamic glacigenic deformation of near-surface metamorphic bedrock: Heath Steele Mines, New Brunswick. *Journal of Geology* 101, 523–530.

Broster, B.E. and Seaman, A.A. 1991. Glacigenic rafting of weathered granite: Charlie Lake, New Brunswick. *Canadian Journal of Earth Sciences* 28, 649–654.

Brotchie, J.F. and Silvester, R. 1969. On crustal flexure. *Journal of Geophysical Research* 74, 5240–5252.

Brown, C.S., Meier, M.F. and Post, A. 1982. Calving speed of Alaskan tidewater glaciers with applications to the Columbia Glacier, Alaska. USGS Professional Paper 1258-C.

Brown, G.H., Sharp, M.J., Tranter, M., Gurnell, A.M. and Nienow, P.W. 1994. Impact of post-mixing chemical reactions on the major ion chemistry of bulk meltwater draining the Haut Glacier d'Arolla. *Hydrological Processes* 8, 465–480.

Brown, I.M. 1993. Pattern of deglaciation of the last (Late Devensian) Scottish ice sheet: evidence from ice-marginal deposits in the Dee valley, northeast Scotland, *Journal of Quaternary Science* 8, 235–250.

Brown, L.F. and Fisher, W.L. 1977. Seismic-stratigraphic interpretation of depositional systems: examples from Brazilian rift and pull-apart basins. In Payton, C.E. (ed.), *Seismic Stratigraphy*. AAPG Memoir 26, 213–248.

Brown, N.E., Hallet, B. and Booth, D.B. 1987. Rapid soft-bed sliding of the Puget glacial lobe. *Journal of Geophysical Research* 92, 8985–8997.

Bruckner, E. 1887. Die Hohe der Schneelinie und ihre Bestimmung. *Meteorologische Zeitschrift* 4, 31–32.

Brunsden, D. 1979. Mass movements. In Embleton, C. and Thornes, J. (eds), *Process in Geomorphology*. Edward Arnold, London, 130–186.

Brunsden, D. and Prior, D.B. 1984. *Slope Instability*. Wiley, Chichester.

Bruun, A.F., Brodie, J.W. and Fleming, C.A. 1955. Submarine geology of Milford Sound, New Zealand. *New Zealand Journal of Science and Technology* B36, 397–410.

Bryan, R.B., Campbell, I.A. and Yair, A. 1987. Postglacial geomorphic development of the Dinosaur Provincial Park badlands, Alberta. *Canadian Journal of Earth Sciences* 24, 135–146.

Budd, W.F. 1966. The dynamics of the Amery Ice Shelf. *Journal of Glaciology* 6, 335–358.

Budd, W.F. 1969. The dynamics of ice masses. Australian National Antarctic Research Expeditions, Scientific Report 108.

Budd, W.F. 1970. Ice flow over bedrock perturbations. *Journal of Glaciology* 9, 29–48.

Budd, W.F. 1975. A first simple model of periodically self-surging glaciers. *Journal of Glaciology* 14, 3–21.

Budd, W.F. and Carter, D.B. 1971. An analysis of the relation between the surface and bedrock profiles of ice caps. *Journal of Glaciology* 10, 197–209.

Budd, W.F., Corry, M.J. and Jacka, T.H. 1982. Results of the Amery Ice Shelf Project. *Annals of Glaciology* 3, 36–41.

Budd, W.F., Jacka, T.H. and Morgan, V.I. 1980. Antarctic iceberg melt rates derived from size distributions and movement rates. *Annals of Glaciology* 1, 103–112.

Budd, W.F., Keage, P.L. and Blundy, N.A. 1979. Empirical studies of ice sliding. *Journal of Glaciology* 23, 157–170.

Budd, W.F. and Smith, I.N. 1981. The growth and retreat of ice sheets in response to orbital radiation changes. In Allison I. (ed.), *Sea Level, Ice and Climatic Change.* IAHS Publication 131, 369–409.

Budd, W.F. and Smith, I.N. 1987. Conditions for growth and retreat of the Laurentide Ice Sheet. *Géographie Physique et Quaternaire* 41, 279–290.

Budel, J. 1948. Die klima-morphologischen Zonen der Polarländer. *Erdkunde* 2, 22–53.

Bull, C., McKelvey, B.C. and Webb, P.N. 1962. Quaternary glaciations in southern Victoria Land, Antarctica. *Journal of Glaciology* 4, 63–78.

Bull, W.B. and Cooper, A.F. 1986. Uplifted marine terraces along the Alpine fault, New Zealand. *Science* 234, 1225–1228.

Burbank, D.W. and Kang, J.C. 1991. Relative dating of Quaternary moraines, Rongbuk Valley, Mount Everest, Tibet, implications for an ice sheet on the Tibetan Plateau. *Quaternary Research* 36, 1–18.

Burrows, C.J. 1977. Late Pleistocene and Holocene glacial episodes in South Island, New Zealand and some climatic implications. *New Zealand Geographer* 33, 34–39.

Burrows, C.J. and Gellatly, A.F. 1982. Holocene glacial activity in New Zealand. *Striae* 18, 41–47.

Byers, A.R. 1959. Deformation of the Whitemud and Eastend Formations near Claybank, Saskatchewan. *Transactions of the Royal Society of Canada* 53, ser. 3, 1–11.

Caldenius, C. 1932. Las glaciaciones cuaternarias en la Patagonia y Tierra del Fuego. Dirección General Minas y Geología, Publication 95, 150 and *Geografiska Annaler* 14, 1–64.

Calkin, P.E. 1964. Geomorphology and glacial geology of the Victoria Valley system, Southern Victoria Land, Antarctica. Institute of Polar Studies, Report 10, 1–66.

Calkin, P.E. 1971. Glacial geology of the Victoria Valley system, Southern Victoria Land, Antarctica. *American Geophysical Union, Antarctic Research Series* 16, 363–412.

Calkin, P.E. and Bull, C. 1972. Interaction of the East Antarctic Ice Sheet, alpine glaciations and sea level in the Wright Valley area, southern Victoria Land. In Adie R.J. (ed.), *Antarctic Geology and Geophysics.* Universitatsforlaget, Oslo, 435–440.

Calkin, P.E. and Ellis, J.M. 1982. Holocene glacial chronology of the Brooks Range, northern Alaska. *Striae* 18, 3–8.

Calkin, P.E., Ellis, J.M., Haworth, L.A. and Burns, P.E. 1985. Cirque glacier regime and neoglaciation, Brooks Range, Alaska. *Zeitschrift für Gletscherkunde und Glazialgeologie* 21, 371–378.

Calkin, P.E. and Rutford, R.H. 1974. The sand dunes of Victoria Valley, Antarctica. *Geographical Review* 64, 189–216.

Calkin, P.E. and Young, G.M. 1995. Global glacial chronologies and causes of glaciation. In Menzies, J. (ed.), *Glacial Environments,* Vol. 1, *Modern Glacial Environments, Processes, Dynamics and Sediments.* Butterworth-Heinemann, Oxford, 9–75.

Cameron, T.D.J., Stoker, M.S. and Long, D. 1987. The history of Quaternary sedimentation in the UK sector of the North Sea basin. *Journal of the Geological Society of London* 144, 43–58.

Campbell, N.J. and Collin, A.E. 1958. The discoloration of Foxe Basin ice. *Journal of the Fisheries Research Board Canada* 15, 1175–1188

Carlson, C.G. and Freers, T.F. 1975. Geology of Benson and Pierce Counties, *North Dakota. North Dakota Geological Survey, Bulletin* 59, part 1.

Carlson, P.R. 1978. Holocene slump on continental shelf off Malaspina Glacier, Gulf of Alaska. *Bulletin of the American Association of Petroleum Geologists* 62, 2412–2426.

Carlson, P.R., Bruns, T.R. and Fisher, M.A. 1990. Development of slope valleys in the glacimarine environment of a complex subduction zone, northern Gulf of Alaska. In Dowdeswell, J.A. and Scourse, J.D. (eds), *Glacimarine Environments, Processes and Sediments.* Geological Society Special Publication 53, 139–153.

Carlson, P.R. and Molnia, B.F. 1978. Submarine faults and slides on the continental shelf, northern Gulf of Alaska. *Marine Geotechnology* 2, 275–290.

Carlson, P.R., Powell, R.D. and Phillips, A.C. 1992. Submarine sedimentary features on a fjord delta front, Queen Inlet, Glacier Bay, Alaska. *Canadian Journal of Earth Sciences* 29, 565–573.

Carlson, P.R., Powell, R.D. and Rearic, D.M. 1989. Turbidity-current channels in Queen Inlet, Glacier Bay, Alaska. *Canadian Journal of Earth Sciences* 26, 807–820.

Carmack, E.C., Gray, C.B.J., Pharo, C.H. and Daley, R.J. 1979. Importance of lake–river interaction on the physical limnology of the Kamloops Lake/Thompson River system. *Limnology and Oceanography* 24, 634–644.

Carney, F. 1910. Glacial erosion on Kelleys Island, Ohio. *Geological Society of America Bulletin* 10, 640–645.

Carol, H. 1947. The formation of roches moutonnées. *Journal of Glaciology* 1, 57–59.

Carr, M.H. 1981. *The Surface of Mars.* Yale University Press, New Haven, Conn.

Carr, M.H. 1986. Mars, a water-rich planet. *Icarus* 68, 187–216.

Carr, M.H. and 12 others 1980. *Viking Orbiter Views of Mars.* US Government Printing Office, Washington DC.

Carroll, J.J. and Fitch, B.W. 1981. Effects of solar elevation and cloudiness on snow albedo at the South Pole. *Journal of Geophysical Research* 86, 5271–5276.

Carson, M.A. and Kirkby, M.J. 1972. *Hillslope Form and Process.* Cambridge University Press, Cambridge.

Carter, L., Mitchell, J.S. and Day, N.J. 1981. Suspended sediment beneath permanent and seasonal ice, Ross Ice Shelf, Antarctica. *New Zealand Journal of Geology and Geophysics* 24, 249–262.

Cas, R.A.F. and Landis, C.A. 1987. A debris-flow deposit with multiple plug-flow channels and associated side accretion deposits. *Sedimentology* 34, 901–910.

Casshyap, S.M. and Qidwai, H.A. 1974. Glacial sedimentation of late Paleozoic Talchir diamictite, Pench Valley coalfield, central India. *Geological Society of America Bulletin* 85, 749–760.

Casshyap, S.M. and Tewari, R.C. 1982. Facies analysis and paleogeographic implications of a late Paleozoic glacial outwash deposit, Bihar, India. *Journal of Sedimentary Petrology* 52, 1243–1256.

Cathles, L.M. 1975. *The Viscosity of the Earth's Mantle*. Princeton University Press, Princeton, NJ.

Chamberlin, T.C. 1888. The rock scourings of the great ice invasions. *USGS Annual Report* 7, 155–248.

Chappell, J. 1974a. Late Quaternary glacio- and hydro-isostasy, on a layered earth. *Quaternary Research* 4, 429–440.

Chappell, J. 1974b. Geology of coral terraces, Huon Peninsula, New Guinea, a study of Quaternary tectonic movements and sea level changes. *Geological Society of America Bulletin* 85, 553–570.

Chappell, J. and Shackleton, N.J. 1986. Oxygen isotopes and sea level. *Nature* 324, 137–140.

Charlesworth, J.K. 1956. The Lateglacial history of the Highlands and islands of Scotland. *Transactions of the Royal Society of Edinburgh* 62, 769–929.

Charlesworth, J.K. 1957. *The Quaternary Era*. Edward Arnold, London.

Cheel, R.J. and Rust, B.R. 1982. Coarse grained facies of glaciomarine deposits near Ottawa, Canada. In Davidson-Arnott, R., Nickling, W. and Fahey, B.D. (eds), *Research in Glaciofluvial and Glaciolacustrine Systems*. Geo Books, Norwich, 279–295.

Cheetham, G.H. 1976. Palaeohydrological investigations of river terrace gravels. In Davidson, D.A. and Shackley, M. (eds), *Geoarchaeology, Earth Science and the Past*. Duckworth, London, 335–343.

Cheetham, G.H. 1979. Flow competence in relation to stream channel form and braiding. *Geological Society of America Bulletin* 90, 877–886.

Cheetham, G.H. 1980. Late Quaternary palaeohydrology, the Kennet valley case-study. In Jones, D.K.C. (ed.), *The Shaping of Southern England*. Institute of British Geographers Special Publication 11, Academic Press, London, 203–223.

Chen, J. and Funk, M. 1990. Mass balance of Rhonegletscher during 1881/83–1986/87. *Journal of Glaciology* 36, 199–209.

Chernicof, S.E. 1983. Glacial characteristics of a Pleistocene ice lobe in east-central Minnesota. *Geological Society of America Bulletin* 94, 1401–1414.

Chinn, T.J. 1980. Glacier balances in the Dry Valleys area, Victoria Land, Antarctica. IAHS Publication No. 125, 237–247.

Chinn, T.J. 1988. The Dry Valleys of Victoria Land. In *Satellite Image Atlas of Glaciers of the World, Antarctica*. USGS Professional Paper 1386-B.

Chinn, T.J., McSaveney, M.J. and McSaveney, E. 1992. The Mt Cook rock avalanche of 14 December 1991. Institute of Geological and Nuclear Sciences Information Leaflet, New Zealand.

Chinn, T.J.H. 1989. Glaciers of New Zealand. In Williams, R.S. and Ferrigno, J.G. (eds), *Satellite Image Atlas of the World*. USGS Professional Paper 1386-H, 21–48.

Chinn, T.J.H. and Dillon, A. 1987. Observations on a debris-covered polar glacier 'Whisky Glacier', James Ross Island, Antarctic Peninsula, Antarctica. *Journal of Glaciology* 33, 1–10.

Chinn, T.J.H. and Kirkbride, M. 1988. *Guidebook for the Post Symposium Glacier Tour of New Zealand*. International Glaciological Society Symposium on Ice Dynamics, Hobart, Australia.

Chizov, O.P. 1964. Precipitation, feeding and melting of ice sheets of northeastern Atlantic in present climatic conditions. *Results of Researches IGY, Glaciology* 13, 30–38.

Chorley, R.J. 1959. The shape of drumlins. *Journal of Glaciology* 3, 339–344.

Chorley, R.J. and Kennedy, B.A. 1971. *Physical Geography, A Systems Approach*. Prentice-Hall, London.

Chorley, R.J., Schumm, S.A. and Sugden, D.E. 1984. *Geomorphology*. Methuen, London.

Chorlton, J.C. and Lister, H. 1968. Snow accumulation over Antarctica. In Gow, A.J. *et al.* (eds), *International Symposium on Antarctic Glaciological Exploration*, IASH Publication 86, 254–263.

Chorlton, J.C. and Lister, H. 1971. Geographical control of glacier budget gradients in Norway. *Norsk Geografisk Tidsskrift* 25, 159–164.

Christian, C.S. 1957. The concept of land units and land systems. *Proceedings of the 9th Pacific Science Conference* 20, 74–81.

Christian, C.S. and Stewart, G.A. 1952. Summary of General Report on Survey of Katherine–Darwin Region, 1946 CSIRO, Australia, Land Research Series, I.

Christiansen, E.A. 1956. Glacial geology of the Moose Mountain area, Saskatchewan. Saskatchewan Department of Mineral Resources, Report 21.

Christiansen, E.A. 1961. Geology and ground-water resources of the Regina Area, Saskatchewan. Saskatchewan Research Council, Geology Division, Report 2.

Christiansen, E.A. 1971a. Tills in southern Saskatchewan, Canada. In Goldthwait, R.P. (ed.), *Till, A Symposium*. Ohio State University Press, Columbus, 167–183.

Christiansen, E.A. 1971b. Geology and groundwater resources of the Melville Area 62K, L. Saskatchewan. Saskatchewan Research Council, Geology Division, Map 12.

Christiansen, E.A. 1979. The Wisconsinan deglaciation of southern Saskatchewan and adjacent areas. *Canadian Journal of Earth Sciences* 16, 913–938.

Christiansen, E.A., Gendzwill, D.J. and Meneley, W.A. 1982. Howe Lake, a hydrodynamic blowout structure. *Canadian Journal of Earth Sciences* 19, 1122–1139.

Christiansen, E.A. and Whitaker, S.H. 1976. Glacial thrusting of drift and bedrock. In Legget, R.F. (ed.), *Glacial Till*. Royal Society of Canada, Special Publication 12, 121–130.

Church, M. 1972. Baffin Island sandurs, a study of arctic fluvial processes. *Geological Survey of Canada Bulletin* 216.

Church, M. 1978. Palaeohydrological reconstructions from a Holocene valley fill. In Miall, A.D. (ed.), *Fluvial Sedimentology*. Canadian Society of Petroleum Geologists Memoir No. 5, 743–772.

Church, M. and Gilbert, R. 1975. Proglacial fluvial and lacustrine sediments. In Jopling, A.V. and McDonald, B.C. (eds), *Glaciofluvial and Glaciolacustrine Sedimentation*. SEPM, Special Publication 23, 22–100.

Church, M. and Jones, D. 1982. Channel bars in gravel-bed rivers. In Hey, R.D., Bathurst, J.C. and Thorne, C.R. (eds), *Gravel-Bed Rivers*. Wiley, London, 291–338.

Church, M. and Ryder, J.M. 1972. Paraglacial sedimentation, a consideration of fluvial processes conditioned by glaciation. *Geological Society of America Bulletin* 83, 3059–3072.

Church, M. and Slaymaker, O. 1989. Disequilibrium of Holocene sediment yield in glaciated British Columbia. *Nature* 337, 452–454.

Churski, Z. 1973. Hydrographic features of the proglacial area of Skeidararjökull. *Geographia Polonica* 26, 209–254.

Clague, J.J. 1975. Glacier flow patterns and the origin of the Late Wisconsinan till in the southern Rocky Mountain trench, British Columbia. *Geological Society of America Bulletin* 86, 721–731.

Clague, J.J. 1984. Quaternary geology and geomorphology, Smithers-Terrace–Prince Rupert area, British Columbia. Geological Survey of Canada Memoir 413.

Clague, J.J. 1986. The Quaternary stratigraphic record of British Columbia, evidence for periodic sedimentation and erosion controlled by glaciation. *Canadian Journal of Earth Sciences* 23, 885–894.

Clague, J.J. 1989. Cordilleran Ice Sheet (Quaternary Geology of the Canadian Cordillera). In Fulton, R.J. (ed.), *Quaternary Geology of Canada and Greenland*. Geological Survey of Canada, Geology of Canada No. 1, 40–42.

Clague, J.J. and Evans, S.G. 1992. A self-arresting moraine dam failure, St Elias Mountains, British Columbia. Geological Survey of Canada Paper 92–1A, 185–188.

Clague, J.J. and Evans, S.G. 1994a. Historic retreat of Grand Pacific and Melbern Glaciers, Saint Elias Mountains, Canada, an analogue for decay of the Cordilleran ice sheet at the end of the Pleistocene? *Journal of Glaciology* 39, 619–624.

Clague, J.J. and Evans, S.G. 1994b. Formation and failure of natural dams in the Canadian Cordillera. *Geological Survey of Canada, Bulletin* 464.

Clague, J.J., Evans, S.G. and Brown, I.G. 1985. A debris flow triggered by the breaching of a moraine-dammed lake, Klattasine Creek, British Columbia. *Canadian Journal of Earth Sciences* 22, 1492–1502.

Clague, J.J. and Mathews, W.H. 1973. The magnitude of jökulhlaups. *Journal of Glaciology* 12, 501–504.

Clapperton, C.M. 1970. The evidence for the Cheviot ice cap. *Transactions of the Institute of British Geographers* 50, 115–127.

Clapperton, C.M. 1971a. Evidence of cirque glaciation in the Falkland Islands. *Journal of Glaciology* 10, 121–125.

Clapperton, C.M. 1971b. The location and origin of subglacial meltwater phenomena in the eastern Cheviot Hills. *Proceedings of the Yorkshire Geological Society* 38, 361–380.

Clapperton, C.M. 1975. The debris content of surging glaciers in Svalbard and Iceland. *Journal of Glaciology* 14, 395–406.

Clapperton, C.M. 1985. Significance of a late-glacial readvance in the Ecuadorian Andes. In Rabassa, J. (ed.), *Quaternary of South America and Antarctic Peninsula*. Balkema, Rotterdam, 149–158.

Clapperton, C.M. 1987. Glacial geomorphology, Quaternary glacial sequence and palaeoclimatic inferences in the Ecuadorian Andes. In Gardiner, V. (ed.), *International Geomorphology*, Part II. Wiley, Chichester, 843–870.

Clapperton, C.M. 1989. Asymmetrical drumlins in Patagonia, Chile. *Sedimentary Geology* 62, 387–398.

Clapperton, C.M. (ed.) 1990a. Quaternary glaciations in the southern hemisphere. *Quaternary Science Reviews* 9.

Clapperton, C.M. 1990b. Quaternary glaciations in the southern hemisphere, an overview. *Quaternary Science Reviews* 9, 299–304.

Clapperton, C.M. 1990c. Quaternary glaciations in the Southern Ocean and Antarctic Peninsula area. *Quaternary Science Reviews* 9, 229–252.

Clapperton, C.M. 1993. *Quaternary Geology and Geomorphology of South America*. Elsevier, Amsterdam.

Clapperton, C.M., Clayton, J.D., Benn, D.I., Marden, C.J. and Argollo, J. 1996. Late Quaternary glacier advances and palaeolake highstands in the Bolivian Altiplano. *Quaternary International* 38/39, 49–59.

Clapperton, C.M. and McEwan, C. 1985. Late Quaternary moraines in the Chimborazo area, Ecuador. *Arctic and Alpine Research* 17, 135–142.

Clapperton, C.M. and Sugden, D.E. 1972. The Aberdeen and Dinnet glacial limits reconsidered. In Clapperton, C.M. (ed.), *North-east Scotland Geographical Essays*. Department of Geography, University of Aberdeen, 19–22.

Clapperton, C.M. and Sugden, D.E. 1990. Late Cenozoic glacial history of the Ross Embayment, Antarctica. *Quaternary Science Reviews* 9, 253–272.

Clapperton, C.M., Sugden, D.E., McCulloch, R.D. and Kaufmann, D.S. 1995. The last glaciation in central Magellan Strait, southernmost Chile. *Quaternary Research*, 44, 133–148.

Clapperton, C.M., Sugden, D.E. and Pelto, M. 1989. Relationship of land terminating and fjord glaciers to Holocene climatic change, South Georgia, Antarctica. In Oerlemans, J. (ed.), *Glacier Fluctuations and Climatic Change*. Kluwer, Dordrecht, 57–75.

Clark, C.D. 1993. Mega-scale lineations and cross-cutting ice-flow landforms. *Earth Surface Processes and Landforms* 18, 1–29.

Clark, C.D. 1994. Large-scale ice-moulding, a discussion of genesis and glaciological significance. *Sedimentary Geology* 91, 253–268.

Clark, D.L. and Hanson, A. 1983. Central Arctic Ocean sediment texture, a key to ice transport mechanisms. In Molnia, B.F. (ed.), *Glacial-Marine Sedimentation*. Plenum Press, New York, 301–330.

Clark, J.A. 1976. Greenland's rapid postglacial emergence, a result of ice–water gravitational attraction. *Geology* 4, 310–312.

Clark, J.A. 1980. A numerical model of worldwide sea level changes on a viscoelastic earth. In Morner, N.A. (ed.), *Earth Rheology, Isostasy and Eustasy*. Wiley, Chichester, 525–534.

Clark, J.A., Farrell, W.E. and Peltier, W.R. 1978. Global changes in postglacial sea level, a numerical calculation. *Quaternary Research* 9, 265–287.

Clark, P.U. 1987. Subglacial sediment dispersal and till composition. *Journal of Geology* 95, 527–541.

Clark, P.U. 1989. Relative differences between glacially crushed quartz transported by mountain and continental ice, some examples from North America and East

Africa, Discussion. *American Journal of Science* 289, 1195–1198.

Clark, P.U. 1988. Glacial geology of the Torngat Mountains, Labrador. *Canadian Journal of Earth Sciences* 25, 1184–1198.

Clark, P.U. 1991a. Striated clast pavements, products of deforming subglacial sediment? *Geology* 19, 530–533.

Clark, P.U. 1991b. Landscapes of glacial erosion, Torngat Mountains, northern Labrador/Ungava. *Canadian Geographer* 35, 208–213.

Clark, P.U. 1994. Unstable behavior of the Laurentide Ice Sheet over deforming sediment and its implications for climate change. *Quaternary Research* 41, 19–25.

Clark, P.U., Clague, J.J., Curry, B.B., *et al.* 1993. Initiation and development of the Laurentide and Cordilleran ice sheets following the last interglaciation. *Quaternary Science Reviews* 12, 79–114.

Clark, P.U. and Hansel, A.K. 1989. Clast ploughing, lodgement and glacier sliding over a soft glacier bed. *Boreas* 18, 201–207.

Clark, P.U. and Lea, P.D. (eds) 1992. The last interglacial–glacial transition in North America. Geological Society of America, Special Paper 270.

Clark, P.U. and Walder, J.S. 1994. Subglacial drainage, eskers, and deforming beds beneath the Laurentide and Eurasian ice sheets. *Bulletin of the Geological Society of America* 106, 304–314.

Clark, M.J. 1987. The alpine sediment system, a context for glacio-fluvial processes. In Gurnell, A.M. and Clark, M.J. (eds), *Glacio-fluvial Sediment Transfer, An Alpine Perspective*. Wiley, London, 9–31.

Clarke, G.K.C. 1982. Glacier outburst flood from 'Hazard lake', Yukon Territory, and the problem of flood magnitude prediction. *Journal of Glaciology* 28, 3–21.

Clarke, G.K.C. 1986. Professor Mathews, outburst floods, and other glaciological disasters. *Canadian Journal of Earth Sciences* 23, 859–868.

Clarke, G.K.C. 1987a. A short history of scientific investigations on glaciers. *Journal of Glaciology*, Special Issue, 4–24.

Clarke, G.K.C. 1987b. Fast glacier flow, ice streams, surging and tidewater glaciers. *Journal of Geophysical Research* 92, 8835–8841.

Clarke, G.K.C. 1987c. Subglacial till, a physical framework for its properties and processes. *Journal of Geophysical Research* 92, 9023–9036.

Clarke, G.K.C. 1991. Length, width and slope influences on glacier surging. *Journal of Glaciology* 37, 236–246.

Clarke, G.K.C. 1996a. Lumped-element model for subglacial transport of solute and suspended sediment. *Annals of Glaciology* 22, 152–159.

Clarke, G.K.C. 1996b. Lumped-element analysis of subglacial hydraulic circuits. *Journal of Geophysical Research* 101, B8, 17547–17559.

Clarke, G.K.C. and Blake, E.W. 1991. Geometric and thermal evolution of a surge-type glacier in its quiescent state, Trapridge Glacier, Yukon Territory, Canada, 1969–89. *Journal of Glaciology* 37, 158–169.

Clarke, G.K.C., Collins, S.G. and Thompson, D.E. 1984a. Flow, thermal structure and subglacial conditions of a surge-type glacier. *Canadian Journal of Earth Sciences* 21, 232–240.

Clarke, G.K.C., Mathews, W.H. and Pack, R.T. 1984b. Outburst floods from glacial Lake Missoula. *Quaternary Research* 22, 289–299.

Clarke, G.K.C., Schmok, J.P., Ommaney, C.S.L. and Collins, S.G. 1986. Characteristics of surge-type glaciers. *Journal of Geophysical Research* 91, 7165–7180.

Clarke, G.K.C. and Waldron, D.A. 1984. Simulation of the August 1979 sudden discharge of glacier-dammed Flood Lake, British Columbia. *Canadian Journal of Earth Sciences* 21, 502–504.

Clarke, T.S. 1991. Glacier dynamics in the Sutsina River basin Alaska, USA. *Journal of Glaciology* 37, 97–106.

Clarke, T.S. and Echelmeyer, K. 1996. Seismic-reflection evidence for a deep subglacial trough beneath Jacobshavns Isbrae, west Greenland. *Journal of Glaciology* 43, 219–232.

Clayton, K.M. 1965. Glacial erosion in the Finger Lakes region, New York State, USA. *Zeitschrift für Geomorphologie* 9, 50–62.

Clayton, K.M. 1974. Zones of glacial erosion. In Brown, E.H. and Waters, R.S. (eds), *Progress in Geomorphology*. Institute of British Geographers, Special Publication 7, 163–176.

Clayton, K.M. and Linton, D.L. 1964. A qualitative scale of intensity of glacial erosion. *20th International Geographical Congress, London. Abstracts of Papers*, 18–19.

Clayton, L. 1964. Karst topography on stagnant glaciers. *Journal of Glaciology* 5, 107–112.

Clayton, L. and Cherry, J.A. 1967. Pleistocene superglacial and ice-walled lakes of west-central North America. *North Dakota Geological Survey, Miscellaneous Series* 30, 47–52.

Clayton, L., Mickelson, D.M. and Attig, J.W. 1989. Evidence against pervasively deformed bed material beneath rapidly moving lobes of the southern Laurentide Ice Sheet. *Sedimentary Geology* 62, 203–208.

Clayton, L. and Moran, S.R. 1974. A glacial process–form model. In Coates, D.R. (ed.), *Glacial Geomorphology*. State University of New York, Binghamton, 89–119.

Clayton, L., Moran, S.R. and Bluemle, J.P. 1980. Explanatory text to accompany the Geologic Map of North Dakota. North Dakota Geological Survey, Report of Investigation No. 69.

Clayton, L., Teller, J.T. and Attig, J.W. 1985. Surging of the southwestern part of the Laurentide Ice Sheet. *Boreas* 14, 235–241.

Clemmensen, L.B. and Houmark-Nielsen, M. 1981. Sedimentary features of a Weichselian glaciolacustrine delta. *Boreas* 10, 229–245.

CLIMAP Project Members 1976. The surface of the ice age earth. *Science* 191, 1131–1136.

Clough, J.W. and Hansen, B.L. 1979. The Ross Ice Shelf Project. *Science* 203, 433–434.

Close, M.H. 1867. Notes on the general glaciation of Ireland. *Royal Geological Society of Ireland Journal* 1, 207–242.

Cohen, J.M. 1979. Deltaic sedimentation in glacial lake Blessington, County Wicklow, Ireland. In Schluchter, C. (ed.), *Moraines and Varves*. Balkema, Rotterdam, 357–367.

Coleman, A.P. 1907. A lower Huronian ice age. *American Journal of Science* 23, 187–192.

Coleman, A.P. 1926. *Ice Ages, Recent and Ancient.* MacMillan, London.

Coleman, J.M. 1982. *Deltas, Processes of Deposition and Models for Exploration.* International Human Resources Development Corporation, Boston.

Colhoun, E.A. 1991. Geological evidence for changes in the East Antarctica ice sheet (60°–120° E) during the last glaciation. *Polar Record* 27, 345–355.

Colhoun, E.A. and Fitzsimons, S.J. 1990. Late Cainozoic glaciation in western Tasmania, Australia. *Quaternary Science Reviews* 9, 199–216.

Collela, A., de Boer, P.L. and Nio, S.D. 1987. Sedimentology of a marine intermontane Pleistocene Gilbert-type fan-delta complex in the Crati Basin southern Italy. *Sedimentology* 34, 721–736.

Collela, A. and Prior, D.B. (eds) 1990. *Coarse-Grained Deltas.* International Association of Sedimentologists, Special Publication 10. Blackwell, Oxford.

Collett, L.W. 1926. The lakes of Scotland and of Switzerland. *Geographical Journal* 67, 193–213.

Collins, D.N. 1978. Hydrology of an Alpine glacier as indicated by the chemical composition of meltwater. *Zeitschrift für Gletscherkunde und Glazialgeologie* 13, 219–238.

Collins, D.N. 1979a. Quantitative determination of the subglacial hydrology of two Alpine glaciers. *Journal of Glaciology* 23, 347–362.

Collins, D.N. 1979b. Hydrochemistry of meltwaters draining from an Alpine glacier. *Arctic and Alpine Research* 11, 307–324.

Collins, D.N. 1979c. Sediment concentration in meltwater as an indicator of erosion processes beneath an Alpine glacier. *Journal of Glaciology* 23, 247–257.

Collins, D.N. 1984. Water and mass balance measurements in glacierized drainage basins. *Geografiska Annaler* 66A, 197–214.

Collinson, J.D. 1970. Deep channels, massive beds and turbidity-current genesis in the Central Pennine Basin. *Proceedings of the Yorkshire Geological Society* 37, 495–519.

Collinson, J.D. 1986. Alluvial sediments. In Reading, H.G. (ed.), *Sedimentary Environments and Facies*, 2nd edition. Blackwell, Oxford, 20–62.

Collinson, J.D. and Thompson, D.B. 1989. *Sedimentary Structures.* Unwin Hyman, London.

Colton, R.B. 1958. Note on the intersecting minor ridges in the Lake Agassiz basin North Dakota. North Dakota Geological Survey, Miscellaneous Series 10, 74–77.

Cook, J.H. 1946a. Ice contacts and the melting of ice below a water level. *American Journal of Science* 244, 502–512.

Cook, J.H. 1946b. Kame complexes and perforation deposits. *American Journal of Science* 244, 573–583.

Costa, J.E. 1988. Floods from dam failures. In Baker, V.R., Kochel, R.G. and Patton, P.C. (eds), *Flood Geomorphology.* Wiley, New York, 439–463.

Costin, A.B., Jennings, J.N., Black, H.P. and Thom, B.G. 1964. Snow action on Mount Twynam, Snowy Mountains, Australia. *Journal of Glaciology* 5, 219–228.

Coude, A. 1989. Comparative study of three drumlin fields in western Ireland, geomorphological data and genetic implications. *Sedimentary Geology* 62, 321–335.

Cowan, E.A. and Powell, R.D. 1991. Suspended sediment transport and deposition of cyclically interlaminated sediment in a temperate glacial fiord, Alaska, USA. In

Dowdeswell, J.A. and Scourse, J.D. (eds), *Glacimarine Environments, Processes and Sediments.* Special Publication of the Geologists' Association 53, 75–89.

Cowan, E.A., Powell, R.D. and Smith, N.D. 1988. Marine event sedimentation from a rainstorm at tidewater front of a temperate glacier. *Geology* 16, 409–412.

Cowan, W.R. 1968. Ribbed moraine, till fabric analysis and origin. *Canadian Journal of Earth Sciences* 5, 1145–1159.

Coxon, P., Owen, L.A. and Mitchell, W.A. (1996) A late Quaternary catastrophic flood in the Lahul Himalayas. *Journal of Quaternary Science* 11, 495–510.

Craig, H., Gordon, L.I. and Horibe, Y. 1963. Isotopic exchange effects in the evaporation of water. *Science* 133, 1833–1834.

Crary, A.P. 1958. Arctic ice islands and ice shelf studies, part 1. *Arctic* 11, 3–42.

Crary, A.P. 1960. Arctic ice islands and ice shelf studies, part 2. *Arctic* 13, 32–50.

Crary, A.P. 1966. Mechanism for fjord formation indicated by studies of an ice-covered inlet. *Geological Society of America Bulletin* 77, 911–929.

Crimes, T.P. and Harper, J.C. (eds) 1970. *Trace Fossils.* Geological Journal, Special Issue 3, 323–330.

Cronin T.M. 1983. Rapid sea level and climate change, evidence from continental and island margins. *Quaternary Science Reviews* 1, 177–214.

Croot, D.G. 1987. Glacio-tectonic structures, a mesoscale model of thin-skinned thrust sheets? *Journal of Structural Geology* 9, 797–808.

Croot, D.G. 1988a. *Glaciotectonics, Forms and Processes.* Balkema, Rotterdam.

Croot, D.G. 1988b. Morphological, structural and mechanical analysis of neoglacial ice-pushed ridges in Iceland. In Croot, D.G. (ed.), *Glaciotectonics, Forms and Processes.* Balkema, Rotterdam, 33–47.

Croot, D.G. 1988c. Glaciotectonics and surging glaciers, a correlation based on Vestspitsbergen, Svalbard, Norway. In Croot, D.G. (ed.), *Glaciotectonics, Forms and Processes.* Balkema, Rotterdam, 49–61.

Crosby, W.O. 1902. Origin of eskers. *American Geology* 30, 1–38.

Crossen, K.J. 1991. Structural control of deposition by Pleistocene tidewater glaciers, Gulf of Maine. In Anderson, J.B. and Ashley, G.M. (eds), *Glacial Marine Sedimentation: Paleoclimatic Significance.* Geological Society of America Special Paper 261, 127–135.

Crowell, J.C. 1957. Origin of pebbly mudstones. *Bulletin of the Geological Society of America* 68, 993–1110.

Crozaz, G., Langway, C.C. and Picciotto, E. 1966. Artificial radioactivity reference horizons in Greenland firn. *Earth and Planetary Science Letters* 1, 42–48.

Crozier, M.J. 1975. On the origin of the Peterborough drumlin field; testing the dilatancy theory. *Canadian Geographer* 14, 181–195.

Cudlip, W. and McIntyre, N.F. 1987. Seasat altimeter observations of an Antarctic 'lake'. *Annals of Glaciology* 9, 55–59.

Curl, J.E. 1980. A glacial history of the South Shetland Islands. Institute of Polar Studies, Ohio, Report 63.

Curray, J.R. 1964. Transgressions and regressions. In Miller, R.L. (ed.), *Papers in Marine Geology.* Macmillan, New York, 175–203.

Currey, D.R. 1974. Probable pre-Neoglacial age of type Temple Lake moraine, Wyoming. *Arctic and Alpine Research* 6, 293–300.

Czechowna, L. 1953. Zagadnienie drumlinów w świetle literatury. *Czasopismo Geograficzne* 23/24, 50–90. English translation, 'The question of drumlins in literature' in Evans, D.J.A. (ed.), *Cold Climate Landforms*. Wiley, Chichester, 269–291.

Dackombe, R.V. and Thomas, G.S.P. 1991. Glacial deposits and Quaternary stratigraphy of the Isle of Man. In Ehlers, J., Gibbard, P.L. and Rose, J. (eds), *Glacial Deposits in Great Britain and Ireland*. Balkema, Rotterdam, 333–344.

Dahl, E. 1947. On the origin of the strand flat. *Norsk Geografisk Tidsskrift* 11, 159–171.

Dahl, E. 1965. Biogeographic and geologic indications of unglaciated areas of Scandianvia during the glacial ages. *Geological Society of America Bulletin* 76, 1499–1520.

Dahl, R. 1965. Plastically sculptured detail forms on rock surfaces in northern Nordland, Norway. *Geografiska Annaler* 47, 83–140.

Dahl-Jensen, D. and Johnsen, S.J. 1986. Palaeotemperatures still exist in the Greenland ice sheet. *Nature* 320, 250–252.

Dalrymple, R.W. 1992. Tidal depositional systems. In Walker, R.G. and James, N.P. (eds), *Facies Models, Response to Sea-Level Change*. Geological Association of Canada, Toronto, 195–218.

Dansgaard, W. 1961. The isotopic composition of natural waters with special reference to the Greenland Ice Cap. *Meddelelser am Grønland* 165, 1–120.

Dansgaard, W., Clausen, H.B., Gundestrup, N., Hammer, C.U., Johnsen, S.J., Kristinsdottir, P.M. and Reeh, N. 1982. A new Greenland deep ice core. *Science* 218, 1273–1277.

Dansgaard, W., Johnsen, S.J., Clausen, H.B. and Gundestrup, N. 1973. Stable isotope glaciology. *Meddelelser am Grønland* 197, 1–53.

Dansgaard, W. and Tauber, H. 1969. Glacier oxygen–18 content and Pleistocene ocean temperatures. *Science* 166, 499–502.

Dardis, G.F. 1985. Till facies associations in drumlins and some implications for their mode of formation. *Geografiska Annaler* 67A, 13–22.

Dardis, G.F. 1987. Sedimentology of late Pleistocene drumlins in south-central Ulster, Northern Ireland. In Menzies, J. and Rose, J. (eds), *Drumlin Symposium*. Balkema, Rotterdam, 215–224.

Dardis, G.F. and Hanvey, P.M. 1994. Sedimentation in a drumlin lee-side wave cavity, northwest Ireland. *Sedimentary Geology* 91, 97–114.

Dardis, G.F. and McCabe, A.M. 1983. Facies of subglacial channel sedimentation in late Pleistocene drumlins, Northern Ireland. *Boreas* 12, 263–278.

Dardis, G.F. and McCabe, A.M. 1987. Subglacial sheetwash and debris flow deposits in late Pleistocene drumlins, Northern Ireland. In Menzies, J. and Rose, J. (eds), *Drumlin Symposium*. Balkema, Rotterdam, 225–240.

Dardis, G.F. and McCabe, A.M. (eds) 1994. *Subglacial Processes, Sediments and Landforms*. *Sedimentary Geology* 91 (special volume).

Dardis, G.F., McCabe, A.M. and Mitchell W.I. 1984. Characteristics and origins of lee-side stratification sequences in late Pleistocene drumlins, Northern Ireland. *Earth Surface Processes and Landforms* 9, 409–424.

Davies, D.A., Berrisford, M.J. and Matthews, J.A. 1990. Boulder-paved river channels, a case study of a fluvioperiglacial landform. *Zeitschrift für Geomorphologie* 34, 213–231.

Davies, J.L. 1969. *Landforms of Cold Climates*. MIT Press, Cambridge, Mass.

Davis, W.M. 1909. Glacial erosion in North Wales. *Quarterly Journal of the Geological Society of London* 65, 281–350.

Davis, W.M. 1920. A roxen lake in Canada. *Scottish Geographical Magazine* 41, 65–74.

Dawson, A.G. 1979. A Devensian medial moraine in Jura. *Scottish Journal of Geology* 15, 43–48.

Dawson, A.G. 1980. Shore erosion by frost, an example from the Scottish Lateglacial. In Lowe, J.J., Gray, J.M. and Robinson, J.E. (eds), *Studies in the Lateglacial of Northwest Europe*. Pergamon, Oxford, 45–54.

Dawson, A.G. 1982. Periglacial shore erosion and Lateglacial sea level changes in western Scotland. In Colquhoun, D.J. (ed.), *Holocene Sea Level Fluctuation, Magnitude and Causes*. INQUA Report, Columbia, South Carolina, 34–41.

Dawson, A.G. 1983. Glacier-dammed lake investigations in the Hullet Lake area, south Greenland. *Meddelelser om Gronland, Geoscience* 11, 24p.

Dawson, A.G. 1992. *Ice Age Earth, Late Quaternary Geology and Climate*. Routledge, London.

Dawson, A.G., Long, D. and Smith, D.E. 1988. The Storegga slides, evidence from eastern Scotland for a possible tsunami. *Marine Geology* 82, 271–276.

Dawson, A.G., Matthews, J.A. and Shakesby, R.A. 1986. A catastrophic landslide (sturzstrom) in Verkilsdalen, Rondane National Park, southern Norway. *Geografiska Annaler* 68A, 77–87.

Dawson, A.G., Matthews, J.A. and Shakesby, R.A. 1987. Rock platform erosion on periglacial shores, a modern analogue for Pleistocene rock platforms in Britain. In Boardman, J. (ed.), *Periglacial Processes and Landforms in Britain and Ireland*. Cambridge University Press, Cambridge, 173–182.

Dawson, G.M. 1895. Report on the area of the Kamloops map sheet. *Geological Society of Canada, Annual Report* 1894, 7, 1–427.

Dawson, M.R. 1985. Environmental reconstructions of a late Devensian terrace sequence: some preliminary findings. *Earth Surface Processes and Landforms* 10, 237–246.

Dawson, M.R. and Bryant, I.D. 1987. Three-dimensional facies geometry in Pleistocene outwash sediments, Worcestershire, UK. In Ethridge, F.G., Flores, R.M. and Harvey, M.D. (eds), *Recent Developments in Fluvial Sedimentology*. SEPM Special Publication 39, Tulsa, Okla., 191–196.

Dawson, M.R. and Gardiner, V. 1987. River terraces, the general model and a palaeohydrological and sedimentological interpretation of the terraces of the River Severn. In Gregory, K.J., Lewin, J. and Thornes, J.B. (eds), *Palaeohydrology in Practice*. Wiley, Chichester, 269–305.

Day, T.E. and Morris, W.A. 1982. Magnetic methods rapid in drift exploration. *The Northern Miner*, June.

Dayton, P.K., Robbiliard, G.A. and De Vries, A.L. 1969. Anchor ice formation in McMurdo Sound, Antarctica, and its biological effects. *Science* 163, 273–274.

Dean, W.G. 1953. The drumlinoid landforms of the 'Barren Grounds', NWT. *Canadian Geographer* 3, 19–30.

Deeley, R.M. and Parr, P.H. 1914. The Hintereis glacier. *Philosophical Magazine*, Ser. 6, 26, 85–111.

Deevey, F.S. and Flint, R.F. 1957. Post-glacial hypsithermal interval. *Science* 125, 182–184.

De Geer, G. 1892. On Pleistocene changes of level in eastern North America. *Proceedings of the Boston Society of Natural History* 25, 454–477.

De Geer, G. 1897. Om rullstensasarnas bildningssatt. *Geologiska Foreningens i Stockholm, Forhandlingar* 19, 366–388.

De Geer, G. 1912. A geochronology of the last 12,000 years. *11th International Geological Congress, Stockholm, Compte Rendu* 1, 241–258.

De Geer, G. 1940. *Geochronologia Suecica, Principles*. Kunglica Svenska Vetenskab Akademiens Handlingar 18, 1–90.

De la Beche, H.T. 1839. Report on the Geology of Cornwall, Devon and west Somerset. Memoirs of the Geological Survey, London.

Dellwig, L.F. and Baldwin, A.D. 1965. Ice-push deformation in northeastern Kansas. *Kansas Geological Survey, Bulletin* 175, part 2.

Demorest, M. 1937. Glaciation of the upper Nugssuak Peninsula, West Greenland. *Zeitschrift für Gletscherkunde* 25, 36–56.

Demorest, M. 1938. Ice flowage as revealed by glacial striae. *Journal of Geology* 46, 700–725.

Demorest, M. 1939. Glacial movement and erosion, a criticism. *American Journal of Science* 237, 594–605.

Denton, G.H. 1975. Glaciers of the Canadian Rocky Mountains. In Field, W.O. (ed.), *Mountain Glaciers of the Northern Hemisphere*. CRREL, Hanover, NH, USA, 603–653.

Denton, G.H. and Armstrong, R.L. 1969. Miocene–Pliocene glaciations in southern Alaska. *American Journal of Science* 267, 1121–1142.

Denton, G.H., Bockheim, J.G., Wilson, S.C. and Stuiver, M. 1989a. Late Wisconsin and early Holocene glacial history, Inner Ross Embayment, Antarctica. *Quaternary Research* 31, 151–182.

Denton, G.H., Bockheim, J.G., Wilson, S.C., Leide, J.E. and Andersen, B.G. 1989b. Late Quaternary ice surface fluctuations of Beardmore Glacier, Transantarctic Mountains. *Quaternary Research* 31, 183–209.

Denton, G.H. and Hughes, T.J. 1981a. The Arctic Ice Sheet, an outrageous hypothesis. In Denton, G.H. and Hughes, T.J. (eds), *The Last Great Ice Sheets*. Wiley, New York.

Denton, G.H. and Hughes, T.J. 1981b. *The Last Great Ice Sheets*. Wiley, New York.

Denton, G.H. and Hughes, T.J. 1983. Milankovitch theory of ice ages, ice sheet link between regional insolation input and global climatic output. *Quaternary Research* 20, 125–144.

Denton, G.H. and Karlen, W. 1973. Holocene climatic variations: their pattern and possible causes. *Quaternary Research* 3, 155–205.

Denton, G.H. and Karlen, W. 1977. Holocene glacial and tree line variations in the White River valley and Skolai Pass, Alaska and Yukon Territory. *Quaternary Research* 7, 63–111.

Denton, G.H. and Porter, S.C. 1970. Neo-glaciation. *Scientific American* 222, 101–110.

Denton, G.H., Sugden, D.E., Marchant, D.R., Hall, B.L. and Wich, T.I. 1993. East Antarctic ice sheet sensitivity to Pliocene climatic change from a Dry Valleys perspective. *Geografiska Annaler* 75A, 155–204.

Derbyshire, E. 1961. Subglacial col gullies and the deglaciation of the north-east Cheviots. *Transactions of the Institute of British Geographers* 29, 31–46.

Derbyshire, E. 1962. Fluvioglacial erosion near Knob Lake, central Quebec–Labrador, Canada. *Geological Society of America Bulletin* 73, 1111–1126.

Derbyshire, E. 1964. Cirques, Australian landform example no. 2. *Australian Geographer* 9, 178–179.

Derbyshire, E. 1968. Cirques. In Fairbridge, R.W. (ed.), *The Encyclopedia of Geomorphology 1*. St Martin's Press, New York.

Derbyshire, E. (ed.) 1995. Wind blown sediments in the Quaternary record. *Quaternary Proceedings 4*, Wiley, Chichester.

Derbyshire, E. and Evans, I.S. 1976. The climatic factor in cirque variation. In Derbyshire, E. (ed.), *Geomorphology and Climate*. Wiley, Chichester, 447–494.

Derbyshire, E. and Owen, L.A. 1990. Quaternary alluvial fans in the Karakoram mountains. In Rachocki, A.H. and Church, M. (eds), *Alluvial Fans, A Field Approach*. Wiley, Chichester, 27–53.

Derbyshire, E. and Owen, L.A. 1996. Glacioaeolian processes, sediments and landforms. In Menzies, J. (ed.), *Glacial Environments 2, Past Glacial Environments – Sediments, Forms and Techniques*. Butterworth-Heinemann, Oxford, 213–237.

Derbyshire, E., Shi, Y., Li, J., Zheng, B., Li, S. and Wang, J. 1991. Quaternary glaciation of Tibet, the geological evidence. *Quaternary Science Reviews* 10, 485–510.

Desloges, J.R., Jones, D.P. and Ricker, K.E. 1989. Estimates of peak discharge from the drainage of ice-dammed Ape Lake, British Columbia, Canada. *Journal of Glaciology* 35, 349–354.

Desloges, J.R. and Ryder, J.M. 1990. Neoglacial history of the Coast Mountains near Bella Coola, British Columbia. *Canadian Journal of Earth Sciences* 27, 281–290.

Dewart, G. 1966. Moulins on Kaskawush Glacier, Yukon Territory. *Journal of Glaciology* 6, 320–331.

Deynoux, M., Miller, J.M.G., Domack, E.W., Eyles, N., Fairchild, I.J. and Young, G.M. (eds) 1993. *Earth's Glacial Record*. Cambridge University Press, Cambridge.

Deynoux, M. and Trompette, R. 1976. Late Precambrian mixtites, glacial and/or non-glacial? Dealing especially with the mixtites of West Africa. *American Journal of Science* 276, 1302–1324.

DiLabio, R.N.W. 1981. Glacial dispersal of rocks and minerals at the south end of Lac Mistassini, Quebec, with special reference to the Icon dispersal train. *Geological Survey of Canada Bulletin* 323.

DiLabio, R.N.W. 1982. Gold and tungsten abundance vs grain size in till at Waverly, Nova Scotia. Geological Survey of Canada, Paper 82–1B, 57–62.

DiLabio, R.N.W. 1990a. Drift prospecting, geologists use glacial sediments to find ore deposits. *Geos* 19, 7–15.

DiLabio, R.N.W. 1990b. Glacial dispersal trains. In Kujan-suu, R. and Saarnisto, M. (eds), *Glacier Indicator Tracing*. Balkema, Rotterdam, 109–122.

DiLabio, R.N.W. and Coker, W.B. (eds) 1989. Drift prospecting. Geological Survey of Canada, Paper 89–20.

DiLabio, R.N.W. and Shilts, W.W. 1979. Composition and dispersal of debris by modern glaciers, Bylot Island, Canada. In Schluchter, C. (ed.), *Moraines and Varves*. Balkema, Rotterdam, 145–155.

Ding, Z., Rutter, N.W. and Liu, T. 1993. Pedostratigraphy of Chinese loess deposits and climatic cycles in the last 2.5 Ma. *Catena* 20, 73–91.

Dionne, J.-C. 1968. Shore morphology on the south shore of the St Lawrence Estuary. *American Journal of Science* 266, 380–388.

Dionne, J.-C. 1973. La notion de pied de glace (icefoot), en particulier dans l'estuaire du Saint-Laurent. Cahiers de Géographie de Québec 17, 221–250. English translation in Evans, D.J.A. (ed.), *Cold Climate Landforms*. Wiley, Chichester, 325–349.

Dionne, J.-C. 1978. Formes et phénomènes périglaciel en Jamesie, Québec subarctique. *Géographie Physique et Quaternaire* 32, 187–247.

Dionne, J.-C. 1984. An estimate of ice-drifted sediments based on the mud content of the ice cover at Mont-magny, middle St Lawrence estuary. *Marine Geology* 47, 149–166.

Dionne, J.-C. 1985. Drift-ice abrasion marks along rocky shores. *Journal of Glaciology* 31, 237–241.

Dionne, J.-C. 1987. Tadpole rock (rocdrumlin), a glacial streamline moulded form. In Menzies, J. and Rose, J. (eds), *Drumlin Symposium*. Balkema, Rotterdam, 149–159.

Doake, C.S.M. 1976. Thermodynamics of the interaction between ice shelves and the sea. *Polar Record* 18, 37–41.

Dolgushin L.D. 1961. Main features of the modern glaciation of the Urals. *IASH* 54, 335–347.

Dolgushin L.D. and Osipova, G.B. 1975. Glacier surges and the problem of their forecast. International Association of Hydrological Sciences Publication 104, 292–304.

Dolgushin L.D. and Osipova, G.B. 1978. Balance of a surging glacier as the basis for forecasting its periodic advances. *Materialny Glyatsiologicheskikh Issledovaniy, Khronica Obsuzhdeniya* 32, 260–265.

Dolgushin L.D., Yevteyev, S.A. and Kotlyakov, V.M. 1962. Current changes in the Antarctic Ice Sheet. *IASH* 58, 286–294.

Domack, E.W. 1982. Sedimentology of glacial and glacial marine deposits on the George V–Adelie continental shelf, East Antarctica. *Boreas* 11, 79–97.

Domack, E.W. 1984. Rhythmically bedded glaciomarine sediments on Whidbey Island, northwestern Washington State and southwestern British Columbia. *Journal of Sedimentary Petrology* 54, 589–602.

Domack, E.W. 1988. Biogenic facies in the Antarctic glacimarine environment, basis for a polar glacimarine summary. *Palaeogeography, Palaeoclimatology, Palaeoecology* 63, 357–372.

Domack, E.W., Anderson, J.B. and Kurtz, D.D. 1980. Clast shape as an indicator of transport and depositional mechanisms in glacial marine sediments, George V con-tinental shelf, Antarctica. *Journal of Sedimentary Petrology* 50, 813–820.

Domack, E.W. and Lawson, D.E. 1985. Pebble fabric in an ice-rafted diamicton. *Journal of Geology* 93, 577–592.

Domaradzki, J. 1951. Blockstrome in Kanton Graubunden. *Ergebnisse der Wissenschaften der Untersuchungen des Schweizerische Nationalparks* 3, 23, and *Arbeiten Geographisches Institut der Universität Zürich*, Ser. A, 54, 173–235.

van Donk, J. 1976. An [18]O record of the Atlantic Ocean for the entire Pleistocene. In Cline, R.M. and Hays, J.D. (eds), *Investigation of Late Quaternary Paleoceanography and Paleoclimatology*. Geological Society of America Memoir 145, 145–163.

Donn, W.L., Farrand, W.R. and Ewing, M. 1962. Pleistocene ice volumes and sea level changes. *Journal of Geology* 70, 206–214.

Donner, J.J. 1969. Land/sea level changes in southern Finland during the formation of the Salpausselkä end moraines. *Bulletin of the Geological Society of Finland* 41, 135–150.

Donner, J.J. 1970. Land/sea level changes in Scotland. In Walker, D. and West, R.G. (eds), *Studies in the Vegetational History of the British Isles*. Cambridge University Press, Cambridge, 23–39.

Donner, J.J. 1978. The dating of the levels of the Baltic Ice Lake and the Salpausselka moraines in south Finland. *Societas Scientiarum Fennica, Commentationes Physico-Mathematicae* 48, 11–38.

Donner, J.J. 1980. The determination and dating of synchronous Late Quaternary shorelines in Fennoscandia. In Morner, N.A. (ed.), *Earth Rheology, Isostasy and Eustasy*. Wiley, Chichester, 285–293.

Donner, J.J. 1982. Fluctuations in water level of the Baltic Ice Lake. *Annales Academiae Scientarum Fennicae AIII* 134, 13–28.

Donner, J.J. 1995. *The Quaternary History of Scandinavia*. Cambridge University Press, Cambridge.

Donovan, D.T. 1973. The geology and origin of the Silver Pit and other closed basins in the North Sea. *Proceedings of the Yorkshire Geological Society* 39, 267–293.

Doornkamp, J.C. and King, C.A.M. 1971. *Numerical Analysis in Geomorphology*. Edward Arnold, London.

Dor, A.G. 1992. The base Tertiary surface of southern Norway and the northern North Sea. *Norsk Geologisk Tidsskrift* 72, 259–265.

Dorrer, E. 1971. Movement of the Ward Hunt Ice Shelf, Ellesmere Island, NWT, Canada. *Journal of Glaciology* 10, 211–225.

Douglas, B.C., Cheney, R.E., Miller, L., Agreen, R.W., Carter, W.E. and Robertson, D.S. 1990. Greenland ice sheet, is it growing or shrinking? *Science* 248, 288.

Dowdeswell, E.K. and Andrews, J.T. 1985. The fiords of Baffin Island, description and classification. In Andrews, J.T. (ed.), *Quaternary Environments, Eastern Canadian Arctic, Baffin Bay and Western Greenland*. Allen & Unwin, Boston, 93–123.

Dowdeswell, J.A. 1987. Processes of glacimarine sedimentation. *Progress in Physical Geography* 11, 52–90.

Dowdeswell, J.A. 1989. On the nature of Svalbard icebergs. *Journal of Glaciology* 35, 224–234.

Dowdeswell, J.A. and Dowdeswell, E.K. 1989. Debris in icebergs and rates of glacimarine sedimentation, observations from Spitsbergen and a simple model. *Journal of Geology* 97, 221–231.

Dowdeswell, J.A. and Drewry, D.J. 1989. The dynamics of Austfonna, Nordaustlandet, Svalbard, surface velocities, mass balance, and subglacial meltwater. *Annals of Glaciology* 12, 37–45.

Dowdeswell, J.A., Drewry, D.J., Cooper, A.P.R., Gorman, M.R., Liestøl, O. and Orheim, O. 1986. Digital mapping of the Nordaustlandet ice caps from airborne geophysical investigations. *Annals of Glaciology* 8, 51–58.

Dowdeswell, J.A., Drewry, D.J., Liestøl, O. and Orheim, O. 1984a. Airborne radio echo sounding of sub-polar glaciers in Spitsbergen. *Norsk Polarinstitutt Skrifter* 182.

Dowdeswell, J.A., Drewry, D.J., Liestøl, O. and Orheim, O. 1984b. Radio echo sounding of Spitsbergen glaciers, problems in the interpretation of layer and bottom returns. *Journal of Glaciology* 30, 16–21.

Dowdeswell, J.A., Hambrey, M.J. and Wu, R. 1985. A comparison of clast fabric and shape in Late Precambrian and modern glacigenic sediments. *Journal of Sedimentary Petrology*, 55, 691–704.

Dowdeswell, J.A., Hamilton, G.S. and Hagen, J.O. 1991. The duration of the active phase of surge-type glaciers: contrasts between Svalbard and other regions. *Journal of Glaciology* 37, 388–400.

Dowdeswell, J.A., Maslin M.A., Andrews, J.T. and McCave, I.N. 1995. Iceberg production, debris rafting, and the extent and thickness of Heinrich layers (H–1, H–2) in North Atlantic sediments. *Geology* 23, 301–304.

Dowdeswell, J.A. and Murray, T. 1990. Modelling rates of sedimentation from icebergs. In Dowdeswell, J.A. and Scourse, J.D. (eds), *Glacimarine Environments, Processes and Products*. Geological Society Special Publication 53, London, 121–137.

Dowdeswell, J.A. and Sharp, M.J. 1986. Characterization of pebble fabrics in modern terrestrial glacigenic sediments. *Sedimentology* 33, 699–710.

Dowdeswell, J.A., Villinger, H., Whittington, R.J. and Marienfeld, P. 1993. Iceberg scouring in Scoresby Sund and the East Greenland continental shelf. *Marine Geology* 111, 37–53.

Dowdeswell, J.A., Whittington, R.J. and Marienfeld, P. 1994. The origin of massive diamicton facies by iceberg rafting and scouring, Scoresby Sund, east Greenland. *Sedimentology* 41, 21–35.

Drake, L.D. 1972. Mechanisms of clast attrition in basal till. *Geological Society of America Bulletin* 83, 2159–2165.

Dredge, L.A. and Cowan, W.R. 1989. Quaternary geology of the southwestern Canadian shield. In Fulton, R.J. (ed.), *Quaternary Geology of Canada and Greenland*. Geological Survey of Canada, Geology of Canada No. 1, 214–235.

Dredge, L.A., Nixon, F.M. and Richardson, R.J. 1986. Quaternary geology and geomorphology of northwestern Manitoba. Geological Survey of Canada, Memoir 418.

Dredge, L.A. and Thorleifson, L.H. 1987. The Middle Wisconsinan history of the Laurentide Ice Sheet. *Géographie Physique et Quaternaire* 41, 215–235.

Dreimanis, A. 1953. Studies of friction cracks along shores of Cirrus Lake and Kasakokwog Lake, Ontario. *American Journal of Science* 251, 769–783.

Dreimanis, A. 1969. Selection of genetically significant parameters for investigation of tills. *Geografia* 8,15–29.

Dreimanis, A. 1976a. Criteria for recognition of various types of till. In Stankowski, W. (ed.), Till, its genesis and diagenesis. *Geografia* 12, 177–178.

Dreimanis, A. 1976b. Tills, their origin and properties. In Leggett, R.F. (ed.), *Glacial Till*. Royal Society of Canada, Special Publication 12, 11–49.

Dreimanis, A. 1979. The problem of waterlain tills. In Schluchter, C. (ed.) *Moraines and Varves*. Balkema, Rotterdam, 167–177.

Dreimanis, A. 1982a. Two origins of the stratified Catfish Creek till at Plum Point, Ontario, Canada. *Boreas* 11, 173–180.

Dreimanis, A. 1982b. Work Group I. Genetic classification of tills and criteria for their differentiation, progress report on activities 1977–1982 and definitions of glacigenic terms. In Schlüchter, C. (ed.), *INQUA Commission on Genesis and Lithology of Quaternary Deposits, Report on Activities 1977–1882*, 12–31. Zurich.

Dreimanis, A. 1984a. Comments on 'Sedimentation in a large lake, an interpretation of the late Pleistocene stratigraphy at Scarborough Bluffs, Ontario, Canada'. *Geology* 12, 185–186.

Dreimanis, A. 1984b. Lithofacies types and vertical profile models, an alternative approach to the description and environmental interpretation of glacial diamict and diamictite sequences. Discussion. *Sedimentology* 31, 885–886.

Dreimanis, A. 1987. Genetic complexity of a subaquatic till tongue at Port Talbot, Ontario, Canada. In Kujansuu, R. and Saarnisto, M. (eds), *INQUA Report on Activities 1982–1987*, ETH, Zurich, 68–78.

Dreimanis, A. 1989. Tills, their genetic terminology and classification. In Goldthwait, R.P. and Matsch, C.L. (eds), *Genetic Classification of Glacigenic Deposits*. Balkema, Rotterdam, 17–84.

Dreimanis, A., Hamilton, J.P. and Kelly, P.E. 1986. Complex subglacial sedimentation of Catfish Creek till at Bradtville, Ontario, Canada. In van der Meer, J.J.M. (ed.), *Tills and Glaciotectonics*. Balkema, Rotterdam, 73–87.

Dreimanis, A. and Lundqvist, J. 1984. What should be called a till? In Konigsson, L.K. (ed.), Ten years of Nordic till research. *Striae* 20, 5–10.

Dreimanis, A. and Vagners, U.J. 1971. Bimodal distribution of rock and mineral fragments in basal tills. In Goldthwait, R.P. (ed.), *Till: A Symposium*. Ohio State University Press, Columbus, 237–250.

Dreimanis, A. and Vagners, U.J. 1972. The effect of lithology on the texture of till. In Yatsu, E. and Falconer, A. (eds), *Research Methods in Pleistocene Geomorphology*. University of Guelph, Guelph, 66–82.

Drew, F. 1873. Alluvial and lacustrine deposits and glacial records of the upper Indus basin. *Geological Society of London Quarterly Journal* 29, 441–471.

Drewes, H., Fraser, G.D., Snyder, G.L. and Barnett, H.F. 1961. Geology of Unalaska Island and adjacent insular shelf, Aleutian Islands, Alaska. *US Geological Survey, Bulletin* 1028–5, 583–676.

Drewry, D.J. 1972. A quantitative assessment of dirt-cone dynamics. *Journal of Glaciology* 11, 431–446.

Drewry, D.J. 1975. Initiation and growth of the East Antarctic Ice Sheet. *Journal of the Geological Society of London* 131, 255–273.

Drewry, D.J. 1978. Aspects of the early evolution of West Antarctic ice. In van Zinderen Bakker, E.M. (ed.), *Antarctic Glacial History and World Palaeoenvironments*. Balkema, Rotterdam, 25–32.

Drewry, D.J. 1979. Late Wisconsin reconstruction for the Ross Sea region, Antarctica. *Journal of Glaciology* 24, 231–244.

Drewry, D.J. 1983. *Antarctica, Glaciological and Geophysical Folio*. University of Cambridge.

Drewry, D.J. 1986. *Glacial Geologic Processes*. Edward Arnold, London.

Drewry, D.J. 1991. The response of the Antarctic ice sheet to climatic change. In Harris, C. and Stonehouse, B. (eds), *Antarctica and Global Climate Change*. SPRI and Belhaven Press, Cambridge, 90–106.

Drewry, D.J. and Cooper, A.P.R. 1981. Processes and models of Antarctic glaciomarine sedimentation. *Annals of Glaciology* 2, 117–122.

Drewry, D.J., Jordan, S.R. and Jankowski, E. 1982. Measured properties of the Antarctic ice sheet, surface configuration, ice thickness, volume and bedrock characteristics. *Annals of Glaciology* 3, 83–91.

Drozdowski, E. 1977. Ablation till and related indicatory forms at the margins of Vestspitsbergen glaciers. *Boreas* 6, 107–114.

Drozdowski, E. 1987. 'Surge moraines'. In Gardiner, V. (ed.), *International Geomorphology*. Wiley, Chichester, 675–692.

Dubois, J.M. and Dionne, J.C. 1985. The Quebec North Shore Moraine system, a major feature of late Wisconsin deglaciation. In Borns, H.W., LaSalle, P. and Thompson, W.B. (eds), *Late Pleistocene History of Northeastern New England and Adjacent Quebec*. Geological Society of America, Special Paper 197, 125–133.

Dubrovin, L.I. 1976. Major types of Antarctic ice shores. *Polar Geography* 3, 67–74.

Dudeney, J.R. 1987. The Antarctic atmosphere. In Walton, D.W.H. (ed.), *Antarctic Science*. Cambridge University Press, Cambridge, 191–247.

Dugmore, A.J. and Sugden, D.E. 1991. Do the anomalous fluctuations of Solheimajokul reflect ice-divide migration? *Boreas* 20, 105–113.

Duk-Rodkin, A. and Hughes, O.L. 1992. Pleistocene montane glaciations in the Mackenzie Mountains, NWT. *Géographie Physique et Quaternaire* 46, 69–83.

Duk-Rodkin, A. and Hughes, O.L. 1994. Tertiary–Quaternary drainage of the pre-glacial Mackenzie basin. *Quaternary International* 22/23, 221–241.

Duk-Rodkin, A., Jackson, L.E. and Rodkin, O. 1986. A composite profile of the Cordilleran ice sheet during McConnell Glaciation, Glenlyon and Tay River map areas, Yukon Territory. Geological Survey of Canada Paper 86–1B, 257–262.

Dunbar, R.B., Anderson, J.B. and Domack, E.W. 1985. Oceanographic influences on sedimentation along the Antarctic continental shelf. In *Oceanology of the Antarctic Continental Shelf*, Antarctic Research Series, 291–312.

Dunbar, R.B., Leventer, A.R. and Stochton, W.L. 1989. Biogenic sedimentation in McMurdo sound, Antarctica. *Marine Geology* 85, 155–180.

Dury, G.H. 1953. A glacial breach in the northwestern highlands. *Scottish Geographical Magazine* 69, 106–117.

Dury, G.H. (ed.) 1970. *Rivers and River Terraces*. Macmillan, London.

Duval, P. 1981. Grain growth and the mechanical behaviour of polar ice. *Annals of Glaciology*, 6, 79–82.

Duynkerke, P.G. and van den Broeke, M.R. 1994. Surface energy balance and katabatic flow over glacier and tundra during GIMEX–91. *Global and Planetary Change* 9, 17–28.

Dyer, K.R. 1973. *Estuaries, A Physical Introduction*. Wiley, New York.

Dyke, A.S. 1979. Glacial and sea level history of the southwestern Cumberland Peninsula, Baffin Island, NWT, Canada. *Arctic and Alpine Research* 11, 179–202.

Dyke, A.S. 1983. Quaternary geology of Somerset Island, District of Franklin. Geological Survey of Canada Memoir 404.

Dyke, A.S. 1984a. Multiple deglaciations of the Hudson Bay lowlands, Canada, since deposition of the Missinaibi (last interglacial?) Formation, discussion. *Quaternary Research* 22, 247–252.

Dyke, A.S. 1984b. Quaternary geology of Boothia Peninsula and northern District of Keewatin, central Canadian arctic. Geological Survey of Canada, Memoir 407.

Dyke, A.S. 1987. A re-interpretation of the ages of glacial and marine limits around the northwestern Laurentide Ice Sheet. *Canadian Journal of Earth Sciences* 24, 591–601.

Dyke, A.S. 1990a. A lichenometric study of Holocene rock glaciers and Neoglacial moraines, Frances Lake Map Area, southeastern Yukon Territory and NWT. Geological Survey of Canada Bulletin 394.

Dyke, A.S. 1990b. Quaternary geology of the Frances Lake map area, Yukon and Northwest Territories. Geological Survey of Canada Memoir 426.

Dyke, A.S. 1993a. Glacial and sea level history of Lowther and Griffith Islands, a hint of tectonics. *Géographie Physique et Quaternaire* 47, 133–145.

Dyke, A.S. 1993b. Landscapes of cold-centred Late Wisconsinan ice caps, Arctic Canada. *Progress in Physical Geography* 17, 223–247.

Dyke, A.S., Andrews, J.T. and Miller, G.H. 1982a. Quaternary geology of Cumberland Peninsula, Baffin Island, District of Franklin. Geological Survey of Canada Memoir 403.

Dyke, A.S. and Dredge, L.A. 1989. Quaternary geology of the northwestern Canadian Shield. In Fulton, R.J. (ed.), *Quaternary Geology of Canada and Greenland*. Geological Survey of Canada, Geology of Canada No. 1, 178–214.

Dyke, A.S., Dredge, L.A. and Vincent, J.S. 1982b. Configuration and dynamics of the Laurentide Ice Sheet during the Late Wisconsinan maximum. *Géographie Physique et Quaternaire* 36, 5–14.

Dyke, A.S. and Morris, T.F. 1988. Drumlin fields, dispersal trains and ice streams in arctic Canada. *Canadian Geographer* 32, 86–90.

Dyke, A.S., Morris, T.F. and Green, D.E.C. 1991. Postglacial tectonic and sea level history of the central

Canadian arctic. Geological Survey of Canada, Bulletin 397.

Dyke, A.S., Morris, T.F., Green, D.E.C. and England, J. 1992. Quaternary geology of Prince of Wales Island, arctic Canada. Geological Survey of Canada, Memoir 433.

Dyke, A.S. and Prest, V.K. 1987. Late Wisconsinan and Holocene history of the Laurentide Ice Sheet. *Géographie Physique et Quaternaire* 41, 237–263.

Dyke, A.S., Vincent, J-S., Andrews, J.T., Dredge, L.A. and Cowan, W.R. 1989. The Laurentide Ice Sheet and an introduction to the Quaternary geology of the Canadian Shield. In Fulton, R.J. (ed.), *Quaternary Geology of Canada and Greenland*. Geological Survey of Canada, Geology of Canada No. 1, 178–189.

Dylik, J. 1961. The Łódź region. *VI INQUA Congress, Guidebook of Excursion*, Łódź.

Dyson, J.L. 1952. Ice-ridged moraines and their relation to glaciers. *American Journal of Science* 250, 204–212.

Dzulynski, S. and Walton, E.K. 1965. *Sedimentary Features of Flysch and Greywackes*. Elsevier, Amsterdam.

Easterbrook, D.J. 1992. Advance and retreat of Cordilleran ice sheets in Washington, USA. *Géographie Physique et Quaternaire* 46, 51–68.

Eberl, B. 1930. *Die Eiszeitenfolge im nordlichen Alpenvorlande*. Benno Filser, Augsberg.

Ebers, E. 1961. Die Gletscherschliffe und -rinnen. In Fehn, H. (ed.), *Der Gletscherschliffe von Fischbach am Inn*. Landeskundliche Forschungen, Munich, 40.

Echelmeyer, K., Butterfield, R. and Cullard, D. 1987. Some observations on a recent surge of Peters Glacier, Alaska, USA. *Journal of Glaciology* 33, 341–345.

Echelmeyer, K., Clarke, T.S. and Harrison, W.D. 1991. Surficial glaciology of Jakobshavns Isbrae, west Greenland, Part I. Surface morphology. *Journal of Glaciology* 37, 368–382.

Echelmeyer, K. and Harrison, W.D. 1990. Jacobshavns Isbrae, west Greenland, seasonal variations in velocity – or lack thereof. *Journal of Glaciology* 36, 82–88.

Echelmeyer, K. and Wang, Z. 1987. Direct observation of basal sliding and deformation of basal drift at sub-freezing temperatures. *Journal of Glaciology* 33, 83–98.

Edelman, N. 1951. Glacial abrasion and ice movement in the area of Rosala-Noto, SW Finland. *Geological Survey of Finland, Bulletin* 154, 157–169.

Edwards, M.B. 1975. Gravel fraction on the Spitsbergen bank NW Barents shelf. *Norges Geologiske Undersoegelse* 315, 205–217.

Ehlers, J. 1978. Fine gravel analysis after the Dutch method tested out on Ristinge Klint, Denmark. *Bulletin of the Geological Society of Denmark* 27, 157–165.

Ehlers, J. 1981. Some aspects of glacial erosion and deposition in north Germany. *Annals of Glaciology* 2, 143–146.

Ehlers, J. 1983a. Different till types in North Germany and their origin. In Evenson, E.B., Schluchter, C. and Rabassa, J. (eds), *Tills and Related Deposits*. Balkema, Rotterdam, 61–80.

Ehlers, J. (ed.) 1983b. *Glacial Deposits of North-West Europe*. Balkema, Rotterdam.

Ehlers, J. 1990. Reconstructing the dynamics of the north-west European Pleistocene ice sheets. *Quaternary Science Review* 9, 71–83.

Ehlers, J. and Gibbard, P.L. 1991. Anglian glacial deposits in Britain and the adjoining offshore regions. In Ehlers, J., Gibbard, P.L. and Rose, J. (eds), *Glacial Deposits in Great Britain and Ireland*. Balkema, Rotterdam, 17–24.

Ehlers, J., Gibbard, P.L. and Rose, J. (eds) 1991. *Glacial Deposits in Great Britain and Ireland*. Balkema, Rotterdam.

Ehlers, J. and Linke, G. 1989. The origin of deep buried channels of Elsterian age in northwest Germany. *Journal of Quaternary Science* 4, 255–265.

Ehlers, J., Meyer, K.D. and Stephan, H.J. 1984. Pre-Weichselian glaciation of North-West Europe. *Quaternary Science Reviews* 3, 1–40.

Ehlers, J. and Stephan, H.J. 1979. Forms at the base of till strata as indicators of ice movement. *Journal of Glaciology* 22, 345–356.

Ehlers, J. and Wingfield, R. 1991. The extension of the Late Weichselian/Late Devensian ice sheets in the North Sea Basin. *Journal of Quaternary Science* 6, 313–326.

Eisbacher, G.H. and Clague, J.J. 1984. Destructive mass movements in high mountains, hazard and management. Geological Survey of Canada Paper 84–16.

Eklund, A. and Hart, J.K. 1996. Glaciotectonic deformation within a flute from the Isfallsglaciären, Sweden. *Journal of Quaternary Science* 11, 299–310.

Elfström, Å. and Rossbacher, L. 1985. Erosional remnants in the Båldakatj area, Lappland, northern Sweden. *Geografiska Annaler* 67A, 167–176.

Elliott, T. 1986a. Siliciclastic shorelines. In Reading, H.G. (ed.), *Sedimentary Environments and Facies*. Oxford, Blackwell, 155–188.

Elliott, T. 1986b. Deltas. In Reading, H.G. (ed.), *Sedimentary Environments and Facies*. Blackwell, Oxford, 113–154.

Ellis, J.M. and Calkin, P.E. 1979. Nature and distribution of glaciers, neoglacial moraines and rock glaciers, east-central Brooks Range, Alaska. *Arctic and Alpine Research* 11, 403–420.

Ellis, J.M. and Calkin, P.E. 1984. Chronology of Holocene glaciation, central Brooks Range, Alaska. *Geological Society of America Bulletin* 95, 897–912.

Ellis-Gruffydd, I.D. 1977. Late Devensian glaciation in the Upper Usk basin. *Cambria* 4, 46–55.

Elliston, G.R. 1973. Water movement through the Gornergletscher. IASH Publication 95, 79–84.

Elson, J.A. 1957. Origin of washboard moraines. *Bulletin of the Geological Society of America* 68, 1721.

Elson, J.A. 1961. The geology of tills. In Penner, E. and Butler, J. (eds), *Proceedings of the 14th Canadian Soil Mechanics Conference*, 5–36. Commission for Soil and Snow Mechanics Technical Memoir 69.

Elson, J.A. 1969. Washboard moraines and other minor moraine types. In Fairbridge, R.W. (ed.), *Encyclopedia of Geomorphology 3*. Reinhold, New York, 1213–1219.

Elson, J.A. 1989. Comment on glacitectonite, deformation till, and comminution till. In Goldthwait, R.P. and Matsch, C.L. (eds), *Genetic Classification of Glacigenic Deposits*. Balkema, Rotterdam, 85–88.

Elverhoi, A., Lonne, O. and Sealand, R. 1983. Glaciomarine sedimentation in a modern fjord environment, Spitsbergen. *Polar Research* 1, 127–149.

Elverhoi, A. and Roaldset, E. 1983. Glaciomarine sediments and suspended particulate matter, Weddell Sea shelf, Antarctica. *Polar Research* 1, 1–21.

Embleton, C. 1964a. Subglacial drainage and supposed ice-dammed lakes in north-east Wales. *Proceedings of the Geologists' Association* 75, 31–38.

Embleton, C. 1964b. The deglaciation of Arfon and southern Anglesey, and the origin of the Menai Straits. *Proceedings of the Geologists' Association* 75, 407–430.

Embleton, C. and King, C.A.M. 1975. *Glacial Geomorphology*. Edward Arnold, London.

Emery, K.O. 1963. Oceanic transportation of marine sediments. In Hill, M.N. (ed.), *The Sea*. Wiley, New York, 776–793.

Emery, K.C. 1968. Relict sediments on continental shelves of the world. *Bulletin of the American Association of Petroleum Geologists* 52, 445–464.

Emiliani, C. 1978. The cause of the ice ages. *Earth and Planetary Science Letters* 37, 349–352.

Engelhardt, H., Humphrey, N., Kamb, B. and Fahnestock, M. 1990. Physical conditions at the base of a fast moving Antarctic ice stream. *Science* 248, 57–59.

Engelhardt, H.F., Harrison, W.D. and Kamb, B. 1978. Basal sliding and conditions at the glacier bed as revealed by bore-hole photography. *Journal of Glaciology* 20, 469–508.

von Engeln, O.D. 1937. Rock sculpture by glaciers, a review. *Geographical Review* 27, 478–482.

von Engeln, O.D. 1938. Glacial geomorphology and glacier motion. *American Journal of Science* 35, 426–440.

England, J. 1976a. Late Quaternary glaciation of the Queen Elizabeth Islands, NWT, Canada, alternative models. *Quaternary Research* 6, 185–202.

England, J. 1976b. Postglacial isobases and uplift curves from the Canadian and Greenland high arctic. *Arctic and Alpine Research* 8, 61–78.

England, J. 1978. The glacial geology of northeastern Ellesmere Island, NWT, Canada. *Canadian Journal of Earth Sciences* 15, 603–617.

England, J. 1982. Postglacial emergence along northern Nares Strait. *Meddelelser om Grønland*, Geoscience 8, 65–75.

England, J. 1983. Isostatic adjustments in a full glacial sea. *Canadian Journal of Earth Sciences* 20, 895–917.

England, J. 1985. The late Quaternary history of Hall Land, northwest Greenland. *Canadian Journal of Earth Sciences* 22, 1394–1408.

England, J. 1986. Glacial erosion of a high arctic valley. *Journal of Glaciology* 32, 60–64.

England, J. 1987. Glaciation and the evolution of the Canadian high arctic landscape. *Geology* 15, 419–424.

England, J. 1990. The late Quaternary history of Greely Fiord and its tributaries, west-central Ellesmere Island. *Canadian Journal of Earth Sciences* 27, 255–270.

England, J. 1992. Postglacial emergence in the Canadian high arctic, integrating glacioisostasy, eustasy, and late deglaciation. *Canadian Journal of Earth Sciences* 29, 984–999.

England, J. and Bednarski, J. 1986. Postglacial isobases from northern Ellesmere Island and Greenland, new data. *Géographie Physique et Quaternaire* 40, 299–305.

England, J. and Bednarski, J. 1989. Northeast Ellesmere Island (Quaternary geology of the Queen Elizabeth Islands). In Fulton, R.J. (ed.), *Quaternary Geology of Canada and Greenland*. Geological Survey of Canada, Geology of Canada No. 1, 459–464.

England, J., Bradley, R.S. and Miller, G.H. 1978. Former ice shelves in the Canadian high Arctic. *Journal of Glaciology* 20, 393–404.

England, J., Bradley, R.S. and Stuckenrath, R. 1981. Multiple glaciations and marine transgressions, western Kennedy Channel, NWT, Canada. *Boreas* 10, 71–89.

England, J., Sharp, M., Lemmen, D.S. and Bednarski, J. 1991. On the extent and thickness of the Innuitian Ice Sheet, a postglacial-adjustment approach, discussion. *Canadian Journal of Earth Sciences* 28, 1689–1695.

Enos, P. 1977. Flow regimes in debris flow. *Sedimentology* 24, 133–142.

Enquist, F. 1917. Der Einfluss des Windes auf die Verteilung der Gletscher. *Bulletin of the Geological Institute, University of Upsala* 14, 1–108.

Epprecht, W. 1987. A major calving event of Jacobshavns Isbrae, West Greenland, on 9 August 1982. *Journal of Glaciology* 33, 169–172.

Erdmann, E. 1873. Iaktagelser ofver moranbildningar och deraf betackta skiktade jordarter i Skane. *Geologiska Foreningens i Stockholm, Forhandlingar* 2, 13–24.

Erikstad, L. and Sollid, J.L. 1986. Neoglaciation in south Norway using lichenometric methods. *Norsk Geografisk Tidsskrift* 40, 85–105.

Eronen, M. 1983. Late Weichselian and Holocene shore displacement in Finland. In Smith, D.E. and Dawson, A.G. (eds), *Shorelines and Isostasy*. Academic Press, London, 183–208.

Escobar, F., Vidal, F., Garin, C. and Naruse, R. 1992. Water balance in the Patagonia Icefield. In Naruse, R. and Aniya, M. (eds), *Glaciological Researches in Patagonia 1990*, 109–119.

Espizua, L.E. 1986. Fluctuations of the Rio del Plomo glaciers. *Geografiska Annaler* 68A, 317–327.

Etzelmüller, B., Hagen, J.O., Vatne, G., Ødegård, R.S. and Sollid, J.L. 1996. Glacier debris accumulation and sediment deformation influenced by permafrost, examples from Svalbard. *Annals of Glaciology* 22, 53–62.

Evans, D.J.A. 1985. Wisconsinan ice dynamics in Alberta, a review of prevalent hypotheses and the development of recent ideas. In Deshaies, L. and Pelletier, R. (eds), *Proceedings of the Canadian Association of Geographers, Trois Rivières*, 1985, 13–33.

Evans, D.J.A. 1989a. The nature of glacitectonic structures and sediments at subpolar glacier margins, northwest Ellesmere Island, Canada. *Geografiska Annaler* 71A, 113–123.

Evans, D.J.A. 1989b. Apron entrainment at the margins of subpolar glaciers, northwest Ellesmere Island, Canadian high arctic. *Journal of Glaciology* 35, 317–324.

Evans, D.J.A. 1990a. The effect of glacier morphology on surficial geology and glacial stratigraphy in a high arctic mountainous terrain. *Zeitschrift für Geomorphologie* 34, 481–503.

Evans, D.J.A. 1990b. The last glaciation and relative sea level history of northwest Ellesmere Island, Canadian high arctic. *Journal of Quaternary Science* 5, 67–82.

Evans, D.J.A. 1991a. A gravel/diamicton lag on the south Albertan prairies, Canada, evidence of bed armoring in early deglacial sheet-flood/spillway courses. *Geological Society of America Bulletin* 103, 975–982.

Evans, D.J.A. 1991b. Glaciated land on the rebound. *Geography Review* 4, 2–6.

Evans, D.J.A. 1993. High latitude rock glaciers, a case study of forms and processes in the Canadian arctic. *Permafrost and Periglacial Processes* 4, 17–35.

Evans, D.J.A. 1994. The stratigraphy and sedimentary structures associated with complex subglacial thermal regimes at the southwestern margin of the Laurentide Ice Sheet, southern Alberta, Canada. In Warren, W.P. and Croot, D.G. (eds), *Formation and Deformation of Glacial Deposits*. Balkema, Rotterdam, 203–220.

Evans, D.J.A. 1996. A possible origin for a mega-fluting complex on the southern Alberta prairies, Canada. *Zeitschrift für Geomorphologie*, Supplement Band 106, 125–148.

Evans, D.J.A., Butcher, C. and Kirthisingha, A.V. 1994. Neoglaciation and an early 'Little Ice Age' in western Norway, lichenometric evidence from the Sandane area. *The Holocene* 4, 278–289.

Evans, D.J.A. and Campbell, I.A. 1992. Glacial and postglacial stratigraphy of Dinosaur Provincial Park and surrounding plains, southern Alberta, Canada. *Quaternary Science Reviews* 11, 535–555.

Evans, D.J.A. and Campbell, I.A. 1995. Quaternary stratigraphy of the buried valleys of the lower Red Deer River, Alberta, Canada. *Journal of Quaternary Science* 10, 123–148.

Evans, D.J.A. and England, J. 1991. Canadian landform examples 19, high arctic thrust block moraines. *Canadian Geographer* 35, 93–97.

Evans, D.J.A. and England, J. 1992. Geomorphological evidence of Holocene climatic change from northwest Ellesmere Island, Canadian high arctic. *The Holocene* 2, 148–158.

Evans, D.J.A. and Fisher, T.G. 1987. Evidence of a periodic ice-cliff avalanche on north-west Ellesmere Island, NWT, Canadian high arctic. *Journal of Glaciology* 33, 68–71.

Evans, D.J.A. and Hansom, J.D. 1996. Scottish landform examples – 15, the Edinburgh Castle crag-and-tail. *Scottish Geographical Magazine* 112, 129–131.

Evans, D.J.A., Owen, L.A. and Roberts, D. 1995. Stratigraphy and sedimentology of Devensian (Dimlington Stadial) glacial deposits, east Yorkshire, England. *Journal of Quaternary Science* 10, 241–265.

Evans, D.J.A., Rea, B.R. and Benn, D.I. 1997. Subglacial deformation and bedrock plucking in areas of hard bedrock. *Glacial Geology and Geomorphology* (in press).

Evans, D.J.A. and Rogerson, R.J. 1986. Glacial geomorphology and chronology in the Selamiut Range/Nachvak Fiord area, Torngat Mountains, Labrador. *Canadian Journal of Earth Sciences* 23, 66–76.

Evans, I.S. 1969. The geomorphology and morphometry of glacial and nival areas. In Chorley, R.J. (ed.), *Water, Earth and Man*. Methuen, London, 369–380.

Evans, I.S. 1977. World-wide variations in the direction and concentration of cirque and glacier aspects. *Geografiska Annaler* 59A, 151–175.

Evans, I.S. 1987. A new approach to drumlin morphometry. In Menzies, J. and Rose, J. (eds), *Drumlin Symposium*. Balkema, Rotterdam, 119–130.

Evans, I.S. 1990. Climatic effects on glacier distribution across the southern Coast Mountains, BC, Canada. *Annals of Glaciology* 14, 58–64.

Evans, I.S. 1994. Lithological and structural effects on forms of glacial erosion, cirques and lake basins. In Robinson, D.A. and Williams, R.B.G. (eds), *Rock Weathering and Landform Evolution*. Wiley, Chichester, 455–472.

Evans, I.S. 1996. Abraded rock landforms (whalebacks) developed under ice streams in mountain areas. *Annals of Glaciology* 22, 9–16.

Evans, I.S. and Cox, N.J. 1974. Geomorphometry and the operational definition of cirques. *Area* 6, 150–153.

Evans, I.S. and Cox, N.J. 1995. The form of glacial cirques in the English Lake District, Cumbria. *Zeitschrift für Geomorphologie* 39, 175–202.

Evans, S.G. and Clague, J.J. 1993. Glacier-related hazards and climate change. In Bras, R. (ed.), *The World at Risk, Natural Hazards and Climate Change*. American Institute of Physics, Conference Proceedings 277, 48–60.

Evans, S.G. and Clague, J.J. 1994. Recent climatic change and catastrophic geomorphic processes in mountain environments. *Geomorphology* 10, 107–128.

Evenson, E.B. 1971. The relationship of macro and microfabrics of till and the genesis of glacial landforms in Jefferson County, Wisconsin. In Goldthwait, R.P. (ed.), *Till: A Symposium*. Ohio State University Press, Columbus, 345–364.

Evenson, E.B. and Clinch, J.M. 1987. Debris transport mechanisms at active glacier margins, Alaskan case studies. In Kujansuu, R. and Saarnisto, M. (eds), *INQUA Till Symposium, Finland, 1985*. Geological Survey of Finland Special Paper 3, 111–136.

Evenson, E.B., Dreimanis, A. and Newsome, J.W. 1977. Subaquatic flow tills, a new interpretation for the genesis of some laminated till deposits. *Boreas* 6, 115–133.

Eybergen, F.A. 1987. Glacier snout dynamics and contemporary push moraine formation at the Turtmannglacier, Wallis, Switzerland. In van der Meer, J.J.M. (ed.), *Tills and Glaciotectonics*. Balkema, Rotterdam, 217–231.

Eyles, C.H. 1987. Glacially-influenced submarine channel sedimentation in the Yakataga Formation, Middleton Island, Alaska. *Journal of Sedimentary Petrology* 57, 1004–1017.

Eyles, C.H. 1988a. Glacially and tidally influenced shallow marine sedimentation of the late Precambrian Port Askaig Formation. *Palaeogeography, Palaeoclimatology, Palaeoecology* 68, 1–25.

Eyles, C.H. 1988b. A model for striated boulder pavement formation on glaciated shallow marine shelves, an example from the Yakataga Formation, Alaska. *Journal of Sedimentary Petrology* 58, 62–71.

Eyles, C.H. and Eyles, N. 1983a. Glaciomarine model for upper Precambrian diamictites of the Port Askaig Formation, Scotland. *Geology* 11, 692–696.

Eyles, C.H. and Eyles, N. 1983b. Sedimentation in a large lake, a reinterpretation of the late Pleistocene stratigraphy at Scarborough Bluffs, Ontario, Canada. *Geology* 11, 146–152.

Eyles, C.H. and Eyles, N. 1984a. Glaciomarine sediments of the Isle of Man as a key to late Pleistocene stratigraphic investigations in the Irish Sea Basin. *Geology* 12, 359–364.

Eyles, C.H. and Eyles, N. 1984b. Sedimentation in a large lake – reply. *Geology* 12, 188–190.

Eyles, C.H., Eyles, N. and Lagoe, M.B. 1991. The Yakataga Formation, a late Miocene to Pleistocene record of temperate glacial marine sedimentation in the Gulf of Alaska. In Anderson, J.B. and Ashley, G.M. (eds), Glacial-marine sedimentation, palaeoclimatic significance. Geological Society of America Special Paper 261, 159–180.

Eyles, C.H., Eyles, N. and Miall, A.D. 1985. Models of glaciomarine sedimentation and their application to the interpretation of ancient glacial sequences. *Palaeogeography, Palaeoclimatology, Palaeoecology* 51, 15–84.

Eyles, C.H. and Lagoe, M.B. 1990. Sedimentation patterns and facies geomoetries on a temperate glacially-influenced continental shelf; the Yakataga Formation, Middleton Island, Alaska. In Dowdeswell, J.A. and Scourse, J.D. (eds), *Glacimarine Environments, Processes and Sediments*. Geological Society of London, Special Publication 53, 363–386.

Eyles, N. 1978a. Scanning electron microscopy and particle size analysis of debris from a British Columbian glacier, a comparative report. In Whalley, W.B. (ed.), *Scanning Electron Microscopy in the Study of Sediments*. Geo Abstracts, Norwich, 227–242.

Eyles, N. 1978b. Rock glaciers in Esjufjoll Nunatak area, southeast Iceland. *Jökull* 28, 53–60.

Eyles, N. 1979. Facies of supraglacial sedimentation on Icelandic and alpine temperate glaciers. *Canadian Journal of Earth Sciences* 16, 1341–1361.

Eyles, N. 1983a. *Glacial Geology*. Pergamon, Oxford.

Eyles, N. 1983b. Modern Icelandic glaciers as depositional models for 'hummocky moraine' in the Scottish highlands. In Evenson, E.B., Schluchter, C. and Rabassa, J. (eds), *Tills and Related Deposits*. Balkema, Rotterdam, 47–59.

Eyles, N. 1983c. Glacial geology, a landsystems approach. In Eyles, N. (ed.), *Glacial Geology*. Pergamon, Oxford, 1–18.

Eyles, N. 1983d. The glaciated valley landsystem. In Eyles, N. (ed.), *Glacial Geology*. Pergamon, Oxford, 91–110.

Eyles, N. 1987. Late Pleistocene debris flow deposits in large glacial lakes in British Columbia and Alaska. *Sedimentary Geology* 53, 33–71.

Eyles, N. 1993. Earth's glacial record and its tectonic setting. *Earth Science Reviews* 35, 1–248.

Eyles, N. and Clague, J.J. 1978. Landsliding caused by glacial lake ponding: Late Pleistocene example from central British Columbia. *Canadian Geotechnical Journal* 24, 656–663.

Eyles, N. and Clark, B.M. 1985. Gravity-induced soft-sediment deformation in glaciomarine sequences of Upper Proterozoic Port Askaig Formation, Scotland. *Sedimentology* 32, 789–814.

Eyles, N. and Clark, B.M. 1988. Storm dominated deltas and ice scouring in a Late Pleistocene glacial lake. *Geological Society of America Bulletin* 100, 793–809.

Eyles, N., Clark, B.M. and Clague, J.J. 1987. Coarse-grained sediment-gravity flow facies in a large supraglacial lake. *Sedimentology* 34, 193–216.

Eyles, N., Clark, B.M. and Clague, J.J. 1988a. Coarse-grained sediment gravity flow facies in a large supraglacial lake – reply. *Sedimentology* 35, 529–530.

Eyles, N. and Dearman, W.R. 1981. A glacial terrain map of Britain for engineering purposes. *Bulletin of the International Association of Engineering Geology* 24, 173–184.

Eyles, N., Dearman, W.R. and Douglas, T.D. 1983. The distribution of glacial landsystems in Britain and North America. In Eyles, N. (ed.), *Glacial Geology*. Pergamon, Oxford, 213–228.

Eyles, N. and Eyles, C.H. 1992. Glacial depositional systems. In Walker, R.G. and James, N.P. (eds), *Facies Models, Response to Sea-Level Change*. Geological Association of Canada, Toronto, 73–100.

Eyles, N., Eyles, C.H. and McCabe, A.M. 1988b. Late Pleistocene subaerial debris flow facies of the Bow Valley near Banff, Canadian Rocky Mountains. *Sedimentology* 35, 465–480.

Eyles, N., Eyles, C.H. and McCabe, A.M. 1989. Sedimentation in an ice-contact subaqueous setting, the mid Pleistocene 'North Sea Drifts' of Norfolk, UK. *Quaternary Science Reviews* 8, 57–74.

Eyles, N., Eyles, C.H. and McCabe, A.M. 1990. Late Pleistocene subaerial debris flow sediments near Banff, Canada – reply. *Sedimentology* 37, 544–547.

Eyles, N., Eyles, C.H. and Miall, A.D. 1983. Lithofacies types and vertical profile models, an alternative approach to the description and environmental interpretation of glacial diamict and diamictite sequences. *Sedimentology* 30, 393–410.

Eyles, N. and Kocsis, S. 1988. Sedimentology and clast fabric of subaerial debris flow facies in a glacially-influenced alluvial fan. *Sedimentary Geology* 59, 15–28.

Eyles, N. and Lagoe, M.B. 1989. Sedimentology of shell-rich deposits (coquinas) in the glaciomarine upper Cenozoic Yakataga Formation, Middleton Island, Alaska. *Geological Society of America Bulletin* 101, 129–142.

Eyles, N. and McCabe, A.M. 1989a. The Late Devensian <22,000 BP Irish Sea Basin: the sedimentary record of a collapsed ice sheet margin. *Quaternary Science Reviews* 8, 307–351.

Eyles, N. and McCabe, A.M. 1989b. Glaciomarine facies within subglacial tunnel valleys, the sedimentary record of glacio-isostatic downwarping in the Irish Sea Basin. *Sedimentology* 36, 431–448.

Eyles, N. and McCabe, A.M. 1991. Glaciomarine deposits of the Irish Sea Basin: the role of glacioisostatic disequilibrium. In Ehlers, J., Gibbard P.L. and Rose, J. (eds), *Glacial Deposits in Great Britain and Ireland*. Balkema, Rotterdam, 311–331.

Eyles, N. and Menzies, J. 1983. The subglacial landsystem. In Eyles, N. (ed.), *Glacial Geology*. Pergamon, Oxford, 19–70.

Eyles, N. and Miall, A.D. 1984. Glacial facies. In Walker, R.G. (ed.), *Facies Models*. Geoscience Canada Reprint Series 1, 15–38.

Eyles, N. and Rogerson, R.J. 1978a. A framework for the investigation of medial moraine formation: Austerdalsbreen, Norway, and Berendon Glacier, British Columbia, Canada. *Journal of Glaciology* 20, 99–113.

Eyles, N. and Rogerson, R.J. 1978b. Sedimentology of medial moraines on Berendon Glacier, British Columbia, Canada: implications for debris transport in a glacierized basin. *Bulletin of the Geological Society of America* 89, 1688–1693.

Eyles, N., Sasseville, D.R., Slatt, R.M. and Rogerson, R.J. 1982a. Geochemical denudation rates and solute transport mechanisms in a maritime temperate glacier basin. *Canadian Journal of Earth Sciences* 19, 1570–1581.

Eyles, N., Sladen, J.A. and Gilroy, S. 1982b. A depositional model for stratigraphic complexes and facies superimposition in lodgement tills. *Boreas* 11, 317–333.

Eyles, N. and Young, G.M. 1993. In Deynoux, M., Miller, J.M.G., Domack, E.W., Eyles, N., Fairchild, I.J. and Young, G.M. (eds), *Earth's Glacial Record*. Cambridge University Press, Cambridge, 1–28.

Eythorsson, J. 1935. On the variations of glaciers in Iceland. Some studies made in 1931. *Geografiska Annaler* 17, 121–137.

Eythorsson, J. 1952. Thaettir ur sogu Breidar. *Jökull* 2, 17–20.

Eythorsson, J. 1963a. Variation of Iceland glaciers 1931–1960. *Jökull* 13, 31–33.

Eythorsson, J. 1963b. Joklabreytingar 1961/62 og 1962/63. *Jökull* 13, 29–31.

Eythorsson, J. 1964. Joklabreytingar 1962/63. *Jökull* 14, 97–99.

Eythorsson, J. 1965. Joklabreytingar 1963/64. *Jökull* 15, 148–150.

Eythorsson, J. 1966. Joklabreytingar 1964/66. *Jökull* 16, 230–231.

Faegri, K. 1952. On the origin of potholes. *Journal of Glaciology* 2, 24–25.

Fahnestock, R.K. 1963. Morphology and hydrology of a glacial stream: White River, Mount Ranier, Washington. USGS Professional Paper 422A, 1–70.

Fahnestock, R.K. 1969. Morphology of the Slims River. *Icefield Ranges Research Project Scientific Results* 1, 161–172.

Fahnestock, R.K. and Bradley, W.C. 1973. Knik and Matanuska Rivers, Alaska: a contrast in braiding. In Morisawa, M. (ed.), *Fluvial Geomorphology, Proceedings of 4th Geomorphology Symposium, Binghamton, New York*, 220–250.

Fairbairn, H.W., Hurley, P.M., Card, K.D. and Knight, C.J. 1969. Correlation of radiometric ages of Nipissing diabase and Huronian metasediments with Proterozoic orogenic events in Ontario. *Canadian Journal of Earth Sciences* 6, 489–497.

Fairbanks, R.G. 1989. A 17,000 year glacio-eustatic sea level record: influence of glacial melting rate on the Younger Dryas event and deep ocean circulation. *Nature* 342, 637–642.

Fairbridge, R.W. 1961. Eustatic changes in sea level. *Physics and Chemistry of the Earth* 5, 99–185.

Fairbridge, R.W. 1970. South Pole reaches the Sahara. *Science* 168, 878–881.

Fairbridge, R.W. 1974. Glacial grooves and periglacial features in the Saharan Ordovician. In Coates, D.R. (ed.), *Glacial Geomorphology*. State University, Binghamton, New York, 317–327.

Fairbridge, R.W. 1979. Traces from the desert: Ordovician. In John, B.S. (ed.), *Winters of the World: Earth under the Ice Ages*. David & Charles, Newton Abbot, 131–153.

Fairbridge, R.W. 1983. Isostasy and eustasy. In Smith, D.E. and Dawson, A.G. (eds), *Shorelines and Isostasy*. Academic Press, London, 3–25.

Fairchild, H.L. 1918. Postglacial uplift of northeastern America. *Geological Society of America Bulletin* 29, 187–234.

Fairchild, H.L. 1929. New York drumlins. *Proceedings of the Rochester Academy of Sciences* 3, 1–37.

Falconer, G., Ives, J.D., Loken, O.H. and Andrews, J.T. 1965. Major end moraines in eastern and central Arctic Canada. *Geographical Bulletin* 7, 137–153.

Farrell, W.E. and Clark, J.A. 1976. On postglacial sea level. *Geophysical Journal of the Royal Astronomical Society* 46, 647–667.

Fastook, J.L. 1984. West Antarctica, the sea level controlled marine instability, past and future. In Hansen, J.T. and Takahashi, T. (eds), *Climate Processes and Climatic Sensitivity*. American Geophysical Union, Geophysical Monograph 29, Washington DC, 275–287.

Fastook, J.L. 1987. Use of a new finite element continuity model to study the transient behaviour of Ice Stream C and the causes of its present low velocity. *Journal of Geophysical Research* 92, 8941–8949.

Fastook, J.L. and Holmlund, P. 1994. A glaciological model of the Younger Dryas event in Scandinavia. *Journal of Glaciology* 40, 125–131.

Fastook, J.L. and Schmidt, W.F. 1982. Finite element anaysis of calving from ice fronts. *Annals of Glaciology* 3, 103–106.

Fecht, K.R. and Tallman, A.M. 1977. Bergmounds along the western margin of the Channeled Scablands, south-central Washington. *Geological Society of America, Abstracts with Programs* 10, 400.

Feenstra, B.H. and Frazer, J.Z. 1988. Fonthill kame-delta. In Barnett, P.J. and Kelly, R.I. (eds), *Quaternary History of Southern Ontario, Guidebook for Field Excursion A–11*, XII INQUA Congress.

Feininger, T. 1971. Chemical weathering and glacial erosion of crystalline rocks and the origin of till. US Geological Survey, Professional Paper 750-C, 65–81.

Fenton, M.M., Trudell, M.R., Pawlowicz, J.G., Jones, C.E., Moran, S.R. and Nikols, D.J. 1986. Glaciotectonic deformation and geotechnical stability in open pit coal mining. In Singhal, R.K. (ed.), *Geotechnical Stability in Surface Mining*. Balkema, Rotterdam, 225–234.

Ferguson, R.I. 1993. Understanding braiding processes in gravel-bed rivers: progress and unsolved problems. In Best, J.L. and Bristow, C.S. (eds), *Braided Rivers*. Geological Society of London Special Publication 75, 73–87.

Ferguson, R.I. and Ashworth, P.J. 1992. Spatial patterns of bedload transport and channel change in braided and near-braided rivers. In Billi, P., Hey, R.D., Thorne, C.R. and Tacconi, P. (eds), *Dynamics of Gravel-Bed Rivers*. Wiley, Chichester, 477–492.

Fernlund, J.M.R. 1988. The Halland Coastal Moraines: are they end moraines or glaciotectonic ridges? In Croot,

D.G. (ed.), *Glaciotectonics: Forms and Processes*. Balkema, Rotterdam, 77–90.

Ferrigno, J.G. and Gould, W.G. 1987. Substantial changes in the coastline of Antarctica revealed by satellite imagery. *Polar Record* 23, 577–583.

Feruglio, E. 1944. Estudios geológicos y glaciológico en la region del Lago Argentino (Patagonia). *Boletín Academia Nacional de Ciencias de Córdoba* 37, 1–208.

Ficker, E., Sonntag, G. and Weber, E. 1980. Ansätze zur mechanischen Deutung der Rissentstehung bei Parabelrissen und Sichelbrüchen auf glazialgeformten Felsoberflächen. *Zeitschrift für Gletscherkunde und Glazialgeologie* 16, 25–43.

Field, W.O. 1947. Glacier recession in Muir Inlet, Glacier Bay, Alaska. *Geographical Review* 37, 369–399.

Field, W.O. 1975. Glaciers of the St Elias Mountains. In Field, W.O. (ed.), *Mountain Glaciers of the Northern Hemisphere*. CRREL, Hanover, NH, USA, 143–297.

Figge, K. 1983. Morainic deposits in the German Bight area of the North Sea. In Ehlers, J. (ed.), *Glacial Deposits in North-West Europe*. Balkema, Rotterdam, 299–304.

Finsterwalder, R. 1959. Chamonix glaciers. *Journal of Glaciology* 3, 547–548.

Firth, C.R. and Haggart, B.A. 1989. Loch Lomond Stadial and Flandrian shorelines in the inner Moray Firth area, Scotland. *Journal of Quaternary Science* 4, 37–50.

Fisher, D. 1973. Subglacial leakage of Summit Lake, British Columbia, by dye determinations. In Glen, J.W., Adie, R.J. and Johnson, D.M. (eds), *Symposium on the Hydrology of Glaciers*, IASH Publication, Cambridge, 95, 111–116.

Fisher, D.A., Reeh, N. and Langley, K. 1985. Objective reconstructions of the Late Wisconsinan Laurentide Ice Sheet and the significance of deformable beds. *Géographie Physique et Quaternaire* 39, 229–238.

Fisher, T.G. and Shaw, J. 1992. A depositional model for Rogen moraine, with examples from the Avalon Peninsula, Newfoundland. *Canadian Journal of Earth Sciences* 29, 669–686.

Fitzsimons, S.J. 1990. Ice-marginal depositional processes in a polar maritime environment, Vestfold Hills, Antarctica. *Journal of Glaciology* 36, 279–286.

Fitzsimons, S.J. 1996. Formation of thrust-block moraines at the margins of dry-based glaciers, south Victoria Land, Antarctica. *Annals of Glaciology* 22, 68–74.

Fitzsimons, S.J. 1997. Depositional models for moraines in East Antarctic coastal oases. *Journal of Glaciology*, in press.

Fleisher, P.J. 1986. Dead-ice sinks and moats, environments of stagnant ice deposition. *Geology* 14, 39–42.

Flinn, D. 1967. Ice front in the North Sea. *Nature* 215, 1151–1154.

Flinn, D. 1978. The most recent glaciation of the Orkney–Shetland Channel and adjacent regions. *Scottish Journal of Geology* 14, 109–123.

Flint, R.F. 1928. Eskers and crevasse fillings. *American Journal of Science* 15, 410–416.

Flint, R.F. 1943. Growth of North American ice sheet during the Wisconsin age. *Geological Society of America Bulletin* 54, 325–362.

Flint, R.F. 1947. *Glacial Geology and the Pleistocene Epoch*. Wiley, New York.

Flint, R.F. 1957. *Glacial and Pleistocene Geology*. Wiley, New York.

Flint, R.F. 1971. *Glacial and Quaternary Geology*. Wiley, New York.

Flint, R.F., Sanders, J.E. and Rodgers, J. 1960a. Symmictite: a name for nonsorted terrigenous sedimentary rocks that contain a wide range of particle sizes. *Geological Society of America Bulletin* 71, 507–510.

Flint, R.F., Sanders, J.E. and Rodgers, J. 1960b. Diamictite: a substitute term for symmictite. *Geological Society of America Bulletin* 71, 1809.

Fluckiger, O. 1934. Glaziale Felsformen. *Petermanns Mitteilungen, Erganzungshefte* 218.

Fogelberg, P. 1970. Geomorphology and deglaciation at the second Salpausselka between Vaaksy and Vierumaki, southern Finland. *Commentationes Physico-Mathematicae Societas Scientarium Fennica, Helsinki* 39.

Foldvik, A. and Kvinge, T. 1974. Conditional instability of sea water at the freezing point. *Deep Sea Research* 21, 169–174.

Foldvik, A. and Kvinge, T. 1977. Thermohaline convection in the vicinity of an ice shelf. In Dunbar, M.J. (ed.), *Polar Oceans*. Arctic Institute of North America, Calgary, 247–255.

Forbes, J.D. 1843. *Travels through the Alps of Savoy*. Edinburgh.

Forbes, J.D. 1846. Illustrations of the viscous theory of glacier motion III. An attempt to establish by observation the plasticity of glacier ice. *Philosophical Transactions* 7, 157–176.

Foster, J.D. and Palmquist, R.C. 1969. Possible subglacial origin for 'minor moraine' topography. *Proceedings of the Iowa Academy of Science* 76, 296–310.

Fountain A.G. 1992. Subglacial water flow inferred from stream measurements at South Cascade Glacier, Washington State, USA. *Journal of Glaciology* 38, 51–64.

Fountain A.G. 1993. Geometry and flow conditions of subglacial water at South Cascade Glacier, Washington State, USA: an analysis of tracer injections. *Journal of Glaciology* 39, 143–156.

Fountain, A.G. 1994. Borehole water-level variations and implications for the subglacial hydraulics run-away: a mechanism for thermally regulated surges of ice sheets. *Journal of Glaciology* 40, 293–304.

Fowler, A.C. 1986. Sub-temperate basal sliding. *Journal of Glaciology* 32, 3–5.

Fowler, A.C. 1987a. Sliding with cavity formation. *Journal of Glaciology* 33, 255–267.

Fowler, A.C. 1987b. A theory of glacier surges. *Journal of Geophysical Research* 92, 9111–9120.

Fowler, A.C. and Ng, F.S.L. 1996. The role of sediment transport in the mechanics of jökulhlaups. *Annals of Glaciology* 22, 255–259.

Fowler A.G. and Johnson, C. 1995. Hydraulic run-away: a mechanism for thermally regulated surges of ice sheets. *Journal of Glaciology* 41, 554–561.

Frakes, L.A. 1979. *Climates through Geologic Time*. Elsevier, Amsterdam.

Frakes, L.A., Amos, A.J. and Crowell, J.C. 1969. Origin and stratigraphy of late Paleozoic diamictites in Argentina and Bolivia. In *Gondwana Stratigraphy, IUGS Symposium Earth Sciences 2, Buenos Aires, October 1967*, UNESCO, Paris, 821–843.

Frakes, L.A. and Crowell, J.C. 1969. Late Paleozoic glaciation: I, South America. *Geological Society of America, Bulletin* 80, 1007–1042.

Frakes, L.A. and Francis, J.E. 1988. A guide to Phanerozoic cold polar climates from high latitude ice-rafting in the Cretaceous. *Nature* 333, 547–549.

Frakes, L.A., Francis, J.E. and Sykes, J.I. 1992. *Climate Modes of the Phanerozoic*. Cambridge University Press, Cambridge.

Francis, P. 1993. *Volcanoes, A Planetary Perspective*. Clarendon Press, Oxford.

Francou, B., Ribstein, P., Saravia, R. and Tiriau, E. 1995. Monthly balance and water discharge on an intertropical glacier: the Zongo Glacier, Cordilleran Real, Bolivia, 16° S. *Journal of Glaciology* 41, 61–67.

Fraser, F.J., McLearn, F.H., Russell, L.S., Warren, P.S. and Wickenden, R.T.D. 1935. *Geology of Saskatchewan*. Geological Survey of Canada, Memoir 176.

Fraser, G.S., Bleuer, N.K. and Smith, N.D. 1983. History of Pleistocene alluviation of the middle and upper Wabash Valley: Field Trip 13. In Shaver, R.H. and Sunderman, J.A. (eds), *Field Trips in Midwestern Geology*, vol. 1. Indiana Geological Survey, Bloomington, 197–224.

Fraser, J.Z. 1982. Derivation of a summary facies sequence based on Markov Chain analysis of the Caledon outwash: a Pleistocene braided glacial fluvial deposit. In Davidson-Arnott, R., Nickling, W. and Fahey, B.D. (eds), *Research in Glacial, Glacio-fluvial, and Glaciolacustrine Systems*. Geo Books, Norwich, 175–202.

Freeze, R.A. and Cherry, J.A. 1979. *Groundwater*. Prentice-Hall, Englewood Cliffs, NJ.

French, H.M. and Harry, D.G. 1990. Observations on buried glacier ice and massive segregated ice, western arctic coast, Canada. *Permafrost and Periglacial Processes* 1, 31–43.

Fritz, W.J. and Moore, J.N. 1988. *Basics of Physical Stratigraphy and Sedimentology*. Wiley, New York.

Frodin G. 1925. Studien über die Eisscheide in Zentralskandinavien. *Bulletin of the Geological Institute, University of Uppsala* 19, 129–214.

Fuller, M.B. 1925. The bearing of some remarkable potholes on the early Pleistocene glaciation of the Front Range, Colorado. *Journal of Geology* 33, 224–235.

Fuller, M.D. 1964. A magnetic fabric in till. *Geology Magazine* 99, 233–237.

Fulton, R.J. 1967. Deglaciation studies in the Kamloops region, an area of moderate relief, British Columbia. Geological Survey of Canada Bulletin 154.

Fulton, R.J. (ed.) 1989. *Quaternary Geology of Canada and Greenland*. Geological Survey of Canada, Geology of Canada, no. 1.

Fulton, R.J. and Andrews, J.T. 1987. The Laurentide Ice Sheet. *Géographie Physique et Quaternaire* 41.

Fulton, R.J. and Pullen, M.J.L.T. 1969. Sedimentation in Upper Arrow Lake, BC. *Canadian Journal of Earth Sciences* 6, 785–791.

Funder, S. 1972. Deglaciation of the Scoresby Sund fjord region, northeast Greenland. In Price, R.J. and Sugden, D.E. (eds), *Polar Geomorphology*. Institute of British Geographers Special Publication 4, 33–42.

Funder, S. 1989. Quaternary geology of the ice-free areas and adjacent shelves of Greenland. In Fulton, R.J. (ed.), *Quaternary Geology of Canada and Greenland*. Geological Survey of Canada, Geology of Canada No. 1, 743–792.

Furbish, D.J. and Andrews, J.T. 1984. The use of hypsometry to indicate long-term stability and response of valley glaciers to changes in mass transfer. *Journal of Glaciology* 30, 199–211.

Fushimi, H. 1977. Structural studies of glaciers in the Khumbu region. *Seppyo* 39, Special Issue, 30–39.

Fushimi, H., Yoshida, M., Watanabe, O. and Upadhyay, B.P. 1980. Distributions and grain sizes of supraglacial debris in the Khumbu Glacier, Khumbu Region, East Nepal. *Seppyo* 42, 18–25.

Fyfe, G.J. 1990. The effect of water depth on ice-proximal glaciolacustrine sedimentation: Salpausselkä I, southern Finland. *Boreas* 19, 147–164.

Fyles, J.G. 1967. Winter Harbour moraine, Melville Island. Geological Survey of Canada, Paper 69–1A, 194–195.

Gade, H.G. 1979. Melting of ice in sea water: a primitive model with applications to the Antarctic ice shelf and icebergs. *Journal of Physical Oceanography* 9, 189–198.

Gade, H.G., Lake, R.A., Lewis, E.L. and Walker, E.R. 1974. Oceanography of an arctic bay. *Deep Sea Research* 21, 547–571.

Gage, M. 1961. On the definition, date and character of the Ross Glaciation. *Transactions of the Royal Society of New Zealand* 88, 631–637.

Gale, S.J. and Hoare, P.G. 1991. *Quaternary Sediments*. Wiley, New York.

Galloway, R.W. 1956. The structure of moraines in Lyngdalen, north Norway. *Journal of Glaciology* 2, 730–733.

Galloway, R.W. 1963. Glaciation in the Snowy Mountains, a re-appraisal. *Proceedings of the Linnean Society of New South Wales* 88, 180–198.

Galloway, W.E. 1989. Genetic stratigraphic sequences in basin analysis I: Architecture and genesis of flooding-surface bounded depositional units. *American Association of Petroleum Geologists Bulletin* 73, 125–142.

Galon, R. 1961. North Poland, area of the last glaciation. *VI INQUA Congress, Guidebook of Excursion, Łódź*.

Galon, R. 1973. Geomorphological and geological analysis of the proglacial area of Skeidarárjökull: central section. *Geographia Polonica* 26, 15–57.

de Gans, W., de Groot, T. and Zwaan, H. 1987. The Amsterdam basin, a case study of a glacial basin in the Netherlands. In van der Meer, J.J.M. (ed.), *Tills and Glaciotectonics*. Balkema, Rotterdam, 205–216.

Gardner, J.S. 1980. Frequency, magnitude and spatial distribution of mountain rockfalls and rockslides in the Highwood Pass area, Alberta, Canada. In Coates, D. and Vitek, J. (eds), *Thresholds in Geomorphology*. Allen & Unwin, London, 267–295.

Gardner, J.S. 1982. Alpine mass wasting in contemporary time: some examples from the Canadian Rocky Mountains. In Thorn, C.E. (ed.), *Space and Time in Geomorphology*. Allen and Unwin, London, 171–192.

Gardner, J.S. 1987. Evidence for headwall weathering zones, Boundary Glacier, Canadian Rocky Mountains. *Journal of Glaciology* 33, 60–67.

Gardner, J.S. and Hewitt, K. 1990. A surge of Bualtar Glacier, Karakoram Range, Pakistan: a possible landslide trigger. *Journal of Glaciology* 36, 159–162.

Gardner, J.S. and Jones, N.K. 1993. Sediment transport and yield at the Raikot Glacier, Nanga Parbat, Punjab Himalaya. In Shroder, J.F. (ed.), *Himalaya to the Sea*. Routledge, London, 184–197.

Garfield, D.E. and Ueda, H.T. 1976. Resurvey of the 'Byrd' station, Antarctica, drill hole. *Journal of Glaciology* 17, 29–34.

Garwood, E.J. and Gregory, J.W. 1898. Contributions to the glacial geology of Spitsbergen. *Quarterly Journal of the Geological Society of London* 5A, 197–227.

Gates, W.L. 1976a. Modelling the ice age climate. *Science* 191, 1138–1144.

Gates, W.L. 1976b. The numerical simulation of ice age climate with a global general circulation model. *Journal of the Atmospheric Sciences* 33, 1844–1873.

Geikie, J. 1877. *The Great Ice Age*, 2nd edition. Daldy, Isbister, London.

Geikie, J. 1894. *The Great Ice Age*, 3rd edition. Edward Stanford, London.

Gellatly, A.F., Gordon, J.E., Whalley, W.B. and Ferguson, R.I. 1986a. Movement of the ice front. *Geographical Magazine* 58, 294–299.

Gellatly, A.F., Gordon, J.E., Whalley, W.B. and Hansom, J.D. 1988. Thermal regime and geomorphology of plateau ice caps in northern Norway, observations and implications. *Geology* 16, 983–986.

Gellatly, A.F., Röthlisberger, F. and Geyh, M.A. 1985. Holocene glacier variations in New Zealand South Island. *Zeitschrift fur Gletscherkunde und Glazialgeologie* 21, 265–273.

Gellatly, A.F., Whalley, W.B. and Gordon, J.E. 1986b. Topographic control over recent changes in southern Lyngen Peninsula, north Norway. *Norsk Geografisk Tidsskrift* 40, 211–218.

Ghibaudo, G. 1992. Subaqueous sediment gravity flow deposits: practical criteria for their field description and classification. *Sedimentology* 39, 423–454.

Giardino, J.R. 1983. Movement of ice-cemented rock glaciers by hydrostatic pressure, an example from Mount Mestas, Colorado. *Zeitschrift für Geomorphologie* 27, 297–310.

Giardino, J.R., Shroder, J.F. and Vitek, J.D. (eds) 1987. *Rock Glaciers*. Allen & Unwin, London.

Giardino, J.R. and Vitek, J.D. 1988. The significance of rock glaciers in the glacial–periglacial landscape continuum. *Journal of Quaternary Science* 3, 97–103.

Gibbard, P.L. 1980. The origin of stratified Catfish Creek till by basal melting. *Boreas* 9, 71–85.

Gibbard, P.L. 1985. *The Pleistocene History of the Middle Thames Valley*. Cambridge University Press, Cambridge.

Gibbard, P.L. 1988. The history of the great northwest European rivers during the past three million years. *Philosophical Transactions of the Royal Society of London* B318, 559–602.

Gibbard, P.L., West, R.G., Andrew, R. and Pettit, M. 1992. The margin of a Middle Pleistocene ice advance at Tottenhill, Norfolk, England. *Geological Magazine* 129, 59–76.

van Gijssel, K. 1987. A lithostratigraphic and glaciotectonic reconstruction of the Lamstedt Moraine, Lower Saxony, FRG. In van der Meer, J.J.M. (ed.), *Tills and Glaciotectonics*. Balkema, Rotterdam, 145–156.

Gilbert, G.K. 1890. *Lake Bonneville*. US Geological Survey Monograph 1, 1–438.

Gilbert, G.K. 1903. *Glaciers and Glaciation of Alaska*. Doubleday, Page & Co., New York.

Gilbert, G.K. 1904a. *Glaciers and Glaciation of Alaska*. In *Harriman Alaska Expedition*, Vol. III. New York.

Gilbert, G.K. 1904b. Systematic asymmetry of crest lines in the High Sierra of California. *Journal of Geology* 12, 579–588.

Gilbert, G.K. 1906a. Crescentic gouges on glaciated surfaces. *Geological Society of America Bulletin* 17, 303–316.

Gilbert, G.K. 1906b. Moulin work under glaciers. *Geological Society of America Bulletin* 17, 317–320.

Gilbert, G.K. 1910. Crescentic gouges on glaciated surfaces. *Geological Society of America Bulletin* 17, 303–316.

Gilbert, R. 1973. Processes of underflow and sediment transport in a British Columbia mountain lake. *Proceedings 9th Canadian Hydrology Symposium, NRC Canada*, 493–507.

Gilbert, R. 1975. Sedimentation in Lillooet Lake, British Columbia. *Canadian Journal of Earth Sciences* 12, 1697–1711.

Gilbert, R. 1982. Contemporary sedimentary environments on Baffin Island, NWT, Canada, glaciomarine processes in fiords of eastern Baffin Island. *Arctic and Alpine Research* 14, 1–12.

Gilbert, R. 1983. Sedimentary processes of Canadian Arctic fjords. *Sedimentary Geology* 36, 147–175.

Gilbert, R. 1984. The movement of gravel by the alga *Fucus vesiculosus L.* on an arctic intertidal flat. *Journal of Sedimentary Petrology* 54, 463–468.

Gilbert, R. 1990. Rafting in glaciomarine environments. In Dowdeswell, J.A. and Scourse, J.D. (eds), *Glacimarine Environments, Processes and Sediments*. Special Publications of the Geologists' Association 53, 105–120.

Gilbert, R. and Church, M. 1983. Contemporary sedimentary environments on Baffin Island, NWT, Canada, reconnaissance of lakes on Cumberland Peninsula. *Arctic and Alpine Research* 15, 321–332.

Gilbert, R. and Shaw, J. 1981. Sedimentation in proglacial Sunwapta Lake, Alberta. *Canadian Journal of Earth Sciences* 18, 81–93.

Gillberg, G. 1965. Till distribution and ice movements on the northern slopes of the south Swedish highlands. *Geologiska Foreningens i Stockholm Forhandlingar* 86, 433–484.

Gillberg, G. 1967. Further discussion of the lithological homogeneity of till. *Geologiska Foreningens i Stockholm Forhandlingar* 89, 29–49.

Gillberg, G. 1968. Lithological distribution and homogeneity of glaciofluvial material. *Geologiska Foreningens i Stockholm Forhandlingar* 90, 189–204.

Gillespie, A. and Molnar, P. 1995. Asynchronous maximum advances of mountain and continental glaciers. *Reviews of Geophysics* 33, 311–364.

Giovinetto, M.B. and Bentley, C.R. 1985. Surface balance in ice drainage systems of Antarctica. *US Antarctic Journal* 20, 6–13.

Gjessing, J. 1960. Isvasmeltningstidens drenering dens forlop og Formdannende virkning i Nordre Atnedalen. *Ad Novas* 3.

Gjessing, J. 1965. On 'plastic scouring' and 'subglacial erosion'. *Norsk Geografisk Tidsskrift* 20, 1–37.

Gjessing, J. 1966. Some effects of ice erosion on the development of Norwegian valleys and fjords. *Norsk Geografisk Tidsskrift* 20, 273–299.

Gjessing, J. 1967a. Potholes in connection with plastic scouring forms. *Geografiska Annaler* 49A, 178–187.

Gjessing, J. 1967b. Norway's paleic surface. *Norsk Geografisk Tidsskrift* 21, 69–132.

Glasser, N.F. 1995. Modelling the effect of topography on ice sheet erosion, Scotland. *Geografiska Annaler* 77A, 67–82.

Glasser, N.F. and Warren, C.R. 1990. Medium scale landforms of glacial erosion in south Greenland: process and form. *Geografiska Annaler* 72A, 211–215.

Glen, J.W. 1954. The stability of ice-dammed lakes and other water-filled holes in glaciers. *Journal of Glaciology* 2, 316–318.

Glen, J.W. 1955. The creep of polycrystalline ice. *Proceedings of the Royal Society*, Series A, 228, 519–538.

Glen, J.W. 1963. Contribution to discussion. *IAHS Bulletin* 8, 68.

Glen, J.W. 1987. Fifty years of progress in ice physics. *Journal of Glaciology* special issue, 52–59.

Glen, J.W., Donner, J.J. and West, R.G. 1957. On the mechanism by which stones in till become oriented. *American Journal of Science*, 255, 194–205.

Gloyne, R.W. 1964. Some characteristics of the natural wind and the modification by natural and artificial obstructions. *Scientific Horticulture* 17, 7–19.

Gluckert, G. 1973. Two large drumlin fields in central Finland. *Fennia* 120.

Gluckert, G. 1974. Map of glacial striation of the Scandinavian ice sheet during the last (Weichsel) glaciation in northern Europe. *Bulletin of the Geological Society of Finland* 46, 1–8.

Gluckert, G. 1977. On the Salpausselkä ice-marginal formations in southern Finland. *Zeitschrift für Geomorphologie* Supp. 27, 79–88.

Gluckert, G. 1986. The first Salpausselkä at Lohja, southern Finland. *Bulletin of the Geological Society of Finland* 58, 45–55.

Godard, A. 1961. L'efficacité de l'érosion glaciaire en Ecosse du Nord. *Revue de Géomorphologie Dynamique* 12, 32–42.

Godard, A. 1989. Les vestiges des manteaux d'altération sur les socles des hautes latitudes, identification, signification. *Zeitschrift für Geomorphologie* 72, 1–20. English translation in Evans, D.J.A. (ed.), *Cold Climate Landforms*. Wiley, Chichester, 397–411.

Goldsmith, R. 1982. Recessional moraines and ice retreat in southeastern Connecticat. In Larson, G.J. and Stone, B.D. (eds), *Late Wisconsinan Glaciation of New England*. Kendall/Hunt, Dubuque, Iowa, 61–76.

Goldstein, B. 1989. Lithology, sedimentology, and genesis of the Wadena drumlin field, Minnesota, USA. *Sedimentary Geology* 62, 241–277.

Goldthwait, J.W. 1908. A reconstruction of water planes of the extinct glacial lakes in the Lake Michigan Basin. *Journal of Geology* 16, 459–476.

Goldthwait, R.P. 1951. Development of end moraines in east-central Baffin Island. *Journal of Geology* 59, 567–577.

Goldthwait, R.P. 1960. Study of ice cliff in Nunatarssuaq, Greenland. Snow, Ice, Permafrost Research Establishment, Technical Report 39, 1–103.

Goldthwait, R.P. 1961. Regimen of an ice cliff on land in northwest Greenland. *Folia Geographica Danica* 9, 107–115.

Goldthwait, R.P. 1971. Introduction to till, today. In Goldthwait, R.P. (ed.), *Till: A Symposium*, 3–26. Ohio State University Press, Columbus.

Goldthwait, R.P. 1979. Giant grooves made by concentrated basal ice streams. *Journal of Glaciology* 23, 297–307.

Goldthwait, R.P. 1989. Classification of glacial morphologic features. In Goldthwait, R.P. and Matsch, C.L. (eds), *Genetic Classification of Glacigenic Deposits*. Balkema, Rotterdam, 267–277.

Gomez, B., Dowdeswell, J.A. and Sharp, M. 1988. Microstructural control of quartz sand grain shape and texture: implications for the discrimination of debris transport pathways through glaciers. *Sedimentary Geology* 57, 119–129.

Gomez, B. and Small, R.J. 1985. Medial moraines of the Haut Glacier d'Arolla, Valais, Switzerland: debris supply and implications for moraine formation. *Journal of Glaciology* 31, 303–307.

Gordon, J. 1976. *The New Science of Strong Materials, or Why You Don't Fall through the Floor*. Penguin, Harmondsworth.

Gordon, J. 1978. *Structures, or Why Things Don't Fall Down*. Penguin, Harmondsworth.

Gordon, J.E. 1977. Morphometry of cirques in the Kintail–Affric–Cannich area of northwest Scotland. *Geografiska Annaler* 59A, 177–194.

Gordon, J.E. 1979. Reconstructed Pleistocene ice sheet temperatures and glacial erosion in northern Scotland. *Journal of Glaciology* 22, 331–344.

Gordon, J.E. 1981. Ice-scoured topography and its relationships to bedrock structure and ice movement in parts of northern Scotland and west Greenland. *Geografiska Annaler* 63A, 55–65.

Gordon, J.E. 1993. The Cairngorms. In Gordon, J.E. and Sutherland, D.G. (eds), *Quaternary of Scotland*. Chapman & Hall, London, 259–276.

Gordon, J.E. and Birnie, R.V. 1986. Production and transfer of subaerially generated rock debris and resulting landforms on South Georgia: an introductory perspective. *British Antarctic Survey Bulletin* 72, 25–46.

Gordon, J.E., Birnie, R.V. and Timmis, R. 1978. A major rockfall and debris slide on the Lyell Glacier, South Georgia. *Arctic and Alpine Research* 10, 49–60.

Gordon, J.E., Darling, W.G., Whalley, W.B. and Gellatly, A.F. 1988. $\partial D - \partial^{18}O$ relationships and the thermal history of the basal ice near the margins of two glaciers in Lyngen, north Norway. *Journal of Glaciology* 34, 265–268.

Gordon, J.E., Whalley, W.B., Gellatly, A.F. and Vere, D.M. 1992. The formation of glacial flutes: assessment of models with evidence from Lyngsdalen, north Norway. *Quaternary Science Reviews* 11, 709–731.

Gorrell, G. and Shaw, J. 1991. Deposition in an esker, bead and fan complex, Lanark, Ontario, Canada. *Sedimentary Geology* 72, 285–314.

Goudie, A. (ed.) 1990. *Geomorphological Techniques*, 2nd edition. Routledge, London.

Gould, S.J. 1980. *The Panda's Thumb*. Norton, New York.

Gow, A.J. and Epstein S. 1972. On the use of stable isotopes to trace the origins of ice in a floating ice tongue. *Journal of Geophysical Research* 77, 6552–6557.

Gow, A.J., Epstein S. and Sheehy, W. 1979. On the origin of stratified debris in ice cores from the bottom of the Antarctic ice sheet. *Journal of Glaciology* 23, 185–192.

Gow, A.J. and Williamson, T. 1976. Rheological implications of the internal structure and crystal fabrics of the West Antarctic ice sheet as revealed by deep core drilling at Byrd Station. Geological Society of America Bulletin 87, 1665–1677.

Graf, W.L. 1970. The geomorphology of the glacial valley cross-section. *Arctic and Alpine Research* 2, 303–312.

Graf, W.L. 1976. Cirques as glacier locations. *Arctic and Alpine Research* 8, 79–90.

Grano, O. 1958. The Vesso esker of south Finland and its economic importance. *Fennia* 82, 3–33.

Grant, D.R. 1975. Glacial style and the Quaternary stratigraphic record in the Atlantic Provinces, Canada. Geological Survey of Canada, Paper 75–1B, 109–110.

Grant, D.R. 1980. Quaternary sea-level change in Atlantic Canada as an indication of crustal delevelling. In Morner, N.A. (ed.), *Earth Rheology, Isostasy and Eustasy*. Wiley, Chichester, 201–214.

Grant, D.R. and King, L.H. 1984. A stratigraphic framework for the Quaternary history of the Atlantic Provinces, Canada. In Fulton R.J. (ed.), *Quaternary Stratigraphy of Canada*. Geological Survey of Canada, Paper 84–10, 173–191.

Grant, U.S. and Higgins, D.F. 1913. Coastal glaciers of Prince William Sound and Kenai Peninsula, Alaska. US Geological Survey, Bulletin 526.

Grass, A.J. 1971. Structural features of turbulent flow over smooth and rough boundaries. *Journal of Fluid Mechanics* 50, 233–255.

Gravenor, C.P. 1953. The origin of drumlins. *American Journal of Science* 251, 674–681.

Gravenor, C.P. 1974. The Yarmouth drumlin field, Nova Scotia, Canada. *Journal of Glaciology* 13, 45–54.

Gravenor, C.P. 1975. Erosion by continental ice sheets. *American Journal of Science* 275, 594–604.

Gravenor, C.P. 1980. Heavy minerals and sedimentological studies on the glaciogenic Late Precambrian Gaskiers Formation of Newfoundland. *Canadian Journal of Earth Sciences* 17, 1331–1341.

Gravenor, C.P., von Brunn, V. and Dreimanis, A. 1984. Nature and classification of waterlain glaciogenic sediments, exemplified by Pleistocene, Late Palaeozoic and Late Precambrian deposits. *Earth Science Reviews* 20, 105–166.

Gravenor, C.P. and Kupsch, W.O. 1959. Ice disintegration features in western Canada. *Journal of Geology* 67, 48–64.

Gravenor, C.P. and Meneley, W.A. 1958. Glacial flutings in central and northern Alberta. *American Journal of Science* 256, 715–728.

Gravenor, C.P. and Monteiro, R. 1983. Ice-thrust features and a possible intertillite pavement in the Proterozoic Macaubas Group, Jequital area, Minas Gerais, Brazil. *Journal of Geology* 91, 113–116.

Gravenor, C.P. and Stupavsky, M. 1974. Magnetic susceptibility of the surface tills of southern Ontario. *Canadian Journal of Earth Sciences* 11, 658–663.

Gravenor, C.P., Stupavsky, M. and Symons, D.T.A. 1973. Paleomagnetism and its relationship to till deposition. *Canadian Journal of Earth Sciences* 10, 1068–1078.

Gray, J.M. 1975. The Loch Lomond Readvance and contemporaneous sea-levels in Loch Etive and neighbouring areas of western Scotland. *Proceedings of the Geologists' Association* 86, 227–238.

Gray, J.M. 1981. P-forms from the Isle of Mull. *Scottish Journal of Geology* 17, 39–47.

Gray, J.M. 1982a. The last glaciers (Loch Lomond Advance) in Snowdonia, north Wales. *Geological Journal* 17, 111–133.

Gray, J.M. 1982b. Unweathered glaciated bedrock on an exposed lake bed in Wales. *Journal of Glaciology* 28, 483–497.

Gray, J.M. 1991. Glaciofluvial landforms. In Ehlers, J., Gibbard, P.L. and Rose, J. (eds), *Glacial Deposits of Britain and Ireland*. Balkema, Rotterdam, 443–453.

Gray, J.M. 1992. Scarisdale: P-forms. In Walker, M.J.C., Gray, J.M. and Lowe, J.J. (eds), *The South-West Scottish Highlands: Field Guide*. Quaternary Research Association, Cambridge, 85–88.

Gray, J.M. 1995. Influence of Southern Upland ice on glacio-isostatic rebound in Scotland, the Main Rock Platform in the Firth of Clyde. *Boreas* 24, 30–36.

Gray, J.M. and Brooks, C.L. 1972. The Loch Lomond Readvance moraines of Mull and Menteith. *Scottish Journal of Geology* 8, 95–103.

Gray, J.M. and Coxon, P. 1991. The Loch Lomond Stadial glaciation in Britain and Ireland. In Ehlers, J., Gibbard, P.L. and Rose, J. (eds), *Glacial Deposits in Great Britain and Ireland*. Balkema, Rotterdam, 89–105.

Gray, J.M. and Lowe, J.J. 1982. Problems in the interpretation of small-scale erosional forms on glaciated bedrock surfaces, examples from Snowdonia, North Wales. *Proceedings of the Geologists' Association* 93, 403–414.

Green, C.P., Hey, R.W. and McGregor, D.F.M. 1980. Volcanic pebbles in Pleistocene gravels of the Thames in Buckinghamshire and Hertfordshire. *Geological Magazine* 117, 59–64.

Greene, D. 1992. Topography and former Scottish tidewater glaciers. *Scottish Geographical Magazine* 108, 164–171.

Gregory, J.W. 1913. *The Nature and Origin of Fiords*. J. Murray, London.

Gregory, J.W. 1927. The fjords of the Hebrides. *Geographical Journal* 69, 193–216.

Gregory, K.H. 1986. Interpretations of striae and roches moutonnees in central Snowdonia, north Wales. *Quaternary Studies* 2, 2–13.

Griffey, N. and Whalley, W.B. 1979. A rock glacier and moraine complex, Lyngen Peninsula, north Norway. *Norsk Geografisk Tidsskrift* 33, 117–124.

Griffiths, J.C. 1967. *Scientific Method in Analysis of Sediments*. McGraw-Hill, New York.

Griggs, G.B. and Kulm, L.D. 1969. Glacial marine sediments from the northern Pacific. *Journal of Sedimentary Petrology* 39, 1142–1148.

Gripp, K. 1929. Glaciologische und geologische Ergebnisse der Hamburgischen Spitzbergen-Expedition 1927.

Naturwissenschaften Verein in Hamburg Abhandlungen aus dem Gebiete der Naturwissenschaften 22, 146–249.

Gripp, K. 1975. 100 Jahre Untersuchungen über das Geschehen am Rande des nordeuropäischen Inlandeises. *Eiszeitalter und Gegenwart* 26, 31–73.

Gripp, K. and Todtmann, E.M. 1925. Die Endmorane des Green-Bay Gletschers auf Spitsbergen, eine studie zum Verstandnis norddeutscher Diluvial-Gebilde. *Mitteilungen aus Geographischen Geseltscahft in Hamburg* 37, 45–75.

Gronlie, A. 1981. The late and postglacial isostatic rebound, the eustatic rise of the sea level and the uncompensated depression in the area of the Blue Road Geotraverse. *Earth Evolution Sciences* 1, 50–57.

Groom, G.E. 1959. Niche glaciers in Bunsow-land, Vestspitsbergen. *Journal of Glaciology* 3, 369–376.

Grootes, P.M., Stuiver, M., White, J.W.C., Johnsen, S. and Jouzel, J. 1993. Comparison of oxygen isotope records from the GISP and GRIP Greenland ice cores. *Nature* 366, 552–554.

Grosswald, M.G. 1980. Late Weichselian ice sheet of northern Eurasia. *Quaternary Research* 13, 1–32.

Grosswald, M.G. 1984. Glaciation of the continental shelves (Parts 1 and 2). *Polar Geography and Geology* 8, 194–258, 287–351.

Grosswald, M.G. 1988. An Antarctic-style ice sheet in the northern hemisphere: toward a new global glacial theory. *Polar Geography and Geology* 12, 239–267.

Grosswald, M.G. and Hughes, T.J. 1995. Paleoglaciology's grand unsolved problem. *Journal of Glaciology* 41, 313–332.

Grousset, F.E., Labeyrie, J.A., Sinks, M., *et al.* 1993. Patterns of ice-rafted detritus in the glacial North Atlantic (40–55° N). *Palaeoceanography* 8, 175–192.

Grove, J.M. 1988. *The Little Ice Age.* Routledge, London.

Grube, F. 1979. Übertiefte Taler im Hamburger Raum. *Eiszeitalter und Gegenwart* 29, 157–172.

Grube, F. 1983. Tunnel valleys. In Ehlers, J. (ed.), *Glacial Deposits in North-West Europe.* Balkema, Rotterdam, 257–258.

Gruell, W. and Oerlemans, J. 1986. Sensitivity studies with a mass-balance model including temperature profile calculations inside the glacier. *Zeitschrift für Gletscherkunde und Glazialgeologie* 22, 101–124.

Gry, H. 1940. De istektoniske forhold i moleromraadet. *Meddelelser Dansk Geologisk Forening* 9, 586–627.

Gudmundsson, M.T., Bjornsson, H. and Palsson, F. 1995. Changes in jökulhlaup sizes in Grimsvötn, Vatnajökull, Iceland, 1934–1991, deduced from in-situ measurements of subglacial lake volume. *Journal of Glaciology* 41, 263–272.

Gudmundsson, G. and Sigbjarnarsson, G. 1972. Analysis of glacier runoff and meteorological observations. *Journal of Glaciology* 11, 303–318.

Guilcher, A. 1969. Pleistocene and Holocene sea level changes. *Earth Science Reviews* 5, 69–98.

Guilcher, A., Bodére, J.-C., Coudé, A., Hansom, J.D., Moign, A. and Peulvast, J.-P. 1986. Le problème des strandflats en cinq pays de hautes latitudes. *Revue de Géologie Dynamique et de Géographie Physique* 27, 47–79. English translation in Evans, D.J.A. (ed.), *Cold Climate Landforms.* Wiley, Chichester, 351–393.

Gurnell, A.M. 1982. The dynamics of suspended sediment concentration in an Alpine pro-glacial stream network. In *Hydrological Aspects of Alpine and High Mountain Areas*, IASH Publication 138, 319–330.

Gurnell, A.M., Clark, M.J. and Hill, C.T. 1992. Analysis and interpretation of patterns within and between hydroclimatological time series in an Alpine glacier basin. *Earth Surface Processes and Landforms* 17, 821–839.

Gurnell, A.M. and Fenn, C.R. 1985. Spatial and temporal variations in electrical conductivity in a proglacial stream. *Journal of Glaciology* 31, 108–114.

Gustavson, T.C. 1974. Sedimentation on gravel outwash fans, Malaspina Glacier foreland, Alaska. *Journal of Sedimentary Petrology* 44, 374–389.

Gustavson, T.C. 1975a. Microrelief gilgai structures on expansive clays of the Texas coastal plain: their recognition and significance in engineering construction. Bureau of Economic Geology, University of Texas, Austin, Geological Circular 75–7.

Gustavson, T.C. 1975b. Sedimentation and physical limnology in proglacial Malaspina Lake, southeastern Alaska. In Jopling, A.V. and McDonald, B.C. (eds), *Glaciofluvial and Glaciolacustrine Sedimentation.* SEPM Special Publication 23, 249–263.

Gustavson, T.C., Ashley, G.M. and Boothroyd, J.C. 1975. Depositional sequences in glaciolacustrine deltas. In Jopling, A.V. and McDonald, B.C. (eds), *Glaciofluvial and Glaciolacustrine Sedimentation.* SEPM Special Publication 23, 264–280.

Gustavson, T.C. and Boothroyd, J.C. 1987. A depositional model for outwash, sediment sources, and hydrologic characteristics, Malaspina Glacier, Alaska: a modern analog of the southeastern margin of the Laurentide Ice Sheet. *Geological Society of America Bulletin* 99, 187–200.

Gwyn, Q.H.J. and Dreimanis, A. 1979. Heavy mineral assemblages in tills and their use in distinguishing glacial lobes in the Great Lakes region. *Canadian Journal of Earth Sciences* 16, 2219–2235.

Gwynne, C.S. 1942. Swell and swale pattern of the Mankato lobe of the Wisconsin drift plain in Iowa. *Journal of Geology* 50, 200–208.

Gwynne, C.S. 1951. Minor moraines in South Dakota and Minnesota. *Bulletin of the Geological Society of America* 62, 233–250.

Haarsted, V. 1956. De kvartaergeologiske og geomorfologiske forhold pa Møn. *Meddelelser Dansk Geologisk Forening* 13, 124–126.

Haavisto-Hyvarinen, M., Kielosto, S. and Niemela, J. 1989. Precrags and drumlin fields in Finland. *Sedimentary Geology* 62, 337–348.

Hack, J.T. 1957. Studies of longitudinal stream profiles in Virginia and Maryland. USGS Professional Paper 294B, 45–94.

Haeberli, W. 1985. Creep of mountain permafrost: internal structure and flow of alpine rock glaciers. *Mitteilungen der Versuchsanstalt für Wasserbau, Hydrologie und Glazialogie*, ETH, Zurich. No. 77.

Haefeli, R. 1961. Contribution to the movement and the form of ice sheets in the Arctic and the Antarctic. *Journal of Glaciology* 3, 1133–1151.

Haefeli, R. 1963. Observations in ice tunnels and the flow law of ice. In Kingery, W.D. (ed.), *Ice and Snow*. MIT Press, Cambridge, Mass.

Haefeli, R. 1968. Gedanken zum Problem der glazialen Erosion. *Felsmechanik und Ingenieurgeologie* 4, 31–51.

Hagen, J.O., Lefauconnier, B. and Liestol, O. 1991. Glacier mass balance in Svalbard since 1912. In Kotlyakov, V.M., Ushakov, A. and Glazovsky, A. (eds), *Glaciers–Ocean–Atmosphere Interactions*. IAHS Publication 208, Great Yarmouth, UK, 313–328.

Hagen, J.O. and Liestøl, O. 1990. Long term glacier mass balance investigations in Svalbard 1950–1988. *Annals of Glaciology* 14, 102–106.

Hagen, J.O., Wold, B., Liestøl, O., Ostrem, G. and Sollid, J.L. 1983. Subglacial processes at Bondhusbreen, Norway. *Annals of Glaciology* 4, 91–98.

Haggart, B.A. 1989. Variations in the pattern and rate of isostatic uplift indicated by a comparison of Holocene sea-level curves from Scotland. *Journal of Quaternary Science* 4, 67–76.

Haldorsen, S. 1981. Grain-size distribution of subglacial till and its relation to subglacial crushing and abrasion. *Boreas* 10, 91–105.

Haldorsen, S. 1983. Mineralogy and geochemistry of basal till and its relationship to till-forming processes. *Norsk Geologisk Tidsskrift* 63, 15–25.

Haldorsen, S. and Shaw, J. 1982. The problem of recognizing melt-out till. *Boreas* 11, 261–277.

Hall, A.M. 1985. Cenozoic weathering covers in Buchan, Scotland, and their significance. *Nature* 315, 392–395.

Hall, A.M. 1986. Deep weathering patterns in north-east Scotland and their geomorphological significance. *Zeitschrift für Geomorphologie* 30, 407–422.

Hall, A.M. and Sugden, D.E. 1987. Limited modification of mid-latitude landscapes by ice sheets, the case of north-east Scotland. *Earth Surface Processes and Landforms* 12, 531–542.

Hall, B.L., Denton, G.H., Lux, D.R. and Bockheim, J.G. 1993. Late Tertiary Antarctic paleoclimate and ice sheet dynamics inferred from surficial deposits in Wright Valley. *Geografiska Annaler* 75A, 239–267.

Hall, K.J. 1990. Quaternary glaciations in the Southern Ocean, sector 0° long.–180° long. *Quaternary Science Reviews* 9, 217–228.

Hallet, B. 1976a. Deposits formed by subglacial precipitation of $CaCO_3$. *Geological Society of America Bulletin* 87, 1003–1015.

Hallet, B. 1976b. The effect of subglacial chemical processes on glacier sliding. *Journal of Glaciology* 17, 209–221.

Hallet, B. 1979a. A theoretical model of glacial abrasion. *Journal of Glaciology* 23, 39–50.

Hallet, B. 1979b. Subglacial regelation water film. *Journal of Glaciology* 23, 321–334.

Hallet, B. 1981. Glacial abrasion and sliding: their dependence on the debris concentration in basal ice. *Annals of Glaciology*, 2, 23–28.

Hallet, B. 1996. Glacial quarrying: a simple theoretical model. *Annals of Glaciology* 22, 1–8.

Hallet, B. and Anderson, R.S. 1982. Detailed glacial geomorphology of a proglacial bedrock area at Castleguard

Glacier, Alberta, Canada. *Zeitschrift für Gletscherkunde und Glazialgeologie* 16, 171–184.

Hallet, B., Hunter, L. and Bogen, J. 1996. Rates of erosion and sediment evacuation by glaciers: a review of field data and their implications. *Global and Planetary Change* 12, 213–235.

Hallet, B., Lorrain, R. and Souchez, R. 1978. The composition of basal ice from a glacier sliding over limestones. *Geological Society of America Bulletin* 89, 314–320.

Hallet, B., Walder, J.S. and Stubbs, C.W. 1991. Weathering by segregation ice growth in microcracks at subfreezing temperatures, verification from an experimental study using acoustic emissions. *Permafrost and Periglacial Processes* 2, 283–300.

Ham, N.R. and Mickelson, D.M. 1994. Basal till fabric and deposition at Burroughs Glacier, Glacier Bay, Alaska. *Geological Society of America Bulletin* 106, 1552–1559.

Hamblin P.F. and Carmack, E.C. 1978. River-induced currents in a fjord lake. *Journal of Geophysical Research* 83, 885–899.

Hambrey, M.J. 1975. The origin of foliation in glaciers: evidence from some Norwegian examples. *Journal of Glaciology* 14, 181–185.

Hambrey, M.J. 1976a. Structure of the glacier Charles Rabots Breen, Norway. *Geological Society of America Bulletin* 87, 1629–1637.

Hambrey, M.J. 1976b. Debris, bubble, and crystal fabric characteristics of foliated glacier ice, Charles Rabots Breen, Okstindan, Norway. *Arctic and Alpine Research* 8, 49–60.

Hambrey, M.J. 1982. Late Proterozoic diamictites of northeastern Svalbard. *Geological Magazine* 119, 527–551.

Hambrey, M.J. 1983. Correlation of Late Proterozoic tillites in the North Atlantic region and Europe. *Geological Magazine* 120, 209–232.

Hambrey, M.J. 1984. Sedimentary processes and buried ice phenomena in the proglacial areas of Spitsbergen glaciers. *Journal of Glaciology* 30, 116–119.

Hambrey, M.J. 1992. Secrets of a tropical ice age. *New Scientist*, 1 February, 42–49.

Hambrey, M.J. 1994. *Glacial Environments*. UCL Press, London.

Hambrey, M.J., Barrett, P.J., Ehrmann, W.U. and Larsen, B. 1992. Cenozoic sedimentary processes on the Antarctic continental margin and the record from deep drilling. *Zeitschrift für Geomorphologie*, Supp. Bd. 86, 77–103.

Hambrey, M.J. and Harland, W.B. (eds) 1981. *Earth's Pre-Pleistocene Glacial Record*. Cambridge University Press, Cambridge.

Hambrey, M.J. and Huddart, D. 1995. Englacial and proglacial glaciotectonic processes at the snout of a thermally complex glacier in Svalbard. *Journal of Quaternary Science* 10, 313–326.

Hambrey, M.J., Larsen, B., Ehrmann, W.U. and Shipboard Scientific Party 1989. Forty million years of Antarctic glacial history revealed by Leg 119 of the Ocean Drilling Program. *Polar Record* 25, 99–106.

Hambrey, M.J. and Milnes, A.G. 1975. Boudinage in glacier ice: some examples. *Journal of Glaciology* 14, 383–393.

Hambrey, M.J. and Milnes, A.G. 1977. Structural geology of an Alpine glacier (Griesgletscher, Valais, Switzerland). *Eclogae Geologicae Helvetiae* 70, 667–684.

Hambrey, M.J., Milnes, A.G. and Siegenthaler, H. 1980. Dynamics and structure of Griesgletscher, Switzerland. *Journal of Glaciology*, 25, 215–228.

Hambrey, M.J. and Müller, F. 1978. Structures and ice deformation in White Glacier, Axel Heiberg Island, NWT, Canada. *Journal of Glaciology* 20, 41–67.

Hamilton, G.S. and Dowdeswell, J.A. 1996. Controls on glacier surging in Svalbard. *Journal of Glaciology* 42, 157–168.

Hamilton, S.J. and Whalley, W.B. 1995. Preliminary results from the lichenometric study of the Nautardalur rock glacier, Trollaskagi, northern Iceland. *Geomorphology* 12, 123–132.

Hamilton, T.D. and Thorson, R.M. 1983. The Cordilleran ice sheet in Alaska. In Porter, S.C. (ed.), *Late Quaternary Environments of the United States*, vol. 1: *The Late Pleistocene*. University of Minnesota Press, Minneapolis, 38–52.

Hamilton, W. and Krinsley, D. 1967. Upper Paleozoic glacial deposits of South Africa and southern Australia. *Geological Society of America Bulletin* 78, 783–800.

Hammer, C.U. 1977. Past volcanism revealed by Greenland ice sheet impurities. *Nature* 270, 482–486.

Hammer, C.U., Clausen, H.B. and Dansgaard, W. 1980. Greenland ice sheet evidence of post-glacial volcanism and its climatic impact. *Nature* 288, 230–235.

Hampton, M.A. 1972. The role of subaqueous debris flow in generating turbidity currents. *Journal of Sedimentary Petrology* 42, 775–793.

Hampton, M.A. 1975. Competence of fine-grained debris flows. *Journal of Sedimentary Petrology* 45, 834–844.

Hansel, A.K. and Johnson, W.H. 1987. Ice marginal sedimentation in a late Wisconsinan end moraine complex, northeastern Illinois, USA. In van der Meer, J.J.M. (ed.), *Tills and Glaciotectonics*. Balkema, Rotterdam, 97–104.

Hansen, K. 1971. Tunnel valleys in Denmark and northern Germany. *Bulletin of the Geological Society of Denmark* 20, 295–306.

Hansen, K. and Nielsen, A.V. 1960. Glacial geology of southern Denmark. *Guide to Excursions Nos A44 and C39, 21st International Geological Congress, Copenhagen.*

Hanshaw, B.B. and Hallet, B. 1978. Oxygen isotope composition of subglacially precipitated calcite: possible palaeoclimatic implications. *Science* 200, 1267–1270.

Hansom, J.D. 1983a. Ice-formed intertidal boulder pavements in the sub-Antarctic. *Journal of Sedimentary Petrology* 53, 135–145.

Hansom, J.D. 1983b. Shore platform development in the South Shetland Islands, Antarctica. *Marine Geology* 52, 211–229.

Hanson, B. 1987. Reconstructing mass balance profiles from climate for an arctic ice cap. IAHS Publication 170, Great Yarmouth, UK, 181–189.

Hantz, D. and Lliboutry, L. 1983. Waterways, ice permeability at depth, and water pressures at Glacier d'Argentière, French Alps. *Journal of Glaciology* 29, 227–239.

Hanvey, P.M. 1987. Sedimentology of lee-side stratification sequences in late Pleistocene drumlins, north-west Ireland. In Menzies, J. and Rose, J. (eds), *Drumlin Symposium*. Balkema, Rotterdam, 241–253.

Hanvey, P.M. 1989. Stratified flow deposits in a late Pleistocene drumlin in northwest Ireland. *Sedimentary Geology* 62, 211–221.

Haq, B.U., Hardenbol, J. and Vail, P.R. 1987. Chronology of fluctuating sea levels since the Triassic. *Science* 235, 1156–1166.

Harbor, J.M. 1992a. Application of a general sliding law to simulating flow in a glacier cross-section. *Journal of Glaciology* 38, 182–190.

Harbor, J.M. 1992b. Numerical modelling of the development of U-shaped valleys by glacial erosion. *Geological Society of America Bulletin* 104, 1364–1375.

Harbor, J.M., Hallet, B. and Raymond, C.F. 1988. A numerical model of landform development by glacial erosion. *Nature* 333, 347–349.

Hardy, L. 1977. La déglaciation et les épisodes lacustre et marin sur le versant québécois des basses-terres de la baie de James. *Géographie Physique et Quaternaire* 31, 261–273.

Hardy, L. 1982. Le Wisconsinien supérieur à l'est de la baie James, Québec. *Naturaliste Canadien* 109, 333–351.

Hardy, R.M. and Leggett, R.F. 1960. Boulders in a varved clay at Steep Rock lake, Ontario, Canada. *Bulletin of the Geological Society of America* 71, 93–94.

Hare, F.K. 1959. A photo-reconnaissance survey of Labrador-Ungava. Geographical Branch, Memoir 6. Mines and Technology Surveys, Ottawa.

Harker, A. 1901. Ice erosion in the Cuillin Hills, Skye. *Transactions of the Royal Society of Edinburgh* 40, 221–252.

Harland, W.B. 1964a. Evidence of late Precambrian glaciation and its significance. In Nairn, A.E.M. (ed.), *Problems in Palaeoclimatology*. Interscience, New York, 119–149.

Harland, W.B. 1964b. Critical evidence for a great infra-Cambrian glaciation. *Geologische Rundschau* 54, 45–61.

Harland, W.B. 1972. The Ordovician ice age. *Geological Magazine* 109, 451–456.

Harland, W.B. and Herod, K.N. 1975. Glaciations through time. In Wright, A.E. and Moseley, F. (eds), *Ice Ages, Ancient and Modern*. Seel House Press, Liverpool, 189–216.

Harland, W.B., Herod, K.N. and Krinsley, D.H. 1966. The definition and identification of tills and tillites. *Earth Science Reviews* 2, 225–256.

Harper, J.T. 1993. Glacier terminus fluctuations on Mount Baker, Washington, USA, 1940–1990, and climatic variations. *Arctic and Alpine Research* 25, 332–340.

Harris, C. 1981. *Periglacial Mass Wasting, A Review of Research*. Geo Abstracts, Norwich.

Harris, C. and Bothamley, K. 1984. Englacial deltaic sediments as evidence for basal freezing and marginal shearing, Leirbreen, Norway. *Journal of Glaciology* 30, 30–34.

Harris, C. and Wright, M.D. 1980. Some last glaciation drift deposits near Pontypridd, south Wales. *Geological Journal* 15, 7–20.

Harris, P.W.V. 1976. The seasonal temperature–salinity structure of a glacial lake, Jökulsarlon, southeast Iceland. *Geografiska Annaler* 58A, 329–336.

Harris, S.A. and Boydell, A.N. 1972. Glacial history of the Bow River and Red Deer River areas and the adjacent foothills. In Slaymaker, O. and McPherson, H.J. (eds), *Mountain Geomorphology*. Tantalus Press, Vancouver, 47–53.

Harris, S.E. 1943. Friction cracks and the direction of glacial movement. *Journal of Geology* 51, 244–258.

Harris, W.K. 1981. Permian diamictites in South Australia. In Hambrey, M.J. and Harland, W.B. (eds), *Earth's Pre-Pleistocene Glacial Record*. Cambridge University Press, Cambridge, 469–473.

Harrison, I.M. and Jolleymore, P.G. 1974. Iceberg furrow marks on the continental shelf northeast of Belle Island, Newfoundland. *Canadian Journal of Earth Sciences* 11, 43–52.

Harrison, S. 1991. A possible paraglacial origin for the drift sheets of upland Britain. *Quaternary Newsletter* 64, 14–18.

Harrison, S. 1993. Solifluction sheets in the Bowmont Valley, Cheviot Hills. Scottish landform examples – 8. *Scottish Geographical Magazine* 109, 119–122.

Harrison, W.D., Echelmeyer, K.A., Chacho, E.F., Raymond, C.F. and Benedict, R.J. 1994. The 1897–88 surge of West Fork Glacier, Susitna Basin, Alaska, USA. *Journal of Glaciology* 40, 241–254.

Harry, D.G., French, H.M. and Pollard, W.H. 1988. Massive ground ice and ice-cored terrain near Sabine Point, Yukon Coastal Plain. *Canadian Journal of Earth Sciences* 25, 1846–1856.

Hart, J.K. 1990. Proglacial glaciotectonic deformation and the origin of the Cromer Ridge push moraine complex, north Norfolk, England. *Boreas* 19, 165–180.

Hart, J.K. 1994a. Till fabric associated with deformable beds. *Earth Surface Processes and Landforms* 19, 15–32.

Hart, J.K. 1994b. Proglacial glaciotectonic deformation at Melabakkar-Ásbakkar, west Iceland. *Boreas* 23, 112–121.

Hart, J.K. 1995a. Subglacial erosion, deposition and deformation associated with deformable beds. *Progress in Physical Geography* 19, 173–191.

Hart, J.K. 1995b. Recent drumlins, flutes and lineations at Vestari-Hagafellsjökull, Iceland. *Journal of Glaciology* 41, 596–606.

Hart, J.K. and Boulton, G.S. 1991. The inter-relation of glaciotectonic and glaciodepositional processes within the glacial environment. *Quaternary Science Reviews* 10, 335–350.

Hart, J.K., Gane, F. and Watts, R.J. 1996. Deforming bed conditions on the Dänischer Wold peninsula, northern Germany. *Boreas* 25, 101–113.

Hart, J.K., Hindmarsh, R.C.A. and Boulton, G.S. 1990. Styles of subglacial glaciotectonic deformation within the context of the Anglian Ice Sheet. *Earth Surface Processes and Landforms* 15, 227–241.

Hart, J.K. and Roberts, D.H. 1994. Criteria to distinguish between subglacial glaciotectonic and glaciomarine sedimentation, I. Deformation styles and sedimentology. *Sedimentary Geology* 91, 191–213.

Hartshorn, J.H. 1957. Flowtill in southeastern Massachusetts. *Geological Society of America Bulletin* 69, 477–482.

Harwood, D.M. 1986. Oldest record of Cainozoic glacial-marine sedimentation in Antarctica (31 Myr), results from MSSTS–1 drill hole. *South African Journal of Science* 82, 516–519.

Haselton, G.M. 1979. Some glaciogenic landforms in Glacier Bay National Monument, southeastern Alaska. In Schluchter, C. (ed.), *Moraines and Varves*. Balkema, Rotterdam, 197–206.

Hastenrath, S. 1971. On the Pleistocene snowline depression in the arid regions of the South American Andes. *Journal of Glaciology* 10, 255–267.

Hastenrath, S. 1981. *The Glaciation of the Ecuadorian Andes*. Balkema, Rotterdam.

Hastenrath, S. 1984. *The Glaciers of Equatorial East Africa*. Reidel, Dordrecht.

Hastenrath, S. 1989. Ice flow and mass changes of Lewis Glacier, Mount Kenya, East Africa, observations 1974–86, modelling, and predictions to the year 2000 AD. *Journal of Glaciology* 35, 325–332.

Hastenrath, S. and Kruss, P.D. 1992. The dramatic retreat of Mount Kenya's glaciers between 1963 and 1987, greenhouse forcing. *Annals of Glaciology* 16, 127–133.

Hastenrath, S. and Kutzbach, J. 1985. Late Pleistocene climate and water budget of the South American Altiplano. *Quaternary Research* 24, 249–256.

Hattersley-Smith, G. 1955. Northern Ellesmere Island 1953 and 1954. *Arctic* 8, 1–36.

Hattersley-Smith, G. 1957a. The Ellesmere ice shelf and the ice islands. *Canadian Geographer* 9, 65–70.

Hattersley-Smith, G. 1957b. The rolls on the Ellesmere ice shelf. *Arctic* 10, 32–44.

Hattersley-Smith, G. 1969. Glacial features of Tanquary Fjord and adjoining areas of northern Ellesmere Island, NWT. *Journal of Glaciology* 8, 23–50.

Hattersley-Smith, G. 1974. Present arctic ice cover. In Ives, J.D. and Barry, R.G. (ed.), *Arctic and Alpine Environments*. Methuen, London, 195–223.

Hattersley-Smith, G., Keys, J.E., Serson, H. and Mielke, J.E. 1970. Density stratified lakes in northern Ellesmere Island. *Nature* 225, 55–56.

Hattersley-Smith, G. and Serson, H.V. 1964. Stratified water of a glacial lake in northern Ellesmere Island. *Arctic* 17, 108–111.

Hattersley-Smith, G. and Serson, H.V. 1970. Mass balance of the Ward Hunt Ice Rise and Ice Shelf: a 10 year record. *Journal of Glaciology* 9, 247–252.

Haughton, P.D.W. 1994. Deposits of deflected and ponded turbidity currents, Sorbas basin, southeast Spain. *Journal of Sedimentary Research* A64, 233–246.

Hawkins, F. 1985. Equilibrium-line altitudes and palaeo-environments in the Merchants Bay area, Baffin Island, NWT, Canada. *Journal of Glaciology* 31, 205–213.

Hay, J.E. and Fitzharris, B.B. 1988. A comparison of the energy balance and bulk-aerodynamic approaches for estimating glacier melt. *Journal of Glaciology* 34, 145–153.

Haynes, V.M. 1968. The influence of glacial erosion and rock structure on corries in Scotland. *Geografiska Annaler* 50A, 221–234.

Haynes, V.M. 1972. The relationship between the drainage areas and sizes of outlet troughs of Sukkertoppen Ice Cap, west Greenland. *Geografiska Annaler* 54A, 66–75.

Haynes, V.M. 1977. The modification of valley patterns by ice-sheet activity. *Geografiska Annaler* 59A, 195–207.

Haynes, V.M. 1983. Scotland's landforms. In Clapperton, C.M. (ed.), *Scotland, A New Study*. David & Charles, Newton Abbot, 28–63.

Haynes, V.M. 1995. Alpine valley heads on the Antarctic Peninsula. *Boreas* 24, 81–94.

Healy, T.R. 1975. Thermokarst: a mechanism of de-icing ice-cored moraines. *Boreas* 4, 19–23.

Hebrand, M. and Åmark, M. 1989. Esker formation and glacier dynamics in eastern Skane and adjacent areas, southern Sweden. *Boreas* 18, 67–81.

Heidenreich, C. 1964. Some observations on the shape of drumlins. *Canadian Geographer* 8, 101–107.

Heikkinen, O. and Tikkanen, M. 1989. Drumlins and flutings in Finland: their relationships to ice movement and to each other. *Sedimentary Geology* 62, 349–355.

Hein, F.J. 1982. Depositional mechanisms of deep-sea coarse clastic sediments, Cap Enragée Formation, Quebec. *Canadian Journal of Earth Sciences* 19, 267–287.

Hein, F.J. 1984. Deep-sea and fluvial braided channel conglomerates: a comparison of two case studies. In Koster, E.H. and Steel, R.J. (eds), *Sedimentology of Gravels and Conglomerates*. Canadian Society of Petroleum Geologists Memoir 10, 33–49.

Hein, F.J. and Walker, R.G. 1977. Bar evolution and development of stratification in the gravelly, braided, Kicking Horse River, British Columbia. *Canadian Journal of Earth Sciences* 14, 562–570.

Heinrich, H. 1988. Origin and consequences of cyclic ice rafting in the northeast Atlantic Ocean during the past 130,000 years. *Quaternary Research* 29, 142–152.

Heinrichs, T.A., Mayo, L.R., Echelmeyer, K.A. and Harrison, W.D. 1996. Quiescent-phase evolution of a surge-type glacier, Black Rapids Glacier, Alaska, USA. *Journal of Glaciology* 42, 110–122.

Hellaakoski, A. 1930. On the transportation of materials in the esker of Laitila. *Fennia* 52, 1–41.

Hendy, C.H., Healy, T.R., Rayner, E.M., Shaw, J. and Wilson, A.T. 1979. Late Pleistocene glacial chronology of the Taylor Valley, Antarctica, and the global climate. *Quaternary Research* 11, 172–184.

Hessell, J.W.D. 1983. Climatic effects on the recession of the Franz Josef Glacier. *New Zealand Journal of Science* 26, 315–320.

Heusser, C.J. 1956. Postglacial environments in the Canadian Rocky Mountains. *Ecological Monographs* 26, 263–302.

Hewitt, K.J. 1964. The great ice dam. *Indus* 5, 18–30.

Hewitt, K.J. 1967. Ice-front deposition and the seasonal effect, a Himalayan example. *Transactions of the Institute of British Geographers* 42, 93–106.

Hewitt, K.J. 1982. Natural dams and outburst floods of the Karakoram Himalaya. *IAHS Publication* 138, 259–269.

Hewitt, K.J. 1988. Catastrophic landslide deposits in the Karakoram Himalaya. *Science* 242, 64–66.

Hewitt, K.J. 1993. Altitudinal organisation of Karakoram geomorphic processes and depositional environments. In Shroder, J.F. (ed.), *Himalaya to the Sea*. Routledge, London, 159–183.

Heyworth, A. 1986. Submerged forests as sea level indicators. In van de Plassche (ed.), *Sea-Level Research*. Geo Books, Norwich, 401–411.

Hicock, S.R. 1990. Genetic till prism. *Geology* 18, 517–519.

Hicock, S.R. 1991. On subglacial stone pavements in till. *Journal of Geology* 99, 607–619.

Hicock, S.R. 1992. Lobal interactions and rheologic superposition in subglacial till near Bradtville, Ontario, Canada. *Boreas* 21, 73–88.

Hicock, S.R. 1993. Glacial octahedron. *Geografiska Annaler* 75A, 35–39.

Hicock, S.R. and Dreimanis, A. 1992a. Deformation till in the Great Lakes region, implications for rapid flow along the south-central margin of the Laurentide Ice Sheet. *Canadian Journal of Earth Sciences*, 29, 1565–1579.

Hicock, S.R. and Dreimanis, A. 1992b. Sunnybrook drift in the Toronto area, Canada: reinvestigation and reinterpretation. In Clark, P.U. and Lea, P.D. (eds), *The Last Interglacial–Glacial Transition in North America*. Geological Society of America Special Paper 270, 139–161.

Hicock, S.R., Dreimanis, A. and Broster, B.E. 1981. Submarine flow tills at Victoria, British Columbia. *Canadian Journal of Earth Sciences* 18, 71–80.

Hicock, S.R. and Fuller, E.A. 1995. Lobal interactions, rheologic superposition, and implications for a Pleistocene ice stream on the continental shelf of British Columbia. *Geomorphology* 14, 167–184.

Higgins, C.G. 1957. Origin of potholes in glaciated regions. *Journal of Glaciology* 3, 11–12.

Higgs, R. 1978. Provenance of Mesozoic and Cenozoic sediments from the Labrador and west Greenland continental margins. *Canadian Journal of Earth Sciences* 15, 1850–1860

Higuchi, K., Ageta, Y., Yasumari, T. and Inoue, J. 1982. Characteristics of precipitation during the monsoon season in high-mountain areas of the Nepal Himalaya. In *Hydrological Aspects of Alpine and High Mountain Areas*. IAHS Publication No. 138, 21–30.

Higuchi, K. and Tanaka, Y. 1982. Flow pattern of meltwater in mountain snow cover. IAHS Publication 138, 63–69.

Hill, A.R. 1971. The internal composition and structure of drumlins in North Down and South Antrim, Northern Ireland. *Geografiska Annaler* 53A, 14–31.

Hill, A.R. 1973. The distribution of drumlins in County Down, Ireland. *Annals of the American Association of Geographers* 63, 226–240.

Hill, P.R., Mudie, P.J., Moran, K. and Blasco, S.M. 1985. A sea level curve for the Canadian Beaufort Shelf. *Canadian Journal of Earth Sciences* 22, 1383–1393.

Hillaire-Marcel, C. and Occhietti, S. 1980. Chronology, palaeogeography, and palaeoclimatic significance of the late and postglacial events in eastern Canada. *Zeitschrift für Geomorphologie* 24, 373–392.

Hillaire-Marcel, C., Occhietti, S. and Vincent, J.-S. 1981. Sakami moraine, Quebec: a 500 km long moraine without climatic control. *Geology* 9, 210–214.

Hillefors, A. 1973. The stratigraphy and genesis of stoss- and lee-side moraines. *Bulletin of the Geological Institute of the University of Uppsala* NS 5, 139–154.

Hinsch, W. 1979. Rinnen an der Basis des glaziären Pleistozäns in Schleswig-Holstein. *Eiszeitalter und Gegenwart* 29, 173–178.

Hirvas, H. and Nenonen, K. 1985. The till stratigraphy of Finland. In Kujansuu, R. and Saarnisto, M. (eds), INQUA Till Symposium. Geological Survey of Finland, Special Paper 3, 49–63.

Hjulström, F. 1935. Studies of the morphological activity of rivers as illustrated by the River Fyris. *Bulletin of the Geological Institute, University of Uppsala* 25, 221–527.

Hjulström, F. 1952. The geomorphology of the alluvial outwash plains (sandurs) of Iceland, and the mechanics of braided rivers. *International Geographical Union, 17th Congress Proceedings, Washington*, 337–342.

Hobbie, J.E. 1961. Summer temperatures in Lake Schrader, Alaska. *Limnology and Oceanography* 6, 326–329.

Hobbs, W.H. 1910. The cycle of mountain glaciation. *Geographical Journal* 35, 146–163 and 268–284.

Hobbs, W.H. 1926. *Earth Features and their Meaning.* Macmillan, New York.

Hock, R. and Hooke, R. le B. 1993. Evolution of the internal drainage system in the lower part of the ablation area of Storglaciären, Sweden. *Geological Society of America Bulletin* 105, 537–546.

Hodge, S.M. 1974. Variations in the sliding of a temperate glacier. *Journal of Glaciology* 13, 349–369.

Hodge, S.M. 1976. Direct measurement of basal water pressures, a pilot study. *Journal of Glaciology* 16, 205–218.

Hodge, S.M. 1979. Direct measurement of basal water pressures, progress and problems. *Journal of Glaciology* 23, 309–319.

Hodgkins, R. and Dowdeswell, J.A. 1994. Tectonic processes in Svalbard tide-water glacier surges, evidence from structural geology. *Journal of Glaciology* 40, 553–560.

Hodgson, D.A. 1985. The last glaciation of west-central Ellesmere Island, arctic archipelago, Canada. *Canadian Journal of Earth Sciences* 22, 347–368.

Hodgson, D.A. 1994. Episodic ice streams and ice shelves during retreat of the northwesternmost sector of the late Wisconsinan Laurentide Ice Sheet over the central Canadian Arctic Archipelago. *Boreas* 23, 14–28.

Hodgson, D.A. and Vincent, J.-S. 1984. A 10,000 yr BP extensive ice shelf over Viscount Melville Sound, Arctic Canada. *Quaternary Research* 22, 18–30.

Hodgson, D.M. 1986. A study of fluted moraines in the Torridon area, NW Scotland. *Journal of Quaternary Science* 1, 109–118.

Hodgson, G.J., Lever, J.H., Woodworth-Lynas, C.M.T. and Lewis, C.F.M. (eds) 1988. The dynamics of iceberg grounding and scouring (DIGS) experiment and repetitive mapping of the eastern Canadian continental shelf. Environmental Studies Research Funds Report No. 094, Ottawa.

Hoinkes, H.C. 1968. Glacier variation and weather. *Journal of Glaciology* 7, 3–19.

Hoinkes, H.C. 1971. Methoden und Möglichkeiten von Massenhaushaltsstudien auf Gletschern. *Zeitschrift für Gletscherkunde und Glazialgeologie* 6, 37–89.

Holdsworth, G. 1973a. Ice calving into the proglacial Generator Lake, Baffin Island, NWT, Canada. *Journal of Glaciology* 12, 235–250.

Holdsworth, G. 1973b. Ice deformation and moraine formation at the margin of an ice cap adjacent to a proglacial lake. In Fahey, B.D. and Thompson, R.D. (eds), *Research in Polar and Alpine Geomorphology.* Geo Abstracts, Norwich, 187–199.

Holdsworth, G. 1974. Erebus Glacier tongue, McMurdo Sound, Antarctica. *Journal of Glaciology* 13, 27–35.

Holdsworth, G. 1977. Tidal interaction with ice shelves. *Annales de Géophysique* 33, 133–146.

Holdsworth, G. 1982. Dynamics of Erebus Glacier tongue. *Annals of Glaciology* 3, 131–137.

Holdsworth, G. and Bull, C. 1970. The flow law of cold ice; investigations on Meserve Glacier, Antarctica. In Gow, A.J. *et al.* (eds), *International Symposium on Antarctic Glaciological Exploration (ISAGE)*, IASH Publication 86, 204–216.

Holdsworth, G. and Glynn, J.E. 1978. Iceberg calving from floating glaciers by a vibrating mechanism. *Nature* 274, 464–466.

Holdsworth, G. and Glynn, J.E. 1981. A mechanism for the formation of large icebergs. *Journal of Geophysical Research* 86, 3210–3222.

Hollick, A. 1894. Dislocations in certain portions of the Atlantic coastal plain strata and their probable causes. *Transactions of the New York Academy of Sciences* 14, 8–20.

Hollin, J.T. 1962. On the glacial history of Antarctica. *Journal of Glaciology* 4, 173–195.

Hollin, J.T. 1969. Ice-sheet surges and the geological record. *Canadian Journal of Earth Sciences* 6, 903–910.

Hollin, J.T. 1970. Is the Antarctic Ice Sheet growing thicker? *IASH* 86, 363–374.

Hollin, J.T. and Schilling, D.H. 1981. Late Wisconsin–Weichselian mountain glaciers and small ice caps. In Denton, G.H. and Hughes, T.J. (eds), *The Last Great Ice Sheets.* Wiley, New York, 179–206.

Hollingworth, S.E. 1931. The glaciation of western Edenside and the Solway Basin. *Quarterly Journal of the Geological Society of London* 87, 281–359.

Holm, L. 1981. Heavy mineral distribution in Weichselian till successions in eastern Denmark. *Bulletin of the Geological Society of Denmark* 30, 1–10.

Holmes, C.D. 1941a. Kames. *American Journal of Science* 245, 240–249.

Holmes, C.D. 1941b. Till fabric. *Geological Society of America Bulletin* 52, 1301–1352.

Holmes, C.D. 1947. Kames. *American Journal of Science* 245, 240–249.

Holmes, C.D. 1960. Evolution of till-stone shapes, New York. *Geological Society of America Bulletin* 71, 1645–1660.

Holmes, G.W. 1955. Morphology and hydrology of the Mint Julep area, southwest Greenland. Mint Julep Reports Part II, Arctic Desert Topic Information Center US Air University Publication A–104-B.

Holmes, J.A. 1993. Present and past patterns of glaciation in the northwest Himalaya, climatic, tectonic and topographic controls. In Shroder, J.F. (ed.), *Himalaya to the Sea.* Routledge, London, 72–90.

Holmgren, B. 1971. Climate and energy exchange on a sub-polar ice cap in summer. Parts A to F. Uppsala Universitet, Meteorologiska Institutionen, Meddelende, Uppsala, Sweden, 107–112.

Holmlund, P. 1988a. Internal geometry and evolution of moulins, Storglaciärien, Sweden. *Journal of Glaciology* 34, 242–248.

Holmlund, P. 1988b. Is the longitudinal profile of Storglaciärien, northern Sweden, in balance with the present climate? *Journal of Glaciology* 34, 269–273.

Holmlund, P. and Hooke, R. Le B. 1983. High water-pressure events in moulins, Storglaciärien, Sweden. *Geografiska Annaler,* 65A, 19–25.

Holmlund, P. and Naslund, J.O. 1994. The glacially sculptured landscape in Dronning Maud Land, Antarctica, formed by wet-based mountain glaciation and not by the present ice sheet. *Boreas* 23, 139–148.

Holtedahl, H. 1959. Den norske strandflate, med saerlig henblikk p dens utvikling i Kystomradene pa Möre. *Norsk Geografisk Tidsskrift* 16, 285–303.

Holtedahl, H. 1960. MountaIn fjord, strandflat, geomorphology and general geology of western Norway. In Dons, J.A. (ed.), *Guide to Excursions, 21st International Geological Congress, Oslo, Norway*, 1–29.

Holtedahl, H. 1967. Notes on the formation of fjords and fjord valleys. *Geografiska Annaler* 49A, 188–203.

Holtedahl, H. 1975. The geology of the Hardangerfjord, west Norway. *Norges Geologiske Undersoegelse Bulletin* 36, 1–87.

Holtedahl, O. 1929. On the geology and physiography of some arctic and antarctic islands, with notes on the character of fjords and strandflats of some northern islands. *Scientific Research of the Norwegian Antarctic Expedition 3*, Oslo.

Hooke, R. Le B. 1970. Morphology of the ice-sheet margin near Thule, Greenland. *Journal of Glaciology* 9, 303–324.

Hooke, R. Le B. 1973a. Structure and flow at the margin of the Barnes Ice Cap, Baffin Island, NWT, Canada. *Journal of Glaciology* 12, 423–438.

Hooke, R. Le B. 1973b. Flow near the margin of the Barnes Ice Cap, and the development of ice-cored moraines. *Geological Society of America Bulletin* 84, 3929–3948.

Hooke, R. Le B. 1984. On the role of mechanical energy in maintaining subglacial water conduits at atmospheric pressure. *Journal of Glaciology* 30, 180–187.

Hooke, R. Le B. 1991. Positive feedbacks associated with erosion of glacial cirques and overdeepenings. *Geological Society of America Bulletin* 103, 1104–1108.

Hooke, R. Le B., Brzozowski, J. and Bronge, C. 1983. Seasonal variations in surface velocity, Storglaciären, Sweden. *Geografiska Annaler* 65A, 263–277.

Hooke, R. Le B., Calla, P., Holmlund, P., Nilsson, M. and Stroeven, A. 1989. A three-year record of seasonal variations in surface velocity, Storglaciären, Sweden. *Journal of Glaciology* 35, 235–247.

Hooke, R. Le B. and Hudleston, P.J. 1978. Origin of foliation in glaciers. *Journal of Glaciology* 20, 285–299.

Hooke, R. Le B. and Iverson, N.R. 1995. Grain-size distribution in deforming subglacial tills, role of grain fracture. *Geology*, 23, 57–60.

Hooke, R. Le B., Laumann, T. and Kohler, J. 1990. Subglacial water pressures and the shape of subglacial conduits. *Journal of Glaciology* 36, 67–71.

Hooke, R. Le B., Miller, S.B. and Kohler, J. 1988. Character of the englacial and subglacial drainage system in the upper part of the ablation area of Storglaciären, Sweden. *Journal of Glaciology* 34, 228–231.

Hope, G.S., Peterson, J.A., Radok, U. and Allison, I. (eds) 1976. *The Equatorial Glaciers of New Guinea*. Balkema, Rotterdam.

Hope, R., Lister, H. and Whitehouse, R. 1972. The wear of sandstone by cold, sliding ice. In Price, R.J. and Sugden, D.E. (eds), *Polar Geomorphology*. Institute of British Geographers, Special Publication 4, 21–31.

Hopkins, O.B. 1923. Some structural features of the plains area of Alberta caused by Pleistocene glaciation. *Bulletin of the Geological Society of America* 34, 419–430.

Hopley, D. 1978. Sea level change on the Great Barrier Reef, an introduction. *Philosophical Transactions of the Royal Society of London* A291, 159–166.

Hopley, D. 1983. Deformation of the north Queensland continental shelf in the late Quaternary. In Smith, D.E. and Dawson, A.G. (eds), *Shorelines and Isostasy*. Academic Press, London, 347–366.

Hopley, D. 1987. Holocene sea-level changes in Australasia and the southern Pacific. In Devoy, R.J.N. (ed.), *Sea Surface Studies*. Croom Helm, London, 375–407.

Hoppe, G. 1948. Isrecessionen fran Norbottens Kustland. *Geographica* 20, 112.

Hoppe, G. 1951. Drumlins in nordostra Norrbotten. *Geografiska Annaler* 33, 157–165.

Hoppe, G. 1952a. Hummocky moraine regions, with special reference to the interior of Norrbotten. *Geografiska Annaler* 34, 1–72.

Hoppe, G. 1952b. Some observations on Icelandic glaciers during the summer of 1952. Ymer, Stockholm, 73e année, 241–265.

Hoppe, G. 1957. Problems of glacial morphology and the ice age. *Geografiska Annaler* 39, 1–18.

Hoppe, G. 1959. Glacial morphology and inland ice recession in northern Sweden. *Geografiska Annaler* 41, 193–212.

Hoppe, G. and Schytt, V. 1953. Some observations on fluted moraine surfaces. *Geografiska Annaler* 35, 105–115.

Hopson, P.M. 1995. Chalk rafts in Anglian till in north Hertfordshire. *Proceedings of the Geologists' Association* 106, 151–158.

Horberg, L. 1951. Intersecting minor ridges and periglacial features in the Lake Agassiz basin North Dakota. *Journal of Geology* 59, 1–18.

Horton, R.E. 1945. Erosional development of streams and their drainage basins: hydrophysical approach to quantitative morphology. *Bulletin of the Geological Society of America* 56, 275–370.

Hoskin, C.M. and Burrell, D.C. 1972. Sediment transport and accumulation in a fjord basin Glacier Bay, Alaska. *Journal of Geology* 80, 539–551.

Hoskin, C.M., Burrell, D.C. and Freitag, G.R. 1978. Suspended sediment dynamics in Blue Fjord, Western Prince William Sound, Alaska. *Estuarine and Coastal Marine Science* 7, 1–16.

Hoskin, C.M. and Valencia, S.M. 1976. Sediment transport by ice rafting in south central Alaska. In Hood, D.W. and Burrell, D.E. (eds), *Assessment of the Arctic Marine Environment, Selected Topics*. Institute of Marine Science, University of Alaska, Fairbanks, 173–185.

Houghton, J.T., Jenkins, G.J. and Ephraums, J.J. (eds) 1990. *Climate Change, The IPCC Scientific Assessment*. Cambridge University Press, Cambridge.

Houghton, J.T., Meira Filho, L.G., Callander, B.A., Harris, N., Kattenberg, A. and Maskell, K. 1996. *Climate Change 1995, The Science of Climate Change*. Cambridge University Press, Cambridge.

Houmark-Nielsen, M. 1987. Pleistocene stratigraphy and glacial history of the central part of Denmark. *Bulletin of the Geological Society of Denmark* 36, 1–189.

Houmark-Nielsen, M. 1988. Glaciotectonic unconformities in Pleistocene stratigraphy as evidence for the behaviour of former Scandinavian ice sheets. In Croot, D.G.

(ed.), *Glaciotectonics, Forms and Processes*. Balkema, Rotterdam, 91–99.

Howard, A.D. 1981. Etched plains and braided ridges of the south polar region of Mars: features produced by basal melting of ground ice? Reports of the Planetary Geology Program, NASA TM 84211, 286–288.

Howarth, P.J. and Price, R.J. 1969. The proglacial lakes of Breidamerkurjökull and Fjallsjökull, Iceland. *Geographical Journal* 135, 573–581.

Hsu, K.J. 1975. Catastrophic debris streams (sturzstroms) generated by rockfalls. *Geological Society of America Bulletin* 86, 129–140.

Huang Maohuan 1992. The movement mechanisms of Urumqi Glacier No. 1, Tien Shan mountains, China. *Annals of Glaciology* 16, 39–44.

Hubbard, B. 1991. Freezing-rate effects on the physical characteristics of basal ice formed by net adfreezing. *Journal of Glaciology* 37, 339–347.

Hubbard, B. and Sharp, M.J. 1989. Basal ice formation and deformation: a review. *Progress in Physical Geography*, 13, 529–558.

Hubbard, B. and Sharp, M.J. 1993. Weertman regelation, multiple refreezing effects and the isotopic evolution of the basal ice layer. *Journal of Glaciology* 39, 275–291.

Hubbard, B. and Sharp, M.J. 1995. Basal ice facies and their formation in the Western Alps. *Arctic and Alpine Research* 27, 301–310.

Hubbard, B., Sharp, M.J. and Lawson, W. 1996. On the sedimentological character of Alpine basal ice facies. *Annals of Glaciology* 22, 187–193.

Hubbard, B.P., Sharp, M.J., Willis, I.C., Nielsen, M.K. and Smart, C.C. 1995. Borehole water-level variations and the structure of the subglacial hydrological system of Haut Glacier d'Arolla, Valais, Switzerland. *Journal of Glaciology* 41, 572–583.

Hubbert, M.K. and Rubey, W.W. 1959. Role of fluid pressure in mechanics of overthrust faulting. *Geological Society of America Bulletin* 70, 115–166.

Huddart, D. 1983. Flow tills and ice-walled lacustrine sediments, the Petteril Valley, Cumbria, England. In Evenson, E.B., Schluchter, C. and Rabassa, J. (eds), *Tills and Related Deposits*. Balkema, Rotterdam, 81–94.

Huddart, D. 1991. The glacial history and glacial deposits of the north and west Cumbrian lowlands. In Ehlers, J., Gibbard, P.L. and Rose, J. (eds), *Glacial Deposits of Britain and Ireland*. Balkema, Rotterdam, 151–168.

Huddart, D. 1994. The late Quaternary glacigenic sequence, landforms and environments in coastal Cumbria. In Boardman, J. and Walden, J. (eds), *The Quaternary of Cumbria, Field Guide*. Quaternary Research Association, Oxford, 59–77.

Huddart, D. and Lister, H. 1981. The origin of ice marginal terraces and contact ridges of east Kangerdluarssuk Glacier, SW Greenland. *Geografiska Annaler* 63A, 31–39.

Hudleston, P.J. 1976. Recumbent folding in the base of the Barnes Ice Cap, Baffin Island, Northwest Territories, Canada. *Geological Society of America Bulletin* 87, 1684–1692.

Hughes, O.L. 1964. Surficial geology, Nichicum–Kaniapiskau map area, Quebec. *Geological Survey of Canada Bulletin* 106.

Hughes, O.L., Rutter, N.W. and Clague, J.J. 1989. Yukon Territory Quaternary geology of the Canadian Cordillera. In Fulton, R.J. (ed.), *Quaternary Geology of Canada and Greenland*. Geological Survey of Canada, Geology of Canada No. 1, 58–61.

Hughes, T.J. 1970. Convection in the Antarctic ice sheet leading to a surge of the ice sheet and possibly to a new ice age. *Science* 170, 630–633.

Hughes, T.J. 1972. Is the West Antarctic Ice Sheet disintegrating? Ice Streamline Co-operative Antarctic Project, Bulletin 1.

Hughes, T.J. 1973. Is the West Antarctic Ice Sheet disintegrating? *Journal of Geophysical Research* 78, 7884–7910.

Hughes, T.J. 1975. The west Antarctic ice sheet: instability, disintegration and initiation of ice ages. *Review of Geophysics and Space Physics* 13, 502–526.

Hughes, T.J. 1986. The marine ice transgression hypothesis. *Geografiska Annaler* 69A, 237–250.

Hughes, T.J. 1987. Ice dynamics and deglaciation models when ice sheets collapsed. In Ruddiman, W.F. and Wright, H.E. (eds), *North America and Adjacent Oceans during the Last Deglaciation*. Geological Society of America, *The Geology of North America*, Vol. K–3, 183–220.

Hughes, T.J. 1989. Calving ice walls. *Annals of Glaciology* 12, 74–80.

Hughes, T.J. 1992. Abrupt climatic change related to unstable ice-sheet dynamics: toward a new paradigm. *Palaeogeography, Palaeoclimatology, Palaeoecology* 97, 203–234.

Hughes, T.J., Denton, G.H., Andersen, G.B., Schilling, D.H., Fastook, J.L. and Lingle, C.S. 1981. The last great ice sheets: a global view. In Denton, G.H. and Hughes, T.J. (eds), *The Last Great Ice Sheets*. Wiley, New York, 263–317.

Hughes, T.J., Denton, G.H. and Fastook, J.L. 1985. The Antarctic Ice Sheet: an analog for northern hemisphere paleo-ice sheets? In Woldenberg, M.J. (ed.), *Models in Geomorphology*. Allen & Unwin, Boston, 25–72.

Hughes, T.J., Denton, G.H. and Grosswald, M.G. 1977. Was there a late-Wurm arctic ice sheet? *Nature* 266, 596–602.

Hughes, T.J. and Grosswald, M.G. 1995. Palaeoglaciology's grand unsolved problem. *Journal of Glaciology* 41, 313–332.

Hughes, T.J. and Nakagawa, M. 1989. Bending shear: the rate-controlling mechanism for calving ice walls. *Journal of Glaciology* 35, 260–266.

Hulton, N., Sugden, D.E., Payne, A.J and Clapperton, C.M. 1994. Glacier modelling and the climate of Patagonia during the last glacial maximum. *Quaternary Research* 42, 1–19.

Humlum, O. 1978. Genesis of layered lateral moraines, implications for palaeoclimatology and lichenometry. *Geografisk Tidsskrift* 77, 65–72.

Humlum, O. 1981. Observations on debris in the basal transport zone of Myrdalsjøkull, Iceland. *Annals of Glaciology* 2, 71–77.

Humlum, O. 1982. Rock glacier types on Disko, central west Greenland. *Norsk Geografisk Tidsskrift* 82, 59–66.

Humlum, O. 1985a. Changes in texture and fabric of particles in glacial traction with distance from source,

Myrdalsjøkull, Iceland. *Journal of Glaciology* 31, 150–156.

Humlum, O. 1985b. Genesis of an imbricate push moraine, Höfdabrekkujøkull, Iceland. *Journal of Geology* 93, 185–195.

Humphrey, N.F., Kamb, B., Fahnestock, M. and Engelhardt, H. 1993. Characteristics of the bed of the lower Columbia Glacier, Alaska. *Journal of Geophysical Research* 98, 837–846.

Humphrey, N.F. and Raymond, C.F. 1994. Hydrology, erosion and sediment production in a surging glacier: Variegated Glacier, Alaska, 1982–82. *Journal of Glaciology* 40, 539–552.

Humphrey, N.F., Raymond, C.F. and Harrison, N. 1986. Discharges of turbid water during mini-surges of Variegated Glacier, Alaska, USA. *Journal of Glaciology* 32, 195–207.

Hunter, L.E., Powell, R.D. and Lawson, D.E. 1996. Morainal-bank sediment budgets and their influence on the stability of tidewater termini of valley glaciers entering Glacier Bay, Alaska, USA. *Annals of Glaciology* 22, 211–216.

Hutchinson, G.E. 1957. *A Treatise on Limnology*, Vol. 1, *Geography, Physics and Chemistry*. Wiley, New York.

Hutter, K. 1983. *Theoretical Glaciology, Material Science of Ice and the Mechanics of Glaciers and Ice Sheets*. Tokyo, Reidel.

Huybrechts, P. 1993. Glaciological and climatological aspects of a stabilist versus dynamic view of the Late Cenozoic glacial history of East Antarctica. *Geografiska Annaler* 75A, 221–238.

Hyvärinen, H. and Eronen, M. 1979. The Quaternary history of the Baltic: the northern part. In Gudelis, V. and Königsson, L.-K. (eds), *The Quaternary History of the Baltic*. Almqvist & Wiksell, Uppsala, 7–27.

Ignatius, H. 1958. On the Late-Wisconsin deglaciation in eastern Canada. *Acta Geographica* 16, 1–34.

Ignatius, H., Korpela, K. and Kujansuu, R. 1980. The deglaciation of Finland after 10,000 BP. *Boreas* 9, 217–228.

Iken, A. 1977. Movement of a large ice mass before breaking off. *Journal of Glaciology* 19, 595–605.

Iken, A. 1981. The effect of subglacial water pressure on the sliding velocity of a glacier in an idealized numerical model. *Journal of Glaciology* 27, 407–421.

Iken, A. and Bindschadler, R.A. 1986. Combined measurements of subglacial water pressure and surface velocity of Findelengletscher, Switzerland, conclusions about drainage system and sliding mechanism. *Journal of Glaciology* 32, 101–119.

Iken, A., Röthlisberger, H., Flotron, A. and Haeberli, W. 1983. The uplift of Unteraargletscher at the beginning of the melt season: a consequence of water storage at the bed? *Journal of Glaciology* 23, 28–47.

Imbrie, J., Berger, A., Boyle, E.A., Clemens, S.C. *et al.* 1993. On the structure and origin of major glaciation cycles 2. The 100,000 year cycle. *Palaeoceanography* 8, 699–735.

Imbrie, J., Boyle, E.A., Clemens, S.C., Duffy, A. *et al.* 1992. On the structure and origin of major glaciation cycles 1. Linear responses to Milankovitch forcing. *Palaeoceanography* 7, 701–738.

Imbrie, J. and Imbrie, K.P. 1979. *Ice Ages: Solving the Mystery*. Macmillan, London.

Innes, M.J.S., Goodacre, A.K., Weston, A., Argun, A. and Weber, J.R. 1968. Gravity and isostasy in the Hudson Bay region. In Beals, C.S. (ed.), *Science, History and Hudson Bay*, Vol. 2, Canada Department of Energy, Mines and Resources, Ottawa, 703–728.

Inoue, J. 1977. Mass budget of Khumbu Glacier. *Seppyo* 39, special issue, 15–19.

Inoue, J. and Yoshida, M. 1980. Ablation and heat exchange over the Khumbu Glacier. *Seppyo* 42, 26–33.

Irwin, J. and Pickrill, R.A. 1982. Water temperature and turbidity in a glacial fed lake: Lake Tekapo. *New Zealand Journal of Freshwater Research* 16, 189–200.

Iverson, N. 1990. Laboratory simulations of glacial abrasion: comparison with theory. *Journal of Glaciology* 36, 304–314.

Iverson, N. 1991. Potential effects of subglacial water pressure fluctuations on quarrying. *Journal of Glaciology* 37, 27–36.

Iverson, N.R. 1993. Regelation of ice through debris at glacier beds: implications for sediment transport. *Geology* 21, 559–562.

Iverson, N.R. 1995. Processes of erosion. In Menzies, J. (ed.), *Modern Glacial Environments: Processes, Dynamics and Sediments*. Butterworth-Heinemann, Oxford, 241–260.

Iverson, N.R., Hanson, B., Hooke, R. Le B. and Jansson, P. 1995. Flow mechanism of glaciers on soft beds. *Science* 267, 80–81.

Iverson, N.R., Hooyer, T.S. and Hooke, R. LeB. 1996. A laboratory study of sediment deformation, stress heterogeneity and grain-size evolution. *Annals of Glaciology* 22, 167–175.

Ives, J.D. 1957. Glaciation of the Torngat Mountains. *Geographical Bulletin* 10, 67–87.

Ives, J.D. 1978. The maximum extent of the Laurentide Ice Sheet along the eastern coast of North America during the last glaciation. *Arctic* 31, 24–53.

Ives, J.D., Andrews, J.T. and Barry, R.G. 1975. Growth and decay of the Laurentide ice sheet and comparisons with Fenno-Scandinavia. *Naturwissenschaften* 62, 118–125.

Ives, J.D. and Kirby, R.P. 1964. Fluvioglacial erosion near Knob Lake, central Quebec–Labrador, Canada: discussion. *Geological Society of America Bulletin* 75, 917–922.

Iwata, S., Watanabe, O. and Fushimi, H. 1980. Surface morphology in the ablation area of the Khumbu Glacier. *Seppyo* 42, 9–17.

Jackson, L.E. 1979. A catastrophic glacial outburst flood (jökulhlaup) mechanism for debris flow generation at the Spiral Tunnels, Kicking Horse River basin, British Columbia. *Canadian Geotechnical Journal* 16, 806–813.

Jackson, L.E. and MacDonald, G.M. 1980. Movement of an ice-cored rock glacier, Tungsten, NWT, Canada, 1963–1980. *Arctic* 33, 842–847.

Jackson, L.E., MacDonald, G.M. and Wilson, M.C. 1982. Paraglacial origin for terraced river sediments in Bow Valley, Alberta. *Canadian Journal of Earth Sciences* 19, 2219–2231.

Jackson, R.G. 1975. Hierarchical attributes and a unifying model of bedform composed of cohesionless material and deposited by shearing flow. *Geological Society of America Bulletin* 86, 1523–1533.

Jacobs, J.J., Heron, R. and Luther, J.E. 1993. Recent changes at the northwest margin of the Barnes Ice Cap, Baffin Island, NWT, Canada. *Arctic and Alpine Research* 25, 341–352.

Jacobs, S.S., Gordon, A.L. and Ardai, J.L. 1979. Circulation and melting beneath the Ross Ice Shelf. *Science* 203, 439–443.

Jacobs, S.S., Helmer, H.H., Doake, C.S.M., Jenkins, A. and Frolich, R.M. 1992. Melting of ice shelves and the mass balance of Antarctica. *Journal of Glaciology* 38, 375–387.

Jacobs, S.S., Huppert, H.E., Holdsworth, G. and Drewry, D.J. 1981. Thermohaline steps induced by melting of the Erebus Glacier tongue. *Journal of Geophysical Research* 86, 6547–6555.

Jacobs, S.S., MacAyeal, D.R. and Ardal, J.L. 1986. The recent advance of the Ross Ice Shelf, Antarctica. *Journal of Glaciology* 32, 464–474.

Jahn, A. 1950. Nowe dane o płożeniu kry jurajskiej w Łukowie. (New facts concerning the ice-transported blocks of the Jurassic at Łuków.) *Annales Societatis Geologorum Poloniae* 19, 372–385.

Jahn, A. 1956. Wyżyna Lubelska, rzezba i czwartorzed (Geomorphology and Quaternary history of Lublin Plateau). *Polska Akademia Nauk, Instytut geografii, Prace Geograficzne* 7.

Jahns, R.H. 1941. Outwash chronology in northeastern Massachusetts (abstract). *Geological Society of America Bulletin* 52, 1910.

Jahns, R.H. 1943. Sheet structure in granites, its origin and use as a measure of glacial erosion in New England. *Journal of Geology* 51, 71–98.

Jamieson, T.F. 1865. On the history of the last geological changes in Scotland. *Quarterly Journal of the Geological Society of London* 21, 161–203.

Jamieson, T.F. 1874. On the last stage of the glacial period in north Britain. *Quarterly Journal of the Geological Society of London* 30, 317–337.

Jamieson, T.F. 1882. On the cause of the depression and re-elevation of the land during the glacial period. *Geological Magazine* 9, 400–407.

Jansen, F. 1976. Late Pleistocene and Holocene history of the northern North Sea, based on acoustic reflection records. *Netherlands Journal of Sea Research* 10, 1–43.

Jansson, P. 1995. Water pressure and basal sliding on Storglaciären, northern Sweden. *Journal of Glaciology* 41, 232–240.

Jauhiainen, E. 1975. Morphometric analysis of drumlin fields in northern central Europe. *Boreas* 4, 219–230.

Jeffries, M.O. 1984. Milne Glacier, northern Ellesmere Island, N.W.T., Canada: a surging glacier? *Journal of Glaciology* 30, 251–253.

Jeffries, M.O. 1986. Glaciers and the morphology and structure of Milne ice shelf. *Arctic and Alpine Research* 18, 397–405.

Jeffries, M.O. 1987. The growth, structure and disintegration of arctic ice shelves. *Polar Record* 23, 631–649.

Jeffries, M.O. 1992. The source and calving of ice island ARLIS-II. *Polar Record* 28, 137–144.

Jeffries, M.O., Reynolds, G.J. and Miller, J.M. 1992. First Landsat multi-spectral images of the Canadian Arctic north of 80° N. *Polar Record* 28, 1–6.

Jeffries, M.O., Sachinger, W.M., Krouse, H.R. and Serson, H.V. 1988. Water circulation and ice accretion beneath Ward Hunt Ice Shelf (northern Ellesmere Island, Canada), deduced from salinity and isotope analysis of ice cores. *Annals of Glaciology* 10, 68–72.

Jeffries, M.O. and Shaw, M.A. 1993. The drift of ice islands from the Arctic Ocean into the channels of the Canadian Arctic archipelago: the history of Hobson's Choice Ice Island. *Polar Record* 29, 305–312.

Jessen, A. 1931. Lonstrup Klint. *Danmarks Geologiske Undersogelse*, II raekke 49, 142 pp.

Jewtuchowicz, S. 1956. Structure des drumlins aux environs de Zbojno. *Acta Geographica Universitatis Lodziensis* 7, 1–74.

Johannesson, M., Dale, T., Gjessing, E.T., Henriksen, A. and Wright, R.F. 1977. Acid precipitation in Norway: the regional distribution of contaminants in snow and the chemical concentration processes during snowmelt. In *Isotopes and Impurities in Snow and Ice*. IASH Publication 118, 116–120.

Johannesson, T., Raymond, C. and Waddington, E. 1989. Timescale for adjustment of glaciers to changes in mass balance. *Journal of Glaciology* 35, 355–369.

Johansson, C.E. 1963. Orientation of pebbles in running water: a laboratory study. *Geografiska Annaler* 45A, 85–112.

Johansson, P. 1994. The subglacially engorged eskers in the Lutto River basin northeastern Finnish Lapland. In Warren, W.P. and Croot, D.G. (eds), *Formation and Deformation of Glacial Deposits*. Balkema, Rotterdam, 89–94.

John, B.S. 1972. A Late Weichselian kame terrace at Mullock Bridge, Pembrokeshire. *Proceedings of the Geologists' Association* 83, 213–229.

John, B.S. (ed.) 1979a. *The Winters of the World: Earth under the Ice Ages*. David & Charles, Newton Abbot.

John, B.S. 1979b. The great ice age: Permo-Carboniferous. In John, B.S. (ed.), *The Winters of the World: Earth under the Ice Ages*. David & Charles, Newton Abbot, 154–172.

John, B.S. and Sugden, D.E. 1962. The morphology of Kaldalon, a recently deglaciated valley in Iceland. *Geografiska Annaler* 44, 347–365.

John, B.S. and Sugden, D.E. 1971. Raised marine features and phases of glaciation in the South Shetland Islands. *British Antarctic Survey Bulletin* 24, 45–111.

John, B.S. and Sugden, D.E. 1975. Coastal geomorphology of high latitudes. *Progress in Geography* 7, 53–132.

Johnson, A.M. 1970. *Physical Processes in Geology*. Freeman, Cooper & Co., San Francisco.

Johnson, A.M. and Rodine, J.R. 1984. Debris flow. In Brunsden, D. and Prior, D.B. (eds), *Slope Instability*. Wiley, Chichester, 257–361.

Johnson, D.W. 1941. Function of meltwater in cirque formation, a criticism. *Journal of Geomorphology* 4, 253–262.

Johnson, H.D. and Baldwin, C.T. 1986. Shallow siliciclastic seas. In Reading, H.G. (ed.), *Sedimentary Environments and Facies*, 2nd edition. Blackwell, Oxford, 229–282.

Johnson, M.D. and Gillam, M.L. 1995. Composition and construction of late Pleistocene end moraines, Durango, Colorado. *Geological Society of America Bulletin* 107, 1241–1253.

Johnson, M.D., Mickelson, D.M., Clayton, L. and Attig, J.W. 1995. Composition and genesis of glacial hummocks, western Wisconsin, USA. *Boreas* 24, 97–116.

Johnson, N., Kargel, J.S., Strom, R.G. and Knight, C. 1991. Chronology of glaciation in the Hellas region of Mars. *Lunar and Planetary Science* 22, 651–652.

Johnson, P.G. 1971. Ice cored moraine formation and degradation, Donjek Glacier, Yukon Territory, Canada. *Geografiska Annaler* 53A, 198–202.

Johnson, P.G. 1972. The morphological effects of surges of the Donjek Glacier, St. Elias Mountains, Yukon. *Journal of Glaciology* 11, 227–234.

Johnson, P.G. 1975. Recent crevasse fillings at the terminus of the Donjek Glacier, St Elias Mountains, Yukon Territory. *Quaestiones Geographicae* 2, 53–59.

Johnson, P.G. 1978. Rock glacier types and their drainage systems, Grizzly Creek, Yukon Territory. *Canadian Journal of Earth Sciences* 15, 1496–1507.

Johnson, P.G. 1980a. Glacier–rock glacier transition in the southwest Yukon Territory, Canada. *Arctic and Alpine Research* 12, 195–204.

Johnson, P.G. 1980b. Rock glaciers: glacial and non-glacial origins. International Association of Scientific Hydrology, Publication 126, 285–293.

Johnson, P.G. 1984a. Paraglacial conditions of instability and mass movement: a discussion. *Zeitschrift für Geomorphologie* 28, 235–250.

Johnson, P.G. 1984b. Rock glacier formation by high-magnitude low-frequency slope processes in the southwest Yukon. *Annals of the Association of American Geographers* 74, 408–419.

Johnson, P.G. 1987. Rock glaciers, glacier debris systems or high-magnitude low-frequency flows. In Giardino, J.R., Shroder, J.F. and Vitek, J.D. (eds), *Rock Glaciers*. Allen & Unwin, London, 175–192.

Johnson, P.G. 1997. Spatial and temporal variability of ice-dammed lake sediments in alpine environments. *Quaternary Science Reviews* 16, 635–647.

Johnson, W.D. 1904. The profile of maturity in Alpine glacial erosion. *Journal of Geology* 12, 569–578.

Johnson, W.H. and Hansel, A.K. 1990. Multiple Wisconsinan glacigenic sequences at Wedron, Illinois. *Journal of Sedimentary Petrology* 60, 26–41.

Johnson, W.H. and Menzies, J. 1996. Pleistocene supraglacial and ice-marginal deposits and landforms. In Menzies, J. (ed.), *Past Glacial Environments: Sediments, Forms and Techniques*. Butterworth-Heinemann, Oxford, 137–160.

Johnsson, G. 1956. Glacialmorfologiska studier i Sodra Sverige med sarskild hansyn till glaciala riktningselement och periglaciala frostfenomen. *Meddelanden fran Lunds Geografiska Institution* 30, 1–407.

Johnstrup, F. 1874. Über die Lagerungsverhaltnisse und die Hebungsphänomene in den Kreidefelsen auf Mön und Rugen. *Zeitschrift der Deutschem Geologischen Gesellschaft* 1874, 533–585.

Jones, D.K.C., Brunsden, D. and Goudie, A.S. 1983. A preliminary geomorphological assessment of part of the Karakoram Highway. *Quarterly Journal of Engineering Geology* 16, 331–355.

Jones, J.G. 1969. Intraglacial volcanoes in the Laugarvatn region, south-west Iceland. *Quarterly Journal of the Geological Society of London* 495, 197–211.

Jones, M.E. and Preston, R.M.F. (eds) 1987. *Deformation of Sediments and Sedimentary Rocks*. Geological Society of London, Special Publication 29.

Jones, M.J. (ed.) 1973. *Prospecting in Areas of Glaciated Terrain*. Institution of Mining and Metallurgy, London.

Jones, N. 1982. The formation of glacial flutings in east-central Alberta. In Davidson-Arnott, R., Nickling, W. and Fahey, B.D. (eds), *Research in Glacial, Glacio-fluvial and Glacio-lacustrine Systems*. Geo Books, Norwich, 49–70.

Jones, R.L. and Keen, D.H. 1993. *Pleistocene Environments in the British Isles*. Chapman & Hall, London.

de Jong, J.D. 1952. On the structure of the pre-glacial Pleistocene of the Archemerberg Province of Overijsel, Netherlands. *Geologie en Mijnbouw* 14, 86.

de Jong, M.G.G. and Rappol, M. 1983. Ice-marginal debris flow deposits in western Allgau, southern West Germany. *Boreas* 12, 57–70.

Jonsson, J. 1982. Notes on the Katla volcanoglacial debris flows. *Jökull* 32, 61–68.

Jopling, A.V. and Walker, R.G. 1968. Morphology and origin of ripple-drift cross-lamination, with examples from the Pleistocene of Massachusetts. *Journal of Sedimentary Petrology* 38, 971–984.

Josenhans, H.W. and Fader, G.B.J. 1989. A comparison of models of glacial sedimentation along the eastern Canadian margin. *Marine Geology* 85, 273–300.

Josenhans, H.W. and Woodworth-Lynas, C.M.T. 1988. Enigmatic linear furrows and pits on the upper continental slope, northwest Labrador Sea: are they sediment furrows or feeding traces? *Maritime Sediments and Atlantic Geology* 24, 149–155.

Jouzel, J. and 16 others. 1993. Extending the Vostok ice core record of palaeoclimate to the penultimate glacial period. *Nature* 364, 407–412.

Jouzel, J. and Souchez, R.A. 1982. Melting–refreezing at the glacier sole and the isotopic composition of the ice. *Journal of Glaciology* 28, 35–42.

Kaldal, I. 1978. The deglaciation of the area north and northeast of Hofsjökull, central Iceland. *Jökull* 28, 18–31.

Kalin, M. 1971. The active push moraine of the Thompson Glacier, Axel Heiberg Island, Canadian Arctic Archipelago. McGill University, Axel Heiberg Island, Research Report, Glaciology 4, McGill University, Montreal.

Kalvoda, J. 1992. *Geomorphological Record of the Quaternary Orogeny in the Himalaya and the Karakoram*. Elsevier, Amsterdam.

Kamb, B. 1970. Sliding motion of glaciers: theory and observation. *Reviews of Geophysics and Space Physics* 8, 673–728.

Kamb, B. 1987. Glacier surge mechanism based on linked cavity configuration of the basal water conduit system. *Journal of Geophysical Research* 92, 9083–9100.

Kamb, B. 1991. Rheological nonlinearity and flow instability in the deforming bed mechanism of ice stream motion. *Journal of Geophysical Research* 96, 585–595.

Kamb, B. and Engelhardt, H. 1987. Waves of accelerated motion in a glacier approaching surge: the mini surges of Variegated Glacier, Alaska, USA. *Journal of Glaciology* 33, 27–46.

Kamb, B. and LaChapelle, E. 1964. Direct observation of the mechanism of glacier sliding over bedrock. *Journal of Glaciology* 5, 159–172.

Kamb, B., Engelhardt, H., Fahnestock, M.A., Humphrey, N., Meier, M. and Stone, D. 1995. Mechanical and hydrologic basis for the rapid motion of a large tidewater glacier 2. Interpretation. *Journal of Geophysical Research* 99, 15231–15244.

Kamb, B., Engelhardt, H.F. and Harrison, W.D. 1988. The ice–rock interface and basal sliding process as revealed by direct observation in bore holes and tunnels. *Journal of Glaciology* 23, 416–419.

Kamb, B., Raymond, C.F., Harrison, W.D., Engelhardt, H., Echelmeyer, K.A., Humphrey, N., Brugman, M.M. and Pfeffer, T. 1985. Glacier surge mechanism: 1982–1983 surge of Variegated Glacier, Alaska. *Science* 227, 469–479.

Karczewski, A. 1974. Structural features of kame forms as an expression of the dynamics of morphogenetic environment. *Quaestiones Geographicae* 1, 53–64.

Kargel, J.S. and Strom, R.G. 1990. Ancient glaciation on Mars. *Lunar and Planetary Science* 21, 598–599.

Kargel, J.S. and Strom, R.G. 1991. Terrestrial glacial eskers: analogs for Martian sinuous ridges. *Lunar and Planetary Science* 21, 598–599.

Kargel, J.S. and Strom, R.G. 1992a. Ancient glaciation on Mars. *Geology* 20, 3–7.

Kargel, J.S. and Strom, R.G. 1992b. The ice ages of Mars. *Astronomy* 20, 40–45.

Kargel, J.S. and Strom, R.G. 1996. Global climatic change on Mars. *Scientific American*, November, 60–68.

Kargel, J.S., Strom, R.G. and Johnson, N. 1991. Glacial geology of the Hellas region on Mars. *Lunar and Planetary Science* 22, 687–688.

Karlen, W. 1973. Holocene glacier and climate variations, Kebnekaise Mountains, Swedish Lappland. *Geografiska Annaler* 55A, 29–63.

Karrow, P.F. 1981. Till texture in drumlins. *Journal of Glaciology* 27, 497–502.

Karrow, P.F. 1984a. Comments on 'Sedimentation in a large lake: a reinterpretation of the late Pleistocene stratigraphy at Scarborough Bluffs'. *Geology* 12, 185.

Karrow, P.F. 1984b. Lithofacies types and vertical profile models: an alternative approach to the description and environmental interpretation of glacial diamict and diamictite sequences. Discussion. *Sedimentology* 31, 883–884.

Kaser, G. 1995. How do tropical glaciers behave? Some comparisons between tropical and mid-latitude glaciers. In Ribstein, P. and Francou, B. (eds), Aguas glaciares y cambios en los Andes tropicales. *Conferencias y posters, Seminaro Internacional, La Paz, 13–16 Junio 1995*, 207–218.

Kaser, G., Ames, A. and Zamora, M. 1990. Glacier fluctuations in the Cordillera Blanca, Peru. *Annals of Glaciology* 14, 136–140.

Kaufman, D.S., Miller, G.H., Stravers, J.A. and Andrews, J.T. 1993. Abrupt early Holocene (9.9–9.6 ka) ice-stream advance at the mouth of Hudson Strait, arctic Canada. *Geology* 21, 1063–1066.

Kaye, C.A. 1964a. Outline of the Pleistocene geology of Martha's Vineyard, Massachusetts. US Geological Survey, Professional Paper 501-C, 134–139.

Kaye, C.A. 1964b. Illinoian and early Wisconsin moraines of Martha's Vineyard, Massachusetts. US Geological Survey, Professional Paper 501-C, 140–143.

Kaye, C.A. 1980. Geologic profile of Gay Head Cliff, Martha's Vineyard, Massachusetts. US Geological Survey, Open File Report 80–148.

Kehew, A.E. 1982. Catastrophic flood hypothesis for the origin of the Souris spillway, Saskatchewan and North Dakota. *Geological Society of America Bulletin* 93, 1051–1058.

Kehew, A.E. and Clayton, L. 1983. Late Wisconsinan floods and development of the Souris–Pembina spillway system in Saskatchewan, North Dakota and Manitoba. In Teller, J.T. and Clayton, L. (eds), *Glacial Lake Agassiz*. Geological Association of Canada Special Paper 26, 187–209.

Kehew, A.E. and Lord, M.L. 1986. Origin and large-scale erosional features of glacial-lake spillways in the northern Great Plains. *Geological Society of America Bulletin* 97, 162–177.

Kehew, A.E. and Lord, M.L. 1987. Glacial lake outbursts along the mid-continent margins of the Laurentide ice-sheet. In Mayer, L. and Nash, D. (eds), *Catastrophic Flooding*. Allen & Unwin, Boston, 95–120.

Kehew, A.E. and Lord, M.L. 1989. Canadian landform examples 12: Glacial lake spillways of the central interior plains, Canada–USA. *Canadian Geographer* 33, 274–277.

Kehew, A. and Teller, J.T. 1994. History and late glacial runoff along the southwestern margin of the Laurentide ice sheet. *Quaternary Science Reviews* 13, 859–879.

Kellogg, T.B., Truesdale, R.S. and Osterman, L.E. 1979. Late Quaternary extent of the West Antarctic ice sheet: new evidence from Ross Sea cores. *Geology* 7, 249–253.

Kemmis, T.J. 1981. Importance of the regelation process to certain properties of basal tills deposited by the Laurentide ice sheet in Iowa and Illinois, USA. *Annals of Glaciology* 2, 147–152.

Kemmis, T.J. and Hallberg, T.J. 1984. Lithofacies types and vertical profile models: an alternative approach to the description and environmental interpretation of glacial diamict and diamictite sequences. Discussion. *Sedimentology* 31, 886–890.

Kemp, D.D. 1994. *Global Environmental Issues: A Climatological Approach*, 2nd edition. Routledge, London.

Kendall, P.F. 1902. A system of glacier lakes in the Cleveland Hills. *Quarterly Journal of the Geological Society of London* 58, 471–571.

Kennedy, J.F. 1963. The mechanics of dunes and antidunes in erodible-bed channels. *Journal of Fluid Mechanics* 16, 521–544.

Kennett, J.P. 1978. Cainozoic evolution of circumantarctic palaeoceanography. In van Zinderen Bakker, E.M. (ed.), *Antarctic Glacial History and World Palaeoenvironments*. Balkema, Rotterdam, 41–56.

Kennett, J.P. and Hodell, D.A. 1993. Evidence for relative climatic stability of Antarctica during the early Pliocene: a marine perspective. *Geografiska Annaler* 75A, 205–220.

Kent, P.E. 1966. The transport mechanism in catastrophic rock falls. *Journal of Geology* 74, 79–83.

Kerr, J.W. 1980. Structural framework of Lancaster aulacogen, arctic Canada. *Geological Survey of Canada Bulletin* 319.

Keys, J.E. 1978. Water regime of Disraeli Fiord, Ellesmere Island, Canada. Department of National Defence, Defence Research Establishment, Report 792.

Keys, J.E., Johannesson, O.M. and Long, A. 1969. The oceanography of Disraeli Fiord, northern Ellesmere Island. Canada Defence Research Board, Geophysics 34.

Kieslinger, A. 1960. Residual stress and relaxation in rocks. *International Geological Congress, Copenhagen, Session 21*, 270–276.

King, C.A.M. 1959. Geomorphology in Austerdalen, Norway. *Geographical Journal* 125, 357–369.

King, C.A.M. 1974. Morphometry in glacial geomorphology. In Coates, D.R. (ed.), *Glacial Geomorphology*, Binghamton, New York State University Press, 147–162.

King, C.A.M. and Buckley, J. 1968. The analysis of stone size and shape in arctic environments. *Journal of Sedimentary Petrology* 38, 200–214.

King, C.A.M. and Gage, M. 1961. Note on the extent of glaciation in part of west Kerry. *Irish Geographer* 4, 202–208.

King, L.H. 1980. Aspects of regional surficial geology related to site investigation requirements – eastern Canadian shelf. In Ardus, D.A. (ed.), *Offshore Site Investigation*. Graham & Trotman, London, 37–57.

King, L.H. 1993. Till in the marine environment. *Journal of Quaternary Science* 8, 347–358.

King, L.H. and Fader, G. 1986. Wisconsinan glaciation of the continental shelf – southeast Atlantic Canada. Geological Survey of Canada, Bulletin 363.

King, L.H., Rokoengen, K., Fader, G.B.J. and Gunleiksrud, T. 1991. Till tongue stratigraphy. *Bulletin of the Geological Society of America* 103, 637–659.

King, L.H., Rokoengen, K. and Gunleiksrud, T. 1987. Quaternary seismostratigraphy of the Mid-Norwegian Shelf, 65°–67°31′N – a till tongue stratigraphy. Institutt for Kontinentals Okkelndersokelser publication 114.

Kirkbride, M.P. 1993. The temporal significance of transitions from melting to calving termini at glaciers in the central Southern Alps of New Zealand. *The Holocene* 3, 232–240.

Kirkbride, M.P. 1995a. Processes of transportation. In Menzies, J. (ed.), *Glacial Environments*, Vol. 1: *Modern Glacial Environments: Processes, Dynamics and Sediments*. Butterworth-Heinemann, Oxford, 261–292.

Kirkbride, M.P. 1995b. Ice flow vectors on the debris-mantled Tasman Glacier, 1957–1986. *Geografiska Annaler* 77A, 147–157.

Kirkbride, M.P. and Matthews, D. 1997. The role of fluvial and glacial erosion in landscape evolution: the Ben Ohau Range, New Zealand. *Earth Surface Processes and Landforms* 22, 317–327.

Kirkbride, M.P. and Spedding, N. 1996. The influence of englacial drainage on sediment-transport pathways and till texture of temperate valley glaciers. *Annals of Glaciology* 22, 160–166.

Kirkbride, M.P. and Sugden, D.E. 1992. New Zealand loses its top. *Geographical Magazine*, July, 30–34.

Kirkbride, M.P. and Warren, C.R. 1997. Calving processes at a grounded ice cliff. *Annals of Glaciology*, in press.

Klassen, R.A. 1982. Glaciotectonic thrust plates, Bylot Island, District of Franklin. Geological Survey of Canada, Paper 82–1A, 369–373.

Klassen, R.A. 1993. Quaternary geology and glacial history of Bylot Island, Northwest Territories. Geological Survey of Canada, Memoir 429.

Klassen, R.A. and Fisher, D.A. 1988. Basal flow conditions at the northeastern margin of the Laurentide Ice Sheet, Lancaster Sound. *Canadian Journal of Earth Sciences* 25, 1740–1750.

Klassen, R.A. and Thompson, F.J. 1989. Ice flow history and glacial dispersal patterns, Labrador. In DiLabio, R.N.W. and Coker, W.B. (eds), *Drift Prospecting*. Geological Survey of Canada, Paper 89–20, 21–29.

Klassen, R.A. and Thompson, F.J. 1993. Glacial history, drift composition, and mineral exploration, central Labrador. Geological Survey of Canada, Bulletin 435.

Klassen, R.W. 1986. Surficial geology of north-central Manitoba. Geological Survey of Canada, Memoir 419.

Klassen, R.W. 1989. Quaternary geology of the southern Canadian interior plains. In Fulton, R.J. (ed.), *Quaternary Geology of Canada and Greenland*. Geological Survey of Canada, Geology of Canada no.1, Ottawa, 138–173.

Kleman, J. 1988. Linear till ridges in the southern Norwegian–Swedish mountains: evidence for a subglacial origin. *Geografiska Annaler* 70A, 35–45.

Kleman, J. 1990. On the use of glacial striae for reconstruction of paleo-ice sheet flow patterns, with application to the Scandinavian ice sheet. *Geografiska Annaler* 72A, 217–236.

Kleman, J. 1992. The palimpsest glacial landscape in northwestern Sweden: Late Weichselian deglaciation landforms and traces of older west-centred ice sheets. *Geografiska Annaler* 74A, 305–325.

Kleman, J. 1994. Preservation of landforms under ice sheets and ice caps. *Geomorphology* 9, 19–32.

Kleman, J. and Bergstrom, I. 1994. Glacial land forms indicative of a partly frozen bed. *Journal of Glaciology* 40, 255–264.

Kleman, J., Bergstrom, I. and Hattestrand, C. 1994. Evidence for a relict glacial landscape in Quebec–Labrador. *Palaeogeography, Palaeoclimatology, Palaeoecology* 111, 217–228.

Kleman, J., Bergstrom, I., Robertsson, A.M. and Lillieskold, M. 1992. Morphology and stratigraphy from several deglaciations in the Transtrand Mountains, western Sweden. *Journal of Quaternary Science* 7, 1–17.

Klemsdal, T. 1982. Coastal classification of the Coast of Norway. *Norsk Geografisk Tidsskrift* 36, 129–152.

Klimaszewski, M. 1964. On the effect of the preglacial relief on the course and the magnitude of glacial erosion in the Tatra Mountains. *Geographica Polonica* 2, 11–21.

Klimek, K. 1973. Geomorphological and geological analysis of the proglacial area of the Skeidararjökull. *Geographia Polonica* 26, 89–113.

Kline, S.J., Reynolds, W.C., Schraub, F.A. and Runstadler, P.W. 1967. The structure of turbulent boundary layers. *Journal of Fluid Mechanics* 30, 741–773.

Knight, P.G. 1987. Observations at the edge of the Greenland ice sheet: boundary condition implications for modellers. In Waddington, E.D. and Walder, J.S. (eds), *The Physical Basis of Ice Sheet Modelling*. IAHS Publication 170, 359–366.

Knight, P.G. 1989. Stacking of basal debris layers without bulk freezing on: isotopic evidence from west Greenland. *Journal of Glaciology* 35, 214–216.

Knight, P.G. 1994. Two-facies interpretation of the basal ice layer of the Greenland ice sheet contributes to a unified model of basal ice formation. *Geology* 22, 971–974.

Knight, P.G., Sugden, D.E. and Minty, C. 1994. Ice flow around large obstacles as indicated by basal ice exposed at the margin of the Greenland ice sheet. *Journal of Glaciology* 40, 359–367.

Knighton, D. 1972. Meandering habit of supraglacial streams. *Geological Society of America Bulletin* 83, 201–204.

Knighton, A.D. 1973. Grain-size characteristics of super-glacial dirt. *Journal of Glaciology* 12, 522–524.

Knudsen, O. 1995. Concertina eskers, Bruarjökull, Iceland: an indicator of surge-type glacier behaviour. *Quaternary Science Reviews* 14, 487–493.

Koch, E. 1924. Die prädiluviale Auflagerungsfläche unter Hamburg und Umgegend. *Mitteilungen aus dem Mineralogisch-Geologischen Staatsinstitut* 6, 31–95.

Koenig, L.S., Greenaway, K.R., Dunbar, M. and Hattersley-Smith, G. 1952. Arctic ice islands. *Arctic* 5, 67–103.

Koerner, R.M. 1970. The mass balance of the Devon Island ice cap, NWT, Canada, 1961–66. *Journal of Glaciology* 9, 325–336.

Koerner, R.M. 1977a. Devon Island Ice Cap: core stratigraphy and paleoclimate. *Science* 196, 15–18.

Koerner, R.M. 1977b. Ice thickness measurements and their implications with respect to past and present ice volumes in the Canadian high arctic ice caps. *Canadian Journal of Earth Sciences* 14, 2697–2705.

Koerner, R.M. 1979. Accumulation, ablation and oxygen isotope variations on the Queen Elizabeth Island ice caps, Canada. *Journal of Glaciology* 22, 25–41.

Koerner, R.M. 1980. Instantaneous glacierization, the rate of albedo change, and feedback effects at the beginning of an ice age. *Quaternary Research* 13, 153–159.

Koerner, R.M. 1989. Ice core evidence for extensive melting of the Greenland ice sheet in the last interglacial. *Science* 244, 964–968.

Koerner, R.M. and Taniguchi, H. 1976. Artificial radioactivity layers in the Devon Island Ice Cap, NWT. *Canadian Journal of Earth Sciences* 13, 1251–1255.

Komar, P.D. 1983. Shapes of streamlined islands on Earth and Mars: experiments and analyses of the minimum-drag form. *Geology* 11, 651–654.

Komar, P.D. 1984. The lemniscate loop: comparisons with the shapes of streamlined landforms. *Journal of Geology* 92, 133–145.

Kondo, H. and Yamada, T. 1988. Some remarks on the mass balance and the lateral-terminal fluctuations of San Rafael Glacier, the Northern Patagonia Icefield. *Bulletin of Glacier Research* 6, 55–63.

Kor, P.S.G., Shaw, J. and Sharpe, D.R. 1991. Erosion of bedrock by subglacial meltwater, Georgian Bay, Ontario: a regional view. *Canadian Journal of Earth Sciences* 28, 623–642.

Kor, P.S.G. and Teller, J.T. 1986. Canadian landform examples 1: Ouimet Canyon, Ontario – deep erosion by glacial meltwater. *Canadian Geographer* 30, 273–276.

Kostaschuk, R.A. and McCann, S.B. 1983. Observations on delta-forming processes in a fjord-head delta, British Columbia, Canada. *Sedimentary Geology* 36, 269–288.

Kostaschuk, R.A. and McCann, S.B. 1987. Subaqueous morphology and slope processes in a fjord delta, Bella Coola, British Columbia. *Canadian Journal of Earth Sciences* 24, 52–59.

Koster, E.A. 1988. Ancient and modern cold climate aeolian sand deposition: a review. *Journal of Quaternary Science* 3, 69–83.

Koster, E.A. and Dijkmans, J.W.A. 1988. Niveo-aeolian deposits and denivation, with special reference to the Great Kobuk Sand Dunes, Northwestern Alaska. *Earth Surface Processes and Landforms* 13, 153–170.

Koteff, C. 1974. The morphologic sequence concept and deglaciation of southern New England. In Coates, D.R. (ed.), *Glacial Geomorphology*, State University of New York, Binghamton, 121–144.

Koteff, C. and Pessl, F. 1981. Systematic ice retreat in New England. USGS Professional Paper 1179.

Kotlyakov, V.M. 1961. The intensity of nourishment of the Antarctic Ice Sheet. *IASH* 55, 100–110.

Kozarski, S. 1959. O genezie chodzieskiej moreny czolowej. *Badania Fizjograficzne nad Polska Zachodnia* 5, 45–69. English translation: 'On the origin of the Chodziez end moraine', in Evans, D.J.A. (ed.), *Cold Climate Landforms*. Wiley, Chichester, 293–312.

Kranck, K. 1973. Flocculation of suspended sediment in the sea. *Nature* 246, 348–350.

Kraus, H. 1973. Energy exchange at air–ice interface. IAHS Publication No. 107, 128–162.

Krause, F.F. and Oldershaw, A.E. 1979. Submarine carbonate breccia beds: a depositional model for two-layer, sediment gravity flows from the Sekwi Formation (Lower Cambrian), Mackenzie Mountains, Northwest Territories, Canada. *Canadian Journal of Earth Sciences* 16, 189–199.

Krenke, A.N. and Khodakov, V.G. 1966. Connection between the surface melting of glaciers and the air temperature. *Materialny Glyatsiologichekikh Issledovaniy, Khronika Obsuzhdeniya* 12, 153–164.

Krigstrom, A. 1962. Geomorphological studies of sandur plains and their braided rivers in Iceland. *Geografiska Annaler* 44, 328–346.

Krimmel, R.M. and Trabant, D.C. 1992. The terminus of Hubbard Glacier, Alaska. *Annals of Glaciology* 16, 151–157.

Krimmel, R.M. and Vaughn, B.H. 1987. Columbia Glacier, Alaska: changes in velocity 1977–1986. *Journal of Geophysical Research* 92, 8961–8968.

Krinsley, D.H. and Doornkamp, J.C. 1973. *Atlas of Quartz Sand Surface Textures*. Cambridge University Press, Cambridge.

Krishnappen, B.G. 1975. Dispersion of Granular Material Dumped in Deep Water. Inland Waters Directorate, Canada Centre for Inland Waters, Scientific Series 55.

Kristensen, M. 1983. Iceberg calving and deterioration in Antarctica. *Progress in Physical Geography* 7, 313–328.

Krüger, J. 1979. Structures and textures in till indicating subglacial deposition. *Boreas* 8, 323–340.

Krüger, J. 1983. Glacial morphology and deposits in Denmark. In Ehlers, J. (ed.), *Glacial Deposits in North-West Europe*. Balkema, Rotterdam, 181–191.

Krüger, J. 1984. Clasts with stoss-lee form in lodgement tills: a discussion. *Journal of Glaciology* 30, 241–243.

Krüger, J. 1985. Formation of a push moraine at the margin of Hofdabrekkujökull, south Iceland. *Geografiska Annaler* 67A, 199–212.

Krüger, J. 1987. Traek af et glaciallandskabs udvikling ved nordranden af Myrdalsjökull, Island. *Dansk Geologisk Foreningens, Arsskrift for 1986*, 49–65.

Krüger, J. 1993. Moraine-ridge formation along a stationary ice front in Iceland. *Boreas* 22, 101–109.

Krüger, J. 1994. Glacial processes, sediments, landforms, and stratigraphy in the terminus region of Myrdalsjökull, Iceland. *Folia Geographica Danica* 21, 1–233.

Krüger, J. and Thomsen, H.H. 1984. Morphology, stratigraphy and genesis of small drumlins in front of the glacier Myrdalsjökull, south Iceland. *Journal of Glaciology* 30, 94–105.

Krumbein, W.C. 1937. Sediments and exponential curves. *Journal of Geology* 45, 577–601.

Kruss, P.D. and Hastenrath, S. 1987. The role of radiation geometry in the climate response of Mount Kenya's glaciers. Part 1. Horizontal reference surfaces. *Journal of Climatology* 7, 493–505.

Kruss, P.D. and Hastenrath, S. 1990. The role of radiation geometry in the climate response of Mount Kenya's glaciers. Part 3. The latitude effect. *Journal of Climatology* 10, 321–328.

Kuhle, M. 1985. Ein subtropisches Inlandeis als Eiszeitauslöser. Sonderdruck aus der Georgia Augusta, Mai 1985, *Nachrichten aus der Universität Göttingen*, 1–17.

Kuhle, M. 1986. The upper limit of glaciation in the Himalayas. *Geojournal* 13, 331–346.

Kuhle, M. 1988a. Topography as a fundamental element of glacial systems. *Geojournal* 17, 545–568.

Kuhle, M. 1988b. Geomorphological findings on the build up of Pleistocene glaciation in southern Tibet and on the problem of inland ice: results of the Shisha Pangma and Mt Everest Expedition 1984. *Geojournal* 17, 581–595.

Kuhle, M. 1990. Ice marginal ramps and alluvial fans in semiarid mountains: convergence and difference. In Richocki, A.H. and Church, M. (eds), *Alluvial Fans: A Field Approach*. Wiley, Chichester, 55–68.

Kuhn, M. 1979. On the computation of heat transfer coefficients from energy-balance gradients on a glacier. *Journal of Glaciology* 22, 263–272.

Kuhn, M. 1981. Climate and glaciers. *Symposium on Sea Level, Ice and Climatic Change, Canberra 1979*. IAHS Publication 131, Great Yarmouth, UK, 3–20.

Kuhn, M. 1984. Mass budget imbalances as criterion for a climatic classification of glaciers. *Geografiska Annaler* 66A, 229–238.

Kuhn, M. 1989a. The role of land ice and snow in climate. *Geophysical Monograph, American Geophysical Union* 52, 17–28.

Kuhn, M. 1989b. The response of the equilibrium line altitude to climate fluctuations: theory and observations. In Oerlemans, J. (ed), *Glacier Fluctuations and Cliamte Change*. Kluwer, Dordrecht, 407–417.

Kujansuu, R. 1967. On the deglaciation of western Finnish Lapland. *Bulletin de la Commission Géologique de Finlande* 232.

Kujansuu, R. and Saarnisto, M. (eds) 1990. *Handbook of Glacial Indicator Tracing*. Balkema, Rotterdam.

Kukla, G. 1987. Loess stratigraphy in central China. *Quaternary Science Reviews* 6, 191–219.

Kulhawy, F.H. 1975. Stress deformation properties of rock and rock discontinuities. *Engineering Geology* 9,

327–350.

Kulkarni, A.V. 1992. Mass balance of Himalayan glaciers using AAR and ELA methods. *Journal of Glaciology* 38, 101–104.

Kupsch, W.O. 1962. Ice-thrust ridges in western Canada. *Journal of Geology* 70, 582–594.

Kurter, A. 1991. Glaciers of Turkey. In Williams, R.S. and Ferrigno, J.G. (eds), *Satellite Image Atlas of Glaciers of the World*. USGA Professional Paper 1386-G, 1–30.

Kurtz, D.D. and Anderson, J.B. 1979. Recognition and sedimentologic description of recent debris flow deposits from the Ross Sea and Weddell Sea, Antarctica. *Journal of Sedimentary Petrology* 49, 1159–1170.

Kuster, H. and Meyer, K.D. 1979. Glaziare Rinnen im mittleren und nordöstlichen Niedersachsen. *Eiszeitalter und Gegenwart* 29, 135–156.

Kutzbach, J.E. and Guetter, P.J. 1986. The influence of changing orbital parameters and surface boundary conditions on climate simulations for the past 18,000 years. *Journal of the Atmospheric Sciences* 43, 1726–1759.

Kutzbach, J.E., Guelter, P.J., Ruddiman, W.F. and Prell, W.L. 1989. Sensitivity of climate to late Cenozoic uplift in southern Asia and the American West: numerical experiments. *Journal of Geophysical Research* 94, 18393–18407.

LaChapelle, E. 1962. Assessing glacier mass budgets by reconnaissance aerial photography. *Journal of Glaciology* 4, 290–297.

Lagerback, R. 1988. The Veiki moraines in northern Sweden: widespread evidence of an Early Weichselian deglaciation. *Boreas* 17, 469–486.

Lagerlund, E. 1983. The Pleistocene stratigraphy of Skane, southern Sweden. In Ehlers J. (ed.), *Glacial Deposits in North-West Europe*. Balkema, Rotterdam, 155–159.

Lagoe, M.B., Eyles, C.H. and Eyles, N. 1989. Paleoenvironmental significance of foraminiferal biofacies in the glaciomarine Yakataga Formation, Middleton Island, Gulf of Alaska. *Journal of Foraminiferal Research* 19, 194–209.

Lahee, F.H. 1912. Crescentic fractures of glacial origin. *American Journal of Science* 33, 41–44.

Laine, E.P. 1980. New evidence from beneath the western North Atlantic for the depth of glacial erosion in Greenland and North America. *Quaternary Research* 14, 188–198.

Laitakari, I. and Aro, K. 1985. The effect of jointing on glacial erosion of bedrock hills in southern Finland. *Fennia* 163, 369–371.

Lake, R.A. and Walker, E.R. 1976. A Canadian Arctic fjord with some comparisons to fjords of the western Americas. *Journal of the Fisheries Research Board Canada* 33, 2272–2285.

Lamb, H.H. and Woodroffe, A. 1970. Atmospheric circulation during the last ice age. *Quaternary Research* 1: 29–58.

Lambeck, K. 1990. Glacial rebound, sea level change and mantle viscosity. *Quarterly Journal of the Royal Astronomical Society* 31, 1–30.

Lambeck, K. 1991. Glacial rebound and sea level change in the British Isles. *Terra Nova* 3, 379–389.

Lambeck, K. 1993a. Glacial rebound and sea level change: an example of a relationship between mantle and surface processes. *Tectonophysics* 223, 15–37.

Lambeck, K. 1993b. Glacial rebound of the British Isles. I: Preliminary model results. *Geophysical Journal International* 115, 941–959.

Lambeck, K. 1993c. Glacial rebound of the British Isles. II: A high resolution, high precision model. *Geophysical Journal International* 115, 960–990.

Lambeck, K. 1995. Constraints on the late Weichselian ice sheet over the Barents Sea from observations of raised shorelines. *Quaternary Science Reviews* 14, 1–16.

Lambeck, K. 1995. Late Devensian and Holocene Shorelines of the British Isles and North Sea from Models of glacio-hydro-isostatic rebound. *Journal of the Geological Society* 152, 437–448.

Lambert, A.M. 1982. Trubeströme des Rheins am Grund des Bodensees (Turbidity currents from the Rhine River on the bottom of Lake Constance). *Wasserwirtschaft* 72, 1–4.

Lambert, A.M. and Hsu, K.J. 1979. Non-annual cycles of varve-like sedimentation in Walensee, Switzerland. *Sedimentology* 26, 453–461.

Lambert, A.M., Kelts, K.R. and Marshall, N.F. 1976. Measurement of density underflows from Walensee, Switzerland. *Sedimentology* 23, 87–105.

Lamplugh, G.W. 1911. On the shelly moraine of the Sefstromglacier and other Spitsbergen phenomena illustrative of the British glacial conditions. *Proceedings of the Yorkshire Geological Society* 17, 216–241.

Landvik, J.Y. and Mangerud, J. 1985. A Pleistocene sandur in western Norway: facies relationships and sedimentological characteristics. *Boreas* 14, 161–174.

Lang, H. 1968. Relations between glacier runoff and meteorological factors observed on and outside the glacier. Union de Géodésie et Géophysique Internationale, *IAHS Assemblée générale de Berne 1967. Commission de Neiges et Glaces, Rapports et discussions*, 429–439.

Langway, C.C. 1967. Stratigraphic analysis of a deep ice core from Greenland. CRREL Research Report 77.

Larsen, E. and Holtedahl, H. 1985. The Norwegian strandflat: a reconsideration of its age and origin. *Norsk Geologisk Tidsskrift* 65, 247–254.

Larsen, E., Longva, O. and Follestad, B.A. 1991. Formation of De Geer moraines and implications for deglaciation dynamics. *Journal of Quaternary Science* 6, 263–277.

Larsen, E. and Mangerud, J. 1981. Erosion rate of a Younger Dryas cirque glacier at Krakanes, western Norway. *Annals of Glaciology* 2, 153–158.

Larsen, H.V. 1959. Runoff studies from the Mitdluagkat Gletcher in SE Greenland during the late summer 1958. *Geografisk Tidsskrift* 58, 54–65.

Laumann, T. and Reeh, N. 1993. Sensitivity to climate change of the mass balance of glaciers in southern Norway. *Journal of Glaciology*, 39, 656–665.

Lauriol, B. and Gray, J.T. 1980. Processes responsible for the concentration of boulders in the intertidal zone in Leaf Basin, Ungava. Geological Survey of Canada Paper 80–10, 281–292.

Laverdiere, C. and Bernard, C. 1970. Bibliographie annotée sur les broutures glaciaires. *Revue de Géographie de Montréal* 24, 79–89.

Laverdiere, C., Guimont, P. and Dionne, J.C. 1985. Les formes et les marques de l'érosion glaciaire du plancher rocheux: signification, terminologie, illustration. *Palaeogeography, Palaeoclimatology, Palaeoecology* 51, 365–387.

Laverdiere, C., Guimont, P. and Pharand, M. 1979. Marks and forms on glacier beds: formation and classification. *Journal of Glaciology* 23, 414–416.

Lavrushin, J.A. 1968. Features of deposition and structures of the glacial-marine deposits under conditions of a fjord coast. *Litologiya: Poloznyye Iskopayemyye* 3, 63–79.

Lavrushin, J.A. 1970. Recognition of facies and subfacies in ground moraine of continental glaciations. *Lithology of Economic Deposits* 6, 684–692.

Lawn, B.R. and Wilshaw, R. 1975. Review of indentation fracture: principles and applications. *Journal of Materials Science* 10, 1049–1081.

Lawrance, C.J. 1972. Terrain evaluation in West Malaysia: Part I: Terrain classification and survey methods. Report LR 506 Transport and Road Research Laboratory, Crowthorne, UK.

Lawrence, D.B. and Elson, J.A. 1953. Periodicity of deglaciation in North America. Part II. Late Wisconsin recession. *Geografiska Annaler* 35, 96.

Lawson, D.E. 1979a. Sedimentological analysis of the western terminus region of the Matanuska Glacier, Alaska. Cold Regions Research and Engineering Laboratory, Report 79–9, Hanover, NH.

Lawson, D.E. 1979b. A comparison of the pebble orientations in ice and deposits of the Matanuska Glacier, Alaska. *Journal of Geology* 87, 629–645.

Lawson, D.E. 1981a. Distinguishing characteristics of diamictons at the margin of the Matanuska Glacier, Alaska. *Annals of Glaciology* 2, 78–84.

Lawson, D.E. 1981b. Sedimentological characteristics and classification of depositional processes and deposits in the glacial environment. Cold Regions Research and Engineering Laboratory, Report 81–27, Hanover, NH.

Lawson, D.E. 1982. Mobilisation, movement and deposition of subaerial sediment flows, Matanuska Glacier, Alaska. *Journal of Geology* 90, 279–300.

Lawson, D.E. 1989. Glacigenic resedimentation: classification concepts and application to mass-movement processes and deposits. In Goldthwait, R.P. and Matsch, C.L. (eds), *Genetic Classification of Glacigenic Deposits*. Balkema, Rotterdam, 147–169.

Lawson, D.E. 1995. Sedimentary and hydrologic processes within modern terrestrial valley glaciers. In Menzies, J. (ed.), *Modern Glacial Environments: Processes, Dynamics and Sediments*. Butterworth-Heinemann, Oxford, 337–363.

Lawson, D.E. and Kulla, J.B. 1978. An oxygen isotope investigation of the origin of the basal zone of the Matanuska Glacier, Alaska. *Journal of Geology*, 86, 673–685.

Lawson, T.J. 1990. Former ice movement in Assynt, Sutherland, as shown by the distribution of glacial erratics. *Scottish Journal of Geology* 26, 25–32.

Lawson, T.J. 1995. Boulder trains as indicators of former ice flow in Assynt, N.W. Scotland. *Quaternary Newsletter* 75, 15–21.

Lawson, T.J. and Ballantyne, C.K. 1995. Late Devensian glaciation of Assynt and Coigach. In Lawson, T.J. (ed.), *The Quaternary of Assynt and Coigach*. Quaternary Research Association Field Guide, Cambridge, 19–34.

Lawson, W. 1996. Structural evolution of Variegated Glacier, Alaska, USA, since 1948. *Journal of Glaciology* 42, 261–270.

Lawson, W., Sharp, M. and Hambrey, M.J. 1994. The

structural geology of a surge-type glacier. *Journal of Structural Geology* 16, 1447–1462.

Laymon, C.A. 1991. Marine episodes in Hudson Strait and Hudson Bay, Canada, during the Wisconsin glaciation. *Quaternary Research* 35, 53–62.

Lee, H.A. 1959. Surficial geology of southern District of Keewatin and the Keewatin ice divide, NWT. Geological Survey of Canada Bulletin 51.

Lee, H.A. 1965. Investigation of eskers for mineral exploration. Geological Survey of Canada, Paper 65–14, 1–17.

Leeder, M.R. 1982. *Sedimentology: Process and Product.* Allen & Unwin, London.

Lefauconnier, B. and Hagen, J.O. 1990. Glaciers and climate in Svalbard, statistical analysis and reconstruction of the Brøgger Glacier mass balance for the last 77 years. *Annals of Glaciology* 14, 148–152.

Lefauconnier, B., Hagen, J.O., Pinglot, J.F. and Pourchet, M. 1994. Mass-balance estimates on the glacier complex Kongsvegen and Sveabreen, Spitsbergen, Svalbard, using radioactive layers. *Journal of Glaciology* 40, 368–376.

Leg 113 Shipboard Scientific Party. 1987. Glacial history of Antarctica. *Nature* 328, 115–116.

Leggett, J. 1990. (ed.) *Global Warming: The Greenpeace Report.* Oxford University Press, Oxford.

Lehmann, R. 1992. Arctic push moraines, a case study of the Thompson Glacier moraine, Axel Heiberg Island, NWT, Canada. *Zeitschrift für Geomorphologie*, Suppl. Bd. 86, 161–171.

Lehr, P. and Horvath, E. 1975. Glaciers of China. In Field, W.O. (ed.), *Mountain Glaciers of the Northern Hemisphere.* Cold Regions Research and Engineering Laboratory, Hanover, NH, 449–476.

Leiviska, I. 1907. Über die Oberflachenformen Mittelostbottniens. *Fennia* 25.

Lemke R.W. 1958. Narrow linear drumlins near Valva, North Dakota. *American Journal of Science* 256, 270–283.

Lemmen, D.S. 1989. The last glaciation of Marvin Peninsula, northern Ellesmere Island, high arctic, Canada. *Canadian Journal of Earth Sciences* 26, 2578–2590.

Lemmen, D.S., Aitken, A.E. and Gilbert, R. 1994a. Early Holocene deglaciation of Expedition and Strand fiords, Canadian high arctic. *Canadian Journal of Earth Sciences* 31, 943–958.

Lemmen, D.S., Duk-Rodkin, A. and Bednarski, J. 1994b. Late glacial drainage systems along the northwestern margin of the Laurentide Ice Sheet. *Quaternary Science Reviews* 13, 805–828.

Lemmen, D.S., Evans, D.J.A. and England, J. 1988. Canadian landform examples 10: Ice shelves of northern Ellesmere Island, NWT. *Canadian Geographer* 32, 363–367.

Lemmens, M., Lorrain, R. and Haren, J. 1982. Isotopic composition of ice and subglacially precipitated calcite in an alpine area. *Zeitschrift für Gletscherkunde und Glazialgeologie* 18, 151–159.

Lemmens, M.M. and Roger, M. 1978 Influence of ion exchange on dissolved load of Alpine meltwaters. *Earth Surface Processes* 3, 179–187.

Leopold, L.B. and Wolman, M.G. 1957. River channel patterns: braided, meandering and straight. US Geological Survey Professional Paper 282-B.

Lerman, A. 1979. *Geochemical Processes, Water and Sediment Environments.* Wiley, New York.

Letreguilly, A. 1988. Relation between the mass balance of western Canadian mountain glaciers and meteorological data. *Journal of Glaciology* 34, 11–18.

Letreguilly, A. and Reynaud, L. 1989. Spatial patterns of mass balance fluctuations of North American glaciers. *Journal of Glaciology* 35, 163–168.

Levson, V.M. and Rutter, N.W. 1986. A facies approach to the stratigraphic analysis of Late Wisconsinan sediments in the Portal Creek area, Jasper National Park, Alberta. *Géographie Physique et Quaternaire* 40, 129–144.

Levson, V.M. and Rutter, N.W. 1989a. A lithofacies analysis and interpretation of depositional environments of montane glacial diamictons, Jasper, Alberta, Canada. In Goldthwait, R.P. and Matsch, C.L. (eds), *Genetic Classification of Glacigenic Deposits.* Balkema, Rotterdam, 117–140.

Levson, V.M. and Rutter, N.W. 1989b. Late Quaternary stratigraphy, sedimentology, and history of the Jasper townsite area, Alberta, Canada. *Canadian Journal of Earth Sciences* 26, 1325–1342.

Lewinski, J. and Rozycki, S.Z. 1929. Dwa profile geologiczne przez Warszawe. *Sprawozdania Towarzystura Naukowego Warszawskiego, Warsaw* 22, 30–50.

Lewis, W.V. 1938. A melt-water hypothesis of cirque formation. *Geological Magazine* 75, 249–265.

Lewis, W.V. 1940a. Dirt cones on the northern margins of Vatnajökull, Iceland. *Journal of Geomorphology* 3, 16–26.

Lewis, W.V. 1940b. The function of meltwater in cirque formation. *Geographical Review* 30, 64–83.

Lewis, W.V. 1947. The formation of roches moutonnées: some comments on Dr H. Carol's article. *Journal of Glaciology* 1, 60–63.

Lewis, W.V. 1949. An esker in process of formation: Boverbreen, Jotunheimen. *Journal of Glaciology* 1, 314–319.

Lewis, W.V. 1954. Pressure release and glacial erosion. *Journal of Glaciology* 2, 417–422.

Lewis, W.V. (ed.) 1960. *Norwegian Cirque Glaciers.* Royal Geographical Society Research Series 4.

Lewkowicz, A.G. 1985. Use of an ablatometer to measure short-term ablation of exposed ground ice. *Canadian Journal of Earth Sciences* 22, 1767–1773.

Liestøl, O. 1955. Glacier-dammed lakes in Norway. *Norsk Geografisk Tidsskrift* 15, 122–149.

Liestøl, O. 1967. Storbreen glacier in Jotunheimen. *Norsk Polarinstittut Skrifter* 141.

Liestøl, O. 1976. Arsmorener foran Nathorstbreen? *Norsk Polarinstittut Årbok 1976*, 361–363.

Liestøl, O. 1977. Pingos, springs and permafrost in Spitsbergen. *Norsk Polarinstittut Årbok 1975*, 7–29.

Lile, R.C. 1978. The effect of anisotropy on the creep of polycrystalline ice. *Journal of Glaciology* 21, 475–483.

Lillieskold, M. 1990. Lithology of some Swedish eskers. University of Stockholm, Department of Quaternary Geology, Report 17.

Linden, A. 1975. Till petrographical studies in an Archaean bedrock area in southern central Sweden. *Striae* 1.

Lindmar-Bergstrom, K. 1988. Exhumed Cretaceous landforms in south Sweden. *Zeitschrift für Geomorphologie* 72, 21–40.

Lindsay, J.F. 1970. Clast fabric of till and its development. *Journal of Sedimentary Petrology* 40, 629–641.

Lindsay, J.F., Summerson, C.H. and Barrett, P.J. 1970. A long axis clast fabric comparison of the 'Squantum' tillite, Massachusetts and the Gowganda Formation, Ontario. *Journal of Sedimentary Petrology* 40, 475–479.

Lindsey, D.A. 1969. Glacial sedimentology of the Precambrian Gowganda Formation, Ontario, Canada. *Geological Society of America Bulletin* 80, 1685–1702.

Lindsey, D.A. 1971. Glacial marine sediments in the Precambrian Gowganda Formation at Whitefish Falls (Ontario, Canada). *Palaeogeography, Palaeoclimatology, Palaeoecology* 9, 7–25.

Lindstrom, E. 1988. Are roches moutonnées mainly preglacial forms? *Geografiska Annaler* 70A, 323–331.

Lingle, C.S. 1984. A numerical model of interactions between a polar ice stream and the ocean: application to Ice Stream E, West Antarctica. *Journal of Geophysical Research* 89, 3523–3549.

Lingle, C.S., Brenner, A.C., Zwally, H.J. and DiMarzio, J.P. 1991. Multiyear elevation changes near the west margin of the Greenland Ice Sheet from satellite radar altimetry. In Weller, G., Wilson, C.L. and Severin, B.A.B. (eds), *International Conference on the Role of the Polar Regions in Global Change*, Vol. 1. University of Alaska, Fairbanks, 35–42.

Lingle, C.S., Hughes, T.J. and Kollmeyer, R.C. 1981. Tidal flexure of Jacobshavns Glacier, west Greenland. *Journal of Geophysical Research* 86, 3960–3968.

Link, P.K. and Gostin, V.A. 1981. Facies and paleogeography of Sturtian glacial strata (late Precambrian), South Australia. *American Journal of Science* 281, 353–374.

Linke, G. 1983. *Geologische Übersichtskarte Raum Hamburg 1:50,000, Quartarbasis*, Blatt 1: *Morphologie*. Geologisches Landesamt, Hamburg.

Linton, D.L. 1962. Glacial erosion on soft rock outcrops in central Scotland. *Biuletyn Peryglacjalny* 11, 247–257.

Linton, D.L. 1963. The forms of glacial erosion. *Transactions of the Institute of British Geographers* 33, 1–28.

Linton, D.L. 1967. Divide elimination by glacial erosion. In Wright H.E. and Osburn, W.H. (eds), *Arctic and Alpine Environments*. Indiana University Press, 241–248.

Linton, D.L. and Moisley, H.A. 1960. The origin of Loch Lomond. *Scottish Geographical Magazine* 76, 26–37.

List, E.J. 1982. Turbulent jets and plumes. *Annual Review of Fluid Mechanics* 14, 189–212.

Lister, H., Pendlington, A. and Chorlton, J. 1968. Laboratory experiments on abrasion of sandstones by ice. *IASH* 79, 98–106.

Liverman, D.G.E., Catto, N.R. and Rutter, N.W. 1989. Laurentide glaciation in west-central Alberta: a single (Late Wisconsinan) event. *Canadian Journal of Earth Sciences* 26, 266–274.

Ljungner, E. 1930. Spaltentektonik und Morphologie der schwedischen Skaggerrack-Kuste. *Bulletin of the Geological Institute*, University of Uppsala 21, 1–478.

Ljunger, E. 1948. East-west balance of the Quaternary ice caps in Patagonia and Scandinavia. *Bulletin of the Geological Institute, University of Uppsala* 33, 12–96.

Lliboutry, L. 1953. Internal moraines and rock glaciers. *Journal of Glaciology* 2, 296.

Lliboutry, L. 1955. Origine et evolution des glaciers rocheux. *Comtes Rendus des Séances de l'Académie des Sciences (Paris)* 240, 1913–1915.

Lliboutry, L. 1958. La dynamique de la Mer de Glace et la vague de 1891–95 d'après les mesures de Joseph Vallot. International Association of Hydrological Sciences Publication 47, 125–138.

Lliboutry, L. 1964, 1965. *Traité de Glaciologie*. Masson, Paris. 2 vols.

Lliboutry, L. 1968. General theory of subglacial cavitation and sliding of temperate glaciers. *Journal of Glaciology* 7, 21–58.

Lliboutry, L. 1971. Les catastrophes glaciaires. *La Recherche* 2, 417–425.

Lliboutry, L. 1974. Multivariate statistical analysis of glacier annual balance. *Journal of Glaciology* 13, 371–392.

Lliboutry, L. 1976. Physical processes in temperate glaciers. *Journal of Glaciology* 16, 151–158.

Lliboutry, L. 1977. Glaciological problems set by the control of dangerous lakes in Cordillera Blanca, Peru. II. Movement of a covered glacier embedded within a rock glacier. *Journal of Glaciology* 18, 255–273.

Lliboutry, L. 1979. Local friction laws for glaciers: a critical review and new openings. *Journal of Glaciology* 23, 67–95.

Lliboutry, L. 1983. Modifications to the theory of intraglacial waterways for the case of subglacial ones. *Journal of Glaciology* 29, 216–226.

Lliboutry, L. 1987. Realistic, yet simple bottom boundary conditions for glaciers and ice sheets. *Journal of Geophysical Research* 92, 9101–9109.

Lliboutry, L. 1993. Internal melting and ice accretion at the bottom of temperate glaciers. *Journal of Glaciology* 39, 50–64.

Lliboutry, L., Arnao, B.M., Pautre, A. and Schneider, B. 1977. Glaciological problems set by the control of dangerous lakes in Cordillera Blanca, Peru. I. Historical failures of morainic dams, their causes and prevention. *Journal of Glaciology* 18, 239–254.

Lobanov, I.N. 1967. Structure and lithology of an esker ridge near Kuznechnaya station on the Karel'skii Isthmus. *Lithologiya i Poleznye Iskopaemye* 4, 76–84.

Loewe, F. 1971. Considerations on the origin of the Quaternary ice sheet of North America. *Arctic and Alpine Research* 3, 331–344.

Løken, O.H. and Andrews, J.T. 1966. Glaciology and chronology of fluctuations of the ice margin at the south end of the Barnes Ice Cap, Baffin Island, NWT. *Geographical Bulletin* 8, 341–359.

Løken, O.H. and Hodgson, D.A. 1971. On the submarine geomorphology along the east coast of Baffin Island. *Canadian Journal of Earth Sciences* 8, 185–195.

Long, D., Laban, C., Streif, H., Cameron, T.D.J. and Schuttenhelm, R.T.E. 1988. The sedimentary record of climatic variation in the southern North Sea. *Philosophical Transactions of the Royal Society of London* B318, 523–537.

Longva, O. and Thoresen, M.K. 1991. Iceberg scours, iceberg gravity craters and current erosion marks from a gigantic Preboreal flood in southeastern Norway. *Boreas* 20, 47–62.

Lønne, I. 1993. Physical signatures of ice advance in a Younger Dryas ice-contact delta, Troms, northern Norway: implications for glacier-terminus history. *Boreas* 22, 59–70.

Loomis, S.R. 1970. Morphology and ablation processes on glacier ice. *Proceedings of the Association of American Geographers* 2, 88–92.

Lord, M.L. and Kehew, A.E. 1987. Sedimentology and palaeohydrology of glacial-lake outbursts in southeastern Saskatchewan and northwestern North Dakota. *Geological Society of America Bulletin* 99, 663–673.

Lorenzo, J.L. 1959. *Los Glaciares de México*. Instituto de Geofísica, Cuidad Universitaria, México.

Lorius, C., Jouzel, J., Raynaud, D., Hansen, J. and Le Treut, H. 1990. The ice-core record: climate sensitivity and future greenhouse warming. *Nature* 347, 139–145.

Lotan, J.E. and Shetron, S.C. 1968. Characteristics of drumlins in Leelanau County, Michigan. *Papers of the Michigan Academy of Science and Arts Letters* 53, 79–89.

Lowe, D.R. 1976a. Grain flow and grain flow deposits. *Journal of Sedimentary Petrology* 46, 188–199.

Lowe, D.R. 1976b. Subaqueous liquefied and fluidised sediment flows and their deposits. *Sedimentology* 23, 285–308.

Lowe, D.R. 1979. Sediment gravity flows: their classification and some problems of application to natural flows and deposits. In Doyle, L.J. and Pilkey, O.H. (eds), *Geology of Continental Slopes*. SEPM Special Publication 27, 75–82.

Lowe, D.R. 1982. Sediment-gravity flows. II. Depositional models with special reference to the deposits of high-density turbidity currents. *Journal of Sedimentary Petrology* 52, 279–297.

Lowe, J.J. and Walker, M.J.C. 1984. *Reconstructing Quaternary Environments*. Longman, London.

Lowell, J.D. 1985. *Structural Styles in Petroleum Exploration*. Oil and Gas Consultants International Publication, Tulsa.

Lowell, T.V., Kite, J.S., Calkin, P.E. and Halter, E.F. 1990. Analysis of small scale erosional data and a sequence of late Pleistocene flow reversal, northern New England. *Geological Society of America Bulletin* 102, 74–85.

Luckman, B.H. 1975. Drop stones resulting from snow avalanche deposition on lake ice. *Journal of Glaciology* 14, 186–188.

Luckman, B.H. 1977. Lichenometric dating of Holocene moraines at Mount Edith Cavell, Jasper, Alberta. *Canadian Journal of Earth Sciences* 14, 1809–1822.

Luckman, B.H. 1978. Geomorphic work of snow avalanches in the Canadian Rocky Mountains. *Arctic and Alpine Research* 10, 261–276.

Luckman, B.H. 1993. Glacier fluctuations and tree-ring records for the last millennium in the Canadian Rockies. *Quaternary Science Reviews* 12, 441–450.

Luckman, B.H., Harding, K.A. and Hamilton, J.P. 1987. Recent glacier advances in the Premier Range, British Columbia. *Canadian Journal of Earth Sciences* 24, 1149–1161.

Luckman, B.H. and Osborn, G.D. 1979. Holocene glacier fluctuations in the middle Canadian Rocky Mountains. *Quaternary Research* 11, 52–77.

Lundqvist, G. 1943. Norrlands jordater. *Sverige Geologiske Undersoegelse*, Series C, 457.

Lundqvist, J. 1958. Beskrivning till jordartskarta over Varmlands lan. *Sverige Geologiske Undersoegelse*, Series Ca, 38.

Lundqvist, J. 1969. Problems of the so-called Rogen moraine. *Sverige Geologiske Undersoegelse*, Series C, 648.

Lundqvist, J. 1970. Studies of drumlin tracts in central Sweden. *Acta Geographica Lodziensia* 24, 317–326.

Lundqvist, J. 1977. Till in Sweden. *Boreas* 6, 73–85.

Lundqvist, J. 1986. Late Weichselian glaciation and deglaciation in Scandinavia. *Quaternary Science Reviews* 5, 269–292.

Lundqvist, J. 1988. Late glacial ice lobes and glacial landforms in Scandinavia. In Goldthwait, R.P. and Matsch, C.L. (eds), *Genetic Classification of Glacigenic Deposits*. Balkema, Rotterdam, 217–225.

Lundqvist, J. 1989a. Rogen (ribbed) moraine: identification and possible origin. *Sedimentary Geology* 62, 281–292.

Lundqvist, J. 1989b. Till and glacial landforms in a dry, polar region. *Zeitschrift für Geomorphologie* 33, 27–41.

Lundqvist, J. 1989c. Late glacial ice lobes and glacial landforms in Scandinavia. In Goldthwait, R.P. and Matsch, C.L. (eds), *Genetic Classification of Glacigenic Deposits*. Balkema, Rotterdam, 217–226.

Lykke-Andersen, H. 1986. On the buried Norrea Valley: a contribution to the geology of the Norrea Valley, Jylland – and other buried overdeepened valleys. *Geogskrifter* 24, 211–223.

Lyons, J.B., Ragle, R.H. and Tamburi, A.J. 1972. Growth and grounding of the Ellesmere Island ice rises. *Journal of Glaciology* 11, 43–52.

Lyons, J.B., Savin, S.M. and Tamburi, A.J. 1971. Basement ice, Ward Hunt Ice Shelf, Ellesmere Island, Canada. *Journal of Glaciology* 10, 93–100.

Maag, H. 1969. Ice dammed lakes and marginal glacial drainage on Axel Heiberg Island. Axel Heiberg Island Research Report, McGill University, Montreal.

Maarleveld, G.C. 1953. Standen van het landijs in Nederland. *Boor en Spade* 4, 95–105.

Mabbutt, J.A. 1968. Review of concepts of land classification. In Stewart, G.A. (ed.), *Land Evaluation*. Macmillan, Melbourne, 11–28.

MacAyeal, D.R. 1993a. A low-order model of the Heinrich Event cycle. *Palaeoceanography* 8, 767–773.

MacAyeal, D.R. 1993b. Binge/purge oscillations of the Laurentide ice sheet as a cause of the North Atlantic's Heinrich Events. *Palaeoceanography* 8, 775–784.

McCabe, A.M. 1985. Geomorphology. In Edwards, K.J. and Warren, W.P. (eds), *The Quaternary History of Ireland*. Academic Press, London, 67–93.

McCabe, A.M. 1986. Glaciomarine facies deposited by retreating tidewater glaciers: an example from the late Pleistocene of Northern Ireland. *Journal of Sedimentary Petrology* 56, 880–894.

McCabe, A.M. 1996. Dating and rhythmicity from the last deglacial cycle in the British Isles. *Journal of the Geological Society, London* 153, 499–502.

McCabe, A.M. and Dardis, G.F. 1989. Sedimentology and depositional setting of late Pleistocene drumlins, Galway Bay, western Ireland. *Journal of Sedimentary Petrology* 59, 944–959.

McCabe, A.M. and Dardis, G.F. 1994. Glaciotectonically induced water-throughflow structures in a Late Pleistocene drumlin, Kanrawer, County Galway, western Ireland. *Sedimentary Geology* 91, 173–190.

McCabe, A.M., Dardis, G.F. and Hanvey, P.M. 1984. Sedimentology of a Late Pleistocene submarine-moraine complex, County Down, Northern Ireland. *Journal of Sedimentary Petrology* 54, 716–730.

McCabe, A.M., Dardis, G.F. and Hanvey, P.M. 1987. Sedimentation at the margins of a late Pleistocene ice lobe terminating in shallow marine environments, Dundalk Bay, eastern Ireland. *Sedimentology* 34, 473–493.

McCabe, A.M. and Haynes, J.R. 1996. A late Pleistocene intertidal boulder pavement from an isostatically emergent coast, Dundalk Bay, eastern Ireland. *Earth Surface Processes and Landforms* 21, 555–572.

McCabe, G.J. and Fountain, A.G. 1995. Relations between atmospheric circulation and mass balance of South Cascade Glacier, Washington, USA. *Arctic and Alpine Research* 27, 226–233.

McCabe, L.H. 1939. Nivation and corrie erosion in west Spitsbergen. *Geographical Journal* 94, 447–465.

McCall, J.G. 1960. The flow characteristics of a cirque glacier and their effect on glacial structure and cirque formation. In W.V. Lewis (ed.), *Investigations on Norwegian Cirque Glaciers*. Royal Geographical Society Research Series 4, London, 39–62.

McCarroll, D. and Harris, C. 1992. The glacigenic deposits of western Lleyn, north Wales: terrestrial or marine? *Journal of Quaternary Science* 7, 19–29.

McCarroll, D., Matthews, J.A. and Shakesby, R.A. 1989. 'Striations' produced by catastrophic subglacial drainage of a glacier-dammed lake, Mjolkedalsbreen, southern Norway. *Journal of Glaciology* 35, 193–196.

MacClintock, P. 1953. Crescentic crack, crescentic gouge, friction crack, and glacier movement. *Journal of Geology* 61, 186.

MacClintock, P. and Dreimanis, A. 1964. Reorientation of till fabric by over-riding glacier in the St Lawrence Valley. *American Journal of Science* 262, 133–142.

McCulloch, J. 1819. *A Description of the Western Islands of Scotland, including the Isle of Man, Comprising an Account of their Geological Structure, with Remarks on their Agriculture, Scenery, and Antiques*. London, 3 vols.

McDonald, B.C. and Bannerjee, I. 1971. Sediments and bedforms on a braided outwash plain. *Canadian Journal of Earth Sciences* 8, 1282–1301.

McDonald, B.C. and Shilts, W.W. 1975. Interpretation of faults in glaciofluvial sediments. In Jopling, A.V. and McDonald, B.C. (eds), *Glaciofluvial and Glaciolacustrine Sedimentation*. SEPM Special Publication 23, 123–131.

MacDonald, G.M., Beukens, R.P., Keiser, W.E. and Vitt, D.H. 1987. Comparative radiocarbon dating of terrestrial plant macrofossils and aquatic moss from the 'ice-free corridor' of western Canada. *Geology* 15, 837–840.

McDougall, D.A. 1995. The identification of plateau glaciers in the geomorphological record: a case study from the Lake District, northwest England. In McLelland, S.J., Skellern, A.R. and Porter, P.R. (eds), *Postgraduate Research in Geomorphology: Selected Papers from the 17th BGRG Postgraduate Symposium*. BGRG, Leeds, 1–8.

McGreevy, J.P. 1981. Perspectives on frost shattering. *Progress in Physical Geography* 8, 413–425.

McIntyre, N.F. 1985. The dynamics of ice sheet outlets. *Journal of Glaciology* 31, 99–107.

McKay, D.S., Gibson, E.F., Thomas-Kaprta, K.L. *et al.* 1996. Search for past life on Mars: possible relic biogenic activity in Martian meteorite ALH84001. *Science* 273, 924–930.

Mackay, J.R. 1959. Glacier ice-thrust features of the Yukon coast. *Geographical Bulletin* 13, 5–21.

Mackay, J.R. 1960. Crevasse fillings and ablation slide moraines, Stopover Lake area, NWT. *Geographical Bulletin* 14, 89–99.

Mackay, J.R. 1972. The world of underground ice. *Annals of the Association of American Geographers* 62, 1–22.

Mackay, J.R. 1973. Problems in the origins of massive icy beds, western arctic, Canada. In *Permafrost: The North American Contribution to the Second International Conference, Washington DC*. National Academy of Sciences, 223–228.

Mackay, J.R., Konischev, V.N. and Popov, A.I. 1979. Geological controls on the origin, characteristics and distribution of ground ice. In *Proceedings, Third International Conference on Permafrost*, vol. 2. National Research Council of Canada, Ottawa, 1–18.

Mackay, J.R. and Mathews, W.H. 1964. The role of permafrost in ice thrusting. *Journal of Geology* 72, 378–380.

Mackay, J.R., Rampton, V.N. and Fyles, J.G. 1972. Relic Pleistocene permafrost, western Arctic, Canada. *Science* 176, 1321–1323.

Mackay, J.R. and Stager, J.K. 1966. Thick tilted beds of segregated ice, Mackenzie delta area, NWT. *Biuletyn Peryglacjalny* 15, 39–43.

McKee, E.D. and Weir, G.W. 1953. Terminology for stratification and cross-stratification in sedimentary rocks. *Geological Society of America Bulletin* 64, 381–390.

McKenna-Neuman, C. and Gilbert, R. 1986. Aeolian processes and landforms in glaciofluvial environments of southeastern Baffin Island, NWT, Canada. In Nickling, W.H. (ed.), *Aeolian Geomorphology*. Allen & Unwin, Boston, 213–235.

Mckenzie, G.D. 1969. Observations on a collapsing kame terrace in Glacier Bay National Monument, south-eastern Alaska. *Journal of Glaciology* 8, 413–425.

Mackiewicz, N.E., Powell, R.D., Carlson, P.R. and Molnia, B.F. 1984. Interlaminated ice-proximal glacimarine sediments in Muir Inlet, Alaska. *Marine Geology* 57, 113–147.

McManus, J. and Duck, R.W. 1988. Localised enhanced sedimentation from icebergs in a proglacial lake in Briksdal, Norway. *Geografiska Annaler* 70A, 215–223.

McManus, J.F., Bond, G.C., Broeker, W.S., Johnsen, S., Labeyrie, L. and Higgins, S. 1994. High resolution climatic records from the North Atlantic during the last interglacial. *Nature* 371, 326–329.

Mader, H. 1992. Observations of the water-vein system in polycrystalline ice. *Journal of Glaciology* 38, 333–347.

Madgett, P.A. and Catt, J.A. 1978. Petrography, stratigraphy and weathering of late Pleistocene tills in east Yorkshire, Lincolnshire and north Norfolk. *Proceedings of the Yorkshire Geological Society* 42, 55–108.

Madsen, V. 1916. Ristinge Klint. *Danmarks Geologiske Undersogelse*, IV raekke 1.

Madsen, V. 1921. Terrainformerne pa Skovbjerg Bakkeo. *Danmark geologische Undersoekelse* 4.

Mahaney, W.C. 1990. Glacially crushed quartz grains in late Quaternary deposits in the Virunga Mountains,

Rwanda: indicators of wind transport from the north? *Boreas* 19, 81–89.

Mahaney, W.C. 1995. Glacial crushing, weathering and diagenetic histories of quartz grains inferred from scanning electron microscopy. In Menzies, J. (ed.), *Modern Glacial Environments: Processes, Dynamics and Sediments*. Butterworth-Heinemann, Oxford, 487–506.

Mahaney, W.C., Vortisch, W.B. and Julig, P. 1988. Relative differences between glacially crushed quartz transported by mountain and continental ice: some examples from North America and East Africa. *American Journal of Science* 288, 810–826.

Mair, R. and Kuhn, M. 1994. Temperature and movement measurements at a bergschrund. *Journal of Glaciology* 40, 561–565.

Maizels, J.K. 1977. Experiments on the origin of kettle holes. *Journal of Glaciology* 18, 291–303.

Maizels, J.K. 1983. Palaeovelocity and palaeodischarge determination for coarse gravel deposits. In Gregory, K.J. (ed.), *Background to Palaeohydrology*. Wiley, Chichester, 101–140.

Maizels, J.K. 1989a. Sedimentology and palaeohydrology of Holocene flood deposits in front of a jökulhlaup glacier, south Iceland. In Bevan, K. and Carling, P.J. (eds), *Floods: The Geomorphological, Hydrological and Sedimentological Consequences*. Wiley, New York, 239–251.

Maizels, J.K. 1989b. Sedimentology, palaeoflow dynamics and flood history of jökulhlaup deposits: palaeohydrology of Holocene sediment sequences in southern Iceland sandur deposits. *Journal of Sedimentary Petrology* 59, 204–223.

Maizels, J.K. 1991. The origin and evolution of Holocene sandur deposits in areas of jökulhlaup drainage, south Iceland. In Maizels, J.K. and Caseldine, C. (eds), *Environmental Change in Iceland: Past and Present*. Kluwer, Dordrecht, 267–302.

Maizels, J.K. 1992. Boulder ring structures produced during jokulhlaup flows: origin and hydraulic significance. *Geografiska Annaler* 74A, 21–33.

Maizels, J.K. 1993. Lithofacies variations within sandur deposits: the role of runoff regime, flow dynamics and sediment supply characteristics. *Sedimentary Geology* 85, 299–325.

Maizels, J.K. 1995. Sediments and landforms of modern proglacial terrestrial environments. In Menzies, J. (ed.), *Modern Glacial Environments*. Butterworth-Heinemann, Oxford, 365–416.

Manabe, S. and Broccoli, A.J. 1984a. Influence of the CLIMAP ice sheet on the climate of a general circulation model: implications for the Milankovitch theory. In Berger, A., Imbrie, J., Hays, J., Kukla, G. and Saltzman, B. (eds), *Milankovitch and Climate*. D. Reidel, Bingham, Mass., 789–799.

Manabe, S. and Broccoli, A.J. 1984b. Ice age climate and continental ice sheets: some experiments with a general circulation model. *Annals of Glaciology* 5, 100–105.

Manabe, S. and Broccoli, A.J. 1985. The influence of continental ice sheets on the climate of an ice age. *Journal of Geophysical Research* 90, 2167–2190.

Mandryk, G.B. and Rutter, N.W. 1990. Discussion: Late Pleistocene subaerial debris flow sediments near Banff, Canada. *Sedimentology* 37, 541–544.

Mangerud, J. 1991a. The last interglacial/glacial cycle in northern Europe. In Shane, L.K.C. and Cushing, E.J. (eds), *Quaternary Landscapes*. University of Minnesota Press, Minneapolis, 38–75.

Mangerud, J. 1991b. The Scandinavian Ice Sheet through the last interglacial/glacial cycle. In Frenzel, B. (ed.), *Klimageschichtliche Probleme der letzten 130,000 Jahre*. Fischer, New York, 307–330.

Mangerud, J., Larsen, E., Longva, O. and Sonstegaard, E. 1979. Glacial history of western Norway 15,000–10,000 BP. *Boreas* 8, 179–187.

Manley, G. 1955. On the occurrence of ice domes and permanently snow-covered summits. *Journal of Glaciology* 2, 453–456.

Manley, G. 1959. The late-glacial climate of north-west England. *Liverpool and Manchester Geological Journal* 2, 188–215.

Mann, D.H. 1986. Reliability of a fjord glacier's fluctuations for palaeoclimatic reconstructions. *Quaternary Research* 25, 10–24.

Mannerfelt, C.M. 1945. Nagra glacialmorfologiska formelement. *Geografiska Annaler* 27, 1–239.

Mannerfelt, C.M. 1949. Marginal drainage channels as indicators of the gradients of Quaternary ice caps. *Geografiska Annaler* 31, 194–199.

Mannerfelt, C.M. 1960. Oviksfjallen: a key glaciomorphological region. *Ymer* 80, 102–113.

Marangunic, C. and Bull, C. 1968. The landslide on the Sherman glacier. In *The Great Alaska Earthquake of 1964*, vol. 3: *Hydrology*. National Acad. Sci. Publication 1603, Washington DC, 383–394.

Marchant, D.R., Denton, G.H., Bockheim, J.G., Wilson, S.C. and Kerr, A.R. 1994. Quaternary changes in level of the upper Taylor Glacier, Antarctica: implications for paleoclimate and East Antarctic Ice Sheet dynamics. *Boreas* 23, 29–43.

Marchant, D.R., Denton, G.H., Sugden, D.E. and Swisher, C.C. 1993a. Miocene glacial stratigraphy and landscape evolution of the western Asgard Range, Antarctica. *Geografiska Annaler* 75A, 303–330.

Marchant, D.R., Denton, G.H. and Swisher, C.C. 1993b. Miocene–Pliocene–Pleistocene glacial history of Arena Valley, Quatermain Mountains, Antarctica. *Geografiska Annaler* 75A, 269–302.

Marcus, M.G. 1960. Periodic drainage of glacier-dammed Tulsequah Lake, British Columbia. *Geographical Review* 50, 89–106.

Marcus, M.G., Chambers, F.B. and Brazel, A.J. 1992. Climate and glacier change. *Zeitschrift für Geomorphologie* 86, 1–14.

Marcus, M.G., Moore, R.D. and Owens, I.F. 1985. Short-term estimates of surface energy transfers and ablation on the Lower Franz Josef Glacier, South Westland, New Zealand. *New Zealand Journal of Geology and Geophysics* 28, 559–567.

Marcussen, I. 1973. Stones in Danish tills as a stratigraphical tool: a review. *Bulletin of the Geological Institutions of Uppsala*, NS 5, 177–181.

Marcussen, I. 1977. Deglaciation landscapes formed during the wasting of the late Middle Weichselian ice sheet in Denmark. *Geological Survey of Denmark*, II Series, 110.

Marienfeld, P. 1992. Recent sedimentary processes in Scoresby Sund, east Greenland. *Boreas* 21, 169–186.

Markgren, M. and Lassila, M. 1980. Problems of moraine morphology: Rogen moraine and Blattnick moraine. *Boreas* 9, 271–274.

Marlowe, J.I. 1968. Unconsolidated marine sediments in Baffin Bay. *Journal of Sedimentary Petrology* 38, 1065–1078.

Marr, J.E. 1926. The Kailpot, Ullswater. *Geological Magazine* 63, 338–341.

Marshall, S.J., Clarke, G.K.C., Dyke, A.S. and Fisher, D.A. 1996. Geologic and topographic controls on fast flow in the Laurentide and Cordilleran Ice Sheets. *Journal of Geophysical Research* 101, B8, 17827–17839.

Martin, H.E. and Whalley, W.B. 1987a. Rock glaciers. Part I: Rock glacier morphology, classification and distribution. *Progress in Physical Geography* 2, 260–282.

Martin, H.E. and Whalley, W.B. 1987b. A glacier ice-covered rock glacier in Trollaskagi, northern Iceland. *Jökull* 37, 49–55.

Martin, S. and Kauffman, P. 1977. An experimental and theoretical study of the turbulent and laminar convection generated under a horizontal ice sheet floating on warm salty water. *Journal of Physical Oceanography* 7, 272–283.

Martini, I.P. 1990. Pleistocene glacial fan deltas in southern Ontario, Canada. In Colella, A. and Prior, D. (eds), *Coarse Grained Deltas*. International Association of Sedimentologists Special Publication 10, 281–295.

Martini, I.P. and Brookfield, M.E. 1995. Sequence analysis of upper Pleistocene (Wisconsinan) glaciolacustrine deposits of the north-shore bluffs of Lake Ontario, Canada. *Journal of Sedimentary Research* B65, 388–400.

de Martonne, E. 1957. *Traité de Géographie Physique*. Armand Colin, Paris.

Mason, K. 1929. Indus floods and Shyok glaciers. *Himalayan Journal* 1, 10–29.

Mason, K. 1930a. The Shyok flood 1929: III – The Shyok flood: a commentary. *Himalayan Journal* 2, 40–47.

Mason, K. 1930b. The glaciers of the Karakoram and neighbourhood. *Records of the Geological Survey of India* 63, 214–279.

Mason, K. 1935. The study of threatening glaciers. *Geographical Journal* 85, 24–41.

Massari, F. 1984. Resedimented conglomerates of a Miocene fan-delta complex, southern Alps, Italy. In Koster, E.H. and Steel, R.J. (eds), *Sedimentology of Gravels and Conglomerates*. Canadian Society of Petroleum Geologists Memoir 10, 259–278.

Mastalerz, K. 1990. Diurnally and seasonally controlled sedimentation on a glaciolacustrine foreset slope: an example from the Pleistocene of eastern Poland. In Colella, A. and Prior, D. (eds), *Coarse Grained Deltas*. International Association of Sedimentologists Special Publication 10, 297–309.

Mather, K.B. and Miller, G.S. 1967. Notes on topographic factors affecting the surface wind in Antarctica, with special reference to katabatic winds. University of Alaska Technical Report UAG-K-189.

Mathews, W.H. 1947. Tuyas. Flat-topped volcanoes in northern British Columbia. *American Journal of Science* 245, 560–570.

Mathews, W.H. 1973. Record of two jökulhlaups. *International Symposium on the Hydrology of Glaciers*, IASH Publication, Cambridge, 95, 99–110.

Mathews, W.H. 1974. Surface profiles of the Laurentide ice sheet in its marginal areas. *Journal of Glaciology* 13, 37–43.

Mathews, W.W. and Mackay, J.R. 1960. Deformation of soils by glacier ice and the influence of pore pressures and permafrost. *Transactions of the Royal Society of Canada*, Section 3, 54, 27–36.

Matsch, C.L. and Ojakangas, R.W. 1991. Comparisons in depositional style of 'polar' and 'temperate' glacial ice; Late Paleozoic Whiteout Conglomerate (West Antarctica) and late Proterozoic Mineral Fork Formation (Utah). In Anderson, J.B. and Ashley, G.M. (eds), *Glacial Marine Sedimentation: Paleoclimatic Significance*. Geological Society of America, Special Paper 261, 191–206.

Matthes, F.E. 1900. Glacial sculpture of the Bighorn Mountains, Wyoming. *USGS 21st Annual Report*, 167–190.

Matthes, F.E. 1930. Geologic history of the Yosemite Valley. USGS Professional Paper 160, 137 pp.

Matthews, J.A., Cornish, R. and Shakesby, R.A. 1979. 'Saw-tooth' moraines in front of Bodalsbreen, southern Norway. *Journal of Glaciology* 22, 535–546.

Matthews, J.A., Dawson, A.G. and Shakesby, R.A. 1986. Lake shoreline development, frost weathering and rock platform erosion in an alpine periglacial environment, Jotunheimen, southern Norway. *Boreas* 15, 33–50.

Matthews, J.A., McCarroll, D. and Shakesby, R.A. 1995. Contemporary terminal-moraine ridge formation at a temperate glacier: Styggedalsbreen, Jotunheimen, southern Norway. *Boreas* 24, 129–139.

Matthews, J.A. and Petch, J.R. 1982. Within-valley asymmetry and related problems of Neoglacial lateral moraine development at certain Jotunheimen glaciers, southern Norway. *Boreas* 11, 225–247.

Matthews, J.A. and Shakesby, R.A. 1984. The status of the 'Little Ice Age' in southern Norway: relative age dating of Neoglacial moraines with Schmidt hammer and lichenometry. *Boreas* 13, 333–346.

Matthews, J.B. and Quinlan, A.V. 1975. Seasonal characteristics of water masses in Muir Inlet, a fjord with tidewater glaciers. *Journal of the Fisheries Research Board Canada* 32, 1693–1703.

Mattsson, A. 1960. Sprickfyllnader och hallskulptur. Nagra iakttagelser fran Aland, Bla Jungfrun och Bornholm. *Svensk Geografisk Årsbok* 36, 85–105.

Mattsson, A. 1962. Morphologische Studien in Sudschweden und auf Bornholm über die nichtglaziale Formenwelt der Felsenkulptur. *Lund Studies in Geography, Physical Geography* 20.

Mawdsley, J.B. 1936. The wash-board moraines of the Opawica–Chibougamau area, Quebec. *Transactions of the Royal Society of Canada*, Series 3, 30, 9–12.

May, R.W. 1977. Facies model for sedimentation in the glaciolacustrine environment. *Boreas* 6, 175–180.

Mayewski, P.A. and Hassinger, J. 1980. Characteristics and significance of rock glaciers in southern Victoria Land, Antarctica. *Antarctic Journal of the United States* 15, 68–69.

Mayewski, P.A. and Jeschke, P.A. 1979. Himalayan and Trans-Himalayan glacier fluctuations since AD 1812. *Arctic and Alpine Research* 11, 267–287.

Mayewski, P.A., Pregent, G.P., Jeschke, P.A. and Ahmad, N. 1980. Himalayan and Trans-Himalayan glacier fluctuations and the south Asian monsoon record. *Arctic and Alpine Research* 12, 171–182.

Mayo, L.R. 1984. Glacier mass balance and runoff research in the USA. *Geografiska Annaler* 66A, 215–227.

Mayo, L.R. 1988. Advance of Hubbard Glacier and closure of Russell Fiord, Alaska: environmental effects and hazards in the Yakutat area. United States Geological Survey Circular 1016.

van der Meer, J.J.M. (ed.) 1987. *Tills and Glaciotectonics*. Balkema, Rotterdam.

van der Meer, J. 1993. Microscopic evidence of subglacial deformation. *Quaternary Science Reviews* 12, 553–587.

Meier, M.F. 1951. Recent eskers in the Wind River mountains of Wyoming. *Iowa Academy of Science* 58, 291–294.

Meier, M.F. 1960. Distribution and variations of glaciers in the United States exclusive of Alaska. *IASH Publication* 54, 420–429.

Meier, M.F. 1961. Mass budget of South Cascade Glacier, 1957–1960. US Geological Survey Professional Paper 424-B, 206–211.

Meier, M.F. 1965. Glaciers and climate. In Wright, H.E. and Frey, D.G. (eds), *The Quaternary of the United States*. Princeton University Press, Princeton, NJ, 795–805.

Meier, M.F. 1974. Flow of Blue Glacier, Olympic Mountains, Washington, USA. *Journal of Glaciology* 13, 187–212.

Meier, M.F., Lundstrom, S., Stone, D. *et al.* 1995. Mechanical and hydrologic basis for the rapid motion of a large tidewater glacier 1. Observations. *Journal of Geophysical Research* 99, 15219–15229.

Meier, M.F. and Post, A.S. 1962. Recent variations in mass net budgets of glaciers in western North America. *IASH* 58, 63–77.

Meier, M.F. and Post, A.S. 1969. What are glacier surges? *Canadian Journal of Earth Sciences* 6, 807–819.

Meier, M.F. and Post, A.S. 1987. Fast tidewater glaciers. *Journal of Geophysical Research* 92, 9051–9058.

Meier, M.F., Rasmussen, L.A., Krimmel, R.M., Olsen, R.W. and Franj, D. 1985. Photogrammetric determination of surface altitude, terminus position and ice velocity of Columbia Glacier, Alaska. US Geological Survey Professional Paper 1258-F.

Meier, M.F. and Tangborn, W.V. 1965. Net budget and flow of South Cascade Glacier, Washington. *Journal of Glaciology* 5, 547–566.

Meier, S. 1983. Portrait of an Antarctic outlet glacier. *Hydrological Sciences Journal* 28, 403–416.

Meierding, T.C. 1982. Late Pleistocene glacial equilibrium-line in the Colorado Front Range: a comparison of methods.*Quaternary Research* 18, 289–310.

Mellor, M. 1967. Antarctic ice budget. In Fairbridge, R.W. (ed.), *The Encyclopedia of Atmospheric Sciences and Astrogeology*. Reinhold, New York, 16–19.

Menzies, J. 1979. A review of the literature on the formation and location of drumlins. *Earth Science Reviews* 14, 315–359.

Menzies, J. 1981. Freezing fronts and their possible influence upon processes of subglacial erosion and deposition. *Annals of Glaciology* 2, 52–56.

Menzies, J. 1984. *Drumlins: A Bibliography*. Geo Books, Norwich.

Menzies, J. 1987. Towards a general hypothesis on the formation of drumlins. In Menzies, J. and Rose, J. (eds), *Drumlin Symposium*. Balkema, Rotterdam, 9–24.

Menzies, J. 1995. Glaciers and ice sheets. In Menzies, J. (ed.), *Modern Glacial Environments*. Butterworth-Heinemann, Oxford, 101–138.

Menzies, J. and Maltman, A.J. 1992. Microstructures in diamictons: evidence of subglacial bed conditions. *Geomorphology* 6, 27–40.

Menzies, J. and Rose, J. 1987. Drumlins: trends and perspectives. *Episodes* 10, 29–31.

Menzies, J. and Rose, J. 1989. Subglacial bedforms: an introduction. *Sedimentary Geology* 62, 117–122.

Mercer, J.H. 1961. The response of fjord glaciers to changes in the firn limit. *Journal of Glaciology* 10, 850–858.

Mercer, J.H. 1965. Glacier variations in southern Patagonia. *Geographical Review* 55, 390–413.

Mercer, J.H. 1968. Variations of some Patagonian glaciers since the late-glacial. *American Journal of Science* 266, 91–109.

Mercer, J.H. 1970. Variations of some Patagonian glaciers since the late-glacial, II. *American Journal of Science* 269, 1–25.

Mercer, J.H. 1972. Some observations on the glacial geology of the Beardmore Glacier area. In Adie, R.J. (ed.), *Antarctic Geology and Geophysics*. International Union of Geological Sciences, Series B, 1, 427–433.

Mercer, J.H. 1976. Glacial history of southernmost South America. *Quaternary Research* 6, 125–166.

Mercer, J.H. 1978. West Antarctic Ice Sheet and CO_2 greenhouse effect: a threat of disaster? *Nature* 271, 321–325.

Mercer, J.H. and Palacios, O. 1977. Radiocarbon dating of the last glaciation in Peru. *Geology* 5, 600–604.

Merrill, F.J.H. 1886a. On the geology of Long Island. *Annals of the New York Academy of Sciences* 3, 341–364.

Merrill, F.J.H. 1886b. On some dynamic effects of the ice sheet. *Proceedings of the American Association for the Advancement of Science* 35, 228–229.

Messerli, B., Messerli, P., Pfister, C. and Zumbuhl, H.J. 1978. Fluctuations of climate and glaciers in the Bernese Oberland, Switzerland, and their geoecological significance, 1600 to 1975. *Arctic and Alpine Research* 10, 247–260.

Messerli, B. and Zurbuchen, M. 1968. Blockgletscher in Weissmies und Aletsch und ihre photogrammetrische Kartierung. *Die Alpen* 44, 139–152.

Metcalf, R.C. 1986. The cationic denudation rate of an alpine glacier catchment: Gornergletscher, Switzerland. *Zeitschrift für Gletscherkunde und Glazialgeologie* 22, 19–32.

Metzger, S.M. 1991. A survey of esker morphologies, the connection to New York State glaciation and criteria for subglacial meltwater channel deposits on the planet Mars. *Lunar and Planetary Science* 22, 891–892.

Miall, A.D. 1977. A review of the braided river depositional environment. *Earth Science Reviews* 13, 1–62.

Miall, A.D. 1978. Lithofacies types and vertical profile models in braided river deposits: a summary. In Miall, A.D. (ed.), *Fluvial Sedimentology*. Canadian Society of Petroleum Geologists Memoir 5, 597–604.

Miall, A.D. 1983a. Glaciomarine sedimentation in the Gowganda Formation (Huronian), northern Ontario. *Journal of Sedimentary Petrology* 53, 477–491.

Miall, A.D. 1983b. Glaciofluvial transport and deposition. In Eyles, N. (ed.), *Glacial Geology*. Pergamon, New York, 168–183.

Miall, A.D. 1984. Deltas. In Walker, R.G. (ed.), *Facies Models*. Geoscience Canada, Reprint Series 1, 105–118.

Miall, A.D. 1985a. Architectural-element analysis: a new method of facies analysis applied to fluvial deposits. *Earth Science Reviews* 22, 261–308.

Miall, A.D. 1985b. Sedimentation on an early Proterozoic continental margin; the Gowganda Formation (Huronian), Elliot Lake area, Ontario, Canada. *Sedimentology* 32, 763–788.

Miall, A.D. 1991. Hierarchies of architectural units in clastic rocks, and their relationship to sedimentation rate. In Miall, A.D. and Tyler, N. (eds), *The Three-dimensional Facies Architecture of Terrigenous Clastic Sediments, and its Implications for Hydrocarbon Discovery and Recovery*. SEPM, Concepts in Sedimentology and Palaeontology 3, 6–12.

Miall, A.D. 1992. Alluvial deposits. In Walker, R.G. and James, N.P. (eds), *Facies Models: Response to Sea-Level Change*. Geological Association of Canada, Toronto, 119–142.

Mickelson, D.M. 1973. Nature and rate of basal till deposition in a stagnating ice mass, Burroughs Glacier, Alaska. *Arctic and Alpine Research* 5, 17–27.

Mickelson, D.M. 1986. Observed processes of glacial deposition in Glacier Bay. In Anderson, P.J., Goldthwait, R.P. and McKenzie, G.D. (eds), *Landform and Till Genesis in the Eastern Burroughs Glacier-Plateau Remnant Area, Glacier Bay, Alaska*. Institute of Polar Studies Miscellaneous Publication 236, Ohio State University, 47–61.

Mickelson, D.M. and Berkson, J.M. 1974. Till ridges presently forming above and below sea level in Wachusett Inlet, Glacier Bay, Alaska. *Geografiska Annaler* 56A, 111–119.

Mickelson, D.M., Ham, N.R. and Ronnert, L. 1992. Comment on 'Striated clast pavements: products of deforming subglacial sediment?'. *Geology* 20, 285.

Middleton, G.V. 1966a. Experiments on density and turbidity currents 1: Motion of the head. *Canadian Journal of Earth Sciences* 3, 523–546.

Middleton, G.V. 1966b. Experiments on density and turbidity currents 2: Uniform flow of density currents. *Canadian Journal of Earth Sciences* 3, 627–637.

Middleton, G.V. 1966c. Experiments on density and turbidity currents 3: Deposition of sediment. *Canadian Journal of Earth Sciences* 4, 475–505.

Middleton, G.V. 1970. Experimental studies related to problems of flysch sedimentation. In Lajoie, J. (ed.), *Flysch Sedimentology in North America*. Geological Association of Canada Special Paper 7, 253–272.

Middleton, G.V. and Hampton, M. 1973. Sediment gravity flows: mechanics of flow and deposition. In Middleton, G.V. and Bouma, A.H. (eds), *Turbidites and Deep Water Sedimentation*. SEPM Short Course Notes, 1–38.

Miller, G.H. 1973. Late Quaternary glacial and climatic history of northern Cumberland Peninsula, Baffin Island, NWT, Canada. *Quaternary Research* 3, 561–583.

Miller, G.H. 1976. Anomalous local glacier activity, Baffin Island, Canada: paleoclimatic implications. *Geology* 4, 502–504.

Miller, G.H., Bradley, R.S. and Andrews, J.T. 1975. The glaciation level and lowest equilibrium line altitude in the high Canadian arctic: maps and climatic interpretation. *Arctic and Alpine Research* 7, 155–168.

Miller, G.H. and Dyke, A.S. 1974. Proposed extent of Late Wisconsin Laurentide ice on eastern Baffin Island. *Geology* 2, 125–130.

Miller, G.H. and Kaufman, D.S. 1990. Rapid fluctuations of the Laurentide Ice Sheet at the mouth of Hudson Strait: new evidence for ocean/ice-sheet interactions as a control on the Younger Dryas. *Paleoceanography* 5, 907–919.

Miller, J.W. 1972. Variations in New York drumlins. *Annals of the American Association of Geographers* 62, 418–423.

Miller, M.M. 1952. Preliminary notes concerning certain glacier structures and glacial lakes on the Juneau Ice Field. American Geographical Society, JIRP Report 6, 49–86.

Mills, H.C. and Wells, P.D. 1974. Ice-shove deformation and glacial stratigraphy of Port Washington, Long Island, New York. *Geological Society of America Bulletin* 85, 357–364.

Mills, H.H. 1980. An analysis of drumlin forms in the northeastern and north-central United States. *Geological Society of America Bulletin* 91, 2214–2289.

Mills, H.H. 1987. Morphometry of drumlins in the northeastern and north-central USA. In Menzies, J. and Rose, J. (eds), *Drumlin Symposium*. Balkema, Rotterdam, 131–147.

Mills, H.H. 1991. Three-dimensional clast orientation in glacial and mass-movement sediments. US Geological Survey, Open File Report 90–128.

Milthers, V. 1948. Det danske Istidslandskabs Terraenformer og deres Opstaaen. *Danmark Geologiske Undersoekelse* 28. Reitzel, Copenhagen.

Mitchell, C.W. 1973 *Terrain Evaluation*. Longman, London.

Mitchell, W.A. 1991. *Western Pennines: Field Guide*. Quaternary Research Association, Cambridge.

Mitchell, W.A. 1994. Drumlins in ice sheet reconstructions, with reference to the western Pennines, northern England. *Sedimentary Geology* 91, 313–331.

Mix, A.C. 1992. The marine oxygen isotope record: constraints on timing and extent of ice-growth events (120–65 ka). In Clark, P.U. and Lea, P.D. (eds), *The Last Interglacial–Glacial Transition in North America*. Geological Society of America, Special Paper 270, 19–30.

Mode, W.M., Nelson, A.R. and Brigham, J.K. 1983. A facies model of Quaternary glacial marine cyclic sedimentation along eastern Baffin Island, Canada. In Molnia, B.F. (ed.), *Glacial-Marine Sedimentation*. Plenum Press, New York, 495–533.

Mojski, J.E. 1982. Outline of the Pleistocene stratigraphy in Poland. *Biuletyn Instytutu Geologicznego* 343, 9–29.

Mollard, J.D. 1983. The origin of reticulate and orbiculate patterns on the floor of the Lake Agassiz basin. In Teller, J.T. and Clayton, L. (eds), *Glacial Lake Agassiz*. Geological Association of Canada Special Paper 26, 355–374.

Mollard, J.D. 1984. Extraordinary landscape patterns: a quest for their origin. *Canadian Journal of Remote Sensing* 10, 121–134.

Mollard, J.D. and Liang, T. 1951. The interpretation of glaciated landscapes from airphotos. Office of Naval Research, NR 257 001, Landform Series 5, Technical Report 3, Cornell University.

Moller, H. 1962. Annuella och interannuella andmoraner. *Geoliska Foreningens i Stockholm. Forhandlingar* 84, 134–143.

Molnar, P. and England, P. 1990. Late Cenozoic uplift of mountain ranges and global climate change: chicken or egg? *Nature* 346, 29–34.

Molnia, B.F. 1983a. Distal glacial marine sedimentation: abundance, composition and distribution of North Atlantic Ocean Pleistocene ice rafted sediment. In Molnia, B.F. (ed.), *Glacial Marine Sedimentation*. Plenum Press, New York, 593–628.

Molnia, B.F. 1983b. Subarctic glacial-marine sedimentation: a model. In Molnia, B.F. (ed.), *Glacial-Marine Sedimentation*. Plenum Press, New York, 95–144.

Molnia, B.F. and Carlson, P.R. 1980. Quaternary sedimentary facies on the continental shelf of the northeast Gulf of Alaska. In Colburn, I., Field, M., Ingle, J. and Douglas, R. (eds), *Quaternary Depositional Environments of the Pacific Coast*. SEPM Symposium volume, 157–168.

Molnia, B.F., Carlson, P.R. and Bruns, T.R. 1977. Large submarine slide in Kayak Trough, Gulf of Alaska. *Geological Society of America Review of Engineering Geology* 3, 137–148.

Moncrieff, A.C.M. and Hambrey, M.J. 1990. Marginal-marine glacial sedimentation in the late Precambrian succession of east Greenland. In Dowdeswell, J.A. and Scourse, J.D. (eds), *Glacimarine Environments: Processes and Sediments*. Geological Society Special Publication 53, 387–410.

Mooers, H.D. 1989a. On the formation of the tunnel valleys of the Superior Lobe, Central Minnesota. *Quaternary Research* 32, 24–35.

Mooers, H.D. 1989b. Drumlin formation: a time transgressive model. *Boreas* 18, 99–107.

Mooers, H.D. 1990a. Ice-marginal thrusting of drift and bedrock: thermal regime, subglacial aquifers, and glacial surges. *Canadian Journal of Earth Sciences* 27, 849–862.

Mooers, H.D. 1990b. A glacial-process model: the role of spatial and temporal variations in glacier thermal regime. *Geological Society of America Bulletin* 102, 243–251.

Moran, S.R. 1971. Glaciotectonic structures in drift. In Goldthwait, R.P. (ed.), *Till: A Symposium*. Ohio State University Press, Columbus, 127–148.

Moran, S.R., Clayton, L., Hooke, R.L., Fenton, M.M. and Andriashek, L.D. 1980. Glacier-bed landforms of the prairie region of North America. *Journal of Glaciology* 25, 457–476.

Morawski, W. 1985. Pleistocene glaciogenic sediments of the water-morainic facies. *Quaternary Studies Poland* 6, 99–115.

Morgan, V.I. 1972. Oxygen isotope evidence for bottom freezing on the Amery Ice Shelf. *Nature* 238, 393–394.

Morgan, V.I. and Budd, W.F. 1975. Radio echo sounding of the Lambert Glacier basin. *Journal of Glaciology* 15, 103–111.

Morgenstern, N.R. 1967. Submarine slumping and the initiation of turbidity currents. In *Marine Geotechnique*. University of Illinois Press, Urbana, 189–230.

Morland, L.W. and Boulton, G.S. 1975. Stress in an elastic hump: the effects of glacier flow over elastic bedrock. *Proceedings of the Royal Society of London Series A*, 344, 157–173.

Morland, L.W. and Morris, E.M. 1977. Stress fields in an elastic bedrock hump due to glacier flow. *Journal of Glaciology* 18, 67–75.

Mörner, N.A. 1976. Eustasy and geoid changes. *Journal of Geology* 84, 123–152.

Mörner, N.A. 1977. Eustasy and instability of the geoid configuration. *Geologiska Foreningens i Stockholm Forhandlingar* 99, 369–376.

Mörner, N.A. 1979. The Fennoscandian uplift and late Cenozoic geodynamics: geological evidence. *Geojournal* 3, 287–318.

Mörner, N.A. 1980a. The Fennoscandian uplift: geological data and their geodynamical implication. In Mörner, N.A. (ed.), *Earth Rheology, Isostasy and Eustasy*. Wiley, Chichester, 251–284.

Mörner, N.A. (ed.) 1980b. *Earth Rheology, Isostasy and Eustasy*. Wiley, Chichester.

Mörner, N.A. 1980c. Eustasy and geoid changes as a function of core/mantle changes. In Mörner, N.A. (ed.), *Earth Rheology, Isostasy and Eustasy*. Wiley, Chichester, 535–553.

Mörner, N.A. 1987. Models of global sea level changes. In Tooley, M.J. and Shennan, I. (eds), *Sea Level Changes*. Blackwell, Oxford, 332–355.

Morris, E.M. and Morland, L.W. 1976. A theoretical analysis of the formation of glacial flutes. *Journal of Glaciology* 17, 311–324.

Mosley, M.P. 1983. The response of braided rivers to changing discharge, New Zealand. *Journal of Hydrology* 22, 18–67.

Mosley, M.P. 1988. Bedload transport and sediment yield in the Onyx River, Antarctica. *Earth Surface Processes and Landforms* 13, 51–67.

Mott, R.J., Grant, D.R., Stea, R.R. and Occhietti, S. 1986. Late-glacial climatic oscillation in Atlantic Canada equivalent to the Allerod–Younger Dryas event. *Nature* 323, 247–250.

Mottershead, D.N. 1975. Observation of a temporary ice-dammed lake, Brimkjelen, southern Norway. *Norsk Geografisk Tidsskrift* 29, 69–74.

Mottershead, D.N. and Collin, R.L. 1976. A study of glacier-dammed lakes over 75 years: Brimkjelen, southern Norway. *Journal of Glaciology* 17, 491–505.

Mountjoy, E.W. 1958. Jasper area, Alberta, a source of the Foothills Erratics Train. *Journal of the Alberta Society of Petroleum Geology* 6, 218–226.

Muhs, D.R. 1992. The last interglacial–glacial transition in North America: evidence from uranium-series dating of coastal deposits. In Clark, P.U. and Lea, P.D. (eds), *The Last Interglacial–Glacial Transition in North America*. Geological Society of America, Special Paper 270, 31–51.

Muller, E. 1983. Dewatering during lodgement of till. In Evenson, E.B., Schluster, C. and Rabassa, J. (eds), *Tills and Related Deposits*. Balkema, Rotterdam, 13–18.

Müller, F. 1962. Zonation in the accumulation area of the glaciers of Axel Heiberg Island, NWT, Canada. *Journal of Glaciology* 4, 302–313.

Müller, F. 1969. Was the Good Friday Glacier on Axel Heiberg Island surging? *Canadian Journal of Earth Sciences* 6, 891–894.

Müller, F. 1980. Present and late Pleistocene equilibrium line altitudes in the Mt Everest region: an application of the glacier inventory. IAHS Publication No. 126, 75–94.

Müller, F. and Iken, A. 1973. Velocity fluctuations and water regime of Arctic valley glaciers. International Association of Hydrological Sciences Publication 95, 165–182.

Müller, F. and Keeler, C.M. 1969. Errors in short term ablation measurements on melting ice surfaces. *Journal of Glaciology* 8, 91–105.

Mullins, H.T. and Hinchey, E.J. 1989. Origin of New York Finger Lakes: a historical perspective on the ice erosion debate. *Northeastern Geology* 11, 166–181.

Mulugeta, G. and Koyi, H. 1987. Three-dimensional geometry and kinematics of experimental piggyback thrusting. *Geology* 15, 1052–1056.

Munro, D.S. 1990. Comparison of melt energy computations and ablatometer measurements on melting ice and snow. *Arctic and Alpine Research* 22, 153–162.

Murray, T. and Dowdeswell, J.A. 1992. Water throughflow and the physical effects of deformation of sedimentary glacier beds. *Journal of Geophysical Research* 97, 8993–9002.

Nakamura, T. and Jones, S.J. 1973. Mechanical properties of impure ice crystals. In Whalley, E., Jones, S.J. and Gold, L.W. (eds), *Physics and Chemistry of Ice*. Royal Society of Canada, Ottawa, 365–369.

Nakawo, M. and Young, G.J. 1981. Field experiments to determine the effect of a debris layer on ablation of glacier ice. *Annals of Glaciology* 2, 85–91.

Nakawo, M. and Young, G.J. 1982. Estimate of glacier ablation under debris layer from surface temperature and meteorological variables. *Journal of Glaciology* 28, 29–34.

Nansen, F. 1904. *The Norwegian North Polar Expedition 1893–1896*. Longman, Green & Co., London.

Nansen, F. 1921. The strandflat and isostasy. *Det Norske Videnskaps-Akademi i Oslo: Mathematics and Natural Science* 11.

Nansen, F. 1922. The strandflat and isostasy. *Videnskabs-selskobets i Kristiania Skrifter I. Mathematisk Naturridenskobelig Klasse II*, Oslo.

Naruse, R., Fukami, H. and Aniya, M. 1992. Short-term variations in flow velocity of Glaciar Soler, Patagonia, Chile. *Journal of Glaciology* 38, 152–156.

Neef, E. 1970. *Das Gesicht der Erde*. Harri Deutsch Verlag, Frankfurt.

Nelson, A.R. 1981. Quaternary glacial and marine stratigraphy of the Qivitu Peninsula, northern Cumberland Peninsula, Baffin Island, Canada: summary. *Bulletin of the Geological Society of America* 92, 512–518.

Nemec, W. 1990. Aspects of sediment movement on steep delta slopes. In Colella, A. and Prior, D. (eds), *Coarse Grained Deltas*. International Association of Sedimentologists Special Publication 10, 29–73.

Nemec, W. and Steel, R.J. 1984. Alluvial and coastal conglomerates: their significant features and some comments on gravelly mass-flow deposits. In Koster, E.H. and Steel, R.J. (eds), *Sedimentology of Gravels and Conglomerates*. Canadian Society of Petroleum Geologists Memoir 10, 1–31.

Nemec, W. and Steel, R.J. (eds) 1988. *Fan Deltas: Sedimentology and Tectonic Settings*. Blackie, London.

Nemec, W., Steel, R.J., Gjelberg, J., Collinson, J.D., Prestholm, E. and Oxnevad, I.E. 1988. Anatomy of collapsed and re-established delta front in Lower Cretaceous of eastern Spitsbergen: gravitational sliding and sedimentation processes. *Bulletin of the American Association of Petroleum Geologists* 72, 454–476.

Nenonen, J. 1994. The Kaituri drumlin and drumlin stratigraphy in the Kangasniemi area, Finland. *Sedimentary Geology* 91, 365–372.

Nesje, A. 1989. Glacier-front variations of outlet galciers from Jostedalsbreen and climate in the Jostedalsbre region of western Norway in the period 1901–80. *Norsk Geografisk Tidsskrift* 43, 3–17.

Nesje, A. 1992. Topographical effects on the equilibrium line altitude on glaciers. *Geojournal* 27, 383–391.

Nesje, A. and Dahl, S.O. 1990. Autochthonous block fields in southern Norway: implications for the geometry, thickness, and isostatic loading of the Late Weichselian Scandinavian ice sheet. *Journal of Quaternary Science* 5, 225–234.

Nesje, A., Dahl, S.O., Anda, E. and Rye, N. 1988. Block fields in southern Norway: significance for the Late Weichselian ice sheet. *Geologisk Tidsskrift* 68, 149–169.

Nesje, A., Dahl, S.O., Valen, V. and Ovstedal, J. 1992. Quaternary erosion in the Sognefjord drainage basin, western Norway. *Geomorphology* 5, 511–520.

Nesje, A. and Kvamme, M. 1991. Holocene glacier and climate variations in western Norway: evidence for early Holocene glacier demise and multiple Neoglacial events. *Geology* 19, 610–612.

Nesje, A. and Sejrup, H.P. 1988. Late Weichselian/Devensian ice sheets in the North Sea and adjacent land areas. *Boreas* 17, 371–384.

Nesje, A. and Whillans, I.M. 1994. Erosion of Sognefjord, Norway. *Geomorphology* 9, 33–45.

Newman, W.A. and Mickelson, D.M. 1994. Genesis of the Boston Harbor drumlins, Masachusetts. *Sedimentary Geology* 91, 333–343.

New Zealand Geological Survey, 1973. Quaternary Geology of New Zealand. Miscellaneous Series Map 6. 1:1,000,000. Department of Scientific and Industrial Research, Wellington.

Nichols, R.L. 1965. Antarctic interglacial features. *Journal of Glaciology* 5, 433–449.

Nichols, R.L. 1971. Glacial geology of the Wright Valley, McMurdo Sound. In Quam, L.O. (ed.), *Research in the Antarctic*. American Association for the Advancement of Science, 293–339.

Nichols, R.L. and Miller, M.M. 1952. The Moreno Glacier, Lago Argentino, Patagonia. *Journal of Glaciology* 2, 41–50.

Nicholson, R. 1963. A note of the relation of rock fracture and fjord direction. *Geografiska Annaler* 45, 303–304.

Nickling, W.G. and Bennett, L. 1984. The shear strength characteristics of frozen coarse granular debris. *Journal of Glaciology* 30, 348–357.

Nielsen, D.N. 1970. Washboard moraines in northeastern North Dakota. *Compass* 47, 154–162.

Nikiforoff, C.C. 1952. Origin of microrelief in the Lake Agassiz basin. *Journal of Geology* 60, 99–103.

Nikkarinen, M., Kallio, E., Lestinen, P. and Ayras, M. 1984. Mode of occurrence of Cu and Zn in till over three mineralized areas in Finland. In Bjorklund, A.J. (ed.), Geochemical exploration – 1983. *Journal of Geochemical Exploration* 21, 239–247.

Nobles, L.H. and Weertman, J. 1971. Influence of irregularities of the bed of an ice sheet on deposition rate of till. In Goldthwait, R.P. (ed.), *Till: A Symposium*. Ohio State University Press, Columbus, 117–126.

Nordkalott Project 1986a. Map of Quaternary Geology, sheet 2: Glacial Geomorphology, Northern Fennoscandia, 1:1,000,000. Geological Surveys of Finland, Norway and Sweden.

Nordkalott Project 1986b. Map of Quaternary Geology, sheet 5: Ice Flow Directions, Northern Fennoscandia, 1:1,000,000. Geological Surveys of Finland, Norway and Sweden.

Norman, G.W.H. 1938. The last Pleistocene ice-front in Chibougamau District, Quebec. *Transactions of the Royal Society of Canada*, Series 3, Section IV, 32, 69–86.

Nougier, J. 1972. Aspects de morpho-tectonique glaciaire aux Îles Kerguelen. *Revue de Géographie Physique et Géologie Dynamique* 14, 499–505.

Nye, J.F. 1951. The flow of glaciers and ice sheets as a problem in plasticity. *Proceedings of the Royal Society of London, Series A* 207, 554–572.

Nye, J.F. 1952. A method of calculating the thickness of ice sheets. *Nature* 169, 529–530.

Nye, J.F. 1957. The distribution of stress and velocity in glaciers and ice sheets. *Proceedings of the Royal Society of London, Series A* 239, 113–133.

Nye, J.F. 1958. Surges in glaciers. *Nature* 181, 1450–1451.

Nye, J.F. 1960. The response of glaciers and ice sheets to seasonal and climatic changes. *Proceedings of the Royal Society of London, Series A* 256, 559–584.

Nye, J.F. 1963. The response of a glacier to changes in the rate of nourishment and wastage. *Proceedings of the Royal Society of London, Series A* 275, 87–112.

Nye, J.F. 1965a. Stability of a circular cylindrical hole in a glacier. *Journal of Glaciology* 5, 505–507.

Nye, J.F. 1965b. The frequency response of glaciers. *Journal of Glaciology* 5, 567–587.

Nye, J.F. 1967. Plasticity solution for a glacier snout. *Journal of Glaciology* 6, 695–715.

Nye, J.F. 1969a. A calculation on the sliding of ice over a wavy surface using a Newtonian viscous approximation. *Proceedings of the Royal Society of London, Series A* 311, 445–467.

Nye, J.F. 1969b. The effect of longitudinal stress on the shear stress at the base of an ice sheet. *Journal of Glaciology* 8, 207–213.

Nye, J.F. 1970. Glacier sliding without cavitation in a linear viscous approximation. *Proceedings of the Royal Society of London, Series A* 315, 381–403.

Nye, J.F. 1973. Water at the bed of a glacier. *Symposium on the Hydrology of Glaciers*, IASH Publication, Cambridge, 95, 189–194.

Nye, J.F. 1976. Water flow in glaciers; jökulhlaups, tunnels and veins. *Journal of Glaciology* 17, 181–207.

Nye, J.F. 1989. The geometry of water veins and nodes in polycrystalline ice. *Journal of Glaciology* 35, 17–22.

Nye, J.F. and Frank, F.C. 1973. Hydrology of the intergranular veins in a temperate glacier. In *Symposium on the Hydrology of Glaciers, Cambridge, 7–13 September 1969*, IASH Publication 95, 157–161.

Nye, J.F. and Martin, P.C.S. 1968. Glacial erosion. IASH Publication 79, 78–86.

Nystuen, J.P. 1976. Facies and sedimentation of the Late Precambrian Moelv Tillite in the eastern part of the Sparagmite region, southern Norway. *Norges Geologiske Undersoekelse* 329, 1–70.

Ó Cofaigh, C. 1996. Tunnel valley genesis. *Progress in Physical Geography* 20, 1–19.

O'Connor, J.E. and Baker, V.R. 1992. Magnitudes and implications of peak discharges from glacial Lake Missoula. *Geological Society of America Bulletin* 104, 267–279.

Odynsky, W., Wynnyk, A. and Newton, J.D. 1952. Soil survey of the High Prairie and McLennan sheets. Alberta Soil Survey Report 17.

Oerlemans, J. 1982. Glacial cycles and ice sheet modelling. *Climatic Change* 4, 353–374.

Oerlemans, J. 1989. On the response of valley glaciers to climatic change. In Oerlemans, J. (ed.), *Glacier Fluctuations and Climate Change*. Kluwer, Dordrecht, 353–371.

Oerlemans, J. and van der Veen, C.J. 1984. *Ice Sheets and Climate*. D. Reidel, Dordrecht.

Oestreich, K. 1906. Die Taler des nordwestlichen Himalaya. *Erganzungscheft zu Petermanns Mitteilungen* 155, Justus Perthes, Gotha.

Ohmura, A. 1982. Climate and energy balance on the arctic tundra. *Journal of Climatology* 2, 65–84.

Ohmura, A., Kasser, P. and Funk, M. 1992. Climate at the equilibrium line of glaciers. *Journal of Glaciology* 38, 397–411.

Ohmura, A. and Reeh, N. 1991. New precipitation and accumulation maps for Greenland. *Journal of Glaciology* 37, 140–148.

Okko, V. 1950. Friction cracks in Finland. *Geological Survey of Finland, Bulletin* 150, 45–50.

Okko, V. 1955. Glacial drift in Iceland: its origin and morphology. *Bulletin de la Commission Géologique de Finlande* 170, 1–133.

Okko, V. 1962. On the development of the first Salpausselkä, west of Lahti. *Bulletin de la Commission Géologique de Finlande* 202, 1–162.

Oldale, R.N. 1982. Pleistocene stratigraphy of Nantuket, Martha's Vineyard, the Elizabeth Islands, and Cape Cod, Massachusetts. In Larson, G.J. and Stone, B.D. (eds), *Late Wisconsinan Glaciation of New England*. Kendall/Hunt, Dubuque, Iowa, 1–34.

Oldale, R.N. and O'Hara, C.J. 1984. Glaciotectonic origin of the Massachusetts coastal end moraines and a fluctuating late Wisconsinan ice margin. *Bulletin of Geological Society of America* 95, 61–74.

Oldfield, F. 1991. Environmental magnetism: a personal perspective. *Quaternary Science Reviews* 10, 73–85.

Oldfield, F., Maher, B.A., Donaghue, J. and Pierce, J. 1985. Particle-size related mineral magnetic source–sediment linkages in the Rhode River Catchment, Maryland, USA. *Journal of the Geological Society of London* 142, 1035–1046.

Oliver, J. 1986. Fluids expelled tectonically from orogenic belts: their role in hydrocarbon migration and other geologic phenomena. *Geology* 14, 99–102.

Ollier, C.D. 1977. Terrain classification, principles and applications. In Hails, J.R. (ed.), *Applied Geomorphology*. Elsevier, Amsterdam, 277–316.

Ommanney, C.S.L. 1969. A study in glacier inventory: the ice masses of Axel Heiberg Island, Canadian Arctic Archipelago. Axel Heiberg Research Report, Glaciology 3, McGill University.

Orheim, O. 1970. Glaciological investigations of Store Supphellebre, west Norway. Norsk Polarinstitutt Skrifter 151.

Orheim, O. 1980. Physical characteristics and life expectancy of tabular Antarctic icebergs. *Annals of Glaciology* 1, 11–18.

Orheim, O. 1985. Iceberg discharge and the mass balance of Antarctica. In *Glaciers, Ice Sheets and Sea Level: Effect of a CO_2 Induced Climatic Change*. US Department of Energy, Washington DC, Publication DOE/ER/60235–1, 210–215.

Orheim, O. and Elverhoi, A. 1981. Model for submarine glacial deposition. *Annals of Glaciology* 2, 123–128.

Orombelli, G. and Porter, S.C. 1982. Late Holocene fluctuations of Brenva Glacier. *Geografia Fisica e Dinamica Quaternaria* 5, 14–37.

Orombelli, G. and Porter, S.C. 1983. Lichen growth curves for the southern flank of the Mont Blanc massif, western Italian Alps. *Arctic and Alpine Research* 15, 193–200.

Osborn, G. 1985. Holocene tephrostratigraphy and glacier fluctuations in Waterton Lakes and Glacier national parks, Alberta and Montana. *Canadian Journal of Earth Sciences* 22, 1093–1101.

Osborn, G. 1986. Lateral moraine stratigraphy and Neoglacial history of Bugaboo Glacier, British Columbia. *Quaternary Research* 26, 171–178.

Osborn, G. and Karlstrom, E.T. 1989. Holocene moraine and paleosol stratigraphy, Bugaboo Glacier, British Columbia. *Boreas* 18, 311–322.

Østrem, G. 1959. Ice melting under a thin layer of moraine and the existence of ice in moraine ridges. *Geografiska Annaler* 41, 228–230.

Østrem, G. 1963. Comparative crystallographic studies on ice from ice-cored moraine, snow banks and glaciers. *Geografiska Annaler* 45, 210–240.

Østrem, G. 1964. Ice-cored moraines in Scandinavia. *Geografiska Annaler* 46, 282–337.

Østrem, G. 1966. The height of the glaciation limit in southern British Columbia and Alberta. *Geografiska Annaler* 48A, 126–138.

Østrem, G. 1971. Rock glaciers and ice-cored moraines, a reply to D. Barsch. *Geografiska Annaler* 53A, 207–213.

Østrem, G. 1972. Height of the glaciation level on northern British Columbia and southeastern Alaska. *Geografiska Annaler* 54A, 76–84.

Østrem, G. 1973. The transient snowline and glacier mass balance in southern British Columbia and Alberta, Canada. *Geografiska Annaler* 55A, 93–106.

Østrem, G. 1974. Present alpine ice cover. In Ives, J.D. and Barry, R.G. (eds), *Arctic and Alpine Environments*. Methuen, London, 226–250.

Østrem, G. 1975a. ERTS data in glaciology: an effort to monitor glacier mass balance from satellite imagery. *Journal of Glaciology* 15, 403–415.

Østrem, G. 1975b. Sediment transport in glacial meltwater streams. In Jopling, A.V. and MacDonald, B.C. (eds), *Glaciofluvial and Glaciolacustrine Sedimentation*. SEPM Special Publication 23, 101–122.

Østrem, G. and Arnold, K. 1970. Ice-cored moraines in southern British Columbia and Alberta, Canada. *Geografiska Annaler* 52A, 120–128.

Østrem, G. and Brugman, M. 1991. *Mass Balance Measurements: A Manual for Field and Office Work*. National Hydrology Research Institute, Scientific Report 4, Environment Canada, Saskatoon and Norges Vassdrags og Elektrisitetsvesen, Oslo.

Østrem, G. and Haakensen, N. 1993. Glaciers of Europe – Glaciers of Norway. In Williams, R.S. and Ferrigno, J.G. (eds), *Satellite Image Atlas of Glaciers of the World*. USGS Professional Paper 1386-E, 63–109.

Østrem, G., Haakensen, N. and Eriksson, T. 1981. The glaciation level in southern Alaska. *Geografiska Annaler* 63A, 251–260.

Østrem, G., Haakensen, N. and Melander, O. 1973. *Atlas over Breer i Nord-Skandinavia*. Norges Vassdrags og Elektrisitetsvesen og Stockholm Univ.

Østrem, G. and Stanley, A. 1969. Glacier mass balance measurements. Canadian Department of Energy, Mines and Resources and the Norwegian Water Resources and Electricity Board.

Oswald, G.K.A. 1975. Investigation of sub-ice bedrock characteristics by radio-echo sounding. *Journal of Glaciology* 15, 75–87.

Oswald, G.K.A. and Robin, G. de Q. 1973. Lakes beneath the Antarctic Ice Sheet. *Nature* 245, 251–254.

Outcalt, S.I. and Benedict, J.B. 1965. Photointerpretation of two types of rock glacier in the Colorado Front Range, USA. *Journal of Glaciology* 5, 849–856.

Ovenshine, A.T. 1970. Observations of iceberg rafting in Glacier Bay, Alaska. *Geological Society of America Bulletin* 81, 891–894.

Overweel, C.J. 1977. Distribution and transport of Fennoscandian indicators. *Scripta Geologica*, Leiden, 43.

Owen, L.A. 1989. Terraces, uplift and climate in the Karakoram Mountains, northern Pakistan: Karakoram intermontane basin evolution. *Zeitschrift für Geomorphologie*, Supp. Bd. 76, 117–146.

Owen, L.A. 1991. Mass movement deposits in the Karakoram Mountains. *Zeitschrift für Geomorphologie* 35, 401–424.

Owen, L.A. 1994. Glacial and non-glacial diamictons in the Karakoram Mountains and Western Himalayas. In Warren, W.P. and Croot, D.G. (eds), *The Formation and Deformation of Glacial Deposits*. Balkema, Rotterdam, 9–28.

Owen, L.A., Benn, D.I., Derbyshire, E., Evans, D.J.A., Mitchell, W., Thompson, D.M., Richardson, S., Lloyd, M. and Holden, C. 1995. The geomorphology and landscape evolution of the Lahul Himalaya, northern India. *Zeitschrift für Geomorphologie* 39, 145–174.

Owen, L.A., Benn, D.I., Derbyshire, E., Evans, D.J.A., Mitchell, W.A. and Richardson, S. 1996. The Quaternary glacial history of the Lahul Himalaya, northern India. *Journal of Quaternary Science* 11, 25–42.

Owen, L.A. and Derbyshire, E. 1988. Glacially-deformed diamictons in the Karakoram Mountains, northern Pakistan. In Croot, D.G. (ed.), *Glacitectonics: Forms and Processes*. Balkema, Rotterdam, 149–176.

Owen, L.A. and Derbyshire, E. 1989. The Karakoram glacial depositional system. *Zeitschrift für Geomorphologie*, Supp. Bd. 76, 33–73.

Owen, L.A. and Derbyshire, E. 1993. Quaternary and Holocene intermontane basin sedimentation in the Karakoram Mountains. In Shroder, J.F. (ed.), *Himalaya to the Sea*. Routledge, London, 108–131.

Owens, I.F., Marcus, M.G. and Moore, R.D. 1984. Temporal variations of energy transfers over the lower part of the Franz Josef Glacier. In *Geography for the 1980s, Proceedings of Twelfth New Zealand Geography Conference, Christchurch*. New Zealand Geographical Society, Conference Proceedings Series 12, 83–87.

Owens, W.H. 1974. Mathematical model studies on factors affecting the magnetic anisotropy of deformed rocks. *Tectonophysics* 24, 115–131.

Pardee, J.T. 1910. The glacial Lake Missoula, Montana. *Journal of Geology* 18, 376–386.

Pardee, J.T. 1922. Glaciation in the Cordilleran region. *Science* 56, 686–687.

Pardee, J.T. 1942. Unusual currents in glacial Lake Missoula, Montana. *Geological Society of America Bulletin* 53, 1569–1600.

Parish, T.R. and Bromwich, D.H. 1989. Instrumented aircraft observations of the katabatic wind regime near Terra Nova Bay. *Monthly Weather Review* 117, 1570–1585.

Parizek, R.R. 1964. Geology of the Willow Bunch Lake Area 72-H. Saskatchewan. Saskatchewan Research Council, Geology Division, Report 4.

Parizek, R.R. 1969. Glacial ice-contact rings and ridges. Geological Society of America Special Paper 123, 49–102.

Park, R.G. 1983. *Foundations of Structural Geology*. Blackie, London.

Parker, T.J., Gorsline, D.S., Saunders, R.S., Pieri, D.C. and Schneeberger, D.M. 1993. Coastal geomorphology of the Martian northern plains. *Journal of Geophysical Research* E98, 11061–11078.

Parkin, G.W. and Hicock, S.R. 1989. Sedimentology of a Pleistocene glacigenic diamicton sequences near Campbell River, Vancouver Island, British Columbia. In Goldthwait, R.P. and Matsch, C.L. (eds), *Genetic Classification of Glacigenic Deposits*. Balkema, Rotterdam, 97–116.

Parson, C.G. 1987. Rock glaciers and site characteristics of the Blanca Massif, Colorado, USA. In Giardino, J.R., Shroder, J.F. and Vitek, J.D. (eds), *Rock Glaciers*. Allen & Unwin, London, 127–143.

Parsch, J. 1882. *Die Gletscher der Vorzeit in den Karpathen und den Mittelgebirgen Deutschlands*. Breslau.

Paschinger, V. 1912. Die Schneegrenze in verschiedenen Klimaten. *Petermanns Mitteilungen. Erganzungshefte* 173.

Paterson, W.S.B. 1972. Laurentide Ice Sheet: estimated volumes during late Wisconsin. *Revue of Geophysics and Space Physics* 10, 885–917.

Paterson, W.S.B. 1980. Ice sheets and ice shelves. In Colbeck, S.C. (ed.), *Dynamics of Snow and Ice Masses*. Academic Press, New York, 1–78.

Paterson, W.S.B. 1981. *The Physics of Glaciers*, 2nd edition. Pergamon, Oxford.

Paterson, W.S.B. 1994. *The Physics of Glaciers*, 3rd edition. Pergamon, Oxford.

Paterson, W.S.B. and Savage, J.C. 1970. Excess pressure observed in a water-filled cavity in Athabasca Glacier, Canada. *Journal of Glaciology* 9, 103–107.

Patterson, C.J. 1994. Tunnel-valley fans of the St Croix moraine, east-central Minnesota, USA. In Warren, W.P. and Croot, D.G. (eds), *Formation and Deformation of Glacial Deposits*. Balkema, Rotterdam, 69–87.

Patterson, C.J. and Hooke, R. Le B. 1996. Physical environment of drumlin formation. *Journal of Glaciology* 41, 30–38.

Patton, P.C. and Schumm, S.A. 1981. Ephemeral-stream processes: implications for studies of Quaternary valley fills. *Quaternary Research* 15, 24–43.

Paul, M.A. 1983. The supraglacial landsystem. In Eyles, N. (ed.), *Glacial Geology*. Pergamon, Oxford, 71–90.

Paul, M.A. and Evans, H. 1974. Observations on the internal structure and origin of some flutes in glaciofluvial sediments, Blomstrandbreen, north-west Spitsbergen. *Journal of Glaciology* 13, 393–400.

Paul, M.A. and Eyles, N. 1990. Constraints on the preservation of diamict facies (melt-out tills) at the margins of stagnant glaciers. *Quaternary Science Reviews* 9, 51–69.

Payne, A. and Sugden, D.E. 1990. Topography and ice sheet growth. *Earth Surface Processes and Landforms* 15, 625–639.

Pedersen, S.A.S. 1989. Glacitectonite: brecciated sediments and cataclastic sedimentary rocks formed subglacially. In Goldthwait, R.P. and Matsch, C.L. (eds), *Genetic Classification of Glacigenic Deposits*. Balkema, Rotterdam, 89–91.

Pedersen, S.A.S. 1993. The glaciodynamic event and glaciodynamic sequence. In Aber, J.S. (ed.), *Glaciotectonics and Mapping Glacial Deposits*. Canadian Plains Research Center, University of Regina, 67–85.

Pedersen, S.A.S., Petersen, K.S. and Rasmussen, L.A. 1988. Observations on glaciodynamic structures at the Main Stationary Line in western Jutland, Denmark. In Croot, D.G. (ed.), *Glaciotectonics: Forms and Processes*. Balkema, Rotterdam, 177–183.

Peixoto, J.P. and Oort, A.H. 1992. *Physics of Climate*. American Institute of Physics, New York.

Pelletier, B.R. 1966. Development of submarine physiography in the Canadian Arctic and its relation to crustal movements. In Garland, G.D. (ed.), *Continental Drift*. University of Toronto Press, Toronto, 77–101.

Peltier, L.C. 1950. The geographic cycle in periglacial regions as it is related to climatic geomorphology. *Annals of the Association of American Geographers* 40, 214–236.

Peltier, W.R. 1987. Glacial isostasy, mantle viscosity, and Pleistocene climatic change. In Ruddiman, W.F. and Wright, H.E. (eds), *North America and Adjacent Oceans during the Last Deglaciation*. Geological Society of America, *The Geology of North America*, Vol. K–3, 155–182.

Peltier, W.R. and Andrews, J.T. 1983. Glacial geology and glacial isostasy of the Hudson Bay region. In Smith, D.E. and Dawson, A.G. (eds), *Shorelines and Isostasy.* Academic Press, London, 285–320.

Pelto, M.S. 1988. The annual balance of North Cascade glaciers, Washington, USA, measured and predicted using an activity-index method. *Journal of Glaciology* 34, 194–199.

Pelto, M.S. 1989. Time-series analysis of mass balance and local climatic records from four northwestern North American glaciers. In Colbeck, S.C. (ed.), *Snow Cover and Glacier Variations.* IAHS Publication 183, Great Yarmouth, UK, 95–102.

Pelto, M.S. and Warren, C. 1991. Relationship between tidewater calving velocity and water depth at the calving front. *Annals of Glaciology* 15, 115–118.

Penck, A. 1905. Glacial features in the surface of the Alps. *Journal of Geology* 13, 1–17.

Perrin, R.M.S., Rose, J. and Davies, H. 1979. The distribution, variation and origins of pre-Devensian tills in eastern England. *Philosophical Transactions of the Royal Society of London, Series B* 287, 535–570.

Perry, W.J., Roeder, D.H. and Lageson, D.R. 1984. North American thrust faulted terrains. American Association of Petroleum Geologists, Reprint 27.

Perttunen, M. 1977. The lithological relation between till and bedrock in the region of Hämeenlinna, southern Finland. Geological Survey of Finland Bulletin 291.

Perttunen, M.1989. Transportation of garnets in glaciofluvial deposits in southeastern Finland. In DiLabio, R.N.W. and Coker, W.B. (eds), *Drift Prospecting.* Geological Survey of Canada, Paper 89–20, 13–20.

Pessl, F. 1971. Till fabrics and till stratigraphy in western Connecticut. In Goldthwait, R.P. (ed.), *Till: A Symposium.* Ohio State University Press, Columbus, 92–105.

Peterson, D.N. 1970. Glaciological investigations on the Casement Glacier, south-east Alaska. Ohio State University, Institute of Polar Studies Report 36.

Petrie, G. and Price, R.J.1966. Photogrammetric measurements of the ice wastage and morphological changes near the Casement Glacier, Alaska. *Canadian Journal of Earth Science* 3, 827–840.

Pettijohn, F.J. and Potter, P.E. 1964. *Atlas and Glossary of Primary Sedimentary Structures.* Springer-Verlag, Berlin.

Peulvast, J.P. 1985. In situ weathered rocks on plateaus, slopes and strandflat areas of the Lofoten-Vesteralen, north Norway. *Fennia* 163, 333–340.

Peuraniemi, V. 1984. Weathering of sulphide minerals in till in some mineralized areas in Finland. In Jones, M.J. (ed.), *Prospecting in Areas of Glaciated Terrain.* Institution of Mining and Metallurgy, London, 127–135.

Péwé, T.L. 1960. Multiple glaciation in the McMurdo Sound region, Antarctica, a progress report. *Journal of Geology* 68, 489–514.

Pfeffer, W.T. 1992. Stress-induced foliation in the terminus of Variegated Glacier, Alaska, USA, formed during the 1982–83 surge. *Journal of Glaciology* 38, 213–222.

Pflug, R. and Scholl, W.U. 1975. Proterozoic glaciations in eastern Brazil: a review. *Geologische Rundschau* 64, 287–299.

Pharo, C.H. and Carmack, E.C. 1979. Sedimentation processes in a short residence-time intermontane lake, Kam-loops, British Columbia. *Sedimentology* 26, 523–541.

Phillips, A.C., Smith, N.D. and Powell, R.D. 1991. Laminated sediments in prodeltaic deposits, Glacier Bay, Alaska. In Anderson, J.B. and Ashley, G.M. (eds), *Glacial Marine Sedimentation: Paleoclimatic Significance.* Geological Society of America Special Paper 261, 51–60.

Pickard, J.1983. Surface lowering of ice-cored moraine by wandering lakes. *Journal of Glaciology* 29, 338–342.

Pickering, K.T. and Owen, L.A. 1994. *An Introduction to Global Environmental Issues.* Routledge, London.

Pickrill, R.A. and Irwin, J. 1982. Predominant headwater inflow and its control of lake-river interactions in Lake Wakatipu. *New Zealand Journal of Freshwater Research* 16, 201–213.

Pickrill, R.A. and Irwin, J. 1983. Sedimentation in a deep glacier-fed lake: Lake Tekapo, New Zealand. *Sedimentology* 30, 63–75.

Pierson, T.C. and Costa, J.E. 1987. A rheologic classification of subaerial sediment–water flows. In Costa, J.E. and Wieczorek, G.F. (eds), *Debris Flows/Avalanches: Process, Recognition, and Mitigation.* Reviews in Engineering Geology 7, 1–12. Geological Society of America.

Piotrowski, J.A. 1994. Tunnel valley formation in northwest Germany: geology, mechanisms of formation and subglacial bed conditions for the Bornhöved tunnel valley. *Sedimentary Geology* 89, 107–141.

Piotrowski, J.A. and Smalley, I.J. 1987. The Woodstock drumlin field, southern Ontario, Canada. In Menzies, J. and Rose, J. (eds), *Drumlin Symposium.* Balkema, Rotterdam, 309–322.

Piper, D.J.W.1976. The use of ice rafted marine sediments in determining glacial conditions. *Revue de Géographie de Montréal* 30, 207–212.

Piper, D.J.W. and Normark, W.R. 1989. Late Cenozoic sea-level changes and the onset of glaciation: impact on continental slope progradation off eastern Canada. *Marine Petroleum Geology* 6, 336–348.

Pirazzoli, P.A. 1991. *World Atlas of Holocene Sea-Level Changes.* Elsevier, Amsterdam.

Plafker, G. 1987. Regional geology and petroleum potential of the northern Gulf of Alaska continental margin. In Scholl, D.W. *et al.* (eds), *Geology and Resource Potential of the Continental Margin of Western North America and Adjacent Ocean Basins.* Circum-Pacific Council for Energy and Mineral Resources, Houston, 229–268.

Plafker, G. and Addicott, W.O.1976. Glaciomarine deposits of Miocene through Holocene age in the Yakataga Formation along the Gulf of Alaska margin, Alaska. In Miller, T.P. (ed.), *Recent and Ancient Sedimentary Environments in Alaska.* Alaska Geological Society Symposium Proceedings, Q1–Q23.

Plassche, van der O. 1986. *Sea-Level Research: A Manual for the Collection and Evaluation of Data.* Geo Books, Norwich.

Pohjola, V.A. 1993. TV-video observations of bed and basal sliding on Storglaciären, Sweden. *Journal of Glaciology* 39, 111–118.

Pohjola, V.A. 1994. TV-video observation of englacial voids in Storglaciären, Sweden. *Journal of Glaciology* 40, 231–240.

Porter, S.C. 1975a. Glaciation limit in New Zealand's southern alps. *Arctic and Alpine Research* 7, 33–37.

Porter, S.C. 1975b. Equilibrium line altitudes of late Quaternary glaciers in the Southern Alps, New Zealand. *Quaternary Research* 5, 27–47.

Porter, S.C. 1978. Glacier Peak tephra in the North Cascade Range, Washington: stratigraphy, distribution, and relationship to Late-Glacial events. *Quaternary Research* 10, 30–41.

Porter, S.C. 1981a. Pleistocene glaciation in the southern Lake District of Chile. *Quaternary Research* 16, 263–292.

Porter, S.C. 1981b. Glaciological evidence of Holocene climatic change. In Wigley, T.M.L., Ingram, M.J. and Farmer, G. (eds), *Climate and History Studies on Past Climates and their Impact on Man*. Cambridge University Press, Cambridge, 82–110.

Porter, S.C. 1981c. Lichenometric studies in the Cascade Range of Washington: establishment of *Rhizocarpon geographicum* growth curves at Mount Ranier. *Arctic and Alpine Research* 13, 11–23.

Porter, S.C. 1989. Some geological implications of average Quaternary glacial conditions. *Quaternary Research* 32, 245–261.

Porter, S.C., Clapperton, C.M. and Sugden, D.E. 1992. Chronology and dynamics of deglaciation along and near the Strait of Magellan, southernmost South America. *Sveriges Geologiska Undersokning Ser. Ca.* 81, 233–239.

Porter, S.C. and Denton, G.H. 1967. Chronology of Neoglaciation in the North American Cordillera. *American Journal of Science* 265, 177–210.

Porter, S.C. and Orombelli, G. 1980. Catastrophic rockfall of September 12, 1717 on the Italian flank of the Mont Blanc massif. *Zeitschrift für Geomorphologie* NF 24, 200–218.

Porter, S.C. and Orombelli, G. 1981. Alpine rockfall hazards. *American Scientist* 69, 67–75.

Porter, S.C. and Orombelli, G. 1982. Late glacial ice advances in the western Italian Alps. *Boreas* 11, 125–140.

Posamentier, H.W., Jervey, M.T. and Vail, P.R. 1988. Eustatic controls on clastic deposition I – conceptual framework. In Wilgus, C.K. *et al.* (eds), *Sea-Level Changes: An Integrated Approach*. SEPM Special Publication 42, 109–124.

Post, A.S. 1967. Effects of the March 1964 Alaska earthquake on glaciers. USGS Professional Paper, 544-D.

Post, A.S. 1969. Distribution of surging glaciers in western North America. *Journal of Glaciology* 8, 229–240.

Postma, G. 1984a. Slumps and their deposits on fan delta front and slope. *Geology* 12, 27–30.

Postma, G. 1984b. Mass-flow conglomerates in a submarine canyon: Abrioja fan-delta, Pliocene, south-east Spain. In Koster, E.H. and Steel, R.J. (eds), *Sedimentology of Gravels and Conglomerates*. Canadian Society of Petroleum Geologists Memoir 10, 237–258.

Postma, G. 1986. Classification for sediment–gravity flow deposits based on flow conditions during sedimentation. *Geology* 14, 291–294.

Postma, G. 1990. Depositional architecture and facies of river and fan deltas: a synthesis. In Colella, A. and Prior, D.B. (eds), *Coarse-Grained Deltas*. International Association of Sedimentologists, Special Publication 10. Blackwell, Oxford, 13–27.

Postma, G., Babic, L., Zupanic, J. and Roe, S.L. 1988a. Delta front failure and associated bottomset deformation in a marine, gravelly Gilbert-type fan delta. In Nemec, W. and Steel, R.J. (eds), *Fan Deltas: Sedimentology and Tectonic Settings*. Blackie, Glasgow, 91–102.

Postma, G. and Cruickshank, C. 1988. Sedimentology of a late Weichselian to Holocene terraced fan delta, Varangerfjord, northern Norway. In Nemec, W. and Steel, R.J. (eds), *Fan Deltas: Sedimentology and Tectonic Settings*. Blackie, London, 144–157.

Postma, G., Nemec, W. and Kleinspehn, K.L. 1988b. Large floating clasts in turbidites: a mechanism for their emplacement. *Sedimentary Geology* 58, 47–61.

Postma, G. and Roep, T.B. 1985. Resedimented conglomerates in the bottomsets of Gilbert-type gravel deltas. *Journal of Sedimentary Petrology* 55, 874–885.

Postma, G., Roep, T.B. and Ruegg, G.H.J. 1983. Sandy gravelly mass flow deposits in an ice-marginal lake (Saalian, Leuvenumsche Beek Valley, Veluwe, The Netherlands) with emphasis on plug flow deposits. *Sedimentary Geology* 34, 59–82.

Potter, N. 1972. Ice-cored rock glacier, Galena Creek, Northern Absaroka Mountains, Wyoming. *Geological Society of America Bulletin* 83, 3025–3058.

Powell, R.D. 1981a. A model for sedimentation by tidewater glaciers. *Annals of Glaciology* 2, 129–134.

Powell, R.D. 1981b. Sedimentation conditions in Taylor Valley, Antarctica, inferred from textural analysis of DVDP cores. In McGinnis, L.D. (ed.), *Dry Valleys Drilling Project*. American Geophysical Union, Antarctic Research Series 33, 331–349.

Powell, R.D. 1983a. Glacial marine sedimentation processes and lithofacies of temperate tidewater glaciers, Glacier Bay, Alaska. In Molnia, B.F. (ed.), *Glacial-Marine Sedimentation*. Plenum Press, New York, 195–232.

Powell, R.D. 1983b. Submarine flow tills at Victoria, British Columbia: discussion. *Canadian Journal of Earth Sciences* 20, 509–510.

Powell, R.D. 1984. Glacimarine processes and inductive lithofacies modelling of ice shelf and tidewater glacier sediments based on Quaternary examples. *Marine Geology* 57, 1–52.

Powell, R.D. 1990. Glacimarine processes at grounding-line fans and their growth to ice-contact deltas. In Dowdeswell, J.A. and Scourse, J.D. (eds), *Glacimarine Enviroments: Processes and Sediments*. Geological Society Special Publication 53, 53–73.

Powell, R.D. 1991. Grounding-line systems as second-order controls on fluctuations of tidewater termini of temperate glaciers. In Anderson, J.B. and Ashley, G.M. (eds), *Glacial Marine Sedimentation: Paleoclimatic Significance*. Geological Society of America Special Paper 261, 75–93.

Powell, R.D., Dawber, M., McInnes, J.N. and Pyne, A.R. 1996. Observations of the grounding-line area at a floating glacier terminus. *Annals of Glaciology* 22, 217–223.

Powell, R.D. and Domack, E. 1995. Modern glaciomarine environments. In Menzies, J. (ed.), *Glacial Environments*, vol. 1: *Modern Glacial Environments: Processes, Dynamics and Sediments*. Butterworth-Heinemann, Oxford, 445–486.

Powell, R.D. and Molnia, B.F. 1989. Glacimarine sedimentary processes, facies and morphology of the south-southeast Alaska shelf and fjords. *Marine Geology* 85, 359–390.

Preller, C.S. duR. 1896. The Merjelen lake (Aletsch glacier). *Geological Magazine* 3, 97–102.

Prest, V.K. 1968. Nomenclature of moraines and ice-flow features as applied to the glacial map of Canada. Geological Survey of Canada, Paper 67–57.

Prest, V.K. 1969. Retreat of Wisconsin and recent ice in North America. Geological Survey of Canada, Map 1257A, scale 1:5,000,000.

Prest, V.K. 1983. Canada's heritage of glacial features. Geological Survey of Canada, Miscellaneous Report 28.

Prest, V.K. 1990. Laurentide ice flow patterns: a historical review, and implications of the dispersal of Belcher Island erratics. *Géographie Physique et Quaternaire* 44, 113–136.

Prest, V.K., Grant D.G. and Rampton V.N. 1968. Glacial Map of Canada. Geological Survey of Canada, Map 1253A, 1:5,000,000.

Price, N.J. and Cosgrove J.W. 1990. *Analysis of Geological Structures*. Cambridge University Press, Cambridge.

Price, R.J. 1966. Eskers near the Casement Glacier, Alaska. *Geografiska Annaler* 48, 111–125.

Price, R.J. 1969. Moraines, sandar, kames and eskers near Breidamerkurjökull, Iceland. *Transactions of the Institute of British Geographers* 46, 17–43.

Price, R.J. 1970. Moraines at Fjallsjökull, Iceland. *Arctic and Alpine Research* 2, 27–42.

Price, R.J. 1973. *Glacial and Fluvioglacial Landforms*. Oliver & Boyd, Edinburgh.

Price, R.J. 1982. Changes in the proglacial area of Breidamerkurjökull, southeastern Iceland: 1890–1980. *Jökull* 32, 29–35.

Price, R.J. 1983. *Scotland's Environment during the Last 30,000 Years*. Scottish Universities Press, Glasgow.

Price, R.J. and Howarth, P.J. 1970. The evolution of the drainage system (1904–1965) in front of Breidamerkurjökull, Iceland. *Jökull* 20, 27–37.

Priestley, R.E. 1923. *Physiography (Robertson Bay and Terra Nova Regions)*. Report of the British Antarctic Expedition 1910–1913. Harrison & Sons, London.

Prior, D.B. and Bornhold, B.D. 1988. Submarine morphology and processes of fjord fan deltas and related high-gradient systems: modern examples from British Columbia. In Nemec, W. and Steel, R.J. (eds), *Fan Deltas: Sedimentology and Tectonic Settings*. Blackie, London, 125–143.

Prior, D.B. and Bornhold, B.D. 1989. Submarine sedimentation on a developing Holocene fan delta. *Sedimentology* 36, 1053–1076.

Prior, D.B. and Bornhold, B.D. 1990. The underwater development of Holocene fan deltas. In Colella, A. and Prior, D.B. (eds), *Coarse-Grained Deltas*. International Association of Sedimentologists, Special Publication 10, 75–90.

Prior, D.B., Bornhold, B.D., Wiseman, W.J. and Lowe, D.R. 1987. Turbidity current activity in a British Columbia fjord. *Science* 237, 1330–1333.

Prior, D.B., Wiseman, W.J. and Bryant, W.R. 1981. Submarine chutes on the slopes of fjord deltas. *Nature* 290, 326–328.

Punkari, M. 1995. Glacial flow systems in the zone of confluence between the Scandinavian and Novaya Zemlya ice sheets. *Quaternary Science Reviews* 14, 589–603.

Puranen, R. 1977. Magnetic susceptibility and its anisotropy in the study of glacial transport in northern Finland. In Jones M.J. (ed.), *Prospecting in Areas of Glaciated Terrain*. Institution of Mining and Metallurgy, London, 111–119.

Pytte, R. and Østrem, G. 1965. Glasio-hydrologiske undersøkelser i Norge 1964. *Norges Vassdrags- og Elektrisitetsvesen, Meddelse fra Hydrologisk Avdeling*, 14 (with English summary).

Quigley, R.M. 1983. Glaciolacustrine and glaciomarine clay deposition: a North American perspective. In Eyles, N. (ed.), *Glacial Geology*. Pergamon, Oxford, 140–167.

Quinlan, G. 1985. A numerical model of postglacial relative sea level change near Baffin Island. In Andrews, J.T. (ed.), *Quaternary Environments: Eastern Canadian Arctic, Baffin Island, and Western Greenland*. Allen & Unwin, London, 560–584.

Quinlan, G. and Beaumont, C. 1981. A comparison of observed and theoretical postglacial relative sea level in Atlantic Canada. *Canadian Journal of Earth Sciences* 18, 1146–1163.

Quinlan, G. and Beaumont, C. 1982. The deglaciation of Atlantic Canada as reconstructed from the postglacial relative sea-level record. *Canadian Journal of Earth Sciences* 19, 2232–2246.

Rabassa, J. and Clapperton, C.M. 1990. Quaternary glaciations of the southern Andes. *Quaternary Science Reviews* 9, 153–174.

Rabassa, J., Rubulis, S. and Suarez, J. 1979. Rate of formation and sedimentology of (1976–1978) push moraines, Frias Glacier, Mount Tronador (41° 10′S; 71° 53′W), Argentina. In Schluchter, C. (ed.), *Moraines and Varves*. Balkema, Rotterdam, 65–79.

Ragotzkie, R.A. 1978. Heat budgets of lakes. In Lerman, A. (ed.), *Lakes: Chemistry, Geology, Physics*. Springer-Verlag, New York, 1–19.

Rains, R.B. and Selby, M.J. 1972. Relatively young moraines of the Webb Glacier, Barwick Valley, Victoria Land, Antarctica. *Proceedings of 7th Geography Conference, New Zealand Geographical Society, Hamilton*, 235–245.

Rains, R.B, Selby, M.J. and Smith, C.J.R. 1980. Polar desert sandar, Antarctica. *New Zealand Journal of Geology and Geophysics* 23, 595–604.

Rains, R.B. and Shaw, J. 1981. Some mechanisms of controlled moraine development, Antarctica. *Journal of Glaciology* 27, 113–128.

Rains, R.B., Shaw, J., Skoye, K.R., Sjogren, D.B. and Kvill, D.R. 1993. Late Wisconsin subglacial megaflood paths in Alberta. *Geology* 21, 323–326.

Raiswell, R. 1984. Chemical models of solute acquisition in glacial melt waters. *Journal of Glaciology* 30, 49–57.

Rampton, V.N. 1974. The influence of ground ice and thermokarst upon the geomorphology of the Mackenzie–Beaufort region. In Fahey, B.D. and Thompson, R.D. (eds), *Research in Polar and Alpine Geomorphology*. Geo Books, Norwich, 43–59.

Rampton, V.N. 1982. Quaternary geology of the Yukon Coastal Plain. Geological Survey of Canada, Bulletin.

Rampton, V.N. 1988. Quaternary Geology of the Tuktoyaktuk coastlands, NWT. Geological Survey of Canada, Memoir 423.

Rampton, V.N., Gauthier, R.C., Thibault, J. and Seaman, A.A. 1984. Quaternary geology of New Brunswick. Geological Survey of Canada Memoir 416.

Ramsay, J.G. 1967. *Folding and Fracturing of Rocks.* McGraw-Hill, New York.

Randall, B.A.O. 1961. On the relationship of valley and fjord directions to the fracture pattern of Lyngen, Troms, N. Norway. *Geografiska Annaler* 43, 336–338.

Rao, U.M. and Carstens, T. 1971. Sediment-laden submerged horizontal jet. *Proceedings 14th Congress International Association of Hydraulic Research*, 135–143.

Rapp, A. 1957. Studien über Schutthalden in Lappland und auf Spitzbergen. *Zeitschrift für Geomorphologie* 1, 179–200. English translation 'Studies of debris cones in Lappland and Spitsbergen', in Evans, D.J.A. (ed.), *Cold Climate Landforms.* Wiley, Chichester, 415–435.

Rapp, A. 1959. Avalanche boulder tongues in Lappland. *Geografiska Annaler* 41, 34–48.

Rapp, A. 1960a. Recent development of mountain slopes in Karkevagge and surroundings, northern Scandinavia. *Geografiska Annaler* 42, 71–200.

Rapp, A. 1960b. Talus slopes and mountain walls at Templefjorden, Spitsbergen. Norsk Polarinstitutt Skrifter 119.

Rappol, M. and Stoltenberg, H.M.P. 1985. Compositional variability of Saalian till in the Netherlands. *Boreas* 14, 33–50.

Rastas, J. and Seppälä, M. 1981. Rock jointing and abrasion forms on roches moutonnées, SW Finland. *Annals of Glaciology* 2, 159–163.

Ray, L.L. 1935. Some minor features of valley glaciers and valley glaciation. *Journal of Geology* 43, 297–322.

Raymo, M.E. and Ruddiman, W.F. 1992. Tectonic forcing of late Cenozoic climate. *Nature* 359, 117–122.

Raymond, C.F. 1971. Flow in a transverse section of Athabasca Glacier, Alberta, Canada. *Journal of Glaciology* 10, 55–84.

Raymond, C.F. 1987. How do glaciers surge? A review. *Journal of Geophysical Research* 92, 9121–9134.

Raymond, C.F. and Harrison, W.D. 1985. Some observations on the behaviour of the liquid and gas phases in temperate ice. *Journal of Glaciology* 14, 213–233.

Raymond, C.F. and Harrison, W.D. 1987. Winter initiation of surges. *Hydrol. Glaciol. Mitt.* 90, 85–86.

Raymond, C.F. and Harrison, W.D. 1988. Evolution of Variegated Glacier, Alaska, USA, prior to its surge. *Journal of Glaciology* 34, 154–169.

Raymond, C.F., Johannesson, T., Pfeffer, T., and Sharp, M. 1987. Propagation of a glacier surge into stagnant ice. *Journal of Geophysical Research* 92, 9037–9049.

Rea, B.R. 1994. Joint control in the formation of rock steps in the subglacial environment. In Robinson, D.A. and Williams, R.B.G. (eds), *Rock Weathering and Landform Evolution.* Wiley, Chichester, 473–486.

Rea, B.R. and Evans, D.J.A. 1996. Landscapes of areal scouring in NW Scotland. *Scottish Geographical Magazine* 112, 47–50.

Rea, B.R. and Whalley, W.B. 1994. Subglacial observations from Oksfjordjøkelen, north Norway. *Earth Surface Processes and Landforms* 19, 659–673.

Rea, B.R., Whalley, W.B., Rainey, M.M. and Gordon, J.E. (1996) Blockfields, old or new? Evidence and implications from plateaus in northern Norway. *Geomorphology* 15, 109–112.

Reading, H.G. (ed.) 1978. *Sedimentary Environments and Facies.* Blackwell, Oxford.

Reading, H.G. 1986. Facies. In Reading, H.G. (ed.), *Sedimentary Environments and Facies*, 2nd edition. Blackwell, Oxford, 4–19.

Reece, A. 1950. The ice of Crown Prince Gustav Channel, Graham Land, Antarctica. *Journal of Glaciology* 1, 404–409.

Reed, B., Galvin, C.J. and Miller, J.P. 1962. Some aspects of drumlin geometry. *American Journal of Science* 260, 200–210.

Reeh, N. 1968. On the calving of ice from floating glaciers and ice shelves. *Journal of Glaciology* 7, 215–232.

Reeh, N. 1969. Calving from floating glaciers. Reply to Professor F. Loewe's comments. *Journal of Glaciology* 8, 322–324.

Reeh, N. 1991. Parameterization of melt rate and surface temperature on the Greenland ice sheet. *Polarforschung* 59, 113–128.

Reeh, N. and Olesen, O.B. 1986. Velocity measurements on Daugaard-Jensen Gletscher, Scoresby Sund, east Greenland. *Annals of Glaciology* 8, 146–150.

Reeves, B.O.K. 1973. The nature and age of the contact between the Laurentide and Cordilleran Ice Sheets in the western interior of North America. *Arctic and Alpine Research* 5, 1–16.

Reid, J.R. 1969. Effects of a debris slide on 'Sioux Glacier', south-central Alaska. *Journal of Glaciology* 8, 353–367.

Reimnitz, E. and Kempema, E.W. 1987. Field observations of slush ice generated during freeze-up in arctic coastal waters. *Marine Geology* 77, 219–231.

Reimnitz, E. and Kempema, E.W. 1988. Ice rafting: an indication of glaciation? *Journal of Glaciology* 34, 254–255.

Reimnitz, E., Kempema, E.W. and Barnes, P.W. 1987. Anchor ice, seabed freezing, and sediment dynamics in shallow arctic seas. *Journal of Geophysical Research* 92, 14671–14678.

Reinick, H.E. and Singh, I.B. 1980. *Depositional Sedimentary Environments.* Springer-Verlag, Berlin.

Rencz, A.N. and Shilts, W.W. 1980. Nickel in soils and vegetation of glaciated terrains. In Nriagu, J.O. (ed.), *Nickel in the Environment.* Wiley, New York, 151–188.

Retelle, M.J. 1986a. Glacial geology and Quaternary marine stratigraphy of the Robeson Channel area, northeastern Ellesmere Island, NWT. *Canadian Journal of Earth Sciences* 23, 1001–1012.

Retelle, M.J. 1986b. Stratigraphy and sedimentology of coastal lacustrine basins, northeastern Ellesmere Island, NWT. *Géographie Physique et Quaternaire* 40, 117–128.

Reusch, H. 1894. Strandflaten, et nyt traek i Norges geografi. *Norges Geologisk Undersoegelse* 14, 1–14.

Reynaud, L. 1980. Can the linear balance model be extended to the whole Alps? In Müller, F. (ed.), *World Glacier Inventory. Proceedings of the Riederalp Workshop, 1978.* IAHS Publication 126, 273–284.

Reynaud, L. 1983. Recent fluctuations of alpine glaciers and their meteorological causes: 1880–1980. In Street-

Perrott, A., Beran, M. and Ratcliff, R. (eds), *Variations in the Global Water Budget*. Reidel, Dordrecht, 195–202.

Reynaud, L. 1984. Mesures des fluctuations glaciaires dans les Alpes françaises: collecte des données et résultats. *La Houille Blanche* 39, 519–525.

Reynaud, L. 1987. The November 1986 survey of the Grand Moulin on the Mer de Glace, Mont Blanc Massif, France. *Journal of Glaciology* 33, 130–131.

Reynolds, R.C. and Johnson, N.M. 1972. Chemical weathering in the temperate glacial environment of the Northern Cascade Mountains. *Geochimica et Cosmochimica Acta* 36, 537–554.

Ribstein, P., Titiau, E., Francou, B. and Saravia, R. 1995. Tropical climate and glacier hydrology: a case study in Bolivia. *Journal of Hydrology* 165, 221–234.

Rice, R.J. and Douglas, T. 1991. Wolstonian glacial deposits and glaciation in Britain. In Ehlers, J., Gibbard, P.L. and Rose, J. (eds), *Glacial Deposits in Great Britain and Ireland*. Balkema, Rotterdam, 25–35.

Rich, J.L. 1943. Buried stagnant ice as a normal product of a progressively retreating glacier in a hilly region. *American Journal of Science* 241, 95–99.

Richards, K., Sharp, M., Arnold, N., Gurnell, A., Clark, M., Tranter, M., Nienow, P., Brown, G., Willis, I. and Lawson, W. 1996. An integrated approach to modelling hydrology and water quality in glacierized catchments. *Hydrological Processes* 10, 479–508.

Richards, M.A. 1988. Seismic evidence for a weak basal layer during the 1982 surge of Variegated glacier, Alaska, USA. *Journal of Glaciology* 34, 111–120.

Richardson, C. and Holmlund, P. 1996. Glacial cirque formation in northern Scandinavia. *Annals of Glaciology* 22, 102–106.

Richardson, D. 1968. Glacier outburst floods in the Pacific Northwest. US Geological Survey, Professional Paper 600D, 79–86.

Richardson, D. 1973. Effect of snow and ice on runoff at Mount Rainier, Washington. International Association of Hydrological Sciences Publication 107, 1172–1185.

Richardson, P.D. 1968. The generation of scour marks near obstacles. *Journal of Sedimentary Petrology* 38, 965–970.

Richmond, G.M. 1965. Glaciation of the Rocky Mountains. In Wright, H.E. and Frey, D.G. (eds), *The Quaternary of the United States*. Princeton University Press, Princeton, NJ, 217–230.

Richmond, G.M. 1986. Tentative correlation of deposits of the Cordilleran ice sheet in the northern Rocky Mountains. In Sibrava, V., Bowen, D.Q. and Richmond, G.M. (eds), *Quaternary Glaciations in the Northern Hemisphere*. Quaternary Science Reviews 5, 129–144.

Ridley, J.K., Cudlip, W. and Laxon, S.W. 1993. Identification of subglacial lakes using ERS-1 radar altimeter. *Journal of Glaciology* 39, 625–634.

Riley, J.M. 1987. Drumlins of the southern Vale of Eden, Cumbria, England. In Menzies, J. and Rose, J. (eds), *Drumlin Symposium*. Balkema, Rotterdam, 323–333.

Riley, N.W. 1979. Discussion comments on Budd *et al.* (1979): Empirical studies of ice sliding. *Journal of Glaciology* 23, 384–385.

Rind, D., Peteet, D., Broecker, W.G., McIntyre, A. and Ruddiman, W. 1986. The impact of North Atlantic sea surface temperatures on climate: implications for the Younger Dryas cooling (11–10 kyr). *Climate Dynamics* 1, 3–33.

Ringberg, B. 1980. *Beskrivning till Jordartskartan Malmö SO* (Description to the Quaternary map Malmö SO). Sveriges Geologiska Undersokning, Ae38.

Ringberg, B. 1983. Till stratigraphy and glacial rafts of chalk at Kvarnby, southern Sweden. In Ehlers, J. (ed.), *Glacial Deposits in North-West Europe*. Balkema, Rotterdam, 151–154.

Ringberg, B. 1988. Late Weichselian geology of southeastern Sweden. *Boreas* 17, 243–263.

Ringberg, B., Holland, B. and Miller, U. 1984. Till stratigraphy and provenance of the glacial chalk rafts at Kvarnby and Angdala, southern Sweden. *Striae* 20, 79–90.

Ringrose, S. 1982. Depositional processes in the development of eskers in Manitoba. In Davidson-Arnott, R., Nickling, W. and Fahey, B.D. (eds), *Research in Glacial, Glacio-fluvial and Glacio-lacustrine Systems*. Geo Books, Norwich, 117–138.

Rist, S. 1955. Skeidararhlaup 1954 (The hlaup of Skeidara 1954). *Jökull* 5, 43–46.

Rist, S. 1984. Jöklabreytingar 1964/65–1973/74 (10 ar), 1974/75–1981/82 (8 ar) og 1982/83. *Jökull* 34, 173–179.

Ritter, D.F. 1978. *Process Geomorphology*. Brown, Dubuque, Iowa.

Roaldset, E., Pettersen, E., Longva, O. and Mangerud, J. 1982. Remnants of preglacial weathering in western Norway. *Norsk Geologisk Tidsskrift* 62, 169–178.

Robe, R.Q. 1980. Iceberg drift and deterioration. In Colbeck, S.C. (ed.), *Dynamics of Snow and Ice Masses*. Academic Press, New York, 211–259.

Roberts, M.C. and Cunningham, F.F. 1992. Post-glacial loess deposition in a montane environment: South Thompson River valley, British Columbia, Canada. *Journal of Quaternary Science* 7, 291–301.

Roberts, M.C. and Rood, R.M. 1984. The role of ice contributing area in the morphology of transverse fiords, British Columbia. *Geografiska Annaler* 66A, 381–393.

Roberts, N. 1989. *The Holocene: An Environmental History*. Blackwell, Oxford.

Robertson-Rintoul, M.S.E. 1986. A quantitative soil-stratigraphic approach to the correlation and dating of postglacial river terraces in Glen Feshie, western Cairngorms. *Earth Surface Processes and Landforms* 11, 605–617.

Robin, G. de Q. 1955. Ice movement and temperature distribution in glaciers and ice sheets. *Journal of Glaciology* 2, 523–532.

Robin, G. de Q. 1975. Ice shelves and ice flow. *Nature* 253, 168–172.

Robin, G. de Q. 1976. Is the basal ice of a temperate glacier at the pressure melting point? *Journal of Glaciology* 16, 183–196.

Robin, G. de Q. 1979. Formation, flow, and disintegration of ice shelves. *Journal of Glaciology* 24, 259–271.

Robin, G. de Q. 1983. *The Climatic Record in Polar Ice Sheets*. Cambridge University Press, Cambridge.

Robin, G. de Q. and Barnes, P. 1969. Propagation of glacier surges. *Canadian Journal of Earth Sciences* 6, 969–977.

Robin, G. de Q., Drewry, D.J. and Meldrum, D.T. 1977. International studies of ice sheet and bedrock. *Philo-*

sophical Transactions of the Royal Society of London series B 279, 185–196.

Robinson, G., Peterson, J.A. and Anderson, P.M. 1971. Trend surface analysis of corrie altitudes in Scotland. *Scottish Geographical Magazine* 87, 142–146.

Robinson, W. and Haskell, T.G. 1992. Calving of Erebus glacier tongue. *Nature* 346, 615–616.

Rocha-Campos, A.C. and Hasui, Y. 1981. Tillites of the Macaubas Group (Proterozoic) in central Minas Gerais and southern Bahia, Brazil. In Hambrey, M.J. and Harland, W.B. (eds), *Earth's Pre-Pleistocene Glacial Record*. Cambridge University Press, Cambridge, 933–939.

Rodahl, K. 1954. Ice islands in the arctic. *Scientific American* 191, 40–45.

Rodine, J.D. and Johnson, A.M. 1976. The ability of debris, heavily freighted with coarse clastic materials, to flow down gentle slopes. *Sedimentology* 23, 213–234.

Roed, M.A. 1975. Cordilleran and Laurentide multiple glaciation, west central Alberta. *Canadian Journal of Earth Sciences* 12, 1493–1515.

Roed, M.A. and Waslyk, D.G. 1973. Age of inactive alluvial fans: Bow River Valley, Alberta. *Canadian Journal of Earth Sciences* 10, 1834–1840.

Rogerson, R.J. and Batterson, M.J. 1982. Contemporary push moraine formation in the Yoho Valley, BC. In Davidson-Arnott, R., Nickling, W. and Fahey, B.D. (eds), *Research in Glacial, Glacio-fluvial and Glaciolacustrine Systems*. Geo Books, Norwich, 71–90.

Ronnert, L. and Mickelson, D.M. 1992. High porosity of basal till at Burroughs Glacier, southeastern Alaska. *Geology* 20, 849–852.

Rose, J. 1981. Field guide to the Quaternary geology of the southern part of the Loch Lomond basin. *Proceedings of the Geological Society of Glasgow* 122/123, 12–28.

Rose, J. (ed.) 1983. *Diversion of the Thames: Field Guide*. Quaternary Research Association, Cambridge.

Rose, J. 1987a. Status of the Wolstonian Glaciation in the British Quaternary. *Quaternary Newsletter* 53, 1–9.

Rose, J. 1987b. Drumlins as part of a glacier bedform continuum. In Menzies, J. and Rose, J. (eds), *Drumlin Symposium*. Balkema, Rotterdam, 103–116.

Rose, J. 1989a. Stadial type sections in the British Quaternary. In Rose, J. and Schlüchter, C. (eds), *Quaternary Type Sections: Imagination or Reality?* Balkema, Rotterdam, 45–67.

Rose, J. 1989b. Glacier stress patterns and sediment transfer associated with the formation of superimposed flutes. *Sedimentary Geology* 62, 151–176.

Rose, J. 1989c. Tracing the Baginton–Lillington sands and gravels from the West Midlands to East Anglia. In Keen, D.H. (ed.), *The Pleistocene of the West Midlands: Field Guide*. Quaternary Research Association, Cambridge, 102–110.

Rose, J. 1992. Boulder clusters in glacial flutes. *Geomorphology* 6, 51–58.

Rose, J. and Letzer, J.M. 1977. Superimposed drumlins. *Journal of Glaciology* 18, 471–480.

Rosqvist, G. 1990. Quaternary glaciations in Africa. *Quaternary Science Reviews* 9, 281–297.

Röthlisberger, F. 1986. *10,000 Jahre Gletschergeschichte der Erde*. Verlag Sauerlander, Aarau.

Röthlisberger, F., Haas, P., Holzhauser, H., Keller, W., Bircher, W. and Renner, F. 1980. Holocene climatic fluc-

tuations: radiocarbon dating of fossil soils (fAh) and woods from moraines and glaciers in the Alps. *Geographica Helvetica* 35, 21–52.

Röthlisberger, H. 1972. Water pressure in intra- and subglacial channels. *Journal of Glaciology* 11, 177–203.

Röthlisberger, H. and Iken, A. 1981. Plucking as an effect of water-pressure variations at the glacier bed. *Annals of Glaciology* 2, 57–62.

Röthlisberger, H. and Lang, H. 1987. Glacial hydrology. In Gurnell, A.M. and Clark, M.J. (eds), *Glacio-fluvial Sediment Transfer*. Wiley, New York, 207–284.

Rotnicki, K. 1976. The theoretical basis for and a model of glaciotectonic deformations. *Quaestiones Geographicae* 3, 103–139.

Rott, H., Skvarca, P. and Nagler, T. 1996. Rapid collapse of Northern Larsen Ice Shelf, Antarctica. *Science* 271, 788–792.

Rudberg, S. 1954. Vasterbottens berggrundsmorfologi. Ett forsok till rekonstruktion av preglaciala erosions- generationer i Sverige. *Geografica* 25.

Rudberg, S. 1973. Glacial erosion forms of medium size: a discussion based on four Swedish case studies. *Zeitschrift für Geomorphologie* 17, 33–48.

Ruddiman, W.F. and Kutzbach, J.E. 1991. Plateau uplift and climatic change. *Scientific American* 264, 42–50.

Ruddiman, W.F. and McIntyre, A. 1979. Warmth of the subpolar North Atlantic Ocean during northern hemisphere ice sheet growth. *Science* 204, 173–175.

Ruddiman, W.F. and McIntyre, A. 1981. The North Atlantic during the last deglaciation. *Palaeogeography, Palaeoclimatology, Palaeoecology* 35, 145–214.

Ruddiman, W.F., Raymo, M.E., Martinson, D.G., Clement, B.M. and Backman, J. 1989. Pleistocene evolution: northern hemisphere ice sheets and North Atlantic Ocean. *Paleoceanography* 4, 353–412.

Ruddiman, W.F. and Wright, H.E. (eds) 1987. *North America and Adjacent Oceans during the Last Deglaciation*. Geological Society of America, Boulder.

Rudoy, A.N. 1990. Ice flow and ice-dammed lakes of the Altay in the Pleistocene. *Izvestiya Vsesoyuznogo Geograficheskogo Obstichestva* 122, 43–52.

Rudoy, A.N. and Baker, V.R. 1993. Sedimentary effects of cataclysmic late Pleistocene glacial outburst flooding, Altay Mountains, Siberia. *Sedimentary Geology* 85, 53–62.

Rudoy, A.N., Galachov, V.P. and Danilin, A.L. 1989. Reconstruction of glacial discharge in the head of the Chuja River and alimentation of ice-dammed lakes in the late Pleistocene. *Izvestiya Vsesoyuznogo Geograficheskogo Obstichestva* 121, 236–244.

Ruff, S.W. and Greeley, R. 1990. Sinuous ridges of the south polar region, Mars: possible origins. *Lunar and Planetary Science* 21, 1047–1048.

Rundle, A.S. 1985. The mechanism of braiding. *Zeitschrift für Geomorphologie*, SB 55, 1–13.

Russell, A.J. 1989. A comparison of two recent jökulhlaups from an ice-dammed lake, Sondre Stromfjord, west Greenland. *Journal of Glaciology* 35, 157–162.

Russell, I.C. 1901. *Glaciers of North America*. Ginn & Co., Boston.

Russell, R.J. 1933. Alpine landforms of western United States. *Geological Society of America Bulletin* 44, 927–949.

Russell-Head, D.S. 1980. The melting of free-drifting icebergs. *Annals of Glaciology* 1, 119–122.

Russell-Head, D.S. and Budd, W.F. 1979. Ice-sheet flow properties derived from bore-hole shear measurements combined with ice-core studies. *Journal of Glaciology* 24, 117–130.

Rust, B.R. 1972a. Pebble orientation in fluvial sediments. *Journal of Sedimentary Petrology* 42, 384–388.

Rust, B.R. 1972b. Structure and process in a braided river. *Sedimentology* 18, 221–245.

Rust, B.R. 1975. Fabric and structure in glaciofluvial gravels. In Jopling, A.V. and McDonald, B.C. (eds), *Glaciofluvial and Glaciolacustrine Sedimentation*. SEPM Special Publication 23, 238–248.

Rust, B.R. 1977. Mass flow deposits in a Quaternary succession near Ottawa, Canada: diagnostic criteria for subaqueous outwash. *Canadian Journal of Earth Sciences* 14, 175–184.

Rust, B.R. 1978. Depositional models for braided alluvium. In Miall, A.D. (ed.), *Fluvial Sedimentology*. Canadian Society of Petroleum Geologists, Memoir 5, 605–625.

Rust, B.R. and Gibling, M.R. 1990. Three-dimensional antidunes as HCS mimics an a fluvial sandstone: the Pennsylvanian South Bar Formation near Sydney, Nova Scotia. *Journal of Sedimentary Petrology* 60, 540–548.

Rust, B.R. and Romanelli, R. 1975. Late Quaternary subaqueous outwash deposits near Ottawa, Canada. In Jopling, A.V. and McDonald, B.C. (eds), *Glaciofluvial and Glaciolacustrine Sedimentation*. SEPM Special Publication 23, 177–192.

Ruszczynska-Szenajch, H. 1976. Glacitektoniczne depresje i kry lodowcowe na tle budowy geologicznej południowo-wschodniego Mazowsza i południowego Podlasia (Glacitectonic depressions and glacial rafts in mid-eastern Poland). *Studia Geologica Polonica* 50, 1–106.

Ruszczynska-Szenajch, H. 1978. Glacitectonic origin of some lake basins in areas of Pleistocene glaciations. *Polskie Archiwum Hydrobiologii* 25, 373–381.

Ruszczynska-Szenajch, H. 1987. The origin of glacial rafts: detachment, transport, deposition. *Boreas* 16, 101–112.

Rutten, M.G. 1960. Ice-pushed ridges, permafrost and drainage. *American Journal of Science* 258, 293–297.

Rutter, N.W. 1980. Late Pleistocene history of the western Canadian ice-free corridor. *Canadian Journal of Anthropology* 1, 1–8.

Rutter, N.W. 1984. Pleistocene history of the western Canadian ice-free corridor. In Fulton, R.J. (ed.), *Quaternary Stratigraphy of Canada*. Geological Survey of Canada, Paper 84–10, 49–56.

Rutter, N.W. 1995. Problematic ice sheets. *Quaternary International* 28, 19–37.

Ryder, J.M. 1971a. The stratigraphy and morphology of paraglacial alluvial fans in south-central British Columbia. *Canadian Journal of Earth Sciences* 8, 279–298.

Ryder, J.M. 1971b. Some aspects of the morphometry of paraglacial alluvial fans in south-central British Columbia. *Canadian Journal of Earth Sciences* 8, 1252–1264.

Ryder, J.M. 1981. Geomorphology of the southern part of the Coast Mountains of British Columbia. *Zeitschrift für Geomorphologie*, Supp. 37, 120–147.

Ryder, J.M. 1987. Neoglacial history of the Stikine–Iskut area, northern Coast Mountains, British Columbia. *Canadian Journal of Earth Sciences* 24, 1294–1301.

Ryder, J.M. 1989. Holocene glacier fluctuations (Canadian Cordillera). In Fulton, R.J. (ed.), *Quaternary Geology of Canada and Greenland*. Geological Survey of Canada, Geology of Canada No. 1, 74–76.

Ryder, J.M. 1995. Recognition and interpretation of flow direction indicators for former glaciers and meltwater streams. In Bobrowsky, P.T., Sibbick, S.J., Newell, J.M. and Matysek, P.F. (eds), *Drift Exploration in the Canadian Cordillera, British Columbia*. Ministry of Energy, Mines and Petroleum Resources, Paper 1995–2, 1–22.

Ryder, J.M., Fulton, R.J. and Clague, J.J. 1991. The Cordilleran ice sheet and the glacial geomorphology of southern and central British Columbia. *Géographie Physique et Quaternaire* 45, 365–377.

Ryder, J.M. and Thomson, B. 1986. Neoglaciation in the southern Coast Mountains of British Columbia: chronology prior to the late Neoglacial maximum. *Canadian Journal of Earth Sciences* 23, 273–287.

Sabadini, R., Lambeck, K. and Boschi, E. (eds) 1991. *Glacial Isostasy, Sea Level and Mantle Rheology*. Kluwer, Dordrecht.

Saettem, J. 1990. Glaciotectonic forms and structures on the Norwegian continental shelf: observations, processes and implications. *Norsk Geologisk Tidsskrift* 70, 81–94.

Saettem, J. 1994. Glaciotectonic structures along the southern Barents shelf margin. In Warren, W.P. and Croot, D.G. (eds), *Formation and Deformation of Glacial Deposits*. Balkema, Rotterdam, 95–113.

Sagar, R.B. 1966. Glaciological and climatological studies on the Barnes Ice Cap, 1962–64. *Geographical Bulletin* 8, 3–47.

Sahlstrom, K.E. 1914. Glacial skulptur i Stockholms yttre skargard. *Sveriges Geologiske Undersoekelse* C258, 1–36.

St Onge, D.A. 1972. Sequence of glacial lakes in north-central Alberta. Geological Survey of Canada Bulletin 213.

St Onge, D.A. and Lajoie, J. 1986. The late Wisconsinan olistostrome of the lower Coppermine River valley, Northwest Territories. *Canadian Journal of Earth Sciences* 23, 1700–1708.

St Onge, D.A. and McMartin, I. 1995. Quaternary geology of the Inman River area, Northwest Territories. Geological Survey of Canada, Bulletin 446.

Salinger, M.J., Heine, M.J. and Burrows, C.J. 1983. Variations of the Stocking (Te Wae Wae) Glacier, Mount Cook, and climatic relationships. *New Zealand Journal of Science* 26, 321–328.

Salisbury, R.D. 1896. Stratified drift. *Journal of Geology* 4, 948–970.

Salonen, V.-P. 1986. Glacial transport distance distributions of surface boulders in Finland. Geological Survey of Finland Bulletin 338.

Salonen, V.-P. 1987. Observations on boulder transport in Finland. Geological Survey of Finland, Special Paper 3, 103–110.

Sammis, C., King, G. and Biegel, R. 1987. The kinematics of gouge deformation. *Pure and Applied Geophysics* 125, 777–812.

Sanderson, T.J.O. 1979. Equilibrium profile of ice shelves. *Journal of Glaciology* 22, 435–460.

Sardeson, F.W. 1905. A particular case of glacial erosion. *Journal of Geology* 13, 351–357.

Sardeson, F.W. 1906. The folding of subjacent strata by glacial action. *Journal of Geology* 14, 226–232.

Sauer, E.K. 1978. The engineering significance of glacier ice-thrusting. *Canadian Geotechnical Journal* 15, 457–472.

Saunderson, H.C. 1975. Sedimentology of the Brampton esker and its associated deposits: an empirical test of theory. In Jopling, A.V. and McDonald, B.C. (eds), *Glaciofluvial and Glaciolacustrine Sedimentation.* SEPM Special Publication 23, 155–176.

Saunderson, H.C. 1977. The sliding bed facies in esker sands and gravels: a criterion for full-pipe (tunnel) flow? *Sedimentology* 24, 623–638.

Sauramo, M. 1923. Studies on the Quaternary varve sediments in southern Finland. *Bulletin de la Commission Géologique de Finlande* 60.

Sauramo, M. 1929. The Quaternary Geology of Finland. *Bulletin de la Commission Géologique de Finlande* 86.

de Saussure H.B. 1786. *Voyages dans les Alpes*, vol II. Barde, Manget & Cie, Geneva.

Savage, W.Z. 1968. Application of plastic flow analyses to drumlin formation. Unpublished MSc thesis, Syracuse University, New York.

Schafer, J.P. and Hartshorn, J.H. 1965. The Quaternary of New England. In Wright, H.E. and Frey, D.G. (eds), *The Quaternary of the United States*. Princeton University Press, Princeton, NJ, 113–127.

Schei, B., Eilertsen, H.G., Falk-Larsen, S., Gullikson, B. and Taasen, J.P. 1979. Marinbiologiske undersokelser i Van Mijenfjorden (Vest Spitsbergen) etter objesolle-kasje ved Sveagruva 1978. Tromsø Museums Rapportserie, Naturvitenshap, no 2. Universitetet i Tromsø 2.

Schermerhorn, L.J.G. 1966. Terminology of mixed coarse–fine sediments. *Journal of Sedimentary Petrology* 36, 831–835.

Schermerhorn, L.J.G. 1974. Late Precambrian mixtites: glacial and/or non-glacial? *American Journal of Science* 274, 673–824.

Schermerhorn, L.J.G. 1975. Tectonic framework of Late Precambrian supposed glacials. In Wright, A.E. and Moseley, F. (eds), *Ice Ages: Ancient and Modern*. Seel House Press, Liverpool, 241–274.

Schlüchter, C. 1986. The Quaternary glaciations of Switzerland, with special reference to the northern Alpine Foreland. *Quaternary Science Reviews* 5, 413–419.

Schneider, T. and Bronge, C. 1996. Suspended sediment transport in the Storglaciären drainage basin. *Geografiska Annaler* 78A, 155–161.

Schumm, S.A. 1977. *The Fluvial System*. Wiley, New York.

Schumm, S.A. and Khan, H.R. 1972. Experimental study of channel patterns. *Geological Society of America Bulletin* 83, 1755–1770.

Schumm, S.A. and Lichty, R.W. 1965. Time, space and causality in geomorphology. *American Journal of Science* 263, 110–119.

Schwab, W.C. and Lee, H.J. 1983. Geotechnical analyses of submarine landslides in glacial-marine sediment, NE Gulf of Alaska. In Molnia, B.F. (ed.), *Glacial-Marine Sedimentation*. Plenum Press, New York, 145–184.

Schwab, W.C. and Lee, H.J. 1988. Causes of two slope failure types in continental shelf sediment, northeast Gulf of Alaska. *Journal of Sedimentary Petrology* 58, 1–11.

Schwab, W.C., Lee, H.J. and Molnia, B.F. 1987. Causes of varied sediment gravity flow types on the Alsek prodelta, northeast Gulf of Alaska. *Marine Geotechnology* 7, 312–342.

Schweizer, J. and Iken, A. 1992. The role of bed separation and friction in sliding over an undeformable bed. *Journal of Glaciology* 38, 77–92.

Schytt, V. 1956. Lateral drainage channels along the northern side of the Moltke Glacier, northwest Greenland. *Geografiska Annaler* 38, 64–77.

Schytt, V. 1959. The glaciers of the Kebnekajse massif. *Geografiska Annaler* 41, 213–227.

Schytt, V. 1962. Mass balance studies in Kebnekajse. *Journal of Glaciology* 4, 281–288.

Schytt, V. 1963. Fluted moraine surfaces: letter. *Journal of Glaciology* 4, 825–827.

Schytt, V. 1967. A study of 'ablation gradient'. *Geografiska Annaler* 49A, 327–332.

Schytt, V. 1981. The net mass balance of Storgläciaren, Kebnekaise, Sweden, related to the height of the equilibrium line and to the height of the 500 mb surface. *Geografiska Annaler* 63A, 219–223.

Scott, D.H. and Underwood, J.R. 1991. Mottled terrain: a continuing Martian enigma. In Ryder, G. and Sharpton, V.L. (eds), *Proceedings of the 21st Lunar and Planetary Science Conference*. Lunar and Planetary Institute, Houston, Texas, 627–634.

Seaberg, S.Z., Seaberg, J.Z., Hooke, R. Le B. and Wiberg, D.W. 1988. Character of the englacial and subglacial drainage system in the lower part of the ablation area of Storglaciären, Sweden, as revealed by dye-trace studies. *Journal of Glaciology* 34, 217–227.

Seddon, B. 1957. Late-glacial cwm glaciers in Wales. *Journal of Glaciology* 3, 94–99.

Seilacher, A. 1953. Studien zur Palichnologie. *Neues Jahrbuch Geologie und Palaeontologie, Abhandlungen* 96, 421–452.

Sejrup, H.P., Aarseth, I., Ellingsen, K.L., Reither, E., Jansen, E., Lovlie, R., Bent, A., Brigham-Grette, J., Larsen, E. and Stoker, M. 1987. Quaternary stratigraphy of the Fladen area, central North Sea: a multidisciplinary study. *Journal of Quaternary Science* 2, 35–58.

Sejrup, H.P., Miller, G.H., Brigham-Grette, J., Lovlie, R. and Hopkins, D. 1984. Amino acid epimerization implies rapid sedimentation rates in Arctic Ocean cores. *Nature* 310, 772–775.

Selby, M.J. 1983. *Hillslope Materials and Processes*. Oxford University Press, Oxford.

Selby, M.J. 1993. *Hillslope Materials and Processes*, 2nd edition. Oxford University Press, Oxford.

Sexton, D.J., Dowdeswell, J.A., Solheim, A. and Elverhøi, A. 1992. Seismic architecture and sedimentation in northwest Spitsbergen fjords. *Marine Geology* 103, 53–68.

Shabtaie, S. and Bentley, C.R. 1987. West Antarctic ice streams draining into the Ross Ice Shelf: configuration and mass balance. *Journal of Geophysical Research* 92, 1311–1336.

Shackleton, N.J. 1987. Oxygen isotopes, ice volume and sea level. *Quaternary Science Reviews* 6, 183–190.

Shackleton, N.J. and Kennett, J.P. 1975. Paleotemperature history of the Cenozoic and the initiation of Antarctic glaciation: oxygen and carbon isotope analysis in DSDP sites 277, 279 and 281. In Kennett, J.P. and Houtz, R.E. (eds), *Initial Reports of the Deep Sea Drilling Project*. US Government Printing Office, Washington DC, 743–755.

Shackleton, N.J. and Opdyke, N.D. 1973. Oxygen isotope and palaeomagnetic stratigraphy of equatorial Pacific core V28–238: oxygen isotope temperatures and ice volumes on a 105 and 106 year scale. *Quaternary Research* 3, 39–55.

Shafer, N.E. and Zare, R.N. 1991. Through a beer glass darkly. *Physics Today* 44, 48–52.

Shakesby, R.A. and Matthews, J.A. 1993. Loch Lomond Stadial glacier at Fan Hir, Mynydd Du (Brecon Beacons), south Wales: critical evidence and palaeoclimatic implications. *Geological Journal* 28, 69–79.

Shaler, N.S. 1874. Preliminary report on the recent changes of level on the coast of Maine. *Memoirs of the Boston Society of Natural History* 2, 321–340.

Shaler, N.S. 1888. Geology of Martha's Vineyard. US Geological Survey, Report for 1886, 3, 297–363.

Shaler, N.S. 1898. Geology of the Cape Cod district. US Geological Survey, Report for 1896–97, 2, 497–593.

Sharp, M.J. 1982. Modification of clasts in lodgement tills by glacial erosion. *Journal of Glaciology* 28, 475–481.

Sharp, M.J. 1984. Annual moraine ridges at Skalafellsjökull, south-east Iceland. *Journal of Glaciology* 30, 82–93.

Sharp, M.J. 1985a. 'Crevasse-fill' ridges: a landform type characteristic of surging glaciers? *Geografiska Annaler* 67A, 213–220.

Sharp, M.J. 1985b. Sedimentation and stratigraphy at Eyjabakkajökull – an Icelandic surging glacier. *Quaternary Research* 24, 268–284.

Sharp, M.J. 1988a. Surging glaciers: behaviour and mechanisms. *Progress in Physical Geography* 12, 349–370.

Sharp, M. 1988b. Surging glaciers: geomorphic effects. *Progress in Physical Geography* 12, 533–559.

Sharp, M.J. 1992. Influence of glacier hydrology on the dynamics of a large Quaternary ice sheet. *Journal of Quaternary Science* 7, 109–124.

Sharp, M.J., Brown, G.H., Tranter, M., Willis, I.C. and Hubbard, B. 1995a. Comments on the use of chemically based mixing models in glacier hydrology. *Journal of Glaciology* 41, 241–246.

Sharp, M.J., Dowdeswell, J.A. and Gemmell, J.C. 1989a. Reconstructing past glacier dynamics and erosion from glacial geomorphic evidence: Snowdon, North Wales. *Journal of Quaternary Science* 4, 115–130.

Sharp, M.J., Gemmell, J.C. and Tison, J.-L. 1989b. Structure and stability of the former subglacial drainage system of the Glacier de Tsanfleuron, Switzerland. *Earth Surface Processes and Landforms* 14, 119–134.

Sharp, M.J. and Gomez, B. 1986. Processes of debris comminution in the glacial environment and implications for quartz sand-grain micromorphology. *Sedimentary Geology* 46, 33–47.

Sharp, M.J., Jouzel, J., Hubbard, B. and Lawson, W. 1994. The character, structure and origin of the basal ice layer of a surge-type glacier. *Journal of Glaciology* 40, 327–340.

Sharp, M.J., Lawson, W. and Anderson, R.S. 1988. Tectonic processes in a surge-type glacier. *Journal of Structural Geology* 10, 499–515.

Sharp, M.J., Tison, J.-L. and Fierens, G. 1990. Geochemistry of subglacial calcites: implications for the hydrology of the basal water film. *Arctic and Alpine Research* 22, 141–152.

Sharp, M.J., Tranter, M., Brown, G.H. and Skidmore, M. 1995b. Rates of chemical denudation and CO_2 drawdown in a glacier-covered catchment. *Geology* 23, 61–64.

Sharp, R.P. 1949. Studies of superglacial debris on valley glaciers. *American Journal of Science* 247, 289–315.

Sharp, R.P. 1951. Features of the firn on Upper seward Glacier, St. Elias Mountains, Canada. *Journal of Geology* 59, 599–621.

Sharp, R.P. 1974. Ice on Mars. *Journal of Glaciology* 13, 173–186.

Sharp, R.P. 1988. *Living Ice: Understanding Glaciers and Glaciation*. Cambridge University Press, Cambridge.

Sharp, R.P. and Malin, M.C. 1975. Channels on Mars. *Geological Society of America Bulletin* 86, 593–609.

Sharpe, D.R. 1982. Allan Park kame-delta. In Karrow, P.F., Jopling, A.V. and Martini, I.P. (eds), *International Association of Sedimentologists 11th International Congress, Excursion 11A – Late Quaternary Sedimentary Environments of a Glaciated Area: Southern Ontario*. McMaster University, Hamilton, 53–60.

Sharpe, D.R. 1984. Comments on 'Sedimentation in a large lake: a reinterpretation of the Late Pleistocene stratigraphy at Scarborough Bluffs, Ontario, Canada'. *Geology* 12, 186–187.

Sharpe, D.R. 1987. The stratified nature of drumlins from Victoria Island and southern Ontario, Canada. In Menzies, J. and Rose, J. (eds), *Drumlin Symposium*. Balkema, Rotterdam, 185–214.

Sharpe, D.R. 1988a. Glaciomarine fan deposits in the Champlain Sea. In Gadd, N.R. (ed.), *The Late Quaternary Development of the Champlain Sea Basin*. Geological Association of Canada Special Paper 35, 63–82.

Sharpe, D.R. 1988b. Late glacial landforms of Wollaston Peninsula, Victoria Island, NWT: product of ice-marginal retreat, surge, and mass stagnation. *Canadian Journal of Earth Sciences* 25, 262–279.

Sharpe, D.R. and Shaw, J. 1989. Erosion of bedrock by subglacial meltwater, Cantley, Quebec. *Geological Society of America Bulletin* 101, 1011–1020.

Shaw, G. 1944. Moraines of Late Pleistocene ice fronts near James Bay, Quebec. *Transactions of the Royal Society of Canada, Section IV*, 79–85.

Shaw, J. 1972. Sedimentation in the ice-contact environment, with examples from Shropshire, England. *Sedimentology* 18, 23–62.

Shaw, J. 1975. Sedimentary successions in Pleistocene ice marginal lakes. In Jopling, A.V. and McDonald, B.C. (eds), *Glaciofluvial and Glaciolacustrine Sedimentation*. SEPM Special Publication 23, 281–303.

Shaw, J. 1977a. Till body morphology and structure related to glacier flow. *Boreas* 6, 189–201.

Shaw, J. 1977b. Sedimentation in an alpine lake during deglaciation, Okanagan Valley, British Columbia, Canada. *Geografiska Annaler* 59A, 221–240.

Shaw, J. 1977c. Tills deposited in arid polar environments. *Canadian Journal of Earth Sciences* 14, 1239–1245.

Shaw, J. 1979. Genesis of the Sveg tills and Rogen moraines of central Sweden: a model of basal melt-out. *Boreas* 8, 409–426.

Shaw, J. 1980. Drumlins and large scale flutings related to glacier folds. *Arctic and Alpine Research* 12, 287–298.

Shaw, J. 1982. Melt-out till in the Edmonton area, Alberta, Canada. *Canadian Journal of Earth Sciences* 19, 1548–1569.

Shaw, J. 1983a. Drumlin formation related to inverted meltwater erosional marks. *Journal of Glaciology* 29, 461–479.

Shaw, J. 1983b. Forms associated with boulders in melt-out till. In Evenson, E.B., Schluchter, C. and Rabassa, J. (eds), *Tills and Related Deposits*. Balkema, Rotterdam, 3–12.

Shaw, J. 1985. Subglacial and ice marginal environments. In Ashley, G.M., Shaw, J. and Smith, N.D. (eds), *Glacial Sedimentary Environments*. SEPM Short Course 16, 7–84.

Shaw, J. 1987. Glacial sedimentary processes and environmental reconstruction based on lithofacies. *Sedimentology* 34, 103–116.

Shaw, J. 1988a. Sublimation till. In Goldthwait, R.P. and Matsch, C.L. (eds), *Genetic Classification of Glacigenic Deposits*. Balkema, Rotterdam, 141–142.

Shaw, J. 1988b. Coarse-grained sediment flow facies in a large supraglacial lake: discussion. *Sedimentology* 35, 527–529.

Shaw, J. 1988c. Subglacial erosional marks, Wilton Creek, Ontario. *Canadian Journal of Earth Sciences* 25, 1256–1267.

Shaw, J. 1989. Drumlins, subglacial meltwater floods, and ocean responses. *Geology* 17, 853–856.

Shaw, J. 1993. Geomorphology. In Godfrey, J.D. (ed.), *Edmonton beneath our Feet: A Guide to the Geology of the Edmonton Region*. Edmonton Geological Society, University of Alberta, 21–32.

Shaw, J. 1994. Hairpin erosional marks, horseshoe vortices and subglacial erosion. *Sedimentary Geology* 91, 269–283.

Shaw, J. and Archer, J. 1978. Winter turbidity current deposits in Late Pleistocene glaciolacustrine varves, Okanagan Valley, British Columbia, Canada. *Boreas* 7, 123–130.

Shaw, J. and Archer, J. 1979. Deglaciation and glaciolacustrine sedimentation conditions, Okanagan Valley, British Columbia, Canada. In Schluchter, C. (ed.), *Moraines and Varves*. Balkema; Rotterdam, 347–355.

Shaw, J. and Carter, R.W.G. 1980. Late Midlandian sedimentation and glaciotectonics of the North Antrim end moraine. *Irish Naturalist Journal* 20, 67–69.

Shaw, J. and Freschauf, R.C. 1973. A kinematic discussion of the formation of glacial flutings. *Canadian Geographer* 17, 19–35.

Shaw, J., Gilbert, R. and Archer, J. 1978. Proglacial lacustrine sedimentation during winter. *Arctic and Alpine Research* 10, 689–699.

Shaw, J. and Healy, T.R. 1980. Morphology of the Onyx River system, McMurdo Sound region, Antarctica. *New Zealand Journal of Geology and Geophysics* 23, 223–238.

Shaw, J. and Kellerhals, R. 1977. Palaeohydraulic interpretation of antidune bedforms with applications to antidunes in gravels. *Journal of Sedimentary Petrology* 47, 257–266.

Shaw, J. and Kvill, D. 1984. A glaciofluvial origin for drumlins of the Livingstone Lake area, Saskatchewan. *Canadian Journal of Earth Sciences* 12, 1426–1440.

Shaw, J., Kvill, D. and Rains, R.B. 1989. Drumlins and catastrophic subglacial floods. *Sedimentary Geology* 62, 177–202.

Shaw, J. and Sharpe, D.R. 1987. Drumlin formation by subglacial meltwater erosion. *Canadian Journal of Earth Sciences* 24, 2316–2322.

Sheldon, P. 1926. Significant characteristics of glacial erosion as illustrated by an erosion channel. *Journal of Geology* 34, 257–265.

Shennan, I. 1989. Holocene crustal movements and sea level changes in Great Britain. *Journal of Quaternary Science* 4, 77–89.

Shennan, L., Innes, J.B., Long, A.J. and Zong, Y. 1993. Late Devensian and Holocene relative sea-level changes at Rumach, near Arisaig, northwest Scotland. *Norsk Geologisk Tidsskrift* 73, 161–174.

Shepard, F.P. 1960. Rise of sea level along north-west Gulf of Mexico: recent sediments. In Shepard, F.P., Phleger, F.B. and van Andel, T.H. (eds), *Recent Sediments, Northwest Gulf of Mexico*. American Association of Petroleum Geologists, Tulsa, 338–344.

Shepard, F.P. 1963. Thirty-five thousand years of sea level. In *Essays in Honor of K.O. Emery*. University of Southern California Press, Los Angeles, 1–10.

Sheridan, M.F., Wohletz, K. and Dehn, J. 1987. Discrimination of grain-size subpopulations in pyroclastic deposits. *Geology* 15, 367–370.

Shetsen, I. 1987. Quaternary Geology, Southern Alberta. Alberta Research Council Map, Edmonton, Alberta.

Shi, Y., Ren, B, Wang, J. and Derbyshire, E. 1986. Quaternary glaciation in China. *Quaternary Science Reviews* 5, 503–507.

Shih Ya-feng, Hsieh Tze-chu, Cheng Pen-hsing and Li Chi-chun 1980. Distribution, features and variations of glaciers in China. In *World Glacier Inventory, Proceedings of the Workshop at Riederalp, 1978*. IAHS Publication 126, 111–116.

Shih Ya-feng and Wang Jingtai 1979. *The Fluctuations of Climate, Glaciers and Sea Level since the Late Pleistocene in China*. Langzhou Institute of Glaciology and Cryopedology and Desert Research.

Shilts, W.W. 1971. Till studies and their application to regional drift prospecting. *Canadian Mining Journal* 92, 45–50.

Shilts, W.W. 1975. Principles of geochemical exploration for sulphide deposits using shallow samples of glacial drift. *Canadian Institute of Mining Bulletin* 68, 73–80.

Shilts, W.W. 1976. Glacial till and mineral exploration. In Leggett, R.F. (ed.), *Glacial Till*. Royal Society of Canada Special Publication 12, 205–224.

Shilts, W.W. 1977. Geochemistry of till in perennially frozen terrain of the Canadian Shield: application to prospecting. *Boreas* 6, 203–212.

Shilts, W.W. 1978. Detailed sedimentological study of till sheets in a stratigraphic section, Samson River, Quebec. Geological Survey of Canada Bulletin 285.

Shilts, W.W. 1980. Flow patterns in the central North American ice sheet. *Nature* 286, 213–218.

Shilts, W.W. 1982a. Glacial dispersal: principles and practical applications. *Geoscience Canada* 9, 42–48.

Shilts, W.W. 1982b. Quaternary evolution of the Hudson/James Bay region. *Naturaliste Canadien* 109, 309–332.

Shilts, W.W. 1984a. Esker sedimentation models, Deep Rose Lake map area, District of Keewatin. Geological Survey of Canada, Paper 84–1B, 217–222.

Shilts, W.W. 1984b. Quaternary events – Hudson Bay lowland and southern District of Keewatin. In Fulton, R.J. (ed.), *Quaternary Stratigraphy of Canada*. Geological Survey of Canada, Paper 84–10, 117–126.

Shilts, W.W. 1984c. Till geochemistry in Finland and Canada. *Journal of Geochemical Exploration* 21, 95–117.

Shilts, W.W. 1991. Principles of glacial dispersal and sedimentation. In Coker, W.B. (ed.), *Exploration Geochemistry Workshop*. Geological Survey of Canada, Open File 2390, 2/1–2/42.

Shilts, W.W. 1993. Geological Survey of Canada's contributions to understanding the composition of glacial sediments. *Canadian Journal of Earth Sciences* 30, 333–353.

Shilts, W.W., Aylsworth, J.M., Kaszycki, C.A. and Klassen, R.A. 1987. Canadian Shield. In Graf, W.L. (ed.), *Geomorphic Systems of North America*. Geological Society of America, Centennial Special Volume 2, 119–161.

Shilts, W.W., Cunningham, C.M. and Kaszycki, C.A. 1979. Keewatin ice sheet: re-evaluation of the traditional concept of the Laurentide ice sheet. *Geology* 7, 537–541.

Shilts, W.W. and Kettles, I.M. 1990. Geochemical/mineralogical profiles through fresh and weathered till. In Kujansuu, R. and Saarnisto, M. (eds), *Handbook of Glacial Indicator Tracing*. Balkema, Rotterdam, 187–216.

Shilts, W.W. and Wyatt, P.H. 1989. Gold and base metal exploration using drift as a sample medium, Kaminak Lake–Turquetal Lake area, District of Keewatin. Geological Survey of Canada, Open File 2132.

Shoemaker, E.M. 1986a. Subglacial hydrology for an ice sheet resting on a deformable aquifer. *Journal of Glaciology* 32, 20–30.

Shoemaker, E.M. 1986b. The formation of fjord thresholds. *Journal of Glaciology* 32, 65–71.

Shoemaker, E.M. 1988. On the formulation of basal debris drag for the case of sparse debris. *Journal of Glaciology* 34, 259–264.

Shoemaker, E.M. 1991. On the formation of large subglacial lakes. *Canadian Journal of Earth Sciences* 28, 1975–1981.

Shoemaker, E.M. 1992a. Subglacial floods and the origin of low-relief ice-sheet lobes. *Journal of Glaciology* 38, 105–112.

Shoemaker, E.M. 1992b. Water sheet outburst floods from the Laurentide Ice Sheet. *Canadian Journal of Earth Sciences* 29, 1250–1264.

Shoemaker, E.M. 1994. Reply to comments on 'Subglacial floods and the origin of low-relief ice-sheet lobes' by J.S. Walder. *Journal of Glaciology* 40, 201–202.

Shoemaker, E.M. and Leung, H.K.N. 1987. Subglacial drainage for an ice sheet resting on a layered deformable bed. *Journal of Geophysical Research* 92, 4936–4946.

Short, N.M. and Blair, R.W. 1986. *Geomorphology from Space*. NASA, Washington.

Shotton, F.W. 1977. *The English Midlands*. INQUA Excursion Guide A2, Xth INQUA Congress, Birmingham, 51 pp.

Shotton, F.W. 1983. The Wolstonian Stage of the British Pleistocene in and around its type area of the English Midlands. *Quaternary Science Reviews* 2, 261–280.

Shreve, R.L. 1966. Sherman landslide, Alaska. *Science* 154, 1639–43.

Shreve, R.L. 1968. Leakage and fluidization in air-layer lubricated landslides. *Geological Society of America Bulletin* 79, 653–658.

Shreve, R.L. 1972. Movement of water in glaciers. *Journal of Glaciology* 11, 205–214.

Shreve, R.L. 1984. Glacier sliding at subfreezing temperatures. *Journal of Glaciology* 30, 341–347.

Shreve, R.L. 1985a. Esker characteristics in terms of glacier physics, Katahdin esker system, Maine. *Geological Society of America Bulletin* 96, 639–646.

Shreve, R.L. 1985b. Late Wisconsin ice-surface profile calculated from esker paths and types, Katahdin esker system, Maine. *Quaternary Research* 23, 27–37.

Shroder, J.F., Owen, L.A. and Derbyshire, E. 1993. Quaternary glaciation of the Karakoram and Nanga Parbat Himalaya. In Shroder, J.F. (ed.), *Himalaya to the Sea*. Routledge, London, 132–158.

Shulmeister, J. 1989. A conceptual model for the deposition of the Dummer Moraine, southern Ontario. *Geomorphology* 2, 385–392.

Shultz, A.W. 1984. Subaerial debris flow deposition in the Upper Palaeozoic Cutler Formation, western Colorado. *Journal of Sedimentary Petrology* 54, 759–772.

Shumsky, P.A. 1950. The energy of glaciation and the life of glaciers. SIPRE Translation 7, Corps of Engineers, US Army.

Sibrava, V., Bowen, D.Q. and Richmond, G.M. 1986. *Quaternary Glaciations in the Northern Hemisphere*. Quaternary Science Reviews 5.

Siddans, A.W.B. 1984. Thrust tectonics: a mechanistic view from West and Central Alps. *Tectonophysics* 104, 257–281.

Siegenthaler, C. and Huggenberger, P. 1993. Pleistocene Rhine gravel: deposits of a braided river system with dominant pool preservation. In Best, J.L. and Bristow, C.S. (eds), *Braided Rivers*. Geological Society of London Special Publication 75, 147–162.

Sirkin, L. 1980. Wisconsinan glaciation of Long Island, New York to Block Island, Rhode Island. In Larson, G.J. and Stone, B.D. (eds), *Late Wisconsin Glaciation of New England*. Kendall/Hunt, Dubuque, Iowa, 35–59.

Sissons, J.B. 1958a. Sub-glacial stream erosion in southern Northumberland. *Scottish Geographical Magazine* 74, 163–174.

Sissons, J.B. 1958b. Supposed ice-dammed lakes in Britain, with particular reference to the Eddleston Valley, southern Scotland. *Geografiska Annaler* 40, 159–187.

Sissons, J.B. 1958c. The deglaciation of part of East Lothian. *Transactions of the Institute of British Geographers* 25, 59–77.

Sissons, J.B. 1960a. Some aspects of glacial drainage systems in Britain, Part I. *Scottish Geographical Magazine* 76, 131–146.

Sissons, J.B. 1960b. Subglacial, marginal and other glacial drainage in the Syracuse–Oneida area, New

York. *Geological Society of America Bulletin* 71, 1575–1588.

Sissons, J.B. 1961a. Some aspects of glacial drainage systems in Britain, Part II. *Scottish Geographical Magazine* 77, 15–36.

Sissons, J.B. 1961b. A subglacial drainage system by the Tinto Hills, Lanarkshire. *Transactions of the Edinburgh Geological Society* 18, 175–193.

Sissons, J.B. 1963. The glacial drainage system around Carlops, Peebleshire. *Transactions of the Institute of British Geographers* 32, 95–111.

Sissons, J.B. 1967. *The Evolution of Scotland's Scenery.* Oliver & Boyd, Edinburgh.

Sissons, J.B. 1971. The geomorphology of central Edinburgh. *Scottish Geographical Magazine* 87, 185–196.

Sissons, J.B. 1972. Dislocation and non-uniform uplift of raised shorelines in the western part of the Forth Valley. *Transactions of the Institute of British Geographers* 55, 149–159.

Sissons, J.B. 1974a. The Quaternary in Scotland: a review. *Scottish Journal of Geology* 10, 311–337.

Sissons, J.B. 1974b. Late-glacial marine erosion in Scotland. *Boreas* 3, 41–48.

Sissons, J.B. 1974c. A Lateglacial ice cap in the central Grampians. *Transactions of the Institute of British Geographers* 62, 95–114.

Sissons, J.B. 1977a. The Loch Lomond Readvance in the northern mainland of Scotland. In Gray, J.M. and Lowe, J.J. (eds), *Studies in the Scottish Lateglacial Environment.* Pergamon, Oxford, 45–59.

Sissons, J.B. 1977b. Former ice-dammed lakes in Glen Moriston, Inverness-shire and their significance in upland Britain. *Transactions of the Institute of British Geographers* 2, 224–242.

Sissons, J.B. 1978. The Parallel Roads of Glen Roy and adjacent glens, Scotland. *Boreas* 7, 229–244.

Sissons, J.B. 1979a. The Loch Lomond Stadial in the British Isles. *Nature* 280, 199–203.

Sissons, J.B. 1979b. The Loch Lomond Advance in the Cairngorm Mountains. *Scottish Geographical Magazine* 95, 66–82.

Sissons, J.B. 1979c. Palaeoclimatic inferences from former glaciers in Scotland and the Lake District. *Nature* 278, 518–521.

Sissons, J.B. 1979d. The limit of the Loch Lomond Advance in Glen Roy and vicinity. *Scottish Journal of Geology* 15, 31–42.

Sissons, J.B. 1980. The Loch Lomond Advance in the Lake District, northern England. *Transactions of the Royal Society of Edinburgh, Earth Sciences* 71, 13–27.

Sissons, J.B. 1981. Ice dammed lakes in Glen Roy and vicinity: a summary. In Neale, J. and Flenley, J. (eds), *The Quaternary in Britain.* Pergamon, Oxford, 174–183.

Sissons, J.B. 1982. A former ice-dammed lake and associated glacier limits in the Achnasheen area, central Ross-shire. *Transactions of the Institute of British Geographers* NS 7, 98–116.

Sissons, J.B. 1983. Shorelines and isostasy in Scotland. In Smith, D.E. and Dawson, A.G. (eds), *Shorelines and Isostasy.* Academic Press, London, 209–225.

Sissons, J.B. and Brooks, C.L. 1971. Dating of early Postglacial land and sea level changes in the western Forth Valley. *Nature* 234, 124–127.

Sissons, J.B. and Cornish, R. 1982a. Differential glacio-isostatic uplift of crustal blocks at Glen Roy, Scotland. *Quaternary Research* 18, 268–288.

Sissons, J.B. and Cornish, R. 1982b. Rapid localized glacio-isostatic uplift at Glen Roy, Scotland. *Nature* 297, 213–214.

Sissons, J.B. and Dawson, A.G. 1981. Former sea levels and ice limits in part of Wester Ross, north-west Scotland. *Proceedings of the Geologists' Association* 12, 115–124.

Sissons, J.B. and Sutherland, D.G. 1976. Climatic inferences from former glaciers in the south-east Grampian Highlands. *Journal of Glaciology* 17, 325–346.

Sjorring, S. 1983. Ristinge Klint. In Ehlers, J. (ed.), *Glacial Deposits in North-West Europe.* Balkema, Rotterdam, 219–226.

Sjorring, S., Nielsen, P.E., Frederiksen, J.K., Hegner, J., Hyde, G., Jensen, J.B., Morgensen, A. and Vortisch, W. 1982. Observationer fra Ristinge Klint, felt- og laboratorie- undersogelser. *Dansk Geologisk Forening, Arsskrift for 1981*, 135–149.

Slater, G. 1927a. The structure of the disturbed deposits in the lower part of the Gipping Valley near Ipswich. *Proceedings of the Geologists' Association* 38, 157–182.

Slater, G. 1927b. The structure of the disturbed deposits of the Hadleigh Road area, Ipswich. *Proceedings of the Geologists' Association* 38, 183–261.

Slater, G. 1927c. The structure of the disturbed deposits of Moens Klint, Denmark. *Transactions of the Royal Society of Edinburgh* 55, 289–302.

Slater, G. 1927d. The disturbed glacial deposits in the neighbourhood of Lonstrup, near Hjorring, north Denmark. *Transactions of the Royal Society of Edinburgh* 55, 303–315.

Slater, G. 1927e. Structure of the Mud Buttes and Tit Hills in Alberta. *Bulletin of the Geological Society of America* 38, 721–730.

Slater, G. 1929. The structure of drumlins exposed on the south shore of Lake Ontario. *New York State Museum Bulletin* 281, 3–19.

Slater, G. 1931. The structure of the Bride Moraine, Isle of Man. *Proceedings of the Liverpool Geological Society* 14, 184–196.

Slatt, R.M. and Eyles, N. 1981. Petrology of glacial sand: implications for the origin and mechanical durability of lithic fragments. *Sedimentology* 28, 171–183.

Small, R.J. 1983. Lateral moraines of Glacier de Tsidjiore Nouve: form, development and implications. *Journal of Glaciology* 29, 250–259.

Small, R.J. 1987a. Englacial and supraglacial sediment: transport and deposition. In Gurnell, A.M. and Clark, M.J. (eds), *Glacio-fluvial Sediment Transfer: An Alpine Perspective.* Wiley, Chichester, 111–145.

Small, R.J. 1987b. Moraine sediment budgets. In Gurnell, A.M. and Clark, M.J. (eds), *Glacio-fluvial Sediment Transfer: An Alpine Perspective.* Wiley, Chichester, 165–197.

Small, R.J., Clark, M.J. and Cawse, T.J.P. 1979. The formation of medial moraines on Alpine glaciers. *Journal of Glaciology* 22, 43–52.

Small, R.J. and Gomez, B. 1981. The nature and origin of debris layers within Glacier de Tsidjiore Nouve, Valais, Switzerland. *Annals of Glaciology* 2, 109–113.

Smalley, I.J. 1966. Drumlin formation: a rheological model. *Science* 151, 1379–1380.

Smalley, I.J. and Piotrowski, J.A. 1987. Critical strength/stress ratios at the ice–bed interface in the drumlin forming process: from 'dilatancy' to 'cross-over'. In Menzies, J. and Rose, J. (eds), *Drumlin Symposium*. Balkema, Rotterdam, 81–86.

Smalley, I.J. and Unwin, D.J. 1968. The formation and shape of drumlins and their distribution and orientation in drumlin fields. *Journal of Glaciology* 7, 377–390.

Smalley, I.J. and Warburton, J. 1994. The shape of drumlins, their distribution in drumlin fields, and the nature of the sub-ice shaping forces. *Sedimentary Geology* 91, 241–252.

Smart, C.C. 1986. Some observations on subglacial groundwater flow. *Journal of Glaciology* 32, 232–234.

Smed, P. 1962. Studier over den fynske ogruppes glaciale landskabsformer. *Meddelelser Dansk Geologisk Forening* 15, 1–74.

Smellie, J.L. and Skilling, I.P. 1994. Products of subglacial volcanic eruptions under different ice thicknesses: two examples from Antarctica. *Sedimentary Geology* 91, 115–129.

Smiraglia, C. 1989. The medial moraines of Ghiacciaio dei Forni, Valtellina, Italy: morphology and sedimentology. *Journal of Glaciology* 35, 81–84.

Smith, A.J. 1963. Evidence for a Talchir (Lower Gondwana) glaciation: striated pavement and boulder bed at Irai, central India. *Journal of Sedimentary Petrology* 33, 739–750.

Smith, D.D. 1964. Ice lithologies and structure of the ice island ARLIS-II. *Journal of Glaciology* 5, 17–38.

Smith, D.E. 1993. Western Forth Valley. In Gordon, J.E. and Sutherland, D.G. (eds), *Quaternary of Scotland*. Chapman & Hall, London, 456–464.

Smith, D.E. and Dawson, A.G. (eds) 1983. *Shorelines and Isostasy*. Academic Press, London.

Smith, D.E., Dawson, A.G., Cullingford, R.A. and Harkness, D.D. 1985. The stratigraphy of Flandrian relative sea-level changes at a site in Tayside, Scotland. *Earth Surface Processes and Landforms* 10, 17–25.

Smith, D.G. 1976. Effect of vegetation on lateral migration of anastomosed channels of a glacier meltwater river. *Geological Society of America Bulletin* 87, 857–860.

Smith, D.G. 1983. Anastomosed fluvial deposits: modern examples from western Canada. In Collinson, J.D. and Lewin, J. (eds), *Modern and Ancient Fluvial Systems*. International Association of Sedimentologists, Special Publication No. 6, 155–168.

Smith, D.G. and Smith, N.D. 1980. Sedimentation in anastomosed river systems: examples from alluvial valleys near Banff, Alberta. *Journal of Sedimentary Petrology* 50, 157–164.

Smith, G.W. 1982. End moraines and the pattern of the last ice retreat from central and south coastal Maine in late Wisconsin glaciation of New England. In Larson, G.J. and Stone, B.D. (eds), *Late Wisconsin Glaciation of New England*. Kendall/Hunt, Dubuque, Iowa, 195–210.

Smith, H.T.U. 1948. Giant glacial grooves in northwest Canada. *American Journal of Science* 246, 503–514.

Smith, N.D. 1971. Transverse bars and braiding in the lower Platte River, Nebraska. *Geological Society of America Bulletin* 82, 3407–3420.

Smith, N.D. 1974. Sedimentology and bar formation in the upper Kicking Horse River, a braided outwash stream. *Journal of Geology* 81, 205–223.

Smith, N.D. 1975 Sedimentary environments and late Quaternary history of a 'low energy' mountain delta. *Canadian Journal of Earth Sciences* 12, 2004–2013.

Smith, N.D. 1978. Sedimentation processes and patterns in a glacier-fed lake with low sediment input. *Canadian Journal of Earth Sciences* 15, 741–756.

Smith, N.D. 1981. The effect of changing sediment supply on sedimentation in a glacier-fed lake. *Arctic and Alpine Research* 13, 75–82.

Smith, N.D. 1985. Proglacial fluvial environment. In Ashley, G.M., Shaw, J. and Smith, N.D. (eds), *Glacial Sedimentary Environments*. SEPM Short Course No. 16, 85–136.

Smith, N.D. 1990. The effects of glacial surging on sedimentation in a modern ice-contact lake, Alaska. *Geological Society of America Bulletin* 192, 1393–1403.

Smith, N.D. and Ashley, G.M. 1985. Proglacial lacustrine environments. In Ashley, G.M., Shaw, J. and Smith, N.D. (eds), *Glacial Sedimentary Environments*. SEPM Short Course 16, 135–215.

Smith, N.D. and Syvitski, J.P.M. 1982. Sedimentation in a glacier-fed lake: the role of pelletisation on deposition of fine-grained suspensates. *Journal of Sedimentary Petrology* 52, 503–513.

Smith, N.D.. Vendl, M.A. and Kennedy, S.K. 1982. Comparison of sedimentation regimes in four glacier-fed lakes of western Alberta. In Davidson-Arnott, R., Nickling, W. and Fahey, B.D. (eds), *Research in Glacial, Glaciofluvial, and Glaciolacustrine Systems*. Geo Books, Norwich, 203–238.

Sneed, E.D. and Folk, R.L. 1958. Pebbles in the lower Colorado River, Texas, a study in particle morphogenesis. *Journal of Geology* 66, 114–150.

Solheim, A. 1986. Submarine evidence of glacier surges. *Polar Research* 4, 91–95.

Solheim, A. 1991. The depositional environment of surging sub-polar tidewater glaciers. Norsk Polarinstitutt Skrifter 194.

Solheim, A. and Pfirman, S.L. 1985. Sea-floor morphology outside a grounded, surging glacier: Brasvellbreen, Svalbard. *Marine Geology* 65, 127–143.

Solheim, A., Russwurm, L., Elverhoi, A. and Nyland Berg, M. 1990. Glacial geomorphic features in the northern Barents Sea: direct evidence for grounded ice and implications for the pattern of deglaciation and late glacial sedimentation. In Dowdeswell, J.A. and Scourse, J.D. (eds), *Glacimarine Environments: Processes and Sediments*. Geological Society Special Publication 53, 253–268.

Sollid, J.L. 1975. Some comments on P-forms. *Norsk Geografisk Tidsskrift* 29, 74–75.

Sollid, J.L. and Carlson, A.B. 1984. De Geer moraines and eskers in Pasvik, north Norway. *Striae* 20, 55–61.

Sollid, J.L. and Reite, A.J. 1983. The last glaciation and deglaciation of central Norway. In Ehlers, J. (ed.), *Glacial Deposits in North-west Europe*. Balkema, Rotterdam, 41–59.

Sollid, J.L. and Sorbel, L. 1984. Distribution and genesis of moraines in central Norway. *Striae* 20, 63–67.

Souchez, R.A. 1966. Sur les mécanismes de l'érosion en Antarctique. *Bulletin de la Société belge Etude Géographique* 35, 25–34.

Souchez, R.A. 1967a. The formation of shear moraines: an example from south Victoria Land, Antarctica. *Journal of Glaciology* 6, 837–843.

Souchez, R.A. 1967b. Le recult des verrous-gradins et les rapports glaciaire–periglaciaire en Antarctique. *Revue de Géomorphologie Dynamique* 17, 49–54.

Souchez, R.A. 1971. Ice-cored moraines in south-western Ellesmere Island, NWT, Canada. *Journal of Glaciology* 10, 245–254.

Souchez, R.A. and de Groote, J.H. 1985. ∂D-$\partial^{18}O$ relationships in ice formed by subglacial freezing: palaeoclimatic implications. *Journal of Glaciology* 31, 229–232.

Souchez, R.A. and Jouzel, J. 1984. On the isotopic composition in ∂D and $\partial^{18}O$ of water and ice during freezing. *Journal of Glaciology* 30, 369–372.

Souchez, R.A. and Lemmens, M. 1985. Subglacial carbonate deposition: an isotopic study of a present-day case. *Palaeogeography, Palaeoclimatology, Palaeoecology* 51, 357–364.

Souchez, R.A. and Lemmens, M.M. 1987. Solutes. In Gurnell, A.M. and Clark, M.J. (eds), *Glacio-fluvial Sediment Transfer: An Alpine Perspective*. Wiley, Chichester, 285–303.

Souchez, R.A., Lemmens, M., Lorrain, R.D., Tison, J.L., Jouzel, J. and Sugden, D.E. 1990. Influence of hydroxyl-bearing minerals on the isotopic composition of ice from the basal zone of an ice sheet. *Nature*, 345, 244–246.

Souchez, R.A. and Lorrain, R.D. 1978. Origin of the basal ice layer from Alpine glaciers indicated by its geochemistry. *Journal of Glaciology* 20, 319–328.

Souchez, R.A. and Lorrain, R.D. 1987. The subglacial sediment system. In Gurnell, A.M. and Clark, M.J. (eds), *Glaciofluvial Sediment Transfer: An Alpine Perspective*. Chichester, Wiley, 147–163.

Souchez, R.A. and Lorrain, R.D. 1991. *Ice Composition and Glacier Dynamics*. Springer-Verlag, Berlin.

Southard, J.B., Smith, N.D. and Kuhnle, R.A. 1984. Chutes and lobes: newly identified elements of braiding in shallow gravelly streams. In Koster, E.H. and Steel, R.J. (eds), *Sedimentology of Gravels and Conglomerates*. Canadian Society of Petroleum Geologists, Memoir 10, 51–59.

Speden, I.G. 1960. Post-glacial terraces near Cape Chocolate, McMurdo Sound, Antarctica. *New Zealand Journal of Geology and Geophysics* 3, 203–217.

Spencer, A.M. 1971. Late Precambrian Glaciation in Scotland. Geological Society of London, Memoir 6.

Spencer, A.M. 1975. Late Precambrian glaciation in the North Atlantic region. In Wright, A.E. and Moseley, F. (eds), *Ice Ages: Ancient and Modern*. Seel House Press, Liverpool, 217–240.

Spencer, A.M. 1981. Late Precambrian Port Askaig tillite in Scotland. In Hambrey, M.J. and Harland, W.B. (eds), *Earth's Pre-Pleistocene Glacial Record*. Cambridge University Press, Cambridge, 632–636.

Spjeldnaes, N. 1973. Moraine stratigraphy, with examples from the basal Cambrian ('Eocambrian') and Ordovi-cian glaciations. *Bulletin of the Geological Institutions of Uppsala*, NS 5, 165–171.

Spring, U. and Hutter, K. 1981. Numerical studies of jökulhlaups. *Cold Regions Science and Technology* 4, 221–244.

Squyres, S.W. 1989. Water on Mars. *Icarus* 79, 229–288.

Squyres, S.W. and Carr, M.H. 1986. Geomorphic evidence for the distribution of ground ice on Mars. *Science* 231, 249–252.

Stalker, A.MacS. 1956. The Erratics Train, Foothills of Alberta. Geological Survey of Canada, Bulletin 37.

Stalker, A.MacS. 1960. Ice-pressed drift forms and associated deposits in Alberta. Geological Survey of Canada, Bulletin 57.

Stalker, A.MacS. 1961. Buried valleys in central and southern Alberta. Geological Survey of Canada Paper 60–32.

Stalker, A.MacS. 1973a. Surficial geology of the Drumheller area. Geological Survey of Canada, Memoir 370.

Stalker, A.MacS. 1973b. The large interdrift bedrock blocks of the Canadian Prairies. Geological Survey of Canada, Paper 75–1A, 421–422.

Stalker, A.MacS. 1976a. Megablocks, or the enormous erratics of the Albertan Prairies. Geological Survey of Canada, Paper 76–1C, 185–188.

Stalker, A.MacS. 1976b. Quaternary stratigraphy of the southwestern Canadian Prairies. In Mahaney, W.C. (ed.), *Quaternary Stratigraphy of North America*. Dowden, Hutchinson & Ross, Stroudsburg, Mass., 381–407.

Stalker, A.MacS. 1980. The geology of the ice-free corridor. *Canadian Journal of Anthropology* 1, 11–13.

Stalker, A.MacS. and Harrison, J.E. 1977. Quaternary glaciation of the Waterton–Castle River region of Alberta. *Canadian Petroleum Geology Bulletin* 25, 882–906.

Stanford, S.D. and Mickelson, D.H. 1985. Till fabric and deformational structures in drumlins near Waukesha, Wisconsin, USA. *Journal of Glaciology* 31, 220–228.

Stanley, D.J. 1968. Reworking of glacial sediments in the northwest arm of a fjord-like inlet on the southeast coast of Nova Scotia. *Journal of Sedimentary Petrology* 38, 1224–1241.

Stea, R.R. 1980. A study of a succession of tills exposed in a wave-cut drumlin, Meisners Reef, Lunenberg County. *Nova Scotia Department of Mines and Energy, Report of Activities* 80–81, 9–20.

Stea, R.R. 1994. Relict and palimpsest glacial landforms in Nova Scotia, Canada. In Warren, W.P. and Croot, D.G. (eds), *Formation and Deformation of Glacial Deposits*. Balkema, Rotterdam, 141–158.

Stea, R.R. and Brown, Y. 1989. Variation in drumlin orientation, form and stratigraphy relating to successive ice flows in southern and central Nova Scotia. *Sedimentary Geology* 62, 223–240.

Stea, R.R. and Mott, R.J. 1989. Deglaciation environments and evidence for glaciers of Younger Dryas age in Nova Scotia, Canada. *Boreas* 18, 169–187.

Stea, R.R., Turner, R.G., Finck, P.W. and Graves, R.M. 1989. Glacial dispersal in Nova Scotia: a zonal concept. In DiLabio, R.N.W. and Coker, W.B. (eds), *Drift Prospecting*. Geological Survey of Canada, Paper 89–20, 155–169.

Steiner, J. and Grillmair, E. 1973. Possible galactic causes for periodic and episodic glaciations. *Geological Society of America Bulletin* 84, 1003–1018.

Stenborg, T. 1969. Studies of the internal drainage of glaciers. *Geografiska Annaler* 51A, 13–41.

Stephan, H.J. 1985. Deformations striking parallel to glacier movement as a problem in reconstructing its direction. *Bulletin of the Geological Society of Denmark* 34, 47–53.

Stephan, H.J. 1987. Form, composition, and origin of drumlins in Schleswig-Holstein. In Menzies, J. and Rose, J. (eds), *Drumlin Symposium*. Balkema, Rotterdam, 335–345.

Stephenson, S.N. and Bindshadler, R.A. 1988. Observed velocity fluctuations on a major Antarctic ice stream. *Nature* 334, 695–697.

Stewart, A.D. 1991. Torridonian. In Craig, G.Y. (ed.), *Geology of Scotland*, 3rd edition. The Geological Society, London, 65–85.

Stewart, G.A. and Perry, R.A. 1953. Survey of Townsville–Bowen Region (1950). Land Research Series 2 CSIRO 1120 (Australia).

Stewart, R.A., Bryant, D. and Sweat, M.J. 1988. Nature and origin of corrugated ground moraine of the Des Moines lobe, Story County, Iowa. *Geomorphology* 1, 111–130.

Stewart, T.G. 1991. Glacial marine sedimentation from tidewater glaciers in the Canadian High Arctic. In Anderson, J.B. and Ashley, G.M. (eds), *Glacial Marine Sedimentation: Palaeoclimatic Significance*. Geological Society of America Special Paper 261, 95–105.

Stewart, T.G. and England, J. 1983. Holocene sea ice variations and paleoenvironmental change, northernmost Ellesmere Island, NWT, Canada. *Arctic and Alpine Research* 15, 1–17.

Stoker, M.S. 1990. Glacially-influenced sedimentation on the Hebridean slope, northwestern United Kingdom continental margin. In Dowdeswell, J.A. and Scourse, J.D. (eds), *Glacimarine Environments: Processes and Sediments*. Geological Society Special Publication 53, 349–362.

Stoker, M.S., Long, D. and Fyfe, J.A. 1985. The Quaternary succession in the central North Sea. *Newsletters on Stratigraphy* 14, 119–128.

Stokes, J.C. 1958. An esker-like ridge in process of formation, Flatisen, Norway. *Journal of Glaciology* 3, 286–290.

Stone, K.H. 1963. Alaskan ice-dammed lakes. *Annals of the Association of American Geographers* 53, 332–349.

Stow, D.A.V. 1986. Deep clastic seas. In Reading, H.G. (ed.), *Sedimentary Environments and Facies*, 2nd edition. Blackwell, Oxford, 399–444.

Stratten, T. 1969. A preliminary report of a directional study of the Dwyka Tillites in the Karroo Basin of South Africa. In *Gondwana Stratigraphy, IUGS Symposium (Earth Sciences 2), Buenos Aires, October 1967*. UNESCO, Paris, 741–761.

Stratten, T. 1971. Tectonic framework of sedimentation during the Dwyka period in South Africa. In Haughton, S.H. (ed.), *2nd IUGS Symposium on Gondwana Stratigraphy, Cape Town and Johannesburg, July 1970*. CSIR, Pretoria, 483–490.

Stravers, J.A., Miller, G.H. and Kaufman, D.S. 1992. Late glacial ice margins and deglacial chronology for southeastern Baffin Island and Hudson Strait, eastern Canadian arctic. *Canadian Journal of Earth Sciences* 29, 1000–1017.

Straw, A. 1968. Late Pleistocene glacial erosion along the Niagara Escarpment of southern Ontario. *Bulletin of the Geological Society of America* 79, 889–910.

Street-Perrot, F.A. and Perrot, R.A. 1990. Abrupt climate fluctuations in the tropics: the influence of Atlantic Ocean circulation. *Nature* 343, 607–612.

Streiff-Becker, R. 1951. Pot-holes and glacier mills. *Journal of Glaciology* 1, 488–490.

Strom, K.M. 1948. The geomorphology of Norway. *Geographical Journal* 112, 19–27.

Strom, R.G., Kargel, J.S., Johnson, N. and Knight, C. 1991. Glacial and marine chronology of Mars. *Lunar and Planetary Science* 22, 1351–1352.

Stromberg, B. 1965. Mappings and geochronological investigations in some moraine areas of south-central Sweden. *Geografiska Annaler* 47A, 73–82.

Stuiver, M., Denton, G.H., Hughes, T.J. and Fastook, J.L. 1981. History of the marine ice sheet in West Antarctica during the last glaciation: a working hypothesis. In Denton, G.H. and Hughes, T.J. (eds), *The Last Great Ice Sheets*. Wiley, New York, 319–436.

Stupavsky, M. and Gravenor, C.P. 1974. Water release from the base of active glaciers. *Geological Society of America Bulletin* 85, 433–436.

Stupavsky, M. and Gravenor, C.P. 1984. Paleomagnetic dating of Quaternary sediments: a review. In Mahaney, W.C. (ed.), *Quaternary Dating Methods*. Elsevier, Amsterdam, 123–140.

Stupavsky, M., Symons, D.T.A. and Gravenor, C.P. 1974. Paleomagnetism of the Port Stanley Till, Ontario. *Geological Society of America Bulletin* 85, 141–144.

Sturm, M., Benson, C.S. and MacKeith, P. 1986. Effects of the 1966–68 eruptions of Mount Redoubt on the flow of Drift Glacier, Alaska, USA. *Journal of Glaciology* 32, 355–362.

Sturm, M., Hall, D.K., Benson, C.S. and Field, W.O. 1991. Non-climatic control of glacier-terminus fluctuations in the Wragell and Chugach Mountains, Alaska, USA. *Journal of Glaciology* 37, 348–356.

Sturm, M. and Matter, A. 1978. Turbidites and varves in Lake Brienz (Switzerland): deposition of clastic detritus by density currents. In Matter, A. and Tucker, M.E. (eds), *Modern and Ancient Lake Sediments*. International Association of Sedimentologists, Special Publication 2, 147–168.

Sugden, D.E. 1968. The selectivity of glacial erosion in the Cairngorm Mountains, Scotland. *Transactions of the Institute of British Geographers* 45, 79–92.

Sugden, D.E. 1969. The age and form of corries in the Cairngorm Mountains, Scotland. *Scottish Geographical Magazine* 85, 34–46.

Sugden, D.E. 1970. Landforms of deglaciation in the Cairngorm Mountains, Scotland. *Transactions of the Institute of British Geographers* 51, 201–219.

Sugden, D.E. 1974. Landscapes of glacial erosion in Greenland and their relationship to ice, topographic and bedrock conditions. Institute of British Geographers, Special Publication 7, 177–195.

Sugden, D.E. 1976. A case against deep erosion of shields by ice sheets. *Geology* 4, 580–582.

Sugden, D.E. 1977. Reconstruction of the morphology, dynamics and thermal characteristics of the Laurentide ice sheet at its maximum. *Arctic and Alpine Research* 9, 27–47.

Sugden, D.E. 1978. Glacial erosion by the Laurentide ice sheet. *Journal of Glaciology* 20, 367–391.

Sugden, D.E. 1988. Will the West Antarctic Ice Sheet collapse? *Geography Review* 2/1, 26–31.

Sugden, D.E. 1993. Ice sheets and global warming. *Geography Review* 6, 2–6.

Sugden, D.E. and Clapperton, C.M. 1981. An ice-shelf moraine, George VI Sound, Antarctica. *Annals of Glaciology* 2, 135–141.

Sugden, D.E., Clapperton, C.M., Gemmell, A.M.D. and Knight, P.G. 1987. Stable isotopes and debris in basal glacier ice, South Georgia, Southern Ocean. *Journal of Glaciology* 33, 324–329.

Sugden, D.E., Clapperton, C.M. and Knight, P.G. 1985. A jökulhlaup near Sondre Stromfjord, west Greenland, and some effects on the ice-sheet margin. *Journal of Glaciology* 31, 366–368.

Sugden, D.E., Denton, G.H. and Marchant, D.R. 1991. Subglacial meltwater channel systems and ice sheet overriding, Asgard Range, Antarctica. *Geografiska Annaler* 73A, 109–121.

Sugden, D.E., Glasser, N.F. and Clapperton, C.M. 1992. Evolution of large roches moutonnées. *Geografiska Annaler* 74A, 253–264.

Sugden, D.E. and John, B.S. 1976. *Glaciers and Landscape*. Edward Arnold, London.

Sugden, D.E., Knight, P.G., Livesey, N., Lorrain, R.D., Souchez, R.A., Tison, J.L. and Jouzel, J. 1987. Evidence for two zones of debris entrainment beneath the Greenland ice sheet. *Nature* 328, 238–241.

Sugden, D.E., Marchant, D.R. and Denton, G.H. 1993. The case for a stable East Antarctic Ice Sheet: the background. *Geografiska Annaler* 75A, 151–154.

Sugden, D.E., Marchant, D.R., Potter, N. Jr, Souchez, R.A., Denton, G.H., Swisher, C.C. III and Tison, J.L. 1995. Preservation of miocene glacier ice in East Antarctica. *Nature* 376, 412–414.

Suggate, R.P. 1990. Late Pliocene and Quaternary glaciations of New Zealand. *Quaternary Science Reviews* 9, 175–197.

Summerfield, M. 1991. *Global Geomorphology*. Longman, Harlow.

Sutherland, D.G. 1984a. The Quaternary deposits and landforms of Scotland and the neighbouring shelves: a review. *Quaternary Science Reviews* 3, 157–254.

Sutherland, D.G. 1984b. Modern glacier characteristics as a basis for inferring former climates with particular reference to the Loch Lomond Stadial. *Quaternary Science Reviews* 3, 291–309.

Sutherland, D.G. 1986. A review of Scottish marine shell radiocarbon dates, their standardisation and interpretation. *Scottish Journal of Geology* 22, 145–164.

Svensson, H. 1959. Is the cross section of a glacial valley a parabola? *Journal of Glaciology* 3, 362–363.

Sverdrup, H.V. 1935. The temperature of the firn on Isachsen's Palteau and general conclusions regarding the temperature of the glaciers on West Spitzbergen. *Geografiska Annaler* 17, 53–88.

Swithinbank, C.W.M. 1950. The origin of dirt cones on glaciers. *Journal of Glaciology* 1, 461–465.

Swithinbank, C.W.M. 1955. Ice shelves. *Geographical Journal* 121, 64–76.

Swithinbank, C.W.M. 1964. To the valley glaciers that feed the Ross ice shelf. *Geographical Journal* 130, 32–48.

Swithinbank, C.W.M. 1969. Giant icebergs in the Weddell Sea, 1967–68. *Polar Record* 14, 477–478.

Swithinbank, C.W.M. 1988. Antarctica. In Williams, R.S. and Ferrigno, J.G. (eds), *Satellite Image Atlas of Glaciers of the World*. USGS Professional Paper 1386-B.

Swithinbank, C.W.M., McClain, P. and Little, P. 1977. Drift tracks of Antarctic icebergs. *Polar Record* 18, 495–501.

Swithinbank, C.W.M. and Zumberge, J.H. 1965. The ice shelves. In Hatherton T. (ed.), *Antarctica*. Methuen, London, 199–220.

Synge, F.M. 1970. The Irish Quaternary: current views (1969). In Stephens, N. and Glasscock, R.E. (eds), *Irish Geographical Studies*. Queen's University, Belfast, 34–48.

Synge, F.M. and Stephens, N. 1960. The Quaternary period in Ireland: an assessment. *Irish Geographer* 4, 121–130.

Syverson, K.M., Gaffield, S.J. and Mickelson, D.M. 1994. Comparison of esker morphology and sedimentology with former ice-surface topography, Burroughs Glacier, Alaska. *Geological Society of America Bulletin* 106, 1130–1142.

Syvitski, J.P.M., Burrell, D.C. and Skei, J.M. 1987. *Fjords: Processes and Products*. Springer, New York.

Syvitski, J.P.M., Fader, G.B., Josenhans, H.W., Maclean, B. and Piper, D.J.W. 1983. Seabed investigations of the Canadian east coast and Arctic using Pisces TV. *Geoscience Canada* 10, 59–68.

Syvitski, J.P.M. and Farrow, G.E. 1983. Structures and processes in bay head deltas: Knight and Bute inlets, British Columbia. *Sedimentary Geology* 30, 217–244.

Syvitski, J.P.M. and Farrow, G.G. 1989. Fjord sedimentation as an analogue for small hydrocarbon-bearing fan deltas. In Whateley, M.K.G. and Pickering, K.T. (eds), *Deltas: Sites and Traps for Fossil Fuels*. Geological Society of London Special Publication 41, 21–43.

Syvitski, J.P.M. and Murray, J.W. 1981. Particle interaction in fjord suspended sediment. *Marine Geology* 39, 215–242.

Tanner, V. 1932. The problems of the eskers. *Fennia* 55, 1–13.

Tanner, V. 1934. The problems of the eskers. *Fennia* 58.

Tanner, V. 1937. The problems of the eskers. *Fennia* 63.

Tarr, R.S. 1894. The origin of drumlins. *American Geologist* 13, 393–407.

Tarr, R.S. and Martin, L. 1914. *Alaskan Glacier Studies of the National Geographical Society in the Yakutat Bay, Prince William Sound and Lower Copper River regions*. National Geographical Society, Washington DC.

Taylor, R.S. 1960. Some Pleistocene lakes of northern Alberta and adjacent areas. *Journal of the Alberta Society of Petroleum Geologists* 8, 167–185.

Tchernia, P. and Jeanin, P.F. 1983. Quelques aspects de la circulation océanique Antartique révélés par l'observation de la dérive d'icebergs (1972–1983). CNRS, Muséum national d'Histoire Naturelle.

Teller, J.T. 1985. Glacial Lake Agassiz and its influence on the Great Lakes. In Karrow, P.F. and Calkin, P.E. (eds), *Quaternary Evolution of the Great Lakes*. Geological Association of Canada Special Paper 30, 1–16.

Teller, J.T. 1987. Proglacial lakes and the southern margin of the Laurentide Ice Sheet. In Ruddiman, W.F. and Wright, H.E. (eds), *North America and Adjacent Oceans during the Last Deglaciation*. Geological Society of America, *The Geology of North America*, Vol. K–3, 39–69.

Teller, J.T. 1990. Meltwater and precipitation runoff to the North Atlantic, Arctic and Gulf of Mexico from the Laurentide Ice Sheet and adjacent regions during the Younger Dryas. *Paleoceanography* 5, 897–905.

Teller, J.T. 1995. History and drainage of large ice-dammed lakes along the Laurentide Ice Sheet. *Quaternary International* 28, 83–92.

Teller, J.T. and Clayton, L. (eds) 1983. *Glacial Lake Agassiz*. Geological Association of Canada Special Paper 26.

Teller, J.T. and Mahnic, P. 1988. History of sedimentation in the northwestern Lake Superior basin and its relation to Lake Agassiz overflow. *Canadian Journal of Earth Sciences* 25, 1660–1673.

Teller, J.T. and Thorleifson, L.H. 1987. Catastrophic flooding into the Great Lakes from Lake Agassiz. In Mayer, L. and Nash, D. (eds), *Catastrophic Flooding*. Allen & Unwin, London, 121–138.

Temple, P.H. 1965. Some aspects of cirque distribution in the west-central Lake District, northern England. *Geografiska Annaler* 47, 185–193.

Terwindt, J.H.J. and Augustinus, P.G.E.F. 1985. Lateral and longitudinal successions in sedimentary structures in the Middle Mause esker, Scotland. *Sedimentary Geology* 45, 161–188.

Tharin, J.C. 1969. The Foothills Erratics Train of Alberta, Canada. *Michigan Academician* 2, 113–124.

Theakstone, W.H. 1967. Basal sliding and movement near the margin of the glacier Osterdalsisen, Norway. *Journal of Glaciology* 6, 805–816.

Theakstone, W.H. 1978. The 1977 drainage of Austre Okstindbreen ice-dammed lake, its causes and consequences. *Norsk Geografisk Tidsskrift* 32, 159–171.

Theakstone, W.H. 1982. Sediment fans and sediment flows generated by snowmelt: observations at Austerdalsbreen, Norway. *Journal of Geology* 90, 583–588.

Thomas, G.S.P. 1984a. A late Devensian glaciolacustrine fan-delta at Rhosesmor, Clwyd, North Wales. *Geological Journal* 19, 125–141.

Thomas, G.S.P. 1984b. Sedimentation of a subaqueous esker-delta at Strathabie, Aberdeenshire. *Scottish Journal of Geology* 20, 9–20.

Thomas, G.S.P. 1984c. On the glacio-dynamic structure of the Bride Moraine, Isle of Man. *Boreas* 13, 355–364.

Thomas, G.S.P. 1989. The Late Devensian glaciation along the western margin of the Cheshire–Shropshire lowland. *Journal of Quaternary Science* 4, 167–181.

Thomas, G.S.P., Connaughton, M. and Dackombe, R.V. 1985. Facies variation in a late Pleistocene supraglacial outwash sandur from the Isle of Man. *Geological Journal* 20, 193–213.

Thomas, G.S.P. and Connell, R.J. 1985. Iceberg drop, dump and grounding structures from Pleistocene glacio-lacustrine sediments, Scotland. *Journal of Sedimentary Petrology* 55, 243–249.

Thomas, G.S.P. and Dackombe, R.V. 1985. Comment on 'Glaciomarine sediments of the Isle of Man as a key to Late Pleistocene stratigraphic investigations in the Irish sea basin' (Eyles and Eyles, 1984). *Geology* 13, 445–446.

Thomas, G.S.P. and Summers, A.J. 1982. Dropstone and allied structures from Pleistocene waterlain till at Ely House, Co. Wexford. *Journal of Earth Science Research, Dublin Society* 4, 109–119.

Thomas, R.H. 1973a. The creep of ice shelves: theory. *Journal of Glaciology* 12, 45–53.

Thomas, R.H. 1973b. The creep of ice shelves: interpretation of observed behaviour. *Journal of Glaciology* 12, 55–70.

Thomas, R.H. 1979a. The dynamics of marine ice sheets. *Journal of Glaciology* 24, 167–177.

Thomas, R.H. 1979b. Ice shelves: a review. *Journal of Glaciology* 24, 273–286.

Thomas, R.H. 1985. Responses of the polar ice sheets to climatic warming. In *Glaciers, Ice Sheets and Sea Level: Effect of CO_2 Induced Climatic Change*. Workshop, US Department of Energy, Seattle, 1984, 301–316.

Thomas, R.H. and Bentley, C.R. 1978a. A model for Holocene retreat of the West Antarctic ice sheet. *Quaternary Research* 10, 150–170.

Thomas, R.H. and Bentley, C.R. 1978b. The equilibrium state of the eastern half of the Ross Ice Shelf. *Journal of Glaciology* 20, 509–518.

Thompson, A. 1988. Historical development of the proglacial landforms of Svinafellsjökull and Skaftafellsjökull, southeast Iceland. *Jökull* 38, 17–31.

Thompson, H.R. and Bonnlander, B.H. 1956. Temperature measurements at a cirque bergschrund in Baffin Island: some results of W.R.B. Battle's work in 1953. *Journal of Glaciology* 2, 762–769.

Thompson, L.G. 1995. Ice core evidence from Peru and China. In Bradley, R.S. and Jones, P.D. (eds), *Climate since AD 1500*. Routledge, London, 517–548.

Thompson, L.G., Mosely-Thompson, E. and Arnao, B.M. 1984. Major El Niño/Southern Oscillation events recorded in stratigraphy of the tropical Quelccaya Ice Cap. *Science* 226, 50–52.

Thompson, L.G., Mosely-Thompson, E., Dansgaard, W. and Grootes, P.M. 1986. The Little Ice Age as recorded in the stratigraphy of the tropical Quelccaya Ice Cap. *Science* 234, 361–364.

Thompson, L.G., Mosely-Thompson, E., Davis, M.E. *et al.* 1989. Pleistocene climatic record from Qinghai–Tibetan Plateau ice cores. *Science* 246, 474–477.

Thompson, L.G., Mosely-Thompson, E., Davis, M.E. *et al.* 1990. Glacial stage ice core records from the subtropical Dunde Ice Cap, China. *Annals of Glaciology* 14, 288–297.

Thompson, R. and Oldfield, F. 1986. *Environmental Magnetism*. Allen & Unwin, London.

Thorarinsson, S. 1939. The ice-dammed lakes of Iceland, with particular reference to their values as indicators of glacier oscillations. *Geografiska Annaler* 21, 216–242.

Thorarinsson, S. 1953. Some new aspects of the Grimsvötn problem. *Journal of Glaciology* 2, 267–274.

Thorarinsson, S. 1956a. On the variations of Svinafells-jökull, Skaftafellsjökull and Kviarjökull in Oraefi. *Jökull* 6, 1–15.

Thorarinsson, S. 1956b. *The Thousand Years Struggle against Ice and Fire*. Bokautgafa Menningarsjods, Reykjavik.

Thorarinsson, S. 1957. The jökulhlaup from the Katla area in 1955 compared with other jökulhlaups in Iceland. *Jökull* 7, 21–25.

Thorarinsson, S. 1967. Forvitnilegir jökulgarder vid jadar Skeidararjökuls. Washboard moraines in front of Skeidararjökull (English abstract). *Jökull* 17, 311–312.

Thorarinsson, S. 1969. Glacier surges in Iceland, with special reference to the surges of Bruarjøkull. *Canadian Journal of Earth Sciences* 6, 875–882.

Thorleifson, L.H., Wyatt, P.H., Shilts, W.W. and Nielsen, E. 1992. Hudson Bay lowland Quaternary stratigraphy: evidence for Early Wisconsinan glaciation centered on Quebec. In Clark, P.U. and Lea, P.D. (eds), *The Last Interglacial–Glacial Transition in North America*. Geological Society of America, Special Paper 270, 207–221.

Thorn, C.E. 1979. Ground temperatures and surficial transport in colluvium during snowpatch meltout: Colorado Front Range. *Arctic and Alpine Research* 11, 41–52.

Thorn, C.E. 1988. Nivation: a geomorphic chimera. In Clark, M.J. (ed.), *Advances in Periglacial Geomorphology*. Wiley, Chichester, 3–31.

Thorn, C.E. and Hall, K. 1980. Nivation: an arctic–alpine comparison and reappraisal. *Journal of Glaciology* 25, 109–124.

Thorp, P.W. 1981. A trimline method for defining the upper limit of Loch Lomond Readvance glaciers: examples from the Loch Leven and Glencoe areas. *Scottish Journal of Geology* 17, 49–64.

Thorp, P.W. 1986. A mountain icefield of Loch Lomond Stadial age, western Grampians, Scotland. *Boreas* 15, 83–97.

Thorp, P.W. 1991. Surface profiles and basal shear stresses of outlet glaciers from a Lateglacial mountain icefield in western Scotland. *Journal of Glaciology* 37, 77–89.

Tilly, G.P. 1969. Erosion caused by impact of solid particles. In Scott, D. (ed.) *Treatise on Materials Science and Technology* 13: *Wear*. Academic Press, New York, 289–319.

Tipper, H.W. 1971. Glacial geomorphology and Pleistocene history of central British Columbia. Geological Survey of Canada Bulletin 196.

Tippett, C.R. 1985. Glacial dispersal train of Paleozoic erratics, central Baffin Island, NWT, Canada. *Canadian Journal of Earth Sciences* 22, 1818–1826.

Tison, J.-L., Petit, J.-R., Barnola, J.-M. and Mahaney, W.C. 1993. Debris entrainment at the ice–bedrock interface in sub-freezing temperatures, Terre Adélie, Antarctica. *Journal of Glaciology* 39, 303–315.

Todd, S.P. 1989. Stream-driven, high-density gravelly traction carpets: possible deposits in the Trabeg Conglomerate Formation, SW Ireland and some theoretical considerations of their origin. *Sedimentology* 36, 513–530.

Tokarsky, O. 1986. Hydrogeologic cross section M–M, Medicine Hat 72L. Alberta Environment, Water Resources Management Services, Edmonton.

Tómasson, H. 1996. The jökulhlaup from Katla in 1918. *Annals of Glaciology* 22, 249–254.

Tooley, M.J. and Shennan, I. (eds) 1987. *Sea Level Changes*. Blackwell, Oxford.

Torell, O. 1872. Undersokningar ofver istiden del I. Aftryck ur Ofversigt af Kungliga Vetenskapsakademiens Forhandlingar 1872. P.A. Nordstedt och Soner, Stockholm.

Torell, O. 1873. Undersokningar ofver istiden del II. Skandinaviska landisens utsrackning under isperioden. *Ofversigt af Kungliga Vetenskapsakademiens Forhandlingar 1873* 1, 47–64.

Torsnes, I., Rye, N. and Nesje, A. 1993. Modern and Little Ice Age equilibrium-line altitudes on outlet valley glaciers from Jostedalsbreen, western Norway: an evaluation of different approaches to their calculation. *Arctic and Alpine Research* 25, 106–116.

Trainer, F.W. 1973. Formation of joints in bedrock by moving glacial ice. *US Geological Survey Journal of Research* 1, 229–235.

Tranter, M. and Raiswell, R. 1991. The composition of the englacial and subglacial component in bulk meltwaters draining the Gornergletscher, Switzerland. *Journal of Glaciology* 37, 59–66.

Trefethen, J. and Trefethen, H. 1945. Lithology of the Kennebec Valley esker. *American Journal of Science* 242, 521–527.

Trenhaile, A.S. 1971. Drumlins: their distribution, orientation and morphology. *Canadian Geographer* 15, 113–126.

Trenhaile, A.S. 1975. The morphology of a drumlin field. *Annals of the Association of American Geographers* 65, 297–312.

Trenhaile, A.S. 1990. *The Geomorphology of Canada: An Introduction*. Oxford University Press, Toronto.

Tricart, J. 1956. Etude expérimentale du problème de la gélivation. *Biuletyn Peryglacjalny* 4, 285–318.

Tricart, J. 1969. *Geomorphology of Cold Environments*, trans. E. Watson. Macmillan, London.

Tricart, J. and Cailleux, A. 1962. *Le Modèle glaciaire et nival*. Sedes, Paris.

Tricart, J. and Cailleux, A. 1965. *Introduction à la géomorphologie climatique*. Sedes, Paris.

Tryggvason, E. 1960. Earthquakes, jökulhlaups and subglacial eruptions. *Jökull* 10, 18–22.

Tsui, P.C., Cruden, D.M. and Thomson, S. 1989. Ice-thrust terrains and glaciotectonic settings in central Alberta. *Canadian Journal of Earth Sciences* 26, 1308–1318.

Tufnell, L. 1984. *Glacier Hazards*. Longman, London.

Turekian, K.K. (ed.) 1971. *Late Cenozoic Glacial Ages*. Yale University Press, New Haven, Conn.

Turner, J.S. 1986. Turbulent entrainment: the development of the entrainment assumption, and its application to geophysical flows. *Journal of Fluid Mechanics* 173, 431–471.

Turner, R.G. and Stea, R.R. 1990. Interpretation of till geochemical data in Nova Scotia, Canada, using mapped till units, multi-element anomaly patterns and the relationship of till clast geology to matrix geochemistry. *Journal of Geochemical Exploration* 37, 225–254.

Tushingham, A.M. 1991. On the extent and thickness of the Innuitian Ice Sheet: a postglacial-adjustment approach. *Canadian Journal of Earth Sciences* 28, 231–239.

Tushingham, A.M. and Peltier, W.R. 1991. Ice–3G: a new global model of late Pleistocene deglaciation based upon geophysical predictions of post-glacial relative sea level change. *Journal of Geophysical Research* 96, 4497–4523.

Twiss, R.J. and Moores, E.M. 1992. *Structural Geology*. New York, Freeman & Co.

Tyndall, J. and Huxley, T.H. 1857. On the structure and motion of glaciers. *Philosophical Transactions of the Royal Society of London* 147, 327–346.

Tyrrell, J.B. 1898a. The glaciation of north central Canada. *Journal of Geology* 6, 147–160.

Tyrrell, J.B. 1898b. Report on the Doobount, Kazan and Ferguson Rivers, and the northwest coast of Hudson Bay; and on two overland routes from Hudson Bay to Lake Winnipeg. *Geological Survey of Canada, Annual Report, Series N* 9, 1–218.

Tyrrell, J.B. 1913. The Patrician glacier south of Hudson Bay. *Congrès géologique international, Compte rendu de la XII session, Canada*, 523–534.

Unwin, D.J. 1973. The distribution and orientation of corries in northern Snowdonia, Wales. *Transactions of the Institute of British Geographers* 58, 85–97.

Upham, W. 1892. Conditions of accumulation of drumlins. *American Geologist* 10, 339–362.

Upham, W. 1894. The Madison type of drumlins. *American Geologist* 14, 69–83.

Upham, W. 1895. Map showing the relationship of Lake Agassiz to the drift bearing area of North America and to lakes Bonneville and Lahontan. In *The Glacial Lake Agassiz*, USGS Monograph XXV, Pl. 2.

Upham, W. 1899. Glacial history of the New England islands, Cape Cod, and Long Island. *American Geologist* 24, 79–92.

Upham, W. 1900. Giants' kettles eroded by moulin torrents. *Geological Society of America Bulletin* 12, 25–44.

Ussing, N.V. 1903. On Jyllands Hedesletter og Teorierne om deres Dannelse: Copenhagen. *Oversigt over det Kongelige Danske Videnskabernes Selskabs Forhandlingar* 2, 99–165.

Ussing, N.V. 1907. Omfloddale og randmoraener i Jylland. *Oversigt over det Kongelige Danske Videnskabernes Selskabs Forhandlingar* 4, 161–213.

Valentin, H. 1957. Glazialmorphologische Untersuchungen in Ostengland. *Abhandlungen des Geographischen Institut der Freien Universität Berlin* 4, 1–86.

Vallot, J. 1898. Exploration des moulins de la Mer de Glace. *Annales de l'Observatoire du Mont Blanc* 3, 183–190.

Van Wagoner, J.C., Posamentier, H.W., Mitchum, R.M., Vail, P.R., Sarg, J.F., Loutit, T.S. and Hardenbol, J. 1988. An overview of the fundamentals of sequence stratigraphy and key definitions. In Wilgus, C.K. *et al.* (eds), *Sea-Level Changes: An Integrated Approach*. SEPM Special Publication 42, 39–45.

Varnes, D.J. 1958. Landslide types and processes. Highway Research Board Special report 29, 20–47 (Washington, D.C.)

van der Veen, C.J. 1986. Numerical modelling of ice shelves and ice tongues. *Annales Geophysicae* 4B, 45–54.

van der Veen, C.J. 1996. Tidewater calving. *Journal of Glaciology* 42, 375–385.

van der Veen, C.J. and Whillans, I.M. 1989. Force budget I. Theory and numerical methods. *Journal of Glaciology* 35, 53–60.

Veillette, J.J. 1986. Former southwesterly ice flows in the Abitibi–Timiskaming region: implications for the configuration of the late Wisconsinan ice sheet. *Canadian Journal of Earth Sciences* 23, 1724–1741.

Velichko, A.A., Isayeva, L.L., Makeyev, V.M., Matishov, G.G. and Faustova, M.A. 1984. Late Pleistocene glaciation of the Arctic Shelf, and the reconstruction of Eurasian icesheets. In Velichko, A.A. (ed.), *Late Quaternary Environments of the Soviet Union*. Longman, London, 35–44.

Velichko, A.A., Isayeva, L.L., Oreshkin, D.B. and Faustova, M.A. 1989. The last glaciation of Eurasia. In Herman, Y. (ed.), *The Arctic Seas*. Reinhold, New York, 729–758.

Vere, D.M. and Benn, D.I. 1989. Structure and debris characteristics of medial moraines in Jotunheimen, Norway: implications for moraine classification. *Journal of Glaciology* 35, 276–280.

Vere, D.M. and Matthews, J.A. 1985. Rock glacier formation from a lateral moraine at Bukkeholsbreen, Jotunheimen, Norway: a sedimentological approach. *Zeitschrift für Geomorphologie* 29, 397–415.

Vernon, P. 1966. Drumlins and Pleistocene ice flow over the Ards peninsula – Strangford Lough area, County Down, Ireland. *Journal of Glaciology* 6, 401–409.

Veyret, P. 1955. Le lit glaciaire: contradiction apparente des formes et logique réelle des processus d'érosion. *Revue de Géographie Alpine* 43, 495–509.

Veyret, P. 1971. Observations recentes sur deux glaciers du massif du Mont Blanc. *Annales der Météorologie* 5, 219–220.

Vialov, S.S. 1958. Regularities of glacial shield movements and the theory of plastic viscous flow. *IASH* 47, 266–275.

Vielette, J.J. 1986. Former southwesterly ice flows in Abitibi–Timiskaming region: implications for the configuration of the Late Wisconsinan ice sheet. *Canadian Journal of Earth Sciences* 23, 1724–1741.

Vielette, J.J. 1988. Déglaciation et évolution des lacs proglaciaires Post-Algonquin et Barlow au Temiscamingue, Québec et Ontario. *Géographie Physique et Quaternaire* 42, 7–31.

Vikingsson, S. 1978. The deglaciation of the southern part of the Skagafjordur District, northern Iceland. *Jökull* 28, 1–17.

Vilborg, L. 1977. The cirque forms of Swedish Lapland. *Geografiska Annaler* 59A, 89–150.

Vincent, J.S. 1977. Le Quaternaire récent de la région du cours inferieur de La Grande Rivière, Quebec. Geological Survey of Canada, Etude 76–19.

Vincent, J.S. 1989. Quaternary geology of the southeastern Canadian shield. In Fulton, R.J. (ed.), *Quaternary Geology of Canada and Greenland*. Geological Survey of Canada, Geology of Canada, No. 1, 249–275.

Vincent, J.S. and Hardy, L. 1979. The evolution of glacial lakes Barlow and Ojibway, Quebec and Ontario. Geological Survey of Canada Bulletin 316.

Vincent, J.S. and Prest, V.K. 1987. The Early Wisconsinan history of the Laurentide Ice Sheet. *Géographie Physique et Quaternaire* 41, 199–213.

Vincent, P.J. 1976. Some periglacial deposits near Aberystwyth, Wales, as seen with a scanning electron microscope. *Biuletyn Periglacjalny* 25, 59–64.

Virkkala, K. 1952. On the bed structure of till in eastern Finland. *Geological Survey of Finland Bulletin* 157, 97–109.

Virkkala, K. 1958. Stone counts in the esker of Hämeenlinna, southern Finland. *Comptes Rendus de la Société Géologique de Finlande* 30, 87–103.

Virkkala, K. 1960. On the striations and glacier movements in the Tampere region, southern Finland. *Geological Society of Finland, Bulletin* 188, 159–176.

Virkkala, K. 1963. On ice-marginal features in southwestern Finland. Bulletin de la Commission Géologique de Finlande 210.

Virkkala, K. 1969. On the lithology and provenance of the till of a gabbro area in Finland. In Ters, M. (ed.), *Etudes sur le Quaternaire dans le monde*. INQUA, VIII Congress, Paris, 711–714.

Visser, J.N.J. 1983a. Glacial-marine sedimentation in the Late Paleozoic Karoo Basin, Southern Africa. In Molnia, B.F. (ed.), *Glacial Marine Sedimentation*. Plenum Press, London, 667–701.

Visser, J.N.J. 1983b. Submarine debris flow deposits from the Upper Carboniferous Dwyka Till Formation in the Kalahari Basin, South Africa. *Sedimentology* 30, 511–524.

Visser, J.N.J. 1983c. The problems of recognizing ancient subaqueous debris flow deposits in glacial sequences. *Transactions of the Geological Society of South Africa* 86, 127–135.

Visser, J.N.J. 1991. The paleoclimatic setting of the late Paleozoic marine ice sheet in the Karoo Basin of southern Africa. In Anderson, J.B. and Ashley, G.M. (eds), *Glacial Marine Sedimentation: Paleoclimatic Significance*. Geological Society of America, Special Paper 261, 181–189.

Visser, J.N.J. 1994. The interpretation of massive rain-out and debris flow diamictites from the glacial marine environment. In Deynoux, M., Miller, J.M.G., Domack, E.W., Eyles, N., Fairchild, I.J. and Young, G.M. (eds), *Earth's Glacial Record*. Cambridge University Press, Cambridge, 83–94.

Visser, J.N.J., Colliston, W.B. and Terblanche, J.C. 1984. The origin of soft sediment deformation structures and related deposits in Permo-Carboniferous glacial and proglacial beds, South Africa. *Journal of Sedimentary Petrology* 54, 1183–1196.

Visser, J.N.J. and Hall, K.J. 1985. Boulder beds in the glaciogenic Permo-Carboniferous Dwyka Formation in South Africa. *Sedimentology* 32, 281–294.

Visser, J.N.J., Loock, J.C. and Colliston, W.P. 1987. Subaqueous outwash fan and esker sandstones in the Permo-Carboniferous Dwyka Formation of South Africa. *Journal of Sedimentary Petrology* 57, 467–478.

Vita-Finzi, C. 1986. *Recent Earth Movements: An Introduction to Neotectonics*. Academic Press, London.

Vivian, R. 1970. Hydrologie et érosion sous-glaciaires. *Revue de Géographie Alpine* 58, 241–264.

Vivian, R. 1974. Les débâcles glaciaires dans les Alpes Occidentales. In Castiglioni, G.B. (ed.), *Le Calamità naturali nelle Alpi*. Istituto di Geografia dell'Università di Padova.

Vivian, R. 1975. *Les Glaciers des Alpes Occidentales*. Allier, Grenoble.

Vivian, R. and Bocquet, G. 1973. Subglacial cavitation phenomena under the Glacier d'Argentière, Mont Blanc, France. *Journal of Glaciology* 12, 439–451.

Vivian, R. and Zumstein, J. 1973. Hydrologie sous-glaciaire au Glacier d'Argentière (Mont Blanc, France). IASH Publication 95, 53–64.

Vorren, T.O. 1977. Weichselian ice movement in south Norway and adjacent areas. *Boreas* 6, 247–257.

Vorren, T.O., Hald, M., Edvardsen, M, and Lind-Hansen, O.W. 1983. Glacigenic sediments and sedimentary environments on continental shelves: general principles with a case study from the Norwegian shelf. In Ehlers, J. (ed.), *Glacial Deposits in North-west Europe*. Balkema, Rotterdam, 61–73.

Vorren, T.O and Kristoffersen, Y. 1986. Late Quaternary glaciation in the south-western Barents Sea. *Boreas* 15, 51–59.

Vorren, T.O., Lebesbye, E., Andreassen, K. and Larsen, K.B. 1989. Glacigenic sediments on a passive continental margin as exemplified by the Barents Sea. *Marine Geology* 85, 251–272.

Vorren, T.O., Lebesbye, E. and Larsen, K.B. 1990. Geometry and genesis of the glacigenic sediments in the southern Barents Sea. In Dowdeswell, J.A. and Scourse, J.D. (eds), *Glacimarine Environments: Processes and Sediments*. Geological Society Special Publication 53, 269–288.

Vuichard, D. and Zimmerman, M. 1986. The Langmoche flash-flood, Khumbu Himal, Nepal. *Mountain Research and Development* 6, 90–94.

Vuichard, D. and Zimmerman, M. 1987. The 1985 catastrophic drainage of a moraine-dammed lake, Khumbu Himal, Nepal: cause and consequences. *Mountain Research and Development* 7, 91–110.

Waddington, E.D. 1986. Wave ogives. *Journal of Glaciology* 32, 325–334.

Wahrhaftig, C. and Cox, A. 1959. Rock glaciers in the Alaska Range. *Geological Society of America Bulletin* 70, 383–436.

Waitt, R.B. 1980. About forty last-glacial Lake Missoula jökulhlaups through southern Washington. *Journal of Geology* 88, 653–679.

Waitt, R.B. 1984. Periodic jökulhlaups from Pleistocene glacial Lake Missoula: new evidence from varved sediment in northern Idaho and Washington. *Quaternary Research* 22, 46–58.

Waitt, R.B. 1985. Case for periodic, colossal jökulhlaups from glacial Lake Missoula. *Geological Society of America Bulletin* 95, 1271–1286.

Waitt, R.B. and Thorson, R.M. 1983. The Cordilleran ice sheet in Washington, Idaho and Montana. In Porter, S.C. (ed.), *Late Quaternary Environments of the United States*, vol. 1: *The Late Pleistocene*. University of Minnesota Press, Minneapolis, 53–70.

van de Wal, R.S.W. and Oerlemans, J. 1994. An energy balance model for the Greenland ice sheet. *Global and Planetary Change* 9, 115–131.

van de Wal, R.S.W. and Oerlemans, J. 1995. Response of valley glaciers to climate change and kinematic waves: a study with a numerical ice-flow model. *Journal of Glaciology* 41, 142–152.

Walcott, R.I. 1970. Isostatic response to loading of the crust in Canada. *Canadian Journal of Earth Sciences* 7, 716–727.

Walcott, R.I. 1972. Past sea levels, eustasy, and deformation of the Earth. *Quaternary Research* 2, 1–14.

Walden, J. 1994. Late Devensian sedimentary environments in the Irish Sea basin: glacioterrestrial or glaciomarine? In Boardman, J. and Walden, J. (eds), *The Quaternary of Cumbria: Field Guide*. Quaternary Research Association, Oxford, 15–18.

Walden, J., Smith, J.P. and Dackombe, R.V. 1987. The use of mineral magnetic analyses in the study of glacial diamicts: a pilot study. *Journal of Quaternary Science* 2, 73–80.

Walden, J., Smith, J.P. and Dackombe, R.V. 1992. Mineral magnetic analyses as a means of lithostratigraphic correlation and provenance indication of glacial diamicts: intra- and inter-unit variation. *Journal of Quaternary Science* 7, 257–270.

Walden, J., Smith, J.P., Dackombe, R.V. and Rose, J. 1995. Mineral magnetic analyses of glacial diamicts from the Midland Valley of Scotland. *Scottish Journal of Geology* 31, 79–89.

Walder, J.S. 1982. Stability of sheet flow of water beneath temperate glaciers and implications for glacier surging. *Journal of Glaciology* 28, 273–293.

Walder, J.S. 1986. Hydraulics of subglacial cavities. *Journal of Glaciology* 32, 439–445.

Walder, J.S. 1994. Comments on 'Subglacial floods and the origin of low-relief ice-sheet lobes' by E.M. Shoemaker. *Journal of Glaciology* 40, 199–200.

Walder, J.S. and Driedger, C.L. 1995. Frequent outburst floods from South Tahoma Glacier, Mount Rainier, USA: relation to debris flows, meteorological origin and implications for subglacial hydrology. *Journal of Glaciology* 41, 1–10.

Walder, J.S. and Fowler, A. 1994. Channelized subglacial drainage over a deformable bed. *Journal of Glaciology* 40, 3–15.

Walder, J.S. and Hallet, B. 1979. Geometry of former subglacial water channels and cavities. *Journal of Glaciology* 23, 335–346.

Walder, J.S. and Hallet, B. 1985. A theoretical model for the fracture of rock during freezing. *Geological Society of America Bulletin* 96, 336–346.

Walder, J.S. and Hallet, B. 1986. The physical basis for frost weathering: toward a more fundamental and unified perspective. *Arctic and Alpine Research* 18, 27–32.

Walker, R.G. 1975. Generalised facies models for resedimented conglomerates of turbidite association. *Geological Society of America Bulletin* 86, 737–748.

Walker, R.G. 1976. Facies models: I. General introduction. *Geoscience Canada* 3, 21–24.

Walker, R.G. (ed.) 1984. *Facies Models*. Geoscience Canada, Reprint Series 1, Geological Association of Canada.

Walker, R.G. 1990. Facies modeling and sequence stratigraphy. *Journal of Sedimentary Petrology* 60, 777–786.

Walker, R.G. 1992a. Facies, facies models and modern stratigraphic concepts. In Walker, R.G. and James, N.P. (eds), *Facies Models: Response to Sea-Level Change*. Geological Association of Canada, Toronto, 1–14.

Walker, R.G. 1992b. Turbidites and submarine fans. In Walker, R.G. and James, N.P. (eds), *Facies Models: Response to Sea-Level Change*. Geological Association of Canada, Toronto, 239–263.

Walker, R.G. and Plint, A.G. 1992. Wave- and storm-dominated shallow marine systems. In Walker, R.G. and James, N.P. (eds), *Facies Models: Response to Sea-Level Change*. Geological Association of Canada, Toronto, 219–238.

Walker, R.J. 1977. Deposition of Upper Mesozoic resedimented conglomerates and associated turbidites in southwestern Oregon. *Geological Society of America Bulletin* 88, 272–285.

Wallen, C.C. 1948. Glacial-meteorological investigations on the Karsa Glacier in Swedish Lappland 1942–48. *Geografiska Annaler* 30, 451–672.

Walters, R.A. and Dunlap, W.W. 1987. Analysis of time series of glacier speed: Columbia Glacier, Alaska. *Journal of Geophysical Research* 92, 8969–8975.

Walther, J. 1894. *Einleitung in die Geologie als historische Wissenschaft*, Bd. 3: *Lithogenesis der Gegenwart*. Fischer Verlag, Jena, 535–1055.

Warburton, J. and Fenn, C.R. 1994. Unusual flood events from an Alpine glacier: observations and deductions on generating mechanisms. *Journal of Glaciology* 40, 176–186.

Warren, C.R. 1992. Iceberg calving and the glacioclimatic record. *Progress in Physical Geography* 16, 253–282.

Warren, C.R. 1993. Rapid recent fluctuations of the calving San Rafael Glacier, Chilean Patagonia: climatic or non-climatic? *Geografiska Annaler* 75A, 111–125.

Warren, C.R. 1995. Glaciers in the greenhouse. *Geography Review* 8, 2–7.

Warren, C.R., Glasser, N.F., Harrison, S., Winchester, V., Kerr, A.R. and Rivera, A. 1995. Characteristics of tidewater calving at Glaciar San Rafael, Chile. *Journal of Glaciology* 41, 273–289.

Warren, C.R. and Hulton, N.R.J. 1990. Topographic and glaciological controls on Holocene ice margin dynamics, central west Greenland. *Annals of Glaciology* 14, 307–310.

Warren, C.R. and Sugden, D.E. 1993. The Patagonian icefields: a glaciological review. *Arctic and Alpine Research* 25, 316–331.

Warren, W.P. 1987. Glaciodiagenesis in gelifluctuated deposits on the south coast of County Cork, Ireland. In van der Meer, J.J.M. (ed.), *Tills and Glaciotectonics*. Balkema, Rotterdam, 105–115.

Warren, W.P. 1988. Protalus till. In Goldthwait, R.P. and Matsch, C.L. (eds), *Genetic Classification of Glacigenic Deposits*. Balkema, Rotterdam, 145–146.

Warren, W.P. and Ashley, G.M. 1994. Origins of the ice-contact stratified ridges (eskers) of Ireland. *Journal of Sedimentary Research* A64, 433–449.

Warren, W.P. and Croot, D.G. (eds) 1994. *Formation and Deformation of Glacial Deposits*. Balkema, Rotterdam.

Warrick, R.A., Le Provost, C., Meier, M.F., Oerlemans, J. and Woodworth, P.L. 1996. Changes in sea level. In Houghton, J.T., Meira Filho, L.G., Callander, B.A., Harris, N., Kattenberg, A. and Maskell, K. (eds), *Climate Change 1995: The Science of Climate Change*. Cambridge University Press, Cambridge, 359–405.

Washburn, A.L. 1979. *Geocryology*. Edward Arnold, London.

van der Wateren, D.F.M. 1981. Glacial tectonics at the Kwintelooijen sandpit, Rhenen, the Netherlands. In Ruegg, G.H.J. and Zandstra, J.G. (eds), *Geology and Archaeology of Pleistocene Deposits in the Ice-Pushed Ridge near Rhenen and Veenendal*. Mededelingen Rijks Geologische Dienst 35–2/7, 252–268.

van der Wateren, D.F.M. 1985. A model of glacial tectonics, applied to the ice-pushed ridges in the central Netherlands. *Bulletin of the Geological Society Denmark* 34, 55–74.

van der Wateren, D.F.M. 1987. Structural geology and sedimentology of the Dammer Berge push moraine, FRG. In van der Meer, J.J.M. (ed.), *Tills and Glaciotectonics*. Balkema, Rotterdam, 157–182.

van der Wateren, D.F.M. 1994. Proglacial subaquatic outwash fan and delta sediments in push moraines: indicators of subglacial meltwater activity. *Sedimentary Geology* 91, 145–172.

van der Wateren, D.F.M. 1995. Processes of glaciotectonism. In Menzies, J. (ed.), *Glacial Environments*, vol. 1: *Modern Glacial Environments: Processes, Dynamics and Sediments*. Butterworth-Heinemann, Oxford, 309–335.

Watson, E. 1966. Two nivation cirques near Aberystwyth, Wales. *Biuletyn Peryglacjalny* 15, 79–101.

Watson, E. and Watson, S. 1967. The periglacial origin of the drifts at Morfa-bychan, near Aberystwyth. *Geological Journal* 5, 419–440.

Wayne, W.J. 1981. Ice segregation as an origin for lenses of non-glacial ice in 'ice-cemented' rock glaciers. *Journal of Glaciology* 27, 506–510.

Wayne, W.J. 1984. Ice segregation as an origin for lenses of non-glacial ice in 'ice-cemented' rock glaciers. *Journal of Glaciology* 29, 524.

Webb, P.N. and Harwood, D.M. 1991. Late Cenozoic glacial history of the Ross Embayment, Antarctica. *Quaternary Science Reviews* 10, 215–224.

Webb, P.N., Harwood, D.M., McKelvey, B.C., Mercer, J.H. and Stott, L.D. 1984. Cenozoic marine sedimentation and ice volume variation on the East Antarctic craton. *Geology* 12, 287–291.

Webb, P.N. and McKelvey, P.C. 1959. Geological investigations in South Victoria land, Antarctica. *New Zealand Journal of Geology and Geophysics* 2, 120–136.

ter Wee, M.W. 1983. The Saalian glaciation in the northern Netherlands. In Ehlers, J. (ed.), *Glacial Deposits in North-West Europe*. Balkema, Rotterdam, 405–412.

Weeks, W.F. and Mellor, M. 1978. Some elements of iceberg technology. In Husseiny, A.A. (ed.), *Iceberg Utilization*. Pergamon, New York, 45–98.

Weertman, J. 1957a. Deformation of floating ice shelves. *Journal of Glaciology* 3, 38–42.

Weertman, J. 1957b. On the sliding of glaciers. *Journal of Glaciology* 3, 33–38.

Weertman, J. 1961a. Stability of ice age ice sheets. *Journal of Geophysical Research* 66, 3783–3792.

Weertman, J. 1961b. Equilibrium profile of ice caps. *Journal of Glaciology* 3, 953–964.

Weertman, J. 1961c. Mechanism for the formation of inner moraines found near the edge of cold ice caps and ice sheets. *Journal of Glaciology* 3, 965–978.

Weertman, J. 1964. The theory of glacier sliding. *Journal of Glaciology* 5, 287–303.

Weertman, J. 1969. Water lubrication mechanism of glacier surges. *Canadian Journal of Earth Sciences* 6, 929–942.

Weertman, J. 1972. General theory of water flow at the base of a glacier or ice sheet. *Reviews of Geophysics and Space Physics* 10, 287–333.

Weertman, J. 1973. Position of ice divides and ice centers on ice sheets. *Journal of Glaciology* 12, 353–360.

Weertman, J. 1979. The unsolved general glacier sliding problem. *Journal of Glaciology* 23, 97–115.

Weertman, J. 1983. Creep deformation of ice. *Annual Review of Earth and Planetary Sciences* 11, 215–240.

Weidick, A. 1985. Review of glacier changes in west Greenland. *Zeitschrift für Gletscherkunde und Glazialgeologie* 21, 301–309.

Weidick, A. 1988. Surging glaciers in Greenland: a status. *Gronl. Geol. Undersolgelse Rapp.* 140, 106–110.

Weirich, F.H. 1984. Turbidity currents: monitoring their occurrence and movement with a three-dimensional sensor network. *Science* 224, 384–387.

Weirich, F.H. 1986. The record of density-induced underflow in a glacial lake. *Sedimentology* 33, 261–277.

Weirich, F.H. 1989. The generation of turbidity currents by subaerial debris flows, California. *Bulletin of the Geological Society of America* 101, 278–291.

Welch, R. and Howarth, P.J. 1968. Photogrammetric measurement of glacial landforms. *Photogrammetric Record* 6, 75–96.

Wendler, G. and Ishikawa, N. 1974. The effect of slope, exposure, and mountain screening on the solar radiation of McCall Glacier, Alaska: a contribution to the International Hydrological Decade. *Journal of Glaciology* 13, 213–226.

Werritty, A. 1992. Downstream fining in a gravel-bed river in southern Poland: lithologic controls and the role of abrasion. In Billi, P., Hey, R.D., Thorne, C.R. and Tacconi, P. (eds), *Dynamics of Gravel-Bed Rivers*. Wiley, London, 333–350.

Werth, E. 1907. Studien zur glazialen Bodengestaltung in den skandinavischen Ländern. *Zeitschrift der Gesellschaft für Erdkunde zu Berlin*, 27–43 and 87–109.

Westgate, J.A. 1968. Linear sole markings in Pleistocene till. *Geological Magazine* 106, 501–505.

Whalley, W.B. 1971. Observations of the drainage of an ice-dammed lake – Strupvatnet, Troms, Norway. *Norsk Geografisk Tidsskrift* 25, 165–174.

Whalley, W.B. 1983. Rock glaciers – permafrost features or glacial relics? *Proceedings of the Fourth International Conference on Permafrost*, vol. 1. National Academy Press, Washington DC 1, 1396–1401.

Whalley, W.B. 1996. Scanning electron microscopy. In Menzies, J. (ed.), *Past Glacial Environmnents: Sediments, Forms and Techniques*. Butterworth-Heinemann, Oxford, 357–375.

Whalley, W.B. and Krinsley, D.H. 1974. A scanning electron microscope study of surface textures of quartz grains from glacial environments. *Sedimentology* 21, 87–105.

Whalley, W.B. and Martin, H.E. 1992. Rock glaciers. II: Models and mechanisms. *Progress in Physical Geography* 16, 127–186.

Whalley, W.B., Palmer, C., Hamilton, S. and Gordon, J. 1994. Ice exposures in rock glaciers. *Journal of Glaciology* 40, 427–429.

Whillans, I.M. 1978. Erosion by continental ice sheets. *Journal of Geology* 86, 516–524.

Whillans, I.M., Bolzan, J. and Shabtaie, S. 1987. Velocity of Ice streams B and C, Antarctica. *Journal of Geophysical Research* 92, 8895–8902.

Whillans, I.M. and van der Veen, C.J. 1993. New and improved determinations of velocity of Ice Streams B and C, West Antarctica. *Journal of Glaciology* 39, 483–490.

White, I.D., Mottershead, D.N. and Harrison, S.J. 1984. *Environmental Systems: An Introductory Text*. Unwin Hyman, London.

White, S.E. 1976. Is frost weathering really only hydration shattering? *Arctic and Alpine Research* 8, 1–6.

White, W.A. 1972. Deep erosion by continental ice sheets. *Geological Society of America Bulletin* 83, 1037–1056.

White, W.A. 1988. More on deep glacial erosion by continental ice sheets and their tongues of distributary ice. *Quaternary Research* 30, 137–150.

Whittecar, G.R. and Mickelson, D.M. 1979. Composition, internal structures, and an hypothesis of formation for drumlins, Waukesha County, Wisconsin, USA. *Journal of Glaciology* 22, 357–370.

Whittington, R.J. 1977. A late-glacial drainage pattern in the Kish Bank area and postglacial sediments in the Central Irish Sea. In Kidson, C. and Tooley, M.J. (eds), *The Quaternary History of the Irish Sea*. Seel House Press, Liverpool, 55–68.

Whittow, J.B. 1960. Some observations on the snowfall of Ruwenzori. *Journal of Glaciology* 3, 765–772.

Whittow, J.B., Shepherd, A., Goldthorpe, J.E. and Temple, P.H. 1963. Observations on the glaciers of the Ruwenzori. *Journal of Glaciology* 4, 581–616.

Wightman, C.A. 1986. Variability of clast size and roundness in contemporary meltwater rivers at Okstindan, north Norway. In Bridgland, D.R. (ed.), *Clast Lithological Analysis*. Quaternary Research Association, Cambridge, 179–192.

Wilch, T.I., Denton, G.H., Lux, D.R. and McIntosh, W.C. 1993. Limited Pliocene glacier extent and surface uplift in Middle Taylor Valley, Antarctica. *Geografiska Annaler* 75A, 331–351.

Wiles, G.C., Calkin, P.E. and Post, A. 1995. Glacier fluctuations in the Kenai Fjords, Alaska, USA: an evaluation of controls on iceberg-calving glaciers. *Arctic and Alpine Research* 27, 234–245.

Williams, G.P. 1988. Palaeofluvial extimates from dimensions of former channels and meanders. In Baker, V.R., Kochel, R.C. and Patton, P.C. (eds), *Flood Geomorphology*. Wiley, Chichester, 321–334.

Williams, H. and King, A.F. 1979. Trepassey map area, Newfoundland. Geological Survey of Canada, Memoir 389.

Williams, J., Barry, R.G. and Washington, W.M. 1974. Simulation of the atmospheric circulation using the NCAR global circulation model with ice age boundary conditions. *Journal of Applied Meteorology* 13, 305–317.

Williams, L.D. 1978. The Little Ice Age glaciation level on Baffin Island, Arctic Canada. *Palaeogeography, Palaeoclimatology, Palaeoecology* 25, 199–207.

Williams, L.D. 1979. An energy balance model of potential glacierization of northern Canada. *Arctic and Alpine Research* 11, 443–456.

Williams, M.A.J., Dunkerley, D.L., De Deckker, P., Kershaw, A.P. and Stokes, T. 1993. *Quaternary Environments*. Edward Arnold, London.

Williams, P.F. and Rust, B.R. 1969. The sedimentology of a braided river. *Journal of Sedimentary Petrology* 39, 649–679.

Williams, R.S. 1986. Glaciers and glacial landforms. In Short, N.M. and Blair, R.W. (eds), *Geomorphology from Space: A Global Overview of Regional Landforms*. NASA Special Publication SP–486, 521–596.

Williams, R.S. and Ferrigno, J.G. (eds) 1988. Satellite image atlas of glaciers of the world: Antarctica. USGS Professional Paper 1386-B.

Williams, R.S. and Ferrigno, J.G. (eds) 1989. Satellite image atlas of glaciers of the world: Irian Jaya, Indonesia and New Zealand. USGS Professional Paper 1386-H.

Williams, R.S. and Ferrigno, J.G. (eds) 1991. Satellite image atlas of glaciers of the world: Middle East and Africa. USGS Professional Paper 1386-G.

Williams, R.S. and Ferrigno, J.G. (eds) 1993. Satellite image atlas of glaciers of the world: Europe. USGS Professional Paper 1386-E.

Williams, R.S. and Hall, D.K. 1993. Glaciers. In Gurney, R.J., Foster, J.L. and Parkinson, C.L. (eds), *Atlas of Satellite Observations Related to Global Change*. Cambridge University Press, Cambridge, 401–422.

Willis, I.C. 1995. Intra-annual variations in glacier motion: a review. *Progress in Physical Geography* 19, 61–106.

Willis, I.C., Sharp, M.J. and Richards, K.S. 1990. Configuration of the drainage system of Mitdalsbreen, Norway, as indicated by dye-tracing experiments. *Journal of Glaciology* 36, 89–101.

Wilson, A.T. 1978. Past surges in the West Antarctic ice sheet and their climatological significance. In van Zinderen Bakker, E.M. (ed.), *Antarctic Glacial History and World Palaeoenvironments*. Balkema, Rotterdam, 33–39.

Wilson, G.S. 1995. The Neogene East Antarctic Ice Sheet: a dynamic or stable feature? *Quaternary Science Reviews* 14, 101–123.

Wilson, J.T. 1938. Glacial geology of part of northwestern Quebec. *Transactions of the Royal Society of Canada, Section 4*, 32, 49–59.

Wingfield, R.T.R. 1989. Glacial incisions indicating Middle and Upper Pleistocene ice limits off Britain. *Terra Nova* 1, 31–52.

Wingfield, R.T.R. 1990. The origin of major incisions within the Pleistocene deposits of the North Sea. *Marine Geology* 91, 31–52.

Wingfield, R.T.R. Eyles, N. and McCabe, A.M. 1992. The Late Devensian (<22,000 BP) Irish Sea Basin: the sedimentary record of a collapsed ice sheet margin: discussion and reply. *Quaternary Science Reviews* 11, 377–379.

Winters, W.A. 1961. Landforms associated with stagnant ice. *Professional Geographer* 13, 19–23.

Wintges, von Th. and Heuberger, H. 1980. Parabelrisse, Sichelbrüche und Sichelwannen im Vereinigungsbereich zweier zillertaler Gletscher (Tirol). *Zeitschrift für Gletscherkunde und Glazialgeologie* 16, 11–23.

Woldstedt, P. 1926. Probleme der Seenbildung in Norddeutschland. *Zeitschrift der Gesellschaft für Erdkunde zu Berlin* 1926, 103–124.

Woldstedt, P. 1952. Die Entstehung der Seen in den ehemals vergletscherten Gebieten. *Eiszeitalter und Gegenwart* 2, 146–153.

Woldstedt, P. 1954. *Das Eiszeitalter: Grundlinien einer Geologie des Quartars*, vol. 1. Ferdinand Enke, Stuttgart.

Woldstedt, P. 1955. *Norddeutschland und angrenzende Gebiete im Eiszeitalter*. Koehler Verlag, Stuttgart.

Woldstedt, P. and Duphorn, K. 1974. *Norddeutschland und angrenzende Gebiete im Eiszeitalter*. Koehler Verlag, Stuttgart.

Woodland, A.W. 1970. The buried tunnel valleys of East Anglia. *Proceedings of the Yorkshire Geological Society* 37, 521–578.

Woodworth, J.B. 1897. Unconformities of Martha's Vineyard and of Block Island. *Bulletin of the Geological Society of America* 8, 197–212.

Woodworth, J.B. and Wigglesworth, E. 1934. Geography and geology of the region including Cape Cod, the Elizabeth Islands, Nantucket, Martha's Vineyard, No Mans Land, and Block Island. Memoirs of the Museum of Comparative Zoology, Harvard College.

Woodworth-Lynas, C.M.T. and Dowdeswell, J.A. 1993. Soft sediment strated surfaces and massive diamicton facies produced by floating ice. In Deynoux *et al.* (eds), *Earth's Glacial Record*. Cambridge University Press, Cambridge, 241–259.

Woodworth-Lynas, C.M.T. and Guigné, J.Y. 1990. Iceberg scours in the geological record: examples from glacial Lake Agassiz. In Dowdeswell, J.A. and Scourse, J.D. (eds), *Glacimarine Environments: Processes and Sediments*. Special Publications of the Geologists' Association 53, 217–223.

World Glacier Monitoring Service 1989. *World Glacier Inventory*. IAHS–UNEP–UNESCO.

Worsley, P. 1969. The Cheshire–Shropshire lowlands. In Lewis, C.A. (ed.), *The Glaciations of Wales and Adjoining Regions*. Longmans, London, 83–106.

Worsley, P. 1974. Recent 'annual' moraine ridges at Austre Okstindbreen, Okstindan, north Norway. *Journal of Glaciology* 13, 265–277.

Wright, H.E. 1957. Stone orientation in Wadena drumlin field, Minnesota. *Geografiska Annaler* 39, 19–31.

Wright, H.E. 1962. Role of the Wadena lobe in the Wisconsin glaciation of Minnesota. *Bulletin of the Geological Society of America* 73, 73–100.

Wright, H.E. 1973. Tunnel valleys, glacial surges and subglacial hydrology of the Superior Lobe, Minnesota. In Black, R.F., Goldthwait, R.P. and Williams, H.B. (eds), *The Wisconsinan Stage*. Geological Society of America, Memoir 36, 251–276.

Wright, H.E. 1989. The amphi-Atlantic distribution of the Younger Dryas paleoclimatic oscillation. *Quaternary Science Reviews* 8, 295–306.

Wright, L.D. 1977. Sediment transport and deposition at river mouths: a synthesis. *Bulletin of the Geological Society of America* 88, 857–868.

Wright, L.D. 1978. River deltas. In Davies, R.A. (ed.), *Coastal Sedimentary Environments*. Springer-Verlag, Heidelberg, 5–68.

Wright, L.D. and Coleman, J.M. 1973. Variations in morphology of major river deltas as functions of ocean wave and river discharge regimes. *Bulletin of the American Association of Petroleum Geologists* 57, 370–398.

Wright, M.D. 1983. The distribution and engineering significance of superficial deposits in the Upper Clydach Valley, south Wales. *Quarterly Journal of Engineering Geology* 16, 319–330.

Wright, M.D. 1991. Pleistocene deposits of the South Wales Coalfield and their engineering significance. In Forster, A., Culshaw, M.G., Cripps, J.C., Little, J.A. and Moon, C.F. (eds), *Quaternary Engineering Geology*. Geological Society, Engineering Geology Special Publication 7, 441–448.

Wright, M.D. and Harris, C. 1980. Superficial deposits in the South Wales Coalfield. In Perkins, J.W. (ed.), *Cliff and Slope Stability*. University College Cardiff, Department of Extra-Mural Studies, 193–205.

Wright, R. and Anderson, J.B. 1982. The importance of sediment gravity flow to sediment transport and sorting in a glacial marine environment: eastern Weddell Sea, Antarctica. *Geological Society of America Bulletin* 93, 951–963.

Wright, R.F. and Nydegger, P. 1980. Sedimentation of detrital particulate matter in lakes: influence of currents produced by inflowing rivers. *Water Resources Research* 16, 597–601.

Wright, W.B. 1914. *The Quaternary Ice Age*. MacMillan, London.

Wu, P. and Peltier, W.R. 1983. Glacial isostatic adjustment and the free air gravity anomaly as a constraint upon deep mantle viscosity. *Geophysical Journal of the Royal Astronomical Society* 74, 377–449.

Wysota, W. 1994. Morphology, internal composition and origin of drumlins in the southeastern part of the Chelmno-Dobrzyn Lakeland, north Poland. *Sedimentary Geology* 91, 345–364.

Yamada, T. 1988. Glaciological characteristics revealed by 37.6 m deep core drilled at the accumulation area of San Rafael Glacier, the Northern Patagonia Icefield. *Bulletin of Glacier Research* 4, 59–8.

Yasunari, T. and Inoue, J. 1978. Characteristics of monsoonal precipitation around peaks and ridges in Shorong and Khumbu Himal. *Seppyo* 40, Special Issue, 26–32.

Yen, Y.C. 1981. Review of thermal properties of snow, ice and sea ice. CRREL Report 81–10.

Young, G.J. 1981. The mass balance of Peyto Glacier, Alberta, Canada, 1965–1978. *Arctic and Alpine Research* 13, 307–318.

Young, G.M. 1970. An extensive early Proterozoic glaciation in North America? *Palaeogeography, Palaeoclimatology, Palaeoecology* 7, 85–101.

Young, G.M. 1973. Tillites and aluminous quartzites as possible time markers for middle Precambrian (Aphebian) rocks of North America. In Young, G.M. (ed.), *Huronian Stratigraphy and Sedimentation*. Geological Association of Canada, Special Paper 12, 97–127.

Young, G.M. 1979. The earliest ice ages: Precambrian. In John, B.S. (ed.), *Winters of the World: Earth under the Ice Ages*. David and Charles, Newton Abbot, 107–130.

Young, G.M. 1991. The geologic record of glaciation: relevance to the climatic history of the Earth. *Geoscience Canada* 18, 100–108.

Young, G.M. and Gostin, V.A. 1991. Late Proterozoic (Sturtian) succession of the North Flinders Basin, South Australia; an example of temperate glaciation in an active rift setting. In Anderson, J.B. and Ashley, G.M. (eds), *Glacial Marine Sedimentation: Paleoclimatic Significance*. Geological Society of America, Special Paper 261: 207–222.

Young, G.M. and Nesbitt, H.W. 1985. The Gowganda Formation in the southern part of the Huronian outcrop belt, Ontario, Canada. Stratigraphy, depositional environments and regional tectonic significance. *Precambrian Research* 29, 265–301.

Young, J.A.T. 1975. Ice wastage in Glen Feshie, Inverness-shire. *Scottish Geographical Magazine* 91, 91–101.

Young, J.A.T. 1978. The landforms of upper Strathspey. *Scottish Geographical Magazine* 94, 76–94.

Young, J.A.T. and Hastenrath, S. 1991. Glaciers of Africa. In Williams, R.S. and Ferrigno, J.G. (eds), *Satellite Image Atlas of Glaciers of the World*. USGS Professional Paper 1386-G, 49–70.

Zhang Xiangsong, Zheng Benxiang and Xie Zichu 1981. Recent variations of existing glaciers on the Qinghai–Xizang Tibet. Plateau. *Geological and Ecological Studies of Qinghai–Xizang Plateau*, vol. 2. Science Press, Beijing, 1625–1629.

Zilliacus, H. 1976. DeGeer-moraner och isrecessionen i sodra Finlands ostra delar. *Terra* 88, 176–184.

Zilliacus, H. 1981. DeGeer-moranerna pa Replot och Bjorkon i Vasa skargard. *Terra* 93, 12–24.

Zilliacus, H. 1987a. The DeGeer moraines in Finland. In Gardiner V. (ed.), *International Geomorphology 1986*, II. *Proceedings of the First International Conference on Geomorphology*. Wiley, Chichester, 711–724.

Zilliacus, H. 1987b. DeGeer moraines in Finland and the annual moraine problem. *Fennia* 165, 145–239.

Zilliacus, H. 1989. Genesis of DeGeer moraines in Finland. *Sedimentary Geology* 62, 309–317.

Zingg, T. 1935. Beiträge zur Schotteranalyse. *Schweizerische Mineralogische und Petrographische Mitteilungen* 15, 38–140.

Zotikov, I.A. 1986. *The Thermophysics of Glaciers*. D. Reidel, Dordrecht.

Zotikov, I.A., Zagorodnov, V.S. and Raikovsky, J.V. 1979. Sea ice on bottom of Ross Ice Shelf. *Antarctic Journal* 14, 65–66.

Zotikov, I.A., Zagorodnov, V.S. and Raikovsky, J.V. 1980. Core drilling through the Ross Ice Shelf (Antarctica) confirmed basal freezing. *Science* 207, 1463–1465.

Zumberge, J.H. 1952. The lakes of Minnesota, their origin and classification. Minnesota Geological Survey Bulletin 35.

Zumberge, J.H. 1955. Glacial erosion in tilted rock layers. *Journal of Geology* 63, 149–158.

Zwally, H.J. 1989. Growth of Greenland ice sheet: interpretation. *Science* 246, 1589–1591.

Zwally, H.J., Brenner, A.C., Major, J.A., Bindschadler, R.A. and Marsh, J.G. 1989. Growth of Greenland Ice Sheet: measurement. *Science* 246, 1587–1589.

Zwally, H.J., Brooks, R.L., Stanley, H.R. and Campbell, W.J. 1979. Ice sheet elevation and changes observable by satellite radar altimetry. *Journal of Glaciology* 24, 213–221.

Zwally, H.J. and Fiegles, S. 1994. Extent and duration of Antarctic surface melting. *Journal of Glaciology* 40, 463–476.

INDEX